WEB TUTORIALS

COMPANION WEBSITE WEB TUTORIALS

104 Web Tutorials review key concepts from the text in an engaging format using animations, instructional narration, and review quizzes.

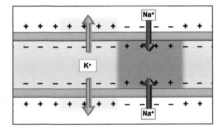

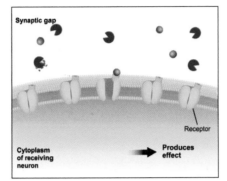

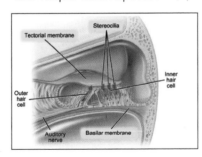

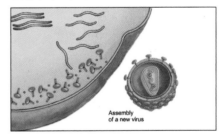

Assembly of a new virus

INSTRUCTOR MEDIA BUILT TO SUPPORT **YOUR** TEACHING APPROACH AND INVIGORATE **YOUR** LECTURES...

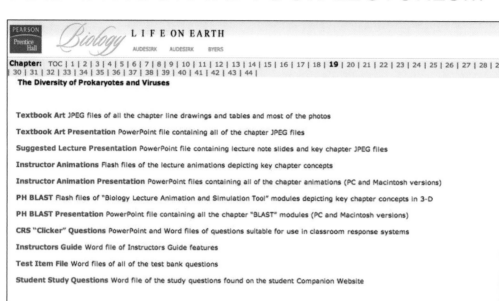

Instructor Resource Center on CD/DVD

This multiple-disc package of lecture and teaching resources is designed to allow you to teach the course and its concepts in the way you choose.

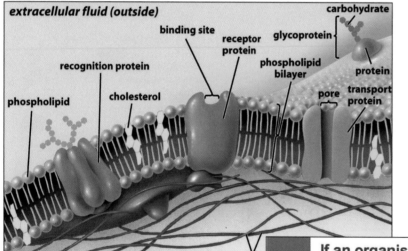

All Textbook Figures, Photos, and Tables, Fully Optimized in JPEG and PowerPoint® Files

- JPEGs of all textbook figures in labeled, unlabeled, and editable label versions plus all text photos and tables
- Every JPEG loaded into PowerPoint®
- Chapter Lecture PowerPoint® Presentations
- Over 250 Instructor Animations and BLAST animations in PowerPoint® in PC and Macintosh versions
- Chapter CRS "Clicker" Question PowerPoint® Presentations

If an organism has cilia, you will most likely find it:

1. in a plant.
2. in an animal.
3. is a bacterial cell.
4. living in a dry environment.
5. living in an watery environment.

Biology: Life on Earth, 8th Edition, CRS "Clicker" Questions
Chapter 4   Question 10.   Answer: 3
Difficulty level: Easy   Skill: Application

Text reference: Section 4.3 What Are the Major Features of Eukaryotic Cells? p. 63

Notes:
1. The goal of this question is to get students to think about what cilia are and then to relate cilia structure to function.

• New Generation CRS "Clicker" Question PowerPoint® Presentations

For the first time: "Clicker" Questions preloaded into PowerPoint® consisting of 15-20 new questions—not repurposed from the Test Bank, but original CRS questions written specifically for this textbook. All questions offer Instructor Notes including insights about how many worked when used in classes with this textbook.

BLAST
BIOLOGY LECTURE ANIMATION & SIMULATION TOOL

Available only with Prentice Hall books—BLAST offers unparalleled vivid 3-D animation and scientific precision to enliven lectures. BLAST is designed to support individual teaching approaches and class needs.

CHEMISTRY

- Covalent Bonds
- Hydrogen Bonds
- Amino Acids
- Building Proteins
- Protein Structure
- Unfolding and Refolding a Protein
- Nucleic Acid Structure and Function
- Osmosis
- Diffusion

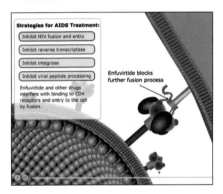

- Passive Diffusion Across a Membrane
- Active Transport: Sodium-Potassium Pump

ENERGY

- Structure of ATP
- Formation and Breakdown of ATP in the Cell
- Anatomy of an Enzyme: Chymotrypsin
- Regulation of Enzyme Activity
- Photosynthesis Overview
- Light-Dependent Reactions
- Light-Independent Reactions
- Harvesting Energy Overview
- Glycolysis
- Krebs Cycle
- Electron Transport Chain

CELLS

- Surface Area and Volume Calculator
- Prokaryotic Cell Size
- Eukaryotic Cell Size
- Bacteria Cell
- Animal Cell
- Plant Cell
- Cell Signaling Types
- Cell Signaling Amplification

GENETICS

- DNA Packaging
- DNA Replication
- Transcription and Translation
- Mitosis
- Meiosis
- Mitosis and Meiosis Compared
- Genetic Variation from Meiosis
- Cell Cycle Control and Cancer
- Single-Trait Crosses
- Two-Trait Crosses
- Stem Cells
- Gel Electrophoresis and DNA Fingerprinting
- Genetic Recombination in Bacteria

EVOLUTION

- Antibiotic Resistance in Bacteria

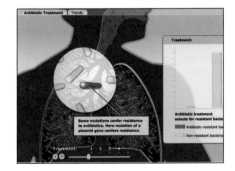

- Homologous Limb Evolution
- Population Dynamics
- Natural Selection

ECOLOGY

- Carbon Cycle
- Nitrogen Cycle
- Energy Flow
- Greenhouse Effect

PLANTS

- Plant Life Cycles
- Plant Pollination and Fertilization

ANIMALS

- Feedback Loops
- Negative Feedback: Body Temperature
- Positive Feedback: Labor
- Action Potential
- Signal Transmission at Synapses
- How Muscle Works

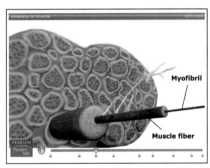

- Cardiac Cycle Overview
- Cardiac Cycle Details
- Electrical Coordination of the Cardiac Cycle
- Kidney Anatomy and Function
- Innate (Nonspecific) Immunity
- Adaptive Immunity
- HIV Structure
- HIV Life Cycle
- AIDS Onset
- AIDS Treatment Strategy

OVER 70 BLAST 3-D ANIMATIONS & 250+ to 2-D INSTRUCTOR ANIMATIONS
ALL ANIMATIONS PRELOADED INTO POWERPOINT® IN PC & MAC VERSIONS

YOUR ACCESS TO SUCCESS

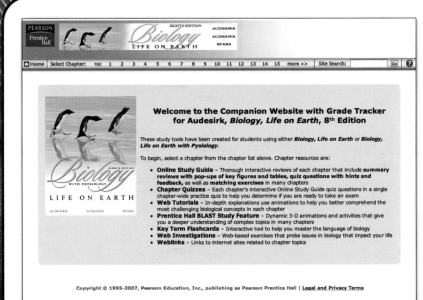

Students with new copies of Audesirk/Audesirk/Byers, *Biology: Life on Earth,* Eighth Edition, have full access to the book's Companion Website with Grade Tracker—a 24/7 study tool with quizzes, animated tutorials, and other features designed to help you make the most of your limited study time.

Just follow the easy website registration steps listed below...

Registration Instructions for
www.prenhall.com/audesirk

1. Go to *www.prenhall.com/audesirk*
2. Click the cover for Audesirk/Audesirk/Byers, *Biology: Life on Earth,* Eighth Edition.
3. Click "Register."
4. Using a coin (not a knife) scratch off the metallic coating below to reveal your Access Code.
5. Complete the online registration form, choosing your own personal Login Name and Password.
6. Enter your pre-assigned Access Code exactly as it appears below.
7. Complete the online registration form by entering your School Location information.
8. After your personal Login Name and Password are confirmed by e-mail, go back to *www.prenhall.com/audesirk*, enter your new Login Name and Password, and click "Log In."

Your Access Code is:

If there is no metallic coating covering the access code above, the code may no longer be valid and you will need to purchase online access using a major credit card to use the website. To do so, go to *www.prenhall.com/audesirk*, click the cover for Audesirk/Audesirk/Byers, *Biology: Life on Earth,* Eighth Edition, and follow the instructions under "Students."

Important: Please read the Subscription and End-User License Agreement located on the "Log In" screen before using the Audesirk/Audesirk/Byers, *Biology: Life on Earth,* Eighth Edition, Companion Website with Grade Tracker. By using the website, you indicate that you have read, understood, and accepted the terms of the agreement.

Minimum system requirements
PC Operating Systems:

Windows 2000/XP

Pentium II 233 MHz processor. 64 MB RAM In addition to the minimum memory required by your OS

Internet Explorer™ 6.0, Netscape Navigator™ 7.2, or Firefox 1.0.x

Macintosh Operating Systems:

Macintosh Power PC with OS X (10.2 and 10.3)

In addition to the RAM required by your OS, this application requires 64 MB RAM, with 40MB Free RAM, with Virtual Memory enabled

Netscape Navigator™ 7.2, Safari 1.0, 1.3, or Firefox 1.0.x

Macromedia Shockwave™

8.50 release 326 plugin

Macromedia Flash Player 6.0.79 & 7.0

Acrobat Reader 6.0.1

800 x 600 pixel screen resolution

Technical Support Call 1-800-677-6337. Phone support is available Monday–Friday, 8am to 8pm and Sunday 5pm to 12am, Eastern time. Visit our support site at *http://247.pearsoned.com*. E-mail support is available 24/7.

Biology
LIFE ON EARTH
with Physiology

EIGHTH EDITION

Teresa Audesirk
University of Colorado at Denver and Health Science Center

Gerald Audesirk
University of Colorado at Denver and Health Science Center

Bruce E. Byers
University of Massachusetts, Amherst

PEARSON

Prentice Hall

Upper Saddle River, NJ 07458

Library of Congress Cataloging-in-Publication Data

Audesirk, Teresa.
 Biology : life on earth with physiology / Teresa Audesirk, Gerald Audesirk, Bruce E. Byers.— 8th ed.
  p. cm.
 Includes bibliographical references and index.
 ISBN 0-13-195766-X
 1. Biology. I. Audesirk, Gerald. II. Byers, Bruce E. III. Title.
 QH308.2.A93 2008b
 570—dc22
                            2006037067

Editor: Jeff Howard
Development Editor: Anne Scanlan-Rohrer
Production Editor: Tim Flem/PublishWare
Media Editor: Patrick Shriner
Executive Managing Editor: Kathleen Schiaparelli
Editor in Chief of Development: Carol Trueheart
Media Production: nSight
Managing Editor, Science Media: Rich Barnes
Director of Marketing: Patrick Lynch
Marketing Assistant: Jessica Muraviov
Director of Creative Services: Paul Belfanti
Creative Director: Juan Lopez
Art Director: John Christiana
Interior Design: Maureen Eide
Cover Designers: Maureen Eide and John Christiana
Page Composition: PublishWare
Manufacturing Manager: Alexis Heydt-Long
Buyer: Alan Fischer
Senior Managing Editor, Art Production and Management: Patricia Burns
Manager, Production Technologies: Matthew Haas

Managing Editor, Art Management: Abigail Bass
Art Development Editor: Jay McElroy
Art Production Editor: Rhonda Aversa
Manager, Art Production: Sean Hogan
Assistant Manager, Art Production: Ronda Whitson
Illustrations: ESM Art Production; Lead Illustrators: Daniel
    Knopsnyder, Stacy Smith, Nathan Storck; Imagineering;
    Stephen Graepel
Cartographer: GeoNova, LLC
Assistant Managing Editor, Science Supplements: Karen Bosch
Editorial Assistant: Gina Kayed
Production Assistant: Nancy Bauer
Director, Image Resource Center: Melinda Reo
Manager, Rights and Permissions: Zina Arabia
Interior Image Specialist: Beth Boyd Brenzel
Cover Image Specialist: Karen Sanatar
Image Permission Coordinator: Debbie Latronica
Photo Researcher: Yvonne Gerin
Cover Photograph: Rockhopper Penguins; The Neck, Saunders Island,
    Falkland Islands, by Laura Crawford Williams

ISBN 0-13-195766-X

Pearson Education Ltd., *London*
Pearson Education Australia Pty., Limited, *Sydney*
Pearson Education Singapore, Pte. Ltd.
Pearson Education North Asia Ltd., *Hong Kong*
Pearson Education Canada, Ltd., *Toronto*
Pearson Educación de Mexico, S.A. de C.V.
Pearson Education—Japan, *Tokyo*
Pearson Education Malaysia, Pte. Ltd.

Brief Contents

This textbook is available in two versions:

Biology: Life on Earth (0-13-238061-7) consists of chapters 1–30, which provide biology students with a thorough overview of cell biology, genetics, evolution, diversity, and ecology.

Biology: Life on Earth with Physiology (0-13-195766-X), an expanded version of the text, contains chapters 1-44: the thirty chapters noted above, plus fourteen chapters covering plant and animal anatomy and physiology.

Essays

Contents

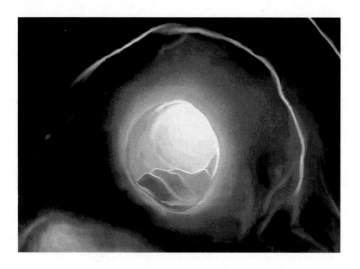

4 Cell Structure and Function 56

5 Cell Membrane Structure and Function 80

UNIT 2

Inheritance   147

9   DNA: The Molecule of Heredity   148

10   Gene Expression and Regulation   166

11 The Continuity of Life: Cellular Reproduction   190

12 Patterns of Inheritance   220

13 Biotechnology 250

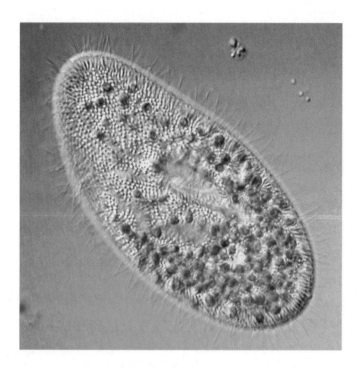

17 The History of Life   330

18 Systematics: Seeking Order Amidst Diversity   356

19 The Diversity of Prokaryotes and Viruses   370

20 The Diversity of Protists   386

21 The Diversity of Plants   402

26 Population Growth and Regulation 512

27 Community Interactions 536

28 How Do Ecosystems Work? 558

29 Earth's Diverse Ecosystems 580

30 Conserving Life on Earth 610

UNIT 5
Animal Anatomy and Physiology 633

31 Homeostasis and the Organization of the Animal Body 634

32 Circulation 648

36 Immunity: Defenses Against Disease   720

37 Chemical Control of the Animal Body: The Endocrine System   740

38 The Nervous System and the Senses   760

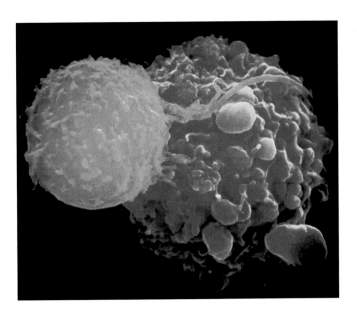

39 Action and Support: The Muscles and Skeleton 796

40 Animal Reproduction 814

41 Animal Development 836

UNIT 6
Plant Anatomy and Physiology   857

42   Plant Anatomy and Nutrient Transport   858

43   Plant Reproduction and Development   886

Preface

Global warming, bioengineered crops, stem cell research, mad cow disease, biodiversity—our students are and will continue to be inundated with scientific information and misinformation throughout their lives. Faced with the rapidly expanding field of biology, how does one decide which concepts and facts to convey? Which types of biological knowledge will best help students to make informed choices relating to their own lives, now and in the future? Which will best prepare students for their upper-level courses? Recognizing that there are no uniquely correct answers, we have revised the eighth edition of *Biology: Life on Earth* to give our users more choices.

In talking to educators engaged in the exciting but challenging mission of introducing students to biology, a single consensus emerged: "We need to help our students become scientifically literate." Scientific literacy endows a student with the mental tools to cope with expanding knowledge. It requires a foundation of factual knowledge that provides a cognitive framework into which new information can be integrated. But scientific literacy also includes the ability to grasp and evaluate new information from the news and popular press. A scientifically literate individual recognizes the interrelatedness of concepts and the need to integrate information from many areas.

BIOLOGY: LIFE ON EARTH EFFECTIVELY MANAGES A WEALTH OF SCIENTIFIC INFORMATION

Our eighth edition of *Biology: Life on Earth* is not only a revised and improved textbook, but a complete package of learning aids for students and teaching aids for instructors. Our major goals are

- To help instructors present biological information in a way that will foster scientific literacy among students
- To help students acquire information according to their own learning styles
- To help students relate this information to their own lives so as to understand its importance and relevance

BIOLOGY: LIFE ON EARTH

. . . Is Organized Clearly and Uniformly

Throughout each chapter, students will find aids that help them navigate through the information.

- Each chapter opens with "At a Glance," presenting the chapter's major sections and essays. Instructors can easily assign—and students can easily locate—key topics within the chapter.
- Major sections are introduced with general questions, while minor subheadings are summary statements reflecting their more specific content. An important pedagogical goal of this organization is its emphasis on biology as a hierarchy of interrelated concepts, rather than merely a compendium of isolated, independent subjects.
- The "Summary of Key Concepts" pulls together important concepts using major subheadings and their numbering system to allow instructors and students to review the chapter efficiently.
- Thought questions are found at the end of each Case Study, incorporated into selected figure captions, and in "Applying the Concepts" sections. These features all stimulate students to think about science rather than merely memorizing facts.

. . . Contains Revitalized Illustrations

Benefiting from the advice of reviewers and careful scrutiny by the authors, we have once again improved the illustration program. For the eighth edition, we have

- **Added and replaced many photographs** to help capture student interest. Our new, more flexible text layout allows us to add "thumbnail" photos illustrating some plants and animals previously described only in words.
- **Continued our emphasis on color consistency** Consistent colors illustrate specific atoms, structures, and processes.
- **Added more key process figures** As well as redrawing many diagrams for greater interest and clarity, we have added new figures that visually illustrate and pull together complex processes such as photosynthesis and cellular respiration.
- **Enhanced label clarity** We have placed additional boxed text within figures to ensure clearer explanations.
- **Continued our use of thought questions in selected figure captions** Answers to these questions are available for the first time in the back of the book.

. . . Has Been Updated and Reorganized

We incorporate scientific discoveries that students might read about in the news, placing them within a scientific context to help foster scientific literacy. Although each chapter has been carefully revised, here are some highlights of the eighth edition:

- **Unit One: The Life of a Cell** New case studies introduce the student to bioengineered skin and the puzzling prions responsible for mad cow disease. In response to reviewers' suggestions, we have reversed the order of presentation in Chapters 4 and 5. Cell structure is now covered before membrane structure, allowing us to describe the details of membrane specializations in the context of cell structures already introduced. In Chapter 5,

"Cell Membrane Structure and Function," our coverage of diffusion, osmosis, and tonicity is enhanced, and we have added a Scientific Inquiry essay on the discovery and structure of aquaporins.

- **Unit Two: Inheritance** In Chapter 9, "DNA: The Molecule of Heredity," we have expanded our discussion of the discovery that DNA is the molecule of heredity with a Scientific Inquiry essay describing the Hershey-Chase experiment as an application of the scientific method. Additional coverage of prokaryotic gene organization, transcription, and translation highlights important similarities and differences between prokaryotes and eukaryotes in Chapter 10, "Gene Expression and Regulation." In Chapter 11, "The Continuity of Life," a new section describes the importance and timing of mitosis and meiosis in eukaryotic life cycles. Also in Chapter 11, the Case Study on skin cancer brings a real-world perspective to a new section on the control of cell division, and a Health Watch essay describes the genetic basis of cancer. We have updated Chapter 13, "Biotechnology," with new essays on biotech applications. As in past editions, the chapter continues its dual focus on technologies and applications, emphasizing both practical and ethical controversies over the medical and agricultural uses of biotechnology.

- **Unit Three: Evolution and Diversity of Life** In response to reviewer requests, we now devote two chapters to enhanced coverage of bacteria, archaea, and viruses as well as the diverse eukaryotic protists. Unit Three includes expanded discussions of several key topics in systematics and evolutionary genetics, and new figures to illustrate these sometimes challenging concepts. The chapters covering the diversity of life include expanded discussion of numerous taxa, within both vertebrate and invertebrate groups. Throughout the unit, we have added or revised topics to reflect new discoveries in evolutionary biology.

- **Unit 4: Behavior and Ecology** In this unit, we have enlivened Chapter 25, "Animal Behavior," with new photographs and examples. Chapter 26, "Population Growth and Regulation," now covers life tables, logistic growth, and demographics. Chapter 28, "How Do Ecosystems Work?" contains updated environmental coverage with sections on clean air and water, and a new Earth Watch essay, "Poles in Peril." Chapter 29, "Earth's Diverse Ecosystems," now describes two new aquatic habitats: freshwater streams and rivers, and marine benthic communities. A highlight of the eighth edition is a new chapter on the emerging field of conservation biology. In Chapter 30, "Conserving Life on Earth," we describe the services provided by ecosystems and attempts to estimate their value to humanity. We explain how human activities can reduce biodiversity, and we discuss how conservation efforts and sustainable use can preserve and restore functional ecosystems.

- **Unit Five: Animal Anatomy and Physiology** This unit begins with revised coverage of homeostasis and thermoregulation. Students will find added and updated coverage on current topics including anorexia and obesity, bird flu, the neurochemistry of love, high-tech reproduction, new contraceptives, sexually transmitted diseases, stem cells, and fetal alcohol syndrome. We have retained our human emphasis while providing additional comparative coverage, adding topics such as countercurrent gas exchange in fishes, Malphigian tubules in insects, and sections on invertebrate hormones and defenses against disease.

- **Unit Six: Plant Anatomy and Physiology** This unit boasts many figure revisions and new photos to better illustrate plant anatomy and physiological processes as well as their fascinating environmental adaptations. We have also expanded coverage of the agricultural uses of plant hormones.

. . . Engages and Motivates Students

Scientific literacy cannot be imposed on students; they must actively participate in acquiring both the necessary information and skills. Thus, it is crucial for students to recognize that biology is about their personal lives as well as the life all around them. To help engage and motivate students, this new edition continues to offer these features:

- **Links to Life** Our short, informally written Links to Life relate to subjects that are both familiar to the student and relevant to the chapter.

- **Case Studies** In the eighth edition, we have retained and updated our more popular case studies while introducing several new ones. Case studies are based on recent news items, situations in which students might find themselves, or particularly fascinating biological topics. At the end of each chapter, the Case Study Revisited feature allows students to explore the topic further in light of what they have learned. Students also will find an in-depth investigation of each Case Study available on the companion Web site.

- **Bioethics** Many topics explored in the text have ethical implications for human life. These include genetic engineering and cloning, the use of animals in research, and human impact on other species. They are now identified with a bioethics icon that alerts students and instructors to the possibility for further discussion and investigation.

- **Essays** We retain our full suite of essays in this edition. Earth Watch essays explore pressing environmental issues, while Health Watch essays cover medical topics. A Closer Look essays allow instructors to explore selected topics in greater detail; Scientific Inquiry essays explain how scientific knowledge is acquired. Evolutionary Connections essays end selected chapters by placing topics in an evolutionary context.

. . . Provides Media and Print Supplements That Support Your Teaching Approach and Focus Student Study Time

- **Instructor Resource Center on CD/DVD (0-13-195773-2)** This multiple-disc instructor tool was built to support your teaching approach and invigorate your lectures. No other textbook for this course offers as many options and as much innovation and quality in instructor support. Resources include the entire art program of the book (in labeled, unlabeled, and editable-label formats) offered as fully optimized JPEGs and in various PowerPoint® files including Chapter Lecture presentations as well as hundreds of 2-D and 3-D animations and simulations preloaded into PowerPoint® presentations. *See the back endpapers for more information on the instructor materials supporting this book.*

- **CRS "Clicker" Questions (0-13-233497-6)** For the first time, a set of 900+ unique and original CRS questions—*not* questions taken from the test bank. Many questions include art, and all offer Instructor Notes that often describe insights relating to the use of the questions in classes where this textbook was assigned and Clickers were used.

- **Test Item File (printed 0-13-195768-6; TestGen® 0-13-233500-X)** The most respected collection of test questions in this market has been revised and thoroughly updated.

- **Instructor Resource Guide (0-13-195767-8)** A collection of printed instructor tools including class-tested lecture activities is offered both as Word files on the IRC on CD/DVD and in this printed version.

- **Companion Web site with Grade Tracker (www.prenhall. com/audesirk/)** This student Web site is a 24/7 focused study tool to help in mastering course concepts. Key features include the Online Study Guide to organize study, Chapter Quizzes to help students determine how well they know information, and 103 Web Tutorials that use animation and activities to help explain the most challenging concepts in each chapter. *See the front endpapers for more information on the student media.*

- **Study Guide (0-13-195769-4)** A traditional printed study guide with helpful materials including practice quizzes and activities.

ACKNOWLEDGMENTS

Biology: Life on Earth is truly a team effort. Our Development Editor Anne Scanlan-Rohrer sought ways to make the text more clear, consistent, and student friendly. Art Director John Christiana developed and executed a fresh design for this new edition, and Art Editor Rhonda Aversa deftly coordinated the art program. The many new and improved illustrations were artfully rendered by Artworks with the help of Jay McElroy. Photo Researcher Yvonne Gerin tirelessly tracked down excellent photos. Christianne Thillen tackled the job of copyediting with meticulous attention to detail. Tim Flem, our Production Editor, brought the art, photos, and manuscript together into a seamless whole, accepting last-minute changes with remarkable good nature. Media Editor Patrick Shriner and Assistant Editor Crissy Dudonis coordinated production of all the media and study aids that contribute so much to the total package that is *Biology: Life on Earth*. Our marketing manager, Mandy Jellerichs, helped create a marketing strategy that will effectively communicate our message to our audience. Editors Teresa Chung and Jeff Howard directed the project with energy and imagination. We thank Teresa for her unfailing faith in the project and for assembling a fantastic team to implement it. We add a special thanks to Jeff for shepherding this enormous endeavor to completion with good-natured patience and skill.

Terry and Gerry Audesirk

Bruce E. Byers

EIGHTH EDITION REVIEWERS

George C. Argyros, *Northeastern University*
Peter S. Baletsa, *Northwestern University*
John Barone, *Columbus State University*
Michael C. Bell, *Richland College*
Melissa Blamires, *Salt Lake Community College*
Robert Boyd, *Auburn University*
Michael Boyle, *Seattle Central Community College*
Matthew R. Burnham, *Jones County Junior College*
Nicole A. Cintas, *Northern Virginia Community College*
Jay L. Comeaux, *Louisiana State University*
Sharon A. Coolican, *Cayuga Community College*
Mitchell B. Cruzan, *Portland State University*
Lewis Deaton, *University of Louisiana—Lafayette*
Dennis Forsythe, *The Citadel*
Teresa L. Fulcher, *Pellissippi State Technical Community College*

Martha Groom, *University of Washington*
Richard Hanke, *Rose State College*
Kelly Hogan, *University of North Carolina—Chapel Hill*
Dale R. Horeth, *Tidewater Community College*
Joel Humphrey, *Cayuga Community College*
James Johnson, *Central Washington University*
Joe Keen, *Patrick Henry Community College*
Aaron Krochmal, *University of Houston—Downtown*
Stephen Lebsack, *Linn-Benton Community College*
David E. Lemke, *Texas State University*
Jason L. Locklin, *Temple College*
Cindy Malone, *California State University—Northridge*
Mark Manteuffel, *St. Louis Community College*
Steven Mezik, *Herkimer County Community College*
Christine Minor, *Clemson University*

Lee Mitchell, *Mt. Hood Community College*
Nicole Moore, *Austin Peay University*
James Mulrooney, *Central Connecticut State University*
Charlotte Pedersen, *Southern Utah University*
Robert Kyle Pope, *Indiana University South Bend*
Kelli Prior, *Finger Lakes Community College*
Jennifer J. Quinlan, *Drexel University*
Robert N. Reed, *Southern Utah University*
Wenda Ribeiro, *Thomas Nelson Community College*
Elizabeth Rich, *Drexel University*
Frank Romano, *Jacksonville State University*
Amanda Rosenzweig, *Delgado Community College*
Marla Ruth, *Jones County Junior College*
Eduardo Salazar, *Temple College*
Brian W. Schwartz, *Columbus State University*
Steven Skarda, *Linn-Benton Community College*
Mark Smith, *Chaffey College*

Dale Smoak, *Piedmont Technical College*
Jay Snaric, *St. Louis Community College*
Phillip J. Snider, *University of Houston*
Gary Sojka, *Bucknell University*
Nathaniel J. Stricker, *Ohio State University*
Martha Sugermeyer, *Tidewater Community College*
Peter Svensson, *West Valley College*
Sylvia Torti, *University of Utah*
Rani Vajravelu, *University of Central Florida*
Lisa Weasel, *Portland State University*
Diana Wheat, *Linn-Benton Community College*
Lawrence R. Williams, *University of Houston*
Michelle Withers, *Louisiana State University*
Taek You, *Campbell University*
Martin Zahn, *Thomas Nelson Community College*
Izanne Zorin, *Northern Virginia Community College—Alexandria*

MEDIA AND SUPPLEMENT CONTRIBUTORS AND REVIEWERS

Tamatha Barbeau, *Francis Marion University*
Linda Flora, *Montgomery County Community College*
Anne Galbraith, *University of Wisconsin–La Crosse*
Christopher Gregg, *Louisiana State University*
Theresa Hornstein, *Lake Superior College*
Dawn Janich, *Community College of Philadelphia*
Steve Kilpatrick, *University of Pittsburgh at Johnstown*
Bonnie L. King, *Quinnipiac University*

Michael Kotarski, *Niagara University*
Nancy Pencoe, *University of West Georgia*
Kelli Prior, *Finger Lakes Community College*
Greg Pryor, *Francis Marion University*
Mark Sugalski, *Southern Polytechnic State University*
Eric Stavney, *DeVry University*
Michelle D. Withers, *Louisiana State University*
Michelle Zurawski, *Moraine Valley Community College*

PREVIOUS EDITIONS REVIEWERS

W. Sylvester Allred, *Northern Arizona University*
Judith Keller Amand, *Delaware County Community College*
William Anderson, *Abraham Baldwin Agriculture College*
Steve Arch, *Reed College*
Kerri Lynn Armstrong, *Community College of Philadelphia*
G. D. Aumann, *University of Houston*
Vernon Avila, *San Diego State University*
J. Wesley Bahorik, *Kutztown University of Pennsylvania*
Bill Barstow, *University of Georgia-Athens*
Colleen Belk, *University of Minnesota, Duluth*
Michael C. Bell, *Richland College*
Gerald Bergtrom, *University of Wisconsin*
Arlene Billock, *University of Southwestern Louisiana*
Brenda C. Blackwelder, *Central Piedmont Community College*
Raymond Bower, *University of Arkansas*
Marilyn Brady, *Centennial College of Applied Arts and Technology*
Virginia Buckner, *Johnson County Community College*
Arthur L. Buikema, Jr., *Virginia Polytechnic Institute*
J. Gregory Burg, *University of Kansas*
William F. Burke, *University of Hawaii*
Robert Burkholter, *Louisiana State University*
Kathleen Burt-Utley, *University of New Orleans*
Linda Butler, *University of Texas-Austin*
W. Barkley Butler, *Indiana University of Pennsylvania*
Jerry Button, *Portland Community College*
Bruce E. Byers, *University of Massachusetts-Amherst*
Sara Chambers, *Long Island University*

Nora L. Chee, *Chaminade University*
Joseph P. Chinnici, *Virginia Commonwealth University*
Dan Chiras, *University of Colorado-Denver*
Bob Coburn, *Middlesex Community College*
Joseph Coelho, *Culver Stockton College*
Martin Cohen, *University of Hartford*
Walter J. Conley, *State University of New York at Potsdam*
Mary U. Connell, *Appalachian State University*
Jerry Cook, *Sam Houston State University*
Joyce Corban, *Wright State University*
Ethel Cornforth, *San Jacinto College-South*
David J. Cotter, *Georgia College*
Lee Couch, *Albuquerque Technical Vocational Institute*
Donald C. Cox, *Miami University of Ohio*
Patricia B. Cox, *University of Tennessee*
Peter Crowcroft, *University of Texas-Austin*
Carol Crowder, *North Harris Montgomery College*
Donald E. Culwell, *University of Central Arkansas*
Robert A. Cunningham, *Erie Community College, North*
Karen Dalton, *Community College of Baltimore County—Catonsville Campus*
Lydia Daniels, *University of Pittsburgh*
David H. Davis, *Asheville-Buncombe Technical Community College*
Jerry Davis, *University of Wisconsin, LaCrosse*
Douglas M. Deardon, *University of Minnesota*
Lewis Deaton, *University of Southwestern Louisiana*
Fred Delcomyn, *University of Illinois-Urbana*

David M. Demers, *University of Hartford*
Lorren Denney, *Southwest Missouri State University*
Katherine J. Denniston, *Towson State University*
Charles F. Denny, *University of South Carolina-Sumter*
Jean DeSaix, *University of North Carolina-Chapel Hill*
Ed DeWalt, *Louisiana State University*
Daniel F. Doak, *University of California-Santa Cruz*
Matthew M. Douglas, *University of Kansas*
Ronald J. Downey, *Ohio University*
Ernest Dubrul, *University of Toledo*
Michael Dufresne, *University of Windsor*
Susan A. Dunford, *University of Cincinnati*
Mary Durant, *North Harris College*
Ronald Edwards, *University of Florida*
Rosemarie Elizondo, *Reedley College*
George Ellmore, *Tufts University*
Joanne T. Ellzey, *University of Texas-El Paso*
Wayne Elmore, *Marshall University*
Thomas Emmel, *University of Florida*
Carl Estrella, *Merced College*
Nancy Eyster-Smith, *Bentley College*
Gerald Farr, *Southwest Texas State University*
Rita Farrar, *Louisiana State University*
Marianne Feaver, *North Carolina State University*
Susannah Feldman, *Towson University*
Linnea Fletcher, *Austin Community College-Northridge*
Charles V. Foltz, *Rhode Island College*
Dennis Forsythe, *The Citadel*
Douglas Fratianne, *Ohio State University*
Scott Freeman, *University of Washington*
Donald P. French, *Oklahoma State University*
Harvey Friedman, *University of Missouri—St. Louis*
Don Fritsch, *Virginia Commonwealth University*
Teresa Lane Fulcher, *Pellissippi State Technical Community College*
Michael Gaines, *University of Kansas*
Irja Galvan, *Western Oregon University*
Gail E. Gasparich, *Towson University*
Farooka Gauhari, *University of Nebraska-Omaha*
John Geiser, *Western Michigan University*
George W. Gilchrist, *University of Washington*
David Glenn-Lewin, *Iowa State University*
Elmer Gless, *Montana College of Mineral Sciences*
Charles W. Good, *Ohio State University-Lima*
Margaret Green, *Broward Community College*
David Grise, *Southwest Texas State University*
Lonnie J. Guralnick, *Western Oregon University*
Martin E. Hahn, *William Paterson College*
Madeline Hall, *Cleveland State University*
Georgia Ann Hammond, *Radford University*
Blanche C. Haning, *North Carolina State University*
Richard Hanke, *Rose State College*
Helen B. Hanten, *University of Minnesota*
John P. Harley, *Eastern Kentucky University*
William Hayes, *Delta State University*
Stephen Hedman, *University of Minnesota*
Jean Helgeson, *Collins County Community College*
Alexander Henderson, *Millersville University*
Timothy L. Henry, *University of Texas-Arlington*
James Hewlett, *Finger Lakes Community College*
Alison G. Hoffman, *University of Tennessee-Chattanooga*
Leland N. Holland, *Paso-Hernando Community College*
Laura Mays Hoopes, *Occidental College*

Michael D. Hudgins, *Alabama State University*
David Huffman, *Southwest Texas State University*
Donald A. Ingold, *East Texas State University*
Jon W. Jacklet, *State University of New York-Albany*
Rebecca M. Jessen, *Bowling Green State University*
J. Kelly Johnson, *University of Kansas*
Florence Juillerat, *Indiana University—Purdue University at Indianapolis*
Thomas W. Jurik, *Iowa State University*
Arnold Karpoff, *University of Louisville*
L. Kavaljian, *California State University*
Jeff Kenton, *Iowa State University*
Hendrick J. Ketellapper, *University of California, Davis*
Jeffrey Kiggins, *Blue Ridge Community College*
Harry Kurtz, *Sam Houston State University*
Kate Lajtha, *Oregon State University*
Tom Langen, *Clarkson University*
Patricia Lee-Robinson, *Chaminade University of Honolulu*
William H. Leonard, *Clemson University*
Edward Levri, *Indiana University of Pennsylvania*
Graeme Lindbeck, *University of Central Florida*
Jerri K. Lindsey, *Tarrant County Junior College-Northeast*
John Logue, *University of South Carolina-Sumter*
William Lowen, *Suffolk Community College*
Ann S. Lumsden, *Florida State University*
Steele R. Lunt, *University of Nebraska-Omaha*
Daniel D. Magoulick, *The University of Central Arkansas*
Paul Mangum, *Midland College*
Richard Manning, *Southwest Texas State University*
Ken Marr, *Green River Community College*
Kathleen A. Marrs, *Indiana University—Purdue University Indianapolis*
Michael Martin, *University of Michigan*
Linda Martin-Morris, *University of Washington*
Kenneth A. Mason, *University of Kansas*
Margaret May, *Virginia Commonwealth University*
D. J. McWhinnie, *De Paul University*
Gary L. Meeker, *California State University, Sacramento*
Thoyd Melton, *North Carolina State University*
Joseph R. Mendelson III, *Utah State University*
Karen E. Messley, *Rockvalley College*
Timothy Metz, *Campbell University*
Glendon R. Miller, *Wichita State University*
Hugh Miller, *East Tennessee State University*
Neil Miller, *Memphis State University*
Jeanne Mitchell, *Truman State University*
Jack E. Mobley, *University of Central Arkansas*
John W. Moon, *Harding University*
Richard Mortenson, *Albion College*
Gisele Muller-Parker, *Western Washington University*
Kathleen Murray, *University of Maine*
Robert Neill, *University of Texas*
Harry Nickla, *Creighton University*
Daniel Nickrent, *Southern Illinois University*
Jane Noble-Harvey, *University of Delaware*
David J. O'Neill, *Community College of Baltimore County-Dundalk Campus*
James T. Oris, *Miami University of Ohio*
Marcy Osgood, *University of Michigan*
C. O. Patterson, *Texas A&M University*
Fred Peabody, *University of South Dakota*
Harry Peery, *Tompkins–Cortland Community College*
Rhoda E. Perozzi, *Virginia Commonwealth University*

Gary B. Peterson, *South Dakota State University*
Bill Pfitsch, *Hamilton College*
Ronald Pfohl, *Miami University of Ohio*
Bernard Possident, *Skidmore College*
Ina Pour-el, *DMACC—Boone Campus*
Elsa C. Price, *Wallace State Community College*
Marvin Price, *Cedar Valley College*
James A. Raines, *North Harris College*
Paul Ramp, *Pellissippi State Technical College*
Mark Richter, *University of Kansas*
Robert Robbins, *Michigan State University*
Jennifer Roberts, *Lewis University*
Chris Romero, *Front Range Community College*
Paul Rosenbloom, *Southwest Texas State University*
K. Ross, *University of Delaware*
Mary Lou Rottman, *University of Colorado-Denver*
Albert Ruesink, *Indiana University*
Connie Russell, *Angelo State University*
Christopher F. Sacchi, *Kutztown University*
Doug Schelhaas, *University of Mary*
Brian Schmaefsky, *Kingwood College*
Alan Schoenherr, *Fullerton College*
Edna Seaman, *University of Massachusetts, Boston*
Patricia Shields, *George Mason University*
Marilyn Shopper, *Johnson County Community College*
Anu Singh-Cundy, *Western Washington University*
Linda Simpson, *University of North Carolina-Charlotte*
Russel V. Skavaril, *Ohio State University*
John Smarelli, *Loyola University*
Shari Snitovsky, *Skyline College*
John Sollinger, *Southern Oregon University*
Sally Sommers Smith, *Boston University*
Jim Sorenson, *Radford University*
Mary Spratt, *University of Missouri, Kansas City*
Bruce Stallsmith, *University of Alabama-Huntsville*
Benjamin Stark, *Illinois Institute of Technology*
William Stark, *Saint Louis University*
Barbara Stebbins-Boaz, *Willamette University*
Kathleen M. Steinert, *Bellevue Community College*
Barbara Stotler, *Southern Illinois University*

Gerald Summers, *University of Missouri-Columbia*
Marshall Sundberg, *Louisiana State University*
Bill Surver, *Clemson University*
Eldon Sutton, *University of Texas-Austin*
Dan Tallman, *Northern State University*
David Thorndill, *Essex Community College*
William Thwaites, *San Diego State University*
Professor Tobiessen, *Union College*
Richard Tolman, *Brigham Young University*
Dennis Trelka, *Washington and Jefferson College*
Sharon Tucker, *University of Delaware*
Gail Turner, *Virginia Commonwealth University*
Glyn Turnipseed, *Arkansas Technical University*
Lloyd W. Turtinen, *University of Wisconsin-Eau Claire*
Robert Tyser, *University of Wisconsin-La Crosse*
Robin W. Tyser, *University of Wisconsin-LaCrosse*
Kristin Uthus, *Virginia Commonwealth University*
F. Daniel Vogt, *State University of New York-Plattsburgh*
Nancy Wade, *Old Dominion University*
Susan M. Wadkowski, *Lakeland Community College*
Jyoti R. Wagle, *Houston Community College-Central*
Lisa Weasel, *Portland State University*
Michael Weis, *University of Windsor*
DeLoris Wenzel, *University of Georgia*
Jerry Wermuth, *Purdue University-Calumet*
Jacob Wiebers, *Purdue University*
Carolyn Wilczynski, *Binghamton University*
P. Kelly Williams, *University of Dayton*
Roberta Williams, *University of Nevada-Las Vegas*
Emily Willingham, *University of Texas–Austin*
Sandra Winicur, *Indiana University-South Bend*
Bill Wischusen, *Louisiana State University*
Chris Wolfe, *North Virginia Community College*
Stacy Wolfe, *Art Institutes International*
Colleen Wong, *Wilbur Wright College*
Wade Worthen, *Furman University*
Robin Wright, *University of Washington*
Brenda L. Young, *Daemen College*
Cal Young, *Fullerton College*
Tim Young, *Mercer University*

About the Authors

TERRY AND GERRY AUDESIRK grew up in New Jersey, where they met as undergraduates. After marrying in 1970, they moved to California, where Terry earned her doctorate in marine ecology at the University of Southern California and Gerry earned his doctorate in neurobiology at the California Institute of Technology. As postdoctoral students at the University of Washington's marine laboratories, they worked together on the neural bases of behavior, using a marine mollusk as a model system.

They are now emeritus professors of biology at the University of Colorado at Denver, where they taught introductory biology and neurobiology from 1982 through 2006. In their research, funded primarily by the National Institutes of Health, they investigated the mechanisms by which neurons are harmed by low levels of environmental pollutants and protected by estrogen.

Terry and Gerry share a deep appreciation of nature and of the outdoors. They enjoy hiking in the Rockies, running near their home in the foothills west of Denver, and attempting to garden at 7000 feet in the presence of hungry deer and elk. They are long-time members of many conservation organizations. Their daughter, Heather, has added another focus to their lives.

BRUCE E. BYERS, a midwesterner transplanted to the hills of western Massachusetts, is a professor in the biology department at the University of Massachusetts, Amherst. He's been a member of the faculty at UMass (where he also completed his doctoral degree) since 1993. Bruce teaches introductory biology courses for both nonmajors and majors; he also teaches courses in ornithology and animal behavior.

A lifelong fascination with birds ultimately led Bruce to scientific exploration of avian biology. His current research focuses on the behavioral ecology of birds, especially on the function and evolution of the vocal signals that birds use to communicate. The pursuit of vocalizations often takes Bruce outdoors, where he can be found before dawn, tape recorder in hand, awaiting the first songs of a new day.

To Heather, Jack, and Lori and in memory of Eve and Joe

T. A. & G. A.

To Bob and Ruth, with gratitude

B. E. B.

1

An Introduction to Life on Earth

Life on Earth is confined to the biosphere, a thin film encompassing Earth's surface. Earth, seen here from the moon, is an oasis of life in our solar system.

AT A GLANCE

CASE STUDY LIFE ON EARTH—AND ELSEWHERE?

"Viewed from the distance of the moon, the astonishing thing about the earth, catching the breath, is that it is alive. The photographs show the dry, pounded surface of the moon in the foreground, dead as an old bone. Aloft, floating free beneath the moist gleaming surface of bright blue sky, is the rising earth, the only exuberant thing in this part of the cosmos."

—Lewis Thomas in *The Lives of a Cell* (1974)

WHEN LEWIS THOMAS, biomedical researcher and physician, viewed the early photographs of Earth taken by astronauts from the surface of the moon (see the photo on the opposite page), he—like most of humanity—felt a sense of awe. The dry and barren surface of the moon in the foreground reminds us of how truly special Earth is—blanketed with green plants, blue oceans, and white clouds. But is Earth itself "alive"? There is no question that life has invaded nearly every nook and cranny of Earth. The toughest life-forms are also the simplest—single-celled organisms collectively described as *extremophiles*. These "survivalist microbes" inhabit the most inhospitable environments on Earth. Some thrive in vents in the deep ocean floor, where the pressure is 30 times that on Earth's surface and which spew water at temperatures over 212°F (100°C). Others have been discovered in ice cores 1200 feet below the surface of an Antarctic lake frozen for hundreds of thousands of years. Extremophiles inhabit the highly acid environments produced by mining wastes and hot springs, and they have been discovered in rock samples taken from 4 miles beneath Earth's surface. These life-forms seem as foreign to us as alien life from another solar system. Indeed, their existence on Earth fuels guarded optimism that life may exist, or may once have existed, in the seemingly hostile conditions found on other planets. What is life? How did it evolve? Could life survive on the barren surface of the moon, or in the harsh environments of other planets?

1.1   HOW DO SCIENTISTS STUDY LIFE?

Life Can Be Studied at Different Levels of Organization

Biology utilizes the same principles and methods as other sciences. In fact, a basic tenet of modern biology is that living things obey the same laws of physics and chemistry that govern nonliving matter. Just as sand can be formed into bricks that provide the building blocks of a wall, and walls in turn provide the basis of a structure, so scientists view the living and nonliving world as a series of *levels of organization*, with each level providing the building blocks for the next level (**FIG. 1-1**).

All matter on Earth is formed of substances called **elements**, each one of them unique. An **atom** is the smallest particle of an element that retains the properties of that element. For example, a diamond is a form of the element carbon. The smallest possible unit of the diamond is an individual carbon atom. Atoms may combine in specific ways to form assemblies called **molecules**; for example, one carbon atom can combine with two oxygen atoms to form

Community	Two or more populations of different species living and interacting in the same area	antelope, hawk, grass
Population	Members of one species inhabiting the same area	herd of pronghorn antelope
Multicellular Organism	An individual living thing composed of many cells	pronghorn antelope
Organ System	Two or more organs working together in the execution of a specific bodily function	the digestive system
Organ	A structure usually composed of several tissue types that form a functional unit	the stomach
Tissue	A group of similar cells that perform a specific function	epithelial tissue
Cell	The smallest unit of life	red blood cells epithelial cells nerve cell
Molecule	A combination of atoms	water glucose DNA
Atom	The smallest particle of an element that retains the properties of that element	hydrogen carbon nitrogen oxygen

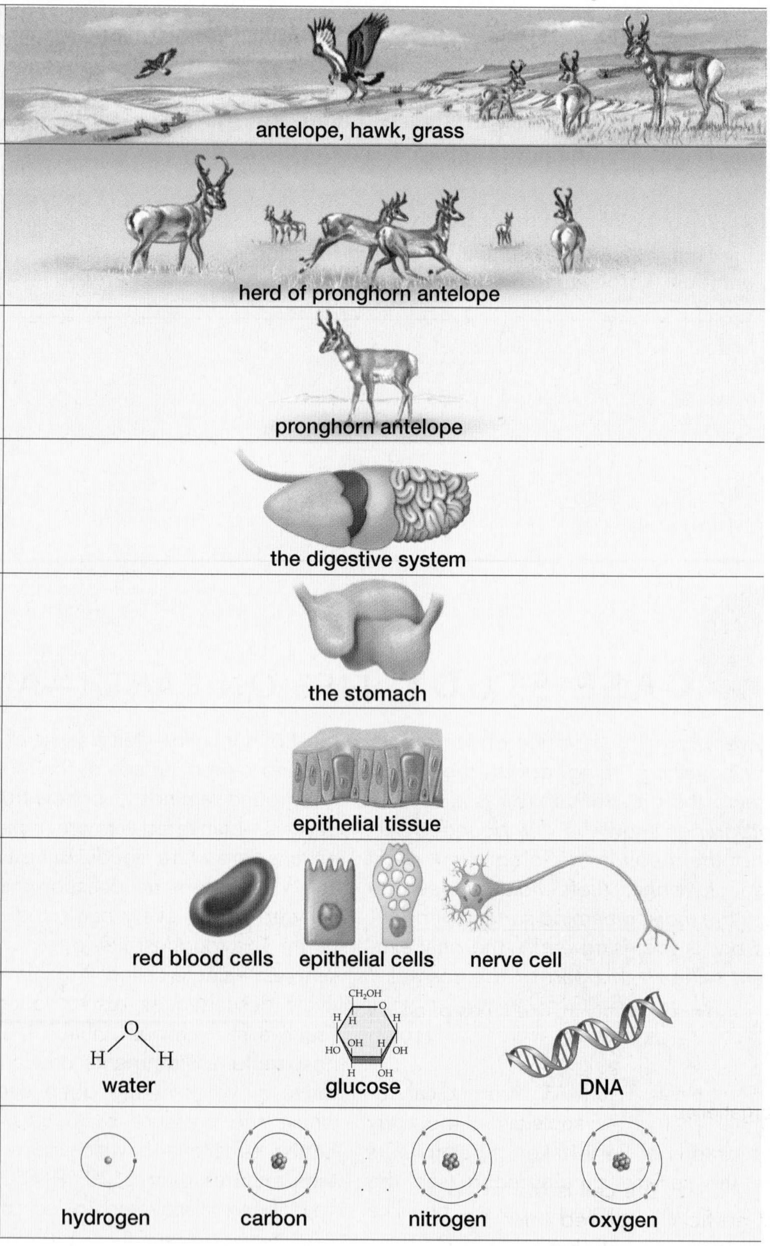

FIGURE 1-1 Levels of organization of matter
All life has a chemical basis, but the quality of life itself emerges on the cellular level. Interactions among the components of each level and the levels below it allow the development of the next-higher level of organization. **EXERCISE** Think of a scientific question that can be answered by investigating at the cell level, but that would be impossible to answer at the tissue level. Then think of one answerable at the tissue level but not the cell level. Repeat the process for two other pairs of adjacent levels of organization.

a molecule of carbon dioxide. Although many simple molecules form spontaneously, only living things manufacture extremely large and complex molecules. The bodies of living things are composed primarily of complex molecules called **organic molecules**, meaning that they contain a framework of carbon, to which at least some hydrogen is bound.

Although atoms and molecules form the building blocks of life, the quality of life itself emerges on the level of the cell. Just as an atom is the smallest unit of an element, so the **cell** is the smallest unit of life (**FIG. 1-2**). Although many forms of life consist of single cells, in multicellular forms, cells of similar type combine to form structures known as **tissues**; for example, muscle is a type of tissue. Different tissues, in turn, can combine to form **organs** (for example, a stomach or a kidney). Organs united by a common overall function are called **organ systems** (for example, the stomach is part of the digestive system, and the kidney is part of the urinary system). A multicellular organism will generally have several organ systems.

Levels of organization extend above the individual. **Organisms** of the same type that are capable of breeding with one another are collectively called a **species**. Within a given area, a group of organisms of the same species constitutes a **population**, and a collection of different populations that interact with one another makes up a **community** (see Fig. 1-1). Note that each level of organization incorporates many members of the previous level; a community contains many populations, a population contains many organisms, and so on.

Biologists work at many different levels of organization, depending on the question they are investigating. For example, to find out how antelope digest their food, a biologist might study the organs of the antelope digestive system, or at a smaller level, the cells that line the digestive tract. Delving deeper, the researcher might investigate the biological molecules secreted into the digestive tract that break down the animal's food. On the other hand, to find out whether habitat destruction is reducing the number of antelope, the scientists would investigate antelope populations as well as the interacting populations of other species that make up the community to which antelope belong. Scientists must recognize and choose the level of organization that is most appropriate to the question at hand.

Scientific Principles Underlie All Scientific Inquiry

All scientific inquiry, including biology, is based on a small set of assumptions. Although these assumptions can never be proven absolutely, they have been so thoroughly tested and validated that we might call them scientific principles. These principles are *natural causality*, *uniformity in space and time*, and *common perception*.

Natural Causality Is the Principle That All Events Can Be Traced to Natural Causes

Over the course of human history, two approaches have been taken to the study of life and other natural phenomena. The first assumes that some events happen through the intervention of supernatural forces beyond our understanding. Through the middle ages, many people believed that life arose spontaneously from nonliving matter. People of the seventeenth century believed that maggots arose from rotting meat (see "Scientific Inquiry: Controlled Experiments, Then and Now"), and that mice could be created from sweaty underwear combined with wheat husks in an open jar. The seizures of epilepsy were once thought to be the result of a visitation from the gods. In contrast, science adheres to the principle of **natural causality**: all events can be traced to natural causes that are potentially within our ability to comprehend. Today, we realize that maggots are the larval form of flies, and that epilepsy is a disease of the brain in which groups of nerve cells are uncontrollably activated. The principle of natural causality has an important corollary: the natural evidence we gather has not been deliberately distorted to fool us. This corollary may seem obvious, yet not so very long ago some people argued that fossils are not evidence of evolution; rather, they were placed on Earth by God to test our faith. The enormous accomplishments of science rely on the premise of natural causality.

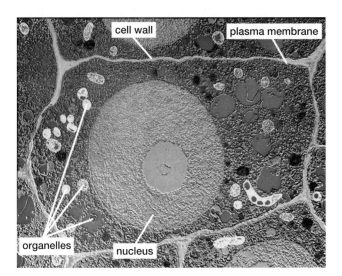

FIGURE 1-2 The cell is the smallest unit of life
This artificially colored micrograph of a plant cell shows the supporting cell wall that surrounds plant (but not animal) cells. Just inside the cell wall, the plasma membrane (found in all cells) has control over which substances enter and leave. The nucleus houses the cell's DNA. The cell also contains several types of specialized organelles. Some store food; some break down food to provide usable energy; and, in plants, some capture light energy.

The Natural Laws That Govern Events Apply Everywhere and for All Time

A second fundamental principle of science is that natural laws—laws derived from the study of nature—are uniform in space and time. The laws of gravity, the behavior of light, and the interactions of atoms, for example, are the same today as they were a billion years ago, and they hold true in Moscow as well as in New York, or even on Mars. Uniformity in space and time is especially vital to biology, because many important biological events, such as the evolution of today's diversity of living things, happened before humans were around to observe them. Some people believe that each of the different types of organisms was individually created at one time in the past by the direct intervention of God, a philosophy called *creationism*. Scientists freely admit that this idea cannot be absolutely disproved, but creationism is contrary to both natural causality and uniformity in time. The overwhelming success of science in explaining natural events through natural causes has led scientists to reject creationism as an explanation for the diversity of life on Earth.

Scientific Inquiry Is Based on the Assumption That People Perceive Natural Events in Similar Ways

A third basic assumption of science is that, generally, all human beings perceive natural events in fundamentally the same way, and that these perceptions provide us with reliable information about the natural world. Common perception is, to some extent, a principle peculiar to science. Value systems, such as those involved in the appreciation of art, poetry, and music, do not assume common perception. We may perceive the colors and shapes in a painting in a similar way (the scientific aspect of art), but we may disagree about the aesthetic value of the painting (the humanistic aspect of art; **FIG. 1-3**). Values may differ among individuals, often as a result of cultural or religious beliefs. Because value systems are subjective and not objective or measurable, science cannot answer certain types of philosophical or moral questions, such as the morality of abortion.

The Scientific Method Is the Basis for Scientific Inquiry

Given these assumptions, how do biologists study the workings of life? Scientific inquiry is a rigorous method for making observations of specific phenomena and searching for the order underlying those phenomena. Biology and other sciences commonly use the **scientific method**, which consists of six interrelated operations: *observation, question, hypothesis, prediction, experiment*, and *conclusion* (**FIG. 1-4a**). All scientific inquiry begins with an **observation** of a specific phenomenon. The observation, in turn, leads to a **question**—"How did this happen?" Then, in a flash of insight—or more typically after long, hard thought—a hypothesis is formulated. A **hypothesis** is a supposition, based on previous observations, that is offered as an answer to the question and a natural explanation for the observed phenomenon. To be useful, the hypothesis must lead to a **prediction**, typically expressed in "If … then" language. The prediction is test-

FIGURE 1-3 Value systems differ
Although people will generally agree about the colors and shapes in this artwork, questions such as "What does this mean?" or "Is this beautiful?" will be answered in different ways by different observers.

ed by carefully controlled observations called **experiments**. These experiments produce results that either support or refute the hypothesis, allowing the scientist to reach a **conclusion** about the validity of the hypothesis. A single experiment is never an adequate basis for a conclusion; the results must be repeatable not only by the original researcher but also by others.

Simple experiments test the assertion that a single factor, or **variable**, is the cause of a single observation. To be scientifically valid, the experiment must rule out other possible variables as the cause of the observation. For this reason, scientists design **controls** into their experiments. Controls, in which all the variables not being tested remain constant, are then compared with the experimental situation, in which only the variable being tested is changed. In the early 1600s, Francesco Redi used the scientific method to test the hypothesis that flies do not arise spontaneously from rotting meat, and this method is still used today, as illustrated by Malte Andersson's experiment to test the hypothesis that female widowbirds prefer to mate with males with long tails (see "Scientific Inquiry: Controlled Experiments, Then and Now").

You probably use some variation of the scientific method to solve everyday problems (**FIG. 1-4b**). For example, late for an important date, you rush to your car, turn the ignition key, and make the *observation* that it won't start. Your *question*: Why won't the car start? immediately leads to a *hypothesis*: The battery is dead. Your hypothesis leads to the *prediction*: If the battery is dead, then a new battery will allow you to start your car. Quickly, you

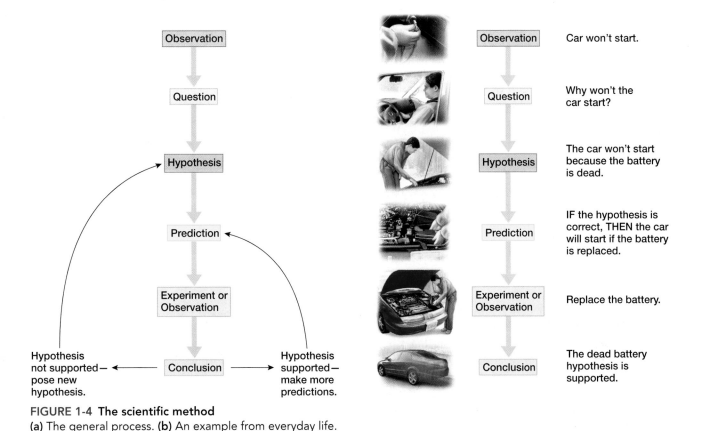

FIGURE 1-4 **The scientific method**
(a) The general process. (b) An example from everyday life.

design an *experiment*: you replace your battery with the battery from your roommate's new car and try to start your car again. The result supports your hypothesis, because your car starts immediately. But wait! You haven't provided controls for several variables. Perhaps the battery cable was loose and simply needed to be tightened. Realizing the need for a good *control*, you replace your old battery, making sure the cables are secured tightly, and attempt to restart the car. If your car repeatedly refuses to start with the old battery and tightened cables but then starts immediately when you put in your roommate's new battery, you have isolated a single *variable*, the battery. Although you are very late for your date, you can now safely draw the *conclusion* that your old battery is dead.

The scientific method is powerful, but it is important to recognize its limitations. In particular, scientists can seldom be sure that they have controlled *all* the variables other than the one they are trying to study. Therefore, scientific conclusions must always remain tentative and are subject to revision if new observations or experiments demand it.

Communication Is Crucial to Science

A final important element of science is *communication*. No matter how well designed an experiment is, it is useless if it is not communicated thoroughly and accurately. Redi's experimental design and conclusions survive today only because he carefully recorded his methods and observations. If experiments are not communicated to other scientists in enough detail, they cannot be repeated to verify the conclusions. Without verification, scientific findings cannot be safely used as the basis for new hypotheses and further experiments.

A fascinating aspect of scientific inquiry is that whenever a scientist reaches a conclusion, the conclusion immediately raises further questions that lead to further hypotheses and more experiments (why did your battery die?). Science is a never-ending quest for knowledge.

Science Is a Human Endeavor

Scientists are real people. They are driven by the same ambitions, pride, and fears as other people, and they sometimes make mistakes. As you will read in Chapter 9, ambition played an important role in the discovery of the structure of DNA by James Watson and Francis Crick. Accidents, lucky guesses, controversies with competing scientists, and, of course, the intellectual powers of individual scientists contribute greatly to scientific advances. To illustrate what we might call "real science," let's consider an actual case.

To study bacteria, microbiologists use pure *cultures*—that is, plates of bacteria that are free from contamination by other bacteria or molds. Only by studying a single type at a time can they learn about the properties of that

A classic experiment by the Italian physician Francesco Redi (1621–1697) beautifully demonstrates the scientific method and helps to illustrate the principle of natural causality, on which modern science is based. Redi investigated why maggots (which are the larval form of flies) appear on spoiled meat. In Redi's time, the appearance of maggots on meat was considered to be evidence of *spontaneous generation*, the production of living things from nonliving matter.

Redi *observed* that flies swarm around fresh meat and that maggots appear on meat left out for a few days. He formed a testable *hypothesis*: The flies produce the maggots. In his *experiment*, Redi wanted to test just one variable—the access of flies to the meat. Therefore, he took two clean jars and filled them with similar pieces of meat. He left one jar open (the *control* jar) and covered the other with gauze to keep out flies (the *experimental* jar). He did his best to keep all the other variables the same (for example, the type of jar, the type of meat, and the temperature). After a few days, he observed maggots on the meat in the open jar, but saw none on the meat in the covered jar. Redi *concluded* that his hypothesis was correct and that maggots are produced by flies, not by the nonliving meat (**FIG. E1-1**). Only through controlled experiments could the age-old hypothesis of spontaneous generation be laid to rest.

More than 300 years after Redi's experiment, today's scientists still use the same approach to design their experiments. Consider the experiment that Malte Andersson designed to investigate the long tails of male widowbirds. Andersson *observed* that male, but not female, widowbirds have extravagantly long tails, which they display while flying across African grasslands (**FIG. E1-2**). This observation led Andersson to ask the *question*: Why do the males, and only the males, have such long tails? His *hypothesis* was that males have long tails because females prefer to mate with long-tailed males, which therefore have more offspring than shorter-tailed males. From this hypothesis, Andersson *predicted* that if his hypothesis were true, then more females would build nests on the territories of males with artificially lengthened tails than would build nests on the territories of males with artificially shortened tails. He then captured some males, trimmed their tails to about half their original length, and released them (*experimental* group 1). Another group of males had the tail feathers that had been removed from the first group glued on as tail extensions (*experimental* group 2). Finally, Andersson had two *control* groups. In one, the tail was cut and then glued back in place (to control for the ef-

FIGURE E1-1 The experiments of Francesco Redi
QUESTION Redi's experiment falsified spontaneous generation, but did his experiment conclusively demonstrate that flies cause maggots? What kind of follow-up experiment would be necessary to better determine the source of maggots?

Observation:	Flies swarm around meat left in the open; maggots appear on meat.
Question:	Where do maggots on meat come from?
Hypothesis:	Flies produce the maggots.
Prediction:	IF the hypothesis is correct, THEN keeping the flies away from the meat will prevent the appearance of maggots.

Experiment

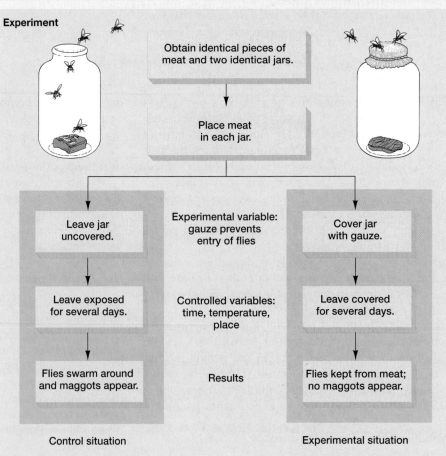

Control situation	**Experimental situation**

Conclusion:	The experiment supports the hypothesis that flies are the source of maggots and that spontaneous generation of maggots does not occur.

fects of capturing the birds and manipulating their feathers). In the other, the birds were simply captured and released. The experimenter was doing his best to make sure that tail length was the only variable that was changed. After a few days, Andersson counted the number of nests that females had built on each male's territory. He found that males with lengthened tails had the most nests on their territories, males with shortened tails had the fewest, and control males (with normal-length tails) had an intermediate number (**FIG. E1-3**). Andersson *concluded* that his hypothesis was correct, and that female widowbirds prefer to mate with males that have long tails.

FIGURE E1-2 A male widowbird

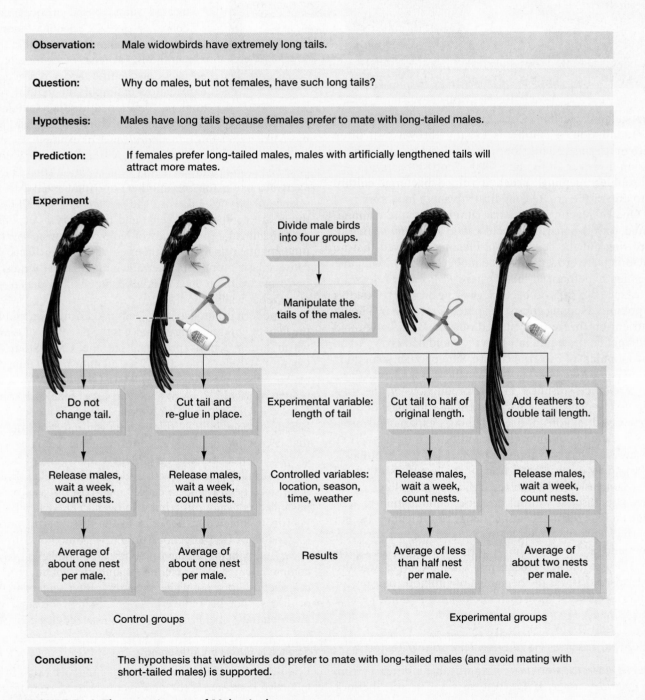

Observation:	Male widowbirds have extremely long tails.
Question:	Why do males, but not females, have such long tails?
Hypothesis:	Males have long tails because females prefer to mate with long-tailed males.
Prediction:	If females prefer long-tailed males, males with artificially lengthened tails will attract more mates.

Experiment

Divide male birds into four groups.

Manipulate the tails of the males.

Experimental variable: length of tail

Controlled variables: location, season, time, weather

Results

Do not change tail.

Cut tail and re-glue in place.

Cut tail to half of original length.

Add feathers to double tail length.

Release males, wait a week, count nests.

Release males, wait a week, count nests.

Release males, wait a week, count nests.

Release males, wait a week, count nests.

Average of about one nest per male.

Average of about one nest per male.

Average of less than half nest per male.

Average of about two nests per male.

Control groups

Experimental groups

Conclusion:	The hypothesis that widowbirds do prefer to mate with long-tailed males (and avoid mating with short-tailed males) is supported.

FIGURE E1-3 The experiments of Malte Andersson

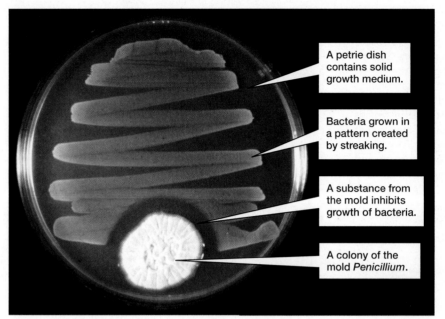

A petrie dish contains solid growth medium.

Bacteria grown in a pattern created by streaking.

A substance from the mold inhibits growth of bacteria.

A colony of the mold *Penicillium*.

FIGURE 1-5 Penicillin kills bacteria
A fuzzy white colony of the mold *Penicillium* has inhibited the growth of colonies of the disease-causing bacteria *Staphlococcus aureus*, which have been smeared back and forth across this plate of jellylike growth medium. Both the mold and the bacteria are visible only when they grow at high densities, as in the colonies seen here. QUESTION Why do some molds produce substances that are toxic to bacteria?

particular bacterium. Consequently, at the first sign of contamination, a culture is usually thrown out, often with mutterings about sloppy technique. On one such occasion however, in the late 1920s, Scottish bacteriologist Alexander Fleming turned a ruined bacterial culture into one of the greatest medical advances in history.

One of Fleming's bacterial cultures became contaminated with a patch of a mold called *Penicillium*. Before throwing out the culture dish, Fleming observed that no bacteria were growing near the mold (**FIG. 1-5**). Why not? Fleming hypothesized that perhaps *Penicillium* releases a substance that kills off bacteria growing nearby. To test this hypothesis, Fleming grew some pure *Penicillium* in a liquid nutrient broth. He then filtered out the *Penicillium* mold and applied the liquid in which the mold had grown to an uncontaminated bacterial culture. Sure enough, something in the liquid killed the bacteria. Further research into these mold extracts resulted in the production of the first *antibiotic*—penicillin, a bacteria-killing substance that has since saved millions of lives. Fleming's experiments are a classic example of the use of scientific methodology. They began with an observation and proceeded to a hypothesis, followed by experimental tests of the hypothesis that led to a conclusion. But the scientific method alone would have been useless without the lucky combination of accident and a brilliant scientific mind. Had Fleming been a "perfect" microbiologist, he would not have had any contaminated cultures. Had he been less observant, the contamination would have been just another spoiled culture dish. Instead, it was the beginning of antibiotic therapy for bacterial diseases. As French microbiologist Louis Pasteur said, "Chance favors the prepared mind."

Scientific Theories Have Been Thoroughly Tested

Scientists use the word *theory* in a way that is different from its everyday usage. If Dr. Watson were to ask Sherlock Holmes, "Do you have a theory as to the perpetrator of this

foul deed?" in scientific terms, he would be asking Holmes for a hypothesis—an "educated guess" based on observable evidence, or clues. A **scientific theory** is far more general and more reliable than a hypothesis. Far from being an educated guess, a scientific theory is a general explanation of important natural phenomena, developed through extensive and reproducible observations. In common English, it is more like a *principle* or a *natural law*. For example, scientific theories such as the atomic theory (that all matter is composed of atoms) and the theory of gravitation (that objects exert attraction for one another) are fundamental to the science of physics. Likewise, the *cell theory* (that all living things are composed of cells) and the *theory of evolution* are fundamental to the study of biology. Scientists describe fundamental principles as "theories" rather than "facts" because a basic premise of scientific inquiry is that it must be performed with an open mind. If compelling evidence arises, a theory will be modified.

A modern example of the need to keep an open mind in the light of new scientific evidence is the discovery of *prions*, which are infectious proteins (see the Case Study for Chapter 3). Before the early 1980s, all known infectious disease agents possessed genetic material—either DNA or the related molecule, RNA. When neurologist Stanley Prusiner from the University of California at San Francisco published evidence in 1982 that scrapie (an infectious disease that causes the brains of sheep to degenerate) is actually caused and transmitted by a protein with no genetic material, his results were met with widespread disbelief. Prions have since been found to cause "mad cow disease," which has killed not only cattle but also over 150 people who ate beef from infected cattle. Prior to the discovery of prions, the concept of an infectious protein was unknown to science. But by being willing to modify accepted beliefs to accommodate new data, scientists maintained the integrity of the scientific process while expanding their understanding of disease. For his pioneering work, Stanley Prusiner was awarded the Nobel Prize in Physiology or Medicine in 1997.

Science Is Based on Reasoning

Scientific theories arise through inductive reasoning. **Inductive reasoning** is the process of creating a generalization as a result of making many observations that support it, and none that contradict it. Simplistically, the theory that Earth exerts gravitational forces on objects arose from repeated observations of objects falling down toward Earth and from a complete lack of observations of objects "falling up." Likewise, the cell theory arises from the observation that all organisms that have the attributes of life are composed of one or more cells, and that nothing that is not composed of cells shares all these attributes.

Once a scientific theory has been formulated, it can be used to support deductive reasoning. In science, **deductive reasoning** is the process of generating hypotheses about how a specific experiment or observation will turn out, based on a well-supported generalization such as a scientific theory. For example, based on the cell theory, if a new organism is found that shares all the attributes of life, scientists can confidently deduce or hypothesize that it will be composed of cells. Of course, the new organism must be carefully scrutinized under the microscope to determine its cellular structure; if compelling new evidence arises, a theory can be modified.

Scientific Theories Are Formulated in Ways That Can Potentially Be Disproved

A major difference between a scientific theory and a belief based on faith is that a scientific theory can be disproved or *falsified*, while a faith-based assertion cannot. The potential to be falsified is why scientists continue to refer to basic precepts of science as "theories." For example, let's look at the existence of elves. The scientific approach to elves is that no solid evidence of their existence can be detected, and therefore elves do not exist. People who have faith in the existence of elves might describe them as creatures so secretive that they can never be captured, observed, or otherwise detected. Alternatively, these believers might claim that elves manifest themselves only to people who believe in them. The scientific theory that elves do not exist could easily be falsified if someone caught one or provided other repeatable, objective evidence of their existence. In contrast, the faith-based assertion that elves exist, as well as other faith-based assertions such as creationism, are formulated in ways that can never be disproved. For this reason, they are articles of faith rather than science.

1.2 EVOLUTION: THE UNIFYING THEORY OF BIOLOGY

In the words of biologist Theodosius Dobzhansky, "Nothing in biology makes sense, except in the light of evolution." Why don't snakes have legs? Why are there dinosaur fossils but no living dinosaurs? Why are monkeys so like us, not only in appearance but also in the structure of their genes and proteins? The answers to those questions, and thousands more, lie in the processes of evolution (examined in detail in Unit Three). Evolution is so vital to our understanding and appreciation of biology that we introduce its major principles here in our opening chapter.

Evolution not only explains the origin of diverse forms of life but accounts for the remarkable similarities among different life-forms as well. Ever since the theory of evolution was formulated in the mid-1800s by two English naturalists, Charles Darwin and Alfred Russel Wallace, it has been supported by fossil finds, geological studies, radioactive dating of rocks, genetics, molecular biology, biochemistry, and breeding experiments. People who refer to evolution as "just a theory" profoundly misunderstand what scientists mean by the word "theory."

Three Natural Processes Underlie Evolution

The scientific theory of **evolution** states that modern organisms descended, with modification, from preexisting life-forms. The most important force in evolution is **natural selection**, the process by which organisms with specific traits that help them cope with the rigors of their environment reproduce more successfully than do others that lack these traits. The changes that occur during evolution are a result of natural selection acting on the inherited variation that occurs among individuals in a population, causing changes in the population over successive generations. The variation upon which natural selection acts is a result of small differences in the genetic makeup of the individuals within the population.

Evolution arises as a consequence of three natural processes: *genetic variation* among members of a population owing to differences in their DNA, *inheritance* of those variations by offspring of parents who carry the variation, and *natural selection*, the enhanced reproduction of organisms with variations that help them cope with their environment.

Genetic Variability Among Organisms Is Inherited

Look around at your classmates and notice how different they are, or go to an animal shelter and observe the differences among the dogs in size, shape, and coat color. Although some of this variation (particularly among your classmates) is due to differences in environment and lifestyles, much of it is influenced by genes. For example, most of us could pump iron for the rest of our lives and never develop a body like that of "Mr. Universe."

But what are genes? The hereditary information of all known forms of life is contained within a type of molecule called **deoxyribonucleic acid**, or **DNA** (FIG. 1-6). An organism's DNA, which is contained in **chromosomes** in each cell, is the cell's genetic blueprint or molecular instruction manual, a guide to the construction and the operation of its body. Genes are segments of DNA; each gene directs the formation of one of the crucial molecular components of the organism's body. When an organism reproduces, it passes a copy of its chromosomes containing DNA to its offspring.

The accuracy of the DNA copying process is astonishingly high; in people only about 25 mistakes, called **mutations**, occur for every billion bits of information copied. Mutations

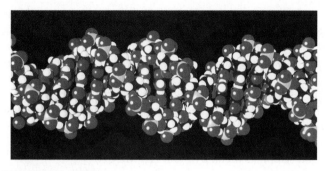

FIGURE 1-6 DNA
A computer-generated model of DNA, the molecule of heredity. As James Watson, its codiscoverer, put it: "A structure this pretty just had to exist."

can also result from damage to DNA, for example by ultraviolet light, radioactive particles, or toxic chemicals such as those in cigarette smoke. These occasional errors alter the information in genes or alter the collections of genes within chromosomes. Most mutations have no effect or are harmful. For example, mutations in skin cells caused by too much ultraviolet light can cause skin cancer. Mutations in lung cells caused by poisons in cigarette smoke can cause lung cancer. On very rare occasions, however, a mutation will occur when a sperm or egg cell is being formed, allowing it to be passed to the organism's offspring. As a result, each cell in the new individual's body will carry this *hereditary mutation*. Some of these will prevent the new organism from developing, while other changes in genetic material cause diseases such as Down syndrome. Still other mutations, many of which occurred millions of years ago and have been passed from parent to offspring through countless generations, account for differences such as height, body proportions, facial features, and the color of skin, hair, and eyes.

Natural Selection Tends to Preserve Genes That Help an Organism Survive and Reproduce

On average, organisms that best meet the challenges of their environment will leave the most offspring; these offspring will inherit the genes that made their parents successful. Thus, natural selection preserves genes that help organisms flourish in their environment. To create a hypothetical example, a mutated gene that caused ancestral beavers to grow larger teeth allowed those with this mutation to chew down trees more efficiently, build bigger dams and lodges, and eat more bark than "ordinary" beavers could. Because these big-toothed beavers obtained more food and better shelter, they were able to raise more offspring who inherited their parents' genes for larger teeth. Over time, less-successful, smaller-toothed beavers became increasingly scarce; after many generations, all beavers had large teeth.

Structures, physiological processes, or behaviors that aid in survival and reproduction in a particular environment are called **adaptations**. Most of the features that we admire so much in our fellow life-forms, such as the long limbs of deer, the wings of eagles, and the mighty trunks of redwood trees, are adaptations molded by millions of years of natural selection acting on random mutations.

Over millennia, the interplay of environment, genetic variation, and natural selection inevitably results in evolution: a change in the genetic makeup of species. This change has been documented innumerable times both in laboratory settings and in the wild. For example, antibiotics have acted as agents of natural selection on bacterial populations, causing the evolution of antibiotic-resistant forms. Lawn mowers have caused changes in the genetic makeup of populations of dandelions, favoring those that produce flowers on very short stems. Scientists have documented the spontaneous emergence of entirely new species of plants due to mutations that alter their chromosome number.

What helps an organism survive today can become a liability tomorrow. If environments change—for example, as global warming occurs—the genetic makeup that best adapts organisms to their environment will also change over time. When random new mutations increase the fitness of an organism in the altered environment, these mutations will spread throughout the population. Populations within a species that live in different environments will be subjected to different types of natural selection. If the differences are great enough and continue for long enough, they may eventually cause the populations to become sufficiently different from one another to prevent interbreeding—a new species will have evolved.

If, however, favorable mutations do not occur, a changing environment may doom a species to extinction. Dinosaurs (**FIG. 1-7**) are extinct not because they were failures—after all, they flourished for 100 million years—but because they could not adapt rapidly enough to changing conditions.

Within particular habitats, diverse organisms have evolved complex interrelationships with one another and with their nonliving surroundings. The diversity of species and the interactions that sustain them are encompassed by the term **biodiversity**. In recent decades, the rate of environmental change has been drastically accelerated by human activities. Many wild species are unable to adapt to this rapid change. In habitats most affected by humans, many species are being driven to extinction. This concept is explored further in "Earth Watch: Why Preserve Biodiversity?"

1.3 WHAT ARE THE CHARACTERISTICS OF LIVING THINGS?

What is life? If you look up *life* in a dictionary, you will find definitions such as "the quality that distinguishes a vital and functioning being from a dead body," but you won't find out what that "quality" is. Life emerges as a result of incredibly complex, ordered interactions among nonliving molecules. How did life originate? Although scientists have several hypotheses as to how life on Earth first arose (see Chapter 17), there are no scientific theories that describe the origin of life. Life is an intangible quality that defies simple definition, but we can describe some of the

FIGURE 1-7 A fossil of *Triceratops*
This *Triceratops* died in what is now Montana about 70 million years ago. No one is certain what caused the extinction of the dinosaurs, but we do know that they were unable to evolve new adaptations to keep up with changes in their habitat.

- Living things grow.
- Living things reproduce themselves, using the molecular blueprint of DNA.
- Living things, as a whole, have the capacity to evolve.

Let's explore these characteristics in more detail.

Living Things Are Complex, Organized, and Composed of Cells

In Chapter 4 you will learn how researchers in the early 1800s, observing life with primitive microscopes, devised the **cell theory**, which states that the cell is the basic unit of life. Even a single cell has an elaborate internal structure (see Fig. 1-2). All cells contain **genes**, units of heredity that provide the information needed to control the life of the cell, and small structures called **organelles** that are specialized to carry out specific functions such as moving the cell, obtaining energy, or synthesizing large molecules. Cells are always surrounded by a thin **plasma membrane** that encloses the **cytoplasm** (organelles and the fluid surrounding them) and separates the cell from the outside world. Some life-forms, mostly invisible to the naked eye, consist of just one cell. Your body—and the bodies of organisms that are most familiar to us—is composed of many cells that are specialized and elaborately organized to perform specific functions. The water flea beautifully illustrates the complexity found in a multicellular form of life far smaller than the letter "o" in this text (**FIG. 1-8**).

characteristics of living things. Taken together, these attributes are not shared by nonliving objects. As you walk outside, you can observe many of these attributes (see "Links to Life: The Life Around Us"). Characteristics of life include the following:

- Living things are composed of cells that have a complex, organized structure.
- Living things respond to stimuli from their environment.
- Living things actively maintain their complex structure and their internal environment, a process called *homeostasis*.
- Living things acquire and use materials and energy from their environment and convert them into different forms.

Living Things Maintain Relatively Constant Internal Conditions Through Homeostasis

Complex, organized structures are not easy to maintain. Whether we consider the molecules of your body or the books and papers on your desk, organization tends to disintegrate into chaos unless energy is used to sustain it (we explore this concept further in Chapter 6). To stay alive and function effectively, organisms must keep the conditions within their bodies fairly constant; in other words, they must maintain **homeostasis** (derived from Greek words meaning "to stay the same"). For example, organisms must precisely regulate the amount of water and salts within their cells. Their bodies must also be maintained at appropriate temperatures for biological functions to occur. Among warm-blooded animals, vital organs such as the brain and heart are kept at a warm, constant temperature despite wide fluctuations in outside temperature. Homeostasis is

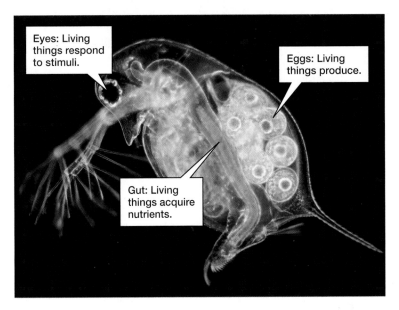

Eyes: Living things respond to stimuli.

Eggs: Living things produce.

Gut: Living things acquire nutrients.

FIGURE 1-8 Life is both complex and organized
The water flea, *Daphnia longispina*, is only 1 millimeter long (1/1000 meter), yet it has legs, a mouth, a digestive tract, reproductive organs, light-sensing eyes, and even a rather impressive brain in relation to its size.

"The loss of species is the folly our descendants are least likely to forgive us."
—E. O. Wilson, Professor, Harvard University

What is biodiversity, and why should we be concerned with preserving it? *Biodiversity* refers to the total number of species within a given region and to the resulting complexity of interactions among them. Over the 3.5-billion-year history of life on Earth, evolution has produced an estimated 8 to 10 million unique and irreplaceable species. Of these, scientists have named only about 1.4 million, and only a tiny fraction of this number have been studied. Evolution has not, however, merely churned out millions of independent species. Over thousands of years, organisms in a given area have been molded by forces of natural selection exerted by other living species as well as by the nonliving environment in which they live. The outcome is the community, a highly complex web of interdependent life-forms whose interactions sustain one another. By participating in the natural cycling of water, oxygen, and other nutrients, and by producing rich soil and purifying wastes, these communities contribute to the sustenance of human life as well. The concept of biodiversity has emerged as a result of our increasing concern over the loss of countless forms of life and the habitats that sustain them.

The tropics are home to the vast majority of all the species on Earth, perhaps 7 or 8 million of them, living in complex communities. The rapid destruction of habitats in the tropics—from rain forests to coral reefs—as a result of human activities is producing high rates of extinction of many species (**FIG. E1-4**). Most of these species have never been named, and others never even discovered. Aside from ethical concerns over eradicating irreplaceable forms of life, as we drive unknown organisms to extinction, we lose potential sources of medicine, food, and raw materials for industry.

For example, a wild relative of corn that is not only very disease-resistant but also *perennial* (that is, lasts more than one growing season) was found growing only on a 25-acre plot of land in Mexico that was scheduled to be cut and burned within a week of the discovery. The genes of this plant might one day enhance the disease resistance of corn or create a perennial corn plant. The rosy periwinkle, a flowering plant found in the tropical forest of the island of Madagascar (off the eastern coast of Africa) produces two substances that are now widely marketed for the treatment of leukemia and Hodgkin's disease, a cancer of the lymphatic organs. Only about 3 percent of the world's flowering plants have been examined for substances that might fight cancer or other diseases. Closer to home, loggers of the Pacific Northwest frequently cut and burned the Pacific yew tree as a "nuisance species" until the active ingredient that has since gone into making the anticancer drug Taxol® was discovered in its bark.

Many conservationists are also concerned that as species are eliminated, either locally or through total extinction, the communities of which they were a part might change, becoming less stable and more vulnerable to damage by diseases or adverse environmental conditions. Some experimental evidence supports this viewpoint, but the interactions within communities are so complex that these hypotheses are difficult to test. Clearly, some species have a much larger role than others in preserving the stability of a given ecosystem. Which species are most crucial in each ecosystem? No one knows. Human activities have increased the natural rate of extinction by a factor of at least 100 and possibly by as much as 1000 times the prehuman rate. By reducing biodiversity to support increasing numbers of humans and wasteful standards of living, we have ignorantly embarked on an uncontrolled global experiment, using planet Earth as our laboratory. In their book *Extinction* (1981), Stanford ecologists Paul and Anne Ehrlich compare the loss of biodiversity to the removal of rivets from the wing of an airplane. The rivet-removers continue to assume that there are far more rivets than needed, until one day, when the airplane takes off, they are proven tragically wrong. As human activities drive species to extinction while we have little knowledge of the role each plays in the complex web of life, we run the risk of removing "one rivet too many."

FIGURE E1-4 Biodiversity threatened
Destruction of tropical rain forests by indiscriminate logging threatens Earth's greatest storehouse of biological diversity. Interrelationships such as those that have evolved between this *Heliconia* flower and its hummingbird pollinator, and this frog and the bromeliad on which it lives, sustain these diverse communities and are threatened by human activities.

maintained by a variety of mechanisms. In the case of temperature regulation, these include sweating during hot weather and exercise, dousing oneself with cool water (**FIG. 1-9**), metabolizing more food in cold weather, basking in the sun, or even adjusting a thermostat.

Of course, not everything stays the same throughout an organism's life. Major changes, such as growth and reproduction, occur; but these are not failures of homeostasis. Rather, they are specific, genetically programmed parts of the organism's life cycle.

Living Things Respond to Stimuli

In order to stay alive, reproduce, and maintain homeostasis, organisms must perceive and respond to stimuli in their internal and external environments. Animals have evolved elaborate sensory organs and muscular systems that allow them to detect and respond to light, sound, touch, chemicals, and many other stimuli from their surroundings. Internal stimuli are perceived by receptors for stretch, temperature, pain, and various chemicals. For example, when you feel hungry, you perceive contractions of your empty stomach and low levels of sugars and fats in your blood. You then respond to external stimuli by choosing appropriate objects to eat, such as a sandwich rather than a plate. Yet animals, with their elaborate nervous systems and motile bodies, are not the only organisms that perceive and respond to stimuli. The plants on your windowsill grow toward light, and even the bacteria in your intestines manufacture different digestive enzymes depending on whether you drink milk, eat candy, or both.

Living Things Acquire and Use Materials and Energy

Organisms need materials and energy to maintain their high level of complexity and organization, to grow, to maintain homeostasis, and to reproduce (see Fig. 1-8). Organisms acquire the materials they need, called **nutrients**, from air, water, or soil, or from other living things. Nutrients include minerals, oxygen, water, and all the other chemical building blocks that make up biological molecules. These nutrients are obtained from the environment, where they are continuously exchanged and recycled among living things and their nonliving surroundings (**FIG. 1-10**).

To sustain life, organisms must obtain **energy**—the ability to do work, such as carrying out chemical reactions, growing leaves in the spring, or contracting a muscle. Ultimately, the energy that sustains nearly all life comes from sunlight. Plants and some single-celled organisms capture the energy of sunlight directly and store it in energy-rich

FIGURE 1-9 Living things maintain homeostasis
Evaporative cooling by water, both from sweat and from a bottle, helps Lance Armstrong (seven-time winner of the Tour de France bicycle race) maintain temperature homeostasis. QUESTION In addition to reducing body temperature, how else does sweating affect homeostasis?

molecules, such as sugars, using a process called **photosynthesis**. These organisms are called **autotrophs**, meaning "self-feeders." Organisms that cannot photosynthesize, such as animals and fungi, must acquire energy prepackaged in the molecules of the bodies of other organisms; hence, these organisms are called **heterotrophs**, meaning "other-feeders." Thus, energy flows in a one-way path from the sun through nearly all forms of life. That energy is eventually released again as heat, which cannot be used to power life (see Fig. 1-10).

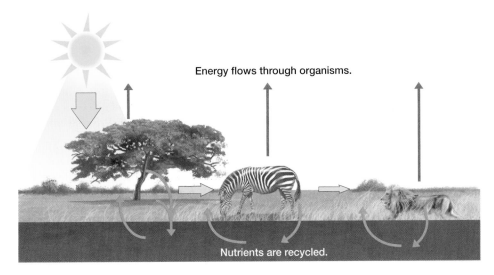

Energy flows through organisms.

Nutrients are recycled.

FIGURE 1-10 The flow of energy and the recycling of nutrients
Nutrients are recycled among organisms and their nonliving environment. In contrast, energy is acquired from sunlight, transferred through heterotrophs (yellow arrows), and lost as heat (red arrows) in a one-way flow. Photosynthetic organisms (autotrophs) capture solar energy and obtain nutrients from soil and water. Other forms of life (heterotrophs) obtain their energy and most of their nutrients from autotrophs either directly (in the case of herbivores) or indirectly by consuming other heterotrophs (in the case of carnivores).

LINKS TO LIFE   The Life Around Us

The next time you walk across campus, look at the astonishing array of creatures thriving in a place as domesticated as a college campus. During the right seasons, you will undoubtedly pass beds of flowers and see honeybees or butterflies flitting among them, gathering the sweet nectar that powers their flight.

As you observe life, think about the "why" behind what you see. The plants' green color is due to a unique molecule, chlorophyll, that absorbs specific wavelengths of solar energy and uses them to power the life of the plant and synthesize the sugar in the nectar gathered by the bees and butterflies. Showy flowers evolved to entice insects to the energy-rich nectar. Why? If you look carefully at a bee, you

may see yellow pollen clinging to its legs or to the hairs coating its body. The plants "use" the insects to fertilize each other, and both benefit. The sugar in nectar is assembled by chemical reactions that combine carbon dioxide and water, releasing oxygen as a waste product. So as you breathe out air rich in carbon dioxide, you are nourishing the plants with your "waste gas." Conversely, with each breath you take, you are inhaling the life-sustaining "waste gas" from the plants around you: oxygen. Wherever you look, if you look in the right way, you'll see evidence of the interdependence of living things, and you will never take life on Earth for granted.

Living Things Grow

At some time in its life cycle, every organism becomes larger—that is, it *grows*. Although this characteristic is obvious in most animals and plants, even single-celled bacteria grow to about double their original size before they divide. In all cases, growth involves the conversion of materials acquired from the environment into the specific molecules of the organism's body.

Living Things Reproduce Themselves

Organisms reproduce, giving rise to offspring of the same type and creating continuity of life. The processes by which this occurs vary, but the result is the same—the perpetuation of the parents' genes.

Living Things, Collectively, Have the Capacity to Evolve

Populations of organisms evolve in response to changing environments. Although the genetic makeup of a single organism remains essentially the same over its lifetime, the genetic makeup of a population will change over time as a result of natural selection.

1.4   HOW DO SCIENTISTS CATEGORIZE THE DIVERSITY OF LIFE?

Although all living things share the general characteristics discussed earlier, evolution has produced an amazing variety of life-forms. Organisms can be grouped into three major categories, called **domains**: Bacteria, Archaea, and Eukarya. This classification reflects fundamental differences among the cell types that compose these organisms. Members of both the Bacteria and the Archaea usually consist of single, simple cells. Members of the Eukarya have bodies composed of one or more highly complex cells. This domain includes three major subdivisions or **kingdoms**: the Fungi, Plantae, and Animalia, as well as a diverse collection

of mostly single-celled organisms collectively known as "protists" (**FIG. 1-11**). There are exceptions to any simple set of criteria used to characterize the domains and kingdoms, but three characteristics are particularly useful: cell type, the number of cells in each organism, and how it acquires energy (Table 1-1).

Categories within the various kingdoms are phylum, class, order, family, genus, and species. These groupings form a hierarchy in which each category includes all of those below it. Within the final category, the species, all members are so similar that they can interbreed. Biologists use a **binomial system** for naming species. As the word "binomial" suggests, each type of organism is assigned a scientific name that consists of two parts: its genus and its species. The genus name is always capitalized, and the species name is not; both are shown in italics. So *Daphnia longispina*, the water flea in Figure 1-8, is in the genus *Daphnia* (which includes many other "water fleas") and the species *longispina* (which refers to the long spine protruding from its tail end). People are classified as *Homo sapiens*; we are the only members of this genus and species. This binomial system of naming organisms allows scientists worldwide to communicate very precisely about any given organism. In the following paragraphs, we provide a brief introduction to the domains and kingdoms of life. You will learn far more about life's incredible diversity and how it evolved in Unit Three.

The Domains Bacteria and Archaea Consist of Prokaryotic Cells; the Domain Eukarya Is Composed of Eukaryotic Cells

There are two fundamentally different types of cells: **prokaryotic** and **eukaryotic**. *Karyotic* refers to the **nucleus** of a cell, a membrane-enclosed sac containing the cell's genetic material (see Fig. 1-2). *Eu* means "true" in Greek; eukaryotic cells possess a "true" membrane-enclosed nucleus. Eukaryotic cells are generally larger than prokaryotic cells and contain a variety of other organelles, many surrounded by membranes. *Pro* means "before" in Greek; prokaryotic

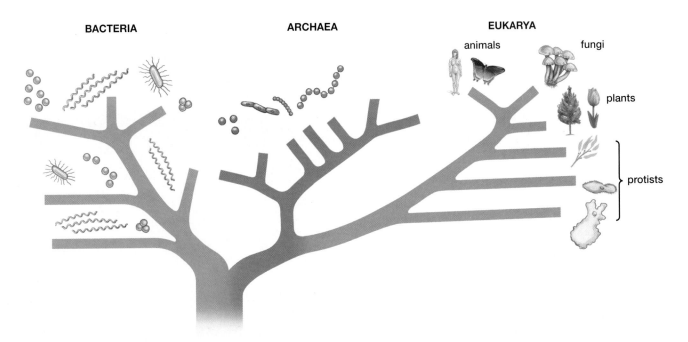

BACTERIA   ARCHAEA   EUKARYA

animals   fungi

plants

protists

FIGURE 1-11 **The domains and kingdoms of life**

cells almost certainly evolved before eukaryotic cells (and, as we will see in Chapter 17, eukaryotic cells almost certainly evolved from prokaryotic cells). Prokaryotic cells do not have a nucleus; their genetic material resides in their cytoplasm. They are usually small—only 1 or 2 micrometers in diameter—and lack membrane-bound organelles. The domains Bacteria and Archaea consist of prokaryotic cells; as its name implies, the cells of Eukarya are eukaryotic.

Bacteria, Archaea, and the Protists Are Mostly Unicellular; Members of the Kingdoms Fungi, Plantae, and Animalia Are Primarily Multicellular

Most members of the domains Bacteria and Archaea and the protists from the domain Eukarya are single-celled, or **unicellular**, although a few live in strands or mats of cells with little communication, cooperation, or organization among them. Most members of the kingdoms Fungi, Plantae, and Animalia are many-celled, or **multicellular**; their lives depend on intimate communication and cooperation among many specialized cells.

Members of the Different Kingdoms Have Different Ways of Acquiring Energy

Photosynthetic organisms—including plants, some bacteria, and some protists—are autotrophic, meaning "self-feeding." Organisms that cannot photosynthesize are heterotrophic, meaning "other-feeding." Many archaea, bacteria, protists, and all fungi and animals are heterotrophs. Heterotrophs differ in the size of the food they eat. Some, such as bacteria and fungi, absorb individual food molecules from outside their bodies; others, including most animals, take in chunks of food (*ingestion*) and break them down to molecules in their digestive tracts.

1.5 HOW DOES KNOWLEDGE OF BIOLOGY ILLUMINATE EVERYDAY LIFE?

Some people regard science as a "dehumanizing" activity, thinking that too deep an understanding of the world robs us of vision and awe. Nothing could be farther from the truth, as we repeatedly discover anew in our own lives.

Table 1-1 Some Characteristics Used in Classification of Organisms

Domain	Kingdom	Cell Type	Cell Number	Energy Acquisition
Bacteria	(Under discussion)	Prokaryotic	Unicellular	Autotrophic or heterotrophic (absorb nutrients)
Archaea	(Under discussion)	Prokaryotic	Unicellular	Heterotrophic (absorb)
Eukarya	Fungi	Eukaryotic	Multicellular	Heterotrophic (absorb)
	Plantae	Eukaryotic	Multicellular	Autotrophic
	Animalia	Eukaryotic	Multicellular	Heterotrophic (ingest)
	"protists"*	Eukaryotic	Uni- and multicellular	Autotrophic or heterotrophic (ingest or absorb)

*The "protists" are a diverse collection of organisms that includes several kingdoms under discussion.

(a)

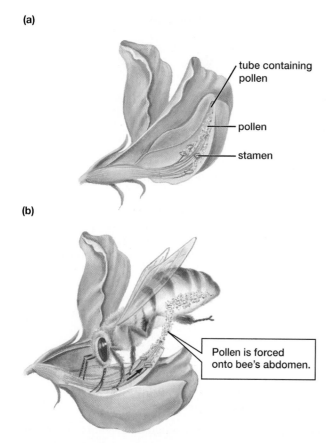

tube containing pollen

pollen

stamen

(b)

Pollen is forced onto bee's abdomen.

FIGURE 1-12 Complex adaptations help ensure pollination
(a) In young lupine flowers, the lower petals form a tube enclosing the reproductive structures, including the male stamens, that shed pollen within the tube. **(b)** The weight of a foraging bee compresses the tube, thrusting the reproductive structures forward and forcing pollen out of the tube onto the bee's abdomen. Some pollen adheres to the abdomen and may come off on the sticky female stigma of the next flower that the bee visits, thus pollinating the flower.

FIGURE 1-13 Wild lupines and subalpine fir trees
Thousands of people visit Hurricane Ridge in Washington State's Olympic National Park each summer to gaze in awe at Mt. Olympus, but few bother to investigate the wonders at their feet.

Years ago, we watched a bee foraging at a spike of lupine flowers. Members of the pea family, lupines have a complicated structure, with two petals on the lower half of the flower enclosing the pollen-laden male reproductive parts (*stamens*) and sticky pollen-capturing female reproductive parts (*stigma*) within a tubelike structure. We had recently learned that in young lupine flowers (**FIG. 1-12a**), the weight of a bee pushing on these petals compresses the stamens, pushing pollen out of the tube and onto the bee's abdomen (**FIG. 1-12b**). In flowers that are ready to be fertilized, the stigma protrudes through the lower petals; when a pollen-dusted bee visits, it usually leaves behind a few grains of pollen.

Did our newfound insights into the functioning of lupine flowers detract from our appreciation of them? Far from it. Rather, we now looked on lupines with new delight, understanding something of the interplay of form and function, bee and flower, that shaped the evolution of the lupine. A few months later we ventured atop Hurricane Ridge in Olympic National Park in Washington State, where the alpine meadows burst with color in August (**FIG.** 1-13). As we crouched beside a wild lupine, an elderly man stopped to ask what we were looking at so intently. He listened with interest as we explained the structure to him; he then went off to another patch of lupines to watch the bees foraging. He too felt the increased sense of wonder that comes with understanding.

Throughout this text, we try to convey to you that sense of understanding and wonder. We also emphasize that biology is not a completed work but an exploration that has really just begun. As Lewis Thomas, a physician and natural philosopher, eloquently stated: "The only solid piece of scientific truth about which I feel totally confident is that we are profoundly ignorant about nature. Indeed, I regard this as the major discovery of the past hundred years of biology … but we are making a beginning."

We cannot urge you strongly enough, even if you are not contemplating a career in biology, to join in the journey of biological discoveries throughout your life. Don't think of biology as just another course to take, just another set of facts to memorize. Biology is a pathway to a new understanding of yourself and of the life on Earth around you.

CASE STUDY REVISITED   LIFE ON EARTH—AND ELSEWHERE?

Is there life on the moon? NASA was not taking any chances. When the *Apollo 11* astronauts who had spent 2.5 hours on the lunar surface splashed down in the ocean on July 24, 1969, a decontamination specialist met them and had them don biological isolation suits while still in the *Apollo 11* module. After the astronauts left the spacecraft, he sterilized the outsides of their isolation suits and the hatch of the spacecraft with disinfectant. The astronauts then stayed in a mobile decontamination unit aboard the recovery ship for four days, until they reached the Johnson Space Center in Houston, Texas. There they remained in quarantine for an additional three weeks.

No foreign microorganisms were found on the astronauts or on the moon rocks they carried back with them. The only microbes found on the moon were discovered by *Apollo 12* astronauts in November 1969. Visiting the unmanned spacecraft *Surveyor 3* that had landed on the moon in 1967, they collected material from inside *Surveyor 3* in a sterile container. From this sample, scientists back on Earth cultured bacteria of the genus *Streptococcus*; ironically, this resident of the human mouth, nose, and throat may have been deposited by a NASA technician who sneezed as he assembled the spacecraft before it was launched. Normally residing in the warm, moist conditions inside the human body, these amazing microbes had, for two years, survived the vacuum of outer space and temperatures as low as –170°F (–110°C).

Astronomers estimate that there may be billions of Earth-like planets in the universe. Thus, the probability is very high that life has evolved elsewhere, although the likelihood of intelligent life is far less certain—and hotly debated. But as an intelligent species, we have hardly begun to understand the diversity, the complexity, and the incredible versatility of life on our own home planet.

Consider This In the late 1970s and 1980s, Dr. James Lovelock, a British chemist, published the controversial and provocative "Gaia hypothesis" (named after the Greek goddess who is said to have brought forth the living world from chaos). Lovelock suggested the living and nonliving components of Earth together constitute a superorganism—an enormous living thing. He noted that the interconnectedness among all forms of life and their environment and the way that living things modify their nonliving surroundings helps maintain conditions conducive to life. Research Lovelock's Gaia hypothesis, either in the library or on the Internet, and discuss how the definition of *life* given in this chapter would need to be changed to accommodate his ideas. Do you believe that Gaia is a useful hypothesis? Is it falsifiable? Should it be elevated to the status of a scientific theory? Explain your answer.

CHAPTER REVIEW

SUMMARY OF KEY CONCEPTS

1.1 How Do Scientists Study Life?

Scientists identify a hierarchy of levels of organizations, as illustrated in Figure 1-1. Biology is based on the scientific principles of natural causality, uniformity in space and time, and common perception. Knowledge in biology is acquired through application of the scientific method, which starts with an observation leading to a question, which leads to a hypothesis. The hypothesis leads to a prediction that is tested by controlled experiments. The experimental results, which must be repeatable, either support or refute the hypothesis, leading to a conclusion about the validity of the hypothesis. A scientific theory is a general explanation of natural phenomena, developed through extensive and reproducible experiments and observations.

Web Tutorial 1.1 Hypothesis Formation and Testing

Web Tutorial 1.2 Spontaneous Generation

1.2 Evolution: The Unifying Theory of Biology

Evolution is the scientific theory that modern organisms descended, with modification, from preexisting life-forms. Evolution occurs as a consequence of genetic variation among members of a population, caused by mutation, inheritance of those variations by offspring, and natural selection of the variations that best adapt an organism to its environment.

1.3 What Are the Characteristics of Living Things?

Organisms possess the following characteristics: their structure is complex and organized; they maintain homeostasis; they acquire energy and materials from the environment; they respond to stimuli; they grow; they reproduce; and they have the capacity to evolve. Most autotrophic organisms capture and store the energy of sunlight in energy-rich molecules by means of photosynthesis. Autotrophs obtain nutrients from their nonliving environment. Heterotrophs obtain all of their energy and most of their nutrients from the bodies of other organisms.

Web Tutorial 1.3 Defining Life

1.4 How Do Scientists Categorize the Diversity of Life?

Organisms can be grouped into three major categories, called domains: Archaea, Bacteria, and Eukarya. Within the Eukarya are three kingdoms, Fungi, Plantae, and Animalia, and unicellular eukaryotes known collectively as "protists." Features used to classify organisms include the cell type (eukaryotic or prokaryotic), cell number (unicellular or multicellular), and energy acquisition (autotrophic or heterotrophic). The genetic material of eukaryotic cells is enclosed within a membrane-bound nucleus. Prokaryotic cells do not have a nucleus. Food obtained by heterotrophic organisms is either ingested in chunks or absorbed molecule by molecule from the environment. The features of the domains and kingdoms are summarized in Table 1-1.

1.5 How Does Knowledge of Biology Illuminate Everyday Life?

The more you know about living things, the more fascinating they become!

KEY TERMS

adaptation *page 10*	deoxyribonucleic acid (DNA) *page 9*	kingdom *page 14*	organ system *page 3*
atom *page 2*		molecule *page 2*	photosynthesis *page 13*
autotroph *page 13*	domain *page 14*	multicellular *page 15*	plasma membrane *page 11*
binomial system *page 14*	element *page 2*	mutation *page 9*	population *page 3*
biodiversity *page 10*	energy *page 13*	natural causality *page 3*	prediction *page 4*
cell *page 3*	eukaryotic *page 14*	natural selection *page 9*	prokaryotic *page 14*
cell theory *page 11*	evolution *page 9*	nucleus *page 14*	question *page 4*
chromosomes *page 9*	experiment *page 4*	nutrient *page 13*	scientific method *page 4*
community *page 3*	gene *page 11*	observation *page 4*	scientific theory *page 8*
conclusion *page 4*	heterotroph *page 13*	organ *page 3*	species *page 3*
control *page 4*	homeostasis *page 11*	organelle *page 11*	tissue *page 3*
cytoplasm *page 11*	hypothesis *page 4*	organic molecule *page 3*	unicellular *page 15*
deoxyribonucleic reasoning *page 9*	inductive reasoning *page 8*	organism *page 3*	variable *page 4*

THINKING THROUGH THE CONCEPTS

1. List the hierarchy of organization of life from an atom to a multicellular organism, briefly explaining each level.

2. What is the difference between a scientific theory and a hypothesis? Explain how scientists use each operation. Why do scientists refer to basic principles as "theories," not "facts"?

3. Explain the differences between inductive and deductive reasoning, and provide an example, real or hypothetical, of each.

4. Describe the scientific method. In what ways do you use the scientific method in everyday life?

5. What are the differences between a salt crystal and a tree? Which is living? How do you know?

6. Define and explain the terms *natural selection*, *evolution*, *mutation*, *creationism*, and *population*.

7. What is evolution? Briefly describe how evolution occurs.

8. Define *homeostasis*. Why must organisms continuously acquire energy and materials from the external environment to maintain homeostasis?

APPLYING THE CONCEPTS

1. Review the properties of life, and then discuss whether humans are unique.

2. Design an experiment to test the effects of a new dog food, "Super Dog," on the thickness and water-shedding properties of the coats of golden retrievers. Include all the parts of a scientific experiment. Design objective methods to assess coat thickness and water-shedding ability.

3. Science is based on principles, including uniformity in space and time and common perception. Assume that humans encounter intelligent beings from a planet in another galaxy where they evolved under very different conditions. Discuss the two principles just mentioned, and explain how they would affect the nature of scientific observations on the different planets as well as communications about these observations.

4. Identify two different types of organisms that you have seen interacting, for example, a caterpillar on a plant such as a milkweed, or a beetle in a flower. Now, form a single, simple hypothesis about this interaction. Use the scientific method and your imagination to design an experiment that tests this hypothesis. Be sure to identify variables and control for them.

5. Explain an instance in which understanding a phenomenon enhances your appreciation of it.

FOR MORE INFORMATION

Dawkins, R. *The Blind Watchmaker*. New York: Norton, 1986. An engagingly written description of the process of evolution, which Dawkins compares to the work of a blind watchmaker.

Leopold, A. *A Sand County Almanac*. New York: Oxford University Press, 1949 (reprinted in 1989). A classic by a natural philosopher; provides an eloquent foundation for the conservation ethic.

Thomas, L. *The Medusa and the Snail*. New York: Bantam Books, 1980, and *The Lives of a Cell*, 1973. The late physician, researcher, and philosopher Lewis Thomas shares his awe of the living world in a series of delightful essays.

Wilson, E. O. *The Diversity of Life*. New York: Norton, 1992. A celebration of the diversity of life, how it evolved, and how humans are impacting it. Wilson's writings have won two Pulitzer prizes.

Zimmer, C. *At the Water's Edge*. New York: The Free Press, 1998. Delightfully written guide to the 4-billion-year journey in time from microbes to people.

The Life of a Cell

Single cells can be complex, independent organisms, such as this protist, a ciliate of the genus *Vorticella*. *Vorticella* consists of a large, round cell body with a mouth at the top. Beating, hairlike cilia protrude from the mouth and create water currents that sweep in food (smaller protists and bacteria). A springy stalk attaches *Vorticella* to objects in its freshwater home. When the cell senses a disturbance, the stalk contracts rapidly and pulls the cell body away from danger.

2

Atoms, Molecules, and Life

Basilisk lizards and ice skaters both exploit unique properties of water.

CASE STUDY   WALKING ON WATER

A YOUNG GIRL IN MEXICO startles a young basilisk lizard as it searches for insects near a small pond. The lizard runs away, moving upright on its strong hind legs. But instead of avoiding the water, it begins to stride across the water's still surface! The young girl gasps—a miracle? Hardly. Natural selection has endowed the basilisk with both speed and specialized feet that allow it to exploit a special property of water: its high surface tension. Put simply, water molecules tend to stick together. With care, you can float a paper clip in a bowl of water, but it will sink instantly in alcohol, which has far less surface tension than water does.

Much farther north, an ice skater whirls at dizzying speed. Frozen water has unique properties that make ice skating both fun and feasible. First, ice is slippery, which allows this athletic feat. Second, ice floats on the top of water, rather than sinking to the bottom. Have you ever wondered why? Most other liquids freeze into denser solids. For example, if the skating pond were filled with oil, the frozen oil would sink to the bottom. Skaters and basilisk lizards are exploiting different and unique properties of water in its liquid and solid phases.

All the diverse molecules that make up living organisms function in a watery environment. But how are water molecules formed? How do water molecules interact with each other and with other forms of matter? What properties give liquid water surface tension, and cause it to expand and become slippery when it freezes?

2.1 WHAT ARE ATOMS?

Atoms, the Basic Structural Units of Matter, Are Composed of Still Smaller Particles

If you cut a diamond (a form of carbon) into pieces, each piece would still be carbon. If you could make finer and finer divisions of these pieces, you would eventually produce a pile of carbon atoms. **Atoms** are the fundamental structural units of matter. Atoms themselves, however, are composed of a central **atomic nucleus** (often called simply the *nucleus*; plural, *nuclei*—don't confuse it with the nucleus of a cell!). The nucleus contains two types of subatomic particles of equal weight: positively charged **protons** and uncharged **neutrons**. Other subatomic particles, called **electrons**, orbit the atomic nucleus (**FIG. 2-1**). Electrons are lighter, negatively charged particles. An atom has an equal number of electrons and protons and is therefore electrically neutral.

There are 92 types of atoms that occur naturally, each forming the structural unit of a different element. An **element** is a substance that can neither be broken down nor converted to other substances by ordinary chemical means. The number of protons in the nucleus—called the **atomic number**—is characteristic of each element. For example, every hydrogen atom has one proton in its nucleus; every carbon atom has six protons; and every oxygen atom has eight. Each element has unique chemical properties based on the number and configuration of its subatomic particles. Some elements, such as oxygen and hydrogen, are gases at room temperature; others, such as lead, are extremely dense solids. Most elements are quite rare, and relatively few are essential to life on Earth. Table 2-1 lists the most common elements in the human body.

Atoms of the same element may have different numbers of neutrons. When this occurs, the atoms are called **isotopes** of each other. Some, but not all, isotopes are **radioactive**; that is, they spontaneously break apart, form-ing different atoms and releasing energy in the process. Radioactive isotopes are extremely useful tools for studying biological processes (see "Scientific Inquiry: Radioactivity in Research").

Electrons Travel Within Specific Regions Called Electron Shells That Correspond to Different Energy Levels

As you may know from experimenting with magnets, like poles repel each other and opposite poles attract each other. In a similar way, electrons, which are negatively charged, repel one another but are attracted to the positively charged protons of the nucleus. However, because of their mutual repulsion, only limited numbers of electrons can occupy the space closest to the nucleus. A large atom can accommodate many electrons because its electrons are found at increasing distances from the nucleus. The electrons move within restricted three-dimensional spaces called **electron shells**. Each electron shell corresponds to a higher energy level as it gets farther from the nucleus. For simplicity, we have drawn these shells as rings around the nucleus (see **FIGS. 2-1** and **2-2**).

The electron shell closest to the atomic nucleus is the smallest and can hold only two electrons. Electrons in this shell are at the lowest energy level. This first shell is the only shell in hydrogen and helium atoms (see Fig. 2-1). The second shell, corresponding to a higher energy level, can hold up to eight electrons. The electrons in an atom fill the shell closest to the nucleus first, and then begin to occupy higher-level shells. Thus, a carbon atom with six electrons has two electrons in the first shell (closest to the nucleus) and four electrons in its second shell (see Fig. 2-2). Although large atoms may have complex energy shells, all atoms important to life (except hydrogen) need (or behave as if they need) eight electrons to fill their outermost shells. This is called the *octet rule*.

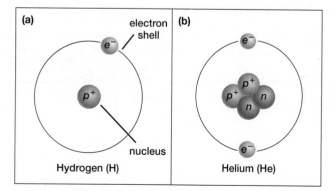

FIGURE 2-1 Atomic models
Structural representations of the two smallest atoms, **(a)** hydrogen and **(b)** helium. In these simplified models, the electrons (pale blue) are represented as miniature planets, circling in specific orbits around a nucleus that contains protons (brown) and neutrons (bluish purple).

Table 2-1 Common Elements in Living Organisms		
Element	Atomic Number[a]	% in Human Body[b]
Hydrogen (H)	1	9.5
Helium (He)	2	Trace
Carbon (C)	6	18.5
Nitrogen (N)	7	3.3
Oxygen (O)	8	65
Sodium (Na)	11	0.2
Magnesium (Mg)	12	0.1
Phosphorus (P)	15	1
Sulfur (S)	16	0.3
Chlorine (Cl)	17	0.2
Potassium (K)	19	0.4
Calcium (Ca)	20	1.5
Iron (Fe)	26	Trace

[a]Atomic number = number of protons in the atomic nucleus.

[b]Approximate percentage of atoms of this element, by weight, in the human body

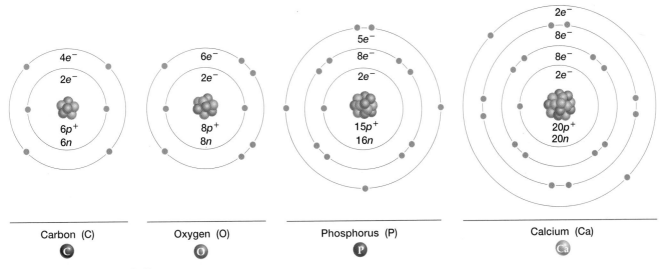

| Carbon (C) | Oxygen (O) | Phosphorus (P) | Calcium (Ca) |

FIGURE 2-2 Electron shells in atoms

Most biologically important atoms have at least two shells of electrons. The first shell, closest to the nucleus, can hold two electrons; the next shell holds a maximum of eight electrons. More-distant shells can hold larger numbers of electrons. QUESTION Why do atoms that tend to react with other atoms have outer shells that are not full?

Nuclei and electron shells play complementary roles in atoms. Nuclei (assuming they are not radioactive) provide stability, while the electron shells allow interactions, or *bonds*, with other atoms. Nuclei resist disturbance by outside forces. Ordinary sources of energy, such as heat, electricity, and light, hardly affect them at all. Because its nucleus is stable, a carbon atom remains carbon whether it is part of a diamond, carbon dioxide, or sugar. Electron shells, however, are dynamic; as you will soon see, atoms bond with one another by gaining, losing, or sharing electrons.

Life Depends on the Ability of Electrons to Capture and Release Energy

Because electron shells correspond to energy levels, when an atom is excited by energy, such as heat or light, this energy causes electrons to jump from a lower-energy electron shell to a higher-energy shell. Soon afterward, the electron spontaneously falls back into its original electron shell, releasing the energy (**FIG. 2-3**).

We make use of this every day. When we switch on a light, electricity flowing through the filament in the bulb heats it, and the heat energy bumps electrons in the metal filament into higher-energy electron shells. As the electrons drop back down into their original electrons' shells, they emit their captured energy as light. Life also depends on the ability of electrons to capture and release energy, as you will learn in Chapters 7 and 8 when we discuss photosynthesis and cellular respiration.

2.2 HOW DO ATOMS INTERACT TO FORM MOLECULES?

Atoms Interact with Other Atoms When There Are Vacancies in Their Outermost Electron Shells

A **molecule** consists of two or more atoms of the same or different elements, held together by interactions among their outermost electron shells. A substance whose molecules are formed of different types of atoms is called a **compound**. Atoms interact with one another according to two basic principles:

- An atom will not react with other atoms when its outermost electron shell is completely full. Such an atom is described as being *inert*.

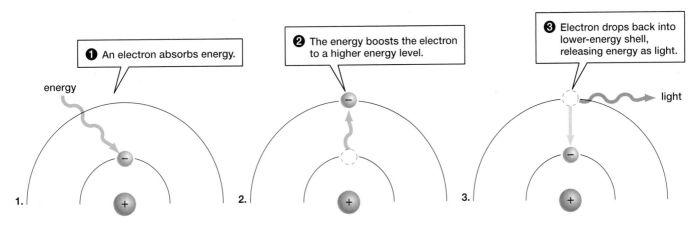

❶ An electron absorbs energy.

❷ The energy boosts the electron to a higher energy level.

❸ Electron drops back into lower-energy shell, releasing energy as light.

energy

light

1.   2.   3.

FIGURE 2-3 Energy capture and release

How do biologists know that DNA is the genetic material of cells (Chapter 9)? How do paleontologists measure the ages of fossils (Chapter 17)? How do botanists know how plants transport sugars made in their leaves during photosynthesis to other parts of the plant (Chapter 42)? These discoveries, and many more, were made possible only through the use of radioactive isotopes. During *radioactive decay*, the process by which a radioactive isotope spontaneously breaks apart, an isotope emits particles that can be detected with devices such as Geiger counters.

A particularly fascinating and medically important use of radioactive isotopes is *positron emission tomography*, or *PET scans* (**FIG. E2-1**). In one common application of PET scans, a subject is given glucose sugar with a harmless radioactive isotope of fluorine attached. As the isotope decays, it emits two bursts of energy that travel in opposite directions. Detectors in a ring around the subject's head capture the emissions, recording the nearly simultaneous arrival times of the two energy bursts from each decaying particle. A powerful computer then calculates the location within the brain where the decay occurred and generates a color-coded map of the frequency of decays within a given

"slice" of the brain. The more active a brain region is, the more glucose it uses as an energy source, and the more radioactivity is concentrated there. For example, tumor cells divide rapidly and use large amounts of glucose; they show up in PET scans as "hot spots" (see Fig. E2-1c). Normal brain regions activated by a specific mental task (for example, a math problem) will also have higher glucose demands that can be detected by PET scans. Thus, physicians can use PET to locate brain problems, while researchers can use it to study what parts of the brain are activated by different mental processes.

The development of PET scans required close cooperation among biologists and physicians (who recognized the need for brain scanning and can interpret the data), chemists (who developed and synthesized the radioactive probes), physicists (who interpreted the nature of isotopes and their energy emissions), and engineers (who designed and built the computers and other electronic components). Continued teamwork among scientists from different fields promises further advances in the fundamental understanding of biological processes as well as more practical applications such as the PET scanner.

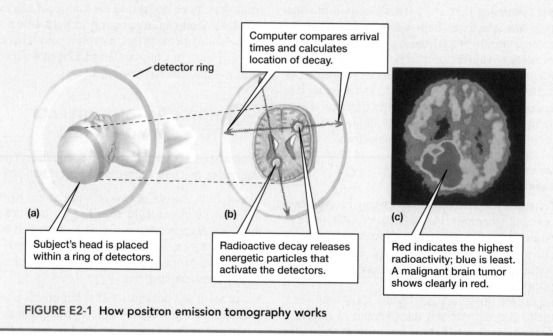

detector ring

Computer compares arrival times and calculates location of decay.

(a) Subject's head is placed within a ring of detectors.

(b) Radioactive decay releases energetic particles that activate the detectors.

(c) Red indicates the highest radioactivity; blue is least. A malignant brain tumor shows clearly in red.

FIGURE E2-1 How positron emission tomography works

- An atom will react with other atoms when its outermost electron shell is only partially full. Such an atom is described as *reactive*.

To demonstrate these principles, consider three types of atoms: hydrogen, helium, and oxygen (see Figs. 2-1 and 2-2). Hydrogen (the smallest atom) has one proton in its nucleus and one electron in its single (and therefore outermost) electron shell, which can hold two electrons. Oxygen has six electrons in its outer shell, which can hold eight. In contrast, helium has two protons in its nucleus, and two electrons fill its single electron shell. Therefore, we can predict that hydrogen and oxygen atoms, both with partially empty outer shells, should be reactive, while helium atoms, with full

shells, should be stable. We might further predict that hydrogen and oxygen atoms could gain stability by reacting with each other. The single electrons from each of two hydrogen atoms would fill the outer shell of an oxygen atom, forming water (H_2O; see Fig. 2-6b). As we predicted, hydrogen reacts vigorously with oxygen. The space shuttle and other rockets use liquid hydrogen as fuel to power liftoff. The hydrogen fuel reacts explosively with oxygen, releasing water as a byproduct as well as tremendous amounts of heat. In contrast, helium, with its full outer shell, is almost completely inert and does not react with other molecules.

An atom with an outermost electron shell that is partially full can gain stability by losing electrons (emptying the shell completely), gaining electrons (filling the shell),

Table 2-2 Common Types of Bonds in Biological Molecules

Type	Interaction	Example
Ionic bond	An electron is transferred, creating positive and negative ions that attract one another.	Occurs between sodium (Na^+) and chloride (Cl^-) ions of table salt (NaCl)
Covalent bond	Electron pairs are shared.	
Nonpolar	Equal sharing	Occurs between the two oxygen atoms in oxygen gas (O_2)
Polar	Unequal sharing	Occurs between the hydrogen and oxygen atoms of a water molecule (H_2O)
Hydrogen bond	The slightly positive charge on a hydrogen atom involved in a polar covalent bond attracts the slightly negative charge on an oxygen or nitrogen atom involved in a polar covalent bond.	Occurs between water molecules; slightly positive charges on hydrogens attract slightly negative charges on oxygens in adjacent molecules

or sharing electrons with another atom (allowing both atoms to behave as though they had full outer shells). The results of losing, gaining, and sharing electrons are **chemical bonds**, which are attractive forces that hold atoms together in molecules. Each element has chemical bonding properties that arise from the configuration of electrons in its outer shell. **Chemical reactions**, the making and breaking of chemical bonds to form new substances, are essential for the maintenance of life and for the working of modern society. Whether they occur in a plant cell as it captures solar energy, your brain as it forms new memories, or your car's engine as it guzzles gas, chemical reactions consist of making new chemical bonds and/or breaking existing ones. There are three major types of chemical bonds: ionic bonds, covalent bonds, and hydrogen bonds (Table 2-2).

Charged Atoms Called Ions Interact to Form Ionic Bonds

Atoms that have an almost empty outermost electron shell, as well as atoms that have an almost full outermost shell, can become stable by losing electrons (emptying their outermost shell) or by gaining electrons (filling their outermost shell). The formation of table salt (sodium chloride) demonstrates this principle. Sodium (Na) has only one electron in its outermost electron shell, and chlorine (Cl) has seven electrons in its outer shell—one electron short of a full shell (**FIG. 2-4a**).

Sodium, therefore, can become stable by losing the electron from its outer shell to chlorine, leaving that shell empty; chlorine then fills its outer shell by gaining the electron. Atoms that have lost or gained electrons, altering the

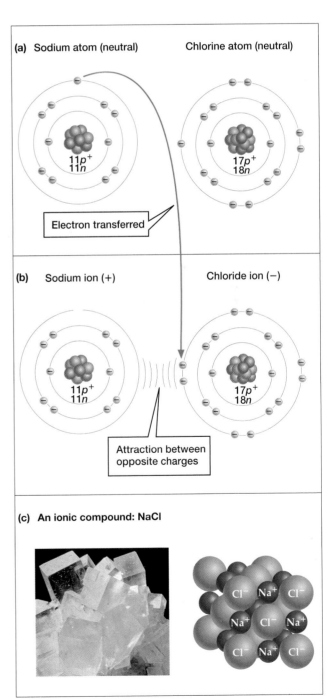

(a) Sodium atom (neutral) Chlorine atom (neutral)

Electron transferred

(b) Sodium ion (+) Chloride ion (−)

Attraction between opposite charges

(c) An ionic compound: NaCl

FIGURE 2-4 The formation of ions and ionic bonds
(a) Sodium has only one electron in its outer electron shell; chlorine has seven. **(b)** Sodium can become stable by losing an electron, and chlorine can become stable by gaining an electron. Sodium then becomes a positively charged ion and chlorine a negatively charged ion. **(c)** Because oppositely charged particles attract one another, the resulting sodium ions (Na^+) and chloride ions (Cl^-) nestle closely together in a crystal of salt, NaCl. (Inset) The organization of ions in salt causes it to form cubic crystals.

balance between protons and electrons, are *charged*. These charged atoms are called **ions**. To form sodium chloride, sodium loses an electron and thereby becomes a positively charged sodium ion (Na^+); chlorine picks up the electron and becomes a negatively charged chloride ion (Cl^-)(**FIG. 2-5**).

FIGURE 2-5 Ionic bond

The two ions are held together by **ionic bonds**: the electrical attraction between positively and negatively charged ions (**FIG. 2-4b**). The ionic bonds between sodium and chloride ions form crystals containing a repeating, orderly arrangement of the two ions; we call this substance "table salt" (**FIG. 2-4c**). As we will see later, water can easily break ionic bonds.

Uncharged Atoms Can Become Stable by Sharing Electrons, Forming Covalent Bonds

An atom with a partially full outermost electron shell can become stable by sharing electrons with another atom, forming a **covalent bond** (**FIG. 2-6**).

Electron Sharing Determines Whether a Covalent Bond Is Nonpolar or Polar

Like two children tugging to gain possession of a teddy bear, electrons in covalent bonds are pulled in opposing directions by the nuclei of the atoms involved. If the children are of equal strength, the teddy bear will remain stretched between them. Likewise, atomic nuclei of equal charge will share electrons equally between them. A covalent bond involving

equal sharing of electrons is called a **nonpolar covalent bond** (**FIG. 2-7**). Consider the hydrogen atom, which has one electron in a shell that can hold two. A hydrogen atom can become reasonably stable if it shares its single electron with another hydrogen atom, forming a molecule of hydrogen gas (H_2) in which each atom behaves almost as if it had two electrons in its outer shell.

(uncharged)

FIGURE 2-7 Nonpolar covalent bond

Two oxygen atoms also share electrons equally, with each atom contributing two electrons to produce a molecule of oxygen gas (O_2) with a double covalent bond. Because the two nuclei in H_2 and in O_2 are identical, their nuclei attract the electron equally, and so the shared electrons spend equal time near each nucleus. Therefore, not only is the molecule as a whole electrically neutral or uncharged, but each end, or *pole*, of the molecule is also uncharged. These molecules and biological molecules such as fats, which are formed with nonpolar covalent bonds, are described as *nonpolar* molecules(see Fig. 2-6a).

In many molecules that form covalent bonds, one nucleus has a larger positive charge than the other and therefore attracts electrons more strongly. Just as a stronger child will be able to pull the teddy bear closer to himself, so the electrons will spend more time near the larger, more positive nucleus and less time near the smaller nucleus. The larger atom thus takes on a slight negative charge (−) from the

(a) Nonpolar covalent bonding

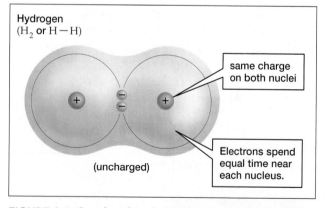

FIGURE 2-6 Covalent bonds involve shared electrons
(a) In hydrogen gas, an electron from each hydrogen atom is shared, forming a single nonpolar covalent bond. **(b)** Oxygen lacks two electrons to fill its outer shell, so oxygen can form polar covalent bonds with two hydrogen atoms, creating water. Oxygen exerts a greater pull on the electrons than does hydrogen, so the "oxygen end" of the molecule has a slight negative charge (−) and the "hydrogen end" has a slight positive charge (+). **QUESTION** In water's polar bonds, why is oxygen's pull on electrons greater than hydrogen's?

(b) Polar covalent bonding

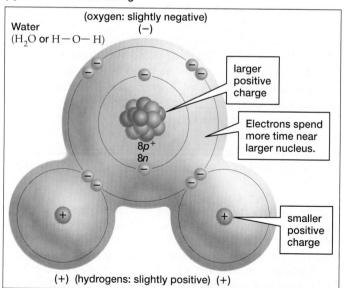

proximity of the electron, and the smaller atom acquires a slight positive charge (+). This situation produces a **polar covalent bond** (**FIG. 2-8**). Although the molecule as a whole is electrically neutral, it has charged poles. In water, for example, oxygen attracts electrons more strongly than does hydrogen, so the oxygen end of a water molecule is slightly negative, and each hydrogen is slightly positive (see Fig. 2-6b). Water is an example of a *polar molecule*.

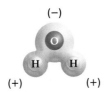

(−)

(+)   (+)

FIGURE 2-8 Polar covalent bonds in water

Most Biological Molecules Utilize Covalent Bonding

Covalent bonds are crucial to life. Since biological molecules must function in a watery environment in which ionic bonds rapidly dissociate, the atoms in most biological molecules, such as those found in proteins, sugars, and cellulose, are joined by covalent bonds. Hydrogen, carbon, oxygen, nitrogen, phosphorus, and sulfur are the most common atoms found in biological molecules. Hydrogen can form a covalent bond with one other atom; oxygen and sulfur with two other atoms; nitrogen with three; and phosphorus and carbon with up to four (Table 2-3). Phosphorus is unusual; although it has only three spaces in its outer shell, it can form up to five covalent bonds with up to four other atoms. This diversity of bonding arrangements allows the construction of biological molecules that have enormous variety and complexity.

Free Radicals Are Highly Reactive and Can Damage Cells

Some reactions, particularly those that occur in cells as they process energy, give rise to molecules that have atoms (often oxygen atoms) with one or more unpaired electrons in their outer shells. This type of molecule, called a **free radical**, is very unstable. Most free radicals react readily with nearby

FIGURE 2-9 Free-radical damage
Aging is partly a result of accumulating free-radical damage to the biological molecules that compose our bodies. For example, sunlight can generate free radicals in skin, damaging molecules that give skin its elasticity and contributing to wrinkles as we age. QUESTION How do free radicals damage biological molecules?

molecules, capturing electrons to fill their outer shells. But when a free radical steals an electron from the molecule it attacks, it creates a new free radical and begins a chain reaction that can lead to the destruction of biological molecules crucial to life. Cell death caused by free radicals contributes to a variety of human ailments, including heart disease and nervous system disorders such as Alzheimer's disease. By damaging genetic material, free radicals may also cause some forms of cancer. The gradual deterioration of the body that accompanies aging is believed by many scientists to result, at least in part, from free-radical damage accumulating over a lifetime of exposure (**FIG. 2-9**). Radiation (such as

Table 2-3 Bonding Patterns of Atoms Commonly Found in Biological Molecules				
Atom	Capacity of Outer Electron Shell	Electrons in Outer Shell	Number of Covalent Bonds Usually Formed	Common Bonding Patterns
Hydrogen	2	1	1	—H
Carbon	8	4	4	—C— —C= =C= —C≡
Nitrogen	8	5	3	—N— —N= —N≡
Oxygen	8	6	2	—O— —O=
Phosphorus	8	5	5	—P—
Sulfur	8	6	2	—S—

from the sun and X-rays), chemicals in automobile exhaust, and industrial metals (such as mercury and lead) can also enter our bodies and generate free radicals. Fortunately, some molecules, called **antioxidants**, react with free radicals and render them harmless. Our bodies synthesize several antioxidants, and others can be obtained from a healthy diet. Vitamins E and C are antioxidants, as are a variety of substances found in fruits and vegetables. To learn about another source of antioxidants, see "Links to Life: Health Food?"

Hydrogen Bonds Are Electrical Attractions Between or Within Molecules with Polar Covalent Bonds

Due to the polar nature of their covalent bonds, nearby polar molecules, such as water molecules, attract one another. The partially negatively charged oxygen atoms of water molecules attract the partially positively charged hydrogen atoms of nearby water molecules. This electrical attraction is called a **hydrogen bond** (FIG. 2-10). Like children holding sweaty hands on a hot day, the individual hydrogen bonds of liquid water are easily broken and reformed, allowing water to flow freely. As we will see shortly, hydrogen bonds between molecules give water several unusual properties that are essential to life on Earth.

Hydrogen bonds are important in biological molecules. They exist in common biological molecules in which hydrogen is bonded to nitrogen or to oxygen, such as occurs in proteins and in DNA. In each case, polar covalent bonds produce a slightly positive charge on a hydrogen atom and a slightly negative charge on the oxygen or nitrogen atom, which both attract electrons more strongly than do hydrogen atoms. The resulting polar parts of the molecules can form hydrogen bonds with water, with other biological molecules, or with polar parts of the same molecule. Although individual hydrogen bonds are quite weak relative to ionic or covalent bonds, many of them working together are quite strong. As we will see in Chapter 3, hydrogen bonds play crucial roles in shaping the three-dimensional structures of proteins. In Chapter 9, you'll discover their importance in DNA.

2.3   WHY IS WATER SO IMPORTANT TO LIFE?

As naturalist Loren Eiseley so eloquently stated, "If there is magic on this planet, it is contained in water." Water is extraordinarily abundant on Earth, has unusual properties, and is so essential to life that it merits special consideration. Life is very likely to have arisen in the waters of the primeval Earth. Living organisms still contain about 60% to 90% water, and all life depends intimately on its properties. Why is water so crucial to life?

Water Interacts with Many Other Molecules

Water is involved in many of the chemical reactions that occur in living cells. The oxygen that green plants release into the air is derived from water during photosynthesis.

FIGURE 2-10 Hydrogen bonds
Like children holding slippery hands, the partial charges on different parts of water molecules produce weak attractive forces called *hydrogen bonds* (dotted lines) between the oxygen and hydrogen atoms in adjacent water molecules. As water flows, these bonds are constantly breaking and re-forming.

When manufacturing a protein, fat, nucleic acid, or sugar, your body produces water; conversely, when you digest proteins, fats, and sugars in the foods you eat, water is used in the reactions. But why is water so important in biological chemical reactions?

Water is an extremely good **solvent**, meaning that it is capable of dissolving a wide range of substances, including protein, salts, and sugars. Water or other solvents containing dissolved substances are called *solutions*. Recall that a crystal of table salt is held together by the electrical attraction between positively charged sodium ions and negatively charged chloride ions (see Fig. 2-4c). Because water is a polar molecule, it has positive and negative poles. If a salt crystal is dropped into water, the positively charged hydrogen ends of the water molecules will be attracted to and will surround the negatively charged chloride ions, and the negatively charged oxygen poles of water molecules will surround the positively charged sodium ions. As water molecules enclose the sodium and chloride ions and shield them from interacting with each other, the ions separate

LINKS TO LIFE   Health Food?

Fruits and vegetables—particularly those with yellow, orange, and red coloring—contain not only vitamins C and E, but other antioxidants as well. But did you know that chocolate (**FIG. E2-2**)—sometimes described as "sinfully delicious" and often a source of guilt for those who indulge in it—contains antioxidants, and so might be a type of health food? Although it is extremely difficult to do controlled studies on the effects of antioxidants in the human diet, there is evidence that diets high in antioxidants may be beneficial. For example, the low incidence of heart disease among the French (many of whom eat a relatively high-fat diet) has been attributed in part to antioxidants in wine, which the French consume regularly. They also eat considerably more fruits and vegetables than Americans do (except for potatoes—fat-laden "french" fries are consumed far more regularly in the United States than in France). Antioxidant supplements abound in nutrition catalogs and in grocery and health food stores. Now, amazingly, researchers have given us an excuse to eat chocolate and feel good about it! Cocoa powder (the dark, bitter powder made from the seeds inside cacao pods; see Fig. E2-2) contains high concentrations of *flavenoids*, which are powerful antioxidants, chemically related to those found in wine. No studies have yet been done to determine whether high consumption of chocolate actually reduces the risk of cancer or heart disease, but there will certainly be no shortage of volunteers for this research. It's important to keep in mind that

the most sinfully delicious chocolates are also high in fat and sugar, and becoming overweight by indulging could counteract any positive effects of the pure cocoa powder. However, slim "chocoholics" have reason to relax and enjoy!

FIGURE E2-2 Chocolate
Cocoa powder is derived from the cacao beans found inside cacao pods (inset), which grow on trees in tropical regions of the Americas.

from the crystal and drift away in the water—thus, the salt dissolves (**FIG. 2-11**).

Water also dissolves molecules held together by polar covalent bonds. The positive and negative poles of water are attracted to oppositely charged regions of dissolving molecules. Ions and polar molecules are termed **hydrophilic** (Greek for "water-loving") because of their electrical attraction for water molecules. Many biological molecules, including sugars and amino acids, are hydrophilic and dissolve readily in water. Water also dissolves nonpolar gases such as oxygen and carbon dioxide. These molecules are small enough that they fit into spaces among the water molecules without disrupting their hydrogen bonds. Fish swimming below the ice on a frozen lake rely on oxygen that dissolved before the ice formed, and the CO_2 they release dissolves in the water. Because it can dissolve such a wide variety of molecules, the watery substance inside a cell provides a suitable environment for the countless chemical reactions essential to life.

Larger molecules with nonpolar covalent bonds usually do not dissolve in water and hence are called **hydrophobic** ("water-fearing"). Nevertheless, water has an important effect on such molecules. Oils, for example, form globules

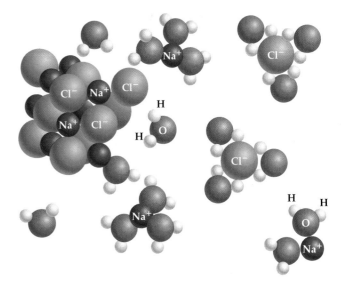

FIGURE 2-11 Water as a solvent
When a salt crystal is dropped into water, the water surrounds the sodium and chloride ions with oppositely charged poles of its molecules. The ions disperse as the surrounding water molecules isolate them from one another, and the crystal gradually dissolves.

FIGURE 2-12 Oil and water don't mix
Yellow oil has just been poured into this beaker of water and is rising to the surface. Oil floats because it is lighter than water, and it forms droplets in water because it is a hydrophobic, nonpolar molecule that is not attracted to the water's polar molecules.

when spilled into water (**FIG. 2-12**)—for an example, look at some chicken soup. Oil molecules in water disrupt the hydrogen bonding among adjacent water molecules. When oil molecules encounter one another in water, their nonpolar surfaces nestle closely together, surrounded by water molecules that form hydrogen bonds with one another but not with the oil. Thus, the oil molecules remain together in droplets. Because oil is lighter than water, these droplets float on the water's surface. The tendency of oil molecules to clump together in water is described as a **hydrophobic interaction**. As we discuss in Chapter 5, the membranes of living cells owe much of their structure to hydrophobic interactions.

Water Molecules Tend to Stick Together

In addition to interacting with other molecules, water molecules interact with each other. Because hydrogen bonds interconnect water molecules, liquid water has high **cohesion**—that is, water molecules have a tendency to stick together. Cohesion among water molecules at the water's surface produces **surface tension**, the tendency for the water surface to resist being broken. If you've ever experienced the slap and sting of a belly flop into a swimming pool, you've discovered firsthand the power of surface tension. Surface tension can support fallen leaves, some spiders and water insects (**FIG. 2-13a**), and even a running basilisk lizard.

Cohesion of water plays a crucial role in the life of land plants. Since a plant absorbs water through its roots, how does the water reach the aboveground parts, especially if the plant is a 100-meter-tall redwood (**FIG. 2-13b**)? As we will see in Chapter 42, water molecules are pulled up by the leaves, filling tiny tubes that connect the leaves, stem, and roots. Water molecules that evaporate from the leaves pull water up the tubes, much like a chain being pulled up from the top. The system works because the hydrogen bonds among water molecules are stronger than the weight of the water in the tubes (even 100 meters' worth); thus, the water "chain" doesn't break. Without the cohesion of water, there would be no land plants as we know them, and terrestrial life would undoubtedly have evolved

FIGURE 2-13 Cohesion among water molecules
(a) Buoyed by surface tension, this fishing spider sprints across the water to capture an insect. **(b)** In giant redwoods, cohesion holds water molecules together in continuous strands from the roots to the topmost leaves as high as 300 feet above the ground.

quite differently. The sting of a belly flop, the ability of a lizard to run on water, and the movement of water up a tree are all made possible by the hydrogen bonds between water molecules.

Water exhibits another property—*adhesion*, or a tendency to stick to polar surfaces having slight charges that attract polar water molecules. Adhesion helps water move within small spaces, such as the thin tubes in plants that carry water from roots to leaves. If you stick the end of a narrow glass tube into water, the water will move a short distance up the tube. Put some water in a narrow glass bud vase or test tube, and you'll see that the upper surface is curved; water pulls itself up the sides of the glass by its adhesion to the surface of the glass and by the cohesion among water molecules.

Water-Based Solutions Can Be Acidic, Basic, or Neutral

Although water is generally regarded as a stable compound, a small fraction of water molecules are ionized. That is, they are broken apart into hydrogen ions (H^+) and hydroxide ions (OH^-)(**FIG. 2-14**).

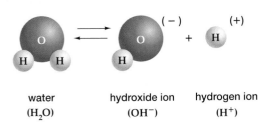

water
(H_2O)

hydroxide ion
(OH^-)

hydrogen ion
(H^+)

FIGURE 2-14 Some water is ionized

A hydroxide ion is negatively charged because it has gained an electron from the hydrogen atom. By losing an electron, the hydrogen atom is converted into a positively charged hydrogen ion. Pure water contains equal concentrations of hydrogen ions and hydroxide ions.

In many solutions, however, the concentrations of H^+ and OH^- are not the same. If the concentration of H^+ exceeds the concentration of OH^-, the solution is **acidic**. An **acid** is a substance that releases hydrogen ions when it is dissolved in water. When hydrochloric acid (HCl), for example, is added to pure water, almost all of the HCl molecules separate into H^+ and Cl^-. Therefore, the concentration of H^+ greatly exceeds the concentration of OH^-, and the resulting solution is acidic. Many acidic substances, such as lemon juice and vinegar, have a sour taste. This is because the sour receptors on your tongue are specialized to respond to the excess of H^+.

If the concentration of OH^- is greater, the solution is **basic**. A **base** is a substance that combines with hydrogen ions, reducing their number. If, for instance, sodium hydroxide (NaOH) is added to water, the NaOH molecules separate into Na^+ and OH^-. Some OH^- ions combine with H^+, reducing their number and creating a basic solution.

The degree of acidity is expressed on the **pH scale** (**FIG. 2-15**), in which neutrality (equal numbers of H^+ and OH^-) is indicated by the number 7. Pure water, with equal concentrations of H^+ and OH^-, has a pH of 7. Acids have a pH below 7; bases have a pH above 7. Each unit on the pH

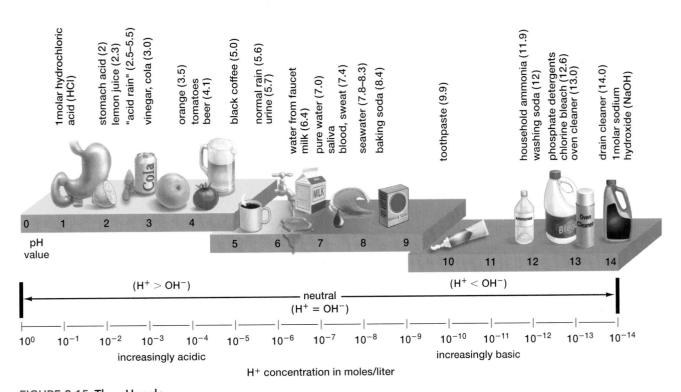

FIGURE 2-15 The pH scale
The pH scale reflects the concentration of hydrogen ions in a solution. The pH (upper scale) is the negative log of the H^+ concentration (lower scale). Each unit on the scale represents a tenfold change. Lemon juice, for example, is about 10 times more acidic than orange juice, and the most severe acid rains in the northeastern United States are almost 1000 times more acidic than normal rainfall. Except for the inside of your stomach, nearly all the fluids in your body are finely adjusted to a pH of about 7.4.

scale represents a tenfold change in the concentration of H^+. Thus, a cola drink with a pH of 3 has a concentration of H^+ 10,000 times that of water, which has a pH of 7.

A Buffer Helps Maintain a Solution at a Relatively Constant pH

In most mammals, including humans, both the cell interior (cytoplasm) and the fluids that bathe the cell are nearly neutral (pH about 7.3 to 7.4). Small increases or decreases in pH may cause drastic changes in both the structure and function of biological molecules, leading to the death of cells or entire organisms. Nevertheless, living cells seethe with chemical reactions that take up or give off H^+. How, then, does the pH remain constant overall? The answer lies in the many buffers found in living organisms. A **buffer** is a compound that tends to maintain a solution at a constant pH by accepting or releasing H^+ in response to small changes in H^+ concentration. If the H^+ concentration rises, buffers combine with them; if the H^+ concentration falls, buffers release H^+. Thus, the original concentration of H^+ is maintained. Common buffers in living organisms include bicarbonate (HCO_3^-) and phosphate ($H_2PO_4^-$ and HPO_4^{2-}), both of which can accept or release H^+ depending on the circumstances. If the blood becomes too acidic, for example, bicarbonate accepts H^+ to form carbonic acid:

$$HCO_3^- \quad + \quad H^+ \quad \rightarrow \quad H_2CO_3$$
$$\text{(bicarbonate)} \quad \text{(hydrogen ion)} \quad \text{(carbonic acid)}$$

If the blood becomes too basic, carbonic acid liberates hydrogen ions, which combine with the excess hydroxide ions, forming water:

$$H_2CO_3 \quad + \quad OH^- \quad \rightarrow \quad HCO_3^- \quad + \quad H_2O$$
$$\text{(carbonic acid)} \quad \text{(hydroxide ion)} \quad \text{(bicarbonate)} \quad \text{(water)}$$

In either case, the result is that the blood pH remains near its normal value.

Water Moderates the Effects of Temperature Changes

Your body and the bodies of other organisms can survive only within a limited temperature range. As we will see in Chapter 6, high temperatures may damage enzymes that guide the chemical reactions essential to life. Low temperatures are also dangerous because enzyme action slows as the temperature drops. Subfreezing temperatures within the body are usually lethal, because spear-like ice crystals can rupture cells. Fortunately, water has important properties that moderate the effects of temperature changes. These properties help keep the bodies of organisms within tolerable temperature limits. Also, large lakes and the oceans have a moderating effect on the climate of nearby land, making it warmer in winter and cooler in summer.

It Takes a Lot of Energy to Heat Water

The energy required to heat a gram of a substance by 1°C is called its *specific heat*. Because of its polar nature and hydrogen bonding, water has a very high specific heat, and therefore water moderates temperature changes. Temperature reflects the speed of molecules; the higher the temperature,

the greater their average speed. Generally, if heat energy enters a system, the molecules of that system move more rapidly, and the temperature of the system rises. Remember that individual water molecules are weakly linked to one another by hydrogen bonds (see Fig. 2-10). When heat enters a watery system such as a lake or a living cell, much of the heat energy initially goes into breaking hydrogen bonds rather than speeding up individual molecules. Thus, it requires more energy to heat water than to heat most other substances by the same amount. For example, 1 **calorie** of energy will raise the temperature of 1 gram of water by 1°C, whereas it takes only 0.02 calorie to heat a gram of rock, such as marble, to that temperature. So the energy required to heat a pound (a pint) of water by 1°C would raise the temperature of a pound of rock by 50°C. For this reason, if a lizard wants to warm up, it will seek out a rock rather than a puddle, because after being exposed to the same amount of heat from the sun, the rock will be a lot warmer. Because the human body is mostly water, a sunbather can absorb a lot of heat energy without sending his or her body temperature soaring (**FIG. 2-16a**).

Water Moderates High and Low Temperatures

Water also moderates the effects of high temperatures because it takes a great deal of heat energy (539 calories per gram) to convert liquid water to water vapor. This, too, is due to the polar nature of water molecules and the hydrogen bonds that interconnect them. For a water molecule to evaporate, it must absorb sufficient energy to make it move quickly enough to break all the hydrogen bonds that hold it to nearby water molecules. Only the fastest-moving water molecules, carrying the most energy, can break their hydrogen bonds and escape into the air as water vapor. The remaining liquid is cooled by the loss of these high-energy molecules. As children romp through a sprinkler on a hot summer day, heat energy is transferred from their skin to the water, which absorbs more energy as it evaporates (**FIG. 2-16b**). When you perspire, the evaporating perspiration produces a considerable loss of heat without much loss of water. The heat required to vaporize water is called its *heat of vaporization*—water's heat of vaporization is one of the highest known.

It Takes a Lot of Energy to Freeze Water

Finally, water moderates the effects of low temperatures because an unusually large amount of energy must be removed from molecules of liquid water before they form the precise crystal arrangement of ice (see the next section). As a result, water freezes more slowly than many other liquids at a given temperature and loses more heat to the environment in the process. This property of a substance is called its *heat of fusion*; water's heat of fusion is very high.

Water Forms an Unusual Solid: Ice

Water will become a solid after prolonged exposure to temperatures below its freezing point. But even solid water is unusual. Most liquids become denser when they solidify; therefore, the solid sinks. Ice is unique because it is less

(a)   (b)

FIGURE 2-16 The high specific heat and heat of vaporization of water affects human behavior
(a) Because our bodies are mostly water, sunbathers can absorb a great deal of heat without sending their body temperatures skyrocketing, a result of water's high specific heat. (b) Water's high heat of vaporization (evaporative cooling) and specific heat together make water a very effective coolant on a hot day.

dense than liquid water. The regular arrangement of water molecules in ice crystals (**FIG. 2-17**) keeps them farther apart than they are in the liquid phase, where they jumble more closely together; thus, ice is less dense than water.

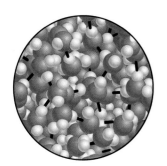 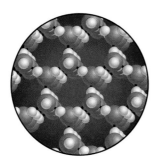

FIGURE 2-17 Water (left) and ice (right)

When a pond or lake starts to freeze in winter, the ice floats on top, forming an insulating layer that delays the freezing of the rest of the water and provides a slippery surface for skaters. This insulation allows fish and other lake residents to survive in the liquid water below. If ice were to sink, many ponds and lakes around the world would freeze solid from the bottom up during the winter, killing plants, fish, and other underwater organisms.

CASE STUDY REVISITED   WALKING ON WATER

Most water-walkers are extremely light-weight insects—a 4-ounce basilisk lizard is probably the heaviest animal able to support itself on its feet while moving across water. When the leaping lizard slaps its foot down hard on the water's surface, resistance caused by surface tension spreads special fringes on the lizard's toes, creating a larger surface area. As the lizard propels itself forward, stroking downward and backward, each of its fringed feet traps and pushes an air bubble behind it. Trapped between the water's surface tension and the lizard's foot, the air pocket acts as a momentary flotation device, providing

support for a fraction of a second before the other foot splashes down ahead and repeats the process.

Ice skaters, meanwhile, exploit the floating property of frozen water. Below their skates, an entire community of lake-dwellers is insulated and protected. But why is ice so slippery? Surprisingly, scientists are not sure. They know that the water molecules in ice crystals are only loosely bonded to one another. Some speculate that the molecules on the icy surface readily move past one another when something solid slides over them, acting like molecular ball bearings. Others hypothesize that slipperiness is due to yet another

unique property of ice: when ice is compressed, it melts. Perhaps under pressure from skates (or shoes, or tires) a microscopically thin layer of water forms and lubricates the surface of the ice.

Consider This Many of water's unique properties are the result of its polar covalent bonds, which allow water molecules to form hydrogen bonds with each other. What if water molecules used nonpolar covalent bonds? What might the implications be? Using the information in this chapter, make a list of ways in which such bonds might affect the properties of water and life on Earth in general.

CHAPTER REVIEW

SUMMARY OF KEY CONCEPTS

2.1   What Are Atoms?

An element is a substance that can neither be broken down nor converted to different substances by ordinary chemical means. The smallest possible particle of an element is the atom, which is itself composed of a central nucleus, containing protons and neutrons, and electrons outside the nucleus. All atoms of a given element have the same number of protons, an amount different from the number of protons in the atoms of every other element. Electrons orbit the nucleus in electron shells within specific regions around the nucleus that correspond to different energy levels, with higher energy levels at increasing distances from the nucleus. Electrons in a lower-energy electron shell can absorb energy from heat, light, or electricity and jump to a higher-energy shell. They then release this energy as light and return to their original shell. Each shell can contain a fixed maximum number of electrons. The chemical reactivity of an atom depends on the number of electrons in its outermost electron shell; an atom is most stable, and therefore least reactive, when its outermost shell is completely full.

Web Tutorial 2.1 Atomic Structure and Chemical Bonding

2.2   How Do Atoms Interact to Form Molecules?

Atoms may combine to form molecules. The forces holding atoms together in molecules are called *chemical bonds*. Atoms

that have lost or gained electrons are negatively or positively charged particles called ions. Ionic bonds are electrical attractions between charged ions, holding them together in crystals. When two atoms share electrons, covalent bonds form. In a nonpolar covalent bond, the two atoms share electrons equally. In a polar covalent bond, one atom may attract the electron more strongly than the other atom does; in this case, the strongly attracting atom bears a slightly negative charge, and the weakly attracting atom bears a slightly positive charge. Some polar covalent bonds give rise to hydrogen bonding, the attraction between charged regions of individual polar molecules or distant parts of a large polar molecule.

2.3   Why Is Water So Important to Life?

Water interacts with many other molecules, and it dissolves many polar and charged substances. Water forces nonpolar substances, such as fat, to assume certain types of physical organization. Water participates in chemical reactions. Water molecules cohere to each other using hydrogen bonds. Because of its high specific heat, heat of vaporization, and heat of fusion, water helps maintain a fairly stable temperature in spite of wide temperature fluctuations in the environment.

Web Tutorial 2.2 Introducing Water's Properties

Web Tutorial 2.3 Specific Heat of Water

KEY TERMS

acid   *page 31*
acidic   *page 31*
antioxidant   *page 28*
atom   *page 22*
atomic nucleus   *page 22*
atomic number   *page 22*
base   *page 31*
basic   *page 31*
buffer   *page 32*
calorie   *page 32*

chemical bond   *page 25*
chemical reaction   *page 25*
cohesion   *page 30*
compound   *page 23*
covalent bond   *page 26*
electron   *page 22*
electron shell   *page 22*
element   *page 22*
free radical   *page 27*

hydrogen bond   *page 28*
hydrophilic   *page 29*
hydrophobic   *page 29*
hydrophobic interaction
   page 30
ion   *page 26*
ionic bond   *page 26*
isotope   *page 22*
molecule   *page 23*

neutron   *page 22*
nonpolar covalent bond
   page 26
pH scale   *page 31*
polar covalent bond   *page 27*
proton   *page 22*
radioactive   *page 22*
solvent   *page 28*
surface tension   *page 30*

THINKING THROUGH THE CONCEPTS

1. What are the six most abundant elements that occur in living organisms?

2. Distinguish among atoms and molecules; elements and compounds; and protons, neutrons, and electrons.

3. Compare and contrast covalent bonds and ionic bonds.

4. Why can water absorb a great amount of heat with little increase in its temperature?

5. Describe how water dissolves a salt. How does this phenomenon compare with the effect of water on a hydrophobic substance such as corn oil?

6. Define *acid*, *base*, and *buffer*. How do buffers reduce changes in pH when hydrogen ions or hydroxide ions are added to a solution? Why is this phenomenon important in organisms?

APPLYING THE CONCEPTS

1. Fats and oils do not dissolve in water; polar and ionic molecules dissolve easily in water. Detergents and soaps help clean by dispersing fats and oils in water so that they can be rinsed away. From your knowledge of the structure of water and the hydrophobic nature of fats, what general chemical structures (for example, polar or nonpolar parts) must a soap or detergent have, and why?

2. What would the effects be for aquatic life if the density of ice were greater than that of liquid water? What would be the impact on terrestrial organisms?

3. How does sweating help you regulate your body temperature? Why do you feel hotter and more uncomfortable on a hot, humid day than on a hot, dry day?

4. Free radicals are commonly formed when animals use oxygen to metabolize sugar to make high-energy molecules. Ross Hardison, a researcher at Pennsylvania State University, stated eloquently: "Keeping oxygen under control while using it in energy production has been one of the great compromises struck in the evolution of life on Earth." What did he mean by this? (You may wish to revisit this question after reading Chapter 8.)

FOR MORE INFORMATION

Eiseley, L. *The Immense Journey*. New York: Vintage Books, 1957. A delightful series of essays by a gifted naturalist and writer.

Glasheen, J. W., and McMahon, T. A. "Running on Water." *Scientific American*, September 1997. Answers the question, "How does the basilisk lizard run on water?"

Matthews, R. "Water: The Quantum Elixir." *New Scientist*, April 8, 2006. What unique properties of water make it so crucial for life on Earth?

Raloff, J. "Chocolate Hearts." *Science News*, March 18, 2000. Describes recent research indicating that chocolate is high in antioxidants.

Storey, K. B., and Storey, J. M. "Frozen and Alive." *Scientific American*, December 1990. By triggering ice formation here, suppressing it there, and loading up their cells with antifreeze molecules, some animals, including certain lizards and frogs, can survive with 60% of their body water frozen solid.

Woodley, R. "The Physics of Ice." *Discover*, June 1999. Ice is such a complicated solid that researchers are still unsure of exactly why it acts in the ways it does.

3

Biological Molecules

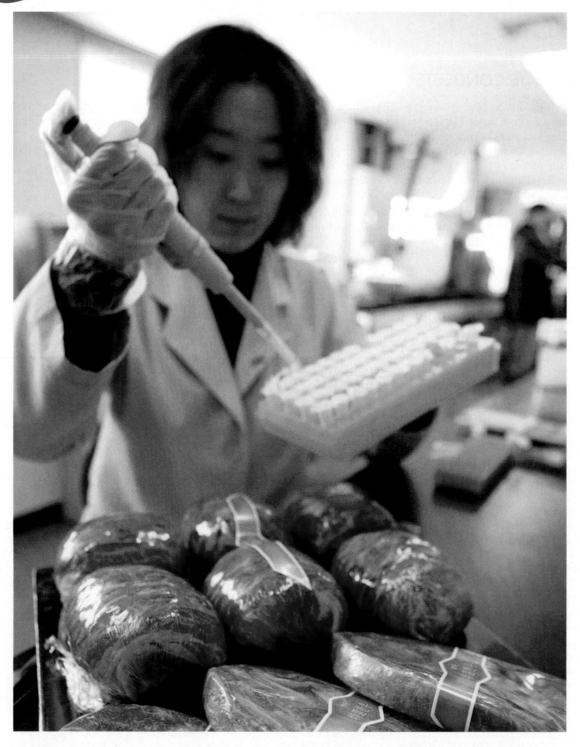

Meat in South Korea is tested to determine its origin after discovery of a case of mad cow disease in a U.S. herd.

AT A GLANCE

CASE STUDY PUZZLING PROTEINS

"YOU KNOW, LISA, I think something is wrong with me," the vibrant, 22-year-old scholarship winner told her sister. It was 2001 and Charlene had been living in the United States for nine years when she began to lose her memory and experience mood swings. During the next year, her symptoms worsened; Charlene's hands shook, she was subject to uncontrollable episodes of biting and hitting, and she became unable to walk. Charlene was a victim of "mad cow disease," almost certainly contracted over 10 years earlier while she was residing in England. In June of 2004, having been bedridden and unable to swallow for two years, Charlene became the first U.S. resident to die of variant Creutzfeldt-Jakob disease (vCJD), the human form of mad cow disease (bovine spongiform encephalitis, or BSE). The daunting scientific name of mad cow disease refers to the spongy appearance of infected cow brains viewed under the microscope. The brains of human vCJD victims, usually young adults, are also riddled with microscopic holes. The human disease is so named because it resembles a long-recognized

malady called Creutzfeldt-Jakob disease. Both CJD and vCJD are always fatal.

Why did cattle suddenly begin to die from BSE? For centuries, it was known that sheep could suffer from a "spongiform encephalitis" called scrapie, which was not transmitted to humans or other livestock. Because the symptoms of BSE strongly resemble scrapie, scientists now believe that a mutated form of scrapie became capable of infecting cattle, perhaps in the early 1980s. The since-discontinued practice of supplementing cattle feed with bone and protein supplements derived from sheep and other livestock probably transmitted the mutated form of scrapie from sheep to cattle. Since BSE was first identified in British cattle in 1986, over 180,000 cattle have been diagnosed with the disease; and millions were slaughtered and their bodies burned as a precaution. It was not until the mid-1990s, about the time that the cattle outbreak was controlled, that officials recognized that the disease could be spread to people who ate meat from infected cattle. Although millions of people are likely to have eaten beef from infected cattle before the danger was

recognized, only about 155 people have died of vCJD worldwide. There is no evidence of transmission between people, except by blood transfusion or organ donation from an infected donor.

Fatal infectious diseases are common—so why does mad cow disease fascinate scientists? In the early 1980s, Dr. Stanley Prusiner, a researcher at the University of California in San Francisco, startled the scientific world by providing evidence that a protein with no genetic material was the cause of scrapie, and that this protein could transmit the disease to experimental animals in the laboratory. He dubbed the infectious proteins "prions" (pronounced PREE-ons), short for "proteinacious infectious particles." Because no infectious agent had ever been identified that lacked all genetic material (DNA or RNA), Prusiner's findings were greeted with enormous skepticism from his fellow scientists.

What are proteins? How do they differ from DNA and RNA? How can a protein with no hereditary material infect another organism and then increase in numbers, resulting in disease?

3.1   WHY IS CARBON SO IMPORTANT IN BIOLOGICAL MOLECULES?

You have probably seen "organic" fruits and vegetables in your grocery store. To a chemist, this phrase is redundant; all produce is organic because it is formed from biological molecules. In chemistry, the term **organic** describes molecules that have a carbon skeleton and contain some hydrogen atoms. The word *organic* is derived from the ability of living *organisms* to synthesize and use this general type of molecule. **Inorganic** molecules include carbon dioxide and all molecules without carbon, such as water and salt.

The versatile carbon atom is the key to the tremendous variety of organic molecules. This variety, in turn, allows the diversity of structures within single organisms and even within individual cells. A carbon atom has four electrons in its outermost shell, with room for eight. Therefore, a carbon atom can become stable by bonding with up to four other atoms, and they can also form double or even triple bonds. Molecules with many carbon atoms can assume complex shapes, including chains, branches, and rings—the basis for an amazing diversity of molecules.

Organic molecules are much more than just complicated skeletons of carbon atoms, however. Attached to the carbon backbone are **functional groups**, groups of atoms that determine the characteristics and chemical reactivity of the molecules. Functional groups are far less stable than the carbon backbone and are more likely to participate in chemical reactions. The common functional groups found in biological molecules are shown in Table 3-1.

The similarity among organic molecules from all forms of life is a consequence of two main features: the use of the same basic set of functional groups in virtually all organic molecules in all types of organisms, and the "modular approach" of synthesizing large organic molecules.

3.2   HOW ARE ORGANIC MOLECULES SYNTHESIZED?

In principle, there are two ways to manufacture a large, complex molecule: by combining atom after atom following an extremely detailed blueprint, or by preassembling smaller molecules and hooking them together. Just as trains are made by coupling engines to various train cars, life also

Table 3-1 Important Functional Groups in Biological Molecules

Group	Structure	Properties	Found In
Hydrogen (—H)	—H	Polar or nonpolar, depending on which atom hydrogen is bonded to; involved in dehydration and hydrolysis reactions	Almost all organic molecules
Hydroxyl (—OH)	—O—H	Polar; involved in dehydration and hydrolysis reactions	Carbohydrates, nucleic acids, alcohols, some acids, and steroids
Carboxylic acid (—COOH)	—C(=O)—O—H	Acidic; involved in peptide bonds	Amino acids, fatty acids
Amino (—NH$_2$)	—N(H)(H)	Basic; may bond an additional H$^+$, becoming positively charged; involved in peptide bonds	Amino acids, nucleic acids
Phosphate (—H$_2$PO$_4$)	—O—P(=O)(O—H)—O—H	Acidic; links nucleotides in nucleic acids; energy-carrier group in ATP	Nucleic acids, phospholipids
Methyl (—CH$_3$)	—C(H)(H)(H)	Nonpolar; tends to make molecules hydrophobic	Many organic molecules; especially common in lipids

takes a "modular approach." Small organic molecules (for example, sugars) are used as subunits that combine to form longer molecules (for example, starches)—like cars in a train. The individual subunits are often called **monomers** (from Greek words meaning "one part"); long chains of monomers are called **polymers** ("many parts").

Biological Molecules Are Joined Together or Broken Apart by Removing or Adding Water

In Chapter 2, you learned some of the reasons that water is so important to life. But water also plays a central role in reactions that break down biological molecules to liberate subunits that the body can use. In addition, when complex biological molecules are synthesized in the body, water is often produced as a by-product.

The subunits that make up large biological molecules almost always join together by means of a chemical reaction called **dehydration synthesis** (literally, "to form by removing water"). In dehydration synthesis, a hydrogen ion (H^+) is removed from one subunit and a hydroxyl ion (OH^-) is removed from a second subunit, creating openings in the outer electron shells of atoms in the two subunits. These openings are filled when the subunits share electrons, creating a covalent bond that links them. The hydrogen ion and the hydroxyl ion then combine to form a molecule of water (H_2O)(**FIG. 3-1**).

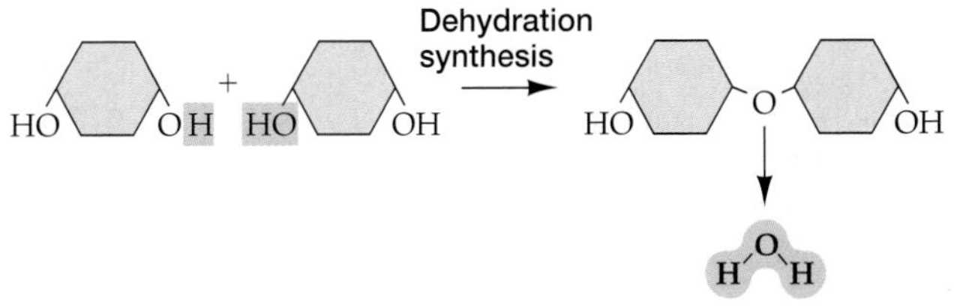

FIGURE 3-1 Dehydration synthesis

The reverse reaction, **hydrolysis** ("to break apart with water"), splits the molecule back into its original subunits (**FIG. 3-2**).

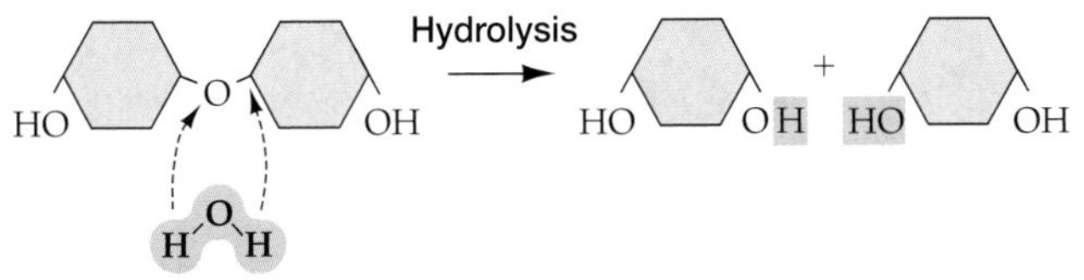

FIGURE 3-2 Hydrolysis

Hydrolysis is the major way our digestive enzymes break down food. For example, the starch in a cracker is composed of a series of glucose (simple sugar) molecules (see Fig. 3-8). Enzymes in our saliva and small intestines promote hydrolysis of the starch into individual sugar molecules that can be absorbed into the body.

Considering how complicated living things are, it might surprise you to learn that nearly all biological molecules fall into one of only four general categories: *carbohydrates, lipids, proteins,* and *nucleic acids* (**Table 3-2**).

3.3 WHAT ARE CARBOHYDRATES?

Carbohydrate molecules are composed of carbon, hydrogen, and oxygen in the approximate ratio of 1:2:1 or CH_2O. This formula explains the origin of the word "carbohydrate," which literally means "carbon plus water." All carbohydrates are either small, water-soluble **sugars** or polymers of sugar, such as starch. If a carbohydrate consists of just one sugar molecule, it is called a **monosaccharide** (Greek for "single sugar"). When two monosaccharides are linked, they form a **disaccharide** ("two sugars"), and a polymer of many monosaccharides is called a **polysaccharide** ("many sugars"). While sugars and starches are used to store energy in many organisms, other carbohydrates provide support. Various types of carbohydrates strengthen plant, fungal, and bacterial cell walls, or form a supportive armor over the bodies of insects, crabs, and their relatives.

The hydroxyl groups of sugars are polar and form hydrogen bonds with water, making sugars water soluble. **FIGURE 3-3** illustrates how a monosaccharide (glucose) forms hydrogen bonds with water molecules.

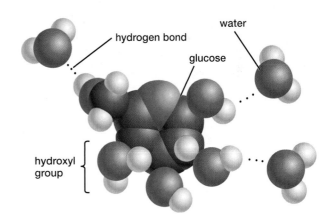

FIGURE 3-3 Sugar dissolving

There Are Several Monosaccharides with Slightly Different Structures

Monosaccharides usually have a backbone of three to seven carbon atoms. Most of these carbon atoms have both a hydrogen (—H) and a hydroxyl group (—OH) attached to them; therefore, carbohydrates generally have the approximate chemical formula $(CH_2O)_n$ (n is the number of carbons in the backbone). When dissolved in water, such as in the cytosol of a cell, the carbon backbone of a sugar usually forms a ring. Sugars in ring form can link together to make disaccharides (see Fig. 3-7) and polysaccharides (see

Table 3-2 The Principal Biological Molecules

Class of Molecule	Principal Subtypes	Example	Function
Carbohydrate: Usually contains carbon, oxygen, and hydrogen, in the approximate formula $(CH_2O)_n$	*Monosaccharide*: Simple sugar with the formula $C_6H_{12}O_6$	Glucose	Important energy source for cells; subunit of polysaccharides
	Disaccharide: Two monosaccharides bonded together	Fructose	Energy-storage molecule in fruits and honey
		Sucrose	Principal sugar transported throughout bodies of land plants
	Polysaccharide: Many monosaccharides (usually glucose) bonded together	Starch	Energy storage in plants
		Glycogen	Energy storage in animals
		Cellulose	Structural material in plants
Lipid: Contains high proportion of carbon and hydrogen; usually nonpolar and insoluble in water	*Triglyceride*: Three fatty acids bonded to glycerol	Oil, fat	Energy storage in animals, some plants
	Wax: Variable numbers of fatty acids bonded to long-chain alcohol	Waxes in plant cuticle	Waterproof covering on leaves and stems of land plants
	Phospholipid: Polar phosphate group and two fatty acids bonded to glycerol	Phosphatidylcholine	Component of cell membranes
	Steroid: Four fused rings of carbon atoms with functional groups attached	Cholesterol	Common component of membranes of eukaryotic cells; precursor for other steroids such as testosterone, bile salts
Protein: Chains of amino acids; contains carbon, hydrogen, oxygen, nitrogen, and sulfur	*Peptide*: Short chain of amino acids	Keratin	Helical protein, principal component of hair
		Silk	Beta-pleated sheet protein produced by silk moths and spiders
	Polypeptide: Long chain of amino acids; also called "protein"	Hemoglobin	Globular protein composed of four subunit peptides; transport of oxygen in vertebrate blood
Nucleic acid: Made of nucleotide subunits containing carbon, hydrogen oxygen, nitrogen, and phosphorus. May consist of a single nucleotide or long chain of nucleotides.	*Long-chain nucleic acids*: polymer of nucleotide subunits	Deoxyribonucleic acid (DNA)	Genetic material of all living cells
		Ribonucleic acid (RNA)	Genetic material of some viruses; in cells, essential in transfer of genetic information from DNA to protein
	Single nucleotides	Adenosine triphosphate (ATP)	Principal short-term energy carrier molecule in cells
		Cyclic adenosine monophosphate (cyclic AMP)	Intracellular messenger

Fig. 3-8). **FIGURES 3-3** and **3-4** show various ways of depicting the chemical structure of the common sugar **glucose**. In many figures, we will use the simplified versions of molecules. Keep in mind that every unlabeled "joint" in a ring is actually a carbon atom.

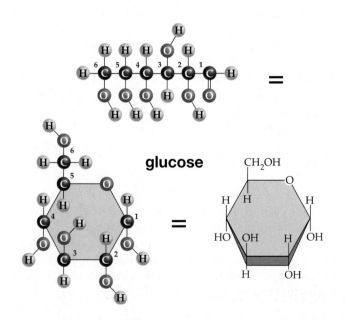

glucose

FIGURE 3-4 Glucose structure
Chemists can represent the same molecule in a variety of ways; glucose is shown here in linear (straight) form and in two different versions of its ring form. Glucose forms a ring when dissolved in water. Notice that each unlabeled joint in a ring structure is a carbon atom.

LINKS TO LIFE   Fake Food?

In societies blessed with an overabundance of food, obesity is a serious health problem. One goal of food scientists is to modify biological molecules to make them noncaloric; sugar and fat are prime candidates. Several artificial sweeteners such as aspartame (Nutrasweet™) and sucralose (Splenda™) add a sweet taste to foods, while providing few or no calories. The artificial oil called olestra is completely indigestible, ensuring that potato chips made with olestra have no fat calories and far fewer total calories than normal chips (**FIG. E3-1**).

How are these "nonbiological molecules" made? Aspartame is a combination of two amino acids: aspartic acid and phenylalanine (see Fig. 3-19). For unknown reasons, aspartame is far more effective than sugar in stimulating the sweet taste receptors on our tongues. Sucralose is a modified sucrose molecule in which three of its hydroxyl groups are replaced with chlorine atoms (**FIG. E3-2**).

Sucralose activates our sweet taste receptors 600 times as effectively as sucrose, but our enzymes cannot digest it, so it provides no calories. Sucralose is gaining in popularity because it is more stable than other artificial sweeteners and can be used in baked goods, as well as in ice cream, diet drinks, and tabletop sweetener packets.

To understand olestra, look at Figure 3-13 and you will see that oils combine a glycerol backbone with three fatty acid chains. Olestra, however, contains a sucrose backbone with six to eight fatty acids attached. Apparently, the large number of fatty acid chains prevents digestive enzymes from reaching the digestible sucrose backbone of the olestra molecule. Since the molecule is not broken into absorbable fragments, it is not digested; but it adds the same appealing feel and flavor to foods that oil does.

FIGURE E3-1 Artificial "foods"
The sucralose in Splenda™ and the olestra in WOW™ potato chips are synthetic, indigestible versions of sugar and oil that are intended to help people lose weight.

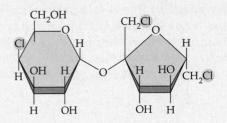

FIGURE E3-2 The structure of Splenda™

Glucose is the most common monosaccharide in living organisms and is a subunit of many polysaccharides. Glucose has six carbons, so its chemical formula is $C_6H_{12}O_6$. Many organisms synthesize other monosaccharides that have the same chemical formula as glucose but slightly different structures. These include *fructose* ("fruit sugar" found in corn syrup, fruits, and honey) and *galactose* (part of lactose, or "milk sugar")(**FIG. 3-5**).

Other common monosaccharides, such as *ribose* and *deoxyribose* (found in DNA and RNA) have five carbons (**FIG. 3-6**).

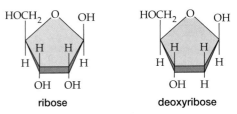

ribose   deoxyribose

FIGURE 3-6 Ribose sugars

Disaccharides Consist of Two Single Sugars Linked by Dehydration Synthesis

Monosaccharides may be broken down in cells to free their chemical energy for use in various cellular activities or may be linked together by dehydration synthesis to

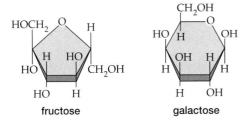

fructose   galactose

FIGURE 3-5 Monosaccharides

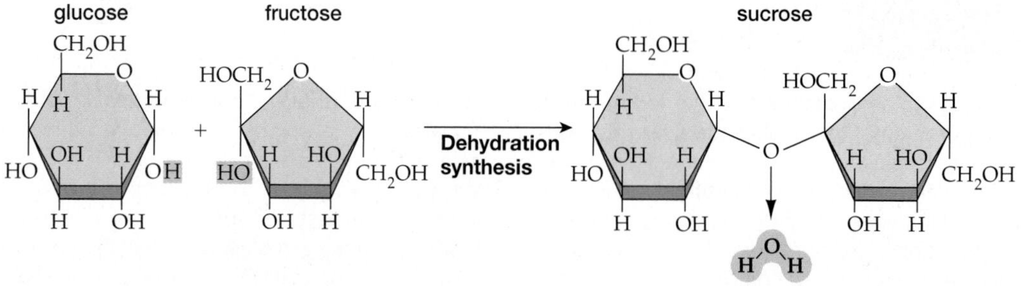

glucose   fructose   sucrose

FIGURE 3-7 Synthesis of a disaccharide
The disaccharide sucrose is synthesized by dehydration synthesis reaction in which a hydrogen (—H) is removed from glucose and a hydroxyl group (—OH) is removed from fructose. A water molecule (H—O—H) is formed in the process, leaving the two monosaccharide rings joined by single bonds to the remaining oxygen atom. Hydrolysis of sucrose is just the reverse of its synthesis—that is, water is split and added to the monosaccharides.

form disaccharides or polysaccharides (**FIG. 3-7**). Disaccharides are often used for short-term energy storage, especially in plants. When energy is required, the disaccharides are broken apart into their monosaccharide subunits by hydrolysis (see Fig. 3-2). Many foods we eat contain disaccharides. Perhaps you had toast and coffee with cream and sugar at breakfast. You stirred **sucrose** (glucose plus fructose, used as an energy storage molecule in sugarcane and sugar beets) into your coffee; added milk containing **lactose** (milk sugar: glucose plus galactose). The disaccharide **maltose** (glucose plus glucose) is rare in nature, but it is formed as enzymes (such as those in your digestive tract) attack and hydrolyze starches such as those in your toast. Digestive enzymes then hydrolyze each maltose into two glucose molecules that

your body can absorb and enzymes in your cells can break down to liberate energy.

If you are on a diet, you may have used an artificial "sugar substitute" such as Splenda™ or Equal™ in your coffee. These interesting molecules are described in "Links to Life: Fake Food?"

Polysaccharides Are Chains of Single Sugars

Try chewing a cracker for a long time. Does it taste sweeter the longer you chew? It should, because over time, enzymes in saliva cause hydrolysis of the **starch** (a polysaccharide) in crackers into its component glucose sugar molecules (monosaccharides), which taste sweet. While plants often

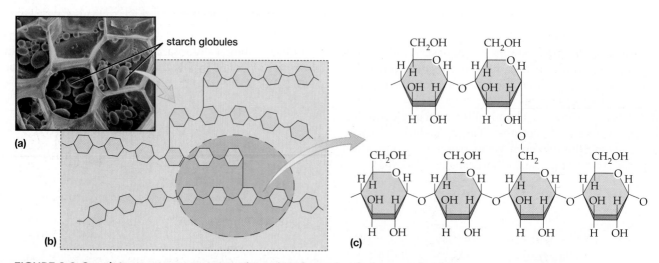

FIGURE 3-8 Starch is an energy-storage polysaccharide made of glucose subunits
(a) Starch globules inside individual potato cells. Most plants synthesize starch, which forms water-insoluble globules consisting of many starch molecules. **(b)** A small section of a single starch molecule, which occurs as branched chains of up to half a million glucose subunits. **(c)** The precise structure of the blue highlighted portion of the starch molecule in (b). Notice the linkage between the individual glucose subunits for comparison with cellulose (see Fig. 3-9).

use starch (**FIG. 3-8**) as an energy-storage molecule, animals commonly store **glycogen**. Both substances consist of polymers of glucose subunits. Starch is commonly formed in roots and seeds—in your cracker, from wheat seeds. Starch often takes the form of branched chains of up to half a million glucose subunits. Glycogen, stored as an energy source in the liver and muscles of animals (including ourselves), is a much shorter chain of glucose subunits with many branches, which probably make it easy to split off the glucose subunits for quick energy release.

Many organisms also use polysaccharides as structural materials. One of the most important structural polysaccharides is **cellulose**, which makes up most of the cell walls of plants, the fluffy white bolls of a cotton plant, and about half the bulk of a tree trunk (**FIG. 3-9**). When you imagine the vast fields and forests that blanket much of our planet, it may not surprise you to learn that there is probably more cellulose on Earth than all other organic molecules combined. Ecologists estimate that about 1 trillion tons of cellulose are synthesized on Earth each year.

Cellulose, like starch, is a polymer of glucose. However, while most animals can easily digest starch, only a few microbes—such as those in the digestive tracts of cows or termites—can digest cellulose. Why is this the case, given that both starch and cellulose consist of glucose? The orientation of the bonds between subunits differs between the two polysaccharides. In cellulose, every other glucose is "upside down" (compare Fig. 3-8c with Fig. 3-9d). This arrangement prevents animals' digestive enzymes from attacking the bonds between glucose subunits. Certain microbes, however, synthesize enzymes that can break these bonds and consume cellulose as food. But for most animals, cellulose passes undigested through the digestive tract. While it serves a valuable function as fiber that prevents constipation, we don't derive any nutrients from it.

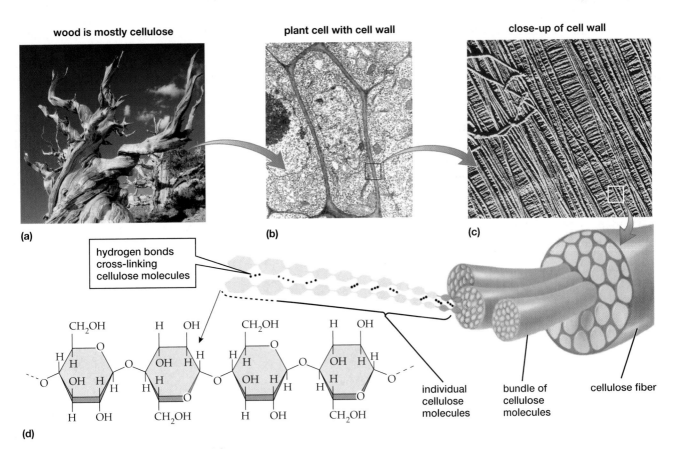

(a)

(b)

(c)

hydrogen bonds cross-linking cellulose molecules

individual cellulose molecules

bundle of cellulose molecules

cellulose fiber

wood is mostly cellulose

plant cell with cell wall

close-up of cell wall

(d)

FIGURE 3-9 Cellulose structure and function
Cellulose can be incredibly tough. **(a)** Wood in this 3000-year-old bristlecone pine is primarily cellulose. **(b)** Cellulose forms the cell wall that surrounds each plant cell. **(c)** Plant cell walls often consist of cellulose fibers in layers that run at angles to each other and resist tearing in both directions. **(d)** Cellulose is composed of glucose subunits. Compare this structure with Fig. 3-8c and notice that every other glucose molecule in cellulose is "upside down." QUESTION Many types of plastic are composed of molecules derived from cellulose, but engineers are working hard to develop plastics based on starch molecules. Why might starch-based plastics be an improvement over existing types of plastic?

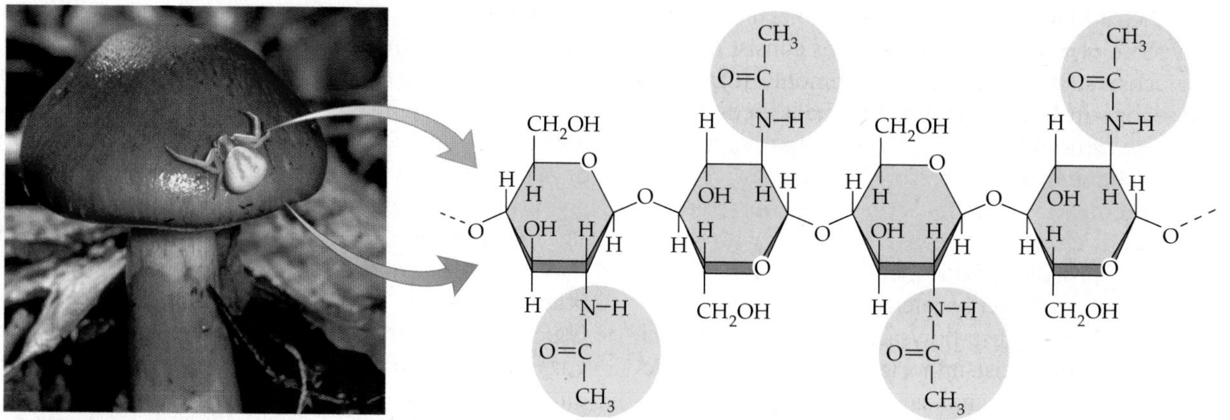

FIGURE 3-10 Chitin: A unique polysaccharide
Chitin has the same bonding configuration of glucose molecules as cellulose does. In chitin, however, the glucose subunits have a nitrogen-containing functional group (yellow) instead of a hydroxyl group. Tough, flexible chitin supports the otherwise soft bodies of arthropods (insects, spiders, and their relatives) and certain fungi, such as this mushroom.

The hard outer coverings (exoskeletons) of insects, crabs, and spiders are made of **chitin**, a polysaccharide in which the glucose subunits bear a nitrogen-containing functional group (**FIG. 3-10**). Interestingly, chitin also stiffens the cell walls of many fungi, including mushrooms. Other types of polysaccharides are found in the cell walls of bacteria, in fluids that lubricate our joints, and in the transparent corneas of our eyes.

A variety of other molecules—including mucus, some chemical messengers called *hormones*, and many molecules in the plasma membrane that surrounds each cell—consist, in part, of carbohydrate. Perhaps the most interesting of these molecules are the nucleic acids (containing sugars) that carry hereditary information. We will discuss these molecules later in this chapter.

3.4   WHAT ARE LIPIDS?

Lipids form a diverse group of molecules with two important features. First, lipids contain large regions composed almost entirely of hydrogen and carbon, with nonpolar carbon–carbon or carbon–hydrogen bonds. Second, these nonpolar regions make lipids hydrophobic and insoluble in water. Lipids serve a wide variety of functions. Some are energy-storage molecules, some form waterproof coverings on plant or animal bodies, some make up the bulk of all the membranes of a cell, and still others are hormones.

Lipids are classified into three major groups: (1) oils, fats, and waxes, which are similar in structure and contain only carbon, hydrogen, and oxygen; (2) phospholipids, which are structurally similar to oils but also contain phosphorus and nitrogen; and (3) the "fused-ring" family of steroids.

Oils, Fats, and Waxes Are Lipids Containing Only Carbon, Hydrogen, and Oxygen

Oils, fats, and waxes are related in three ways. First, they contain only carbon, hydrogen, and oxygen. Second, they contain one or more **fatty acid** subunits, which are long

chains of carbon and hydrogen with a *carboxylic acid group* (—COOH) at one end. Finally, most do not have ring structures. **Fats** and **oils** are formed by dehydration synthesis from three fatty acid subunits and one molecule of **glycerol**, a short, three-carbon molecule (**FIG. 3-11**). This structure gives fats and oils their chemical name: **triglycerides**. Notice that a double bond between two carbons in the fatty acid subunit creates a kink in the chain.

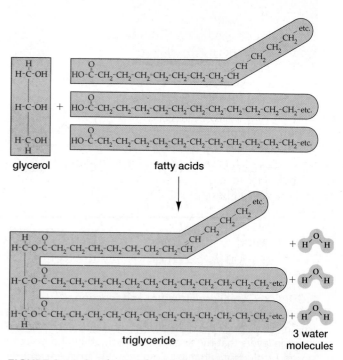

FIGURE 3-11 Synthesis of a triglyceride
Dehydration synthesis links a single glycerol molecule with three fatty acids to form a triglyceride and three water molecules.

(a) Fat

(b) Wax

FIGURE 3-12 Lipids
(a) A fat European brown bear ready to hibernate. If this bear stored the same amount of energy in carbohydrates instead of fat, she probably would be unable to walk! **(b)** Wax is a highly saturated lipid that
remains very firm at normal outdoor temperatures. Its rigidity allows it to be used to form the strong but
thin-walled hexagons of this honeycomb.

Fats and oils contain over twice as many calories per gram as do sugars and proteins, making them very efficient energy-storage molecules for both plants and animals, such as the bear in **FIGURE 3-12a**. People who want to avoid looking like this bear may turn to foods manufactured with fat substitutes such as olestra, described in "Links to Life: Health Food?" Most of the saturated fat in the human diet, such as butter and bacon fat, comes from animals. The difference between a fat (such as beef fat), which is a solid at room temperature, and an oil (such as those used to make potato chips or french fries) lies in their fatty acids. Fats have fatty acids with all single bonds in their carbon chains. Hydrogens occupy all the other bond positions on the carbons. The resulting fatty acid is said to be **saturated**, because it has as many hydrogen atoms as possible. Lacking double bonds between carbons, the carbon chain of the fatty acid is straight. The straight carbon chains of saturated fatty acids of fats (such as the beef fat molecule illustrated below) can nestle closely together, forming solid lumps at room temperature (**FIG. 3-13**).

If there are double bonds between some of the carbons, and consequently fewer hydrogens, the fatty acid is said to be **unsaturated**. Oils have mostly unsaturated fatty acids. We get most of our unsaturated oils from the seeds of plants, where they are stored for the plants' developing embryos. Corn oil, peanut oil, and canola oil are all examples. The double bonds in the unsaturated fatty acids of oils produce kinks in the fatty acid chains, as illustrated by the linseed oil molecule (**FIG. 3-14**).

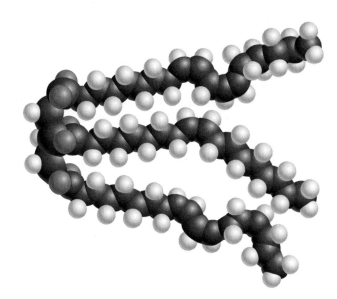

FIGURE 3-14 Linseed oil is unsaturated

The kinks caused by the double bonds in unsaturated fatty acids keep oil molecules apart; as a result, oil is liquid at room temperature. An oil can be converted to a fat by

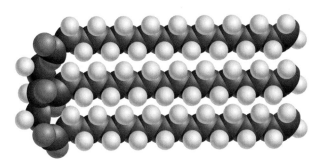

FIGURE 3-13 Beef fat is saturated

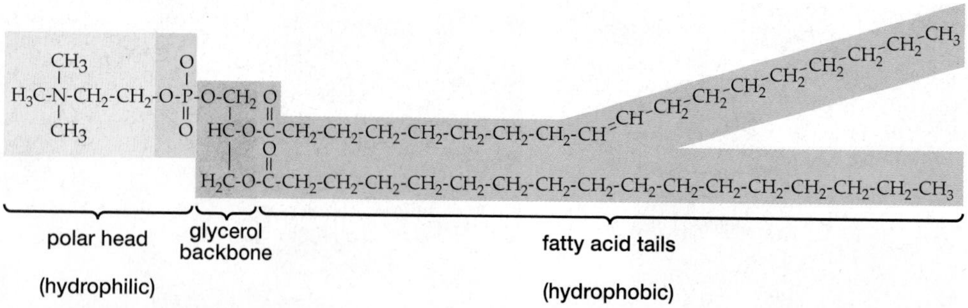

$$CH_3 \qquad O$$
$$H_3C-N-CH_2-CH_2-O-\overset{\overset{O}{\|}}{P}-O-CH_2 \quad O$$
$$CH_3 \qquad O \quad HC-O-C-CH_2-CH_2-CH_2-CH_2-CH_2-CH_2-CH_2-CH \overset{CH-CH_2-CH_2-CH_2-CH_2-CH_2-CH_2-CH_2-CH_3}{}$$
$$\underset{O}{\overset{O}{\|}}$$
$$H_2C-O-C-CH_2-CH_2-CH_2-CH_2-CH_2-CH_2-CH_2-CH_2-CH_2-CH_2-CH_2-CH_2-CH_2-CH_2-CH_3$$

polar head | glycerol backbone | fatty acid tails

(hydrophilic) | (hydrophobic)

FIGURE 3-15 Phospholipids
Phospholipids have two fatty acid tails attached to the glycerol backbone. The third position on glycerol is occupied by a polar "head" consisting of a phosphate group to which a second (usually nitrogen-containing) functional group is attached. The phosphate group bears a slight negative charge, and the nitrogen-containing group bears a slight positive charge, making the heads hydrophilic.

breaking some of the double bonds between carbons, replacing them with single bonds, and adding hydrogens to the remaining bond positions. The resulting substance is the "hydrogenated oil" that causes the oil to be solid at room temperature. The partial hydrogenation process results in a configuration of double and single bonds, called the *trans* configuration, that is rare in nature. In this configuration, the carbon chain bends in a zigzag shape that allows adjacent fatty acids to stack together: the "zigs" of one strand nestle within the "zags" of adjacent strands. This resembles the packing that occurs between the straight-chain fatty acids in saturated fats, and it allows the *trans fats* to assemble into a solid, as do saturated fats. These artificially produced trans fats are found in many commercial food products, including margarine, cookies, crackers, and french fries. Recently, however, researchers have become concerned about our consumption of trans fats (see "Health Watch: Cholesterol—Friend and Foe"). As a result, many manufacturers have made efforts to reduce the use of these substances in processed food.

Although **waxes** are chemically similar to fats, they are not a food source; people and most other animals do not have the appropriate enzymes to break them down. Waxes are highly saturated and therefore solid at normal outdoor temperatures. Waxes form a waterproof coating over the leaves and stems of land plants. Animals synthesize waxes as waterproofing for mammalian fur and insect exoskeletons and, in a few cases, use waxes to build elaborate structures such as beehives (see Fig. 3-12b).

Phospholipids Have Water-Soluble "Heads" and Water-Insoluble "Tails"

The plasma membrane that surrounds each cell contains several types of **phospholipids**. A phospholipid is similar to an oil, except that one of the three fatty acids is replaced by a phosphate group with a short, polar functional group (typically containing nitrogen) attached to the end (**FIG. 3-15**). A phospholipid has two dissimilar ends: two nonpolar fatty acid "tails," which are insoluble in water, and one phosphate–nitrogen "head" that is polar and water soluble.

As you will learn in Chapter 5, the dual nature of phospholipids is crucial to the structure and function of the plasma membrane.

Steroids Consist of Four Carbon Rings Fused Together

Steroids are structurally different from all other lipids. In contrast to other lipids that lack rings, all steroids are composed of four rings of carbon, fused together, with various functional groups protruding from them (**FIG. 3-16**). One type of steroid is *cholesterol*. Cholesterol is a vital component of the membranes of animal cells and is also used by cells to synthesize other steroids, including the male sex

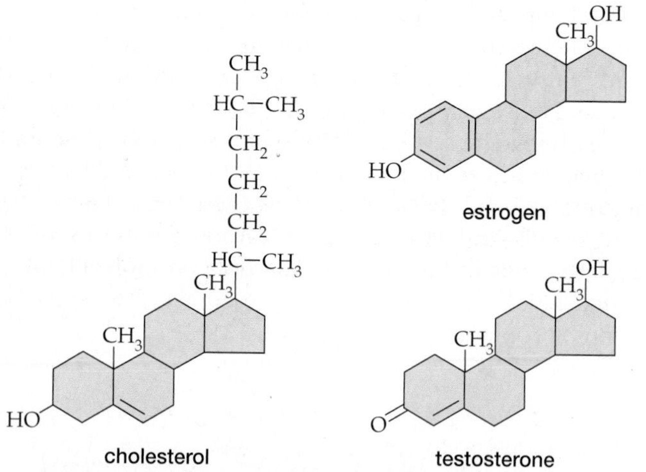

FIGURE 3-16 Steroids
Steroids are synthesized from cholesterol. All steroids have a similar, nonpolar molecular structure (compare the carbon rings). Differences in steroid function result from differences in functional groups attached to the rings. Notice the similarity between the male sex hormone testosterone and the female sex hormone estradiol (estrogen). QUESTION Why are steroid hormones, after traveling in the bloodstream, able to penetrate plasma membranes and nuclear membranes of cells to exert their effects?

HEALTH WATCH   Cholesterol—Friend and Foe

Why are so many foods advertised as "cholesterol free" or "low in cholesterol"? Although cholesterol is crucial to life, medical researchers have found that individuals with high levels of cholesterol in their blood are at increased risk for heart attacks and strokes. Cholesterol contributes to the formation of obstructions in arteries, called *plaques* (**FIG. E3-3**), which in turn can promote the formation of blood clots. If a clot breaks loose and blocks an artery supplying the heart muscle, it can cause a heart attack. If the clot blocks an artery to the brain, it may cause stroke.

Cholesterol comes from animal-derived foods: egg yolks are a particularly rich source; sausages, bacon, whole milk, and butter contain it as well. Perhaps you have heard of "good" and "bad" cholesterol. Because cholesterol molecules are nonpolar, they do not dissolve in blood (which is mostly water). Thus, clusters of cholesterol molecules, surrounded by polar protein carrier molecules and phospholipids, are transported in the blood. These packets of cholesterol plus carriers are called *lipoproteins* (lipids plus proteins). If these lipoproteins have more protein and less lipid, they are described as "high-density lipoproteins," or HDL, because protein is denser than lipid. The HDL form of cholesterol packet is the good kind; these are transported to the liver, where the cholesterol is removed from circulation and further metabolized (used in bile synthesis, for example). In contrast, bad cholesterol is transported in low-density lipoprotein packets ("LDL cholesterol"), which have less protein and more cholesterol. The LDL cholesterol circulates to cells throughout the body and can be deposited in artery walls. A high ratio of HDL ("good") to LDL ("bad") is correlated with a reduced risk of heart disease. A complete cholesterol screening test will distinguish between these two forms in your blood.

Perhaps you've also heard about trans fatty acids as dietary villains. These are not found in nature, but are produced when oils are artificially hardened to make them solid at room temperature. Research has revealed that these trans fatty acids are not metabolized normally and can both increase LDL cholesterol and decrease HDL cholesterol, suggesting that they may place consumers at a higher risk of heart disease. The U.S. Food and Drug Administration now requires food labels to include the trans fat content. In response to health concerns, many food manufacturers and fast-food chains are reducing or eliminating trans fats in their products.

Animals, including people, can synthesize all the cholesterol their bodies require. Roughly 85 percent of the cholesterol in human blood is synthesized by the body, and the other 15 percent comes from diet. Lifestye choices also play a role; exercise tends to increase HDL cholesterol, while obesity and smoking increase LDL levels. Because of genetic differences, some people's bodies manufacture more cholesterol than do others. Studies of identical twins indicate that genetics also influences how much effect diet has on cholesterol levels. Some people's bodies can compensate for a high-cholesterol diet by manufacturing less. Other people compensate poorly, and their diet strongly influences cholesterol levels. Pairs of identical twins share these compensating or non-compensating traits.

People with high cholesterol (about 25 percent of all adults in the United States) are advised to switch to a diet low in cholesterol and saturated fats, to maintain a healthy weight, and to exercise. For people with dangerous levels of cholesterol who are unable to reduce it adequately through lifestyle changes, doctors often prescribe cholesterol-reducing drugs.

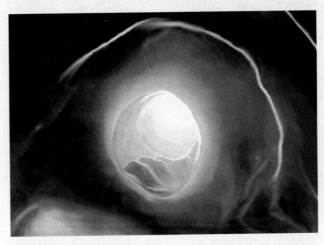

FIGURE E3-3 Plaque
A plaque deposit (ripply structure) partially blocks a carotid artery.

---

hormone testosterone, the female sex hormone estrogen, and bile that assists in fat digestion. However, cholesterol can also contribute to heart disease, as explained in "Health Watch: Cholesterol—Friend and Foe."

3.5   WHAT ARE PROTEINS?

Proteins are molecules composed of one or more chains of *amino acids*. Proteins perform many functions; this diversity of function is made possible by the diversity of protein structures (Table 3-3). Most cells contain hundreds of different **enzymes**, which are proteins that guide

Table 3-3 **Functions of Proteins**	
Function	**Example**
Structure	Collagen in skin; keratin in hair, nails, horns
Movement	Actin and myosin in muscle
Defense	Antibodies in bloodstream
Storage	Albumin in egg white
Signaling	Growth hormone in bloodstream
Catalyzing reactions	Enzymes (Ex.: amylase digests carbohydrates; ATP synthase makes ATP)

(a) Hair

(b) Horn

(c) Silk

FIGURE 3-17 Structural proteins
Common structural proteins include keratin, which is the predominant protein found in **(a)** hair, **(b)** horn, and **(c)** the silk of a spider web.

almost all the chemical reactions occurring inside cells, as you will learn in Chapter 6. Other types of proteins are used for structural purposes. These include *elastin*, which gives skin its elasticity; *keratin*, the principal protein of hair, horns, nails, scales, and feathers; and the silk

of spider webs and silk moth cocoons (**FIG. 3-17**). Still other proteins provide a source of amino acids for developing young animals, such as *albumin* protein in egg white and *casein* protein in milk. The protein *hemoglobin* transports oxygen in the blood, while contractile proteins, such as those in muscle, allow both individual cells and entire animal bodies to move. Some hormones, including insulin and growth hormone, are proteins, as are *antibodies* (which help fight disease and infection) and many poisons (such as rattlesnake venom) produced by animals.

Proteins Are Formed from Chains of Amino Acids

Proteins are polymers of **amino acids**. All amino acids have the same fundamental structure (**FIG. 3-18**), consisting of a central carbon bonded to four different functional groups: a nitrogen-containing amino group ($-NH_2$); a carboxylic acid group ($-COOH$); a hydrogen; and a group that varies among different amino acids (R).

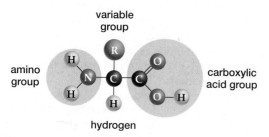

FIGURE 3-18 Amino acid structure

The R group gives each amino acid its distinctive properties (**FIG. 3-19**). Twenty amino acids are commonly found in the proteins of organisms. Some amino acids are hydrophilic and water soluble because their R groups are polar. Others are hydrophobic, with nonpolar R groups that are insoluble in water. The R group of one amino acid, cysteine (Fig. 3-19c), contains sulfur that can form covalent bonds with the sulfur in other cysteines; these bonds are called **disulfide bridges**. Disulfide bridges can link different chains of amino acids to one another or connect different parts of the same amino acid chain, causing the protein to bend or fold. For example, disulfide bridges link chains in the keratin protein of hair, making it curly or straight (see "A Closer Look: A Hairy Subject").

Amino acids differ in their chemical and physical properties—size, water solubility, electrical charge—because of their different R groups. Therefore, the sequence of amino acids largely determines the properties and function of each protein—whether it is water soluble and whether it is an enzyme, a hormone, or a structural protein. In some cases, just one wrong amino acid can cause a protein to function incorrectly.

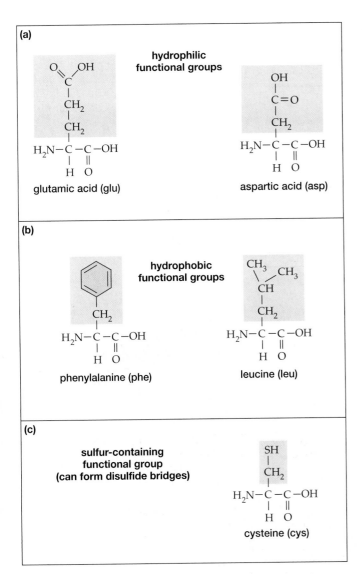

FIGURE 3-19 Amino acid diversity
The diversity of amino acids is a consequence of the variable R group (colored blue), which may be hydrophilic or hydrophobic. The R group of cysteine is unique in possessing a sulfur atom, which can form covalent bonds with the sulfur in other cysteines, creating disulfide bridges that can bend a protein or link adjacent polypeptide chains.

Amino Acids Are Joined to Form Chains by Dehydration Synthesis

Like lipids and polysaccharides, proteins are formed by dehydration synthesis. The nitrogen in the amino group ($-NH_2$) of one amino acid is joined to the carbon in the carboxylic acid group ($-COOH$) of another amino acid by a single covalent bond (**FIG. 3-20**). This bond is called a **peptide bond**, and the resulting chain of two amino acids is called a **peptide**. More amino acids are added, one by one, until the protein is complete. Amino acid chains in living cells vary in length from three to thousands of amino acids. The term *protein* or *polypeptide* is often reserved for long chains—say, 50 or more amino acids in length—and the term *peptide* refers to shorter chains.

A Protein Can Have Up to Four Levels of Structure

Proteins come in many shapes, and biologists recognize four levels of organization in protein structure. A single molecule of hemoglobin, the oxygen-carrying protein in red blood cells, illustrates all four structural levels (**FIG. 3-21**). The **primary structure** refers to the sequence of amino acids that make up the protein (see Fig. 3-21a). This sequence is specified by genes within molecules of DNA. Different types of proteins have different sequences of amino acids.

Polypeptide chains can exhibit two types of simple, repeating **secondary structures**. You may recall that hydrogen bonds can form between parts of polar molecules that have slight negative and slight positive charges, which attract one another (see Chapter 2). Hydrogen bonds between amino acids produce the secondary structures of proteins. Many proteins, such as keratin in hair and the subunits of the hemoglobin molecule (see Fig. 3-21b) have a coiled, springlike secondary structure called a **helix**. Hydrogen bonds that form between the oxygens of $-C=O$ in the carboxylic acid groups (which have a partial negative charge) and the hydrogens of the $-N-H$ in the amino groups (which have a partial positive charge) hold the turns of the coils together. Other proteins, such as silk, consist of polypeptide chains that repeatedly fold back

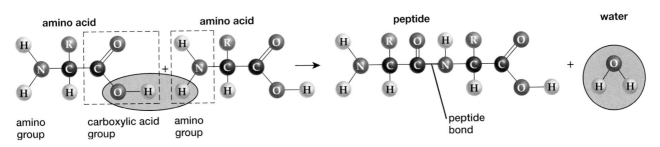

FIGURE 3-20 Protein synthesis
In protein synthesis, a dehydration reaction joins the carbon of the carboxylic acid group of one amino acid to the nitrogen of the amino group of a second amino acid, releasing water. The resulting covalent bond between amino acids is called a *peptide bond*.

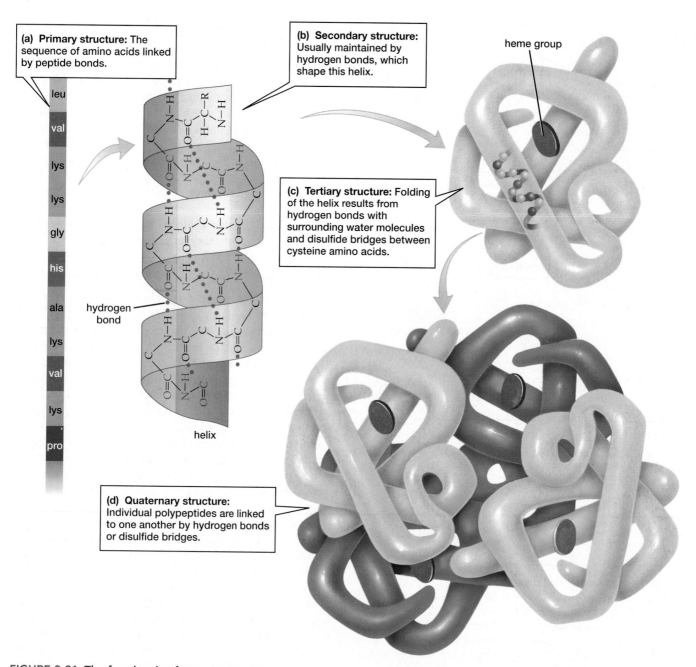

(a) Primary structure: The sequence of amino acids linked by peptide bonds.

(b) Secondary structure: Usually maintained by hydrogen bonds, which shape this helix.

(c) Tertiary structure: Folding of the helix results from hydrogen bonds with surrounding water molecules and disulfide bridges between cysteine amino acids.

(d) Quaternary structure: Individual polypeptides are linked to one another by hydrogen bonds or disulfide bridges.

heme group

hydrogen bond

helix

leu
val
lys
lys
gly
his
ala
lys
val
lys
pro

FIGURE 3-21 The four levels of protein structure
Levels of protein structure are represented here by hemoglobin, the oxygen-carrying protein in red blood cells (red discs represent the iron-containing heme group that binds oxygen). Levels of protein structure are generally determined by the amino acid sequence of the protein, interactions among the R groups of the amino acids, and interactions between the R groups and their surroundings. QUESTION Why do most proteins, when heated, lose their ability to function?

upon themselves, with hydrogen bonds holding adjacent segments of the polypeptide together in a **pleated sheet** arrangement (**FIG. 3-22**).

In addition to their secondary structures, proteins assume complex, three-dimensional **tertiary structures** that determine the final configuration of the polypeptide (see Fig. 3-21c). Probably the most important influence on the tertiary structure of a protein is its cellular environment—specifically, whether the protein is dissolved in the watery cytoplasm within a cell, in the lipids of cellular membranes, or spanning a cell membrane and thus straddling the two environments. Hydrophilic amino acids can form hydrogen bonds with nearby water molecules, whereas hydrophobic amino acids cannot. Therefore, a protein dissolved in water folds in a way that exposes its hydrophilic amino acids to the watery environment outside and causes its hydrophobic amino acids to cluster together in the center of the molecule. Disulfide bridges can also contribute to tertiary

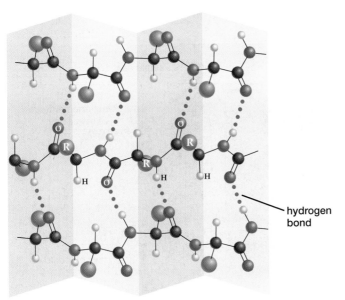

pleated sheet

FIGURE 3-22 The pleated sheet is an example of protein secondary structure
In a pleated sheet, a single polypeptide chain is folded back upon itself repeatedly (connecting portions not shown). Adjacent segments of the folded polypeptide are linked by hydrogen bonds (dotted lines), creating a sheetlike configuration. The R groups (green) project alternately above and below the sheet. Despite its accordion-pleated appearance, produced by bonding patterns between adjacent amino acids, each peptide chain is in a fully extended state and cannot easily be stretched farther. For this reason, pleated sheet proteins such as silk are not elastic.

structure by linking cysteine amino acids from different regions of the polypeptide. In keratin (**FIG. 3-23**), disulfide bridges within individual helical polypeptides can distort them, creating a tertiary structure that makes hair kinky or curly (see "A Closer Look: A Hairy Subject").

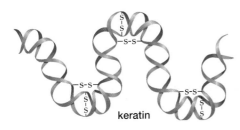

FIGURE 3-23 Keratin structure

Individual polypeptides are sometimes linked together, forming a fourth level of protein organization called **quaternary structure**. Hemoglobin consists of four polypeptide chains (two pairs of very similar peptides) held together by hydrogen bonds (see Fig. 3-21d). Each of the four peptides holds an iron-containing organic molecule called a *heme* (red disks in Fig. 3-21c, d), which can bind one molecule (two atoms) of oxygen.

The Functions of Proteins Are Linked to Their Three-Dimensional Structures

Within a protein, the exact type, position, and number of amino acids bearing specific R groups determine both the structure of the protein and its biological function. In hemoglobin, for example, certain amino acids bearing specific R groups must be present in precisely the right places to hold the iron-containing heme group that binds oxygen. In contrast, the amino acids on the outside of a hemoglobin molecule serve mostly to keep it dissolved in the cytoplasm of a red blood cell. Therefore, as long as they are hydrophilic, changes in these amino acids will not alter the protein's function. As you will see in Chapter 12, replacing a hydrophilic amino acid with a hydrophobic amino acid can have catastrophic effects on the solubility of the hemoglobin molecule. Such a substitution is the molecular cause of a painful and sometimes life-threatening disorder called sickle-cell anemia.

For an amino acid to be in the proper location within a protein, the amino acids must be in their proper sequence, and the protein must have the correct secondary and tertiary structures. For example, enzymes, such as those in your digestive system that break down starch into glucose molecules, are proteins that rely on a precise three-dimensional shape to function properly. The infectious prion protein described in our Case Study has a tertiary structure different from the normal, noninfectious version. If the secondary and tertiary structures of a protein are altered (leaving the peptide bonds between amino acids intact), the protein is said to be **denatured**, and it will no longer perform its function. Although scientists do not yet know what causes the shape change of infectious prions, there are many ways to denature proteins in everyday life. In a fried egg, for example, the heat of the frying pan causes so much movement of the atoms in the albumin protein that hydrogen bonds are ripped apart. Due to the loss of its secondary structure, the egg white changes from clear to white and its texture changes from liquid to solid. Sterilization using heat or ultraviolet rays denatures the proteins of bacteria or viruses and causes them to lose their function. Salty, acidic solutions can denature bacterial proteins, killing the bacteria—dill pickles are preserved from bacterial attack in this way. When hair is permed, the disulfide bridges of keratin are altered, denaturing the natural protein.

A CLOSER LOOK   A Hairy Subject

Pull out a strand of hair, and notice the root or follicle that was embedded in the scalp. Hair is composed mostly of a helical protein called *keratin*. Living cells in the hair follicle produce new keratin at the rate of 10 turns of the protein helix every second. The keratin proteins in a hair entwine around one another, held together by disulfide bridges (**FIG. E3-4**). If you pull gently on the hair, you will find it to be both strong and stretchy. When hair stretches, hydrogen bonds that create the helical structure of keratin are broken, allowing the strand of protein to be extended. Most of the covalent disulfide bonds, in contrast, are distorted by stretching but do not break. When tension is released, these disulfide bridges return the hair to its normal length, and the hydrogen bonds reform. Now wet the hair and notice how limp it becomes. In wet hair, the hydrogen bonds of the helices are broken and replaced by hydrogen bonds between the hair amino acids and the water molecules surrounding them, so the protein is denatured and the helices collapse. Notice that the hair is now both longer and easier to stretch, as well. If you roll the wet hair around a curler and allow it to dry, the hydrogen bonds will re-form in slightly different places, holding the hair in a curl. However, the slightest moisture (even humid air) allows these hydrogen bonds to rearrange into their natural configuration.

If your hair is naturally curly (because of the particular sequence of amino acids specified by your genes), the disulfide bridges within and between the individual keratin helices form at locations that bend the keratin molecules, producing a curl (**FIG. E3-5**).

In straight hair, the disulfide bridges occur in places that do not distort the keratin (as shown in Fig. E3-4). When straight hair is given a "permanent wave," two lotions are applied. The first lotion breaks disulfide bonds, denaturing the protein. After the hair is rolled tightly onto curlers, a second solution is applied that re-forms the disulfide bridges.

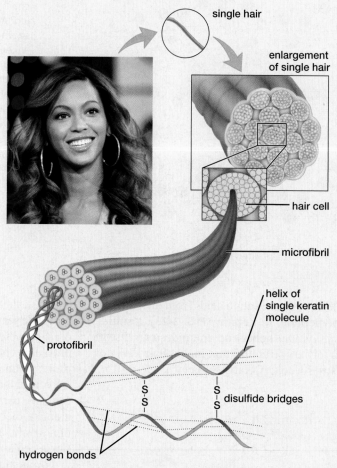

FIGURE E3-4 The structure of hair
At the microscopic level, a single hair is organized into bundles of "protofibrils" within larger bundles called "microfibrils." Each protofibril consists of keratin molecules held in a helical shape by hydrogen bonds, with different keratin strands cross-linked by disulfide bridges. These bonds give hair both elasticity and strength.

The new disulfide bridges reconnect the keratin helices at new positions determined by the curler, as in the curly hair shown in Figure E3-5. These new bridges are permanent, transforming genetically straight hair into "biochemically curly" hair.

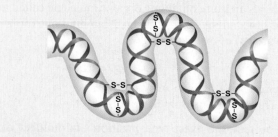

FIGURE E3-5 Curly hair

3.6 WHAT ARE NUCLEIC ACIDS?

Nucleic acids are long chains of similar subunits called **nucleotides**. All nucleotides have a three-part structure: a five-carbon sugar (ribose or deoxyribose), a phosphate group, and a nitrogen-containing base that differs among nucleotides; the base adenine is illustrated below (**FIG. 3-24**).

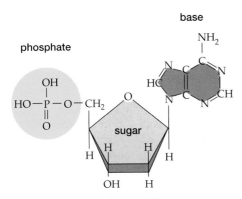

FIGURE 3-24 Deoxyribose nucleotide

There are two types of nucleotides: ribose nucleotides (containing the 5-carbon sugar ribose) and deoxyribose nucleotides (containing the sugar deoxyribose, which has one fewer oxygen than ribose has). The base component of a deoxyribose nucleotide can be adenine, guanine, cytosine, or thymine. Like the adenine molecule in Fig. 3-24, these all have carbon- and nitrogen-containing rings. In adenine and guanine the rings are double, while cytosine and thymine have single-ring structures.

Nucleotides may be strung together in long chains (**FIG. 3-25**), forming nucleic acids. In nucleic acids, the phosphate group of one nucleotide is covalently bonded to the sugar of another.

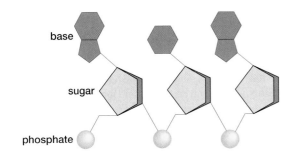

FIGURE 3-25 Nucleotide chain

DNA and RNA, the Molecules of Heredity, Are Nucleic Acids

Deoxyribose nucleotides form chains millions of units long, called **deoxyribonucleic acid**, or **DNA**. DNA is found in the chromosomes of all living things. Its sequence of nucleotides, like the dots and dashes of a biological Morse code, spells out the genetic information needed to construct

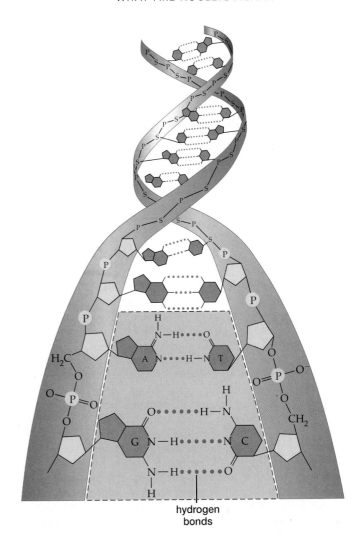

FIGURE 3-26 DNA
Like a twisted ladder, the double helix of DNA is formed by helical strands of nucleotides that spiral around one another. The two strands are held together by hydrogen bonds linking the bases of nucleotides in opposing strands, which form the "rungs" of the ladder.

the proteins of each organism. Each DNA molecule consists of two chains of nucleotides entwined in the form of a double helix. The nucleotides on the opposing strands form hydrogen bonds with one another, linking the two strands (**FIG. 3-26**). Chains of ribose nucleotides, called **ribonucleic acid**, or **RNA**, are copied from the DNA in the nucleus of each cell. The RNA carries DNA's genetic code into the cell's cytoplasm and directs the synthesis of proteins. DNA and RNA are described in detail in Chapters 9 and 10.

Other Nucleotides Act as Intracellular Messengers and Energy Carriers

Not all nucleotides are part of nucleic acids. Some exist singly in the cell or occur as parts of other molecules. Some, such as the nucleotide *cyclic AMP*, are intracellular messengers that carry chemical signals within a cell.

Other nucleotides have additional phosphate groups. These diphosphate and triphosphate nucleotides, such as **adenosine triphosphate (ATP)**, are unstable molecules that carry energy from place to place within a cell, storing the energy in bonds between the phosphate groups (**FIG. 3-27**).

Nucleotides such as ATP can release energy to drive energy-demanding reactions (to synthesize a protein, for example). Other nucleotides (NAD^+ and FAD) are known as "electron carriers" that transport energy in the form of high-energy electrons. You will learn more about these nucleotides in Chapters 6, 7, and 8.

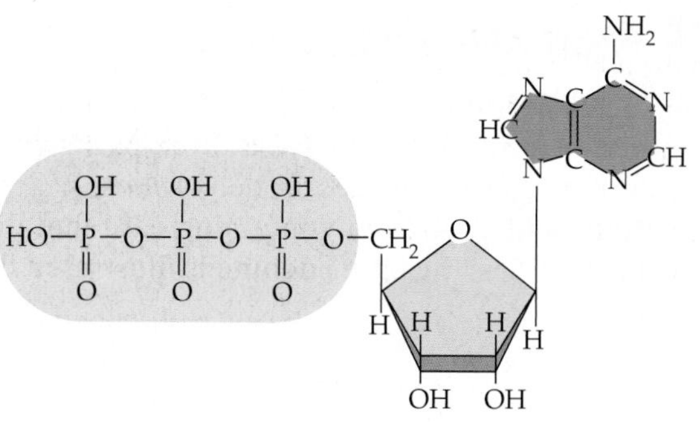

FIGURE 3-27 The energy-carrier molecule ATP

CASE STUDY REVISITED   PUZZLING PROTEINS

Prusiner and his associates identified a protein found normally throughout the animal kingdom as the culprit in scrapie, and now in mad cow disease. But the infectious prions, with the same amino acid sequence, were different. You now know that the three-dimensional folding of a protein is crucial to its proper functioning. Infectious prions are folded very differently from normal prion proteins, which do not cause disease. Like Transformer™ toys, the same structure can take on extremely different appearances. Unfortunately, the infectious prions resist attack by both heat and enzymes that break down their normal counterparts. How does the infectious prion "reproduce" itself? Prusiner and other researchers have strong experimen-

tal evidence supporting a radical hypothesis: the misfolded protein interacts with normal proteins and causes them to change configuration, misfolding into the infectious form. The new misfolded "converts" then go on to transform additional normal proteins in an ever-expanding chain reaction. As in Charlene's case, it can be years after infection that enough of the proteins have been transformed to cause disease symptoms. Scientists are investigating how the abnormal folding occurs, how it can induce more misfolding, and why misfolded prions cause disease.

Stanley Prusiner's efforts led to the recognition of a totally new disease process. A major strength of the scientific method is that hypotheses can be tested experimentally. If repeated experiments

support the hypothesis, then even well-established scientific principles—for example, that an infectious agent must have genetic material—must be redefined. Although some scientists still insist that proteins can't be infectious, Prusiner's efforts have been so convincing to the scientific community that he was awarded a Nobel Prize in 1997.

Consider This A disorder called "chronic wasting disease" has been reported among both wild and captive elk and deer populations in several western states in the United States. Like scrapie and BSE, wasting disease is a fatal brain disorder caused by prions. No human cases have been confirmed. If you were a hunter in an affected region, would you eat deer or elk meat? Explain why or why not.

CHAPTER REVIEW

SUMMARY OF KEY CONCEPTS

3.1 Why Is Carbon So Important in Biological Molecules?

Organic molecules are so diverse because the carbon atom is able to form many types of bonds. This ability, in turn, allows organic molecules (molecules with a backbone of carbon) to form many complex shapes, including chains, branches, and rings. The presence of functional groups, shown in Table 3-1, produces further diversity among biological molecules.

3.2 How Are Organic Molecules Synthesized?

Most large biological molecules are polymers synthesized by linking together many smaller subunits, or monomers. The subunits are connected by covalent bonds through dehydration synthesis; the chains may be broken apart by hydrolysis reactions. The most important organic molecules fall into one of four classes: carbohydrates, lipids, proteins, and nucleic acids. Their major characteristics are summarized in Table 3-2.

Web Tutorial 3.1 Dehydration Synthesis and Hydrolysis

3.3 What Are Carbohydrates?

Carbohydrates include sugars, starches, chitin, and cellulose. Sugars (monosaccharides and disaccharides) are used for temporary

storage of energy and for the construction of other molecules. Starches and glycogen are polysaccharides that provide longer-term energy storage in plants and animals, respectively. Cellulose forms the cell walls of plants, and chitin strengthens the exoskeletons of many invertebrates and many types of fungi. Other types of polysaccharides form the cell walls of bacteria.

Web Tutorial 3.2 Carbohydrate Structure and Function

3.4 What Are Lipids?

Lipids are nonpolar, water-insoluble molecules of diverse chemical structure. Lipids include oils, fats, waxes, phospholipids, and steroids. Lipids are used for energy storage (oils and fats), as waterproofing for the outside of many plants and animals (waxes), as the principal component of cellular membranes (phospholipids), and as hormones (steroids).

Web Tutorial 3.3 Lipid Structure and Function

3.5 What Are Proteins?

Proteins are chains of amino acids that possess primary, secondary, tertiary, and sometimes quaternary structure. Both the structure and the function of a protein are determined by the sequence of amino acids in the chain as well as by how these amino acids interact with their surroundings and with each other. Proteins can be enzymes (which promote and guide chemical reactions), structural molecules (hair, horn), hormones (insulin), or transport molecules (hemoglobin).

Web Tutorial 3.4 Protein Structure

3.6 What Are Nucleic Acids?

The nucleic acid molecules deoxyribonucleic acid (DNA) and ribonucleic acid (RNA) are chains of nucleotides. Each nucleotide is composed of a phosphate group, a sugar group, and a nitrogen-containing base. Molecules formed from single nucleotides include intracellular messengers, such as cyclic AMP, and energy-carrier molecules, such as ATP.

Web Tutorial 3.5 The Structure of DNA

KEY TERMS

adenosine triphosphate (ATP) *page 54*
amino acid *page 48*
carbohydrate *page 39*
cellulose *page 43*
chitin *page 44*
dehydration synthesis *page 39*
denatured *page 51*
deoxyribonucleic acid (DNA) *page 53*
disaccharide *page 39*
disulfide bridge *page 48*
enzyme *page 47*

fat *page 44*
fatty acid *page 44*
functional group *page 38*
glucose *page 40*
glycerol *page 44*
glycogen *page 43*
helix *page 49*
hydrolysis *page 39*
inorganic *page 38*
lactose *page 42*
lipid *page 44*
maltose *page 42*
monomer *page 39*
monosaccharide *page 39*

nucleic acid *page 53*
nucleotide *page 53*
oil *page 44*
organic *page 38*
peptide *page 49*
peptide bond *page 49*
phospholipid *page 46*
pleated sheet *page 50*
polymer *page 39*
polysaccharide *page 39*
primary structure *page 49*
protein *page 47*
quaternary structure *page 51*

ribonucleic acid (RNA) *page 53*
saturated *page 45*
secondary structure *page 49*
starch *page 42*
steroid *page 46*
sucrose *page 42*
sugar *page 39*
tertiary structure *page 50*
triglyceride *page 44*
unsaturated *page 45*
wax *page 46*

THINKING THROUGH THE CONCEPTS

1. Which elements are common components of biological molecules?

2. List the four principal types of biological molecules and give an example of each.

3. What roles do nucleotides play in living organisms?

4. One way to convert corn oil to margarine (a solid at room temperature) is to add hydrogen atoms, decreasing the number of double bonds in the molecules of oil. What is this process called? Why does it work?

5. Describe and compare dehydration synthesis and hydrolysis. Give an example of a substance formed by each chemical reaction, and describe the specific reaction in each instance.

6. Distinguish among the following: monosaccharide, disaccharide, and polysaccharide. Give two examples of each, and list their functions.

7. Describe the synthesis of a protein from amino acids. Then describe primary, secondary, tertiary, and quaternary structures of a protein.

8. Most structurally supportive materials in plants and animals are polymers of special sorts. Where would we find cellulose? Chitin? In what way(s) are these two polymers similar? Different?

9. Which kinds of bonds or bridges between keratin molecules are altered when hair is (a) wet and allowed to dry on curlers and (b) given a permanent wave?

APPLYING THE CONCEPTS

1. A preview question for Chapter 4: In Chapter 2, you learned that hydrophobic molecules tend to cluster when immersed in water. In this chapter, you read that a phospholipid has a hydrophilic head and hydrophobic tails. What do you think would be the configuration of phospholipids that are immersed in water?

2. Fat contains twice as many calories per unit weight as carbohydrate does, so fat is an efficient way for animals, who must move about, to store energy. Compare the way fat and carbohydrates interact with water, and explain why this interaction also gives fat an advantage for weight-efficient energy storage.

FOR MORE INFORMATION

Burdick, A., "Cement on the Half Shell." *Discover*, February 2003. Mussels produce a protein polymer that is waterproof and incredibly strong.

Gorman J. "Trans Fat." *Science News*, November 10, 2001. Reviews the structure and origin of trans fats and studies that implicate it in heart disease.

Hill, J. W., and Kolb, D. K. *Chemistry for Changing Times*. 10th ed. Upper Saddle River, NJ: Prentice Hall, 2004. A chemistry textbook for nonscience majors that is both clearly readable and thoroughly enjoyable.

King, J., Haase-Pettingell, C., and Gossard, D. "Protein Folding and Misfolding." *American Scientist*, September–October 2002. Protein folding holds the key to diverse functions.

Kunzig, R., "Arachnomania." *Discover*, September 2001. Researchers work to unravel the mystery of spider silk, and develop a process to synthesize it.

Prusiner, S. B. "Detecting Mad Cow Disease." *Scientific American*, July, 2004. A clear discussion of prions, mad cow disease, and techniques for detection and possible future treatments, written by the discoverer of prions.

4 Cell Structure and Function

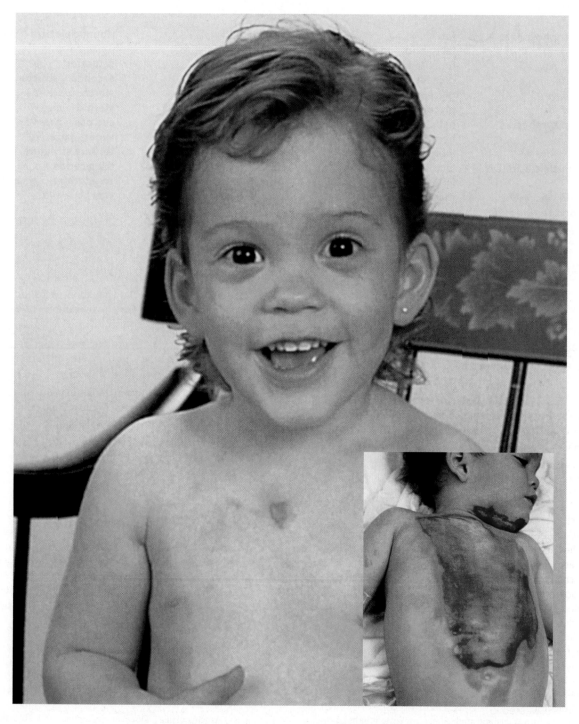

Just six months before this photo was taken, the child's chest was severely burned (inset). Today, healing time has been radically reduced and scarring almost eliminated by bioartificial skin.

CASE STUDY SPARE PARTS FOR HUMAN BODIES

"I DON'T THINK I've ever screamed so loud in my life"—a mother looks back on the terrible day when bubbling oil from a deep-fat fryer spilled from the stove onto her 10-month-old baby, burning over 70% of his body. "The 911 operator said to get his clothes off, but they'd melted onto him. I pulled his socks off and the skin came right with them." A few decades ago, this child's burns would have been fatal. Now, the only evidence of the burn on his chest is slightly crinkled skin. Zachary was saved by the bioengineering marvel of artificial skin.

Skin consists of several specialized cell types with complex interactions. The outer (epithelial) cells of skin are masters of multiplication, so minor burns heal without a trace. However, if the deeper (dermal) layers of the skin are completely destroyed, healing occurs only very slowly, from the edges of the burn. Deep burns are often treated by grafting skin, including some dermis taken from other

sites on the body; but for extensive burns, the lack of healthy skin makes this approach impossible. Until recently, the only alternative was to use skin from human cadavers or pigs. At best, these tissues serve as temporary "biological bandages" because the victim's body eventually rejects both of them. Extensive and disfiguring scars are a common legacy.

The availability of bioengineered skin has radically changed the prognosis for burn victims. The child in this chapter's opening photograph was treated with a bioengineered skin that contains living skin cells. The cells are obtained from the donated foreskins of infants who were circumcised at birth. After being cultured in the laboratory, a single square inch of tissue can provide enough cells to produce 250,000 square feet of artificial skin. The cells are grown under exacting conditions and seeded onto sponge-like, degradable polyester scaffolds. When complete, the artificial skin is frozen at −94°F (−70°C), a

temperature that allows cells to survive. The skin is shipped in dry ice to hospitals treating burn patients.

The living cells in bioengineered skin produce a variety of proteins, including fibrous proteins that form outside the cells in normal deep layers of skin, and protein growth factors that stimulate regeneration of deeper tissue layers and encourage growth of new blood vessels to nourish the tissue. As new tissue forms within the scaffold, the polyester breaks down into carbon dioxide, oxygen, and water.

Creation of artificial skin demonstrates our increasing power to manipulate cells, the fundamental units of life. All living things are constructed of cells, including tissues and organs that can be damaged by injury or disease. If scientists can shape cells into artificial, but living, skin, might they someday be able to sculpt cells into working bones, livers, kidneys, and lungs?

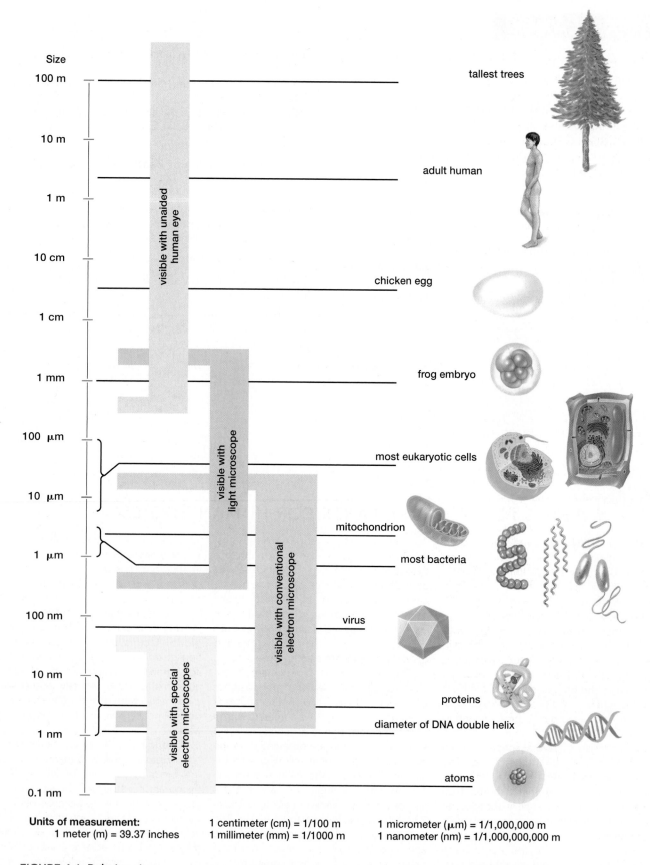

Size

100 m — tallest trees

10 m

1 m — adult human

10 cm

1 cm — chicken egg

1 mm — frog embryo

100 μm — most eukaryotic cells

10 μm

mitochondrion

1 μm — most bacteria

100 nm — virus

10 nm — proteins

1 nm — diameter of DNA double helix

0.1 nm — atoms

visible with unaided human eye

visible with light microscope

visible with conventional electron microscope

visible with special electron microscopes

Units of measurement:
1 meter (m) = 39.37 inches
1 centimeter (cm) = 1/100 m
1 millimeter (mm) = 1/1000 m
1 micrometer (μm) = 1/1,000,000 m
1 nanometer (nm) = 1/1,000,000,000 m

FIGURE 4-1 Relative sizes
Dimensions commonly encountered in biology range from about 100 meters (the height of the tallest red-woods) through a few stricrometers (the diameter of most cells) to a few nanometers (the diameter of many large molecules). Note that in the metric system (used almost exclusively in science and in many regions of the world), separate names are given to dimensions that differ by factors of 10, 100, and 1000.

4.1 WHAT IS THE CELL THEORY?

In the late 1850s, Austrian pathologist Rudolf Virchow wrote, "Every animal appears as a sum of vital units, each of which bears in itself the complete characteristics of life." Furthermore, Virchow predicted, "All cells come from cells." Virchow's insights were based on foundations laid by earlier microscopists, as you will learn later in "Scientific Inquiry: The Search for the Cell." The three principles of modern cell theory, a fundamental precept of biology, echo Virchow's statements:

- Every living organism is made up of one or more cells.
- The smallest living organisms are single cells, and cells are the functional units of multicellular organisms.
- All cells arise from preexisting cells.

All living things, from microscopic bacteria to a mighty oak tree to the human body, are composed of cells. Whereas each bacterium consists of a single, relatively simple cell, the human body consists of trillions of complex cells, specialized to perform an enormous variety of functions. To survive, all cells must obtain energy and nutrients from their environment, synthesize a variety of proteins and other molecules necessary for their growth and repair, and eliminate wastes. Many cells need to interact with other cells. To ensure the continuity of life, cells must also reproduce. These activities are accomplished by specialized parts of each cell, described later in this chapter.

4.2 WHAT ARE THE BASIC ATTRIBUTES OF CELLS?

Cell Function Limits Cell Size

Most cells range in size from about 1 to 100 micrometers (millionths of a meter) in diameter (FIG. 4-1). Because cells are so small, they were not discovered until the invention of the microscope. Ever since seeing the first cells in the late 1600s, scientists have devised increasingly sophisticated ways of observing them, as described in "Scientific Inquiry: The Search for the Cell."

Why are most cells small? The answer lies in the need for cells to exchange nutrients and wastes with their external environment through the plasma membrane. As you will learn in Chapter 5, many nutrients and wastes move into, through, and out of cells by *diffusion*, the movement of molecules from places of high concentration of those molecules to places of low concentration. This relatively slow process requires that no part of the cell be too far away from the external environment (look ahead to Fig. 5-17).

All Cells Share Common Features

Despite their diversity, all cells—from the prokaryotic bacteria and archaea to the eukaryotic protists, fungi, plants, and animals—share common features, as described in the following sections.

The Plasma Membrane Encloses the Cell and Mediates Interactions Between the Cell and Its Environment

Each cell is surrounded by an extremely thin, rather fluid membrane called the **plasma membrane** (FIG. 4-2). As you will learn in Chapter 5, the plasma membrane and the other membranes within cells consist of a double layer of phospholipids (see Chapter 3) in which a variety of proteins are embedded. Important functions of the plasma membrane include:

- It isolates the cell's contents from the external environment.
- It regulates the flow of materials into and out of the cell.
- It allows interaction with other cells and with the extracellular environment.

The phospholipid and protein components of cellular membranes have very different roles. Each phospholipid has a hydrophilic ("water-loving") head that faces the watery interior or watery exterior of the cell; it also has a pair of hydrophobic ("water-fearing") tails that face the interior of the membrane. Although some small molecules including oxygen, carbon dioxide, and water are able to diffuse through it, the phospholipid bilayer (referring to

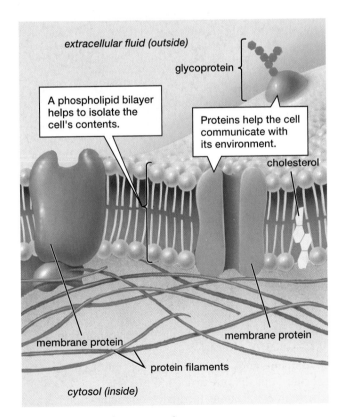

extracellular fluid (outside)

glycoprotein

A phospholipid bilayer helps to isolate the cell's contents.

Proteins help the cell communicate with its environment.

cholesterol

membrane protein

membrane protein

protein filaments

cytosol (inside)

FIGURE 4-2 The plasma membrane
The plasma membrane encloses the cell. Its structure, similar to that of all cellular membranes, consists of a double layer of phospholipid molecules in which a variety of proteins are embedded.

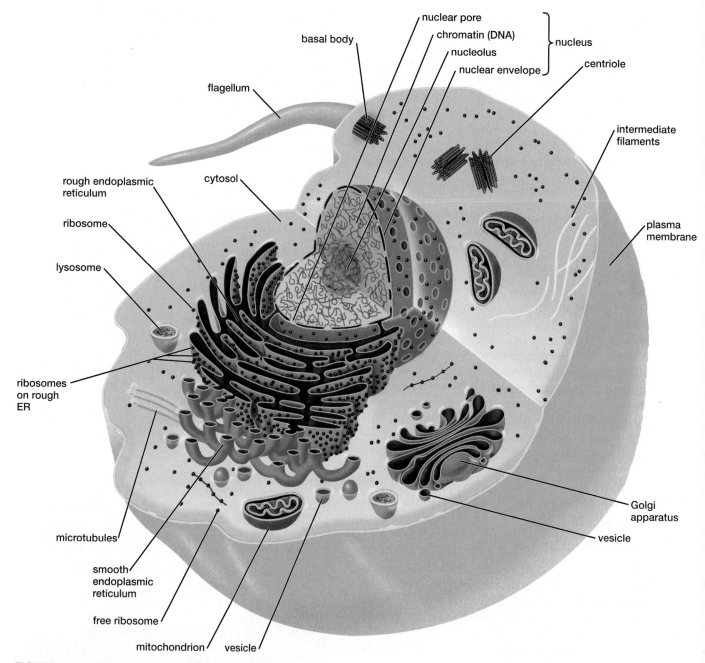

FIGURE 4-3 **A generalized animal cell**

the double layer of molecules) forms a barrier to most hydrophilic molecules and ions. The phospholipid bilayer helps isolate the cell from its surroundings, allowing the cell to maintain differences in concentrations of materials inside and out that are crucial to life.

In contrast, proteins facilitate communication between the cell and its environment. Some allow specific molecules or ions to move through the plasma membrane, while others promote chemical reactions inside the cell. Some membrane proteins attach cells together, and others

receive and respond to signals from molecules (such as hormones) in the fluid surrounding the cell (see Fig. 4-2). In Chapter 5, we discuss the plasma membrane in detail.

All Cells Contain Cytoplasm

The **cytoplasm** consists of all the material and structures that lie inside the plasma membrane, but outside the region of the cell that contains DNA (**FIGS. 4-3** and **4-4**). The fluid portion of the cytoplasm in both prokaryotic and eukaryotic cells, called the **cytosol**, contains water, salts, and an as-

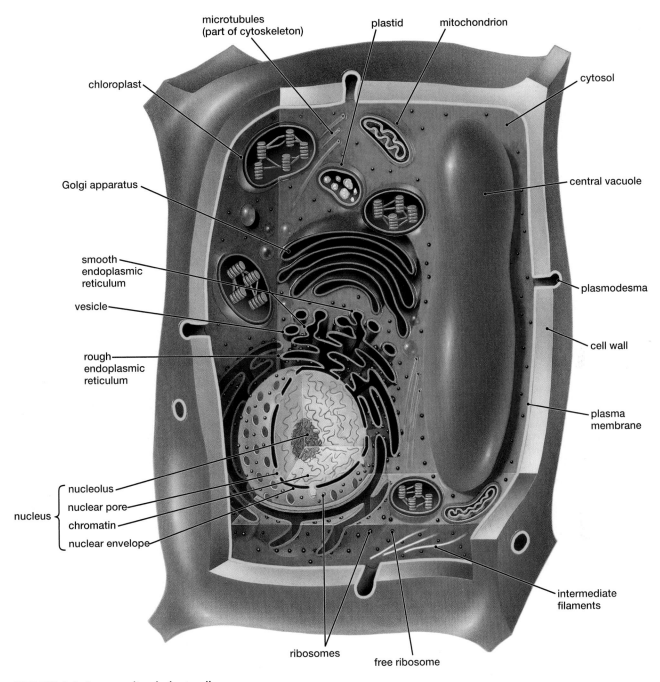

FIGURE 4-4 A generalized plant cell

sortment of organic molecules, including proteins, lipids, carbohydrates, sugars, amino acids, and nucleotides (see Chapter 3). The cytoplasm includes the cytosol and a variety of structures. Most of the cell's metabolic activities—the biochemical reactions that support life—occur in the cell cytoplasm. Protein synthesis is one example. This complex process takes place on special structures called *ribosomes*, located in the cytoplasm of all cells. The many types of proteins synthesized by cells include all those found in cell membranes and enzymes that allow metabolic reactions to occur, as we will see in Chapter 6.

All Cells Use DNA as a Hereditary Blueprint and RNA to Copy the Blueprint and Implement Construction

Each cell contains genetic material, an inherited blueprint that stores the instructions for making the other parts of the cell and for producing new cells. The genetic material in cells is **deoxyribonucleic acid (DNA)**. This fascinating molecule, described in detail in Chapter 9, contains genes consisting of precise sequences of nucleotides (see Chapter 3). During cell division, the original or "parent cells" pass exact copies of their DNA to their newly formed offspring or "daughter cells." **Ribonucleic acid (RNA)** is chemically related to DNA and comes in various forms that copy the "blueprint" of genes on DNA and help construct proteins based on this blueprint. All cells contain RNA.

All Cells Obtain Energy and Nutrients from Their Environment

To maintain their incredible complexity, cells must continuously acquire and expend energy. As we explain in Chapters 6, 7, and 8, essentially all of the energy powering life on Earth originates in sunlight. Cells that can harness this energy directly provide energy for nearly all other forms of life. The building blocks of biological molecules, such as carbon, nitrogen, oxygen, and a variety of minerals, ultimately come from the environment—the air, water, rocks, and other forms of life. All cells obtain the materials to generate the molecules of life, and the energy to power this synthesis, from their living and nonliving environment.

There Are Two Basic Types of Cells: Prokaryotic and Eukaryotic

All forms of life are composed of only two fundamentally different types of cells. **Prokaryotic** (Greek for "before the nucleus"; see Fig. 4-20a) cells form the bodies of **bacteria** and **archaea**, the simplest forms of life on Earth. **Eukaryotic** (Greek for "true nucleus"; see Figs. 4-3 and 4-4) cells are

Table 4-1 Functions and Distribution of Cell Structures

Structure	Function	Prokaryotes	Eukaryotes: Plants	Eukaryotes: Animals
Cell surface				
Cell wall	protects, supports cell	present	present	absent
Cilia	move cell through fluid or move fluid past cell surface	absent	absent	present
Flagella	move cell through fluid	present[1]	present[2]	present
Plasma membrane	isolates cell contents from environment; regulates movement of materials into and out of cell; communicates with other cells	present	present	present
Organization of genetic material				
Genetic material	encodes information needed to construct cell and control cellular activity	DNA	DNA	DNA
Chromosomes	contain and control use of DNA	single, circular, no proteins	many, linear, with proteins	many, linear, with proteins
Nucleus	membrane-bound container for chromosomes	absent	present	present
Nuclear envelope	encloses nucleus; regulates movement of materials into and out of nucleus	absent	present	present
Nucleolus	synthesizes ribosomes	absent	present	present
Cytoplasmic structures				
Mitochondria	produce energy by aerobic metabolism	absent	present	present
Chloroplasts	perform photosynthesis	absent	present	absent
Ribosomes	provide site of protein synthesis	present	present	present
Endoplasmic reticulum	synthesizes membrane components, proteins, and lipids	absent	present	present
Golgi apparatus	modifies and packages proteins and lipids; synthesizes some carbohydrates	absent	present	present
Lysosomes	contain intracellular digestive enzymes	absent	present	present
Plastids	store food, pigments	absent	present	absent
Central vacuole	contains water and wastes; provides turgor pressure to support cell	absent	present	absent
Other vesicles and vacuoles	transport secretory products; contain food obtained through phagocytosis	absent	present	present
Cytoskeleton	gives shape and support to cell; positions and moves cell parts	absent	present	present
Centrioles	produce the microtubules of cilia and flagella, and those that form the spindle during animal cell division.	absent	absent (in most)	present

[1]Some prokaryotes have structures called flagella, which lack microtubules and move in a fundamentally different way than do eukaryotic flagella.
[2]A few types of plants have flagellated sperm.

far more complex and comprise the bodies of animals, plants, fungi, and protists. As their names imply, one striking difference between prokaryotic cells and eukaryotic cells is that the genetic material of eukaryotic cells is contained within a membrane-enclosed nucleus. In contrast, the genetic material of prokaryotic cells is not enclosed within a membrane. Other membrane-enclosed structures, called *organelles*, contribute to the far greater structural complexity of eukaryotic cells. Table 4-1 summarizes the features of prokaryotic and eukaryotic cells, which we discuss in the sections that follow.

4.3 WHAT ARE THE MAJOR FEATURES OF EUKARYOTIC CELLS?

Eukaryotic cells comprise the bodies of animals, plants, protists, and fungi, so as you might imagine, these cells are extremely diverse. Within the body of any multicellular organism there exists a variety of eukaryotic cells specialized to perform different functions. In contrast, the single-celled bodies of protists and some fungi must be sufficiently complex to perform all the necessary activities needed to sustain life, grow, and reproduce independently. Here we emphasize plant and animal cells; the specialized structures of protists and fungi are covered in more detail in Chapters 20 and 22, respectively.

Eukaryotic cells differ from prokaryotic cells in many ways. For example, they are usually larger than prokaryotic cells—typically more than 10 micrometers in diameter. The cytoplasm of eukaryotic cells includes a variety of **organelles**—membrane-enclosed structures that perform specific functions within the cell—such as the nucleus and mitochondria. The **cytoskeleton**, a network of protein fibers, gives shape and organization to the cytoplasm of eukaryotic cells. Many of the organelles are attached to the cytoskeleton.

Figures 4-3 and 4-4 illustrate the structures that are found in animal and plant cells, respectively, although not all individual cells possess all the features shown in either drawing. Each type of cell has a few unique organelles not found in the other. Plant cells, for example, are surrounded by a *cell wall*, and they contain chloroplasts, plastids, and a central vacuole. Only animal cells possess centrioles. You may want to refer to these illustrations as we describe the structures of the cell in more detail. The major components of eukaryotic cells (see Table 4-1) are described in more detail in the following sections.

Some Eukaryotic Cells Are Supported by Cell Walls

The outer surfaces of plants, fungi, and some protists are covered with nonliving, relatively stiff coatings—called cell walls—that support and protect the delicate plasma membrane. Single-celled protists that live in the ocean may have cell walls made of cellulose, protein, or glassy silica (see Chapter 20). Plant cell walls are composed of cellulose and other *polysaccharides*, whereas fungal cell walls are made

of polysaccharides and chitin (a modified polysaccharide; described in Chapter 3.) Prokaryotic cells also have cell walls, made of a chitin-like framework to which short chains of amino acids and other molecules are attached.

Cell walls are produced by the cells they surround. Plant cells secrete cellulose through their plasma membranes, forming the *primary cell wall*. Many plant cells, when they mature and stop enlarging, secrete more cellulose and other polysaccharides beneath the primary wall to form a *secondary cell wall*, pushing the primary cell wall away from the plasma membrane. The primary cell walls of adjacent cells are joined by the *middle lamella*, a layer made primarily of the polysaccharide *pectin* (**FIG. 4-5**). If you've ever made or enjoyed fruit jelly, you might be interested to know that pectin from the cell walls of fruit is what makes it gel.

Cell walls support and protect otherwise fragile cells. For example, cell walls allow plants and mushrooms to resist the forces of gravity and wind and to stand erect on land. Tree trunks, composed almost entirely of cellulose and other materials laid down over the years and capable of supporting impressive loads, are the ultimate proof of cell wall strength.

Cell walls are usually porous, allowing oxygen, carbon dioxide, and water with dissolved molecules to move through easily. The structure governing the interactions between a cell and its external environment is the plasma membrane, located just beneath the cell wall (when a wall is present). The plasma membrane is described earlier in this chapter and covered in detail in Chapter 5.

The Cytoskeleton Provides Shape, Support, and Movement

Organelles and other structures within eukaryotic cells do not drift about the cytoplasm haphazardly; most are attached to the network of protein fibers that comprises the

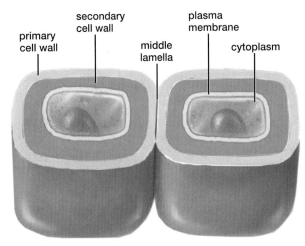

FIGURE 4-5 Plant cell walls
Primary and secondary cells walls are made primarily of cellulose. Growing cells have only a flexible primary cell wall. The more rigid secondary cell wall is secreted by some plant cells when they reach maturity. Adjacent cells are linked by a middle lamella made of pectin.

Human understanding of the cellular nature of life came slowly. In 1665, English scientist and inventor Robert Hooke reported observations with a primitive microscope. He aimed his instrument at an "exceeding thin . . . piece of Cork" and saw "a great many little Boxes" (**FIG. E4-1a**). Hooke called the boxes "cells," because he thought they resembled the tiny rooms, or cells, occupied by monks. Cork comes from the dry outer bark of the cork oak, and we now know that he was looking at the nonliving cell walls that surround all plant cells. Hooke wrote that in the living oak and other plants, "These cells [are] fill'd with juices."

In the 1670s, Dutch microscopist Anton van Leeuwenhoek was constructing his own simple microscopes and observing a previously unknown world. A self-taught amateur scientist, his descriptions of myriad "animalcules" (his term for protists) in rain, pond, and well water caused quite an uproar because in those days, water was consumed without

being treated. Eventually, van Leeuwenhoek made careful observations of an enormous range of microscopic specimens, including blood cells, sperm, and the eggs of small insects such as weevils, aphids, and fleas. His discoveries struck a blow to the then common belief in spontaneous generation; at that time, fleas were believed to emerge spontaneously from sand or dust, as were weevils from grain! Although they appeared much more primitive than Hooke's microscopes, Leeuwenhoek's microscopes provided much clearer images and higher magnification (**FIG. E4-1b**).

More than a century passed before biologists began to understand the role of cells in life on Earth. Microscopists first noted that many plants consist entirely of cells. The thick wall surrounding all plant cells, first observed by Hooke, made their observations easier. Animal cells, however, escaped notice until the 1830s, when German zoologist Theodor Schwann saw that cartilage contains cells that "ex-

(a) 17ᵗʰ century microscope and cork cells

(b) Leeuwenhoek's microscope

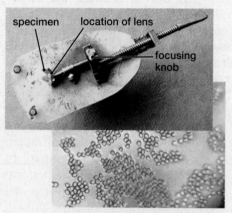

specimen   location of lens

focusing knob

blood cells photographed through Leeuwenhoek's microscope

(c) Electron microscope

FIGURE E4-1 Microscopes yesterday and today
(a) Robert Hooke's drawings of cork cells, as he viewed them with an early light microscope similar to the one shown here. Only the cell walls remain. **(b)** One of Leeuwenhoek's microscopes, and a photograph of blood cells taken through a Leeuwenhoek microscope. The specimen is viewed through a tiny hole just underneath the lens. **(c)** This electron microscope is capable of performing both scanning and transmission electron microscopy.

actly resemble [the cells of] plants." In 1839, after studying cells for years, Schwann was confident enough to publish his *cell theory*, calling cells the elementary particles of both plants and animals. By the mid-1800s, German botanist Matthias Schleiden further refined science's view of cells when he wrote: "It is . . . easy to perceive that the vital process of the individual cells must form the first, absolutely indispensable fundamental basis [of life]."

Ever since the pioneering efforts of Robert Hooke and Anton van Leeuwenhoek, biologists, physicists, and engineers have collaborated in the development of a variety of advanced microscopes to view the cell and its components:

Light microscopes use lenses, usually made of glass, to focus light rays that either pass through or bounce off a specimen, thereby magnifying its image. Light microscopes provide a wide range of images, depending on how the specimen is illuminated and whether it has been stained

(**FIG. E4-2a**). The *resolving power* of light microscopes— that is, the smallest structure that can be seen—is about 1 micrometer (a millionth of a meter).

Electron microscopes (**FIG. E4-1c**) use beams of electrons instead of light, which are focused by magnetic fields rather than by lenses. Some types of electron microscopes can resolve structures as small as a few nanometers (billionths of a meter). *Transmission electron microscopes* (TEMs) pass electrons through a thin specimen and can reveal the details of interior cell structure, including organelles and plasma membranes (**FIG. E4-2b**). *Scanning electron microscopes* (SEMs) bounce electrons off specimens that have been coated with metals and provide three-dimensional images. SEMs can be used to view the surface details of structures that range in size from entire insects down to cells and even organelles (**FIG. E4-2c, d**).

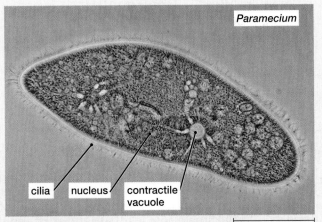

(a) Light microscope — 60 micrometers

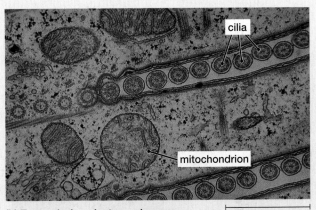

(b) Transmission electron microscope — 1.5 micrometers

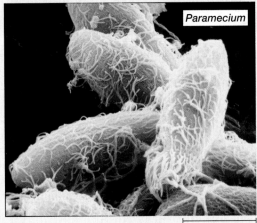

(c) Scanning electron microscope — 70 micrometers

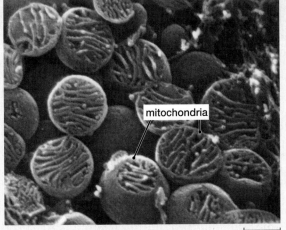

(d) Scanning electron microscope — 0.5 micrometers

FIGURE E4-2 A comparison of microscope images
(a) A living *Paramecium* (a single-celled freshwater protist) viewed through a light microscope. **(b)** A false-color TEM photo of a *Paramecium* showing sections of mitochondria and of the bases of the cilia that cover this amazing cell. **(c)** An SEM photo of cilia-covered *Paramecia*. **(d)** An SEM photo at much higher magnification, showing mitochondria (many of which are sliced open) within the cytoplasm.

(a)

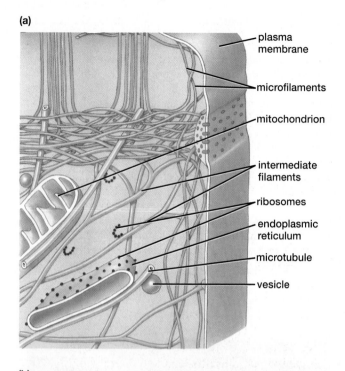

plasma membrane

microfilaments

mitochondrion

intermediate filaments

ribosomes

endoplasmic reticulum

microtubule

vesicle

(b)

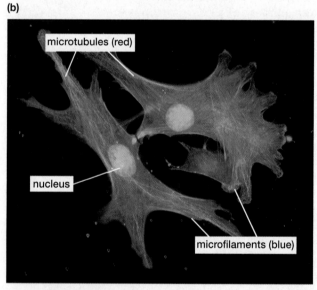

microtubules (red)

nucleus

microfilaments (blue)

FIGURE 4-6 The cytoskeleton
(a) Eukaryotic cells are given shape and organization by the cytoskeleton, which consists of three types of proteins: microtubules, intermediate filaments, and microfilaments. **(b)** This cell from the lining of a cow artery has been treated with fluorescent stains to reveal microtubules and microfilaments, as well as the nucleus.

cytoskeleton (**FIG. 4-6**). Even individual enzymes, which are often a part of complex metabolic pathways, may be fastened in sequence to the cytoskeleton, so that molecules can be passed from one enzyme to the next in the correct order for a particular chemical transformation. Several types of protein fibers, including thin **microfilaments**, medium-sized **intermediate filaments**, and thick **microtubules**, make up the cytoskeleton.

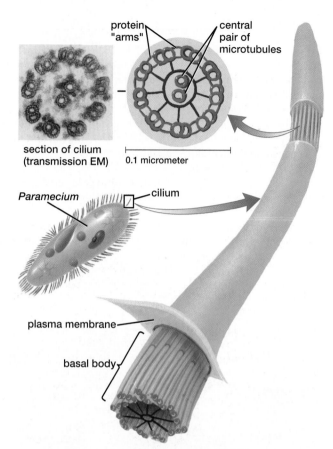

protein "arms"

central pair of microtubules

section of cilium (transmission EM)

0.1 micrometer

Paramecium

cilium

plasma membrane

basal body

FIGURE 4-7 Cilia and flagella
Both cilia and flagella contain microtubules arranged in an outer ring of nine fused pairs of microtubules surrounding a central unfused pair. The outer pairs have "arms" made of protein that interact with adjacent pairs to provide the force for bending. Cilia and flagella arise from basal bodies located just beneath the plasma membrane.

The cytoskeleton performs the following important functions:

- *Cell shape.* In cells without cell walls, the cytoskeleton, especially networks of intermediate filaments, determines the shape of the cell.
- *Cell movement.* Cell movement occurs as microfilaments or microtubules assemble, disassemble, or slide past one another. Examples of moving cells include single-celled protists propelled by cilia, swimming sperm, and contracting muscle cells.
- *Organelle movement.* Microtubules and microfilaments move organelles from place to place within a cell. For example, microfilaments attach to vesicles formed during *endocytosis* when large particles are engulfed by the plasma membrane, and pull the vesicles into the cell (see Chapter 5). Vesicles budded off the *endoplasmic reticulum* (ER) and *Golgi apparatus* are guided by the cytoskeleton as well.

- *Cell division*. Microtubules and microfilaments are essential to cell division in eukaryotic cells. First, when eukaryotic nuclei divide, microtubules move the *chromosomes* (packets of genetic material) into the daughter nuclei. Second, animal cells divide when a ring of microfilaments contracts, pinching the "parent" cell inward around the middle to form two new "daughter" cells. **Centrioles** (see Fig. 4-3), which form the spindle that helps apportion the genetic material during animal cell division, are composed of microtubules. Cell division is covered in detail in Chapter 11.

Cilia and Flagella Move the Cell Through Fluid or Move Fluid Past the Cell

Both **cilia** (Latin for "eyelash") and **flagella** ("whip") are slender extensions of the plasma membrane, supported internally with microtubules of the cytoskeleton. Each cilium and flagellum contains a ring of nine pairs of microtubules, with another pair in the center (**FIG. 4-7**). These microtubules, which extend the entire length of the cilium or flagellum, extend upward from a **basal body** (derived from a centriole; see Fig. 4-3) anchored just beneath the plasma membrane.

Tiny "arms" of protein attach neighboring pairs of microtubules in cilia and flagella. When these arms flex, they slide one pair of microtubules past the adjacent pairs, causing the cilia or flagellum to move. Energy released from adenosine triphosphate (ATP) powers the movement of the protein arms during microtubule sliding. Cilia and flagella often move almost continuously; the energy to power

this motion is supplied by mitochondria, usually found in abundance near the basal bodies.

Cilia and flagella differ in their length, number, and the direction of the force they generate. In general, cilia are shorter and more numerous than flagella. Like the oars in a rowboat, cilia provide force in a direction parallel to the plasma membrane. This is accomplished by means of a "rowing" motion (**FIG. 4-8a**, left). Flagella are longer and fewer in number; they provide force perpendicular to the plasma membrane, more like the engine on a motorboat (**FIG. 4-8b**, left).

Some unicellular organisms, such as *Paramecium* (see Fig. E4-2a, c), use cilia to swim through water; others use flagella. Some small aquatic invertebrates swim by beating rows of cilia, like the oars of an ancient Roman galley ship. In animals cilia usually move fluids and suspended particles past a surface. Ciliated cells line such diverse structures as the gills of oysters (where they move water rich in food and oxygen over the gills), the oviducts of female mammals (where they move an egg through fluid from the ovary to the uterus), and the respiratory tracts of most land vertebrates (where they clear mucus that carries debris and microorganisms from the trachea and lungs; Fig. 4-8a, right). Nearly all animal sperm and a few types of plant sperm cells rely on flagella for movement (Fig. 4-8b, right).

The Nucleus Is the Control Center of the Eukaryotic Cell

A cell's DNA stores all the information needed to construct the cell and direct the countless chemical reactions necessary for life and reproduction. The genetic information in

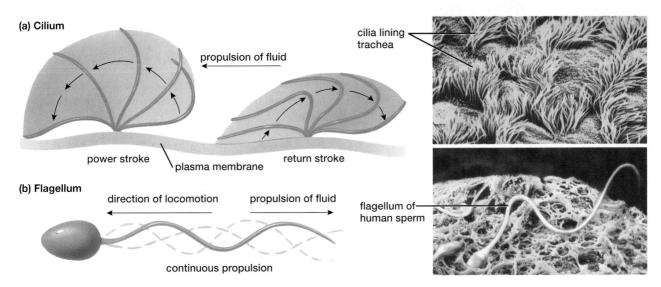

FIGURE 4-8 How cilia and flagella move
(a) (left) Cilia usually "row," providing a force of movement parallel to the plasma membrane. Their movement resembles the arms of a swimmer doing the breast stroke. (right) SEM photo of cilia lining the trachea (which conducts air to the lungs); these cilia sweep out mucus and trapped particles. **(b)** (left) Flagella move in a wavelike motion, providing continuous propulsion perpendicular to the plasma membrane. In this way, a flagellum attached to a sperm can move the sperm straight ahead. (right) A human sperm cell on the surface of a human egg cell.

(a)

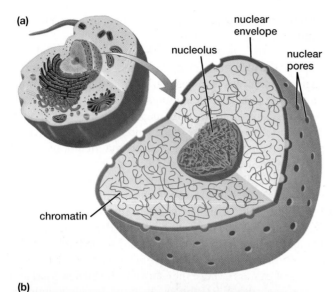

(b)

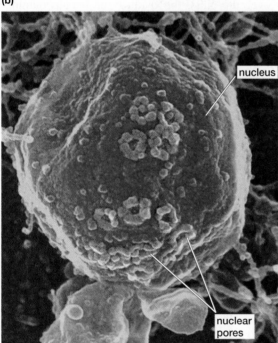

FIGURE 4-9 The nucleus
(a) The nucleus is bounded by a double outer membrane. Inside are chromatin and a nucleolus. **(b)** An electron micrograph of a yeast cell that has been frozen and broken open to reveal its internal structures. The large nucleus, with nuclear pores penetrating its nuclear membrane, is clearly visible. The pink structures are the "gatekeeper proteins" that line the pores.

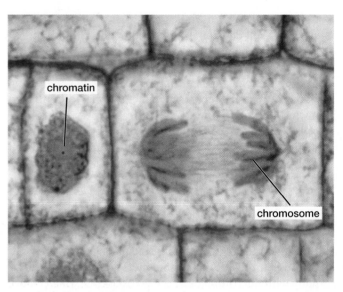

FIGURE 4-10 Chromosomes
Chromosomes, seen here in a light micrograph of a dividing cell (on the right) in an onion root tip, contain the same material (DNA and proteins) as the chromatin seen in adjacent non-dividing cells, but in a more compact state.

The Nuclear Envelope Allows Selective Exchange of Materials

The nucleus is isolated from the rest of the cell by a **nuclear envelope** that consists of a double membrane. The membrane is perforated with tiny membrane-lined channels called *nuclear pores*. Water, ions, and small molecules such as ATP can pass freely through the pores, but the passage of large molecules—particularly proteins, pieces of ribosomes, and RNA—is regulated by specialized "gatekeeper proteins" that line each nuclear pore. Ribosomes stud the outer nuclear membrane, which is continuous with membranes of the rough ER described later (see Figs. 4-3 and 4-4).

Chromatin Contains DNA, Which Codes for the Synthesis of Proteins

Because the nucleus is highly colored by stains used in light microscopy, early microscopists, not knowing its function, named the nuclear material **chromatin**, meaning "colored substance." Biologists have since learned that chromatin consists of DNA associated with proteins. Eukaryotic DNA and its associated proteins form long strands called **chromosomes** ("colored bodies"). When cells divide, each chromosome coils upon itself, becoming thicker and shorter. The resulting "condensed" chromosomes are easily visible even with light microscopes (**FIG. 4-10**).

The genes on DNA provide a blueprint or a "molecular code" for a huge variety of proteins. Some of these proteins form structural components of the cell. Others regulate the movement of materials through cell membranes, and still

DNA is used selectively by the cell, depending on its stage of development, its environment, and the function of the cell in a multicellular body. In eukaryotic cells, DNA is housed within the nucleus.

The **nucleus** is an organelle (usually the largest in the cell) consisting of three major parts: the *nuclear envelope*, *chromatin*, and the *nucleolus*, shown in **FIGURE 4-9** and described in the following sections.

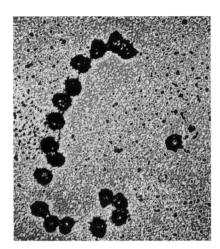

FIGURE 4-11 Ribosomes
Ribosomes may be found free in the cytoplasm either singly or strung along messenger RNA molecules as they participate in protein synthesis, as seen in this electron micrograph. Ribosomes also stud the rough endoplasmic reticulum (see **FIG. 4-12**).

accomplish this, genetic information is copied from DNA into molecules of RNA (called *messenger RNA*, or *mRNA*; seen in **FIG. 4-11** linking a series of ribosomes), which move through the pores of the nuclear envelope into the cytoplasm. This information, coded by the sequence of nucleotides in mRNA, is then used to direct the synthesis of cellular proteins, a process that occurs on ribosomes, which are composed of *ribosomal RNA* and protein. We take a closer look at these processes in Chapter 10.

The Nucleolus Is the Site of Ribosome Assembly

Eukaryotic nuclei have one or more darkly staining regions called *nucleoli* ("little nuclei"; see Fig. 4-9a). Nucleoli are the sites of ribosome synthesis. The **nucleolus** consists of ribosomal RNA, proteins, ribosomes in various stages of synthesis, and DNA (bearing genes that specify the blueprint for ribosomal RNA).

A **ribosome** is a small particle composed of RNA and proteins that serves as a kind of "workbench" for the synthesis of proteins within the cell cytoplasm. Just as a workbench can be used to construct many different objects, any ribosome can be used to synthesize any of the thousands of proteins made by a cell. In electron micrographs, ribosomes appear as dark granules, either distributed in the cytoplasm (Fig. 4-11) or clustered along the membranes of the nuclear envelope and the endoplasmic reticulum (**FIG. 4-12**).

others are enzymes that promote chemical reactions within the cell that are responsible for growth and repair, nutrient and energy acquisition and use, and reproduction.

Because proteins are synthesized in the cytoplasm, copies of the protein blueprints on DNA must be ferried out through the nuclear membrane into the cytoplasm. To

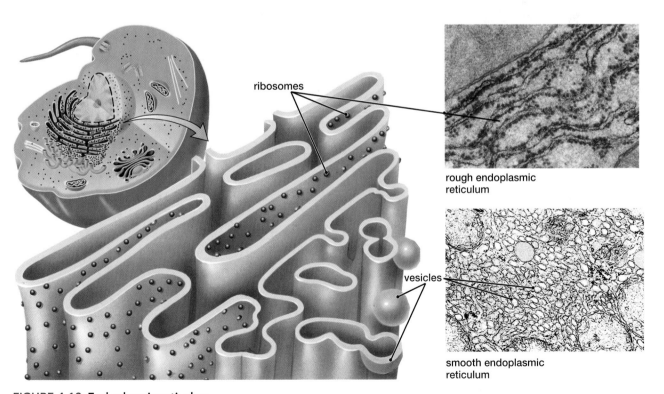

FIGURE 4-12 Endoplasmic reticulum
There are two types of endoplasmic reticulum: rough ER and smooth ER. In some cells, the rough and smooth ER are thought to be linked, as depicted in the drawing. In others, smooth ER is likely separate. Ribosomes (black) stud the cytosolic face of the rough ER membrane.

Eukaryotic Cytoplasm Includes an Elaborate System of Membranes

All eukaryotic cells have an elaborate system of membranes that enclose the cell and create internal compartments within the cytoplasm. Imagine a series of rooms within a large factory. Each room houses specialized machinery. The rooms are often interconnected to allow a complex product to be manufactured in stages. Some products must be moved between buildings before they are finished. The factory must import raw materials, but it makes and repairs its own machinery and exports some of its products. In a comparable way, specialized regions within the cytoplasm separate a variety of biochemical reactions from one another and process different types of molecules in specific ways. The fluid property of membranes allows them to fuse with one another, so that these internal compartments can interconnect, exchange fragments of membrane among themselves, and transfer their contents to different compartments for various types of processing. Membrane sacs called **vesicles** ferry membranes and specialized contents among the separate regions of the membrane system. Vesicles also fuse with the plasma membrane, exporting their contents outside the cell (see Fig. 4-14). How do the vesicles know where to go within the complex membrane system? Researchers have discovered that various proteins embedded in membranes serve as "mailing labels" that specify the address to which the sac and its contents are being sent.

The cell's membrane system includes the plasma membrane, nuclear membrane, endoplasmic reticulum, Golgi apparatus, lysosomes, vesicles, and vacuoles, which we explore further in the sections below.

The Endoplasmic Reticulum Forms Membrane-Enclosed Channels Within the Cytoplasm

The **endoplasmic reticulum (ER)** is a series of interconnected membrane-enclosed tubes and channels in the cytoplasm (*reticulum* means "network" and *endoplasmic* means "within the cytoplasm"; Fig. 4-12). Eukaryotic cells have two forms of ER: rough and smooth. Parts of the rough ER are continuous with the nuclear membrane (see Fig. 4-3). Numerous ribosomes stud the outside of the *rough endoplasmic reticulum*, giving it a rough appearance. In contrast, *smooth endoplasmic reticulum* lacks ribosomes. The membranes of both rough and smooth ER contain enzymes that can synthesize the various lipids, such as phospholipids and cholesterol, which are needed to produce the lipid portions of cell membranes.

Smooth Endoplasmic Reticulum

Smooth ER has a variety of functions and is specialized for different activities in different cells. In some cells, smooth ER manufactures large quantities of lipids such as steroid hormones made from cholesterol. For example, sex hormones are produced by smooth ER in mammalian reproductive organs. Smooth ER is also abundant in liver cells,

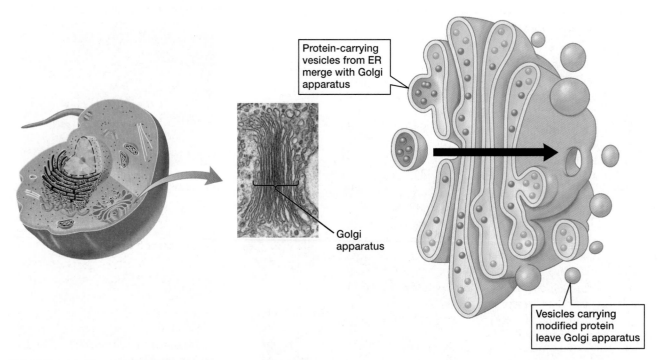

Protein-carrying vesicles from ER merge with Golgi apparatus

Golgi apparatus

Vesicles carrying modified protein leave Golgi apparatus

FIGURE 4-13 The Golgi apparatus
The Golgi apparatus is a stack of flattened membranous sacs derived from the endoplasmic reticulum. Vesicles constantly bud off from and fuse with the Golgi and ER, transporting both material and cell membrane from the ER to the Golgi and back again. The large arrow indicates the direction of movement of materials within the Golgi as they are modified and sorted. Vesicles bud from the face of the Golgi opposite the ER; some produce lysosomes, and others transport substances to the plasma membrane for exocytosis.

where it contains enzymes that detoxify harmful drugs such as alcohol and metabolic by-products such as ammonia. Other enzymes in the smooth ER of the liver break down glycogen (a carbohydrate stored in the liver) into glucose molecules that provide energy. Smooth ER stores calcium in all cells, but in skeletal muscles it is enlarged and specialized to store large amounts of calcium that are required for muscle contraction.

Rough Endoplasmic Reticulum

The ribosomes on rough ER are sites of protein synthesis. For example, the various proteins embedded in cellular membranes are manufactured here, so rough ER is capable of producing all the components of new membranes. Continuous production of new membrane is important because ER membrane is constantly being budded off and transported to the Golgi apparatus, lysosomes, and the plasma membrane.

Ribosomes on rough ER are also sites for manufacturing proteins such as digestive enzymes and protein hormones (for example, insulin) that some cells export into their surroundings. As these proteins are synthesized, they are inserted through the ER membrane into the interior compartment. Proteins synthesized either for secretion outside the cell or use elsewhere within the cell move through the ER channels. Here they are modified chemically and folded into their proper three-dimensional structures. Eventually the proteins accumulate in pockets of membrane that bud off as vesicles that carry their protein cargo to the Golgi apparatus.

The Golgi Apparatus Sorts, Chemically Alters, and Packages Important Molecules

The **Golgi apparatus** (or **Golgi**; named for the Italian physician and cell biologist Camillo Golgi, who discovered it in the late 1800s) is a specialized set of membranes, derived from the endoplasmic reticulum, that looks like a stack of flattened and interconnected sacs (**FIG. 4-13**). Its main purpose is to modify, sort, and package proteins produced by the rough ER. The compartments of the Golgi act like the finishing rooms of a factory, where the final touches are put on products and they are packaged and exported. Vesicles from the rough ER fuse with one face of the Golgi apparatus, adding their membranes to the Golgi and emptying their contents into the Golgi sacs. Within the flattened Golgi compartments, protein molecules synthesized in the rough ER are modified further. Carbohydrates are added to form glycoproteins. Some large proteins are cleaved into smaller fragments. Finally, vesicles bud off from the opposite face of the Golgi, carrying away finished products for use or export.

The Golgi apparatus performs the following:

- The Golgi modifies some molecules; an important role is to add carbohydrates to proteins to make glycoproteins. It also breaks up some proteins into smaller peptides.
- The Golgi synthesizes some polysaccharides, such as cellulose and pectin used in plant cell walls.

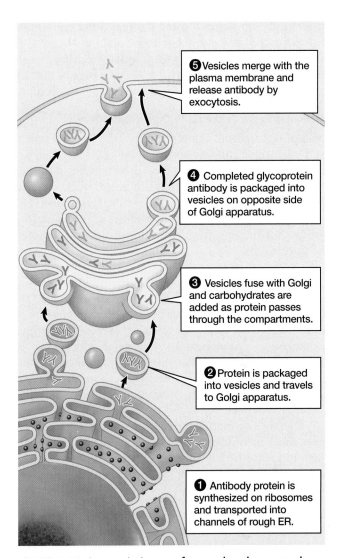

5 Vesicles merge with the plasma membrane and release antibody by exocytosis.

4 Completed glycoprotein antibody is packaged into vesicles on opposite side of Golgi apparatus.

3 Vesicles fuse with Golgi and carbohydrates are added as protein passes through the compartments.

2 Protein is packaged into vesicles and travels to Golgi apparatus.

1 Antibody protein is synthesized on ribosomes and transported into channels of rough ER.

FIGURE 4-14 A protein is manufactured and exported

- The Golgi separates proteins and lipids received from the ER according to their destinations; for example, it separates digestive enzymes, which are bound for lysosomes, from cholesterol used in new membrane synthesis, from the protein hormones that the cell will secrete.
- The Golgi packages finished molecules into vesicles that are then transported to other parts of the cell or to the plasma membrane for export.

Secreted Proteins Travel Through the Cell for Export

To understand how some of the components of the membrane system work together, let's look at the manufacture and export of an extremely important protein called an *antibody* (**FIG. 4-14**). Antibodies, produced by white blood cells, are glycoproteins that bind to foreign invaders (such as disease-causing bacteria) and help destroy them. Antibody proteins are synthesized on ribosomes of the rough ER within the white blood cell, and they are then packaged into vesicles formed from ER membrane. These vesicles travel to the Golgi, where the membranes fuse, releasing the protein into the Golgi apparatus. Within the Golgi, carbohydrates are attached to the protein, which is then repackaged into vesicles formed from Golgi membrane. The vesicle containing the completed antibody then travels to the plasma membrane and fuses with it, releasing the antibody outside the cell, where it will make its way into the bloodstream to help defend the body against infection.

Lysosomes Serve as the Cell's Digestive System

Some of the proteins manufactured in the ER and sent to the Golgi apparatus are intracellular digestive enzymes that can break proteins, fats, and carbohydrates into their component subunits. In the Golgi, these enzymes are packaged in membrane-enclosed vesicles called **lysosomes** (FIG. 4-15). One major function of lysosomes is to digest food particles, which range from individual proteins to complete microorganisms.

As you will learn in Chapter 5, many cells "eat" by *phagocytosis*—that is, by engulfing particles just outside the cell using extensions of the plasma membrane. The plasma membrane with its enclosed food then pinches off inside the cytosol and forms a **food vacuole**. Lysosomes recognize these food vacuoles and merge with them. The contents of the two vesicles mix, and lysosomal enzymes digest the food into small molecules such as amino acids, monosaccharides, and fatty acids that can be used within the cell. Lysosomes also digest excess cellular membranes and defective or malfunctioning organelles. The cell encloses them in vesicles made of membrane from the ER, which then fuse with lysosomes. Digestive enzymes within the lysosome enable the cell to recycle valuable molecules from the defunct organelles.

Membrane Flows Through the Membrane System of the Cell

The nuclear envelope, rough and smooth ER, Golgi apparatus, lysosomes, food vacuoles, and the plasma membrane all form an integrated membrane system. By reviewing Figures 4-14 and 4-15, you can get an idea of how the membranes themselves interconnect. The ER synthesizes the phospholipids and proteins that make up the plasma membrane and buds off some of this membrane in vesicles, which then fuse with the Golgi membranes. Some of the ER membrane that fuses with the Golgi carries protein "address labels" that direct it to return to the ER, restoring crucial proteins (such as some enzymes) to the ER membrane. Other parts of the ER membrane are modified by the Golgi; for example, carbohydrates may be added to make membrane glycoproteins. Eventually, this membrane leaves the Golgi as a vesicle that may fuse with the plasma membrane, replenishing or enlarging it.

Vacuoles Serve Many Functions, Including Water Regulation, Support, and Storage

Most cells contain one or more **vacuoles**—sacs of cell membrane filled with fluid containing various molecules. Some, such as the food vacuoles that form during phagocytosis (see Fig. 4-15), are only temporary. However, many cells contain permanent vacuoles that have important roles in maintaining the integrity of the cell, most notably by regulating the cell's water content.

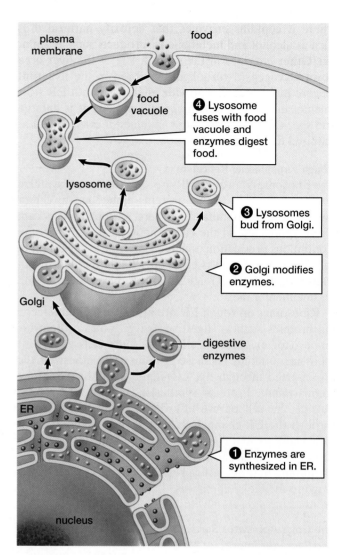

FIGURE 4-15 Formation and function of lysosomes and food vacuoles

Freshwater Microorganisms Have Contractile Vacuoles

Freshwater protists such as *Paramecium* consist of a single eukaryotic cell. Many of these organisms possess **contractile vacuoles** composed of collecting ducts, a central reservoir, and a tube leading to a pore in the plasma membrane (FIG. 4-16). These complex cells live in fresh water, which constantly leaks into them through their plasma membranes (we describe this process, called *osmosis*, in Chapter 5). The influx of water would soon burst the fragile creature if it did not have a mechanism to excrete the water. Cellular energy is used to pump salts from the cytoplasm of the protist into collecting ducts. Water follows by osmosis and drains into the central reservoir. When the reservoir of the contractile vacuole is full, it contracts, squirting the water out through a pore in the plasma membrane.

Plant Cells Have Central Vacuoles

Three-quarters or more of the volume of many plant cells is occupied by a large **central vacuole** (see Fig. 4-4). The central vacuole has several functions. Filled mostly with

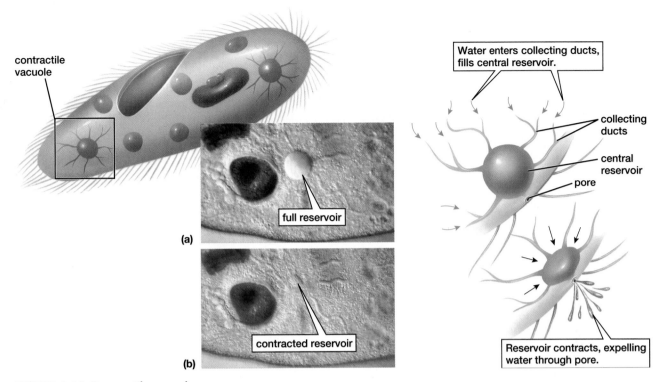

FIGURE 4-16 Contractile vacuoles
Many freshwater protists contain contractile vacuoles. **(a)** Water constantly enters the cell by osmosis. In the cell, water is taken up by collecting ducts and drains into the central reservoir of the vacuole. **(b)** When full, the reservoir contracts, expelling the water through a pore in the plasma membrane.

water, the central vacuole is involved in the cell's water balance. It also provides a "dump site" for hazardous wastes, which plant cells often cannot excrete. Some plant cells store extremely poisonous substances, such as sulfuric acid, in their vacuoles. These poisons deter animals from munching on the otherwise tasty leaves. Vacuoles may also store sugars and amino acids not immediately needed by the cell for later use. Blue or purple pigments stored in central vacuoles are responsible for the colors of many flowers. As you will learn in Chapter 5, dissolved substances attract water into the vacuole. The water pressure within the vacuole, called *turgor pressure*, pushes the fluid portion of the cytoplasm up against the cell wall with considerable force. Cell walls are usually somewhat flexible, so both the overall shape and the rigidity of the cell depend on turgor pressure within the cell. Turgor pressure thus provides support for the non-woody parts of plants (look ahead to Fig. 5-11 to see what happens when you forget to water your houseplants).

Mitochondria Extract Energy from Food Molecules, and Chloroplasts Capture Solar Energy

Every cell requires a continuous supply of energy to manufacture complex molecules and structures, to acquire nutrients from the environment and excrete waste materials, to move, and to reproduce. All eukaryotic cells have *mitochondria* that convert energy stored in sugar to ATP. The cells of plants (and some protists) also have *chloroplasts*, which can capture energy directly from sunlight and store it in sugar molecules.

Most biologists accept the hypothesis that both mitochondria and chloroplasts evolved from prokaryotic bacteria that took up residence long ago within the cytoplasm of other prokaryotic cells, by a process called *endosymbiosis* (literally, "living together inside"). Mitochondria and chloroplasts are similar to each other and to prokaryotic cells in several ways. Both are about the size of some prokaryotic cells (1 to 5 micrometers in diameter). Both are surrounded by a double membrane; the outer membrane may have come from the original host cell and the inner membrane from the guest cell. Both have assemblies of enzymes that synthesize ATP, as would have been needed by an independent cell. Finally, both possess their own DNA and ribosomes that more closely resemble prokaryotic than eukaryotic ribosomes and DNA. The **endosymbiont hypothesis** of mitochondrial and chloroplast evolution is discussed in more detail in Chapter 17.

Mitochondria Use Energy Stored in Food Molecules to Produce ATP

All eukaryotic cells have **mitochondria**, which are sometimes called the "powerhouses of the cell" because they extract energy from food molecules and store it in the high-energy bonds of ATP. As you will see in Chapter 8, different amounts of energy can be released from a food molecule, depending on how it is metabolized. The breakdown

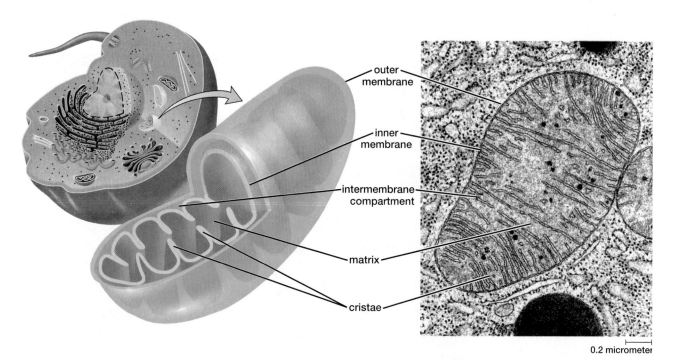

outer membrane

inner membrane

intermembrane compartment

matrix

cristae

0.2 micrometer

FIGURE 4-17 A mitochondrion
Mitochondria consist of a pair of membranes enclosing two fluid compartments: the intermembrane compartment between the outer and inner membranes, and the matrix within the inner membrane. The outer membrane is smooth, but the inner membrane forms deep folds called cristae.

of food molecules begins with enzymes in the cytosol and does not use oxygen. This **anaerobic** (without oxygen) metabolism does not convert much food energy into ATP energy. Mitochondria enable a eukaryotic cell to use oxygen to break down high-energy molecules even further. These **aerobic** (with oxygen) reactions generate energy much more effectively; 16 times more ATP is generated by aerobic metabolism in the mitochondria than by anaerobic metabolism in the cytosol. Not surprisingly, mitochondria are found in large numbers in metabolically active cells, such as muscle, and they are less abundant in cells that are less active, such as those of bone and cartilage.

Mitochondria are round, oval, or tubular organelles that possess a pair of membranes (**FIG. 4-17**). Although the outer mitochondrial membrane is smooth, the inner membrane forms deep folds called *cristae* (singular, *crista*, meaning "crest"). The mitochondrial membranes enclose two fluid-filled spaces: the *intermembrane compartment* between the inner and outer membranes and the *matrix*, or *inner compartment*, within the inner membrane. Some of the reactions that break down high-energy molecules occur in the fluid of the matrix inside the inner membrane; the rest are conducted by a series of enzymes attached to the membranes of the cristae within the intermembrane compartment. The role of mitochondria in energy production is described in detail in Chapter 8.

Chloroplasts Are the Sites of Photosynthesis

If chloroplasts didn't exist, you wouldn't be reading this; none of the eukaryotic life-forms that dominate Earth today would exist, as you will learn in Chapter 7. Photosyn-

thesis in the eukaryotic cells of plants and photosynthetic protists occurs in chloroplasts. **Chloroplasts** (**FIG. 4-18**) are specialized organelles surrounded by a double membrane. The inner membrane of the chloroplast encloses a fluid called the *stroma*. Within the stroma are interconnected stacks of hollow, membranous sacs. The individual sacs are called *thylakoids*, and a stack of sacs is a *granum* (plural, *grana*).

The thylakoid membranes contain the green pigment molecule **chlorophyll** (which gives plants their green color) as well as other pigment molecules. During photosynthesis, chlorophyll captures the energy of sunlight and transfers it to other molecules in the thylakoid membranes. These molecules in turn transfer the energy to ATP and other energy-carrier molecules. The energy carriers diffuse into the stroma, where their energy is used to drive the synthesis of sugar from carbon dioxide and water.

Plants Use Plastids for Storage

Chloroplasts are highly specialized **plastids**, which are organelles found only in plants and photosynthetic protists. Plastids are surrounded by a double membrane and serve a variety of functions. Plants and photosynthetic protists use nonchloroplast types of plastids as storage containers for various molecules, including pigments that give ripe fruits their yellow, orange, or red colors. In plants that continue growing from one year to the next, plastids store photosynthetic products from the summer for use during the following winter and spring. Most plants convert the sugars made during photosynthesis into starch, which is

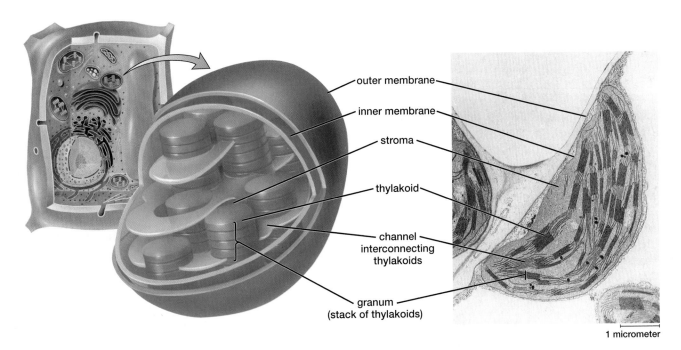

FIGURE 4-18 A chloroplast
Chloroplasts are surrounded by a double membrane, although the inner membrane is not usually visible in electron micrographs. The fluid stroma is enclosed by the inner membrane; within the stroma are stacks of thylakoid sacs called grana. Chlorophyll is embedded in the membranes of the thylakoids.

also stored in plastids (**FIG. 4-19**). Potatoes, for example, are composed almost entirely of cells stuffed with starch-filled plastids.

4.4 WHAT ARE THE MAJOR FEATURES OF PROKARYOTIC CELLS?

Prokaryotic Cells Are Small and Possess Specialized Surface Features

Most prokaryotic cells are very small (less than 5 micrometers in diameter), with a simple internal structure compared to eukaryotic cells (**FIG. 4-20**; compare this to Figs. 4-3 and 4-4). Nearly all prokaryotic cells are surrounded by a stiff cell wall, which protects the cell and confers its characteristic shape. Most prokaryotes take the form of rods (bacilli; **FIG. 4-20a**), spheres (cocci; **FIG. 4-20b**), or helices that resemble "squiggles" (spirilla; **FIG. 4-20b**). Several types of antibiotics, including penicillin, fight bacterial infections by interfering with cell wall synthesis, causing the bacteria to rupture. Some bacteria and archaea can move, propelled by flagella (different from eukaryotic flagella). Prokaryotes lack cilia.

Bacteria that infect other organisms—such as the bacteria that cause tooth decay, diarrhea, pneumonia, or urinary tract infections—possess surface features that help them adhere to specific host tissues, such as the surface of a tooth or the lining of the small intestine, lungs, or bladder. These surface features include *capsules* and *slime layers*, which are polysaccharide coatings that some bacteria secrete outside their cell walls, as well as *pili* (literally,

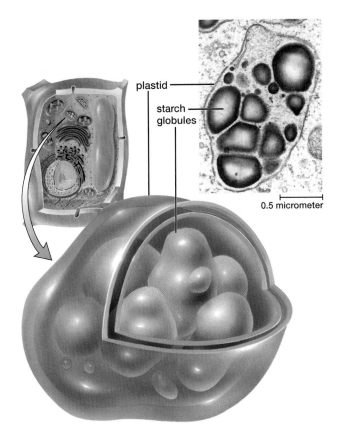

FIGURE 4-19 A plastid
Plastids, found in the cells of plants and photosynthetic protists, are organelles surrounded by a double outer membrane. Chloroplasts are the most familiar plastids; other types store various materials, such as the starch filling these plastids in potato cells.

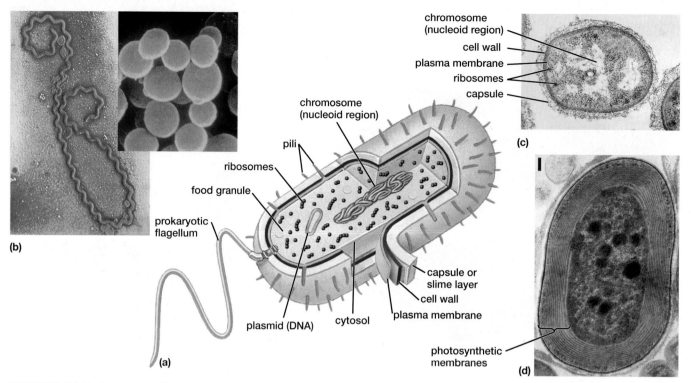

FIGURE 4-20 Prokaryotic cells
(a) Prokaryotic cells are simpler than eukaryotic cells. Some, such as in this diagram, are rod shaped. **(b)** Others take the form of spheres or helices. **(c)** A TEM photo of a spherical bacterium with a capsule. **(d)** Some photosynthetic bacteria have internal membranes where photosynthesis occurs.

"hairs"), which are proteins jutting outward from the prokaryotic cell wall. When Leeuwenhoek observed material scraped from his teeth under his simple microscope, he observed many bacteria that adhere using slime layers (see "Links to Life: Unwanted Guests"). Capsules and slime layers also help some prokaryotic cells avoid drying out. Some types of bacteria form sex pili, which are hollow protein tubes used to exchange genetic material (DNA) between bacterial cells. The features of prokaryotic cells are discussed and illustrated in more detail in Chapter 19.

Prokaryotic Cells Have Fewer Specialized Structures Within Their Cytoplasm

The cytoplasm of most prokaryotic cells is rather homogeneous in appearance when compared to eukaryotic cells. Prokaryotic cells generally have a single, circular *chromosome* consisting of a long strand of DNA that carries essential genetic information for the cell. The chromosome is usually coiled and localized in the central region of the cell; this region is called the **nucleoid** (Fig. 4-20). The nucleoid is not separated from the rest of the cytoplasm by a membrane. Most prokaryotic cells also contain small rings of DNA called *plasmids*, which are located outside the nucleoid. Plasmids usually carry genes that give the cell special properties; for example, some disease-causing bacteria possess plasmids that allow them to inactivate antibiotics, making them much more difficult to kill.

Prokaryotic cells lack nuclei and the other membrane-enclosed organelles (such as chloroplasts, mitochondria, ER, Golgi, and other components of the membrane system) that eukaryotic cells possess. Nonetheless, some prokaryotic cells do use membranes to organize the enzymes responsible for a series of biochemical reactions. The enzymes are situated in a particular sequence along the membrane to promote reactions in the necessary order. For example, photosynthetic bacteria possess internal membranes in which light-capturing proteins and enzymes that catalyze the synthesis of high-energy molecules are embedded in a specific order (Fig. 4-20d). In prokaryotic cells, reactions that harvest energy from the breakdown of sugars are catalyzed by enzymes that may be either localized along the inner plasma membrane or floating free in the cytosol.

Bacterial cytoplasm contains ribosomes (see Fig. 4-20a). Although their function is similar to that of eukaryotic ribosomes, they are smaller and contain different proteins. These ribosomes resemble those found in eukaryotic mitochondria and chloroplasts, providing support for the endosymbiotic hypothesis described earlier. Prokaryotic cytoplasm also may contain *food granules* that store energy-rich molecules, such as glycogen, but these are not enclosed by membranes.

You might wish to go back and look at Table 4-1 at this point to review the differences between prokaryotic and eukaryotic cells. The diversity and specialized structures of bacteria and archaea are covered in more detail in Chapter 19.

LINKS TO LIFE Unwanted Guests

In the late 1600s, Anton van Leeuwenhoek scraped white matter from between his teeth and viewed it through a microscope that he had constructed himself. To his consternation, he saw millions of cells that he called "animalcules": microscopic single-celled organisms that we now recognize as bacteria. Annoyed at the presence of these life-forms in his mouth, he attempted to kill them with vinegar and hot coffee—with little success. The warm, moist environment of the human mouth, particularly the crevices of the teeth and gums, is an ideal habitat for a variety of bacteria. Some forms of bacteria produce slime layers that help them and others adhere to the tooth. Each divides repeatedly to form a colony of offspring. Thick layers of bacteria, slime, and glycoproteins from saliva make up the white substance—called plaque—that Leeuwenhoek scraped from his teeth. Sugar in foods and beverages nourishes the bacteria, which break down the sugar into lactic acid. The acid eats away at the tooth enamel, producing tiny crevices in which the bacteria multiply further, eventually producing a cavity. Fluoride in toothpaste and drinking water can help prevent cavities by incorporating fluoride into the enamel, helping it resist attack by acid. So, although he didn't know why, Leeuwenhoek was right to be concerned by the "animalcules" in his mouth!

CASE STUDY REVISITED SPARE PARTS FOR HUMAN BODIES

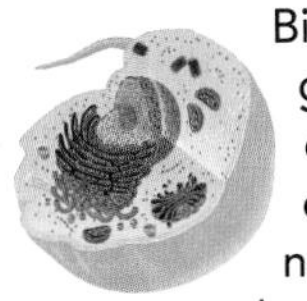

BIOETHICS

Bioengineering tissues and organs such as skin requires the coordinated efforts of biochemists, biomedical engineers, cell biologists, and physicians. To heal broken bones, teams of researchers are working to use degradable plastics, incorporating protein growth factors into the material. These growth factors would encourage nearby bone cells and tiny blood vessels to invade the plastic as it breaks down, eventually replacing it with the patient's own bone.

In laboratories across the world, teams of scientists are working to grow not only skin and bone but cartilage, heart valves, bladders, and breast tissue on plastic scaffolds, and they are implanting some of these artificial tissues into experimental animals. The mouse shown in **FIGURE 4-21** is incubating an ear-shaped scaffold seeded with human cartilage cells (cartilage supports the natural ear). In the future, artificial ears might be grown directly on people with missing or deformed ears.

Researchers continue to refine tissue-culturing techniques and to develop better scaffolding materials with the goal of duplicating entire organs. Bioartificial bladders have been created using muscle and bladder-lining cells from patients with improperly functioning bladders. The cells were seeded onto a bladder-shaped scaffold composed of collagen and transplanted into the patients. Seven recipients of these engineered bladders continue to report improved bladder function after about four years.

A major challenge in growing new organs is that, unlike the bladder, most organs are relatively thick, and it's difficult to deliver nutrients to the innermost cells. To solve this problem, Dr. Joseph Vacanti of Massachusetts General Hospital in Boston has teamed up with a microengineering expert to devise a bioengineered liver with its own blood supply. The team created a plastic cast of a liver's blood vessels by injecting liquid plastic into the vessels and (after the plastic hardened) dissolving away the surrounding tissue. They then produced a three-dimensional computer image of the plastic blood vessel network and used it to create a mold for scaffolding material. This material will be seeded with at least seven different types of cells that form the bulk of the liver. The network of blood vessels is represented by channels of varying size penetrating the scaffold framework. The researchers will inject these channels with blood vessel cells, which will hopefully line the channels and eventually form new vessels. Since the complexity of the project is staggering, it is unlikely that any of the over 17,000 people currently awaiting liver transplants in the United States will benefit from this research. In the future, however, bioengineered organs could save hundreds of thousands of lives worldwide each year.

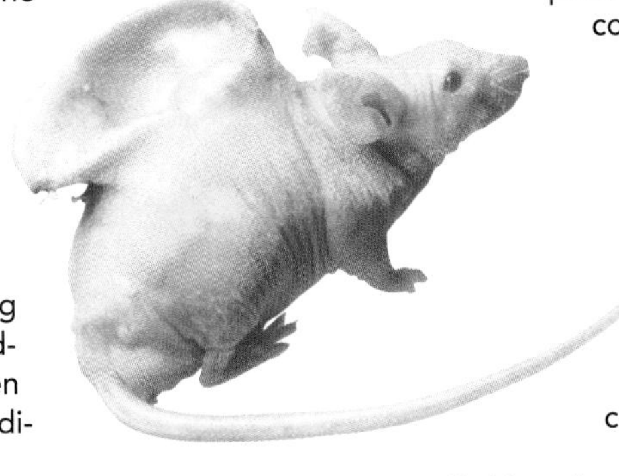

FIGURE 4-21 Ear-shaped scaffold under the skin of a mouse

Consider This The mouse photo below created controversy, angering some individuals who believe that this was an inappropriate use of laboratory animals. But virtually all modern medicinal drugs and medical procedures were developed using animal research. Do you believe that using certain types of animals or performing some types of animal experimentation is unethical and should be prohibited? If so, explain your position. If you oppose all use of any animals in research, what techniques do you believe medical researchers should use to develop better treatments for human health problems?

CHAPTER REVIEW

SUMMARY OF KEY CONCEPTS

4.1   What Is the Cell Theory?

The principles of the cell theory are as follows:
- Every living organism is made up of one or more cells.
- The smallest living organisms are single cells, and cells are the functional units of multicellular organisms.
- All cells arise from preexisting cells.

4.2   What Are the Basic Attributes of Cells?

Cells are limited in size because they must exchange materials with their surroundings by diffusion, a slow process that requires that the interior of the cell must never be too far from the plasma membrane. All cells are surrounded by a plasma membrane that regulates the interchange of materials between the cell and its environment. Cells contain cytoplasm that consists of a watery cytosol and various organelles, not including the nucleus. All cells use DNA as a genetic blueprint and RNA to assist in protein sysnthesis based on this blueprint. All cells obtain the materials to generate the molecules of life, and the energy to power this synthesis, from their living and nonliving environment. There are two fundamentally different types of cells: prokaryotic and eukaryotic.

Web Tutorial 4.1 Cell Structure

4.3   What Are the Major Features of Eukaryotic Cells?

Cells of plants, fungi, and some protists are supported by porous cell walls outside the plasma membrane. All eukaryotic cells have an internal cytoskeleton of protein filaments that organizes and gives shape to eukaryotic cells and moves and anchors organelles. Some eukaryotic cells have cilia or flagella, extensions of the plasma membrane that contain microtubules in a characteristic pattern. These structures move fluids past the cell or move the cell through its fluid environment.

Genetic material (DNA) is contained within the nucleus, which is bounded by the double membrane of the nuclear envelope. Pores in the nuclear envelope regulate the movement of molecules between nucleus and cytoplasm. The genetic material is organized into strands called *chromosomes*, which consist of DNA and proteins. The nucleolus consists of ribosomal RNA and ribosomal proteins, as well as the genes that code for ribosome synthesis. Ribosomes are particles of ribosomal RNA and protein that are the sites of protein synthesis.

The membrane system of a cell consists of the plasma membrane, endoplasmic reticulum (ER), Golgi apparatus, vacuoles, and vesicles derived from these membranes. Endoplasmic reticulum consists of a series of interconnected compartments whose membranes have enzymes to produce more membrane lipids. The ER is a major site of membrane synthesis within the cell. Rough ER, which bears ribosomes, manufactures many cellular proteins. Smooth ER, lacking ribosomes, manufactures lipids such as steroid hormones, detoxifies drugs and metabolic wastes, breaks glycogen into glucose, and stores calcium. The Golgi apparatus is a series of membranous sacs derived from the ER. The Golgi apparatus processes and modifies materials synthesized in the rough ER. Substances modified in the Golgi are packaged into vesicles for transport elsewhere in the cell. Lysosomes are vesicles that contain digestive enzymes, which digest food particles and defective organelles.

All eukaryotic cells contain mitochondria—organelles that use oxygen to complete the metabolism of food molecules, capturing much of their energy as ATP. Cells of plants and some protists contain plastids—including chloroplasts that capture the energy of sunlight during photosynthesis—that enable the cells to manufacture organic molecules, particularly sugars, from simple inorganic molecules. Both mitochondria and chloroplasts probably originated from bacteria. Storage plastids store pigments or starch.

Many eukaryotic cells contain sacs, called *vacuoles*, that are bounded by a single membrane and function to store food or wastes, excrete water, or support the cell. Some protists have contractile vacuoles, which collect and expel water. Plants use central vacuoles to support the cell as well as to store wastes and toxic materials.

Web Tutorial 4.2 Membrane Traffic

4.4   What Are the Major Features of Prokaryotic Cells?

Prokaryotic cells are generally very small and have a simple internal structure. Most are surrounded by a relatively stiff cell wall. The cytoplasm of prokaryotic cells lacks membrane-enclosed organelles (although some photosynthetic bacteria have extensive internal membranes). A single, circular strand of DNA is found in the nucleoid. Table 4-1 compares prokaryotic cells to the eukaryotic cells of plants and animals.

Study Note

Figures 4-3, 4-4, and 4-20 illustrate the overall structure of animal, plant, and prokaryotic cells, respectively. Table 4-1 lists the principal organelles, their functions, and their occurrence in animals, plants, and prokaryotes.

KEY TERMS

THINKING THROUGH THE CONCEPTS

1. Diagram "typical" prokaryotic and eukaryotic cells, and describe their important similarities and differences.

2. Which organelles are common to both plant and animal cells, and which are unique to each?

3. Define *stroma* and *matrix*.

4. Describe the nucleus, including the nuclear envelope, chromatin, chromosomes, DNA, and nucleolus.

5. What are the functions of mitochondria and chloroplasts? Why do scientists believe that these organelles arose from prokaryotic cells?

6. What is the function of ribosomes? Where in the cell are they typically found? Are they limited to eukaryotic cells?

7. Describe the structure and function of the endoplasmic reticulum and Golgi apparatus.

8. How are lysosomes formed? What is their function?

9. Diagram the structure of cilia and flagella.

APPLYING THE CONCEPTS

1. If samples of muscle tissue were taken from the legs of a world-class marathon runner and a sedentary individual, which would you expect to have a higher density of mitochondria? Why?

2. One of the functions of the cytoskeleton in animal cells is to give shape to the cell. Plant cells have a fairly rigid cell wall surrounding the plasma membrane. Does this mean that a cytoskeleton is unnecessary for a plant cell?

3. Most cells are very small. What physical and metabolic constraints limit cell size? What problems would an enormous cell encounter? What adaptations might help a very large cell survive?

FOR MORE INFORMATION

de Duve, C. "The Birth of Complex Cells." *Scientific American*, April 1996. A famous cell biologist describes the mechanisms by which the first eukaryotic cells were produced from prokaryotic ancestors.

Ford, B. J. "The Earliest Views." *Scientific American*, April 1998. The author used the original microscopes of Antoni van Leeuwenhoek to see the microscopic world as Leeuwenhoek saw it. Photographic images taken through these early and very primitive instruments reveal remarkable detail.

Hoppert, M., and Mayer, F. "Prokaryotes." *American Scientist*, November–December 1999. These relatively simple cells actually possess a great deal of internal organization.

Ingber, D. E. "The Architecture of Life." *Scientific American*, January 1998. Counteracting forces stabilize the design of organic structures, from carbon compounds to the cytoskeleton-reinforced architecture of the cell.

CHAPTER 5

Membrane Structure and Function

A rattlesnake prepares to strike. (Inset) A brown recluse spider.

CASE STUDY VICIOUS VENOMS

EAGER TO EXPLORE their new environs, Karl and Mark, freshmen roommates at a university in southern California, drove to a trailhead in the Mojave Desert. Karl kidded Mark about his cell phone—what kind of a wilderness experience could they have with a phone along? Mark joked about the large field guide, *Desert Flora and Fauna*, weighing down Karl's pack. Competitive and athletic, they spotted a rocky bluff and raced each other to the top. As Karl reached for a rock outcropping to pull himself upward, he gasped at the feel of smooth, scaly coils writhing under his hand. A sudden unmistakable warning rattle was followed almost immediately by an intense burning pain at the base of his thumb. Seeing a huge snake slithering back into a crevice, Mark quickly dialed 911. By the time they

heard the medical evacuation helicopter, they had used Karl's field guide to identify the rattler as a western diamondback (see the chapter-opening photo). Before he reached the hospital, a large bruised-looking area was spreading over Karl's hand, his blood pressure had dropped, and the paramedics were administering oxygen because he was gasping for air.

Meanwhile, in rural Kentucky, Melissa was anticipating a romantic dinner with her fiancé in front of a cozy fire. As she loaded her arms with wood from the dim old storage shed behind her home, she was unaware of the leggy brown spider, clinging to a small web deep in the pile, that now found itself pressed against her skin (see the chapter-opening photo, inset). She never felt the bite of the brown recluse spider. Hours later, upon experiencing a

stinging sensation, she noticed a small red swelling on her arm. Melissa had trouble sleeping as the pain increased. The next morning, alarmed at the spreading purplish welt, Melissa sought medical help. After a series of tests to rule out other causes, the doctor said she suspected a brown recluse bite. In many cases, the doctor warned, such bites kill both the surrounding skin and underlying tissue, resulting in an open wound that could become quite large and might take months to heal. When an anguished Melissa asked if there wasn't some medication to prevent this, the doctor regretfully shook her head.

How do rattlesnake and brown recluse spider venoms cause leaky blood vessels, disintegrating skin, and other potentially life-threatening symptoms throughout the body? Can venoms attack cell membranes?

5.1 HOW IS THE STRUCTURE OF A MEMBRANE RELATED TO ITS FUNCTION?

Cell Membranes Isolate the Cell Contents While Allowing Communication with the Environment

As you now know, all cells, as well as many of the organelles within eukaryotic cells, are surrounded by membranes. Cell membranes perform several crucial functions:

- They selectively isolate the cell's contents from the external environment, allowing concentration gradients of dissolved substances to be produced across the membrane.
- They regulate the exchange of essential substances between the cell and the extracellular fluid, or between membrane-enclosed organelles and the surrounding cytosol.
- They communicate with other cells.
- They create attachments within and between cells.
- They regulate many biochemical reactions.

These are formidable tasks for a structure so thin that 10,000 membranes stacked atop one another would scarcely equal the thickness of this page. The key to membrane function lies in its structure. Membranes are not simply uniform sheets; they are complex, heterogeneous structures whose different parts perform very distinct functions, and they change dynamically in response to their surroundings.

All the membranes of a cell have a similar basic structure: proteins floating in a double layer of phospholipids, (see Chapter 3). Phospholipids are responsible for the isolating function of membranes, whereas proteins are responsible for selectively exchanging substances and communicating with the environment, controlling biochemical reactions associated with the cell membrane, and forming attachments.

Membranes Are "Fluid Mosaics" in Which Proteins Move Within Layers of Lipids

Prior to the 1970s, although cell biologists knew that cell membranes contained proteins and lipids, microscopes lacked the resolution to determine their exact structure. In 1972, cell researchers S. J. Singer and G. L. Nicolson developed the **fluid mosaic model** of cell membranes, which is

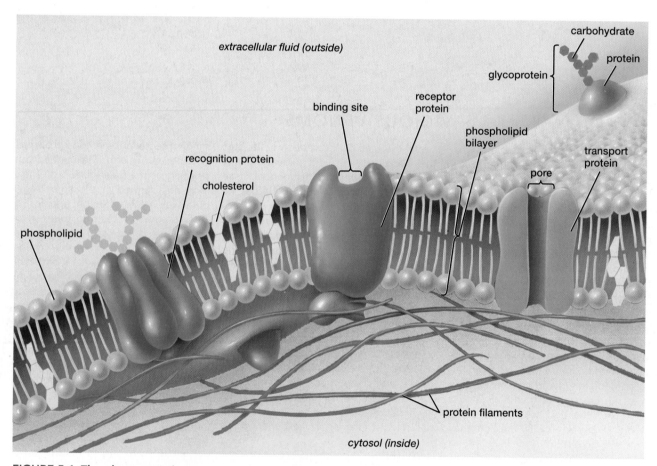

FIGURE 5-1 The plasma membrane
The plasma membrane is a bilayer of phospholipids that form a fluid matrix in which various proteins (blue) are embedded. Many proteins have carbohydrates attached to them, forming glycoproteins. Three of the five major types of membrane proteins are illustrated here: recognition, receptor, and transport proteins.

now known to be accurate. According to this model, each membrane consists of a mosaic or "patchwork" of different proteins that constantly shift and flow within a viscous fluid formed by a double layer of phospholipids (**FIG. 5-1**). Although the components of the plasma membrane remain relatively constant, the overall distribution of proteins and various types of phospholipids can change over time. Let's look more closely at the structure of membranes.

The Phospholipid Bilayer Is the Fluid Portion of the Membrane

As you learned in Chapter 3, a phospholipid consists of two very different parts: a polar, hydrophilic head (attracted to water) and a pair of nonpolar, hydrophobic fatty acid tails (repelled by water). Membranes contain many different phospholipids of the general type shown in **FIGURE 5-2**. Notice that in this particular phospholipid, a double bond (which makes the lipid unsaturated) creates a kink in the fatty acid tail that helps keep the membrane fluid.

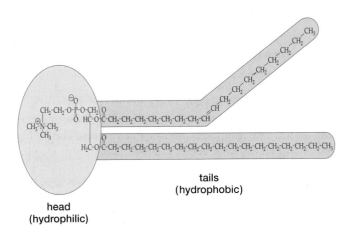

FIGURE 5-2 Phospholipid

All cells are surrounded by a watery medium. Single-celled organisms may live in fresh water or in the ocean, while animal cells are bathed in a weakly salty *extracellular fluid* that filters out of the blood. The cytosol (the fluid inside the cell within which all the organelles are suspended; see Chapter 4) is mostly water. Thus, plasma membranes separate the watery cytosol from its watery external environment, and similar membranes surround watery compartments within the cell. Under these conditions, phospholipids spontaneously arrange themselves into a double layer called a **phospholipid bilayer** (**FIG. 5-3**). Hydrogen bonds can form between water and the phospholipid heads, so the hydrophilic heads face both the watery cytosol and the extracellular fluid, forming the inner- and outermost portions of the bilayer. Hydrophobic interactions (see Chapter 2) cause the phospholipid tails to hide inside the bilayer.

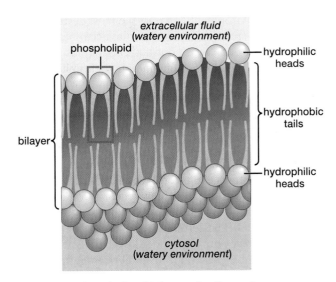

FIGURE 5-3 Phospholipid bilayer of cell membrane

Individual phospholipid molecules are not bonded to one another, and membranes include phospholipids with unsaturated fatty acids whose double bonds introduce "kinks" into their "tails" (see Chapter 3). These features allow phospholipids to move about easily within each layer, and they make the bilayer quite fluid. So, the more double bonds there are to form kinks in the lipid tails, the more fluid the membrane will be (**FIG. 5-4**).

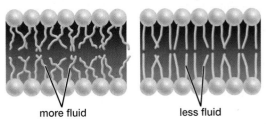

FIGURE 5-4 Kinks in phospholipid tails increase membrane fluidity

Cells may have different degrees of saturation in the lipid bilayer, and these differences in membrane fluidity allow them to perform different functions or to function well in different environments. For example, membranes tend to become more fluid at high temperatures (because molecules move faster) and less fluid at low temperatures (molecules move more slowly). Cell membranes of organisms living in low-temperature environments are therefore likely to be rich in unsaturated phospholipids to allow the membrane to retain the necessary fluidity (see "Evolutionary Connections: Caribou Legs and Membrane Diversity" later in this chapter).

Most biological molecules, including salts, amino acids, and sugars, are polar and water soluble, and so they are hydrophilic. In fact, most substances that contact a cell are water soluble—hydrophilic—and so cannot easily pass through the nonpolar, hydrophobic fatty acid tails of the phospholipid bilayer. The phospholipid bilayer is largely responsible for the first of the three membrane functions listed earlier: selectively isolating the cell's contents from the external environment. Some of the most devastating effects of certain snake and spider venoms occur because they contain enzymes that break down phospholipids and thereby destroy cell membranes, eliminating their ability to confine the contents of the cell. The isolation provided by the plasma membrane is not complete, however. As we will describe later, very small molecules—such as water, oxygen, and carbon dioxide—as well as larger uncharged, lipid-soluble molecules can pass through the phospholipid bilayer.

In most animal cells, the phospholipid bilayer of membranes also contains cholesterol (see Fig. 5-1). Some cellular membranes have just a few cholesterol molecules; others have as many cholesterol molecules as they do phospholipids. Cholesterol affects membrane structure and function in several ways: it makes the bilayer stronger and more flexible, less fluid at higher temperatures, less solid at lower temperatures, and less permeable to water-soluble substances such as ions or monosaccharides.

The flexible, fluid nature of the bilayer is very important for membrane function. As you breathe, move your eyes, and turn the pages of this book, cells in your body change shape. If plasma membranes were stiff instead of flexible, cells would break open and die. Further, as you learned in Chapter 4, membranes within eukaryotic cells are in constant motion. Membrane-enclosed compartments ferry substances into the cell, carry materials within the cell, and expel them to the outside, merging membranes in the process. This flow and merger of membranes is made possible by the fluid nature of the phospholipid bilayer.

A Variety of Proteins Form a Mosaic Within the Membrane

Thousands of proteins are embedded within or attached to the surface of a membrane's phospholipid bilayer. Many of the proteins in plasma membranes have carbohydrate groups attached to the portion that is exposed on the outer membrane surface (see Fig. 5-1). These are called **glycoproteins** ("glyco" is from the Greek word for "sweet" and refers to the carbohydrate portion with its sugar-like subunits; see Chapter 3).

Membrane proteins may be grouped into five major categories based on their function: receptor proteins, recognition proteins, enzymatic proteins, attachment proteins, and transport proteins.

Most cells bear dozens of types of **receptor proteins** spanning their plasma membranes. Each receptor protein has a binding site for a specific molecule—a hormone, for example. When the appropriate molecule binds to the receptor, the receptor is activated (often by changing shape) and this in turn may trigger a sequence of chemical reactions within the cell that results in changes in cellular activities (**FIG. 5-5**).

A hormone produced by the adrenal gland, for example, causes stronger heart muscle contractions when it binds the appropriate receptors. Other molecules binding to various receptors may initiate cell division, movement toward a nutrient source, or secretion of hormones. Some receptor proteins act like gates on channel proteins; specific chemicals

binding these receptors cause the gates to open and allow ions to flow through the channels. Receptors allow cells of the immune system to recognize and attack foreign invaders that can cause disease. They also allow nerve cells to communicate with one another, and cells throughout the body to respond to hormones.

Recognition proteins are glycoproteins located on the surfaces of cells that serve as identification tags (see Fig. 5-1). The cells of your immune system, for example, can recognize a bacterium or virus as a foreign invader and target it for destruction, in part by responding to its unique glycoproteins. These same immune cells ignore the trillions of cells in your own body because your body cells have different identification glycoproteins on their surfaces. Glycoproteins on the surfaces of red blood cells bear different sugar groups and determine whether your blood is type A, B, AB, or O (see Chapter 12). Human sperm recognize the unique glycoproteins on human egg cells, thus allowing fertilization to occur.

Enzymes are proteins that are often attached to the inner surfaces of membranes. These promote chemical reactions that synthesize or break apart biological molecules without being changed themselves. We discuss enzymes in detail in Chapter 6.

Attachment proteins anchor the cell membrane in various ways. Some attachment proteins bind the plasma membrane to the network of protein filaments within the cytoplasm called the cytoskeleton (see Fig. 4-6). The attachments between plasma membrane proteins and the underlying protein filaments produce the characteristic shapes of animal cells, from the dimpled discs of red blood cells to the elaborate branching of nerve cells. Other membrane attachment proteins bind the cell to a matrix of other protein fibers that exist outside the cell. Still other attachment proteins form junctions between adjacent cells, described later in this chapter.

Transport proteins regulate the movement of hydrophilic molecules through the plasma membrane. Some transport proteins, called *channel proteins*, form channels whose central pores allow specific ions or water molecules to pass through the membrane along their concentration gradients (see Fig. 5-1). Other transport proteins, called *carrier proteins*, have binding sites that can temporarily attach to specific molecules on one side of the membrane. The protein then changes shape (in some cases using cellular energy), moves the molecule across the membrane, and releases it on the other side. In the next section, you will learn more about transport proteins.

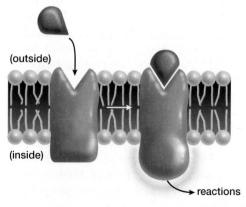

(outside)

(inside)

reactions

FIGURE 5-5 Receptor activation

5.2 HOW DO SUBSTANCES MOVE ACROSS MEMBRANES?

Molecules in Fluids Move in Response to Gradients

You now know that substances move directly across membranes by diffusion through the phospholipid bilayer, or by traveling through specialized transport proteins. To understand this process more fully requires some background knowledge and definitions. Because the plasma membrane separates the fluid in the cell's cytosol from its fluid extracellular environment, let's begin our study of membrane transport with a brief look at the characteristics of fluids, starting with a few definitions:

- **Fluid:** any substance whose molecules move freely past one another; as a result, fluids have no defined shape. Both liquids and gases are fluids.

- **Solutes** and **solvents:** A solute is a substance that can be dissolved (dispersed into individual atoms, molecules, or ions) in a solvent, which is a fluid (usually a liquid) capable of dissolving the solute. Water, in which all biological processes occur, dissolves more different substances than any other solvent and is sometimes called the "universal solvent."

- The **concentration** of a substance in a fluid is a measurement of the number of molecules of that substance contained in a given volume of the fluid. The term can refer to molecules in a gas; for example, the concentration of oxygen in air. The concentration of a substance defines the amount of solute in a given amount of solvent.

- A **gradient** is a physical difference in properties such as temperature, pressure, electrical charge, or concentration of a particular substance in a fluid between two adjoining regions of space. Basic principles of physics tell us that it requires energy to create gradients and that, over time, gradients tend to break down unless energy is supplied to maintain them or unless a barrier separates them. For example, gradients in temperature cause a flow of energy from the higher-temperature region to the lower-temperature region. Electrical gradients can drive the movement of ions. Gradients of concentration or pressure cause molecules or ions to move from one region to the other in a manner that tends to equalize the difference. Cells use energy and the unique properties of their cell membranes to generate **concentration gradients** of ions and various molecules in solution within their cytosol relative to their watery surroundings.

It is also important to be aware that, at temperatures above absolute zero (−459.4°F, or −273°C), atoms, molecules, and ions are in constant random motion. As the temperature rises, their rate of motion increases, and at temperatures that support life, these particles are moving quite rapidly indeed. So molecules and ions in solution are continuously bombarding and shifting past one another. Over time, random movements produce a net movement of molecules from regions of high concentration to regions of low concentration, by a process called **diffusion**. If there are no factors opposing this movement, such as electrical charge, pressure differences, or physical barriers, the random movement of molecules will continue until the substance is evenly dispersed throughout the fluid.

To envision how the random movement of molecules or ions within a fluid equalizes concentration gradients, consider a sugar cube dissolving in coffee, or perfume molecules moving from an open bottle into the air. In each of these examples you have a concentration gradient. If you leave the perfume jar open long enough or forget your coffee, eventually you will have an empty perfume bottle and a very fragrant room, or cold, but uniformly sweet, coffee. In an analogy with gravity, we refer to such movements as going "down" the concentration gradient.

To watch diffusion in action, place a drop of food coloring in a glass of water. With time, the drop will spread and become paler, until eventually, even without being stirred, the entire glass of water will be uniformly faintly colored. Random motion propels dye molecules both into and out of the dye droplet. However, because there is much more water than dye, dye molecules have a greater chance of moving randomly into the water than back into the dye droplet (**FIG. 5-6**). Simultaneously, random motion causes some water molecules to enter the dye droplet. So there is a net movement of dye into the water, and of water into the dye. At first there is a very steep concentration gradient, and the dye diffuses rapidly. As the concentration differences

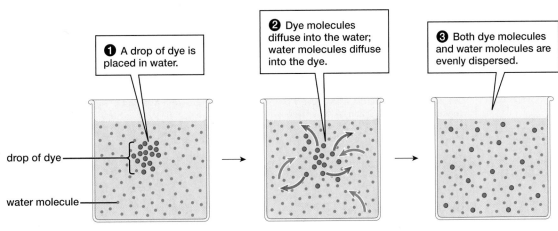

❶ A drop of dye is placed in water.

❷ Dye molecules diffuse into the water; water molecules diffuse into the dye.

❸ Both dye molecules and water molecules are evenly dispersed.

drop of dye

water molecule

FIGURE 5-6 Diffusion of a dye in water

lessen, the dye diffuses more and more slowly. In other words, the greater the concentration gradient, the faster the rate of diffusion will be. The net movement of dye will continue until the dye is uniformly dispersed in the water. Then, with no concentration gradient of either dye or water, diffusion stops. Individual molecules still move about randomly, but no overall changes occur in concentration of either water or dye.

If you compare diffusion of dye in hot and cold water, you will see that heat increases the diffusion rate. This is because heat increases the random movement rate of molecules. But even at body temperature, diffusion cannot move molecules rapidly over long distances. As you learned in Chapter 4, the slow rate of diffusion over long distances is one of the reasons that most cells are extremely small, and it is why larger cells tend to be very thin.

SUMMING UP
The Principles of Diffusion

- Diffusion is the net movement of molecules down a gradient from high to low concentration.
- The greater the concentration gradient, the faster the rate of diffusion.
- The higher the temperature, the faster the rate of diffusion.
- If no other processes intervene, diffusion will continue until the concentration gradient is eliminated.
- Diffusion cannot move molecules rapidly over long distances.

Movement Across Membranes Occurs by Both Passive and Active Transport

There are significant concentration gradients of ions and molecules across the plasma membranes of every cell. This occurs because proteins in the cell membrane expend energy to create these gradients, and the selective permeability of the plasma membrane helps maintain them. In its role as gatekeeper of the cell, the plasma membrane provides for two types of movement: *passive transport* and *energy-requiring transport* (Table 5-1). The movement of molecules directly through a cell membrane using energy is described as *active transport*.

Passive transport can be described as diffusion of substances across cell membranes. Because diffusion always occurs down concentration gradients, passive transport requires no expenditure of energy. Concentration gradients drive the movement and determine the direction of movement across the membrane. The phospholipids and the protein channels of the plasma membrane regulate which ions or molecules can cross, but they do not influence the direction of movement.

During **active transport**, the cell uses energy to move substances across a membrane against a concentration gradient. In this case, transport proteins control the direction of movement. The difference between passive and active transport can be compared to riding a bike. If you don't pedal, you can go only downhill, as in passive transport. However, if you put enough energy into pedaling, you can go uphill as well, as in active transport. Active transport that uses energy to create a concentration gradient can be compared to using your muscular energy to pedal a bike up a hill. Passive transport by diffusion that reduces concentration gradients is like coasting down the hill—no energy is expended. Keep in mind that both passive transport and coasting down a hill require an initial investment of energy, either by active transport to create the concentration gradient, or by muscular effort to move your body and bike up to the hilltop.

Passive Transport Includes Simple Diffusion, Facilitated Diffusion, and Osmosis

Diffusion can occur within a fluid or across a membrane separating two fluid compartments. Many molecules cross plasma membranes by diffusion, driven by differences between their concentration in the cytosol and in the external environment. Due to the properties of the plasma membrane, different molecules cross the plasma membrane at different locations and at different rates.

Table 5-1 Transport Across Membranes

Passive transport	Diffusion of substances across a membrane down a gradient of concentration, pressure, or electrical charge. Does not require the cell to expend energy.
Simple diffusion	Diffusion of water, dissolved gases, or lipid-soluble molecules through the phospholipid bilayer of a membrane.
Facilitated diffusion	Diffusion of water, ions, or water-soluble molecules through a membrane *via* a channel or carrier protein.
Osmosis	Diffusion of water across a selectively permeable membrane from a region of higher free water concentration to a region of lower free water concentration.
Energy-requiring transport	Movement of substances into or out of a cell using cellular energy, usually ATP.
Active transport	Movement of individual small molecules or ions against their concentration gradients through membrane-spanning proteins.
Endocytosis	Movement of large particles, including large molecules or entire microorganisms, into a cell; occurs as the plasma membrane engulfs the particle in a membranous sac that enters the cytosol.
Exocytosis	Movement of materials out of a cell; occurs as the plasma membrane encloses the material in a membranous sac that moves to the cell surface, fuses with the plasma membrane, and opens to the outside, allowing its contents to diffuse out.

(a) Simple diffusion through the phospholipid bilayer

(b) Facilitated diffusion through a channel protein

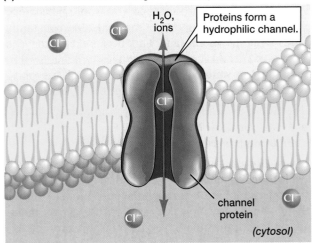

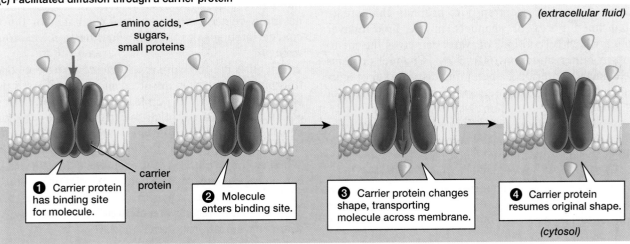

(c) Facilitated diffusion through a carrier protein

FIGURE 5-7 Diffusion through the plasma membrane
(a) Simple diffusion: gases such as oxygen and carbon dioxide and lipid-soluble molecules can diffuse directly through the phospholipids. **(b)** Facilitated diffusion through a channel protein: channels (pores) allow passage of some water-soluble molecules, principally ions, that cannot diffuse directly through the bilayer. **(c)** Facilitated diffusion through a carrier protein. EXERCISE Imagine an experiment that measures the initial rate of diffusion into cells placed in sucrose solutions of various concentrations. Sketch a graph (initial diffusion rate versus solution concentration) that shows the result expected if diffusion is simple, and a graph that shows the result expected for facilitated diffusion.

Therefore, plasma membranes are said to be **selectively permeable**—that is, they *selectively* allow some molecules to pass through, or *permeate*, but prevent other molecules from passing.

Some Molecules Move Across Membranes by Simple Diffusion

Lipid-soluble molecules such as alcohol, some vitamins (A, D, and E), and steroid hormones diffuse directly across the phospholipid bilayer in either direction, as do very small molecules, including water and dissolved gases such as oxygen and carbon dioxide. This process is called **simple diffusion** (**FIG. 5-7a**). Generally, the rate of simple diffusion is a function of the concentration gradient across the mem-

brane, the temperature, the size of the molecule, and how easily it dissolves in lipids (its lipid *solubility*). A larger concentration gradient, higher temperature, smaller molecular size, and higher lipid solubility all increase the rate of simple diffusion.

Right now you might be asking yourself, "How can water—a polar molecule—diffuse across the hydrophobic (literally "water-fearing") phospholipid bilayer?" The answer is that enormous numbers of water molecules bump randomly against the cell membrane all the time. Because there is no bonding between the phospholipids, a relatively small number of water molecules stray into the thicket of phospholipid tails. As their random motion continues, some reach the far side of the membrane. You might predict that diffusion of water through the bilayer would be a relatively slow and inefficient process, and you would be correct. However, because so many water molecules are constantly colliding with the membrane, and because cells have a large area of membrane relative to their volume, significant amounts of water manage to leak though the phospholipid bilayer.

Other Molecules Cross Membranes by Facilitated Diffusion Using Membrane Transport Proteins

Most ions (for example, K^+, Na^+, Ca^{2+}), and water-soluble molecules such as amino acids and monosaccharides (simple sugars), cannot move through the phospholipid bilayer on their own. These molecules can diffuse across membranes only with the aid of specific transport proteins, either *channel proteins* or *carrier proteins*. This process is called **facilitated diffusion**.

Channel proteins are transport proteins that create protein-lined pores, or channels, in the lipid bilayer through which certain ions or water can cross the membrane in either direction (**FIG. 5-7b**). Channel proteins have a specific interior diameter and distribution of electrical charges lining the pore, so that only certain types of ions can pass through them. Nerve cells, for example, have separate channels for sodium ions, potassium ions, and calcium ions. Although water can diffuse directly through the phospholipid bilayer in all cells, many cells have specialized water channels called **aquaporins** (literally "water pores"). Aquaporins allow water to cross membranes by facilitated diffusion, which is faster than simple diffusion (see "Scientific Inquiry: The Discovery of Aquaporins").

Carrier proteins are transport proteins with distinct regions called *active sites* that bind specific molecules from the cytosol or extracellular fluid, such as particular amino acids, sugars, or small proteins. The binding triggers a change in the shape of the carrier that allows the molecules to pass through the protein and across the membrane. The carrier proteins that make facilitated diffusion possible do not use cellular energy, and they can move molecules only down their concentration gradients (**FIG. 5-7c**).

Osmosis Is the Diffusion of Water Across Selectively Permeable Membranes

The diffusion of water through membranes from regions of high water concentration to regions of lower water concentration has such dramatic and important effects on cells that we refer to it by a special name: **osmosis**. (Although water movement across membranes is increased by greater pressures and temperatures, here we focus on its movement in response to concentration gradients.)

What do we mean when we describe a solution as having a "high water concentration" or a "low water concentration"? The answer is simple: pure water has the highest possible water concentration. Any substance that dissolves in water (any solute) displaces some of the water molecules, and it also forms hydrogen bonds with many more of the water molecules, preventing them from moving across a water-permeable membrane. So, the higher the concentration of dissolved substances, the lower the concentration of "free" water that is available to move across the membrane. As you might predict, the higher the concentration of dissolved substances in a solution, the greater the tendency of water to move across a water-permeable membrane into that solution.

For example, when sugar solutions are separated by a membrane that is selectively permeable to water, water will move by osmosis from the solution with a lower sugar concentration into the solution with a higher sugar concentration. This occurs because there are more free water molecules in the less-concentrated sugar solution, so more water molecules will collide with—and move through—the water-permeable membrane on that side. Because it appears to draw water across the membrane, the more concentrated sugar solution is described as having a *higher osmotic strength* than that of the less-concentrated sugar solution, which has a *lower osmotic strength*.

Scientists use the word *tonicity* to compare the concentrations of substances dissolved in water across a membrane that is selectively permeable to water. Solutions with equal concentrations of dissolved substances (and thus equal concentrations of water) are described as being **isotonic** to one another (the prefix "iso" means "same"). When isotonic solutions (for example, two solutions each containing 20 percent sugar) are separated by a water-permeable membrane, such as the bag in **FIGURE 5-8**, there is no net movement of water between them, because their water concentrations are the same.

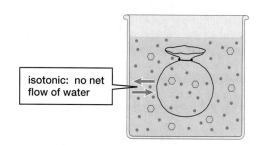

isotonic: no net flow of water

FIGURE 5-8 Isotonic solution

What if a water-permeable membrane separates a solution with a higher concentration of dissolved substances from another with a lower concentration of solutes? In this case scientists describe the more concentrated solution as having a higher tonicity, or being **hypertonic** to the other solution (the prefix "hyper" means "greater than"), and the more dilute solution as having lower tonicity or being **hypotonic** ("hypo" means "below"). Because they have more solute molecules, hypertonic solutions (with higher osmotic strength) have fewer free water molecules to bombard the membrane, and so water moves into them. Hypotonic solutions have a higher free water content (and a lower osmotic strength), so they lose water across

Louis Pasteur's observation that "Chance favors the prepared mind" is as true today as it was when he first expressed it in the 1800s. Scientists have long realized that osmosis directly through the phospholipid bilayer is much too slow to account for water movement across some cell membranes, including those of kidney tubules (which must reabsorb tremendous quantities of water that the kidney filters from the blood each day) and those of red blood cells (see Fig. 5-10). Partly because water is abundant on both sides of the membrane, and partly because water can move directly through the bilayer, attempts to identify selective transport proteins for water repeatedly met with failure.

Then, as often happens in science, chance and "prepared minds" met. In the mid-1980s, Dr. Peter Agre (**FIG. E5-1**), then at the Johns Hopkins School of Medicine in Maryland, was attempting to determine the structure of a recognition glycoprotein on red blood cells. The protein he isolated was contaminated, however, with large quantities of another protein. Instead of discarding the unknown protein, he and his coworkers collaborated to identify its structure. They found that it was similar to previously identified membrane proteins that were hypothesized to be channels but whose function was unknown. Agre and colleagues investigated the protein's function by forcing frog eggs (which are only slightly permeable to water) to insert the protein into their plasma membranes. While eggs without the mystery protein swelled only slightly when placed in a hypotonic solution, those with the protein swelled rapidly and exploded in the same solution (**FIG E5-2a**). Further studies showed that no other ions or molecules traversed this channel, which was dubbed "aquaporin." In 2000, Agre and other research teams reported the three-dimensional structure of aquaporin and described how specific amino acids in its interior allow billions of water molecules to move through the channel in single file every second, while repelling other ions or molecules (**FIG. E5-2b**).

Many types of aquaporins have now been identified (including at least 11 different versions in the human body), and they have been found in all forms of life that have been investigated. For example, the plasma membrane of the central vacuole of plant cells is rich in aquaporins, which allow it to fill rapidly when water is available (see Fig. 5-11). Because aquaporins are so widely distributed in human tissues including the brain, lungs, muscles, and kidneys, and because aquaporin mutations have now been linked to several human disorders, the medical implications of these "water pores" are enormous. In 2003, Agre shared the Nobel Prize in Chemistry for his discovery—the result of both persistence and chance, or what Agre himself described as "sheer blind luck."

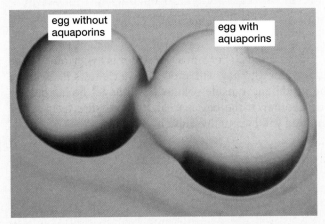

(a)

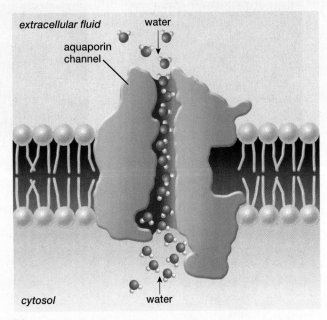

(b)

FIGURE E5-2 Function and structure of aquaporin
(a) The frog egg on the right has aquaporin in its plasma membrane, while the egg on the left does not. They have both been sitting in a hypotonic solution for 30 seconds. The egg on the right has burst, while that on the left has swollen only slightly. **(b)** Aquaporin consists of proteins that form a narrow pore (seen here in cross-section) in which charged amino acids interact with water molecules, promoting their movement through in either direction while repelling other substances.

FIGURE E5-1 Peter Agre

water-permeable membranes (**FIG. 5-9**). What will happen if we place our water-permeable bag containing a 20 percent sugar solution into a beaker of pure water?

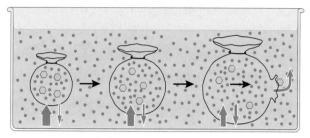

FIGURE 5-9 Hypotonic solution

Because the 20 percent sugar water is hypertonic to the pure water, the bag will swell as water enters it. Because the inside and outside solutions will never be isotonic to one another, if the bag is weak, the pressure from water entering will eventually cause it to burst.

SUMMING UP
The Principles of Osmosis

- Osmosis is the movement of water across a selectively water-permeable membrane by simple diffusion or facilitated diffusion through aquaporins.
- Water moves across a selectively water-permeable membrane down its concentration gradient from the side with a higher concentration of free water molecules to the side with a lower concentration of free water molecules.
- Dissolved substances reduce the concentration of free water molecules in a solution.
- When comparing solutions separated by a membrane that is selectively permeable to water, scientists describe the solution with a higher concentration of dissolved materials as hypertonic and as having a higher osmotic strength (ability to pull water across) than the other solution. The solution with the lower concentration of dissolved materials is described as hypotonic and as having a lower osmotic strength than the other solution.

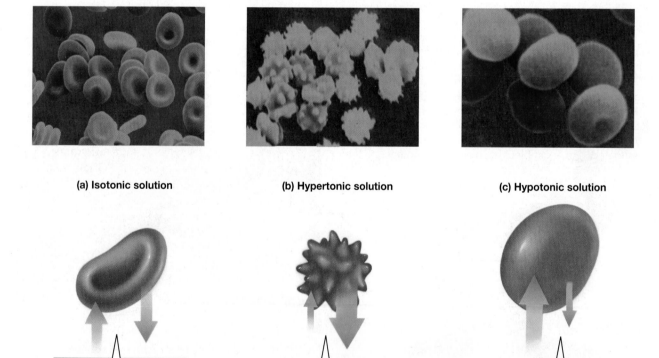

(a) Isotonic solution   **(b) Hypertonic solution**   **(c) Hypotonic solution**

Equal movement of water into and out of cells.

Net water movement out of cells.

Net water movement into cells.

FIGURE 5-10 The effects of osmosis
(a) If red blood cells are immersed in an isotonic salt solution, there is no net movement of water across the plasma membrane. The red blood cells keep their characteristic dimpled disk shape. **(b)** A hypertonic solution, with more salt than is in the cells, causes water to leave the cells and the cells to shrivel. **(c)** A hypotonic solution, with less salt than is in the cells, causes water to enter; the cells swell, and may burst. QUESTION All freshwater fishes swim in a solution that is hypotonic to the fluid inside their bodies. Why don't freshwater fishes swell up and burst?

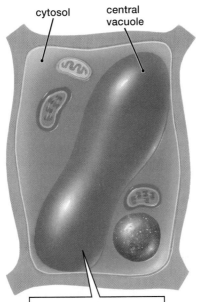

cytosol   central
           vacuole

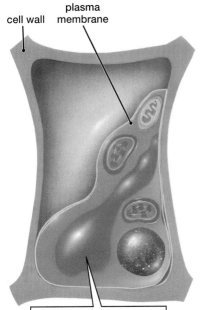

                    plasma
        cell wall   membrane

When water is plentiful, it fills the central vacuole, pushes the cytosol against the cell wall, and helps maintain the cell's shape.

When water is scarce, the central vacuole shrinks and the cell wall is unsupported.

FIGURE 5-11 Turgor pressure in plant cells Aquaporins allow water to move rapidly in and out of the central vacuoles of plant cells. The cell (top) and plant (bottom) on the right are supported by water turgor pressure, whereas those on the left have lost pressure due to dehydration. QUESTION If a plant cell is placed in water containing no solutes, will the cell eventually burst? Explain.

Water pressure supports the leaves of this impatiens plant.

Deprived of the support from water, the plant wilts.

Osmosis Across the Plasma Membrane Plays an Important Role in the Lives of Cells

The extracellular fluid of animals is usually isotonic to the cytosol of their cells; that is, the concentration of water inside is the same as that outside, so there is no net tendency for water to enter or leave the cells. Although the types of dissolved particles are seldom the same inside and outside the cells, the total concentration of all dissolved particles is equal; therefore, the water concentration inside is equal to that outside the cells.

If you immerse red blood cells in salt solutions of varying concentrations, you can observe the effects of water movement across the cell membranes. In an isotonic salt solution, the size of the cell remains constant (**FIG. 5-10a**). If the salt solution is hypertonic to the cytosol of the red

blood cells, water will leave the cells by osmosis, and the cells will shrivel (**FIG. 5-10b**). Conversely, if the salt solution is very dilute and hypotonic to the red blood cell cytosol, water will enter the cells, causing them to swell (**FIG. 5-10c**). If red blood cells are placed in pure water, they will continue to swell until they burst.

Osmosis helps explain why protists that live in fresh water, such as *Paramecium*, have special structures called *contractile vacuoles* to eliminate the water that continuously leaks into the cytosol, which is hypertonic to the fresh water in which the protist lives. Cellular energy is used to pump salts from the cytosol into collecting ducts of the contractile vacuole. Water follows by osmosis, filling the central reservoir. When the reservoir is full, it contracts, squirting the water out through a pore in the plasma membrane (see Fig. 4-16).

Osmosis across plasma membranes is crucial to many biological processes, including water uptake by the roots of plants, absorption of dietary water from the intestine, and the reabsorption of water in kidneys.

Nearly every living plant cell is supported by water that enters through osmosis. As you learned in Chapter 4, most plant cells have a large, membrane-enclosed compartment called the central vacuole, which is filled with dissolved substances that are stored there. These dissolved substances make the vacuole contents hypertonic to the cell cytosol, which in turn is usually hypertonic to the extracellular fluid that bathes the cells. Water therefore enters the cytosol and then the vacuole by osmosis. The water pressure within the vacuole, called **turgor pressure**, pushes the cytosol up against the cell wall with considerable force (**FIG. 5-11**, top). Cell walls are usually somewhat flexible, so both the overall shape and the rigidity of the cell depend on turgor pressure. Turgor pressure thus provides support for the non-woody parts of plants. If you forget to water your houseplants, the central vacuole and cytosol of each cell loses water and the plasma membrane shrinks away from its cell wall as the vacuole collapses, in a process called

plasmolysis. Just as a balloon goes limp when its air leaks out, so too the plant droops as its cells lose turgor pressure and plasmolysis occurs (**FIG. 5-11**, bottom).

Active Transport Uses Energy to Move Molecules Against Their Concentration Gradients

All cells need to move some materials "uphill" across their plasma membranes against concentration gradients. For example, every cell requires some nutrients that are less concentrated in the environment than in the cell's cytosol; diffusion would cause the cell to lose, not gain, these nutrients. Other substances, such as sodium and calcium ions, are maintained at much lower concentrations inside the cell than in the extracellular fluid. When these ions diffuse into the cell, they must be pumped out again against their concentration gradients.

In active transport, membrane proteins use cellular energy to move molecules or ions across the plasma membrane against their concentration gradients (**FIG. 5-12**). These active-transport proteins span the width of the membrane and have two active sites. One active site (which may face the inside or outside of the plasma membrane, depending on the transport protein) binds a particular molecule or ion, such as a calcium ion. The second site (always on the inside of the membrane) binds an energy-carrier molecule, usually adenosine triphosphate (ATP; introduced in Chapter 3). The ATP donates energy to the protein, causing it to change shape and move the calcium ion across the membrane (in the process, it releases one of its phosphate groups, becoming adenosine diphosphate [ADP]). Active-transport proteins are often called *pumps*—in an analogy to water pumps—because they use energy to move ions or molecules uphill against a concentration gradient. As we will see, plasma membrane pumps are vital in mineral uptake by plants, mineral absorption in the intestines of animals, and maintaining concentration gradients essential to nerve cell function.

Cells Engulf Particles or Fluids by Endocytosis

Cells have evolved several processes that use cellular energy to acquire or expel particles or substances that are too large to be transported directly across the membrane. Cells can acquire fluids or particles from their extracellular environment, especially large proteins or entire microorganisms such as bacteria, by a process called **endocytosis** (Greek for "into the cell"). During endocytosis, the plasma membrane engulfs the fluid droplet or particle and pinches off a membranous sac called a *vesicle* into the cytosol with the fluid or particle inside. We can distinguish three types of endocytosis, based on the size and type of material acquired and the method of acquisition: *pinocytosis, receptor-mediated endocytosis,* and *phagocytosis.*

Pinocytosis Moves Liquids into the Cell

In **pinocytosis** ("cell drinking"), a very small patch of plasma membrane dimples inward as it surrounds extracellular fluid and buds off into the cytosol as a tiny vesicle (**FIG. 5-13**). Pinocytosis moves a droplet of extracellular fluid, contained within the dimpling patch of membrane, into the cell. Therefore, the cell acquires materials in the same concentration as in the extracellular fluid.

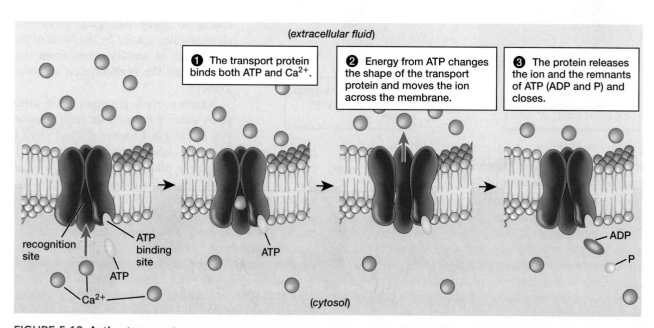

FIGURE 5-12 Active transport
Active transport uses cellular energy to move molecules across the plasma membrane against a concentration gradient. A transport protein (blue) has an ATP binding site and a recognition site for the molecules to be transported, in this case calcium ions (Ca^{2+}). Notice that when ATP donates its energy, it loses its third phosphate group and becomes ADP + P.

(a) Pinocytosis

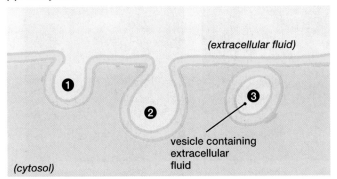

❶ A dimple forms in the plasma membrane, which **❷** deepens and surrounds the extracellular fluid. **❸** The membrane encloses the extracellular fluid, forming a vesicle.

(b) Pinocytosis in a smooth muscle cell.

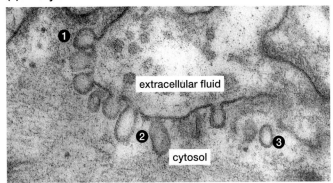

FIGURE 5-13 Pinocytosis
The circled numbers correspond to both the diagram and the electron micrograph.

Receptor-Mediated Endocytosis Moves Specific Molecules into the Cell

Cells take up certain molecules or complexes of molecules (packets containing protein and cholesterol, for example) by the process of **receptor-mediated endocytosis** (see **FIG. 5-14**). This process can selectively concentrate specific molecules inside a cell. Most plasma membranes bear many receptor proteins on their outside surfaces, each with a binding site for a particular molecule. In some cases, the receptors accumulate in depressions on the plasma membrane called *coated pits*. If the right molecule contacts a receptor protein in one of these coated pits, the molecule attaches to the binding site. The coated pit deepens into a U-shaped pocket that eventually pinches off into the cytosol as a vesicle. Both the receptor with its bound molecules and a bit of extracellular fluid move into the cytosol encased within the vesicle.

FIGURE 5-14 Receptor-mediated endocytosis
The circled numbers correspond to both the diagram and the electron micrograph.

Receptor-mediated endocytosis

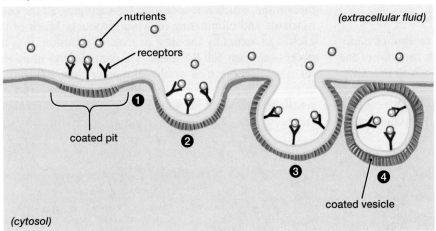

❶ Receptor proteins for specific molecules or complexes of molecules are localized at coated pit sites.

❷ The receptors bind the molecules and the membrane dimples inward.

❸ The coated pit region of the membrane encloses the receptor-bound molecules.

❹ A vesicle ("coated vesicle") containing the bound molecules is released into the cytosol.

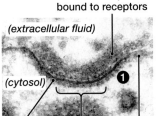

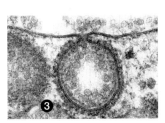

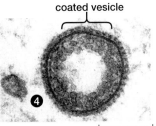

0.1 micrometer

(a) Phagocytosis

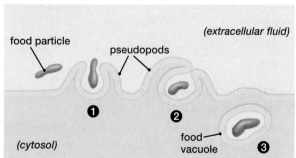

food particle

pseudopods

(extracellular fluid)

food
vacuole

(cytosol)

❶ The plasma membrane extends pseudopods toward an extracellular particle (for example, food). ❷ The ends of the pseudopods fuse, encircling the particle. ❸ A vesicle called a food vacuole is formed containing the engulfed particle.

FIGURE 5-15 Phagocytosis

(b) *Amoeba*

An *Amoeba* (a freshwater protist), engulfs a *Paramecium* using phagocytosis.

(c) White blood cell

A white blood cell ingests bacteria using phagocytosis.

Phagocytosis Moves Large Particles into the Cell

Cells use **phagocytosis** (which means "cell eating") to pick up large particles, including whole microorganisms (**FIG. 5-15a**). When the freshwater protist *Amoeba*, for example, senses a tasty *Paramecium*, *Amoeba* extends parts of its surface membrane. These membrane extensions are called *pseudopods* (Latin for "false feet"). The pseudopods fuse around the prey, which is enclosed inside a vesicle called a *food vacuole* for digestion (**FIG. 5-15b**). Like *Amoeba*, white blood cells also use phagocytosis and intracellular digestion to engulf and destroy invading bacteria (**FIG. 5-15c**) in a drama that occurs frequently within your body.

Exocytosis Moves Material Out of the Cell

Cells often use energy to accomplish the reverse of endocytosis, a process called **exocytosis** (Greek for "out of the cell"), to dispose of unwanted materials such as the waste products of digestion, or to secrete materials such as hormones into the extracellular fluid (**FIG. 5-16**). During exocytosis, a membrane-enclosed vesicle carrying material to

be expelled moves to the cell surface, where the vesicle's membrane fuses with the cell's plasma membrane. The vesicle then opens to the extracellular fluid, and its contents diffuse outward.

Exchange of Materials Across Membranes Influences Cell Size and Shape

As you learned in Chapter 4, most cells are too small to be seen with the naked eye; they range from about 1 to 100 micrometers (millionths of a meter) in diameter (see Fig. 4-1). Why? As a roughly spherical cell becomes larger, its innermost regions become farther away from the plasma membrane, which is responsible for supplying all the cells' nutrients and eliminating its waste products. Much of the exchange occurs by the slow process of diffusion. In a hypothetical giant cell 8.5 inches (20 centimeters) in diameter, oxygen molecules would take more than 200 days to diffuse to the center of the cell, by which time the cell would be long dead for lack of oxygen. As a spherical cell enlarges, its volume increases more rapidly than does its

FIGURE 5-16 Exocytosis
Exocytosis is functionally the reverse of endocytosis. QUESTION How does exocytosis differ from diffusion of materials out of a cell?

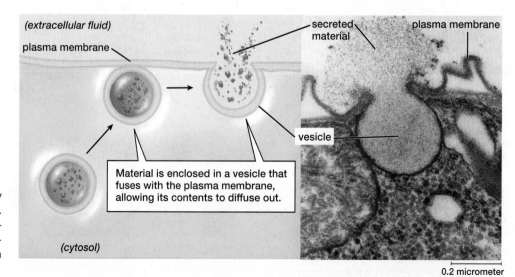

(extracellular fluid)

plasma membrane

secreted material

plasma membrane

vesicle

Material is enclosed in a vesicle that fuses with the plasma membrane, allowing its contents to diffuse out.

(cytosol)

0.2 micrometer

surface area. So a larger cell, which requires more nutrients and produces more wastes, has a relatively smaller area of membrane to accomplish this exchange than does a small cell (**FIG. 5-17**).

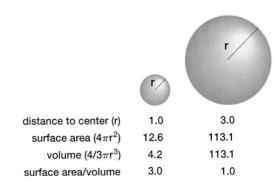

distance to center (r)	1.0	3.0
surface area ($4\pi r^2$)	12.6	113.1
volume ($4/3\pi r^3$)	4.2	113.1
surface area/volume	3.0	1.0

FIGURE 5-17 Surface area and volume relationships

In a very large, roughly spherical cell, the surface area of the plasma membrane would be too small to keep up with the cell's metabolic needs. This constraint limits the size of most cells. However, some cells, such as neurons and muscle cells, can grow much larger because they have an elongated shape that increases their membrane surface area and keeps the ratio of surface area to volume relatively high. For example, cells lining the small intestine have plasma membranes that project out in fringe-like folds, called *microvilli* (**FIG. 5-18a**, middle). These structures create an enormous surface area of membrane to absorb nutrients from digested food.

5.3 HOW DO SPECIALIZED JUNCTIONS ALLOW CELLS TO CONNECT AND COMMUNICATE?

In multicellular organisms, plasma membranes hold together clusters of cells and provide avenues through which cells communicate with their neighbors. Depending on the organism and the cell type, four types of connection may occur between cells: *desmosomes*, *tight junctions*, *gap junctions*, and *plasmodesmata*. While plasmodesmata are restricted to plant cells, some animal cells possess all three of the other types of junctions.

Desmosomes Attach Cells Together

As you know, animals tend to be flexible, mobile organisms. Many of an animal's tissues are stretched, compressed, and bent as the animal moves. Cells in the skin, intestine, urinary bladder, and other organs must adhere firmly to one another to avoid tearing under the stresses of movement. Such animal tissues have junctions called **desmosomes**, which hold adjacent cells together (Fig. 5-18a). In a desmosome, the membranes of adjacent cells are held together by proteins and carbohydrates. Protein filaments attached to the insides of the desmosomes extend into the interior of each cell, further strengthening the attachment.

Tight Junctions Make Cell Attachments Leakproof

The animal body contains many tubes and sacs that must hold their contents without leaking; for example, leaky skin or a leaky urinary bladder would spell

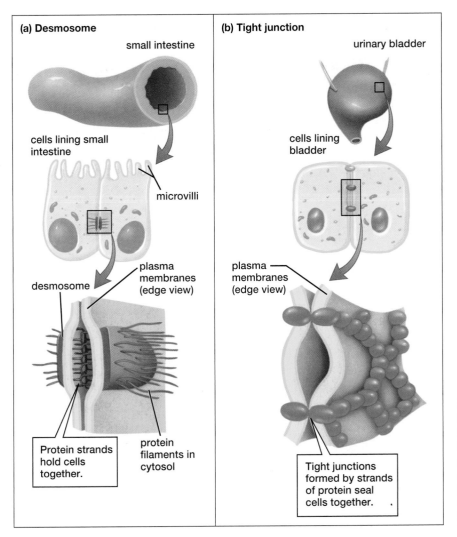

FIGURE 5-18 Cell attachment structures (a) Cells lining the small intestine are attached by desmosomes. Protein filaments bound to the inside surface of each desmosome extend into the cytosol and attach to other filaments inside the cell, strengthening the connection between cells. (b) Tight junctions prevent leakage between cells, as between cells of the urinary bladder.

(a) Desmosome

small intestine

cells lining small intestine

microvilli

desmosome

plasma membranes (edge view)

Protein strands hold cells together.

protein filaments in cytosol

(b) Tight junction

urinary bladder

cells lining bladder

plasma membranes (edge view)

Tight junctions formed by strands of protein seal cells together.

disaster for the rest of the body. When cells must create a reasonably waterproof barrier, the spaces between them are blocked with strands of protein that form **tight junctions** (**FIG. 5-18b**). These protein "gaskets" prevent fluid from moving between adjacent cells.

Gap Junctions and Plasmodesmata Allow Direct Communication Between Cells

Multicellular organisms must coordinate the actions of their component cells. In animals, most cells that contact other cells—in other words, nearly all cells in the body—communicate through protein channels that connect the insides of adjacent cells. These cell-to-cell channels are called **gap junctions** (**FIG. 5-19a**). Hormones, nutrients, ions, and even electrical signals can pass through the channels at gap junctions.

Virtually all of the living cells of plants are connected to one another by **plasmodesmata**. Plasmodesmata are openings in the walls of adjacent plant cells, lined with plasma membrane and filled with cytosol. Plasmodesmata create continuous cytosolic bridges between the insides of adjacent cells (**FIG. 5-19b**). Many plant cells have thousands of plasmodesmata, allowing water, nutrients, and hormones to pass quite freely from one cell to another.

EVOLUTIONARY CONNECTIONS

Caribou Legs and Membrane Diversity

The membranes of all cells are similar in structure, reflecting the common evolutionary heritage of all life on Earth. Membrane function varies tremendously from organism to organism, however, and even from cell to cell within a single organism.

Our discussion of membranes emphasized the unique functions of membrane proteins. Consequently, you may think that the phospholipids are just a waterproof place for the proteins to reside. This isn't quite true, as we can see by examining the plasma membrane phospholipids in cells of the legs of caribou, animals that live in very cold regions of North America (**FIG. 5-20**). During the long arctic winters, temperatures plummet far below freezing, and keeping legs and feet really warm would waste precious energy. Fortunately, these cold conditions have favored the evolution of specialized arrangements of arteries and veins in caribou legs that allow the temperature of their lower legs to drop almost to freezing (32°F, or 0°C), thus conserving body heat. The upper legs and body, in contrast, remain at

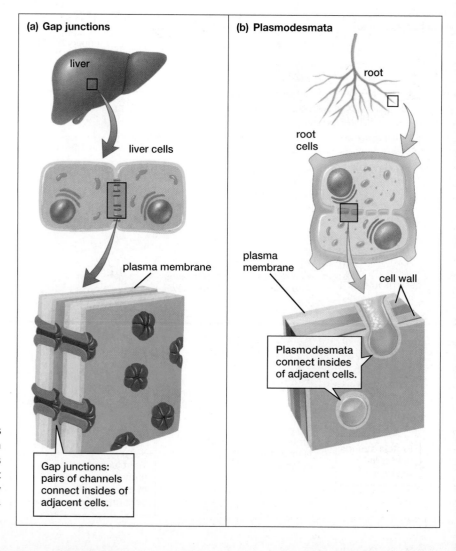

FIGURE 5-19 Cell communication structures (a) Gap junctions, such as those between cells of the liver, contain cell-to-cell channels that interconnect the cytosol of adjacent cells. **(b)** Plant cells are interconnected by plasmodesmata, which form cytosolic connections through the walls of adjacent cells.

(a) Gap junctions

liver

liver cells

plasma membrane

Gap junctions: pairs of channels connect insides of adjacent cells.

(b) Plasmodesmata

root

root cells

plasma membrane

cell wall

Plasmodesmata connect insides of adjacent cells.

about 105°F (40.5°C). How can cell membranes remain fluid at these radically different temperatures so that proteins can move freely within the membrane to sites where they are needed?

Recall that the fluidity of a membrane is a function of the fatty acid tails of its phospholipids. Unsaturated fatty acids remain more fluid at lower temperatures than do saturated fatty acids. In caribou legs, the membranes of cells near the chilly hoof have lots of unsaturated fatty acids and kinky tails, whereas the membranes of cells near the warmer trunk have more saturated fatty acids and fewer kinky tails. This arrangement gives the plasma membranes throughout the leg the proper fluidity despite great differences in temperature.

Throughout this text, we will return many times to the concepts of membrane structure and transport described in this chapter. Understanding the diversity of membrane lipids and proteins is the key to understanding not just the isolated cell but also entire organs, which could not perform as they do without the specialized membrane properties of their component cells.

FIGURE 5-20 **Caribou browse on the frozen Alaskan tundra** The lipid composition of the membranes in the cells of a caribou's legs varies with distance from the animal's trunk. Unsaturated phospholipids predominate in the lower leg; more saturated phospholipids are found in the upper leg.

CASE STUDY REVISITED VICIOUS VENOMS

Both rattlesnake and brown recluse spider venoms are complex mixtures of poisonous proteins. In each case, the proteins responsible for the worst symptoms are enzymes. As you will learn in Chapter 6, enzymes cause the breakdown of biological molecules while remaining unchanged themselves. Enzymes are often named after the molecules they break apart, with the suffix "–ase" added to identify the protein as an enzyme. Several of the toxic enzymes in snake and spider venoms are phospholipases; the name tells us that they break down phospholipids. You now know that within cell membranes, the fluid bilayer portion—which allows the membrane to maintain the gradients that are crucial to life—consists of phospholipids. Although the phospholipases and the other toxic proteins that form the "witches brew" of spider and rattlesnake venoms differ, in both cases (as you could now predict) the venom attacks cell membranes, causing cells to rupture and die. Cell death can destroy the tissue around a rattlesnake or brown recluse spider bite (**FIG. 5-21**). The phospholipases of these venoms also attack the membranes of red blood cells (which carry oxygen throughout the body), and so both venoms can cause anemia (an inadequate number of red blood cells). The rattlesnake can inject far more venom far more deeply, so it is more likely to reduce the oxygen-carrying capacity of the blood, causing the victim to become short of breath, as happened to Karl. Both venoms break down the membranes of cells that form the tiny blood vessels called capillaries, causing bleeding under the skin around the bite or, in severe cases, internally as well.

(a)

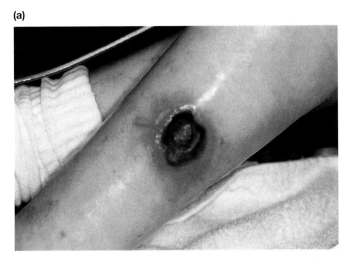

(b)

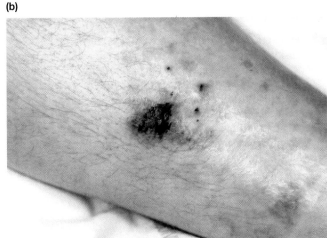

FIGURE 5-21 **Phospholipases in venoms can destroy cells**
(a) A brown recluse spider bite on a person's forearm. (b) A rattlesnake bite on the forearm. Both show extensive tissue destruction caused by phospholipases.

Karl was extremely lucky that Mark had brought his cell phone. Trying to hike back to the car would have quickly spread the venom throughout Karl's body, and the added delay would have decreased his chances of survival. Because the young men had identified the snake, the hospital had the correct antivenin waiting. Antivenin contains proteins that bind and neutralize the toxins in the snake venom. Unfortunately, no antivenin is available for brown recluse bites, and treatment generally consists of preventing infection, controlling pain and swelling, and waiting

patiently to heal. Melissa was glad her fiancé was not squeamish and could help her care for the wound.

Although both snake and spider bites can have serious consequences, it's important to realize that only a tiny fraction of the large number of species of spiders and snakes found in the Americas are dangerous to people. The best defense is to learn which poisonous snakes and spiders live in your area and where they prefer to hang out. If your activities bring you to such places, wear protective clothing—and always look before you reach! With

education and appropriate care, we humans can comfortably coexist with spiders and snakes, avoid their bites, and keep our cell membranes intact.

Consider This Phospholipases and other digestive enzymes are found in animal (including human, snake, and spider) digestive tracts as well as in snake and spider venom. What different roles do phospholipases play in snake and spider venom as compared to the roles of snake and spider digestive enzymes?

CHAPTER REVIEW

SUMMARY OF KEY CONCEPTS

5.1 How Is the Structure of a Membrane Related to Its Function?

The plasma membrane has three major functions: it selectively isolates the cytoplasm from the external environment; it regulates the flow of materials into and out of the cell; and it communicates with other cells. The membrane consists of a bilayer of phospholipids in which a variety of proteins are embedded. There are five major categories of membrane proteins: receptor proteins, which bind molecules and trigger changes in the cellular metabolism; recognition proteins, which serve as identification tags and attachment sites; enzymatic proteins, which promote chemical reactions without being changed themselves; attachment proteins, which bind the plasma membrane to protein filaments inside or outside the cell as well as bind cells to one another; and finally, transport proteins, which regulate the movement of most water-soluble substances through the membrane.

Web Tutorial 5.1 Plasma Membrane Structure and Transport

5.2 How Do Substances Move Across Membranes?

Diffusion is the movement of particles from regions of higher concentration to regions of lower concentration. In simple diffusion, water, dissolved gases, and lipid-soluble molecules diffuse through the phospholipid bilayer. In facilitated diffusion, water-soluble molecules cross the membrane through protein channels or with the assistance of protein carriers. In both cases, molecules move down their concentration gradients, and cellular energy is not required.

Osmosis is the diffusion of water across a selectively permeable membrane down its concentration gradient, from solutions with higher free water concentration (lower concentration of solutes) to solutions with lower free water concentration (with higher concentrations of solutes). Water can diffuse directly through the phospholipid bilayer. In many cells it also moves by facilitated diffusion through membrane water channels called aquaporins.

Several types of transport require energy. In active transport, protein carriers in the membrane use cellular energy (ATP) to drive the movement of molecules across the plasma membrane, usually against concentration gradients. Large molecules (for example, proteins), particles of food, microorganisms, and extracellular fluid may be acquired by endocytosis—either by pinocytosis, receptor-mediated endocytosis, or phagocytosis. The secretion of substances such as hormones and the excretion of wastes from a cell are accomplished by exocytosis.

Web Tutorial 5.2 Osmosis

5.3 How Do Specialized Junctions Allow Cells to Connect and Communicate?

Cells may be connected to one another by a variety of junctions. Desmosomes attach cells firmly to one another, preventing a tissue from tearing during movement or stretching. Tight junctions seal off the spaces between adjacent cells, leakproofing organs such as the skin and urinary bladder. Gap junctions in animals and plasmodesmata in plants interconnect the cytosol of adjacent cells.

KEY TERMS

active transport   *page 86*
aquaporin   *page 88*
attachment protein
   page 84
carrier protein   *page 88*
channel protein   *page 88*
concentration   *page 85*
concentration gradient
   page 85
desmosome   *page 95*
diffusion   *page 85*

endocytosis   *page 92*
enzyme   *page 84*
exocytosis   *page 94*
facilitated diffusion   *page 88*
fluid   *page 85*
fluid mosaic model   *page 82*
gap junction   *page 96*
glycoprotein   *page 84*
gradient   *page 85*
hypertonic   *page 88*
hypotonic   *page 88*

isotonic   *page 88*
osmosis   *page 88*
passive transport   *page 86*
phagocytosis   *page 94*
phospholipid bilayer
   page 83
pinocytosis   *page 92*
plasmodesmata   *page 96*
plasmolysis   *page 92*
receptor-mediated
   endocytosis   *page 93*

receptor protein   *page 84*
recognition protein   *page 84*
selectively permeable
   page 87
simple diffusion   *page 87*
solute   *page 85*
solvent   *page 85*
tight junction   *page 96*
transport protein   *page 84*
turgor pressure   *page 91*

THINKING THROUGH THE CONCEPTS

1. Describe and diagram the structure of a plasma membrane. What are the two principal types of molecules in plasma membranes? What are the five principal functions of plasma membranes?

2. What are the five categories of proteins commonly found in plasma membranes, and what is the function of each one?

3. Define *diffusion*, and compare that process to osmosis. How do these two processes help plant leaves remain firm?

4. Define *hypotonic, hypertonic,* and *isotonic.* What would be the fate of an animal cell immersed in each of the three types of solution?

5. Describe the following types of transport processes in cells: simple diffusion, facilitated diffusion, active transport, pinocytosis, receptor-mediated endocytosis, phagocytosis, and exocytosis.

6. Name the protein that allows facilitated diffusion of water. What experiment demonstrated the function of this protein?

7. Imagine a container of glucose solution, divided into two compartments (A and B) by a membrane that is permeable to water and glucose but not to sucrose. If some sucrose is added to compartment A, how will the contents of compartment B change?

8. Name four types of cell-to-cell junctions, and explain the function of each. Which of them function in plants, and which in animals?

APPLYING THE CONCEPTS

1. Different cells have somewhat different plasma membranes. The plasma membrane of a *Paramecium,* for example, is only about 1% as permeable to water as the plasma membrane of a human red blood cell. Referring to our discussion of the effects of osmosis on red blood cells and the role of contractile vacuoles in *Paramecium,* what do you think is the function of the low water permeability of *Paramecium*? Is *Paramecium* likely to have aquaporins in its plasma membrane? Explain your answer.

2. You have been introduced to active and passive transport and their associated proteins, to receptor proteins, and to concentration gradients. Describe how a hypothetical cell could use ions in solution along with these proteins to generate a flow of ions in response to a chemical stimulus.

3. Red blood cells will swell up and burst when placed in a hypotonic solution such as pure water. Why don't we swell up and burst when we swim in water that is hypotonic to our cells and body fluids?

FOR MORE INFORMATION

Davis, K., "'Ghost Bugs' Could Help Cut Pesticide Use." *NewScientist,* September 19, 2004. Bacterial cells with their cytoplasm removed could be used as containers for delivering pesticides to plants.

Kunzig, R., "They Love the Pressure." *Discover,* August 2001. Living at depths that exert a pressure of 15,000 pounds per square inch necessitates alterations in the membranes of deep-sea dwellers.

Martindale, D. "The Body Electric." *New Scientist,* May 15, 2004. Electrical gradients created by cells govern many critical biological processes, from wound healing to developmental events.

Rothman, J. E., and Orci, L. "Budding Vesicles in Living Cells." *Scientific American,* March 1996. Membranes within cells form small containers called vesicles, which transport materials inside the cell. Researchers are discovering the mechanisms by which these containers are formed.

Energy Flow in
the Life of a Cell

The bodies of these runners convert energy stored in fats and carbohydrates to the energy of movement and heat. Their pounding footsteps noticeably shake the Verrazano Narrows Bridge during the New York Marathon.

AT A GLANCE

CASE STUDY ENERGY UNLEASHED

PICTURE A WOMAN battling cancer; a supporter of wildlife conservation dressed as a rhino; a 91-year-old man proceeding at a slow shuffle; a fireman wearing full firefighting gear to honor his fallen comrades; a man without a leg, using crutches; and blind people guided by the sighted. All these people participated in a 26-mile journey; a personal odyssey for each individual, and a collective testimony to human persistence and endurance.

The 20,000-plus runners in the New York Marathon collectively expend more than 50 million Calories, travelling a total of 520,000 miles, and shaking the Verrazano Narrows Bridge. Once finished, they douse their overheated bodies with water and refuel on high-energy snacks. Finally, cars, buses, and airplanes—burning vast quantities of fuel and releasing enormous amounts of heat—carry the runners back to their homes throughout the world.

What exactly is energy? Do our bodies use it according to the same principles that govern energy use in the engines of cars and airplanes? Why do our bodies generate heat, and why do we give off more heat when exercising than when watching TV?

We often talk about "burning" Calories. How does setting sugar on fire compare to the set of reactions that allow your body to "burn" sugar that you've eaten? In both cases, oxygen combines with sugar to produce carbon dioxide, water, and heat. Why aren't our bodies roasted when we metabolize food? How do we capture energy in molecules that power muscle movement and the huge variety of metabolic processes within our cells? How do we control the breakdown of high-energy molecules to produce useful energy?

6.1   WHAT IS ENERGY?

Energy is defined as simply the capacity to do work. Scientists define *work* as a force acting upon an object that causes the object to move. The objects upon which energy acts are not always easy to see or even to measure. It is obvious that the marathoners in our Case Study are expending energy and moving things—their chests heave, their arms pump, and their legs stride. It is less obvious where that energy comes from, but we know that it originates in molecules stored in the runners' bodies: sugars, glycogen, and fats. In fact, *chemical energy*, the subject of this chapter, powers all of life on Earth. The "objects" in chemical energy are electrons. Energy forces determine the position of electrons in atoms and their interactions with other atoms that allow molecules to be formed and transformed. As these positions and interactions among electrons change, molecules are formed or broken down, and energy is stored or released. Specialized protein molecules synthesized by cells can lengthen or shorten, causing the cell to move. Muscle cells can contract powerfully as a result of interactions among specialized proteins that are driven by chemical energy released from molecules of ATP. Synchronized contractions of muscle cells move runners' bodies—and collectively can shake one of the world's great bridges.

There are two fundamental types of energy: *kinetic energy* and *potential energy*, each of which takes several forms. **Kinetic energy** is the energy of movement. It includes light (movement of photons), heat (movement of molecules), electricity (movement of electrically charged particles), and any movement of larger objects—such as your eyes as you scan this page, and marathon runners as they strive to complete the grueling race. **Potential energy**, or stored energy, includes chemical energy stored in the bonds that hold atoms together in molecules, electrical energy stored in a battery, and positional energy stored in a penguin poised to spring (**FIG. 6-1**). Under the right conditions, kinetic energy can be transformed into potential energy, and vice versa. For example, a penguin converts kinetic energy of movement into potential energy of position when it climbs up from the water onto the ice. When it jumps off, the potential energy is converted back into kinetic energy. During this process, the potential energy stored in chemical bonds of molecules in the penguin's body is transformed into the kinetic energy of movement.

To understand energy flow and change, we need to know more about the properties and "behavior" of energy, described by the *laws of thermodynamics*.

The Laws of Thermodynamics Describe the Basic Properties of Energy

The **laws of thermodynamics** describe the quantity (the total amount) and the quality ("usefulness") of energy. The **first law of thermodynamics** states that energy can nei-

FIGURE 6-1 From potential to kinetic energy
Perched atop an ice floe, the body of the penguin has potential energy, because the ice is much higher than the ocean. As it dives, the potential energy is converted to the kinetic energy of motion of the penguin's body. Finally, some of this kinetic energy transferred to the water causes the water to splash and ripple.

ther be created nor destroyed by ordinary processes (outside of nuclear reactions). Energy can, however, change form—for example, from chemical energy to the energy of heat or movement. If you had a *closed system*, where neither energy nor matter could enter or leave, and if you could measure energy in all its forms both before and after a particular process occurred, you would find the total energy before and after the process to be unchanged. Therefore, the first law is often called the *law of conservation of energy*.

To illustrate the first law, consider your car. Before you turn the ignition key, the energy in the car is all potential energy, stored in the chemical bonds of its fuel. As you drive, about 25 percent of this potential energy is converted into the kinetic energy of motion. According to the first law of thermodynamics, however, no energy is created or destroyed. So where is the "missing" energy? The burning gas not only moves the car but also heats up the engine, the exhaust system, and the air around the car. The friction of tires on the pavement slightly heats the road. So, as the first law dictates, no energy is lost. The total amount of energy remains the same, although it has changed in form. Likewise, a runner converts the potential chemical energy stored in food molecules into the same total amount of kinetic energy of movement plus heat.

The **second law of thermodynamics** states that when energy is converted from one form to another, the amount of useful energy decreases. Put another way, the second law states that all reactions or physical changes cause energy to be converted from more-useful into less-useful forms. Again, consider the examples just presented. The 75 percent of the energy stored in gasoline that did not go into moving the car was converted into heat energy (FIG. 6-2). Heat is a less-usable form of energy because it merely increases the random movement of molecules in the car, the road, and the air.

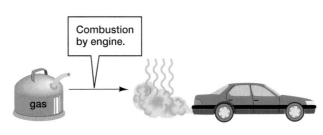

100 units chemical energy
(concentrated)

75 units heat + 25 units kinetic energy
energy (motion)

FIGURE 6-2 **Energy conversions result in a loss of useful energy**

Similarly, the energy of heat that runners liberate to the air when food "burns" in their bodies cannot be harnessed to allow them to run farther or faster. Thus, the second law tells us that no energy conversion process, including those that occur in the body, is 100 percent efficient in using energy to achieve a specific outcome.

The second law of thermodynamics also tells us something about the organization of matter. Useful energy tends to be stored in highly ordered matter, and whenever energy is used within a closed system, there is an overall increase in randomness and disorder of matter. We all experience this in our homes. Without energy-demanding cleaning and organizing efforts, dirty dishes accumulate; books, newspapers, and clothes collect in confusion on the floor; and the bed remains rumpled.

In the case of chemical energy, the eight carbon atoms in a single molecule of gasoline have a much more orderly arrangement than do the carbon atoms of the eight separate, randomly moving molecules of carbon dioxide and the nine molecules of water that are formed when the gasoline burns. The same is true for the glycogen molecules stored in a runner's muscles, which are converted from highly organized chains of sugar molecules into simpler water and carbon dioxide as they are used by the muscles. This tendency toward loss of complexity, orderliness, and useful energy—and the concurrent increase in randomness, disorder, and less-useful energy—is called **entropy**. To counteract entropy requires that energy be infused into the system from an outside source.

When the eminent Yale scientist Evelyn Hutchinson stated, "Disorder spreads through the universe, and life alone battles against it," he made an eloquent reference to entropy and the second law of thermodynamics. Fortunately,

Earth is not a closed system, for life as we know it depends on a constant infusion of energy from a source that is 93 million miles away.

Living Things Use the Energy of Sunlight to Create the Low-Entropy Conditions of Life

If you think about the second law of thermodynamics, you may wonder how life can exist at all. If chemical reactions, including those inside living cells, cause the amount of unusable energy to increase, and if matter tends toward increasing randomness and disorder, how can organisms accumulate the usable energy and precisely ordered molecules that characterize living things? The answer is that nuclear reactions in the sun produce energy in the form of sunlight, a process that also produces vast increases in entropy in the form of heat. Living things use a continuous input of solar energy to synthesize complex molecules and maintain orderly structures—to "battle against disorder." The highly organized, low-entropy systems that characterize life do not violate the second law of thermodynamics, since they are achieved through a continuous influx of usable energy from the sun. The solar reactions that provide this usable energy here on Earth cause a far greater loss of usable energy in the sun, which will eventually burn out. Because the solar energy that powers life on Earth occurs with an enormous net increase in solar entropy, life does not violate the second law of thermodynamics.

6.2 HOW DOES ENERGY FLOW IN CHEMICAL REACTIONS?

A **chemical reaction** is a process that forms or breaks the chemical bonds that hold atoms together. Chemical reactions convert one set of chemical substances, the **reactants**, into another set, the **products**. All chemical reactions either require an overall (net) input of energy or produce a net release of energy. A reaction is **exergonic** (Greek for "energy out"; the prefix "exo-" means "out") if it releases energy; that is, if the reactants contain more energy than the products. Exergonic reactions give off some of their energy as heat (FIG. 6-3).

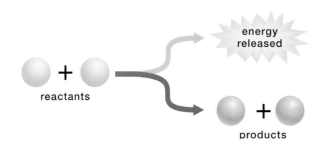

energy released

reactants

products

FIGURE 6-3 **Exergonic reaction**

A reaction is **endergonic** (endergonic is Greek for "energy in"; the prefix "endo–" means "in") if it requires a net input of energy; that is, if the products contain more energy than the reactants. According to the second law of thermodynamics, endergonic reactions require a net influx of energy from an outside source (**FIG. 6-4**).

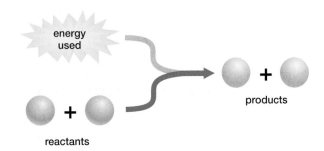

FIGURE 6-4 Endergonic reaction

Let's look at two processes that illustrate each type of reaction: burning sugar and photosynthesis.

Exergonic Reactions Release Energy

In an exergonic reaction, the reactants contain more energy than the products. The sugar that the runners' bodies use as fuel contains more energy than the carbon dioxide and water that are produced when that sugar breaks down. The extra energy is liberated as muscular movement and heat. Sugar can also be burned, as any cook can tell you. As it is burned, a sugar (glucose for example) undergoes the same basic reactions as sugar in a runner's body: sugar ($C_6H_{12}O_6$) is combined with oxygen (O_2) to produce carbon dioxide (CO_2) and water (H_2O) and release energy, as shown below (**FIG. 6-5**).

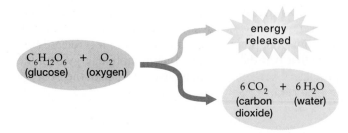

FIGURE 6-5 Burning glucose

Because molecules of sugar contain much more energy than do molecules of carbon dioxide and water, the reaction releases energy. Once ignited, the sugar will continue to burn spontaneously. It may be helpful to think of exergonic reactions as running "downhill," from high energy to low energy, as shown in **FIGURE 6-6**.

All Chemical Reactions Require Activation Energy to Begin

Although burning sugar releases energy overall, a spoonful of sugar doesn't burst into flames by itself. This observation leads to an important concept: all chemical reactions, even those that can continue spontaneously, require energy to get started. Think of a rock sitting at the top of a hill. It will remain there indefinitely unless something gives it a push to start it rolling down. In chemical reactions, the energy "push" is called the **activation energy** (Fig. 6-6). Chemical reactions require activation energy to get started because shells of negatively charged electrons surround all atoms and molecules. For two molecules to react with each other, their electron shells must be forced together, despite their mutual electrical repulsion. Forcing the electron shells together requires activation energy.

The usual source of activation energy is the kinetic energy of moving molecules. Molecules moving with sufficient speed collide hard enough to force their electron shells to mingle and react. Because molecules move faster as the temperature increases, most chemical reactions occur more readily at high temperatures. The initial heat provided by a match setting sugar on fire allows these reactions to begin. The combination of sugar with oxygen then releases enough of its own heat to sustain the reaction, so the reaction continues spontaneously. Now, think about lighting a match; where does the heat to start that reaction come from? How is adequate activation energy generated in the body to allow it to "burn" sugar? Keep this question in mind—you'll find the answer a bit later in the chapter.

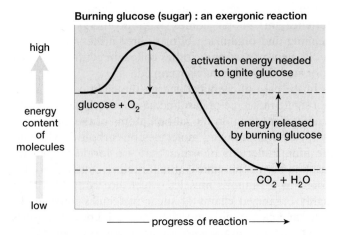

FIGURE 6-6 Energy relations in exergonic reactions
An exergonic ("downhill") reaction, such as burning sugar, proceeds from high-energy reactants (here, glucose and O_2) to low-energy products (CO_2 and H_2O). The energy difference between the chemical bonds of the reactants and products is released as heat. To start the reaction, however, an initial input of energy—the activation energy—is required. QUESTION In addition to heat and sunlight, what are some other potential sources of activation energy?

Endergonic Reactions Require a Net Input of Energy

In contrast to what happens when sugar or a match is burned, many reactions in living systems result in products that contain more energy than the reactants do. Sugar, produced by photosynthetic organisms such as plants, contains far more energy than does the carbon dioxide and water from which it was formed. The protein in a muscle cell contains more energy than the individual amino acids that were joined together to synthesize it. In other words, synthesizing complex biological molecules requires an input of energy. As we will see in Chapter 7, photosynthesis in green plants takes low-energy water and carbon dioxide and produces oxygen and high-energy sugar from them (**FIG. 6-7**).

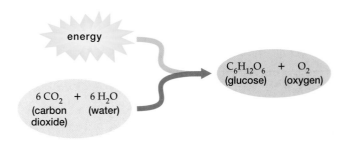

energy

6 CO_2 + 6 H_2O
(carbon (water)
dioxide)

$C_6H_{12}O_6$ + O_2
(glucose) (oxygen)

FIGURE 6-7 Photosynthesis

Endergonic reactions are not spontaneous; we might call them "uphill" reactions, because the reactants contain less energy than the products do. Going from low energy to high energy is like pushing a rock up to the top of the hill. The overall reactions of photosynthesis are endergonic, requiring a net input of energy, which photosynthetic organisms (plants, some protists, and some bacteria) obtain from sunlight. But where do we and other animals get the energy to synthesize muscle protein and other complex biological molecules?

Coupled Reactions Link Exergonic with Endergonic Reactions

Because endergonic reactions require a net input of energy, they must obtain this energy from exergonic, or energy-releasing, reactions. In a **coupled reaction**, an exergonic reaction provides the energy needed to drive an endergonic reaction. When you drive a car, the exergonic reaction of burning gasoline provides the energy for the endergonic reaction of starting the car moving and keeping it going; in the process, much energy is lost as heat. Photosynthesis is another coupled reaction. In photosynthesis, the exergonic reaction occurs in the sun, and the endergonic reaction occurs in the plant. Most of the energy liberated by the sun is lost as heat, so the second law of thermodynamics still applies: the net usable energy (in this case, in the solar system) decreases, and entropy increases.

Essentially all organisms depend on solar energy. This may be trapped directly by photosynthesis, or obtained from breaking down high-energy molecules that were derived from the bodies of other organisms. The energy in these molecules also came, ultimately, from photosynthesis. Within their bodies, organisms constantly use the energy given off by exergonic reactions (such as the chemical breakdown of sugar) to drive essential endergonic reactions (such as brain activity, muscular contraction, and other types of movement) or to synthesize complex molecules. Since some energy is lost as heat every time it is transformed, the energy provided by exergonic reactions must exceed the energy needed to drive endergonic reactions. The exergonic and endergonic parts of coupled reactions often occur in different places within cells, so there must be some way to transfer the energy from the exergonic reactions that release energy to the endergonic reactions that require it. In coupled reactions within cells, energy is usually transferred from place to place by *energy-carrier* molecules such as ATP.

6.3 HOW IS CELLULAR ENERGY CARRIED BETWEEN COUPLED REACTIONS?

As we saw earlier, cells couple reactions so that energy released by exergonic reactions drives endergonic reactions. In a runner, the breakdown of a sugar (glucose) releases energy, and this energy release is coupled to energy-consuming reactions, such as those powering muscle contraction. But glucose cannot be used directly for muscle contraction. Instead, the energy from glucose must first be transferred to an **energy-carrier molecule**, which provides the muscle with energy to contract. Energy carriers work something like rechargeable batteries; they pick up an energy charge at an exergonic reaction, move to another location within the cell, and release the energy to drive an endergonic reaction. Because energy-carrier molecules are unstable, they are used for short-term energy transfers within cells. They are not used to transport energy from cell to cell, nor are they used for long-term energy storage. Muscles store energy in glycogen, a stable carbohydrate molecule that consists of chains of glucose molecules, described in Chapter 3. When energy is needed—for example, when the marathon begins—the glycogen in the body is broken down by enzymes first to glucose and then to carbon dioxide and water. The energy is captured and transferred to muscle protein molecules by ATP.

ATP Is the Principal Energy Carrier in Cells

Several exergonic reactions in cells produce **adenosine triphosphate**, or **ATP**, the most common energy-carrier molecule in cells. By providing energy for a wide variety of endergonic reactions, ATP serves as a common currency of

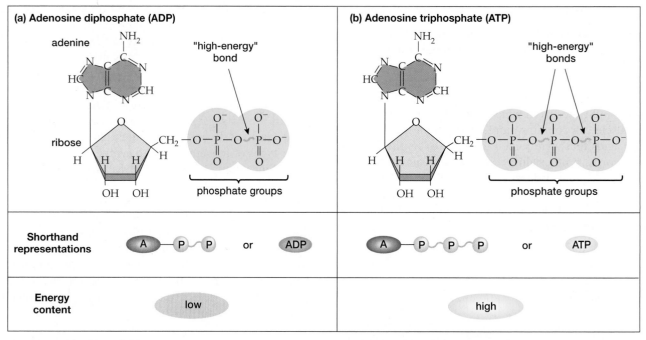

FIGURE 6-8 ADP and ATP
A phosphate group is added to **(a)** ADP (adenosine diphosphate) to make **(b)** ATP (adenosine triphosphate). In most cases, only the last phosphate group and its high-energy bond are used to carry energy and transfer it to endergonic reactions within a cell. QUESTION Why does conversion of ATP to ADP release energy for cellular work?

energy transfer, and it is sometimes called the "energy currency" of cells. As you learned in Chapter 3, ATP is a nucleotide composed of the nitrogen-containing base adenine, the sugar ribose, and three phosphate groups (**FIG. 6-8**).

Energy released in cells during glucose breakdown is used to synthesize ATP from **adenosine diphosphate (ADP)** and phosphate (**FIG. 6-9**).

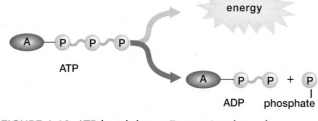

FIGURE 6-10 ATP breakdown: Energy is released

FIGURE 6-9 ATP synthesis: Energy is stored in ATP

ATP stores energy within chemical bonds and carries the energy to sites where energy-requiring reactions, such as the synthesis of proteins or muscle contraction, occur. The ATP is then broken down to form ADP and phosphate (**FIG. 6-10**).

During these energy transfers, some heat is given off at each stage, and there is an overall loss of usable energy (**FIG. 6-11**). Heat, generated as a by-product of every biochemical transformation, is used by warm-blooded animals to maintain a high body temperature. By speeding up biochemical reactions, this elevated body temperature allows animals to move faster and respond more quickly to stimuli than if their body temperatures were lower.

ATP is admirably suited to carrying energy within cells. The bonds joining the last two phosphate groups of ATP to the rest of the molecule (sometimes called *high-energy bonds*) require a large amount of energy to form, so considerable energy can be trapped from exergonic reactions by synthesizing ATP molecules. ATP is also unstable; it readily releases its energy in the presence of appropriate enzymes. Under most circumstances, only the bond joining the last phosphate group (the one joining phosphate to

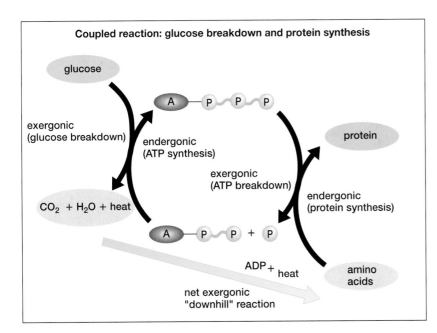

FIGURE 6-11 Coupled reactions within living cells
Exergonic reactions (such as glucose breakdown) drive the endergonic reaction that synthesizes ATP from ADP. The ATP molecule can carry its chemical energy to a part of the cell where the energy from the breakdown of ATP is needed to drive an essential endergonic reaction (such as protein synthesis). The ADP and phosphate are recycled back to ATP via endergonic reactions. The overall reaction is exergonic, or downhill: more energy is released by the exergonic reaction than is needed to drive the endergonic reaction.

ADP to form ATP) carries energy from exergonic to endergonic reactions.

The life span of an ATP molecule in a living cell is very short because this energy carrier is continuously formed, broken down to ADP and phosphate, and resynthesized. If the molecules of ATP that you use just sitting at your desk all day could be captured (instead of recycled), they would weigh 40 kg—nearly 90 pounds! A marathon runner may recycle the equivalent of a pound of ATP every minute. (The ADP must be quickly converted back to ATP, or it would be a very brief run.) So as you can see, ATP is *not* a long-term energy-storage molecule. More stable molecules, such as glycogen or fat, can store energy for hours, days, or (in the case of fat) even years.

Electron Carriers Also Transport Energy Within Cells

In addition to ATP, other energy-carrier molecules can also transport energy within a cell. In some exergonic reactions, including both glucose metabolism and the light-capturing stage of photosynthesis, some energy is transferred to electrons. These energetic electrons (in some cases, along with hydrogen atoms) are captured by **electron carriers** (FIG. 6-12). Common electron carriers include *nicotinamide adenine dinucleotide* (NAD^+) and its relative *flavin adenine dinucleotide* (FAD). Loaded electron carriers then donate the electrons, along with most of their energy, to other molecules. You will learn more about electron carriers and their role in cellular metabolism in Chapters 7 and 8.

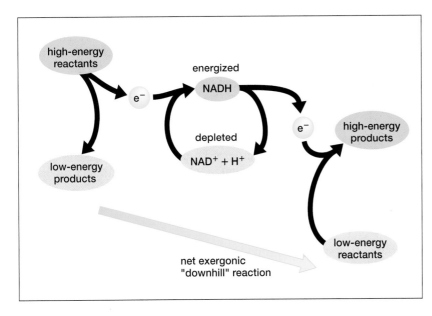

FIGURE 6-12 Electron carriers
Low-energy electron-carrier molecules such as NAD^+ pick up electrons generated by exergonic reactions and hold them in high-energy outer electron shells. Hydrogen ions are often picked up simultaneously. The electron is then transferred, with most of its energy, to another molecule to drive an endergonic reaction, often the synthesis of ATP.

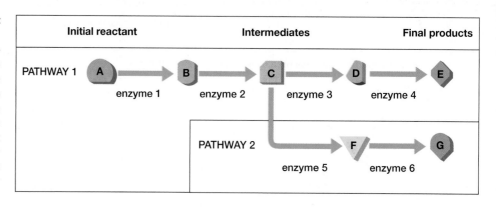

FIGURE 6-13 Simplified metabolic pathways
The initial reactant molecule, A, undergoes a series of reactions, each catalyzed by a specific enzyme. The product of each reaction serves as the reactant for the next reaction in the pathway. Metabolic pathways are commonly interconnected such that the product of a step in one pathway may serve as a reactant for the next reaction or for a reaction in another pathway.

6.4 HOW DO CELLS CONTROL THEIR METABOLIC REACTIONS?

Cells are miniature, incredibly complex chemical factories. The **metabolism** of a cell is the sum of all its chemical reactions. Many of these reactions are linked in sequences called **metabolic pathways** (**FIG. 6-13**). Within metabolic pathways, molecules are both synthesized and broken down. Photosynthesis (Chapter 7) is a metabolic pathway that results in the synthesis of high-energy molecules, including glucose. The metabolic pathway of glycolysis begins the breakdown of glucose (Chapter 8). Different metabolic pathways may use the same molecules; as a result, all the metabolic reactions within a cell are directly or indirectly interconnected.

The chemical reactions in a cell are governed by the same laws of thermodynamics that control other reactions. How, then, do orderly metabolic pathways arise? The biochemistry of cells is finely tuned in three ways:

- Cells couple reactions together, powering energy-requiring endergonic reactions with the energy released by exergonic reactions.
- Cells synthesize energy-carrier molecules that capture energy from exergonic reactions and transport it to endergonic reactions.
- Cells regulate chemical reactions through the use of proteins called *enzymes*, which are biological catalysts that help overcome activation energy.

At Body Temperatures, Spontaneous Reactions Proceed Too Slowly to Sustain Life

In general, the speed at which a reaction occurs is determined by its activation energy; that is, how much energy is required to start the reaction (see Fig. 6-6). Reactions with low activation energies can proceed swiftly at body temperature, whereas reactions with high activation energies, such as gasoline combining with oxygen, are practically nonexistent at similar temperatures. Most reactions can be accelerated by raising the temperature, thereby increasing the speed of molecules.

The reaction of sugar with oxygen to yield carbon dioxide and water is exergonic, but it has a high activation energy. The heat of a match flame can start sugar and oxygen molecules moving and colliding violently enough to cause them to react. The energy then released from this exergonic reaction is sufficient to cause more of the sugar molecules to combine with oxygen, and the sugar burns spontaneously. At the temperatures found in living organisms, however, sugar and many other energy-containing molecules would almost never break down spontaneously and give up their energy. But enzymes, which are biological catalysts produced by cells, allow sugar to be an important energy source. Let's see how enzymes and some nonbiological catalysts promote chemical reactions.

Catalysts Reduce Activation Energy

Catalysts are molecules that speed up the rate of a reaction without themselves being used up or permanently altered. A catalyst speeds up a reaction by reducing its activation energy (**FIG. 6-14**). As an example of catalytic action, let's consider the catalytic converters on automobile exhaust systems. When gasoline is burned completely, the final products are carbon dioxide and water:

$$2\ C_8H_{18} + 25\ O_2 \rightarrow 16\ CO_2 + 18\ H_2O + energy$$
(octane)

However, flaws in the combustion process generate other substances, including poisonous carbon monoxide (CO). Carbon monoxide reacts spontaneously but slowly with oxygen in the air to form carbon dioxide:

$$2\ CO + O_2 \rightarrow 2\ CO_2 + energy$$

In heavy traffic, however, the spontaneous reaction of CO with O_2 can't keep pace with the vast amount of CO emitted, and unhealthy levels of carbon monoxide accumulate. Enter the catalytic converter. Catalysts such as platinum in the converter provide a specialized surface upon which oxygen and CO combine more readily, hastening the conversion of CO to CO_2 and reducing air pollution.

Here are three important features of all catalysts:

- Catalysts speed up reactions.
- Catalysts can speed up only those reactions that would occur spontaneously if activation energy could be surmounted.
- Catalysts are not consumed or permanently changed by the reactions they promote.

Enzymes Are Biological Catalysts

Enzymes are biological catalysts, composed primarily of protein and synthesized by living organisms. Some enzymes require small nonprotein helper molecules called

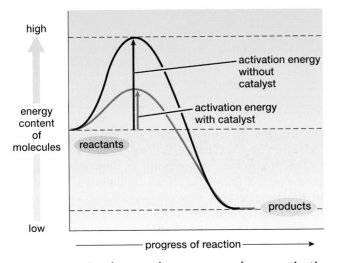

FIGURE 6-14 Catalysts such as enzymes lower activation energy
A high activation energy (black curve) means that reactant molecules must collide very forcefully in order to react. Catalysts lower the activation energy of a reaction (red curve), so a much higher proportion of molecules move fast enough to react when they collide. Therefore, the reaction proceeds much more rapidly. Enzymes are protein catalysts for biological reactions. QUESTION Can a catalyst make a non-spontaneous reaction occur spontaneously?

coenzymes in order to function. Many water-soluble vitamins (such as the B vitamins) are essential to humans because they are used by the body to synthesize coenzymes.

Enzymes, which can catalyze several million reactions per second, use their precise chemical structures to orient, distort, and reconfigure other molecules while emerging unchanged themselves. Besides the characteristics of catalysts that we just described, enzymes have two additional attributes that set them apart from nonbiological catalysts:

- Enzymes are usually very specific, catalyzing at most only a few types of chemical reactions. Generally, an enzyme catalyzes a single type of reaction involving specific molecules but leaves similar molecules unchanged.

- Enzyme activity is often regulated—that is, enhanced or suppressed—by negative feedback that controls the rate at which enzymes synthesize or break down biological molecules.

Enzyme Structures Allow Them to Catalyze Specific Reactions

Enzyme function is intimately related to enzyme structure. Each enzyme has a "pocket," called the **active site**, into which one or more reactant molecules, called **substrates**, can enter. You may recall from Chapter 3 that proteins have complex three-dimensional shapes. Their primary structure is determined by the precise order in which amino acids are linked together. The chain of amino acids then folds around itself in a configuration (often a helix or pleated sheet) called secondary structure. The protein then acquires the additional twists and bends of tertiary structure. In enzyme proteins, the order of amino acids and the precise way in which they are folded creates a distinctive

shape and distribution of electrical charges that are complementary to the enzyme's substrate. Some enzymes make use of quaternary protein structure, linking chains of amino acids together to create the necessary shape and arrangement of charges within the active site.

Because the enzyme and its substrate must fit together precisely, only certain molecules can enter the active site. Take the enzyme *amylase*, for example. Amylase breaks down starch molecules by hydrolysis but leaves cellulose molecules intact, even though both substances consist of chains of glucose. A different bonding pattern between the glucose molecules in cellulose prevents them from fitting into the enzyme's active site. If you chew a soda cracker long enough, you'll notice a sweet flavor caused by the release of sugar molecules from the starch in the cracker by amylase in your saliva. The stomach enzyme *pepsin* is selective for proteins, attacking them at many sites along their amino acid chains. Certain other protein-digesting enzymes (*trypsin*, for example) will break bonds only between specific amino acids. Digestive systems manufacture several different enzymes that work together to completely break down dietary protein into its individual amino acids.

How does an enzyme catalyze a reaction? First, both the shape and the charge of the active site allow substrates to enter the enzyme only in specific orientations (**FIG. 6-15**, step ①). Second, when substrates enter the active site, both the substrate and active site change shape (step ②). Certain amino acids within the enzyme's active site may temporarily bond with atoms of the substrates, or electrical

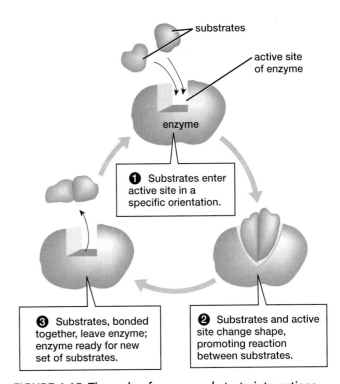

FIGURE 6-15 The cycle of enzyme-substrate interactions
As you look at this figure, imagine the opposite type of reaction as well, in which an enzyme binds to a single molecule and causes it to dissociate into two smaller molecules. QUESTION How would you modify reaction conditions if you wanted to increase the rate at which an enzyme-catalyzed reaction produced its product?

interactions between the amino acids in the active site and substrates may distort chemical bonds within the substrates. The combination of substrate selectivity, substrate orientation, temporary chemical bonds, and bond distortion promotes the specific chemical reaction catalyzed by a particular enzyme. When the final reaction between the substrates is finished, the product(s) no longer fit properly into the active site and drift away (step ③). The enzyme reverts to its original configuration, and it is ready to accept another set of substrates (back to step ①).

How do enzymes speed up the rate of chemical reactions? The breakdown or synthesis of a molecule within a cell usually occurs in many discrete steps, each catalyzed by a different enzyme (see Fig. 6-13). Each of these enzymes lowers the activation energy for its particular reaction (see Fig. 6-14), allowing the reaction to occur readily at body temperature. For example, a spoonful of sugar at body temperature would remain sugar indefinitely because of its high activation energy. But within a cell, sugar is readily combined with oxygen (oxidized) to water and carbon dioxide in the process of cellular respiration (described in detail in Chapter 8). How? Think of a rock climber ascending a steep cliff by finding a series of hand- and footholds that allow her to scale it one small step at a time. In a similar manner, a series of reaction steps, each catalyzed by an enzyme that lowers activation energy, allows the overall reaction—in this case, oxidizing sugar—to surmount its high activation energy "cliff," and proceed at body temperature.

Cells Regulate Metabolism by Controlling Enzymes

To be useful, the metabolic reactions within cells must be carefully controlled; they must occur at the proper rate and with the proper timing. This fine-tuning of metabolic reactions is accomplished by regulating the enzymes that promote the reactions, as we describe in the following paragraphs.

Cells Regulate Enzyme Synthesis

Cells exert tight control over all the types of proteins they produce. Genes that code for specific proteins are turned on or off depending on the need for that protein, a process described in detail in Chapter 10. Enzyme proteins govern all the cell's metabolic activities, and these activities must be able to change to meet the cell's changing needs. Some enzymes, therefore, are synthesized in larger quantities when more of their substrate is available. For example, the liver produces more of an enzyme (alcohol dehydrogenase) that breaks down alcohol in people who consume large amounts of this drug. Unfortunately, alcohol is converted to other toxic substances by this and other liver enzymes, and so the livers of alcoholic individuals are often seriously damaged as a result.

Accidental changes in genes can result in a lack of specific enzymes, sometimes with severe consequences. For example, people with phenylketonuria don't make the enzyme

that begins the breakdown of the amino acid phenylalanine, common in proteins. A buildup of phenylalanine is toxic to developing infants, and can cause mental retardation.

Cells Regulate Enzyme Activity

Some enzymes are synthesized in inactive forms. Another mechanism by which cells exert control over enzymes is by synthesizing some enzymes in an inactive form that becomes active under the proper conditions. Examples are the protein-digesting enzymes *pepsin* and *trypsin*, mentioned earlier. Cells synthesize and release these enzymes in inactive forms, preventing self-digestion. In the stomach where pepsin works, stomach acid removes the block from the pepsin active site, allowing the enzyme to function. Trypsin, in contrast, works best in the low-acid conditions of the small intestine, where it is activated by another enzyme.

Some enzymes are controlled by regulator molecules. Certain enzymes have their activity enhanced or inhibited by regulator molecules, in a process called **allosteric regulation**. In allosteric regulation the molecule that acts as regulator is neither the substrate nor the product of the enzyme it regulates.

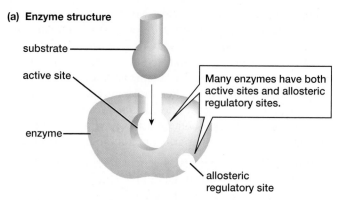

(a) **Enzyme structure**

substrate

active site

enzyme

Many enzymes have both active sites and allosteric regulatory sites.

allosteric regulatory site

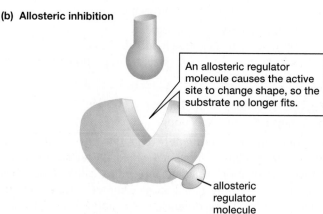

(b) **Allosteric inhibition**

An allosteric regulator molecule causes the active site to change shape, so the substrate no longer fits.

allosteric regulator molecule

FIGURE 6-16 Some enzymes are controlled by allosteric regulation
(a) Many enzymes have an active site and an allosteric regulatory site on different parts of the molecule. (b) When enzymes are inhibited by allosteric regulation, binding by a regulator molecule alters the active site so the enzyme is less compatible with its substrate.

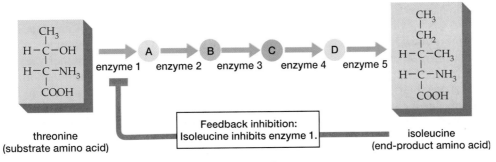

FIGURE 6-17 Enzyme regulation by feedback inhibition
In this example, the first enzyme in the metabolic pathway that converts threonine (an amino acid substrate) to isoleucine (an amino acid product) is inhibited by high concentrations of isoleucine, which acts as a regulator molecule. If a cell lacks isoleucine, the reactions proceed. As isoleucine builds up, it binds to the allosteric regulatory site of enzyme 1, blocking the pathway. When concentrations of isoleucine drop and fewer isoleucine molecules are available to inhibit the enzyme, production resumes.

The regulator molecule binds in a reversible manner to a special *allosteric regulatory site* on the enzyme, which is separate from the enzyme's active site (**FIG. 6-16a**). This temporary binding of the *regulator molecule* modifies the active site of the enzyme ("allosteric" literally means "other shape"), and the enzyme may become either more or less able to bind its substrate as a result (**FIG. 6-16b**). The specific enzyme and specific regulator molecule determine whether allosteric regulation increases or decreases enzyme activity.

One important type of allosteric regulation is **feedback inhibition**. This is a type of negative feedback that causes a metabolic pathway to stop producing a product when quantities reach a desired level, just as a thermostat turns off a heater when a room becomes warm enough. In feedback inhibition, the activity of an enzyme is inhibited by a regulator molecule that is the end product of a metabolic pathway. The regulator molecule typically inhibits an enzyme early in the series of reactions that produced it, as illustrated in **FIGURE 6-17**. For example, assume that a series of reactions, each catalyzed by a different enzyme, converts one amino acid to another. When enough of the product amino acid is present, the series of reactions is halted because the product amino acid binds to an allosteric regulatory site on an enzyme early in the pathway, inhibiting it.

Poisons, Drugs, and Environmental Conditions Influence Enzyme Activity

Drugs and poisons that act on enzymes usually inhibit them. Both competitive and noncompetitive forms of inhibition are exhibited by poisons and drugs.

Some Inhibitors Compete with Substrate for the Active Site of the Enzyme

Some drugs and poisons bind to the active site of an enzyme in a reversible way, so that both the normal substrate and the foreign substance compete for the site. Sometimes the foreign substance is broken down by the enzyme; sometimes it merely gets in the way of the normal substrate. This process is called **competitive inhibition** (**FIG. 6-18**). Methanol, for example, is a highly toxic form of alcohol used as a solvent. It competes for the active site of the enzyme alcohol dehydrogenase, whose normal substrate is ethanol (found in fermented fruit and alcoholic beverages). Alcohol dehydrogenase can break down methanol, but in the process it produces formaldehyde, which can cause blindness. Taking advantage of competitive inhibition, doctors administer ethanol to victims of methanol poisoning. By competing with methanol for the active site of alcohol dehydrogenase, ethanol blocks formaldehyde production. This example illustrates an important property of competitive inhibition: the normal substrate or the inhibitor can each displace the other if its concentration is high enough.

Some anticancer drugs are competitive inhibitors of enzymes. Because cancers consist of rapidly dividing cells, they generate large quantities of DNA. Some anticancer drugs resemble the subunits that comprise DNA. These

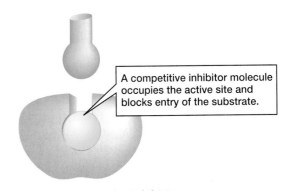

A competitive inhibitor molecule occupies the active site and blocks entry of the substrate.

FIGURE 6-18 Competitive inhibition
A drug or poison reversibly blocks the active site, competing with the normal substrate.

drugs compete with the normal subunits, tricking enzymes into building defective DNA, which in turn prevents the cancer cells from proliferating. Unfortunately, these drugs also interfere with the growth of other rapidly dividing cells, including those in hair follicles and lining the digestive tract. This explains why hair loss and nausea are side effects of some types of cancer chemotherapy drugs.

Some Inhibitors Bind Permanently to the Enzyme

Some poisons or drugs bind irreversibly to the enzyme. These irreversible inhibitors may enter and permanently block the enzyme's active site, or they may attach to another part of the enzyme, changing the enzyme's shape or charge so that it can no longer bind its substrates properly.

For example, some nerve gases and insecticides permanently block the active site of the enzyme *acetylcholinesterase* that breaks down acetylcholine (a substance that neurons release to activate muscles). This allows acetylcholine to build up and overstimulate muscles, causing paralysis. Death ensues because victims are unable to breathe. Other poisons, including arsenic, mercury, and lead, are toxic because they bind permanently to other parts of various enzymes, inactivating them.

The Activity of Enzymes Is Influenced by the Environment

The complex three-dimensional structures of enzymes are also sensitive to environmental conditions. You may recall from Chapter 3 that much of the three-dimensional structure of proteins is produced by hydrogen bonds between partially charged amino acids. These bonds rely for their existence on a narrow range of chemical and physical conditions, including the proper pH, temperature, and salt concentration. Most enzymes have a very narrow range of conditions in which they function optimally (**FIG. 6-19**).

Although the protein-digesting enzyme *pepsin* requires the acidic conditions of the stomach (pH = 2), most enzymes, including the starch-digesting enzyme amylase, function optimally at a pH between 6 and 8—the level found in most body fluids and maintained within living cells (Fig. 6-19a). An acid pH alters the charges on amino acids by adding hydrogen ions to them. Stomach acid kills many bacteria by inactivating their enzymes.

Temperature also affects the rate of enzyme-catalyzed reactions, which are slowed by lower temperatures and accelerated by moderately higher temperatures because the rate of movement of molecules determines how likely they are to encounter the active site of an enzyme (Fig. 6-19b). Cooling the body can drastically slow human metabolic reactions. In one true example, a boy who fell through the ice on a lake was rescued and survived unharmed after 20 minutes under water. Although at normal body temperature the brain dies after about 4 minutes without oxygen, the boy's body temperature and metabolic rate were lowered by the icy water, drastically reducing his need for oxygen. In contrast, when temperatures rise too high, the hydrogen bonds that regulate protein shape may be broken apart by the excessive molecular motion. Think of the

protein in egg white, and how its color and texture are completely altered by cooking. Far lower temperatures than those required to fry an egg can still be too hot to allow enzymes to function properly. Excessive heat may be fatal, partly because the greater motion of atoms at high temperatures breaks hydrogen bonds and distorts the three-dimensional structure of enzymes and other proteins necessary to life. Every summer, dozens of children in the United States die from heat stroke when left unattended in overheated cars.

Bacteria and fungi, which exist in nearly all our food, are responsible for food spoiling. Food remains fresh in the refrigerator or freezer because cooling slows the enzyme-catalyzed reactions upon which these microorganisms rely to grow and reproduce. Before the advent of refrigeration, meat was commonly preserved by using

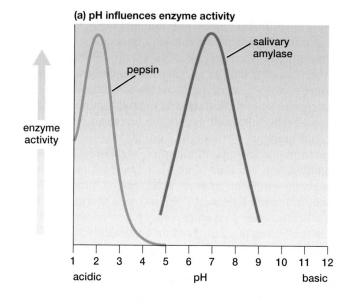

(a) pH influences enzyme activity

pepsin

salivary amylase

enzyme activity

1   2   3   4   5   6   7   8   9   10 11 12
acidic   pH   basic

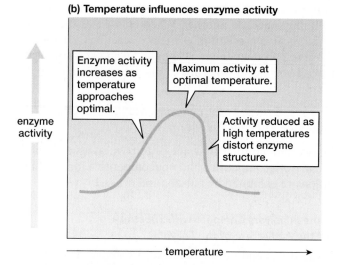

(b) Temperature influences enzyme activity

Enzyme activity increases as temperature approaches optimal.

Maximum activity at optimal temperature.

Activity reduced as high temperatures distort enzyme structure.

enzyme activity

temperature

FIGURE 6-19 Enzymes function best within narrow ranges of pH and temperature

LINKS TO LIFE   Lack of an Enzyme Leads to Lactose Intolerance

Is it difficult for you to imagine life without milk, ice cream, or even pizza? Although some people consider these foods to be staples of the U.S. diet, they are not enjoyed by most of the world's population. Why? About 75% of people worldwide, including 25% of people in the United States, lose the ability to digest lactose, or "milk sugar," in early childhood. Roughly 75% of African Americans, Hispanics, and Native Americans, as well as 90% of Asian Americans, are *lactose intolerant*. From an evolutionary perspective, this makes perfect sense. The lactose enzyme, called lactase, is found in the small intestines of all normal young children. After being weaned in early childhood, our early ancestors no longer consumed milk—the main source of lactose. Because it takes energy to synthesize enzymes, losing the ability to synthesize an enzyme that is no longer needed provides an adaptive advantage. However, a relatively small proportion of people, primarily those of northern European descent, retained the ability to digest lactose, raised cattle for their milk, and made dairy products a regular part of their diet.

When people who lack the enzyme lactase consume milk products, the undigested lactose draws water into the intestines by osmosis and also feeds intestinal bacteria that produce gas. The combination of excess water and gas leads to abdominal pain, bloating, diarrhea, and flatulence—a high price to pay for indulging in pizza or ice cream! Most people who are lactose intolerant do not need to avoid milk products altogether. Some produce enough lactase to tolerate a few servings daily. Aged cheeses (such as cheddar) and yogurt with live bacteria have relatively little lactose because the bacteria break it down. Lactase supplements can also be consumed along with dairy products. But compared to other consequences of enzyme deficiency, the inability to tolerate milk is a relatively minor inconvenience. Because these biological catalysts we know as enzymes are crucial to all aspects of life, mutations that render certain enzymes nonfunctional may prevent an embryo from developing at all, or they may cause life-threatening disorders.

concentrated salt solutions (think of bacon or salt pork), which kill most bacteria. Salts dissociate into ions, which form bonds with amino acids in enzyme proteins. Too much (or too little) salt interferes with the normal three-dimensional structure of enzymes, destroying their activity. Dill pickles are very well preserved in a vinegar-salt solution, which combines both highly salty and acidic conditions. Organisms that live in saline environments, as you might predict, have enzymes whose configuration depends on the presence of salt ions.

CASE STUDY REVISITED   ENERGY UNLEASHED

Although runners and all other organisms use sugar as fuel, we life-forms burn it in a controlled fashion, using enzymes rather than flames to overcome the activation energy. Like a climber descending a mountain in a series of small steps rather than jumping off the peak, enzymes allow our cells to break down sugar in many steps, each liberating a small, safe quantity of energy. Key steps in the pathway liberate just enough energy to be captured in energy-carrier molecules and used in energy-consuming reactions.

Life, with its constant demand for usable energy, generates heat, as dictated by the laws of thermodynamics. In marathoners, for example, as ATP is broken down to power muscle contraction, some of the chemical energy is converted to the kinetic energy of movement, and some is lost as heat. You learned in Chapter 2 that because water has one of the highest heats of vaporization of any molecule, we use sweat (which is almost entirely water) to cool our bodies. Marathon runners lose large amounts of water through sweating during the race, and they risk overheating if they don't replenish it.

Consider This When a runner's body temperature begins to rise, the body activates several mechanisms, including sweating and circulating more blood to the skin. Compare this response to overheating with feedback inhibition in enzymes.

CHAPTER REVIEW

SUMMARY OF KEY CONCEPTS

6.1 What Is Energy?

Energy is the capacity to do work. Kinetic energy is the energy of movement (light, heat, electricity, movement of large particles). Potential energy is stored energy (chemical energy, positional energy). The first law of thermodynamics, the law of conservation of energy, states that in a closed system the total amount of energy remains constant, although it may change in form. The second law of thermodynamics states that any use of energy causes a decrease in the quantity of useful energy and an increase in randomness and disorder, or entropy, within a system. The highly organized, low-entropy systems that characterize life do not violate the second law of thermodynamics, because they are achieved through a continuous influx of usable energy from the sun, accompanied by an enormous net increase in solar entropy.

Web Tutorial 6.1 Energy and Coupled Reactions

6.2 How Does Energy Flow in Chemical Reactions?

Chemical reactions fall into two categories. In exergonic reactions, the reactant molecules have more energy than do the product molecules, so the reaction releases energy. In endergonic reactions, the reactants have less energy than do the products, so the reaction requires an input of energy. Exergonic reactions can occur spontaneously; but all reactions, including exergonic ones, require an initial input of energy (the activation energy) to overcome electrical repulsions between reactant molecules. Exergonic and endergonic reactions may be coupled such that the energy liberated by an exergonic reaction drives the endergonic reaction. Organisms couple exergonic reactions, such as light-energy capture or sugar metabolism, with endergonic reactions, such as the synthesis of organic molecules.

6.3 How Is Cellular Energy Carried Between Coupled Reactions?

Energy released by chemical reactions within a cell is captured and transported within the cell by energy-carrier molecules, such as ATP and electron carriers. These molecules are the major means by which cells couple the exergonic and endergonic reactions occurring at different places in the cell.

6.4 How Do Cells Control Their Metabolic Reactions?

Cellular reactions are linked in interconnected sequences called metabolic pathways. The biochemistry of cells is regulated in three ways: first, through the use of protein catalysts called enzymes; second, by coupling endergonic with exergonic reactions; and third, through the use of energy-carrier molecules that transfer energy within cells.

High activation energies slow many reactions, even exergonic ones, to an imperceptible rate under normal environmental conditions. Catalysts lower the activation energy and thereby speed up chemical reactions without being permanently changed themselves. Organisms synthesize enzyme catalysts that promote one or a few specific reactions. The reactants temporarily bind to the active site of the enzyme, making it easier to form the new chemical bonds of the products. Enzyme action is regulated in many ways; these include altering the rate of enzyme synthesis, activating previously inactive enzymes, feedback inhibition, allosteric regulation, and competitive inhibition. Environmental conditions including pH, salt concentration, and temperature can promote or inhibit enzyme function by altering the three-dimensional structure of an enzyme.

Web Tutorial 6.2 Enzymes

KEY TERMS

activation energy *page 104*
active site *page 109*
adenosine diphosphate (ADP) *page 106*
adenosine triphosphate (ATP) *page 105*
allosteric regulation *page 110*
catalyst *page 108*
chemical reaction *page 103*

coenzyme *page 109*
competitive inhibition *page 111*
coupled reaction *page 105*
electron carrier *page 107*
endergonic *page 103*
energy *page 102*
energy-carrier molecule *page 105*

entropy *page 103*
enzyme *page 108*
exergonic *page 103*
feedback inhibition *page 111*
first law of thermodynamics *page 102*
kinetic energy *page 102*
laws of thermodynamics *page 102*

metabolic pathway *page 108*
metabolism *page 108*
potential energy *page 102*
product *page 103*
reactant *page 103*
second law of thermodynamics *page 102*
substrate *page 109*

THINKING THROUGH THE CONCEPTS

1. Explain why organisms do not violate the second law of thermodynamics. What is the ultimate energy source for most forms of life on Earth?

2. Define *metabolism*, and explain how reactions can be coupled to one another.

3. What is activation energy? How do catalysts affect activation energy? How does this change the rate of reactions?

4. Describe some exergonic and endergonic reactions that occur regularly in plants and animals.

5. Describe the structure and function of enzymes. How is enzyme activity regulated?

APPLYING THE CONCEPTS

1. One of your nerdiest friends walks in while you are vacuuming. Trying to impress her, you casually mention that you are infusing energy into your room to create a lower-entropy state, and the energy is coming from electricity. As she looks at you in bemusement, you add that this doesn't violate the second law of thermodynamics because a lot of heat is released at the power plant where the electricity is generated, and the air coming out of the vacuum is warmer, too. Irritatingly, she adds that ultimately you are actually taking advantage of the increase in entropy of the sun to clean your room. What is she talking about? Hint: Look for clues in Chapter 7.

2. As you learned in Chapter 3, the subunits of virtually all organic molecules are joined by condensation reactions and can be broken apart by hydrolysis reactions. Why, then, does your digestive system produce separate enzymes to digest proteins, fats, and carbohydrates—in fact, several of each type?

3. A preview question for evolution (Unit Three): Suppose someone tried to refute evolution using the following argument: "According to evolutionary theory, organisms have increased in complexity through time. However, the increase in complexity contradicts the second law of thermodynamics. Therefore, evolution is impossible." Is this a valid statement?

4. When a brown bear eats a salmon, does the bear acquire all the energy contained in the body of the fish? Why or why not? What implications do you think this answer would have for the relative abundance (by weight) of predators and their prey?

FOR MORE INFORMATION

Collins, T. J., and Walter, C. "Little Green Molecules." *Scientific American,* March, 2006. Small, non-protein molecules are being designed by chemists to act like enzymes and degrade man-made persistent toxic chemicals in the environment.

Farid, R. S. "Enzymes Heat Up." *Science News*, May 9, 1998. Scientists explore new ways to synthesize enzymes that will function at high temperatures.

Madigan, M. T., and Narrs, B. L. "Extremophiles." *Scientific American*, April 1997. Industrial processes are benefiting from an understanding of the molecules, particularly enzymes, that allow certain microbes to thrive under highly acidic, salty, or extremely hot conditions that would denature most proteins.

Wu, C. "Hot-Blooded Proteins." *Science News*, May 9, 1998. Bacteria that thrive in near-boiling conditions have special enzymes that allow them to function at these extreme temperatures.

CHAPTER 7

Capturing Solar Energy: Photosynthesis

A doomed dinosaur watches a giant meteor streak toward Earth.
Some scientist believe that a meteor impact may have caused
a massive extinction about 65 million years ago.

AT A GLANCE

 CASE STUDY **DID THE DINOSAURS DIE FROM LACK OF SUNLIGHT?**

IT IS SUMMER in the year 65,000,000 B.C. The Cretaceous period is about to reach an abrupt and catastrophic end. On an Earth where the continents we know as the Americas are largely submerged by shallow seas, an 80-foot-long, 35-ton *Apatosaurus* grazes on lush, tropical vegetation in what is now southern California. Suddenly, a deafening roar startles the animal, and it gazes upward to see an immense fireball ripping through the blue sky. A meteorite six miles in diameter has entered the atmosphere and is about to irrevocably alter life on Earth. Although any creatures that witnessed the meteorite strike Earth were immediately incinerated by the blast wave from the impact, the aftereffects were experienced by plants and animals all over the world. As it plowed into the ocean at the tip of the Yucatán Peninsula, the meteorite dug a crater 1 mile deep and 120 miles wide. The force of its impact sent trillions of tons of debris from Earth's crust and from the meteorite itself rocketing into the atmosphere. The heat of the blast caused fires that may have charred 25 percent of all the vegetation on land. Ashes, smoke, and dust obliterated the sun. Earth was plunged into a night that lasted for months. What would happen if the sun were obscured for months on end? Why is sunlight so important? Could a meteorite have been responsible for ending the rule of the dinosaurs?

7.1   WHAT IS PHOTOSYNTHESIS?

At least 2 billion years ago, some cells, through chance changes (mutations) in their genetic makeup, acquired the ability to harness the energy in sunlight. These cells combined simple inorganic molecules—carbon dioxide and water—into more complex organic molecules such as glucose. In the process of *photosynthesis*, the cells captured a small fraction of the sunlight's energy, storing it as chemical energy in complex organic molecules. Exploiting this new source of energy without competition, early photosynthetic cells filled the seas, releasing oxygen as a by-product. A new element in the atmosphere, free oxygen was harmful to many organisms. But the endless variation produced by random genetic mistakes eventually gave rise to some cells that could survive in oxygen and, later, to cells that made use of oxygen to break down glucose in a new, more efficient process: *cellular respiration*. Today, most forms of life on Earth—including you—depend on the sugar produced by photosynthetic organisms as an energy source and release the energy from that sugar through cellular respiration, using the photosynthetic by-product oxygen (**FIG. 7-1**). In Chapter 8, we will examine the processes by which nearly all living things break down the sugary energy-storage molecules produced by photosynthesis and reclaim the energy to power other metabolic reactions. Sunlight powers practically all life on Earth and is captured only by photosynthesis.

Starting with the simple molecules of carbon dioxide (CO_2) and water (H_2O), **photosynthesis** converts the energy of sunlight into chemical energy stored in the bonds of glucose ($C_6H_{12}O_6$) and releases oxygen (O_2). The simplest overall chemical reaction for photosynthesis is:

$$6\,CO_2 + 6\,H_2O + \text{light energy} \rightarrow C_6H_{12}O_6 + 6\,O_2$$

Photosynthesis occurs in eukaryotic plants, algae, and certain types of prokaryotes, all of which are described as *autotrophs* (literally, "self-feeders"). In this chapter, we will limit our discussion of photosynthesis to plants, particularly land plants. Photosynthesis in plants takes place within chloroplasts, most of which are contained within leaf cells. Let's begin, then, with a brief look at the structures of both leaves and chloroplasts.

Leaves and Chloroplasts Are Adaptations for Photosynthesis

The leaves of most land plants are only a few cells thick; their structure is elegantly adapted to the demands of photosynthesis (**FIG. 7-2**). The flattened shape of leaves exposes a large surface area to the sun, and their thinness ensures that sunlight can penetrate them to reach the light-trapping chloroplasts inside. Both the upper and lower surfaces of a leaf consist of a layer of transparent cells, the *epidermis*. The outer surface of both epidermal layers is covered by the *cuticle*, a waxy, waterproof covering that reduces the evaporation of water from the leaf (Fig. 7-2b).

A leaf obtains CO_2 for photosynthesis from the air; adjustable pores in the epidermis, called **stomata** (singular, *stoma*; Greek for "mouth," shown in **FIG. 7-3**), open and close at appropriate times to admit air carrying CO_2.

Inside the leaf are layers of cells collectively called *mesophyll* (which means "middle of the leaf"). The mesophyll cells contain most of a leaf's chloroplasts (see Fig. 7-2b, c), and consequently photosynthesis occurs principally in these cells. *Vascular bundles*, or veins (see Fig. 7-2b), supply water and minerals to the mesophyll cells and carry the sugars they produce to other parts of the plant.

A single mesophyll cell can have from 40 to 200 chloroplasts, which are sufficiently small that 2000 of them lined up would just span your thumbnail. As described in Chapter 4, chloroplasts are organelles that consist of a double outer membrane enclosing a semifluid medium, the **stroma** (see Fig. 7-2d). Embedded in the

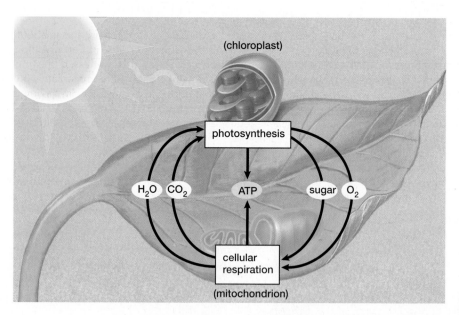

FIGURE 7-1 Interconnections between photosynthesis and cellular respiration Chloroplasts in green plants use the energy of sunlight to synthesize high-energy carbon compounds such as glucose from low-energy molecules of water and carbon dioxide. Plants themselves, and other organisms that eat plants or one another, extract energy from these organic molecules by cellular respiration, yielding water and carbon dioxide once again. This energy in turn drives all the reactions of life.

(a) Leaves

(b) Internal leaf structure

(d) Chloroplast

(c) Mesophyll cell containing chloroplasts

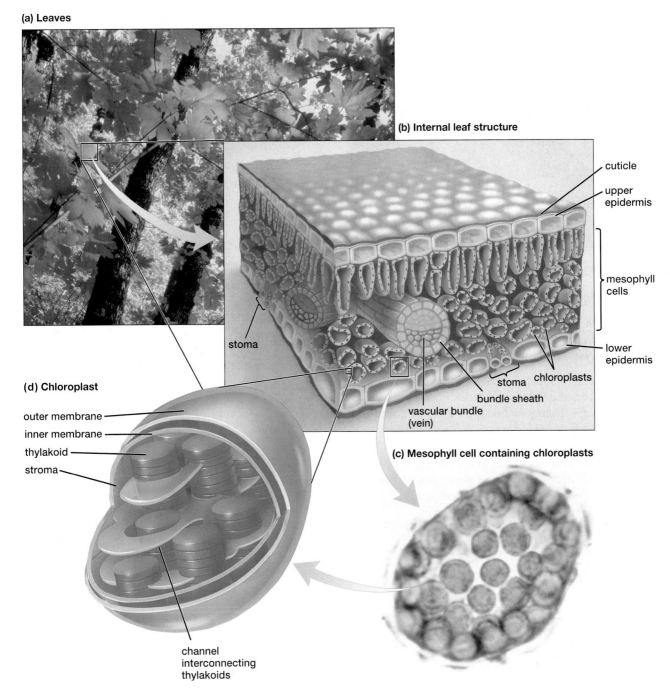

FIGURE 7-2 **An overview of photosynthetic structures**
(a) Photosynthesis occurs primarily in the leaves of land plants. (b) A section of a leaf, showing mesophyll cells where chloroplasts are concentrated and the waterproof cuticle that coats the leaf on both surfaces. (c) A mesophyll cell packed with green chloroplasts. (d) A single chloroplast, showing the stroma and thylakoids where photosynthesis occurs.

stroma are disk-shaped, interconnected membranous sacs called **thylakoids**. The chemical reactions of photosynthesis that depend on light (*light-dependent reactions*) occur within the membranes of the thylakoids, while the photosynthetic reactions that can continue for a time in darkness (*light-independent reactions*) occur in the surrounding stroma.

Photosynthesis Consists of Light-Dependent and Light-Independent Reactions

The simple chemical summary of photosynthesis obscures the fact that photosynthesis actually involves dozens of enzymes catalyzing dozens of individual reactions. These reactions can be classified as either light-dependent reactions

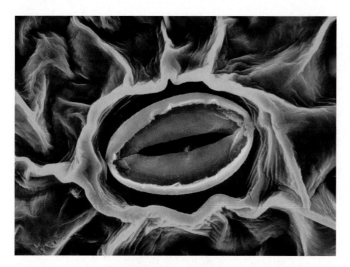

FIGURE 7-3 Stoma in leaf of pea plant

or light-independent reactions. Each group of reactions occurs within a different region of the chloroplast; the two are linked by energy-carrier molecules.

- In **light-dependent reactions**, chlorophyll and other molecules embedded in the membranes of the thylakoids capture sunlight energy and convert some of it into the chemical energy stored in energy-carrier molecules (ATP and NADPH). Oxygen gas is released as a by-product.
- In **light-independent reactions**, enzymes in the stroma use the chemical energy of the carrier molecules to drive the synthesis of glucose or other organic molecules.

The relationship of light-dependent and light-independent reactions is illustrated in **FIGURE 7-4**.

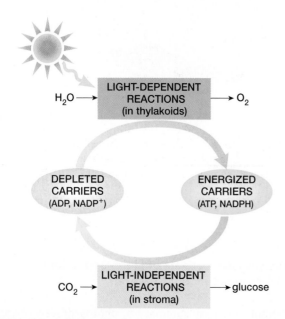

FIGURE 7-4 **Relationship between the light-dependent and light-independent reactions**

7.2 LIGHT-DEPENDENT REACTIONS: HOW IS LIGHT ENERGY CONVERTED TO CHEMICAL ENERGY?

The light-dependent reactions capture the energy of sunlight, storing it as chemical energy in two different energy-carrier molecules: the familiar energy carrier ATP (*adenosine triphosphate*) and the high-energy electron carrier *NADPH* (*nicotinamide adenine dinucleotide phosphate*). The chemical energy stored in these carrier molecules will then be used to power the synthesis of high-energy storage molecules, such as glucose, during the light-independent reactions.

During Photosynthesis, Light Is First Captured by Pigments in Chloroplasts

The sun emits energy in a broad spectrum of electromagnetic radiation. The *electromagnetic spectrum* ranges from short-wavelength gamma rays, through ultraviolet, visible, and infrared light, to very long-wavelength radio waves (**FIG. 7-5**). Light and the other types of radiation are composed of individual packets of energy called **photons**. The

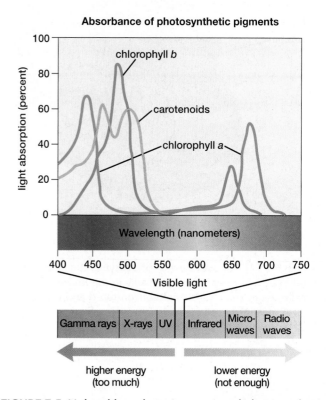

FIGURE 7-5 **Light, chloroplast pigments, and photosynthesis** Visible light, a small part of the electromagnetic spectrum, consists of wavelengths that correspond to the colors of the rainbow. Chlorophyll (blue and green curves) strongly absorbs violet, blue, and red light. Carotenoids (orange curve) absorb blue and green wavelengths.

energy of a photon corresponds to its wavelength: short-wavelength photons are very energetic, whereas longer-wavelength photons have lower energies. Visible light consists of wavelengths with energies that are strong enough to alter the shape of certain pigment molecules (such as those in chloroplasts) but weak enough not to damage crucial molecules such as DNA. It is no coincidence that these wavelengths—with "just the right amount" of energy—not only power photosynthesis but also stimulate the pigments in our eyes and allow us to see the world around us.

One of three events occurs when light strikes an object such as a leaf: the light is either *absorbed* (captured), *reflected* (bounced back), or *transmitted* (passed through). Light that is absorbed can heat up the object or drive biological processes, such as photosynthesis. Light that is reflected or transmitted is not captured by the object and can reach the eyes of an observer, giving an object its color.

Chloroplasts contain various pigment molecules that absorb different wavelengths of light. **Chlorophyll**, the key light-capturing pigment molecule in chloroplasts, strongly absorbs violet, blue, and red light but reflects green, thus giving green leaves their color (see Fig. 7-5). Chloroplasts also contain other molecules, called *accessory pigments*, that absorb additional wavelengths of light energy and transfer them to *chlorophyll a*. Some accessory pigments are actually slightly different forms of green chlorophyll; in land plants, *chlorophyll a* is the main light-capturing pigment, while *chlorophyll b* serves as an accessory pigment. **Carotenoids** are accessory pigments found in all chloroplasts. They absorb blue and green light and appear mostly yellow or orange because they reflect these wavelengths to our eyes (see Fig. 7-5).

Although carotenoids (particularly yellow and orange forms) are present in leaves, their color is usually masked by more abundant green chlorophyll. In the autumn, as leaves begin to die, chlorophyll breaks down before carotenoids do, revealing the bright yellow and orange carotenoids as fall colors. (Red and purple fall leaf colors are primarily pigments that are not involved in photosynthesis.) The aspen leaves in **FIGURE 7-6** show green chlorophyll fading, revealing yellow carotenoids.

You may have heard of the carotenoid beta-carotene. This pigment helps capture light in chloroplasts, and it produces the orange color of vegetables such as carrots. Beta-carotene is the principal source of vitamin A for animals. In a beautiful symmetry, vitamin A is used to form the visual pigment that captures light in animal (including human) eyes. Thus, carotenoids capture light energy in plants and (indirectly) in animals as well.

The Light-Dependent Reactions Occur in Association with the Thylakoid Membranes

The thylakoid membranes contain highly organized clusters of proteins, chlorophyll, and accessory pigment molecules including carotenoids; these clusters are called **photosystems**. Each thylakoid contains thousands of copies of each of two photosystems—*photosystem I* (PS I) and *photosystem II* (PS II). Both photosystems are activated by light, and they work simultaneously. Each photosystem contains roughly 250 to 400 chlorophyll and carotenoid molecules. These pigments absorb light and pass its energy to a pair of specialized chlorophyll molecules within a small region of the photosystem called the **reaction center**. The reaction-center chlorophyll molecules are located

FIGURE 7-6 Loss of chlorophyll reveals yellow carotenoids

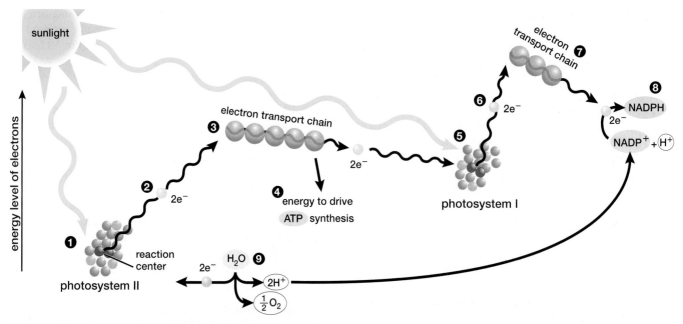

FIGURE 7-7 The light-dependent reactions of photosynthesis
① Light is absorbed by photosystem II, and the energy is passed to electrons in the reaction-center chlorophyll molecules. ② Energized electrons leave the reaction center. ③ The electrons move into the adjacent electron transport chain. ④ The chain passes the electrons along, and some of their energy is used to drive ATP synthesis by chemiosmosis. Energy-depleted electrons replace those lost by photosystem I. ⑤ Light strikes photosystem I, and the energy is passed to electrons in the reaction-center chlorophyll molecules. ⑥ Energized electrons leave the reaction center. ⑦ The electrons move into the electron transport chain. ⑧ The energetic electrons from photosystem I are captured in molecules of NADPH. ⑨ The electrons lost from the reaction center of photosystem II are replaced by electrons obtained from splitting water, a reaction that also releases oxygen, and H⁺ used to form NADPH. QUESTION If these reactions produce ATP and NADPH, then why do plant cells need mitochondria?

adjacent to an **electron transport chain (ETC)**, which is a series or "chain" of electron carrier molecules embedded in the thylakoid membrane. As you will see in **FIGURES 7-7** and **7-8**, each photosystem is associated with a different electron transport chain.

When the reaction-center chlorophyll molecules receive energy from the surrounding carotenoid molecules, an electron from each of the two reaction-center chlorophylls absorbs the energy. These "energized electrons" leave the chlorophyll molecules and "jump" over to the electron transport chain, where they are passed along from one carrier molecule to the next, losing energy as they go. At certain transfer points along the electron transport chain, the energy released from the electrons is captured and used to synthesize ATP from ADP plus phosphate or NADPH from NADP⁺ plus H⁺. (*NADP* is the electron carrier nicotinamide adenine dinucleotide [NAD], described in Chapter 6, plus a phosphate group.)

The light-dependent reactions in many ways resemble a pinball game. Energy (light) is transferred to a ball (electron) by spring-driven pistons (chlorophyll molecules). The ball is propelled upward (enters a higher energy level). As the ball travels downhill, the energy it releases can be used to turn a wheel (generate ATP), or ring a bell (generate NADPH). With this overall scheme in mind, let's look more closely at the actual sequence of events in the light-dependent reactions. These are illustrated diagrammatically in Figure 7-7, where each step is numbered, and more realistically within the chloroplast membrane in Figure 7-8. As you follow the numbered steps in Figure 7-7, find the same events in Figure 7-8 within the membrane.

Photosystem II Generates ATP

For historical reasons, the photosystems are numbered "backward." The process of capturing light energy is most easily understood by starting with photosystem II and following the events initiated by the capture of two photons of light. The light-dependent reactions begin when photons of light are absorbed by photosystem II (step ① in Fig. 7-7; far left in Fig. 7-8). Light energy passes from molecule to molecule until it reaches the reaction center, where it boosts an electron out of each of the two chlorophyll molecules (step ②). The first electron carrier of the adjacent electron transport chain instantly accepts these two energized electrons

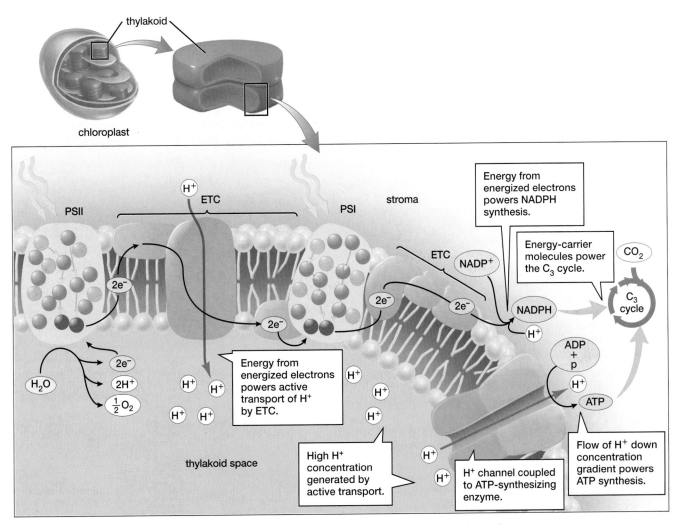

FIGURE 7-8 Events of the light-dependent reactions occur in and near the thylakoid membranes

(step ③). The electrons move along the chain from one carrier molecule to the next, releasing energy as they go; since no energy transfer is 100 percent efficient, some of the energy is lost as heat at each step. But some of the energy released as the electron travels through the electron transport chain is captured and used to pump hydrogen ions (H^+) across the thylakoid membrane into the compartment within the thylakoid, creating an H^+ ion concentration gradient across the thylakoid membrane, shown in Figure 7-8 (left). The energy used to create this gradient is then harnessed to drive the synthesis of ATP, in a process called **chemiosmosis** (step ④). In Figure 7-8 (right), you will see H^+ flowing back down its concentration gradient through a special channel that generates ATP as H^+ flows through. See "A Closer Look: Chemiosmosis—ATP Synthesis in Chloroplasts" for a more detailed description of this process.

Photosystem I Generates NADPH

Meanwhile, light rays have also been striking the pigment molecules of photosystem I (step ⑤ in Fig. 7-7, middle of Fig. 7-8). Energy from the light photons is captured by these

pigment molecules and funneled to the two reaction center chlorophyll molecules, which then eject high-energy electrons (step ⑥). These electrons jump to the electron transport chain associated with photosystem I (step ⑦). The energized electrons ejected from photosystem I move through the adjacent, shorter electron transport chain and are finally transferred to the electron carrier $NADP^+$. The energy carrier molecule NADPH is formed when each $NADP^+$ molecule picks up two energetic electrons and one hydrogen ion (step ⑧; Fig. 7-8, right); the hydrogen ion is obtained from splitting water (step ⑨; Fig. 7-8, left). Both $NADP^+$ and NADPH are water-soluble molecules dissolved in the chloroplast stroma. The reaction-center chlorophylls of photosystem I immediately replace their lost electrons by obtaining the energy-depleted electrons from the final electron carrier of the electron transport chain fed by photosystem II.

Splitting Water Maintains the Flow of Electrons Through the Photosystems

Overall, electrons flow from the reaction center of photosystem II, through its nearby electron transport chain, to

A CLOSER LOOK   Chemiosmosis—ATP Synthesis in Chloroplasts

In the light-dependent reactions of photosynthesis (see Fig. 7-7, Fig. 7-8, and Fig. E7-2), photons energize electrons in photosystem II. In the electron transport chain associated with photosystem II, these energetic electrons lose energy as they move from carrier to carrier. The electron transfers do not directly drive ATP synthesis; rather, the energy they release is used to pump hydrogen ions (H^+) from the stroma across the thylakoid membrane into the thylakoid space. Like charging a battery, active transport of (H^+) stores energy by creating a concentration gradient of (H^+) across the thylakoid membrane. Then, in a separate reaction, the energy stored in this gradient powers ATP synthesis.

How is a gradient of (H^+) used to synthesize ATP? Compare the (H^+) gradient to water stored behind a dam at a hydroelectric plant (**FIG. E7-1**). The water flows out through turbines, rotating them. The turbines convert the energy of moving water into electrical energy. Hydrogen ions in the thylakoid interior (like water stored behind a dam) can move down their gradients into the stroma only through special (H^+) channels linked to ATP-synthesizing enzymes. Like turbines generating electricity, the enzymes linked to (H^+) channels capture the energy liberated by the flow of (H^+) and use it to drive ATP synthesis from ADP plus phosphate (**FIG. E7-2**). About one ATP molecule is synthesized for every three hydrogen ions that pass through the channel.

Scientists are still investigating exactly how the ATP-synthesizing proton channel works. However, this general mechanism of ATP synthesis was first proposed in 1961 by British biochemist Peter Mitchell, who called it *chemiosmosis*. Chemiosmosis has been shown to be the mechanism of ATP generation in chloroplasts, mitochondria (as we will see in Chapter 8), and bacteria. For his brilliant hypothesis, Mitchell was awarded the Nobel Prize for Chemistry in 1978.

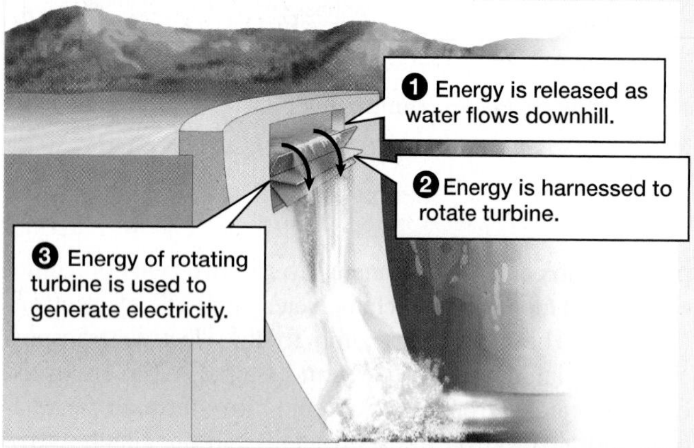

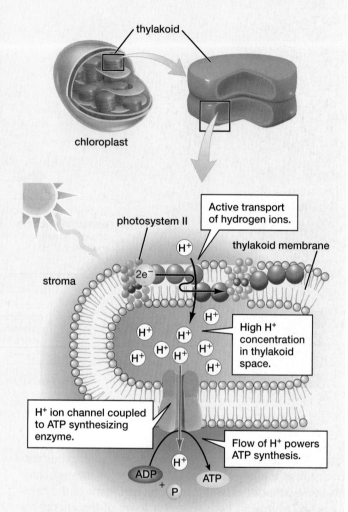

FIGURE E7-2 Chemiosmosis in chloroplasts creates an H^+ gradient and generates ATP by capturing the energy stored in this gradient

❶ Energy is released as water flows downhill.

❷ Energy is harnessed to rotate turbine.

❸ Energy of rotating turbine is used to generate electricity.

FIGURE E7-1 Energy stored in a water "gradient" can be used to generate electricity

the reaction center of photosystem I, and through its nearby electron transport chain; at that point, they finally form NADPH. To sustain this one-way flow of electrons, the reaction center of photosystem II must be continuously supplied with new electrons to replace the ones it gives up. These replacement electrons come from water (step ⑨ in Fig. 7-7; Fig. 7-8, left). In a series of reactions, the reaction center chlorophylls of photosystem II attract electrons from water molecules within the thylakoid compartment, causing the water molecules to split apart:

$$H_2O \rightarrow \tfrac{1}{2}O_2 + 2H^+ + 2e^-$$

For every two photons captured by photosystem II, two electrons are boosted out of the reaction-center chloro-

FIGURE 7-9 Oxygen is a by-product of photosynthesis
The bubbles released by the leaves of this aquatic plant (*Elodea*) are composed of oxygen, a by-product of photosynthesis.

phyll and are replaced by the two electrons obtained by splitting one water molecule. The loss of two electrons from water generates two hydrogen ions (H^+), which are used to form NADPH. As water molecules are split, their oxygen atoms combine to form molecules of oxygen gas (O_2). The oxygen may be used directly by the plant in its own cellular respiration (see Chapter 8) or released to the atmosphere (**FIG. 7-9**).

SUMMING UP

Light-Dependent Reactions

- Chlorophyll and carotenoid pigments of photosystem II absorb light that is used to energize and eject electrons from the reaction-center chlorophyll molecules.
- The electrons are passed along the adjacent electron transport chain, where they release energy. Some of the energy is used to create a hydrogen ion gradient across the thylakoid membrane that is used to drive ATP synthesis.
- The "electron-deprived" reaction-center chlorophylls of photosystem II replace their electrons by splitting water molecules. The resulting H^+ is used in NADPH, and oxygen gas is generated as a by-product.
- Light is also absorbed by photosystem I, which ejects energized electrons from its reaction-center chlorophylls.
- The electron transport chain picks up these energized electrons, and their energy is captured in NADPH.
- Electrons lost from the reaction center of photosystem I are replaced by those from the electron transport chain associated with photosystem II.
- The products of the light-dependent reactions are NADPH, ATP, and O_2.

7.3 LIGHT-INDEPENDENT REACTIONS: HOW IS CHEMICAL ENERGY STORED IN GLUCOSE MOLECULES?

The ATP and NADPH synthesized during the light-dependent reactions are dissolved in the fluid stroma that surrounds the thylakoids. There, these substances provide the energy to power the synthesis of glucose from carbon dioxide and water—a process that requires enzymes, which are also dissolved in the stroma. The reactions that eventually produce glucose are called the light-independent reactions because they can occur independently of light as long as ATP and NADPH are available. However, these high-energy molecules required for glucose synthesis are available only if they have been recharged by light. Thus, any event that reduces light availability (such as the dust, smoke, and ash that would accompany a meteorite collision with Earth) also reduces the availability of these high-energy compounds and consequently decreases the plant's ability to synthesize its food.

The C_3 Cycle Captures Carbon Dioxide

The process of capturing six carbon dioxide molecules from the air and using them to synthesize the six-carbon sugar glucose occurs in a set of reactions known as either the **Calvin-Benson cycle** (after its discoverers) or the **C_3 cycle**. The C_3 cycle requires CO_2 (usually from the air); the sugar, *ribulose bisphosphate* (RuBP); enzymes to catalyze each of its many reactions; and energy in the form of ATP and NADPH, which are provided by the light-dependent reactions.

FIGURE 7-10 The C$_3$ cycle of carbon fixation
① Six molecules of RuBP react with 6 molecules of CO$_2$ to form 12 molecules of PGA. This reaction is carbon fixation—the capture of carbon from CO$_2$ into organic molecules. ② The energy of 12 ATPs and the electrons and hydrogens of 12 NADPHs are used to convert the 12 PGA molecules to 12 G3Ps. ③ Energy from 6 ATPs is used to rearrange 10 G3Ps into 6 RuBPs, completing one turn of the C$_3$ cycle. ④ Two G3P molecules are available to synthesize glucose or other organic molecules. The process in ④ occurs outside the chloroplast and is not part of the C$_3$ cycle.

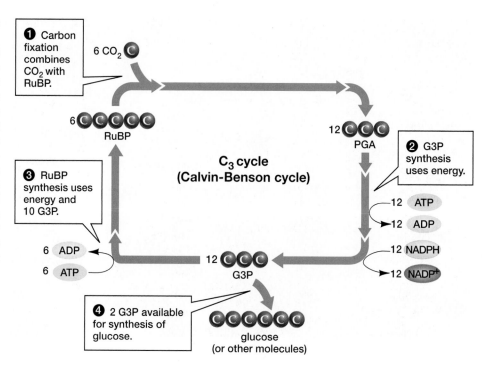

The C$_3$ cycle is best understood if we mentally divide it into three parts: carbon fixation, the synthesis of *glyceraldehyde-3-phosphate* (G3P, which is used to synthesize sugar), and finally the regeneration of RuBP. Keep track of the number of carbon atoms as you follow the process in **FIGURE 7-10**. Also keep in mind that all the energy used in this cycle was captured from sunlight during the light-dependent reactions of photosynthesis.

1. *Carbon fixation.* During **carbon fixation,** plants capture carbon dioxide and incorporate (fix) its carbon atoms into a larger organic molecule. The C$_3$ cycle uses an enzyme called *rubisco* to combine the carbon from carbon dioxide with the five-carbon sugar molecule RuBP, forming an unstable six-carbon molecule that immediately splits in half to form two, three-carbon molecules of PGA (*phosphoglyceric acid*). The three carbons of PGA give the C$_3$ cycle its name (step ① in Fig. 7-10).
2. *Synthesis of G3P.* In a series of enzyme-catalyzed reactions, energy donated by ATP and NADPH (generated during the light-dependent reactions) is used to convert PGA to G3P (step ②).
3. *Regeneration of RuBP.* Through a series of enzyme-catalyzed reactions requiring ATP energy, G3P is used to regenerate RuBP (step ③) used at the start of the cycle. The remaining two molecules of G3P will be used to synthesize glucose and other molecules needed by the plant (step ④).

Carbon Fixed During the C$_3$ Cycle Is Used to Synthesize Glucose

Because the C$_3$ cycle starts with RuBP, adds carbon from CO$_2$, and ends each "cycle" with RuBP, there is carbon left over from the captured CO$_2$. Using the simplest "carbon accounting" numbers shown in Figure 7-10, if you start and end one passage through the cycle with six molecules of RuBP, two molecules of G3P are left over. In light-independent reactions that occur outside the C$_3$ cycle, these two G3P molecules (three carbons each) are combined to form one molecule of glucose (six carbons). Most of these are then used to form sucrose (table sugar; a disaccharide storage molecule consisting of a glucose linked to a fructose), or linked together in long chains to form starch (another storage molecule) or cellulose (a major component of plant cell walls). Most of the synthesis of glucose from G3P and the subsequent synthesis of more-complex molecules from glucose occurs outside the chloroplast. Later, glucose molecules may be broken down during cellular respiration to provide energy for the plant.

SUMMING UP
Light-Independent Reactions

- For the synthesis of one molecule of glucose through the C$_3$ cycle, six molecules of RuBP capture six molecules of CO$_2$. A series of reactions driven by energy from ATP and NADPH (obtained from the light-dependent reactions) produces 12 molecules of G3P.
- Two G3P molecules join to become one molecule of glucose.
- ATP energy is used to regenerate six RuBP molecules from the remaining 10 G3P molecules.
- Light-independent reactions generate glucose and depleted energy carriers (ADP and NADP$^+$) that will be recharged during light-dependent reactions.

FIGURE 7-11 A summary diagram of photosynthesis

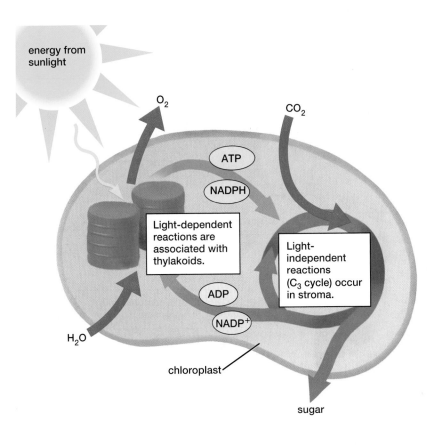

energy from sunlight

O₂

CO₂

ATP

NADPH

Light-dependent reactions are associated with thylakoids.

Light-independent reactions (C₃ cycle) occur in stroma.

ADP

NADP⁺

H₂O

chloroplast

sugar

7.4 WHAT IS THE RELATIONSHIP BETWEEN LIGHT-DEPENDENT AND LIGHT-INDEPENDENT REACTIONS?

FIGURE 7-11 illustrates the relationship between light-dependent and light-independent reactions, placing each in its appropriate location within the chloroplast. Both Figure 7-11 and Fig 7-4 illustrate the interdependence of these two sets of reactions in the overall process of photosynthesis. Simply put, the "photo" part of photosynthesis refers to the capture of light energy by the light-dependent reactions. The "synthesis" part of photosynthesis refers to the synthesis of glucose that occurs during the light-independent reactions, using the energy captured by the light-dependent reactions. Stated in more detail, the light-dependent reactions in the membranes of the thylakoids use light energy to "charge up" the energy-carrier molecules ADP and NADP⁺ to form ATP and NADPH. During light-independent reactions, the energized carriers move to the stroma, where their energy drives the C₃ cycle. This produces G3P, which is used to synthesize glucose and other carbohydrates. The depleted carriers ADP and NADP⁺ are then recharged by the light-dependent reactions into ATP and NADPH.

7.5 WATER, CO₂, AND THE C₄ PATHWAY

Photosynthesis requires light and carbon dioxide. Therefore, an ideal leaf should have a large surface area to intercept lots of sunlight and be very porous to allow lots of CO₂ to enter the leaf from the air. For land plants, howev-er, being porous to air also allows water to evaporate easily from the leaf. Loss of water from leaves is a prime cause of stress to land plants and may even be fatal.

Many plants have evolved leaves with features that represent a compromise between obtaining adequate light energy and CO₂ and reducing water loss. These leaves have a large surface area for intercepting light, a water-proof coating to reduce evaporation, and leaves with adjustable pores (the stomata) through which air carrying CO₂ can diffuse. In most plant leaves, chloroplasts are found in mesophyll cells and in stomata cells (see Fig. 7-3 and **FIG. 7-12**). When water supplies are adequate, the stomata open, letting in CO₂. If the plant is in danger of drying out, the stomata close. Closing the stomata reduces evaporation but has two disadvantages: it reduces CO₂ intake, and it also limits the ability of the leaf to release O₂, a by-product of photosynthesis.

When Stomata Are Closed to Conserve Water, Wasteful Photorespiration Occurs

What happens to carbon fixation when the stomata close, CO₂ levels drop, and O₂ levels rise? Unfortunately, the enzyme rubisco that catalyzes the reaction of RuBP with CO₂ is not very selective; it can cause either CO₂ or O₂ to combine with RuBP (Fig. 7-12a), an example of competitive inhibition. When O₂ (rather than CO₂) is combined with RuBP, a wasteful process called **photorespiration** occurs. During photorespiration (as in cellular respiration), O₂ is used up and CO₂ is generated. But unlike cellular respiration, photorespiration does not produce any useful cellular energy. It also prevents the light-independent reactions from synthesizing glucose. Thus, photorespiration undermines the plant's ability to fix carbon.

Some photorespiration occurs all the time, even under the best of conditions. But during hot, dry weather, the stomata seldom open; CO₂ from the air can't get in, and the O₂ generated by photosynthesis can't get out. In this situation, O₂ outcompetes CO₂ for the active site, and photorespiration dominates (see Fig. 7-12a). Plants, especially fragile seedlings, may die during hot, dry weather because they are unable to capture enough energy to meet their metabolic needs.

You could argue that rubisco is one of the most important enzymes on Earth, because it catalyzes the reaction by which carbon enters the biosphere, and all life is based on carbon. Why do plants have such a nonselective and

(a)   C₃ plants use the C₃ pathway

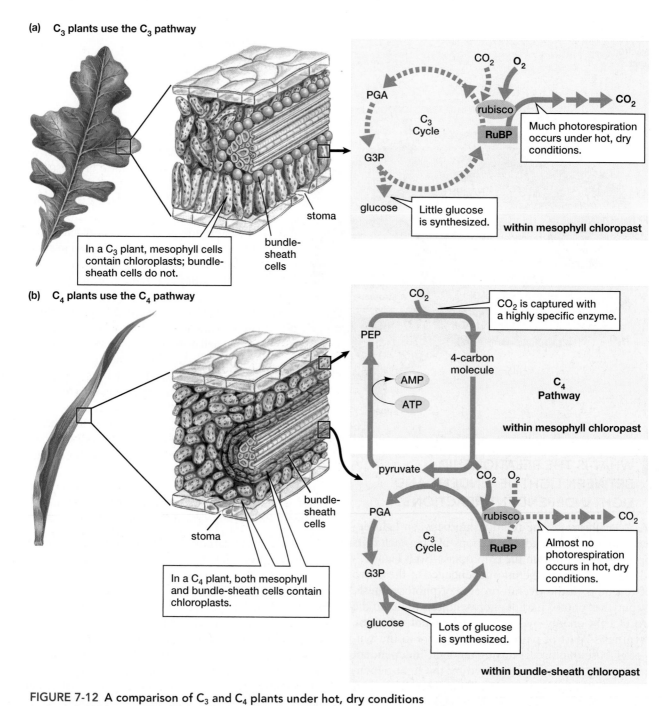

(b)   C₄ plants use the C₄ pathway

FIGURE 7-12 A comparison of C₃ and C₄ plants under hot, dry conditions
(a) With low CO_2 and high O_2 levels, photorespiration dominates in C₃ plants, because the enzyme rubisco causes RuBP to combine with O_2 instead of CO_2. **(b)** In C₄ plants, CO_2 is combined with PEP by a more selective enzyme found in mesophyll cells, and the carbon is shuttled into bundle-sheath cells by a four-carbon molecule, which releases CO_2 there. Higher CO_2 levels allow the C₃ pathway to work efficiently in the bundle-sheath cells. Notice that it takes energy from ATP to regenerate the PEP. QUESTION Why do C₃ plants have an advantage over C₄ plants under conditions that are not hot and dry?

inefficient enzyme? In Earth's early atmosphere, when photosynthesis first evolved, there was far less oxygen and far more carbon dioxide. Because oxygen was rare, there was very little selection pressure for the active site of the enzyme to favor carbon dioxide over oxygen. Although in today's atmosphere this would be a highly adaptive mutation, apparently it has never occurred. Instead, over evolutionary time, plants have developed mechanisms to circumvent wasteful photorespiration, although these involve several additional steps and utilize energy.

LINKS TO LIFE   You Owe Your Life to Plants

As you study the details of photosynthesis, it is easy to become bogged down in the complexity of it all, losing sight of why photosynthesis is worth studying. The bottom line is that, without photosynthesis, you wouldn't be here to be perplexed by it—and neither would any other living thing that you encounter in a typical day. Over 2 billion years ago, when, in the words of poet Robinson Jeffers, the first bacteria "invented chlorophyll and ate sunlight," they sparked a revolution in the evolution of life on Earth. Capturing solar energy and using water as a source of electrons liberated oxygen into the primordial atmosphere for the first time. For many non-photosynthetic organisms, this was a disaster. Oxygen is a highly reactive molecule that readily combines with and destroys biological molecules. The single-celled organisms that first encountered an oxygen-containing atmosphere had three "choices": die, hide, or evolve protective mechanisms. Descendants of those bacteria that hid from oxygen in the primordial ooze still survive today, and oxygen is still deadly to

them. The others, through chance mutations, evolved cellular machinery to harness the reactive power of oxygen, using it to derive more energy from food molecules such as the glucose produced during photosynthesis. These efficient, oxygen-loving organisms quickly dominated Earth and gradually evolved into the myriad creatures that inhabit it today, most of which would quickly die if deprived of oxygen.

Not only do we rely on the oxygen produced by photosynthesis, but all the energy in the food we eat originates in plants, which captured it from sunlight. Even if you revel in a diet of double bacon cheeseburgers and fried chicken, the energy stored in these animal fats and proteins ultimately came from the animals' food—plants. Even if you eat a carnivorous tuna fish, you can still trace the food chain (and the energy) that supported the tuna back to photosynthetic organisms in the ocean. So photosynthesis gives us both our food and the oxygen that we need to "burn" it. Have you thanked a plant today?

C₄ Plants Reduce Photorespiration by Means of a Two-Stage Carbon-Fixation Process

One adaptation to circumvent photorespiration is the **C₄ cycle**, a two-stage carbon-fixation pathway. Plants that use this pathway, called *C₄ plants*, thrive in relatively hot, dry weather. In these C₄ plants, which include corn and crabgrass, the bundle-sheath cells (in addition to mesophyll and stomatal cells) also contain chloroplasts (Fig. 7-12b).

The chloroplasts within the mesophyll cells of C₄ plants contain a three-carbon molecule called *phosphoenolpyruvate* (PEP) instead of RuBP. The CO₂ reacts with PEP to form four-carbon intermediate molecules for which C₄ plants are named. The reaction between CO₂ and PEP is catalyzed by an enzyme that, unlike rubisco, is highly specific for CO₂ and is not hindered by high O₂ concentrations. A 4-carbon molecule is used to shuttle carbon from mesophyll cells into bundle-sheath cells, where it breaks down, releasing CO₂. The high CO₂ concentration created in the bundle-sheath cells (up to 10 times higher than atmospheric CO₂) now allows the regular C₃ cycle to proceed with less competition from oxygen. The remnant of the shuttle molecule (a 3-carbon molecule called pyruvate) returns to the mesophyll cells. Back in the mesophyll cells, ATP energy is used to regenerate the PEP molecule from pyruvate, allowing the cycle to continue.

C₃ and C₄ Plants Are Each Adapted to Different Environmental Conditions

Plants using the C₄ process to fix carbon are locked into this pathway, which uses up more energy to produce glucose than does the C₃ pathway. When light energy is abundant but water is not, C₄ plants have an advantage. However, if water is plentiful, allowing the stomata of C₃ plants to stay open and let in lots of CO₂, or if light levels are low, the more efficient C₃ carbon fixation pathway is advantageous to the plant.

Consequently, C₄ plants thrive in deserts and in hotter, drier regions of temperate climates, where light energy is plentiful but water is scarce. Plants using C₄ photosynthesis include corn, sugarcane, sorghum, some grasses (including crabgrass), and some types of thistles. The C₃ plants—which include most trees; grains such as wheat, oats, and rice; and grasses such as Kentucky bluegrass—have the advantage in cool, wet, cloudy climates, because the C₃ pathway is more energy efficient. These differing adaptations explain why your spring lawn of lush Kentucky bluegrass (a C₃ plant) may be taken over by spiky crabgrass (a C₄ plant) during a long, hot, dry summer.

CASE STUDY REVISITED DID THE DINOSAURS DIE FROM LACK OF SUNLIGHT?

Paleontologists (scientists who study fossils) have cataloged the extinction of approximately 70 percent of all living species by the disappearance of their fossils at the end of the Cretaceous period. In sites from around the globe, researchers have found a thin layer of clay deposited around 65 million years ago; the clay has about 30 times the typical levels of a rare element called *iridium*, which is found in high concentrations in some meteorites. The clay also contains soot, such as would have been deposited in the aftermath of massive fires. Did a meteorite wipe out the dinosaurs? Many scientists believe it did. Certainly the evidence of an enormous me-

teorite impact, dated to 65 million years ago, is clear on the Yucatán Peninsula. But other scientists believe that more gradual climate changes, possibly triggered by intense volcanic activity, produced conditions that would no longer support the enormous reptiles. Volcanoes also spew out soot and ash, and iridium is found in higher levels in Earth's molten mantle than on its surface, so furious volcanic activity could also explain the iridium layer.

Either scenario would significantly reduce the amount of sunlight and immediately impact the rate of photosynthesis. Large herbivores (plant eaters), such as *Triceratops*, which might have needed to consume hundreds of pounds of vegeta-

tion daily, would suffer if plant growth slowed significantly. Predators such as *Tyrannosaurus*, which fed on herbivores, would also suffer. In the Cretaceous as now, sunlight captured by photosynthesis powers all the dominant forms of life on Earth—interrupting this vital flow of energy would be catastrophic.

Consider This Design an experiment to test the effects of light-blocking soot (such as may have filled Earth's atmosphere after a huge meteor strike) on photosynthesis. What might you measure to determine the relative amounts of photosynthesis that occurred under normal versus sooty conditions?

CHAPTER REVIEW

SUMMARY OF KEY CONCEPTS

7.1 What Is Photosynthesis?

Photosynthesis captures the energy of sunlight and uses it to convert the inorganic molecules of carbon dioxide and water into high-energy organic molecules such as glucose. In plants, photosynthesis takes place in the chloroplasts, using two major reaction sequences: the light-dependent reactions and the light-independent reactions.

Web Tutorial 7.1 Photosynthesis

7.2 Light-Dependent Reactions: How Is Light Energy Converted to Chemical Energy?

The light-dependent reactions occur in the thylakoids. Light excites electrons in chlorophyll molecules and transfers the energetic electrons to electron transport chains. The energy of these electrons drives three processes:

- *Photosystem II generates ATP.* Some of the energy from the electrons is used to pump hydrogen ions into the thylakoids. The hydrogen ion concentration is therefore higher inside the thylakoids than in the stroma outside. Hydrogen ions move down this concentration gradient through ATP-synthesizing enzymes in the thylakoid membranes, providing the energy to drive ATP synthesis.
- *Photosystem I generates NADPH.* Some of the energy, in the form of energetic electrons, is added to electron-carrier molecules of $NADP^+$ to make the highly energetic carrier NADPH.
- *Splitting water maintains the flow of electrons through the photosystems.* Some of the energy is used to split water, generating electrons, hydrogen ions, and oxygen.

Web Tutorial 7.2 Properties of Light

Web Tutorial 7.3 Chemiosmosis

7.3 Light-Independent Reactions: How Is Chemical Energy Stored in Glucose Molecules?

In the stroma of the chloroplasts, both ATP and NADPH provide the energy that drives the the synthesis of GP3, which is

used to make glucose from CO_2 and H_2O. The light-independent reactions begin with a cycle of chemical reactions called the Calvin-Benson, or C_3, cycle. The C_3 cycle has three major parts: (1) *Carbon fixation*—carbon dioxide and water combine with ribulose bisphosphate (RuBP) to form phosphoglyceric acid (PGA). (2) *Synthesis of G3P*—PGA is converted to glyceraldehyde-3-phosphate (G3P), using energy from ATP and NADPH. G3P may be used to synthesize glucose and other important molecules, such as starch and cellulose. (3) *Regeneration of RuBP*—Ten molecules of G3P are used to regenerate six molecules of RuBP, again using ATP energy. Light-independent reactions continue with the synthesis of glucose and other carbohydrates including sucrose, starch, and cellulose. These reactions occur primarily outside of the chloroplast.

7.4 What Is the Relationship Between Light-Dependent and Light-Independent Reactions?

The light-dependent reactions produce the energy carrier ATP and the electron carrier NADPH. Energy from these carriers is used in the synthesis of organic molecules during the light-independent reactions. The depleted carriers, ADP and $NADP^+$, return to the light-dependent reactions for recharging.

7.5 Water, CO_2, and the C_4 Pathway

The enzyme rubisco that catalyzes the reaction between RuBP and CO_2 also catalyzes a reaction, called photorespiration, between RuBP and O_2. If CO_2 concentrations drop too low or if O_2 concentrations rise too high, wasteful photorespiration, which prevents carbon fixation and does not generate ATP, may exceed carbon fixation. C_4 plants have evolved an additional step for carbon fixation that minimizes photorespiration. In the mesophyll cells of these C_4 plants, CO_2 combines with phosphoenolpyruvic acid (PEP) to form a four-carbon molecule, which is modified and transported into adjacent bundle-sheath cells, where it releases CO_2, thereby maintaining a high CO_2 concentration in those cells. This CO_2 is then fixed using the C_3 cycle.

KEY TERMS

C_3 cycle *page 125*
C_4 cycle *page 129*
Calvin-Benson cycle *page 125*
carbon fixation *page 126*
carotenoids *page 121*
chemiosmosis *page 123*

chlorophyll *page 121*
electron transport chain
  (ETC) *page 122*
light-dependent reactions
  page 120

light-independent reactions
  page 120
photon *page 120*
photorespiration *page 127*
photosynthesis *page 118*

photosystems *page 121*
reaction center *page 121*
stomata *page 118*
stroma *page 118*
thylakoid *page 119*

THINKING THROUGH THE CONCEPTS

1. Write the overall equation for photosynthesis. Does the overall equation differ between C_3 and C_4 plants?

2. Draw a simplified diagram of a chloroplast, and label it. Explain specifically how chloroplast structure is related to its function.

3. Briefly describe the light-dependent and light-independent reactions. In what part of the chloroplast does each occur?

4. What is the difference between carbon fixation in C_3 and in C_4 plants? Under what conditions does each mechanism of carbon fixation work most effectively?

5. Describe the process of chemiosmosis in chloroplasts, tracing the flow of energy from sunlight to ATP.

APPLYING THE CONCEPTS

1. Many lawns and golf courses are planted with bluegrass, a C_3 plant. In the spring, the bluegrass grows luxuriously. In the summer, crabgrass, a weed and a C_4 plant, often appears and spreads rapidly. Explain this sequence of events, given the normal weather conditions of spring and summer and the characteristics of C_3 versus C_4 plants.

2. Suppose an experiment is performed in which plant I is supplied with normal carbon dioxide but with water that contains radioactive oxygen atoms. Plant II is supplied with normal water but with carbon dioxide that contains radioactive oxygen atoms. Each plant is allowed to perform photosynthesis, and the oxygen gas and sugars produced are tested for radioactivity. Which plant would you expect to produce radioactive sugars, and which plant would you expect to produce radioactive oxygen gas? Why?

3. You continuously monitor the photosynthetic oxygen production from the leaf of a plant illuminated by white light. Explain what will happen (and why) if you place filters in front of the light source that transmit (a) red, (b) blue, and (c) green light onto the leaf.

4. A plant is placed in a CO_2-free atmosphere in bright light. Will the light-dependent reactions continue to generate ATP and NADPH indefinitely? Explain how you reached your conclusion.

5. You are called before the Ways and Means Committee of the House of Representatives to explain why the U.S. Department of Agriculture should continue to fund photosynthesis research. How would you justify the expense of producing, by genetic engineering, the enzyme that catalyzes the reaction of RuBP with CO_2 and prevents RuBP from reacting with oxygen as well as CO_2? What are the potential applied benefits of this research?

FOR MORE INFORMATION

Bazzazz, F. A., and Fajer, E. D. "Plant Life in a CO_2-Rich World." *Scientific American*, January 1992. Burning fossil fuels is increasing CO_2 levels in the atmosphere. This increase could tip the balance between C_3 and C_4 plants.

George, A. "Photosynthesis." *American Scientist*, April 9, 2005. One of life's greatest inventions.

Grodzinski, B. "Plant Nutrition and Growth Regulation by CO_2 Enrichment." *BioScience*, 1992. The author discusses how higher CO_2 levels influence plant metabolism.

Kring, D. A., and Durda, D. D. "The Day the World Burned." *Scientific American,* December, 2003. Describes the wildfires that would have followed the meteorite that may have killed the dinosaurs.

Monastersky, R. "Children of the C_4 World." *Science News*, January 3, 1998. What role did a shift in global vegetation toward C_4 photosynthesis play in the evolution of humans?

Mooney, H. A., Drake, B. G., Luxmoore, R. J., Oechel, W. C., and Pitelka, L. F. "Predicting Ecosystems' Responses to Elevated CO_2 Concentrations." *BioScience*, 1994. What effects will CO_2 enrichment of the atmosphere due to human activities have on ecosystems?

Robbins, M. W. "The Promise of Pond Scum." *Discover,* October, 2005. Can we tap the energy stored by photosynthetic algae as a substitute for fossil fuels?

8 Harvesting Energy: Glycolysis and Cellular Respiration

The leg muscles of these racing cyclists require both glucose and oxygen to obtain the energy they need. (inset) Johann Mühlegg is among the elite athletes penalized for artificially boosting the oxygen supply to their cells to increase athletic performance.

CASE STUDY

WHEN ATHLETES BOOST THEIR BLOOD COUNTS: DO CHEATERS PROSPER?

THOUSANDS OF SPECTATORS cheered wildly as the leaders of the 50-km cross-country ski race entered the home stretch at the 2002 Winter Olympics. As the grueling race drew to its conclusion, the skiers were clearly exhausted, struggling to find the energy for a final burst. One skier, however, came on strong. Johann Mühlegg, competing for Spain, raced to the front of the pack and pulled away, finishing almost 15 seconds ahead of the second-place skier and claiming the gold medal. Mühlegg's triumph was short lived. Shortly after the race, he was stripped of his medals and expelled from the Games. His offense? Blood doping.

Blood doping increases a person's physical endurance by increasing the blood's ability to carry oxygen. Mühlegg accomplished this by injecting the drug darbepoetin. This drug mimics the effect of the natural hormone erythropoietin (Epo), which is also used as a blood-doping substance. Erythropoietin is present in normal human bodies, where it stimulates bone marrow to produce more red blood cells. A healthy body produces just enough Epo to ensure that red blood cells are replaced as they age and die. An injection of extra Epo, however, can stimulate the production of a huge number of extra red blood cells. The extra cells greatly increase the oxygen-carrying capacity of the blood.

Do Epo injections really improve endurance? In one study, researchers divided 20 human subjects into two groups, one of which received injections of Epo. After 4 weeks, the subjects were tested for endurance and for oxygen consumption during exercise. Individuals in the Epo group had much better endurance and consumed significantly more oxygen than the control group did. The researchers concluded that Epo injections improve endurance and increase the body's capacity to carry oxygen.

Why is endurance improved by extra oxygen molecules in the bloodstream? Think about this question as we examine the role of oxygen in supplying energy to muscle cells.

8.1 HOW DO CELLS OBTAIN ENERGY?

Cells require a continuous supply of energy to power the multitude of metabolic reactions required to sustain life. To drive a reaction, however, energy must be in a usable form; generally, this means it must be stored in the bonds of energy-carrier molecules, especially **adenosine triphosphate (ATP)**. Some of the most important reactions in cells are the ones that transfer energy from energy-storage molecules, such as glucose, to energy-carrier molecules, such as ATP.

Photosynthesis Is the Ultimate Source of Cellular Energy

As you learned in Chapter 7, photosynthetic organisms capture and store the energy of sunlight in glucose. While photosynthesis produces some ATP, plants store much of the energy from photosynthesis as sugar. Like all eukaryotic cells, plants have mitochondria, and they rely on glucose breakdown to supply the energy they need to sustain life. During glucose breakdown, all cells release the solar energy that was originally captured by plants through photosynthesis and use it to make ATP. The chemical equations for glucose formation by photosynthesis and for the complete metabolism of glucose back to CO_2 and H_2O (the original reactants in photosynthesis) are almost perfectly symmetrical:

Photosynthesis:

$6\,CO_2 + 6\,H_2O + \text{light energy} \rightarrow C_6H_{12}O_6 + 6\,H_2O + \text{heat energy}$

Complete Glucose Metabolism:

$C_6H_{12}O_6 + 6\,H_2O \rightarrow 6\,CO_2 + 6\,H_2O +$
$\text{chemical energy (ATP)} + \text{heat energy}$

As you may recall from our discussion of the second law of thermodynamics (Chapter 6), with each reaction that occurs, there is a decrease in useful energy, and heat is generated. Although over half of the energy produced by glucose breakdown is liberated as heat, cells are extremely efficient at capturing chemical energy, trapping about 40 percent of the energy in glucose as ATP. If cells were as inefficient as our gasoline engines (25 percent or less), animals would need to eat voraciously to remain active. Athletes participating in long-distance events might need to stop for meals!

Glucose Is a Key Energy-Storage Molecule

Most cells can metabolize a variety of organic molecules to produce ATP. In this chapter, we focus on the breakdown of glucose for three reasons. First, virtually all cells metabolize glucose for energy at least part of the time. Some, such as the nerve cells of the brain, rely almost entirely on glucose as a source of energy. Second, glucose metabolism is less complex than the metabolism of most other organic molecules. Finally, when using other organic molecules as energy sources, cells usually first convert the molecules to glucose or other compounds that enter the pathways of glucose metabolism (see "Health Watch: Why Can You Get Fat by Eating Sugar?" later in the chapter).

An Overview of Glucose Breakdown

FIGURE 8-1 summarizes the major steps of glucose metabolism in eukaryotic cells. The initial reactions are collectively called **glycolysis** (Greek, "to break apart a sweet"). Glycolysis, which occurs in the cytosol and does not require oxygen, breaks glucose into pyruvate, capturing energy in two molecules of ATP. If oxygen is absent (anaerobic conditions), glycolysis is followed by fermentation, which does not produce any additional chemical energy. During fermentation, pyruvate is converted into either lactate or into ethanol and CO_2.

If oxygen is present (aerobic conditions), most forms of life use a process called *cellular respiration* to break the pyruvate down further into carbon dioxide and water. In eukaryotic cells (fungi, protists, plants, and animals), cellular respiration occurs in mitochondria. As in photosynthesis, cellular respiration produces both ATP and high-energy electrons that travel through an *electron transport chain (ETC)*. In cellular respiration, oxygen acts as the final electron acceptor, combining with the electrons and hydrogen ions to form water. Cellular respiration captures far more energy than does glycolysis, producing an additional 34 or 36 ATP molecules, depending on the cell type.

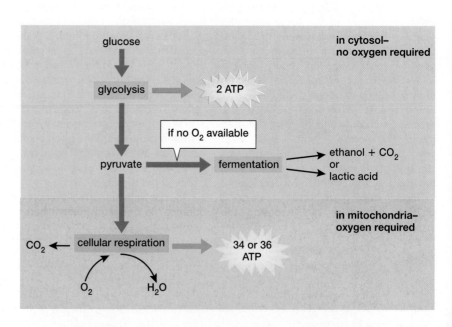

FIGURE 8-1 A summary of glucose metabolism

8.2 HOW IS THE ENERGY IN GLUCOSE CAPTURED DURING GLYCOLYSIS?

Glycolysis Breaks Down Glucose to Pyruvate, Releasing Chemical Energy

Reduced to its essentials, glycolysis consists of two major parts (each with several steps): ① glucose activation and ② energy harvest (**FIG. 8-2**). Before glucose is broken down, it must be activated—an energy-demanding process. During glucose activation, one molecule of glucose undergoes two enzyme-catalyzed reactions, each of which uses ATP energy. These reactions convert a relatively stable glucose molecule into a highly unstable "activated" molecule of *fructose bisphosphate* (Fig. 8-2, left). Fructose is a sugar molecule similar to glucose; *bisphosphate* refers to the two phosphate groups acquired from the ATP molecules. Although forming fructose bisphosphate costs the cell two ATP molecules, this initial investment of energy is necessary to produce greater energy returns later. Because much of the energy from ATP is stored in the bonds linking the phosphate groups to the sugar, fructose bisphosphate is an unstable molecule.

In the energy-harvesting steps, fructose bisphosphate splits apart into two 3-carbon molecules of glyceraldehyde 3-phosphate (G3P—see Fig. 8-2, right; recall that G3P was also formed during the C_3 cycle of photosynthesis). Each G3P molecule, which retains one phosphate with its high-energy bond, then goes through a series of reactions that convert it to pyruvate. During these reactions, two ATPs are generated for each G3P, for a total of four ATPs. But because two ATPs were used up to activate the glucose molecule in the first place, there is a net gain of only two ATPs per glucose molecule. At another step along the way from G3P to pyruvate, two high-energy electrons and a hydrogen ion are added to the "empty" electron-carrier NAD^+ to make the high-energy electron-carrier molecule NADH. Because two G3P molecules are produced per glucose molecule, two NADH carrier molecules are formed when those G3P molecules are converted to pyruvate. For the complete reactions of glycolysis, see "A Closer Look: Glycolysis."

SUMMING UP
Glycolysis

- Each molecule of glucose is broken down to two molecules of pyruvate.
- During these reactions, a net of two ATP molecules and two NADH (high-energy electron carriers) are formed.

In the Absence of Oxygen, Fermentation Follows Glycolysis

Because every living creature on the planet uses it, glycolysis is believed to be one of the most ancient of all biochemical pathways. Scientists hypothesize that the earliest forms of life appeared under anaerobic conditions (before the evolution of oxygen-liberating photosynthesis) and probably relied on glycolysis for energy production. Many microorganisms still thrive in places where oxygen is rare or absent, such as in the stomach and intestines of animals (including people), deep in soil, or in bogs and marshes. Some microorganisms are poisoned by oxygen and rely entirely on the inefficient process of glycolysis to meet their energy needs. Even some of our own body cells—and those of other animals—must cope without oxygen for brief periods of time. Under anaerobic conditions, pyruvate is converted into lactate or ethanol through a process called **fermentation**.

Fermentation does not produce more ATP, but it is necessary to regenerate the high-energy electron carrier molecule NAD^+, which is reused during glycolysis and must be available for glycolysis to continue. Electron carrier molecules such as NAD^+ capture energy by accepting high-energy electrons. One major difference between anaerobic and aerobic glucose breakdown is in the way these high-energy

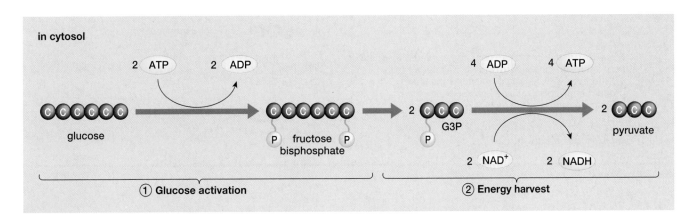

FIGURE 8-2 The essentials of glycolysis
① Glucose activation: The energy of two ATP molecules is used to convert glucose to the highly reactive fructose bisphosphate, which splits into two reactive molecules of G3P. ② Energy harvest: The two G3P molecules undergo a series of reactions that generate four ATP and two NADH molecules. Thus, glycolysis results in a net production of two ATP and two NADH molecules per glucose molecule.

Glycolysis

Glycolysis is a series of enzyme-catalyzed reactions that break down a single molecule of glucose into two molecules of pyruvate. To help you follow the reactions in **FIGURE E8-1**, we show only the "carbon skeletons" of glucose and the molecules produced during glycolysis. Each blue arrow represents a reaction catalyzed by at least one enzyme.

❶ A glucose molecule is energized by the addition of a high-energy phosphate from ATP.

❷ The molecule is slightly rearranged, forming fructose-6-phoshate.

❸ A second phosphate is added from another ATP.

❹ The resulting molecule, fructose-1,6-bisphosphate, is split into two three-carbon molecules, one DHAP (dihydroxacetone phosphate) and one G3P. Each has one phosphate attached.

❺ DHAP rearranges into G3P. From now on, there are two molecules of G3P going through the identical reactions.

❻ Each G3P undergoes two almost-simultaneous reactions. Two electrons and a hydrogen ion are donated to NAD⁺ to make the energized carrier NADH, and an inorganic phosphate (P) is attached to the carbon skeleton with a high-energy bond. The resulting molecules of 1,3-bisphosphoglycerate have two high-energy phosphates.

❼ One phosphate from each bisphosphoglycerate is transferred to ADP to form ATP, for a net of two ATPs. This transfer compensates for the initial two ATPs used in glucose activation.

❽ After another rearrangement, the second phosphate from each phosphoenolpyruvate is transferred to ADP to form ATP, leaving pyruvate as the final product of glycolysis. There is a net profit of two ATPs from each glucose molecule.

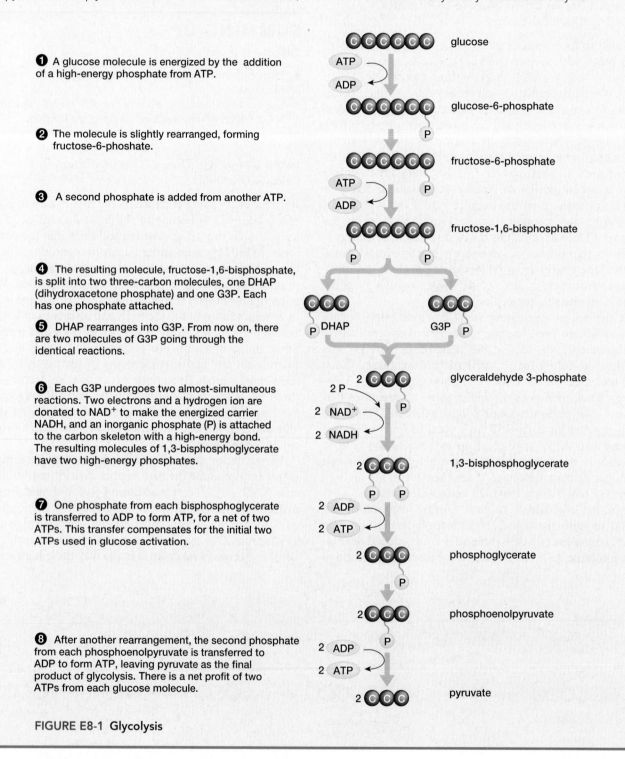

FIGURE E8-1 Glycolysis

electrons are used. During cellular respiration in the presence of oxygen (described later), electron carriers ferry these electrons to the electron transport chain, which requires oxygen to accept the electrons as they leave the chain. This process results in the production of a large quantity of ATP. In the absence of oxygen, however, pyruvate accepts electrons from NADH, producing ethanol or lactate by fermentation.

Under anaerobic conditions, NADH is not used to produce ATP. Rather, converting NAD⁺ to NADH is actually a way of getting rid of the hydrogen ions and electrons produced as glucose is broken down into pyruvate. But as it accepts electrons and hydrogen ions to become NADH, NAD$^+$ is used up. Without a way to regenerate NAD$^+$, when the supply of NAD$^+$ was exhausted, glycolysis would stop, energy production would cease, and the organism would almost instantly die. Fermentation solves this problem by enabling pyruvate to act as the final acceptor of electrons and hydrogen ions from NADH, regenerating NAD$^+$ for use in further glycolysis. Some microorganisms lack the enzymes for cellular respiration; these will ferment glucose even when oxygen is present, and some of them are actually poisoned by oxygen.

There are two main fermentation pathways; one converts pyruvate to lactate, and the other converts pyruvate to ethanol and carbon dioxide.

Some Cells Ferment Pyruvate to Form Lactate

Fermentation of pyruvate to lactate is called *lactic acid fermentation*; in the cytosol, the lactic acid is ionized to form lactate. Lactic acid fermentation occurs in muscles when people and other animals exercise vigorously, such as oc-

curs when a deer runs from a wolf, a runner sprints through the finish line (**FIG. 8-3a**), or when you race to class after you've overslept.

Even though working muscles need lots of ATP and cellular respiration generates much more ATP than does glycolysis, cellular respiration is limited by the organism's ability to provide oxygen (by breathing, for example). When animals exercise vigorously, they may not be able to get enough air into their lungs and enough oxygen into their blood to supply their muscles with sufficient oxygen to allow cellular respiration to meet all their energy needs. This is why athletes, desperate for a competitive edge, may turn to illegal blood doping to increase the ability of their blood to carry oxygen.

When deprived of adequate oxygen, muscles do not immediately stop working. After all, most animals exercise most vigorously when fighting, fleeing, or pursuing prey—and in these activities, the ability to continue just a little bit longer can make the difference between life and death. So glycolysis continues for a short time, providing its meager two ATP molecules per glucose and generating both pyruvate and NADH. Then, to regenerate NAD$^+$, muscle cells

(a)

(b)

FIGURE 8-3 Fermentation
(a) During a sprint, a runner's respiratory and circulatory systems cannot supply oxygen to her leg muscles fast enough to keep up with the demand for energy, so glycolysis must provide some of the ATP. In muscles, lactic acid fermentation follows glycolysis when oxygen is unavailable. **(b)** Bread rises as CO_2 is liberated by fermenting yeast, which converts glucose to ethanol. The dough on the left rose to the level on the right in a few hours. QUESTION Some species of bacteria use aerobic respiration and other species use anaerobic (fermenting) respiration. In an oxygen-rich environment, would either type be at a competitive advantage? What about in an oxygen-poor environment?

ferment pyruvate molecules to lactate, using electrons and hydrogen ions from NADH (**FIG. 8-4**).

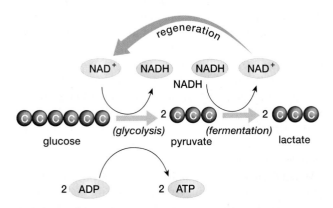

FIGURE 8-4 Glycolysis followed by lactate fermentation

If you're breathing hard after racing to arrive in class on time, your lungs are working to obtain enough oxygen so that your muscles can revert to cellular respiration. As oxygen is replenished, the lactate produced by sprinting is transported in the bloodstream to the liver, where it is again converted to pyruvate. Some of this pyruvate is then broken down by cellular respiration into carbon dioxide and water, capturing additional energy.

Various microorganisms also use lactic acid fermentation, including the bacteria that convert milk into yogurt, sour cream, and cheese. As you may know, acids taste sour, so the lactic acid contributes to the distinctive tastes of these foods. (The acid also denatures milk protein, altering its three-dimensional structure and causing the milk to thicken.)

Other Cells Ferment Pyruvate to Alcohol

Many microorganisms use another type of fermentation to regenerate NAD^+ under anaerobic conditions: *alcoholic fermentation*. These organisms produce ethanol and CO_2 (rather than lactate) from pyruvate, using hydrogen ions and electrons from NADH (**FIG. 8-5**).

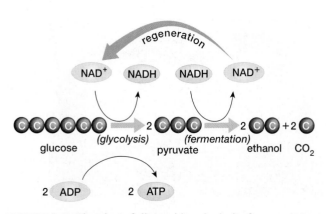

FIGURE 8-5 Glycolysis followed by alcoholic fermentation

Sparkling wines, such as champagne, are bottled while the yeasts are still alive and fermenting, trapping both the alcohol and the CO_2. Yeast in bread dough produces CO_2, making the bread rise; the alcohol evaporates while the bread is baking (see Fig. 8-3b). For more on alcoholic fermentation, see "Links to Life: A Jug of Wine, a Loaf of Bread, and a Nice Bowl of Sauerkraut."

8.3 HOW DOES CELLULAR RESPIRATION CAPTURE ADDITIONAL ENERGY FROM GLUCOSE?

Cellular respiration is a series of reactions that occur under aerobic conditions and produce a large quantity of ATP while breaking down the pyruvate generated by glycolysis into carbon dioxide and water. The reactions of cellular respiration require oxygen because oxygen acts as the final acceptor of electrons from the electron transport chain.

Cellular Respiration in Eukaryotic Cells Occurs in Mitochondria

In eukaryotic cells, cellular respiration occurs within *mitochondria*, organelles that are sometimes called the "powerhouses of the cell." A mitochondrion has two membranes that produce two compartments. The inner membrane encloses a central compartment containing the fluid **matrix**, and the outer membrane surrounds the organelle, producing an **intermembrane space** between the membranes (**FIG. 8-6**).

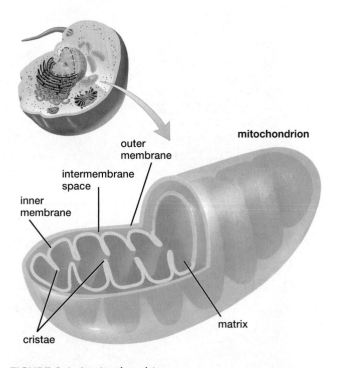

FIGURE 8-6 A mitochondrion
The outer and inner mitochondrial membranes enclose two spaces within the mitochondrion.

LINKS TO LIFE   A Jug of Wine, a Loaf of Bread, and a Nice Bowl of Sauerkraut

Life would be a little less interesting without fermentation. Persian poet Omar Khayyam (1048–1122) described his vision of paradise on Earth as "A Jug of Wine, a Loaf of Bread—and Thou Beside Me." In fact, people have long exploited yeast's ability to ferment the sugars in fruit to form alcohol; historical evidence suggests that wine and beer were commercially produced at least 5000 years ago. Yeasts (single-celled fungi) will engage in cellular respiration if oxygen is available, but switch to alcoholic fermentation if they run out of oxygen. But as you now know, carbon dioxide is also a by-product of alcoholic fermentation; thus, wine must be fermented in containers that allow the carbon dioxide to leave (so the containers don't explode) but prevent the entry of air (so cellular respiration doesn't occur). Sparking (fizzy) wines and champagne are made by adding more yeast and sugar just before the wine is bottled, so that final fermentation occurs in the sealed bottle, trapping carbon dioxide.

FIGURE E8-2 Without fermentation, there would be no cheese, bread, or wine

Fermentation also gives bread its airy texture. All breads contain yeast, flour, and water. Dry yeast is awakened from its dormant state by water, and it multiplies rapidly while metabolizing the sugars present in flour. The carbon dioxide released during fermentation is trapped within the bread dough, where it forms tiny pockets of gas. Kneading distributes the multiplying yeast cells evenly throughout the bread and makes the dough stretchy and resilient so it traps the gas, resulting in an evenly porous texture.

While wine and bread are produced by alcoholic fermentation, bacteria that produce lactic acid by fermentation are responsible for other culinary staples. For thousands of years, people have relied on microorganisms that produce lactic acid to convert milk into sour cream, yogurt, and a wide variety of cheeses (**FIG. E8-2**). In addition, lactate fermentation by salt-loving bacteria converts the sugars in cucumbers and cabbage to lactic acid. The result: dill pickles and sauerkraut, excellent partners for other fermented foods.

Now let's look a little more closely at the processes of cellular respiration in the mitochondria.

Pyruvate Is Broken Down in the Mitochondrial Matrix, Releasing More Energy

Recall that pyruvate is the end product of glycolysis and that it is synthesized in the cytosol. The pyruvate diffuses through the mitochondrial membranes until it reaches the mitochondrial matrix, where it is used in cellular respiration.

The reactions that occur in the mitochondrial matrix occur in two stages: the formation of acetyl CoA from pyruvate (part ① in **FIG. 8-7**) and the Krebs cycle (part ② in Fig. 8-7). First, pyruvate reacts with a molecule called

coenzyme A (CoA). Each pyruvate is split into CO_2 and a two-carbon molecule called an *acetyl group*, which immediately attaches to coenzyme A and forms *acetyl CoA*. During this reaction, two high-energy electrons and a hydrogen ion are transferred to NAD^+, forming NADH.

The next stages of the reaction form a cyclic pathway known as the **Krebs cycle**, named after its discoverer Hans Krebs, a biochemist who won the Nobel Prize for this work in 1953. The Krebs cycle is also called the *citric acid cycle* because citrate (the ionized form of citric acid) is the first molecule produced in the cycle. During the Krebs cycle, each acetyl CoA combines with the four-carbon molecule *oxaloacetate* to form the six-carbon *citrate*. Coenzyme A is released; like an enzyme, it is not permanently altered during

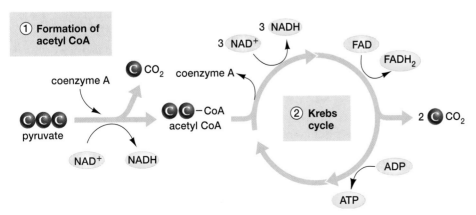

FIGURE 8-7 The reactions in the mitochondrial matrix ① Pyruvate reacts with CoA, forming CO_2 and acetyl CoA. During this reaction, an energetic electron is added to NAD^+ to form NADH. ② When acetyl CoA enters the Krebs cycle, coenzyme A is released. The Krebs cycle produces one ATP, three NADH, one $FADH_2$, and two CO_2 for each acetyl CoA. Because each glucose molecule yields two pyruvates, the total energy harvest per glucose molecule in the matrix is two ATP, eight NADH, and two $FADH_2$.

these reactions and is reused many times. Mitochondrial enzymes then promote several rearrangements that regenerate the oxaloacetate and release two CO_2 molecules. During this sequence of reactions, chemical energy from each acetyl group is captured in one ATP and four high-energy electron carrier molecules: three NADH and one $FADH_2$ (flavin adenine dinucleotide, a related molecule).

To review the complete set of reactions that occur in the mitochondrial matrix, see "A Closer Look: The Mitochondrial Matrix Reactions."

SUMMING UP
The Mitochondrial Matrix Reactions

- The synthesis of acetyl CoA produces one CO_2 and one NADH per pyruvate.

- The Krebs cycle produces two CO_2, one ATP, three NADH, and one $FADH_2$ per acetyl CoA.
- Therefore, at the conclusion of the matrix reactions, the two pyruvates that are produced from a single glucose molecule have been completely broken down, forming six CO_2 molecules.
- During the process, two ATPs and 10 high-energy electron carriers—eight NADH and two $FADH_2$—have been produced from a single glucose molecule.

High-Energy Electrons Travel Through the Electron Transport Chain

At this point, the cell has gained only four ATP molecules from the original glucose molecule: two during glycolysis and two during the Krebs cycle. The cell has, however, captured many high-energy electrons in carrier molecules: 2 NADH during glycolysis, plus 8 more NADH and 2 $FADH_2$ from the matrix reactions, for a total of 10 NADH and 2 $FADH_2$ for each glucose molecule. The carriers deposit their electrons in the **electron transport chain (ETC)**, many copies of which are embedded in the inner mitochondrial membrane (**FIG. 8-8**). These are similar in structure and function to those embedded in the thylakoid membrane of chloroplasts. Energetic electrons move from molecule to molecule along the chain, losing small amounts of energy at each transfer. At certain points along

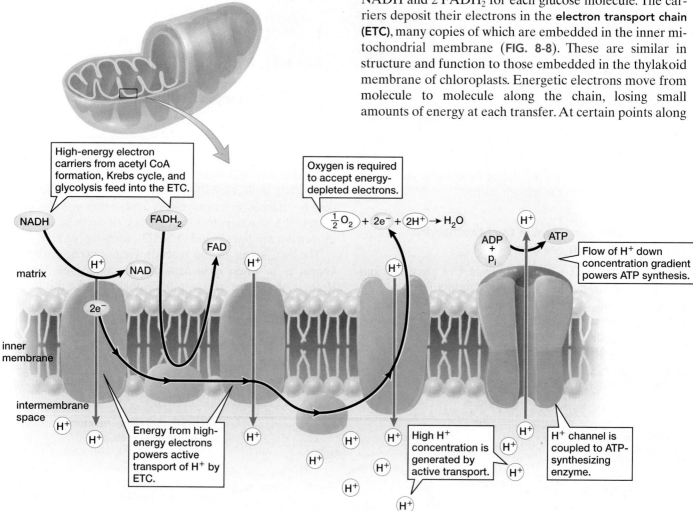

FIGURE 8-8 The electron transport chain of mitochondria
NADH and $FADH_2$ donate their energetic electrons to the carriers of the transport chain. As the electrons pass through the transport chain, some of their energy is used to pump hydrogen ions from the matrix into the intermembrane space. This creates a hydrogen ion gradient that is used to drive ATP synthesis. At the end of the electron transport chain, the energy-depleted electrons combine with oxygen and hydrogen ions in the matrix to form water. QUESTION How would the rate of ATP production be affected by the absence of oxygen?

A CLOSER LOOK The Mitochondrial Matrix Reactions

Mitochondrial matrix reactions occur in two stages: the formation of acetyl coenzyme A and the Krebs cycle (**FIG. E8-3**). Recall that glycolysis produces two pyruvates from each glucose molecule, so each set of matrix reactions occurs twice during the metabolism of a single glucose molecule.

FIRST STAGE: FORMATION OF ACETYL COENZYME A

Pyruvate is split to form CO_2 and an acetyl group. The acetyl group attaches to CoA to form acetyl CoA. Simultaneously, NAD^+ receives two electrons and a hydrogen ion to make NADH. The acetyl CoA enters the second stage of the matrix reactions.

SECOND STAGE: THE KREBS CYCLE

① Acetyl CoA donates its acetyl group to oxaloacetate to make citrate. CoA is released.

② Citrate is rearranged to form isocitrate.

③ Isocitrate loses a carbon to CO_2, forming α-ketoglutarate: NADH is formed from NAD^+.

④ Alpha-ketoglutarate loses a carbon to CO_2, forming succinate; NADH is formed from NAD^+ and additional energy is stored in ATP. (By this stage in the mitochondrial matrix reactions, all three carbons of the original pyruvate have been released as CO_2.)

⑤ Succinate is converted to fumarate, and the electron carrier FAD is charged up to $FADH_2$.

⑥ Fumarate is converted to malate.

⑦ Malate is converted to oxaloacetate, and NADH is formed from NAD^+.

The Krebs cycle produces two CO_2, three NADH, one $FADH_2$, and one ATP per acetyl CoA. The formation of each acetyl CoA generates an additional CO_2 and an NADH. Overall, the mitochondrial matrix reactions produce four NADH, one $FADH_2$, and three CO_2 for each pyruvate supplied by glycolysis. Since each glucose molecule produces two pyruvates, the mitochondrial matrix reactions will generate a total of eight NADH and two $FADH_2$ per glucose molecule. These high-energy electron carriers will release their high-energy electrons to the electron transport chain of the inner membrane, where the energy of the electrons will be used to synthesize more ATP by chemiosmosis.

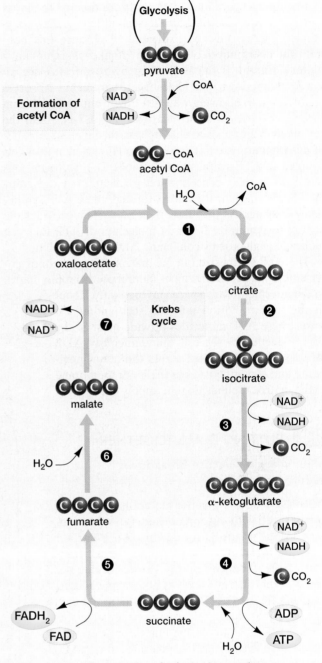

FIGURE E8-3 The mitochondrial matrix reactions

the chain, just enough energy is liberated to pump hydrogen ions from the matrix across the inner membrane and into the intermembrane space during *chemiosmosis* (discussed in the next section).

Finally, at the end of the electron transport chain, oxygen and hydrogen ions (H^+) accept the energy-depleted electrons: two electrons, one oxygen atom, and two hydrogen ions combine to form water (see Fig. 8-8). This step clears out the transport chain, leaving it ready to carry more electrons. Without oxygen, the electrons would be

unable to move through the ETC, and H^+ would not be pumped across the inner membrane. The H^+ gradient would soon dissipate, and ATP synthesis would stop.

Chemiosmosis Captures Energy Stored in a Hydrogen Ion Gradient and Produces ATP

Why pump hydrogen ions across a membrane? As you may recall from Chapter 7, **chemiosmosis** is the process by which a gradient of hydrogen ions (H^+) is produced and then allowed to run down, capturing energy in the bonds of ATP molecules. Pumping H^+ across the inner membrane by the ETC produces a high concentration of H^+ in the intermembrane space and a low concentration in the matrix (see Fig. 8-8). According to the second law of thermodynamics, energy must be expended to produce this nonuniform distribution of H^+, sort of like charging up a battery. Energy is released when hydrogen ions are allowed to flow down their concentration gradient—a process comparable to allowing water to flow from behind a dam through hydroelectric turbines (see Fig. E7-1). As in the thylakoid membranes of chloroplasts, the inner membranes of mitochondria are impermeable to H^+ except at channels coupled to ATP-synthesizing enzymes. The hydrogen ions move down their concentration gradient from the intermembrane space to the matrix through these ATP-synthesizing enzymes. As they flow, their movement provides the energy to synthesize 32 or 34 molecules of ATP for each molecule of glucose by combining ADP and phosphate.

The ATP synthesized in the matrix during chemiosmosis is pumped across the inner membrane from the matrix to the intermembrane space and then diffuses out of the mitochondrion into the surrounding cytosol. These ATP molecules provide most of the energy needed by the cell. Simultaneously, ADP diffuses from the cytosol across the outer membrane and is pumped across the inner membrane to the matrix, replenishing the supply of ADP.

8.4   PUTTING IT ALL TOGETHER

A Summary of Glucose Breakdown in Eukaryotic Cells

FIGURE 8-9 shows glucose metabolism in a eukaryotic cell with oxygen present. Glycolysis occurs in the cytosol, producing two 3-carbon pyruvate molecules and releasing a small fraction of the chemical energy stored in glucose. Some of this energy is lost as heat, some is used to generate two ATP molecules, and some is captured in two NADH (high-energy electron carriers). Under anaerobic conditions, fermentation then occurs and regenerates the NAD, producing either lactate or ethanol and carbon dioxide.

During cellular respiration, the pyruvate enters the mitochondria. First, it reacts with coenzyme A (CoA). This reaction liberates CO_2, captures a high-energy electron in NADH, and produces acetyl CoA (a two-carbon molecule). The acetyl CoA then enters a series of enzyme-

catalyzed reactions called the Krebs cycle (citric acid cycle). The Krebs cycle liberates more CO_2, produces 1 ATP for every acetyl CoA molecule, and captures high-energy electrons in electron carrier molecules: 3 NADH and 1 $FADH_2$ (a related molecule), for every acetyl CoA. These electrons are transferred by their carriers into the ETC. In the process of chemiosmosis, the ETC uses the energy in the high-energy electrons to generate a gradient of H^+. The energy stored in this gradient is then harnessed to generate ATP as the hydrogen ions flow down their gradient through channels coupled to an ATP-synthesizing enzyme. At the completion of the process, oxygen combines with the H^+, forming water. Chemiosmosis in mitochondria generates an additional 32 or 34 ATP molecules for each molecule of glucose (the amount of ATP differs from cell to cell; see the legend for Figure 8-10). The energy produced by each stage of glucose breakdown is shown in FIGURE 8-10.

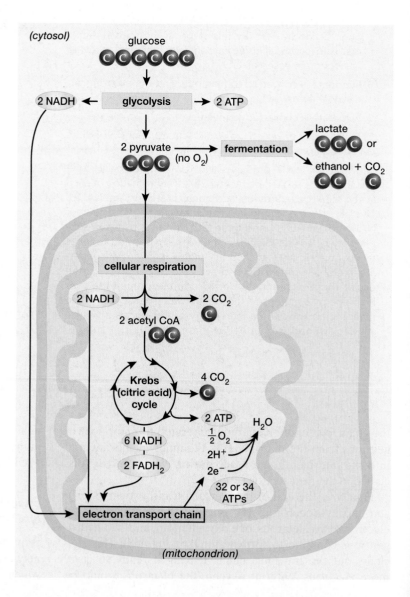

FIGURE 8-9 A summary of glycolysis and cellular respiration

Glycolysis and Cellular Respiration Influence the Way Organisms Function

Many students believe that the details of glycolysis and cellular respiration are hard to learn and don't really help them understand the living world around them. But have you ever read a murder mystery and wondered how cyanide could kill a person almost instantly? Cyanide reacts with the final protein in the ETC more strongly than oxygen does; but unlike oxygen, cyanide will not accept electrons. By preventing oxygen from accepting electrons, cyanide brings cellular respiration to a screeching halt. So dependent are we on cellular respiration that blocking it with cyanide can kill a person within minutes. For your heart to continue beating, your brain to process the information you are reading, and your hand to turn the pages of this book, your cells require a continuous supply of energy. The bodies of most animals store energy in molecules such as glycogen (long chains of glucose molecules) and fat. When food is abundant, sugar and even protein can be converted to fat (as described in "Health Watch: Why Can You Get Fat by Eating Sugar?"). When energy demands are high, glycogen is broken into glucose molecules, which are broken down by glycolysis, which is in turn followed by cellular respiration. But high energy demands produce high oxygen demands. What happens if oxygen becomes limiting? As an extreme example, let's consider Olympic track events.

Why is the average speed of the 5000-meter run in the Olympics slower than that of the 100-meter dash? During the dash, or during the sprint across the finish line of a marathon, runners' leg muscles use more ATP than cellular respiration can supply, because their bodies cannot deliver enough oxygen to keep up with the demand. Glycolysis and lactate fermentation can keep the muscles supplied with ATP for a short time, but soon the effects of lactate buildup (along with several other factors) cause discomfort, fatigue, and cramps. Although runners can do a 100-meter dash without adequate oxygen, distance runners and other distance athletes such as cross-country skiers and cyclists must pace themselves, using cellular respiration to power their muscles for most of the race and saving the anaerobic sprint for the finish. Training for distance events focuses on building up the capacity of the athletes' respiratory and circulatory systems to deliver enough oxygen to their muscles. For this reason, the athletes who engage in distance events are most often those who resort to blood doping.

As you can see, sustaining life depends on efficiently obtaining, storing, and using energy. By gaining an understanding of the principles of cellular respiration, you can more fully appreciate the energy-related adaptations of people and other living organisms.

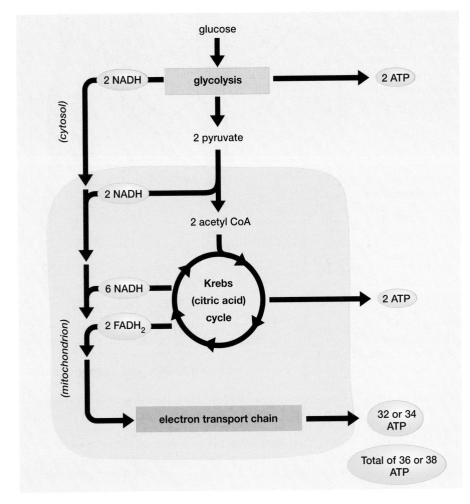

FIGURE 8-10 Energy harvest from the breakdown of glucose
Why do we say that glucose breakdown releases "36 *or* 38 ATP molecules," rather than one specific number? Glycolysis produces two NADH molecules in the cytosol. The electrons from these two NADH molecules must be transported into the matrix before they can enter the electron transport chain. In most eukaryotic cells, the energy of one ATP molecule is used to transport the electrons from each NADH molecule into the matrix. Thus, the two "glycolytic NADH" molecules net only two ATPs, not the usual three, during electron transport. The heart and liver cells of mammals, however, use a different transport mechanism, one that does not consume ATP to transport electrons. In these cells, the two NADH molecules produced during glycolysis net three ATPs each, just as the "mitochondrial NADH" molecules do.

HEALTH WATCH   Why Can You Get Fat by Eating Sugar?

As you know, humans do not live by glucose alone. Nor does the typical diet contain exactly the required amounts of each nutrient. Accordingly, the cells of the human body seethe with biochemical reactions, synthesizing one amino acid from another, making fats from carbohydrates, and channeling surplus organic molecules of many types into energy storage or release. Let's look at two examples of these metabolic transformations: the production of ATP from fats and proteins, and the synthesis of fats from sugars.

HOW ARE FATS AND PROTEINS METABOLIZED?

Even lean people have some fat in their bodies. During fasting or starvation, the body mobilizes these fat reserves for ATP synthesis; even the bare maintenance of life requires a continuous supply of ATP, and seeking out new food sources demands even more energy. Fat metabolism flows directly into the pathways of glucose metabolism.

In Chapter 3 we described the structure of a fat: three fatty acids connected to a glycerol backbone. In fat metabolism, the bonds between the fatty acids and glycerol are hydrolyzed (broken into subunits by the addition of water). The glycerol part of a fat, after activation by ATP, feeds directly into the middle of the glycolysis pathway (**FIG. E8-4**). The fatty acids are transported into the mitochondria, where enzymes in the inner membrane and matrix chop them up into acetyl groups. These groups attach to CoA to form acetyl CoA, which enters the Krebs cycle.

In individuals who are starving (a situation in which muscle protein is broken down to provide energy) or who are on a high-protein diet, amino acids can be used to produce energy. First, the amino acids are converted to pyruvate, acetyl CoA, or the compounds of the Krebs cycle. These molecules then proceed through the remaining stages of cellular respiration, yielding amounts of ATP that vary with their point of entry into the pathway.

HOW IS FAT SYNTHESIZED FROM SUGAR?

The body, in addition to having developed ways of coping with fasting or starvation, has also evolved strategies for coping with situations in which food intake exceeds current energy needs. The sugars and starches in corn, candy bars, and potatoes can all be converted into fats for energy storage. Complex sugars, such as starches and sucrose, are first hydrolyzed into their monosaccharide subunits (see Chapter 3). The monosaccharides are broken down to pyruvate and converted to acetyl CoA. If the cell needs ATP, the acetyl CoA will enter the Krebs cycle. If the cell has plenty of ATP, acetyl CoA will be used to make fatty acids by a series of reactions that are essentially the reverse of fatty acid breakdown. In humans, the liver synthesizes fatty acids; but fat storage is relegated to fat cells, with their all-too-familiar distribution in the body, particularly around the waist and hips. Acetyl CoA and other intermediate molecules of glucose breakdown can also be used in the synthesis of amino acids.

Energy use, fat storage, and nutrient intake are usually precisely balanced. Where the balance point lies, however, varies from person to person. Some people seem able to eat excessively without ever storing much fat; other people

crave high-calorie foods even when they have an excess of fat stores. From an evolutionary perspective, overeating during times of easy food availability is a highly adaptive behavior. During times of famine—which were common during our evolutionary history—heavier people are more likely to survive, while leaner individuals succumb to starvation. It is only recently (from an evolutionary standpoint) that people in societies such as ours have had continuous access to high-calorie food. Under these conditions, the drive to eat and the adaptation of storing excess food as fat leads to obesity, a growing health problem in the United States and many other countries.

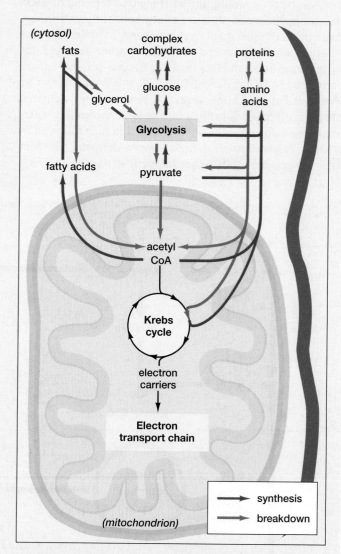

FIGURE E8-4 How various nutrients yield energy and can be interconverted
Metabolic pathways allow interconversion of fats, proteins, and carbohydrates via intermediate molecules formed along the same pathways that break down glucose. Blue arrows show breakdown of these substances to provide energy. Red arrows show that these molecules may also be synthesized when there is an excess of the intermediates.

CASE STUDY REVISITED WHEN ATHLETES BOOST THEIR BLOOD COUNTS: DO CHEATERS PROSPER?

As you have seen, human cells most efficiently extract energy from glucose when an ample supply of oxygen is available to them. The aim of blood-doping athletes, then, is to extend as long as possible the period in which muscle cells have access to oxygen. During a difficult hill climb, a skier who has doped his blood with erythropoietin may be able to ski efficiently, his muscle cells using cellular respiration to churn out abundant ATP. At the same time, his "clean" competitor may labor painfully, leg muscles laden with lactate from fermentation. Because Epo forms naturally in the human body, its abuse is hard to detect. Sports officials assert that the difficulty of detecting Epo has made it the drug of choice among blood-doping skiers, cyclists, distance runners, and other competitive athletes.

Evidence in support of the hypothesis that Epo abuse is widespread includes a study of blood samples taken from participants in the Nordic Ski World Championships. Researchers predicted that if blood doping with Epo were common among top skiers, contestants' blood would contain abnormally high levels of red blood cells (**FIG. 8-11**).

They found that 36 percent of the tested skiers had high red blood cell counts and concluded that many skiers are blood doping. Contestants at the Olympics are now routinely tested for Epo, but the available tests are not considered to be completely reliable. Meanwhile, researchers continue to explore the chemistry of Epo metabolism in hopes of discovering a definitive test for blood doping.

Consider This Some athletes move to high-altitude locations to train for races run at lower altitudes. Is this cheating? Explain your reasoning. Advances in gene therapy may one day make it possible to modify athletes' kidney cells so that they have extra copies of the genes that produce Epo. Is this cheating?

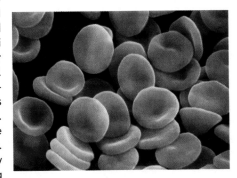

FIGURE 8-11 Red blood cells

CHAPTER REVIEW

SUMMARY OF KEY CONCEPTS

8.1 How Do Cells Obtain Energy?

Cells produce usable energy by breaking down glucose into lower-energy compounds and capturing some of the released energy as ATP. In glycolysis, glucose is metabolized in the cytosol to form two molecules of pyruvate, generating two ATP molecules. In the absence of oxygen, pyruvate is converted by fermentation to lactate or ethanol and CO_2. If oxygen is available, the pyruvate molecules are metabolized to release CO_2 and H_2O through cellular respiration in the mitochondria, generating much more ATP than does fermentation.

Web Tutorial 8.1 Overview of Glucose Metabolism

8.2 How Is the Energy in Glucose Captured During Glycolysis?

During glycolysis, a molecule of glucose is activated by the addition of phosphates from two ATP molecules to form fructose bisphosphate. In a series of reactions, the fructose bisphosphate is broken down into two molecules of pyruvate. These reactions produce a net yield of two ATP and two NADH. Glycolysis, in addition to providing a small yield of ATP, uses up NAD^+ to produce NADH. Once the cell's supply of NAD^+ is consumed, glycolysis must stop. Under anaerobic conditions, NAD^+ is regenerated by fermentation, with no additional ATP gain. If oxygen is present, most cells will regenerate the NAD using cellular respiration, which also produces additional ATP.

Web Tutorial 8.2 Glycolysis and Fermentation

8.3 How Does Cellular Respiration Capture Additional Energy from Glucose?

If oxygen is available, cellular respiration can occur. Pyruvate is transported into the matrix of the mitochondria. Here, it reacts with coenzyme A to form acetyl CoA plus CO_2. One NADH is also formed at this step. The two-carbon acetyl group of acetyl CoA enters the Krebs cycle, which releases the remaining two carbons as CO_2. One ATP, three NADH, and one $FADH_2$ are also formed for each acetyl group that goes through the cycle. At this point, each glucose molecule has produced 4 ATP (2 from glycolysis and 1 from each acetyl CoA during the Krebs cycle), 10 NADH (2 from glycolysis, 1 from each pyruvate during the formation of acetyl CoA, and 3 from each acetyl CoA during the Krebs cycle), and 2 $FADH_2$ (1 from each acetyl CoA during the Krebs cycle).

The NADH and $FADH_2$ deliver their high-energy electrons to the proteins of the electron transport chain (ETC) embedded in the inner mitochondrial membrane. The energy of the electrons is used to pump hydrogen ions across the inner membrane from the matrix to the intermembrane space. At the end of the ETC, the depleted electrons combine with hydrogen ions and oxygen to form water. This is the oxygen-requiring step of cellular respiration. During chemiosmosis, the hydrogen ion gradient created by the electron transport chain is used to produce ATP, as the hydrogen ions diffuse back across the inner membrane through channels in ATP-synthesizing enzymes. Electron transport and chemiosmosis yield 32 or 34 additional ATP, for a net yield of 36 or 38 ATP per glucose molecule.

Web Tutorial 8.3 Cellular Respiration in Mitochondria

8.4 Putting It All Together

Figures 8-1, 8-9, and 8-10 summarize the locations, major mechanisms, and overall energy harvest for the complete metabolism of glucose from glycolysis through cellular respiration.

KEY TERMS

adenosine triphosphate (ATP)   *page 134*
cellular respiration   *page 138*
chemiosmosis   *page 141*

electron transport chain   *page 140*
fermentation   *page 135*

glycolysis   *page 134*
intermembrane space   *page 138*

Krebs cycle   *page 139*
matrix   *page 138*

THINKING THROUGH THE CONCEPTS

1. Starting with glucose $(C_6H_{12}O_6)$, write the overall reactions for (a) aerobic respiration and (b) fermentation in yeast.

2. Draw a labeled diagram of a mitochondrion, and explain how its structure relates to its function.

3. What role do the following play in respiratory metabolism: (a) glycolysis, (b) mitochondrial matrix, (c) inner membrane of mitochondria, (d) fermentation, and (e) NAD^+?

4. Outline the major steps in (a) aerobic and (b) anaerobic respiration, indicating the sites of ATP production. What is the overall energy harvest (in terms of ATP molecules generated per glucose molecule) for each?

5. Describe the Krebs cycle. In what form is most of the energy captured?

6. Describe the mitochondrial electron transport chain and the process of chemiosmosis.

7. Why is oxygen necessary for cellular respiration to occur?

8. Compare the structure of chloroplasts (described in Chapter 7) to that of mitochondria, and describe how the similarities in structure relate to similarities in function. Also describe any differences in structure and function between chloroplasts and mitochondria.

APPLYING THE CONCEPTS

1. Some years ago a freight train overturned, spilling a load of grain. Because the grain was unusable, it was buried in the railroad embankment. Although there is no shortage of other food, the local bear population has become a nuisance by continually uncovering the grain. Yeasts are common in the soil. What do you think has happened to the grain to make the bears keep digging it up, and how is their behavior related to human cultural evolution?

2. In detective novels, "the odor of bitter almonds" is the telltale clue to murder by cyanide poisoning. Cyanide works by attacking the enzyme that transfers electrons from the respiratory electron transport chain to O_2. Why can't the victim survive by using anaerobic respiration? Why is cyanide poisoning almost immediately fatal?

3. Some species of bacteria that live at the surface of sediment on the bottom of lakes are capable of using either glycolysis plus fermentation or cellular respiration to generate ATP. There is very little circulation of water in lakes during the summer. Predict and explain what will happen to the bottom-most water of lakes as the summer progresses, and describe how this situation will effect energy production by bacteria.

4. Dumping large amounts of raw sewage into rivers or lakes typically leads to massive fish kills, although sewage itself is not toxic to fish. Similar fish kills also occur in shallow lakes that become covered with ice during the winter. What kills the fish? How might you reduce fish mortality after raw sewage is accidentally released into a small pond containing large bass?

5. Different cells respire at different rates. Explain why. How could you predict the relative respiratory rates of different tissues in a fish by examining cells through a microscope?

6. Imagine a hypothetical situation in which a starving cell reaches the stage where every bit of its ATP has been depleted and converted to ADP plus phosphate. If at this point you place the cell in a solution containing glucose, will it recover and survive? Explain your answer based on what you know about glucose breakdown.

FOR MORE INFORMATION

Aschwanden, C. "No Cheating in the Blood Test." *New Scientist,* October 2, 2004. Users of Epo can no longer escape detection.

Lovett, R. "Runner's High." *New Scientist,* November 2, 2002. How trainers at the Oregon Project hope to combine traditional training with strategies that explore the limits of human endurance physiology.

Roth, M. R., and Ntstul, T. "Buying Time in Suspended Animation." *Scientific American,* June, 2005. Although hydrogen sulfide is poisonous at high level, our own cells produce small quantities of it. In mice, hydrogen sulfide blocks oxygen use and can put the animal into a state of suspended animation. Could this work in people as well?

Inheritance

Inheritance provides for both similarity and difference. All dogs share many similarities because their genes are nearly identical. The enormous variety of size, fur length and color, and bodily proportions results from tiny differences in their genes.

DNA: The Molecule of Heredity

Ordinary bull or incredible hulk?
A tiny change in DNA makes all
the difference.

CASE STUDY MUSCLES, MUTATIONS, AND MYOSTATIN

NO, THE BULL in the top photo hasn't been pumping iron—he's a Belgian Blue, and they always have bulging muscles. What makes a Belgian Blue look like a bodybuilder gone wild, compared to an ordinary bull, such as the Hereford in the bottom photo?

When any mammal develops, its cells divide many times, enlarge, and become specialized for a specific function. The size, shape, and cell types in any organ are precisely regulated during development, so that you don't wind up with a head the size of a basketball, or have hair growing on your liver. Muscle development is no exception. When you were very young, cells destined to form your muscles multiplied, fused together to form long, relatively thick cells with multiple nuclei, and synthe-sized the specialized proteins that cause muscles to contract and thereby move your skeleton. A protein called *myostatin*, found in all mammals, puts the brakes on this process. The word "myostatin" literally means "to make muscles stay the same," and that is exactly what myostatin does. As muscles develop, myostatin slows down, and eventually stops, the multiplication of these pre-muscle cells. A bodybuilder can bulk up by lifting weights (and by taking so-called anabolic steroids, although that's not a very smart thing to do), which *enlarges* the muscle cells, but doesn't usually add many *more* cells.

Belgian Blues have *more* muscle cells than ordinary cattle. Why? You may already have guessed—they don't produce normal myostatin. And why not? As you will learn in this chapter, proteins are synthesized from the genetic directions contained in deoxyribonucleic acid, or DNA for short. The DNA of a Belgian Blue is very slightly different from the DNA of normal cattle—it has a change, or *mutation*, in the DNA of its myostatin gene. As a result, it produces defective myostatin. Belgian Blue pre-muscle cells multiply more than normal, producing remarkably buff cattle.

In this chapter, we will follow the scientific paths that led to our modern understanding of the structure of DNA. We will look at how it can contain the directions for traits such as muscle development, how those directions might stay the same—or change—from generation to generation, and what might happen if the directions are changed.

9.1 HOW DID SCIENTISTS DISCOVER THAT GENES ARE MADE OF DNA?

By the late 1800s, scientists had learned that genetic information exists in discrete units that they called **genes**. However, they didn't really know what a gene was. Scientists merely knew that genes determine many of the heritable differences among individuals within a species. For example, genes for flower color determine whether roses are red, pink, yellow, or white. By the early 1900s, studies of dividing cells provided strong evidence that genes are parts of **chromosomes** (see Chapters 5, 11, and 12). Soon, biochemists found that eukaryotic chromosomes are composed only of protein and DNA. One of these substances must carry the cell's hereditary blueprint, but which one?

Transformed Bacteria Revealed the Link Between Genes and DNA

In the late 1920s, a British researcher named Frederick Griffith was trying to make a vaccine to prevent bacterial pneu-

monia, a major cause of death at that time. Making vaccines against many infectious bacteria is very difficult (for example, modern vaccines against anthrax are neither completely safe nor completely effective), but this was not known back in the 1920s. Some antibacterial vaccines consist of a weakened strain of the bacteria that doesn't cause illness. Injecting this weakened but living strain into an animal may stimulate immunity against the disease-causing strains. Other vaccines use disease-causing (virulent) bacteria that have been killed by exposure to heat or chemicals. Griffith was trying to make a vaccine using two strains of the *Streptococcus pneumoniae* bacterium. One strain, R, did not cause pneumonia when injected into mice (**FIG. 9-1a**). The other strain, S, was deadly when injected, causing pneumonia and killing the mice in a day or two (**FIG. 9-1b**). As expected, when the S-strain was killed and injected into mice, it did not cause disease (**FIG. 9-1c**). Unfortunately, neither the live R-strain nor the killed S-strain provided immunity against live S-strain bacteria.

Griffith also tried mixing living R-strain bacteria together with heat-killed S-strain bacteria and injecting the

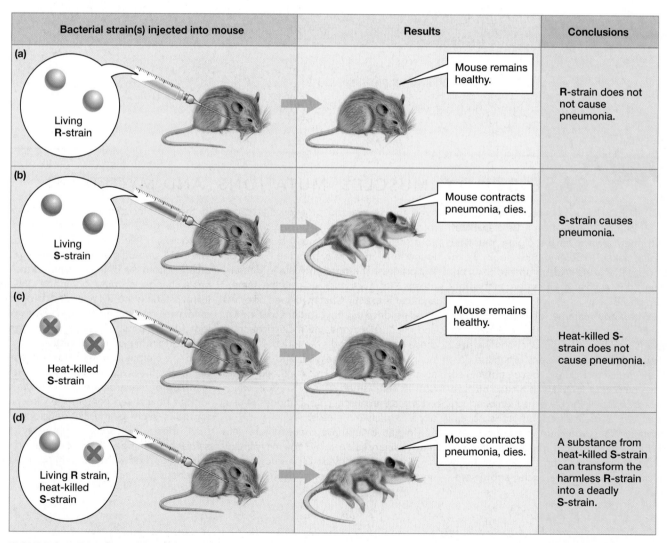

FIGURE 9-1 Transformation in bacteria
Griffith's discovery that bacteria can be transformed from harmless to deadly laid the groundwork for the discovery that DNA contains genes.

mixture into mice (**FIG. 9-1d**). Because neither of these bacterial strains causes pneumonia on its own, he expected the mice to remain healthy. To his surprise, the mice sickened and died. When he autopsied the mice, he recovered *living* S-strain bacteria from them. The simplest interpretation of these results is that some substance in the heat-killed S-strain changed the living but harmless R-strain bacteria into the deadly S-strain, a process he called *transformation*. The transformed S-strain cells then multiplied and caused pneumonia.

Griffith never discovered an effective pneumonia vaccine, so in that sense his experiments were a failure (in fact, an effective and safe vaccine against most forms of *Streptococcus pneumoniae* was not developed until a few years ago). However, Griffith's experiments marked a turning point in our understanding of genetics because other researchers suspected that the substance that causes transformation might be the long-sought molecule of heredity.

The Transforming Molecule Is DNA

In 1933, J. L. Alloway discovered that the mice had no role in transformation, which occurred just as well when live R-strain bacteria were mixed with dead S-strain bacteria in culture dishes. A decade later, Oswald Avery, Colin MacLeod, and Maclyn McCarty discovered that the transforming molecule is **DNA**. Avery, MacLeod, and McCarty isolated DNA from S-strain bacteria, mixed it with live R-strain bacteria, and produced live S-strain bacteria. To show that transformation was caused by DNA, and not by traces of protein contaminating the DNA, they treated some samples with protein-destroying enzymes. These enzymes did not prevent transformation. However, treating samples with DNA-destroying enzymes did prevent transformation.

This discovery helps us interpret the results of Griffith's experiments. Heating S-strain cells killed them but did not completely destroy their DNA. When killed S-strain bacteria were mixed with living R-strain bacteria, fragments of DNA from the dead S-strain cells entered into some of the R-strain cells and became incorporated into the chromosome of the R-strain bacteria (**FIG. 9-2**). If these fragments of DNA contained the genes needed to cause disease, an R-strain cell would be transformed into an S-strain cell. Thus, Avery, MacLeod, and McCarty concluded that genes are made of DNA.

DNA, Not Protein, Is the Molecule of Heredity

Nevertheless, not everyone in the scientific community was persuaded. Some still thought that genes are made of protein, and that the transforming DNA molecules from S-strain bacteria caused a mutation in the genes of R-strain bacteria. Others hypothesized that DNA might be the hereditary molecule of bacteria, but not of other organisms. However, evidence continued to accumulate that DNA is the genetic material in many, or perhaps all,

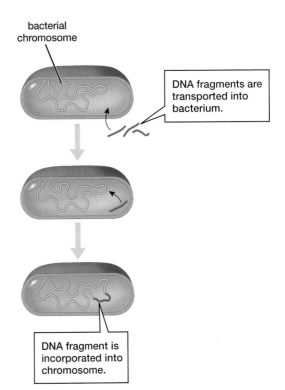

FIGURE 9-2 Molecular mechanism of transformation
Most bacteria have a single large, circular chromosome made of DNA. Transformation may occur when a living bacterium takes up pieces of DNA from its environment and incorporates those fragments into its chromosome.

organisms. For example, before dividing, a eukaryotic cell duplicates its chromosomes (see Chapter 11) and exactly doubles its DNA content—just what would be expected if genes are made of DNA. Finally, virtually all of the remaining skeptics were convinced by a superb set of experiments by Alfred Hershey and Martha Chase, conclusively showing that DNA is the heredity molecule of certain viruses (see "Scientific Inquiry: DNA Is the Hereditary Molecule of Bacteriophages").

9.2 WHAT IS THE STRUCTURE OF DNA?

Knowing that genes are made of DNA does not answer critical questions about inheritance: How does DNA encode genetic information? How is DNA duplicated so that information can be accurately passed from one cell to its daughter cells? (See Chapter 11 for more information about cell reproduction.) The secrets of DNA function, and therefore of heredity itself, are found in the three-dimensional structure of the DNA molecule.

Certain viruses infect only bacteria and are called **bacteriophages**, meaning "bacteria eaters" (**FIG. E9-1**). A bacteriophage (phage for short) depends on its host bacterium for every aspect of its life cycle (Fig. E9-1b). When a phage encounters a bacterium, it attaches to the bacterial cell wall and injects its genetic material into the bacterium. The outer coat of the phage remains outside the bacterium. The bacterium cannot distinguish phage genes from its own genes, so it "reads" the phage genes and uses that information to produce more phages. Finally, one of the phage genes directs the synthesis of an enzyme that ruptures the bacterium, freeing the newly manufactured phages.

Even though many bacteriophages have intricate structures (see Fig. E9-1a), they are chemically very simple, containing only DNA and protein. Therefore, one of these two molecules must be the phage genetic material. In the early 1950s, Alfred Hershey and Martha Chase used the chemical simplicity of bacteriophages to deduce that their genetic material is DNA.

Hershey and Chase knew that infected bacteria should contain phage genetic material, so if they could "label" phage DNA and protein, and separate the infected bacteria from the phage coats left outside, they could see which molecule entered the bacteria (**FIG. E9-2**). As you learned in Chapter 3, DNA and protein both contain atoms of carbon, oxygen, hydrogen, and nitrogen. However, DNA also contains phosphorus but not sulfur, whereas proteins contain sulfur (in the amino acids methionine and cysteine) but not phosphorus. Hershey and Chase forced one population of phages to synthesize DNA using radioactive phosphorus, thereby labeling their DNA. Another population was forced to synthesize protein using radioactive sulfur, labeling their protein. When bacteria were infected by phages containing radioactively labeled protein, the bacteria did not become radioactive. However, when bacteria were infected by phages containing radioactive DNA, the bacteria did become radioactive. Hershey and Chase concluded that DNA, and not protein, was the genetic material of phages.

Hershey and Chase also reasoned that some of the labeled genetic material from the "parental" phages might be incorporated into the genetic material of the "offspring" phages (you will learn more about this in section 9.3). In a second set of experiments, the researchers again labeled DNA in one phage population and protein in another phage population, and allowed the phages to infect bacteria. After enough time had passed so that the phages had reproduced, the bacteria were broken open, and the offspring phages were separated from the bacterial debris. Radioactive DNA, but not radioactive protein, was found in the offspring phages. This second experiment confirmed the results of the first: DNA is the hereditary molecule.

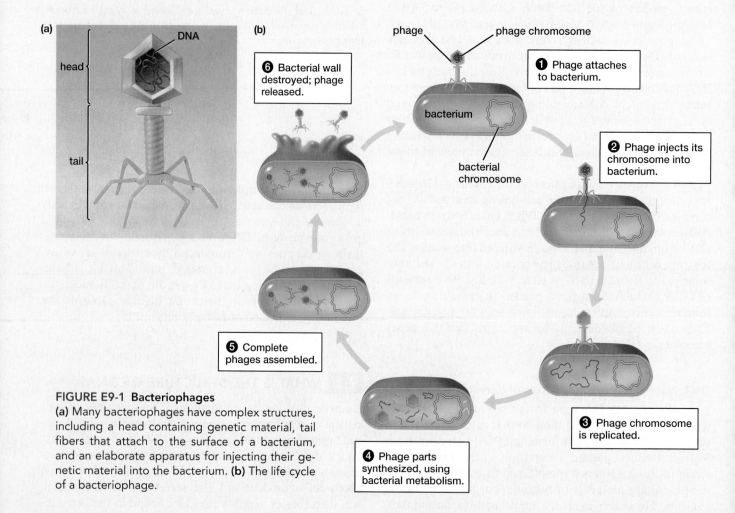

FIGURE E9-1 Bacteriophages
(a) Many bacteriophages have complex structures, including a head containing genetic material, tail fibers that attach to the surface of a bacterium, and an elaborate apparatus for injecting their genetic material into the bacterium. (b) The life cycle of a bacteriophage.

Observations:
1. Bacteriophage viruses consist of only DNA and protein.
2. Bacteriophage inject their genetic material into bacteria, forcing the bacteria to synthesize more phages.
3. The outer coat of bacteriophages stays outside of the bacteria.
4. DNA contains phosphorus but not sulfur.
   a. DNA can be "labeled" with radioactive phosphorus.
5. Protein contains sulfur but not phosphorus.
   a. Protein can be "labeled" with radioactive sulfur.

Question: Is DNA or protein the genetic material of bacteriophages?

Hypothesis: DNA is the genetic material.

Prediction:
1. If bacteria are infected with bacteriophages containing radioactively labeled DNA, the bacteria will be radioactive.
2. If bacteria are infected with bacteriophages containing radioactively labeled protein, the bacteria will not be radioactive.

Experiment:

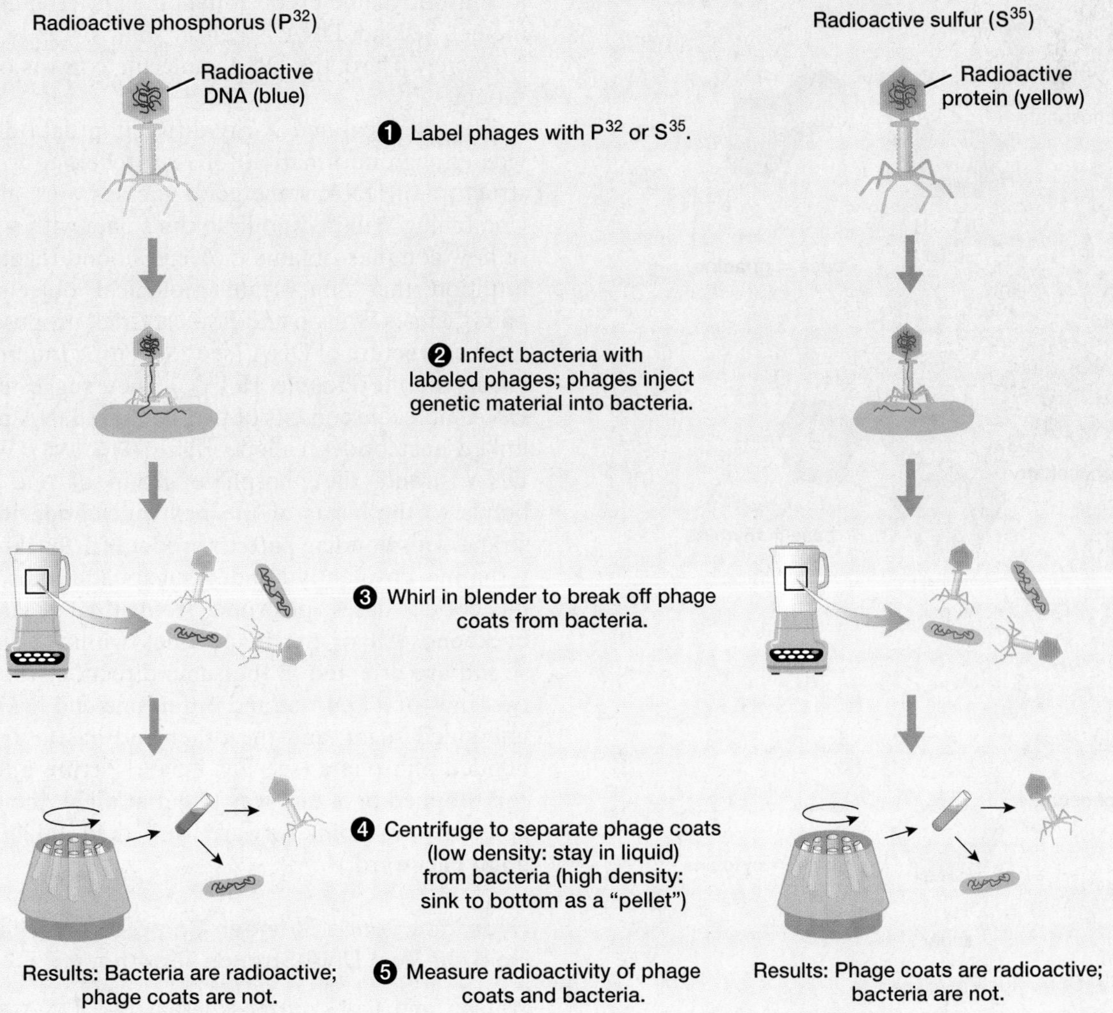

Radioactive phosphorus (P^{32})

Radioactive DNA (blue)

Radioactive sulfur (S^{35})

Radioactive protein (yellow)

❶ Label phages with P^{32} or S^{35}.

❷ Infect bacteria with labeled phages; phages inject genetic material into bacteria.

❸ Whirl in blender to break off phage coats from bacteria.

❹ Centrifuge to separate phage coats (low density: stay in liquid) from bacteria (high density: sink to bottom as a "pellet")

❺ Measure radioactivity of phage coats and bacteria.

Results: Bacteria are radioactive; phage coats are not.

Results: Phage coats are radioactive; bacteria are not.

Conclusion: Infected bacteria are labeled with radioactive phosphorus but not with radioactive sulfur, supporting the hypothesis that the genetic material of bacteriophages is DNA, not protein.

FIGURE E9-2 The Hershey-Chase experiment

DNA Is Composed of Four Nucleotides

As you learned in Chapter 3, DNA consists of four small subunits called **nucleotides**. Each nucleotide in DNA has three parts (**FIG. 9-3**): a phosphate group, a sugar called *deoxyribose*, and one of four possible nitrogen-containing bases—adenine (A), guanine (G), thymine (T), or cytosine (C).

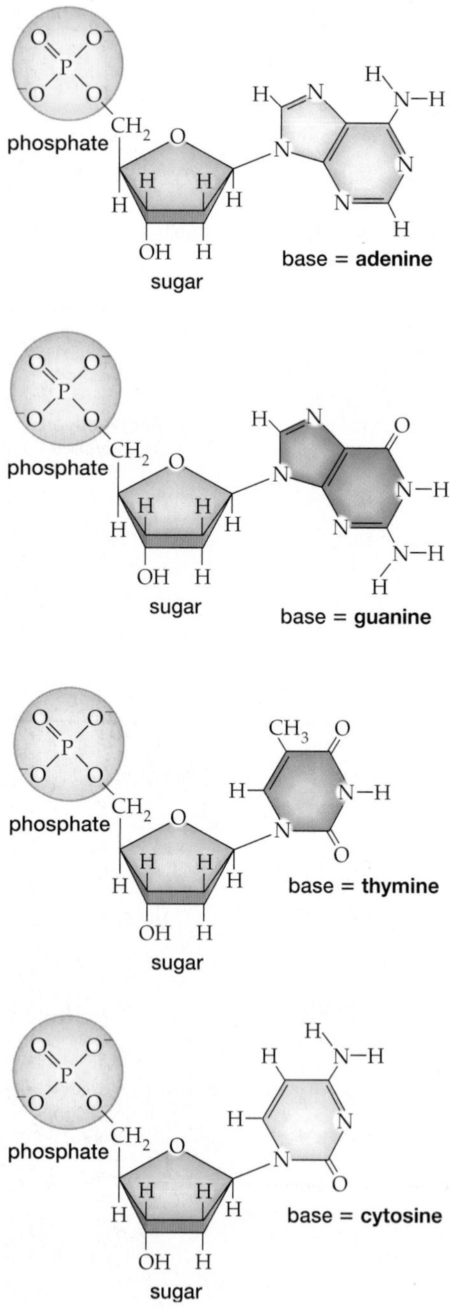

FIGURE 9-3 DNA nucleotides

In the 1940s, when biochemist Erwin Chargaff of Columbia University analyzed the amounts of the four bases in DNA from organisms as diverse as bacteria, sea urchins, fish, and humans, he found a curious consistency. The DNA of any given species contains *equal amounts of adenine and thymine*, as well as *equal amounts of guanine and cytosine*.

This consistency, often called "Chargaff's rule," certainly seemed significant, but it would be almost another decade before anyone figured out what it meant about DNA structure.

DNA Is a Double Helix of Two Nucleotide Strands

Determining the structure of any biological molecule is no simple task, even for scientists today. Nevertheless, in the late 1940s, several scientists began to investigate the structure of DNA. British scientists Maurice Wilkins and Rosalind Franklin used X-ray diffraction to study the DNA molecule. They bombarded crystals of purified DNA with X-rays and recorded how the X-rays bounced off the DNA molecules (**FIG. 9-4a**). As you can see, the resulting "diffraction" pattern does not provide a direct picture of DNA structure. However, experts like Wilkins and Franklin (**FIG. 9-4b, c**) could extract a lot of information about DNA from the pattern. First, a molecule of DNA is long and thin, with a uniform diameter of 2 nanometers (2 billionths of a meter). Second, DNA is helical; that is, it is twisted like a corkscrew. Third, the DNA molecule consists of repeating subunits.

The chemical and X-ray diffraction data did not provide enough information for researchers to work out the structure of DNA; some good guesses were also needed. Combining Wilkins and Franklin's data with a knowledge of how complex organic molecules bond together and an intuition that "important biological objects come in pairs," James Watson and Francis Crick proposed a model for the structure of DNA (see "Scientific Inquiry: The Discovery of the Double Helix.") They suggested that the DNA molecule consists of two separate DNA polymers of linked nucleotides, called *strands* (**FIG. 9-5**). Within each DNA strand, the phosphate group of one nucleotide bonds to the sugar of the next nucleotide in the same strand. This bonding pattern produces a "backbone" of alternating, covalently bonded sugars and phosphates. The nucleotide bases protrude from this **sugar-phosphate backbone**. All of the nucleotides within a single DNA strand are oriented in the same direction. Therefore, the two ends of a DNA strand differ; one end has a "free" or unbonded sugar, and the other end has a "free" or unbonded phosphate (see Fig. 9-5a). (Picture a long line of cars stopped on a one-way street at night; the cars' headlights always point forward, and their taillights always point backward.)

Hydrogen Bonds Between Complementary Bases Hold the Two DNA Strands Together

Watson and Crick proposed that two DNA strands are held together by hydrogen bonds that form between the protruding bases of the individual DNA strands (see Fig. 9-5a). These bonds give DNA a ladder-like structure, with the sugar-phosphate backbones on the outside (forming the uprights of the ladder) and the nucleotide bases on the inside (forming the rungs of the ladder). However, the DNA strands are not straight. Instead, they are twisted

(a)   (b)   (c)

FIGURE 9-4 X-ray diffraction studies of DNA taken by Rosalind Franklin
(a) The X formed of dark spots is characteristic of helical molecules such as DNA. Measurements of various aspects of the pattern indicate the dimensions of the DNA helix; for example, the distance between the dark spots corresponds to the distance between turns of the helix. Maurice Wilkins **(b)** and Rosalind Franklin **(c)** discovered many of the features of DNA by carefully examining such X-ray diffraction patterns. Wilkins shared the Nobel Prize in Physiology or Medicine with Watson and Crick in 1962. However, Franklin died in 1958. Because Nobel Prizes are not awarded posthumously, her contributions often do not receive the recognition they deserve.

about each other to form a **double helix**, resembling a ladder twisted lengthwise into the shape of a circular staircase (see Fig. 9-5b). Further, the two strands in a DNA double helix are oriented in opposite directions, or are *antiparallel*. (Again imagine evening traffic, this time on a two-lane,

north-south highway. All the cars in one lane are going north, and all the cars in the other lane are going south. Therefore, a traffic helicopter pilot overhead would see only the headlights on cars in one lane, and only the taillights of cars in the other lane).

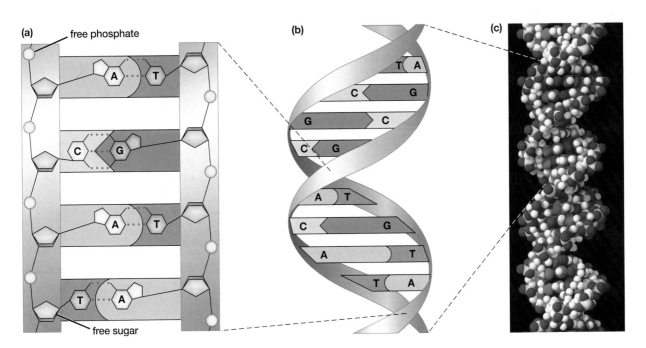

FIGURE 9-5 The Watson-Crick model of DNA structure
(a) Hydrogen bonding between complementary base pairs holds the two strands of DNA together. Three hydrogen bonds (red dotted lines) hold guanine to cytosine, and two hydrogen bonds hold adenine to thymine. Note that each strand has a free phosphate (yellow ball) on one end and a free sugar (blue pentagon) on the opposite end. Further, the two strands run in opposite directions. **(b)** Strands of DNA wind about each other in a double helix, like a twisted ladder, with the sugar-phosphate backbone forming the uprights and the complementary base pairs forming the rungs. **(c)** A space-filling model of DNA structure. QUESTION Which do you think would be more difficult to break apart: an A–T base pair or a C–G base pair?

SCIENTIFIC INQUIRY   The Discovery of the Double Helix

In the early 1950s, many biologists realized that the key to understanding inheritance lay in the structure of DNA. They also knew that whoever deduced the correct structure of DNA would receive recognition, probably including a Nobel Prize. Linus Pauling of Caltech was considered the person most likely to solve the mystery of DNA structure. Pauling probably knew more about the chemistry of large organic molecules than any person alive. Like Rosalind Franklin and Maurice Wilkins, Pauling was an expert in X-ray diffraction techniques. In 1950, he used these techniques to show that many proteins were coiled into single-stranded helices (see Chapter 3). Pauling, however, had two important handicaps. First, he had concentrated on protein research for years and therefore had little data about DNA. Second, he was active in the peace movement. At that time, some government officials, including Senator Joseph McCarthy, considered such activity to be potentially subversive and threatening to national security. This latter handicap may have proved decisive.

The second most likely competitors were Wilkins and Franklin, the British scientists who had set out to determine the structure of DNA by using X-ray diffraction patterns. In fact, they were the only scientists who had good data about the general shape of the DNA molecule. Unfortunately for them, their methodical approach was also slow.

The door was open for the eventual discoverers of the double helix—James Watson and Francis Crick, two scientists with neither Pauling's tremendous understanding of chemical bonds nor Franklin and Wilkins' expertise in X-ray analysis. Watson and Crick did no experiments in the ordinary sense of the word; instead, they spent their time thinking about DNA, trying to construct a molecular model that made sense and fit the data. Because they were working in England and because Wilkins was very open about his and Franklin's data, Watson and Crick were familiar with all the X-ray information relating to DNA. This information was just what Pauling lacked. Because of Pauling's presumed subversive tendencies, the U.S. State Department refused to issue him a passport to leave the United States, so he could neither attend the meetings at which Wilkins presented the X-ray data nor visit England to talk with Franklin and Wilkins directly. Watson and Crick knew that Pauling was working on DNA structure and were terrified that he would beat them to it. In his book, *The Double Helix*, Watson recounts his belief that, had Pauling seen the X-ray pictures, "in a week at most, Linus would have [had] the structure."

You might be thinking, "But wait just a minute! That's not fair. If the goal of science is to advance knowledge, then everyone should have access to all the data. If Pauling was the best, he should have discovered the double helix first." Maybe so. But after all, scientists are people, too. Although virtually all scientists want to see the advancement and benefit of humanity, each of them also wants to be the one responsible for that advancement and receive the credit and glory. And so, Linus Pauling remained in the dark about the X-ray data and was beaten to the correct structure (**FIG. E9-3**). Soon after Watson and Crick deciphered the structure of DNA, Watson described it in a letter to Max Delbruck, a friend and adviser at Caltech. When Delbruck told Pauling about the double helix model for DNA, Pauling graciously congratulated Watson and Crick on their brilliant solution. The race was over.

FIGURE E9-3 The discovery of DNA
James Watson and Francis Crick with a model of DNA structure.

Take a closer look at the pairs of hydrogen-bonded bases that form each rung of the double helix ladder. Notice that adenine forms hydrogen bonds only with thymine, and that guanine forms hydrogen bonds only with cytosine (see Fig. 9-5a, b). These A–T and G–C pairs are called **complementary base pairs**, and they explain "Chargaff's rule"—that the DNA of a given species contains equal amounts of adenine and thymine, as well as equal amounts of cytosine and guanine. Because an A in one DNA strand always pairs with a T in the other strand, the amount of A always equals the amount of T. Similarly, because a G in one strand always pairs with a C in the other DNA strand, the amount of G always equals the amount of C. Finally, look at the sizes of the bases: adenine and guanine are large, whereas thymine and cytosine are small. Because the double helix has only A–T and G–C pairs, all the rungs of the DNA ladder are the same width. Therefore, the double helix has a constant diameter, just as the X-ray diffraction pattern predicted.

The structure of DNA was solved. On March 7, 1953, at the Eagle Pub in Cambridge, England, Francis Crick proclaimed to the lunchtime crowd, "We have discovered the secret of life." This claim was not far from the truth. Although further data would be needed to confirm the details, within just a few years, their DNA model revolutionized biology, from genetics to medicine. As we will see in later chapters, the revolution continues today.

9.3 HOW DOES DNA ENCODE INFORMATION?

Look again at the structure of DNA shown in Figure 9-5. Can you see why many scientists had trouble believing that DNA could be the carrier of genetic information? Consider the many characteristics of just one organism. How can the color of a bird's feathers, the size and shape of its beak, its ability to make a nest, its song, and its ability to migrate all be determined by a molecule with just four simple parts?

The answer is that it's not the *number* of different subunits but their *sequence* that's important. Within a DNA strand, the four types of bases can be arranged in any order, and this sequence is what encodes genetic information. An analogy might help: You don't need a lot of unique letters to make up a language. English has 26 letters, but Hawaiian has only 12, and the binary language of computers uses only two "letters" (0 and 1, or "on" and "off"). Nevertheless, all three languages can spell out thousands of different words. A stretch of DNA that is just 10 nucleotides long can have more than a million possible sequences of the four bases. Because an organism has millions (in bacteria) to billions (in plants or animals) of nucleotides, DNA molecules can encode a staggering amount of information.

Of course, to make sense, words must have the right letters in the right sequence. Similarly, a gene must have the right bases in the right sequence. Just as "friend" and "fiend" mean different things, and "fliend" doesn't mean anything, different sequences of bases in DNA may encode very different pieces of information, or no information at all. Think back to the Case Study at the beginning of this chapter. All "normal" mammals have a DNA sequence that encodes a functional myostatin protein, which limits their muscle growth. Belgian Blue cattle have a mutation that changes a "friendly" gene to a nonsensical "fliendly" one that no longer codes for a functional protein, so they have excessive muscle development.

In Chapter 10, we will discover how the information in DNA is used to produce the structures of living cells. In the remainder of this chapter, we will examine how DNA is replicated during cell division to ensure accurate copying of this genetic information.

9.4 HOW DOES DNA REPLICATION ENSURE GENETIC CONSTANCY DURING CELL DIVISION?

Replication of DNA Is a Critical Event in a Cell's Life

In the 1850s, Austrian pathologist Rudolf Virchow realized that "all cells come from [preexisting] cells." All of the trillions of cells of your body are the offspring (usually called *daughter cells*) of other cells, going all the way back to when you were a fertilized egg. Moreover, nearly every cell of your body contains identical genetic information—the same genetic information present in that fertilized egg. To accomplish this, cells reproduce by a complex process in which one parental cell divides in half, forming two daughter cells (we will learn more about cell division in Chapter 11). Each daughter cell receives a nearly perfect copy of the parent cell's genetic information. Consequently, at an early stage of cell division, the parent cell must synthesize two exact copies of its DNA through a process known as **DNA replication**. Many cells in an adult human never divide at all and therefore do not replicate their DNA. In most of the millions of cells that *do* divide, starting DNA replication irreversibly commits the cell to division. If a cell tries to replicate its DNA without stockpiling enough raw materials or energy to complete the process, it can die. Therefore, the timing of replication is carefully regulated, ensuring that DNA replication does not begin unless the cell is ready to divide. These controls also ensure that the cell's DNA is replicated *exactly* one time prior to each cell division.

Through a complex mechanism involving many other molecules, myostatin prevents pre-muscle cells from replicating their DNA. Therefore, the cells stop dividing, and the number of mature muscle cells is limited. The mutated myostatin of Belgian Blue cattle fails to inhibit DNA replication, so the pre-muscle cells continue to divide, producing more muscle cells.

Once the "decision" is made to divide, the cell replicates its DNA. Recall that DNA is a component of chromosomes. Each chromosome contains a single DNA double helix. DNA replication produces two identical double helices, one of which will be passed to each of the new daughter cells, as we will see in Chapter 11.

DNA Replication Produces Two DNA Double Helices, Each with One Original Strand and One New Strand

How does a cell accurately copy its DNA? In their paper describing DNA structure, Watson and Crick included one of the greatest understatements in all of science: "It has not escaped our notice that the specific [base] pairing we have postulated immediately suggests a possible copying mechanism for the genetic material." In fact, base pairing is the foundation of DNA replication. Remember, the rules for base pairing are that an adenine on one strand must pair with a thymine on the other strand, and a cytosine must pair with a guanine. If one strand reads ATG, for

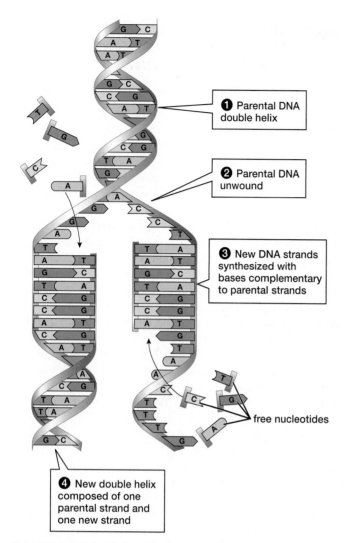

① Parental DNA double helix

② Parental DNA unwound

③ New DNA strands synthesized with bases complementary to parental strands

free nucleotides

④ New double helix composed of one parental strand and one new strand

FIGURE 9-6 Basic features of DNA replication
During replication, the two strands of the parental DNA double helix separate. Free nucleotides that are complementary to those in each strand are joined to make new daughter strands. Each parental strand and its new daughter strand then form a new double helix.

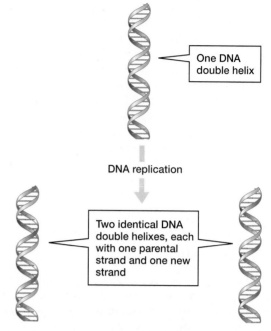

One DNA double helix

DNA replication

Two identical DNA double helixes, each with one parental strand and one new strand

FIGURE 9-7 Semiconservative replication of DNA

example, then the other strand must read TAC. Therefore, the base sequence of each strand contains all the information needed to replicate the other strand.

Conceptually, DNA replication is quite simple (**FIG. 9-6**). Enzymes called **DNA helicases** pull apart the parental DNA double helix, so that the bases of the two DNA strands no longer form base pairs with one another. Now DNA strands complementary to the two parental strands must be synthesized. Other enzymes, called **DNA polymerases**, move along each separated parental DNA strand, matching bases on the strand with complementary **free nucleotides**, previously synthesized in the cytoplasm. DNA polymerase also connects these free nucleotides with one another to form two new DNA strands, each complementary to one of the parental DNA strands. Thus, if a parental DNA strand reads TAG, DNA polymerase will synthesize

a new DNA strand with the complementary sequence ATC. For more information on how DNA is replicated, refer to "A Closer Look: DNA Structure and Replication."

When replication is complete, one parental DNA strand and its newly synthesized, complementary daughter DNA strand wind together into one double helix. At the same time, the other parental strand and its daughter strand wind together into a second double helix. In forming a new double helix, the process of DNA replication conserves one parental DNA strand and produces one newly synthesized strand. Hence, the process is called **semiconservative replication** (**FIG. 9-7**).

The base sequences of both new DNA double helices are identical to the base sequence of the parental DNA double helix and, of course, to each other.

At this point, the two new double helices are still part of a single chromosome while the cell prepares for division. The DNA of every chromosome in the cell replicates in the same manner, so that all of the chromosomes contain two double helices. When the cell divides, one double helix of every chromosome is delivered to each daughter cell. Thus, the two daughter cells normally receive exactly the same genetic information as the original parent cell.

9.5 HOW DO MUTATIONS OCCUR?

Nothing alive is perfect, including the DNA in your cells. Changes in the sequence of bases in DNA, often resulting in a defective gene, are called **mutations**. In most cells, mutations are minimized by highly accurate DNA replication, "proofreading" the newly synthesized DNA, and repairing any changes in DNA that may occur even when the DNA is not being replicated.

DNA Structure and Replication

DNA STRUCTURE

To understand DNA replication, we must first return to the structure of DNA. Recall that the two strands of a double helix run in opposite directions, or are *antiparallel*. Biochemists keep track of the atoms in a complex molecule by numbering them. In the case of a nucleotide, the atoms that form the "corners" of the base are numbered 1 through 6 for the single-ringed cytosine and thymine, or 1 through 9 for the double-ringed adenine and guanine. The carbon atoms of the sugar are numbered 1' through 5'. The prime symbol (') is used to distinguish atoms in the sugar from atoms in the base. The carbons in the sugar are called "1-prime" through "5-prime" (**FIG. E9-4**).

The sugar of a nucleotide has two "ends" that can be involved in synthesizing the sugar-phosphate backbone of a DNA strand: a 3' end, which has a free —OH (hydroxyl) group attached to the 3' carbon, and a 5' end, which has a phosphate group attached to the 5' carbon. When a DNA strand is synthesized, the phosphate of one nucleotide bonds with the hydroxyl group of the next nucleotide (**FIG. E9-5**).

This, of course, still leaves a free hydroxyl group on the 3' carbon of one nucleotide, and a free phosphate group on the 5' carbon of the other nucleotide. This pattern continues, no matter how many nucleotides are strung together.

The sugar-phosphate backbones of the two strands of a double helix are antiparallel. Therefore, at one end of a double helix, one strand has a free sugar group, or 3' end, while the other strand has a free phosphate group, or 5' end. At the other end of the double helix, the strand ends are reversed (**FIG. E9-6**).

DNA REPLICATION

DNA replication involves three major actions (**FIG. E9-7**). First, the DNA double helix must be opened up so that the base sequence can be "read." Then new DNA strands with base sequences complementary to the two original strands must be synthesized. In eukaryotic cells, these new DNA strands are synthesized in fairly short pieces. Therefore, the third step in DNA replication is to stitch the pieces together to form a continuous strand of DNA. Each step is carried out by a distinct set of enzymes.

DNA Helicase Separates the Parental DNA Strands Acting in concert with several other enzymes, *DNA helicase*

("an enzyme that breaks apart the double helix") breaks the hydrogen bonds between complementary base pairs that hold the two parental DNA strands together. This separates and unwinds the parental double helix, forming a replication "bubble" (Fig. E9-7a, b). Within the replication bubble, the nucleotide bases of the parental DNA strands are no longer paired with one another. Each replication bubble contains two replication "forks" where the two parental DNA strands have not yet been unwound.

DNA Polymerase Synthesizes New DNA Strands Replication bubbles are essential because they allow a second enzyme, *DNA polymerase* ("an enzyme that makes a DNA polymer"), to gain access to the bases of each DNA strand (Fig. E9-7c). At each replication fork, a complex of DNA polymerase and other proteins binds to *each parental strand*. Therefore, there will be two DNA polymerase complexes, one on each parental strand. DNA polymerase recognizes an unpaired base in the parental strand and matches it up with a complementary base in a free nucleotide. For example, DNA polymerase pairs up an exposed adenine base in the parental strand with a thymine base in a free nucleotide. Then, DNA polymerase catalyzes the formation of new covalent bonds, linking the phosphate of the incoming free nucleotide (the 5' end) to the sugar of the most recently added nucleotide (the 3' end) of the growing daughter strand. In this way, DNA polymerase synthesizes the sugar-phosphate backbone of the daughter strand.

DNA polymerase always moves away from the 3' end of a parental DNA strand (this is the end with a free sugar group) toward the 5' end (with a free phosphate group); new nucleotides are always added to the 3' end of the daughter strand. In other words, DNA polymerase moves 3' → 5' on a parental strand, and simultaneously moves 5' → 3' on the daughter strand. Finally, because the two strands of the parental DNA double helix are oriented in opposite directions,

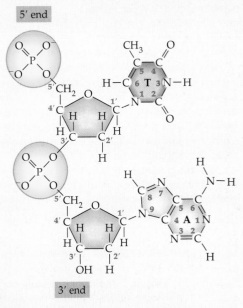

FIGURE E9-5 Numbering of carbon atoms in a dinucleotides

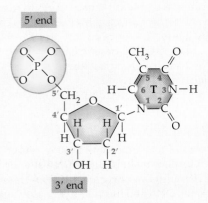

FIGURE E9-4 Numbering of carbon atoms in a nucleotide

the DNA polymerase molecules move in opposite directions on the two parental strands (Fig. E9-7c).

Why make replication bubbles, rather than simply starting at one end of a double helix and letting one DNA polymerase molecule copy the DNA in one continuous piece all the way to the other end? Well, eukaryotic chromosomes are *very* long: human chromosomes range from "only" 23 million bases for the relatively tiny Y chromosome, to 246 million bases for chromosome 1. Eukaryotic DNA is copied at a rate of about 50 nucleotides per second; this seems pretty fast, but it would take about 5 to 57 days to copy human chromosomes in one continuous piece. To replicate an entire chromosome in a reasonable time, many DNA helicase enzymes open up many replication bubbles, allowing many DNA polymerase enzymes to copy the parental strands in fairly small pieces. The bubbles grow as DNA replication progresses, and they merge when they contact one another.

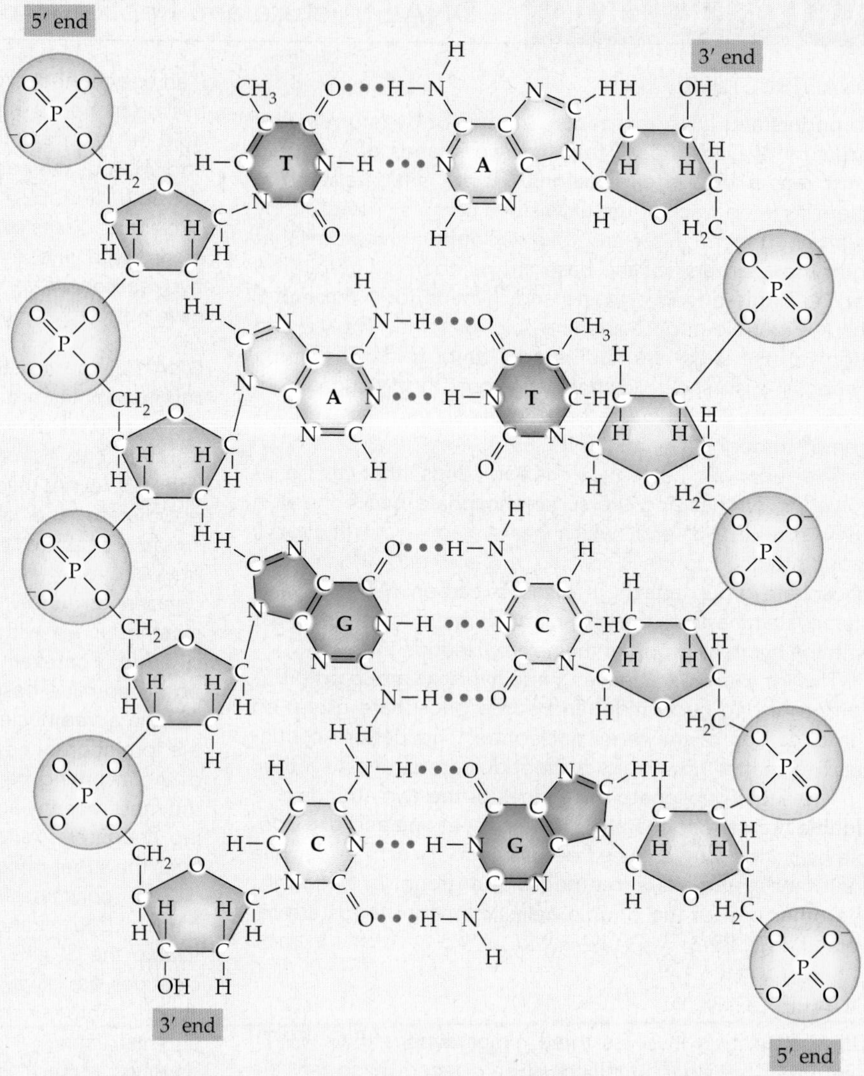

Segments of DNA Are Joined Together by DNA Ligase Now picture DNA helicase and DNA polymerase working together (Fig. E9-7d). DNA helicase "lands" on the double helix and moves along, unwinding the double helix and separating the strands. Because the two DNA strands run in opposite directions, as a DNA helicase enzyme moves toward the 5′ end of one parental strand, it is simultaneously moving toward the 3′ end of the other parental strand. Now visualize two DNA polymerases "landing" on the two separated strands of DNA. One DNA polymerase (call it polymerase #1) can follow behind the helicase toward the 5′ end of the parental strand and can synthesize a continuous, complete daughter DNA strand. This continuous daughter DNA strand is called the *leading strand*. On the other parental strand, however, DNA polymerase #2 moves *away from* the helicase, and therefore can synthesize only part of a new DNA strand, called the *lagging strand*. As the helicase continues to unwind more of the double helix, additional DNA polymerases (#3, #4, and so on) must land on this strand and synthesize more pieces of DNA.

FIGURE E9-6 **The two strands of a DNA double helix are antiparallel**

In this way, multiple DNA polymerases synthesize pieces of DNA of varying lengths. Each chromosome may form hundreds of replication bubbles. Within each bubble, there will be one leading strand, tens to hundreds of thousands of nucleotide pairs long, and dozens to thousands of lagging strands, each perhaps 100 to 200 nucleotide pairs long. Therefore, a cell might synthesize millions of pieces of DNA while replicating a single chromosome. How are all of these pieces sewn together? This is the job of the third major enzyme, **DNA ligase** ("an enzyme that ties DNA together"; Fig. E9-7e). Many DNA ligase enzymes stitch the fragments of DNA together until each daughter strand consists of one long, continuous DNA polymer.

FIGURE E9-7 **DNA replication**
(a) DNA helicase enzymes separate the parental strands of a chromosome to form replication bubbles. (b) Each replication bubble consists of two replication forks, with "unwound" DNA strands between the forks. (c) DNA polymerase synthesizes new pieces of DNA. (d) DNA helicase and DNA polymerase move along a replication bubble. (e) DNA ligase joins the small DNA segments into a single daughter strand. QUESTION During synthesis, why doesn't DNA polymerase move away from the replication fork on both strands?

160

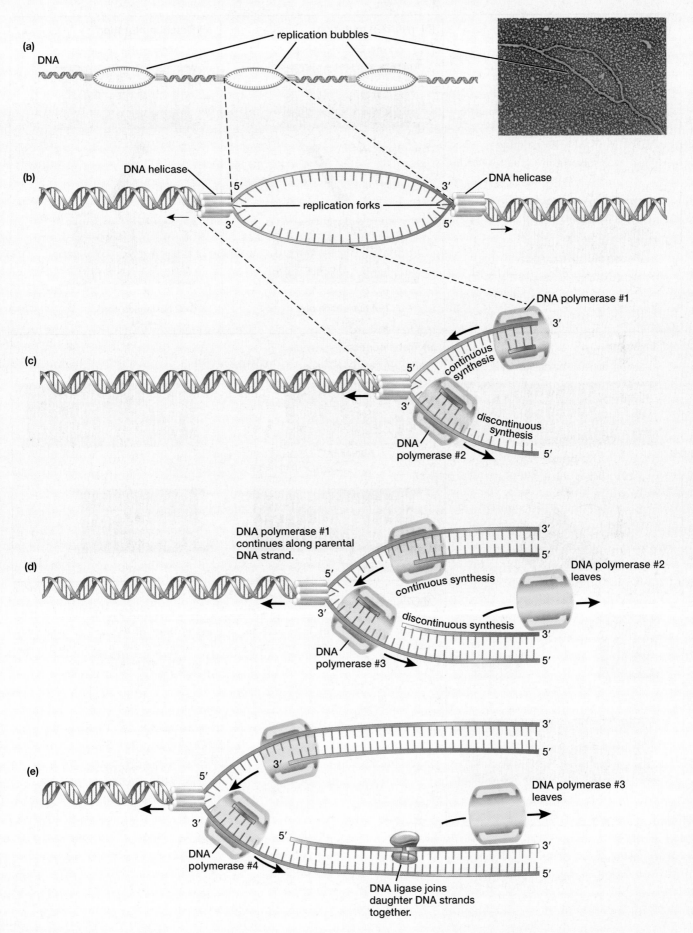

(a)

DNA

replication bubbles

(b)

DNA helicase

5′

replication forks

3′

DNA helicase

3′

5′

DNA polymerase #1

3′

(c)

5′

continuous
synthesis

3′

discontinuous
synthesis

DNA
polymerase #2

5′

(d)

DNA polymerase #1
continues along parental
DNA strand.

3′

5′

5′

continuous synthesis

DNA polymerase #2
leaves

3′

discontinuous synthesis

DNA
polymerase #3

3′

5′

(e)

3′

5′

5′

3′

DNA
polymerase #4

DNA polymerase #3
leaves

3′

5′

DNA ligase joins
daughter DNA strands
together.

161

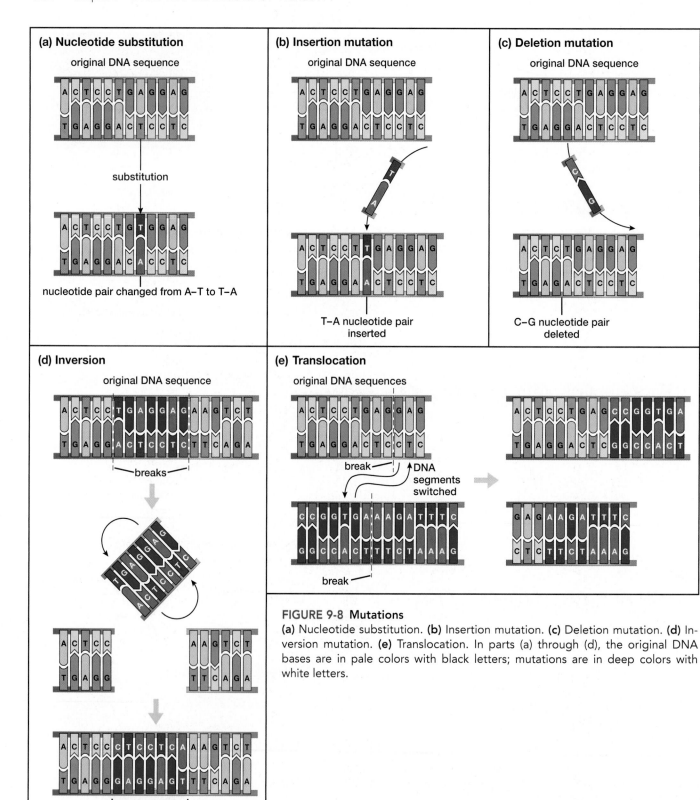

FIGURE 9-8 Mutations
(a) Nucleotide substitution. **(b)** Insertion mutation. **(c)** Deletion mutation. **(d)** Inversion mutation. **(e)** Translocation. In parts (a) through (d), the original DNA bases are in pale colors with black letters; mutations are in deep colors with white letters.

Accurate Replication and Proofreading Produce Almost Error-Free DNA

The specificity of hydrogen bonding between complementary base pairs makes DNA replication highly accurate. Nevertheless, DNA replication isn't perfect. DNA polymerase matches bases incorrectly about once in every 1000 to 100,000 base pairs, partly because replication is so fast (from about 50 nucleotides per second in humans to 1000 per second in some bacteria). However, completed DNA strands contain only about one mistake in every 100 million to 1 billion base pairs (in humans, usually less than one per chromosome per replication). This phenomenally low error rate is ensured by a variety of DNA repair enzymes that proofread each daughter strand during and after its synthesis. For example, some forms of DNA polymerase recognize a base-pairing mistake as it is made. This type of DNA polymerase will pause, fix the mistake, and then continue synthesizing more DNA.

Mistakes Do Happen

Despite this amazing accuracy, neither we nor any other life-forms have perfect, error-free DNA. In addition to rare mistakes made during normal DNA replication, a variety of environmental conditions can damage DNA. For example, certain chemicals (such as various components of cigarette smoke) and some types of radiation (such as X-rays and ultraviolet rays in sunlight) increase the frequency of base-pairing errors during replication, or even induce changes in DNA composition between replications. Most of these changes in DNA sequence are fixed by a variety of repair enzymes in the cell. However, some inevitably remain.

Mutations Range from Changes in Single Nucleotide Pairs to Movements of Large Pieces of Chromosomes

During replication, a pair of bases is occasionally mismatched. Usually, repair enzymes recognize the mismatch, cut out the incorrect nucleotide, and replace it with a nucleotide that bears a complementary base. Sometimes, however, the enzymes replace the *correct* nucleotide instead of the incorrect one. The resulting base pair is complementary, but it is incorrect. These **nucleotide substitutions** are also called **point mutations**, because individual nucleotides in the DNA sequence are changed (FIG. 9-8a). An **insertion mutation** occurs when one or more nucleotide pairs are inserted into the DNA double helix (FIG. 9-8b). A **deletion mutation** occurs when one or more nucleotide pairs are removed from the double helix (FIG. 9-8c).

Pieces of chromosomes ranging in size from a single nucleotide pair to massive pieces of DNA are occasionally rearranged. An **inversion** occurs when a piece of DNA is cut out of a chromosome, turned around, and reinserted into the gap (FIG. 9-8d). Finally, a **translocation** results when a chunk of DNA, often very large, is removed from one chromosome and attached to another one (FIG. 9-8e).

Mutations May Have Varying Effects on Function

Mutations are often harmful, much as randomly changing words in the middle of Shakespeare's *Hamlet* would probably interrupt the flow of the play. If they are really damaging, a cell or an organism inheriting such a mutation may quickly die. Some mutations, however, have no effect or, in very rare instances, are even beneficial, as you will learn in Chapter 10. Mutations that are beneficial, at least in certain environments, may be favored by natural selection, and are the basis for the evolution of life on Earth (see Unit Three).

CASE STUDY REVISITED MUSCLES, MUTATIONS, AND MYOSTATIN

Belgian Blue cattle have a *deletion mutation* in their myostatin gene. The result is that their cells stop synthesizing the myostatin protein about halfway through (in Chapter 10, we'll explain why some mutations cause short, or truncated, proteins to be synthesized). No one knows how this particular mutation arose.

Humans have myostatin, too; not surprisingly, mutations can occur in the human myostatin gene. As you probably know, a child inherits two copies of most genes, one from each parent. Recently, a child was born in Germany who inherited a *substitution mutation* in his myostatin gene, from both parents. This particular substitution mutation also results in short, inactive, myostatin proteins. Even at 7 months, this boy had well-developed calf,

thigh, and buttock muscles (FIG. 9-9). At 4 years, he could hold a 7-pound (3.18-kilogram) dumbbell in each hand, with his arms fully extended horizontally out to his sides (try it—it's not that easy for many adults).

Consider This Mutations may be neutral, harmful, or beneficial. Into which category do myostatin mutations fall? Well, Belgian Blue cattle are born so muscular, and consequently so large, that they usually must be delivered by caesarian section. A few become so muscular that their muscles get in the way, and they can hardly walk. So far, the German boy appears to be healthy. What will happen as he grows up? Will he become a super-athlete, or suffer debilitating health effects as he ages, or both? Only time will tell.

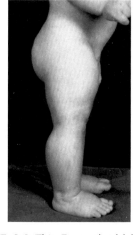

FIGURE 9-9 This 7-month-old boy has remarkable muscle development in his legs, caused by a mutation in his myostatin gene.

CHAPTER REVIEW

SUMMARY OF KEY CONCEPTS

9.1 How Did Scientists Discover That Genes Are Made of DNA?

By the turn of the century, scientists knew that genes must be made of either protein or DNA. Studies by Griffith showed that genes can be transferred from one bacterial strain into another. This transfer could transform the bacterial strain from harmless to deadly. Avery, MacLeod, and McCarty showed that DNA was the molecule that could transform bacteria. Thus, genes must be made of DNA.

9.2 What Is the Structure of DNA?

DNA is composed of subunits called nucleotides, linked together into long strands. Each nucleotide consists of a phosphate group, the five-carbon sugar deoxyribose, and a nitrogen-containing base. Four types of bases occur in DNA: adenine, guanine, thymine, and cytosine. Within DNA, two nucleotide strands wind about one another to form a double helix. Within each strand, the sugar of one nucleotide is linked to the phosphate of the next nucleotide, forming a sugar-phosphate "backbone" on each side of the double helix. The nucleotide bases of each strand pair up in the middle of the helix, held together by hydrogen bonds. Only specific pairs of bases, called complementary base pairs, can bond together in the helix: adenine bonds with thymine, and guanine bonds with cytosine.

Web Tutorial 9.1 DNA Structure

9.3 How Does DNA Encode Information?

Information in DNA is encoded in the sequence of its nucleotides. Just as a language can form thousands of words from a small number of letters by varying the sequence and number of letters in each word, so too DNA can encode large amounts of information with various sequences and numbers of nucleotides in different genes.

9.4 How Does DNA Replication Ensure Genetic Constancy During Cell Division?

When cells reproduce, they must replicate their DNA so that each daughter cell receives all of the original genetic information. During DNA replication, enzymes unwind the two parental DNA strands. The enzyme DNA polymerase binds to each parental DNA strand, selects free nucleotides with complementary bases, and links the nucleotides together to form new DNA strands. The sequence of nucleotides in each newly formed strand is complementary to the sequence of a parental strand. Replication is semiconservative because, when DNA replication is complete, both new DNA double helices consist of one parental DNA strand and one newly synthesized, complementary strand. The two new DNA double helices are therefore duplicates of the parental DNA double helix.

Web Tutorial 9.2 DNA Replication

9.5 How Do Mutations Occur?

Mutations are changes in nucleotide sequence in DNA. DNA polymerase and other repair enzymes "proofread" the DNA, minimizing the number of mistakes during replication, but mistakes do occur. Other nucleotide changes occur as a result of radiation and damage from certain chemicals. Mutations include substitutions, insertions, deletions, inversions, and translocations. Most mutations are harmful or neutral, but some are beneficial and may be favored by natural selection.

KEY TERMS

adenine (A)   *page 154*
bacteriophage   *page 152*
bases   *page 154*
chromosome   *page 150*
complementary base pairs
 page 156
cytosine (C)   *page 154*
deletion mutation   *page 163*

DNA   *page 151*
DNA helicase   *page 158*
DNA ligase   *page 160*
DNA polymerase   *page 158*
DNA replication   *page 157*
double helix   *page 155*
free nucleotides   *page 158*
gene   *page 150*

guanine (G)   *page 154*
insertion mutation   *page 163*
inversion   *page 163*
mutation   *page 158*
nucleotide substitution
 page 163
nucleotides   *page 154*
point mutation   *page 163*

semiconservative replication
 page 158
sugar-phosphate backbone
 page 154
thymine (T)   *page 154*
translocation   *page 163*

THINKING THROUGH THE CONCEPTS

1. Draw the general structure of a nucleotide. Which parts are identical in all nucleotides, and which can vary?

2. Name the four types of nitrogen-containing bases found in DNA.

3. Which bases are complementary to one another? How are they held together in the double helix of DNA?

4. Describe the structure of DNA. Where are the bases, sugars, and phosphates in the structure?

5. Describe the process of DNA replication.

6. How do mutations occur? Describe the principal types of mutations.

APPLYING THE CONCEPTS

1. As you learned in "Scientific Inquiry: The Discovery of the Double Helix," scientists in different laboratories often compete with one another to make new discoveries. Do you think this competition helps promote scientific discoveries? Sometimes researchers in different laboratories collaborate with one another. What advantages does collaboration offer over competition? What factors might provide barriers to collaboration and lead to competition?

2. Genetic information is encoded in the sequence of nucleotides in DNA. Let's suppose that the nucleotide sequence on one strand of a double helix encodes the information needed to synthesize a hemoglobin molecule. Do you think that the sequence of nucleotides on the other strand of the double helix also encodes useful information? Why? (An analogy might help. Suppose that English were a "complementary language," with letters at opposite ends of the alphabet complementary to one another; that is, A is complementary to Z, B to Y, C to X, etc. Would a sentence composed of letters complementary to "To be or not to be?" make sense?) Finally, why do you think DNA is double-stranded?

3. Today, scientific advances are being made at an astounding rate, and nowhere is this more evident than in our understanding of the biology of heredity. Using DNA as a starting point, do you believe that there are limits to the knowledge that people should acquire? Defend your answer.

FOR MORE INFORMATION

Crick, F. *What Mad Pursuit: A Personal View of Scientific Discovery*. New York: Basic Books, 1998. Another view of the race to determine the structure of DNA, by Francis Crick himself.

Gibbs, W. W. "Peeking and Poking at DNA." *Scientific American (Explorations)*, March 31, 1997. An update of techniques for studying the DNA molecule, such as atomic force microscopy.

Judson, H. F. *The Eighth Day of Creation*. Cold Spring Harbor, NY: Cold Spring Harbor Laboratory Press, 1993. A very readable historical perspective on the development of genetics.

Radman, M., and Wagner, R. "The High Fidelity of DNA Duplication." *Scientific American*, August 1988. Faithful duplication of chromosomes requires both reasonably accurate initial replication of DNA sequences and final proofreading.

Rennie, J. "DNA's New Twists." *Scientific American*, March 1993. A reprise of information on DNA structure and function.

Watson, J. D. *The Double Helix*. New York: Atheneum, 1968. If you still believe the Hollywood images that scientists are either maniacs or cold-blooded, logical machines, be sure to read this book. Although hardly models for the behavior of future scientists, Watson and Crick are certainly human enough!

Weinberg, R. "How Cancer Arises." *Scientific American*, September 1996. An overview of the molecular basis of cancer: mutations in DNA.

Wheelwright, J. "Bad Genes, Good Drugs." *Discover*, April 2002. The human genome project provides insights into genetic disorders and possible treatments.

Gene Expression and Regulation

Many of the differences in the body structures of males and females can ultimately be traced to the activity of a single gene.

CASE STUDY VIVE LA DIFFÉRENCE!

MALE AND FEMALE—so alike, yet so different. The physical differences between men and women are pretty obvious, but for a very long time, biologists had only the vaguest ideas about the genetic bases of those differences. It's been less than a century since Theophilus Painter discovered the Y chromosome. Several more decades passed before it was generally accepted that the Y chromosome usually determines maleness in humans and other mammals. But how?

One hypothesis might be that genes on the Y chromosome encode male genitalia, so we would predict that anyone with a Y chromosome will have testes and a penis. But males also have all of the other chromosomes that females have (although males have only one X chromosome, rather than the two that females have). Why, then, don't boys develop both male *and* female genitalia? Furthermore, most of the genes needed to produce male sexual characteristics, including the genitalia, are *not* on the Y chromosome. Girls therefore possess these genes, so why don't girls develop both female and male genitalia?

In boys, the action of a single gene located on the Y chromosome activates the male developmental pathway and deactivates the female developmental pathway. Without this gene we would all be physically female. How can a single gene determine something as complex as the sex of a human being? In this chapter, we will examine the flow of information from an organism's genes to its physical characteristics. Just as information in a book remains hidden until someone opens its cover and reads the text, so too the information in genes may, or may not, be used in different organisms, in the various cells of an individual organism, and at different times during the organism's life.

10.1 HOW ARE GENES AND PROTEINS RELATED?

Information, by itself, doesn't *do* anything. For example, a blueprint may describe the structure of a house in great detail, but unless that information is translated into action, no house will ever be built. Likewise, although the base sequence of DNA, the "molecular blueprint" of every cell, contains an incredible amount of information, DNA cannot carry out any action on its own. So how does DNA determine whether you are male or female, or have brown or blue eyes?

Proteins are a cell's "molecular workers." Every cell contains a particular set of proteins. The activities of these proteins control the cell's shape, function, and reproduction, as well as the synthesis of lipids, carbohydrates, and nucleic acids. Therefore, there must be a flow of information from the DNA of a cell's genes to the proteins that actually carry out the cell's functions.

Most Genes Contain the Information for the Synthesis of a Single Protein

Cells synthesize molecules in a series of linked steps called *biochemical pathways*. Each step in a biochemical pathway is catalyzed by an enzyme. (Recall from Chapters 3 and 6 that enzymes are proteins that catalyze a specific chemical reaction.) Within a biochemical pathway, the product produced by one enzyme becomes the substrate of the next enzyme in the pathway, like a molecular assembly line (look back to Fig. 6-13). How do genes encode the information needed to produce these pathways?

The first hint came from children who are born with a defect in one or more biochemical pathways. For example, defects in the metabolism of two amino acids, phenylalanine and tyrosine, can cause albinism (no pigmentation in skin or hair; see Chapter 12), some types of mental retardation, or phenylketonuria (PKU). In the early 1900s, an English physician named Archibald Garrod studied the inheritance of these *inborn errors of metabolism*. He hypothesized that

(a) Growth characteristics of normal and mutant *Neurospora* on simple medium with different supplements show that defects in a single gene lead to defects in a single enzyme.

		Supplements Added to Medium			Conclusions
		none	ornithine	citrulline	arginine	
Normal *Neurospora*						Normal *Neurospora* can synthesize arginine, citrulline, and ornithine.
Mutants with single gene defect	A					Mutant A grows only if arginine is added. It cannot synthesize arginine because it has a defect in enzyme 2; gene *A* is needed for synthesis of arginine.
	B					Mutant B grows if either arginine or citrulline are added. It cannot synthesize arginine because it has a defect in enzyme 1. Gene *B* is needed for synthesis of citrulline.

(b) The biochemical pathway for synthesis of the amino acid arginine involves two steps, each catalyzed by a different enzyme.

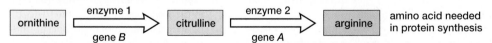

ornithine → (enzyme 1 / gene *B*) → citrulline → (enzyme 2 / gene *A*) → arginine (amino acid needed in protein synthesis)

FIGURE 10-1 Beadle and Tatum's experiments with *Neurospora* mutants
QUESTION What result would you expect for a mutant that lacks an enzyme needed to produce ornithine?

(1) each inborn error of metabolism is caused by a defective version of a specific enzyme, (2) each defective enzyme is caused by a defective version of a single gene, and (3) therefore, at least some genes must encode the information needed for the synthesis of enzymes.

Given the technology of his time, and the obvious limitations to human genetic studies, Garrod couldn't definitively test his hypotheses, and they were largely ignored. However, in the early 1940s, geneticists George Beadle and Edward Tatum used the biochemical pathways of a common bread mold, *Neurospora crassa*, to show that Garrod was correct.

Although we normally encounter *Neurospora* growing on stale bread, it can survive on a much simpler diet. All it needs is an energy source such as sugar, a few minerals, and vitamin B_6. Therefore, *Neurospora* manufactures the enzymes needed to make virtually all of its own organic molecules, including the amino acids. (In contrast, we humans cannot synthesize many vitamins or 9 of the 20 common amino acids; we must obtain these from our food.) Individual *Neurospora*, like any organism, may have mutations in some of its genes. Beadle and Tatum used mutant *Neurospora* to test the hypothesis that many of an organism's genes encode the information needed to synthesize enzymes. If this hypothesis is true, a mutation in a specific gene might disrupt the synthesis of a specific enzyme. Without this enzyme, one of the mold's biochemical pathways wouldn't function properly. The mold would be unable to synthesize some of the organic molecules, such as certain amino acids, that it needs to survive. These mutant *Neurospora* would be able to grow on a simple medium of sugar, minerals, and vitamin B_6 only if the missing organic molecules were added to the medium.

Beadle and Tatum induced mutations in *Neurospora* by exposing them to X-rays. Some of these mutants could grow if the simple medium were supplemented with the amino acid arginine. Arginine is synthesized from citrulline, which in turn is synthesized from ornithine (FIG. 10-1b). Mutant A could grow only if supplemented with arginine, but not if supplemented with citrulline or ornithine (FIG. 10-1a). Therefore, this strain had a defect in the enzyme that converts citrulline to arginine. Mutant B could grow if supplemented with either arginine or citrulline, but not with ornithine (see Fig. 10-1a). This mutant had a defect in the enzyme that converts ornithine to citrulline. Because a mutation in a single gene affected only a single enzyme within a single biochemical pathway, Beadle and Tatum concluded that one gene encodes the information for one enzyme. The importance of this observation was recognized in 1958 with a Nobel Prize, which was shared by Joshua Lederberg, one of Tatum's students.

Almost all enzymes are proteins, but many of the proteins in a cell are not enzymes. For example, keratin is a structural protein in hair and nails, but it does not catalyze any chemical reactions. In addition, many enzymes are composed of more than one protein subunit. For example, DNA polymerase is composed of more than a dozen proteins. Thus, the "one gene, one enzyme" relationship proposed by Beadle and Tatum was later clarified to "one gene, one protein." (As you know from Chapter 3, a protein is a chain of amino acids joined by peptide bonds. Depending on the length of the chain, proteins may be called peptides [short chains] or polypeptides [long chains]. In this text, we usually call any chain of amino acids, regardless of length, a protein.) There are exceptions to the "one gene, one protein" rule, including several in which the final product of a gene isn't protein, but a nucleic acid—called *ribonucleic acid*—described in the next section. Nevertheless, as a generalization, most genes encode the information for the amino acid sequence of a protein.

DNA Provides Instructions for Protein Synthesis via RNA Intermediaries

The DNA of a eukaryotic cell is housed in the nucleus, but protein synthesis occurs on ribosomes in the cytoplasm (see Chapter 5). Therefore, DNA cannot directly guide protein synthesis. There must be an intermediary, a molecule that carries the information from DNA in the nucleus to the ribosomes in the cytoplasm. This molecule is **ribonucleic acid**, or **RNA**.

RNA is similar to DNA but differs structurally in three respects: (1) RNA is normally single-stranded; (2) RNA has the sugar ribose (instead of deoxyribose) in its backbone; and (3) RNA has the base uracil instead of the base thymine found in DNA (Table 10-1).

Table 10-1	A Comparison of DNA and RNA	
	DNA	**RNA**
Strands	2	1
Sugar	Deoxyribose	Ribose
Types of Bases	adenine (A), thymine (T) cytosine (C), guanine (G)	adenine (A), uracil (U) cytosine (C), guanine (G)
Base Pairs	DNA–DNA A–T T–A C–G G–C	RNA–DNA RNA–RNA A–T A–U U–A U–A C–G C–G G–C G–C
Function	Contains genes; sequence of bases in most genes determines the amino acid sequence of a protein	**Messenger RNA (mRNA):** carries the code for a protein-coding gene from DNA to ribosomes **Ribosomal RNA (rRNA):** combines with proteins to form ribosomes, the structures that link amino acids to form a protein **Transfer RNA (tRNA):** carries amino acids to the ribosomes

(a) Messenger RNA (mRNA)

| A | U | G | U | G | C | G | A | G | U | U | A | U | G | G |

The base sequence of mRNA carries the information for the amino acid sequence of a protein.

(b) Ribosome: contains ribosomal RNA (rRNA)

large subunit

catalytic site

1   2

tRNA/amino acid binding sites

small subunit

rRNA combines with proteins to form ribosomes. The small subunit binds mRNA. The large subunit binds tRNA and catalyzes peptide bond formation between amino acids during protein synthesis.

(c) Transfer RNA (tRNA)

tyr

attached amino acid

anticodon

Each tRNA carries a specific amino acid to a ribosome during protein synthesis. The anticodon of tRNA pairs with a codon of mRNA, ensuring that the correct amino acid is incorporated into the protein.

FIGURE 10-2 Cells synthesize three major types of RNA

DNA codes for the synthesis of three major types of RNA: **messenger RNA (mRNA)**, **ribosomal RNA (rRNA)**, and **transfer RNA (tRNA)** (FIG. 10-2). All these RNA molecules are involved in converting the nucleotide sequence of genes into the amino acid sequence of proteins. We will examine their functions in more detail shortly.

Overview: Genetic Information Is Transcribed into RNA and Then Translated into Protein

Information from DNA is used to direct the synthesis of proteins in a two-step process (**FIG. 10-3** and Table 10-2):

1. During RNA synthesis, or **transcription** (see Fig. 10-3a), the information contained in the DNA of a specific gene is copied into messenger RNA (mRNA), transfer RNA (tRNA), or ribosomal RNA (rRNA). Thus, a gene is a segment of DNA that can be copied, or transcribed, into RNA. Transcription is catalyzed by an enzyme, RNA polymerase. In eukaryotic cells, transcription occurs in the nucleus.

2. As we will see shortly, the nucleotide sequence of mRNA encodes the amino acid sequence of a protein. During protein synthesis, or **translation** (see Fig. 10-3b), this mRNA nucleotide sequence is decoded. Ribosomal RNA combines with dozens of proteins to form a complex structure called a **ribosome**. Transfer RNA molecules bring individual amino acids to the ribosome. Messenger RNA binds to the ribosome, where base pairing between mRNA and tRNA converts the nucleotide sequence of mRNA into the amino acid sequence of the protein. In eukaryotic cells, ribosomes are found in the cytoplasm, so translation occurs there as well.

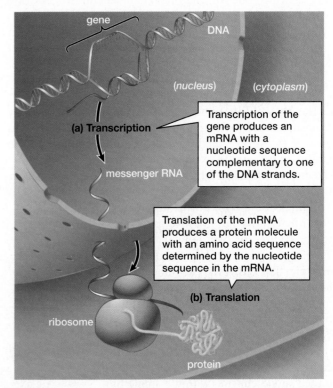

gene

DNA

(nucleus)   (cytoplasm)

(a) Transcription

Transcription of the gene produces an mRNA with a nucleotide sequence complementary to one of the DNA strands.

messenger RNA

Translation of the mRNA produces a protein molecule with an amino acid sequence determined by the nucleotide sequence in the mRNA.

(b) Translation

ribosome

protein

FIGURE 10-3 Genetic information flows from DNA to RNA to protein
(a) During transcription, the nucleotide sequence in a gene specifies the nucleotide sequence in a complementary RNA molecule. For protein-encoding genes, the product is an mRNA molecule that exits from the nucleus and enters the cytoplasm. (b) During translation, the sequence in an mRNA molecule specifies the amino acid sequence in a protein.

Table 10-2 Processes Involved in the Use and Inheritance of Genetic Information

Process	Information for Process	Product	Major Enzyme or Structure Involved in Process	Type of Base Pairing Required
Transcription (synthesis of RNA)	Short segment of one DNA strand	One RNA molecule (mRNA, tRNA, rRNA)	RNA polymerase	DNA with RNA: DNA bases pair with RNA bases in new RNA molecule.
Translation (synthesis of protein)	mRNA	One protein molecule	Ribosome (also requires tRNA)	mRNA with tRNA: Codon in mRNA forms base pairs with anticodon in tRNA.
Replication (synthesis of DNA; occurs only before cells divide)	Entire length of both DNA strands	Two DNA double helices (each with one old and one new strand)	DNA polymerase	DNA with DNA: DNA bases of each parental strand pair with DNA bases in the newly synthesized strands.

It's easy to confuse the terms *transcription* and *translation*. Comparing their common English meanings with their biological meanings may help you understand the difference. In everyday English, to *transcribe* means to make a written copy of something, almost always in the same language. In an American courtroom, for example, verbal testimony is transcribed into a written copy, and both the testimony of the witnesses and the transcriptions are in English. In biology, *transcription* is the process of copying information from DNA to RNA using the common "language" of nucleotides. In contrast, the common English meaning of *translation* is to convert words from one language to a different language. Similarly, in biology, *translation* means to convert information from the "nucleotide language" of RNA to the "amino acid language" of proteins.

The Genetic Code Uses Three Bases to Specify an Amino Acid

We will investigate both transcription and translation in more detail in sections 10.2 and 10.3. First, however, let's see how geneticists broke the language barrier: namely, how does the language of nucleotide sequences in DNA and messenger RNA translate into the language of amino acid sequences in proteins? This translation relies on a "dictionary" called the genetic code.

The **genetic code** translates the sequence of bases in nucleic acids into the sequence of amino acids in proteins. But which combinations of bases code for which amino acids? Both DNA and RNA contain four different bases: A, T (or U in RNA), G, and C (see Table 10-1). However, proteins are made of 20 different amino acids. Therefore, one base cannot code for one amino acid, because there are simply not enough different types of bases. The genetic code must rely on a short sequence of bases to encode each amino acid. If a sequence of two bases codes for an amino acid, there would be 16 possible combinations, which still isn't enough to code for all 20 amino acids. A three-base sequence, however, gives 64 possible combinations of bases, which is more than enough. Under the assumption that nature operates as eco-

nomically as possible, biologists hypothesized that the genetic code must be triplet; that is, three bases specify a single amino acid. In 1961, Francis Crick and three coworkers demonstrated that this hypothesis is correct.

For any language to be understood, its users must know what the words mean, where words start and stop, and where sentences begin and end. To decipher the "words" of the genetic code, researchers ground up bacteria and isolated the components needed to synthesize proteins. To this mixture, they added artificial mRNA, allowing them to control which "words" were to be translated. They could then see which amino acids were incorporated into the resulting proteins. For example, an mRNA strand composed entirely of uracil (UUUUUUUU ...) directed the mixture to synthesize a protein composed solely of the amino acid phenylalanine. Therefore, the triplet UUU must specify phenylalanine. Because the genetic code was deciphered by using these artificial mRNAs, it is usually written in terms of the base triplets in mRNA (rather than in DNA) that code for each amino acid (Table 10-3). These mRNA triplets are called **codons**.

What about punctuation? Given that one mRNA molecule may contain hundreds or even thousands of bases, how does a cell recognize where codons start and stop, and where the code for an entire protein starts and stops? All proteins originally begin with the same amino acid, methionine (though it may be removed after the protein is synthesized). Methionine is specified by the codon AUG, which is known as the **start codon**. Three codons—UAG, UAA, and UGA—are **stop codons**. When the ribosome encounters a stop codon, it releases both the newly synthesized protein and the mRNA. Because all codons consist of three bases, and the beginning and end of a protein are specified, then punctuation ("spaces") between codons is unnecessary. Why? Consider what would happen if English used only three-letter words: a sentence such as THEDOGSAWTHECAT would be perfectly understandable, even without spaces between the words.

Because the genetic code has three stop codons, 61 nucleotide triplets remain to specify only 20 amino acids.

Table 10-3 The Genetic Code (Codons of mRNA)

First Base	Second Base — U	Second Base — C	Second Base — A	Second Base — G	Third Base
U	UUU Phenylalanine (Phe) UUC Phenylalanine UUA Leucine (Leu) UUG Leucine	UCU Serine (Ser) UCC Serine UCA Serine UCG Serine	UAU Tyrosine (Tyr) UAC Tyrosine UAA Stop UAG Stop	UGU Cysteine (Cys) UGC Cysteine UGA Stop UGG Tryptophan (Trp)	U C A G
C	CUU Leucine CUC Leucine CUA Leucine CUG Leucine	CCU Proline (Pro) CCC Proline CCA Proline CCG Proline	CAU Histidine (His) CAC Histidine CAA Glutamine (Gln) CAG Glutamine	CGU Arginine (Arg) CGC Arginine CGA Arginine CGG Arginine	U C A G
A	AUU Isoleucine (Ile) AUC Isoleucine AUA Isoleucine AUG Methionine (Met) Start	ACU Threonine (Thr) ACC Threonine ACA Threonine ACG Threonine	AAU Asparagine (Asp) AAC Asparagine AAA Lysine (Lys) AAG Lysine	AGU Serine (Ser) AGC Serine AGA Arginine (Arg) AGG Arginine	U C A G
G	GUU Valine (Val) GUC Valine GUA Valine GUG Valine	GCU Alanine (Ala) GCC Alanine GCA Alanine GCG Alanine	GAU Aspartic acid (Asp) GAC Aspartic acid GAA Glutamic acid (Glu) GAG Glutamic acid	GGU Glycine (Gly) GGC Glycine GGA Glycine GGG Glycine	U C A G

Therefore, most amino acids are specified by several different codons. For example, six different codons specify leucine (see Table 10-3), so whether UUA or CUG is present in the mRNA sequence, ribosomes add leucine to the growing amino acid chain. However, each codon specifies one, and only one, amino acid.

10.2 HOW IS INFORMATION IN A GENE TRANSCRIBED INTO RNA?

We can view transcription as a process consisting of (1) *initiation*, (2) *elongation*, and (3) *termination*. These three steps correspond to the three major parts of most genes in both eukaryotes and prokaryotes: (1) a *promoter* region at the beginning of the gene, where transcription is started, or initiated; (2) the "body" of the gene where elongation of the RNA strand occurs; and (3) a termination signal at the end of the gene, where RNA synthesis ceases, or terminates.

Initiation of Transcription Occurs When RNA Polymerase Binds to the Promoter of a Gene

The enzyme **RNA polymerase** synthesizes RNA. To initiate transcription, RNA polymerase must first locate the beginning of a gene. Near the beginning of every gene is an untranscribed segment of DNA called the **promoter**. In eukaryotic cells, a promoter consists of two main parts: (1) a short sequence of bases, often TATAAA, that binds RNA polymerase, and (2) one or more other sequences, often called transcription factor binding sites or response elements. When specific cellular proteins, appropriately called transcription factors, attach to one of these response elements, they enhance or suppress binding of RNA polymerase to the promoter, and consequently they enhance or

suppress transcription of the gene. We will return to the important topic of gene regulation in the last section of this chapter.

When RNA polymerase binds to the promoter region of a gene, the DNA double helix at the beginning of the gene unwinds and transcription begins (**FIG. 10-4a**).

Elongation Proceeds Until RNA Polymerase Reaches a Termination Signal

RNA polymerase then travels down one of the DNA strands, called the **template strand**, synthesizing a single strand of RNA with bases complementary to those in the DNA (**FIG. 10-4b**). Like DNA polymerase (see Chapter 9), RNA polymerase always travels along the DNA template strand starting at the 3′ end of a gene and moving toward the 5′ end. Base pairing between RNA and DNA is the same as between two strands of DNA, except that uracil in RNA pairs with adenine in DNA (see Table 10-1).

After about 10 nucleotides have been added to the growing RNA chain, the first nucleotides in the RNA molecule separate from the DNA template strand. This separation allows the two DNA strands to rewind into a double helix (**FIG. 10-4b,c**). Thus, as transcription continues to elongate the RNA molecule, one end of the RNA drifts away from the DNA; RNA polymerase keeps the other end temporarily attached to the DNA template strand (**FIG. 10-4c** and **FIG. 10-5**).

RNA polymerase continues along the template strand of the gene until it reaches a sequence of DNA bases known as the *termination signal*. At this point, RNA polymerase releases the completed RNA molecule and detaches from the DNA (**FIG. 10-4c, d**). The RNA polymerase is then free to bind to another promoter and synthesize another RNA molecule.

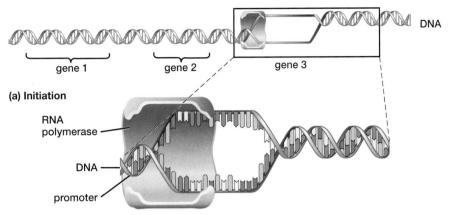

DNA

gene 1   gene 2   gene 3

(a) Initiation

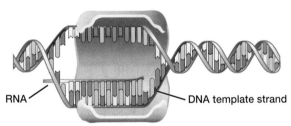

RNA polymerase

DNA

promoter

RNA polymerase binds to the promoter region of DNA near the beginning of a gene, separating the double helix near the promoter.

(b) Elongation

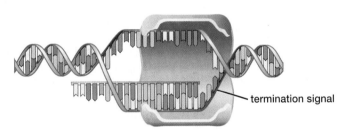

RNA

DNA template strand

RNA polymerase travels along the DNA template strand (blue), catalyzing the addition of ribose nucleotides into an RNA molecule (pink). The nucleotides in the RNA are complementary to the template strand of the DNA.

(c) Termination

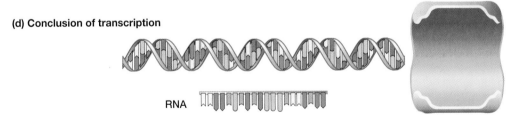

termination signal

At the end of a gene, RNA polymerase encounters a DNA sequence called a termination signal. RNA polymerase detaches from the DNA and releases the RNA molecule.

(d) Conclusion of transcription

RNA

After termination, the DNA completely rewinds into a double helix. The RNA molecule is free to move from the nucleus to the cytoplasm for translation, and RNA polymerase may move to another gene and begin transcription once again.

FIGURE 10-4 Transcription is the synthesis of RNA from instructions in DNA
A gene is a segment of a chromosome's DNA. One of the DNA strands will serve as the template for the synthesis of an RNA molecule with bases complementary to the bases in the DNA strand. QUESTION If the other DNA strand of this molecule were the template strand, in which direction would the RNA polymerase travel?

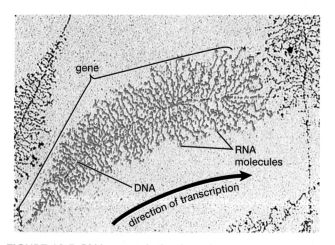

FIGURE 10-5 RNA transcription in action
This colorized electron micrograph shows the progress of RNA transcription in the egg of an African clawed toad. In each tree-like structure, the central "trunk" is DNA (blue) and the "branches" are RNA molecules (red). A series of RNA polymerase molecules (too small to be seen in this micrograph) are traveling down the DNA, synthesizing RNA as they go. The beginning of the gene is on the left. The short RNA molecules on the left have just begun to be synthesized; the long RNA molecules on the right are almost finished.

10.3 HOW IS THE BASE SEQUENCE OF A MESSENGER RNA MOLECULE TRANSLATED INTO PROTEIN?

As their names suggest, each type of RNA has a specific role in protein synthesis.

Messenger RNA Carries the Code for Protein Synthesis from DNA to Ribosomes

All RNA is produced by transcription of DNA, but only mRNA carries the code for the amino acid sequence of a protein. Prokaryotic and eukaryotic cells differ considerably in how they produce a functional mRNA molecule from the instructions in their DNA.

Messenger RNA Synthesis in Prokaryotes

Prokaryotic genes are typically compact: all the nucleotides of a gene code for the amino acids in a protein. What's more, most or all of the genes for a complete metabolic pathway sit end to end on the chromosome (**FIG. 10-6a**). Therefore, prokaryotic cells commonly transcribe a single, very long mRNA from a series of adjacent genes. Because prokaryotic cells do not have a nuclear membrane separating their DNA from the cytoplasm (see Chapter 5), transcription and translation are usually not separated, either in space or in time. In most cases, as an mRNA molecule begins to separate from the DNA during transcription, ribosomes immediately begin translating the mRNA into protein (**FIG. 10-6b**).

Messenger RNA Synthesis in Eukaryotes

In contrast, the DNA of eukaryotic cells is sequestered in the nucleus, while the ribosomes reside in the cytoplasm. Further, the organization of DNA in eukaryotes differs tremendously from the DNA of prokaryotes. In eukaryotes, the genes that encode the proteins needed for a biochemical pathway are not clustered together as they are in prokaryotes, but may be dispersed among several chromosomes. Further, each eukaryotic gene typically consists of two or more segments of DNA with nucleotide sequences that encode for a protein, interrupted by other nucleotide sequences that are not translated into protein. The coding segments are called **exons**, because they are **ex**pressed in protein, and the noncoding segments are called **introns**, because they are "**int**ragenic," meaning "within a gene" (**FIG. 10-7a**). Most eukaryotic genes have introns; in fact, the gene that codes for a type of connective tissue protein in chickens has about 50 introns!

Transcription of a eukaryotic gene produces a very long RNA strand, which starts before the first exon and ends after the last exon (**FIG. 10-7b**). More nucleotides are added at the beginning and end of the mRNA molecule, forming a "cap" and "tail." These nucleotides will help to move the mRNA through the nuclear envelope to the cytoplasm, to bind the mRNA to a ribosome, and to prevent cellular enzymes from breaking down the mRNA molecule before it can be translated. Finally, to convert this RNA molecule into true mRNA, enzymes in the nucleus precisely cut the RNA molecule apart at the junctions between introns and exons, splice together the protein-coding exons, and discard the introns.

Why are eukaryotic genes split up into introns and exons? Gene fragmentation appears to serve at least two functions. The first function is to allow a cell to produce multiple proteins from a single gene by splicing exons together in different ways. Rats, for example, have a gene that is transcribed in both the thyroid and the brain. In the thyroid, one splicing arrangement results in the synthesis of a hormone called calcitonin, which helps regulate calcium concentrations in the blood. In the brain, a different splicing arrangement results in the synthesis of a short protein that is used as a chemical messenger for communication between brain cells. Alternative splicing may occur in the RNA transcribed from over half of all human genes. Therefore, in eukaryotes, the "one gene, one protein" rule must be reworded as "one gene, one *or more* proteins."

The second function of interrupted genes is more speculative, but is supported by some good experimental evidence: fragmented genes may provide a quick and efficient way for eukaryotes to evolve new proteins with new functions. Chromosomes sometimes break apart, and their parts may reattach to different chromosomes. If the breaks occur within the noncoding introns of genes, exons may be moved intact from one chromosome to another. Most such errors would be harmful. But some of these shuffled exons might code for a protein subunit that has a specific function (binding ATP, for example). In rare instances, adding this subunit to an existing gene may cause the gene to code

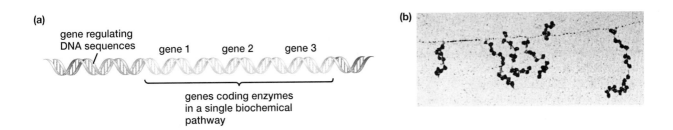

(a)
gene regulating
DNA sequences

gene 1   gene 2   gene 3

genes coding enzymes
in a single biochemical
pathway

(b)

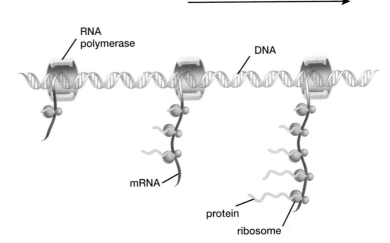

direction of transcription

FIGURE 10-6 Messenger RNA synthesis in prokaryotic cells
(a) In prokaryotes, many or all of the genes for a complete metabolic pathway lie side by side on the chromosome. **(b)** Transcription and translation are simultaneous in prokaryotes. In this colorized electron micrograph, RNA polymerase (not visible at this magnification) travels from left to right on a strand of DNA (blue). As it synthesizes a messenger RNA molecule (red), ribosomes (dark polygons) bind to the mRNA and immediately begin synthesizing a protein (not visible). The diagram below the micrograph shows all of the key molecules involved.

RNA
polymerase

DNA

mRNA

protein

ribosome

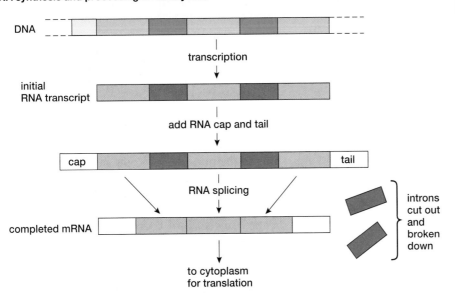

(a) Eukaryotic gene structure

exons

DNA

promoter

introns

A typical eukaryotic gene consists of sequences of DNA called exons, which code for the amino acids of a protein (medium blue), and intervening sequences called introns (dark blue), which do not. The promoter (light blue) determines where RNA polymerase will begin transcription.

(b) RNA synthesis and processing in eukaryotes

DNA

transcription

initial
RNA transcript

add RNA cap and tail

cap   tail

RNA splicing

introns
cut out
and
broken
down

completed mRNA

to cytoplasm
for translation

RNA polymerase transcribes both the exons and introns, producing a long RNA molecule. Enzymes in the nucleus then add further nucleotides at the beginning (cap) and end (tail) of the RNA transcript. Other enzymes cut out the RNA introns and splice together the exons to form the true mRNA, which moves out of the nucleus and is translated on the ribosomes.

FIGURE 10-7 Messenger RNA synthesis in eukaryotic cells

for a new protein with useful functions. The accidental exchange of exons among genes produces new eukaryotic genes that will, on occasion, enhance the survival and reproduction of the organism that carries them.

The final mRNA molecules then leave the nucleus and enter the cytoplasm through pores in the nuclear envelope. In the cytoplasm, mRNA binds to ribosomes, which synthesize a protein specified by the mRNA base sequence. The gene itself remains safely stored in the nucleus, like a valuable document in a library, while mRNA, like a "molecular photocopy," carries the information to the cytoplasm to be used in protein synthesis.

Ribosomes Consist of Two Subunits, Each Composed of Ribosomal RNA and Protein

Ribosomes, the structures that carry out translation, are composed of rRNA and many different proteins. Each ribosome is composed of two subunits—one small and one large. The small subunit has binding sites for mRNA, a "start" (methionine) tRNA, and several other proteins that collectively make up the "initiation complex." The large subunit has binding sites for two tRNA molecules and a catalytic site for joining together the amino acids attached to the tRNA molecules. Unless they are actively synthesizing proteins, the two subunits remain separate (see Fig. 10-2b). During protein synthesis, the small and large subunits come together and sandwich an mRNA molecule between them.

Transfer RNA Molecules Decode the Sequence of Bases in mRNA into the Amino Acid Sequence of a Protein

Delivery of the appropriate amino acids to the ribosome for incorporation into the growing protein chain depends on the activity of tRNA. Each cell synthesizes many different types of tRNA, at least one (and sometimes several) for each amino acid. Twenty enzymes in the cytoplasm, one for each amino acid, recognize the tRNA molecules and use the energy of ATP to attach the correct amino acid to one end (see Fig. 10-2c).

The ability of tRNA to deliver the proper amino acid depends on specific base pairing between tRNA and mRNA. Each tRNA has three exposed bases, called the **anticodon**, which form base pairs with the mRNA codon. For example, the mRNA codon AUG forms base pairs with the anticodon UAC of a tRNA that has the amino acid methionine attached to its end. The ribosome can then incorporate methionine into a growing protein chain.

During Translation, mRNA, tRNA, and Ribosomes Cooperate to Synthesize Proteins

Now that we have introduced the major molecules involved in translation, let's look at the actual events. Protein synthesis differs slightly between eukaryotes and prokaryotes. We will describe translation only in eukaryotic cells

(FIG. 10-8), but the differences between eukaryotes and prokaryotes are crucial to the action of many antibiotics commonly used to treat bacterial infections (see "Links to Life: Genetics, Evolution, and Medicine").

Like transcription, translation has three steps: (1) *initiation*, (2) *elongation* of the protein chain, and (3) *termination*.

Initiation: Protein Synthesis Begins When tRNA and mRNA Bind to a Ribosome

The first AUG codon in a eukaryotic mRNA sequence specifies the start of translation. Because AUG also codes for methionine, all newly synthesized proteins begin with this amino acid. An "initiation complex," which contains a small ribosomal subunit, a methionine tRNA, and several other proteins (Fig. 10-8a), binds to the beginning of an mRNA molecule (Fig. 10-8b). The AUG codon in the mRNA forms base pairs with the UAC anticodon of the methionine tRNA. A large ribosomal subunit then attaches to the small subunit, sandwiching the mRNA between the two subunits and holding the methionine tRNA in its first tRNA binding site (Fig. 10-8c). The ribosome is now fully assembled and ready to begin translation.

Elongation and Termination: Protein Synthesis Proceeds One Amino Acid at a Time Until a Stop Codon Is Reached

The assembled ribosome covers about 30 nucleotides of the mRNA. It holds two mRNA codons in alignment with the two tRNA binding sites of the large subunit. A second tRNA, with an anticodon complementary to the second codon of the mRNA, moves into the second tRNA binding site on the large subunit (Fig. 10-8d). The amino acids attached to the two tRNAs are now side by side. The catalytic site of the large subunit breaks the bond holding the first amino acid (methionine) to its tRNA and forms a peptide bond between this amino acid and the amino acid attached to the second tRNA (Fig. 10-8e). Interestingly, ribosomal RNA, and not one of the proteins of the large subunit, catalyzes the formation of the peptide bond. Therefore, this "enzymatic RNA" is often called a "ribozyme."

After the peptide bond is formed, the first tRNA is "empty," and the second tRNA carries a two-amino-acid chain. The ribosome then releases the empty tRNA and shifts to the next codon on the mRNA molecule (Fig. 10-8f). The tRNA holding the elongating chain of amino acids also shifts, moving from the second to the first binding site of the ribosome. A new tRNA, with an anticodon complementary to the third codon of the mRNA, binds to the empty second site (Fig. 10-8g). The catalytic site on the large subunit now links the third amino acid onto the growing protein chain (Fig. 10-8h). The "empty" tRNA leaves the ribosome, the ribosome shifts to the next codon on the mRNA, and the process repeats, one codon at a time.

A stop codon in the mRNA molecule signals the ribosome to terminate protein synthesis. Stop codons do not bind to tRNA. Instead, proteins called "release factors"

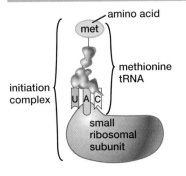

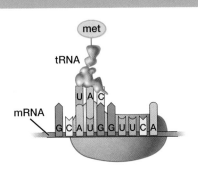

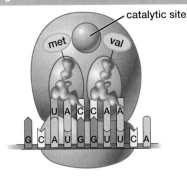

(a) A tRNA with an attached methionine amino acid binds to a small ribosomal subunit, forming an initiation complex.

(b) The initiation complex binds to an mRNA molecule. The methionine (met) tRNA anticodon (UAC) base-pairs with the start codon (AUG) of the mRNA.

(c) The large ribosomal subunit binds to the small subunit. The methionine tRNA binds to the first tRNA site on the large subunit.

Elongation:

(d) The second codon of mRNA (GUU) base-pairs with the anticodon (CAA) of a second tRNA carrying the amino acid valine (val). This tRNA binds to the second tRNA site on the large subunit.

(e) The catalytic site on the large subunit catalyzes the formation of a peptide bond linking the amino acids methionine and valine. The two amino acids are now attached to the tRNA in the second binding position.

(f) The "empty" tRNA is released and the ribosome moves down the mRNA, one codon to the right. The tRNA that is attached to the two amino acids is now in the first tRNA binding site and the second tRNA binding site is empty.

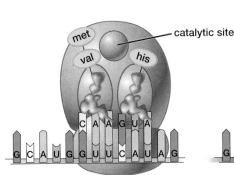

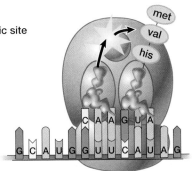

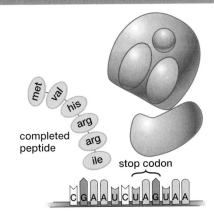

Termination:

(g) The third codon of mRNA (CAU) base-pairs with the anticodon (GUA) of a tRNA carrying the amino acid histidine (his). This tRNA enters the second tRNA binding site on the large subunit.

(h) The catalytic site forms a new peptide bond between valine and histidine. A three-amino-acid chain is now attached to the tRNA in the second binding site. The tRNA in the first site leaves, and the ribosome moves one codon over on the mRNA.

(i) This process repeats until a stop codon is reached; the mRNA and the completed peptide are released from the ribosome, and the subunits separate.

FIGURE 10-8 Translation is the process of protein synthesis
Protein synthesis, or translation, decodes the base sequence of an mRNA into the amino acid sequence of a protein. QUESTION Examine panel (i). If mutations changed all of the guanine molecules visible in the mRNA sequence shown here to uracil, how would translated peptide differ from the one shown?

LINKS TO LIFE   Genetics, Evolution, and Medicine

All life on Earth is related through evolution, sometimes closely (dogs and foxes), sometimes distantly (bacteria and people). As you know, mutations occur constantly, usually at a very low rate. Distantly related organisms may have shared a common ancestor millions of years ago. A lot of mutations may have occurred since then, so that the genes of these organisms may now differ by many nucleotides. Medicine takes advantage of these differences to develop antibiotics to treat bacterial infections.

Streptomycin and neomycin, commonly prescribed antibiotics, kill certain bacteria by binding to a specific sequence of RNA in the small subunits of the bacterial ribosomes, thereby inhibiting protein synthesis. Without adequate protein synthesis, the bacteria die. Patients infected by these bacteria don't die, however, because the small subunits of human eukaryotic ribosomes have a different nucleotide sequence than the bacteria's prokaryotic ribosomes do.

You have probably heard of *antibiotic resistance*, in which bacteria that are frequently exposed to antibiotics evolve defenses against those antibiotics. Bacteria evolve resistance against neomycin and related antibiotics rather rapidly. Why? It's actually pretty straightforward. If eukaryotic ribosomes are insensitive to neomycin, then eukaryotic ribosomes must function perfectly well with a different RNA sequence than prokaryotic ribosomes have. Bacteria that are resistant to neomycin and its relatives have a mutation that changes just a single nucleotide in their ribosomal RNA from adenine to guanine, which is precisely the nucleotide found at the comparable position in eukaryotic ribosomal RNA.

Genetics, mutations, the mechanisms of protein synthesis, and evolution are important not only to biologists, but to physicians, too. In fact, a whole discipline called evolutionary medicine has arisen that uses the evolutionary relationships between people and microbes to help fight disease.

---

bind to the ribosome when it encounters a stop codon, forcing the ribosome to release the finished protein chain and the mRNA (Fig. 10-8i). The ribosome disassembles into its large and small subunits, which can then be used to translate another mRNA.

None of the steps in protein synthesis are "free": they all require considerable amounts of cellular energy, as we explain in "A Closer Look at Protein Synthesis: A High-Energy Business."

Recap: Decoding the Sequence of Bases in DNA into the Sequence of Amino Acids in Protein Requires Transcription and Translation

We can now understand how a cell decodes the genetic information stored in its DNA to synthesize a protein. Each step involves the pairing of complementary bases and requires the action of a variety of proteins and enzymes. Follow these steps in **FIGURE 10-9**:

a. With some exceptions, such as the genes for tRNA and rRNA, each gene codes for the amino acid sequence of a protein.

b. Transcription of a protein-coding gene produces an mRNA molecule that is complementary to one DNA strand of the gene. Starting from the first AUG, each codon within the mRNA is a sequence of three bases that specifies an amino acid or a "stop."

c. Enzymes in the cytoplasm attach the appropriate amino acid to each tRNA based on the tRNA's anticodon.

d. During translation, tRNAs carry their attached amino acids to the ribosome. The appropriate amino acid is selected based on the complementary base pairs formed between the bases in the mRNA codon and the bases in the tRNA anticodon. The ribosome then links the amino acids together in sequence to form a protein.

This "decoding chain," moving from DNA bases to mRNA codons to tRNA anticodons to amino acids, results in the synthesis of a protein with a specific amino acid sequence. The amino acid sequence is ultimately determined by the base sequence within a gene.

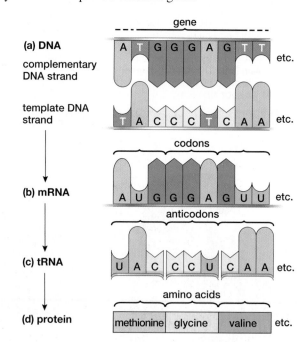

FIGURE 10-9 Complementary base pairing is critical to decode genetic information
(a) DNA contains two strands: the template strand is used by RNA polymerase to synthesize an RNA molecule. **(b)** Bases in the template strand of DNA are transcribed into a complementary mRNA. Codons are sequences of three bases that specify an amino acid or a stop during protein synthesis. **(c)** Unless it is a stop codon, each mRNA codon forms base pairs with the anticodon of a tRNA molecule that carries a specific amino acid. **(d)** The amino acids borne by the tRNAs are joined together to form a protein.

Table 10-4 Effects of Mutations in the Hemoglobin Gene

	DNA (template strand)	mRNA	Amino Acid	Properties of Amino Acid	Functional Effect on Protein	Disease
Original codon 6	CTC	GAG	Glutamic acid	Hydrophilic	Normal protein function	None
Mutation 1	CT**T**	GA**A**	Glutamic acid	Hydrophilic	Neutral; normal protein function	None
Mutation 2	**G**TC	**C**AG	Glutamine	Hydrophilic	Neutral; normal protein function	None
Mutation 3	**CAC**	**GUG**	Valine	Hydrophobic	Loses water solubility; compromises protein function	Sickle-cell anemia
Original codon 17	TTC	AAG	Lysine	Hydrophilic	Normal protein function	None
Mutation 4	**A**TC	**U**AG	Stop codon	Ends translation after amino acid 16	Synthesizes only part of protein; eliminates protein function	Beta-thalassemia

10.4 HOW DO MUTATIONS IN DNA AFFECT THE FUNCTION OF GENES?

As we saw in Chapter 9, mistakes during DNA replication, ultraviolet rays in sunlight, chemicals in cigarette smoke, and a host of other environmental factors may change the sequence of bases in DNA. These changes are called **mutations**. What are the consequences of a mutation on an organism's structure and function? That depends on how the mutation affects the functioning of the protein encoded by a mutated gene.

Mutations May Have a Variety of Effects on Protein Structure and Function

Most mutations may be categorized as substitutions, deletions, insertions, inversions, or translocations (see Chapter 9).

Inversions and Translocations

Inversions and translocations occur when pieces of DNA—sometimes most or even all of a chromosome—are broken apart and reattached, sometimes within a single chromosome, sometimes to a different chromosome. These mutations may be relatively benign, if entire genes, including their promoters, are merely moved from one place to another. However, if a gene is split in two, it will no longer code for a complete, functional protein. For example, almost half the cases of severe hemophilia are caused by an inversion in the gene that encodes a protein required for blood clotting.

Deletions and Insertions

The effects of **deletions** and **insertions** usually depend on how many nucleotides are removed or added. Why? Think back to the genetic code: three nucleotides encode a single amino acid. Therefore, adding or deleting three nucleotides will add or delete a single amino acid to the encoded protein. In most cases, this doesn't alter the function of the protein very much. In contrast, deletions and insertions of one or two nucleotides, or any deletion or insertion that isn't a multiple of 3 nucleotides, can have particularly catastrophic effects, because all of the codons that follow the deletion or insertion will be altered. Recall our English sentence with

all three-letter words: THEDOGSAWTHECAT. Deleting or inserting a letter (deleting the first E, for example) means that all of the following three-letter words will be nonsense, such as THD OGS AWT HEC AT. Similarly, most—and possibly all—of the amino acids of the protein synthesized from an mRNA containing such a *frameshift mutation* will be the wrong ones. Sometimes, one of the new codons following an insertion or deletion will be a stop codon, cutting the protein short. Such proteins will almost always be nonfunctional. Remember the Belgian Blue cattle in Chapter 9? The defective myostatin gene of a Belgian Blue has an 11-nucleotide deletion, generating a "premature" stop codon that terminates translation before the myostatin protein is complete.

Substitutions

Nucleotide substitutions (also called **point mutations**) within a protein-coding gene can produce at least four different outcomes (Table 10-4). As a concrete example, let's consider mutations that occur in the gene encoding beta-globin, one of the subunits of hemoglobin, the oxygen-carrying protein in red blood cells. The other type of subunit in hemoglobin is called alpha-globin; a normal hemoglobin molecule consists of two alpha and two beta subunits. In all but the last example, we will consider the results of mutations that occur in the sixth codon (CTC in DNA, GAG in mRNA), which specifies glutamic acid—a charged, hydrophilic, water-soluble amino acid.

- *The protein may be unchanged.* Remember that most amino acids can be encoded by several different codons. If a mutation changes the beta-globin DNA base sequence from CTC to CTT, this sequence still codes for glutamic acid. Therefore, the protein synthesized from the mutated gene remains the same even though the DNA sequence is different.

- *The new protein may be functionally equivalent to the original one.* Many proteins have regions whose exact amino acid sequence is relatively unimportant. For example, in beta-globin, the amino acids on the outside of the protein must be hydrophilic to keep the protein dissolved in the cytoplasm of red blood cells. Exactly *which* hydrophilic amino acids are on the outside doesn't matter

A CLOSER LOOK AT PROTEIN SYNTHESIS   A High-Energy Business

An old expression says that the good things in life are free. Maybe so, but protein synthesis certainly isn't. At least six different steps in protein synthesis require energy:

1. **Transcription:** RNA polymerase uses free *trinucleotides*—adenosine triphosphate (ATP), guanosine triphosphate (GTP), cytosine triphosphate (CTP), and uracil triphosphate (UTP)—to synthesize a strand of RNA. Like the familiar ATP, the last two phosphates of all trinucleotides are attached by high-energy bonds (see Chapter 6). These two phosphates are split off from the trinucleotide, releasing energy that is used to form the bond between the remaining phosphate and the sugar of the previous nucleotide in the growing RNA chain.

2. **Charging tRNAs:** The energy of ATP is used to attach an amino acid to its tRNA. Much of this energy remains in the tRNA–amino acid bond and is then used to form the peptide bond between amino acids during translation.

3. **Scanning mRNA:** In eukaryotes, mRNA binds to the small ribosomal subunit upstream of the start codon. ATP energy is used to "scan" the mRNA to find the start codon.

4. **Loading tRNA–Amino Acid Complexes:** The energy of one guanosine triphosphate (GTP) is used each time a new tRNA–amino acid complex is loaded onto a ribosome.

5. **Translocation:** Energy from one GTP is also used each time the ribosome moves one codon down the mRNA molecule.

6. **Termination:** One GTP is used to release the finished protein from the ribosome.

Therefore, each amino acid in a protein requires one trinucleotide for mRNA synthesis, one ATP to charge tRNA, one GTP to load the tRNA onto a ribosome, and one GTP to move one codon along mRNA. Starting and stopping translation uses more ATP and GTP. Protein synthesis uses up about 90% of all the energy expended by some cells, such as the common intestinal bacterium, *Escherichia coli.*

---

much. For example, a family in the Japanese town of Machida was found to contain a mutation from CTC to GTC, replacing glutamic acid (hydrophilic) with glutamine (also hydrophilic). Hemoglobin containing this mutant beta-globin protein—known as *Hemoglobin Machida*—appears to function well. Mutations such as the ones in Hemoglobin Machida and in the previous example are called **neutral mutations** because they do not detectably change the function of the encoded protein.

- *Protein function may be changed by an altered amino acid sequence.* A mutation from CTC to CAC replaces glutamic acid (hydrophilic) with valine (hydrophobic). This substitution is the genetic defect that causes sickle-cell anemia (see Chapter 12, page 239). The valines on the outside of the hemoglobin molecules cause them to clump together, distorting the shape of the red blood cells. These changes can cause serious illness.

- *Protein function may be destroyed by a premature stop codon.* A particularly catastrophic mutation occasionally occurs in the 17th codon of the beta-globin gene (TTC in DNA, AAG in mRNA). This codon specifies the amino acid lysine. A mutation from TTC to ATC (UAG in mRNA) results in a stop codon, halting translation of beta-globin mRNA before the protein is completed. People who inherit this mutant gene from both their mother and their father do not synthesize any functional beta-globin protein; they manufacture hemoglobin consisting entirely of alpha-globin subunits. This "pure alpha" hemoglobin does not bind oxygen very well. This condition, called beta-thalassemia, can be fatal unless treated with regular blood transfusions throughout life.

Mutations Provide the Raw Material for Evolution

Mutations in gametes (sperm or eggs) may be passed on to future generations. In humans, mutation rates range from about 1 in every 100,000 gametes to 1 in 1,000,000 gametes. For reference, a man releases about 300 to 400 million sperm per ejaculation; each ejaculate contains about 600 sperm with new mutations. Although most mutations are neutral or potentially harmful, they are essential for evolution, because these random changes in DNA sequence are the ultimate source of all genetic variation. New base sequences undergo natural selection as organisms compete to survive and reproduce. Occasionally, a mutation proves beneficial in an organism's interactions with its environment. Through reproduction over time, the mutant base sequence may spread throughout the population and become common, as organisms that possess it may outcompete rivals bearing the original, unmutated base sequence. This process is described in detail in Unit Three.

10.5   HOW ARE GENES REGULATED?

The complete human genome contains about 21,000 genes. Each of these genes is present in most of your body cells, but any individual cell *expresses* (transcribes and, if the final gene product is a protein, translates) only a small fraction of them. Some genes are expressed in all cells, because they encode proteins or RNA molecules that are essential for the life of any cell. For example, all cells need to synthesize proteins, so they all transcribe tRNA genes, rRNA genes, and genes for ribosomal proteins. Other genes are expressed exclusively in certain types of cells, at certain times in an organism's life, or under specific environmental conditions. For example, even though every cell in your body contains the gene for casein, the major protein in milk, that gene is expressed only in mature women, only in certain cells in the breast, and only when a woman is breast-feeding.

Regulation of gene expression may occur at the level of transcription (which genes are used to make mRNA in a given cell), translation (how much protein is made from a particular type of mRNA), and the activity of proteins (how long the protein lasts in a cell and how rapidly protein enzymes catalyze specific reactions).

Gene Regulation in Prokaryotes

Prokaryotic DNA is often organized in coherent packages called **operons**, in which the genes for related functions lie close to one another (**FIG. 10-10a**). An operon consists of four parts: (1) a **regulatory gene**, which controls the timing or rate of transcription of other genes; (2) a **promoter**, which RNA polymerase recognizes as the place to start transcribing; (3) an **operator**, which governs access of RNA polymerase to the promoter or to the (4) **structural genes**, which actually encode the related enzymes or other proteins. Whole operons are regulated as units, so that functionally related proteins are synthesized simultaneously when the need arises.

Prokaryotic operons may be regulated in a variety of ways, depending on the functions that they control. Some operons synthesize enzymes that are needed by the cell just about all the time, such as the enzymes that synthesize many amino acids. These operons are usually transcribed continuously, except under unusual circumstances when the bacterium encounters a vast surplus of a particular amino acid. Other operons synthesize enzymes that are needed only occasionally, for instance to digest a relatively rare food substance. They are transcribed only when the bacterium encounters the rare food.

As an example of the latter type of operon, consider the common intestinal bacterium *Escherichia coli (E. coli)*. This bacterium must live on whatever types of nutrients its host eats, and it can synthesize a variety of enzymes to metabolize a potentially wide variety of foods. The genes that code for these enzymes are transcribed only when the enzymes are needed. The enzymes that metabolize lactose, the principal sugar in milk, are a case in point. The **lactose operon** contains three structural genes, each coding for an enzyme that aids in lactose metabolism (Fig. 10-10a).

The lactose operon is shut off, or repressed, unless specifically activated by the presence of lactose. The regulatory gene of the lactose operon directs the synthesis of a protein, called a **repressor protein**, which binds to the operator site. RNA polymerase, although still able to bind to the promoter, cannot get past the repressor protein to transcribe the structural genes. Consequently, the lactose-metabolizing enzymes are not synthesized (**FIG. 10-10b**).

When *E. coli* colonize the intestines of a newborn mammal, however, they find themselves bathed in a sea of lac-

(a) Structure of the lactose operon

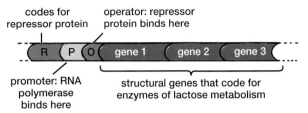

The lactose operon consists of a regulatory gene, a promoter, an operator, and three structural genes that code for enzymes involved in lactose metabolism. The regulatory gene codes for a protein, called a repressor, which can bind to the operator site under certain circumstances.

(b) Lactose absent

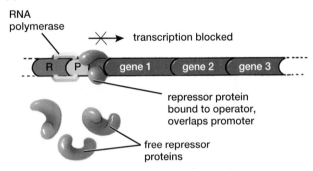

When lactose is not present, repressor proteins bind to the operator of the lactose operon. When RNA polymerase binds to the promoter, the repressor protein blocks access to the structural genes, which therefore cannot be transcribed.

(c) Lactose present

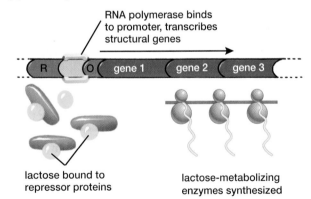

When lactose is present, it binds to the repressor protein. The lactose-repressor complex cannot bind to the operator, so RNA polymerase has free access to the promoter. The RNA polymerase transcribes the three structural genes coding for the lactose-metabolizing enzymes.

FIGURE 10-10 Regulation of the lactose operon

tose whenever the host nurses from its mother. Lactose molecules enter the bacteria and bind to the repressor proteins, changing their shape (**FIG. 10-10c**). The lactose-repressor complex cannot attach to the operator site. Therefore, when RNA polymerase binds to the promoter of the lactose operon, it can transcribe the structural genes.

Lactose-metabolizing enzymes are synthesized, allowing the bacteria to use lactose as an energy source. After the young mammal is weaned, it usually never consumes milk again. The intestinal bacteria no longer encounter lactose, the repressor proteins are free to bind to the operator, and the genes for lactose metabolism are shut down.

Gene Regulation in Eukaryotes

Eukaryotic gene regulation is similar to prokaryotic regulation in some respects. In both, not all genes are transcribed and translated all the time. Further, controlling the rate of transcription is probably the principal mechanism of gene regulation in both. However, the compartmentalization of the DNA in a membrane-bound nucleus, the variety of cell types in multicellular eukaryotes, a very different organization of the genome, and complex processing of RNA transcripts all distinguish gene regulation in eukaryotes from regulation in prokaryotes.

Expression of genetic information by a eukaryotic cell is a multistep process, beginning with transcription of DNA and commonly ending with a protein that performs a particular function. Regulation of gene expression can occur at any of these steps, which are illustrated in **FIGURE 10-11**:

1. *Cells can control the frequency at which an individual gene is transcribed.* The rate at which cells transcribe specific genes depends on the demand for the protein (or RNA) product that they encode. Gene transcription differs between organisms, between cell types in a given organism, and within a given cell at different stages in the organism's life or when it is stimulated by different environmental conditions (see the following section, "Eukaryotic Cells May Regulate the Transcription of Individual Genes, Regions of Chromosomes, or Entire Chromosomes").

2. *The same gene may be used to produce different mRNAs and protein products.* As we described earlier in this chapter, in eukaryotes, the same gene may be used to produce several different protein products, depending on how its RNA transcript is spliced to form the actual mRNA that will be translated on the ribosomes. For example, in the fruit fly *Drosophila*, alternative splicing of the pre-mRNA from a gene called *doublesex* produces a long protein in male flies and a short protein in female flies. The long protein in males suppresses transcription of other genes required for female sexual development

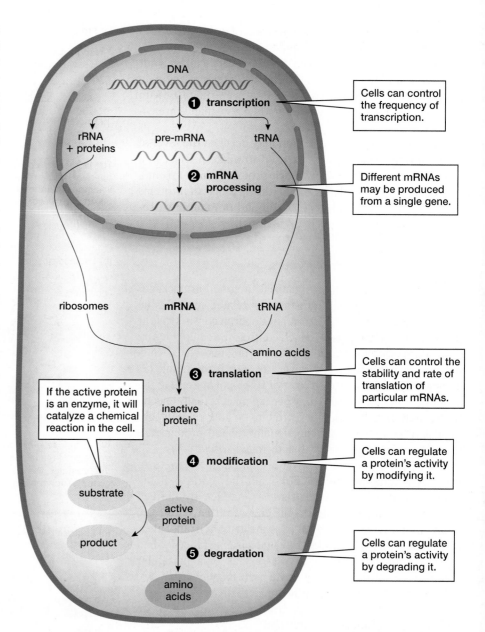

FIGURE 10-11 An overview of information flow in a eukaryotic cell, from gene transcription to structural proteins and chemical reactions catalyzed by enzymes

and enhances transcription of genes required for male sexual development. The short protein in female flies has the opposite effect, often on the same genes.

3. *Cells may control the stability and translation of messenger RNAs.* Some mRNAs are long lasting and are translated into protein many times. Others are translated only a few times before they are degraded. Recently, molecular biologists have discovered that "small regulatory RNA" molecules may block translation of some mRNAs, or may even target some mRNAs for destruction (see "Scientific Inquiry: RNA: It's Not Just a Messenger Any More").

4. *Proteins may require modification before they can carry out their functions.* Many proteins must be modified before they become active. For instance, the protein-digesting enzymes produced by cells in your stomach wall and

In recent years, molecular biologists have discovered an entirely new class of genes in eukaryotic cells—genes that code for "regulatory RNA." Biologists suspect that future research will discover many different types of regulatory RNA molecules, with many different functions. Here, we will describe just one function, called RNA interference, or RNAi. RNA interference is so important to cellular function that its discoverers, Andrew Fire and Craig Mello, shared the Nobel Prize in Physiology or Medicine in 2006.

As you know, messenger RNA is transcribed from DNA and then translated into protein. It's usually the protein that actually performs cellular functions, such as catalyzing a reaction or forming part of the cytoskeleton. How much protein is synthesized depends both on how much mRNA is made *and* on how rapidly and for how long the mRNA is translated. Enter RNAi. Many organisms, as diverse as roundworms, plants, and humans, synthesize small RNA molecules called "micro RNAs." After processing by cellular enzymes, micro RNAs give rise to small regulatory RNA molecules, usually about 20 to 25 nucleotides in length, which are complementary to short stretches of mRNA. In some cases, these small regulatory RNA molecules base-pair with the mRNA, forming a little section of double-stranded RNA that cannot be translated by a ribosome. In other cases, the short RNA strands combine with protein enzymes to form what are called "RNA-induced silencing complexes" or RISC for short. When an interfering RNA strand encounters an mRNA with a complementary sequences of bases, the RISC cuts up the mRNA, which obviously also prevents translation.

Why would a cell want to do that? In the roundworm *Caenorhabditis elegans*, where RNAi was first discovered, RNAi is required during normal development. A single protein is required for the development of juvenile body structures, but must be absent for the worm to mature into an adult. It turns out that the gene encoding this protein is transcribed into mRNA all the time. However, early in development, the mRNA is translated into protein, but later in development, interfering RNA binds to the mRNA, preventing translation. The result is that protein levels decline, and the worm matures.

Some organisms use RNAi to defend against disease. Many plants produce interfering RNA that is complementary to the nucleic acids (usually RNA) of plant viruses. When the interfering RNA finds complementary viral RNA molecules, the RISC cuts up the viral RNA, thereby preventing viral reproduction.

RNAi shows great promise in medicine, too. For example, macular degeneration, which is one of the leading causes of blindness in the elderly, results from the development of weak, leaky blood vessels in the retina of the eye. RNAi can prevent the overproduction of a key growth factor that stimulates the development of these abnormal blood vessels. In 2005, at least two pharmaceutical companies started clinical trials of synthetic micro RNA treatments for macular degeneration. This technology may reach patients within a few years.

---

pancreas are initially synthesized in an inactive form, which prevents the enzymes from digesting the very cells that produce them. After these inactive forms are secreted into the digestive tract, portions of the enzymes are snipped out to unveil the enzyme's active site. Other modifications, such as adding or removing phosphate groups, can temporarily activate or inactivate a protein, allowing second-to-second control of the protein's activity. Similar regulation of protein structure and function occurs in prokaryotic cells.

5. *The life span of a protein can be regulated.* Most proteins have a limited life span within the cell. By preventing or promoting a protein's degradation, a cell can rapidly adjust the amount of a particular protein within it. Protein life span may also be regulated in prokaryotic cells.

Eukaryotic Cells May Regulate the Transcription of Individual Genes, Regions of Chromosomes, or Entire Chromosomes

In eukaryotic cells, transcriptional regulation can operate on at least three levels: the individual gene, regions of chromosomes, or entire chromosomes.

Regulatory Proteins That Bind to the Gene's Promoter Alter the Transcription of Individual Genes

The promoter regions of virtually all genes contain several different response elements. Therefore, whether these genes are transcribed depends on which specific transcription factors are synthesized by the cell and whether those transcription factors are active or not. For example, when cells are exposed to free radicals (see Chapter 2), a protein transcription factor binds to antioxidant response elements in the promoters of several genes. As a result, the cell produces enzymes that break down free radicals to harmless substances.

Many transcription factors require activation before they can affect gene transcription. One of the best-known examples is the role that the sex hormone, estrogen, plays in controlling egg production in birds. The gene for albumin, the major protein in egg whites, is not transcribed in the winter when birds are not breeding and estrogen levels are low. During the breeding season, the ovaries in female birds release estrogen, which enters the cells in the oviduct and binds to a protein (usually called an estrogen receptor, but it is also a transcription factor). The estrogen–receptor complex then attaches to an estrogen response element in the promoter of the albumin gene. This attachment makes it easier for RNA polymerase to bind to the promoter and initiate transcription of mRNA. The mRNA is then translated into large amounts of albumin. Similar activation of gene transcription by steroid hormones occurs in other animals, including humans. The importance of hormonal regulation of transcription during development is illustrated by genetic defects in which receptors for sex hormones are nonfunctional (see "Health Watch: Sex, Aging, and Mutations"). In such cases, cells are unable to respond to the hormone, short-circuiting critical events in sexual development.

HEALTH WATCH   Sex, Aging, and Mutations

Sometime in her early to mid-teens, a girl usually goes through puberty: her breasts swell, her hips widen, and she begins to menstruate. In rare instances, however, a girl may develop all of the outward signs of womanhood, but does not menstruate. Eventually, when it becomes clear that she isn't merely developing a bit late, she reports this symptom to her physician, who may take a blood sample to do a chromosome test. In some cases, the chromosome test gives what might seem to be an impossible result: the girl's sex chromosomes are XY, a combination that would normally give rise to a boy. The reason she has not begun to menstruate is that she lacks ovaries and a uterus, but instead has testes that have remained inside her abdominal cavity. She has about the same concentrations of *androgens* (male sex hormones, such as testosterone) circulating in her blood as would be found in a boy her age. In fact, androgens, produced by the testes, have been present since early in her development. The problem is that her cells cannot respond to them—a rare condition called *androgen insensitivity*. This condition was a problem for Maria Jose Martinez Patino, an outstanding Spanish athlete who reached the Olympics several years ago, only to be barred from the hurdles competition because her cells lacked Barr bodies, which are normally present in females. After three years of struggle, the fact that she had developed as a female was finally recognized, and she was allowed to compete with others of her gender.

Many features of male development, including the formation of a penis, the descent of the testes into sacs outside the body cavity, and sexual characteristics that develop at puberty, such as a beard and increased muscle mass, occur because various body cells are responding to male sex hormones produced by the testes. In normal males, many body cells have androgen receptor proteins. When these proteins bind male hormones such as testosterone, the hormone-receptor complex attaches to androgen response elements in the promoters of specific genes and influences their transcription into

mRNA. The mRNA molecules are translated into proteins that contribute to maleness. In different cells, the testosterone-androgen receptor complex influences gene transcription in different ways, producing a wide range of male characteristics. Like other proteins, androgen receptors are coded by specific genes (interestingly, the gene encoding the androgen receptor protein is on the X chromosome). There are over 200 mu-

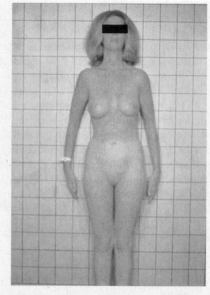

FIGURE E10-1 Androgen insensitivity leads to female features
This individual has an X and a Y chromosome. She has testes that produce testosterone, but a mutation in her androgen receptor genes makes her cells unable to respond to testosterone, resulting in her female appearance.

Some Regions of Chromosomes Are Condensed and Not Normally Transcribed

Certain parts of eukaryotic chromosomes are in a highly condensed, compact state in which most of the DNA seems to be inaccessible to RNA polymerase. Some of these regions are structural parts of chromosomes that don't contain genes. Other tightly condensed regions contain functional genes that are not currently being transcribed. When the product of a gene is needed, the portion of the chromosome containing that gene becomes "decondensed"—loosened so that the nucleotide sequence is accessible to RNA polymerase and transcription can occur.

Large Parts of Chromosomes May Be Inactivated, Thereby Preventing Transcription

In some cases most of a chromosome may be condensed, making it largely inaccessible to RNA polymerase. An exam-

ple occurs in the sex chromosomes of female mammals. Male mammals usually have an X and a Y chromosome (XY), and females usually have two X chromosomes (XX). As a consequence, females have the capacity to synthesize mRNA from genes on their two X chromosomes, while males, with only one X chromosome, may produce only half as much. In 1961, the geneticist Mary Lyon hypothesized that perhaps one of the two X chromosomes in women was inactivated in some way, so that its genes were not expressed. This hypothesis was soon proved correct. More recently, X chromosome inactivation has been found to be another case of "regulatory RNA" controlling gene expression. Early in development (about the 16th day in humans), by a mechanism that is not understood, one X chromosome begins to produce large amounts of a specific RNA molecule, called Xist, which coats the chromosome and causes it to condense into a tight mass. In a light microscope, this condensed X chromosome shows up in the nucleus as a dark spot called a **Barr body** (FIG. 10-12),

tant forms of the androgen receptor gene. The most serious are usually insertions, deletions, or point mutations creating a premature stop codon—and as you know, these types of mutations are likely to have catastrophic effects on protein structure and function.

Even though genetically a male with both X and Y chromosomes, a person with a mutant androgen receptor gene will be unable to make functional androgen receptor proteins, and therefore will be unable to respond to the testosterone that the testes produce. Thus, a change in the nucleotide sequence of a single gene, causing a single type of defective protein to be produced, can cause a person who is genetically male to look and feel like a woman (**FIG. E10-1**).

A second type of mutation provides clues to why people age. Why will your hair whiten, skin wrinkle, joints ache, and eyes cloud as you become elderly? A small number of individuals carry a defective gene that causes *Werner syndrome*, which causes a type of premature aging (**FIG. E10-2**). People with this disorder typically die of aging-related conditions by age 50. Recent research has localized the mutations in most people with Werner syndrome to a gene that codes for an enzyme involved in DNA replication. As you have seen, the accurate replication of DNA is crucial to the production of normally functioning cells. If a mutation interferes with the ability of enzymes to replicate DNA accurately and to proofread and repair errors in DNA, then mutations will accumulate rapidly in cells throughout the body.

The fact that an overall increase in mutations caused by defective replication enzymes produces symptoms of old age provides support for one hypothesis to explain many of the symptoms of normal aging. During a typical long (say, 80-year) life span, mutations gradually accumulate because of mistakes in DNA replication and environmentally induced DNA damage. Eventually, these mutations interfere with nearly every aspect of body functioning and contribute to death from "old age."

Disorders such as androgen insensitivity and Werner syndrome provide profound insights into the impact of mutations, the function of specific genes and their protein products, the ways hormones regulate gene transcription, and even the mystery of aging.

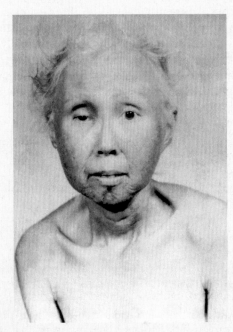

FIGURE E10-2 A 48-year-old woman with Werner syndrome
This condition, most common among people of Japanese ancestry, is the result of a mutation that interferes with proper DNA replication, increasing the incidence of mutations throughout the body.

named after its discoverer, Murray Barr. About 85% of the genes on an inactivated X chromosome are not transcribed.

Up until a few years ago, Olympic officials attempted to verify that athletes who compete in women's events were truly female by performing a gene-based sex test. Women who "passed" the test were given a gender certification card, a requirement for a female athlete's participation in many competitions. One type of sex test, used as recently as the 1996 Olympics in Atlanta, checked the athlete's cells for the presence of Barr bodies. This test caused major problems for a female hurdler from Spain, Maria Jose Martinez Patino, when no Barr bodies were found in her cells. Learn more about her story in "Health Watch: Sex, Aging, and Mutations."

Usually, fairly large clusters of cells (all descended from a common "ancestral" cell during development) have the same X chromosome inactivated. As a result, the bodies of female mammals (including women) are composed of

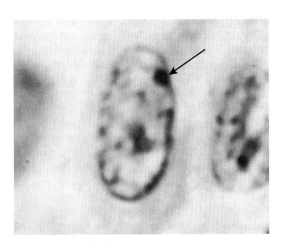

FIGURE 10-12 Barr bodies
The black spot on the upper right side of the nucleus is an inactivated X chromosome called a Barr body, usually found only in the cells of female mammals.

patches of cells in which one of the X chromosomes is fully active, and patches of cells in which the other X chromosome is active. The results of this phenomenon are easily observed in calico cats (**FIG. 10-13**). The X chromosome of a cat contains a gene encoding an enzyme that produces fur pigment. This gene comes in two versions, one producing orange fur and the other producing black fur. If one X chromosome in a female cat has the orange version of the fur-color gene and the other X chromosome has the black version, the cat will have patches of orange and black fur. These patches represent areas of skin that developed from cells in the early embryo in which different X chromosomes were inactivated. Calico coloring is almost exclusively found in female cats. Because male cats usually have only one X chromosome, which is active in all of their cells, normal male cats usually have black fur or orange fur, but not both.

FIGURE 10-13 Inactivation of the X chromosome regulates gene expression
This female calico cat carries a gene for orange fur on one X chromosome and a gene for black fur on her other X chromosome. Inactivation of different X chromosomes produces the black and orange patches. The white color is due to an entirely different gene that prevents pigment formation altogether.

CASE STUDY REVISITED   VIVE LA DIFFÉRENCE!

How does knowing about transcription and translation help us understand the physical differences between males and females? By the 1930s, biologists knew that one or more genes on the Y chromosome were essential for determining whether a mammal would develop into a male or a female. In 1990, a search for this gene led to discovery of the *SRY* gene, for "sex-determining region on the Y chromosome." The *SRY* gene is found in all male mammals, including humans. Experiments with mice demonstrated its importance in sex determination. If a mouse embryo with two X chromosomes is given a copy of *SRY*, but not the rest of the Y chromosome, the embryo develops male characteristics: it will have testes and a penis, and behave like a male mouse. (These XX male mice are sterile, however, because other genes located on the Y chromosome are needed for production of functional sperm.) Mouse embryos that lack an *SRY* gene develop as females, regardless of whether they have two X chromosomes or an X and a Y. The conclusion: male (XY) mammals have all the genes needed to be female but usually aren't,

because they have an *SRY* gene. Likewise, female (XX) mammals have all the genes needed to be male, but because they *don't* have an *SRY* gene, they are usually female.

How does the *SRY* gene exert such an enormous effect on a mammal's characteristics? Based on what you've learned in this chapter, you probably won't be surprised to learn that SRY encodes a transcription factor. The *SRY* gene is transcribed only for a short time in embryonic development, and only in cells that will become the testes. It is then permanently inactivated for the rest of the animal's life. However, in the short time that it is produced, the transcription factor encoded by the *SRY* gene stimulates the expression of many other genes, whose protein products are essential to testicular development. Once formed, the testes in the embryo secrete testosterone, which binds to androgen receptors and activates still other genes, leading to development of the penis and scrotum. The physical expression of gender, therefore, depends on the carefully regulated expression of many genes, with a single gene, *SRY*, serving as the initial switch to activate male development.

Consider This We have briefly described two different ways in which a person with XY sex chromosomes can develop as a female: the Y chromosome may have a defective *SRY* gene, or the X chromosome may have a defective androgen receptor gene. Suppose a 16-year-old girl is tearful and frightened because she has never menstruated, and she asks a physician what is wrong with her. The doctor orders a chromosome test, and perhaps hormone tests as well, and discovers that, in fact, she has both X and Y chromosomes, but either has androgen insensitivity or lacks a functional *SRY* gene. What should the physician tell her? Clearly, she has to be told that she has no uterus, will never menstruate, and cannot bear children. But beyond that, what? To most people, a person with two X chromosomes is a female, and one with one X and one Y chromosome is a male, and that's that. Should the doctor tell her that she is genetically male although physiologically female? What will this do to her self-image and her psychological health? What would *you* do? To see how one physician handled this dilemma, see "The Curse of the Garcias," by Robert Marion, in *Discover* magazine, December 2000.

CHAPTER REVIEW

SUMMARY OF KEY CONCEPTS

10.1 How Are Genes and Proteins Related?

Genes are segments of DNA that can be transcribed into RNA and, for most genes, translated into protein. Transcription produces the three types of RNA needed for translation: messenger RNA (mRNA), transfer RNA (tRNA), and ribosomal RNA (rRNA). During translation, tRNA and rRNA collaborate with enzymes and other proteins to decode the sequence of bases in mRNA and produce a protein with the amino acid sequence specified by the gene. The genetic code consists of codons, sequences of three bases in mRNA that specify either an amino acid in the protein chain or the end of protein synthesis (stop codons).

10.2 How Is Information in a Gene Transcribed into RNA?

Within an individual cell, only certain genes are transcribed. When the cell requires the product of a gene, RNA polymerase binds to the promoter region of the gene and synthesizes a single strand of RNA. This RNA is complementary to the template strand in the gene's DNA double helix. Cellular proteins called transcription factors may bind to parts of the promoter and enhance or suppress transcription of a given gene.

Web Tutorial 10.1 Transcription

10.3 How Is the Base Sequence of a Messenger RNA Molecule Translated into Protein?

In prokaryotic cells, all of the nucleotides of a protein-coding gene code for amino acids, and therefore the RNA transcribed from the gene is the mRNA that will be translated on a ribosome. In eukaryotic cells, protein-coding genes consist of two parts: exons, which code for amino acids in a protein, and introns, which do not. Therefore, the introns of the initial RNA transcript must be cut out and the exons spliced together to produce a true mRNA.

In eukaryotes, mRNA carries the genetic information from the nucleus to the cytoplasm, where ribosomes can use this information to synthesize a protein. Ribosomes contain rRNA and proteins organized into large and small subunits. These subunits come together at the first AUG codon of the mRNA molecule to form the complete protein-synthesizing machine. tRNAs deliver the appropriate amino acids to the ribosome for incorporation into the growing protein. Which tRNA binds, and consequently which amino acid is delivered, depends on

base pairing between the anticodon of the tRNA and the codon of the mRNA. Two tRNAs, each carrying an amino acid, bind simultaneously to the ribosome; the large subunit catalyzes the formation of peptide bonds between the amino acids. As each new amino acid is attached, one tRNA detaches, and the ribosome moves over one codon, binding to another tRNA that carries the next amino acid specified by mRNA. Addition of amino acids to the growing protein continues until a stop codon is reached, signaling the ribosome to disassemble and to release both the mRNA and the newly formed protein.

Web Tutorial 10.2 Translation

10.4 How Do Mutations in DNA Affect the Function of Genes?

A mutation is a change in the nucleotide sequence of a gene. Mutations can be caused by mistakes in base pairing during replication, by chemical agents, and by environmental factors such as radiation. Common types of mutations include inversions, translocations, insertions, deletions, and substitutions (point mutations). Mutations may be neutral or harmful, but in rare cases a mutation will promote better adaptation to the environment and thus will be favored by natural selection.

10.5 How Are Genes Regulated?

The expression of a gene requires that it be transcribed and translated, and the resulting protein must perform some action within the cell. Which genes are expressed in a cell at any given time is regulated by the function of the cell, the developmental stage of the organism, and the environment. Control of gene regulation can occur at many steps. The amount of mRNA synthesized from a particular gene can be regulated by increasing or decreasing the rate of its transcription, as well as by changing the stability of the mRNA itself. Rates of translation of mRNAs can also be regulated. Regulation of transcription and translation affects how many protein molecules are produced from a particular gene. Even after they are synthesized, many proteins must be modified before they can function. In addition to regulation of individual genes, cells can regulate transcription of groups of genes. For example, entire chromosomes or parts of chromosomes may be condensed and inaccessible to RNA polymerase, whereas other portions may be expanded, allowing transcription to occur.

KEY TERMS

anticodon *page 176*
Barr body *page 184*
codon *page 171*
deletion mutation *page 179*
exon *page 174*
genetic code *page 171*
insertion mutation *page 179*
intron *page 174*
lactose operon *page 181*

messenger RNA (mRNA) *page 170*
mutation *page 179*
neutral mutation *page 180*
nucleotide substitution *page 179*
operator *page 181*
operon *page 181*
point mutation *page 179*

promoter *page 172*
regulatory gene *page 181*
repressor protein *page 181*
ribonucleic acid (RNA) *page 169*
ribosomal RNA (rRNA) *page 170*
ribosome *page 170*
RNA polymerase *page 172*

start codon *page 171*
stop codon *page 171*
structural gene *page 181*
template strand *page 172*
transcription *page 170*
transfer RNA (tRNA) *page 170*
translation *page 170*

THINKING THROUGH THE CONCEPTS

1. How does RNA differ from DNA?

2. What are the three types of RNA? What is the function of each?

3. Define the following terms: *genetic code*; *codon*; *anticodon*. What is the relationship among the bases in DNA, the codons of mRNA, and the anticodons of tRNA?

4. How is mRNA formed from a eukaryotic gene?

5. Diagram and describe protein synthesis.

6. Explain how complementary base pairing is involved in both transcription and translation.

7. Describe some mechanisms of gene regulation.

8. Define *mutation*. Are most mutations likely to be beneficial or harmful? Explain your answer.

APPLYING THE CONCEPTS

1. As you have learned in this chapter, many factors influence gene expression, including hormones. The use of anabolic steroids and growth hormones among athletes has created controversy in recent years. Hormones certainly affect gene expression, but, in the broadest sense, so do vitamins and foods. What do you think are appropriate guidelines for the use of hormones? Should athletes take steroids or growth hor-

mones? Should children at risk of being unusually short be given growth hormones? Should parents be allowed to request growth hormones for a child of normal height in the hope of producing a future basketball player?

2. About 40 years ago, some researchers reported that they could transfer learning from one animal (a flatworm) to another by feeding trained animals to untrained animals. Fur-

ther, they claimed that RNA was the active molecule of learning. Given your knowledge of the roles of RNA and protein in cells, do you think that a *specific* memory (for example, remembering the base sequences of codons of the genetic code) could be encoded by a *specific* molecule of RNA and that this RNA molecule could transfer that mem- ory to another person? In other words, in the future, could you learn biology by popping an RNA pill? If so, how would this work? If not, can you propose a reasonable hypothesis for the results with flatworms? How would you test your hypothesis?

FOR MORE INFORMATION

Gibbs, W. W. "The Unseen Genome: Beyond DNA." *Scientific American*, December 2003. Gene expression can be regulated across generations by modifying the nucleotides of DNA.

Grunstein, M. "Histones as Regulators of Genes." *Scientific American*, October 1992. Histones are proteins associated with DNA in eukaryotic chromosomes. Once thought to be merely a scaffold for DNA, they are actually important in gene regulation.

Marion, R. "The Curse of the Garcias." *Discover*, December 2000. How one physician diagnosed and counseled a patient with androgen insensitivity.

Mattick, J. S. "The Hidden Genetic Program of Complex Organisms." *Scientific American*, October 2004. "Advanced" organisms, such as humans, have scarcely any more genes than worms, but they have much more DNA that does *not* code for proteins. Some of this DNA codes for regulatory RNA that may be crucial for the development of complex bodies.

Nirenberg, M. W. "The Genetic Code: II." *Scientific American*, March 1963. Nirenberg describes some of the experiments in which he deciphered much of the genetic code.

Tjian, R. "Molecular Machines That Control Genes." *Scientific American*, February 1995. Complexes of proteins regulate which genes are transcribed in a cell and therefore help determine the cell's structure and function.

The Continuity of Life: Cellular Reproduction

Sunburns are not merely painful; they may ultimately cause skin cancer.

CASE STUDY HOW MUCH IS A GREAT TAN WORTH?

RACHEL LIKED THE OUTDOORS, and she liked the sun. Even when she wasn't swimming on the varsity team, she was often out running, playing volleyball, or just sunbathing. Her friends admired her great tan. That spring, when a teammate noticed that she had a bumpy, black mole on her back, she simply shrugged it off with "I've always had a mole there." Rachel would have ignored it completely, but her swim coach asked her to have it checked by a physician. So, she scheduled an appointment with her family doctor, who removed the mole in his office.

After the minor surgery, Rachel didn't think about the mole at all. The cut healed in time for the next swim meet, she won the 100-meter butterfly, and she was on top of the world. However, her doctor called back a few days later. Following his general policy, he had sent the tissue to a laboratory for examination. The diagnosis was a type of cancer called melanoma.

Melanoma is a skin cancer that usually begins in pigmented cells in the deeper parts of the skin. The cancer may then spread to other parts of the body, including internal organs. The resulting disease is challenging to treat and frequently deadly.

The American Dermatology Association estimates that more than 54,000 people in the United States will be diagnosed with melanoma this year, with about 8000 deaths. It is now the most common cancer in people between 25 and 29 years of age. Melanoma is mostly caused by exposure to ultraviolet rays in sunlight.

Why do cancers form? How can sunlight cause cancer? To answer these questions, we need to understand how cells divide, how they control their rate of cell division, and how cancerous cells escape these controls.

11.1 WHAT IS THE ROLE OF CELLULAR REPRODUCTION IN THE LIVES OF INDIVIDUAL CELLS AND ENTIRE ORGANISMS?

The **cell cycle** is the sequence of activities that occurs from one cell division to the next. When a cell divides, it must provide its offspring (usually called "daughter cells") with genetic information (DNA) and the other cellular components it needs, such as mitochondria, ribosomes, and endoplasmic reticulum. Much of this text is devoted to the activities of cells when they are not dividing. In this chapter, however, we will focus on the mechanisms of cell division and the role of cell division in the lives of individual cells and multicellular organisms.

Reproduction in which offspring are formed from a single parent, rather than through the union of gametes (sperm and egg) from two parents, is called **asexual reproduction**. Single-celled organisms, including *Paramecium* in ponds (**FIG. 11-1a**) and yeast in rising bread (**FIG. 11-1b**), reproduce asexually by cell division—each cell cycle results in two new organisms from each preexisting cell. Asexual reproduction isn't confined to single-celled organisms, however. You, too, have been involved in asexual reproduction throughout your life—or at least your cells have. Since your conception as a single fertilized egg, asexual reproduction via cell division has produced all of the trillions of cells in your body, and continues every day in many organs, such as your skin and intestines.

Entire multicellular organisms may also reproduce by asexual reproduction. Like its relative the sea anemone, a

FIGURE 11-1 Cell division in eukaryotes enables asexual reproduction
(a) In unicellular microorganisms, such as the protist *Paramecium,* cell division produces two new, independent organisms. **(b)** Yeast, a unicellular fungus, reproduces by cell division. **(c)** *Hydra*, a freshwater relative of the sea anemone, grows a miniature replica of itself (a bud) on its side. When fully developed, the bud breaks off and assumes independent life. **(d)** Trees in an aspen grove are often genetically identical. Each tree grows up from the roots of a single ancestral tree. This photo shows three separate groves near Aspen, Colorado. In fall, the appearance of their leaves shows the genetic identity within a grove and the genetic difference between groves.

Hydra reproduces by growing a small replica of itself, called a bud, on its body (**FIG. 11-1c**). Eventually, the bud is able to live independently and separates from its parent. Many plants and fungi reproduce both asexually and sexually. The beautiful aspen groves of Colorado, Utah, and New Mexico (**FIG. 11-1d**) develop asexually from shoots growing up from the root system of a single parent tree. Although the grove seems to be a population of separate trees, it can be considered to be a single individual whose multiple trunks are interconnected by a common root system. Aspen can also reproduce by seeds, which are made through sexual reproduction.

Both prokaryotic and eukaryotic cells have cell cycles that include growth, DNA replication, and cell division. Because of the structural and functional differences between these two cell types, prokaryotic and eukaryotic cell cycles differ considerably.

The Prokaryotic Cell Cycle Consists of Growth and Binary Fission

With enough nutrients and favorable temperatures, many prokaryotic cells are usually either dividing or preparing to divide. The cell cycle consists of a relatively long period of growth—during which the cell also replicates its DNA—followed by rapid cell division (**FIG. 11-2a**).

Cell division in prokaryotic cells is known as **binary fission**, which means "splitting in two." The prokaryotic chromosome is usually a circle of DNA, attached at one point to the plasma membrane (**FIG. 11-2b**, ①). During the long "growth phase" of the prokaryotic cell cycle, the DNA is replicated, producing two identical chromosomes which become attached to the plasma membrane at nearby, but separate, points (Fig. 11-2b, ②). The cell increases in size both during and after DNA replication. As the cell grows, the plasma membrane between the attachment points of the chromosomes enlarges, pushing them apart (Fig. 11-2b, ③). When the cell has approximately doubled in size, the plasma membrane around the middle of the cell rapidly grows inward between the two DNA attachment sites (Fig. 11-2b, ④). Fusion of the plasma membrane along the equator of the cell completes binary fission, producing two daughter cells, each containing one of the chromosomes (Fig. 11-2b, ⑤). Because DNA replication produces two identical DNA molecules (except for the occasional mutation), the two daughter cells are genetically identical to one another and to the parent cell.

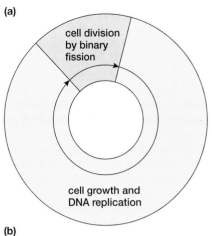

(a)

cell division by binary fission

cell growth and DNA replication

(b)

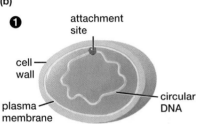

①   attachment site

cell wall

plasma membrane

circular DNA

The circular DNA double helix is attached to the plasma membrane at one point.

②

The DNA replicates and the two DNA double helices attach to the plasma membrane at nearby points.

③

New plasma membrane is added between the attachment points, pushing them further apart.

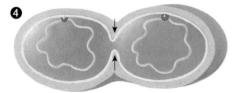

④

The plasma membrane grows inward at the middle of the cell.

⑤

The parent cell divides into two daughter cells.

FIGURE 11-2 The prokaryotic cell cycle
(a) The prokaryotic cell cycle consists of growth and DNA replication, followed by binary fission. **(b)** Binary fission in prokaryotic cells.

FIGURE 11-3 The eukaryotic cell cycle
The eukaryotic cell cycle consists of interphase and mitotic cell division. Some cells enter the G_0 phase and may not divide again.

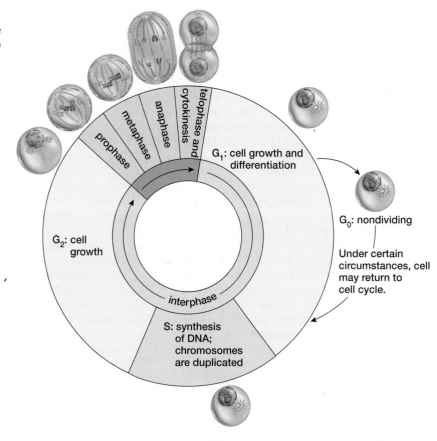

Under ideal conditions, binary fission in prokaryotes occurs rapidly. For example, the common intestinal bacterium *Escherichia coli* can grow, replicate its DNA, and divide in about 20 minutes. Luckily, the environment in our intestines is not ideal for bacteria growth; otherwise, the bacteria would soon outweigh the rest of our bodies!

The Eukaryotic Cell Cycle Consists of Interphase and Cell Division

The eukaryotic cell cycle (**FIG. 11-3**) is somewhat more complex than the prokaryotic cell cycle. Newly formed cells usually acquire nutrients from their environment, synthesize additional cellular components, and grow larger. After a variable amount of time—depending on the organism, the type of cell, and the nutrients available—the cell may divide. Each daughter cell may then enter another cell cycle, producing additional cells. Many newly formed cells, however, divide only if they receive signals, such as growth hormones, that cause them to enter another cell cycle. Still other cells may exit the cell cycle completely and never divide again. In humans, cells in the bone marrow and skin may divide as frequently as once a day. At the other extreme, most nerve and muscle cells never divide after they mature; if one of these cells dies, it is not replaced.

During Interphase, the Eukaryotic Cell Grows in Size and Replicates Its DNA

The eukaryotic cell cycle is divided into two major phases: interphase and cell division (see Fig. 11-3). During **interphase**, the cell acquires nutrients from its environment, grows, and duplicates its chromosomes. With the exception of meiotic cell division (described below), **cell division** parcels one copy of each chromosome and usually about half the cytoplasm (including mitochondria, ribosomes, and other organelles) into each of the two daughter cells.

Most eukaryotic cells spend the majority of their time in interphase, preparing for cell division. For example, some cells in human skin, which divide about once a day, spend roughly 22 hours in interphase. Interphase itself contains three subphases: G_1 (**g**ap or **g**rowth phase 1), S (DNA **s**ynthesis), and G_2 (**g**ap or **g**rowth phase 2).

To explore these stages, let's consider a newly formed daughter cell. This cell enters the G_1 portion of interphase, during which it acquires or synthesizes the materials needed for cell division. During the G_1 phase, the cell is sensitive to internal and external signals that help the cell "decide" whether to divide. If that decision is positive, the cell enters the S phase, which is when DNA synthesis occurs. After replicating its DNA, the cell completes its growth in the G_2 phase before dividing.

Alternatively, if the "division decision" during the G_1 phase is negative, the cell may exit from the cell cycle during G_1 and enter into a phase known as G_0. Cells in G_0 are alive and metabolically active. They may even grow in size, but they do not replicate their DNA or divide. This phase is also the time when many cells specialize, or **differentiate**. Muscle cells fill with the contractile proteins myosin and actin; some cells of the immune system become packed with endoplasmic reticulum to produce massive amounts of antibodies; and nerve cells grow long strands, called axons, that allow them to connect with other cells. Many differentiated cells, including most of those in your heart muscle, eyes, and brain, remain in G_0 for your entire life.

As this discussion suggests, the cell cycle is carefully controlled throughout the life of an organism. Without enough cell divisions at the right time and in the right organs, development falters or body parts fail to replace worn-out or damaged cells. With too many cell divisions, cancers may form. We will investigate how the cell cycle is controlled in section 11.4.

There Are Two Types of Cell Division in Eukaryotic Cells: Mitotic Cell Division and Meiotic Cell Division

Eukaryotic cells may undergo one of two evolutionarily related, but very different, types of cell division: mitotic cell division and meiotic cell division. **Mitotic cell division** consists of nuclear division (called **mitosis**) followed by cytoplasmic division (called **cytokinesis**). The word *mitosis* comes from the Greek for "thread"; during mitosis, chromosomes condense and appear as thin, thread-like structures when viewed through a light microscope. Cytokinesis (from the Greek words for "cell movement") is the process by which the cytoplasm is divided between the two daughter cells. As we will see later in this chapter, mitosis gives each daughter nucleus one copy of the parent cell's replicated chromosomes, and cytokinesis usually places one of these nuclei into each daughter cell. Hence, mitotic cell division typically produces two daughter cells that are genetically identical to each other and to the parent cell, and usually contain about equal amounts of cytoplasm.

Mitotic cell division takes place in all types of eukaryotic organisms. It is the mechanism of asexual reproduction in eukaryotic cells—including unicellular organisms such as yeast, *Amoeba*, and *Paramecium*—and in multicellular organisms such as *Hydra* and aspens. Finally, mitotic cell division is crucially important in multicellular organisms, even when the entire organism does not reproduce asexually.

In the life of any multicellular organism, mitotic cell division followed by differentiation of the daughter cells allows a fertilized egg to grow into an adult with perhaps trillions of specialized cells. Mitotic cell division also allows an organism to maintain its tissues, many of which require frequent replacement. For example, the cells of your stomach lining, which are constantly exposed to acid and digestive enzymes, survive only about three days. Without mitotic cell division to replace these short-lived cells, your body would soon be unable to function properly. Mitotic cell divisions also allow the body to repair itself, or sometimes even regenerate parts following injury.

Mitotic cell division also plays a role in biotechnology. Mitosis produces the nuclei used in cloning, which you will read about in "Scientific Inquiry: Carbon Copies—Cloning in Nature and the Lab" later in this chapter. Because mitosis usually produces daughter cells that are genetically identical to the parent cell, clones are genetically identical to their respective "nuclear donors" (the organisms that provided the nuclei for each cloning procedure). Finally, mitotic cell division may give rise to *stem cells*. These cells, which are found in both embryos and adults, may produce a wide variety of differentiated cell types, such as nerve cells, immune system cells, or muscle cells.

Meiotic cell division is a prerequisite for **sexual reproduction** in all eukaryotic organisms. In animals, meiotic cell division occurs only in ovaries and testes. The process of meiotic cell division involves a specialized nuclear division called **meiosis** and two rounds of cytokinesis to produce four daughter cells that can become **gametes** (eggs or sperm). Gametes carry half of the genetic material of the parent. Thus, the cells produced by meiotic cell division are not genetically identical to each other *or* to the original cell. During sexual reproduction, fusion of two gametes, one from each parent, reconstitutes a full complement of genetic material, forming a genetically unique offspring that is similar to both parents, but identical to neither (see section 11.6).

We will examine the events of mitosis and meiosis shortly. However, to understand the mechanisms of mitosis and meiosis, and their genetic and evolutionary significance, we first need to explore how DNA is packaged into eukaryotic chromosomes.

11.2   HOW IS DNA IN EUKARYOTIC CELLS ORGANIZED INTO CHROMOSOMES?

The Eukaryotic Chromosome Consists of a Linear DNA Double Helix Bound to Proteins

Fitting all the DNA of a eukaryotic cell into the nucleus is no trivial task. If it were laid end to end, the total DNA in a single cell in your body would be about 6 feet (1.83 meters) long, and this DNA must fit into a nucleus that is at least a million times smaller! The degree of DNA compaction, or condensation, varies at each stage of the cell cycle. During most of a cell's life, much of the DNA is extended, making it readily accessible for transcription. In this extended state, individual **chromosomes**, which consist of a single DNA double helix and many associated

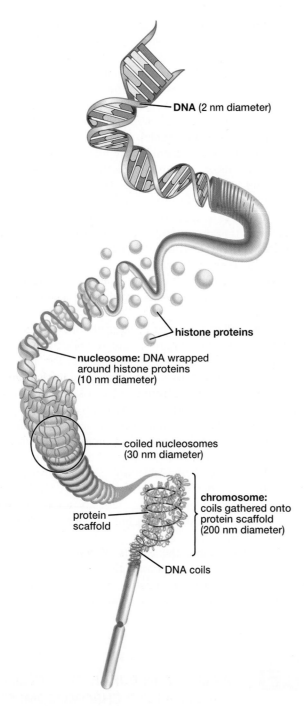

FIGURE 11-4 Chromosome structure
A eukaryotic chromosome contains a single, linear DNA double helix (top), which in humans is about 14 to 73 millimeters (mm) long and 2 nanometers (nm) in diameter. The DNA is wound around proteins called histones, forming nucleosomes (middle); this reduces the length by about a factor of 6. Other proteins coil up adjacent nucleosomes, much like a Slinky toy, reducing the length by another factor of 6 or 7. The coils of DNA and their associated proteins are attached in loops to still larger coils of protein "scaffolding" to complete the chromosome (bottom). All of this wrapping, coiling, and looping makes the extended interphase chromosome roughly 1000 times shorter than the DNA molecule it contains. Still other proteins produce about another tenfold condensation during cell division (see Fig. 11-6).

may contain hundreds or even thousands of genes, arranged in a particular linear order along the DNA strands. Each gene occupies a specific place, or **locus**, on a specific chromosome.

Chromosomes vary in length, and therefore in the number of genes they contain. The largest human chromosome, chromosome 1, contains approximately 3000 genes, whereas one of the smallest human chromosomes, chromosome 22, contains only about 600 genes.

In addition to genes, every chromosome has specialized regions that are crucial to its structure and function: two telomeres and one centromere (**FIG. 11-5**). The two ends of a chromosome consist of repeated nucleotide sequences called **telomeres** ("end body" in Greek), which are essential for chromosome stability. Without telomeres, the ends of chromosomes might be removed by DNA repair enzymes, or the ends of two or more chromosomes might become connected, forming long, unwieldy structures that probably could not be distributed properly to the daughter nuclei during cell division.

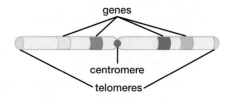

FIGURE 11-5 Principal features of a eukaryotic chromosome

At the time it condenses, the DNA within each chromosome has already replicated, forming two DNA double helices that remain attached to each other at the **centromere** (**FIG. 11-6**). Although *centromere* means "middle body," a chromosome's centromere can be located almost anywhere along the DNA double helix. While the two chromosomes are attached at their centromeres, we refer to

proteins (**FIG. 11-4**), are too thin to be visible in light microscopes. Cell division, however, requires that the chromosomes be sorted out and moved into two daughter nuclei. Just as thread is easier to organize when it is wound onto spools, sorting and transporting chromosomes is easier when they are condensed and shortened. During cell division, proteins fold up the DNA of each chromosome into compact structures that can be seen in a light microscope.

How are chromosomes and genes related? Recall that genes are sequences of DNA from hundreds to thousands of nucleotides long. A single DNA double helix

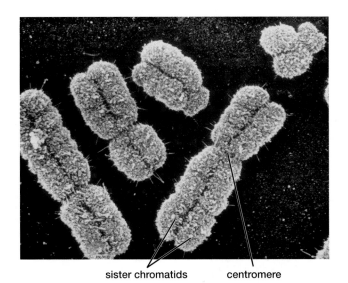

sister chromatids   centromere

FIGURE 11-6 Human chromosomes during mitosis
The DNA and associated proteins in these duplicated human chromosomes have coiled up into thick, short sister chromatids attached at the centromere. Each visible strand of "texture" is a loop of DNA. During cell division, the condensed chromosomes are about 5 to 20 micrometers long.

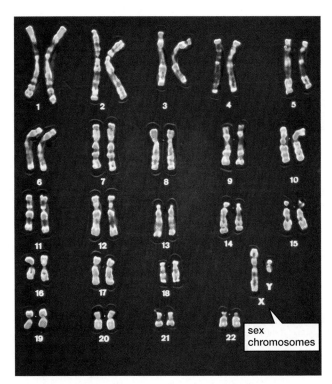

sex chromosomes

FIGURE 11-9 The karyotype of a human male
Staining and photographing the entire set of duplicated chromosomes within a single cell produces a karyotype. Pictures of the individual chromosomes are cut out and arranged in descending order of size. The chromosome pairs (homologues) are similar in both size and staining pattern and have similar genetic material. Chromosomes 1 through 22 are the autosomes; the X and Y chromosomes are the sex chromosomes. Notice that the Y chromosome is much smaller than the X chromosome. If this were a female karyotype, it would have two X chromosomes.

each attached chromosome as a sister **chromatid**. Thus, DNA replication produces a **duplicated chromosome** with two identical sister chromatids (**FIG. 11-7**):

sister chromatids — duplicated chromosome (2 DNA double helices)

FIGURE 11-7 A duplicated chromosome consists of two sister chromatids

During mitotic cell division, the two sister chromatids separate, and each chromatid becomes an independent chromosome that is delivered to one of the two daughter cells (**FIG. 11-8**):

independent daughter chromosomes, each with one identical DNA double helix

FIGURE 11-8 Sister chromatids separate to become two independent chromosomes

Eukaryotic Chromosomes Usually Occur in Homologous Pairs with Similar Genetic Information

The chromosomes of each eukaryotic species have characteristic shapes, sizes, and staining patterns (**FIG. 11-9**). When we view an entire set of stained chromosomes from a single cell (its **karyotype**), it becomes clear that the non-reproductive cells of many organisms, including humans, contain pairs of chromosomes. With one exception that we will discuss shortly, both members of each pair are the same length and have the same staining pattern. This similarity in size, shape, and staining occurs because each chromosome in a pair carries the same genes arranged in the same order. Chromosomes that contain the same genes are called *homologous chromosomes* or **homologues**, from Greek words that mean "to say the same thing." Cells with pairs of homologous chromosomes are called **diploid**, meaning "double."

Let's consider a human skin cell. Although it has 46 chromosomes, it does not have 46 completely *different* chromosomes. The cell has two copies of chromosome 1, two copies of chromosome 2, and so on, up through chromosome 22. These chromosomes, which have similar appearance, similar genetic composition, and are paired in diploid cells of both sexes, are called **autosomes**. The cell also has two **sex chromosomes**: two X chromosomes or an X and a Y chromosome. The X and Y chromosomes are quite different in size (see Fig. 11-9) and genetic composition.

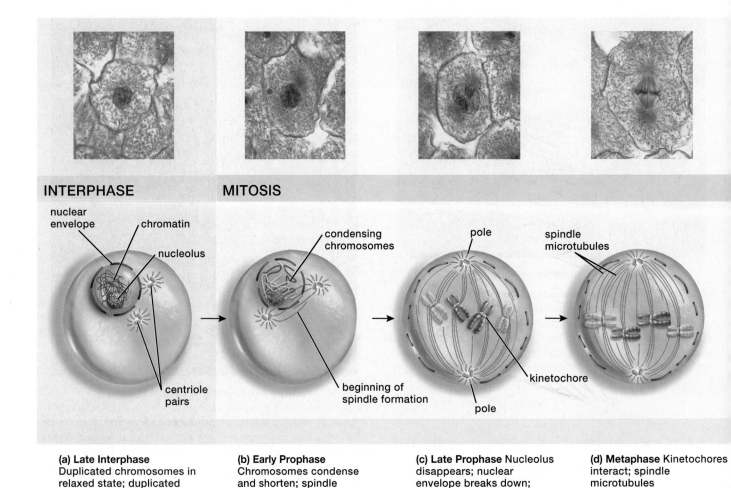

INTERPHASE   MITOSIS

(a) Late Interphase
Duplicated chromosomes in
relaxed state; duplicated
centrioles remain clustered.

(b) Early Prophase
Chromosomes condense
and shorten; spindle
microtubules begin to form
between separating
centriole pairs.

(c) Late Prophase Nucleolus
disappears; nuclear
envelope breaks down;
spindle microtubules attach
to the kinetochore of each
sister chromatid.

(d) Metaphase Kinetochores
interact; spindle
microtubules
line up chromosomes
at cell's equator.

FIGURE 11-10 Mitotic cell division in an animal cell
QUESTION What would the consequences be if one set of sister chromatids failed to separate at anaphase?

Thus, sex chromosomes are an exception to the rule that ho-mologous chromosomes contain the same genes. However, as we will see later on, the X and Y chromosomes behave like homologues during meiotic cell division, and so the X and Y are considered as a pair in our "chromosomal book-keeping."

Most cells within our bodies are diploid. However, during sexual reproduction, cells in the ovaries or testes undergo meiotic cell division to produce gametes (sperm or eggs) that contain only one member of each pair of autosomes and one of the two sex chromosomes. Cells that contain only one of each type of chromosome are called **haploid** (mean-ing "half"). In humans, a haploid cell contains one each of the 22 autosomes, plus either an X or Y sex chromosome, for a total of 23 chromosomes. (Think of a haploid cell as one

that contains *half* the diploid number of chromosomes, or one of each type of chromosome. A diploid cell contains two of each type of chromosome.) When a sperm fertilizes an egg, fusion of the two haploid cells produces a diploid cell with two copies of each type of chromosome.

In biological shorthand, the number of different types of chromosomes in a species is called the *haploid number* and is designated n. For humans, $n = 23$ because we have 23 dif-ferent types of chromosomes (autosomes 1 to 22 plus one sex chromosome). Diploid cells contain $2n$ chromosomes. Thus, each human nonreproductive cell has 46 (2×23) chromosomes.

Every species has a characteristic number of chromo-somes in its cells, but the number differs tremendously be-tween species.

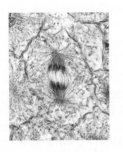
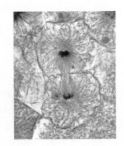
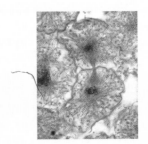

INTERPHASE

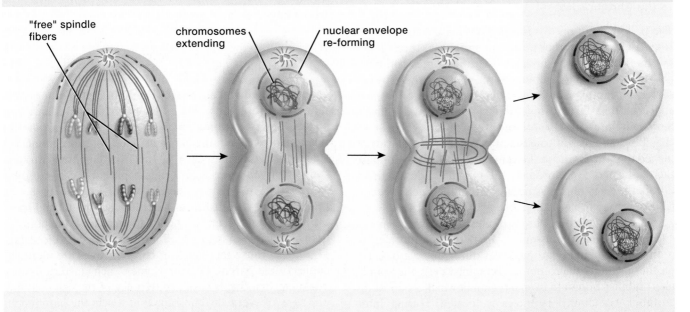

"free" spindle fibers

chromosomes extending

nuclear envelope re-forming

(e) Anaphase Sister chromatids separate and move to opposite poles of the cell; spindle microtubules push poles apart.

(f) Telophase One set of chromosomes reaches each pole and relaxes into extended state; nuclear envelopes start to form around each set; spindle microtubules begin to disappear.

(g) Cytokinesis Cell divides in two; each daughter cell receives one nucleus and about half of the cytoplasm.

(h) Interpase of daughter cells Spindles disappear, intact nuclear envelopes form, chromosomes extend completely, and the nucleolus reappears.

Organism	*n* (haploid number)	*2n* (diploid number)
Human	23	46
Gorilla, chimpanzee	24	48
Dog	39	78
Cat	19	38
Shrimp	127	254
Fruit fly	4	8
Pea	7	14
Potato	24	48
Sweet potato	45	90

Not all organisms are diploid. The bread mold *Neurospora*, for example, has haploid cells for most of its life cycle. Some plants, on the other hand, have more than two copies of each type of chromosome, with 4*n*, 6*n*, or even more chromosomes per cell.

11.3 HOW DO CELLS REPRODUCE BY MITOTIC CELL DIVISION?

As we described earlier, mitotic cell division (**FIG. 11-10**) consists of mitosis (nuclear division) and cytokinesis (cytoplasmic division). After interphase (Fig. 11-10a), when the cell's chromosomes have been replicated and all other necessary preparations for division have been made, mitotic cell division can occur. We will discuss mitosis and cytokinesis separately, even though they may overlap in time.

For convenience, biologists divide mitosis into four phases, based on the appearance and behavior of the chromosomes: (1) *prophase*, (2) *metaphase*, (3) *anaphase*, and (4) *telophase*. As with most biological processes, however, these phases are not really discrete events. Rather, they form a continuum, each phase merging into the next.

During Prophase, the Chromosomes Condense and the Spindle Microtubules Form and Attach to the Chromosomes

The first phase of mitosis is called **prophase** (meaning "the stage before" in Greek). During prophase, three major events occur: (1) the duplicated chromosomes condense, (2) the spindle microtubules form, and (3) the chromosomes are captured by the spindle (Fig. 11-10b, c).

Recall that chromosome duplication occurs during the S phase of interphase. Therefore, when mitosis begins, each chromosome already consists of two sister chromatids attached to one another at the centromere. During prophase, the duplicated chromosomes coil up and condense. In addition, the nucleolus, a structure within the nucleus where ribosomes assemble, disappears.

After the duplicated chromosomes condense, the **spindle microtubules** begin to assemble. In all eukaryotic cells, the proper movement of chromosomes during mitosis depends on these spindle microtubules. In animal cells, the spindle microtubules originate from a region in which a pair of microtubule-containing **centrioles** is located. During interphase, a new pair of centrioles forms near the previously existing pair. During prophase, the centriole pairs migrate to opposite sides of the nucleus. Each centriole pair serves as a central point from which the spindle microtubules radiate, both inward toward the nucleus and outward toward the plasma membrane. These points are called *spindle poles*. Though the cells of plants, fungi, and many algae do not contain centrioles, they nevertheless form functional spindles during mitotic cell division.

As the spindle microtubules form into a complete basket around the nucleus, the nuclear envelope disintegrates, releasing the duplicated chromosomes. Each sister chromatid has a protein-containing structure at its centromere called a **kinetochore**, which serves as the attachment site for the spindle microtubules. In each duplicated chromosome, the kinetochore of one sister chromatid binds to the ends of spindle microtubules leading to one pole of the cell, while the kinetochore of the other sister chromatid binds to spindle microtubules leading to the opposite pole of the cell (Fig. 11-10c). When the sister chromatids separate later in mitosis, the newly independent chromosomes will move along the spindle microtubules to opposite poles. Some spindle microtubules do not attach to chromosomes; rather, they have free ends that overlap along the cell's equator. As we will see, these unattached spindle microtubules will push the two spindle poles apart later in mitosis.

During Metaphase, the Chromosomes Align Along the Equator of the Cell

At the end of prophase, the two kinetochores of each duplicated chromosome are connected to spindle microtubules leading to opposite poles of the cell. As a result, each duplicated chromosome is connected to both spindle poles. During **metaphase** (the "middle stage"), the two kinetochores on a duplicated chromosome engage in a "tug of war." During this process, the microtubules lengthen or shorten, until each chromosome lines up along the equator of the cell, with one kinetochore facing each pole (Fig. 11-10d).

During Anaphase, Sister Chromatids Separate and Are Pulled to Opposite Poles of the Cell

At the beginning of **anaphase** (Fig. 11-10e), the sister chromatids separate, becoming independent daughter chromosomes. This separation allows "motor proteins" in the kinetochores to pull the chromosomes poleward along the spindle microtubules. One of the two daughter chromosomes derived from each original parental chromosome moves to each pole of the cell. As the kinetochores tow their chromosomes toward the poles, the unattached spindle microtubules interact and lengthen to push the poles of the cell apart, forcing the cell into an oval shape (see Fig. 11-10e). Because the daughter chromosomes are identical copies of the parental chromosomes, each cluster of chromosomes that forms at opposite poles of the cell contains one copy of every chromosome that was in the parent cell.

During Telophase, Nuclear Envelopes Form Around Both Groups of Chromosomes

When the chromosomes reach the poles, **telophase** (the "end stage") begins (Fig. 11-10f). The spindle microtubules disintegrate, and a nuclear envelope forms around each group of chromosomes. The chromosomes revert to their extended state, and the nucleoli reappear. In most cells, cytokinesis occurs during telophase, separating each daughter nucleus into a separate cell (Fig. 11-10g).

During Cytokinesis, the Cytoplasm Is Divided Between Two Daughter Cells

In animal cells, microfilaments attached to the plasma membrane form a ring around the equator of the cell. During cytokinesis, the ring contracts and constricts the cell's equator, much like the drawstring on a pair of sweatpants tightens the waist when pulled. Eventually the "waist" constricts completely, dividing the cytoplasm into two new daughter cells (**FIG. 11-11**).

Cytokinesis in plant cells is quite different, perhaps because their stiff cell walls make it impossible to divide one cell into two by pinching at the waist. Instead, carbohydrate-filled vesicles, which bud off the Golgi apparatus, line up along the cell's equator between the two nuclei

(a)

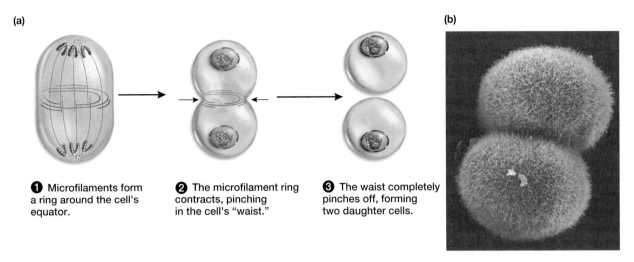

❶ Microfilaments form a ring around the cell's equator.

❷ The microfilament ring contracts, pinching in the cell's "waist."

❸ The waist completely pinches off, forming two daughter cells.

(b)

FIGURE 11-11 Cytokinesis in an animal cell
(a) A ring of microfilaments just beneath the plasma membrane contracts around the equator of the cell, pinching it in two. (b) This scanning electron micrograph of cytokinesis shows the two daughter cells nearly separated.

(FIG. 11-12). The vesicles fuse, producing a structure called the **cell plate**, which is shaped like a flattened sac, surrounded by plasma membrane, and filled with sticky carbohydrates. When enough vesicles have fused, the edges of the cell plate merge with the original plasma membrane around the circumference of the cell. The carbohydrate formerly contained in the vesicles remains between the plasma membranes as part of the cell wall.

Following cytokinesis, eukaryotic cells enter G_1 of interphase, thus completing the cell cycle (Fig. 11-10h).

11.4 HOW IS THE CELL CYCLE CONTROLLED?

As you know, some cells, such as those of the stomach lining, divide frequently throughout the life of an organism. Others divide more or less often, depending on various conditions. For example, liver and skin cells are stimulated to divide by damage, which results in repair and regrowth. Still other cells, such as most cells in the brain, heart, and skeletal muscles, never divide in an adult. Cell division is regulated by a

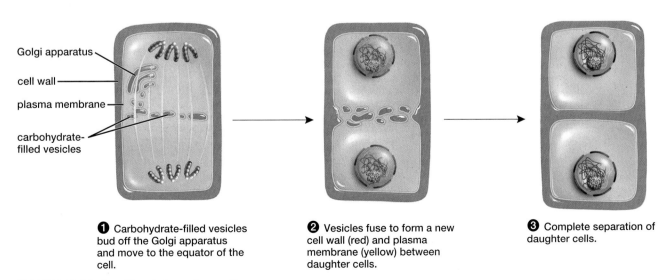

Golgi apparatus
cell wall
plasma membrane
carbohydrate-filled vesicles

❶ Carbohydrate-filled vesicles bud off the Golgi apparatus and move to the equator of the cell.

❷ Vesicles fuse to form a new cell wall (red) and plasma membrane (yellow) between daughter cells.

❸ Complete separation of daughter cells.

FIGURE 11-12 Cytokinesis in a plant cell

The word "cloning" usually brings to mind images of Dolly the sheep or even *Star Wars: Attack of the Clones*, but nature has been quietly cloning for hundreds of millions of years. Everyone knows what **cloning** is: the creation of one or more individual organisms (**clones**) that are genetically identical to a preexisting individual. How are clones produced, either in nature or in the lab? Why is cloning such a hot—and controversial—topic in the news? And why is cloning included in a chapter on cell division?

CLONING IN NATURE: THE ROLE OF MITOTIC CELL DIVISION

Let's address the last question first. As you know, there are two types of cell division: mitotic division and meiotic division. Sexual reproduction relies on meiotic cell division, the production of gametes, and fertilization, and usually produces genetically unique offspring. In contrast, asexual reproduction (see Fig. 11-1) relies on mitotic cell division. Because mitotic cell division creates daughter cells that are genetically identical to the parent cell, offspring produced by asexual reproduction are genetically identical to their parents—clones.

CLONING PLANTS: A FAMILIAR APPLICATION IN AGRICULTURE

Humans have been in the cloning business a lot longer than you might think. For example, consider navel oranges,

which don't produce seeds. Without seeds, how do they reproduce? Navel orange trees are propagated by cutting a piece of stem from an adult navel tree and grafting it onto the top of the root of a seedling orange tree, usually of a different type. (Why would the seedling usually not be a navel?) Therefore, the cells of the aboveground, fruit-bearing parts of the resulting tree are clones of the original navel orange stem. All navel oranges apparently originated from a single mutant bud of an orange tree discovered in Brazil in the early 1800s, and propagated asexually ever since. Three navel orange trees were brought from Brazil to Riverside, California, in the 1870s. (One of them is still there!) All American navels are clones of these three trees.

CLONING ADULT MAMMALS

Animal cloning isn't a recent development either. In the 1950s, John Gurdon and his colleagues inserted a nucleus from early frog embryos into eggs, and some of the resulting cells developed into complete frogs. By the 1990s, several labs had been able to clone mammals using embryonic nuclei, but it wasn't until 1996 that Dr. Ian Wilmut of the Roslin Institute in Edinburgh, Scotland, cloned the first adult mammal, the famous Dolly (**FIG. E11-1**).

Why is it important to clone an adult animal? In agriculture, it's usually worthwhile to clone only adults, because only in adults can we see the traits that we wish to propagate (such as milk and meat production in cows, or speed

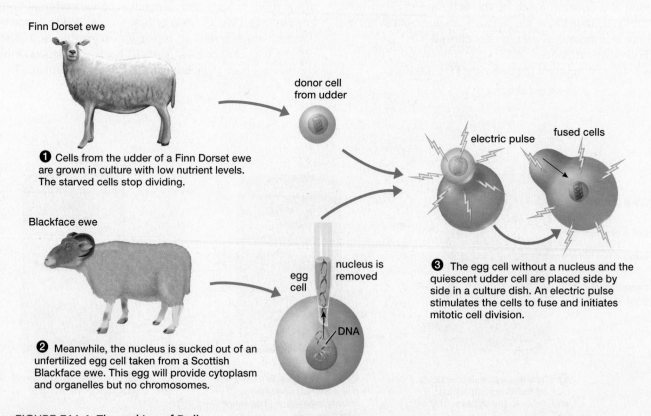

Finn Dorset ewe

donor cell from udder

electric pulse   fused cells

❶ Cells from the udder of a Finn Dorset ewe are grown in culture with low nutrient levels. The starved cells stop dividing.

Blackface ewe

egg cell

nucleus is removed

DNA

❷ Meanwhile, the nucleus is sucked out of an unfertilized egg cell taken from a Scottish Blackface ewe. This egg will provide cytoplasm and organelles but no chromosomes.

❸ The egg cell without a nucleus and the quiescent udder cell are placed side by side in a culture dish. An electric pulse stimulates the cells to fuse and initiates mitotic cell division.

FIGURE E11-1 The making of Dolly

and strength in horses). Cloning an adult would produce "offspring" that are genetically identical to the adult. Therefore, any valuable traits of the adult that are genetically determined would also be expressed in all of its clones. Cloning of embryos would usually not be useful, because the embryonic cells would have been produced by sexual reproduction in the first place, and normally no one could tell if the embryo had any especially desirable traits.

For some medical applications, too, cloning adults is essential. Suppose that a pharmaceutical company genetically engineered (see Chapter 13) a cow that secreted a valuable molecule, such as an antibiotic, in its milk. These techniques are extremely expensive and somewhat hit or miss, so the company might successfully produce only one profitable cow. This cow could then be cloned, creating a whole herd of antibiotic-producing cows. Cloned cows that produce more milk or meat, and pigs tailored to be organ donors for humans, already exist.

Cloning might also help rescue critically endangered species, many of which don't reproduce well in zoos. As Richard Adams of Texas A&M University put it, "You could repopulate the world [with an endangered species] in a matter of a couple of years. Cloning is not a trivial pursuit."

CLONING: AN IMPERFECT TECHNOLOGY

Unfortunately, cloning mammals is inefficient and beset with difficulties. An egg is subjected to severe trauma when its nucleus is sucked out or destroyed, and a new nucleus is inserted (see Fig. E11-1). Often, the egg may simply die. Molecules in the cytoplasm that are needed to control development may be lost or moved to the wrong places, so that even if the egg survives and divides, it may not develop properly. If the eggs develop into viable embryos, the embryos must then be implanted into the uterus of a surrogate mother. Many clones die or are aborted during gestation, often with serious or fatal consequences for the surrogate mother. Even if the clone survives gestation and birth, it may have defects, commonly a deformed head, lungs, or heart. Given the high failure rate—it took 277 tries to produce Dolly—cloning mammals is an expensive proposition.

To make things even more problematic, "successful" clones may have hidden defects. Dolly, for example, had "middle-aged" chromosomes. Remember the telomeres on the ends of chromosomes? At each mitotic cell division, the telomeres get a little shorter, and it appears that cells may die—or at least no longer divide—when their telomeres get too short. Dolly was born with short telomeres, as if she were already over 3 years old. On the other hand, not all cloned mammals have short telomeres; the right cloning techniques, using the optimal adult tissue (skin cells seem to be better than mammary gland cells, from which Dolly was cloned) may eliminate problems with shortened telomeres. However, Dolly seemed to have other troubles, too. She developed arthritis when she was $5\frac{1}{2}$ and was euthanized with a serious lung disease when she was $6\frac{1}{2}$, so her problems occurred at a relatively young age (the typical life span of a sheep is 11 to 16 years), although no one knows if these health problems occurred because she was a clone.

THE FUTURE OF CLONING

A new technology, called *chromatin transfer*, appears to reduce the likelihood of defective clones. Many researchers think that DNA in "old" cells is in a different chemical state than the DNA of a newly fertilized egg. Although inserting

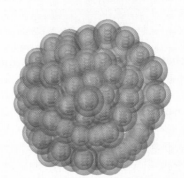

❹ The cell divides, forming an embryo that consists of a hollow ball of cells.

❺ The ball of cells is implanted into the uterus of another Blackface ewe.

❻ The Blackface ewe gives birth to Dolly, a female Finn Dorset lamb, a genetic twin of the Finn Dorset ewe.

(a) CC as a kitten in 2001

(b) Little Nicky, the $50,000 kitten, in 2004

FIGURE E11-2 Cloned cats
(a) "Standard" technology was used to make CC, the first cloned cat, in 2001. CC was the only successful birth out of 87 cloned embryos. **(b)** Little Nicky, the first cat cloned for a paying customer, was cloned using the chromatin transfer technique, which has a much higher success rate.

an old nucleus into an enucleated egg helps to rejuvenate the DNA, it doesn't always do the trick. In chromatin transfer, the membranes of donor cells are made leaky. The permeable cells are then incubated with a "mitotic extract" derived from rapidly dividing, and therefore "young," cells. This remodels the DNA of the older cell and causes it to condense, much like DNA does during prophase of mitotic cell division. The rejuvenated cell is then fused with an enucleated egg, as with more traditional cloning procedures. A company wittily called Genetic Savings and Clone, which cloned the first cat, CC (**FIG. E11-2a**), by conventional methods, now uses chromatin transfer, with a much higher success rate, to clone pet cats (**FIG. E11-2b**). (In 2005, the price to clone your cat was reduced to a mere $32,000!)

Modern cloning technology has now successfully cloned cows, cats, sheep, horses, and a variety of other animals. As the process becomes more routine, it also brings ethical questions. While hardly anyone objects to cloning navel oranges, and few would refuse antibiotics or other medicinal products from cloned livestock, some people think that cloning pets is a frivolous luxury—especially when you consider that every 9 seconds, an unwanted dog or cat is euthanized in the United States. And what about human cloning? Early in 2003, there were claims that two cloned children had been born (although this has never been confirmed). Assuming that the technology existed to clone people, would it be a good idea? What about therapeutic cloning, in which a person's DNA might be used to start a cloned embryo, and some of its young, undifferentiated cells would be used to treat the donor's disease or regenerate an organ, without fear of transplant rejection? What do you think?

---

bewildering array of molecules, not all of which have been identified and studied. Nevertheless, several general principles are common to most eukaryotic cells.

Checkpoints Regulate Progress Through the Cell Cycle

There are three major **checkpoints** in the eukaryotic cell cycle (**FIG. 11-13**). At each checkpoint, protein complexes in the cell determine whether the cell has successfully completed a specific phase of the cycle and regulate the activities of other proteins that move the cell to the next phase:

- G_1 to S: Is the cell's DNA suitable for replication?

- G_2 to mitosis: Has the DNA been completely and accurately replicated?

- Metaphase to anaphase: Are the chromosomes aligned properly at the metaphase plate?

The Activities of Specific Enzymes Drive the Cell Cycle

The cell cycle is controlled by a family of proteins called cyclin-dependent kinases, or Cdk's for short. These proteins get their name from two features: First, a kinase is an enzyme that phosphorylates (adds a phosphate group to) other proteins, stimulating or inhibiting the activity of the target protein. Second, these are "cyclin-dependent" because they are active only when they bind still another protein, called a cyclin. The name "cyclin" tells you a lot about these proteins: their abundance changes during the cell cycle, and in fact helps to regulate the cell cycle.

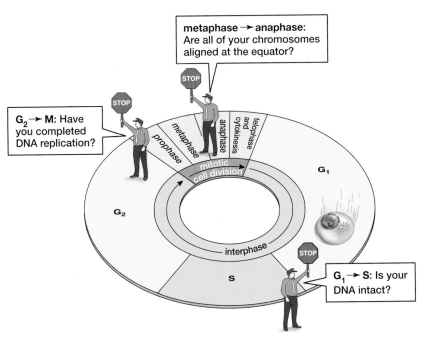

FIGURE 11-13 Control of the cell cycle
Three major "checkpoints" regulate a cell's transitions from one phase of the cell cycle to the next: (1) G₁ to S, (2) G₂ to mitosis (M), and (3) metaphase to anaphase.

Normal cell-cycle control works like this (**FIG. 11-14**). In most cases, a cell will divide only if it receives signals from hormone-like molecules called *growth factors*. For example, if you cut your skin, platelets (cell fragments in the blood that are involved in blood clotting) accumulate at the wound site and release growth factors, including the appropriately named platelet-derived growth factor and epidermal growth factor. These growth factors bind to receptors on the surfaces of cells deep in the skin, triggering a cascade of molecular interactions in which the activity of one molecule stimulates the activity of the next, again and again, ultimately culminating in progression through the cell cycle. When a skin cell in G₁ is stimulated by these growth factors, it synthesizes cyclin proteins that bind to, and activate, specific Cdk's. These Cdk's then stimulate the synthesis and activity of proteins that are required for DNA synthesis to occur. The cell thus enters the S phase and replicates its DNA. After DNA replication is complete, other Cdk's become activated, causing chromosome condensation, breakdown of the nuclear envelope, formation of the spindle, and

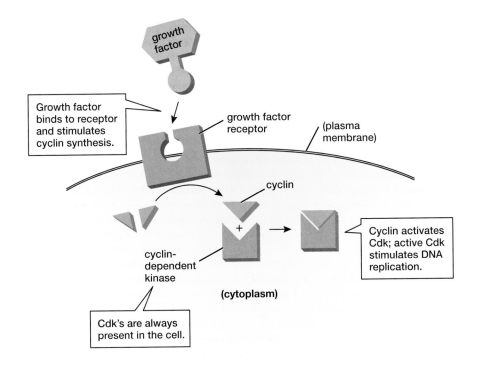

FIGURE 11-14 The G₁ to S checkpoint
Progress through the cell-cycle checkpoints is under the overall control of cyclin and cyclin-dependent kinases (Cdk's). In the G₁ to S checkpoint illustrated here, growth factors stimulate synthesis of cyclin proteins, which activate Cdk's, starting a cascade of events that lead to DNA replication.

attachment of the chromosomes to the spindle micro-tubules. Finally, still other Cdk's stimulate processes that allow the sister chromatids to separate into individual chromosomes, and move to opposite poles of the cell during anaphase.

Controls over the Checkpoints

Many things can go wrong during the cell cycle. For example, the DNA may have suffered mutations, or the cell may not have accumulated enough nutrients. Therefore, there are a variety of controls that regulate movement through the checkpoints.

G_1 to S Checkpoint

Because of its importance in preventing cancer, we'll look at the G_1 to S checkpoint in some detail (**FIG. 11-15**). One of the proteins that is regulated by phosphorylation by Cdk's is called Rb (which stands for retinoblastoma, because defective Rb proteins cause cancer of the retina). Rb inhibits transcription of several genes whose protein products are required for DNA synthesis. Phosphorylation of Rb by Cdk's relieves this inhibition, allowing DNA replication to proceed (Fig. 11-15a).

Another protein, called p53 (which merely means "a *p*rotein with a molecular weight of *53*,000"), indirectly regulates Rb activity (Fig. 11-15b). There is little p53 protein in a healthy cell. However, when DNA has been damaged

(for example, by ultraviolet light in sunlight), p53 levels rise. The p53 protein then stimulates the expression of proteins that inhibit Cdk's. If Cdk's are inhibited, then Rb is not phosphorylated, so DNA synthesis is blocked. p53 also stimulates the synthesis of DNA repair enzymes. After the DNA has been repaired, p53 levels decline, Cdk's become active, Rb becomes phosphorylated, and the cell enters the S phase. If the DNA cannot be repaired, p53 triggers a special form of cell death called *apoptosis*, in which the cell cuts up its DNA and effectively commits suicide.

Remember the muscle-bound bull in Chapter 9? Like p53, myostatin triggers a chain of protein interactions that block phosphorylation of Rb, thereby preventing DNA replication and cell division. Defective myostatin causes excessive activation of Rb, so pre-muscle cells divide more than they normally would, producing extra-muscular cattle.

G_2 to Mitosis Checkpoint

The p53 protein is also involved in controlling the progression from G_2 to mitosis. Increased levels of p53 caused by defective DNA (for example, mismatched base pairs as a result of faulty replication) decrease the synthesis and activity of an enzyme that helps to cause chromosome condensation. Therefore, the chromosomes remain extended and accessible to DNA repair enzymes, while the cell "waits" to enter mitosis until after the DNA has been fixed.

Metaphase-to-Anaphase Checkpoint

Although the mechanisms are not entirely understood, a cell also monitors the attachment of the chromosomes to the spindle, and if the chromosomes are aligned at the equator during metaphase. If even a single chromosome is not attached to the spindle, or the spindle microtubules attaching a chromosome to opposite poles of the cell aren't pulling equally hard (which probably means that the chromosome isn't at the equator), a variety of proteins prevent separation of the sister chromatids and hence prevent progression to anaphase.

As you can see, growth factor stimulation ensures that a cell divides only when it should. Multiple checkpoints ensure that the cell successfully completes DNA synthesis during interphase and proper chromosome movements during mitotic cell division. Of course, sometimes the cell cycle does not progress properly. Defects in either growth factor stimulation or checkpoint function can allow cells to divide without control, forming a cancer. We will explore the mechanisms that subvert the control of the cell cycle in "Health Watch: Cancer—Mitotic Cell Division Run Amok."

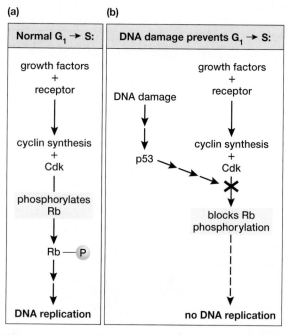

FIGURE 11-15 Controlling the transition from G_1 to S
(a) The Rb protein inhibits DNA synthesis. Toward the end of the G_1 phase, cyclin levels rise. These activate Cdk, which then adds a phosphate group to the Rb protein. Phosphorylated Rb no longer inhibits DNA synthesis, so the cell enters the S phase. **(b)** Damaged DNA stimulates increased levels of the p53 protein, which triggers a cascade of events that inhibit Cdk, thereby preventing entry into the S phase until the DNA has been repaired.

11.5 WHY DO SO MANY ORGANISMS REPRODUCE SEXUALLY?

The largest organism known on Earth is a mushroom whose underground, branching filaments extend through 2200 acres of soil in eastern Oregon. This organism was produced almost entirely by mitotic cell division. Clearly, asexual reproduction via mitotic cell division must work pretty well! Why, then, have nearly all known forms of life, including mushrooms, evolved ways of sexual reproduction? Mitosis can only produce clones of genetically identical offspring. In contrast, sexual reproduction shuffles genes to produce genetically unique offspring. The nearly universal presence of sexual reproduction provides evidence for the tremendous evolutionary advantage that DNA exchange among individuals confers on a species.

Mutations in DNA Are the Ultimate Source of Genetic Variability

As we saw in Chapter 10, the fidelity of DNA replication and proofreading minimizes errors during DNA replication, but changes in DNA base sequences do occur, producing mutations. Although most mutations are either neutral or harmful, they are also the raw material for evolution. Bacteria are different from bison, and you are different from your ancestors, because of differences in DNA sequence that originally arose as mutations. Mutations in gametes may be passed to offspring and become a part of the genetic makeup of the species. Such mutations form **alleles**, alternate forms of a given gene that may produce differences in structure or function—such as black, brown, or blond hair in humans, or different mating calls in frogs. As we saw earlier, most eukaryotic organisms are diploid, containing pairs of homologous chromosomes. Homologous chromosomes have the same genes, but each homologue may have the same alleles of some genes and different alleles of other genes (**FIG. 11-16**).

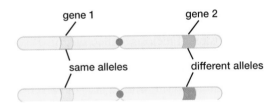

FIGURE 11-16 Homologous chromosomes may have the same (left) or different (right) alleles of individual genes

We'll explore the consequences of having paired genes—and more than one allele of each gene—in the next chapter.

Sexual Reproduction May Combine Different Parental Alleles in a Single Offspring

To illustrate how sexual reproduction promotes genetic variability, let's consider a familiar organism: the domestic cat. As a result of generations of selective breeding by cat fanciers, a lot is known about cat genetics. For example, cat fur comes in two basic lengths, long and short, controlled by two alleles of a single gene. Cats may also have the typical number of toes (four on each paw) or have extra toes (a condition called polydactyly). Toe number is also controlled by two alleles of a single gene (a different gene than the one that determines fur length). Let's suppose

you have short-haired, four-toed cats and long-haired, polydactylous cats, but you'd like to breed short-haired, polydactylous cats. You could breed each type of cat only to similar cats, wait for a mutation in either the fur-length gene or the toe-number gene, and hope that you live for 10,000 years and get lucky. Or, you could mate short-haired, 4-toed cats with long-haired, polydactylous cats. After a few generations, sexual recombination will produce some short-haired, polydactylous cats.

From an evolutionary perspective, making short-haired, polydactylous cats isn't very useful, but you can probably imagine many traits in wild animals or plants that would be useful only when combined. For example, camouflage coloration can help an animal avoid predation only if it stays still when it sees a predator. Both camouflaged animals that constantly jump around and brightly colored animals that freeze when a predator appears will probably be eaten. Let's suppose that a ground-nesting bird has better-than-average camouflage color, while another bird of the same species has more effective "freezing" behavior. Combining the two through sexual reproduction might produce offspring that are able to avoid predation better than either parent. Combining useful, genetically determined traits is one reason sexual reproduction is so nearly ubiquitous in nature.

How does sexual reproduction combine traits from two parents in a single offspring? The first eukaryotic cells to evolve, about 1 billion to 1.5 billion years ago, were probably haploid, with only one copy of each chromosome. Relatively early on, two evolutionary events probably occurred in single-celled eukaryotic organisms that allowed them to shuffle and recombine genetic information. First, two haploid (parental) cells fused, resulting in a diploid cell with two copies of each chromosome. This cell could reproduce by mitotic cell division, producing diploid daughter cells. Second, this population of diploid cells evolved a variation in the process of cell division called meiotic cell division. Meiotic cell division produces haploid cells, each containing one copy of each chromosome. In animals, these haploid cells usually become gametes. A haploid sperm from animal A might contain alleles contributing to camouflage coloration, and a haploid egg from animal B might contain alleles that favor freezing at the first sign of a predator. Fusion of these gametes would produce an animal with camouflage coloration that also becomes motionless when a predator approaches.

11.6 HOW DOES MEIOTIC CELL DIVISION PRODUCE HAPLOID CELLS?

Meiosis Separates Homologous Chromosomes, Producing Haploid Daughter Nuclei

The key to sexual reproduction in eukaryotes is meiosis, the production of haploid nuclei with unpaired chromosomes from diploid parent nuclei with paired chromosomes. In meiotic cell division (meiosis followed by cytokinesis), each daughter cell receives one member of each pair of homologous chromosomes. Therefore, meiosis

Mitotic cell division is essential for the development of multicellular organisms from single fertilized eggs, as well as for routine maintenance of body parts such as the skin and the lining of the digestive tract. Unfortunately, uncontrolled cell division is a menace to life: cancer. How do cancers escape the complex processes that normally regulate the cell cycle? There are many mechanisms, but almost all have two features in common: (1) mutations in DNA leading to (2) overactive *oncogenes* or inactive *tumor suppressor genes* (**FIG. E11-3**).

ONCOGENES

The term "oncogene" literally means "a gene that causes cancer." How can a gene cause cancer? Any gene whose activity tends to promote mitotic cell division, such as growth factor receptors and some of the cyclins and cyclin-dependent kinases, is called a proto-oncogene. By themselves, proto-oncogenes are harmless, and indeed are essential to properly controlled cell division. However, a mutation may convert a proto-oncogene to an oncogene. For example, mutated receptors for growth factors may be "turned on" all the time, regardless of the presence or absence of a growth factor (see Fig. E11-3a). Certain mutations in cyclin genes cause the cyclins to be synthesized at a high rate, regardless of growth factor activity. In either case, a cell may skip through some of the checkpoints that are normally regulated by fluctuating cyclin concentrations.

TUMOR SUPPRESSOR GENES

Although we didn't call them by that name, we have already met two tumor suppressor genes: *Rb* and *p53* (see section 11.4). Recall that *Rb* inhibits the synthesis of proteins needed for DNA replication, unless the Rb protein is phosphory-

lated by cyclin-dependent kinases (Cdk's). Normally, damaged DNA increases *p53* levels, which indirectly inhibit Cdk activity, so the Rb protein cannot be phosphorylated. The result is that the cell doesn't replicate defective DNA. Many carcinogens mutate the *p53* and *Rb* genes, so that the proteins can't do their jobs (Fig. E11-3b). Mutated *p53* is inactive, so Cdk's are overactive, phosphorylating Rb and allowing DNA to be replicated. Mutated *Rb* mimics phosphorylated Rb, which also permits unregulated DNA synthesis. With either mutation, replication proceeds, whether or not the DNA has been damaged. Even if the DNA is otherwise intact, the cell skips right through the G_1 to S checkpoint and may divide much more frequently than it should. Not surprisingly, about half of all cancers—including tumors of the breast, lung, brain, pancreas, bladder, stomach, and colon—have mutations in *p53*. Many others, including tumors of the eye (retinoblastoma), lung, breast, and bladder have mutated *Rb*.

FROM MUTATED CELL TO CANCER

In most cases, a mutation in a cell is quickly fixed by DNA repair enzymes. If a little extra time is needed, *p53* activity blocks the transition from G_1 to S or from G_2 to mitosis until the DNA has been fixed. If the mutation is extensive (such as a translocation or inversion) or cannot be fixed, then prolonged, high *p53* activity usually causes the cell to kill itself by apoptosis. But what if *p53* is mutated as well? Does that doom a person to malignant cancer? Not necessarily. Many mutations cause the surface of a cell to "look different" to the cells of the immune system, which then kills the mutated cell. Occasionally, however, a renegade cell survives and reproduces. Because mitotic cell division faithfully transmits

(from a Greek word meaning "to diminish") reduces the number of chromosomes in a diploid cell by half. For example, each diploid cell in your body contains 23 *pairs* of chromosomes; meiotic cell division produces sperm or eggs with 23 chromosomes, one from each pair.

Meiosis evolved from mitosis, so many of the structures and events of meiosis are similar or identical to those of mitosis. However, meiosis differs from mitosis in a major way: during meiosis, the cell undergoes *one* round of DNA replication followed by *two* nuclear divisions. One round of DNA replication produces two chromatids in each duplicated chromosome. Because diploid cells have pairs of homologous chromosomes—with two chromatids per homologue—a single round of DNA replication creates four chromatids for each type of chromosome (**FIG. 11-17**).

The first division of meiosis (called *meiosis I*) separates the pairs of homologues and sends one of each pair into each of two daughter nuclei, producing two haploid nuclei. Each homologue, however, still consists of two chromatids (**FIG. 11-18**).

FIGURE 11-18 During meiosis I, each daughter cell receives one member of each pair of homologous chromosomes.

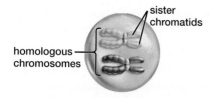

sister chromatids

homologous chromosomes

FIGURE 11-17 Both members of a pair of homologous chromosomes are replicated prior to meiosis

A second division (called *meiosis II*) separates the chromatids and parcels one chromatid into each of two

genetic information from cell to cell, all daughter cells of the original cancerous cell will themselves be cancerous.

Why does medical science, which has conquered smallpox, measles, and a host of other diseases, have such a difficult time curing cancer? Both normal and cancerous cells use the same machinery for cell division, so treatments that slow down the multiplication of cancer cells also inhibit the maintenance of essential body parts, such as the stomach, intestine, and blood cells. Truly effective and *selective* treatments for cancer must target cell division only in cancerous cells. Although great strides have been made in the fight to cure cancer, much remains to be done.

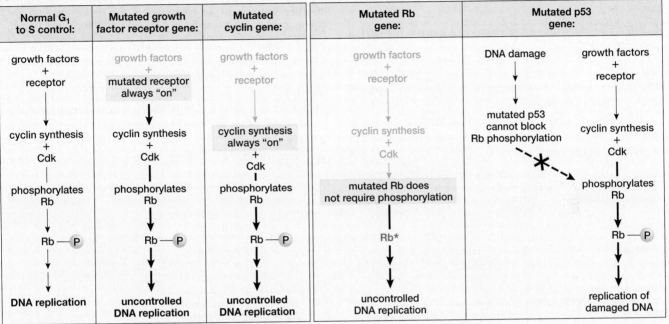

(a) Actions of oncogenes

Normal G$_1$ to S control:	Mutated growth factor receptor gene:	Mutated cyclin gene:
growth factors + receptor	growth factors + mutated receptor always "on"	growth factors + receptor
↓	↓	↓
cyclin synthesis + Cdk	cyclin synthesis + Cdk	cyclin synthesis always "on" + Cdk
↓	↓	↓
phosphorylates Rb	phosphorylates Rb	phosphorylates Rb
↓	↓	↓
Rb — P	Rb — P	Rb — P
↓	↓	↓
DNA replication	**uncontrolled DNA replication**	**uncontrolled DNA replication**

(b) Actions of mutated tumor suppressor genes

Mutated Rb gene:	Mutated p53 gene:	
growth factors + receptor	DNA damage	growth factors + receptor
↓	↓	↓
cyclin synthesis + Cdk	mutated p53 cannot block Rb phosphorylation	cyclin synthesis + Cdk
↓	⟋*	↓
mutated Rb does not require phosphorylation		phosphorylates Rb
↓		↓
Rb*		Rb — P
↓		↓
uncontrolled DNA replication		**replication of damaged DNA**

FIGURE E11-3 Actions of oncogenes and tumor suppressor genes

more daughter nuclei. Therefore, at the end of meiosis, there are four haploid daughter nuclei, each with one copy of each homologous chromosome. Because each nucleus is usually contained in a different cell, meiotic cell division normally produces four haploid cells from a single diploid parent cell (**FIG. 11-19**).

FIGURE 11-19 During meiosis II, sister chromatids separate into independent chromosomes. Each daughter cell receives one of these independent chromosomes.

We'll explore the stages of meiosis in more detail in the following sections.

Meiotic Cell Division Followed by Fusion of Gametes Keeps the Chromosome Number Constant from Generation to Generation

Why is meiotic cell division so important to sexual reproduction? Consider what would happen if gametes were diploid, like the rest of the cells of the parent organism, with two copies of each homologous chromosome. Fertilization would result in a cell with four copies of each homologue, giving the offspring twice as many chromosomes as its parents. After a few generations, the cells of the offspring would have enormous amounts of DNA. On the other hand, when a haploid sperm fuses with a haploid egg, the resulting offspring are diploid, just like their parents (**FIG. 11-20**).

Meiosis I Separates Homologous Chromosomes into Two Haploid Daughter Nuclei

The phases of meiosis have the same names as the roughly equivalent phases in mitosis, followed by I or II to distinguish the two nuclear divisions that occur in meiosis

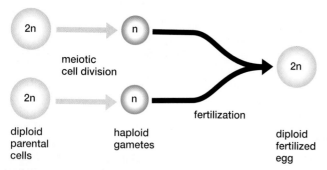

FIGURE 11-20 Meiotic cell division is essential for sexual reproduction

(FIG. 11-21). In the descriptions that follow, we assume that cytokinesis accompanies the nuclear divisions. Meiosis begins with chromosome duplication. As in mitosis, the sister chromatids of each chromosome remain attached to one another at the centromere.

During Prophase I, Homologous Chromosomes Pair Up and Exchange DNA

During mitosis, homologous chromosomes move completely independently of each other. In contrast, during *prophase I* of meiosis, homologous chromosomes line up side by side and exchange segments of DNA (Fig. 11-21a and **FIG. 11-22a**). We'll call one homologue the "maternal chromosome" and the other the "paternal chromosome," because one was originally inherited from the organism's mother and the other from the organism's father. During prophase I, proteins bind the maternal and paternal homologues together so that they match up exactly along their entire length, much like closing a zipper (**FIG. 11-22b**). In addition, enzyme complexes assemble at several places along the paired chromosomes (**FIG. 11-22c**). These enzymes cut through the DNA backbones within the chromosomes and graft the broken DNA ends together again, usually joining the maternal DNA to the paternal DNA and vice versa. This joining creates crosses, or **chiasmata**

MEIOSIS I

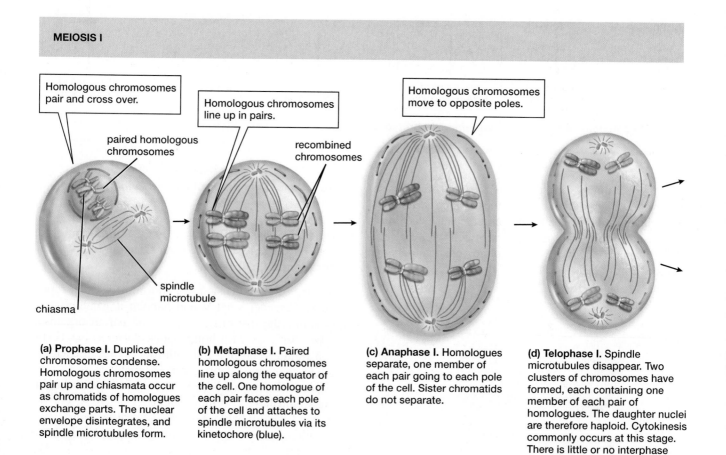

(a) Prophase I. Duplicated chromosomes condense. Homologous chromosomes pair up and chiasmata occur as chromatids of homologues exchange parts. The nuclear envelope disintegrates, and spindle microtubules form.

(b) Metaphase I. Paired homologous chromosomes line up along the equator of the cell. One homologue of each pair faces each pole of the cell and attaches to spindle microtubules via its kinetochore (blue).

(c) Anaphase I. Homologues separate, one member of each pair going to each pole of the cell. Sister chromatids do not separate.

(d) Telophase I. Spindle microtubules disappear. Two clusters of chromosomes have formed, each containing one member of each pair of homologues. The daughter nuclei are therefore haploid. Cytokinesis commonly occurs at this stage. There is little or no interphase between meiosis I and meiosis II.

FIGURE 11-21 Meiotic cell division in an animal cell
In meiotic cell division (meiosis and cytokinesis), the homologous chromosomes of a diploid cell are separated, producing four haploid daughter cells. Each daughter cell contains one member of each pair of parental homologous chromosomes. In these diagrams, two pairs of homologous chromosomes are shown, large and small. The yellow chromosomes are from one parent (for example, the father), and the violet chromosomes are from the other parent (for example, the mother). QUESTION What would the consequences be (for the resulting gametes) if one pair of homologues failed to separate at anaphase I?

(singular, chiasma), where the maternal and paternal chromosomes intertwine (**FIG. 11-22d**). In human cells, each pair of homologues usually forms two or three chiasmata in prophase I. Eventually, the enzyme complexes detach from the chromosomes, and the protein zippers that held homologues together disassemble. Nevertheless, the homologues remain together, held by the chiasmata (**FIG. 11-22e**).

This exchange of DNA between maternal and paternal chromosomes at chiasmata is called **crossing over**. If the chromosomes have different alleles, then the formation of the chiasmata creates slight genetic differences in both chromosomes (see Chapter 12). The result of crossing over, then, is genetic **recombination**: the formation of new combinations of alleles on a chromosome.

As in mitosis, the spindle microtubules begin to assemble outside the nucleus during prophase I. Near the end of prophase I, the nuclear envelope breaks down and the spindle microtubules capture the chromosomes by attaching to their kinetochores.

During Metaphase I, Paired Homologous Chromosomes Line Up at the Equator of the Cell

During *metaphase I*, interactions between the kinetochores and the spindle microtubules move the paired homologues to the equator of the cell (**FIG. 11-21b**). Unlike mitosis, in which *individual* duplicated chromosomes line up along the equator, *homologous pairs* of duplicated chromosomes line up along the equator during metaphase I of meiosis.

The key to understanding meiosis lies in the way the duplicated chromosomes line up during metaphase I. Before going further, then, let's look more closely at the differences in chromosome attachment to spindle microtubules in mitosis versus meiosis I. First, in mitosis, the homologues attach independently to the spindle. In meiosis I, the homologues remain associated with each other via chiasmata, attaching to the spindle as a unit containing both the maternal and paternal homologues. Second, in mitosis, the duplicated chromosome has two functional kinetochores,

MEIOSIS II

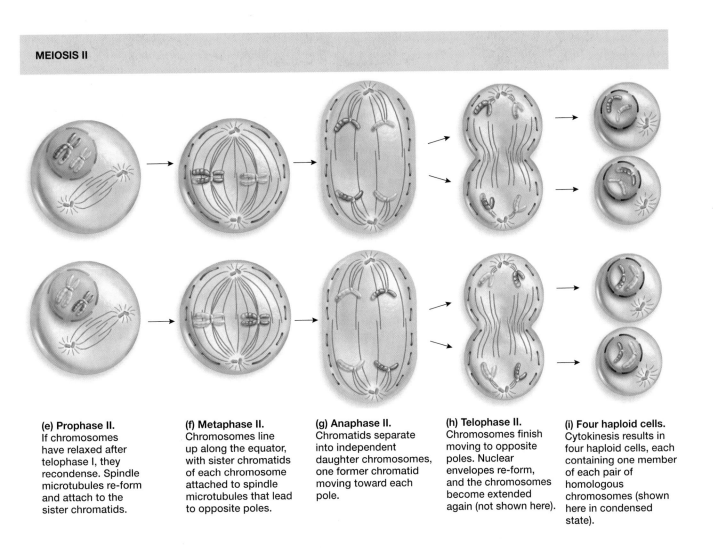

(e) Prophase II.
If chromosomes have relaxed after telophase I, they recondense. Spindle microtubules re-form and attach to the sister chromatids.

(f) Metaphase II.
Chromosomes line up along the equator, with sister chromatids of each chromosome attached to spindle microtubules that lead to opposite poles.

(g) Anaphase II.
Chromatids separate into independent daughter chromosomes, one former chromatid moving toward each pole.

(h) Telophase II.
Chromosomes finish moving to opposite poles. Nuclear envelopes re-form, and the chromosomes become extended again (not shown here).

(i) Four haploid cells.
Cytokinesis results in four haploid cells, each containing one member of each pair of homologous chromosomes (shown here in condensed state).

FIGURE 11-22 The mechanism of crossing over

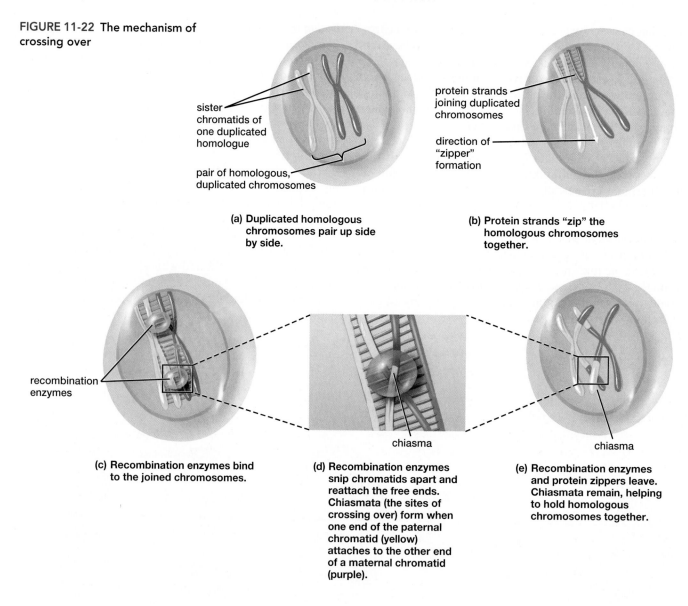

(a) Duplicated homologous chromosomes pair up side by side.

sister chromatids of one duplicated homologue

pair of homologous, duplicated chromosomes

(b) Protein strands "zip" the homologous chromosomes together.

protein strands joining duplicated chromosomes

direction of "zipper" formation

recombination enzymes

(c) Recombination enzymes bind to the joined chromosomes.

chiasma

(d) Recombination enzymes snip chromatids apart and reattach the free ends. Chiasmata (the sites of crossing over) form when one end of the paternal chromatid (yellow) attaches to the other end of a maternal chromatid (purple).

chiasma

(e) Recombination enzymes and protein zippers leave. Chiasmata remain, helping to hold homologous chromosomes together.

one on each sister chromatid. Both kinetochores attach to spindle microtubules, so that each sister chromatid is attached to microtubules that pull toward opposite poles (FIG. 11-23).

spindle microtubules leading to the same pole. However, the chromosomes of a homologous pair attach to spindle microtubules that pull them toward opposite poles (FIG. 11-24).

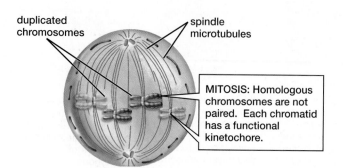

duplicated chromosomes

spindle microtubules

MITOSIS: Homologous chromosomes are not paired. Each chromatid has a functional kinetochore.

FIGURE 11-23 Chromosome attachment to the spindle in mitosis

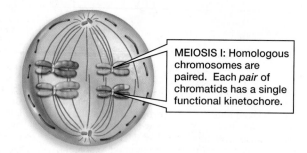

MEIOSIS I: Homologous chromosomes are paired. Each *pair* of chromatids has a single functional kinetochore.

FIGURE 11-24 Chromosome attachment to the spindle in meiosis I

In meiosis I, each duplicated chromosome has only one functional kinetochore, so both sister chromatids attach to

These differences in attachment explain what will happen at anaphase. In mitosis, the *sister chromatids separate* and move to opposite poles. In contrast, in meiosis I the sister chromatids of each duplicated chromosome remain attached to each other and move to the same pole, but the *homologues separate* and move to opposite poles.

During meiosis I, which member of a pair of homologous chromosomes faces which pole of the cell is random. The maternal chromosome may face "north" for some pairs and "south" for other pairs. This randomness (also called *independent assortment*), together with genetic recombination caused by crossing over, is responsible for the genetic diversity of the haploid cells produced by meiosis.

During Anaphase I, Homologous Chromosomes Separate

In *anaphase I*, the homologues separate from one another and are towed by their kinetochores to opposite poles of the cell (Fig. 11-21c). One duplicated chromosome of each homologous pair (still consisting of two sister chromatids) moves to each pole of the dividing cell. At the end of anaphase I, the cluster of chromosomes at each pole contains one member of each pair of homologous chromosomes. Therefore, each cluster contains the haploid number of chromosomes.

During Telophase I, Two Haploid Clusters of Duplicated Chromosomes Form

In *telophase I*, the spindle microtubules disappear. Cytokinesis commonly occurs during telophase I (Fig. 11-21d), and nuclear envelopes may reappear. Telophase I is usually followed immediately by meiosis II, with little or no intervening interphase. It is important to remember that the chromosomes do not replicate between meiosis I and meiosis II.

Meiosis II Separates Sister Chromatids into Four Daughter Nuclei

During meiosis II, the sister chromatids of each duplicated chromosome separate in a process that is virtually identical to mitosis, though it takes place in haploid cells. During *prophase II*, the spindle microtubules re-form (Fig. 11-21e). The duplicated chromosomes attach individually to spindle microtubules as they do in mitosis. Each chromatid contains a functional kinetochore, allowing each sister chromatid in a duplicated chromosome to attach to spindle microtubules extending to opposite poles of the cell. During *metaphase II*, the duplicated chromosomes line up at the cell's equator (Fig. 11-21f). During *anaphase II*, the sister chromatids separate and are towed to opposite poles (Fig. 11-21g). *Telophase II* and cytokinesis conclude meiosis II as nuclear envelopes re-form, the chromosomes relax into their extended state, and the cytoplasm divides (Fig. 11-21h). Commonly, both daughter cells produced in meiosis I undergo meiosis II, producing a total of four haploid cells from the original parental diploid cell (Fig. 11-21i).

Now that we have covered all of the processes in detail, examine Table 11-1 to review and compare mitotic and meiotic cell division.

11.7 WHEN DO MITOTIC AND MEIOTIC CELL DIVISION OCCUR IN THE LIFE CYCLES OF EUKARYOTES?

The life cycles of almost all eukaryotic organisms have a common overall pattern (**FIG. 11-25**). First, two haploid cells fuse during the process of fertilization, bringing together genes

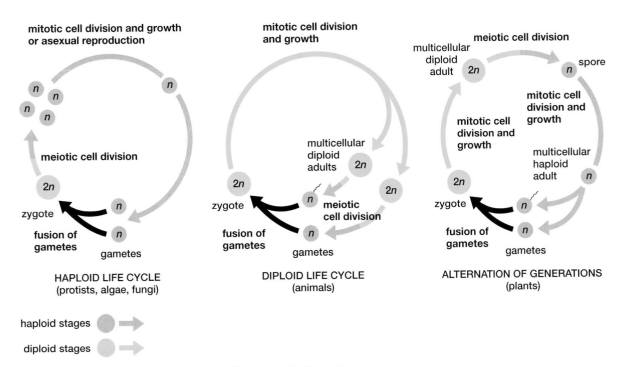

FIGURE 11-25 The three major types of eukaryotic life cycles
The lengths of the arrows correspond roughly to the proportion of the life cycle spent in each stage.

Table 11-1 A Comparison of Mitotic and Meiotic Cell Divisions in Animal Cells

Feature	Mitotic Cell Division	Meiotic Cell Division
Cells in which it occurs	Body cells	Gamete-producing cells
Final chromosome number	Diploid—2n; two copies of each type of chromosome (homologous pairs)	Haploid—1n; one member of each homologous pair
Number of daughter cells	Two, identical to the parent cell and to each other	Four, containing recombined chromosomes due to crossing over
Number of cell divisions per DNA replication	One	Two
Function in animals	Development, growth, repair, and maintenance of tissues; asexual reproduction	Gamete production for sexual reproduction

MITOSIS

no stages comparable to meiosis I

interphase   prophase   metaphase   anaphase   telophase   2 diploid cells

MEIOSIS

Recombination occurs.   Homologues pair.   Sister chromatids remain attached.

interphase prophase metaphase anaphase telophase   prophase metaphase anaphase telophase   4 haploid cells

MEIOSIS I   MEIOSIS II

In these diagrams, comparable phases are aligned. In both mitosis and meiosis, chromosomes are replicated during interphase. Meiosis I, with the pairing of homologous chromosomes, formation of chiasmata, exchange of chromosome parts, and separation of homologues to form haploid daughter nuclei, has no counterpart in mitosis. Meiosis II, however, is similar to mitosis.

from different parental organisms and endowing the resulting diploid cell with new gene combinations. Second, at some point in the life cycle, meiotic cell division occurs, recreating haploid cells. Third, at some time in the life cycle, mitotic cell division of either haploid or diploid cells, or both, results in the growth of multicellular bodies or asexual reproduction.

The seemingly vast differences between the life cycles of, say, ferns and humans are caused by variations in three aspects: (1) the interval between meiotic cell division and the fusion of haploid cells; (2) at what points in the life cycle mitotic and meiotic cell division occur; and (3) the relative proportions of the life cycle spent in the diploid and haploid states. These aspects of life cycles are interrelated, and we can conveniently label life cycles according to the relative dominance of diploid and haploid stages.

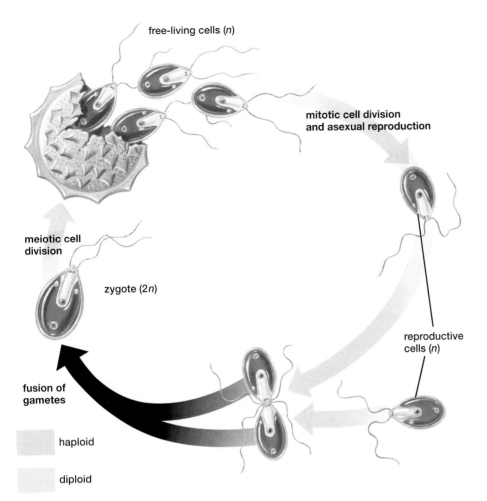

free-living cells (*n*)

mitotic cell division
and asexual reproduction

meiotic cell
division

zygote (2*n*)

reproductive
cells (*n*)

fusion of
gametes

☐ haploid

☐ diploid

FIGURE 11-26 The life cycle of the unicellular alga, *Chlamydomonas* *Chlamydomonas* reproduces asexually by mitotic cell division of haploid cells. When nutrients are scarce, specialized haploid cells (usually from genetically different populations) fuse to form a diploid cell. Meiotic cell division then immediately produces four haploid cells, usually with different genetic compositions than either of the parental strains.

11-26). Asexual reproduction by mitotic cell division produces a population of identical, haploid cells. Under certain environmental conditions, specialized "sexual" haploid cells are produced. Two of these sexual haploid cells fuse, forming a diploid cell. This cell immediately undergoes meiosis, producing haploid cells again. In organisms with haploid life cycles, mitotic cell division never occurs in diploid cells.

In Diploid Life Cycles, the Majority of the Cycle Consists of Diploid Cells

Most animals have life cycles that are just the reverse of the haploid cycle. Virtually the entire animal life cycle is spent in the diploid state (Fig. 11-25b and **FIG. 11-27**). Haploid gametes (sperm in males, eggs in females) are formed by meiotic cell division. These fuse to form a diploid fertilized egg, the zygote. Growth and development of the zygote to the adult organism result from mitotic cell division and differentiation of diploid cells.

In Alternation of Generations Life Cycles, There Are Both Diploid and Haploid Multicellular Stages

The life cycle of plants is called alternation of generations, because it includes both multicellular diploid and

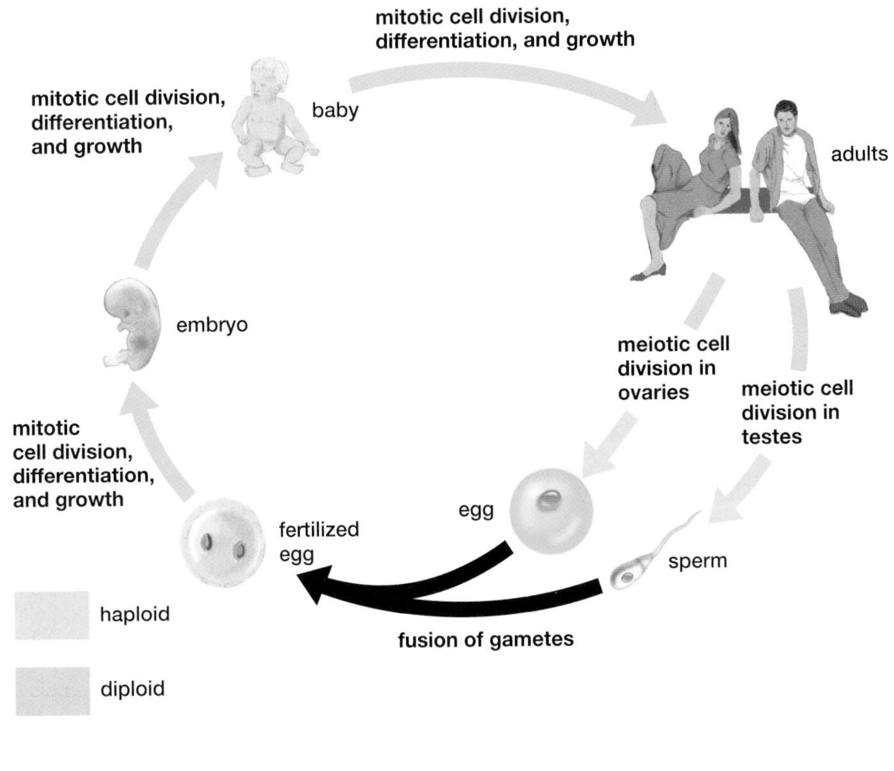

In Haploid Life Cycles, the Majority of the Cycle Consists of Haploid Cells

Some eukaryotes, such as many fungi and unicellular algae, spend most of their life cycles in the haploid state, with single copies of each type of chromosome (Fig. 11-25a and **FIG.**

mitotic cell division,
differentiation, and growth

mitotic cell division,
differentiation,
and growth

baby

adults

mitotic
cell division,
differentiation,
and growth

embryo

meiotic cell
division in
ovaries

meiotic cell
division in
testes

egg

fertilized
egg

sperm

☐ haploid

fusion of gametes

☐ diploid

FIGURE 11-27 The human life cycle Through meiotic cell division, the two sexes produce gametes—sperm in males and eggs in females—that fuse to form a diploid zygote. Mitotic cell division and differentiation of the daughter cells produce an embryo, child, and ultimately a sexually mature adult. The haploid stages last only a few hours to a few days; the diploid stages may survive for a century.

FIGURE 11-28 Alternation of generations in plants

In plants, such as this fern, specialized cells in the multicellular diploid stage undergo meiotic cell division to produce haploid spores. The spores undergo mitotic cell division and differentiation of the daughter cells to produce a multicellular haploid stage. Sometime later, perhaps many weeks later, some of these haploid cells differentiate into sperm and eggs. These fuse to form a diploid zygote. Mitotic cell division and differentiation once again give rise to a multicellular diploid stage.

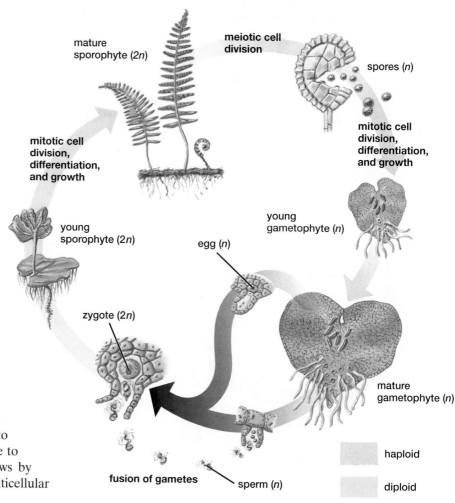

multicellular haploid body forms. In the typical pattern (Fig. 11-25c and **FIG. 11-28**), a multicellular diploid body gives rise to haploid cells, called spores, through meiotic cell division. These spores then undergo mitotic cell division and differentiation of the daughter cells to produce a multicellular haploid stage (the "haploid generation"). At some point, certain haploid cells differentiate into haploid gametes. Two gametes then fuse to form a diploid zygote. The zygote grows by mitotic cell division into a diploid multicellular body (the "diploid generation").

In "primitive" plants, such as ferns, both the haploid and diploid stages are free-living, independent plants. The flowering plants, however, have reduced haploid stages, represented only by the pollen grain and a small cluster of cells in the ovary of the flower.

11.8 HOW DO MEIOSIS AND SEXUAL REPRODUCTION PRODUCE GENETIC VARIABILITY?

Shuffling of Homologues Creates Novel Combinations of Chromosomes

Genetic variability among organisms is essential for survival and reproduction in a changing environment, and therefore for evolution. Mutations occurring randomly over millions of years are the original sources of genetic variability within the populations of organisms that exist today. However, mutations are rare events. Therefore, the genetic variability that occurs from one generation to the next results almost entirely from meiosis and sexual reproduction.

How does meiosis produce genetic diversity? One mechanism is the random distribution of maternal and paternal homologues to the daughter cells at meiosis I. Remember, at metaphase I the paired homologues line up at the cell's equator. In each pair of homologues, the maternal chromosome faces one pole and the paternal chromosome faces the opposite pole, but which homologue faces which pole is random.

Let's consider meiosis in mosquitoes, which have three pairs of homologous chromosomes ($n = 3, 2n = 6$). For simplicity, we'll represent these chromosomes as large, medium, and small. To keep track of the homologues, let's color-code the paternal chromosomes yellow and the maternal chromosomes violet. At metaphase I, the chromosomes can align in 4 configurations (**FIG. 11-29**).

FIGURE 11-29 Possible chromosome arrangements at metaphase of meiosis I

Therefore, anaphase I can produce 8 possible sets of chromosomes ($2^3 = 8$), as shown in **FIGURE 11-30**.

FIGURE 11-30 Possible sets of chromosomes after meiosis I

When each of these chromosome clusters undergoes meiosis II, it produces two gametes. Therefore, a single mosquito, with 3 pairs of homologous chromosomes, can produce gametes with 8 different chromosome sets. A single human, with 23 pairs of homologous chromosomes, can theoretically produce gametes with more than 8 million (2^{23}) different combinations of maternal and paternal chromosomes.

Crossing Over Creates Chromosomes with Novel Combinations of Genes

In addition to the genetic variation resulting from the random assortment of parental chromosomes, crossing over during meiosis produces chromosomes with combinations of alleles that differ from those of either parent. In fact, some of these new combinations may never have existed before, because homologous chromosomes cross over in new and different places at each meiotic division. In humans, therefore, although 1 in 8 million gametes should have the same combination of maternal and paternal chromosomes, in reality, none of those chromosomes will be purely maternal or purely paternal. Even though a man produces about

100 million sperm each day, he may never produce two that carry exactly the same combinations of alleles. In essence, every sperm and every egg is genetically unique.

Fusion of Gametes Adds Further Genetic Variability to the Offspring

At fertilization, two gametes, each probably containing unique combinations of alleles, fuse to form a diploid offspring. Even if we ignore crossing over, every human can produce about 8 million different gametes based solely on the random separation of the homologues. Therefore, fusion of gametes from just two people could produce 8 million × 8 million, or 64 trillion, genetically different children, which is far more people than have ever existed on Earth! Put another way, the chances that your parents could produce another child that is genetically the same as you are about 1/8,000,000 × 1/8,000,000, or about 1 in 64 trillion! When we factor in the almost endless variability produced by crossing over, we can confidently say that (except for identical twins) there never has been, and never will be, anyone just like you.

CASE STUDY REVISITED   HOW MUCH IS A GREAT TAN WORTH?

When ultraviolet (UV) rays in sunlight penetrate into the skin, they may strike the bases of DNA, causing mutations in oncogenes or tumor suppressor genes. These mutations may lead to one of the three common types of skin cancer, each involving a different cell type: basal cell carcinoma, squamous cell carcinoma, and melanoma. About 80% of skin cancers are basal cell carcinomas, 16% are squamous cell carcinomas, and 4% are melanomas. As is usual with cancers, multiple mutations are required to allow unregulated cell multiplication. In both basal cell and squamous cell carcinomas, one of these mutations is in p53, the tumor suppressor protein that halts cell division or even kills cells if DNA damage is present. Ultraviolet light frequently causes mutations in p53, so that the protein permits division even in cells with multiple other mutations, allowing a cancer to form. Fortunately, basal cell and squamous cell carcinomas tend to grow slowly and not to invade distant parts of the body very rapidly.

Not so with melanoma. Although much less common than the other two types of

skin cancer, melanoma is much more likely to spread to other tissues, and to kill. Melanomas are cancers of pigment-forming cells in the skin. About one-third of melanomas arise from preexisting moles, but about two-thirds start in otherwise normal-looking skin. Melanomas usually have multiple mutations, usually caused by UV light. Some of these mutations stimulate the synthesis of cyclin proteins. Cyclins stimulate cyclin-dependent kinases, which phosphorylate the Rb protein, which in turn allows a cell to pass through the G_1 to S checkpoint, replicate its DNA, and divide (**FIG. 11-31**).

As a child, Rachel had many sunburns. Being badly sunburned as a kid about triples the risk of developing melanoma. People with darker skin—which burns less easily—are about 15 times less likely to develop melanomas, though they are not totally protected. Because her melanoma was caught at an early stage, Rachel's prognosis for total recovery is good.

You can learn to spot potential melanomas before they become deadly, because virtually all melanomas either

begin with a preexisting mole or begin as a dark spot on the skin that looks like a new mole forming. Recognizing a possible melanoma is as easy as "ABCD": Examine your moles for Asymmetry, irregular Border, irregular Color, or a Diameter larger than the eraser on the end of a pencil. Check out this text's Web site to learn some tips for recognizing possible melanomas, and have any suspicious spots promptly examined by your doctor.

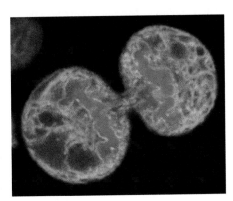

FIGURE 11-31 Dividing melanoma cells

CHAPTER REVIEW

SUMMARY OF KEY CONCEPTS

11.1 What Is the Role of Cellular Reproduction in the Lives of Individual Cells and Entire Organisms?
The prokaryotic cell cycle consists of growth, DNA replication, and division by binary fission. The eukaryotic cell cycle con-

sists of interphase and cell division. During interphase, the cell grows and duplicates its chromosomes. Interphase is divided into G_1 (growth phase 1), S (DNA synthesis), and G_2 (growth phase 2). During G_1, some cells may exit the cell cycle to enter a nondividing state called G_0. Cells may remain permanently in G_0, or they may be induced to reenter the cell cycle. Eukaryotic cells can divide by mitotic or meiotic cell division.

Mitotic cell division consists of two processes: (1) mitosis (nuclear division) and (2) cytokinesis (cytoplasmic division). Mitosis parcels out one copy of every chromosome into two separate nuclei, and cytokinesis subsequently encloses each nucleus in a separate cell, producing two genetically identical daughter cells. Mitotic cell division of a fertilized egg produces genetically identical cells that grow and differentiate into an embryo and, eventually, an adult. Mitotic cell division also maintains body tissues and repairs damage to some organs. Asexual reproduction is based on mitotic cell division, resulting in formation of clones that are genetically identical to the parent.

Meiotic cell division produces haploid cells, which have only half of the parent's DNA. Fusion of haploid gametes creates a fertilized egg that has a different genetic makeup from either parent, which then grows and develops via mitotic cell division.

11.2 How Is DNA in Eukaryotic Cells Organized into Chromosomes?

Each chromosome in a eukaryotic cell consists of a single DNA double helix and proteins that organize the DNA. During cell growth, the chromosomes are extended and accessible for use by enzymes that read their genetic instructions. During cell division, the chromosomes condense into short, thick structures. Eukaryotic cells typically contain pairs of chromosomes called homologues, which appear virtually identical because they carry the same genes with similar nucleotide sequences. Cells with pairs of homologous chromosomes are diploid. Cells with only one member of each chromosome pair are haploid.

11.3 How Do Cells Reproduce by Mitotic Cell Division?

The chromosomes are duplicated during interphase, prior to mitosis. The two identical copies, called chromatids, remain attached to one another at the centromere during the early stages of mitosis. Mitosis consists of four phases (see Fig. 11-10), usually followed by cytokinesis.

1. **Prophase:** The chromosomes condense and their kinetochores attach to the spindle microtubules that form at this time.
2. **Metaphase:** The chromosomes move to the equator of the cell.
3. **Anaphase:** The two chromatids of each duplicated chromosome separate and are pulled along the spindle microtubules to opposite poles of the cell.
4. **Telophase:** The chromosomes relax into their extended state, and nuclear envelopes re-form around each new daughter nucleus.
5. **Cytokinesis:** Cytokinesis normally occurs at the end of telophase and divides the cytoplasm into approximately equal halves, each containing a nucleus. In animal cells, a ring of microfilaments pinches the plasma membrane in along the equator. In plant cells, new plasma membrane forms along the equator by the fusion of vesicles produced by the Golgi apparatus.

Web Tutorial 11.1 Mitosis

11.4 How Is the Cell Cycle Controlled?

Complex interactions among many proteins, particularly cyclins and cyclin-dependent protein kinases, drive the cell cycle. There are three major checkpoints at which progress through the cell cycle is regulated: between G_1 and S, between G_2 and mitosis, and between metaphase and anaphase.

11.5 Why Do So Many Organisms Reproduce Sexually?

Genetic differences among organisms originate as mutations, which, when preserved within a species, produce alternate forms of genes, called alleles. Alleles in different individuals of a species may be combined in offspring through sexual reproduction, creating variation among the offspring and potentially improving their likelihood of surviving and reproducing in their turn.

11.6 How Does Meiotic Cell Division Produce Haploid Cells?

Meiosis separates homologous chromosomes and produces haploid cells with only one homologue from each pair. During interphase before meiosis, chromosomes are duplicated. The cell then undergoes two specialized cell divisions—meiosis I and meiosis II—to produce four haploid daughter cells (see Fig. 11-21).

Meiosis I: During prophase I, homologous duplicated chromosomes, each consisting of two chromatids, pair up and exchange parts by crossing over. During metaphase I, homologues move together as pairs to the cell's equator, one member of each pair facing opposite poles of the cell. Homologous chromosomes separate during anaphase I, and two nuclei form during telophase I. Each daughter nucleus receives only one member of each pair of homologues, and is therefore haploid. The sister chromatids remain attached to each other throughout meiosis I.

Meiosis II: Meiosis II usually occurs in both daughter nuclei and resembles mitosis in a haploid cell. The duplicated chromosomes move to the cell's equator during metaphase II. The two chromatids of each chromosome separate and move to opposite poles of the cell during anaphase II. This second division produces four haploid nuclei. Cytokinesis normally occurs during or shortly after telophase II, producing four haploid cells.

Web Tutorial 11.2 Meiosis

Web Tutorial 11.3 Comparing Mitosis and Meiosis

11.7 When Do Mitotic and Meiotic Cell Division Occur in the Life Cycles of Eukaryotes?

Most eukaryotic life cycles have three parts: (1) Sexual reproduction combines haploid gametes to form a diploid cell. (2) At some point in the life cycle, diploid cells undergo meiotic cell division to produce haploid cells. (3) At some point in the life cycle, mitosis of either a haploid cell, or a diploid cell, or both, results in the growth of multicellular bodies. When these stages occur, and what proportion of the life cycle is occupied by each stage, varies greatly among different species.

Web Tutorial 11.4 The Human Life Cycle

11.8 How Do Meiosis and Sexual Reproduction Produce Genetic Variability?

The random shuffling of homologous maternal and paternal chromosomes creates new chromosome combinations. Crossing over creates chromosomes with allele combinations that may never have occurred before on single chromosomes. Because of the separation of homologues and crossing over, a parent probably never produces any two gametes that are completely identical. The fusion of two such genetically unique gametes adds further genetic variability to the offspring.

KEY TERMS

allele   *page 207*
anaphase   *page 200*
asexual reproduction   *page 192*
autosome   *page 197*
binary fission   *page 193*
cell cycle   *page 192*
cell division   *page 194*
cell plate   *page 201*
centriole   *page 200*
centromere   *page 196*
checkpoint   *page 204*

chiasma (chiasmata)   *page 210*
chromatid   *page 197*
chromosome   *page 195*
clone   *page 202*
cloning   *page 202*
crossing over   *page 211*
cytokinesis   *page 195*
differentiation   *page 195*
diploid   *page 197*
duplicated chromosome
    page 197

gamete   *page 195*
haploid   *page 198*
homologue   *page 197*
interphase   *page 194*
karyotype   *page 197*
kinetochore   *page 200*
locus   *page 196*
meiosis   *page 195*
meiotic cell division   *page 195*
metaphase   *page 200*

mitosis   *page 195*
mitotic cell division   *page 195*
prophase   *page 200*
recombination   *page 211*
sex chromosome   *page 197*
sexual reproduction   *page 195*
spindle microtubule   *page 200*
telomere   *page 196*
telophase   *page 200*

THINKING THROUGH THE CONCEPTS

1. Diagram and describe the eukaryotic cell cycle. Name the various phases, and briefly describe the events that occur during each.

2. Define *mitosis* and *cytokinesis*. What changes in cell structure would result when cytokinesis does not occur after mitosis?

3. Diagram the stages of mitosis. How does mitosis ensure that each daughter nucleus receives a full set of chromosomes?

4. Define the following terms: *homologous chromosome, centromere, kinetochore, chromatid, diploid, haploid*.

5. Describe and compare the process of cytokinesis in animal cells and in plant cells.

6. How is the cell cycle controlled? Why is it essential that cells cannot simply progress through the cell cycle without regulation?

7. Diagram the events of meiosis. At which stage do homologous chromosomes separate?

8. Describe homologue pairing and crossing over. At which stage of meiosis do they occur? Name two functions of chiasmata.

9. In what ways are mitosis and meiosis similar? In what ways are they different?

10. Describe or diagram the three main types of eukaryotic life cycles. When do meiotic cell division and mitotic cell division occur in each?

11. Describe how meiosis provides for genetic variability. If an animal had a haploid number of 2 (no sex chromosomes), how many genetically different gametes could it produce? (Assume no crossing over.) If it had a haploid number of 5?

APPLYING THE CONCEPTS

1. Most nerve cells in the adult human central nervous system, as well as heart muscle cells, remain in the G_0 portion of interphase. In contrast, cells lining the inside of the small intestine divide frequently. Discuss this difference in terms of why damage to the nervous system and heart muscle cells (such as caused by a stroke or heart attack) is so dangerous. What do you think might happen to tissues such as the intestinal lining if some disorder blocked mitotic cell division in all cells of the body?

2. Cancer cells divide out of control. Side effects of chemotherapy and radiation therapy that fight cancers include loss of hair

and of the intestinal lining, producing severe nausea. Note that cells in hair follicles and intestinal lining divide frequently. What can you infer about the mechanisms of these treatments? What would you look for in an improved cancer therapy?

3. Some animal species can reproduce either asexually or sexually, depending on the state of the environment. Asexual reproduction tends to occur in stable, favorable environments; sexual reproduction is more common in unstable or unfavorable circumstances. Discuss the advantages and disadvantages of sexual and asexual reproduction.

FOR MORE INFORMATION

Axtman, K. "Quietly, Animal Cloning Speeds Onward." *Christian Science Monitor*, October 23, 2001. A discussion of the successes and failures of mammalian cloning.

Gibbe, W. W. "Untangling the Roots of Cancer." *Scientific American*, July 2003. Cancerous cells arise by numerous mechanisms. Many involve mutations in the molecules that control the cell cycle.

Grant, M. C. "The Trembling Giant." *Discover*, October 1993. Aspen groves are really single individuals: huge, slowly spreading from the roots of the original parent tree, and potentially almost immortal.

Lanza, R. P., Dresser, B. L., and Damiani, P. "Cloning Noah's Ark." *Scientific American*, November 2000. Cloning rare and endangered species may offer hope of preventing extinction.

Leutwyler, K. "Turning Back the Strands of Time." *Scientific American (Explorations)*, February 2, 1998. A brief discussion of telomeres, the repeating DNA regions at the ends of chromosomes.

Travis, J. "A Fantastical Experiment." *Science News*, April 5, 1997. A clear description of the cloning of Dolly the sheep and some of its implications.

Wilmut, I. "Cloning for Medicine." *Scientific American*, December 1998. Explanation of why cloning experiments might have medical applications.

12 Patterns of Inheritance

Olympic silver medalist Flo Hyman was struck down
by Marfan syndrome at the height of her career.

CASE STUDY   SUDDEN DEATH ON THE COURT

FLO HYMAN, graceful, athletic, and over 6 feet tall, was one of the best woman volleyball players of all time. A star of the 1984 silver medal American Olympic volleyball team, Hyman later joined a professional Japanese team. In 1986, taken out of a game for a short breather, she died while sitting quietly on the bench. How could this happen to someone only 32 years old and in superb physical condition?

Flo Hyman had a genetic disorder called Marfan syndrome. Marfan syndrome is surprisingly common, affecting about one in 5000 people. People with Marfan syndrome are typically tall and slender, with unusually long limbs and large hands and feet. These characteristics helped Flo Hyman to become an outstanding volleyball player. Unfortunately, Marfan syndrome can also be deadly.

An autopsy showed that Hyman died from a ruptured aorta, the massive artery that carries blood from the heart to most of the body. Why did Hyman's aorta break? What does a weak aorta have in common with tallness and large hands? Marfan syndrome is caused by a mutation in the gene that encodes a protein called fibrillin, which forms long fibers that give elasticity and strength to connective tissue. Many parts of the body contain connective tissue, including tendons, ligaments, and artery walls. Defective fibrillin molecules weaken connective tissue, sometimes with tragic consequences. Fibrillin mutations apparently also stimulate growth, causing people with Marfan syndrome to grow tall and lanky.

How did Flo Hyman get this disorder? Did she inherit it from her parents? Or was it a new mutation (most likely in the DNA of either her mother's egg or her father's sperm that fertilized it)? Because new mutations are rare, let's hypothesize that Hyman inherited a defective gene from her parents. Geneticists can't do experiments, in the usual sense, on people; but they can gather other evidence to help them determine modes of inheritance. As you read this chapter, ask yourself a few questions: What evidence would you need to decide if Hyman's case of Marfan syndrome was a new mutation or inherited from her parents? If it was inherited, did it come from both parents, or could she inherit it from just one? If Hyman would have had children before she died, would they be likely to have Marfan syndrome?

12.1 WHAT IS THE PHYSICAL BASIS OF INHERITANCE?

Inheritance is the process by which the characteristics of individuals are passed to their offspring. As you learned in previous chapters, DNA carries genetic information, in the form of sequences of nucleotides. In most cases, segments of DNA ranging from a few hundred to many thousands of nucleotides are the genes that encode the information needed to synthesize a specific protein. Chromosomes are made up of DNA together with various proteins. **Genes**, therefore, are parts of chromosomes. Finally, chromosomes are passed from cell to cell and from organism to organism during reproduction. Inheritance, then, occurs when genes are transmitted from parent to offspring.

We will begin our exploration with a brief overview of the structures—genes and chromosomes—that form the physical basis of inheritance. In this chapter, we will confine our discussion to diploid organisms, including most plants and animals, that reproduce sexually by the fusion of haploid gametes.

Genes Are Sequences of Nucleotides at Specific Locations on Chromosomes

A gene's physical location on a chromosome is called its **locus** (plural, **loci**; FIG. 12-1). Each member of a pair of homologous chromosomes carries the same genes, located at the same loci. Will the nucleotide sequences at the same locus of a pair of homologues always be identical? Think back to Chapters 9 and 10. Errors in DNA replication, certain chemicals, radiation—all of these can cause mutations that change the nucleotide sequence of DNA. *Different* nucleotide sequences at the *same* locus on two homologous chromosomes are called **alleles**. The human A, B, and O blood types, for example, are produced by three different alleles of the blood-type gene.

An Organism's Two Alleles May Be the Same or Different

If both homologous chromosomes have the *same* allele at a given gene locus, the organism is said to be **homozygous** at that locus. (*Homozygous* comes from Greek words meaning "same pair.") For example, the chromosomes in Figure 12-1 are homozygous at the loci for the *M* and *D* genes. If two homologous chromosomes have *different* alleles at a locus, the organism is **heterozygous** ("different pair") at that locus and is sometimes called a **hybrid**. The chromosomes in Figure 12-1 are heterozygous at the locus for the *Bk* gene.

Recall from Chapter 11 that, during meiosis, homologous chromosomes are separated, so each gamete receives one member of each pair of homologous chromosomes. As a result, every gamete has only one allele for each gene. Therefore, all the gametes produced by an organism that is homozygous at a particular gene locus contain the same allele. Gametes produced by an organism that is heterozygous at the same gene locus are of two kinds: half of the gametes contain one allele, and half contain the other allele.

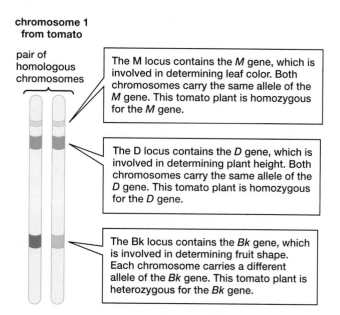

FIGURE 12-1 The relationships among genes, alleles, and chromosomes
Each homologous chromosome carries the same set of genes. Each gene is located at the same relative position, or locus, on its chromosome. Differences in nucleotide sequences at the same gene locus produce different alleles of the gene. Diploid organisms have two alleles of each gene.

The common patterns of inheritance and many essential facts about genes, alleles, and the distribution of alleles in gametes and zygotes during sexual reproduction were deduced in the mid-1800s by an Austrian monk, Gregor Mendel (FIG. 12-2), long before DNA, chromosomes, or meiosis had been discovered. Because his experiments are succinct, elegant examples of science in action, let's follow Mendel's paths of discovery.

12.2 HOW DID GREGOR MENDEL LAY THE FOUNDATIONS FOR MODERN GENETICS?

Before settling down as a monk in the monastery of St. Thomas in Brünn (now Brno, in the Moravian part of the Czech Republic), Gregor Mendel attended the University of Vienna for two years. He studied many subjects, including botany and mathematics. At St. Thomas, Mendel used this training to carry out a groundbreaking series of experiments on inheritance in the common edible pea.

Doing It Right: The Secrets of Mendel's Success

There are three key steps to any successful experiment in biology: choosing the right organism with which to work, designing and performing the experiment correctly, and analyzing the data properly. Mendel was the first geneticist to complete all three steps.

Mendel's choice of the edible pea as an experimental subject was critical to the success of his experiments. Sta-

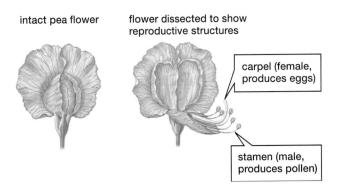

FIGURE 12-3 Flowers of the edible pea
In the intact pea flower (left), the lower petals form a container enclosing the reproductive structures—the stamens (male) and carpel (female). Pollen normally cannot enter the flower from outside, so peas usually self-fertilize. If the flower is opened (right), it can be cross-pollinated by hand.

of inheritance became clear. Numerical analysis was something of an innovation in Mendel's time, but today, statistics is an essential tool in virtually every field of biology.

12.3 HOW ARE SINGLE TRAITS INHERITED?

In the edible pea, flower color is controlled by a single gene. If a pea plant is homozygous for this gene, all of the offspring produced by self-fertilization will have the same flower color, which will be the same as the parent plant. Such plants are called **true-breeding**. Even in Mendel's time, commercial seed dealers sold many types of true-breeding pea varieties. Mendel raised pea plants that were true-breeding for different forms of a single trait, such as flower color, and cross-fertilized them. He saved the resulting hybrid seeds and grew them the following year to determine their characteristics.

In one of these experiments, Mendel cross-fertilized a white-flowered pea plant with a purple-flowered one. This was the *parental generation*, denoted by the letter P. When he grew the resulting seeds, he found that all the first-generation offspring (the "first filial," or F_1, generation) produced purple flowers (**FIG. 12-4**):

FIGURE 12-2 Gregor Mendel
A portrait of Mendel, painted in about 1888, after he had completed his pioneering genetics experiments.

mens, the male reproductive structures of a flower, produce pollen. Each pollen grain contains sperm. Pollination allows sperm to fertilize the female gamete, or *egg*, located within the ovary at the base of the carpel, which is the female reproductive structure of the flower. The petals of a pea flower enclose all of the flower structures, preventing another flower's pollen from entering (**FIG. 12-3**). Instead, each pea flower normally supplies its own pollen, so the egg cells in each flower are fertilized by sperm from the pollen of the same flower, in a process called **self-fertilization**.

Although peas normally self-fertilize, plant breeders can mate two plants by hand, causing **cross-fertilization**. Breeders pull apart the petals and remove the stamens, preventing self-fertilization. By dusting the sticky end of the carpel with pollen from plants they have selected, breeders can control fertilization. In this way, two plants can be mated to see what types of offspring they produce.

Mendel's experimental design was simple, but brilliant. Rather than looking at the entire plant in all of its complexity, Mendel chose to study individual characteristics (usually called *traits*) that had unmistakably different forms, such as white versus purple flowers. He also worked with one trait at a time.

Mendel followed the inheritance of these traits for several generations, counting the numbers of offspring with each type of trait. By analyzing these numbers, the basic patterns

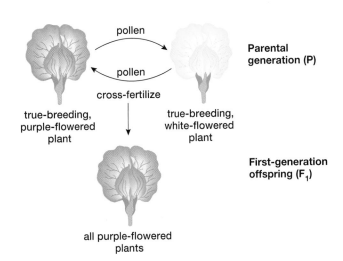

FIGURE 12-4 Cross of white and purple pea flowers

What had happened to the white color? The flowers of the hybrids were just as purple as their parent. The white color seemed to have disappeared in the F_1 offspring.

Mendel then allowed the F_1 flowers to self-fertilize, collected the seeds, and planted them the next spring. In the second generation (F_2), about three-fourths of the plants had purple flowers and one-fourth had white flowers (**FIG. 12-5**):

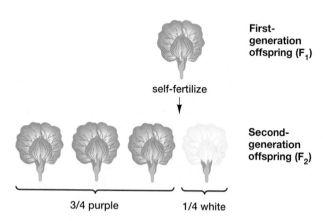

First-generation offspring (F_1)

self-fertilize

Second-generation offspring (F_2)

3/4 purple   1/4 white

FIGURE 12-5 Cross of F_1 purple pea flowers

The exact numbers were 705 purple and 224 white, or a ratio of about 3 purple to 1 white. This result showed that the capacity to produce white flowers had not disappeared in the F_1 plants, but had only been "hidden."

Mendel allowed the F_2 plants to self-fertilize and produce yet a third (F_3) generation. He found that all the white-flowered F_2 plants produced white-flowered offspring; that is, they were true-breeding. For as many generations as he had time and patience to raise, white-flowered parents always gave rise to white-flowered offspring. In contrast, the purple-flowered F_2 plants were of two types: About $\frac{1}{3}$ of these were true-breeding for purple; the remaining $\frac{2}{3}$ were hybrids that produced both purple- and white-flowered offspring, again in the ratio of 3 to 1. Therefore, the F_2 generation included $\frac{1}{4}$ true-breeding purple plants, $\frac{1}{2}$ hybrid purple, and $\frac{1}{4}$ true-breeding white.

The Inheritance of Dominant and Recessive Alleles on Homologous Chromosomes Can Explain the Results of Mendel's Crosses

Mendel's results, supplemented by our knowledge of genes and homologous chromosomes, allow us to develop a five-part hypothesis to explain the inheritance of single traits:

- Each trait is determined by pairs of discrete physical units that we now call genes. Each organism has two alleles for each gene, such as the gene that determines flower color. One allele of the gene is present on each homologous chromosome. True-breeding, white-flowered peas have

different alleles of the "flower-color" gene than true-breeding, purple-flowered peas do.

- When two different alleles are present in an organism, one—the **dominant** allele—may mask the expression of the other—the **recessive** allele. The recessive allele, however, is still present. In the edible pea, the allele for purple flowers is dominant, and the allele for white flowers is recessive.

- The pairs of genes on homologous chromosomes separate from each other during gamete formation, so each gamete receives only one allele of each pair. This is known as Mendel's **law of segregation**: the two alleles of a gene segregate (separate) from one another during meiosis. When a sperm fertilizes an egg, the resulting offspring receives one allele from the father and one from the mother.

- Chance determines which allele is included in a given gamete. Because homologous chromosomes separate at random during meiosis, the distribution of alleles to the gametes is also random.

- True-breeding (homozygous) organisms have two copies of the same alleles for a given gene. Therefore, all of the gametes from a homozygous individual have the same allele for that gene (**FIG. 12-6**):

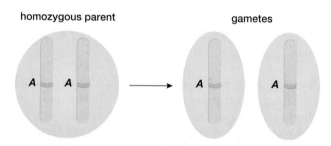

homozygous parent   gametes

FIGURE 12-6 Chromosomes in the gametes of a homozygous parent

Hybrid (heterozygous) organisms have two different alleles for a given gene. Half of the organism's gametes will contain one allele for that gene and half will contain the other allele (**FIG. 12-7**):

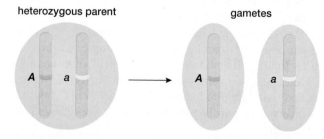

heterozygous parent   gametes

FIGURE 12-7 Chromosomes in the gametes of a heterozygous parent

Let's see how this hypothesis explains the results of Mendel's experiments with flower color. Using letters to represent the different alleles, we will assign the uppercase letter P to the allele for purple (dominant) and the lowercase letter p to the allele for white (recessive). (By Mendel's convention, the dominant allele is represented by a capital letter.) A true-breeding (homozygous) purple-flowered plant has two alleles for purple flowers (PP), whereas a white-flowered plant has two alleles for white flowers (pp). All the sperm and eggs produced by a PP plant carry the P allele; all the sperm and eggs of a pp plant carry the p allele (**FIG. 12-8**):

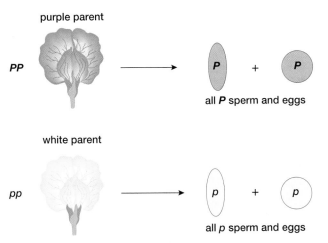

FIGURE 12-8 Gametes from peas homozygous for purple and white flowers

The F$_1$ hybrid offspring are produced when P sperm fertilize p eggs or when p sperm fertilize P eggs. In either case, the F$_1$ offspring are Pp. Because P is dominant to p, all the offspring are purple (**FIG. 12-9**):

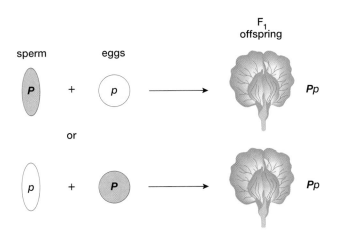

FIGURE 12-9 Combinations of gametes that produce F$_1$ offspring flowers

Each gamete produced by a heterozygous Pp plant has an equal chance of receiving either the P allele or the p al-

lele. That is, the hybrid plant produces equal numbers of P and p sperm and equal numbers of P and p eggs. When a Pp plant self-fertilizes, each type of sperm has an equal chance of fertilizing each type of egg (**FIG. 12-10**):

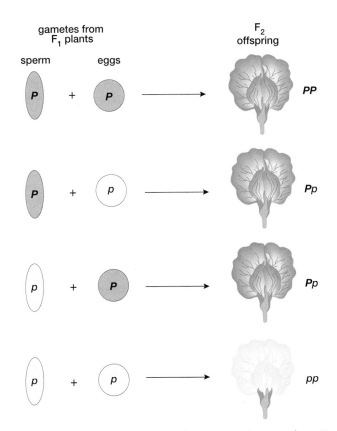

FIGURE 12-10 Combinations of gametes that produce F$_2$ offspring flowers

Therefore, three types of offspring can be produced: PP, Pp, and pp. The three types occur in the approximate proportions of $\frac{1}{4}PP$, $\frac{1}{2}Pp$, and $\frac{1}{4}pp$.

The actual combination of alleles carried by an organism (for example, PP or Pp) is its **genotype**. The organism's traits, including its outward appearance, behavior, digestive enzymes, blood type, or any other observable or measurable feature, make up its **phenotype**. As we have seen, plants with either the PP or the Pp genotype make purple flowers. Thus, even though they have different genotypes, they have the same phenotype. Therefore, the F$_2$ generation consists of three genotypes ($\frac{1}{4}PP$, $\frac{1}{2}Pp$, and $\frac{1}{4}pp$) but only two phenotypes ($\frac{3}{4}$ purple and $\frac{1}{4}$ white).

Simple "Genetic Bookkeeping" Can Predict Genotypes and Phenotypes of Offspring

The **Punnett square method**, named after R. C. Punnett, a famous geneticist of the early 1900s, is a convenient way to predict the genotypes and phenotypes of offspring.

FIGURE 12-11 Determining the outcome of a single-trait cross
(a) The Punnett square allows you to predict both genotypes and phenotypes of specific crosses; here we use it for a cross between plants that are heterozygous for a single trait, flower color.
(1) Assign letters to the different alleles; use uppercase for dominant and lowercase for recessive.
(2) Determine all the types of genetically different gametes that can be produced by the male and female parents.
(3) Draw the Punnett square, with each row and column labeled with one of the possible genotypes of sperm and eggs, respectively. (We have included the fractions of these genotypes with each label.)
(4) Fill in the genotype of the offspring in each box by combining the genotype of sperm in its row with the genotype of the egg in its column. (Multiply the fraction of sperm of each type in the row headers by the fraction of eggs of each type in the column headers.)
(5) Count the number of offspring with each genotype. (Note that *Pp* is the same as *pP*.)
(6) Convert the number of offspring of each genotype to a fraction of the total number of offspring. In this example, out of four fertilizations, only one is predicted to produce the *pp* genotype, so $\frac{1}{4}$ of the total number of offspring produced by this cross is predicted to be white. To determine phenotypic fractions, add the fractions of genotypes that would produce a given phenotype. For example, purple flowers are produced by $\frac{1}{4}PP + \frac{1}{4}Pp + \frac{1}{4}pP$, for a total of $\frac{3}{4}$ of the offspring.
(b) Probability theory can also be used to predict the outcome of a single-trait cross. Determine the fractions of eggs and sperm of each genotype, and multiply these fractions together to calculate the fraction of offspring of each genotype. When two genotypes produce the same phenotype (e.g., *Pp* and *pP*), add the fractions of each genotype to determine the phenotypic fraction.

(a)

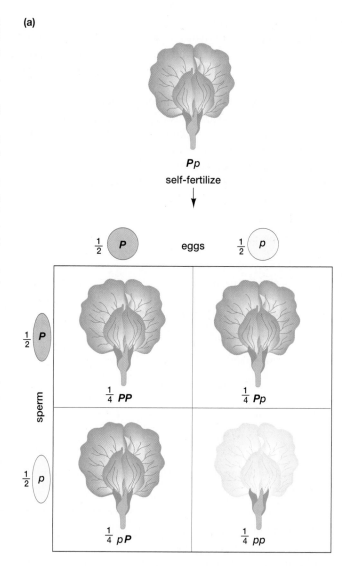

(b)

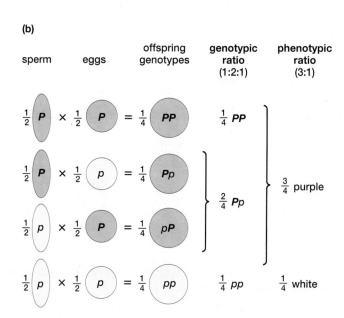

FIGURE 12-11 shows how to use a Punnett square to determine the proportions of offspring that arise from the self-fertilization of a flower that is heterozygous for color (or the proportions of offspring produced by breeding any two organisms that are heterozygous for a single trait). This figure also provides the fractions that allow you to calculate the same outcomes based on probability theory. As you use these "genetic bookkeeping" techniques, keep in mind that in a real experiment, the offspring will occur only in *approximately* the predicted proportions. Let's consider an example. Each time a baby is conceived, it has a 50:50 chance of becoming a boy or a girl. However, many families with two children do not have one girl and one boy. The 50:50 ratio of girls to boys occurs only if we average the genders of the children in many families.

Mendel's Hypothesis Can Be Used to Predict the Outcome of New Types of Single-Trait Crosses

You have probably recognized that Mendel used the scientific method: observing results and formulating a hypothesis based on them. The scientific method has another crucial step: using the hypothesis to predict the results of other experiments and seeing if those experiments support or refute the hypothesis. For example, if the hybrid F_1 flowers have one allele for purple and one for white (*Pp*), then Mendel could predict the outcome of cross-fertilizing these *Pp* plants with homozygous recessive white plants (*pp*). Can you? Mendel predicted that there would be equal numbers of *Pp* (purple) and *pp* (white) offspring, and this is indeed what he found.

This type of experiment also has practical uses. Cross-fertilization of an organism with a dominant phenotype (in this case, a purple flower) but an unknown genotype with a homozygous recessive organism (a white flower) tests whether the organism with the dominant phenotype is homozygous or heterozygous; logically enough, this is called a **test cross** (**FIG. 12-12**). When crossed with a homozygous recessive (*pp*), a homozygous dominant (*PP*) produces all phenotypically dominant offspring, whereas a heterozygous dominant (*Pp*) yields offspring with both dominant and recessive phenotypes in a 1:1 ratio:

12.4 HOW ARE MULTIPLE TRAITS INHERITED?

Mendel Hypothesized That Traits Are Inherited Independently

Having determined the inheritance modes for single traits, Mendel then turned to the more complex question of multiple traits in peas (**FIG. 12-13**). He began by crossbreeding plants that differed in two traits—for example, seed color (yellow or green) and seed shape (smooth or wrinkled). From other crosses of plants with these traits, Mendel already knew that the smooth allele of the seed shape gene (*S*) is dominant to the wrinkled allele (*s*). In addition, the yellow allele of the seed color gene (*Y*) is dominant to the green allele (*y*). He crossed a true-breeding plant with smooth, yellow seeds (*SSYY*) to a true-breeding plant with wrinkled, green seeds (*ssyy*). All the F_1 offspring, therefore, were genotypically *SsYy*. They also all had the same phenotype: smooth, yellow seeds. Allowing these F_1 plants to self-fertilize, Mendel found that the F_2 generation consisted of 315 plants with smooth, yellow seeds; 101 with wrinkled, yellow seeds; 108 with smooth, green seeds; and 32 with wrinkled, green seeds—a ratio of about 9:3:3:1. The F_2 generations produced from other crosses of gametes that were heterozygous for two traits had similar phenotypic ratios.

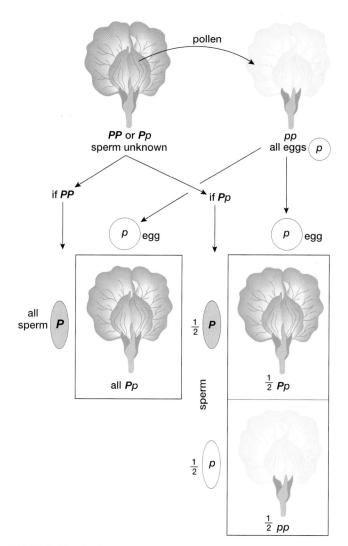

FIGURE 12-12 The test cross

Trait	Dominant form	Recessive form
Seed shape	smooth	wrinkled
Seed color	yellow	green
Pod shape	inflated	constricted
Pod color	green	yellow
Flower color	purple	white
Flower location	at leaf junctions	at tips of branches
Plant size	tall (1.8 to 2 meters)	dwarf (0.2 to 0.4 meters)

FIGURE 12-13 Traits of pea plants that Mendel studied

FIGURE 12-14 Predicting genotypes and phenotypes for a cross between gametes that are heterozygous for two traits In pea seeds, yellow (*Y*) is dominant to green (*y*), and smooth (*S*) is dominant to wrinkled (*s*). **(a)** Punnett square analysis. In this cross, an individual heterozygous for both traits self-fertilizes. Note that the Punnett square predicts both the frequencies of combinations of traits ($\frac{9}{16}$ smooth yellow, $\frac{3}{16}$ smooth green, $\frac{3}{16}$ wrinkled yellow, and $\frac{1}{16}$ wrinkled green) and the frequencies of individual traits ($\frac{3}{4}$ yellow, $\frac{1}{4}$ green, $\frac{3}{4}$ smooth, and $\frac{1}{4}$ wrinkled). **(b)** Probability theory states that the probability of two independent events is the product (multiplication) of their individual probabilities. Seed *shape* is independent of seed *color*. Therefore, multiplying the independent probabilities of the genotypes or phenotypes for each trait produces the predicted frequencies for the combined genotypes or phenotypes of the offspring. These ratios are identical to those generated by the Punnett square. EXERCISE Use Punnett squares to determine if the genotype of a plant bearing smooth yellow seeds can be revealed by a test cross with a plant bearing wrinkled green seeds.

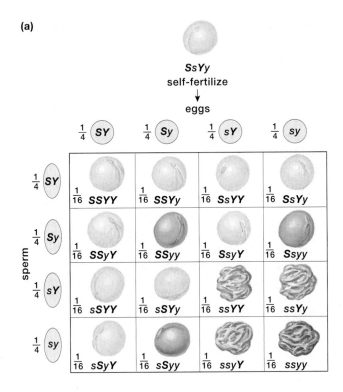

(a)

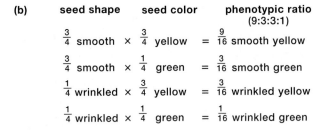

(b)

	seed shape		seed color		phenotypic ratio (9:3:3:1)
$\frac{3}{4}$ smooth	×	$\frac{3}{4}$ yellow	=	$\frac{9}{16}$ smooth yellow
$\frac{3}{4}$ smooth	×	$\frac{1}{4}$ green	=	$\frac{3}{16}$ smooth green
$\frac{1}{4}$ wrinkled	×	$\frac{3}{4}$ yellow	=	$\frac{3}{16}$ wrinkled yellow
$\frac{1}{4}$ wrinkled	×	$\frac{1}{4}$ green	=	$\frac{1}{16}$ wrinkled green

We can explain these results if the genes for seed color and seed shape are inherited independently of each other and do not influence each other during gamete formation. If so, then for each trait $\frac{3}{4}$ of the offspring should show the dominant phenotype and $\frac{1}{4}$ should show the recessive phenotype. This result is just what Mendel observed. He found 423 plants with smooth seeds (of either color) and 133 with wrinkled seeds (a ratio of about 3:1); in this same group of plants, 416 produced yellow seeds (of either shape) and 140 produced green seeds (also about 3:1). **FIGURE 12-14** shows how a Punnett square or probability calculation can be used to determine the outcome of a cross between organisms that are heterozygous for two traits, and how two independent 3:1 ratios combine to produce an overall 9:3:3:1 ratio.

The independent inheritance of two or more distinct traits is called the **law of independent assortment**, which states that the alleles of one gene may be distributed to gametes independently of the alleles for other genes. Independent assortment will occur when the traits being studied are controlled by genes on different pairs of homologous chromosomes. Why? From Chapter 11, recall the movement of chromosomes during meiosis. When paired homologous chromosomes line up during metaphase I, which homologue faces which pole of the cell is random, and the orientation of one homologous pair does not influence other pairs. Therefore, when the homologues separate during anaphase I, which allele of a gene on homologous pair 1 moves "north" does not affect which allele of a gene on homologous pair 2 moves "north"; that is, the alleles of genes on different chromosomes are distributed, or "assorted," independently (**FIG. 12-15**).

In an Unprepared World, Genius May Go Unrecognized

In 1865, Gregor Mendel presented the results of his experiments on inheritance to the Brünn Society for the Study of Natural Science, and they were published the following year. His paper did not mark the beginning of genetics. In fact, it didn't make any impression at all on the study of biology during his lifetime. Mendel's experiments, which eventually spawned one of the most important scientific theories in all of biology, simply vanished from the scene. Apparently, very few biologists read his paper, and those who did failed to recognize its significance.

It was not until 1900 that three biologists—Carl Correns, Hugo de Vries, and Erich Tschermak—working independently and knowing nothing of Mendel's work, rediscovered the principles of inheritance. No doubt to their intense disappointment, when they searched the scientific literature before publishing their results, they found that Mendel had scooped them more than 30 years earlier. To their credit, they graciously acknowledged the important work of the Augustinian monk, who had died in 1884.

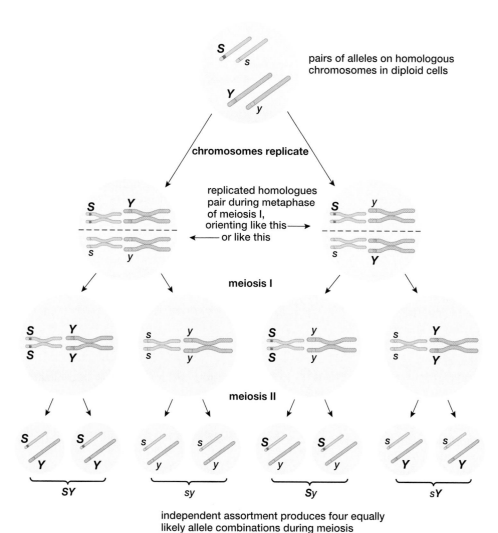

FIGURE 12-15 Independent assortment of alleles
Chromosome movements during meiosis produce independent assortment of alleles of two different genes. Each combination of alleles is equally likely to occur. Therefore, an F_1 plant would produce gametes in the predicted proportions $\frac{1}{4}SY$, $\frac{1}{4}sy$, $\frac{1}{4}sY$, and $\frac{1}{4}Sy$.

pairs of alleles on homologous chromosomes in diploid cells

chromosomes replicate

replicated homologues pair during metaphase of meiosis I, orienting like this⟶ ⟵ or like this

meiosis I

meiosis II

SY   sy   Sy   sY

independent assortment produces four equally likely allele combinations during meiosis

12.5 HOW ARE GENES LOCATED ON THE SAME CHROMOSOME INHERITED?

Gregor Mendel knew nothing about the physical nature of genes or chromosomes. Much later, when scientists discovered that chromosomes are the vehicles of inheritance, it became obvious that there are many more traits (and therefore many more genes) than there are chromosomes. As you may recall from Chapter 11, genes are parts of chromosomes, and each chromosome contains many genes. These facts have important implications for inheritance.

Genes on the Same Chromosome Tend to Be Inherited Together

When chromosomes assort independently during meiosis I, only genes located on *different chromosomes* assort independently into gametes. In contrast, genes on the *same chromosome* tend to be inherited together. Genetic **linkage** is the inheritance of certain genes as a group because they are on the same chromosome. One of the first pairs of linked genes to be discovered was found in the sweet pea,

a different species from Mendel's garden pea. In sweet peas, the gene for flower color and the gene for pollen grain shape are carried on the same chromosome. Thus, the alleles for these genes normally assort *together* into gametes during meiosis and are inherited together.

Consider a heterozygous sweet pea plant with purple flowers and long pollen. Its chromosomes are as shown in **FIGURE 12-16**:

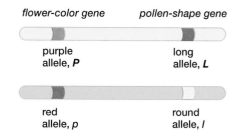

flower-color gene   *pollen-shape gene*

purple allele, **P**   long allele, **L**

red allele, *p*   round allele, *I*

FIGURE 12-16 Homologous chromosomes of the sweet pea, showing genes for flower color and pollen shape

Note that the purple allele of the flower-color gene and the long allele of the pollen-shape gene are located on one homologous chromosome. The red allele of the flower-color gene and the round allele of the pollen-shape gene are located on the other homologue. Therefore, the gametes produced by this sweet pea plant are likely to have *either* purple and long alleles *or* red and round alleles. This pattern of inheritance breaks the law of independent assortment, because the alleles for flower color and pollen shape do *not* segregate independently of one another into the gametes, but tend to stay together during meiosis.

Recombination Can Create New Combinations of Linked Alleles

Although they *tend* to be inherited together, genes on the same chromosome do not *always* stay together. In the sweet pea cross just described, for example, the F$_2$ generation usually includes a few plants in which the genes for flower color and pollen shape have been inherited as if they were not linked. That is, some of the offspring will have purple flowers and round pollen, and some will have red flowers and long pollen. How can this be?

As you learned in Chapter 11, during prophase I of meiosis, homologous chromosomes sometimes exchange parts, a process called **crossing over** (see Fig. 11-22). In most chromosomes, at least one exchange between each homologous pair occurs during each meiotic cell division. The exchange of corresponding segments of DNA during crossing over produces new allele combinations on both homologous chromosomes. Then, when homologues separate at anaphase I, the chromosomes that each haploid daughter cell receives will have different sets of alleles than the chromosomes of the parent cell did.

Crossing over during meiosis explains the appearance of new combinations of alleles that were previously linked. Let's revisit the sweet pea during the early stages of meiosis I, when the chromosomes have duplicated and homologous chromosomes are paired up (**FIG. 12-17**):

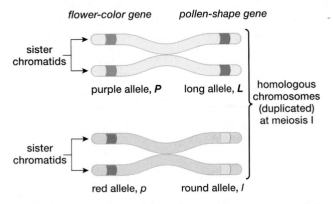

FIGURE 12-17 Replicated homologous chromosomes of the sweet pea

Each homologous chromosome will have one or more regions where crossing over will occur. Imagine that one crossover occurs between the genes for flower color and pollen shape (**FIG. 12-18**):

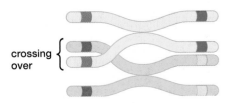

FIGURE 12-18 Crossing over between homologous chromosomes of the sweet pea

At anaphase I, the separated homologous chromosomes will have this gene composition (**FIG. 12-19**):

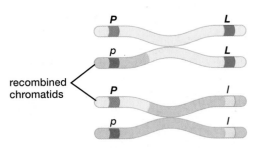

FIGURE 12-19 The results of crossing over in replicated, homologous chromosomes of the sweet pea

Four types of chromosomes are then distributed to the haploid daughter cells during meiosis II (**FIG. 12-20**):

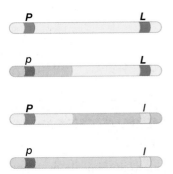

FIGURE 12-20 Homologous chromosomes of the sweet pea, after separation in anaphase II of meiosis

Therefore, some gametes will be produced with each of four chromosome configurations: *PL* and *pl* (the original parental types), and *Pl* and *pL* (recombined chromosomes). By exchanging DNA between homologous chromosomes, this **genetic recombination** creates new combinations of alleles. If a sperm with a *Pl* chromosome fertilizes an egg with a *pl* chromosome, the offspring plant will have purple flowers (*Pp*) and round pollen (*ll*). If a sperm with a *pL* chromosome fertilizes an egg with a *pl* chromosome, then the offspring will have red flowers (*pp*) and long pollen (*Ll*).

Not surprisingly, the farther apart the genes are on a chromosome, the more likely it is that crossing over will occur between them. In fact, if two genes are really far apart, crossing over occurs so often that they seem to be independently assorted, just as if they were on different

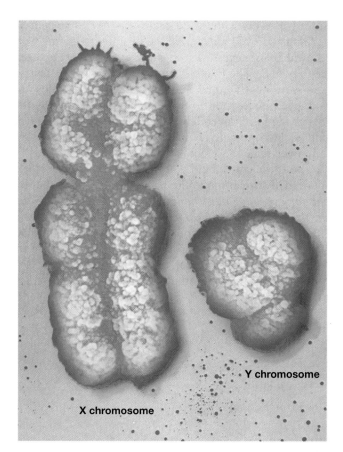

FIGURE 12-21 Photomicrograph of human sex chromosomes Notice the small size of the Y chromosome, which carries relatively few genes.

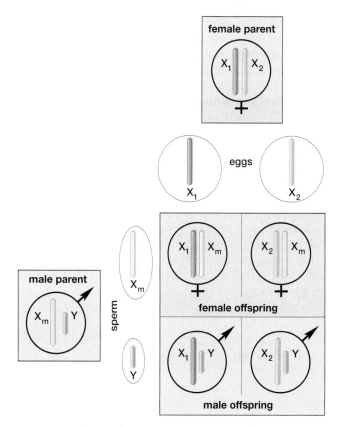

FIGURE 12-22 Sex determination in mammals Male offspring receive their Y chromosome from the father; female offspring receive the father's X chromosome (labeled X_m). Both male and female offspring receive an X chromosome (either X_1 or X_2) from the mother.

chromosomes. When Gregor Mendel discovered independent assortment, he was not only clever and careful, he was also lucky. The seven traits that he studied were controlled by genes on only four different chromosomes; he observed independent assortment because the genes that were on the same chromosome were far apart.

12.6 HOW IS SEX DETERMINED, AND HOW ARE SEX-LINKED GENES INHERITED?

In mammals and many insects, males have the same number of chromosomes as females do, but one "pair," the **sex chromosomes**, is very different in appearance and genetic composition. Females have two identical sex chromosomes, called *X chromosomes*, whereas males have one X chromosome and one *Y chromosome* (**FIG. 12-21**). Although the Y chromosome normally carries far fewer genes than does the X chromosome, a small part of both sex chromosomes is homologous. As a result, the X and Y chromosomes pair up during prophase of meiosis I and separate during anaphase I. The other chromosomes, which occur in pairs that have identical appearance in both males and females, are called **autosomes**. The total number of chromosomes varies tremendously among species, but in all cases there is only one pair of sex chromosomes.

For organisms in which males are XY and females are XX, the sex chromosome carried by the sperm determines the sex of the offspring (**FIG. 12-22**). During sperm formation, the sex chromosomes segregate, and each sperm receives either the X or the Y chromosome (plus one member of each pair of the autosomes). The sex chromosomes also segregate during egg formation, but because females have two X chromosomes, every egg receives one X chromosome (along with one member of each pair of the autosomes). An offspring is male if an egg is fertilized by a Y-bearing sperm or female if an egg is fertilized by an X-bearing sperm.

Sex-Linked Genes Are Found Only on the X or Only on the Y Chromosome

Genes that are on one sex chromosome but not on the other are said to be **sex-linked**. In many animals, the Y chromosome carries only a few genes. In humans, the Y chromosome contains a few dozen genes, many of which play a role in male reproduction. In contrast, the X chromosome contains over 1000 genes, few of which have a specific role in female reproduction. Most, which have no counterpart on the Y chromosome, encode traits that are important in both sexes, such as color vision, blood clotting, and certain structural proteins in muscles. Because they have two X chromosomes, females can be either homozygous or heterozygous for genes on the X chromosome, and dominant versus recessive relationships among alleles will be expressed. Males, in contrast, fully express all the alleles they have on their single X chromosome, whether those alleles are otherwise dominant or recessive. For this reason, in humans, most cases of recessive traits encoded by genes on the X chromosome, such as color blindness, hemophilia, and certain types of muscular dystrophy, occur in males. We will return to this concept later in the chapter.

How does sex linkage affect inheritance? Let's look at the first example of sex linkage to be discovered, the inheritance of eye color in the fruit fly *Drosophila*. Because these flies are small, reproduce rapidly, are easy to raise in the laboratory, and have few chromosomes, *Drosophila* have been favored subjects for genetics studies for more than a century. Normally, *Drosophila* have red eyes. In the early 1900s, researchers in the laboratory of Thomas Hunt Morgan at Columbia University discovered a male fly with white eyes. This white-eyed male was mated to a true-breeding, red-eyed female. All the resulting offspring were red-eyed flies, suggesting that white eye color (*r*) is recessive to red (*R*). The F$_2$ generation, however, was a surprise: there were nearly equal numbers of red-eyed males and white-eyed males, but no females had white eyes! A test cross of the F$_1$ red-eyed females to the original white-eyed male yielded roughly equal numbers of red-eyed and white-eyed males and females.

From these data, could you figure out how eye color is inherited? Morgan made the brilliant hypothesis that *the gene for eye color must be located on the X chromosome and that the Y chromosome has no corresponding gene* (**FIG. 12-23**). In the F$_1$ generation, both male and female offspring received an X chromosome, with its *R* allele for red eyes, from their mother. The F$_1$ males received a Y chromosome from their father with no allele for eye color, and so the males had an *R*- genotype and red-eyed pheno-

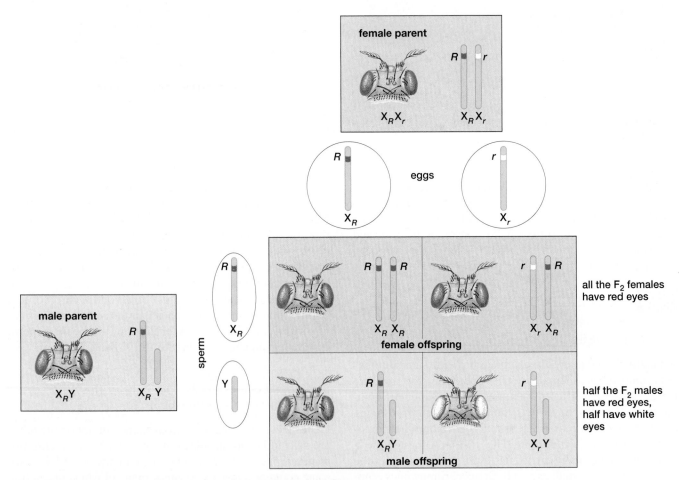

FIGURE 12-23 Sex-linked inheritance of eye color in fruit flies
The gene for eye color is located on the X chromosome; the Y chromosome does *not* contain an eye-color gene. Red (*R*) is dominant to white (*r*). When a white-eyed male is mated to a homozygous red-eyed female, all the offspring have red eyes: F$_1$ females are heterozygous, receiving the *r* allele from their father and the *R* allele from their mother, while male offspring receive only the *R* allele from their mother. In the F$_2$ generation, the single *R* allele of the F$_1$ male parent is passed on to his daughters, so all the F$_2$ daughters have red eyes. The F$_2$ sons receive a Y chromosome from their father, and either the *R* or *r* allele on the X chromosome from their mother, so half the sons have white eyes and half have red eyes.

type. (The "-" means that the Y chromosome doesn't have an eye-color gene.) The F_1 females received the father's X chromosome with its r allele, so the females had an Rr genotype and red-eyed phenotype. Thus, both male and female F_1 offspring had red eyes.

Crossing two F_1 flies, R- × Rr, resulted in an F_2 generation with the chromosome distribution shown in Figure 12-23. All the F_2 females received one X chromosome from their F_1 male parent, with its R allele, and therefore had red eyes. All the F_2 males inherited their single X chromosome from their F_1 female parent, which was heterozygous for eye color (Rr). So the F_2 males had a 50:50 chance of receiving an X chromosome with the R allele or one with the r allele. *With no corresponding gene on the Y chromosome, the F_2 males displayed the phenotype determined by the allele on the X chromosome.* Therefore, half the F_2 males had red eyes, and half had white eyes.

12.7 DO THE MENDELIAN RULES OF INHERITANCE APPLY TO ALL TRAITS?

In our discussion of patterns of inheritance thus far, we have made some major simplifying assumptions: that each trait is completely controlled by a single gene, that there are only two possible alleles of each gene, and that one allele is completely dominant to the other, recessive, allele. Most traits, however, are influenced in more varied and subtle ways.

Incomplete Dominance: The Phenotype of Heterozygotes Is Intermediate Between the Phenotypes of the Homozygotes

When one allele is completely dominant over a second allele, heterozygotes with one dominant allele have the same phenotype as homozygotes with two dominant alleles. However, relationships between alleles are not always this simple. When the heterozygous phenotype is intermediate between the two homozygous phenotypes, the pattern of inheritance is called **incomplete dominance**. In humans, hair texture is influenced by a gene with two incompletely dominant alleles, which we will call C_1 and C_2 (**FIG. 12-24**). A person with two copies of the C_1 allele has curly hair; two copies of the C_2 allele produce straight hair. Heterozygotes, with the C_1C_2 genotype, have wavy hair. (See Chapter 3 for more information about how differences in a protein called keratin determine how curly your hair is.) If two wavy-haired people marry, they might have children with any of the three hair types, with probabilities of $\frac{1}{4}$ curly (C_1C_1), $\frac{1}{2}$ wavy (C_1C_2), and $\frac{1}{4}$ straight (C_2C_2); see Figure 12-24.

A Single Gene May Have Multiple Alleles

Alleles arise through mutation, and the same gene in different individuals may have different mutations, each producing a new allele. Therefore, although an *individual* can

have at most two different alleles, a *species* may have **multiple alleles** of many of its genes. One eye-color gene in *Drosophila*, for example, has more than a thousand alleles! Depending on how they are combined, eye color in fruit flies can be white, yellow, orange, pink, brown, or red. There are hundreds of alleles for both Marfan syndrome and cystic fibrosis (see "Scientific Inquiry: Cystic Fibrosis"), each of which arose as a new mutation.

Human blood types are examples of multiple alleles of a single gene, with an added twist to the pattern of inheritance. The blood types A, B, AB, and O arise as a result of three different alleles (for simplicity, we will designate them A, B, and o) of a single gene on chromosome 9. This gene codes for an enzyme responsible for adding sugar molecules to the ends of glycoproteins that protrude from the surfaces of red blood cells. Alleles A and B code for enzymes that add different sugars to the glycoproteins (we'll call the resulting molecules glycoproteins A and B, respectively). Allele o codes for a nonfunctional enzyme that doesn't add any sugar molecule. A person may have one of

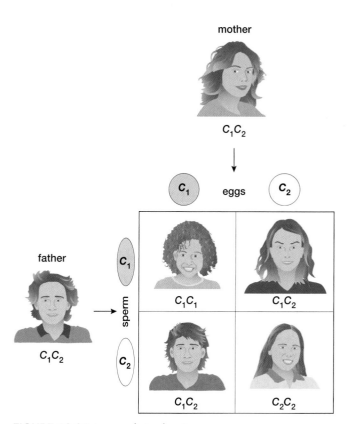

FIGURE 12-24 Incomplete dominance
The inheritance of hair texture in humans is an example of incomplete dominance. In such cases, we use capital letters for both alleles, here C_1 and C_2. Homozygotes may have curly hair (C_1C_1) or straight hair (C_2C_2). Heterozygotes (C_1C_2) have wavy hair. The children of a man and a woman, both with wavy hair, may have curly, straight, or wavy hair, in the approximate ratio of $\frac{1}{4}$ curly: $\frac{1}{2}$ wavy: $\frac{1}{4}$ straight.

"Woe to that child which when kissed on the forehead tastes salty. He is bewitched and soon must die."
—**17th-century English saying**

This adage is based on a remarkably accurate diagnostic tool for the most common recessive genetic disorder in the United States and Europe—cystic fibrosis. About 30,000 Americans, 3,000 Canadians, and 20,000 Europeans have cystic fibrosis. The story of this disease is a blend of physiology, medicine, and both Mendelian and molecular genetics.

Let's begin with a child's salty forehead. Sweat cools the body by evaporating from the skin, and is mostly water. However, sweat also contains a lot of salt (sodium chloride) when it is first secreted, about as much as in blood and extracellular fluid. As sweat moves through tubes connecting the secreting cells with the surface of the skin, most of the salt is reclaimed if you are perspiring slowly enough. How? Transport proteins in the plasma membrane of the cells lining the tubes move negatively charged chloride ions out of the sweat and return them to the extracellular fluid. Positively charged sodium ions follow along by electrical attraction. Cystic fibrosis is caused by defective transport proteins: salt stays in the sweat, so the skin tastes salty.

Salty sweat isn't very harmful, but unfortunately the cells lining the lungs have the same transport proteins. In the lungs, these proteins move chloride onto the surface of the airways. As you may recall from Chapter 4, water "follows" ions by osmosis, so the chloride and sodium ions cause water to move to the airway surfaces. Some cells of the airways also secrete mucus. Ideally, the water dilutes the mucus, so that the fluid on the airway surfaces is thin and watery. Why does this matter? The mucus traps bacteria and debris. The resulting "soup" is swept out of the lungs by cilia on the cells. In cystic fibrosis, reduced chloride transport means that not much water reaches the airway surfaces, so the mucus is thick and the cilia can't move it very well. The mucus clogs the airways, and bacteria remain in the lungs, causing frequent infections. Even if a person survives the infections, the lungs usually become permanently damaged. Mucus also builds up in the stomach and intestines, reducing the absorption of nutrients and causing malnutrition. Before modern medical care, most people with cystic fibrosis died by age 4 or 5; even now, the average life span is only about 35 to 40 years.

Mutations in the *CFTR* gene, which encodes the chloride transport protein, cause cystic fibrosis. Researchers have identified nearly 1000 mutations in this gene. Some mutations introduce a stop codon into the middle of the mRNA, which cuts off translation before the transport protein is completed; others change the amino acid sequence in ways that reduce the transport rate. The most common mutation prevents the protein from moving through the endoplasmic reticulum and Golgi apparatus to the plasma membrane. Overall, about 1 American in 30 carries one of these mutations.

Why is cystic fibrosis a recessive trait? People who are heterozygous, with one normal *CFTR* allele and one copy of any of these mutations, produce enough CFTR proteins to provide adequate chloride transport. Therefore, they are phenotypically normal—that is, they produce watery secretions in their lungs and do not develop cystic fibrosis. Someone with two defective alleles will not have functioning chloride transport proteins and will develop the disease.

Can anything be done to prevent, cure, or control the symptoms of cystic fibrosis? Because it is a genetic disorder, the only way to prevent the disease is to prevent the birth of affected infants. However, people usually don't know whether they are carriers, and therefore don't know whether their children may inherit the disease. Treatments that reduce lung damage include physical manipulation to drain the lungs, medicines that open the airways (similar to those used by people with asthma), and frequent, even continuous, administration of antibiotics (**FIG. E12-1**). Unfortunately, these treatments merely postpone inevitable damage to the lungs, intestine, pancreas, and other organs.

This situation may change in the next few years. Today, medical labs can identify carriers by a blood test and homozygous recessive embryos by prenatal diagnosis. Soon, children with cystic fibrosis may be cured or helped by one of several gene therapies now under development. We will explore these applications of biotechnology in the next chapter.

FIGURE E12-1 Cystic fibrosis
A child is treated for cystic fibrosis. Gentle pounding on the chest and back while the child is held upside-down helps dislodge mucus from the lungs. A device on the child's wrist injects antibiotics into a vein. These treatments combat the lung infections to which cystic fibrosis patients are vulnerable.

six genotypes: *AA, BB, AB, Ao, Bo,* or *oo* (**Table 12-1**). Alleles *A* and *B* are dominant to *o*. Therefore, people with genotypes *AA* or *Ao* have only type A glycoproteins and have type A blood. Those with genotypes *BB* or *Bo* synthesize only type B glycoproteins and have type B blood.

Homozygous recessive *oo* individuals lack both types of glycoproteins and have type O blood. In people with type AB blood, both enzymes are present, so the plasma membranes of their red blood cells have both A and B glycoproteins. When heterozygotes express phenotypes of *both*

Table 12-1 Human Blood Group Characteristics

Blood Type	Genotype	Red Blood Cells	Has Plasma Antibodies to:	Can Receive Blood from:	Can Donate Blood to:	Frequency in U.S.
A	AA or Ao	A glycoprotein	B glycoprotein	A or O (no blood with B glycoprotein)	A or AB	40%
B	BB or Bo	B glycoprotein	A glycoprotein	B or O (no blood with A glycoprotein)	B or AB	10%
AB	AB	Both A and B glycoproteins	Neither A nor B glycoprotein	AB, A, B, O (universal recipient)	AB	4%
O	oo	Neither A nor B glycoprotein	Both A and B glycoproteins	O (no blood with A or B glycoprotein)	O, AB, A, B (universal donor)	46%

of the homozygotes (in this case, both A and B glycoproteins), the pattern of inheritance is called **codominance**, and the alleles are said to be *codominant* to one another.

People make antibodies to the type of glycoprotein(s) that they lack. These antibodies are proteins in blood plasma that bind to foreign glycoproteins by recognizing different end-sugar molecules. The antibodies cause red blood cells that bear foreign glycoproteins to clump together and to rupture. The resulting clumps and fragments can clog small blood vessels and damage vital organs such as the brain, heart, lungs, or kidneys. This means that blood type must be determined and matched carefully before a blood transfusion is made.

Type O blood, lacking any end sugars, is not attacked by antibodies in A, B, or AB blood, so it can be transfused safely to all other blood types. (The antibodies present in transfused blood become too diluted to cause problems.) People with type O blood are called "universal donors." But O blood carries antibodies to both A and B glycoproteins, so type O individuals can receive transfusions of only type O blood. Can you predict the blood type of people called "universal recipients"? Table 12-1 summarizes blood types and transfusion characteristics.

Many Traits Are Influenced by Several Genes

If you look around your class, you are likely to see people of varied heights, skin colors, and body builds. Traits such as these are not governed by single genes but are influenced by interactions among two or more genes, as well as by interactions with the environment. Many traits, such as human height, weight, eye color, and skin color, may have several phenotypes or even seemingly continuous variation that cannot be split up into convenient, easily defined categories. This is an example of **polygenic inheritance**, a form of inheritance in which the interaction of two or more genes contributes to a single phenotype.

Although no one fully understands the inheritance of human skin color, it is probably controlled by at least three genes, each with pairs of incompletely dominant alleles (**FIG. 12-25a**). As you might imagine, the more genes that contribute to a single trait, the greater the number of phenotypes and the finer the distinctions among them. When three or more pairs of genes contribute to a trait, differences between phenotypes are fairly small. If the environment also contributes significantly to the trait, as exposure to sunlight alters skin color, there can be virtually continuous variation in phenotype (**FIG. 12-25b**).

Single Genes Typically Have Multiple Effects on Phenotype

We have just seen that a single phenotype may result from the interaction of several genes. The reverse is also true: single genes commonly have multiple phenotypic effects, a phenomenon called **pleiotropy**. A good example is the *SRY* gene, discovered on the Y chromosome in 1990. The *SRY* gene (short for "sex-determining region of the Y chromosome") codes for a protein that activates other genes; those genes in turn code for proteins that switch on male development in an embryo. Under the influence of the genes activated by the SRY protein, sex organs develop into testes. The testes, in turn, secrete sex hormones that stimulate the development of both internal and external male reproductive structures, such as the epididymis, seminal vesicles, prostate gland, penis, and scrotum. The SRY gene is described more fully in the Chapter 10 Case Study, "*Vive la Différence!*"

(a)

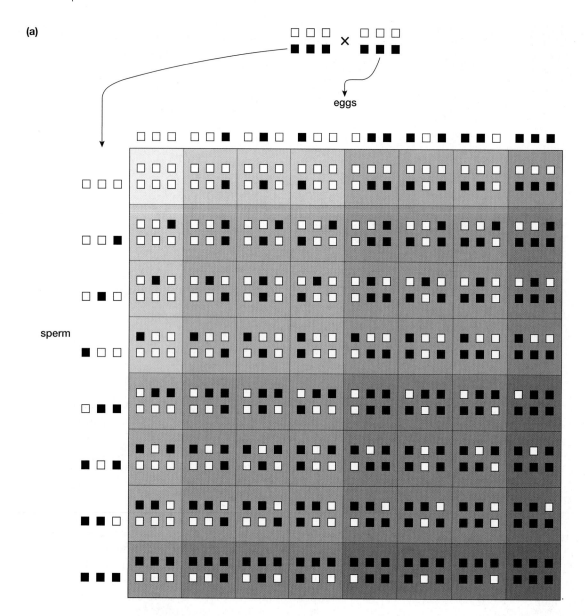

(b)

FIGURE 12-25 Polygenic inheritance of skin color in humans
(a) At least three separate genes, each with two incompletely dominant alleles, determine human skin color (the inheritance is actually much more complex than this). The backgrounds of each box indicate the depth of skin color expected from each genotype. **(b)** The combination of complex polygenic inheritance and environmental effects (especially exposure to sunlight) produces an almost infinite gradation of human skin colors.

FIGURE 12-26 Environmental influence on phenotype
The expression of the gene for black fur in the Himalayan rabbit is a simple case of interaction between genotype and environment producing a particular phenotype. The gene for black fur is expressed in cool areas (nose, ears, and feet).

The Environment Influences the Expression of Genes

An organism is not just the sum of its genes. In addition to its genotype, the environment in which an organism lives profoundly affects its phenotype. A striking example of environmental effects on gene action occurs in the Himalayan rabbit, which, like the Siamese cat, has pale body fur but black ears, nose, tail, and feet (**FIG. 12-26**). The Himalayan rabbit actually has the genotype for black fur all over its body. However, the enzyme that produces the black pigment is inactive at temperatures above about 34°C (93°F). At typical ambient temperatures, extremities such as the ears, nose, and feet are cooler than the rest of the body, so black pigment can be produced there. Because the rabbit's main body surface is warmer than 34°C, the fur there is pale.

Most environmental influences are more complicated and subtle. The complexity of environmental influences is particularly true of human characteristics. The polygenic trait of skin color is modified by the environmental effects of sun exposure (see Fig. 12-25). Height, another polygenic trait, is influenced by nutrition.

The interactions between complex genetic systems and varied environmental conditions can create a continuum of phenotypes that defies analysis into genetic and environmental components. The human generation time is long, and the number of offspring per couple is small. Add to these factors the countless subtle ways in which people respond to their environments, and you can see why it may be exceedingly difficult to determine the precise genetic basis of complex human traits such as intelligence or musical or athletic abilities.

12.8 HOW ARE HUMAN GENETIC DISORDERS INVESTIGATED?

Because experimental crosses with humans are out of the question, human geneticists search medical, historical, and family records to study past crosses. Records extending across several generations can be arranged in the form of family **pedigrees**, diagrams that show the genetic relationships among a set of related individuals (**FIG. 12-27**). Careful

(a) A pedigree for a dominant trait

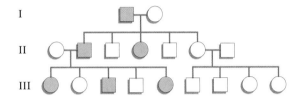

(b) A pedigree for a recessive trait

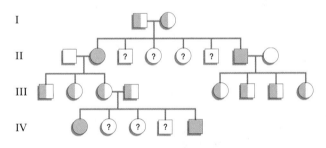

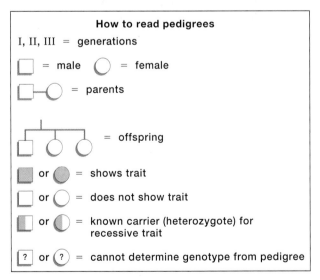

FIGURE 12-27 Family pedigrees
(a) A pedigree for a dominant trait. Note that any offspring showing a dominant trait *must* have at least one parent with the trait (see Figs. 12-11 and 12-14). **(b)** A pedigree for a recessive trait. Any individual showing a recessive trait must be homozygous recessive. If that person's parents did not show the trait, then *both* of the parents must be heterozygotes (carriers). Note that the genotype cannot be determined for some offspring, who may be either carriers or homozygous dominants.

analysis of pedigrees can reveal whether a particular trait is inherited in a dominant, recessive, or sex-linked pattern. Since the mid-1960s, analysis of human pedigrees, combined with molecular genetic technology, has produced great strides in understanding human genetic diseases. For instance, geneticists now know the genes responsible for dozens of inherited diseases, such as sickle-cell anemia, Marfan syndrome, and cystic fibrosis. Research in molecular genetics has increased our ability to predict genetic diseases and perhaps even to cure them, a topic we explore further in Chapter 13.

12.9 HOW ARE HUMAN DISORDERS CAUSED BY SINGLE GENES INHERITED?

Many common human traits, such as freckles, long eyelashes, cleft chin, and widow's peak hairline, are inherited in a simple Mendelian fashion; that is, each trait appears to be controlled by a single gene with a dominant and a recessive allele. Here, we will concentrate on a few examples of medically important genetic disorders and the ways in which they are transmitted from one generation to the next.

Some Human Genetic Disorders Are Caused by Recessive Alleles

The human body depends on the integrated actions of thousands of enzymes and other proteins. A mutation in an allele of the gene coding for one of these enzymes can impair or destroy its function. However, the presence of one normal allele may generate enough functional enzyme or other protein to enable heterozygotes to be phenotypically indistinguishable from homozygotes with two normal alle-

les. Therefore, for many genes, a normal allele that encodes a functional protein is dominant to a mutant allele that encodes a nonfunctional protein. Put another way, a mutant allele of these genes is recessive to a normal allele. Thus, an abnormal phenotype occurs only in individuals who inherit two copies of the mutant allele. Cystic fibrosis, which affects about 30,000 Americans, is a recessive disease of this type (see "Scientific Inquiry: Cystic Fibrosis").

Heterozygous individuals are **carriers** of a recessive genetic trait: they are phenotypically dominant but can pass on their recessive allele to their offspring. Geneticists estimate that each of us carries recessive alleles of 5 to 15 genes that would cause serious genetic defects in homozygotes. Every time we have a child, there is a 50:50 chance that we will pass on the defective allele. This is usually harmless, because an unrelated man and woman are unlikely to possess a defective allele of the *same* gene, so they are unlikely to produce a child who is homozygous recessive for a genetic disease. Related couples, however (especially first cousins or closer), have inherited some of their genes from recent common ancestors, and so are more likely to carry a defective allele of the same gene. If they are heterozygous for the *same* defective recessive allele, such couples have a 1 in 4 chance of having a child affected by the genetic disorder (see Fig. 12-27).

Albinism Results from a Defect in Melanin Production

An enzyme called tyrosinase is needed to produce melanin—the dark pigment in skin, hair, and the iris of the eye. The gene that encodes tyrosinase is called *TYR*. If an individual is homozygous for a mutant *TYR* allele that encodes a defective tyrosinase enzyme, albinism results (**FIG. 12-28**). Albinism in humans and other mammals is mani-

(a) Human   (b) Rattlesnake   (c) Wallaby

FIGURE 12-28 Albinism
Albinism is controlled by a single, recessive allele. Melanin is found throughout the animal kingdom, so albinos of many species have been observed. The female wallaby, having mated with a normally pigmented male, carries a normally colored offspring in her pouch.

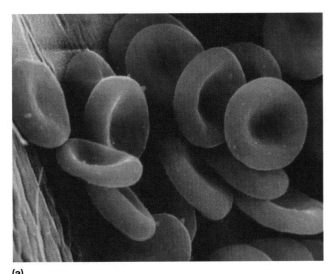

(a)   (b)

FIGURE 12-29 Sickle-cell anemia
(a) Normal red blood cells are disk-shaped with indented centers. **(b)** Sickled red blood cells in a person with sickle-cell anemia occur when blood oxygen is low. In this shape they are fragile and tend to clump together, clogging capillaries.

fested as white skin and hair and pink eyes (without melanin in the iris, you can see the color of the blood vessels in the retina).

Sickle-Cell Anemia Is Caused by a Defective Allele for Hemoglobin Synthesis

Sickle-cell anemia, a recessive disease in which defective hemoglobin is produced, results from a specific mutation in the hemoglobin gene. The hemoglobin protein, which gives red blood cells their color, transports oxygen in the blood. In sickle-cell anemia, the substitution of one nucleotide results in a single incorrect amino acid at a crucial position in hemoglobin, altering the properties of the hemoglobin molecule (see section 10.4 in Chapter 10). Under conditions of low oxygen (such as in muscles during exercise), masses of hemoglobin molecules in each red blood cell clump together. The clumps force the red blood cell out of its normal disk shape (**FIG. 12-29a**) into a longer, sickle shape (**FIG. 12-29b**). The sickled cells are more fragile than normal red blood cells, making them likely to break; they also tend to aggregate, clogging capillaries. Tissues "downstream" of the block do not receive enough oxygen or have their wastes removed. This lack of blood flow can cause pain, especially in joints. Paralyzing strokes can result if blocks occur in blood vessels in the brain. The condition can also cause anemia because so many red blood cells are destroyed. Although heterozygotes have about half normal and half abnormal hemoglobin, they usually have few sickled cells and are not disabled by the disease; in fact, many world-class athletes are heterozygous for the sickle-cell allele.

About 8% of the African American population is heterozygous for sickle-cell anemia, reflecting a genetic legacy of African origins. In some regions of Africa, 15% to 20% of the population carries the allele. The prevalence of the sickle-cell allele in Africa is explained by the fact that heterozygotes have some resistance to the parasite that causes malaria. We will explore this benefit further in Chapter 15.

If two heterozygous carriers have children, every conception will have a 1 in 4 chance of producing a child who is homozygous for the sickle-cell allele and consequently will have sickle-cell anemia. Modern DNA techniques can distinguish the normal hemoglobin allele from the sickle-cell allele, and analysis of fetal cells allows medical geneticists to diagnose sickle-cell anemia in fetuses. These methods are described in Chapter 13.

Some Human Genetic Disorders Are Caused by Dominant Alleles

Many normal physical traits, including cleft chin and freckles, are inherited as dominant traits. Many serious genetic disorders, such as Huntington's disease, are also caused by dominant alleles. For dominant diseases to be passed on to offspring, at least one parent must have the disease; thus, at least some people with dominant diseases must remain healthy enough to grow up and have children. Alternately, the dominant allele may result from a new mutation in a normal individual's eggs or sperm. In this situation, neither parent has the disease.

How can a mutant allele be dominant to the normal allele? Some dominant alleles produce an abnormal protein that interferes with the function of the normal one. For example, some proteins must link together into long chains in order to perform their function in the cell. The abnormal protein may enter into a chain but prevent addition of new protein "links." These shortened chain fragments may be unable to properly perform a needed function. Other dominant alleles may encode proteins that carry out new, toxic reactions. Finally, dominant alleles may encode a protein that is overactive, performing its function at inappropriate times and places.

Some Human Genetic Disorders Are Sex-Linked

As we described earlier, the X chromosome contains many genes that have no counterpart on the Y chromosome. Because males have only one X chromosome, they have only one allele for each of these genes. This single allele will be expressed with no possibility of its activity being "hidden" by expression of another allele.

A son receives his X chromosome from his mother and passes it only to his daughters. Thus, sex-linked disorders caused by a recessive allele have a unique pattern of inheritance. Such disorders appear far more frequently in males and typically skip generations: an affected male passes the trait to a phenotypically normal, carrier daughter, who in turn bears affected sons. The most familiar genetic defects due to recessive alleles of X-chromosome genes are red-green color blindness (**FIG. 12-30**) and **hemophilia** (**FIG. 12-31**). Hemophilia is caused by a recessive allele on the X chromosome that results in a deficiency in one of the proteins needed for blood clotting. People with hemophilia bruise easily and may bleed extensively from minor in-

juries. Hemophiliacs often have anemia due to blood loss. Nevertheless, even before modern treatment with clotting factors, some hemophiliac males survived to pass on their defective allele to their daughters, who carried the allele and could pass it to their sons.

12.10 HOW DO ERRORS IN CHROMOSOME NUMBER AFFECT HUMANS?

In Chapter 11, we examined the intricate mechanisms of meiosis, which ensure that each sperm and egg receive only one chromosome from each homologous pair. Not surprisingly, this elaborate dance of the chromosomes occasionally misses a step, resulting in gametes that have too many or too few chromosomes (**FIG. 12-32**). Such errors in meiosis, called **nondisjunction**, can affect the number of either sex chromosomes or autosomes. Most embryos that arise from the fusion of gametes with abnormal chromosome numbers spontaneously abort, accounting for 20% to 50% of all miscarriages, but some embryos with abnormal chromosome number survive to birth or beyond.

Some Genetic Disorders Are Caused by Abnormal Numbers of Sex Chromosomes

Because the X and Y chromosomes pair up during meiosis, sperm usually carry either an X or a Y chromosome. Nondisjunction of sex chromosomes in males produces

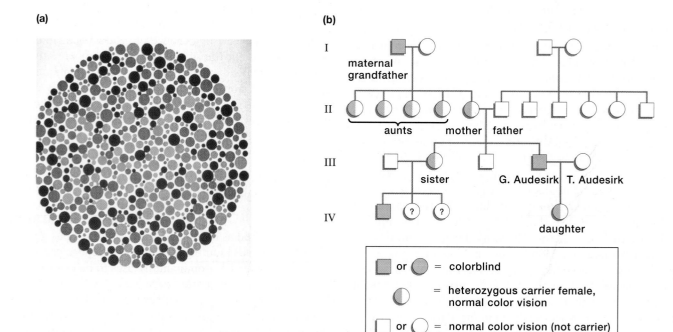

(a)   **(b)**

FIGURE 12-30 Color blindness, a sex-linked recessive trait
(a) This figure, called an *Ishihara chart* after its inventor, distinguishes color-vision defects. People with red-deficient vision see a 6, and those with green-deficient vision see a 9. People with normal color vision see 96. **(b)** Pedigree of one of the authors (G. Audesirk, who sees only a 6 in the Ishihara chart), showing sex-linked inheritance of red-green color blindness. Both the author and his maternal grandfather are color deficient; his mother and her four sisters carry the trait but have normal color vision. This pattern of more-common phenotypic expression in males and transmission from affected male to carrier female to affected male is typical of sex-linked recessive traits.

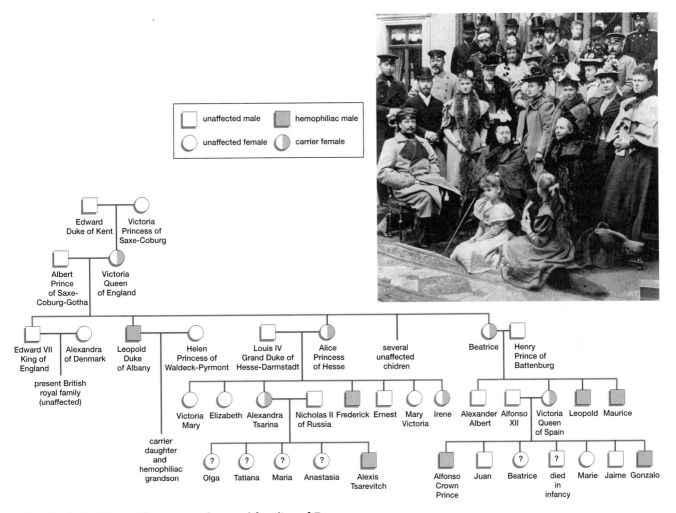

FIGURE 12-31 Hemophilia among the royal families of Europe
A famous genetic pedigree involves the transmission of sex-linked hemophilia from Queen Victoria of England (seated center front, with cane, 1885) to her offspring and eventually to virtually every royal house in Europe. Because Victoria's ancestors were free of hemophilia, the hemophilia allele must have arisen as a mutation either in Victoria herself or in one of her parents (or as a result of marital infidelity). Extensive intermarriage among royalty spread Victoria's hemophilia allele throughout Europe. Her most famous hemophiliac descendant was great-grandson Alexis, Tsarevitch (crown prince) of Russia. The Tsarina Alexandra (Victoria's granddaughter) believed that only the monk Rasputin could control Alexis's bleeding. Rasputin may actually have used hypnosis to cause Alexis to cut off circulation to bleeding areas by muscular contraction. The influence that Rasputin had over the imperial family may have contributed to the downfall of the tsar during the Russian Revolution. In any event, hemophilia was not the cause of Alexis's death; he was killed with the rest of his family by the Bolsheviks (Communists) in 1918.

sperm with 22 autosomes, but either no sex chromosome (often called "O" sperm), or two sex chromosomes (the sperm may be XX, YY, or XY, depending on whether the nondisjunction occurred in meiosis I or II). Nondisjunction of the sex chromosomes in females produces O or XX eggs instead of eggs with one X chromosome. When normal gametes fuse with these defective sperm or eggs, the zygotes have normal numbers of autosomes but abnormal numbers of sex chromosomes (Table 12-2). The most common abnormalities are XO, XXX, XXY, and XYY. (Genes on the X chromosome are essential to survival, so any embryo without at least one X chromosome spontaneously aborts very early in development.)

Turner Syndrome (XO)

About one in every 3000 phenotypically female babies has only one X chromosome, a condition known as **Turner syndrome**. At puberty, hormone deficiencies prevent XO females from menstruating or developing secondary sexual characteristics, such as enlarged breasts. Treatment with estrogen promotes physical development. However, because most women with Turner syndrome lack mature eggs, hormone treatment does not make it possible for them to bear children. Additional characteristics of women with Turner syndrome include short stature, folds of skin around the neck, and increased risk of cardiovascular disease, kidney defects, and hearing loss. Because

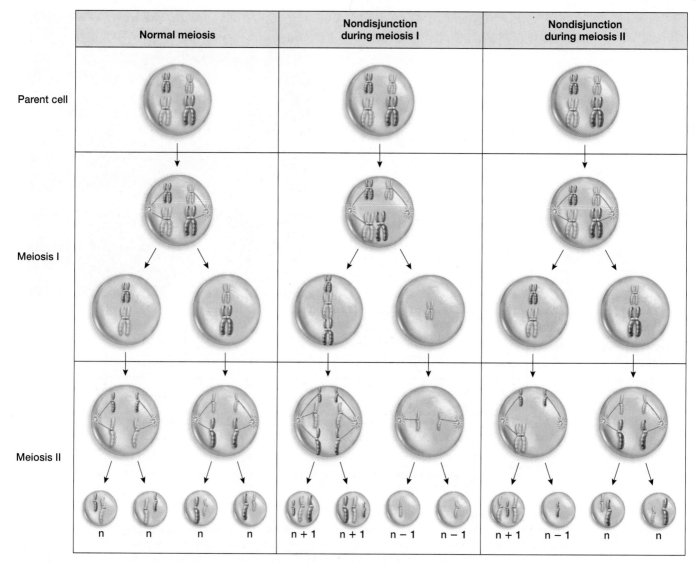

FIGURE 12-32 Nondisjunction during meiosis
Nondisjunction may occur either during meiosis I (left) or meiosis II (right), resulting in gametes with too many ($n + 1$) or too few ($n - 1$) chromosomes.

Table 12-2 Effects of Nondisjunction of the Sex Chromosomes During Meiosis			
Nondisjunction in Father			
Sex Chromosomes of Defective Sperm	Sex Chromosomes of Normal Egg	Sex Chromosomes of Offspring	Phenotype
0 (none)	X	X0	Female—Turner syndrome
XX	X	XXX	Female—Trisomy X
YY	X	XYY	Male—Jacob syndrome
XY	X	XXY	Male—Klinefelter syndrome
Nondisjunction in Mother			
Sex Chromosomes of Normal Sperm	Sex Chromosomes of Defective Egg	Sex Chromosomes of Offspring	Phenotype
X	0 (none)	X0	Female—Turner syndrome
Y	0 (none)	Y0	Dies as embryo
X	XX	XXX	Female—Trisomy X
Y	XX	XXY	Male—Klinefelter syndrome

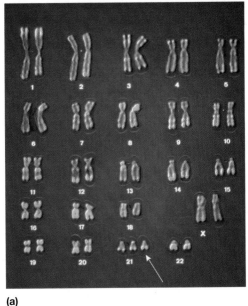

(a) **(b)**

FIGURE 12-33 Trisomy 21, or Down syndrome
(a) This karyotype of a Down syndrome child reveals three copies of chromosome 21 (arrow). **(b)** These girls have the relaxed mouth and distinctively shaped eyes typical of Down syndrome.

women with Turner syndrome have only one X chromosome, they display X-linked recessive disorders, such as hemophilia and color blindness, much more frequently than do XX women.

The differences between XO and XX women suggest that the inactivation of one X chromosome in XX females (see Chapter 10) is not complete. Otherwise, XX women, with only one "active" X chromosome, and XO women, with only one X chromosome, should have identical features. In fact, some 200 genes on the inactivated X chromosome are functional in XX females, preventing the development of Turner syndrome.

Trisomy X (XXX)

About one in every 1000 women has three X chromosomes. Most such women have no detectable defects, except for a tendency to be tall and a higher incidence of below-normal intelligence. Unlike women with Turner syndrome, most **trisomy X** women are fertile and, interestingly enough, almost always bear normal XX and XY children. Some unknown mechanism must operate during meiosis to prevent an extra X chromosome from being included in their eggs.

Klinefelter Syndrome (XXY)

About one male in every 1000 is born with two X chromosomes and one Y chromosome. Most of these men go through life never realizing that they have an extra X chromosome. However, at puberty, some show mixed secondary sexual characteristics, including partial breast development, broadening of the hips, and small testes. These symptoms are known as **Klinefelter syndrome.** XXY men are often infertile because of low sperm count but are not impotent. They are usually diagnosed when the man and his partner seek medical help when they are unable to conceive a baby.

Jacob Syndrome (XYY)

Another common type of sex chromosome abnormality is XYY, occurring in about one male in every 1000. You might expect that an extra Y chromosome, which has few active genes, would not make very much difference, and this seems to be true in most cases. However, XYY males usually have high levels of testosterone, often have severe acne, and are tall (about two-thirds of XYY males are over 6 feet tall, compared with the average male height of 5 feet 9 inches).

Some Genetic Disorders Are Caused by Abnormal Numbers of Autosomes

Nondisjunction of the autosomes may also occur, producing eggs or sperm missing an autosome or with two copies of an autosome. Fusion with a normal gamete (bearing one copy of each autosome) leads to an embryo with either one or three copies of the affected autosome. Embryos that have only one copy of any of the autosomes abort so early in development that the woman never knows she was pregnant. Embryos with three copies of an autosome (trisomy) also usually spontaneously abort. However, a small fraction of embryos with three copies of chromosomes 13, 18, or 21 survive to birth. In the case of trisomy 21, the child may live into adulthood.

Trisomy 21 (Down Syndrome)

In about 1 of every 900 births, the child inherits an extra copy of chromosome 21, a condition called **trisomy 21** or **Down syndrome.** Children with Down syndrome have several distinctive physical characteristics, including weak muscle tone, a small mouth held partially open because it cannot accommodate the tongue, and distinctively shaped eyelids (**FIG. 12-33**). Much more serious defects include

FIGURE 12-34 Down syndrome frequency increases with maternal age The increase in frequency of Down syndrome after maternal age 35 is quite dramatic.

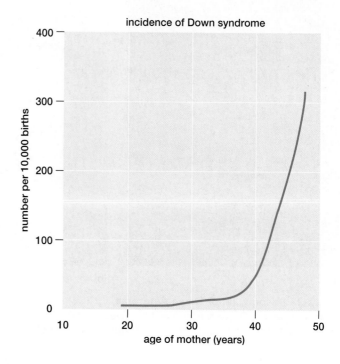

incidence of Down syndrome

low resistance to infectious diseases, heart malformations, and varying degrees of mental retardation, often severe.

The frequency of nondisjunction increases with the age of the parents, especially the mother (**FIG. 12-34**). Nondisjunction in sperm accounts for about a quarter of the cases of Down syndrome, and there is a small increase in these defective sperm with increasing age of the father. Since the 1970s, it has become more common for couples to delay having children, increasing the probability of trisomy 21. Trisomy can be diagnosed before birth by examining the chromosomes of fetal cells (see "Health Watch: Prenatal Genetic Screening" in Chapter 13).

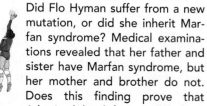

BIOETHICS CASE STUDY REVISITED SUDDEN DEATH ON THE COURT

Did Flo Hyman suffer from a new mutation, or did she inherit Marfan syndrome? Medical examinations revealed that her father and sister have Marfan syndrome, but her mother and brother do not. Does this finding prove that Hyman inherited the defective allele from her father? As you learned in this chapter, diploid organisms, including people, generally have two alleles of each gene, one on each homologous chromosome. It has been known for many years that one defective fibrillin allele is enough to cause Marfan syndrome. Further, the children of a person with Marfan syndrome have a 50% chance of inheriting the disease. What can we conclude from these data?

First, if even one defective fibrillin allele produces Marfan syndrome, then Hyman's mother must carry two normal alleles, because she does not have Marfan syndrome. Second, because new mutations are rare and Hyman's father has Marfan syndrome, it is almost certain that Hyman inherited a defective fibrillin allele from her father.

Third, is Marfan syndrome inherited as a dominant or a recessive condition? Once again, if only one defective allele is enough to cause Marfan syndrome, then this allele must be dominant and the normal allele must be recessive. Finally, if Hyman had borne children, could they have inherited Marfan syndrome from her? For a dominant disorder, any children who inherited her defective allele would develop Marfan syndrome. Therefore, on average, half of her children would have had Marfan syndrome (try working this out with a Punnett square).

Consider This Marfan syndrome can't be detected in an embryo by a simple biochemical test (yet). Most cystic fibrosis mutations, however, can easily be detected in both heterozygotes and homozygotes, in adults, children, and embryos. A few years ago, some states considered making mandatory testing of couples for cystic fibrosis part of the application for a marriage license. If two heterozygotes marry, each of their children has a 25% chance of having the disorder. Although there is no cure, there will probably be better treatments within a few years. Do you think that carrier screening should be mandatory? If you and your spouse were both heterozygotes, would you seek prenatal diagnosis of an embryo? What would you do if your embryo were destined to be born with cystic fibrosis?

CHAPTER REVIEW

SUMMARY OF KEY CONCEPTS

12.1 What Is the Physical Basis of Inheritance?

Homologous chromosomes carry the same genes located at the same loci, but the genes at a particular locus can exist in alternate forms called alleles. An organism whose homologous chromosomes carry the same allele at a locus is homozygous for that particular gene. If the alleles at a locus differ, the organism is heterozygous for that gene.

12.2 How Did Gregor Mendel Lay the Foundation for Modern Genetics?

Gregor Mendel deduced many principles of inheritance in the mid-1800s, before the discovery of DNA, genes, chromosomes, or meiosis. He did this by choosing an appropriate experimental subject, designing his experiments carefully, following progeny for several generations, and analyzing his data statistically.

Web Tutorial 12.1 Self- and Cross-Pollination of Pea Plants

12.3 How Are Single Traits Inherited?

A trait is an observable or measurable feature of the organism's phenotype, such as eye color or blood type. Traits are inherited in particular patterns that depend on the types of alleles that parents pass on to their offspring. Each parent provides its offspring with one copy of every gene, so that the offspring inherits a pair of alleles for every gene. The combination of alleles in the offspring determines whether it displays a particular trait. Dominant alleles mask the expression of recessive alleles. The masking of recessive alleles can result in organisms with the same phenotype but different genotypes. That is, organisms with two dominant alleles (homozygous dominant) have the same phenotype as do organisms with one dominant and one recessive allele (heterozygous). Because each allele segregates randomly during meiosis, we can use the laws of probability to predict the relative proportions of offspring with a particular trait.

Web Tutorial 12.2 The Inheritance of Single Traits

12.4 How Are Multiple Traits Inherited?

If the genes for two traits are located on separate chromosomes, they will be assorted independently of one another into the egg or sperm. Thus, crossing two organisms that are heterozygous at two loci on separate chromosomes produces offspring with 10 different genotypes. If the alleles are typical dominant and recessive alleles, these progeny will display only 4 different phenotypes.

Web Tutorial 12.3 The Inheritance of Multiple Traits

12.5 How Are Genes Located on the Same Chromosome Inherited?

Genes located on the same chromosome are linked to one another (encoded on the same DNA double helix) and they tend to be inherited together. Unless the alleles are separated by chromosomal recombination, the two alleles will be passed together to the offspring.

12.6 How Is Sex Determined, and How Are Sex-Linked Genes Inherited?

In many animals, sex is determined by sex chromosomes, often designated X and Y. The rest of the chromosomes, identical in the two sexes, are called autosomes. In many animals, females have two X chromosomes, whereas males have one X and one Y chromosome. The Y chromosome has many fewer genes than the X chromosome. Because males have only one copy of most X chromosome genes, recessive traits on the X chromosome are more likely to be phenotypically expressed in males.

12.7 Do the Mendelian Rules of Inheritance Apply to All Traits?

Not all inheritance follows the simple dominant-recessive pattern:

- In incomplete dominance, heterozygotes have a phenotype that is intermediate between the two homozygous phenotypes.
- Codominance occurs when two types of proteins, each encoded by a different allele at a single locus, both contribute to the phenotype.
- Many traits are determined by several different genes at different loci that contribute to the phenotype, a phenomenon called polygenic inheritance.
- Many genes have multiple effects on the organism's phenotype (pleiotropy).
- The environment influences the phenotypic expression of most, if not all, traits.

12.8 How Are Human Genetic Disorders Investigated?

The genetics of humans is similar to the genetics of other animals, except that experimental crosses are not feasible. Analysis of family pedigrees and, more recently, molecular genetic techniques must be used to determine the mode of inheritance of human traits.

12.9 How Are Human Disorders Caused by Single Genes Inherited?

Some genetic disorders are inherited as recessive traits; therefore, only homozygous recessive persons show symptoms of the disease. Heterozygotes are called carriers; they carry the recessive allele but do not express the trait. Some other diseases are inherited as simple dominant traits. In such cases, only one copy of the dominant allele is needed to cause disease symptoms. The human Y chromosome bears few genes other than those that determine male reproductive attributes; therefore, men phenotypically display whichever allele they carry on their single X chromosome, a phenomenon called sex-linked inheritance.

12.10 How Do Errors in Chromosome Number Affect Humans?

Errors in meiosis can result in gametes with abnormal numbers of sex chromosomes or autosomes. Many people with abnormal numbers of sex chromosomes have distinguishing physical characteristics. Abnormal numbers of autosomes typically lead to spontaneous abortion early in pregnancy. In rare instances, the fetus may survive to birth, but mental and physical deficiencies always occur, as is the case with Down syndrome (trisomy 21). The likelihood of abnormal numbers of chromosomes increases with increasing age of the mother, and, to a lesser extent, the father.

KEY TERMS

allele   *page 222*
autosome   *page 231*
carrier   *page 238*
codominance   *page 235*
cross-fertilization   *page 223*
crossing over   *page 230*
dominant   *page 224*
Down syndrome   *page 243*
gene   *page 222*
genetic recombination
  page 230
genotype   *page 225*

hemophilia   *page 240*
heterozygous   *page 222*
homozygous   *page 222*
hybrid   *page 222*
incomplete dominance
  page 233
inheritance   *page 222*
Klinefelter syndrome
  page 243
law of independent
  assortment   *page 228*
law of segregation   *page 224*

linkage   *page 229*
locus   *page 222*
multiple alleles   *page 233*
nondisjunction   *page 240*
pedigree   *page 237*
phenotype   *page 225*
pleiotropy   *page 235*
polygenic inheritance
  page 235
Punnett square method
  page 225

recessive   *page 224*
self-fertilization   *page 223*
sex chromosome   *page 231*
sex-linked   *page 231*
sickle-cell anemia   *page 239*
test cross   *page 227*
trisomy 21   *page 243*
trisomy X   *page 243*
true-breeding   *page 223*
Turner syndrome   *page 241*

THINKING THROUGH THE CONCEPTS

1. Define the following terms: *gene, allele, dominant, recessive, true-breeding, homozygous, heterozygous, cross-fertilization,* and *self-fertilization.*

2. Explain why genes located on the same chromosome are said to be linked. Why do alleles of linked genes sometimes separate during meiosis?

3. Define *polygenic inheritance.* Why does polygenic inheritance sometimes allow parents to produce offspring that are notably different in eye or skin color than either parent?

4. What is sex linkage? In mammals, which sex would be most likely to show recessive sex-linked traits?

5. What is the difference between a phenotype and a genotype? Does knowledge of an organism's phenotype always allow you to determine the genotype? What type of experiment would you perform to determine the genotype of a phenotypically dominant individual?

6. In the pedigree of part (a) of Figure 12-27, do you think that the individuals showing the trait are homozygous or heterozygous? How can you tell from the pedigree?

7. Define *nondisjunction,* and describe the common syndromes caused by nondisjunction of sex chromosomes and autosomes.

APPLYING THE CONCEPTS

1. Sometimes the term *gene* is used rather casually. Compare the terms *allele* and *gene.*

2. Mendel's numbers seemed almost too perfect to be real; some believe he may have cheated a bit on his data. Perhaps he continued to collect data until the numbers matched his predicted ratios, then stopped. Recently, there has been much publicity over violations of scientific ethics, including researchers' plagiarizing others' work, using other scientists' methods to develop lucrative patents, or just plain fabricating data. How important is this issue for society? What are the boundaries of

ethical scientific behavior? How should the scientific community or society "police" scientists? What punishments would be appropriate for violations of scientific ethics?

3. Although American society has been described as a "melting pot," people often engage in "assortative mating," in which they marry others of similar height, socioeconomic status, race, and IQ. Discuss the consequences to society of assortative mating among humans. Would society be better off if people mated more randomly? Explain.

GENETICS PROBLEMS

(Note: An extensive group of genetics problems, with answers, can be found in the Study Guide.)

1. In certain cattle, hair color can be red (homozygous R_1R_1), white (homozygous R_2R_2), or roan (a mixture of red and white hairs, heterozygous R_1R_2).

   a. When a red bull is mated to a white cow, what genotypes and phenotypes of offspring could be obtained?

   b. If one of the offspring bulls in (a) were mated to a white cow, what genotypes and phenotypes of offspring could be produced? In what proportion?

2. The palomino horse is golden in color. Unfortunately for horse fanciers, palominos do not breed true. In a series of matings between palominos, the following offspring were obtained:

   65 palominos, 32 cream-colored,
   34 chestnut (reddish brown)

   What is the probable mode of inheritance of palomino coloration?

3. In the edible pea, tall (*T*) is dominant to short (*t*), and green pods (*G*) are dominant to yellow pods (*g*). List the types of

gametes and offspring that would be produced in the following crosses:

a. $TtGg \times TtGg$
b. $TtGg \times TTGG$
c. $TtGg \times Ttgg$

4. In tomatoes, round fruit (R) is dominant to long fruit (r), and smooth skin (S) is dominant to fuzzy skin (s). A true-breeding round, smooth tomato ($RRSS$) was crossbred with a true-breeding long, fuzzy tomato ($rrss$). All the F_1 offspring were round and smooth ($RrSs$). When these F_1 plants were bred, the following F_2 generation was obtained:

Round, smooth: 43   Long, fuzzy: 13

Are the genes for skin texture and fruit shape likely to be on the same chromosome or on different chromosomes? Explain your answer.

5. In the tomatoes of problem 4, an F_1 offspring ($RrSs$) was mated with a homozygous recessive ($rrss$). The following offspring were obtained:

Round, smooth: 583   Round, fuzzy: 21
Long, fuzzy: 602   Long, smooth: 16

What is the most likely explanation for this distribution of phenotypes?

6. In humans, hair color is controlled by two interacting genes. The same pigment, melanin, is present in both brown-haired and blond-haired people, but brown hair has much more of it. Brown hair (B) is dominant to blond (b). Whether any melanin can be synthesized depends on another gene. The dominant form (M) allows melanin synthesis; the recessive form (m) prevents melanin synthesis. Homozygous recessives (mm) are albino. What will be the expected proportions of phenotypes in the children of the following parents?

a. $BBMM \times BbMm$
b. $BbMm \times BbMm$
c. $BbMm \times bbmm$

7. In humans, one of the genes determining color vision is located on the X chromosome. The dominant form (C) produces normal color vision; red-green color blindness (c) is recessive. If a man with normal color vision marries a color-blind woman, what is the probability of their having a color-blind son? A color-blind daughter?

8. In the couple described in problem 7, the woman gives birth to a color-blind but otherwise normal daughter. The husband sues for a divorce on the grounds of adultery. Will his case stand up in court? Explain your answer.

ANSWERS TO GENETICS PROBLEMS

1. a. A red bull (R_1R_1) is mated to a white cow (R_2R_2). The bull will produce all R_1 sperm; the cow will produce all R_2 eggs. All the offspring will be R_1R_2 and will have roan hair (codominance).

b. A roan bull (R_1R_2) is mated to a white cow (R_2R_2). The bull produces half R_1 and half R_2 sperm; the cow produces R_2 eggs. Using the Punnett square method:

eggs
R_2

R_1	R_1R_2
R_2	R_2R_2

sperm

Using probabilities:

sperm	egg	offspring
$\frac{1}{2}R_1$	R_2	$\frac{1}{2}R_1R_2$
$\frac{1}{2}R_2$	R_2	$\frac{1}{2}R_2R_2$

The predicted offspring will be $\frac{1}{2}R_1R_2$ (roan) and $\frac{1}{2}R_2R_2$ (white).

2. The offspring occur in three types, classifiable as dark (chestnut), light (cream), and intermediate (palomino). This distribution suggests incomplete dominance, with the alleles for chestnut (C_1) combining with the allele for cream (C_2) to produce palomino heterozygotes (C_1C_2). We can test this hypothesis by examining the offspring numbers. There are approximately $\frac{1}{4}$ chestnut (C_1C_1), $\frac{1}{2}$ palomino (C_1C_2), and $\frac{1}{4}$

cream (C_2C_2). If palominos are heterozygotes, we would expect the cross $C_1C_2 \times C_1C_2$ to yield $\frac{1}{4}C_1C_1$, $\frac{1}{2}C_1C_2$, and $\frac{1}{4}C_2C_2$. Our hypothesis is supported

3. a. $TtGg \times TtGg$. This is a "standard" cross for differences in two traits. Both parents produce TG, Tg, tG, and tg gametes. The expected proportions of offspring are $\frac{9}{16}$ tall green, $\frac{3}{16}$ tall yellow, $\frac{3}{16}$ short green, $\frac{1}{16}$ short yellow.

b. $TtGg \times TTGG$. In this cross, the heterozygous parent produces TG, Tg, tG, and tg gametes. However, the homozygous dominant parent can produce only TG gametes. Therefore, all offspring will receive at least one T allele for tallness and one G allele for green pods, and thus all the offspring will be tall with green pods.

c. $TtGg \times Ttgg$. The second parent will produce two types of gametes, Tg and tg. Using a Punnett square:

eggs

	Tg	tg
TG	$TTGg$	$TtGg$
Tg	$TTgg$	$Ttgg$
tG	$TtGg$	$ttGg$
tg	$Ttgg$	$ttgg$

sperm

The expected proportions of offspring are $\frac{3}{8}$ tall green, $\frac{3}{8}$ tall yellow, $\frac{1}{8}$ short green, $\frac{1}{8}$ short yellow.

4. If the genes are on separate chromosomes—that is, assort independently—then this would be a typical two-trait cross with expected offspring of all four types (about $\frac{9}{16}$ round smooth, $\frac{3}{16}$

round fuzzy, $\frac{3}{16}$ long smooth, and $\frac{1}{16}$ long fuzzy). However, only the parental combinations show up in the F_2 offspring, indicating that the genes are on the same chromosome.

5. The genes are on the same chromosome and are quite close together. On rare occasions, crossing over occurs between the two genes, producing recombination of the alleles.

6. a. *BBMM* (brown) $\times$ *BbMm* (brown). The first parent can produce only *BM* gametes, so all offspring will receive at least one dominant allele for each gene. Therefore, all offspring will have brown hair.

   b. *BbMm* (brown) $\times$ *BbMm* (brown). Both parents can produce four types of gametes: *BM, Bm, bM,* and *bm.* Filling in the Punnett square:

eggs

	BM	Bm	bM	bm
BM	BBMM	BBMm	BbMM	BbMm
Bm	BBMm	BBmm	BbMm	Bbmm
bM	BbMM	BbMm	bbMM	bbMm
bm	BbMm	Bbmm	bbMm	bbmm

(sperm)

All *mm* offspring are albino, so we get the expected proportions $\frac{9}{16}$ brown-haired, $\frac{3}{16}$ blond-haired, $\frac{4}{16}$ albino.

   c. *BbMm* (brown) $\times$ *bbmm* (albino):

eggs

	bm
BM	BbMm
Bm	Bbmm
bM	bbMm
bm	bbmm

(sperm)

The expected proportions of offspring are $\frac{1}{4}$ brown-haired, $\frac{1}{4}$ blond-haired, $\frac{1}{2}$ albino.

7. A man with normal color vision is *CY* (remember, the Y chromosome does not have the gene for color vision). His color-blind wife is *cc*. Their expected offspring will be:

eggs

	c
C	Cc
Y	cY

(sperm)

We therefore expect that all the daughters will have normal color vision and all the sons will be color-blind.

8. The husband should win his case. All his daughters must receive one X chromosome, with the *C* allele, from him and therefore should have normal color vision. If his wife gives birth to a color-blind daughter, her husband cannot be the father (unless there was a new mutation for color blindness in his sperm line, which is very unlikely).

FOR MORE INFORMATION

Cattaneo, E., Rigamonti, D., and Zuccato, C. "The Enigma of Huntington's Disease." *Scientific American*, December 2002. Although the dominant allele causing Huntington's disease has been discovered, researchers still don't know how it causes the disorder.

McGue, M. "The Democracy of the Genes." *Nature*, July 1997. The environment has a more important role in the development of intelligence than previously thought.

Mendel Museum of Genetics, http://www.mendel-museum.org/. The monastery where Mendel lived and worked is located in what is now the Czech Republic. The Mendel Museum in Brno sponsors the Web site, which describes Mendel's contributions to the discovery of the principles of inheritance.

National Institutes of Health, http://www.nlm.nih.gov/medlineplus/cysticfibrosis.html. "Medline Plus: Cystic Fibrosis." Numerous links to causes and treatment of cystic fibrosis; regularly updated.

National Institutes of Health, http://www.nlm.nih.gov/medlineplus/marfansyndrome.html. "Medline Plus: Marfan Syndrome." Links to causes and treatment of Marfan syndrome; regularly updated.

Sapienza, C. "Parental Imprinting of Genes." *Scientific American*, October 1990. It is not quite true that all genes are equal, regardless of whether they have been inherited from mother or father. In some cases, which parent a gene comes from greatly alters its expression in the offspring.

Stern, C., and Sherwood, E. R. *The Origin of Genetics: A Mendel Source Book*. San Francisco: Freeman, 1966. There is no substitute for the real thing, in this case a translation of Mendel's original paper to the Brünn Society.

13 Biotechnology

DNA profiling proved that Earl Ruffin, shown here with
some of his grandchildren, was innocent of the rape and assault
for which he spent 21 years in prison.

AT A GLANCE

CASE STUDY GUILTY OR INNOCENT?

IT WAS A HORROR STORY straight out of a TV crime show. At about 2 a.m. on December 5, 1981, in Norfolk, Virginia, a divorced mother of three was awakened by a stranger in her bedroom. "You scream, you are dead," he told her. In an interval between repeated rapes, she caught a glimpse of the man's face, lit by the streetlamp outside her bedroom window. Finally finished, he ordered the woman to take a shower. Despite her desperate fear, she carefully avoided washing away evidence that might be used to find her assailant.

A few weeks later, she accidentally encountered a man in an elevator. She believed that she had found her rapist. Julius Earl Ruffin, 28, was arrested and tried for rape and assault. The victim testified that she recognized Ruffin's face. Further, his blood type, called B secretor, matched the semen sample taken from the victim on the night of the rape (secretors "secrete"

their blood antigens into bodily fluids, including semen). About 8% of men are type B secretors. Despite major inconsistencies between Ruffin's appearance and the victim's initial descriptions of her assailant—height, complexion, Ruffin's two gold front teeth—and despite testimony by Ruffin's girlfriend and brother that they had been with Ruffin on the night of the assault, Ruffin was convicted and sentenced to life in prison on each of multiple counts. Sentencing Ruffin put an innocent man behind bars and left the real rapist to roam the streets and strike again.

Although he was terribly wronged, fate eventually smiled upon Ruffin. Remember, the victim had not washed away her rapist's semen when she showered. Although this contributed to Ruffin's initial conviction, it also proved to be the essential ingredient for his eventual exoneration. A second stroke of good fortune for

Ruffin was that Mary Jane Burton was the forensic scientist assigned to his case, and that she routinely violated normal procedures: instead of returning all of the evidence to police investigators (usually to be destroyed), she kept biological material in her case files. And finally, Ruffin was fortunate that the Innocence Project, founded in 1992 by Barry Scheck and Peter Neufeld of the Benjamin Cardozo School of Law at Yeshiva University, turned loose the power of biotechnology in cases like his.

You've probably already guessed how Ruffin's innocence was proved: DNA evidence. In this chapter, we'll investigate the techniques of biotechnology that have started to pervade so much of modern life, including forensics in the courtroom, prenatal diagnosis and treatment of inherited disorders, and genetically modified crops and livestock.

13.1 WHAT IS BIOTECHNOLOGY?

In its broadest sense, **biotechnology** is any use or alteration of organisms, cells, or biological molecules to achieve specific practical goals. Therefore, some aspects of biotechnology are ancient. For example, people have used yeast cells to produce bread, beer, and wine for the past 10,000 years. Selective breeding of plants and animals has an equally long history: 8000- to 10,000-year-old fragments of squash, found in a dry cave in Mexico, have larger seeds and thicker rinds than wild squash, suggesting selective breeding for higher nutritional content. Prehistoric art and animal remains indicate that dogs, sheep, goats, pigs, and camels were domesticated and selectively bred beginning at least 10,000 years ago.

Even today, selective breeding remains an important tool of biotechnology. Modern biotechnology, however, also frequently uses **genetic engineering**, a term that refers to more direct methods for modifying genetic material. Genetically engineered cells or organisms may have had genes deleted, added, or changed. Genetic engineering can be used to learn more about how cells and genes work, to develop better treatments for diseases, to produce valuable biological molecules, and to improve plants and animals for agriculture.

A key tool in genetic engineering is **recombinant DNA**, which is DNA that has been altered to contain genes or portions of genes from different organisms. Large amounts of recombinant DNA can be grown in bacteria, viruses, or yeast, and then transferred into other species. Plants and animals that express DNA that has been modified or derived from other species are called **transgenic**, or **genetically modified organisms (GMOs)**.

Since its development in the 1970s, recombinant DNA technology has grown explosively, providing new methods, applications, and possibilities for genetic engineering. Today, researchers in almost every area of biology routinely use recombinant DNA technology in their experiments. In the pharmaceutical industry, genetic engineering has become the preferred way to manufacture many products, including several human hormones, such as insulin, and some vaccines, such as the vaccine against hepatitis B.

Modern biotechnology also includes many methods of manipulating DNA, whether or not the DNA is subsequently put into a cell or an organism. For example, determining the nucleotide sequence of specific pieces of DNA is crucial to forensic science and the diagnosis of inherited disorders.

In this chapter, we will provide an overview of modern biotechnology. Our emphasis will be on applications of biotechnology and their impacts on society, but we will also briefly describe some of the important methods used in those applications. We will organize our discussion around five major themes: (1) recombinant DNA mechanisms found in nature, mostly in bacteria and viruses; (2) biotechnology in criminal forensics, principally DNA matching; (3) biotechnology in agriculture, specifically the production of transgenic plants and animals; (4) the

Human Genome Project and its applications; and (5) biotechnology in medicine, focusing on the diagnosis and treatment of inherited disorders.

13.2 HOW DOES DNA RECOMBINE IN NATURE?

Most people think that a species' genetic makeup is constant, except for the occasional mutation, but genetic reality is far more fluid. Many natural processes can transfer DNA from one organism to another, sometimes even to organisms of different species. Recombinant DNA technologies used in the laboratory are often based on these naturally occurring processes.

Sexual Reproduction Recombines DNA

Sexual reproduction literally recombines DNA from two different organisms. As we saw in Chapter 11, homologous chromosomes exchange DNA by crossing over during meiosis I. Thus, each chromosome in a gamete usually contains a mixture of alleles from the two parental chromosomes. In this sense, every egg and every sperm contain recombinant DNA, derived from the organism's two parents. When a sperm fertilizes an egg, the resulting offspring also contain recombinant DNA.

Transformation May Combine DNA from Different Bacterial Species

Bacteria can undergo several types of recombination (**FIG. 13-1**). **Transformation** enables bacteria to pick up DNA from the environment (Fig. 13-1b). The DNA may be part of the chromosome from another bacterium, even from another species. You may recall from Chapter 9 that living, non-virulent pneumonia bacteria can pick up genes from dead virulent bacteria, allowing the formerly harmless bacteria to cause pneumonia (see Fig. 9-1). Unraveling the mechanism of bacterial transformation was an important step leading to the discovery that DNA is the genetic material.

Transformation may also occur when bacteria pick up tiny circular DNA molecules called **plasmids** (Fig. 13-1c). Many types of bacteria contain plasmids, ranging in size from about 1000 to 100,000 nucleotides long. For comparison, the *E. coli* chromosome is around 4,600,000 nucleotides long. A single bacterium may contain dozens or even hundreds of copies of a plasmid. When the bacterium dies, it releases these plasmids into the environment, where they may be picked up by other bacteria of the same or different species. In addition, living bacteria can often pass plasmids directly to other living bacteria. Passing plasmids from bacteria to yeast may also occur, moving genes from a prokaryotic cell to a eukaryotic cell.

What use are plasmids? A bacterium's chromosome contains all the genes the cell normally needs for basic survival. However, genes carried by plasmids may permit the bacteria to thrive in novel environments. Some plasmids contain genes that allow bacteria to metabolize unusual energy sources, such as petroleum. Other plasmids carry

(a) Bacterium

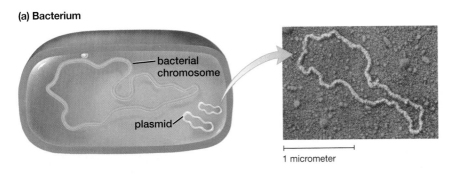

FIGURE 13-1 Recombination in bacteria
(a) In addition to their large circular chromosome, bacteria commonly possess small rings of DNA called plasmids, which often carry additional useful genes. Bacterial transformation occurs when living bacteria take up (b) fragments of chromosomes or (c) plasmids.

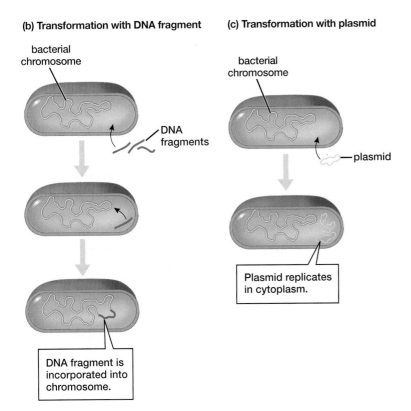

genes that cause disease symptoms, such as diarrhea, in the animal or other organism that the bacterium infects. (Diarrhea may benefit the bacterium, enabling it to spread and infect new hosts.) Still other plasmids carry genes that enable bacteria to grow in the presence of antibiotics, such as penicillin. In environments where antibiotic use is high, particularly in hospitals, bacteria carrying antibiotic-resistance plasmids can quickly spread among patients and health care workers, making antibiotic-resistant infections a serious problem.

Viruses May Transfer DNA Between Species

Viruses, which are often little more than genetic material encased in a protein coat, transfer their genetic material into cells during infection. Within the infected cell, viral genes replicate. Unable to distinguish which genetic information is its own, and which is from the virus, the host cell's enzymes and ribosomes then synthesize viral proteins. The replicated genes and viral proteins assemble in-side the cell, forming new viruses that are released and may infect new cells (FIG. 13-2).

Some viruses can transfer genes from one organism to another. In these instances, the virus inserts its DNA into a host cell's chromosome. The viral DNA may remain there for days, months, or even years. Every time the cell divides, it replicates the viral DNA along with its own DNA. When new viruses are finally produced, some of the host's genes may be incorporated into the viral DNA. If such recombinant viruses infect another cell and insert their DNA into the new host cell's chromosomes, pieces of the previous host cell's DNA will also be inserted.

Most viruses infect and replicate only in the cells of specific bacterial, animal, or plant species. For example, the canine distemper virus, which causes a frequently fatal disease in dogs, usually only infects dogs, raccoons, otters, and related species (although in the 1990s it jumped the "species barrier" and killed thousands of lions in Africa). Therefore, most of the time, viruses move host DNA between different individuals of a single, or fairly closely related, species. However,

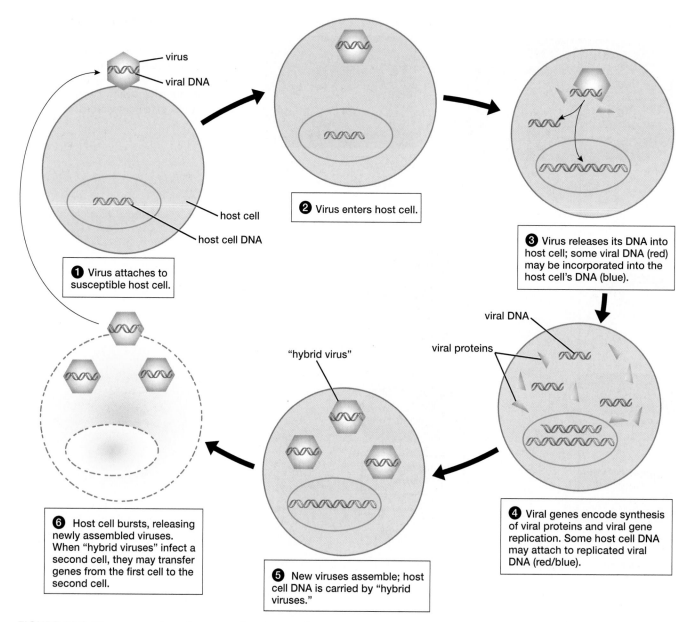

1 Virus attaches to susceptible host cell.

2 Virus enters host cell.

3 Virus releases its DNA into host cell; some viral DNA (red) may be incorporated into the host cell's DNA (blue).

4 Viral genes encode synthesis of viral proteins and viral gene replication. Some host cell DNA may attach to replicated viral DNA (red/blue).

5 New viruses assemble; host cell DNA is carried by "hybrid viruses."

6 Host cell bursts, releasing newly assembled viruses. When "hybrid viruses" infect a second cell, they may transfer genes from the first cell to the second cell.

virus
viral DNA
host cell
host cell DNA
"hybrid virus"
viral DNA
viral proteins

FIGURE 13-2 Viruses may transfer genes between cells

some viruses may infect species quite unrelated to each other; for example, influenza infects birds, pigs, and humans. In these cases, viruses may transfer genes from one species to another.

13.3 HOW IS BIOTECHNOLOGY USED IN FORENSIC SCIENCE?

As with any technology, the applications of DNA biotechnology vary, depending on the goals of those who use it. Forensic scientists need to identify victims and criminals; biotechnology firms need to identify specific genes and insert them into organisms such as bacteria, cattle, or crop plants; and biomedical firms and physicians need to detect defective alleles and, ideally, devise ways to fix them or to insert normally functioning alleles into patients. We will

begin by describing a few common methods of manipulating DNA, using their application to forensic DNA analysis as a specific example. Later, we will investigate how biotechnology is used in agriculture and medicine.

In 2002, when investigators located the semen samples from the Earl Ruffin case, they needed to determine if the samples collected from the rape victim in 1981 came from Ruffin. Now, the bits of DNA left in a 20-year-old sample were probably not in perfect shape. Even if intact DNA could be obtained, how could forensic scientists determine if the DNA samples matched? The technicians used two techniques that have become commonplace in virtually all DNA labs. First, they amplified the DNA so that they had enough material to analyze. Then, they determined whether the DNA from the semen samples matched Ruffin's own DNA. Let's look briefly at these two techniques.

FIGURE 13-3 PCR copies a specific DNA sequence
The polymerase chain reaction consists of a series of 20 to 30 cycles of heating and cooling. After each cycle, the amount of target DNA doubles. After just 20 cycles, a million copies of the target DNA have been synthesized. QUESTION Why are primers necessary for PCR?

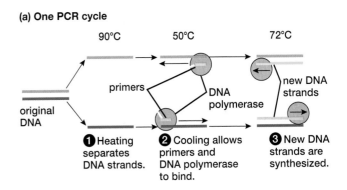

(a) One PCR cycle

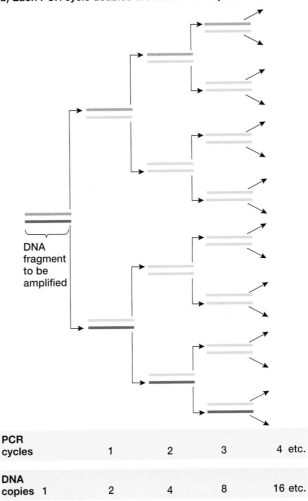

(b) Each PCR cycle doubles the number of copies of the DNA.

PCR cycles	1	2	3	4 etc.
DNA copies	1	2	4	8	16 etc.

The Polymerase Chain Reaction Amplifies DNA

Developed by Kary B. Mullis of the Cetus Corporation in 1986, the **polymerase chain reaction (PCR)** produces virtually unlimited amounts of DNA. Further, PCR can be used to amplify selected pieces of DNA, if desired. PCR is so crucial to molecular biology that it earned Mullis a share in the Nobel Prize for Chemistry in 1993. Let's see how PCR amplifies a specific piece of DNA (**FIG. 13-3**).

When we described DNA replication in Chapter 10, we omitted some of its real-life complexity. One of the things we did not discuss is crucial to PCR: by itself, DNA polymerase doesn't know where to start copying a strand of DNA. When a DNA double helix is unwound, enzymes put a little piece of complementary RNA, called a *primer*, on each strand. DNA polymerase recognizes this "primed" region of DNA as the place to start replicating the rest of the DNA strand.

In PCR, the nucleotide sequence of the beginning and the end of the DNA segment to be amplified must be known. A DNA synthesizer is used to make two sets of DNA fragments, one complementary to the beginning of one strand of the DNA segment and one complementary to the beginning of the other strand. These primers will be used to "tell" DNA polymerase where to start copying.

In a small test tube, DNA is mixed with primers, free nucleotides, and a special DNA polymerase, isolated from microbes that live in hot springs (see "Scientific Inquiry: Hot Springs and Hot Science"). PCR involves the following steps, which are repeated for as many cycles as are needed to generate enough copies of the DNA segment. PCR synthesizes DNA in a geometric progression ($1 \rightarrow 2 \rightarrow 4 \rightarrow 8$ etc.), so 20 PCR cycles make about a million copies, and a little over 30 cycles make a billion copies.

1. The test tube is heated to 194–203°F (90–95°C). High temperatures break the hydrogen bonds between complementary bases, separating the DNA into single strands.
2. The temperature is lowered to about 122°F (50°C), which allows the two primers to form complementary base pairs with the original DNA strands.
3. The temperature is raised to 158–161.6°F (70–72°C). DNA polymerase, directed by the primers, uses the free nucleotides to make copies of the DNA segment bounded by the primers.
4. This cycle is repeated as many times as desired.

Using appropriate mixtures of primers, free nucleotides, and DNA polymerase, a PCR machine automatically runs heating and cooling cycles over and over again. Each cycle takes only a few minutes, so PCR can produce billions of copies of a gene or DNA segment in a single afternoon, starting, if necessary, from a single molecule of DNA. The DNA is then available for forensics, cloning, making transgenic organisms, or many other purposes.

The Choice of Primers Determines Which Segments of DNA Are Amplified

How would a forensics laboratory know which primers to use? After years of painstaking work, forensics experts have found that small, repeating segments of DNA, called *short tandem repeats* (STRs), can be used to identify people

SCIENTIFIC INQUIRY   Hot Springs and Hot Science

At a hot spring, such as those found in Yellowstone National Park, water literally boils out of the ground, gradually cooling as it flows to the nearest stream (**FIG. E13-1**). You might think that such springs, scalding hot and often containing poisonous metals and sulfur compounds, must be lifeless. However, closer examination often reveals a diversity of microorganisms, each adapted to a different temperature zone in the spring. Back in 1966, in a Yellowstone hot spring, Thomas Brock of the University of Wisconsin discovered *Thermus aquaticus*, a bacterium that lives in water as hot as 176°F (80°C).

When Kary Mullis first developed the polymerase chain reaction, he encountered a major technical difficulty. The DNA solution must be heated almost to boiling to separate the double helix into single strands, then cooled so DNA polymerase can synthesize new DNA, and this process must be repeated again and again. "Ordinary" DNA polymerase, like most proteins, is ruined, or *denatured*, by high temperatures. Therefore, new DNA polymerase had to be added after every heat cycle, which was expensive and labor intensive. Enter *Thermus aquaticus*. Like other organisms, it replicates its DNA when it reproduces. But because it lives in hot springs, it has a particularly heat-resistant DNA polymerase. When DNA polymerase from *T. aquaticus* is used in PCR, it needs to be added to the DNA solution only once, at the start of the reaction.

FIGURE E13-1 Thomas Brock surveys Mushroom Spring The colors in hot springs arise from minerals dissolved in the water and from various types of microbes that live at different temperatures.

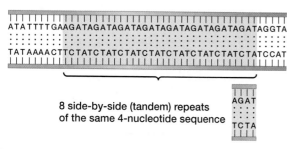

8 side-by-side (tandem) repeats
of the same 4-nucleotide sequence

FIGURE 13-4 Short tandem repeats are common in noncoding regions of DNA This STR, called D5, is not part of any known gene. The sequence AGAT may be repeated from 7 to 13 times in different individuals.

with astonishing accuracy. Think of STRs as very short, stuttering genes (**FIG. 13-4**). Each STR is *short* (consisting of 2 to 5 nucleotides), *repeat*ed (about 5 to 15 times), and *tandem* (having all the repetitions right alongside one another). As with any gene, different people may have different alleles of the STRs. In the case of an STR, each allele is simply a different number of repeats of the same few nucleotides.

In 1999, British and American law enforcement agencies agreed to use a set of 10 to 13 STRs, each 4 nucleotides long, that vary greatly among individuals. A perfect match of 10 STRs in a suspect's DNA and the DNA found at a crime scene means that there is less than one chance in a

trillion that the two DNA samples did not come from the same person. What's more, the DNA around STRs doesn't seem to degrade very fast, so even old DNA samples, such as those in the Ruffin case, usually have STRs that are mostly intact.

Forensics labs use PCR primers that amplify only the DNA immediately surrounding the STRs. Because STR alleles vary in how many times they repeat, they vary in size: an STR with more repeats has more nucleotides and is larger. Therefore, a forensic lab needs to identify each STR in a DNA sample and to determine its size.

Gel Electrophoresis Separates DNA Segments

Modern forensics labs use sophisticated and expensive machines to determine the number of times STRs repeat in their samples. Most of these machines, however, are based on two methods that are used in molecular biology labs around the world: first, separating the DNA by size, and second, labeling specific DNA segments of interest.

The mixture of DNA pieces is separated by a technique called **gel electrophoresis** (**FIG. 13-5**). First, the mixture of DNA fragments is loaded into shallow grooves, or wells, in a slab of agarose, a carbohydrate purified from certain types of seaweed (Fig. 13-5a). Agarose is one of several materials that can form a *gel*, which is simply a meshwork of fibers with holes of various sizes between the fibers. The gel is put into a chamber with electrodes connected to each end. One electrode is made positive and the other negative; therefore, current will flow between the electrodes *through* *the gel.* How does this process separate pieces of DNA? Remember, the phosphate groups in the backbones of DNA are negatively charged. When electrical current flows through the gel, the negatively charged DNA fragments move toward the positively charged electrode. Because smaller fragments slip through the holes in the gel more easily than larger fragments, they move more rapidly toward the positively charged electrode. Eventually the DNA fragments are separated by size, forming distinct bands on the gel (Fig. 13-5b).

DNA Probes Are Used to Label Specific Nucleotide Sequences

Unfortunately, the DNA bands are invisible. There are several dyes that stain DNA, but these are often not very useful in either forensics or medicine. Why not? Because there may be many DNA fragments of approximately the same size; for example, five or six STRs with the same numbers of repeats might be mixed together in the same band. How can a technician identify a *specific* STR? Well, how does *nature* identify sequences of DNA? Right, by base-pairing! Usually, the two strands of the DNA double helix are separated during gel electrophoresis; this allows pieces of synthetic DNA, called **DNA probes**, to base-pair with specific DNA fragments in the sample. DNA probes are short pieces of single-stranded DNA that are complementary to the nucleotide sequence of a given STR (or any other DNA of interest in the gel). The DNA probes are labeled, either by radioactivity or by attaching one of several colored molecules to them. Therefore, a given DNA probe will label certain DNA sequences, and not others (**FIG. 13-6**):

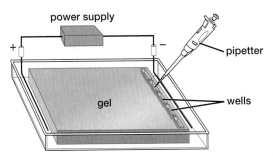

(a) DNA samples are pipetted into wells (shallow slots) in the gel. Electrical current is sent through the gel (negative at end with wells, positive at opposite end).

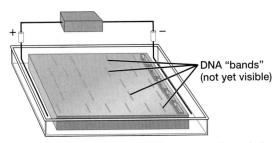

(b) Electrical current moves DNA segments through the gel. Smaller pieces of DNA move farther toward the positive electrode.

(c) Gel is placed on special nylon "paper." Electrical current drives DNA out of gel onto nylon.

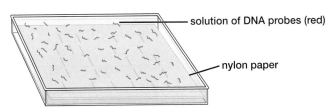

(d) Nylon paper with DNA is bathed in a solution of labeled DNA probes (red) that are complementary to specific DNA segments in the original DNA sample.

(e) Complementary DNA segments are labeled by probes (red bands).

FIGURE 13-5 Gel electrophoresis is used to separate and identify segments of DNA

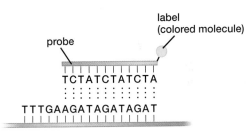

STR 1: probe base-pairs and binds.

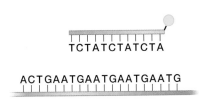

STR 2: probe cannot base-pair; does not bind.

FIGURE 13-6 DNA probes base-pair with complementary DNA segments

When the gel is finished running, the technician transfers the single-stranded DNA segments out of the gel and onto a piece of paper made of nylon (see Fig. 13-5c). Then, the paper is bathed in a solution containing a specific DNA probe (see Fig. 13-5d), which will base-pair with, and therefore bind to, only a specific STR, making this STR visible (see Fig. 13-5e). (Labeling the DNA fragments with radioactive or colored DNA probes is standard procedure in most research applications. In modern forensic applications, however, the STRs are usually directly labeled with colored molecules during the PCR reaction. Therefore, the STRs are immediately visible in the gel and do not have to be stained with DNA probes.)

Every Person Has a Unique DNA Profile

Until the early 1990s, forensic technicians ran DNA samples from a crime scene and from various suspects side by side on a gel, to see which suspect, if any, had DNA matching that found at the scene. In modern STR analysis, however, the suspect and crime scene DNA samples can be run on different gels, in different states or countries, and even years apart. Why? DNA samples run on STR gels produce a pattern, called a **DNA profile (FIG. 13-7)**, which is coded by recording the number of repeats for all the STR genes. The numbers and positions of the bands on the gel are determined by the numbers of repeats of each STR. Because an STR is part of a gene, each person has two copies of each STR—one on each homologous chromosome in every pair. Each of the two copies of the "STR genes" might have the same number of repeats (the person would be homozygous for that STR gene) or different numbers of repeats (the person would be heterozygous). For example, the first person in Fig. 13-7 is heterozygous for Penta D: the gel has two bands, with 9 repeats in one allele and 14 repeats in the other allele. That same person is homozygous for CSF and D16, and the gel has single bands of 11 and 12 repeats, respectively.

In many states, anyone convicted of certain crimes (assault, burglary, attempted murder, etc.) must give a blood sample. Using the standard array of STRs, technicians then determine the criminal's DNA profile. This DNA profile is coded (by the number of repeats of each STR found in the criminal's DNA) and stored in computer files at a state agency, at the FBI, or both. (On *CSI* and other TV crime

shows, when you hear the actors refer to "CODIS," that acronym stands for "Combined DNA Index System," a DNA profile database kept on FBI computers.) Because all forensic labs use the same STRs, computers can easily determine if DNA left behind at another crime scene, even years before or years later (after the criminal may have been released from prison) matches one of the millions of profiles stored in the CODIS database. If the STRs match, then the odds are overwhelming that the crime scene DNA was left by the person with the matching CODIS profile. If there aren't any matches, the crime scene DNA profile will remain on file. Sometimes, years later, a newly convicted criminal's DNA profile will match an archived crime-scene profile, and a "cold case" will be solved (see "Case Study Revisited" at the end of this chapter).

13.4 HOW IS BIOTECHNOLOGY USED IN AGRICULTURE?

The main goals of agriculture are to grow as much food as possible, as cheaply as possible, with minimal loss from pests such as insects and weeds. Many commercial farmers and seed suppliers have turned to biotechnology to achieve these goals.

Many Crops Are Genetically Modified

Currently, almost all of the genetically modified organisms used in agriculture are plants. According to the U.S. Department of Agriculture (USDA), in 2005, about 52% of the corn, 79% of the cotton, and 87% of the soybeans grown in the United States were transgenic; that is, they contained genes from other species (see Table 13-1). Globally, about 200 million acres of land were planted with transgenic crops in 2004, an increase of 20% over the pre-

FIGURE 13-7 DNA profiling
The lengths of short tandem repeats of DNA form characteristic patterns on a gel; this gel displays six different STRs (Penta D, CSF, etc.). The evenly spaced yellow-green bands on the far left and far right sides of the gel show the number of repeats of the individual STRs. DNA samples from 13 different people were run between these standards, resulting in one or two bands per vertical lane. For example, in the enlargement of the D16 STR on the right, the first person's DNA has 12 repeats, the second person's has 13 and 12, the third has 11, and so on. Although some people have the same number of repeats of some STRs, no one has the same number of repeats of all the STRs. (Photo courtesy of Dr. Margaret Kline, National Institute of Standards and Technology.) QUESTION On any individual's DNA profile, a given STR always displays either one or two bands. Further, single bands are always about twice as bright as each band of a pair. For example, in the D16 STR on the right, the single bands of the first and third DNA samples are twice as bright as the pairs of bands of the second, fourth, and fifth samples. Why?

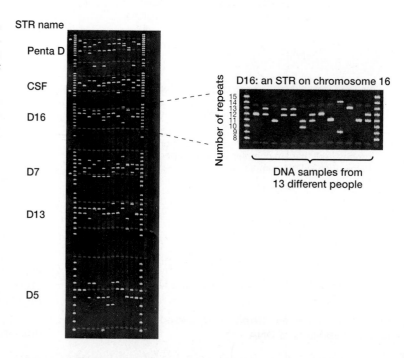

Table 13-1 Genetically Engineered Crops with USDA Approval

Genetically Engineered Trait	Potential Advantage	Examples of Bioengineered Crops with USDA Approval
Resistance to herbicide	Application of herbicide kills weeds, but not crop plants, producing higher crop yields.	beet, canola, corn, cotton, flax, potato, rice, soybean, tomato
Resistance to pests	Crop plants suffer less damage from insects, producing higher crop yields.	corn, cotton, potato, rice, soybean
Resistance to disease	Plants are less prone to infection by viruses, bacteria, or fungi, producing higher crop yields.	papaya, potato, squash
Sterile	Transgenic plants cannot cross with wild varieties, making them safer for the environment and more economically productive for the seed companies that produce them.	chicory, corn
Altered oil content	Oils can be made healthier for human consumption or can be made similar to more expensive oils (such as palm or coconut).	canola, soybean
Altered ripening	Fruits can be more easily shipped with less damage, producing higher returns for the farmer.	tomato

vious year. Crops are most commonly modified to improve their resistance to insects and herbicides.

Many herbicides kill plants by inhibiting an enzyme that is used by plants, fungi, and some bacteria—but not animals—to synthesize amino acids such as tyrosine, tryptophan, and phenylalanine. Without these amino acids, the plants cannot synthesize proteins, and they die. Many herbicide-resistant transgenic crops have been given a bacterial gene that encodes an enzyme that functions even in the presence of these herbicides, so the plants continue to synthesize normal amounts of amino acids and proteins. Herbicide-resistant crops allow farmers to kill weeds without harming their crops. Less competition from weeds means more water, nutrients, and light for the crops, hence larger harvests.

To promote insect resistance, many crops have been given a gene, called *Bt*, from the bacterium *Bacillus thuringiensis*. The protein encoded by the *Bt* gene damages the digestive tract of insects (but not mammals). Transgenic Bt crops often suffer far less damage from insects (**FIG. 13-8**), and farmers can apply less pesticide to their fields.

How would a seed company go about making a transgenic plant? Let's examine the process, using insect-resistant Bt plants as an example.

The Desired Gene Is Cloned

Cloning a gene usually involves two tasks: (1) obtaining the gene and (2) inserting it into a plasmid so that huge numbers of copies of the gene can be made.

There are two common ways of obtaining a gene. For a long time, the only practical method was to isolate the gene from the organism that makes it. Now, biotechnologists can often synthesize the gene—or a modified version of it—in the lab, using PCR or DNA synthesizers.

Once the gene has been obtained, why insert it into a plasmid? Plasmids, small circles of DNA in bacteria (see Fig. 13-1), are replicated when the bacteria multiply. Therefore, once the desired gene has been inserted into a

FIGURE 13-8 Bt plants resist insect attack
Transgenic cotton plants expressing the Bt gene (right) resist attack by bollworms, which eat cotton seeds. The transgenic plants therefore produce far more cotton than non-transgenic plants (left). QUESTION How might herbicide-resistant crops reduce topsoil erosion?

plasmid, producing huge numbers of copies of the gene is as simple as raising lots of bacteria. Inserting the gene into a plasmid also allows it to be easily separated from the bacteria, achieving partial purification of the gene, free of the DNA of the bacterial chromosome. Finally, plasmids may then be taken up by other bacteria (this is important when making transgenic Bt plants) or injected directly into animal eggs.

Restriction Enzymes Cut DNA at Specific Nucleotide Sequences

Genes are inserted into plasmids through the action of **restriction enzymes**, isolated from a wide variety of bacteria. Each restriction enzyme cuts DNA at a specific nucleotide sequence. Many restriction enzymes cut "straight across" a double helix of DNA. Others make a "staggered" cut, snipping the DNA in a different location on each of the two strands, so that single-stranded sections hang off the ends of the DNA. Because these single-stranded regions can base-pair with, and thus stick to, other single-stranded pieces of DNA with complementary bases, they are commonly called "sticky ends" (**FIG. 13-9**):

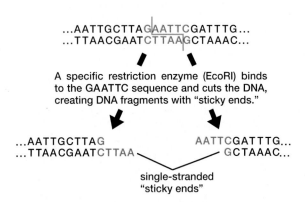

...AATTGCTTAG|AATTC GATTTG...
...TTAACGAATCTTAA|GCTAAAC...

A specific restriction enzyme (EcoRI) binds to the GAATTC sequence and cuts the DNA, creating DNA fragments with "sticky ends."

...AATTGCTTAG   AATTCGATTTG...
...TTAACGAATCTTAA   GCTAAAC...

single-stranded "sticky ends"

FIGURE 13-9 Some restriction enzymes leave "sticky ends" when they cut DNA

Cutting Two Pieces of DNA with the Same Restriction Enzyme Allows the Pieces to Be Joined Together

To insert the *Bt* gene into a plasmid, the same restriction enzyme is used to cut the DNA on either side of the *Bt* gene and to split open the circle of plasmid DNA (**FIG. 13-10a**). As a result, the ends of the *Bt* gene and the opened-up plasmid both have complementary nucleotides in their sticky ends. When the cut *Bt* genes and plasmids are mixed together, base-pairing between their sticky ends allows some of the *Bt* genes to fill in the circle of plasmid DNA (**FIG. 13-10b**). DNA ligase (see Chapter 9) is added to the mix, to permanently bond the *Bt* genes into the plasmid. Bacteria are then transformed with the plasmids (**FIG. 13-10c**). By manipulating the plasmids and bacteria appropriately, biotechnologists can isolate and grow only the bacteria with the desired plasmid.

Plasmids Are Used to Insert the Bt Gene into a Plant

The bacterium *Agrobacterium tumefaciens*, which contains a specialized plasmid called the Ti (tumor-inducing) plasmid, can infect many plant species. When the bacterium infects a plant cell, the Ti plasmid inserts its DNA into one of the plant cell's chromosomes. Thereafter, anytime that plant cell divides, it replicates the Ti plasmid DNA as well, and all of its daughter cells inherit the Ti DNA. (Genes on the Ti plasmid cause plant tumors; however, biotechnologists have learned how to make "disabled" Ti plasmids that are harmless.) To make insect-resistant plants, *Bt* genes are inserted into disabled Ti plasmids. *A. tumefaciens* bacteria are allowed to take up the plasmids and infect plant cells in culture (**FIG. 13-10d**). The modified Ti plasmids insert the *Bt* gene into the plant cells' chromosomes, so that the plant cells now permanently carry the *Bt* gene (**FIG. 13-10e**). Appropriate hormonal treatments stimulate the transgenic plant cells to divide

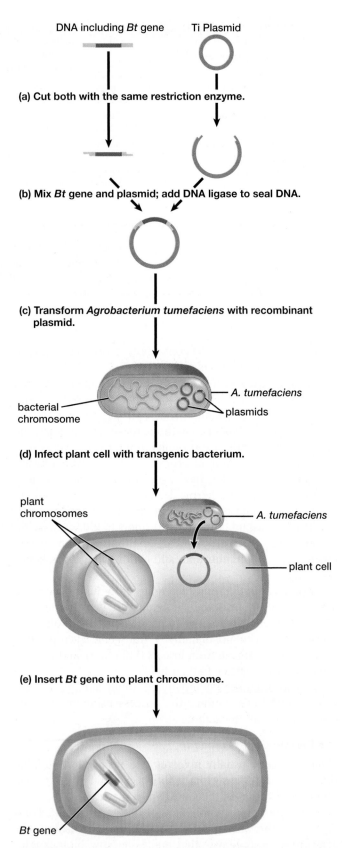

DNA including *Bt* gene   Ti Plasmid

(a) Cut both with the same restriction enzyme.

(b) Mix *Bt* gene and plasmid; add DNA ligase to seal DNA.

(c) Transform *Agrobacterium tumefaciens* with recombinant plasmid.

bacterial chromosome   *A. tumefaciens*
                                        plasmids

(d) Infect plant cell with transgenic bacterium.

plant chromosomes   *A. tumefaciens*

                                        plant cell

(e) Insert *Bt* gene into plant chromosome.

Bt gene

FIGURE 13-10 Using *Agrobacterium tumefaciens* to insert the *Bt* gene into plants

and differentiate into entire plants. These plants are bred to one another, or to other plants, to create commercially valuable crop plants that resist insect attack.

Genetically Modified Plants May Be Used to Produce Medicines

Similar techniques can be used to insert medically useful genes into plants, producing medicines down on the "pharm." For example, a plant could be engineered to produce harmless proteins that are normally found in disease-causing bacteria or viruses. If these proteins resisted digestion in the stomach and small intestine, simply eating such plants could act as a vaccination against the disease organisms. Several years ago, such "edible vaccines" were touted as a great way to provide vaccinations—no need to produce purified vaccines, no refrigeration needed, and, of course, no needles. Recently, however, many biomedical researchers have warned that edible vaccine plants are not really a good idea, because there is no good way to control the dose: too little and the user doesn't develop decent immunity; too much, and the vaccine proteins might be harmful. Nevertheless, producing vaccine proteins in plants is still worthwhile. Pharmaceutical companies just have to extract and purify the proteins before use. Plant-produced vaccines against hepatitis B, rabies, and certain types of diarrhea are now in clinical trials.

Molecular biologists can also engineer plants to produce human antibodies that would combat various diseases. When a microbe invades your body, it takes several days for your immune system to respond and produce enough antibodies to overcome the infection. Meanwhile, you feel terrible, and might even die if the disease is serious enough. A direct injection of large quantities of the right antibodies might be able to cure the disease almost instantly. Although none have yet entered medical practice, plant-derived antibodies against bacteria that cause tooth decay and non-Hodgkin's lymphoma (a cancer of the lymphatic system) are in clinical trials. Ideally, such "plantibodies" can be produced very cheaply, making such therapies available to rich and poor alike.

Genetically Modified Animals May Be Useful in Agriculture and Medicine

Unlike plants, animals—especially vertebrates—are very difficult to make from single cells in culture dishes. Therefore, making transgenic animals usually involves injecting the desired DNA, often incorporated into a disabled virus, into a fertilized egg. The egg is usually allowed to divide a few times in culture before being implanted into a surrogate mother. If the offspring are healthy and express the foreign gene, they are then bred together to produce homozygous transgenic organisms. So far, it has proven difficult to produce commercially valuable transgenic livestock, but several companies in countries around the world are working on it.

One example is Nexia Biotech, which has engineered a herd of goats to carry the genes for spider silk, and to secrete the resulting silk protein into their milk. The resulting Bio-Steel® can be spun into silk that is five times stronger than steel and twice as strong as Kevlar®, the fiber usually used in bulletproof vests. Several types of fish with added growth-hormone genes grow much faster than wild-type fish, and they display no obvious ill effects. However, whether "fish farms" should be allowed to grow these fish remains controversial, principally because of concerns about what would happen if they escaped into the wild.

Biotechnologists are also developing animals that will produce medicines, such as human antibodies or other essential proteins. For example, there are sheep whose milk contains a protein, alpha-1-antitrypsin, that may prove valuable in treating cystic fibrosis. Other transgenic livestock have been engineered so that their milk contains erythropoietin (a hormone that stimulates red blood cell synthesis), clotting factors (for treatment of hemophilia), or clot-busting proteins (to treat heart attacks caused by blood clots in the coronary arteries).

13.5 HOW IS BIOTECHNOLOGY USED TO LEARN ABOUT THE HUMAN GENOME?

Genes influence virtually all the traits of human beings, including gender, size, hair color, intelligence, and susceptibility to disease organisms and toxic substances in the environment. To begin to understand how our genes influence our lives, the Human Genome Project was launched in 1990, with the goal of determining the nucleotide sequence of all the DNA in our entire set of genes, called the human genome.

By 2003, this joint project of molecular biologists in several countries sequenced the human genome with an accuracy of about 99.99%. To many people's surprise, the human genome contains only about 21,000 genes, comprising approximately 2% of the DNA. Some of the other 98% consists of promoters and regions that regulate how often individual genes are transcribed, but it's not really known what most of the DNA does.

What good is it to sequence the human genome? First, many genes were discovered whose functions are completely unknown. Now that these genes have been identified and sequenced, the genetic code allows biologists to predict the amino acid sequences of the proteins they encode. Comparing these proteins to familiar proteins whose functions are already known will enable us to find out what many of these genes do.

Second, knowing the nucleotide sequences of human genes will have an enormous impact on medical practice. In 1990, fewer than 100 genes known to be associated with human diseases had been discovered. By 2003, this number had jumped to over 1400, mostly because of the Human Genome Project.

Third, there is no single "human genome" (or else all of us would be identical twins). *Most* of the DNA of everyone on the planet is the same, but we each also carry our own unique set of alleles. Some of those alleles can cause or predispose people to develop various medical conditions, including Marfan syndrome, sickle-cell anemia, cystic fibrosis (all described in earlier chapters), breast cancer, alcoholism, schizophrenia, heart disease, Huntington's disease, Alzheimer's disease, and many others. A major impact of the Human Genome Project will be to help diagnose genetic disorders or predispositions, and hopefully to devise treatments or even cures in the future, as we describe in the following sections.

Fourth, the Human Genome Project, along with numerous companion projects that have sequenced the genomes of organisms as diverse as bacteria, mice, and chimpanzees, help us to appreciate our place in the evolution of life on Earth. For example, the DNA of humans and chimps differs by only about 1.2%. Comparing the similarities and differences may help biologists to understand what genetic differences help to make us human, and why we are susceptible to certain diseases that chimps are not.

13.6   HOW IS BIOTECHNOLOGY USED FOR MEDICAL DIAGNOSIS AND TREATMENT?

Many people suffer from inherited disorders: sickle-cell anemia, Marfan syndrome, and cystic fibrosis, to name a few that we have discussed earlier in this text. For over a decade, biotechnology has been routinely used to diagnose some inherited disorders. Potential parents can learn if they are carriers of a genetic disorder, and an embryo can be diagnosed early in a pregnancy (see "Health Watch: Prenatal Genetic Screening" later in this chapter). More recently, medical researchers have begun using biotechnology in an attempt to cure, or at least treat, genetic diseases.

DNA Technology Can Be Used to Diagnose Inherited Disorders

A person inherits a genetic disease because he or she inherits one or more dysfunctional alleles. Defective alleles differ from normal, functional alleles because of differences in nucleotide sequence. Two methods are currently used to find out if a person carries a normal allele or a malfunctioning allele.

Restriction Enzymes May Cut Different Alleles of a Gene at Different Locations

Remember that restriction enzymes cut DNA only at specific nucleotide sequences. Because chromosomes are so large, any given restriction enzyme usually cuts the DNA of a chromosome in many places, producing many *restriction fragments*. What if two homologous chromosomes have different alleles of several genes, and some alleles have nucleotide sequences that *can* be cut by a restriction enzyme, while others have nucleotide sequences that *cannot* be cut by the enzyme? The result will be a mixture of DNA segments of various lengths. These are called **restriction fragment length polymorphisms** (**RFLPs**; pronounced "riff-lips"). This rather daunting phrase simply means that *restriction*

enzymes have cut DNA into *fragments* that vary in *length*, and that homologous chromosomes (from the same person or from different people) may differ (or be *polymorphic*) in the lengths of the fragments. Why is this useful? First, if different people have different RFLPs, this can be used to identify DNA samples. In fact, in the early 1990s, before STRs became the gold standard in DNA forensics, RFLPs were used to determine if DNA from a crime scene matched the DNA of a suspect. Second, with diligent research and a little luck, medically important alleles can sometimes be identified by differences in the lengths of the restriction fragments produced by cutting with a specific restriction enzyme.

RFLP analysis has become a standard technique to diagnose sickle-cell anemia, even in an embryo. You may recall that sickle-cell anemia is caused by a point mutation in which thymine replaces adenine near the beginning of the globin gene. This causes a hydrophobic amino acid (valine) to be placed in the globin protein instead of a hydrophilic amino acid (glutamic acid; see p. 172 in Chapter 10). The hydrophobic valines cause hemoglobin molecules to clump together, distorting and weakening the red blood cells.

One restriction enzyme, called MstII, cuts DNA in about the middle of both the normal and sickle-cell alleles. It also cuts DNA just outside of both alleles. However, the normal globin allele, but *not* the sickle-cell allele, is also cut in a third location (**FIG. 13-11a**). MstII also cuts the rest of the chromosome in many other places that have nothing to do with sickle-cell anemia. How can the one unique cut be identified? A DNA probe is synthesized that is complementary to the part of the globin allele spanning the unique cut site. When sickle-cell DNA is cut with MstII and run on a gel, this probe labels a single large band (**FIG. 13-11b**). When normal DNA is cut with MstII, the probe labels two bands, one small and one not quite as large as the sickle-cell band. Someone who is homozygous for the normal globin allele will have two bands; someone who is homozygous for the sickle-cell allele will have one band, and a heterozygote will have three bands. The genotypes of parents, children, and fetuses can be determined by this simple test.

Different Alleles May Bind to Different DNA Probes

In Chapter 12, we briefly discussed cystic fibrosis, a disease caused by a defect in a protein that normally transports chloride ions across cell membranes. There are over 1000 different alleles, all at the same gene locus, each encoding a slightly different defective chloride transport protein. People with either one or two normal alleles synthesize enough functioning chloride transport proteins so that they do not develop cystic fibrosis. People with two defective alleles (they may be the same or different alleles) do not synthesize fully functional transport proteins and develop cystic fibrosis. Therefore, the disease is inherited as a simple recessive trait.

How can anyone hope to diagnose a disorder with a thousand different alleles? Most of these alleles are extremely rare; only 32 alleles account for about 90% of the cases of cystic fibrosis. Still, 32 alleles is a lot. Although researchers could probably find restriction enzymes that would cut most of these alleles differently from the normal allele, testing would involve dozens of different enzymes, producing dozens of different patterns of DNA pieces that would need to be run on dozens of different gels. The cost would be astronomical.

However, each allele has a different nucleotide sequence. Therefore, one strand of each allele will form perfect base-pairs only with its own complementary strand, not with any of others. Several companies now produce cystic fibrosis "arrays," which are pieces of specialized filter paper to which segments of single-stranded DNA are bound. Each piece of DNA is complementary to a different cystic fibrosis allele (**FIG. 13-12a**). A person's DNA is cut into small pieces, separated into single strands, and labeled. The array is then bathed in the resulting solution of labeled DNA fragments. Under the right conditions, only a perfect complementary strand of the person's DNA will bind to any given spot of DNA on the array; even a single

"wrong" base will keep the person's DNA from binding. Depending on the number of different alleles represented on the array, up to 95% of all cases of cystic fibrosis can be diagnosed by this method. In 2005, slightly more complex methods were developed that test for 97 different cystic fibrosis alleles.

Although not yet practical for routine medical use, an expanded version of this type of DNA analysis may one

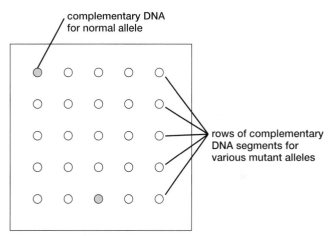

(a) A cystic fibrosis diagnostic array.

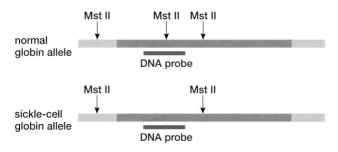

(a) Mst II cuts a normal globin allele in 2 places, but cuts the sickle-cell allele in 1 place.

(b) Gel electrophoresis of globin alleles

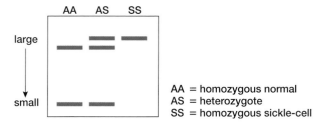

AA = homozygous normal
AS = heterozygote
SS = homozygous sickle-cell

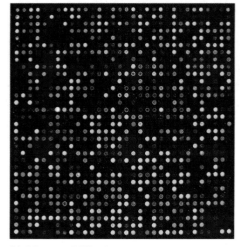

(b) A human DNA microarray.

FIGURE 13-11 Diagnosing sickle-cell anemia with restriction enzymes
(a) The normal globin allele and the sickle-cell allele (both shown in red) are cut in half by the restriction enzyme MstII (far right arrow). The normal allele is also cut in another, unique location (middle arrow). Finally, regardless of which allele is present, the chromosome is cut somewhat ahead of the globin gene locus (far left arrow). A DNA probe (blue) is synthesized that is complementary to DNA on both sides of the unique cut site. Therefore, the probe will label two pieces of DNA from the normal allele but only a single piece of the sickle-cell allele. **(b)** The cut DNA is run on a gel and made visible with the DNA probe. The large piece of DNA of the sickle-cell allele is close to the beginning of the gel, while the smaller pieces of the normal allele run further into the gel.

FIGURE 13-12 DNA arrays in medicine and research
(a) DNA from a patient is cut into small pieces, separated into single strands, and labeled (blue, in this diagram). A cystic fibrosis screening array is bathed in this solution of labeled DNA. Each cystic fibrosis allele can bind to only one specific piece of complementary DNA on the array. In this simplified diagram, the patient has one normal allele (upper left) and one defective allele (middle bottom). **(b)** Each spot contains a DNA probe for a specific human gene. In most research applications, messenger RNA is isolated from a subject (for example, from a human cancer), and labeled with a fluorescent dye. The mRNA is then poured onto the array, and each base-pairs with its complementary template DNA probe. Genes that are particularly active in the cancer will "light up" the corresponding DNA probe.

Table 13-2 Examples of Medical Products Produced by Recombinant DNA Methods

Type of Product	Purpose	Product	Genetic Engineering
Human hormones	Used in treatment of diabetes, growth deficiency	Humulin™ (human insulin)	Human gene inserted into bacteria
Human cytokines (regulate immune system function)	Used in bone marrow transplants and to treat cancers and viral infections, including hepatitis and genital warts	Leukine™ (granulocyte-macrophage colony stimulating factor)	Human gene inserted into yeast
Antibodies (immune system proteins)	Used to fight infections, cancers, diabetes, organ rejection, and multiple sclerosis	Herceptin™ (antibodies to a protein expressed in some breast cancer cells)	Recombinant antibody genes inserted into cultured hamster cell line
Viral proteins	Used to generate vaccines against viral diseases and for diagnosing viral infections	Energiz-B™ (Hepatitis B vaccine)	Viral gene inserted into yeast
Enzymes	Used in treatment of heart attacks, cystic fibrosis, and other diseases, and production of cheeses and detergents	Activase™ (tissue plasminogen activator)	Human gene inserted into cultured hamster cell line

day offer customized medical care. Different people have slightly different alleles of hundreds of genes; these may cause them to be more or less susceptible to many diseases, or to respond more or less well to various treatments. Someday instead of spotting an array with probes for a only few dozen cystic fibrosis genes, biomedical scientists could make a *microarray* containing hundreds, even thousands, of probes for hundreds of disease-related alleles to determine which susceptibility alleles each patient carries. By knowing exactly which alleles a patient has, a physician might be able to tailor her medical care accordingly. Sound too much like science fiction? Well, microarrays containing probes for thousands of human genes are already manufactured (**FIG. 13-12b**). Although not a trivial obstacle, all that stands between these research microarrays and practical medical microarrays is to identify enough disease-related alleles and make DNA probes to match. See "The Magic of Microarrays," listed in this chapter's "For More Information" section, for an especially clear description of how microarrays work and how they may be used in medicine.

DNA Technology Can Help to Treat Disease

Several therapeutically important proteins are now routinely made in bacteria, using technology similar to the first half of making a transgenic plant: restriction enzymes are used to splice appropriate genes into plasmids, and bacteria are then transformed with these plasmids. The first human protein made by recombinant DNA technology was insulin. Prior to 1982, when recombinant human insulin was first licensed for use, the insulin needed by diabetics was extracted from the pancreases of cattle or pigs slaughtered for

meat. Although the insulin from these animals is very similar to human insulin, the slight differences cause an allergic reaction in about 5% of diabetics. Recombinant human insulin does not cause allergic reactions.

Other human proteins, such as growth hormone and clotting factors, can also be also produced in transgenic bacteria. Before recombinant DNA technology, some of these proteins were obtained from either human blood or human cadavers; these sources are expensive and sometimes dangerous. As you know, blood can be contaminated by HIV, the virus that causes AIDS. Cadavers may also contain several hard-to-diagnose infectious diseases, such as Creutzfeld-Jacob syndrome, in which an abnormal protein can be passed from the tissues of an infected cadaver to a patient and cause irreversible, fatal brain degeneration. Engineered proteins grown in bacteria or other cultured cells circumvent these dangers. Some of the categories of human proteins produced by recombinant DNA technology are listed in Table 13-2.

These proteins, although tremendously helpful and often lifesaving, do not *cure* inherited disorders; they merely treat the symptoms. Often, as in the case of insulin-dependent diabetes, a patient may need to take the protein for his entire life. Note that these proteins are soluble molecules, usually found dissolved in the blood, that often act as signaling molecules that tell cells how to regulate their metabolism. Insulin, for example, is released into the bloodstream and travels all over the body, instructing cells, such as liver and muscle cells, to take up glucose from the blood. Imagine how much better it would be if a diabetic could regain the ability to synthesize and release his *own* insulin, rather than taking daily injections. Nevertheless, diabetes sufferers are comparatively fortunate; in some diseases, such as cystic fibrosis, the defective molecule is an integral part of the patient's cells and cannot be replaced by a simple pill or injection.

Biotechnology offers the potential to treat diseases such as cystic fibrosis and possibly cure diseases such as diabetes, although progress has been painfully slow thus far. Let's look at two specific examples of how these advances may treat, or even cure, life-threatening illnesses.

Using Biotechnology to Treat Cystic Fibrosis

Cystic fibrosis causes devastating effects in the lungs, where the lack of chloride transport causes the usually thin, watery fluid lining the airways to become thick and clogged with mucus (see "Scientific Inquiry: Cystic Fibrosis" in Chapter 12). Several research groups are developing methods to deliver the allele for normal chloride transport proteins to the cells of the lungs, get them to synthesize functioning transport proteins, and insert these proteins into their plasma membranes. Although different laboratories use slightly different methods, all involve inserting the DNA of the normal allele into a virus. When a virus infects a cell, it releases its genetic material into the cytoplasm of the cell and uses the cell's own metabolism to transcribe the viral genes and make new viral proteins (see Fig. 13-2).

To treat cystic fibrosis, researchers first disable a suitable virus, so that the treatment doesn't cause yet another disease. Cold viruses are often used, because they normally infect cells of the respiratory tract. The DNA of the normal chloride transport allele is then inserted into the DNA of the virus. The recombinant viruses are suspended in a solution and sprayed into the patient's nose or dripped directly into the lungs through a nasal tube. If all goes well, the viruses enter cells of the lungs and release the normal chloride transport allele into the cells. The cells then manufacture normal proteins, insert them into their plasma membranes, and transport chloride into the fluid lining the lungs. The clinical trials under way for such treatments have been reasonably successful, but for only a few weeks. In all likelihood, the patients' immune systems see the viruses as undesirable invaders and mount an attack that eliminates them—and the helpful genes they carry—from the patients' bodies. Because lung cells are continually replaced over time, a single dose "wears off" as the modified cells die. Several research groups are now trying both to increase the expression of the chloride transport genes in the viruses and to extend the effective lifetime of a single treatment.

Using Biotechnology to Cure Severe Combined Immune Deficiency

Like the cells of the lung, the vast majority of cells in the human body eventually die and are replaced by new cells. In many cases, the new cells come from special populations of cells called **stem cells**; when they divide, stem cells give rise to daughter cells that can differentiate into several different types of mature cells. In the brain, for example, stem cells give rise to several types of nerve cells and multiple types of non-nervous, support cells as well. It's possible that some stem cells, under the right conditions in the laboratory, might be able to give rise to *any* cell type of the entire body! For now, we will look at a more limited function of stem cells in the body: producing or replacing cells of just one or two types.

All the cells of the immune system (mostly white blood cells) originate in the bone marrow. Some go on to produce antibodies, others kill cells that have been infected by viruses, and still others regulate the actions of these other cells. As mature cells die, they are replaced by new cells that arise from division of stem cells in the bone marrow. Severe combined immune deficiency (SCID) is a rare disorder in which a child fails to develop an immune system. About 1 in 80,000 children is born with some form of SCID. Infections that would be trivial in a normal child become life threatening. In some cases, if the child has an unaffected relative with a similar genetic makeup, a bone marrow transplant from the healthy relative can give the child functioning stem cells, so that he or she can develop a working immune system. Most SCID victims, however, die before their first birthday.

Although there are several forms of SCID, most are recessive, single-gene defects. In some cases, children are homozygous recessive for a defective allele that normally codes for an enzyme called adenosine deaminase. In 1990, the first test of human gene therapy was performed on such a SCID patient, 4-year-old Ashanti DeSilva. Some of her white blood cells were removed, genetically altered with a virus containing a functional version of her defective allele, and then returned to her bloodstream. Now, Ashanti is a healthy adult, with a reasonably functional immune system. However, as the altered white blood cells die, they must be replaced with new ones; therefore, Ashanti needs repeated treatments. She is also given regular injections of a form of adenosine deaminase. Although she's now an adult, Ashanti receives only a 4-year-old's dose of adenosine deaminase; so the gene therapy, although not perfect, is certainly making a major difference.

As of 2005, Italian researchers appear to have completely cured Ashanti's type of SCID in six children. Instead of inserting a normal copy of the adenosine deaminase gene into mature white blood cells, the Italian team inserted the gene into stem cells. Because the "cured" stem cells should continue to multiply and churn out new white blood cells, these children will probably have functioning immune systems for the rest of their lives. (In 1990, when Ashanti received her pioneering treatments, stem cell research was in its infancy. At that time, it would not have been possible to isolate her stem cells and correct their adenosine deaminase genes.)

13.7 WHAT ARE THE MAJOR ETHICAL ISSUES OF MODERN BIOTECHNOLOGY?

Modern biotechnology offers the promise—some would say the threat—of greatly changing our lives, and the lives of many other organisms on Earth. As Spider-Man noted, "With great power comes great responsibility." Is humanity capable of handling the responsibility of biotechnology?

LINKS TO LIFE   Biotechnology—From the Sublime to the Ridiculous

Nearly everyone applauds many applications of biotechnology—freeing innocent men from prison, or diagnosing and curing inherited diseases, for example. Many other applications are greeted with a bit of uncertainty, but generally accepted. Most of us, like it or not, have eaten transgenic corn or soybeans and haven't gone out of our way to avoid it. But what about growing GMOs just for fun?

A few years ago, scientists in Singapore thought that transgenic fish might be used as pollution monitors. They inserted the gene for a fluorescent protein from sea anemones into zebra-fish eggs. The fluorescent protein gene was linked to a promoter that would be activated under stressful conditions, such as polluted water. That application hasn't yet proven useful, but a tropical fish breeder saw the fish and decided that fluorescent fish would make great specimens in home aquaria. The result was the GloFish® (FIG. E13-2). GloFish are available throughout the United States except in California, which banned the sale of GloFish basically on the principle that such a trivial application of GMO technology was inappropriate. (California's streams and lakes are almost certainly too cold in the winter for zebra fish to survive, so escape into the wild is probably not an important concern.) As one member of the California Fish and Game Commission put it, "No matter how low the risk is, there needs to be a public benefit that is higher than this." What do you think? Are "useful" applications OK, but not "trivial" applications?

FIGURE E13-2 GloFish®
Red zebra fish that glow under "black lights"? Although this application of transgenic technology seems trivial, similar fish are used in research labs to investigate the mechanisms of development. Transgenic mice "tagged" with fluorescent proteins are used in toxicology, development, and cancer studies.

Controversy swirls around many applications of biotechnology (see "Links to Life: Biotechnology—From the Sublime to the Ridiculous"). Here we will explore two important debates about biotechnology: the use of genetically modified organisms in agriculture and prospects for genetically modifying human beings.

Should Genetically Modified Organisms Be Permitted in Agriculture?

The aims of "traditional" and "modern" agricultural biotechnology are the same: to modify the genetic makeup of living organisms to make them more useful. However, there are three principal differences. First, traditional biotechnology is usually slow; many generations of selective breeding are necessary before significantly useful new traits appear in plants or animals. Genetic engineering, in contrast, can potentially introduce massive genetic changes in a single generation. Second, traditional biotechnology almost always recombines genetic material from the same, or at least very closely related, species, while genetic engineering can recombine DNA from very different species in one organism. Finally, traditional biotechnologists had no way to manipulate the DNA sequence of genes themselves. Genetic engineering, however, can produce new genes never before seen on Earth.

The best transgenic crops have clear advantages for farmers. Herbicide-resistant crops allow farmers to rid their fields of weeds by applying powerful, non-selective herbicides at virtually any stage of crop growth. Insect-resistant crops decrease the need to apply synthetic pesticides, saving the cost of the pesticides themselves, tractor fuel, and labor. Therefore, transgenic crops may produce larger harvests at less cost. These savings may be passed along to the consumer. Transgenic crops also have the potential to be more nutritious than "standard" crops (see "Biotechnology Watch: Golden Rice").

Regardless of potential monetary or health benefits, many people strenuously object to transgenic crops or livestock. For example, in November 2005, voters in Switzerland voted to ban the cultivation of transgenic crops (although food made from transgenic crops, grown elsewhere, can still be imported and sold). There are two principal scientific objections to the use of genetically modified organisms (GMOs) in agriculture: (1) they may be hazardous to human health, and (2) they may be dangerous to the environment.

Are Foods from GMOs Dangerous to Eat?

The first argument against transgenic foods is that they may be dangerous to the people who eat them. In most cases, this is not a major concern. For example, tests have shown that the Bt protein is not toxic to mammals, and should not prove a danger to human health. The Flavr Savr™ tomato, which lacked an enzyme that makes tomatoes get soft as they ripen (and therefore bruise easily during shipment), didn't taste very good and soon disappeared from grocery shelves, but it didn't make people sick. Transgenic fish that

Rice is the principal food for about two-thirds of the people on Earth (**FIG. E13-3**). A bowl of rice provides a good supply of carbohydrates and some protein but is a poor source of many vitamins, including vitamin A. Unless people eat enough fruits and vegetables along with the rice, they often suffer from vitamin A deficiency. According to the World Health Organization, over 100 million children suffer from vitamin A deficiency; each year 250,000 to 500,000 children become blind, principally in Asia, Africa, and Latin America. Especially in Asia, vitamin A deficiency typically strikes the poor, because a bowl of rice may be all they can afford to eat. In 1999 biotechnology provided a possible remedy: rice that was genetically engineered to contain high levels of beta-carotene, a pigment that gives daffodils their bright yellow colors and that the human body easily converts into vitamin A.

Creating a rice strain with high levels of beta-carotene wasn't simple. However, funding from the Rockefeller Institute, the European Community Biotech Program, and the Swiss Federal Office for Education and Science enabled the European molecular biologists Ingo Potrykus and Peter Beyer to tackle the task. They inserted three genes into the rice genome, two from daffodils and one from a bacterium. Regulatory DNA sequences were included with the genes to control their expression, so that the genes would be turned on in the rice grains. As a result, these "Golden Rice" grains synthesize beta-carotene (**FIG. E13-4**, upper right).

Trouble was, the original Golden Rice had several drawbacks. First, it didn't make enough beta-carotene, so people would have had to eat enormous amounts of Golden Rice to get a day's supply of vitamin A. Second, the original Golden Rice strains grow well only in certain areas; for worldwide use, the genes must be inserted into local varieties of rice. Finally, people who might benefit the most from transgenic technologies are often too poor to afford it.

However, the Golden Rice community didn't give up. The first, and perhaps most important, breakthrough was made not by scientists but by businessmen. The biotech company Syngenta and multiple patent holders have given the technology—free—to research centers in the Philippines, India, China, and Vietnam, with the hope that they will modify native rice varieties for local use. Further, any individual farmer who produces less than about $10,000 worth of Golden Rice per year doesn't have to pay any fees to Syngenta or the other patent holders.

Meanwhile, Syngenta biotechnologists set about increasing the beta-carotene levels. Although daffodils are an *obvious* choice from which to get the genes directing beta-carotene synthesis, it turns out that they aren't the *best* choice. Genes from tomatoes, peppers, and especially corn cause rice to produce more beta-carotene. Golden Rice 2, with genes from corn, produces over 20 times more beta-carotene than the original Golden Rice (compare the rice in the upper right and left-hand sections of Fig. E13-4). About 3 cups of cooked Golden Rice 2 should provide enough beta-carotene to equal the full recommended daily amount of vitamin A. Syngenta has donated Golden Rice 2 to the Humanitarian Rice Board for experiments and planting in Southeast Asia.

Is Golden Rice 2 the best way, or the only way, to solve the problems of malnutrition in poor people? Perhaps not. For one thing, many poor people's diets are deficient in many nutrients, not just vitamin A. To help solve that problem, the Bill and Melinda Gates Foundation is funding research by Peter Beyer, one of the originators of Golden Rice, to increase its levels of vitamin E, iron, and zinc. Further, not all poor people have access to any kind of rice, let alone Golden Rice. In parts of Africa, sweet potatoes, not rice, are the staple source of starches. Recent efforts to persuade these people to eat orange, instead of white, sweet potatoes have dramatically increased their vitamin A levels. Finally, in many parts of the world, government and humanitarian organizations are implementing massive vitamin A supplementation programs. In some parts of Africa and Asia, as many as 80% of the children receive large doses of vitamin A a few times when they are very young. Some day, the combination of these efforts may result in a world in which no children suffer blindness from the lack of a simple nutrient in their diets.

FIGURE E13-3 A Field of Dreams?
For hundreds of millions of people, rice provides the major source of calories, but not enough vitamins and minerals. Can biotechnology improve the quality of rice, and hence the quality of life for these people?

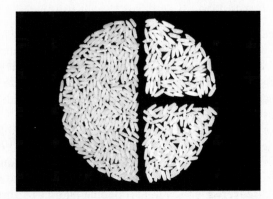

FIGURE E13-4 Golden Rice
Conventional milled rice is white or very pale tan (lower right). The original Golden Rice (upper right) was pale golden-yellow because of its increased beta-carotene content. Second-generation Golden Rice 2 (left) is much deeper yellow, because it contains about 20 times more beta-carotene than original Golden Rice does.

HEALTH WATCH   Prenatal Genetic Screening

Prenatal diagnosis of a variety of genetic disorders, including cystic fibrosis, sickle-cell anemia, and Down syndrome, requires samples of fetal cells or chemicals produced by the fetus. Presently, two main techniques are used to obtain these samples: *amniocentesis* and *chorionic villus sampling*. Techniques for analyzing fetal cells and other fetal substances in maternal blood are also under development. Several types of diagnostic tests can be performed on the samples.

AMNIOCENTESIS

The human fetus, like all animal embryos, develops in a watery environment. A waterproof membrane called the *amnion* surrounds the fetus and holds the fluid. As the fetus develops, it sheds some of its own cells into the fluid, which is called *amniotic fluid*. When a fetus is 16 weeks or older, amniotic fluid can be collected safely by a procedure called **amniocentesis**. A physician determines the position of the fetus by ultrasound scanning, inserts a sterilized needle through the abdominal wall, the uterus, and amnion, and withdraws 10 to 20 milliliters of fluid (**FIG. E13-5**). Biochemical analysis may be performed on the fluid immediately, but there are very few cells in the sample. For most analyses, such as karyotyping for Down syndrome, the cells must first be allowed to multiply in culture. After a week or two, there are normally enough cells to work with.

CHORIONIC VILLUS SAMPLING

The *chorion* is a membrane that is produced by the fetus and becomes part of the placenta. The chorion produces many small projections, called *villi*. In **chorionic villus sampling (CVS)**, a physician inserts a small tube into the uterus through the mother's vagina and suctions off a few fetal villi for analysis (see Fig E13-5); the loss of a few villi does not harm the fetus. CVS has two major advantages over amniocentesis. First, it can be done much earlier in pregnancy—as early as the eighth week. This is especially important if the woman is contemplating a therapeutic abortion if her fetus

has a major defect. Second, the sample contains a much higher concentration of fetal cells than can be obtained by amniocentesis, so analyses can be performed immediately. However, chorionic cells tend to be more likely to have abnormal numbers of chromosomes (even when the fetus is normal), which complicates karyotyping. CVS also appears to have a slightly greater risk of causing miscarriages than amniocentesis does. Finally, CVS cannot detect certain disorders, such as spina bifida. For these reasons, CVS is much less commonly performed than amniocentesis.

MATERNAL BLOOD

A tiny number of fetal cells cross the placenta and enter the mother's bloodstream as early as the 6th week of pregnancy. Separating fetal cells (perhaps as few as one per milliliter of blood) from the huge numbers of maternal cells is challenging, but it can be done. Several companies now offer paternity testing based on fetal cells in maternal blood. A variety of proteins and other chemicals produced by the fetus may also cross into the mother's bloodstream. The presence or concentration of these chemicals in the maternal blood may indicate whether the fetus has certain genetic disorders, such as Down syndrome, or some nongenetic disorders, such as spina bifida or anencephaly (these are both serious defects of the nervous system in which the inside of the nervous system is connected to the skin, with the result that cerebrospinal fluid leaks out of the fetal nervous system). So far, except for paternity tests, examining fetal cells or chemicals in the mother's blood may provide some evidence for fetal defects, but the results are far from certain, so that other tests—such as amniocentesis, chorionic villus sampling, or ultrasound—must still be performed.

ANALYZING THE SAMPLES

Several types of analyses can be performed on amniotic fluid or fetal cells (see Fig E13-5). Biochemical analysis is used to determine the concentration of chemicals in the amniotic

---

produce extra growth hormone are also unlikely to be hazardous to eat, because growth hormone is also made in the human body. If growth-enhanced livestock are ever developed, they will simply have more meat, composed of exactly the same proteins that exist in non-transgenic animals, so they shouldn't be dangerous either.

Another possible danger is that people might be allergic to genetically modified plants. In the 1990s, a gene from Brazil nuts was inserted into soybeans in an attempt to improve the balance of amino acids in soybean protein. It was discovered, however, that people allergic to Brazil nuts would probably also be allergic to the transgenic soybeans and eat them without suspecting that they might cause an allergic reaction. A more unexpected result occurred when researchers inserted a protein that kills pea weevils, a major insect pest, from beans into peas. No allergic responses have been reported when people or animals eat beans that synthesize this protein. However, in

2005, Australian researchers discovered that when this protein is made in peas, it can cause allergic reactions in mice. As you learned in Chapters 4 and 12, sugars are often attached to proteins, to make *glycoproteins*. Apparently, peas and beans attach sugars to different places on the weevil-killing protein, making the pea protein allergenic, although the bean protein is not. (That sugar differences on glycoproteins may provoke different allergic responses is not really surprising; recall from Chapter 12 that human blood types are caused by differences in sugars attached to otherwise identical proteins.) Needless to say, these transgenic plants will never make it to the farm. Because of findings such as these, the U.S. Food and Drug Administration now monitors all new transgenic crop plants for allergenic potential.

In 2003, the U.S. Society of Toxicology studied the risks of genetically modified plants and concluded that the current transgenic plants pose no significant dangers to

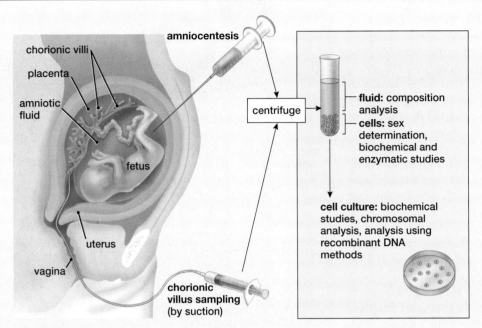

FIGURE E13-5 Prenatal cell sampling techniques
Two methods of obtaining fetal cell samples—amniocentesis and chorionic villus sampling—and some of the tests performed on the fetal cells.

fluid. For example, many metabolic disorders can be detected by a low concentration of enzymes that normally catalyze specific metabolic pathways or by the abnormal accumulation of precursors or by-products. Analysis of the chromosomes of the fetal cells can show if all the chromosomes are present, if there are too many or too few of some, and if any chromosomes show structural abnormalities.

Recombinant DNA techniques can be used to analyze the DNA of fetal cells to detect many defective alleles, such as those for cystic fibrosis or sickle-cell anemia. Prior to the development of PCR, fetal cells typically had to be grown in culture for as long as two weeks before cells had multiplied sufficiently. Now, the second step in prenatal diagnosis is to extract the DNA from a few cells and to use PCR to amplify the region containing the gene of interest. After a few hours, enough DNA is available for techniques such as RFLP analysis, which is used to detect the allele that causes sickle-cell anemia (see Fig. 13-11). If the infant is homozygous for the sickle-cell allele, some therapeutic measures can be taken. In particular, regular doses of penicillin greatly reduce bacterial infections that otherwise kill about 15% of homozygous children. Further, knowing that a child has the disorder ensures correct diagnosis and rapid treatment during a "sickling crisis," when malformed red blood cells clump and block blood flow.

human health. The Society also recognized that past safety does not guarantee future safety, and recommended continued testing and evaluation of all new genetically modified plants. Similar conclusions were drawn by the U.S. National Academy of Sciences in 2004, which found that "the process of genetic engineering has not been shown to be inherently dangerous but … any technique, including genetic engineering, carries the potential to result in unintended changes in the composition of the food."

Are GMOs Hazardous to the Environment?

The environmental effects of GMOs are much more debatable. One clear positive effect of Bt crops is that farmers apply less pesticide to their fields. This should translate into less pollution of the environment, and of the farmers, too. For example, in 2002 and 2003, Chinese farmers planting Bt rice reduced pesticide use by 80% compared to farmers planting conventional rice. Further, they suffered no instances of pesticide poisoning, compared to about 5% of farmers planting conventional rice.

On the other hand, Bt or herbicide-resistance genes might spread outside of the farmer's fields. Because these genes are incorporated into the genome of the transgenic crop, these genes will be in its pollen, too. A farmer cannot control where pollen from a transgenic crop will go; wind might carry pollen miles from the farmer's field. In some instances, this probably doesn't matter very much. In the United States, for example, there are no wild relatives of wheat, so pollen from transgenic wheat probably wouldn't spread resistance genes to wild plants. In Eastern Europe and the Middle East, however, where many crops originated, including oats, wheat, and barley, there are many weedy relatives in the wild. Suppose these plants interbred with transgenic crops and became resistant to herbicides or pests. Would they become significant weed problems for agriculture, because they would not be susceptible to herbicides? Would they displace other native plants in the wild, because they would be less likely to be eaten by insects? Even if transgenic crops have no close relatives in the wild, bacteria and viruses can carry genes from one plant to another, even between unrelated species. Might such "lateral transfer" spread unwanted genes into wild plant populations? No one really knows the answers to these questions.

In 2002, a committee of the U.S. National Academy of Sciences studied the potential impact of transgenic crops on the environment. The committee pointed out that crops modified by both traditional breeding methods and recombinant DNA technologies have the potential to cause major changes in the environment. In addition, the committee found that the United States does not have an adequate system for monitoring changes in ecosystems that might be caused by transgenic crops. It recommended more thorough screening of transgenic plants before they are used commercially, and sustained ecological monitoring of both the agricultural and natural environments after commercialization.

What about transgenic animals? Unlike pollen, most domesticated animals, such as cattle or sheep, are relatively immobile. Further, most have few wild relatives with which they might exchange genes, so the dangers to natural ecosystems appear minimal. However, some transgenic animals, especially fish, have the potential to pose more significant threats, because they can disperse rapidly and are nearly impossible to recapture.

Should the Genome of Humans Be Changed by Biotechnology?

Many of the ethical implications of human applications of biotechnology are fundamentally the same as those connected with other medical procedures. For example, long before biotechnology enabled prenatal testing for cystic fibrosis or sickle-cell anemia, trisomy 21 (Down syndrome) could be diagnosed in embryos by simply counting the chromosomes in cells taken from the amniotic fluid (see "Health Watch: Prenatal Genetic Screening"). Whether parents should use such information as a basis for therapeutic abortion or to prepare to care for the affected child is an ethical issue that generates considerable debate.

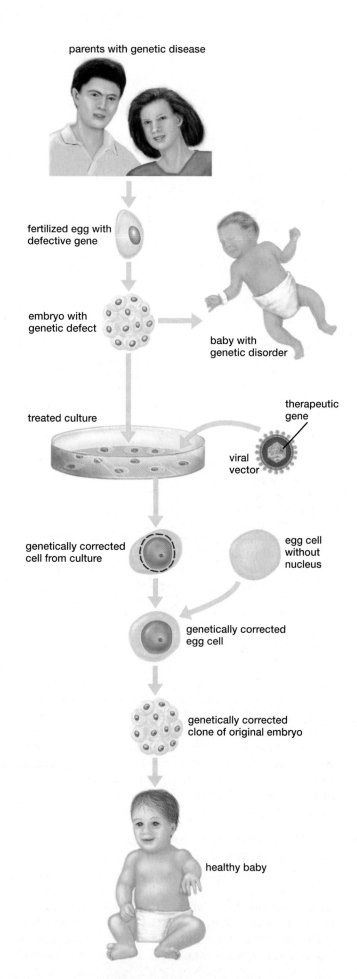

FIGURE 13-13 Human cloning technology might allow permanent correction of genetic defects
In this process, human embryos are derived from eggs fertilized in culture dishes using sperm and eggs from a man and woman, one or both of whom have a genetic disorder. When an embryo containing a defective gene grows into a small cluster of cells, a single cell would be removed from the embryo and the defective gene replaced using an appropriate vector. The repaired nucleus could then be implanted into another egg (taken from the same woman) whose nucleus had been removed. The repaired egg cell would then be implanted in the woman's uterus for normal development.

Other ethical concerns, however, have arisen purely as a result of advances in biotechnology; for example, should people be allowed to select the genomes of their offspring, or even more controversially, should they be allowed to *change* the genomes?

On July 4, 1994, a girl in Colorado was born with Fanconi anemia, a genetic disorder that causes not only anemia but also skeletal abnormalities, such as missing thumbs. It is fatal without a bone marrow transplant. Her parents wanted another child—a very special child. They wanted one without Fanconi anemia, of course, but they also wanted a child who could serve as a donor for their daughter. They went to Yury Verlinsky of the Reproductive Genetics Institute for help. Verlinsky used the parents' sperm and eggs to create dozens of embryos in culture. The embryos were then tested both for the genetic defect and for tissue compatibility with the couple's daughter. Verlinsky chose an embryo with the desired genotype and implanted it into the mother's uterus. Nine months later, a son was born. Blood from his umbilical cord provided cells to transplant into his sister's bone marrow. Today, the girl's bone marrow failure has been cured, although she will always have ane-

mia and many accompanying symptoms. In 2003 her parents had another girl, conceived by in vitro fertilization with one of the remaining healthy embryos. Was this an appropriate use of genetic screening? Should dozens of embryos be created, knowing that the vast majority will never be implanted? Is this ethical if it is the only way to save the life of another child? Assuming that it's possible someday, would this be an ethical method of selecting embryos that would grow up to be bigger or stronger football players?

Today's technology allows physicians only to select among existing embryos, not to change their genomes. But technologies do exist to alter the genomes of, for example, bone marrow stem cells to cure SCID. What if biotechnology could change the genes of fertilized eggs (**FIG. 13-13**)? This is not possible yet, but surely with enough research, the time will come, eventually. If such techniques were used to fix SCID or cystic fibrosis, would they be ethical? What about improving prospective athletes? If and when the technology is developed to cure diseases, it will be difficult to prevent it from being used for nonmedical purposes. Who will determine which uses are appropriate and which are trivial vanity?

CASE STUDY REVISITED GUILTY OR INNOCENT?

Earl Ruffin's innocence could be proven only if the semen collected from the rape victim could be located and if its DNA could be compared against Ruffin's. Beginning in 1989, the Innocence Project and Ruffin filed request after request, trying to find out if the rape kit had been preserved. Finally, in 2002, a Virginia state attorney located the rape kit that had been preserved, along with hundreds of other samples, by Mary Jane Burton. Analysis of STRs showed that Ruffin was not the rapist. The DNA profile indicated that another man, by that time in prison for rape, was the real perpetrator. On February 12, 2003, after 21 years in prison, Earl Ruffin was freed.

What of the other people involved in the Ruffin case? Many people find it almost impossible to admit that they were wrong. To her enormous credit, the woman who was raped that December night is not one of them. She wrote to Ruffin, "I thank God for DNA testing. I do not know how to express my sorrow and devastation." She has asked Virginia legislators to support a bill that would pay Ruffin monetary compensation for his time spent in prison.

Forensic scientist Mary Jane Burton didn't live to see the fruits of her labors, because she died in 1999. However, two other innocent men, Arthur Whitfield and Marvin Anderson, are now free because of Burton's meticulous work. Although Burton is not identified, the forensic lab in Patricia Cornwell's best-selling novel, *Postmortem*, is apparently based partly on Burton's, with whom Cornwell briefly worked.

Police and district attorneys, of course, also use DNA technology as an investigative tool. In 1990, three elderly women in Goldsboro, North Carolina, were raped; two of them were murdered. DNA evidence indicated that all three crimes were committed by the same assailant, known only as the "Night Stalker." Over the years, the FBI and many states have slowly built up DNA databases of criminals, each identified by their DNA profiles of short tandem repeats. In 2001, the Goldsboro police created a DNA profile of the Night Stalker from the evidence they had carefully stored for over a decade. They sent the profile to the North Carolina DNA database and discovered a match. Faced with indisputable DNA evidence, the Stalker confessed. He is now in prison.

Consider This Who are the "heroes" in these stories? There are the obvious ones, of course—Mary Jane Burton, the professors and law students of the Innocence Project, and the members of the Goldsboro Police Department. But what about Thomas Brock, who discovered *Thermus aquaticus* and its unusual lifestyle in Yellowstone hot springs (see Scientific Inquiry: "Hot Springs and Hot Science")? Or molecular biologist Kary Mullis, who discovered PCR? Or the hundreds of biologists, chemists, and mathematicians who, over many decades, developed procedures for gel electrophoresis, labeling DNA, and statistical analysis of sample matching?

Scientists often say that science is worthwhile for its own sake, and that it is difficult or impossible to predict which discoveries will lead to the greatest benefits for humanity. Nonscientists, when asked to pay the costs of scientific projects, are sometimes skeptical of such claims. How do you think that public support of science should be allocated? Forty years ago, would *you* have voted to give Thomas Brock public funds to see what types of organisms lived in hot springs?

CHAPTER REVIEW

SUMMARY OF KEY CONCEPTS

13.1 What Is Biotechnology?

Biotechnology is any industrial or commercial use or alteration of organisms, cells, or biological molecules to achieve specific practical goals. Modern biotechnology generates altered genetic material via genetic engineering. Genetic engineering frequently involves the production of recombinant DNA by combining DNA from different organisms. When DNA is transferred from one organism to another, the recipients are called transgenic or genetically modified organisms (GMOs). Major applications of modern biotechnology include increasing our understanding of gene function, treating disease, improving agriculture, and solving crimes.

13.2 How Does DNA Recombine in Nature?

DNA recombination occurs naturally through processes such as sexual reproduction; bacterial transformation, in which bacteria acquire DNA from plasmids or other bacteria; and viral infection, in which viruses incorporate fragments of DNA from their hosts and transfer the fragments to members of the same or other species.

Web Tutorial 13.1 Genetic Recombination in Nature

13.3 How Is Biotechnology Used in Forensic Science?

Small quantities of DNA, such as might be obtained at a crime scene, can be amplified by the polymerase chain reaction (PCR) technique. The DNA can then be cut into specific, reproducible fragments using restriction enzymes. The most common fragments used in forensics are short tandem repeats (STR). The STRs are separated by gel electrophoresis and made visible with DNA probes. The pattern of STRs is unique to each individual, and can be used to match DNA found at a crime scene with DNA from suspects.

Web Tutorial 13.2 Polymerase Chain Reaction (PCR)

13.4 How Is Biotechnology Used in Agriculture?

Many crop plants have been modified by the addition of genes that promote herbicide resistance or pest resistance. The most common procedure uses restriction enzymes to insert the gene into a plasmid from the bacterium *Agrobacterium tumefaciens*. The genetically modified plasmid is then used to transform the bacteria, which are allowed to infect plant cells. The plasmid inserts the new gene into one of the plant chromosomes. Using cell culture, entire plants are grown from the transgenic cells and eventually planted commercially. Plants may also be modified to produce human proteins, vaccines, or antibodies. Transgenic animals may be produced as well, with properties such as faster growth, increased production of valuable products such as milk, or the ability to produce human proteins, vaccines, or antibodies.

13.5 How Is Biotechnology Used to Learn About the Human Genome?

Techniques of biotechnology were used to discover the complete nucleotide sequence of the human genome. This knowledge will be used to learn the identities and functions of new genes, to discover medically important genes, to explore genetic variability among individuals, and to better understand the evolutionary relationships between humans and other organisms.

13.6 How Is Biotechnology Used for Medical Diagnosis and Treatment?

Biotechnology may be used to diagnose genetic disorders such as sickle-cell anemia or cystic fibrosis. For example, in the diagnosis of sickle-cell anemia, restriction enzymes cut normal and defective globin alleles in different locations. The resulting DNA fragments of different lengths may then be separated and identified by gel electrophoresis. In the diagnosis of cystic fibrosis, DNA probes complementary to various cystic fibrosis alleles are placed on a DNA array. Base-pairing of a patient's DNA to specific probes on the array identifies which alleles are present in the patient.

Inherited diseases are caused by defective alleles of crucial genes. Genetic engineering may be used to insert functional alleles of these genes into normal cells, stem cells, or even into eggs to correct the genetic disorder.

Web Tutorial 13.3 Manufacturing Human Growth Hormone

13.7 What Are the Major Ethical Issues Surrounding Modern Biotechnology?

The use of genetically modified organisms in agriculture is controversial for two major reasons: consumer safety and environmental protection. In general, GMOs contain proteins that are harmless to mammals, are readily digested, or are already found in other foods. The transfer of potentially allergenic proteins to normally nonallergenic foods can be avoided by thorough testing. Environmental effects of GMOs are more difficult to predict. It is possible that foreign genes, such as those for pest resistance or herbicide resistance, might be transferred to wild plants, with resulting damage to agriculture and/or disruption of ecosystems. If they escape, highly mobile transgenic animals might displace their wild relatives.

Genetically selecting or modifying human embryos is highly controversial. As technologies improve, society may be faced with decisions about the extent to which parents should be allowed to correct or enhance the genomes of their children.

KEY TERMS

amniocentesis   *page 268*
biotechnology   *page 252*
chorionic villus sampling
  (CVS)   *page 268*
DNA profile   *page 258*
DNA probe   *page 257*
gel electrophoresis   *page 256*

genetic engineering
  page 252
genetically modified
  organism (GMO)   *page 252*
plasmid   *page 252*
polymerase chain reaction
  (PCR)   *page 255*

recombinant DNA   *page 252*
restriction enzyme   *page 259*
restriction fragment length
  polymorphism (RFLP)   *page 262*

stem cell   *page 265*
transformation   *page 252*
transgenic   *page 252*

THINKING THROUGH THE CONCEPTS

1. Describe three natural forms of genetic recombination, and discuss the similarities and differences between recombinant DNA technology and these natural forms of genetic recombination.

2. What is a plasmid? How are plasmids involved in bacterial transformation?

3. What is a restriction enzyme? How can restriction enzymes be used to splice a piece of human DNA into a plasmid?

4. What is a short tandem repeat? How are short tandem repeats used in forensics?

5. Describe several uses of genetic engineering in agriculture.

6. Describe several uses of genetic engineering in human medicine.

7. Describe amniocentesis and chorionic villus sampling, including the advantages and disadvantages of each. What are their medical uses?

APPLYING THE CONCEPTS

1. Discuss the ethical issues that surround the release of genetically modified organisms (plants, animals, or bacteria) into the environment. What could go wrong? What precautions might prevent the problems you listed from occurring? What benefits do you think would justify the risks?

2. Do you think that using recombinant DNA technologies to change the genetic composition of a human egg cell is ever justified? If so, what restrictions should be placed on such a use?

3. If you were contemplating having a child, would you want both yourself and your spouse tested for the cystic fibrosis gene? If both of you were carriers, how would you deal with this decision?

4. As you may know, many insects have evolved resistance to common pesticides. Do you think that insects might evolve resistance to Bt crops? If this is a risk, do you think that Bt crops should be planted anyway? Why or why not?

FOR MORE INFORMATION

Brownlee, C. "Gene Doping: Will Athletes Go for the Ultimate High?" *Science News*, October 30, 2004. Sometime, probably not too many years in the future, athletes may be able to enhance performance by altering their genes.

Friend, S. H., and Stoughton, R. B. "The Magic of Microarrays." *Scientific American*, February 2002. How microarrays are made, how they work, and how they may be used to provide customized medical care are all clearly explained.

Gura, T. "New Genes Boost Rice Nutrients." *Science*, August 1999. Explanation of how rice was genetically engineered to produce vitamin A precursors.

Hoplin, K. "The Risks on the Table." *Scientific American*, April 2001. Hoplin describes the controversies over whether GM crops are safe to eat.

Langridge, W. H. R. "Edible Vaccines." *Scientific American*, September 2000. Plants may be developed to produce vaccines or treatments for diseases.

Martindale, D. "Pink Slip in Your Genes." *Scientific American*, January 2001. Should employers have access to genetic information about their employees? If so, what should they be allowed to do about it?

Marvier, M. "Ecology of Transgenic Crops." *American Scientist*, March/April 2001. A thoughtful article that weighs the benefits, risks, and uncertainties of bioengineered crops.

Miller, R. V. "Bacterial Gene Swapping in Nature." *Scientific American*, January 1998. How likely is it that genes introduced into bioengineered organisms might be inadvertently transferred to wild organisms?

Palevitz, B. A. "Society Honors Golden Rice Inventor." *The Scientist*, August 2001. Ingo Potrykus, one of the key researchers involved in making golden rice, describes the motives, triumphs, and turmoils.

Scientific American, June 1997. A special issue devoted to the prospects of human gene therapy.

Weidensaul, S. "Raising the Dead." *Audubon*, May–June 2002. Can cloning be used to resurrect extinct species from museum specimens? Don Colgan is trying to recreate the Tasmanian marsupial wolf.

Wheelwright, J. "Body, Cure Thyself." *Discover*, March 2002. The promise of gene therapy is vast, but so far, the results have been decidedly mixed.

Wheelwright, J. "Bad Genes, Good Drugs." *Discover*, April 2002. Researchers are using the results of the Human Genome Project to identify genes that predispose people to diseases, such as Alzheimer's disease, and to develop new drugs to combat those diseases.

Evolution and Diversity of Life

The ghostly grandeur of ancient bones evokes images of a lost world. Fossil remnants of extinct creatures, such as this *Triceratops* dinosaur skeleton, provide clues for the biologists who attempt to reconstruct the history of life.

14

Principles of Evolution

This massive, earthbound ostrich
has useless wings, a legacy of its
evolutionary heritage.

CASE STUDY WHAT GOOD ARE WISDOM TEETH?

HAVE YOU HAD your wisdom teeth removed yet? If not, it's probably only a matter of time. Almost all of us will visit an oral surgeon to have our wisdom teeth extracted. There's just no room in our jaws for these rearmost molars, and removing them is the best way to prevent dental disasters. And removal is harmless, because we don't really need wisdom teeth. They're pretty much useless.

If you've already suffered through a wisdom tooth extraction, you may have found yourself wondering why we even *have* these useless molars. Biologists hypothesize that we have them because our apelike ancestors had them and we inherited them, even though we don't need them. The presence in a living species of structures that have no current function, but that *are* useful in other species, demonstrates the shared ancestry of different species.

Some excellent evidence of the connection between useless traits and evolutionary ancestry is provided by flightless birds. Consider the ostrich, a bird that can be 8 feet tall and weigh 300 pounds. These massive creatures cannot fly. Nonetheless, they have wings, just as sparrows and ducks do. Why do ostriches have useless wings that serve no function? Because the common ancestor of sparrows, ducks, and ostriches had wings. And so do all of its descendants, even the flightless descendants that don't need wings. The bodies of today's organisms may contain now-useless hand-me-downs from their ancestors.

FIGURE 14-1 A timeline of the roots of evolutionary thought
Each bar represents the life span of a key figure in the development of modern evolutionary biology.

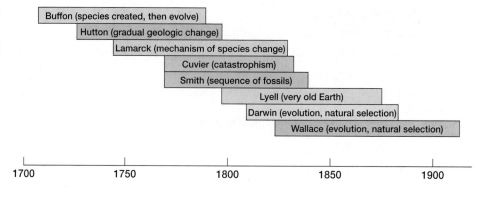

14.1 HOW DID EVOLUTIONARY THOUGHT EVOLVE?

When you began studying biology, you may not have seen a connection between your wisdom teeth and an ostrich's wings. But the connection is there, provided by the concept that unites all of biology: **evolution**, or change over time in the characteristics of populations.

Modern biology is based on our understanding that life has evolved, but early scientists did not recognize this fundamental principle. The main ideas of evolutionary biology were widely accepted only after the publication of Charles Darwin's work in the late nineteenth century. Nonetheless, the intellectual foundation on which these ideas rest developed gradually over the centuries before Darwin's time. (You may wish to refer to the timeline in **FIGURE 14-1** as you read the following historical account.)

Early Biological Thought Did Not Include the Concept of Evolution

Pre-Darwinian science, heavily influenced by theology, held that all organisms were created simultaneously by God and that each distinct life-form remained fixed and unchanging from the moment of its creation. This explanation of how life's diversity arose was elegantly expressed by the ancient Greek philosophers, especially Plato and Aristotle. Plato (427–347 B.C.) proposed that each object on Earth was merely a temporary reflection of its divinely inspired "ideal form." Plato's student Aristotle (384–322 B.C.) categorized all organisms into a linear hierarchy that he called the "ladder of Nature" (**FIG. 14-2**).

These ideas formed the basis of the view that the form of each type of organism is permanently fixed. This view reigned unchallenged for nearly 2000 years. By the eighteenth century, however, several lines of newly emerging evidence began to erode the dominance of this static view of creation.

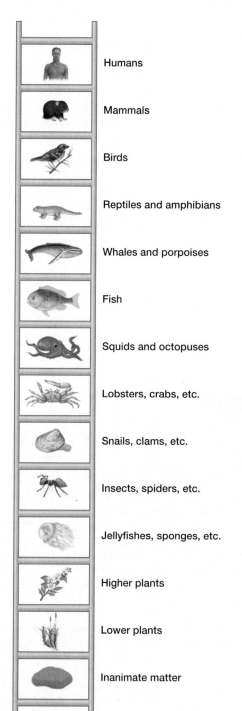

FIGURE 14-2 Aristotle's Ladder of Nature
In Aristotle's view, fixed, unchanging species could be arranged in order of increasing closeness to perfection, with inferior types at the bottom and superior types above.

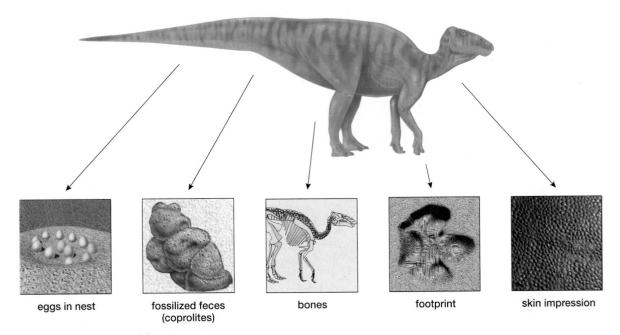

eggs in nest | fossilized feces (coprolites) | bones | footprint | skin impression

FIGURE 14-3 Types of fossils
Any part or trace of an organism that is preserved in rock or sediments is a fossil.

Exploration of New Lands Revealed a Staggering Diversity of Life

The Europeans who explored and colonized Africa, Asia, and the Americas were often accompanied by naturalists who observed and collected the plants and animals of these previously unknown (to Europeans) lands. By the 1700s, the accumulated observations and collections of the naturalists had begun to reveal the true scope of life's variety. The number of species, or different types of organisms, was much greater than anyone had suspected.

Stimulated by the new evidence of life's incredible diversity, some eighteenth-century naturalists began to take note of some fascinating patterns. They noticed, for example, that the species found in one place were different from those found in other places, so that each area had its own distinctive set of species. In addition, the naturalists saw that some of the species in a given location closely resembled one another, yet differed in some characteristics. To some scientists of the day, the differences between the species of different geographical areas and the existence of clusters of similar species within areas seemed inconsistent with the idea that species were fixed and unchanging.

A Few Scientists Speculated That Life Had Evolved

A few eighteenth-century scientists went so far as to speculate that species had, in fact, changed over time. For example, the French naturalist Georges Louis LeClerc (1707–1788), known by the title Comte de Buffon, suggested that perhaps the original creation provided a relatively small number of founding species and that some modern species had been "conceived by Nature and produced by Time"—that is, they had evolved through natural processes.

Fossil Discoveries Showed That Life Has Changed over Time

As Buffon and his contemporaries pondered the implications of new biological discoveries, developments in geology cast further doubt on the idea of permanently fixed species. Especially important was the discovery, during excavations for roads, mines, and canals, of rock fragments that resembled parts of living organisms. People had known of such **fossils** since the fifteenth century, but most thought they were ordinary rocks that wind, water, or people had worked into lifelike forms. As more and more fossils were discovered, however, it became obvious that they were the remains or impressions of plants or animals that had died long ago and had been changed into or in some way preserved in rock (**FIG. 14-3**).

By the beginning of the nineteenth century, some pioneering investigators realized that how fossils were distributed in rock was also significant. Many rocks occur in layers, with newer layers positioned over older layers. The British surveyor William Smith (1769–1839), who studied rock layers and the fossils embedded in them, recognized that certain fossils were always found in the same layers of rock. Further, the organization of fossils and rock layers was consistent: Fossil type A could always be found in a rock layer resting beneath a younger layer containing fossil type B, which in turn rested beneath a still-younger layer containing fossil type C, and so on.

Scientists of the period also discovered that fossil remains showed a remarkable progression. Most fossils found in the oldest layers were very different from modern organisms, and the resemblance to modern organisms gradually increased in progressively younger rocks. Many of the fossils were from plant or animal species that had

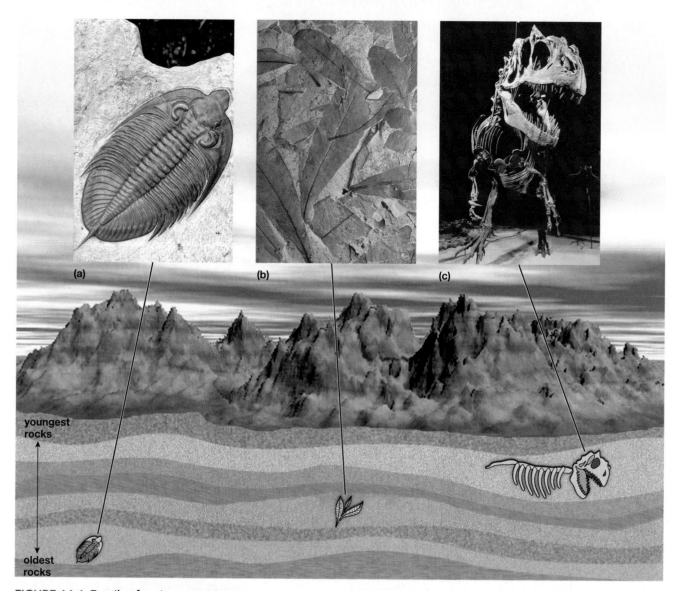

FIGURE 14-4 Fossils of extinct organisms
Fossils provide strong support for the idea that today's organisms were not created all at once, but arose over time by the process of evolution. If all species were created simultaneously, we would not expect **(a)** trilobites to be found in older rock layers than **(b)** seed ferns, which in turn would not be found deeper than **(c)** dinosaurs, such as *Allosaurus*. Trilobites became extinct about 230 million years ago, seed ferns about 150 million years ago, and dinosaurs 65 million years ago.

gone *extinct;* that is, no members of the species still lived on Earth (**FIG. 14-4**).

Putting all of these facts together, some scientists came to an inescapable conclusion. Different types of organisms had lived at different times in the past.

Some Scientists Devised Nonevolutionary Explanations for Fossils

Despite the growing fossil evidence, many scientists of the period did not accept the proposition that species changed and new ones had arisen over time. To account for extinct species while preserving the notion of a single creation by God, Georges Cuvier (1769–1832) proposed the theory of **catastrophism**. Cuvier, a French paleontologist, hypothesized that a vast supply of species was created initially. Successive catastrophes (such as the Great Flood described in the Bible) produced layers of rock and destroyed many species, fossilizing some of their remains in the process. The organisms of the modern world, he theorized, are the species that survived the catastrophes.

Geology Provided Evidence That Earth Is Exceedingly Old

Cuvier's hypothesis of a world shaped by successive catastrophes was challenged by the work of the geologist Charles Lyell (1797–1875). Lyell, building on the earlier thinking of James Hutton (1726–1797), considered the forces of wind, water, and volcanoes and concluded that there was no need to invoke catastrophes to explain the findings of geology. Don't flooding rivers lay down layers of sediment? Don't lava flows produce layers of basalt? Shouldn't we conclude, then, that layers of rock are evidence of ordinary natural processes, occurring repeatedly over long periods of time? This concept, called **uniformitarianism**, had profound

FIGURE 14-5 Darwin's finches, residents of the Galápagos Islands
Each species specializes in eating a different type of food and has a beak of characteristic size and shape, because natural selection has favored the individuals best suited to exploit each food source efficiently. Aside from the differences in their beaks, the finches are quite similar.

(a) Large ground finch, beak suited to large seeds

(b) Small ground finch, beak suited to small seeds

(c) Warbler finch, beak suited to insects

(d) Vegetarian tree finch, beak suited to leaves

implications, because it asserts that Earth is very old.

Before the 1830 publication of Lyell's evidence in support of uniformitarianism, few scientists suspected that Earth could be more than a few thousand years old. Counting generations in the Old Testament, for example, yields a maximum age of 4000 to 6000 years. An Earth this young poses problems for the idea that life has evolved. For example, ancient writers such as Aristotle described wolves, deer, lions, and other organisms that were identical to those present in Europe more than 2000 years later. If organisms had changed so little over that time, how could whole new species possibly have arisen if Earth was created only a couple of thousand years before Aristotle's time?

But if, as Lyell suggested, rock layers thousands of feet thick were produced by slow, natural processes, then Earth must be old indeed—many millions of years old. Lyell, in fact, concluded that Earth was eternal. (Modern geologists estimate that Earth is about 4.5 billion years old; see "Scientific Inquiry: How Do We Know How Old a Fossil Is?" in Chapter 17.)

Lyell (and his intellectual predecessor Hutton) showed that there was enough time for evolution to occur. But what was the mechanism? What process could cause evolution?

Some Pre-Darwin Biologists Proposed Mechanisms for Evolution

One of the first scientists to propose a mechanism for evolution was the French biologist Jean Baptiste Lamarck (1744–1829). Lamarck was impressed by the sequences of organisms in the rock layers. He observed that older fossils tend to be simple, whereas younger fossils tend to be more complex and more like existing organisms. In 1801, Lamarck hypothesized that organisms evolved through the **inheritance of acquired characteristics**, a process in which the bodies of living organisms are modified through the use or disuse of parts, and these modifications are inherited by offspring. Why would bodies be modified? Lamarck proposed that all organisms possess an innate drive for perfection. For example, if ancestral giraffes tried to improve their lot by stretching upward to feed on leaves that grow high up in trees, their necks became slightly longer as a result. Their offspring would inherit these longer necks and

then stretch even farther to reach still higher leaves. Eventually, this process would produce modern giraffes with very long necks indeed.

Today, we understand how inheritance works and can see that Lamarck's proposed evolutionary process could not work as he described it. Acquired characteristics are not inherited. The fact that a prospective father pumps iron doesn't mean that his children will look like Arnold Schwarzenegger. Remember, though, that in Lamarck's time the principles of inheritance had not yet been discovered (Mendel was born a few years before Lamarck's death). In any case, Lamarck's insight that inheritance plays an important role in evolution was an important influence on the later biologists who discovered the key mechanism of evolution.

Darwin and Wallace Proposed a Mechanism of Evolution

By the mid-1800s, a growing number of biologists had concluded that present-day species had evolved from earlier ones. But how? In 1858 Charles Darwin and Alfred Russel Wallace, working separately, provided convincing evidence that evolution was driven by a simple yet powerful process.

Although their social and educational backgrounds were very different, Darwin and Wallace were quite similar in some respects. Both had traveled extensively in the tropics and had studied the plants and animals living there. Both found that some species differed in only a few features (**FIG. 14-5**). Both were familiar with the fossils that

SCIENTIFIC INQUIRY Charles Darwin—Nature Was His Laboratory

Like many students, Charles Darwin excelled only in subjects that intrigued him. Although his father was a physician, Darwin was uninterested in medicine and unable to stand the sight of surgery. He eventually obtained a degree in theology from Cambridge University, although theology too was of minor interest to him. What he really liked to do was to tramp over the hills, observing plants and animals, collecting new specimens, scrutinizing their structures, and categorizing them.

In 1831, when Darwin was 22 years old (**FIG. E14-1**), he secured a position as "gentleman companion" to Captain Robert Fitzroy of the HMS *Beagle*. The *Beagle* soon embarked on a 5-year surveying expedition along the coastline of South America and then around the world.

Darwin's voyage on the *Beagle* sowed the seeds for his theory of evolution. In addition to his duties as companion to the captain, Darwin served as the expedition's official naturalist, whose task was to observe and collect geological and biological specimens. The *Beagle* sailed to South America and made many stops along its coast. There Darwin observed the plants and animals of the Tropics and was stunned by the greater diversity of species compared with that of Europe.

Although he had boarded the *Beagle* convinced of the permanence of species, Darwin's experiences soon led him to doubt it. He discovered a snake with rudimentary hindlimbs, calling it "the passage by which Nature joins the lizards to the snakes" (**FIG. E14-2**). Another snake he encountered vibrated its tail like a rattlesnake but had no rattles and therefore made no noise. Similarly, Darwin noticed that penguins used their wings to paddle through the water rather than fly through the air. If a creator had individually created each animal in its present form, to suit its present environment, what could be the purpose behind these makeshift arrangements?

FIGURE E14-1 A painting of Charles Darwin as a young man

Perhaps the most significant stopover of the voyage was the month spent on the Galápagos Islands off the northwestern coast of South America. There, Darwin found huge tortoises. Different islands were home to distinctively different types of tortoises. Darwin also found several types of finches and noticed that, as with the tortoises, different islands had slightly different finches. Could the differences in these organisms have arisen after they became isolated from one another on separate islands? The diversity of tortoises and finches haunted him for years afterward.

---

had been discovered, which showed a trend of increasing complexity through time. Finally, both were aware of the studies of Hutton and Lyell, who had proposed that Earth is extremely ancient. These facts suggested to both Darwin and Wallace that species change over time. Both men sought a mechanism that might cause such evolutionary change.

Of the two, Darwin was the first to put in writing a proposed mechanism for evolution, which he described in a paper he wrote in 1842. But he did not submit the paper for publication, perhaps because he was fearful of the controversy that publication would cause. Some historians wonder if Darwin would ever have publicized his ideas had he not received, 16 years later, a draft of a paper by Wallace, which outlined ideas remarkably similar to Darwin's own. Darwin realized that he could delay no longer.

In separate but similar papers that were presented to the Linnaean Society in London in 1858, Darwin and Wallace each described the same mechanism for evolution. Initially, their papers had little impact. The secretary of the society, in fact, wrote in his annual report that nothing very interesting happened that year. Fortunately, the next year Darwin published his monumental *On the Origin of Species by Means of Natural Selection*, which attracted a great deal of attention to the new theory.

14.2 HOW DO WE KNOW THAT EVOLUTION HAS OCCURRED?

Today, virtually all biologists consider evolution to be a fact. Why? Because an overwhelming body of evidence permits no other conclusion. The key lines of evidence come from fossils, comparative anatomy (the study of how body structures differ among species), embryology, and biochemistry and genetics.

FIGURE E14-2 The vestigial remnants of hind legs in a snake Some snakes have small "spurs" (large photo, at arrow) where their distant ancestors had rear legs. In some species, these vestigial structures even retain claws (inset).

In 1836, Darwin returned to England after 5 years on the *Beagle* and became established as one of the foremost naturalists of his time. But the problem of how isolated populations come to differ from each other gnawed constantly at his mind. Part of the solution came to him from an unlikely source: the writings of an English economist and clergyman, Thomas Malthus. In his *Essay on Population*, Malthus wrote, "It may safely be pronounced, therefore, that [human] population, when unchecked, goes on doubling itself every 25 years, or increases in a geometrical ratio."

Darwin realized that a similar principle holds true for plant and animal populations. In fact, most organisms can reproduce much more rapidly than can humans (consider rabbits, dandelions, and houseflies) and consequently could produce overwhelming populations in short order. Nonetheless, the world is not chest-deep in rabbits, dandelions, or flies: natural populations do not grow "unchecked" but tend to remain approximately constant in size. Clearly, vast numbers of individuals must die in each generation, and most must not reproduce.

From his experience as a naturalist, Darwin realized that members of a species typically differ from one another. Further, which individuals die without reproducing in each generation is not arbitrary, but depends to some extent on the structures and abilities of the organisms. This observation was the source of the theory of evolution by natural selection. As Darwin's colleague Alfred Wallace put it, "Those which, year by year, survived this terrible destruction must be, on the whole, those which have some little superiority enabling them to escape each special form of death to which the great majority succumbed." Here you see the origin of the expression "survival of the fittest." That "little superiority" that confers greater success might be better resistance to cold, more efficient digestion, or any of hundreds of other advantages, some very subtle.

Everything now fell into place. Darwin wrote, "It at once struck me that under these circumstances favorable variations would tend to be preserved, and unfavorable ones to be destroyed." If a favorable variation were inheritable, then the entire species would eventually consist of individuals possessing the favorable trait. With the continual appearance of new variations (due, as we now know, to mutations), which in turn are subject to further natural selection, "the result...would be the formation of new species. Here, then, I had at last got a theory by which to work."

When Darwin finally published *On the Origin of Species* in 1859, his evidence had become truly overwhelming. Although its full implications would not be realized for decades, Darwin's theory of evolution by natural selection has become a unifying concept for virtually all of biology.

Fossils Provide Evidence of Evolutionary Change over Time

If it is true that many fossils are the remains of species ancestral to modern species, we might expect to find progressive series of fossils that start with an ancient, primitive organism, progress through several intermediate stages, and culminate in a modern species. Such series have indeed been found. For example, fossils of the ancestors of modern whales illustrate stages in the evolution of an aquatic species from land-dwelling ancestors (**FIG. 14-6**). Series of fossil giraffes, elephants, horses, and mollusks also show the evolution of body structures over time. These fossil series suggest that new species evolved from, and replaced, previous species. Certain sequences of fossil snails have such slight gradations in body structures between successive rock layers that paleontologists cannot easily decide where one species leaves off and the next one begins.

Comparative Anatomy Gives Evidence of Descent with Modification

Fossils provide snapshots of the past that allow biologists to trace evolutionary changes, but careful examination of today's organisms can also uncover evolution's story. Comparing the bodies of organisms of different species can reveal similarities that can be explained only by shared ancestry and differences that could result only from evolutionary change during descent from a common ancestor. In this way, the study of comparative anatomy has supplied strong evidence that different species are linked by a common evolutionary heritage.

Homologous Structures Provide Evidence of Common Ancestry

The same body structure may be modified by evolution to serve different functions in different species. The forelimbs

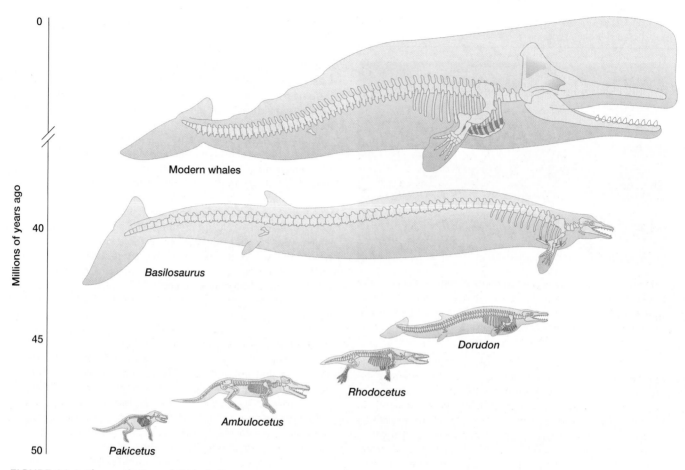

FIGURE 14-6　The evolution of the whale
Over the past 50 million years, whales have evolved from four-legged land dwellers, to semiaquatic pad-dlers, to fully aquatic swimmers with shrunken hind legs, to today's sleek ocean dwellers. **QUESTION** The fossil history of some kinds of modern organisms, such as sharks and crocodiles, shows that their structure and appearance have changed very little over hundreds of millions of years. Is this evidence that such or-ganisms have not evolved over that time?

of birds and mammals, for example, are variously used for flying, swimming, running over several types of terrain, and grasping objects, such as branches and tools. Despite this enormous diversity of function, the internal anatomy of all bird and mammal forelimbs is remarkably similar (**FIG. 14-7**). It seems inconceivable that the same bone arrange-ments would be used to serve such diverse functions if each animal had been created separately. Such similarity is exactly what we would expect, however, if bird and mam-mal forelimbs were derived from a common ancestor. Through natural selection, each has been modified and now performs a particular function. Such internally similar structures are called **homologous structures**, meaning that they have the same evolutionary origin despite any differ-ences in current function or appearance.

Functionless Structures Are Inherited from Ancestors

Evolution by natural selection also helps explain the curi-ous circumstance of **vestigial structures** that serve no ap-parent purpose. Examples include such things as molar teeth in vampire bats (which live on a diet of blood and therefore don't chew their food) and pelvic bones in whales

and certain snakes (**FIG. 14-8**). Both of these vestigial struc-tures are clearly homologous to structures that are found in—and used by—other vertebrates (animals with a back-bone). Their continued existence in animals that have no use for them is best explained as a sort of "evolutionary baggage." For example, the ancestral mammals from which whales evolved had four legs and a well-developed set of pelvic bones (see Fig. 14-7). Whales do not have hind legs, yet they have small pelvic and leg bones embedded in their sides. During whale evolution, losing the hind legs provided an advantage, better streamlining the body for movement through water. The result is the modern whale with small, useless, and unused pelvic bones.

Some Anatomical Similarities Result from Evolution in Similar Environments

The study of comparative anatomy has demonstrated the shared ancestry of life by identifying a host of homologous structures that different species have inherited from com-mon ancestors, but comparative anatomists have also iden-tified many anatomical similarities that do not stem from common ancestry. Instead, these similarities stem from

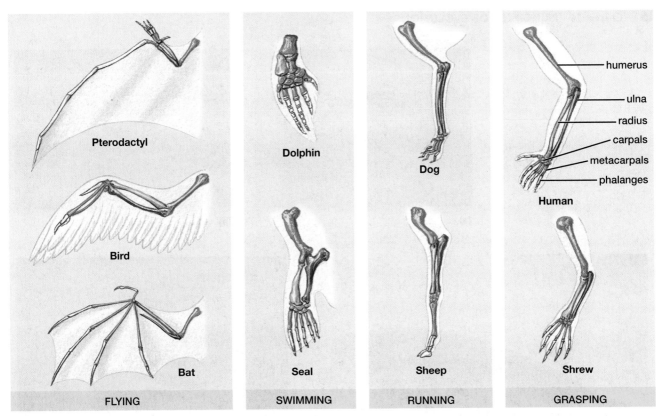

FIGURE 14-7 Homologous structures
Despite wide differences in function, the forelimbs of all these animals contain the same set of bones, inherited through evolution from a common ancestor. The different colors of the bones highlight the correspondences among the various species.

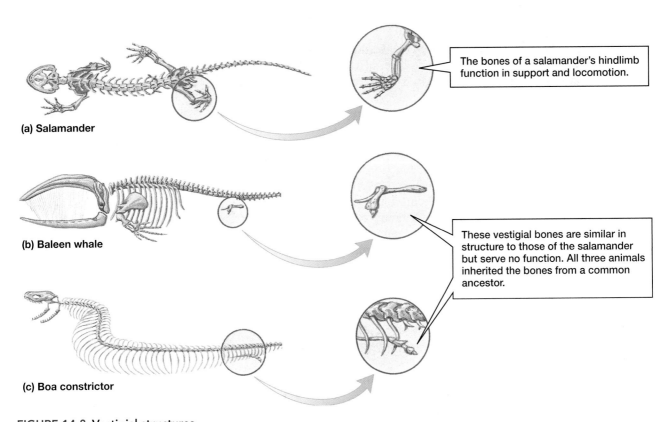

The bones of a salamander's hindlimb function in support and locomotion.

(a) Salamander

(b) Baleen whale

These vestigial bones are similar in structure to those of the salamander but serve no function. All three animals inherited the bones from a common ancestor.

(c) Boa constrictor

FIGURE 14-8 Vestigial structures
Many organisms have vestigial structures that serve no apparent function. The **(a)** salamander, **(b)** whale, and **(c)** snake all inherited hindlimb bones from a common ancestor; the bones remain functional in the salamander but are vestigial in the whale and snake. **EXERCISE** Compile a list of human vestigial structures. For each structure, name the corresponding homologous structure in a nonhuman species.

(a)

(b)

(c)

(d)

FIGURE 14-9 Analogous structures
Convergent evolution can produce outwardly similar structures that differ anatomically. The wings of **(a)** insects and **(b)** birds and the sleek, streamlined shapes of **(c)** seals and **(d)** penguins are examples of such analogous structures. **QUESTION** Are a peacock's tail (see Fig. 15-12) and a dog's tail homologous structures or analogous structures?

convergent evolution, in which natural selection causes nonhomologous structures that serve similar functions to resemble one another. For example, both birds and insects have wings, but this similarity did not arise from evolutionary modification of a structure that both birds and insects inherited from a common ancestor. Instead, the similarity arose from modification of two different, nonhomologous structures that eventually gave rise to superficially similar structures. Because natural selection favored flight in both birds and insects, the two groups evolved superficially similar structures—wings—that are useful for flight. Such outwardly similar but nonhomologous structures are called **analogous structures** (FIG. 14-9). Analogous structures are typically very different in internal anatomy, because the parts are not derived from common ancestral structures.

Embryological Similarity Suggests Common Ancestry

In the early 1800s, German embryologist Karl von Baer noted that all vertebrate embryos (developing organisms in the period from fertilization to birth or hatching) look quite similar to one another early in their development

(FIG. 14-10). In their early embryonic stages, fish, turtles, chickens, mice, and humans all develop tails and gill slits. But of this group of animals, only fish retain gills as adults, and only fish, turtles, and mice retain substantial tails.

Why do vertebrates that are so different have similar developmental stages? The only plausible explanation is that ancestral vertebrates possessed genes that directed the development of gills and tails. All of their descendants still have those genes. In fish, these genes are active throughout development, resulting in adults with fully developed tails and gills. In humans and chickens, these genes are active only during early developmental stages, and the structures are lost or inconspicuous in adults.

Modern Biochemical and Genetic Analyses Reveal Relatedness Among Diverse Organisms

Biologists have been aware of anatomical and embryological similarities among organisms for centuries, but it took the emergence of modern technology to uncover similarity at the molecular level. An especially powerful tool is the ability to quickly determine the sequence of nucleotides in a

(a)

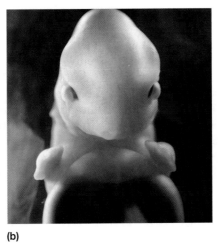
(b)

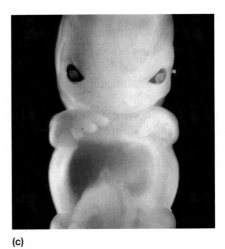
(c)

FIGURE 14-10 Embryological stages reveal evolutionary relationships
The early embryonic stages of a (a) lemur, (b) pig, and (c) human show strikingly similar anatomical features.

DNA molecule. Biologists can now compare the DNA of different organisms. These comparisons have uncovered biochemical similarities that provide perhaps the most striking evidence of the evolutionary relatedness of organisms. Just as relatedness is revealed by homologous anatomical structures, it is revealed by homologous molecules.

A particularly useful feature of molecular comparisons is that they can reveal the relatedness of organisms that have no anatomical features in common. For example, the protein cytochrome c is present in the cells of all plants, all animals, and many single-celled organisms and performs the same function in all of them. This widespread presence of a particular protein is excellent evidence that these diverse organisms share a common ancestor that had cytochrome c in its cells.

A deeper examination, at the level of the DNA that encodes cytochrome c, shows that the differences among organisms can be as revealing as the similarities (see Chapters 9 and 10 for information on DNA and how it encodes proteins). For example, the DNA nucleotide sequence of the gene for human cytochrome c is very similar to the sequence for mouse cytochrome c, but a few nucleotides (about 10% of the total) differ between the two species (FIG. 14-11). These differences in an otherwise identical sequence show that humans and mice share a common ancestor, but that the cytochrome c gene that each inherited from that common ancestor has changed a bit during the time that the two species have been evolving separately. In more distantly related species, the number of differences is greater. For example, in a comparison of the

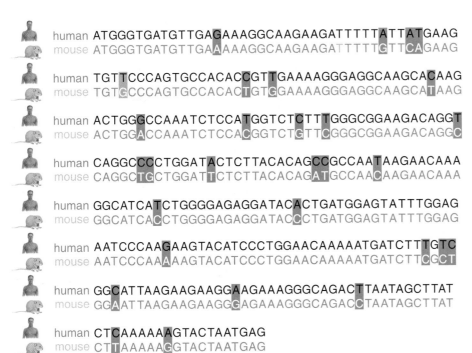

human ATGGGTGATGTTGAGAAAGGCAAGAAGATTTTTATTATGAAG
mouse ATGGGTGATGTTGAAAAAGGCAAGAAGATTTTTGTTCAGAAG

human TGTTCCCAGTGCCACACCGTTGAAAAGGGAGGCAAGCACAAG
mouse TGTGCCCAGTGCCACACTGTGGAAAAGGGAGGCAAGCATAAG

human ACTGGGCCAAATCTCCATGGTCTCTTTGGGCGGAAGACAGGT
mouse ACTGGACCAAATCTCCACGGTCTGTTCGGGCGGAAGACAGGC

human CAGGCCCCTGGATACTCTTACACAGCCGCCAATAAGAACAAA
mouse CAGGCTGCTGGATTCTCTTACACAGATGCCAACAAGAACAAA

human GGCATCATCTGGGGAGAGGATACACTGATGGAGTATTTGGAG
mouse GGCATCACCTGGGGAGAGGATACCCTGATGGAGTATTTGGAG

human AATCCCAAGAAGTACATCCCTGGAACAAAAATGATCTTTGTC
mouse AATCCCAAAAAGTACATCCCTGGAACAAAAATGATCTTCGCT

human GGCATTAAGAAGAAGGAAGAAAGGGCAGACTTAATAGCTTAT
mouse GGAATTAAGAAGAAGGGAGAAAGGGCAGACCTAATAGCTTAT

human CTCAAAAAAGTACTAATGAG
mouse CTTAAAAAGGTACTAATGAG

FIGURE 14-11 Molecular similarity shows evolutionary relationships
The DNA sequences of the genes that code for cytochrome c in a human and a mouse. Of the 315 nucleotides in the gene, 30 (shaded in blue) differ between the two species.

cytochrome genes of humans and corn, about one-third of the nucleotides differ.

Some molecules (and the genes that encode them) are so widespread that they occur in all living things, from bacteria to blue whales, and serve as evidence of the common ancestry of all life. For example, a particular RNA molecule that forms part of the ribosome (the cellular structure on which proteins are assembled) is present in the cells of every organism. As with cytochrome *c*, the degree of similarity of the DNA sequences coding for the ribosomal RNA of two organisms can tell us how recently the common ancestor of the organisms lived.

In addition to molecules that are common to all living things, certain universally shared biochemical processes demonstrate the common heritage of all organisms:

- All cells use DNA as the carrier of genetic information.
- All cells use RNA and approximately the same genetic code to translate genetic information into proteins.
- All cells use roughly the same set of 20 amino acids to build proteins.
- All cells use ATP as a cellular energy carrier.

14.3   HOW DOES NATURAL SELECTION WORK?

The evidence that evolution has occurred does not tell us *how* life evolves. Darwin and Wallace proposed that life's huge variety of excellent designs arose by a process of descent with modification, in which individuals in each generation differ slightly from the members of the preceding generation. Over long stretches of time, these small differences accumulate to produce major transformations.

Darwin and Wallace's Theory Rests on Four Postulates

The chain of logic that led Darwin and Wallace to their proposed process of evolution turns out to be surprisingly simple and straightforward. It is based on four postulates about **populations**, or all the individuals of one species in a particular area.

Postulate 1: Individual members of a population differ from one another in many respects.

Postulate 2: At least some of the differences among members of a population are due to characteristics that may be passed from parent to offspring.

Postulate 3: In each generation, some individuals in a population survive and reproduce successfully but others do not.

Postulate 4: The fate of individuals is not determined entirely by chance or luck. Instead, an individual's likelihood of survival and reproduction depends on its characteristics. Individuals with advantageous traits survive longest and leave the most offspring, a process known as **natural selection**.

Darwin and Wallace understood that if all four postulates were true, populations would inevitably change over time. If members of a population have different traits (postulate 1), and the ones that are best suited to their environment leave more offspring (postulates 3 and 4), and those individuals pass their favorable traits to the next generation (postulate 2), then the favorable traits will be more common in subsequent generations. The characteristics of the population will change slightly with each generation. This process is evolution by natural selection.

Are the four postulates true? Darwin thought so, and devoted much of *On the Origin of Species* to describing supporting evidence. Let's briefly examine each postulate, in some cases with the advantage of knowledge that was not available to Darwin and Wallace.

Postulate 1: Populations Vary

The accuracy of postulate 1 is apparent to anyone who has glanced around a crowded room. People differ in size, eye color, skin color, and many other physical features. Similar variability is present in populations of other organisms, although it may be less obvious to the casual observer (**FIG. 14-12**). We now know that the variations in natural populations arise purely by chance, as a result of random mutations in DNA (see Chapters 9 and 10). Thus, the differences among individuals extend to the molecular level. The reason that DNA tests can match blood from a crime scene to a suspect is that each person's exact DNA sequence is unique.

FIGURE 14-12 Variation in a population of snails Although these snails are all members of the same population, no two are exactly alike.

Postulate 2: Traits Are Inherited

The principles of genetics had not yet been discovered when Darwin published *On the Origin of Species*. Therefore, although observation of people, pets, and farm animals seemed to show that offspring generally resemble their parents, Darwin and Wallace did not have scientific evidence in support of postulate 2. Mendel's later work, though, demonstrated conclusively that particular traits can be passed to offspring. Since Mendel's time, genetics researchers have produced an incredibly detailed picture of how inheritance works.

Postulate 3: Some Individuals Fail to Survive and Reproduce

Darwin and Wallace's formulation of postulate 3 was heavily influenced by Thomas Malthus's *Essay on the Principle of Population* (1798), which described the perils of unchecked growth of human populations. Darwin was keenly aware that organisms can produce far more offspring than are required merely to replace the parents. He calculated, for example, that a single pair of elephants would multiply to a population of 19 million in 750 years if each descendant had six offspring.

But we aren't overrun with elephants. The number of elephants, like the number of individuals in most natural populations, tends to remain relatively constant. Therefore, more organisms must be born than survive long enough to reproduce. In each generation, many individuals must die young. Even among those that survive, many must fail to reproduce, produce few offspring, or produce less-vigorous offspring that, in turn, fail to survive and reproduce. As you might expect, whenever biologists have measured reproduction in a population, they have found that some individuals have more offspring than others.

Postulate 4: Reproductive Success Is Not Random

If unequal reproduction is the norm in populations, what determines which individuals leave the most offspring? A large amount of scientific evidence has shown that reproductive success depends on an individual's characteristics. For example, larger male elephant seals in a California population had more offspring than did smaller males. In a Colorado population, snapdragon plants with white flowers had more offspring than did plants with yellow flowers. In a laboratory population of flour beetles, pesticide-resistant beetles had more offspring than did pesticide-sensitive ones. These results, and hundreds of other similar ones, show that in the competition to survive and reproduce, winners are determined not by chance but by the traits they possess.

Natural Selection Modifies Populations over Time

Observation and experiment suggest that the four postulates of Darwin and Wallace are sound. Logic suggests that the resulting consequence ought to be change over time in the characteristics of populations. In *On the Origin of Species*, Darwin proposed the following example:

> Let us take the case of a wolf, which preys on various animals, securing [them] by...fleetness....The swiftest and slimmest wolves would have the best chance of surviving, and so be preserved or selected.... Now if any slight innate change of habit or structure benefited an individual wolf, it would have the best chance of surviving and of leaving offspring. Some of its young would probably inherit the same habits or structure, and by the repetition of this process, a new variety might be formed.

The same logic applies to the wolf's prey; the fastest or most alert would be most likely to avoid predation and would pass on these traits to its offspring.

Notice that natural selection acts on individuals within a population. Over generations, the population changes as the percentage of individuals inheriting favorable traits increases. An individual cannot evolve, but a population can.

Although it is easier to understand how natural selection would cause changes *within* a species, under the right circumstances, the same principles might produce entirely *new* species. We will discuss the circumstances that might give rise to new species in Chapter 16.

14.4 WHAT IS THE EVIDENCE THAT POPULATIONS EVOLVE BY NATURAL SELECTION?

Darwin and Wallace's description of the process of natural selection is logical and persuasive. But what is the evidence that evolution occurs by this process?

Controlled Breeding Modifies Organisms

One line of evidence supporting evolution by natural selection is **artificial selection**, the breeding of domestic plants and animals to produce specific desirable features. The various breeds of dog provide a striking example of artificial selection (**FIG. 14-13**). Dogs descended from wolves, and even today, the two will readily crossbreed. With few exceptions, however, modern dogs do not resemble wolves. Some breeds are so different from one another that they would be considered separate species if they were found in the wild. Humans produced these radically different dogs in a few thousand years by doing nothing more than repeatedly selecting individuals with desirable traits for breeding. Therefore, it is quite plausible that natural selection could, by an analogous process acting over hundreds of millions of years, produce the spectrum of living organisms. Darwin was so impressed by the connection between artificial selection and natural selection that he devoted a chapter of *Origin of Species* to the topic.

(a)

(b)

FIGURE 14-13 **Dog diversity illustrates artificial selection**
A comparison of **(a)** the ancestral dog (the gray wolf, *Canis lupus*) and **(b)** various breeds of modern dogs. Artificial selection by humans has caused a great divergence in size and shape of dogs in only a few thousand years.

Evolution by Natural Selection Occurs Today

The logic of natural selection gives us no reason to believe that evolutionary change is limited to the past. After all, inherited variation and competition for access to resources are certainly not limited to the past. If Darwin and Wallace were correct that those conditions lead inevitably to evolution by natural selection, then scientific observers and experimenters ought to be able to detect evolutionary change as it occurs. And they have. We consider next some examples that give us a glimpse of natural selection at work.

Brighter Coloration Can Evolve When Fewer Predators Are Present

On the island of Trinidad, guppies live in streams that are also inhabited by several species of larger, predatory fish that frequently dine on guppies (**FIG. 14-14**). In upstream portions of these streams, however, the water is too shallow for the predators, and guppies are free of danger from predators. When scientists compared male guppies in an upstream area with ones in a downstream area, they found that the upstream guppies were much more brightly colored than the downstream guppies. The scientists knew that the source of the upstream population was guppies that had found their way up into the shallower waters many generations earlier.

The explanation for the difference in coloration between the two populations stems from the sexual preferences of female guppies. The females prefer to mate with the most brightly colored males, so the brightest males have a large advantage when it comes to reproduction. In predator-free areas, male guppies with the bright colors that females prefer have more offspring than duller

males do. Bright color, however, makes guppies more conspicuous to predators and therefore more likely to be eaten. Thus, where predators are common, they act as agents of natural selection by eliminating the bright-colored males before they can reproduce. In these areas, the duller males have the advantage and produce more offspring. The color difference between the upstream and downstream guppy populations is a direct result of natural selection.

FIGURE 14-14 **Guppies evolve to become more colorful in predator-free environments**
Male guppies (top) are more brightly colored than females (bottom). Some male guppies are more colorful than others. In some environments, brighter males are naturally selected; in other environments, duller males are selected.

Natural Selection Can Lead to Pesticide Resistance

Natural selection is also evident in numerous instances of insect pests evolving resistance to the pesticides with which we try to control them. For example, a few decades ago, Florida homeowners were dismayed to realize that roaches were ignoring a formerly effective poison bait called Combat®. Researchers discovered that the bait had acted as an agent of natural selection. Roaches that liked it were consistently killed; those that survived inherited a rare mutation that caused them to dislike glucose, a type of sugar found in the corn syrup used as bait in Combat. By the time researchers identified the problem in the early 1990s, the formerly rare mutation had become common in Florida's urban roach population.

Unfortunately, the evolution of pesticide resistance in insects is a common example of natural selection in action. Such resistance has been documented in more than 500 species of crop-damaging insects, and virtually every pesticide has fostered the evolution of resistance in at least one insect species. We pay a heavy price for this evolutionary phenomenon. The additional pesticides that farmers apply in their attempts to control resistant insects cost almost $2 billion each year in the United States alone and add millions of tons of poisons to Earth's soil and water.

FIGURE 14-15 Anole leg size evolves in response to a changed environment

Experiments Can Demonstrate Natural Selection

In addition to observing natural selection in the wild, scientists have also devised numerous experiments that confirm the action of natural selection. For example, one group of evolutionary biologists released small groups of *Anolis sagrei* lizards onto 14 small Bahamian islands that were previously uninhabited by lizards (**FIG. 14-15**). The original lizards came from a population on Staniel Cay, an island with tall vegetation, including plenty of trees. In contrast, the islands to which the small colonial groups were introduced had few or no trees and were covered mainly with small shrubs and other low-growing plants.

The biologists returned to those islands 14 years after releasing the colonists and found that the original small groups of lizards had given rise to thriving populations of hundreds of individuals. On all 14 of the experimental islands, lizards had legs that were shorter and thinner than the legs of lizards from the original source population on Staniel Cay. In just over a decade, it appeared, the lizard populations had changed in response to new environments.

Why had the new lizard populations evolved shorter, thinner legs? Long legs allow greater speed for escaping predators, but shorter legs allow for more agility and maneuverability on narrow surfaces. So, natural selection favors legs that are as long and thick as possible while still allowing sufficient maneuverability. When the lizards were moved from an environment with thick-branched trees to an environment with only thin-branched bushes, the individuals with formerly favorable long legs were at a disadvantage. In the new environment, more agile, short-er-legged individuals were better able to escape predators and survive to produce a greater number of offspring. Thus, members of subsequent generations had shorter legs on average.

Selection Acts on Random Variation to Favor the Phenotypes That Work Best in Particular Environments

Two important points underlie the evolutionary changes just described:

- *The variations on which natural selection works are produced by chance mutations.* The bright coloration in Trinidadian guppies, distaste for glucose in Florida cockroaches, and shorter legs in Bahamian lizards were not *produced* by the female mating preferences, poisoned corn syrup, or thinner branches. The mutations that produced each of these beneficial traits had arisen spontaneously.

- *Natural selection selects for organisms that are best adapted to a particular environment.* Natural selection is not a process for producing ever-greater degrees of perfection. Natural selection does not select for the "best" in any absolute sense, but only for what is best in the context of a particular environment, which varies from place to place and which may change over time. A trait that is advantageous under one set of conditions may become disadvantageous if conditions change. For example, in the presence of poisoned corn syrup, distaste for glucose yields an advantage to a cockroach, but under natural conditions, avoiding glucose would cause the insect to bypass good sources of food.

14.5 A POSTSCRIPT BY CHARLES DARWIN

These are the concluding sentences of Darwin's *On the Origin of Species*:

> It is interesting to contemplate an entangled bank, clothed with many plants of many kinds, with birds singing on the bushes, with various insects flitting about, and with worms crawling through the damp earth, and to reflect that these elaborately constructed forms...have all been produced by laws acting around us. These laws, taken in the highest sense, being Growth with Reproduction; Inheritance [and] Variability; a Ratio of Increase so high as to lead to a Struggle for Life, and as a consequence to Natural Selection, entailing Divergence of Character and Extinction of less-improved forms.... There is grandeur in this view of life, with its several powers, having been originally breathed into a few forms or into one; and that, whilst this planet has gone cycling on according to the fixed law of gravity, from so simple a beginning endless forms most beautiful and most wonderful have been, and are being, evolved.

CASE STUDY REVISITED   WHAT GOOD ARE WISDOM TEETH?

Wisdom teeth are but one of dozens of human anatomical structures that appear to serve no function. Darwin himself noted many of these "useless, or nearly useless" traits in the very first chapter of *Origin* and declared them to be prime evidence that humans had evolved from earlier species.

One vestigial structure is the appendix, a narrow tube attached to the large intestine. Although the appendix produces some white blood cells, a person clearly does not need one. Each year, about 300,000 diseased appendixes are surgically removed from Americans, who get along just fine without them. The appendix is probably homologous with the cecum, an extension of the large intestine used for food storage in many plant-eating mammals.

Body hair is another functionless human trait. It seems to be an evolutionary relic of the fur that kept our distant ancestors warm (and that still warms our closest evolutionary relatives, the great apes). Not only do we retain useless body hair, we also still have erector pili, the muscle fibers that allow other mammals to puff up their fur for better insulation. In humans, these vestigial structures just give us goose bumps.

Though humans don't have and don't need a tail, we nonetheless have a tailbone. The tailbone consists of a few tiny vertebrae fused into a small structure at the base of the backbone, where a tail would be if we had one. Some small muscles are attached to the tailbone, but people who are born without one or have theirs surgically removed suffer no ill effects.

Consider This Advocates of creationism argue that if a structure can do anything, it cannot be considered functionless, even if its removal has no effect. Thus, according to this view, wisdom teeth are not evidence of evolution, because they *can* be used to chew if not removed. Do you find this argument persuasive?

CHAPTER REVIEW

SUMMARY OF KEY CONCEPTS

14.1 How Did Evolutionary Thought Evolve?

Historically, the most common explanation for the origin of species was the divine creation of each species in its present form; species were believed to remain unchanged after their creation. This view was challenged by evidence from fossils, geology, and biological exploration of the Tropics. Since the middle of the nineteenth century, scientists have realized that species originate and evolve by the operation of natural processes that change the genetic makeup of populations.

14.2 How Do We Know That Evolution Has Occurred?

Many lines of evidence indicate that evolution has occurred, including the following:

- Fossils of ancient species tend to be simpler in form than modern species. Sequences of fossils have been discovered that show a graded series of changes in form. Both of these observations would be expected if modern species evolved from older species.

- Species thought to be related to a common ancestor through evolution show many similar anatomical structures. Examples include the forelimbs of amphibians, reptiles, birds, and mammals.
- Stages in early embryological development are quite similar among very different types of vertebrates.
- Similarities in such biochemical traits as the use of DNA as the carrier of genetic information support the notion of descent of related species through evolution from common ancestors.

14.3 How Does Natural Selection Work?

Charles Darwin and Alfred Russel Wallace independently proposed the theory of evolution by natural selection. Their theory expresses the logical consequences of four postulates about populations. If populations are variable and the variable traits can be inherited, and if there is differential (unequal) reproduction based on the traits of individuals, the characteristics of successful individuals will be "naturally selected" and become more common over time.

Web Tutorial 14.1 Natural Selection for Antibiotic Resistance

Web Tutorial 14.2 Natural Selection in Alpine Skypilots

14.4 What Is the Evidence That Populations Evolve by Natural Selection?

Natural selection is the chief mechanism driving changes in the characteristics of species over time, as indicated by many lines of evidence including the following:

- Inheritable traits have been changed rapidly in populations of domestic animals and plants by selectively breeding organisms with desired features (artificial selection). The immense variations in species produced in a few thousand years of artificial selection by humans makes it seem likely that much larger changes could be wrought by hundreds of millions of years of natural selection.

- Evolution can be observed today. Both natural and human activities drastically change the environment over short periods of time. Characteristics of species have been observed to change significantly in response to such environmental changes.

KEY TERMS

analogous structures *page 286*

artificial selection *page 289*

catastrophism *page 280*

convergent evolution *page 286*

evolution *page 278*

fossil *page 279*

homologous structures *page 284*

inheritance of acquired characteristics *page 281*

natural selection *page 288*

population *page 288*

uniformitarianism *page 280*

vestigial structure *page 284*

THINKING THROUGH THE CONCEPTS

1. Selection acts on individuals, but only populations evolve. Explain why this is true.

2. Distinguish between catastrophism and uniformitarianism. How did these hypotheses contribute to the development of modern evolutionary theory?

3. Describe Lamarck's theory of inheritance of acquired characteristics. Why is it invalid?

4. What is natural selection? Describe how natural selection might have caused differential reproduction among the ancestors of a fast-swimming predatory fish (such as the barracuda).

5. Describe how evolution occurs through the interactions among the following: the reproductive potential of a species, the normally constant size of natural populations, variation among individuals of a species, natural selection, and inheritance.

6. What is convergent evolution? Give an example.

7. How do biochemistry and molecular genetics contribute to the evidence that evolution occurred?

APPLYING THE CONCEPTS

1. Does evolution through natural selection produce "better" organisms in an absolute sense? Are we climbing the "ladder of Nature"? Defend your answer.

2. Both the study of fossils and the idea of divine creation have had an impact on evolutionary thought. Discuss why one is considered scientific endeavor and the other is not.

3. In evolutionary terms, "success" can be defined in many different ways. What are the most successful organisms you can think of in terms of (a) persistence over time, (b) sheer numbers of individuals alive now, (c) numbers of species, and (d) geographical range?

4. In what sense are humans currently acting as agents of selection on other species? Name some organisms that are *favored* by the environmental changes humans cause.

5. Darwin and Wallace's discovery of natural selection is one of the great revolutions in scientific thought. Some scientific revolutions spill over and affect the development of philosophy and religion. Is this true of evolution? Does (or should) the idea of evolution by natural selection affect the way humans view their place in the world?

FOR MORE INFORMATION

Appleman, P. (ed.) *Darwin, a Norton Critical Edition.* New York: Norton, 2001. An excellent collection of early and modern writing about evolution, including excerpts from *The Origin of Species.*

Darwin, C. *On the Origin of Species by Means of Natural Selection.* Garden City, NY: Doubleday, 1960 (originally published in 1859). An impressive array of evidence amassed to convince a skeptical world.

Dennet, D. *Darwin's Dangerous Idea.* New York: Simon & Schuster, 1995. A philosopher's view of Darwinian ideas and their application to the world outside biology. A thought-provoking book that seems to have inspired admiration and condemnation in roughly equal proportions.

Eiseley, L. C. "Charles Darwin." *Scientific American*, February 1956. An essay on the life of Darwin by one of his foremost American biographers. Even if you need no introduction to Darwin, read this as an introduction to Eiseley, author of many marvelous essays.

Gould, S. J. *Ever Since Darwin*, 1977; *The Panda's Thumb*, 1980; and *The Flamingo's Smile*, 1985. New York: Norton. A series of witty, imaginative, and informative essays about evolution, mostly from *Natural History* magazine.

Zimmer, C. *Evolution: The Triumph of an Idea.* New York: HarperCollins, 2001. A well-written, beautifully illustrated survey of evolutionary biology.

How Organisms Evolve

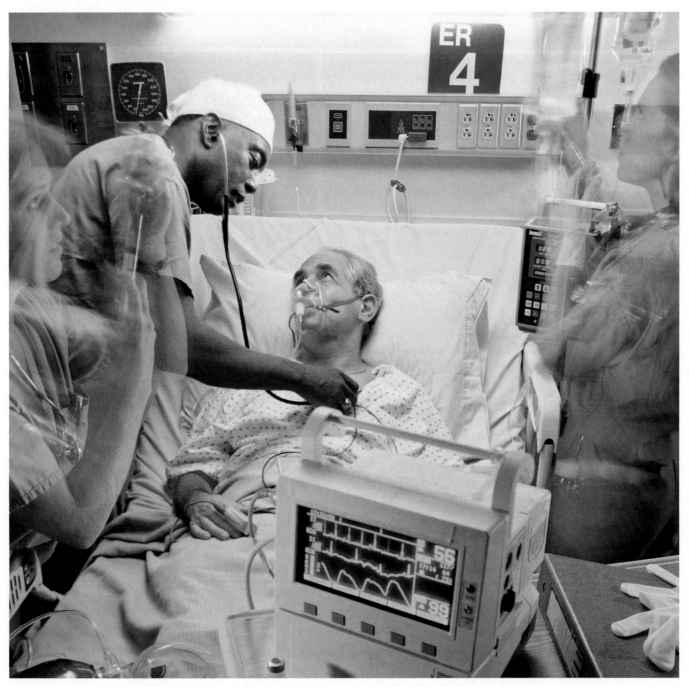

We think of hospitals as places in which to seek protection from disease, but they also foster the evolution of drug-resistant supergerms.

CASE STUDY EVOLUTION OF A MENACE

AS SHE WATCHED Jim leave the clinic, Dr. Lawson sighed and shook her head sadly. Jim, a middle-aged homeless man who was being treated for tuberculosis, was unlikely to recover. Lab tests had revealed that the bacteria infecting Jim were resistant to four different antibiotic drugs commonly used to treat the disease. Multidrug-resistant tuberculosis is extremely difficult to treat; it can be cured only by long-term treatment with a combination of several drugs. Even after such treatment, some cases prove to be incurable, and Dr. Lawson knew that Jim's treatment was especially unlikely to suc-

ceed. Jim lived on the streets, often struggling to survive, and was not likely to complete the treatment: five pills each day for two years. To make matters worse, Jim would no doubt sleep in overpopulated shelters, keep warm in crowded subway stations, and perhaps even spend time in packed city jails. In such places, he would likely pass tuberculosis bacteria to others, thereby hastening the spread of his multidrug-resistant strain.

Multidrug-resistant tuberculosis is a frightening and increasingly widespread threat to public health in many parts of the world, including the United States.

Drug resistance is also becoming common in other dangerous bacteria as well, including those that cause food poisoning, blood poisoning, dysentery, pneumonia, gonorrhea, meningitis, and urinary tract infections. We are experiencing a global onslaught of resistant "supergerms," and are facing the specter of diseases that cannot be cured, even by our best medicines. In order to understand how this crisis arose and to devise a strategy to resolve it, we must have a clear understanding of the mechanisms by which populations evolve.

15.1   HOW ARE POPULATIONS, GENES, AND EVOLUTION RELATED?

If you live in an area with a seasonal climate and you own a dog or cat, you have probably noticed that your pet's fur gets thicker and heavier as winter approaches. Has the animal evolved? No. The changes that we see in an individual organism over the course of its lifetime are not evolutionary changes. Instead, evolutionary changes occur from generation to generation, causing descendants to be different from their ancestors.

Furthermore, we can't detect evolutionary change across generations by looking at a single set of parents and offspring. For example, if you observed that a 6-foot-tall man had an adult son who stood 5 feet tall, could you conclude that humans were evolving to become shorter? Obviously not. Rather, if you wanted to learn about evolutionary change in human height, you would begin by measuring many humans of many generations to see if the average height is changing over time. Clearly, evolution is a property not of individuals but of populations (a **population** is a group that includes all the members of a species living in a given area).

The recognition that evolution is a population-level phenomenon was one of Darwin's key insights. But populations are composed of individuals, and the actions and fates of individuals determine which characteristics will be passed to descendant populations. In this fashion, inheritance provides the link between the lives of individual organisms and the evolution of populations. We will therefore begin our discussion of the processes of evolution by reviewing some principles of genetics as they apply to individuals. We will then extend those principles to the genetics of populations.

Genes and the Environment Interact to Determine Traits

Each cell of every organism contains genetic information encoded in the DNA of its chromosomes. Recall from Chapter 9 that a *gene* is a segment of DNA located at a particular place on a chromosome.

The sequence of nucleotides in a gene encodes the sequence of amino acids in a protein, usually an enzyme that catalyzes a particular reaction in the cell. At a given gene's location, different members of a species may have slightly different nucleotide sequences, called *alleles*. Different alleles generate different forms of the same enzyme. In this way, various alleles of the gene that influences eye color in humans, for example, help produce eyes that are brown, or blue, or green, and so on.

In any population of organisms, there are usually two or more alleles of each gene. An individual of a diploid species whose alleles of a particular gene are both the same is *homozygous* for that gene, and an individual with different alleles for that gene is *heterozygous*. The specific alleles borne on an organism's chromosomes (its *genotype*) interact with the environment to influence the development of its physical and behavioral traits (its *phenotype*).

Let's illustrate these principles with an example. A black hamster's coat is colored black because a chemical reaction in its hair follicles produces a black pigment. When we say that a hamster has the allele for a black coat, we mean that a particular stretch of DNA on one of its chromosomes contains a sequence of nucleotides that codes for the enzyme that catalyzes this reaction. A hamster with the allele for a brown coat has a different sequence of nucleotides at the corresponding chromosomal position. That different sequence codes for an enzyme that cannot produce black pigment. If a hamster is homozygous for the black allele or is heterozygous (one black allele and one brown allele), its fur contains the pigment and is black. But if a hamster is homozygous for the brown allele, its hair follicles produce no black pigment and its coat is brown (**FIG. 15-1**). Because the hamster's coat is black even when only one copy of the black allele is present, the black allele is considered *dominant* and the brown allele *recessive*.

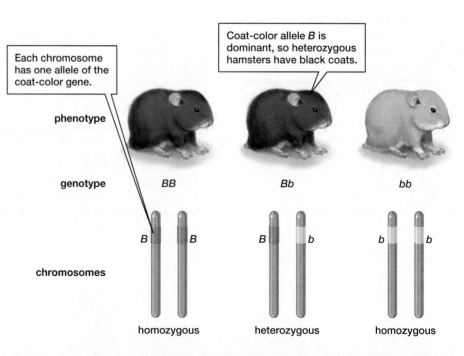

FIGURE 15-1 Alleles, genotype, and phenotype in individuals
An individual's particular combination of alleles is its genotype. The word "genotype" can refer to the alleles of a single gene (as shown here), to a set of genes, or to all of an organism's genes. An individual's phenotype is determined by its genotype and environment. Phenotype can refer to a single trait, a set of traits, or all of an organism's traits.

The Gene Pool Is the Sum of the Genes in a Population

We can often deepen our understanding of a subject by looking at it from more than one perspective. In studying evolution, looking at the process from the viewpoint of a gene has proven to be an enormously effective tool. In particular, evolutionary biologists have made excellent use of the tools of a branch of genetics called population genetics, which deals with the frequency, distribution, and inheritance of alleles in populations. To take advantage of this powerful aid to understanding evolution, you will need to learn a few of the basic concepts of population genetics.

Population genetics defines the **gene pool** as the sum of all the genes in a population. In other words, the gene pool consists of all the alleles of all the genes in all the individuals of a population. Each particular gene can also be considered to have its own gene pool, which consists of all the alleles of that specific gene in a population (**FIG. 15-2**). If we added up all the copies of each allele of that gene in all the individuals in a population, we could determine the relative proportion of each allele, a number called the **allele frequency**. For example, the population of 25 hamsters portrayed in Figure 15-2 contains 50 alleles of the gene that controls coat color (because hamsters are diploid and each hamster thus has two copies of each gene). Twenty of those 50 alleles are of the type that codes for black coats, so the frequency of that allele in the population is 0.40 (or 40%), because 20/50 = 0.40.

Evolution Is the Change of Allele Frequencies Within a Population

A casual observer might define evolution on the basis of changes in the outward appearance or behaviors of the members of a population. A population geneticist, however, looks at a population and sees a gene pool that just happens to be divided into the packages that we call individual organisms. So any outward changes that we observe in the individuals that make up the population can also be viewed as the visible expression of underlying changes to the gene pool. A population geneticist therefore defines evolution as the changes in allele frequencies that occur in a gene pool over time. Evolution is a change in the genetic makeup of populations over generations.

The Equilibrium Population Is a Hypothetical Population in Which Evolution Does Not Occur

It is easier to understand what causes populations to evolve if we begin by considering the characteristics of a population that would *not* evolve. In 1908, English mathematician Godfrey H. Hardy and German physician Wilhelm

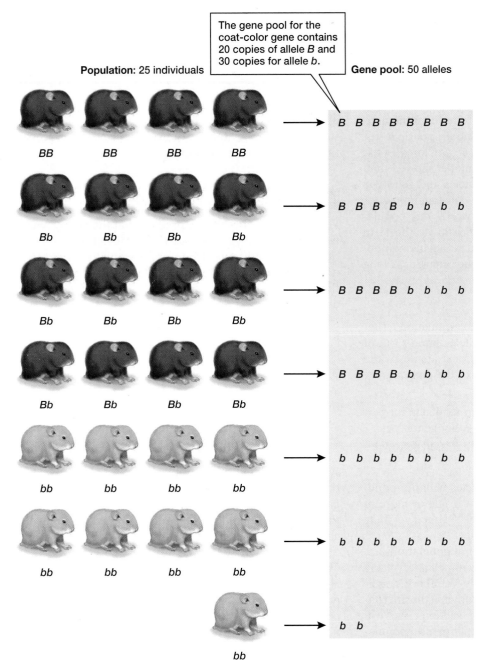

Population: 25 individuals

The gene pool for the coat-color gene contains 20 copies of allele *B* and 30 copies for allele *b*.

Gene pool: 50 alleles

BB BB BB BB → B B B B B B B B

Bb Bb Bb Bb → B B B B b b b b

Bb Bb Bb Bb → B B B B b b b b

Bb Bb Bb Bb → B B B B b b b b

bb bb bb bb → b b b b b b b b

bb bb bb bb → b b b b b b b b

bb → b b

FIGURE 15-2 A gene pool
In diploid organisms, each individual in a population contributes two alleles of each gene to the gene pool.

The Hardy-Weinberg principle states that allele frequencies will remain constant over time in the gene pool of a large population in which there is random mating but no mutation, no gene flow, and no natural selection. In addition, Hardy and Weinberg showed that if allele frequencies do not change in an equilibrium population, the proportion of individuals with a particular genotype will also remain constant.

To better understand the relationship between allele frequencies and the occurrence of genotypes, picture an equilibrium population whose members carry a gene that has two alleles, A_1 and A_2. Note that each individual in this population must carry one of three possible diploid genotypes (combinations of alleles): A_1A_1, A_1A_2, or A_2A_2.

Suppose that in our population's gene pool the frequency of allele A_1 is p and the frequency of allele A_2 is q. Hardy and Weinberg demonstrated that if allele frequencies are given as p and q, then the proportions of the different genotypes in the population can be calculated as

Proportion of individuals with genotype $A_1A_1 = p^2$

Proportion of individuals with genotype $A_1A_2 = 2pq$

Proportion of individuals with genotype $A_2A_2 = q^2$

For example, if, in our population's gene pool, 70% of the alleles of a gene are A_1 and 30% are A_2 (that is, $p = 0.7$ and $q = 0.3$), then genotype proportions would be

Proportion of individuals with genotype $A_1A_1 = 49\%$ (because $p^2 = 0.7 \times 0.7 = 0.49$)

Proportion of individuals with genotype $A_1A_2 = 42\%$ (because $2pq = 2 \times 0.7 \times 0.3 = 0.42$)

Proportion of individuals with genotype $A_2A_2 = 9\%$ (because $q^2 - 0.3 \times 0.3 = 0.09$)

Because every member of the population must possess one of the three genotypes, the three proportions must always add up to one. For this reason, the expression that relates allele frequency to genotype proportions may written as

$$p^2 + 2pq + q^2 = 1$$

where the three terms on the left side of the equation represent the three genotypes.

Weinberg independently developed a simple mathematical model now known as the **Hardy-Weinberg principle** (for more information about the model, see "A Closer Look: The Hardy-Weinberg Principle"). This model showed that, under certain conditions, allele frequencies and genotype frequencies in a population will remain constant no matter how many generations pass. In other words, this population will not evolve. Population geneticists use the term **equilibrium population** for this idealized, nonevolving population in which allele frequencies do not change, as long as the following conditions are met:

- There must be no mutation.
- There must be no **gene flow** between populations. That is, there must be no movement of alleles into or out of the population (as would be caused, for example, by the movement of organisms into or out of the population).
- The population must be very large.
- All mating must be random, with no tendency for certain genotypes to mate with specific other genotypes.
- There must be no natural selection. That is, all genotypes must reproduce with equal success.

Under these conditions, allele frequencies within a population will remain the same indefinitely. If one or more of these conditions is violated, then allele frequencies may change; that is, the population will evolve.

As you might expect, few if any natural populations are truly in equilibrium. What, then, is the importance of the Hardy-Weinberg principle? The Hardy-Weinberg conditions are useful starting points for studying the mechanisms of evolution. In the following sections, we will examine some of the conditions, show that natural populations often fail to meet them, and illustrate the consequences of such failures. In this way, we can better understand both the inevitability of evolution and the processes that drive evolutionary change.

15.2 WHAT CAUSES EVOLUTION?

Population genetics theory predicts that the Hardy-Weinberg equilibrium can be disturbed by deviations from any of its five conditions. Therefore, we might predict five major causes of evolutionary change: mutation, gene flow, small population size, nonrandom mating, and natural selection.

Mutations Are the Original Source of Genetic Variability

A population remains in genetic equilibrium only if there are no **mutations** (changes in DNA sequence). Most mutations occur during cell division, when a cell must make a copy of its DNA. Sometimes, there are errors in the copying process and the copied DNA does not match the original. Most such errors are quickly corrected by cellular systems that identify and repair DNA copying mistakes, but some changes in nucleotide sequence slip past the repair systems. An unrepaired mutation in a cell that gives rise to gametes may be passed to offspring and enter the gene pool of a population.

Inherited Mutations Are Rare but Important

How significant is mutation in changing the gene pool of a population? For any given gene, only a tiny proportion of a population inherits a mutation from the previous generation. For example, a mutant version of a typical human gene will appear in only about one out of every 100,000 ga-

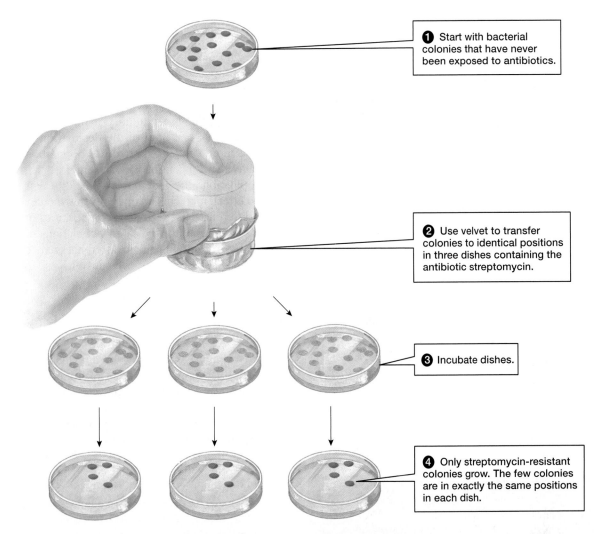

❶ Start with bacterial colonies that have never been exposed to antibiotics.

❷ Use velvet to transfer colonies to identical positions in three dishes containing the antibiotic streptomycin.

❸ Incubate dishes.

❹ Only streptomycin-resistant colonies grow. The few colonies are in exactly the same positions in each dish.

FIGURE 15-3 Mutations occur spontaneously
This experiment demonstrates that mutations occur spontaneously and not in response to environmental pressures. When bacterial colonies that have never been exposed to antibiotics are exposed to the antibiotic streptomycin, only a few colonies grow. The observation that these surviving colonies grow in exactly the same positions in all dishes shows that the mutations for resistance to streptomycin were present in the original dish before exposure to the environmental pressure, streptomycin. **QUESTION** If it were true that mutations *do* occur in response to the presence of antibiotic, how would the result of this experiment have differed from the actual result?

metes produced and, because new individuals are formed by the fusion of two gametes, in only about one of every 50,000 newborns. Therefore, mutation by itself generally causes only very small changes in the frequency of any particular allele.

Despite the rarity of inherited mutations of any particular gene, the cumulative effect of mutations is essential to evolution. Most organisms have a large number of different genes, so even if the rate of mutation is low for any one gene, the sheer number of possibilities means that each new generation of a population is likely to include some mutations. For example, humans have about 30,000 different genes, so each person carries about 60,000 alleles. Thus, even if each allele has, on average, only a one in 100,000 chance of mutation, most newborn individuals will probably carry one or two mutations overall. These mutations

are the source of new alleles, new variations on which other evolutionary processes can work. As such, they are the foundation of evolutionary change. Without mutations there would be no evolution.

Mutations Are Not Goal Directed

A mutation does not arise as a result of, or in anticipation of, environmental necessities. A mutation simply happens and may in turn produce a change in a structure or function of an organism. Whether that change is helpful or harmful or neutral, now or in the future, depends on environmental conditions over which the organism has little or no control (**FIG. 15-3**). The mutation provides a potential for evolutionary change. Other processes, especially natural selection, may act to spread the mutation through the population or to eliminate it from the population.

Gene Flow Between Populations Changes Allele Frequencies

The movement of alleles between populations, known as gene flow, changes how alleles are distributed among populations. When individuals move from one population to another and interbreed at the new location, alleles are transferred from one gene pool to another. In baboons, for example, individuals routinely move to new populations. Baboons live in social groupings called *troops*. Within each troop, all the females mate with a handful of dominant males. Juvenile males usually leave the troop. If they are lucky, they join and perhaps even become dominant in another troop. Thus, the male offspring of one troop carry alleles to the gene pools of other troops.

Although movement of individuals is a common cause of gene flow, alleles can move between populations even if organisms do not. Flowering plants, for example, cannot move, but their seeds and pollen can (**FIG. 15-4**). Pollen, which contains sperm cells, may be carried long distances by wind or by animal pollinators. If the pollen ultimately reaches the flowers of a different population of its species, it may fertilize eggs and add its collection of alleles to the local gene pool. Similarly, seeds may be borne by wind or water or animals to distant locations where they can germinate to become part of a population far from their place of origin.

FIGURE 15-4 Pollen can be an agent of gene flow
Pollen, drifting on the wind, can carry alleles from one population to another.

The main evolutionary effect of gene flow is to increase the genetic similarity of different populations of a species. To see why, picture two glasses, one containing fresh water and the other salt water. If a few spoonfuls are transferred from the seawater glass to the freshwater glass, the liquid in the freshwater glass becomes saltier. In the same way, movement of alleles from one population to another tends to change the gene pool of the destination population so that it is more similar to the source population.

If alleles move continually back and forth between different populations, the gene pools of the different populations will, in effect, mix. This mixing prevents the development of large differences in the genetic compositions of the populations. But if gene flow between populations of a species is blocked, the resulting genetic differences may grow so large that one of the populations becomes a new species, a process we will discuss in Chapter 16.

Allele Frequencies May Drift in Small Populations

Allele frequencies in populations can be changed by chance events. For example, if bad luck prevents some members of a population from reproducing, their alleles will ultimately be removed from the gene pool, altering its makeup. What kinds of bad-luck events can randomly prevent some individuals from reproducing? Seeds can fall into a pond and never sprout, flowers can be destroyed by a hailstorm, organisms can be killed by a fire or by a volcanic eruption. Any event that cuts lives short at random or otherwise allows only a random subset of a population to reproduce can cause random changes in allele frequencies (**FIG. 15-5**). The process by which chance events change allele frequencies is called **genetic drift**.

To see how genetic drift works, imagine a population of 20 hamsters in which the frequency of the black coat-color allele B is 0.50 and the frequency of the brown coat-color allele b is 0.50 (Fig. 15-5, top). If all of the hamsters in the population were to interbreed to yield another population of 20 animals, the frequencies of the two alleles would not change in the next generation. But if we instead allow only two, randomly chosen hamsters to breed and become the parents of the next generation of 20 animals, allele frequencies might be quite different in generation 2 (Fig. 15-5, center). And if breeding in the second generation were again restricted to two randomly chosen hamsters, allele frequencies might change again in the third generation (Fig. 15-5, bottom). Allele frequencies will continue to change in random fashion for as long as reproduction is restricted to a random subset of the population. Notice that the changes caused by genetic drift can include the disappearance of an allele from the population.

Population Size Matters

Genetic drift occurs to some extent in all populations, but it occurs more rapidly and has a bigger effect in small populations than in large ones. If a population is sufficiently large, chance events are unlikely to significantly alter its genetic composition, because random removal of a few individuals'

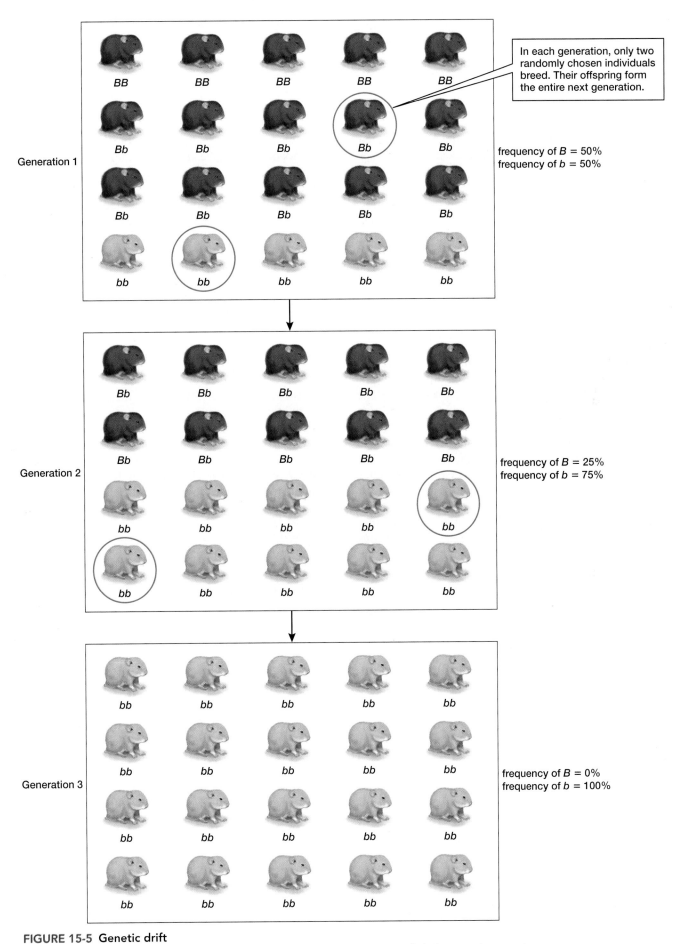

FIGURE 15-5 Genetic drift

If chance events prevent some members of population from reproducing, allele frequencies can change randomly. **QUESTION** Explain why the distribution of genotypes in Generation 2 is as shown.

alleles won't have a big impact on allele frequencies in the population as a whole. In a small population, however, a particular allele may be carried by only a few organisms. Chance events could eliminate most or all examples of such an allele from the population.

To see how population size affects genetic drift, imagine two populations of amoebas in which each amoeba is either red or blue, and color is controlled by two alleles (*A* and *a*) of a gene. Half of the amoebas in each of our two populations are red and half are blue. One population, however, has only four individuals in it, whereas the other has 10,000.

Now let's picture reproduction in our imaginary populations. Let's select at random half of the individuals in each population and allow them to reproduce by binary fission. To do so, each reproducing amoeba splits in half to yield two amoebas, each of which is the same color as the parent. In the large population, 5000 amoebas reproduce, yielding a new generation of 10,000. What are the chances that all 10,000 members of the new generation will be red? Just about nil. In fact, it would be extremely unlikely for even 3000 amoebas to be red or for 7000 to be red. The most likely outcome is that about half will be red and half blue, just as in the original population. In this large population, then, we would not expect a major change in allele frequencies from generation to generation.

One way to test this prediction is to write a computer program that simulates how the allele frequencies of the alleles could change over generations. **FIGURE 15-6a** shows the results from four runs of such a simulation. Notice that the frequency of allele *A*, encoding red color, remains

close to 0.5, consistent with the expectation that half of the amoebas would be red.

In the small population, the situation is different. Only two amoebas reproduce, and there is a 25% chance that both reproducers will be red. (This outcome is as likely as flipping two coins and having both come up heads.) If only red amoebas reproduce, then the next generation will consist entirely of red amoebas—a relatively likely outcome. It is thus possible, within a single generation, for the allele for blue color to disappear from the population.

FIGURE 15-6b shows the fate of allele *A* in four runs of a simulation of our small population. In one of the four runs (red line), allele *A* reaches a frequency of 1.0 (100%) in the second generation, meaning that all the amoebas in the third and following generations are red. In another run, the frequency of *A* drifts to 0.0 in the third generation (blue line), and the population subsequently is all blue. Thus, one of the two amoeba phenotypes disappeared in half of the simulations.

A Population Bottleneck Is an Example of Genetic Drift

Two causes of genetic drift, called the *population bottleneck* and the *founder effect*, further illustrate the effect that small population size may have on the allele frequencies of a species. In a **population bottleneck**, a population is drastically reduced—because of a natural catastrophe or overhunting, for example. Then, only a few individuals are available to contribute genes to the next generation. Population bottlenecks can rapidly change allele frequencies and can reduce genetic variability by eliminating alleles (**FIG. 15-7a**). Even if the population later increases, the ge-

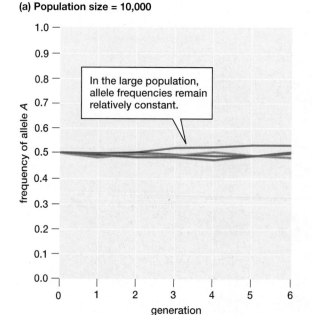

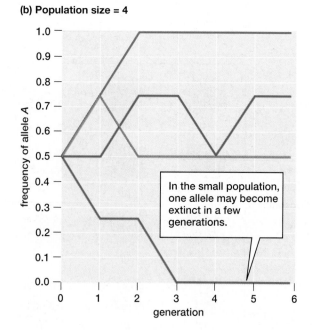

FIGURE 15-6 The effect of population size on genetic drift
Each colored line represents one computer simulation of the change over time in the frequency of allele *A* in a **(a)** large or **(b)** small population in which two alleles, *A* and *a*, were initially present in equal proportions, and in which randomly chosen individuals reproduced. **EXERCISE** Sketch a graph that shows the result you would predict if the simulation were run four times with a population size of 20.

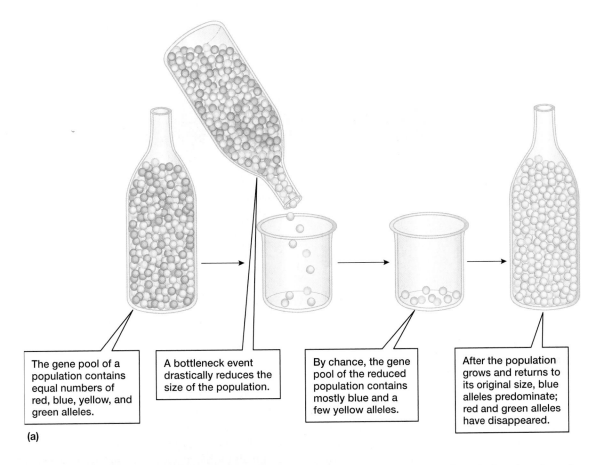

The gene pool of a population contains equal numbers of red, blue, yellow, and green alleles.

A bottleneck event drastically reduces the size of the population.

By chance, the gene pool of the reduced population contains mostly blue and a few yellow alleles.

After the population grows and returns to its original size, blue alleles predominate; red and green alleles have disappeared.

(a)

(b)

FIGURE 15-7 Population bottlenecks reduce variation
(a) A population bottleneck may drastically reduce genetic and phenotypic variation because the few organisms that survive may carry similar sets of alleles. (b) The northern elephant seal passed through a population bottleneck in the recent past, resulting in an almost total loss of genetic diversity. **QUESTION** If a population grows large again after a bottleneck, genetic diversity will eventually increase. Why?

netic effects of the bottleneck may remain for hundreds or thousands of generations.

Loss of genetic variability due to bottlenecks has been documented in numerous species, including the northern elephant seal (**FIG. 15-7b**), The elephant seal was hunted almost to extinction in the 1800s; by the 1890s, only about 20 still survived. Dominant male elephant seals typically monopolize breeding; therefore, with a single male mating with a stable group of females, one male may have fathered all the offspring at this extreme bottleneck point. Since then, elephant seals have increased in number to about 30,000 individuals, but biochemical analysis shows that all northern elephant seals are genetically almost identical. Other species of seals, whose populations have

always remained large, exhibit much more genetic variability. The rescue of the northern elephant seal from extinction is rightly regarded as a triumph of conservation. With very little genetic variation, however, the elephant seal has much less potential to evolve in response to environmental changes. No matter how many elephant seals there are, the species must be considered to be threatened with extinction.

Isolated Founding Populations May Produce Bottlenecks

A special case of a population bottleneck is the **founder effect**, which occurs when isolated colonies are founded by a small number of organisms. A flock of birds, for instance, that becomes lost during migration or is blown off course by a storm may settle on an isolated island. The small founder group may, by chance, have allele frequencies that are very different from the frequencies of the parent population. If they do, the gene pool of the future population in the new location will be quite unlike that of the larger population from which it sprang. For example, a set of genetic defects known as Ellis–van Creveld syndrome (FIG. 15-8) is far more common among the Amish inhabitants of Lancaster County, Pennsylvania, than among the general population. Today's Lancaster County Amish are descended from only 200 or so eighteenth-century immigrants, and one couple among these immigrants is known to have carried the Ellis–van Creveld allele. In such a small founder population, this single occurrence meant that the allele was carried by a comparatively high proportion of the Amish founder population (1 or 2 carriers out of 200, versus perhaps 1 in 1000 in the general population). This high initial allele frequency, combined with subsequent genetic drift, has led to extraordinarily high levels of Ellis–van Creveld syndrome among this Amish group.

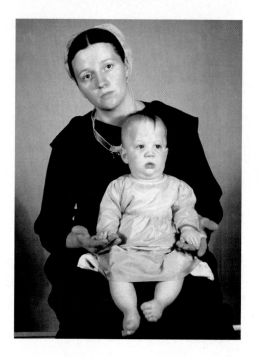

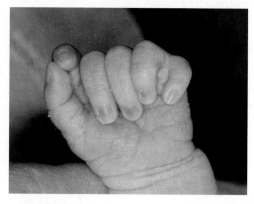

FIGURE 15-8 A human example of the founder effect
An Amish woman with her child (left), who suffers from a set of genetic defects known as Ellis–van Creveld syndrome. Symptoms include short arms and legs, extra fingers (right), and, in some cases, heart defects. The founder effect accounts for the prevalence of Ellis–van Creveld syndrome among the Amish residents of Lancaster County, Pennsylvania.

Mating Within a Population Is Almost Never Random

Nonrandom mating by itself will not alter allele frequencies in a population. Nonetheless, it can have large effects on the distribution of different genotypes, and thus on the distribution of phenotypes, in the population. Certain genotypes may become more common, which can affect the direction of natural selection.

The effects of nonrandom mating can be important, because organisms seldom mate strictly randomly. For example, many organisms have limited mobility and tend to remain near their place of birth, hatching, or germination. In such species, most of the offspring of a given parent live in the same area and thus, when they reproduce, there is a good chance that they will be related to their reproductive partners. Sexual reproduction among relatives is call *inbreeding*.

Because relatives are genetically similar, inbreeding tends to increase the number of individuals that inherit the same alleles from both parents and are therefore homozygous for many genes. This increase in homozygotes can have harmful effects, such as increased occurrence of genetic diseases or defects. Many gene pools include harmful recessive alleles that persist in the population because their negative effects are masked in heterozygous carriers (which have only a single copy of the harmful allele). In-

breeding, however, increases the odds of producing homozygous offspring with two copies of the harmful allele.

In animals, nonrandom mating can also arise if they choose to mate with certain individuals of their species rather than with others. The snow goose is a case in point. Individuals of this species come in two "color phases"; some snow geese are white, while others are blue-gray (FIG. 15-9). Although both white and blue-gray geese belong to the same species, mate choice is not random with respect to color. The birds exhibit a strong tendency to mate with a partner of the same color. This preference for mates that are similar is known as *assortative mating*.

All Genotypes Are Not Equally Beneficial

In a hypothetical equilibrium population, individuals of all genotypes survive and reproduce equally well; that is, no genotype has any advantage over the others. This condi-

FIGURE 15-9 Nonrandom mating among snow geese
Snow geese, which have either white plumage or blue-gray plumage, are most likely to mate with other birds of the same color.

tion, however, is probably met only rarely, if ever, in real populations. Even though some alleles are neutral, in the sense that organisms possessing any of several alleles are equally likely to survive and reproduce, clearly not all alleles are neutral in all environments. Any time an allele provides, in Alfred Russel Wallace's words, "some little superiority," natural selection favors the individuals who possess it. That is, those individuals have higher reproductive success. This phenomenon is illustrated by an example concerning an antibiotic drug.

Antibiotic Resistance Evolves by Natural Selection

The antibiotic penicillin first came into widespread use during World War II, when it was used to combat infections in wounded soldiers. Suppose that an infantryman, brought to a field hospital after suffering a gunshot wound in his arm, develops a bacterial infection in that arm. A medic sizes up the situation and resolves to treat the wounded soldier with an intravenous drip of penicillin. As the antibiotic courses through the soldier's blood vessels, millions of bacteria die before they can reproduce. A few bacteria, however, carry a rare allele that codes for an enzyme that destroys any penicillin that comes into contact with the bacterial cell. (This allele is a variant of a gene that normally codes for an enzyme that breaks down the bacterium's waste products.) The bacteria carrying this rare allele are able to survive and reproduce, and their offspring inherit the penicillin-destroying allele. After a few generations, the frequency of the penicillin-destroying allele has soared to nearly 100%, and the frequency of the normal, waste-processing allele has declined to near zero. As a result of natural selection imposed by the antibiotic's killing power, the population of bacteria within the soldier's body has evolved. The gene pool of the population has changed, and natural selection, in the form of bacterial destruction by penicillin, has caused the change.

Penicillin Resistance Illustrates Key Points About Evolution

The example of penicillin resistance highlights some important features of natural selection and evolution.

Natural selection does not cause genetic changes in individuals. The allele for penicillin resistance arose spontaneously, long before penicillin was dripped into the soldier's vein. Penicillin did not cause resistance to appear; its presence merely favored the survival of bacteria with penicillin-destroying alleles over that of bacteria with waste-processing alleles.

(a)

(b)

FIGURE 15-10 A compromise between opposing pressures (a) A male giraffe with a long neck is at a definite advantage in combat to establish dominance. (b) But a giraffe's long neck forces it to assume an extremely awkward and vulnerable position when drinking. Thus, drinking and male-male contests place opposing evolutionary pressures on neck length.

Natural selection acts on individuals, but it is populations that are changed by evolution. The agent of natural selection, penicillin, acted on individual bacteria. As a result, some individuals reproduced and some did not. However, it was the population as a whole that evolved as its allele frequencies changed.

Evolution is change in allele frequencies of a population, owing to unequal success at reproduction among organisms bearing different alleles. In evolutionary terminology, the **fitness** of an organism is measured by its reproductive success. In our example, the penicillin-resistant bacteria had greater fitness than the normal bacteria did, because the resistant bacteria produced greater numbers of viable (able to survive) offspring.

Evolution is not progressive; it does not make organisms "better." The traits favored by natural selection change as the environment changes. The resistant bacteria were favored only because of the presence of penicillin in the soldier's body. At a later time, when the environment of the soldier's body no longer contained penicillin, the resistant bacteria may have been at a disadvantage relative to other bacteria that could process waste more effectively. Similarly, the long necks of male giraffes are helpful when the animals battle to establish dominance, but are a hindrance to drinking (**FIG. 15-10**). The length of male giraffe necks represents an evolutionary compromise between the advantage of being able to win contests with other males and the disadvantage of vulnerability while drinking water. (The

Table 15-1 **Causes of Evolution**	
Process	**Consequence**
Mutation	Creates new alleles; increases variability
Gene flow	Increases similarity of different populations
Genetic drift	Causes random change of allele frequencies; can eliminate alleles
Nonrandom mating	Changes genotype frequencies but not allele frequencies
Natural and sexual selection	Increases frequency of favored alleles; can produce adaptations

necks of female giraffes are long—though not as long as male necks—because successful males pass the alleles for long necks to daughters as well as to sons.)

Table 15-1 summarizes the different causes of evolution.

15.3 HOW DOES NATURAL SELECTION WORK?

Natural selection is not the *only* evolutionary force. As we have seen, mutation provides variability in heritable traits, and the chance effects of genetic drift may change allele frequencies. Further, evolutionary biologists are now beginning to appreciate the power of random catastrophes in shaping the history of life on Earth; massively destructive events may exterminate thriving and failing species alike. Nevertheless, it is natural selection that shapes the evolution of populations as they adapt to their changing environment. For this reason, we will examine natural selection in more detail.

Natural Selection Stems from Unequal Reproduction

To most people, the words **natural selection** are synonymous with the phrase "survival of the fittest." Natural selection evokes images of wolves chasing caribou, of lions snarling angrily in competition over a zebra carcass. Natural selection, however, is not about survival alone. It is also about reproduction. It is certainly true that if an organism is to reproduce, it must survive long enough to do so. In some cases, it is also true that a longer-lived organism has more chances to reproduce. But no organism lives forever, and the only way that its genes can continue into the future is through successful reproduction. When an organism dies without reproducing, its genes die with it. An organism that reproduces lives on, in a sense, through the genes that it has passed to its offspring. Therefore, although evolutionary biologists often discuss survival, partly because survival is usually easier to observe than reproduction, the main issue of natural selection is *differences in reproduction*: individuals bearing certain alleles leave more offspring (who inherit those alleles) than do other individuals with different alleles.

Natural Selection Acts on Phenotypes

Although we have defined evolution as changes in the genetic composition of a population, it is important to recognize that natural selection cannot act directly on the genotypes of individual organisms. Rather, natural selection acts on phenotypes, the structures and behaviors displayed by the members of a population. This selection of phenotypes, however, inevitably affects the genotypes present in a population, because phenotypes and genotypes are closely tied. For example, we know that a pea plant's height is strongly influenced by the plant's alleles of certain genes. If a population of pea plants were to encounter environmental conditions that favored taller plants, then taller plants would leave more offspring. These offspring would carry the alleles that contributed to their parents' height. Thus, if natural selection favors a particular phenotype, it will necessarily also favor the underlying genotype.

Some Phenotypes Reproduce More Successfully Than Others

As we have seen, natural selection simply means that some phenotypes reproduce more successfully than others do. This simple process is such a powerful agent of change because only the "best" phenotypes pass traits to subsequent generations. But what makes a phenotype the best? Successful phenotypes are those that have the best adaptations to their particular environment. **Adaptations** are characteristics that help an individual survive and reproduce.

An Environment Has Nonliving and Living Components

Individual organisms must cope with an environment that includes not only physical factors but also the other organisms with which the individual interacts. The nonliving (*abiotic*) component of the environment includes such factors as climate, availability of water, and minerals in the soil. The abiotic environment establishes the "bottom line" requirements that an organism must meet to survive and reproduce. However, many of the adaptations that we see in modern organisms have arisen because of interactions with other organisms—the living (*biotic*) component of the environment. As Darwin wrote, "The structure of every organic being is related … to that of all other organic beings, with which it comes into competition for food or residence, or from which it has to escape, or on which it preys." A simple example illustrates this concept.

A buffalo grass plant sprouts in a small patch of soil in the eastern Wyoming plains. Its roots must be able to take up enough water and minerals for growth and reproduction, and to that extent it must be adapted to its abiotic environment. But even in the dry prairies of Wyoming, this requirement is relatively trivial, provided that the plant is alone and protected in its square yard of soil. In reality, however, many other plants—other buffalo grass plants as well as other grasses, sagebrush bushes, and annual wildflowers—also sprout in that same patch of soil. If our buffalo grass is to

FIGURE 15-11 Competition between males favors evolution of structures for ritual combat Two male bighorn sheep spar during the fall mating season. In many species, the losers of such contests are unlikely to mate, while winners enjoy tremendous reproductive success. **QUESTION** Imagine that you studied a population of bighorn sheep and were able to identify the father and mother of each lamb born. Would the difference in number of offspring between the most reproductively successful adult and the least successful one be greater for males or for females?

survive, it must compete with the other plants for resources. Its long, deep roots and efficient methods of mineral uptake have evolved not so much because the plains are dry as because the buffalo grass must share the dry prairies with other plants. Further, buffalo grass must also coexist with animals that wish to eat it, such as the cattle that graze the prairie (and the bison that grazed it in the past). As a result, buffalo grass is extremely tough. Silica compounds reinforce its leaves, an adaptation that discourages grazing. Over time, tougher, hard-to-eat plants survived better and reproduced more than did less-tough plants—another adaptation to the biotic environment.

Competition Acts as an Agent of Selection

As the buffalo grass example shows, one of the major agents of natural selection in the biotic environment is **competition** with other organisms for scarce resources. Competition for resources is most intense among members of the same species. As Darwin wrote in *On the Origin of Species*, "The struggle almost invariably will be most severe between the individuals of the same species, for they frequent the same districts, require the same food, and are exposed to the same dangers." In other words, no competing organism has such similar requirements for survival as does another member of the same species. Different species may also compete for the same resources, although generally to a lesser extent than do individuals within a species.

Both Predator and Prey Act as Agents of Selection

When two species interact extensively, each exerts strong selection on the other. When one evolves a new feature or modifies an old one, the other typically evolves new adaptations in response. This constant, mutual feedback between two species is called **coevolution**. Perhaps the most familiar form of coevolution is found in predator-prey relationships.

Predation includes any situation in which one organism eats another. In some instances, coevolution between predators (those who do the eating) and prey (those who are eaten) is a sort of "biological arms race," with each side evolving new adaptations in response to "escalations" by the other. Darwin used the example of wolves and deer: Wolf predation selects against slow or careless deer, thus leaving faster, more-alert deer to reproduce and continue the species. In their turn, alert, swift deer select against slow, clumsy wolves, because such predators cannot acquire enough food.

Sexual Selection Favors Traits That Help an Organism Mate

In many animal species, males have conspicuous features such as bright colors, long feathers or fins, or elaborate antlers. Males may also exhibit bizarre courtship behaviors or sing loud, complex songs. Although these extravagant features typically play a role in mating, they also seem to be at odds with efficient survival and reproduction. Exaggerated ornaments and displays may help males gain access to females, but they also make the males more vulnerable to predators. Darwin was intrigued by this apparent contradiction. He coined the term **sexual selection** to describe the special kind of selection that acts on traits that help an animal acquire a mate.

Darwin recognized that sexual selection could be driven either by sexual contests among males or by female preference for particular male phenotypes. Male-male competition for access to females can favor the evolution of features that provide an advantage in fights or ritual displays of aggression (**FIG. 15-11**). Female mate choice provides a second source of sexual selection. In animal species in which females actively choose their mates from among males, females often

Many of Earth's species are in danger. According to the World Conservation Union, more than 15,000 species of plants and animals alone are currently threatened with extinction. For most of these endangered species, the main threat is habitat destruction. When a species' habitat shrinks, its population size almost invariably follows suit.

Many people, organizations, and governments are concerned about the plight of endangered species and are working to protect them and their habitats. The hope is that these efforts will not only protect endangered species but also restore their numbers, so that they are no longer in danger of extinction. Unfortunately, however, a population that has already become small enough to warrant endangered status is likely to undergo evolutionary changes that increase its chances of going extinct. The principles of evolutionary genetics that we've explored in the chapter can help us understand these changes.

One problem is that in small populations, mating choices are limited, and a high proportion of matings may be between close relatives. This inbreeding increases the odds that offspring will be homozygous for harmful recessive alleles. These less-fit individuals may die before reproducing, further reducing the size of the population.

The greatest threat to small populations, however, stems from their inevitable loss of genetic diversity (**FIG. E15-1**). From our discussion of population bottlenecks, it is apparent that, when populations shrink to very small sizes, many of the alleles that were present in the original population will not be represented in the gene pool of the remnant population. Furthermore, we have seen that genetic drift in small populations will cause many of the surviving alleles to subsequently disappear permanently from the population (see Fig. 15-6b). Because genetic drift is a random process, many of the lost alleles will be advantageous ones that were previously favored by natural selection. Inevitably, the number of different alleles in the population grows ever smaller. As ecologist Thomas Foose aptly put it, "gene pools are being converted into gene puddles." Even if the size of an endangered population eventually begins to grow, the damage has already been done; lost genetic diversity is regained only very slowly.

Why does it matter if a population's genetic diversity is low? There are two main risks. First, the fitness of the population as a whole is reduced by the loss of advantageous alleles that underlie adaptive traits. A less-fit population is unlikely to thrive. Second, a genetically impoverished population lacks the variation that will allow it to adapt when environmental conditions change. When the environment changes, as it inevitably will, a genetically uniform species is less likely to contain individuals well suited to survive and reproduce under the new conditions. A species unable to adapt to changing conditions is at very high risk of extinction.

What can be done to preserve the genetic diversity of endangered species? The best solution, of course, is to preserve plenty of diverse types of habitat so that species never become endangered. The human population, however, has grown so large and has appropriated so large a share of Earth's resources that this solution is impossible in many places. For many species, the only solution is to ensure that areas of preserved habitat are large enough to hold populations of sufficient size to contain most of a threatened species' total genetic diversity. If, however, circumstances dictate that preserved areas are small, it is important that the small areas be linked by corridors of the appropriate habitat, so that gene flow among populations in the small preserves can increase the spread of new and beneficial alleles.

ETHICAL CONSIDERATIONS

BIOETHICS Does it matter that human activities are causing species to go extinct? Some bioethecists argue that, because humans have the power to extinguish species, we have an ethical obligation to protect the interests of all of the planet's inhabitants. In this view, it is unethical to allow any species to go extinct. For those who believe in the sanctity of other species, the biodiversity crisis poses profound ethical dilemmas. In many cases, the habitat destruction that endangers other species also helps make space for the farmland, housing, and workplaces needed by our growing human population. How can we reconcile the conflict between valid human needs and the needs of endangered species? Furthermore, it is becoming clear that, even with the best of intentions, we cannot save all of the species currently threatened with extinction. The resources available to preserve and manage protected habitats are limited, and we must make choices that will allow some species to survive while others perish. If all species are precious, how can we make such terrible choices? Who should decide which species will live and which will die, and what criteria should be used?

FIGURE E15-1 Endangered by habitat destruction and loss of genetic diversity
Only a few hundred Sumatran rhinoceroses remain.

seem to prefer males with the most elaborate ornaments or most extravagant displays (**FIG. 15-12**). Why?

A popular hypothesis is that male structures, colors, and displays that do not enhance survival might instead provide a female with an outward sign of a male's condition.

Only a vigorous, energetic male can survive when burdened with conspicuous coloration or a large tail that might make him more vulnerable to predators. Conversely, males that are sick or under parasitic attack are dull and frumpy compared with healthy males. A female that chooses the brightest, most ornamented male is also choosing the healthiest, most vigorous male. By doing so, she gains fitness if, for example, the most vigorous male provides superior parental care to offspring or if he carries alleles for disease resistance that will be inherited by offspring and help ensure their survival. Females thus gain a reproductive advantage by choosing the most highly ornamented males, and the traits (including the exaggerated male ornament) of these flashy males will be passed to subsequent generations.

Selection Can Influence Populations in Three Ways

Natural selection and sexual selection can lead to different patterns of evolutionary change. Evolutionary biologists group these patterns into three categories (**FIG. 15-13**).

- **Directional selection** favors individuals with an extreme value of a trait and selects against both average individuals

FIGURE 15-12 Peahens are attracted to the peacock's showy tail
The ancestors of today's peahens were apparently picky when deciding on males with which to mate, favoring males with longer and more colorful tails.

FIGURE 15-13 Three ways that selection affects a population over time
A graphical illustration of three ways natural and/or sexual selection, acting on a normal distribution of phenotypes, can affect a population over time. In all graphs, the beige areas represent individuals that are selected against—that is, they do not reproduce as successfully as do the individuals in the purple range. **QUESTION** When selection is directional, is there any limit to how extreme the trait under selection will become? Why or why not?

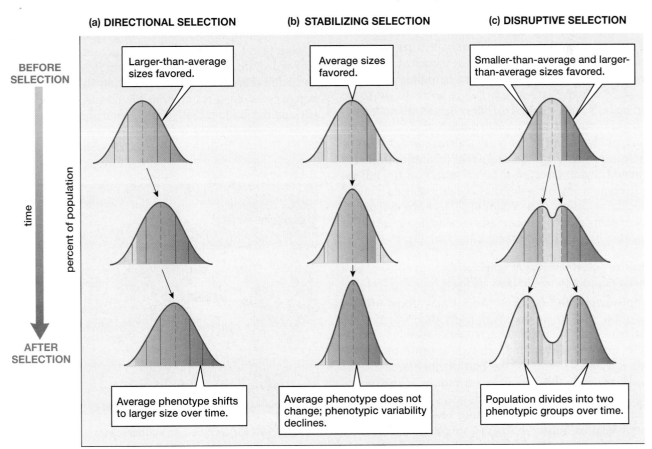

(a) DIRECTIONAL SELECTION

(b) STABILIZING SELECTION

(c) DISRUPTIVE SELECTION

BEFORE SELECTION

Larger-than-average sizes favored.

Average sizes favored.

Smaller-than-average and larger-than-average sizes favored.

time

percent of population

AFTER SELECTION

Average phenotype shifts to larger size over time.

Average phenotype does not change; phenotypic variability declines.

Population divides into two phenotypic groups over time.

range of a particular characteristic (size, color, etc.)

and individuals at the opposite extreme. For example, directional selection might favor small size and select against both average and large individuals in a population.

- **Stabilizing selection** favors individuals with the average value of a trait (for example, intermediate body size) and selects against individuals with extreme values.
- **Disruptive selection** favors individuals at both extremes of a trait (for example, both large and small body sizes) and selects against individuals with intermediate values.

Directional Selection Shifts Character Traits in a Specific Direction

If environmental conditions change in a consistent way, a species may respond by evolving in a consistent direction. For example, if the climate becomes colder, mammal species may evolve thicker fur. The evolution of antibiotic resistance in bacteria is an example of directional selection (see Fig. 15-13a): when antibiotics are present in a bacterial species' environment, individuals with greater resistance reproduce more prolifically than do individuals with less resistance.

Stabilizing Selection Acts Against Individuals Who Deviate Too Far from the Average

Directional selection can't go on forever. What happens once a species is well adapted to a particular environment? If the environment is unchanging, most new variations that appear will be harmful. Under these conditions, we expect species to be subject to stabilizing selection, which favors the survival and reproduction of average individuals (see Fig. 15-13b). Stabilizing selection commonly occurs when a trait is under opposing environmental pressures from two different sources. For example, among lizards of the genus *Aristelliger*, the smallest lizards have a hard time defending territories, but the largest lizards are more likely to be eaten by owls. As a result, *Aristelliger* lizards are under stabilizing selection that favors intermediate body size.

It is widely assumed that many traits are under stabilizing selection. Although the long necks of giraffes probably originated under directional sexual selection for advantage in combat among males, they are probably now under stabilizing selection, as a compromise between the advantages of being able to win contests and the disadvantage of being vulnerable while drinking water (see Fig. 15-10).

Disruptive Selection Adapts Individuals Within a Population to Different Habitats

Disruptive selection (see Fig. 15-13c) may occur when a population inhabits an area with more than one type of useful resource. In this situation, the most adaptive characteristics may be different for each type of resource. For example, the food source of the black-bellied seedcracker (**FIG. 15-14**), a small, seed-eating bird found in the forests of Africa, includes both hard seeds and soft seeds. Cracking hard seeds requires a large, stout beak, but a smaller, pointier beak is a more efficient tool for processing soft seeds. Consequently, black-bellied seedcrackers have beaks in one of two sizes. A bird may have a large beak or small beak, but very few birds have a medium-sized beak; individuals with intermediate-sized beaks have a lower survival rate than individuals with either large or small beaks. Disruptive selection in black-bellied seedcrackers thus favors birds with large beaks and birds with small beaks, but not those with medium-sized beaks.

Black-bellied seedcrackers represent an example of *balanced polymorphism*, in which two or more phenotypes are maintained in a population. In many cases of balanced polymorphism, multiple phenotypes persist because each is favored by a separate environmental factor. For example, consider two different forms of hemoglobin that are present in some African human populations. In these populations, the hemoglobin molecules of people who are homozygous for a particular allele produce defective hemoglobin that clumps up into long chains, which distort and weaken red blood cells. This distortion causes a serious illness known as sickle-cell anemia, which can kill its victims. Before the advent of modern medicine, people homozygous for the sickle-cell allele were unlikely to survive long enough to reproduce. So why hasn't natural selection eliminated the allele?

Far from being eliminated, the sickle-cell allele is present in nearly half the population in some areas of Africa. The persistence of the allele seems to be the result of counterbalancing selection that favors heterozygous carriers of the allele. Heterozygotes, who have one allele for defective hemoglobin and one allele for normal hemoglobin, suffer from mild anemia but they also exhibit increased resistance to malaria, a deadly disease affecting red blood cells, which is widespread in equatorial Africa. In areas of Africa with high risk of malaria infection, heterozygotes must have survived and reproduced more successfully than either type of homozygote. As a result, both the normal hemoglobin allele and the sickle-cell allele have been preserved.

FIGURE 15-14 Black-bellied seedcrackers

CASE STUDY REVISITED   EVOLUTION OF A MENACE

The evolution of antibiotic resistance in populations of bacteria—such as the bacteria that cause multidrug-resistant tuberculosis—is a direct consequence of natural selection applied by antibiotic drugs. When a population of disease-causing bacteria begins to grow in a human body, physicians try to halt population growth by introducing an antibiotic drug to the bacteria's environment. Although many bacteria are killed, some surviving bacteria have genomes with a mutant allele that confers resistance. Bacteria carrying the "resistance allele" produce a disproportionately large share of offspring, which inherit the allele. Soon, resistant bacteria predominate within the population. The resistant bacteria get even more of a boost when the presence of antibiotics is inconsistent, as occurs when a tuberculosis patient neglects to take his or her medicine. During antibiotic-free periods, the populations of resistant bacteria can grow very rapidly and spread to new hosts.

By introducing massive quantities of antibiotics into the bacteria's environment, humans have accelerated the pace of the evolution of antibiotic resistance. Each year, U.S. physicians write more than 100 million prescriptions for antibiotics; the Centers for Disease Control estimates that about half of these prescriptions are unnecessary.

Although medical use and misuse of antibiotics is the most important source of natural selection for antibiotic resistance, antibiotics also pervade the environment outside our bodies. Our food supply, especially meat, contains a portion of the 20 million pounds of antibiotics that are fed to farm animals each year. In addition, Earth's soils and water are laced with antibiotics that enter the environment through human and animal wastes, and from the antibacterial soaps and cleansers that are now routinely used in many households and workplaces. As a result of this massive alteration of the environment, resistant bacteria are now found not only in hospitals and the bodies of sick people but are also widespread in our food, water, and soil. Susceptible bacteria are under constant attack, and resistant strains have little competition. In our fight against disease, we have rashly overlooked some basic principles of evolutionary biology and are now paying a heavy price.

Consider This Because natural selection acts only on existing variation among phenotypes, antibiotic resistance could not evolve if bacteria in natural populations did not already carry alleles that help them resist attack by antibiotic chemicals. Why are such alleles present (albeit at low levels) in bacterial populations? Conversely, if resistance alleles are beneficial, why are they rare in natural populations of bacteria?

CHAPTER REVIEW

SUMMARY OF KEY CONCEPTS

15.1 How Are Populations, Genes, and Evolution Related?

Evolution is change in frequencies of alleles in a population's gene pool. Allele frequencies in a population will remain constant over generations only if the following conditions are met: there is no mutation; there is no gene flow; the population is very large; all mating is random; and all genotypes reproduce equally well (that is, there is no natural selection). These conditions are rarely, if ever, met in nature. Understanding what happens when they are not met helps reveal the mechanisms of evolution.

15.2 What Causes Evolution?

- Mutations are random, undirected changes in DNA composition. Although most mutations are neutral or harmful to the organism, some prove advantageous in certain environments. Inherited mutations are uncommon and do not by themselves change allele frequencies very much, but they provide the raw material for evolution.
- Gene flow is the movement of alleles between different populations of a species. Gene flow tends to reduce differences in the genetic composition of different populations.
- In any population, chance events kill or prevent reproduction by some of the individuals. If the population is small, chance events may eliminate a disproportionate number of individuals who bear a particular allele, thereby greatly changing the allele frequency in the population: this is genetic drift.

- Nonrandom mating, such as assortative mating and inbreeding, can change the distribution of genotypes in a population, in particular by increasing the proportion of homozygotes.
- The survival and reproduction of organisms are influenced by their phenotypes. Because phenotype depends at least partly on genotype, natural selection tends to favor the reproduction of certain alleles at the expense of others.

Web Tutorial 15.1 Agents of Change

Web Tutorial 15.2 The Bottleneck Effect

15.3 How Does Natural Selection Work?

Natural selection is driven by differences in reproductive success among different genotypes. Natural selection proceeds from the interactions of organisms with both the biotic and abiotic parts of their environments. When two or more species exert mutual environmental pressures on each other for long periods of time, both of them evolve in response. Such coevolution can result from any type of relationship between organisms, including competition and predation. Phenotypes that help organisms mate can evolve by sexual selection.

Web Tutorial 15.3 Three Modes of Natural Selection

KEY TERMS

adaptation   *page 306*
allele frequency   *page 297*
coevolution   *page 307*
competition   *page 307*
directional selection   *page 309*
disruptive selection   *page 310*

equilibrium population
    page 298
fitness   *page 305*
founder effect   *page 303*
gene flow   *page 298*
gene pool   *page 297*

genetic drift   *page 300*
Hardy-Weinberg principle
    page 298
mutation   *page 298*
natural selection   *page 306*
population   *page 296*

population bottleneck
    page 302
predation   *page 307*
sexual selection   *page 307*
stabilizing selection   *page 310*

THINKING THROUGH THE CONCEPTS

1. What is a gene pool? How would you determine the allele frequencies in a gene pool?

2. Define *equilibrium population*. Outline the conditions that must be met for a population to stay in genetic equilibrium.

3. How does population size affect the likelihood of changes in allele frequencies by chance alone? Can significant changes in allele frequencies (that is, evolution) occur as a result of genetic drift?

4. If you measured the allele frequencies of a gene and found large differences from the proportions predicted by the Hardy-Weinberg principle, would it prove that natural selection is occurring in the population you are studying? Review the conditions that lead to an equilibrium population, and explain your answer.

5. People like to say that "you can't prove a negative." Study the experiment in Figure 15-3 again, and comment on what it demonstrates.

6. Describe the three ways in which natural selection can affect a population over time. Which way(s) is (are) most likely to occur in stable environments, and which way(s) might occur in rapidly changing environments?

7. What is sexual selection? How is sexual selection similar to and different from other forms of natural selection?

APPLYING THE CONCEPTS

1. In North America, the average height of adult humans has been increasing steadily for decades. Is directional selection occurring? What data would justify your answer?

2. Malaria is rare in North America. In populations of African Americans, what would you predict is happening to the frequency of the hemoglobin allele that leads to sickling in red blood cells? How would you go about determining whether your prediction is true?

3. By the 1940s, the whooping crane population had been reduced to fewer than 50 individuals. Thanks to conservation measures, its numbers are now increasing. What special evolutionary problems do whooping cranes have now that they have passed through a population bottleneck?

4. In many countries, conservationists are trying to design national park systems so that "islands" of natural area (the big parks) are connected by thin "corridors" of undisturbed habitat. The idea is that this arrangement will allow animals and plants to migrate between refuges. Why would such migration be important?

5. A preview question for Chapter 16: A *species* is all the populations of organisms that potentially interbreed with one another but that are reproductively isolated from (cannot interbreed with) other populations. Using the five conditions of the Hardy-Weinberg principle as a starting point, what factors do you think would be important in the splitting of a single ancestral species into two modern species?

FOR MORE INFORMATION

Allison, A. C. "Sickle Cells and Evolution." *Scientific American*, August 1956. The story of the interaction between sickle-cell anemia and malaria in Africa.

Dawkins, R. *Climbing Mount Improbable*. New York: Norton, 1996. An eloquent book-length tribute to the power of natural selection to design intricate adaptations. The chapter on the evolution of the eye is an instant classic.

Dugatkin, L. A., and Godin, J. J. "How Females Choose Their Mates." *Scientific American*, April 1998. A discussion of the role of female mate choice in sexual selection.

Levy, S. B. "The Challenge of Antibiotic Resistance." *Scientific American*, March 1998. An excellent summary of the public health implications of antibiotic resistance. Also discusses some strategies for ameliorating the problem.

Palumbi, S. R. *The Evolution Explosion*. New York: Norton, 2001. An evolutionary biologist explores cases of rapid evolution caused by humans, including antibiotic resistance, pesticide resistance, and the evolution of the virus that causes AIDS.

Rennie, J. "Fifteen Answers to Creationist Nonsense." *Scientific American*, July 2002. A summary of some common misconceptions espoused by creationists, and the scientific response to them.

16

The Origin of Species

The saola, unknown to science until 1992, is one of a number of
previously undiscovered species recently found in the mountains of Vietnam.
The area's distinctive assemblage of species probably arose
during a past period of geographic isolation.

CASE STUDY LOST WORLD

THE STEEP, RAIN-DRENCHED SLOPES of Vietnam's Annamite Mountains are remote and forbidding, cloaked in tropical mists that lend an air of mystery and concealment to the forested mountains. As it turns out, this remote refuge conceals a most astonishing biological surprise: the saola, a hoofed, horned mammal that was unknown to science until the early 1990s. The discovery of a new species of large mammal at this late date was a complete shock. After centuries of human exploration and exploitation in every corner of the world's forests, deserts, and savannas, scientists were certain that no large-sized mammal species could have escaped detection. As long ago as 1812, French naturalist Georges Cuvier wrote that "there is

little hope of discovering new species of large quadrupeds." And yet, the saola, 3 feet high at the shoulder, weighing up to 200 pounds and sporting 20-inch black horns, remained hidden in Annamite Mountain forests, outside the realm of scientific knowledge until 1992 (though local tribespeople had apparently been hunting the creature for some time).

Since the discovery of the saola, scientists have described several additional new (if smaller-sized) mammal species in the same area, including the giant muntjac (also known as the barking deer) and a strange rabbit that has short ears and a brown-striped coat. Recent investigations have also revealed several dozen new species of other vertebrates, including

birds, reptiles, amphibians, and fish. This wave of discoveries has revealed the Vietnamese mountains to be a kind of lost world of animals. Isolated by inhospitable terrain and the wars fought in Vietnam during the last century, the animals of the Annamite Mountains remained unknown to scientists. In the face of increased scientific attention, however, this lost world has become increasingly well known, and the curious biologist may wonder why these wonderfully unfamiliar species are concentrated in this particular part of the world. But before we can fully consider that question, we will need to explore the evolutionary process by which new species arise.

16.1   WHAT IS A SPECIES?

Although Darwin brilliantly explained how evolution shapes complex, amazingly well-designed organisms, his ideas did not fully explain life's diversity. In particular, the process of natural selection cannot by itself explain how living things came to be divided into groups, with each group distinctly different from all other groups. When we look at big cats, we don't see a continuous array of different tiger phenotypes that gradually grades into a lion phenotype. We see lions and tigers as separate, distinct types with no overlap. Each distinct type is known as a species.

Biologists Need a Clear Definition of Species

Before we can study the origin of species, we must first clarify our definition of the term. Throughout most of human history, "species" was a poorly defined concept. In pre-Darwinian Europe, the word "species" simply referred to one of the "kinds" produced by the biblical creation. In this view, humans could not possibly know the criteria of the creator, but could only attempt to distinguish among species on the basis of visible differences in structure. In fact, species is Latin for "appearance."

On a coarse scale, it is easy to use quick visual comparisons to distinguish species. For example, warblers are clearly different from eagles, which are obviously different from ducks. But it is far more difficult to distinguish among different species of warblers, eagles, or ducks. How do scientists make these finer distinctions?

Species Are Groups of Interbreeding Populations

Today, biologists define a **species** as a group of populations that evolves independently. Each species follows a separate evolutionary path because alleles do not move between the gene pools of different species. This definition, however, does not clearly state the standard by which such evolutionary independence is judged. The most widely used standard defines species as "groups of actually or potentially interbreeding natural populations, which are reproductively isolated from other such groups." This definition, known as the *biological species concept*, is based on the observation that **reproductive isolation** (no successful breeding outside the group) ensures evolutionary independence.

The biological species concept has at least two major limitations. First, because the definition is based on patterns of sexual reproduction, it does not help us determine species boundaries among asexually reproducing organisms. Second, it is not always practical or even possible to directly observe whether members of two different groups interbreed. Thus, a biologist who wishes to determine if a group of organisms is a separate species must often make the determination without knowing for sure if group members breed with organisms outside the group.

Despite the limitations of the biological species concept, most biologists accept it for identifying species of sexually reproducing organisms. Nonetheless, scientists who study bacteria and other organisms that mainly reproduce asexually must use alternative definitions. Even some biologists who study sexually reproducing organisms prefer species definitions that do not depend on a property (reproductive isolation) that can be difficult to measure. Several such alternatives to the biological species concept have been proposed; one that has gained many adherents is described in Chapter 18 (page 365).

Appearance Can Be Misleading

Biologists have found that some organisms with very similar appearances belong to different species. For example, the cordilleran flycatcher and the Pacific-slope flycatcher are so similar that even experienced birdwatchers cannot tell them apart (**FIG. 16-1**). Until recently, these birds were

(a)

(b)

FIGURE 16-1 Members of different species may be similar in appearance (a) The cordilleran flycatcher and (b) Pacific-slope flycatcher are different species.

considered to be a single species. However, research revealed that the two kinds of birds do not interbreed and are in fact two different species.

Superficial similarity can sometimes hide multiple species. Researchers recently discovered that the species known until now as the two-barred flasher butterfly is actually a group of at least 10 different species. The caterpillars of the different species do differ in appearance, but the adult butterflies are all so similar that their species identities went undetected for more than two centuries after the butterfly was first described and named.

Conversely, differences in appearance do not always mean that two populations belong to different species. For example, bird field guides published in the 1970s list the myrtle warbler and Audubon's warbler (**FIG. 16-2**) as distinct species. These birds differ in geographical range and in the color of their throat feathers. More recently, scientists decided that these birds are local varieties of the same species. The reason: where their ranges overlap, these warblers interbreed, and the offspring are just as vigorous and fertile as the parents.

16.2 HOW IS REPRODUCTIVE ISOLATION BETWEEN SPECIES MAINTAINED?

What prevents different species from interbreeding? The traits that prevent interbreeding and maintain reproductive isolation are called **isolating mechanisms**. Isolating mechanisms give a clear benefit to individuals. Any individual that mates with a member of another species will probably produce no offspring (or unfit or sterile off-spring), thereby wasting its reproductive effort and contributing nothing to future generations. Thus, natural selection favors traits that prevent mating across species boundaries. Mechanisms that prevent mating between species are called **premating isolating mechanisms**.

When premating isolating mechanisms fail or have not yet evolved, members of different species may mate. If, however, all resulting hybrid offspring die during development, then the two species are still reproductively isolated from one another. Even if viable hybrid offspring are produced, if these hybrids are less fit than their parents or are themselves infertile, the two species may still remain separate, with little or no gene flow between them. Mechanisms that prevent the formation of vigorous, fertile hybrids between species are called **postmating isolating mechanisms**.

Premating Isolating Mechanisms Prevent Mating Between Species

Reproductive isolation can be maintained by a variety of mechanisms, but those that prevent mating attempts are especially effective. We next describe the most important types of such premating isolating mechanisms.

Members of Different Species May Be Prevented from Meeting

Members of different species cannot mate if they never get near one another. *Geographical isolation* prevents interbreeding between populations that do not come into contact because they live in different, physically separated

(a)

(b)

FIGURE 16-2 Members of a species may differ in appearance
(a) The myrtle warbler and (b) Audubon's warbler are members of the same species.

FIGURE 16-3 Geographical isolation
To determine if these two squirrels are members of different species, we must know if they are "actually or potentially interbreeding." Unfortunately, it is hard to tell, because (a) the Kaibab squirrel lives only on the north rim of the Grand Canyon and (b) the Abert squirrel lives only on the south rim. The two populations are geographically isolated but still quite similar. Have they diverged enough since their separation to become reproductively isolated? Because they remain geographically isolated, we cannot say for sure.

(a)

(b)

places (**FIG. 16-3**). However, we cannot determine if geographically separated populations are actually distinct species. Should the physical barrier separating the two populations disappear (a new channel might connect two previously isolated lakes, for example), the reunited populations might interbreed freely and not be separate species after all. If they cannot interbreed, then other mechanisms, such as those considered below, must have developed during their isolation. Geographical isolation, therefore, is usually considered to be a mechanism that allows new species to form rather than a mechanism that maintains reproductive isolation between species.

Different Species May Occupy Different Habitats

Two populations that use different resources may spend time in different habitats within the same general area and thus exhibit *ecological isolation*. White-crowned and white-throated sparrows, for example, have extensively overlapping ranges. The white-throated sparrow, however, frequents dense thickets, whereas the white-crowned sparrow inhabits fields and meadows, seldom penetrating far into dense growth. The two species may coexist within a few hundred yards of one another and yet seldom meet during the breeding season. A more dramatic example is provided by the more than 750 species of fig wasp (**FIG. 16-4**). Each species of fig wasp breeds in (and pollinates) the fruits of a particular species of fig, and each fig species hosts one and only one species of pollinating wasp.

Although ecological isolation may slow down interbreeding, it seems unlikely that it could prevent gene flow entirely. Other mechanisms normally also contribute to reproductive isolation.

Different Species May Breed at Different Times

Even if two species occupy similar habitats, they cannot mate if they have different breeding seasons, a phenomenon called *temporal (time-related) isolation*. For example, the spring field cricket and the fall field cricket both occur in many areas of North America but, as their names suggest, the former species breeds in spring and the latter in autumn. As a result, the two species do not interbreed. In plants, the reproductive structures of different species may

FIGURE 16-4 Ecological isolation
This female fig wasp is carrying fertilized eggs from a mating that took place within a fig. She will find another fig of the same species, enter it through a pore, lay eggs, and die. Her offspring will hatch, develop, and mate within the fig. Because each species of fig wasp reproduces only in its own particular fig species, each wasp species is reproductively isolated.

FIGURE 16-5 Temporal isolation
Bishop pines, such as these, and Monterey pines coexist in nature. In the laboratory they produce fertile hybrids. In the wild, however, they do not interbreed, because they release pollen at different times of the year.

mature at different times. For example, Bishop pines and Monterey pines grow together near Monterey on the California coast (**FIG. 16-5**), but the two species release their sperm-containing pollen (and have eggs ready to receive the pollen) at different times: the Monterey pine releases pollen in early spring, the Bishop pine in summer. For this reason, the two species never interbreed under natural conditions.

Different Species May Have Different Courtship Signals

Among animals, the elaborate courtship colors and behaviors that so enthrall human observers not only serve as recognition and evaluation signals between male and female but also prevent mating with members of other species. Signals and behaviors that differ from species to species create *behavioral isolation*. The distinctive plumage and vocalizations of a male bird, for example, may attract females of his own species, but females of other species treat these displays with indifference. For example, the extravagant plumes and arresting pose of a courting male greater bird of paradise are conspicuous indicators of his species, and there is little chance that females of another species will be mistakenly attracted (**FIG. 16-6**). Among frogs, males are often impressively indiscriminate, jumping on every female in sight, regardless of the species, when the spirit moves them. Females, however, approach only male frogs that utter the call appropriate to their species. If they do find themselves in an unwanted embrace, they utter the "release call," which causes the male to let go. As a result, few hybrids are produced.

Species' Differing Sexual Organs May Foil Mating Attempts

In rare instances, a male and a female of different species attempt to mate. Their attempt is likely to fail. Among animal species with internal fertilization (in which the sperm

is deposited inside the female's reproductive tract), the male's and female's sexual organs simply may not fit together. Incompatible body shapes may also make copulation between species impossible. For example, snails of species whose shells have left-handed spirals may be unable to successfully copulate with snails whose shells have right-handed spirals (**FIG. 16-7**). Among plants, differences in flower size or structure may prevent pollen transfer between species because the differing flowers may attract different pollinators. Isolating mechanisms of this type are called *mechanical incompatibilities*.

Postmating Isolating Mechanisms Limit Hybrid Offspring

Premating isolation sometimes fails. When it does, members of different species may mate, and the sperm of one species may reach the egg of another species. Such matings, however, often fail to produce vigorous, fertile hybrid offspring, owing to postmating isolating mechanisms.

One Species' Sperm May Fail to Fertilize Another Species' Eggs

Even if a male inseminates a female of a different species, his sperm may not be able to fertilize her eggs, an isolating mechanism called *gametic incompatibility*. For example, in

(a)

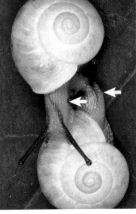

(b)   (c)

FIGURE 16-7 Mechanical isolation
(a) The shells of different snail species may coil in different directions. Among the three closely related species shown, two have shells that coil left and one has a shell that coils right. **(b)** Two snails with matching coils can mate, but **(c)** snails of different species with mismatched coiling cannot mate because the mismatch keeps their genitals (arrow) apart.

FIGURE 16-6 Behavioral isolation
The mate-attraction display of a male greater bird of paradise includes distinctive posture, movements, plumage, and vocalizations that do not resemble those of other bird-of-paradise species.

animals with internal fertilization, fluids in the female reproductive tract may weaken or kill sperm of other species. Gametic incompatibility may be an especially important isolating mechanism in species, such as marine invertebrate animals and wind-pollinated plants, that reproduce by scattering gametes in the water or in the air. For example, sea urchin sperm cells contain a protein that allows them to bind to eggs. The structure of the protein differs among species so that sperm of one sea urchin species cannot bind to the eggs of another species. In abalones (a type of mollusk), eggs are surrounded by a membrane that can be penetrated only by sperm containing a particular enzyme. Each abalone species has a distinctive version of the enzyme, so hybrids are rare, even though several species of abalones coexist in the same waters and spawn during the same period. Among plants, similar chemical incompatibility may prevent the germination of pollen from one species that lands on the stigma (pollen-catching structure) of the flower of another species.

Hybrid Offspring May Survive Poorly

If cross-species fertilization does occur, the resulting hybrid may be unable to survive, a situation called *hybrid inviability*. The genetic instructions directing development of the two species may be so different that hybrids abort early in development. For example, captive leopard frogs can be induced to mate with wood frogs, and the matings generally yield fertilized eggs. The resulting embryos, however, inevitably fail to survive more than a few days.

In other animal species, a hybrid might survive but display behaviors that are mixtures of the two parental types. In attempting to do some things the way species A does them and other things the way species B does them, the hybrid may be hopelessly uncoordinated and therefore unable to reproduce. Hybrids between certain species of lovebirds, for example, have great difficulty learning to carry nest materials during flight and probably could not reproduce in the wild.

FIGURE 16-8 Hybrid infertility
This liger, the hybrid offspring of a lion and a tiger, is sterile. The gene pools of its parent species remain separate.

Hybrid Offspring May Be Infertile

Most animal hybrids, such as the mule (a cross between a horse and a donkey) and the liger (a zoo-based cross between a male lion and a female tiger), are sterile (**FIG. 16-8**). *Hybrid infertility* prevents hybrids from passing on their genetic material to offspring, which blocks gene flow between the two parent populations. A common reason for hybrid infertility is the failure of chromosomes to pair properly during meiosis, so that eggs and sperm fail to develop.

Table 16-1 summarizes the different types of isolating mechanisms.

16.3 HOW DO NEW SPECIES FORM?

Despite his exhaustive exploration of the process of natural selection, Charles Darwin never proposed a complete mechanism of **speciation**, the process by which new species form. One scientist who did play a large role in describing the process of speciation was Ernst Mayr of Harvard University, an ornithologist (expert on birds) and a pivotal figure in the history of evolutionary biology. Mayr developed the biological species concept discussed above. He was also among the first to recognize that speciation depends on two factors acting on a pair of populations: isolation and genetic divergence.

- *Isolation of populations.* If individuals move freely between two populations, interbreeding and the resulting gene flow will cause changes in one population to soon become widespread in the other as well. Thus, two populations cannot grow increasingly different unless something happens to block interbreeding between them. Speciation depends on isolation.

Table 16-1 Mechanisms of Reproductive Isolation

Premating Isolating Mechanisms: factors that prevent organisms of two populations from mating

- **Geographical isolation:** The populations cannot interbreed because a physical barrier separates them.
- **Ecological isolation:** The populations do not interbreed, even if they are within the same area, because they occupy different habitats.
- **Temporal isolation:** The populations cannot interbreed because they breed at different times.
- **Behavioral isolation:** The populations do not interbreed because they have different courtship and mating rituals.
- **Mechanical incompatibility:** The populations cannot interbreed because their reproductive structures are incompatible.

Postmating Isolating Mechanisms: factors that prevent organisms of two populations from producing vigorous, fertile offspring after mating

- **Gametic incompatibility:** Sperm from one population cannot fertilize eggs of another population.
- **Hybrid inviability:** Hybrid offspring fail to survive to maturity.
- **Hybrid infertility:** Hybrid offspring are sterile or have reduced fertility.

• *Genetic divergence of populations.* It is not sufficient for two populations simply to be isolated. They will become separate species only if, during the period of isolation, they evolve sufficiently large genetic differences. The differences must be large enough that, if the isolated populations are reunited, they can no longer interbreed and produce vigorous, fertile offspring. That is, speciation is complete only if divergence results in evolution of an isolating mechanism. Such differences can arise by chance (genetic drift), especially if at least one of the isolated populations is small (see Chapter 15). Large genetic differences can also arise through natural selection, if the isolated populations experience different environmental conditions.

Speciation always requires isolation followed by divergence, but these steps can take place in several different ways. Evolutionary biologists group the different pathways to speciation into two broad categories: **allopatric speciation**, in which two populations are geographically separated from one another, and **sympatric speciation**, in which two populations share the same geographical area.

Geographical Separation of a Population Can Lead to Allopatric Speciation

New species can arise by allopatric speciation when an impassible barrier physically separates different parts of a population.

Organisms May Colonize Isolated Habitats

A small population can become isolated if it moves to a new location (**FIG. 16-9**). For example, some members of a population of land-dwelling organisms might colonize an oceanic island. The colonists might be birds, flying insects, fungal spores, or wind-borne seeds blown by a storm. More earthbound organisms could reach the island on a drifting "raft" of vegetation torn from the mainland coast. Whatever the means, such colonization must occur regularly, given the presence of living things on even the remotest islands.

Isolation by colonization need not be limited to islands. For example, different coral reefs may be separated by miles of open ocean, so any reef-dwelling sponges, fishes, or algae that were carried by ocean currents to a distant reef would be effectively isolated from their original populations. Any bounded habitat, such as a lake, a mountaintop, or a parasite's host can isolate arriving colonists.

Geological and Climate Changes May Divide Populations

Isolation can also result from landscape changes that divide a population. For example, rising sea levels might transform a coastal hilltop into an island, isolating the residents. New rock from a volcanic eruption can divide a previously continuous sea or lake, splitting populations. A river that changes course can also divide populations, as can a newly formed mountain range. Climate shifts, such as those that happened in past ice ages, can change the distribution of vegetation and strand portions of populations in isolated patches of suitable habitat. You can probably

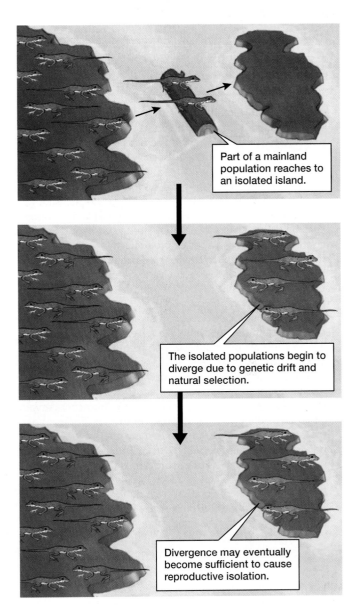

FIGURE 16-9 Allopatric isolation and divergence
In allopatric speciation, some event causes a population to be divided by an impassable geographical barrier. One way the division can occur is by colonization of an isolated island. The two now-separated populations may diverge genetically. If the genetic differences between the two populations become large enough to prevent interbreeding, then the two populations constitute separate species. **EXERCISE** Make a list of events or processes that could cause geographical subdivision of a population. Are the items on your list sufficient to account for formation of the millions of species that have inhabited Earth?

imagine many other scenarios that could lead to the geographical subdivision of a population.

Over the history of Earth, many populations have been divided by continental drift. Earth's continents float on molten rock and slowly move about the surface of the planet. On a number of occasions during Earth's long history, continental landmasses have broken into pieces that

The main cause of extinction is environmental change, especially habitat destruction. Some species with small populations, however, are also threatened by a less obvious danger: hybridization. Although premating isolating mechanisms ensure that, for the most part, members of one species cannot interbreed with members of a different species, matings between members of different species are nonetheless possible. Between-species matings and the resulting hybrid offspring are especially common in birds and plants.

How can hybrid mating be dangerous to endangered species? Recall that postmating isolating mechanisms ensure that, in most cases, hybrid offspring will survive poorly and may even be sterile. Now picture what happens when contact between two species produces hybrids, and one of the species has a much smaller population than the other. If the hybrid offspring fail to survive and reproduce, the numbers of each species will decline, but the decline will have a proportionally larger impact on the small population. When the more abundant species moves into the range of the rare species, the impact on the rare species can be severe. Even if the hybrid offspring survive well, high numbers of hybrids can overwhelm the rare species, essentially absorbing the rare species into the abundant species.

Damage from hybridization is most likely to occur when formerly isolated small populations come into contact with larger populations of a closely related species. For example, the plant *Clarkia lingulata* is extremely rare, known to exist only in two sites in the Sierra Nevada mountains of California. Unfortunately, it readily hybridizes with its abundant relative *Clarkia biloba* to produce sterile hybrid offspring. Because several populations of *biloba* grow near the *lingulata* populations, extinction by hybridization is a real possibility for this rare species.

Human activities often cause contact between an endangered species and a more abundant species with which it can hybridize. For example, the rare Hawaiian duck, found only on the Hawaiian Islands, hybridizes freely with mallard ducks, a nonnative species introduced to Hawaii by hunters in search of new game species. Similarly, the endangered Ethiopian wolf (**FIG. E16-1**) is threatened by interbreeding with feral domestic dogs, and the endangered European wildcat is at risk from hybridization with domestic cats. In these cases and others, a species first declined in number due to habitat destruction and then became vulnerable to further damage by hybridization with a more numerous species that was present as a result of human activities.

FIGURE E16-1 Ethiopian wolves
Fewer than 500 Ethiopian wolves remain. Among the threats to their continued existence is hybridization with wild dogs.

---

subsequently moved apart (see Fig. 17-11 on page 345). Each of these breakups must have split a multitude of populations. The bird group known as the ratites provides evidence of such a split. Ratites are large, flightless birds, including the ostrich of Africa, the rhea of South America, and the emu of Australia. The ancestor of all the ratite species lived on the ancient supercontinent of Gondwana. When Gondwana broke apart, different portions of the ancestral ratite population were isolated on separate drifting continents.

Natural Selection and Genetic Drift May Cause Isolated Populations to Diverge

If two populations become geographically isolated for any reason, there will be no gene flow between them. If the pressures of natural selection differ in the separate locations, then the populations may accumulate genetic differences. Alternatively, genetic differences may arise if one or more of the separated populations is small enough for genetic drift to occur, which may be especially likely in the aftermath of a founder event (in which a few individuals become isolated from the main body of the species). In either case, genetic differences between the separated popu-

lations may eventually become large enough to make interbreeding impossible. At that point, the two populations will have become separate species. Most evolutionary biologists believe that geographical isolation followed by allopatric speciation has been the most common source of new species, especially among animals.

Ecological Isolation of a Population Can Lead to Sympatric Speciation

Only genetic isolation—limited gene flow—is required for speciation, so new species can arise by sympatric speciation when populations become genetically isolated without geographic separation (**FIG. 16-10**). If, for example, a geographical area contains two distinct types of habitats (each with distinct food sources, places to raise young, and so on), different members of a single species may begin to specialize in one habitat or the other. If conditions are right, natural selection in the two different habitats may lead to the evolution of different traits in the two groups. Eventually, these differences may become large enough to prevent members of the two groups from interbreeding, and the formerly single species will have split into two species. Such a split

FIGURE 16-10 Sympatric isolation and divergence

In sympatric speciation, some event blocks gene flow between two parts of a population that remains in a single geographic area. One way in which this genetic isolation can occur is if a portion of a population begins to use a previously unexploited resource, as when some members of an insect population shift to a new host plant species (as has occurred in the fruit fly species *Rhagoletis pomonella*). The two now-isolated populations may diverge genetically. If the genetic differences between the two populations become large enough to prevent interbreeding, then the two populations constitute separate species. QUESTION How might future scientists test whether the current *R. pomonella* has become two species?

seems to be taking place right before biologists' eyes, so to speak, in the case of the fruit fly *Rhagoletis pomonella*.

Rhagoletis is a parasite of the American hawthorn tree. This fly lays its eggs in the hawthorn's fruit; when the maggots hatch, they eat the fruit. About 150 years ago,

entomologists (scientists who study insects) noticed that *Rhagoletis* had begun to infest apple trees, which were introduced into North America from Europe. Today, it appears that *Rhagoletis* is splitting into two species, one that breeds on apples and one that sticks to hawthorns. The

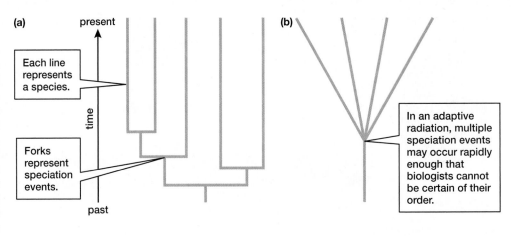

FIGURE 16-11 Interpreting evolutionary trees
Evolutionary history is often represented by **(a)** an evolutionary tree, a graph in which the vertical axis plots time. In **(b)**, an evolutionary tree representing an adaptive radiation, many lines may branch from a single point. This pattern reflects biologists' uncertainty about the order in which the multiple speciation events of the radiation took place. With more research, it may be possible to replace the "starburst" pattern with a more informative tree.

(a) present · time · past
(b)

Each line represents a species.

Forks represent speciation events.

In an adaptive radiation, multiple speciation events may occur rapidly enough that biologists cannot be certain of their order.

two groups have evolved substantial genetic differences, some of which—such as those that affect the timing of emergence of the adult flies—are important for survival on a particular host plant.

The two kinds of flies will become two species only if they maintain reproductive separation. Apple trees and hawthorns typically grow in the same areas, and flies, after all, can fly. So why don't apple flies and hawthorn flies interbreed and cancel out any genetic differences between them? First, female flies usually lay their eggs in the same type of fruit in which they developed. Males also tend to prefer the same type of fruit in which they developed. Therefore, apple-liking males will encounter and mate with apple-liking females. Second, apples mature two or three weeks later than does hawthorn fruit, and the two types of flies emerge with timing appropriate for their chosen host fruit. Thus, the two varieties of flies have very little chance of meeting. Although some interbreeding between the two types of flies occurs, it seems they are well on their way to speciation. Will they make it? Entomologist Guy Bush suggests, "Check back with me in a few thousand years."

The story of *Rhagoletis* illustrates how shifts in habitat or resource use might foster sympatric speciation. To explore another mechanism of sympatric speciation, see "A Closer Look: Speciation by Mutation."

Under Some Conditions, Many New Species May Arise

The mechanisms of speciation and reproductive isolation that we have described lead to forking branches in the *evolutionary tree* of life, as one species splits into two (**FIG. 16-11a**). In some cases, many new species arise in a relatively short time (**FIG 16-11b**). This process, called **adaptive radiation**, can occur when populations of one species invade a variety of new habitats and evolve in response to the differing environmental pressures in those habitats.

Adaptive radiation has occurred many times and in many groups of organisms, typically when species encounter a wide variety of unoccupied habitats. For example, episodes of adaptive radiation took place when some way-

ward finches colonized the Galápagos Islands, when a population of cichlid fish reached isolated Lake Malawi in Africa, and when an ancestral tarweed plant species arrived at the Hawaiian Islands (**FIG. 16-12**). These events gave rise to adaptive radiations of 13 species of Darwin's finches in the Galápagos, more than 300 species of cichlids in Lake Malawi, and 30 species of silversword plants in Hawaii. In these examples, the invading species faced no competitors except other members of their own species, and all the available habitats and food sources were rapidly exploited by new species that evolved from the original invaders.

16.4 WHAT CAUSES EXTINCTION?

Every living organism must eventually die, and the same is true of species. Just like individuals, species are "born" (through the process of speciation), persist for some period of time, and then perish. The ultimate fate of any species is **extinction**, the death of the last of its members. In fact, at least 99.9% of all the species that have ever existed are now extinct. The natural course of evolution, as revealed by fossils, is continual turnover of species as new ones arise and old ones go extinct.

The immediate cause of extinction is probably always environmental change, in either the living or the nonliving parts of the environment. Two major environmental factors that may drive a species to extinction are competition among species and habitat destruction.

Localized Distribution and Overspecialization Make Species Vulnerable in Changing Environments

Species vary widely in their range of distribution and, hence, in their vulnerability to extinction. Some species, such as herring gulls, white-tailed deer, and humans, inhabit entire continents or even the whole Earth; others, such as the Devil's Hole pupfish (**FIG. 16-13**), have extremely limited ranges. Obviously, if a species inhabits only a very small area, any disturbance of that area could easily result in extinction. If Devil's Hole dries up due to a drought or well drilling nearby, its pupfish will immediately vanish. Conversely, wide-ranging species normally do not succumb to local environmental catastrophes.

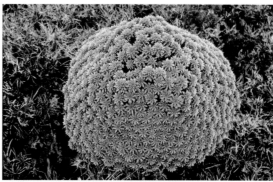

FIGURE 16-12 Adaptive radiation
About 30 species of silversword plants inhabit the Hawaiian Islands. These species are found nowhere else, and all of them descended from a single ancestral population within a few million years. This adaptive radiation has led to a collection of closely related species of diverse form and appearance, with an array of adaptations for exploiting the many different habitats in Hawaii, from warm, moist rain forests to cool, barren volcanic mountaintops. QUESTION Did the Hawaiian silverswords arise by allopatric or sympatric speciation?

Another factor that may make a species vulnerable to extinction is overspecialization. Each species evolves adaptations that help it survive and reproduce in its environment. In some cases, these adaptations include specializations that favor survival in a particular and limited set of environmental conditions. The Karner blue butterfly, for example, feeds only on the blue lupine plant (FIG. 16-14). The butterfly is therefore found only where the plant thrives. But the blue lupine has become quite rare as its habitat of sandy, open woods and clearings in northeast North America has been largely replaced by farms and development. If the lupine disappears, the Karner blue butterfly will surely go extinct along with it.

Interactions with Other Species May Drive a Species to Extinction

As described earlier, interactions such as competition and predation serve as agents of natural selection. In some cases, these same interactions can lead to extinction rather than to adaptation.

FIGURE 16-14 Extreme specialization places species at risk
The Karner blue butterfly feeds exclusively on the blue lupine, found in dry forests and clearings in the northeastern United States. Such behavioral specialization renders the butterfly extremely vulnerable to any environmental change that may exterminate its single host plant species. QUESTION If specialization puts a species at risk for extinction, how could this hazardous trait have evolved?

FIGURE 16-13 Very localized distribution can endanger a species
The Devil's Hole pupfish is found in only one spring-fed water hole in the Nevada desert. This and other isolated small populations are at high risk of extinction.

In some instances, new species can arise nearly instantaneously because of mutations that change the number of chromosomes in an organism's cells. The acquisition of multiple copies of each chromosome is known as **polyploidy** and has been a frequent cause of sympatric speciation in plants (**FIG. E16-2**). As you may remember from Chapter 11, most plants and animals have paired chromosomes and are described as diploid. Occasionally, especially in plants, a fertilized egg duplicates its chromosomes but doesn't divide into two daughter cells. The resulting cell thus becomes *tetraploid*, with four copies of each chromosome. If all of the subsequent cell divisions are normal, this tetraploid zygote will develop into a plant with tetraploid cells. Most tetraploid plants are vigorous and healthy, and many can successfully complete meiosis to form viable gametes. The gametes, however, are diploid (meiosis normally produces haploid gametes from diploid cells). These diploid gametes can fuse with other diploid gametes to produce new tetraploid offspring, so it is no problem for tetraploids to interbreed with other tetraploids of that species or to self-fertilize (as many plants do).

If, however, a tetraploid interbreeds with a diploid individual from the "parental" species, the outcome is not so successful. For example, if a diploid sperm from a tetraploid plant fertilizes a haploid egg cell of the parental species, the offspring will be *triploid*, with three copies of each chromosome. Many triploid individuals experience problems during growth and development. Even if the triploid offspring develops normally, it will be sterile: when a triploid cell attempts to undergo meiosis, the odd number of chromosomes makes chromosome pairing impossible. Meiosis fails, and viable gametes are not formed. Because the offspring of diploid–tetraploid matings are inevitably sterile, tetraploid plants and their diploid parents form distinct reproductive communities that cannot interbreed successfully. A new species can form in a single generation.

Why is speciation by polyploidy common in plants but not in animals? Many plants can either self-fertilize or reproduce asexually, or both. If a tetraploid plant self-fertilizes, then its offspring will also be tetraploid. Asexual offspring, of course, are genetically identical to the parent and are also tetraploid. In either case, the new tetraploid plant may perpetuate itself and form a new species. Most animals, however, cannot self-fertilize or reproduce asexually. Therefore, if an animal produced a tetraploid offspring, the offspring would have to mate with a member of the diploid parental species, producing all triploid offspring. The triploid offspring would almost certainly be sterile. Speciation by polyploidy is extremely common in plants; in fact, nearly half of all species of flowering plants are polyploid, and many of them are tetraploid.

Organisms compete for limited resources in all environments. If a species' competitors evolve superior adaptations and the species doesn't evolve fast enough to keep up, it may become extinct. A particularly striking example of extinction through competition occurred in South America, beginning about 2.5 million years ago. At that time, the isthmus of Panama rose above sea level and formed a land bridge between North America and South America. After the previously separated continents were connected, the mammal species that had evolved in isolation on each continent were able to mix. Many species did indeed expand their ranges, as North American mammals moved southward and South American mammals moved northward. As they moved, each species encountered resident species that occupied the same kinds of habitats and exploited the same kinds of resources. The ultimate result of the ensuing competition was that the North American species diversified and underwent an adaptive radiation that displaced the vast majority of the South American species, many of which went extinct. Clearly, evolution had bestowed on the North American species some (as yet unknown) set of adaptations that enabled their descendants to exploit resources more efficiently and effectively than their South American counterparts could.

Habitat Change and Destruction Are the Leading Causes of Extinction

Habitat change, both contemporary and prehistoric, is the single greatest cause of extinctions. Present-day habitat destruction due to human activities is proceeding at a rapid pace. Many biologists believe that we are presently in the midst of the fastest-paced and most widespread episode of species extinction in the history of life. Loss of tropical forests is especially devastating to species diversity. As many as half the species presently on Earth may be lost over the next 50 years as the tropical forests that contain them are cut for timber and to clear land for cattle and crops. In Chapter 17, we will discuss extinctions due to prehistoric habitat change.

EVOLUTIONARY CONNECTIONS

Scientists Don't Doubt Evolution

In the popular press, conflicts among evolutionary biologists are sometimes seen as conflicts about evolution itself. We occasionally read statements implying that new theories are overthrowing Darwin's and casting doubt on the reality of evolution. Nothing could be farther from the truth. Despite some disagreements about the details of the evolutionary process, biologists unanimously agree that evolution occurred in the past and is still occurring today. The only argument is about the relative importance of the various mechanisms of evolutionary change in the history of life on Earth, their pace, and which factors were most important in shaping the evolution of particular species. Meanwhile, wolves still tend to catch the slowest caribou, small populations still undergo genetic drift, and habitats still change or disappear. Evolution continues, still generating, in Darwin's words, "endless forms most beautiful."

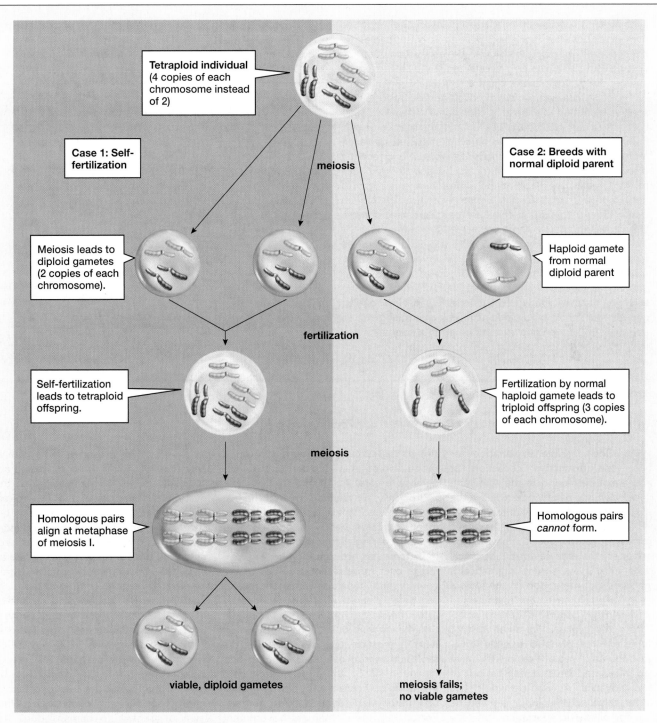

Tetraploid individual (4 copies of each chromosome instead of 2)

Case 1: Self-fertilization

meiosis

Case 2: Breeds with normal diploid parent

Meiosis leads to diploid gametes (2 copies of each chromosome).

Haploid gamete from normal diploid parent

fertilization

Self-fertilization leads to tetraploid offspring.

Fertilization by normal haploid gamete leads to triploid offspring (3 copies of each chromosome).

meiosis

Homologous pairs align at metaphase of meiosis I.

Homologous pairs *cannot* form.

viable, diploid gametes

meiosis fails; no viable gametes

FIGURE E16-2 Speciation by polyploidy
A tetraploid mutant can successfully self-fertilize (or can interbreed with other tetraploid individuals) to yield a new generation of tetraploids, but mating between tetraploids and normal diploid individuals will yield only sterile offspring. Tetraploid mutants are thus reproductively isolated from their diploid ancestors and may constitute a new species.

Looking for a special gift for a friend or loved one? Why not name a species after him or her? Or, for that matter, name one after yourself! Thanks to the BIOPAT project (www.biopat.de), anyone with $3000 can be immortalized in the Latin name of a newly discovered plant or animal.

Typically, the scientist who discovers and describes a new species is entitled to choose its Latin name. Scientists usually choose a name that describes a trait of the species or perhaps the location where it was found. Sometimes, however, more whimsical choices are made. For example, a recently discovered snail was named *Bufonaria borisbeckeri*, in honor of the German tennis player Boris Becker, and a frog was named *Hyla stingi* after the British rock star. *Agathidium bushi* and *Agathidium cheneyi* are beetles named for the U.S. president and vice president.

If you donate money to the BIOPAT project, the name of a new species will be entirely up to you. In return for a contribution that supports efforts to discover and conserve endangered species, the people at BIOPAT will offer you a selection of newly discovered but unnamed species. You can choose your species and pick a name, which is then given an appropriate Latin ending and published in a scientific journal. Your chosen name becomes the official, recognized scientific name of the new species.

In perhaps the most extraordinary example of purchased species naming rights, a newly discovered monkey was named after an online casino (**FIG. E16-3**). In return for a contribution of $650,000, the new species was named *Callicebus aureipalatii*; the species name is Latin for golden palace. The payment will be used for management of the Madidi National Park in Bolivia, where the new species was discovered.

FIGURE E16-3 The Golden Palace monkey is named after a casino

CASE STUDY REVISITED LOST WORLD

One possible explanation for the distinctive collection of species found in the Annamite Mountains of Vietnam lies in the geological history of the region. During the ice ages that have occurred repeatedly over the past million years or so, the area covered by tropical forests must have shrunk dramatically. Organisms that depended on the forests for survival would have been restricted to any remaining "islands" of forest, isolated from their fellows in other, distant patches of forest. What is now the Annamite Mountain region may well have been an isolated forest during periods of glacial advance. As we learned in this chapter, this kind of isolation can set the stage for allopatric speciation and may have created the conditions that gave

rise to the saola, giant muntjac, striped rabbit, and other unique denizens of Vietnamese forests.

Ironically, we have discovered the lost world of Vietnamese animals at a moment when that world is in grave danger of disappearing. Economic development in Vietnam has brought logging and mining to ever more remote regions of the country, and Annamite Mountain forests are being cleared at an unprecedented rate. The increasing local human population means that local animals are hunted heavily; most of our knowledge of the saola comes from carcasses found in local markets. All of the newly discovered mammals of Vietnam are quite rare, seen only infrequently even by local hunters. Fortunately, the Vietnamese government has established a number of national

parks and nature preserves in key areas. Only time will tell if these measures are sufficient to ensure the survival of the mysterious mammals of the Annamites.

Consider This The All Species Foundation is a nonprofit organization that promotes the goal of finding and naming all of Earth's undiscovered species within the next 25 years. According to the foundation, this task "deserves to be one of the great scientific goals of the new century." The foundation estimates the cost of the job at between $700 and $2000 per species, with perhaps millions of undiscovered species remaining to be found. Do you think the search for undiscovered species should continue? What value or benefit to humans does the search for new species provide?

CHAPTER REVIEW

SUMMARY OF KEY CONCEPTS

16.1 What Is a Species?

According to the biological species concept, a species is defined as all the populations of organisms that are potentially capable of interbreeding under natural conditions and that are reproductively isolated from other populations.

16.2 How Is Reproductive Isolation Between Species Maintained?

Reproductive isolation between species may be maintained by one or more of several mechanisms, known as premating isolating mechanisms and postmating isolating mechanisms. Premating isolating mechanisms include geographical isolation, ecological isolation, temporal isolation, behavioral isolation, and mechanical incompatibility. Postmating isolating mechanisms include gametic incompatibility, hybrid inviability, and hybrid infertility.

16.3 How Do New Species Form?

Speciation, the formation of new species, takes place when gene flow between two populations is reduced or eliminated and the populations diverge genetically. Most commonly, speciation follows geographical isolation and subsequent genetic divergence of the separated populations through genetic drift or natural selection.

Web Tutorial 16.1 The Process of Speciation

Web Tutorial 16.2 Allopatric Speciation

Web Tutorial 16.3 Speciation by Polyploidy

16.4 What Causes Extinction?

Factors that can lead to extinction, or the death of all the members of a species, include overspecialization, competition among species, and habitat destruction.

KEY TERMS

adaptive radiation *page 324*
allopatric speciation *page 321*
extinction *page 324*
isolating mechanism *page 317*
polyploidy *page 326*
postmating isolating mechanism *page 317*
premating isolating mechanism *page 317*
reproductive isolation *page 316*
speciation *page 320*
species *page 316*
sympatric speciation *page 321*

THINKING THROUGH THE CONCEPTS

1. Define the following terms: *species, speciation, allopatric speciation,* and *sympatric speciation*. Explain how allopatric and sympatric speciation might work, and give a hypothetical example of each.

2. Many of the oak tree species in central and eastern North America hybridize (interbreed). Are they "true species"?

3. Review the material on the possibility of sympatric speciation in *Rhagoletis* varieties that breed on apples or hawthorns. What types of genotypic, phenotypic, or behavioral data would convince you that the two forms have become separate species?

4. A drug called colchicine affects the mitotic spindle fibers and prevents cell division after the chromosomes have doubled at the start of meiosis. Describe how you would use colchicine to produce a new polyploid species of your favorite garden flower.

5. What are the two major types of reproductive isolating mechanisms? Give examples of each type, and describe how they work.

APPLYING THE CONCEPTS

1. The biological species concept has no meaning with regard to asexual organisms, and it is difficult to apply to extinct organisms that we know only as fossils. Can you devise a meaningful, useful species definition that would apply in all situations?

2. Seedless varieties of fruits and vegetables, created by breeders, are triploid. Explain why they are seedless.

3. Why do you suppose there are so many *endemic* species—that is, species found nowhere else—on islands? Why have the overwhelming majority of recent extinctions occurred on islands?

4. A biologist you've met claims that the fact that humans are pushing other species into small, isolated populations is good for biodiversity because these are the conditions that lead to new speciation events. Comment.

5. Southern Wisconsin is home to several populations of gray squirrels (*Sciurus carolinensis*) with black fur. Design a study to determine if they are actually a separate species.

6. It is difficult to gather data on speciation events in the past or to perform interesting experiments about the process of speciation. Does this difficulty make the study of speciation "unscientific"? Should we abandon the study of speciation?

FOR MORE INFORMATION

Eldredge, N. *Fossils: The Evolution and Extinction of Species*. New York: Abrams, 1991. A nicely illustrated exploration of a paleontologist's approach to examining and interpreting the past, including speciation events.

Levin, D. A. "Hybridization and Extinction." *American Scientist*, May–June 2002. A discussion of the impact of interbreeding on the conservation of rare species.

Quammen, D. *The Song of the Dodo*. New York: Scribner, 1996. Beautifully written exposition of the biology of islands. Read this book to understand why islands are known as "natural laboratories of speciation."

Schilthuizen, M. *Frogs, Flies, and Dandelions: Speciation—The Evolution of New Species*. Oxford: Oxford University Press, 2001. A readable and entertaining summary of the latest biological thought on species and speciation.

Sterling, E., Hurley, M., and Bain, R. "Vietnam's Secret Life." *Natural History*, March 2003. Nicely illustrated account of the distinctive species recently found in Vietnam's mountain forests.

Wilson, E. O. *The Diversity of Life*. New York: Norton, 1992. Elegant description of how species arise, how they disappear, and why we should preserve them.

17

The History of Life

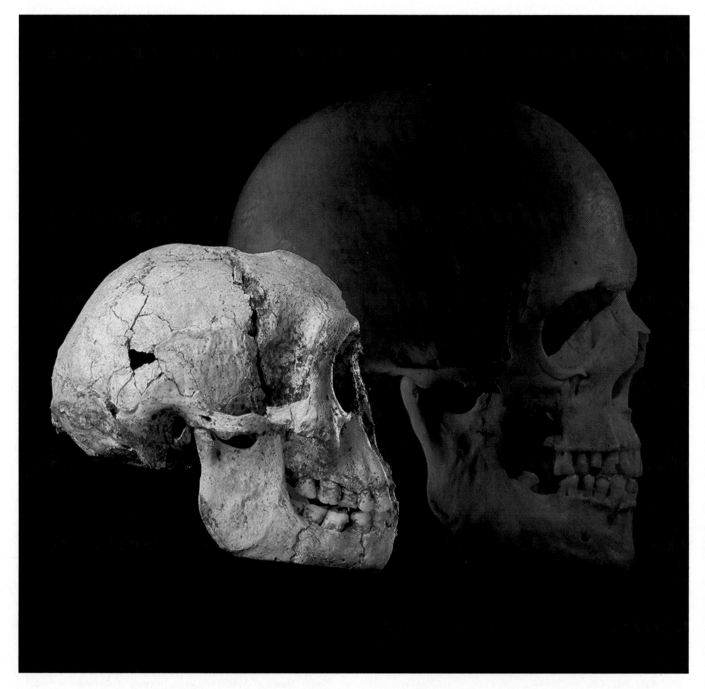

The skull of *Homo floresiensis*, a recently discovered diminutive human relative, is dwarfed by the skull of a modern *Homo sapiens*.

AT A GLANCE

 CASE STUDY LITTLE PEOPLE, BIG STORY

THE WORK OF PALEONTOLOGISTS, the scientists who study fossils, can be rather dull. A paleontologist might spend months, even years, digging very slowly in the soil of some remote location, carefully and laboriously separating small objects from dirt and debris. If all goes well, this labor will produce new information that adds to our understanding of the evolutionary history of life. And every once in a while, a diligent paleontologist will uncover something wonderfully surprising.

One such discovery was made by a small group of paleontologists searching beneath the floor of a cave on the Indonesian island of Flores. In the midst of an ongoing excavation, the researchers were pleased to find the fossilized skeleton of what they at first believed was a human child. Closer examination of the skeleton, however, revealed that it instead appeared to belong to a fully grown adult, but one that was no more than three feet tall. The researchers gave this extraordinary creature the nickname "Hobbit," and transported the skeleton to their lab for closer examination.

Unlike today's small humans, such as pygmies or pituitary dwarves, Hobbit had a very small brain, smaller even than the brain of a typical chimpanzee. So Hobbit was not simply a small *Homo sapiens*. Further tests ruled out the possibility that Hobbit's small size was due to an illness or genetic defect. The researchers concluded that Hobbit was a human relative of a previously unknown species, which they dubbed *Homo floresiensis*.

The bones of the *H. floresiensis* specimen are about 18,000 years old. Scientists had previously believed that, by well before 18,000 years ago, we were the only surviving member of the human family tree. Now, however, it seems clear that we shared Earth with close relatives until fairly recently. It is possible that, in the forests of Flores, people not too long ago encountered members of another, tiny human species. Perhaps still other recent human relatives await discovery.

Although the tale of *H. floresiensis* is especially significant in our human-centered view of the world, it is but one thread among the millions that together make up the story of life's evolution. We therefore turn our attention from our hobbit-like cousin to a brief tour of some of the highlights of life's history.

17.1 HOW DID LIFE BEGIN?

Pre-Darwinian thought held that all species were simultaneously created by God a few thousand years ago. Further, until the nineteenth century most people thought that new members of speceies sprang up all the time, through **spontaneous generation** from both nonliving matter and other, unrelated forms of life. In 1609 a French botanist wrote, "There is a tree … frequently observed in Scotland. From this tree leaves are falling; upon one side they strike the water and slowly turn into fishes, upon the other they strike the land and turn into birds." Medieval writings abound with similar observations. Microorganisms were thought to arise spontaneously from broth, maggots from meat, and mice from mixtures of sweaty shirts and wheat.

Experiments Refuted Spontaneous Generation

You may recall from Chapter 1 that in 1668, the Italian physician Francesco Redi disproved the maggots-from-meat hypothesis simply by keeping flies (whose eggs hatch into maggots) away from uncontaminated meat. In the mid-1800s, Louis Pasteur in France and John Tyndall in England disproved the broth-to-microorganism idea (**FIG. 17-1**). Although their work effectively demolished the notion of spontaneous generation, it did not address the question of how life on Earth originated in the first place. Or, as the biochemist Stanley Miller put it, "Pasteur never proved it didn't happen once, he only showed that it doesn't happen all the time."

The First Living Things Arose from Nonliving Ones

For almost half a century, the subject lay dormant. Eventually, biologists returned to the question of the origin of life. In the 1920s and 1930s, Alexander Oparin in Russia and John B. S. Haldane in England noted that today's oxygen-rich atmosphere would not have permitted the spontaneous formation of the complex organic molecules necessary for life.

Oxygen reacts readily with other molecules, disrupting chemical bonds. Thus, an oxygen-rich environment tends to keep molecules simple.

Oparin and Haldane speculated that the atmosphere of the young Earth must have contained very little oxygen and that, under such atmospheric conditions, complex organic molecules could have arisen through ordinary chemical reactions. Some kinds of molecules could persist in the lifeless environment of early Earth better than others, and would therefore become more common over time. This chemical version of the "survival of the fittest" is called *prebiotic* (meaning "before life") evolution. In the scenario envisioned by Oparin and Haldane, prebiotic chemical evolution gave rise to increasingly complex molecules and eventually to living organisms.

Organic Molecules Can Form Spontaneously Under Prebiotic Conditions

Inspired by the ideas of Oparin and Haldane, Stanley Miller and Harold Urey set out in 1953 to simulate prebiotic evolution in the laboratory. They knew that, based on the chemical composition of the rocks that formed early in Earth's history, geochemists had concluded that the early atmosphere probably contained virtually no oxygen gas, but did contain other substances, including methane, ammonia, hydrogen, and water vapor. Miller and Urey simulated the oxygen-free atmosphere of early Earth by mixing these components in a flask. An electrical discharge mimicked the intense energy of early Earth's lightning storms. In this experimental microcosm, the researchers found that simple organic molecules appeared after just a few days (**FIG. 17-2**). Similar experiments by Miller and others have produced amino acids, short proteins, nucleotides, adenosine triphosphate (ATP), and other molecules characteristic of living things.

In recent years, new evidence has convinced most geochemists that the actual composition of Earth's early atmosphere probably differed from the mixture of gases

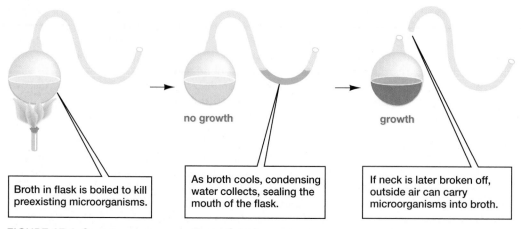

no growth

growth

Broth in flask is boiled to kill preexisting microorganisms.

As broth cools, condensing water collects, sealing the mouth of the flask.

If neck is later broken off, outside air can carry microorganisms into broth.

FIGURE 17-1 Spontaneous generation refuted
Louis Pasteur's experiment disproving the spontaneous generation of microorganisms in broth.

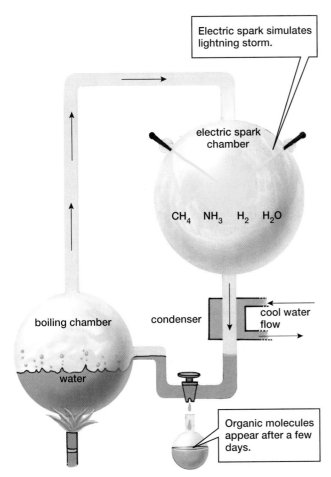

Electric spark simulates lightning storm.

electric spark chamber

CH_4 NH_3 H_2 H_2O

boiling chamber

condenser

cool water flow

water

Organic molecules appear after a few days.

FIGURE 17-2 The experimental apparatus of Stanley Miller and Harold Urey
Life's earliest stages left no fossils, so evolutionary historians have pursued a strategy of re-creating in the laboratory the conditions that may have prevailed on early Earth. The mixture of gases in the spark chamber simulates Earth's early atmosphere. QUESTION How would the experiment's result change if O_2 were included in the spark chamber?

surface. Analysis of present-day meteorites recovered from impact craters on Earth has revealed that some meteorites contain relatively high concentrations of amino acids and other simple organic molecules. Laboratory experiments suggest that these molecules could have formed in interstellar space before plummeting to Earth. When small molecules known to be present in space were placed under space-like conditions of very low temperature and pressure and bombarded with UV light, larger organic molecules were produced.

Organic Molecules Can Accumulate Under Prebiotic Conditions

Prebiotic synthesis was neither very efficient nor very fast. Nonetheless, in a few hundred million years, large quantities of organic molecules accumulated in the early Earth's oceans. Today, most organic molecules have a short life because they are either digested by living organisms or they react with atmospheric oxygen. Early Earth, however, lacked both life and free oxygen, so molecules would not have been exposed to these threats.

Still, the prebiotic molecules must have been threatened by the sun's high-energy UV radiation, because early Earth lacked an ozone layer. The ozone layer is a region high in today's atmosphere that is enriched with ozone (O_3) molecules, which absorb some of the sun's UV light before it reaches Earth's surface. Before the ozone layer formed, UV bombardment must have been fierce. UV radiation, as we have seen, can provide energy for the formation of organic molecules but it can also break them apart. Some places, however, such as those beneath rock ledges or at the bottoms of even fairly shallow seas, would have been protected from UV radiation. In these locations, organic molecules may have accumulated.

Clay May Have Catalyzed the Formation of Larger Organic Molecules

In the next stage of prebiotic evolution, simple molecules combined to form larger molecules. The chemical reactions that formed the larger molecules required that the reacting molecules be packed closely together. Scientists have proposed several processes by which the required high concentrations might have been achieved on early Earth. One possibility is that small molecules accumulated on the surfaces of clay particles, which can have a small electrical charge that attracts dissolved molecules with the opposite charge. Clustered on a clay particle, small molecules would have been sufficiently close together to allow chemical reactions between them. Researchers have demonstrated the plausibility of this scenario with experiments in which adding clay to solutions of dissolved small biological molecules catalyzed the formation of larger, more complex molecules. Such molecules might have formed on clay at the bottom of early Earth's oceans or lakes and gone on to become the building blocks of the first living organisms.

used in the pioneering Miller-Urey experiment. This improved understanding of the early atmosphere, however, has not undermined the basic finding of the Miller-Urey experiment. Additional experiments with more realistic (but still oxygen-free) simulated atmospheres have also yielded organic molecules. In addition, these experiments have shown that electricity is not the only suitable energy source. Other energy sources that were available on early Earth, such as heat or ultraviolet (UV) light, have also been shown to drive the formation of organic molecules in experimental simulations of prebiotic conditions. Thus, even though we may never know exactly what the earliest atmosphere was like, we can be confident that organic molecules formed spontaneously on early Earth.

Additional organic molecules probably arrived from space when meteorites and comets crashed into Earth's

RNA May Have Been the First Self-Reproducing Molecule

Although all living organisms use DNA to encode and store genetic information, it is unlikely that DNA was the earliest informational molecule. DNA can reproduce itself only with the help of large, complex protein enzymes, but the instructions for building these enzymes are encoded in DNA itself. For this reason, the origin of DNA's role as life's information storage molecule poses a "chicken and egg" puzzle: DNA requires proteins, but those proteins require DNA. It is thus difficult to construct a plausible scenario for the origin of self-replicating DNA from prebiotic molecules. It is therefore likely that the current DNA-based system of information storage evolved from an earlier system.

A prime candidate for the first self-replicating informational molecule is RNA. In the 1980s, Thomas Cech and Sidney Altman, working with the single-celled organism *Tetrahymena*, discovered a cellular reaction that was catalyzed not by a protein, but by a small RNA molecule. Because this special RNA molecule performed a function previously thought to be performed only by protein enzymes, Cech and Altman decided to give their catalytic RNA molecule the name **ribozyme.**

In the years since the discovery of these molecules, researchers have found dozens of naturally occurring ribozymes that catalyze a variety of reactions, including cutting other RNA molecules and splicing together different RNA fragments. Ribozymes have also been found in the protein-manufacturing machinery of cells, where they help catalyze the attachment of amino acid molecules to growing proteins. In addition, researchers have been able to synthesize different ribozymes in the laboratory, including some that can catalyze the replication of small RNA molecules.

The discovery that RNA molecules can act as catalysts for diverse reactions, including RNA replication, provides support for the hypothesis that life arose in an "RNA world." According to this view, the current era of DNA-based life was preceded by one in which RNA served as both the information-carrying genetic molecule and the enzyme catalyst for its own replication. This RNA world may have emerged after hundreds of millions of years of prebiotic chemical synthesis, during which RNA nucleotides would have been among the molecules synthesized. After reaching a sufficiently high concentration, perhaps on clay particles, the nucleotides probably bonded together to form short RNA chains.

Let's suppose that, purely by chance, one of these RNA chains was a ribozyme that could catalyze the production of copies of itself. This first self-reproducing ribozyme probably wasn't very good at its job and produced copies with lots of errors. These mistakes were the first mutations. Like modern mutations, most undoubtedly ruined the catalytic abilities of the "daughter molecules," but a few may have been improvements. Such improvements set the stage for the evolution of RNA molecules, as variant ribozymes with increased speed and accuracy of replication repro-

duced faster, making more copies of themselves and displacing less-efficient molecules. Molecular evolution in the RNA world proceeded until, by some still unknown chain of events, RNA gradually receded into its present role as an intermediary between DNA and protein enzymes.

Membrane-Like Vesicles May Have Enclosed Ribozymes

Self-replicating molecules alone do not constitute life; these molecules must be contained within some kind of enclosing membrane. The precursors of the earliest biological membranes may have been simple structures that formed spontaneously from purely physical, mechanical processes. For example, chemists have shown that if water containing proteins and lipids is agitated to simulate waves beating against ancient shores, the proteins and lipids combine to form hollow structures called *vesicles*. These hollow balls resemble living cells in several respects. They have a well-defined outer boundary that separates their internal contents from the external solution. If the composition of the vesicle is right, a "membrane" forms that is remarkably similar in appearance to a real cell membrane. Under certain conditions, vesicles can absorb material from the external solution, grow, and even divide.

If a vesicle happened to surround the right ribozymes, it would form something resembling a living cell. We could call it a **protocell**, structurally similar to a cell but not a living thing. In the protocell, ribozymes and any other enclosed molecules would have been protected from free-roaming ribozymes in the primordial soup. Nucleotides and other small molecules might have diffused across the membrane and been used to synthesize new ribozymes and other complex molecules. After sufficient growth, the vesicle may have divided, with a few copies of the ribozymes becoming incorporated into each daughter vesicle. If this process occurred, the path to the evolution of the first cells would be nearly at its end.

Was there a particular moment when a nonliving protocell gave rise to something alive? Probably not. Like most evolutionary transitions, the change from protocell to living cell was a continuous process, with no sharp boundary between one state and the next.

But Did All This Happen?

The above scenario, although plausible and consistent with many research findings, is by no means certain. One of the most striking aspects of origin-of-life research is a great diversity of assumptions, experiments, and contradictory hypotheses. (Iris Fry's *The Emergence of Life on Earth*, cited in the "For More Information" section at the end of this chapter, offers a taste of these controversies.) Researchers disagree as to whether life arose in quiet pools, in the sea, in moist films on the surfaces of crystals, or in furiously hot deep-sea vents. A few researchers even argue that life arrived on Earth from space. Can we draw any conclusions from the research conducted so far? No one knows for sure, but we can make a few observations.

First, the experiments of Miller and others show that amino acids, nucleotides, and other organic molecules, along with simple membrane-like structures, would have formed in abundance on the early Earth. Second, chemical evolution had long periods of time and huge areas of the Earth available to it. Given sufficient time and a sufficiently large pool of reactant molecules, even extremely rare events can occur many times. So, even if prebiotic evolution yielded only simple molecules, the earliest catalysts were not very efficient, and the earliest membranes were simple, the vast expanses of available time and space would have increased the likelihood of each small step on the path from primordial soup to living cell.

Most biologists accept that the origin of life was probably an inevitable consequence of the working of natural laws. We should emphasize, however, that this proposition cannot be definitively tested. The origin of life left no record, and researchers exploring this mystery can proceed only by developing a hypothetical scenario and then conducting laboratory investigations to determine if the scenario's steps are chemically and biologically possible and plausible.

17.2 WHAT WERE THE EARLIEST ORGANISMS LIKE?

When Earth first formed about 4.5 billion years ago, it was quite hot (FIG. 17-3). A multitude of meteorites smashed into the forming planet, and the kinetic energy of these extraterrestrial rocks was converted into heat on impact. Still more heat was released by the decay of radioactive atoms. The rock composing Earth melted, and heavier ele-ments such as iron and nickel sank to the center of the planet, where they remain molten even today. It must have taken hundreds of millions of years for Earth to cool enough to allow water to exist as a liquid. Nonetheless, it appears that life arose in fairly short order once liquid water was available.

The oldest fossil organisms found so far are in rocks that are about 3.5 billion years old. (Their age was determined using radiometric dating techniques; see "Scientific Inquiry: How Do We Know How Old a Fossil Is?".) Chemical traces in older rocks have led some paleontologists to believe that life is even older, perhaps as old as 3.9 billion years.

The period in which life began is known as the Precambrian era. This interval was designated by geologists and paleontologists, who have devised a hierarchical naming system of eras, periods, and epochs to delineate the immense span of geological time (Table 17-1).

The First Organisms Were Anaerobic Prokaryotes

The first cells to arise in Earth's oceans were **prokaryotes**, cells whose genetic material was not contained within a nucleus, separate from the rest of the cell. These cells probably obtained nutrients and energy by absorbing organic molecules from their environment. There was no oxygen gas in the atmosphere, so the cells must have metabolized the organic molecules anaerobically. You will recall from Chapter 8 that anaerobic metabolism yields only small amounts of energy.

Thus, the earliest cells were primitive anaerobic bacteria. As these bacteria multiplied, they must have eventually used up the organic molecules produced by prebiotic chemical reactions. Simpler molecules, such as carbon dioxide

FIGURE 17-3 Early Earth
Life began on a planet characterized by abundant volcanic activity, frequent electrical storms, repeated meteorite strikes, and an atmosphere that lacked oxygen gas.

Table 17-1   The History of Life on Earth

Era	Period	Epoch	Millions of Years Ago*	Major Events
Cenozoic	Quaternary	Recent	0.01–present	Evolution of genus *Homo*; repeated glaciations in Northern Hemisphere; extinction of many giant mammals.
		Pleistocene	1.8–0.01	
	Tertiary	Pliocene	5–1.8	Widespread flourishing of birds, mammals, insects, and flowering plants; shifting of continents into modern postions; mild climate at beginning of period, with extensive mountain building and cooling toward end.
		Miocene	23–5	
		Oligocene	38–23	
		Eocene	54–38	
		Paleocene	65–54	
Mesozoic	Cretaceous		146–65	Flowering plants appear and become dominant; mass extinctions of marine life and some terrestrial life, including last dinosaurs; modern continents well separated.
	Jurassic		208–146	Dominance of dinosaurs and conifers; first birds; continents partially separated.
	Triassic		245-208	First mammals and dinosaurs; forests of gymnosperms and tree ferns; beginning of breakup of Pangaea.
Paleozoic	Permian		286–245	Massive marine extinctions, including last of trilobites; flourishing of reptiles and decline of amphibians; aggregation of continents into one land mass, Pangaea.
	Carboniferous		360–286	Swamp forests of tree ferns and club mosses; first conifers; dominance of amphibians; numerous insects; first reptiles.
	Devonian		410–360	Fishes and trilobites flourish in sea; first amphibians and insects; first seeds and pollen.
	Silurian		440–410	Many fishes, trilobites, mollusks in sea; first vascular plants; invasion of land by plants; invasion of land by arthropods.
	Ordovician		505–440	Invertebrates, especially arthropods and mollusks, dominate in sea; first fungi.
	Cambrian		544–505	Primitive marine algae flourish; origin of most marine invertebrate types; first fish.
Precambrian			About 1000	First animals (soft-bodied marine invertebrates).
			1200	First multicellular organisms.
			2000	First eukaryotes.
			2200	Accumulation of free oxygen in atmosphere.
			3500	Origin of photosynthesis (in cyanobacteria).
			3900–3500	First living cells (prokaryotes).
			4000–3900	Appearance of first rocks on Earth.
			4600	Origin of solar system and Earth.

and water, would still have been very abundant, as was energy in the form of sunlight. What was lacking, then, was not materials or energy itself, but energetic molecules—molecules in which energy is stored in chemical bonds.

Some Organisms Evolved the Ability to Capture the Sun's Energy

Eventually, some cells evolved the ability to use the energy of sunlight to drive the synthesis of complex, high-energy molecules from simpler molecules: In other words, photosynthesis appeared. Photosynthesis requires a source of hydrogen, and the very earliest photosynthetic bacteria probably used hydrogen sulfide gas dissolved in water for this purpose (much as today's purple photosynthetic bacteria do). Eventually, however, Earth's supply of hydrogen sulfide (which is produced mainly by volcanoes) must have run low. The shortage of hydrogen sulfide set the stage for the evolution of photosynthetic bacteria that were able to use the planet's most abundant source of hydrogen: water (H_2O).

Photosynthesis Increased the Amount of Oxygen in the Atmosphere

Water-based photosynthesis converts water and carbon dioxide to energetic molecules of sugar, releasing oxygen as a by-product. The emergence of this new method for capturing energy introduced significant amounts of free oxygen to the atmosphere for the first time. At first, the newly liberated oxygen was quickly consumed by reactions with other molecules in the atmosphere and in Earth's crust, or surface layer. One especially common reactive atom in the crust was iron, and much of the new oxygen combined with iron atoms to form huge deposits of iron oxide (also known as rust).

After all the accessible iron had turned to rust, the concentration of oxygen gas in the atmosphere began to increase. Chemical analysis of rocks suggests that significant amounts of oxygen first appeared in the atmosphere about 2.2 billion years ago, produced by bacteria that were probably very similar to modern cyanobacteria. (You will undoubtedly breathe in some oxygen molecules today that were expelled 2 billion years ago by one of these early cyanobacteria.) Atmospheric oxygen levels increased steadily until they reached a stable level about 1.5 billion years ago. Since that time, the proportion of oxygen in the atmosphere has been nearly constant, as the amount of oxygen released by photosynthesis worldwide is neatly balanced by the amount that is consumed by aerobic respiration.

Aerobic Metabolism Arose in Response to the Oxygen Crisis

Oxygen is potentially very dangerous to living things, because it reacts with organic molecules, destroying them. Many of today's anaerobic bacteria perish when exposed to what is for them a deadly poison, oxygen. The accumulation of oxygen in the atmosphere of early Earth probably exterminated many organisms and fostered the evolution of cellular mechanisms for detoxifying oxygen. This crisis for evolving life also provided the environmental pressure for the next great advance in the Age of Microbes: the ability to use oxygen in metabolism. This ability not only provides a defense against the chemical action of oxygen, but actually channels oxygen's destructive power through aerobic respiration to generate useful energy for the cell. Because the amount of energy available to a cell is vastly increased when oxygen is used to metabolize food molecules, aerobic cells had a significant selective advantage.

Some Organisms Acquired Membrane-Enclosed Organelles

Hordes of bacteria would offer a rich food supply to any organism that could eat them. There are no fossils of the first predatory cells to roam the seas, but paleobiologists speculate that once a suitable prey population (such as these bacteria) appeared, predation would have evolved quickly. According to the most widely accepted hypothesis, these early predators were prokaryotes that had evolved to become larger than typical bacteria. In addition, they had lost the rigid cell wall that surrounds most bacterial cells, so that their flexible plasma membrane was in contact with the surrounding environment. Thus, the predatory cells were able to envelop smaller bacteria in an infolded pouch of membrane and in this fashion engulf whole bacteria as prey.

These early predators were probably capable of neither photosynthesis nor aerobic metabolism. Although they could capture large food particles, namely bacteria, they metabolized them inefficiently. By about 1.7 billion years ago, however, one predator probably gave rise to the first eukaryotic cell.

The Internal Membranes of Eukaryotes May Have Arisen Through Infolding of the Plasma Membrane

As you know, the cells of **eukaryotes** differ from prokaryotic cells in having an elaborate system of internal membranes, including a nucleus that contains the cell's genetic material. These internal membranes may have originally arisen through inward folding of the cell membrane of a single-celled predator. If, as it is in most of today's bacteria, the DNA of the eukaryotes' ancestor was attached to the inside of its cell membrane, an infolding of the membrane near the point of DNA attachment may have pinched off and become the precursor of the cell nucleus.

In addition to the nucleus, other key eukaryotic structures include the organelles used for energy metabolism: mitochondria and (in plants and algae) chloroplasts. How did these organelles evolve?

Mitochondria and Chloroplasts May Have Arisen from Engulfed Bacteria

The **endosymbiont hypothesis** proposes that early eukaryotic cells acquired the precursors of mitochondria and chloroplasts by engulfing certain types of bacteria.

SCIENTIFIC INQUIRY   How Do We Know How Old a Fossil Is?

Early geologists could date rock layers and their accompanying fossils only in a *relative* way: fossils found in deeper layers of rock were generally older than those found in shallower layers. With the discovery of radioactivity, it became possible to determine *absolute* dates, within certain limits of uncertainty. The nuclei of radioactive elements spontaneously break down, or decay, into other elements. For example, carbon-14 (usually written ^{14}C) decays by emitting an electron to become nitrogen-14 (^{14}N). Each radioactive element decays at a rate that is independent of temperature, pressure, or the chemical compound of which the element is a part. The time it takes for half of a radioactive element's nuclei to decay at this characteristic rate is called its *half-life*. The half-life of ^{14}C, for example, is 5730 years.

How are radioactive elements used in determining the age of rocks? If we know the rate of decay and measure the proportion of decayed nuclei to undecayed nuclei, we can estimate how much time has passed since these radioactive elements became trapped in the rock. This process is called *radiometric dating*. A particularly straightforward dating technique measures the decay of potassium-40 (^{40}K), which has a half-life of about 1.25 billion years, into argon-40 (^{40}Ar). Potassium is a very reactive element commonly found in volcanic rocks such as granite and basalt. Argon, however, is an unreactive gas. Let's suppose that a volcano erupts with a massive lava flow, covering the countryside. All the ^{40}Ar, being a gas, will bubble out of the molten lava, so when the lava first cools and solidifies into rock, it will not contain any ^{40}Ar. Meanwhile, any ^{40}K present in the hardened lava will decay to ^{40}Ar, with half of the ^{40}K decaying every 1.25 billion years. This ^{40}Ar gas will be trapped in the rock. A geologist could take a sample of the rock and determine the proportion of ^{40}K to ^{40}Ar (**FIG. E17-1**). If the analysis finds equal amounts of the two elements, the geologist will conclude that the lava hardened 1.25 billion years ago. With appropri-

ate care, such age estimates are quite reliable. If a fossil is found beneath a lava flow dated at, say, 500 million years, then we know that the fossil is at least that old.

As some radioactive elements decay, they can even give an estimate of the age of the solar system. Analysis of uranium, which decays to lead, has shown that the oldest meteorites and moon rocks collected by astronauts are about 4.6 billion years old.

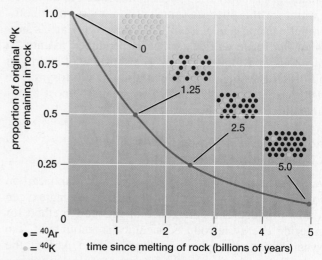

FIGURE E17-1 The relationship between time and the decay of radioactive ^{40}K to ^{40}Ar

EXERCISE Uranium-235 decays to lead-207 with a half-life of 713 million years. If you analyze a rock and find that it contains uranium-235 and lead-207 in a ratio of 3:1, how old is the rock?

These cells and the bacteria trapped inside them (*endo* means "within") gradually entered into a *symbiotic* relationship, a close association between different types of organisms over an extended time. How might this have happened?

Let's suppose that an anaerobic predatory cell captured an aerobic bacterium for food, as it often did, but for some reason failed to digest this particular prey. The aerobic bacterium remained alive and well. In fact, it was better off than ever, because the cytoplasm of its predator-host was chock-full of half-digested food molecules, the remnants of anaerobic metabolism. The aerobe absorbed these molecules and used oxygen to metabolize them, thereby gaining enormous amounts of energy. So abundant were the aerobe's food resources, and so bountiful its energy production, that the aerobe must have leaked energy, probably as ATP or similar molecules, back into its host's cytoplasm.

The anaerobic predatory cell with its symbiotic bacteria could now metabolize food aerobically, gaining a great advantage over other anaerobic cells and leaving a greater number of offspring. Eventually, the endosymbiotic bacterium lost its ability to live independently of its host, and the mitochondrion was born (**FIG. 17-4, ① and ②**).

One of these successful new cellular partnerships must have managed a second feat: it captured a photosynthetic cyanobacterium and similarly failed to digest its prey. The cyanobacterium flourished in its new host and gradually evolved into the first chloroplast (**FIG. 17-4, ③ and ④**). Other eukaryotic organelles may have also originated through endosymbiosis. Many biologists believe that cilia, flagella, centrioles, and microtubules may all have evolved from a symbiosis between a spirilla-like bacterium (a form of bacterium with an elongated corkscrew shape) and an early eukaryotic cell.

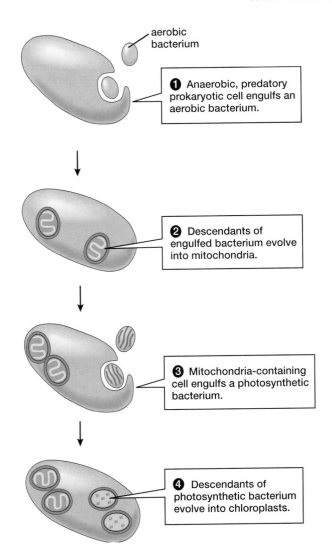

1 Anaerobic, predatory prokaryotic cell engulfs an aerobic bacterium.

aerobic bacterium

2 Descendants of engulfed bacterium evolve into mitochondria.

3 Mitochondria-containing cell engulfs a photosynthetic bacterium.

4 Descendants of photosynthetic bacterium evolve into chloroplasts.

FIGURE 17-4 The probable origin of mitochondria and chloroplasts in eukaryotic cells
QUESTION Scientists have identified a living bacterium believed to be descended from the endosymbiont that gave rise to mitochondria. Would you expect the DNA sequence of this modern bacterium to be most similar to the sequence of DNA from a plant chloroplast, an animal cell nucleus, or a plant mitochondrion?

Evidence for the Endosymbiont Hypothesis Is Strong

Several types of evidence support the endosymbiont hypothesis. A particularly compelling line of evidence is the many distinctive biochemical features shared by eukaryotic organelles and living bacteria. In addition, mitochondria, chloroplasts, and centrioles each contain their own minute supply of DNA, which many researchers interpret as remnants of the DNA originally contained within the engulfed bacteria.

Another kind of support comes from *living intermediates*, organisms alive today that are similar to hypothetical ancestors and thus help show that a proposed evolutionary pathway is plausible. For example, the amoeba *Pelomyxa palustris* lacks mitochondria but hosts a permanent population of aerobic bacteria that carry out much the same

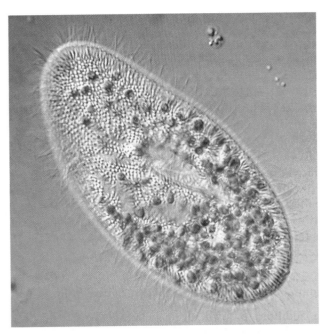

FIGURE 17-5 Symbiosis within a modern cell
The ancestors of the chloroplasts in today's plant cells may have resembled *Chlorella*, the green, photosynthetic, single-celled algae living symbiotically within the cytoplasm of the *Paramecium* pictured here.

role. Similarly, a variety of corals, some clams, a few snails, and at least one species of *Paramecium* harbor a permanent collection of photosynthetic algae in their cells (**FIG. 17-5**). These examples of modern cells that host bacterial endosymbionts suggest that we have no reason to doubt that similar symbiotic associations could have occurred almost 2 billion years ago and led to the first eukaryotic cells.

17.3 WHAT WERE THE EARLIEST MULTICELLULAR ORGANISMS LIKE?

Once predation had evolved, increased size became an advantage. In the marine environments to which life was restricted, a larger cell could easily engulf a smaller cell and would also be difficult for other predatory cells to ingest. Larger organisms can also generally move faster than small ones, making successful predation and escape more likely. But enormous single cells have problems. Oxygen and nutrients going into the cell and waste products going out must diffuse through the plasma membrane. The larger a cell becomes, the less surface membrane is available per unit volume of cytoplasm.

There are only two ways that an organism larger than a millimeter or so in diameter can survive. First, it can have a low metabolic rate so that it doesn't need much oxygen or produce much carbon dioxide. This strategy seems to work for certain very large single-celled algae. Alternatively, an organism can be multicellular; that is, it may consist of many small cells packaged into a larger, unified body.

Some Algae Became Multicellular

The oldest fossils of multicellular organisms are about 1.2 billion years old and include impressions of the first multi-cellular algae, which arose from single-celled eukaryotic cells containing chloroplasts. Multicellularity would have provided at least two advantages for these seaweeds. First, large, many-celled algae would have been difficult for sin-gle-celled predators to engulf. Second, specialization of cells would have provided the potential for staying in one place in the brightly lit waters of the shoreline, as rootlike structures burrowed in sand or clutched onto rocks, while leaflike structures floated above in the sunlight. The green, brown, and red algae lining our shores today—some, such as the brown kelp, more than 200 feet long—are the de-scendants of these early multicellular algae.

Animal Diversity Arose in the Precambrian Era

In addition to fossil algae, billion-year-old rocks have yielded fossil traces of animal tracks and burrows. This ev-idence of early animal life notwithstanding, fossils of ani-mal bodies first appear in Precambrian rocks laid down between 610 million and 544 million years ago. Some of these ancient invertebrate animals (animals lacking a backbone) are quite different in appearance from any ani-mals that appear in later fossil layers and may represent types of animals that left no descendants. Others fossils in these rock layers, however, appear to be ancestors of today's animals. Ancestral sponges and jellyfish appear in the oldest layers, followed later by ancestors of worms, mollusks, and arthropods.

The full range of modern invertebrate animals, howev-er, does not appear in the fossil record until the Cambrian period, marking the beginning of the Paleozoic era, about 544 million years ago. (The phrase "fossil record" is a shorthand reference to the entire collection of all fossil ev-idence that has been found to date.) These Cambrian fos-sils reveal an adaptive radiation (see Chapter 16) that had already yielded a diverse array of complex body plans. Al-most all of the major groups of animals on Earth today were already present in the early Cambrian. The apparent-ly sudden appearance of so many different kinds of ani-mals suggests that the earlier evolutionary history that produced such an impressive range of different animal forms is not preserved in the fossil record.

The early diversification of animals was probably dri-ven in part by the emergence of predatory lifestyles. Co-evolution of predator and prey led to the evolution of new features in many kinds of animals. By the Silurian period (440 million to 410 million years ago), mud-skimming, ar-mored trilobites were preyed on by ammonites and the chambered nautilus, which still survives in almost un-changed form in deep Pacific waters (**FIG. 17-6**).

Many animals of the Paleozoic era were more mobile than their evolutionary predecessors. Predators gain an ad-vantage by being able to travel over wide areas in search of suitable prey, and the ability to make a speedy escape is an advantage for prey. The evolution of efficient movement was often associated with the evolution of greater sensory capabilities and more complex nervous systems. Senses for detecting touch, chemicals, and light became highly devel-oped, along with nervous systems capable of handling the sensory information and directing appropriate behaviors.

About 530 million years ago, one group of animals—the fishes—developed a new form of support for the body: an internal skeleton. These early fishes were inconspicuous members of the ocean community, but by 400 million years ago fishes were a diverse and prominent group. By and large, the fishes proved to be faster than the invertebrates, with more-acute senses and larger brains. Eventually, they became the dominant predators of the open seas.

17.4 HOW DID LIFE INVADE THE LAND?

A compelling subplot in the long tale of life's history is the story of life's invasion of land after more than 3 billion years of a strictly watery existence. In moving to solid ground, organisms had many obstacles to overcome. Life in the sea provides buoyant support against gravity, but on land an organism must bear its weight against the crushing force of gravity. The sea provides ready access to life-sus-taining water, but a terrestrial organism must find ade-quate water. Sea-dwelling plants and animals can reproduce by means of mobile sperm or eggs, or both, which swim to each other through the water, but the ga-metes of land dwellers must be protected from drying out.

Despite the obstacles to life on land, the vast empty spaces of the Paleozoic landmass represented a tremen-dous evolutionary opportunity. The potential rewards of terrestrial life were especially great for plants. Water strongly absorbs light, so even in the clearest water, photo-synthesis is limited to the upper few hundred meters of depth, and usually much less. Out of the water, the dazzling brightness of the sun permits rapid photosynthesis. Fur-thermore, terrestrial soils are rich storehouses of nutrients, whereas seawater tends to be low in certain nutrients, par-ticularly nitrogen and phosphorus. Finally, the Paleozoic sea swarmed with plant-eating animals, but the land was devoid of animal life. The plants that first colonized the land would have had ample sunlight, untouched nutrient sources, and no predators.

Some Plants Became Adapted to Life on Dry Land

In moist soils at the water's edge, a few small green algae began to grow, taking advantage of the sunlight and nu-trients. They didn't have large bodies to support against the force of gravity, and by living right in the film of water on the soil, they could easily obtain water. About 475 million years ago, some of these algae gave rise to the first multicellular land plants. Initially simple, low-growing forms, land plants rapidly evolved solutions to two of the main difficulties of plant life on land: obtain-ing and conserving water and staying upright despite

FIGURE 17-6 Diversity of ocean life during the Silurian period
(a) Life characteristic of the oceans during the Silurian period, 440 million to 410 million years ago. Among the most common fossils from that time are **(b)** trilobites and their predators, such as nautiloids and **(c)** ammonites. This **(d)** living *Nautilus* is very similar in structure to the Silurian nautiloids, showing that a successful body plan may exist virtually unchanged for hundreds of millions of years.

gravity and winds. Waterproof coatings on aboveground parts reduced water loss by evaporation, and rootlike structures delved into the soil, mining water and minerals. Specialized cells formed tubes called vascular tissues to conduct water from roots to leaves. Extra-thick walls surrounding certain cells enabled stems to stand erect.

Primitive Land Plants Retained Swimming Sperm and Required Water to Reproduce

Reproduction out of water presented challenges. As do animals, plants produce sperm and eggs, which must be able to meet in order to produce the next generation. The first land plants had swimming sperm, presumably much like those of some of today's marine algae (some of which have swimming eggs as well). Consequently, the earliest plants were restricted to swamps and marshes, where the sperm and eggs could be released into the water, or to areas with abundant rainfall, where the ground would occasionally be covered with water. Later, plants with swimming sperm prospered during periods in which the climate was warm and moist. For example, the Carboniferous period (360 million to 286 million years ago) was characterized by vast forests of giant tree ferns and club mosses (**FIG. 17-7**). The coal we mine today is derived from the fossilized remains of those forests.

Seed Plants Encased Sperm in Pollen Grains

Meanwhile, some plants inhabiting drier regions had evolved a means of reproduction that no longer depended on water. The eggs of these plants were retained on the parent plant, and the sperm were encased in drought-resistant pollen grains that traveled on the wind from plant to plant. When the pollen grains landed near an egg, they released sperm cells directly into living tissue, eliminating the need for a surface film of water. The fertilized egg remained on the parent plant, where it developed inside a

FIGURE 17-7 The swamp forest of the Carboniferous period
The treelike plants in this artist's reconstruction are tree ferns and giant club mosses, most species of which are now extinct. QUESTION Why are today's ferns and club mosses so small in comparison to their giant ancestors?

seed, which provided protection and nutrients for the developing embryo within.

The earliest seed-bearing plants appeared in the late Devonian period (375 million years ago) and produced their seeds along branches, without any specialized structures to hold them. By the middle of the Carboniferous period, however, a new kind of seed-bearing plant had arisen. These plants, called **conifers**, protected their developing seeds inside cones. Conifers, which did not depend on water for reproduction, flourished and spread during the Permian period (286 to 245 million years ago) when mountains rose, swamps drained, and the climate became much drier. The good fortune of the conifers, however, was not shared by the tree ferns and giant club mosses, which, with their swimming sperm, largely went extinct.

Flowering Plants Enticed Animals to Carry Pollen

About 140 million years ago, during the Cretaceous period, the flowering plants appeared, having evolved from a group of conifer-like plants. Many flowering plants are pollinated by insects and other animals, and this mode of pollination seems to have conferred an evolutionary advantage. Flower pollination by animals can be far more efficient than pollination by wind. Wind-pollinated plants must produce an enormous amount of pollen because the vast majority of pollen grains fail to reach their target. Flowering plants also evolved other advantages, including more-rapid reproduction and, in some cases, much more rapid growth. Today, flowering plants dominate the land, except in cold northern regions, where conifers still prevail.

Some Animals Became Adapted to Life on Dry Land

Soon after land plants evolved, providing potential food sources for other organisms, animals emerged from the sea. The first animals to move onto land were **arthropods** (the group that today includes insects, spiders, scorpions, centipedes, and crabs). Why arthropods? The answer seems to be that they already possessed certain structures that, purely by chance, were suited to life on land. Foremost among these structures was the external skeleton, or **exoskeleton**, a hard covering surrounding the body, such as the shell of a lobster or crab. Exoskeletons are both waterproof and strong enough to support a small animal against the force of gravity.

For millions of years, arthropods had the land and its plants to themselves, and for tens of millions of years more, they were the dominant land animals. Dragonflies with a wingspan of 28 inches (70 centimeters) flew among the Carboniferous tree ferns, while millipedes 6.5 feet (2 meters) long munched their way across the swampy forest floor. Eventually, however, the arthropods' splendid isolation came to an end.

Amphibians Evolved from Lobefin Fishes

About 400 million years ago, a group of Silurian fishes called the lobefins appeared, probably in freshwater. **Lobefins** had two important features that would later enable their descendants to colonize land: (1) stout, fleshy fins with which they crawled about on the bottoms of shallow, quiet waters, and (2) an outpouching of the digestive tract that could be filled with air, like a primitive lung. One group of lobefins colonized very shallow ponds and streams, which shrank during droughts and often became oxygen poor. By taking air into their lungs, these lobefins could obtain oxygen anyway. Some began to use their fins to crawl from pond to pond in search of prey or water, as some modern fish do today (FIG. 17-8).

The benefits of feeding on land and moving from pool to pool favored the evolution of a group of animals that could stay out of water for longer periods and that could

FIGURE 17-8 A fish that walks on land
Some modern fishes, such as this mudskipper, walk on land. Like the ancient lobefin fishes that gave rise to amphibians, mudskippers use their strong pectoral fins to move across dry areas in their swampy habitats. QUESTION Does the mudskipper's ability to walk on land constitute evidence that lobefin fishes were the ancestors of amphibians?

move about more effectively on land. With improvements in lungs and legs, **amphibians** evolved from lobefins, first appearing in the fossil record about 350 million years ago. To an amphibian, the Carboniferous swamp forests were a kind of paradise: no predators to speak of, abundant prey, and a warm, moist climate. As had the insects and millipedes, some amphibians evolved gigantic size, including salamanders more than 10 feet (3 meters) long.

Despite their success, the early amphibians were not fully adapted to life on land. Their lungs were simple sacs without very much surface area, so they had to obtain some of their oxygen through their skin. Therefore, their skin had to be kept moist, a requirement that restricted

them to swampy habitats where they wouldn't dry out. Further, amphibian sperm and eggs could not survive in dry surroundings and had to be deposited in watery environments. So, although amphibians could move about on land, they could not stray too far from the water's edge. Along with the tree ferns and club mosses, amphibians declined when the climate turned dry at the beginning of the Permian period about 286 million years ago.

Reptiles Evolved from Amphibians

As the conifers were evolving on the fringes of the swamp forests, a group of amphibians was also evolving adaptations to drier conditions. These amphibians ultimately gave rise to the **reptiles**, which had three major adaptations to life on land. First, reptiles evolved shelled, waterproof eggs that enclosed a supply of water for the developing embryo. Thus, eggs could be laid on land without the reptiles' having to venture back to the dangerous swamps full of fish and amphibian predators. Second, ancestral reptiles evolved scaly, waterproof skin that helped prevent the loss of body water to the dry air. Finally, reptiles evolved improved lungs that were able to provide the entire oxygen supply for an active animal. As the climate dried during the Permian period, reptiles became the dominant land vertebrates, relegating amphibians to the swampy backwaters where most of them remain today.

A few tens of millions of years later, the climate returned to more moist and stable conditions. This period saw the evolution of some very large reptiles, in particular the dinosaurs. The variety of dinosaur forms was enormous—from predators (**FIG. 17-9**) to plant eaters, from those that dominated the land to others that took to the air, to still others that returned to the sea. Dinosaurs were

FIGURE 17-9 A reconstruction of a Cretaceous forest
By the Cretaceous period, flowering plants dominated terrestrial vegetation. Dinosaurs, such as the predatory pack of 6-foot-long *Velociraptors* shown here, were the preeminent land animals. Although small by dinosaur standards, velociraptors were formidable predators with great running speed, sharp teeth, and deadly, sickle-like claws on their hind feet.

among the most successful animals ever, if we consider persistence as a measure of success. They flourished for more than a hundred million years, until about 65 million years ago, when the last dinosaurs went extinct. No one is certain why they died out, but the aftereffects of a gigantic meteorite's impact with Earth seem to have been the final blow (as discussed in the following section).

Even during the age of dinosaurs, many reptiles remained quite small. One major difficulty faced by small reptiles is maintaining a high body temperature. Being active on land is helped by a warm body that maximizes the efficiency of the nervous system and muscles. But a warm body loses heat to the environment unless the air is also warm. Heat loss is a big problem for small animals, which have a larger surface area per unit of weight than do larger animals. Many species of small reptiles have retained slow metabolisms and have coped with the heat-loss problem by developing lifestyles in which they remain active only when the air is sufficiently warm. Two groups of small reptiles, however, independently followed a different evolutionary pathway: they developed insulation. One group evolved feathers, and another evolved hair.

Reptiles Gave Rise to Both Birds and Mammals

In ancestral birds, insulating feathers helped retain body heat. Consequently, these animals could be active in cool habitats and during the night, when their scaly relatives became sluggish. Later, some ancestral birds evolved longer, stronger feathers on their forelimbs, perhaps under selection for better ability to glide from trees or to jump after insect prey. Ultimately, feathers evolved into structures capable of supporting powered flight. Fully developed, flight-capable feathers are present in 150-million-year-old fossils, so the earlier insulating structures that eventually developed into flight feathers must have been present well before that time.

The earliest fossil **mammal** so far unearthed is almost 200 million years old. Early mammals thus coexisted with the

dinosaurs. They were mostly small creatures. The largest known mammal from the dinosaur era was about the size of a modern raccoon, but most early mammal species were far smaller than that. When the dinosaurs went extinct, however, mammals colonized the habitats left empty by the extinctions. Mammal species prospered, diversifying into the array of modern forms.

Unlike birds, which retained the reptilian habit of laying eggs, mammals evolved live birth and the ability to feed their young with secretions of the mammary (milk-producing) glands. Ancestral mammals also developed hair, which provided insulation. Because the uterus, mammary glands, and hair do not fossilize, we may never know when these structures first appeared, or what their intermediate forms looked like. Recently, however, a team of paleontologists found bits of fossil hair preserved in coprolites, which are fossilized animal feces. These coprolites, found in the Gobi Desert of China, were deposited by an anonymous predator 55 million years ago, so mammals have presumably had hair at least that long.

17.5 WHAT ROLE HAS EXTINCTION PLAYED IN THE HISTORY OF LIFE?

If there is a lesson in the great tale of life's history, it is that nothing lasts forever. The story of life can be read as a long series of evolutionary dynasties, with each new dominant group rising, ruling the land or the seas for a time and, inevitably, falling into decline and extinction. Dinosaurs are the most famous of these fallen dynasties, but the list of extinct groups known only from fossils is impressively long. Despite the inevitability of extinction, however, the overall trend has been for species to arise at a faster rate than they disappear, so the number of species on Earth has tended to increase over time.

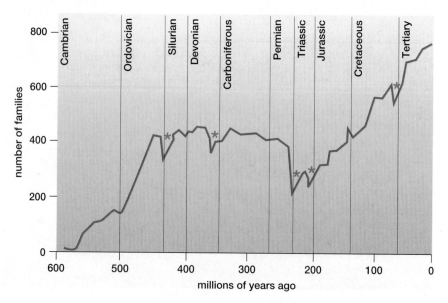

FIGURE 17-10 Mass extinctions
This graph plots the number of marine-animal groups against time, as reconstructed from the fossil record. Notice the general trend toward an increasing number of groups, punctuated by periods of sometimes rapid extinction. Five of these declines, marked by asterisks, are so steep that they qualify as catastrophic mass extinctions. QUESTION If extinction is the ultimate fate of all species, how can the total number of species have increased over time?

Evolutionary History Has Been Marked by Periodic Mass Extinctions

Over much of life's history, dynastic succession has proceeded in a steady, relentless manner. This slow and steady turnover of species, however, has been interrupted by episodes of **mass extinction** (FIG. 17-10). These mass extinctions are characterized by the relatively sudden disappearance of a wide variety of species over a large part of Earth. In the most catastrophic episodes of extinction, more than half of the planet's species disappeared. The worst episode of all, which occurred 245 million years ago at the end of the Permian period, wiped out more than 90% of the world's species, and life came perilously close to disappearing altogether.

Climate Change Contributed to Mass Extinctions

Mass extinctions have had a profound impact on the course of life's history, repeatedly redrawing the picture of life's diversity. What could have caused such dramatic changes in the fortunes of so many species? Many evolutionary biologists believe that changes in climate must have played an important role. When the climate changes, as it has done many times over the course of Earth's history, organisms that are adapted for survival in one climate may be unable to survive in a drastically different climate. In particular, at times when warm climates gave way to drier, colder climates with more variable temperatures, species may have gone extinct after failing to adapt to the harsh new conditions.

One cause of climate change is the changing positions of continents. Earth's surface is divided into portions called plates, which include the continents and the seafloor. The solid plates slowly move above a viscous but fluid layer. This movement is called **plate tectonics**. As the plates wander, their positions may change in latitude (FIG. 17-11). For example, 350 million years ago much of North America was located at or near the equator, an area characterized by consistently warm and wet tropical weather. But plate tectonics carried the continent up into temperate and arctic regions. As a result, the tropical climate was replaced by a regime of seasonal changes, cooler temperatures, and less rainfall. Plate tectonics continues today; the Atlantic Ocean, for example, widens by a few centimeters each year.

FIGURE 17-11 Continental drift from plate tectonics
The continents are passengers on plates moving on Earth's surface as a result of plate tectonics. **(a)** About 340 million years ago, much of what is now North America was positioned at the equator. **(b)** All the plates eventually fused together into one gigantic landmass, which geologists call Pangaea. **(c)** Gradually Pangaea broke up into Laurasia and Gondwanaland, which itself eventually broke up into West and East Gondwana. **(d)** Further plate motion eventually resulted in the modern positions of the continents.

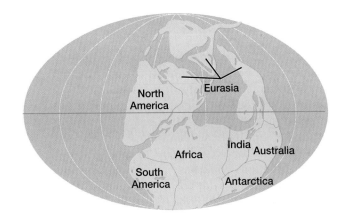

(a) 340 million years ago

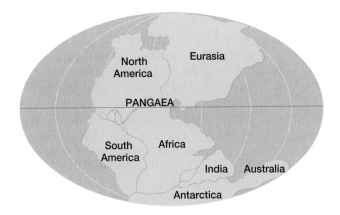

(b) 225 million years ago

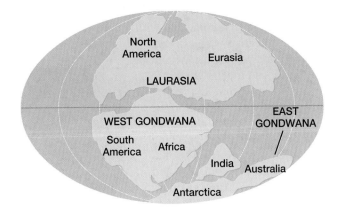

(c) 135 million years ago

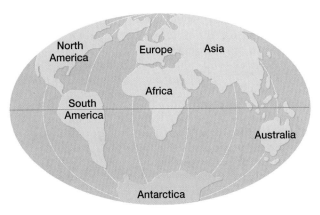

(d) Present

Catastrophic Events May Have Caused the Worst Mass Extinctions

Geological data indicate that most mass extinction events coincided with periods of climatic change. To many scientists, however, the rapidity of mass extinctions suggests that the slow process of climate change could not, by itself, be responsible for such large-scale disappearances of species. Perhaps more sudden events also play a role. For example, catastrophic geological events, such as massive volcanic eruptions, could have had devastating effects. Geologists have found evidence of past volcanic eruptions so huge that they make the 1980 Mount St. Helens explosion look like a firecracker by comparison. Even such gigantic eruptions, however, would directly affect only a relatively small portion of Earth's surface.

The search for the causes of mass extinctions took a fascinating turn in the early 1980s when Luis and Walter Alvarez proposed that the extinction event of 65 million years ago, which wiped out the dinosaurs and many other species, was caused by the impact of a huge meteorite. The Alvarezes' idea was met with great skepticism when it was first introduced, but geological research since that time has generated a great deal of evidence that a massive impact did indeed occur 65 million years ago. In fact, researchers have identified the Chicxulub crater, a 100-mile-wide crater buried beneath the Yucatán Peninsula of Mexico, as the impact site of a giant meteorite—10 miles (16 kilometers) in diameter—that collided with Earth just at the time that dinosaurs disappeared.

Could this immense meteorite strike have caused the mass extinction that coincided with it? No one knows for sure, but scientists suggest that such a massive impact would have thrown so much debris into the atmosphere that the entire planet would have been plunged into darkness for a period of years. With little light reaching the planet, temperatures would have dropped precipitously and the photosynthetic capture of energy (upon which all life ultimately depends) would have declined drastically. The worldwide "impact winter" would have spelled doom for the dinosaurs and a host of other species.

17.6 HOW DID HUMANS EVOLVE?

Scientists are intensely interested in the origin and evolution of humans, especially in the evolution of the gigantic human brain. The outline of human evolution that we present in this section is a synthesis of current thought on the subject, but we note that it is speculative, because the fossil evidence of human evolution is comparatively scarce. Paleontologists disagree about the interpretation of the fossil evidence, and many ideas may have to be revised as new fossils are found.

Humans Inherited Some Early Primate Adaptations for Life in Trees

Humans are members of a mammal group known as **primates**, which also includes lemurs, monkeys, and apes. The oldest primate fossils are 55 million years old, but because primate fossils are relatively rare compared with those of many other animals, the first primates probably arose considerably earlier but left no fossil record. Early primates probably fed on fruits and leaves and were adapted for life in the trees. Many modern primates retain the tree-dwelling lifestyle of their ancestors (**FIG. 17-12**). The common heritage of humans and other primates is reflected in a set of physical characteristics that was present in the earliest primates and that persists in many modern primates, including humans.

(b)

(c)

FIGURE 17-12 Representative primates
The **(a)** tarsier, **(b)** lemur, and **(c)** liontail macaque monkey all have relatively flat faces, with forward-looking eyes providing binocular vision. All also have color vision and grasping hands. These features, retained from the earliest primates, are shared by humans. **(a)**

Binocular Vision Provided Early Primates with Accurate Depth Perception

One of the earliest primate adaptations seems to have been large, forward-facing eyes (see Fig. 17-12). Jumping from branch to branch is risky business unless an animal can accurately judge where the next branch is located. Accurate depth perception was made possible by binocular vision, provided by forward-facing eyes with overlapping fields of view. Another key adaptation was color vision. We cannot, of course, tell if a fossil animal had color vision, but since modern primates have excellent color vision, it seems reasonable to assume that earlier primates did too. Many primates feed on fruit, and color vision helps to detect ripe fruit among a bounty of green leaves.

Early Primates Had Grasping Hands

Early primates had long, grasping fingers that could wrap around and hold onto tree limbs. This adaptation to tree dwelling was the basis for later evolution of human hands that could perform both a *precision grip* (used by modern humans for delicate maneuvers such as manipulating small objects, writing, and sewing) and a *power grip* (used for powerful actions, such as swinging a club or thrusting with a spear).

A Large Brain Facilitated Hand-Eye Coordination and Complex Social Interactions

Primates have brains that are larger, relative to their body size, than the brains of almost all other animals. No one really knows for certain which environmental forces favored the evolution of large brains. It seems reasonable, however, that controlling and coordinating rapid locomotion through trees, dexterous movements of the hands, and binocular, color vision would be facilitated by increased brain power. Most primates also have fairly complex social systems, which probably require relatively high intelligence. If sociality promoted increased survival and reproduction, then there would have been environmental pressures for the evolution of larger brains.

The Oldest Hominid Fossils Are from Africa

On the basis of comparisons of DNA from modern chimps, gorillas, and humans, researchers estimate that the **hominid** line (humans and their fossil relatives) diverged from the ape lineage sometime between 5 million and 8 million years ago. The fossil record, however, suggests that the split must have occurred at the early end of that range. Paleontologists working in the African country of Chad in 2002 discovered fossils of a hominid, *Sahelanthropus tchadensis*, that lived more than 6 million years ago (**FIG. 17-13**). *Sahelanthropus* is clearly a hominid, since it shares several anatomical features with later members of the group. But because this oldest known member of our family also exhibits other features that are more characteristic of apes, it may represent a point on our family tree that is close to the split between apes and hominids.

In addition to *Sahelanthropus*, two other hominid species, *Ardipithecus ramidus* and *Orrorin tugenensis*, are

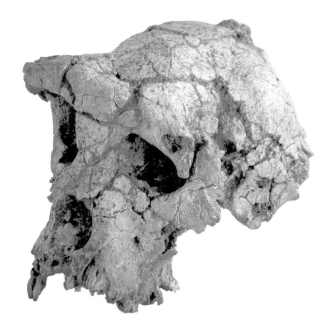

FIGURE 17-13 The earliest hominid
This nearly complete skull of *Sahelanthropus tchadensis*, which is more than 6 million years old, is the oldest hominid fossil yet found.

known from fossils appearing in rocks that are between 4 million and 6 million years old. Our knowledge of these hominids is limited, because only a few specimens have been found so far, mostly in recent discoveries that typically include only small portions of skeletons. A more extensive record of early hominid evolution does not begin until about 4 million years ago. That date marks the beginning of the fossil record of the genus *Australopithecus* (**FIG. 17-14**), a group of African hominid species with brains larger than those of their prehominid forebears but still much smaller than those of modern humans.

The Earliest Hominids Could Stand and Walk Upright

The earliest australopithecines (as the various species of *Australopithecus* are collectively known) had legs that were shorter, relative to their height, than those of modern humans, but their knee joints allowed them to straighten their legs fully, permitting efficient bipedal (upright, two-legged) locomotion. Footprints almost 4 million years old, discovered in Tanzania by anthropologist Mary Leakey, show that even the earliest australopithecines could, and at least sometimes did, walk upright. Upright posture may have evolved even earlier. The discoverers of *Sahelanthropus* and *Orrorin* argue that the leg and foot bones of these earliest hominids have characteristics that indicate bipedal locomotion, but this conclusion will remain speculative until more complete skeletons of these species are found.

The reasons for the evolution of bipedal locomotion among the early hominids remain poorly understood. Perhaps hominids that could stand upright gained an advantage in gathering or carrying food in their forest habitat. Whatever its cause, the early evolution of upright posture was extremely important in the evolutionary history of hominids,

FIGURE 17-14 A possible evolutionary tree for humans
This hypothetical family tree shows facial reconstructions of representative specimens. Although many paleontologists consider this to be the most likely human family tree, there are several alternative interpretations of the known hominid fossils. Fossils of the earliest hominids are scarce and fragmentary, so the evolutionary relationship of these species to later hominids remains unknown.

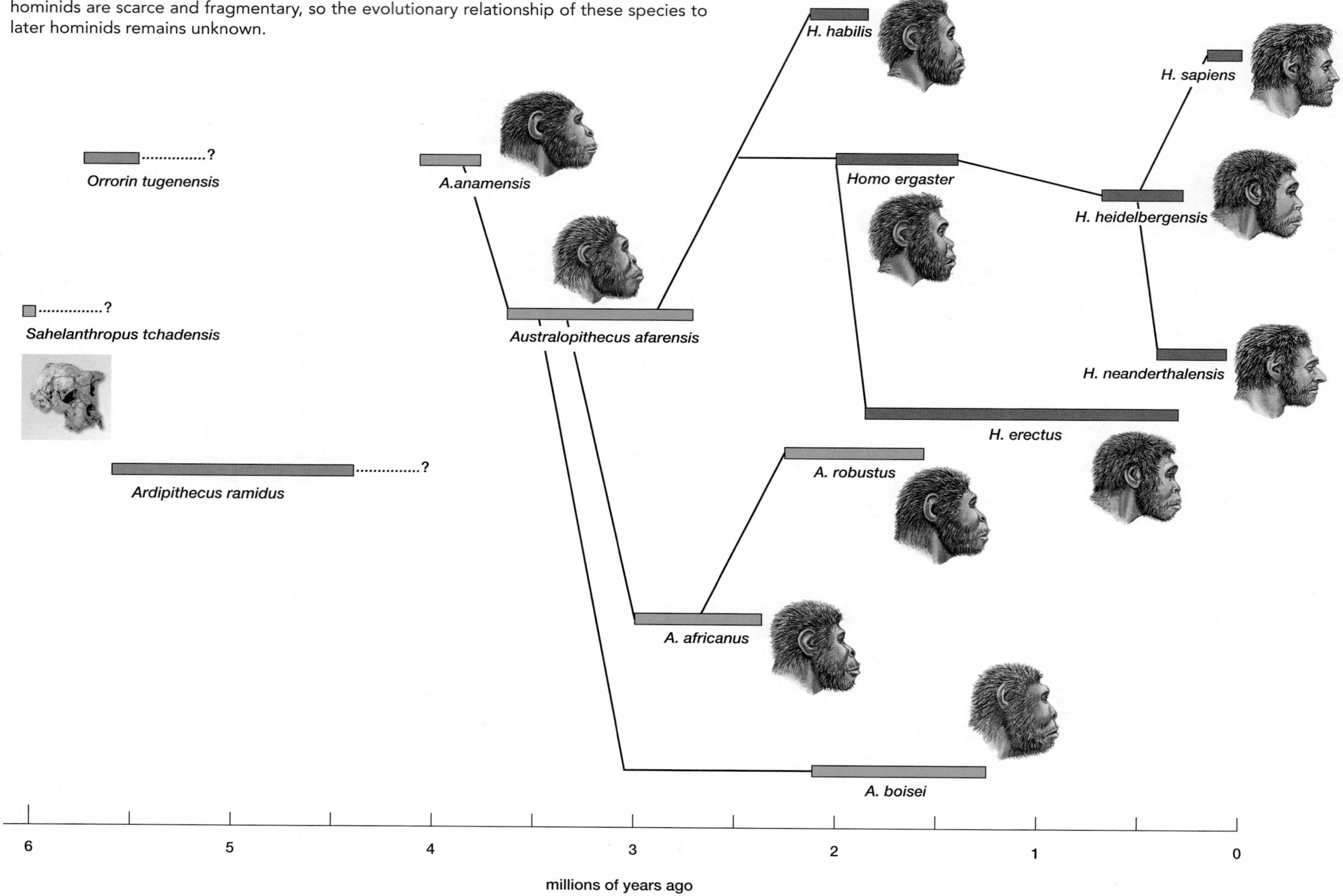

348

because it freed their hands from use in walking. Later hominids were thus able to carry weapons, manipulate tools, and eventually achieve the cultural revolutions produced by modern *Homo sapiens*.

Several Species of *Australopithecus* Emerged in Africa

The oldest australopithecine species, represented by fossilized teeth, skull fragments, and arm bones, was unearthed near an ancient lake bed in Kenya from sediments that were dated, by means of radioactive isotopes, at between 3.9 million and 4.1 million years old (see "Scientific Inquiry: How Do We Know How Old a Fossil Is?"). It was named *Australopithecus anamensis* by its discoverers (*anam* means "lake" in the local Ethiopian language). The second-most-ancient australopithecine, called *Australopithecus afarensis*, was discovered in the Afar region of Ethiopia. Fossil remains of this species as old as 3.9 million years have been unearthed. The *A. afarensis* line apparently gave rise to at least two distinct forms: small, omnivorous species such as *A. africanus* (which was similar to *A. afarensis* in size and eating habits), and larger, herbivorous species such as *A. robustus* and *A. boisei*. All of the australopithecine species had apparently gone extinct by 1.2 million years ago, but one of them (*A. afarensis* in the interpretation shown in Fig. 17-14) first gave rise to a new branch of the hominid family tree, the genus *Homo*.

The Genus Homo Diverged from the Australopithecines 2.5 Million Years Ago

Hominids that are sufficiently similar to modern humans to be placed in the genus *Homo* first appear in African fossils that are about 2.5 million years old. Among the earliest African *Homo* fossils are *H. habilis* (see Fig. 17-14), a species whose body and brain were larger than those of the australopithecines although they retained the apelike long arms and short legs of their australopithecine ancestors. In contrast, the skeletal anatomy of *H. ergaster*, a species whose fossils first appear 2 million years ago, has limb proportions more like those of modern humans. This species is believed by many paleoanthropologists (scientists who study human origins) to be on the evolutionary branch that led ultimately to our own species, *H. sapiens*. In this view, *H. ergaster* was the common ancestor of two distinct branches of hominids. The first branch led to *H. erectus*, which was the first hominid species to leave Africa. The second branch from *H. ergaster* ultimately led to *H. heidelbergensis*, some of which migrated to Europe and gave rise to the Neanderthals, *H. neanderthalensis*. Meanwhile, back in Africa, another branch split off from the *H. heidelbergensis* lineage. This branch ultimately became *H. sapiens*, modern humans.

The Evolution of *Homo* Was Accompanied by Advances in Tool Technology

Hominid evolution is closely tied to the development of tools, a hallmark of hominid behavior. The oldest tools discovered so far were found in 2.5-million-year-old East African rocks, concurrent with the early emergence of the genus *Homo*. Early *Homo*, whose cheek teeth were much smaller than those of the genus's australopithecine ancestors, might first have used stone tools to break and crush tough foods that were hard to chew. Hominids constructed their earliest tools by striking one rock with another to chip off fragments and leave a sharp edge behind. Over the next several hundred thousand years, toolmaking techniques in Africa gradually became more advanced. By 1.7 million years ago, tools had become more sophisticated. Flakes were chipped symmetrically from both sides of a rock to form double-edged tools ranging from hand axes, used for cutting and chopping, to points, probably used on spears (**FIG. 17-15a, b**). *Homo*

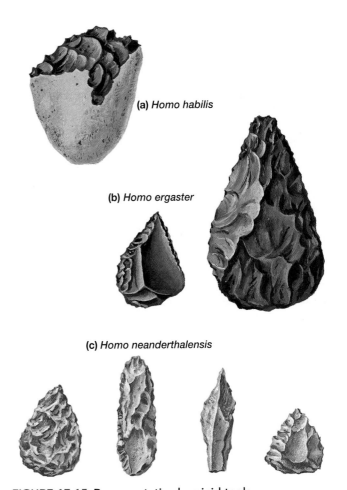

(a) *Homo habilis*

(b) *Homo ergaster*

(c) *Homo neanderthalensis*

FIGURE 17-15 Representative hominid tools
(a) *Homo habilis* produced only fairly crude chopping tools called *hand axes*, usually unchipped on one end to hold in the hand. (b) *Homo ergaster* manufactured much finer tools. The tools were typically sharp all the way around the stone; at least some of these blades were probably tied to spears rather than held in the hand. (c) Neanderthal tools were works of art, with extremely sharp edges made by flaking off tiny bits of stone. In comparing these weapons, note the progressive increase in the number of flakes taken off the blades and the corresponding decrease in flake size. Smaller, more numerous flakes produce a sharper blade and suggest more insight into toolmaking, more patience, finer control of hand movements, or perhaps all three.

ergaster and other bearers of these weapons presumably ate meat, probably acquired from both hunting and scavenging for the remains of prey killed by other predators. Double-edged tools were carried to Europe at least 600,000 years ago by migrating populations of *H. heidelbergensis*, and the Neanderthal descendants of these emigrants took stone-tool construction to new heights of skill and delicacy (**FIG. 17-15c**).

Neanderthals Had Large Brains and Excellent Tools

Neanderthals first appeared in the European fossil record about 150,000 years ago. By about 70,000 years ago, they had spread throughout Europe and western Asia. By 30,000 years ago, however, Neanderthals were extinct.

Contrary to the popular image of a hulking, stoop-shouldered "caveman," Neanderthals were quite similar to modern humans in many ways. Although more heavily muscled, Neanderthals walked fully erect, were dexterous enough to manufacture finely crafted stone tools, and had brains that, on average, were slightly larger than those of modern humans. Many European Neanderthal fossils show heavy brow ridges and a broad, flat skull, but others, particularly from areas around the eastern shores of the Mediterranean Sea, are somewhat more physically similar to *H. sapiens*.

Despite the physical and technological similarities between *H. neanderthalensis* and *H. sapiens*, there is no solid archeological evidence that Neanderthals ever developed an advanced culture that included such characteristically human endeavors as art, music, and rituals. Some anthropologists argue that, because their skeletal anatomy shows that they were physically capable of making the sounds required for speech, Neanderthals might have acquired language. This interpretation of Neanderthal anatomy, however, is not unanimously accepted. In general, the available evidence of the Neanderthal way of life is limited and open to different interpretations, and anthropologists are engaged in a sometimes heated debate about how advanced Neanderthal culture became.

Though some anthropologists argue that Neanderthals were simply a variety of *H. sapiens*, most agree that Neanderthals were a separate species. Dramatic evidence in support of this hypothesis has come from researchers who have isolated DNA from Neanderthal and *H. sapiens* skeletons that were more than 20,000 years old. These extractions of ancient DNA have allowed researchers to compare the nucleotide sequences of Neanderthal genes with the sequences of the same genes in both fossil and modern humans. The comparisons have shown that Neanderthal sequences are very different from those of both modern and fossil humans, but that modern and fossil humans share similar sequences. These findings indicate that the evolutionary branch leading to Neanderthals diverged from the ancestral human line hundreds of thousands of years before the emergence of modern *H. sapiens*.

Modern Humans Emerged Less than 200,000 Years Ago

The fossil record shows that anatomically modern humans appeared in Africa at least 160,000 years ago and possibly as long as 195,000 years ago. The location of these fossils suggests that *Homo sapiens* originated in Africa, but most of our knowledge about our own early history comes from European and Middle Eastern fossil *H. sapiens* collectively known as Cro-Magnons (after the district in France where their remains were first discovered). Cro-Magnons appeared about 90,000 years ago. They had domed heads, smooth brows, and prominent chins (just like us). Their tools were precision instruments similar to the stone tools used until recently in many parts of the world.

Behaviorally, Cro-Magnons seem to have been similar to, but more sophisticated than, Neanderthals. Artifacts from 30,000-year-old Cro-Magnon archeological sites include elegant bone flutes, graceful carved ivory sculptures, and evidence of elaborate burial ceremonies (**FIG. 17-16**). Perhaps the most remarkable accomplishment of Cro-Magnons is the magnificent art left in caves in places such as Altamira in Spain and Lascaux and Chauvet in France (**FIG. 17-17**). The oldest cave paintings so far found are more than 30,000 years old, and even the oldest ones make use of sophisticated artistic techniques. No one knows exactly why these paintings were made, but they attest to minds as fully human as our own.

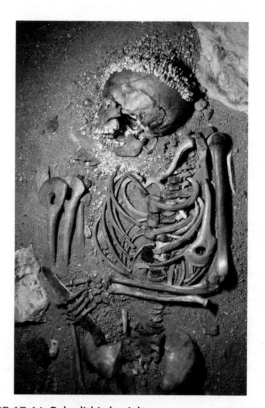

FIGURE 17-16 Paleolithic burial
This 24,000-year-old grave shows evidence that Cro-Magnon people ritualistically buried their dead. The body was covered with a dye known as red ocher, then buried wearing a headdress made of snail shells and with a flint tool in its hand.

Cro-Magnons and Neanderthals Lived Side by Side

Cro-Magnons coexisted with Neanderthals in Europe and the Middle East for perhaps as many as 50,000 years before the Neanderthals disappeared. Some researchers believe that Cro-Magnons interbred extensively with Neanderthals, so Neanderthals were essentially absorbed into the human genetic mainstream. Other scientists disagree, citing mounting evidence such as the fossil DNA described earlier, and suggest that later-arriving Cro-Magnons simply overran and displaced the less well-adapted Neanderthals.

Neither hypothesis does a good job of explaining how the two kinds of hominids managed to occupy the same geographical areas for such a long time. The persistence in one area of two similar but distinct groups for tens of thousands of years seems inconsistent with both interbreeding and direct competition. Perhaps the competition between *H. neanderthalensis* and *H. sapiens* was indirect, so that the two species were able to coexist for a time in the same habitat, until the superior ability of *H. sapiens* to exploit the available resources slowly drove Neanderthals to extinction.

FIGURE 17-17 The art of Cro-Magnon people
Cave paintings by Cro-Magnons have been remarkably preserved by the relatively constant underground conditions of a cave in Lascaux, France.

Several Waves of Hominids Emigrated from Africa

The human family tree is rooted in Africa, but hominids found their way out of Africa on numerous occasions. For example, *H. erectus* reached tropical Asia almost 2 million years ago and apparently thrived there, eventually spreading across Asia. Similarly, *H. heidelbergensis* made it to Europe at least 780,000 years ago. It is increasingly clear that the genus *Homo* made repeated long-distance emigrations, beginning as soon as sufficiently capable limb anatomy evolved. What is less clear is how all this wandering is related to the origin of modern *H. sapiens*. According to the "African replacement" hypothesis (the basis of the scenario outlined earlier), *H. sapiens* emerged in Africa and dispersed less than 150,000 years ago, spreading into the Near East, Europe, and Asia and replacing all other hominids (**FIG. 17-18a**). But some paleoanthropologists believe that populations of *H. sapiens* evolved simultaneously in many regions from the already widespread populations of *H. erectus*. According to this "multiregional origin" hypothesis, continued migrations and interbreeding among *H. erectus* populations in different regions of the world maintained them as a single species as they gradually evolved into *H. sapiens* (**FIG. 17-18b**). Although an increasing number of studies of modern human DNA support the African replacement model of the origin of our species, both hypotheses are consistent with the fossil record. Therefore, the question remains unsettled.

The Evolutionary Origin of Large Brains May Be Related to Meat Consumption

The main physical features that distinguish us from our closest relatives, the apes, are our upright posture and large, highly developed brains. As described earlier, upright posture arose very early in hominid evolution, and hominids walked upright for several million years before large-brained *Homo* species arose. What circumstances might have caused the evolution of increased brain size? Many explanations have been proposed, but little direct evidence is available; hypotheses about the evolutionary origins of large brains are necessarily speculative.

One proposed explanation for the origin of large brains suggests that they evolved in response to increasingly complex social interactions. In particular, fossil evidence suggests that, beginning about 2 million years ago, hominid social life began to include a new type of activity: the cooperative hunting of large game. The resulting access to significant amounts of meat must have fostered a need to develop methods for distributing this valuable, limited resource among group members. Some anthropologists hypothesize that the individuals best able to manage this social interaction would have been more successful at gaining a large share of meat and using their share to their own advantage. Perhaps this social management was best accomplished by individuals with larger, more powerful brains, and natural selection therefore favored such individuals. Observations of chimpanzee societies have shown that the distribution of group-hunted meat often involves

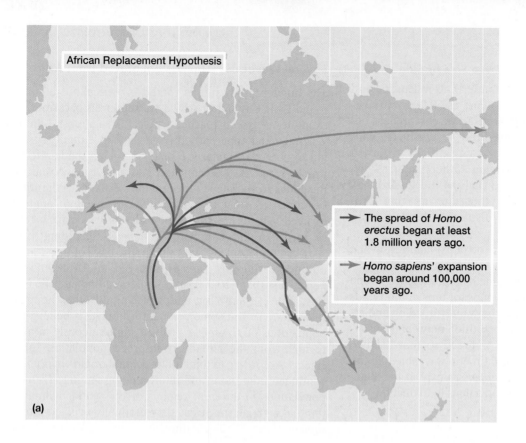

(a)

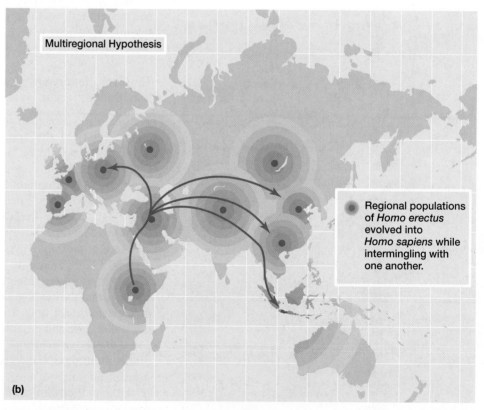

(b)

FIGURE 17-18 Competing hypotheses for the evolution of *Homo sapiens*
(a) The "African replacement" hypothesis suggests that *H. sapiens* evolved in Africa, then migrated throughout the Near East, Europe, and Asia, displacing the other hominid species that were present in those regions. **(b)** The "multiregional" hypothesis suggests that populations of *H. sapiens* evolved in many regions simultaneously from the already widespread populations of *H. erectus*. QUESTION Paleontologists recently discovered fossil hominids with features characteristic of modern humans in 160,000-year-old sediments in Africa. Which hypothesis does this new evidence support?

intricate social interactions in which meat is used to form alliances, repay favors, gain access to sexual partners, placate rivals, and so on. Perhaps the mental skill required to plan, assess, and remember such interactions was the driving force behind the evolution of our large, clever brains.

The Evolutionary Origin of Human Behavior Is Highly Speculative

Even after the evolution of comparatively large brains in species such as *H. erectus*, more than a million years passed before the origin of modern humans and their extremely large brains. And even after the first appearance of modern *H. sapiens*, more than 100,000 years passed before the appearance of any archeological evidence of the distinctively human characteristics that were made possible by a large brain: language, abstract thought, and advanced culture. The evolutionary origin of these human traits is another unresolved question, in part because direct evidence of the transition to advanced culture may never be found. Early humans capable of language and symbolic thought would not necessarily have created artifacts that indicated these capabilities. We can uncover some clues by studying our ape relatives, which possess less-complex versions of many human behaviors and mental processes. Their behavior might resemble that of ancestral hominids. Nonetheless, the late, seemingly rapid origin of advanced human culture remains a puzzle.

The Cultural Evolution of Humans Now Far Outpaces Biological Evolution

In recent millennia, human evolution has come to be dominated by *cultural evolution*, the transmission of learned behaviors from generation to generation. Our recent evolutionary success, for example, was engendered not so much by new physical adaptations as by a series of cultural and technological revolutions. The first such revolution was the development of tools, which began with the early hominids. Tools increased the efficiency with which food and shelter could be acquired and thus increased the number of individuals that could survive within a given ecosystem. About 10,000 years ago, human culture underwent a second revolution as people discovered how to grow crops and domesticate animals. This agricultural revolution dramatically increased the amount of food that could be extracted from the environment, and the human population surged, increasing from about 5 million at the dawn of agriculture to around 750 million by 1750. The subsequent industrial revolution gave rise to the modern economy and its attendant improvements in public health. Longer lives and lower infant mortality led to truly explosive population growth, and today we are more than 6 billion strong.

Human cultural evolution and the accompanying increases in human population have had profound effects on the continuing biological evolution of other life-forms. Our agile hands and minds have transformed much of Earth's terrestrial and aquatic habitats. Humans have become the single overwhelming agent of natural selection. In the words of the late evolutionary biologist Stephen Jay Gould, "We have become, by the power of a glorious evolutionary accident called intelligence, the stewards of life's continuity on Earth. We did not ask for this role, but we cannot renounce it. We may not be suited for it, but here we are."

CASE STUDY REVISITED   LITTLE PEOPLE, BIG STORY

The discovery of the hobbit-sized *Homo floresiensis* was exciting to many people in part because it suggested that our species may have more close relatives than we previously suspected and that at least some of them lived tantalizingly close to the present. Plus, the idea of a society of tiny humans seems to have an inherent appeal. But the discovery also raises a host of fascinating evolutionary questions.

Many of these questions are related to the ancestry of *H. floresiensis*. Some clues point toward an intriguing scenario. First, the only other evidence of early hominid habitation on Flores consists of stone tools found at an 840,000-year-old site. The age and relative crudeness of the tools suggest that they were probably left by *H. erectus*, the only hominid known to be present in Asia that long ago. So, *H. floresiensis* may have descended from a population of *H. erectus* that became isolated on Flores. This con-

clusion is supported by some anatomical similarities between *H. floresiensis* and *H. erectus*. Oddly, *H. floresiensis* is more similar to *H. erectus* specimens from a distant, 1.8-million-year-old site in Central Asia than it is to much younger *H. erectus* specimens found in relatively nearby sites on other Indonesian islands. Perhaps *H. floresiensis* descended from a very early wave of *erectus* migrants.

Regardless of which hominid group gave rise to *H. floresiensis*, it is not obvious how the ancestral population arrived on the island. Unlike some islands, Flores was never connected to the mainland. Archaeologists generally agree that hominids did not construct boats until 60,000 years ago at the earliest. So how did *H. erectus* get to Flores nearly 800,000 years before the invention of boats? Perhaps they drifted across on clumps of floating vegetation.

Another interesting question about *H. floresiensis* is the cause of their small size.

Large animal species that are isolated on islands sometimes evolve smaller body size. For example, the now-extinct elephants of Flores were only about 4 feet tall. Biologists suggest that the absence of large predators on most islands eliminates much of the benefit of large size, conferring an advantage on smaller individuals that require less food. Did this kind of dynamic foster the evolution of small stature in *H. floresiensis*? Are the bodies of hominids, which have weapons to defend against predators and tools to aid in acquiring food, subject to the same evolutionary pressures that shape the bodies of other animals?

Consider This *H. floresiensis* was found on an island. If you were searching for evidence of other undiscovered recent hominid species, would you concentrate your search on islands? Why or why not? In which regions of the world would you search?

CHAPTER REVIEW

SUMMARY OF KEY CONCEPTS

17.1 How Did Life Begin?

Before life arose, lightning, ultraviolet light, and heat formed organic molecules from water and the components of primordial Earth's atmosphere. These molecules probably included nucleic acids, amino acids, short proteins, and lipids. By chance, some molecules of RNA may have had enzymatic properties, catalyzing the assembly of copies of themselves from nucleotides in Earth's waters. These may have been the forerunners of life. Protein-lipid vesicles enclosing these ribozymes may have formed the first protocells.

17.2 What Were the Earliest Organisms Like?

The oldest fossils, about 3.5 billion years old, are of prokaryotic cells that fed by absorbing organic molecules that had been synthesized in the environment. Because there was no free oxygen in the atmosphere, their energy metabolism must have been anaerobic. As the cells multiplied, they depleted the organic molecules that had been formed by prebiotic synthesis. Some cells developed the ability to synthesize their own food molecules by using simple inorganic molecules and the energy of sunlight. These earliest photosynthetic cells were probably ancestors of today's cyanobacteria.

Photosynthesis releases oxygen as a by-product; by about 2.2 billion years ago, significant amounts of free oxygen were accumulating in the atmosphere. Aerobic metabolism, which generates more cellular energy than does anaerobic metabolism, probably arose about this time.

Eukaryotic cells had evolved by about 1.7 billion years ago. The first eukaryotic cells probably arose as symbiotic associations between predatory prokaryotic cells and other bacteria. Mitochondria may have evolved from aerobic bacteria engulfed by predatory cells. Similarly, chloroplasts may have evolved from photosynthetic cyanobacteria.

Web Tutorial 17.1 The Endosymbiont Hypothesis

17.3 What Were the Earliest Multicellular Organisms Like?

Multicellular organisms evolved from eukaryotic cells and first appeared, in the seas, about 1 billion years ago. Multicellularity offers several advantages, including greater size. In plants, increased size offered some protection from predation. Specialization of cells allowed plants to anchor themselves in the nutrient-rich, well-lit waters of the shore. For animals, multicellularity allowed more-efficient predation and more-effective escape from predators. These in turn provided environmental pressures for faster locomotion, improved senses, and greater intelligence.

17.4 How Did Life Invade the Land?

The first land organisms were probably algae. The first multicellular land plants appeared about 400 million years ago. Although life on land required special adaptations for support of the body, reproduction, and the acquisition, distribution, and retention of water, the land also offered abundant sunlight and protection from aquatic herbivores. Soon after land plants evolved, arthropods invaded the land. Absence of predators and abundant land plants for food probably facilitated the invasion of the land by animals.

The earliest land vertebrates evolved from lobefin fishes, which had leg-like fins and a primitive lung. A group of lobefins evolved into the amphibians about 350 million years ago. Reptiles evolved from amphibians, with several further adaptations for life on land: waterproof eggs that could be laid on land, waterproof skin, and better lungs. Birds and mammals evolved independently from separate groups of reptiles. A major advance in the evolution of both birds and mammals was body insulation: feathers and hair.

17.5 What Role Has Extinction Played in the History of Life?

The history of life has been characterized by the constant turnover of species as some species go extinct and are replaced by new ones. Mass extinctions, in which large numbers of species disappear within a relatively short time, have occurred periodically. Mass extinctions were probably caused by some combination of climate changes and catastrophic events, such as volcanic eruptions and meteorite impacts.

Web Tutorial 17.2 Continental Drift from Plate Tectonics

17.6 How Did Humans Evolve?

One group of mammals evolved into the tree-dwelling primates, which were the ancestors of apes and humans. The human and ape lines diverged 7 million to 8 million years ago; the oldest known hominid fossils are between 6 million and 7 million years old and were found in Africa. The first well-known hominid line, the australopithecines, arose in Africa about 4 million years ago. These hominids walked erect, had larger brains than did their forebears, and fashioned primitive tools. One group of australopithecines gave rise to a line of hominids in the genus *Homo*, which in turn gave rise to modern humans.

KEY TERMS

amphibian *page 343*
arthropod *page 342*
conifer *page 342*
endosymbiont hypothesis *page 337*

eukaryote *page 337*
exoskeleton *page 342*
hominid *page 347*
lobefin *page 342*
mammal *page 344*

mass extinction *page 345*
plate tectonics *page 345*
primate *page 346*
prokaryote *page 335*
protocell *page 334*

reptile *page 343*
ribozyme *page 334*
spontaneous generation *page 332*

THINKING THROUGH THE CONCEPTS

1. What is the evidence that life might have originated from non-living matter on early Earth? What kind of evidence would you like to see before you would accept this hypothesis?

2. If the first cells with aerobic metabolism were so much more efficient at producing energy, why didn't they drive cells with only anaerobic metabolism to extinction?

3. Explain the endosymbiont hypothesis for the origin of chloroplasts and mitochondria.

4. Name two advantages of multicellularity for plants and two advantages for animals.

5. What advantages and disadvantages would terrestrial existence have had for the first plants to invade the land? For the first land animals?

6. Outline the major adaptations that emerged during the evolution of vertebrates, from fishes to amphibians to reptiles to birds and mammals. Explain how these adaptations increased the fitness of the various groups for life on land.

7. Outline the evolution of humans from early primates. Include in your discussion such features as binocular vision, grasping hands, bipedal locomotion, social living, toolmaking, and brain expansion.

APPLYING THE CONCEPTS

1. What is cultural evolution? Is cultural evolution more or less rapid than biological evolution? Why?

2. Do you think that studying our ancestors can shed light on the behavior of modern humans? Why?

3. A biologist would probably answer the age-old question, "What is life?" by saying, "The ability to self-replicate." Do you agree with this definition? If so, why? If not, how would you define life in biological terms?

4. Traditional definitions of humans have emphasized "the uniqueness of humans" because we possess language and use tools. But most animals can communicate with other individuals in sophisticated ways, and many vertebrates use tools to accomplish tasks. Pretend that you are a biologist from Mars, and write a taxonomic description of the species *Homo sapiens*.

5. Extinctions have occurred throughout the history of life on Earth. Why should we care if humans are causing a mass extinction event now?

6. The "African replacement" and "multiregional origin" hypotheses of the evolution of *Homo sapiens* make contrasting predictions about the extent and nature of genetic divergence among human races. One predicts that races are old and highly diverged genetically; the other predicts that races are young and little diverged genetically. What data would help you determine which hypothesis is closer to the truth?

7. In biological terms, what do you think was the most significant event in the history of life? Explain your answer.

FOR MORE INFORMATION

de Duve, C. "The Birth of Complex Cells." *Scientific American*, April 1996. Narrative describing the origin of complex eukaryotic cells by repeated instances of endosymbiosis.

Fry, I. *The Emergence of Life on Earth: A Historical and Scientific Overview*. Brunswick, NJ: Rutgers University Press, 2000. A thorough review of research and hypotheses concerning the origin of life.

Maynard Smith, J., and Szathmary. E. *The Origins of Life: From the Birth of Life to the Origin of Language*. New York: Oxford University Press, 1999. A thought-provoking review of the major shifts that have occurred over the 3.5-billion-year history of life.

Monastersky, R. "The Rise of Life on Earth." *National Geographic*, March 1998. A beautifully illustrated and engaging description of current ideas and evidence of how life arose.

Morwood, M., Sutkina, T., and Roberts, R. "The People Time Forgot." *National Geographic*, April 2005. An account of the discovery of *Homo floresiensis* and the implications of the find, written by the paleontologists who made the discovery.

Tattersall, I. "Once We Were Not Alone." *Scientific American*, January 2000. A overview of the evolutionary history that led to modern *Homo sapiens*, with illustrations of some of the hominids that preceded us.

Ward, P. D. *The End of Evolution: On Mass Extinctions and the Preservation of Biodiversity*. New York: Bantam Books, 1994. An engaging first-person account of a paleontologist's investigation of the causes of mass extinction.

Zimmer, C. "What Came Before DNA." *Discover*, June 2004. An overview of recent research on prebiotic evolution.

18 Systematics: Seeking Order Amidst Diversity

Biologists studying the evolutionary history of type 1 human
immunodeficiency virus (HIV-1) discovered that the virus, which
causes AIDS, probably originated in chimpanzees.

CASE STUDY   ORIGIN OF A KILLER

ONE OF THE WORLD'S most frightening diseases is also one of its most mysterious. Acquired immune deficiency syndrome (AIDS) appeared seemingly out of nowhere, and when it was first recognized in the early 1980s, no one knew what caused it or where it came from. Scientists raced to solve the mystery and, within a few years, had identified the infectious agent that caused AIDS: human immunodeficiency virus (HIV). Once HIV had been identified, researchers turned their attention to the question of its origin.

Finding the source of HIV required an evolutionary approach. To ask, "Where did HIV come from?" is really to ask, "What kind of virus was the ancestor of HIV?" Biologists who examine questions of ancestry are known as *systematists*. Systematists strive to categorize organisms according to their evolutionary history, building classifications that accurately reflect the structure of the tree of life. When a systematist concludes that two species are closely related, it means that the two species share a recent common ancestor from which both species evolved.

The systematists who explored the ancestry of HIV discovered that its closest relatives are found not among other viruses that infect humans, but among those that infect monkeys and apes. In fact, the latest research on HIV's evolutionary history has concluded that the closest relative of HIV-1 (the type of HIV that is most responsible for the worldwide AIDS epidemic) is a virus strain that infects a particular chimp subspecies that inhabits a limited range in West Africa. Therefore, the ancestor of the virus that we now know as HIV-1 did not evolve from a preexisting human virus but must have somehow jumped from West African chimpanzees to humans.

18.1 HOW ARE ORGANISMS NAMED AND CLASSIFIED?

Systematics is the science of reconstructing **phylogeny**, or evolutionary history. As part of their effort to reveal the tree of life, systematists name organisms and place them into categories on the basis of their evolutionary relationships. There are eight major categories: **domain, kingdom, phylum, class, order, family, genus,** and **species.** These categories form a nested hierarchy in which each level includes all of the other levels below it. Each domain contains a number of kingdoms; each kingdom contains a number of phyla; each phylum includes a number of classes; each class includes a number of orders; and so on. As we move down the hierarchy, smaller and smaller groups are included. Each category is increasingly narrow and specifies groups whose common ancestor is increasingly recent. Table 18-1 describes some examples of classifications of specific organisms.

Each Species Has a Unique, Two-Part Name

The **scientific name** of an organism is formed from the two smallest categories, the genus and the species. Each genus includes a group of very closely related species, and each species within a genus includes populations of organisms that can potentially interbreed under natural conditions. Thus, the genus *Sialia* (bluebirds) includes the eastern bluebird (*Sialia sialis*), the western bluebird (*Sialia mexicana*), and the mountain bluebird (*Sialia currucoides*)—very similar birds that normally do not interbreed (**FIG. 18-1**).

Each two-part scientific name is unique, so referring to an organism by its scientific name rules out any chance of ambiguity or confusion. For example, the bird *Gavia immer* is commonly known in North America as the common loon, in Great Britain as the northern diver, and by still other names in non-English-speaking countries. But the Latin scientific name *Gavia immer* is recognized by biologists worldwide, overcoming language barriers and allowing precise communication.

Note that, by convention, scientific names are always underlined or *italicized*. The first letter of the genus name is always capitalized, and the first letter of the species name is always lowercase. The species name is never used alone but is always paired with its genus name.

Classification Originated as a Hierarchy of Categories

Aristotle (384–322 B.C.) was among the first to attempt to formulate a logical, standardized language for naming living things. Based on characteristics such as structural complexity, behavior, and degree of development at birth, he classified about 500 organisms into 11 categories. Aristotle's categories formed a hierarchical structure, with each category more inclusive than the one beneath it, a concept that is still used today.

Building on this foundation more than 2000 years later, Swedish naturalist Carl von Linné (1707–1778)—who called himself Carolus Linnaeus, a latinized version of his given name—laid the groundwork for the modern classification system. He placed each organism into a series of hierarchically arranged categories on the basis of its resemblance to other life-forms, and he also introduced the scientific name composed of genus and species.

Nearly 100 years later, Charles Darwin (1809–1882) published *On the Origin of Species*, which demonstrated that all organisms are connected by common ancestry. Biologists then began to recognize that the categories ought to reflect the pattern of evolutionary relatedness among organisms. The more categories two organisms share, the closer their evolutionary relationship.

Systematists Identify Features That Reveal Evolutionary Relationships

Systematists seek to reconstruct the tree of life, but they must do so without much direct knowledge of evolutionary history. Because they can't see into the past, they must infer the past as best they can, on the basis of similarities among living organisms. Not just any similarity will do, however. Some observed similarities stem from convergent evolution in organisms that are not closely related, and such similarities are not useful for inferring evolutionary history. Instead, systematists value the similarities that exist because two kinds of organisms both inherited the characteristic from a common ancestor. Therefore, an important task

Table 18-1 Classification of Selected Organisms, Reflecting Their Degree of Relatedness*

	Human	Chimpanzee	Wolf	Fruit Fly	Sequoia Tree	Sunflower
Domain	**Eukarya**	**Eukarya**	**Eukarya**	**Eukarya**	**Eukarya**	**Eukarya**
Kingdom	**Animalia**	**Animalia**	**Animalia**	**Animalia**	**Plantae**	**Plantae**
Phylum	**Chordata**	**Chordata**	**Chordata**	Arthropoda	Coniferophyta	Anthophyta
Class	**Mammalia**	**Mammalia**	**Mammalia**	Insecta	Coniferosida	Dicotyledoneae
Order	**Primates**	**Primates**	Carnivora	Diptera	Coniferales	Asterales
Family	Hominidae	Pongidae	Canidae	Drosophilidae	Taxodiaceae	Asteraceae
Genus	*Homo*	*Pan*	*Canis*	*Drosophila*	*Sequoiadendron*	*Helianthus*
Species	*sapiens*	*troglodytes*	*lupus*	*melanogaster*	*giganteum*	*annuus*

*Boldface categories are those that are shared by more than one of the organisms classified.
Genus and species names are always italicized or underlined.

FIGURE 18-1 Three species of bluebird
Despite their obvious similarity, these three species of bluebird—from left to right, the eastern bluebird *(Sialia sialis)*, the western bluebird *(Sialia mexicana)*, and the mountain bluebird *(Sialia currucoides)*—remain distinct because they do not interbreed.

is to distinguish informative similarities caused by common ancestry from similarities that result from convergent evolution. In the search for informative similarities, biologists look at many kinds of characteristics.

Anatomy Plays a Key Role in Systematics

Historically, the most important and useful distinguishing characteristics have been anatomical. Systematists look carefully at similarities in external body structure (see Fig. 18-1) and in internal structures, such as skeletons and muscles. For example, homologous structures, such as the finger bones of dolphins, bats, seals, and humans (see Fig. 14-7), provide evidence of a common ancestor. To detect relationships between more closely related species, biologists may use microscopes to discern finer details—the number and shape of the "teeth" on the tonguelike radula of a snail, the shape and position of the bristles on a marine worm, or the external structure of pollen grains of a flowering plant (**FIG. 18-2**).

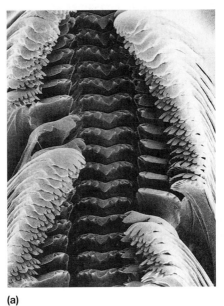
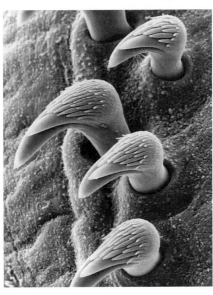
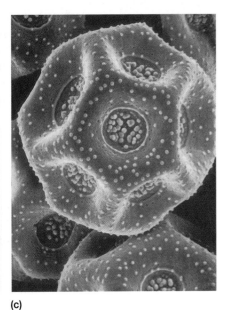

(a)   (b)   (c)

FIGURE 18-2 Microscopic structures may be used to classify organisms
(a) The "teeth" on a snail's tonguelike radula (a structure used in feeding), (b) the bristles on a marine worm, and (c) the shape and surface features of pollen grains are characteristics that are potentially useful in classification. Such finely detailed structures can reveal similarities between species that are not apparent in larger and more obvious structures.

Molecular Similarities Are Also Useful for Reconstructing Phylogeny

The anatomical characteristics shared by related organisms are expressions of underlying genetic similarities, so it stands to reason that evolutionary relationships among species must also be reflected in genetic similarities. Of course, direct genetic comparisons were not possible for most of the history of biology. Since the 1980s, however, advances in the techniques of molecular genetics have revolutionized studies of evolutionary relationships.

As a result of these technical advances, today's systematists can use the nucleotide sequences of DNA (that is, genotype) to investigate relatedness among different types of organisms. Closely related species have similar DNA sequences. In some cases, similarity of DNA sequences will be reflected in the structure of chromosomes. For example, both the DNA sequences and the chromosomes of chimpanzees and humans are extremely similar, showing that these two species are very closely related (**FIG. 18-3**). Some of the key methods of genetic analysis are explored in "Scientific Inquiry: Molecular Genetics Reveals Evolutionary Relationships." The process by which systematists use genetic and anatomical similarities to reconstruct evolutionary history are discussed in "A Closer Look: Reconstructing Phylogenetic Trees."

18.2   WHAT ARE THE DOMAINS OF LIFE?

Before 1970, all forms of life were classified into two kingdoms: Animalia and Plantae. All bacteria, fungi, and photosynthetic eukaryotes were considered to be plants, and all other organisms were classified as animals. As scientists learned more about fungi and microorganisms, however, it became apparent that the two-kingdom system oversimplified evolutionary history. To help rectify this problem, Robert H. Whittaker in 1969 proposed a five-kingdom classification that was eventually adopted by most biologists.

The Five-Kingdom System Improved Classification

Whittaker's five-kingdom system placed all prokaryotic organisms into a single kingdom and divided the eukaryotes into four kingdoms. The designation of a separate kingdom (called Monera) for the prokaryotes reflected growing recognition that the evolutionary pathway of these tiny, single-celled organisms had diverged from that of the eukaryotes early in the history of life. Among the eukaryotes, the five-kingdom system recognized three kingdoms of multicellular organisms (Plantae, Fungi, and Animalia) and placed all of the remaining, mostly single-celled eukaryotes in a single kingdom (Protista).

Because it more accurately reflected current understanding of evolutionary history, the five-kingdom system was an improvement over the old two-kingdom system. As understanding continued to grow, however, our view of life's most fundamental categories needed yet another re-

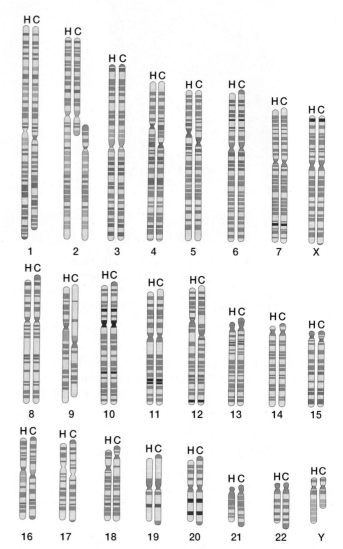

FIGURE 18-3 Human and chimp chromosomes are similar Chromosomes from different species can be compared by means of banding patterns that are revealed by staining. The comparison illustrated here, between human chromosomes (left member of each pair; H) and chimpanzee chromosomes (C), reveals that the two species are genetically very similar. In fact, the entire genomes of both species have been sequenced and are 96% identical. The numbering system shown is that used for human chromosomes; note that human chromosome 2 corresponds to a combination of two chimp chromosomes.

vision. The pioneering work of microbiologist Carl Woese has shown that biologists had overlooked a fundamental event in the early history of life, one that demands a new and more-accurate classification of life.

A Three-Domain System More Accurately Reflects Life's History

Woese and other biologists interested in the evolutionary history of microorganisms have studied the biochemistry of prokaryotic organisms. These researchers, focusing on

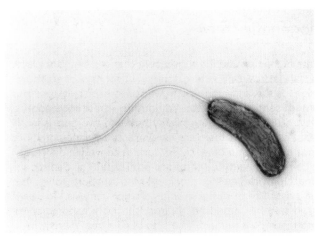

(a)

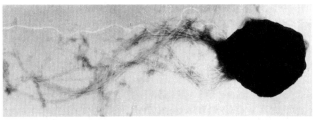

(b)

FIGURE 18-4 Two domains of prokaryotic organisms
Although similar in appearance, **(a)** *Vibrio cholerae* and **(b)** *Methanococcus jannaschi* are less closely related than a mushroom and an elephant. *Vibrio* is in the domain Bacteria, and *Methanococcus* is in Archaea.

radically different. These two groups are no more closely related to one another than either one is to any eukaryote. The tree of life split into three parts very early in the history of life, long before the appearance of plants, animals, and fungi. As a result of this new understanding, the five-kingdom system has been replaced by a classification that divides life into three domains: **Bacteria**, **Archaea**, and **Eukarya** (FIG. 18-5).

Kingdom-Level Classification Remains Unsettled

The move to a three-domain classification system has required systematists to reexamine the kingdoms within each domain, and the process of establishing kingdoms is still under way. If we accept that the striking differences between plants, animals, and fungi demand that each of these evolutionary lineages retains its kingdom status, then the logic of classification requires that we also assign kingdom status to groups that branch off of the tree of life earlier than these three groups of multicellular eukaryotes. Following this logic, prokaryote systematists recognize about 15 kingdoms among the Bacteria and three or so kingdoms among the Archaea. Systematists also recognize additional kingdoms within the Eukarya, reflecting a number of very early evolutionary splits within the diverse array of single-celled eukaryotes formerly lumped together in the kingdom Protista. Systematists, however, have yet to reach a consensus about the precise definitions of new prokaryotic and eukaryotic kingdoms, though new information about the evolutionary history of single-celled organisms is emerging rapidly. Thus, kingdom-level classification is in a state of transition as systematists strive to incorporate the latest information.

This text's descriptions of the diversity of life—which appear in Chapters 19 through 24—sidestep the unsettled state of life's kingdoms. The prokaryotic domains Archaea and Bacteria are discussed without reference to kingdom-level relationships. Among the eukaryotes, fungi, plants, and animals are treated as distinct evolutionary units, and the generic term "protist" designates the diverse collection

nucleotide sequences of the RNA that is found in the organisms' ribosomes, discovered that the supposed kingdom Monera actually included two very different kinds of organisms. Woese has dubbed these two groups the Bacteria and the Archaea (FIG. 18-4).

Despite superficial similarities in their appearance under the microscope, the Bacteria and the Archaea are

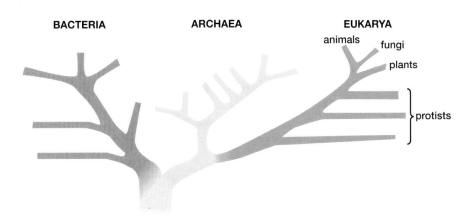

BACTERIA   ARCHAEA   EUKARYA
animals fungi
plants
protists

FIGURE 18-5 The tree of life
The three domains of life represent the three main "trunks" on the tree of life.

Systematists strive to develop a system of classification that reflects the phylogeny (evolutionary history) of organisms. Thus, the main job of systematists is to reconstruct phylogeny. Reconstructing the evolutionary history of all of Earth's organisms is, of course, a huge task, so each systematist typically chooses to work on some particular portion of the history.

The result of a phylogenetic reconstruction is usually represented by a diagram. These diagrams can take a number of different forms, but all of them show the sequence of branching events in which ancestral species split to give rise to descendant species. For this reason, diagrams of phylogeny are generally treelike.

These trees can present the phylogeny of any specified set of *taxa* (singular, *taxon*). A taxon is a named species, such as *Homo sapiens*, or a named group of species, such as primates or beetles or ferns. Thus, phylogenetic trees can show evolutionary history at different scales. Systematists might reconstruct, for example, a tree of 10 species in a particular genus of clams, or a tree of 25 phyla of animals, or a tree of the three domains of life.

After selecting the taxa to include, a systematist is ready to begin building a tree. Most systematists use the *cladistic* approach to reconstruct phylogenetic trees. Under the cladistic approach, relationships among taxa are revealed by the occurrence of similarities known as *synapomorphies*. A synapomorphy is a trait that is similar in two or more taxa because these taxa inherited a "derived" version of the trait that had changed from its original state in a common ancestor. The formation of synapomorphies is illustrated in FIGURE E18-1.

In the imaginary scenario illustrated in Figure E18-1, we can easily identify synapomorphies because we know the ancestral state of the trait and the subsequent changes that took place. In real life, however, systematists would not have direct knowledge of the ancestor, which lived in the distant past and whose identity is unknown. Without this direct knowledge, a systematist observing a similarity between two taxa is faced with a challenge. Is the observed similarity a synapomorphy, or does it have some other cause, such as convergent evolution or common inheritance of the ancestral state? The cladistic approach provides methods for identifying synapomorphies, but the possibility of mistaken interpretation remains. To guard against errors introduced by misidentified synapomorphies, systematists use numerous traits to build a tree, thereby minimizing the influence of any single trait.

In the last phase of the tree-building process, the systematist compares different possible trees. For example, three taxa can be arranged in three different branching patterns (FIG. E18-2). Each branching pattern represents a different hypothesis about the evolutionary history of taxa A, B, and C. Imagine, for example, that a systematist identifies a number of synapomorphies that link taxa A and B but are not shared by C, but has found few synapomorphies that link taxa B and C or taxa A and C. In this case, tree 1 is the best-supported hypothesis.

With large numbers of taxa, the number of possible trees grows dramatically. Similarly, a large number of traits also complicates the problem of identifying the tree best supported by the data. Fortunately, however, systematists have developed sophisticated computer programs to help cope with these complications.

Under the cladistic approach, phylogenetic trees play a key role in classification. Each designated group in a classification should contain only organisms that are more closely

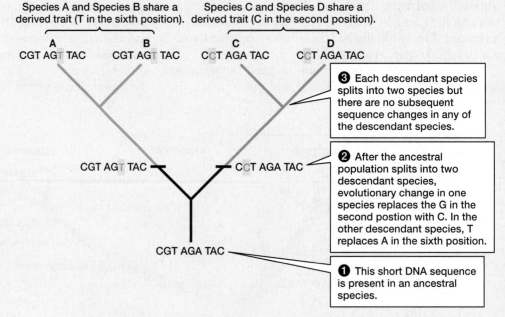

Species A and Species B share a derived trait (T in the sixth position).

Species C and Species D share a derived trait (C in the second position).

A
CGT AGT TAC

B
CGT AGT TAC

C
CCT AGA TAC

D
CCT AGA TAC

❸ Each descendant species splits into two species but there are no subsequent sequence changes in any of the descendant species.

CGT AGT TAC

CCT AGA TAC

❷ After the ancestral population splits into two descendant species, evolutionary change in one species replaces the G in the second position with C. In the other descendant species, T replaces A in the sixth position.

CGT AGA TAC

❶ This short DNA sequence is present in an ancestral species.

FIGURE E18-1 Related taxa are linked by shared derived traits (synapomorphies)

A derived trait is one that has been modified from the ancestral version of the trait. When two or more taxa share a derived trait, the shared trait is said to be a synapomorphy. The hypothetical scenario illustrated here shows how synapomorphies arise.

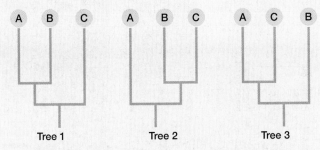

FIGURE E18-2 The three possible trees for three taxa

related to one another than to any organisms outside the group. So, for example, the members of family Canidae (which includes dogs, wolves, foxes, and coyotes) are more closely related to each other than to any member of any other family. Another way to state this principle is to say that each designated group should contain *all* of the living descendants of a common ancestor (**FIG. E18-3a**). In the terminology of cladistic systematics, such groups are said to be *monophyletic*.

Some names, especially names that predate the cladistic approach, designate groups that contain some, but not all, of the descendants of a common ancestor. Such groups are *paraphyletic*. One well-known paraphyletic group is the reptiles (**FIG. E18-3b**). As historically defined, the reptiles exclude the birds, which are now known to belong squarely within the reptile family tree. The reptiles, therefore, do not include all living descendants of the common ancestor that gave rise to snakes, lizards, turtles, crocodilians, and birds. So systematists would prefer to dispense with the former Class Reptilia in favor of a scheme that names only monophyletic groups. The word "reptiles," however, will most likely be with us for some time to come, if only because so many people (including systematists) are accustomed to using it. The word does, after all, provide a convenient way to describe a group of animals that share some interesting adaptations, even if that group isn't monophyletic.

(a)

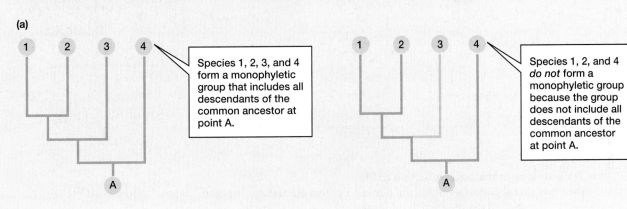

(b)

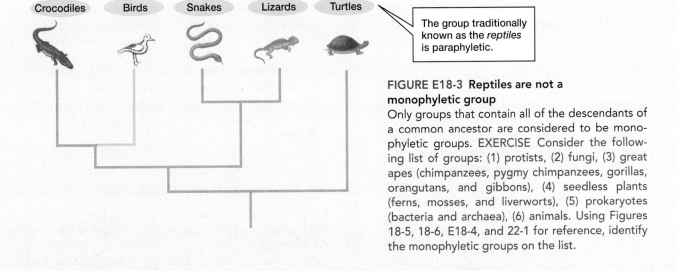

FIGURE E18-3 Reptiles are not a monophyletic group
Only groups that contain all of the descendants of a common ancestor are considered to be monophyletic groups. EXERCISE Consider the following list of groups: (1) protists, (2) fungi, (3) great apes (chimpanzees, pygmy chimpanzees, gorillas, orangutans, and gibbons), (4) seedless plants (ferns, mosses, and liverworts), (5) prokaryotes (bacteria and archaea), (6) animals. Using Figures 18-5, 18-6, E18-4, and 22-1 for reference, identify the monophyletic groups on the list.

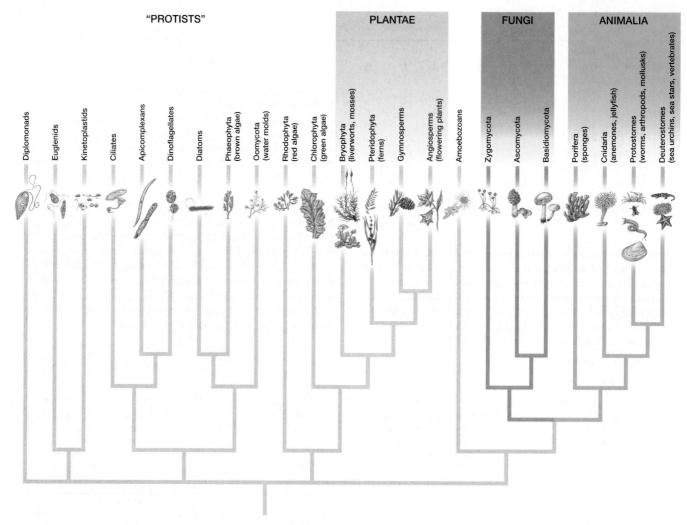

FIGURE 18-6 A closer look at the eukaryotic tree of life
Some of the major evolutionary lineages within the domain Eukarya are shown. The term "protist" refers to the many eukaryotes that are not plants, animals, or fungi.

of eukaryotes that are not members of these three kingdoms. **FIGURE 18-6** shows the evolutionary relationships among some members of the domain Eukarya.

18.3 WHY DO CLASSIFICATIONS CHANGE?

As the emergence of the three-domain system shows, the hypotheses of evolutionary relationships on which classification is based are subject to revision as new data emerge. Even domains and kingdoms, which represent deep, old branchings of the tree of life, must sometimes be rearranged. Such changes at the top levels of classification occur only rarely, but at the other end of the classification hierarchy, among species designations, revisions are more frequent.

Species Designations Change When New Information Is Discovered

As researchers uncover new information, systematists regularly propose changes in species-level classification. For example, until recently, systematists recognized two species of elephant, the African elephant and the Indian elephant. Now, however, they recognize three elephant species; the former African elephant is now divided into two species, the savanna elephant and the forest elephant. Why the change? Genetic analysis of elephants in Africa revealed that there is little gene flow between forest-dwelling and savanna-dwelling elephants. The two groups of elephants are no more genetically similar than lions and tigers.

The Biological Species Definition Can Be Difficult or Impossible to Apply

In some cases, systematists find themselves unable to say with certainty where one species ends and another begins. As discussed in Chapter 16, asexually reproducing organisms pose a particular challenge to systematists, because the criterion of interbreeding (the basis of the biological species definition that we have used in this text) cannot be used to distinguish among species. The irrelevance of this criterion in studies of asexual organisms leaves plenty of

SCIENTIFIC INQUIRY   Molecular Genetics Reveals Evolutionary Relationships

Evolution results from the accumulation of inherited changes in populations. Because DNA is the molecule of heredity, evolutionary changes must be reflected in changes in DNA. Systematists have long known that comparing DNA within a group of species would be a powerful method for inferring evolutionary relationships, but for most of the history of systematics, direct access to genetic information was nothing more than a dream. Today, however, **DNA sequencing**—determining the sequence of nucleotides in segments of DNA—is comparatively cheap, easy, and widely available. The *polymerase chain reaction* (PCR; see Chapter 13) allows systematists to easily accumulate large samples of DNA from organisms, and automated machinery makes sequence determination a comparatively simple task. Sequencing has rapidly become one of the primary tools for uncovering phylogeny.

The logic underlying molecular systematics is straightforward. It is based on the observation that when a single species divides into two species, the gene pool of each resulting species begins to accumulate mutations. The particular mutations in each species, however, will differ because the species are now evolving independently, with no gene flow between them. As time passes, more and more genetic differences accumulate. So, a systematist who has obtained DNA sequences from representatives of both species can compare the two species' nucleotide sequences at any given location in the genome. Fewer differences indicate more closely related organisms.

Putting the simple principles outlined above into practice usually involves more sophisticated thinking. For example, sequence comparisons become far more complex when a researcher wishes to assess relationships among, say, 20 or 30 species. Fortunately, mathematicians and computer programmers have devised some very clever computer-assisted methods for comparing large numbers of sequences and deriving the phylogeny that is most likely to account for the observed sequence differences.

Molecular systematists must also use care in choosing which segment of DNA to sequence. Different parts of the genome evolve at different rates, and it is crucial to sequence a DNA segment whose rate of change is well matched to the phylogenetic question at hand. In general,

slowly evolving genes work best for comparing distantly related organisms, and rapidly changing portions of the genome are best for analyzing closer relationships. It is sometimes difficult to find any single gene that will yield sufficient information to provide an accurate picture of evolutionary change across the genome, so sequences from several different genes are often needed to construct reliable phylogenies, such as the one shown in **FIG. E18-4**.

Today, sequence data are piling up with unprecedented rapidity, and systematists have access to sequences from an ever-increasing number of species. The entire genomes of more than 180 species have been sequenced, and this number is expected to reach 1000 within a decade. The Human Genome Project has been completed, and our own DNA sequences are now a matter of public record. The revolution in molecular biology has fostered a great leap forward in our understanding of evolutionary history.

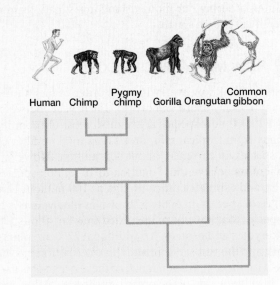

FIGURE E18-4 Relatedness can be determined by comparing DNA sequences
This evolutionary tree was derived from the nucleotide sequences of several different genes that are common to humans and apes.

room for investigators to disagree about which asexual populations constitute a species, especially when comparing groups with similar phenotypes. For instance, some systematists recognize 200 species of the British blackberry (a plant that can produce seeds parthenogenetically—that is, without fertilization), but others recognize only 20 species.

The difficulty of applying the biological species definition to asexual organisms is a serious problem for systematists. After all, a significant portion of Earth's organisms reproduces without sex. Most bacteria, archaea, and protists, for example, reproduce asexually most of the time. Some systematists argue that we need a more universally

applicable definition of species, one that won't exclude asexual organisms and that doesn't depend on the criterion of reproductive isolation.

The Phylogenetic Species Concept Offers an Alternative Definition

A number of alternative species definitions have been proposed over the history of evolutionary biology, but none has been sufficiently compelling to displace the biological species definition. One alternative definition, however, has been gaining adherents in recent years. The *phylogenetic species concept* defines a species as "the smallest diagnosable group that contains all the descendants of a single common ancestor." In other words, if we draw an evolutionary tree that describes the pattern of ancestry among a collection of organisms, each distinctive branch on the tree constitutes a separate species, regardless of whether the individuals represented by that branch can interbreed with individuals from other branches. As you might suspect, rigorous application of the phylogenetic species concept would vastly increase the number of different species recognized by systematists.

Proponents and opponents of the phylogenetic species concept are currently engaged in a vigorous debate about its merits. Perhaps one day it will replace the biological species concept as the "textbook definition" of species. In the meantime, classifications will continue to be debated and revised as systematists learn more and more about evolutionary relationships, particularly with the application of techniques derived from molecular biology. Although the precise evolutionary relationships of many organisms continue to elude us, classification is enormously helpful in ordering our thoughts and investigations into the diversity of life on Earth.

18.4 HOW MANY SPECIES EXIST?

Scientists do not know even within an order of magnitude how many species share our world. Each year, between 7000 and 10,000 new species are named, most of them insects, many from tropical rain forests. The total number of named species is currently about 1.5 million. However, many scientists believe that 7 million to 10 million species may exist, and estimates range as high as 100 million. This total range of species diversity is known as **biodiversity**. Of all the species that have been identified thus far, about 5% are prokaryotes and protists. An additional 22% are plants and fungi, and the rest are animals. This distribution has little to do with the actual abundance of these organisms and a lot to do with the size of the organisms, how easy they are to classify, how accessible they are, and the number of scientists studying them. Historically, systematists have chiefly focused on large or conspicuous organisms in temperate regions, but biodiversity is greatest among small, inconspicuous organisms in the Tropics. In addition to the overlooked species on land and in shallow waters, an entire "continent" of species lies largely unexplored on the deep-sea floor. From the limited samples available, scientists estimate that hundreds of thousands of unknown species may reside there.

Although about 5000 species of prokaryotes have been described and named, prokaryotic diversity remains largely unexplored. Consider a study by Norwegian scientists, who analyzed DNA to count the number of different bacteria species present in a small sample of forest soil. To distinguish among species, the researchers arbitrarily defined bacterial DNA as coming from separate species if it differed by at least 30% from that of any other bacterial DNA in the sample. Using this criterion, they reported more than 4000 types of bacteria in their soil sample and an equal number of forms in a sample of shallow marine sediment.

Our ignorance of the full extent of life's diversity adds a new dimension to the tragedy of the destruction of tropical rain forests. Although these forests cover only about 6% of Earth's land area, they are believed to be home to two-thirds of the world's existing species, most of which have never been studied or named. Because these forests are being destroyed so rapidly, Earth is losing many species that we will never even know existed! For example, in 1990, a new species of primate, the black-faced lion tamarin, was discovered in a small patch of dense rain forest on an island just off the east coast of Brazil (**FIG. 18-7**). Had the patch of forest been cut before this squirrel-sized monkey was discovered, its existence would have remained undocumented. At current rates of deforestation, most of the tropical rain forests, with their undescribed wealth of life, will be gone within the next century.

FIGURE 18-7 The black-faced lion tamarin
Researchers estimate that no more than 260 individuals remain in the wild; captive breeding may be the black-faced lion tamarin's only hope for survival.

LINKS TO LIFE   Small World

In light of humankind's intense curiosity about the origin of our species, it is not surprising that systematists have devoted a great deal of attention to uncovering the evolutionary history of *Homo sapiens*. Though much of this inquiry has focused on revealing the evolutionary connections between modern humans and the species to which we are most closely related, the techniques and methods of systematics have also been used to assess the evolutionary relationships among different populations within our species. Biologists have compared DNA sequences from human populations in many different parts of the world; different investigators have compared different portions of the human genome. As a result, a large amount of data has been gathered, and some interesting findings have emerged.

First, genetic divergence among human populations is very low compared to that of other animal species. For example, the range of genetic differences among all of Earth's humans is only one-tenth the size of the differences among the deer mice of North America (and many other species have even more genetic variability than deer mice). Clearly, all humans are genetically very similar, and the differences among different human populations are tiny.

It is also increasingly apparent that most of the human genetic variability that does exist can be found in African populations. The range of genetic differences found *within* sub-Saharan African populations is greater than the range between African populations and any non-African population. For many genes, all known variants are found in Africa, and no non-African population contains any distinctive variants; rather, non-African populations contain subsets of the African set. This finding strongly suggests that *Homo sapiens* originated in Africa, and that we have not lived anywhere else long enough to evolve very many differences from our African ancestors.

CASE STUDY REVISITED   ORIGIN OF A KILLER

What evidence has persuaded evolutionary biologists that HIV originated in apes and monkeys? To understand the evolutionary thinking behind this conclusion, examine the evolutionary tree shown in **FIGURE 18-8**. This tree illustrates the phylogeny of HIV and its close relatives, as revealed by a comparison of RNA sequences among different viruses. Notice the positions on the tree of the four human viruses (two strains of HIV-1 and two of HIV-2). One strain of HIV-1 is more closely related to a chimpanzee virus than to the other strain of HIV-1. Similarly, one strain of HIV-2 is more closely related to pig-tailed macaque simian immunodeficiency virus (SIV) than to the other strain of HIV-2. Both HIV-1 and HIV-2 are more closely related to ape or monkey viruses than to one another.

The only way for the evolutionary history shown in the tree to have emerged is if viruses jumped between host species. If HIV had evolved strictly within human hosts, the human viruses would be each other's closest relatives. Because the human viruses do not cluster together on the phylogenetic tree, we can infer that cross-species infection occurred, probably on multiple occasions. The most likely means of transmission is human consumption of monkeys (HIV-2) and chimpanzees (HIV-1).

Consider This Can understanding the evolutionary origin of HIV help researchers devise better ways to treat and control the spread of AIDS? How might such understanding influence strategies for treatment and prevention? More generally, how can evolutionary thinking help advance medical research?

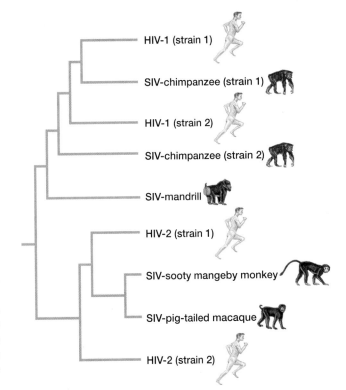

FIGURE 18-8 Evolutionary analysis helps reveal the origin of HIV
In this phylogeny of some immunodeficiency viruses, the viruses with human hosts do not cluster together. This lack of congruence between the evolutionary histories of the viruses and their host species suggests that the viruses must have jumped between host species. (Note that SIV stands for *simian immunodeficiency virus*.)

CHAPTER REVIEW

SUMMARY OF KEY CONCEPTS

18.1   How Are Organisms Named and Classified?

Organisms are classified and placed into hierarchical categories that reflect their evolutionary relationships. The eight major categories, in order of decreasing inclusiveness, are domain, kingdom, phylum, class, order, family, genus, and species. The scientific name of an organism is composed of its genus name and species name. Anatomical and molecular similarities among organisms are a measure of evolutionary relatedness.

Web Tutorial 18.1 Taxonomic Classification

18.2   What Are the Domains of Life?

The three domains of life, each representing one of three main branches of the tree of life, are Bacteria, Archaea, and Eukarya. Each domain contains a number of kingdoms, but the details of kingdom-level classification are in a period of transition and remain unsettled. Within the domain Eukarya, how-ever, the kingdoms Fungi, Plantae, and Animalia are universally accepted as valid, monophyletic groups.

Web Tutorial 18.2 The Tree of Life

18.3   Why Do Classifications Change?

Classifications are subject to revision as new information is discovered. Species boundaries may be hard to define, particularly in the case of asexually reproducing species. However, systematics is essential for precise communication and contributes to our understanding of the evolutionary history of life.

18.4   How Many Species Exist?

Although only about 1.5 million species have been named, estimates of the total number of species range up to 100 million. New species are being identified at the rate of 7000 to 10,000 annually, mostly in tropical rain forests.

KEY TERMS

Archaea   *page 361*
Bacteria   *page 361*
biodiversity   *page 366*
class   *page 358*

DNA sequencing   *page 365*
domain   *page 358*
Eukarya   *page 361*
family   *page 358*

genus   *page 358*
kingdom   *page 358*
order   *page 358*
phylogeny   *page 358*

phylum   *page 358*
scientific name   *page 358*
species   *page 358*
systematics   *page 358*

THINKING THROUGH THE CONCEPTS

1. What contributions did Aristotle, Linnaeus, and Darwin each make to modern taxonomy?

2. What features would you study to determine whether a dolphin is more closely related to a fish or to a bear?

3. What techniques might you use to determine whether the extinct cave bear is more closely related to a grizzly bear or to a black bear?

4. Only a small fraction of the total number of species on Earth has been scientifically described. Why?

5. In England, "daddy longlegs" refers to a long-legged fly, but the same name refers to a spider-like animal in the United States. How do scientists attempt to avoid such confusion?

APPLYING THE CONCEPTS

1. There are many areas of disagreement about the classification of organisms. For example, there is no consensus about whether the red wolf is a distinct species or about how many kingdoms are within the domain Bacteria. What difference does it make whether biologists consider the red wolf a species, or into which kingdom a bacterial species falls? As Shakespeare put it, "What's in a name?"

2. The pressures created by human population growth and economic expansion place storehouses of biological diversity such as the Tropics in peril. The seriousness of the situation is clear when we consider that probably only 1 out of every 20 tropical species is known to science at present. What arguments can you make for preserving biological diversity in poor and developing countries, such as those in many areas of the Tropics? Does such preservation require that these countries sacrifice economic development? Suggest some solutions to the conflict between the growing demand for resources and the importance of conserving biodiversity.

3. During major floods, only the topmost branches of submerged trees may be visible above the water. If you were asked to sketch the branches below the surface of the water solely on the basis of the positions of the exposed tips, you would be attempting a reconstruction somewhat similar to the "family tree" by which taxonomists link various organisms according to their common ancestors (analogous to branching points). What sources of error do both exercises share? What advantages do modern taxonomists have?

4. The Florida panther, found only in the Florida Everglades, is currently classified as an endangered species, protecting it from human activities that could lead to its extinction. It has long been considered a subspecies of cougar (mountain lion), but recent mitochondrial DNA studies have shown that the Florida panther may actually be a hybrid between American and South American cougars. Should the Florida panther be protected by the Endangered Species Act?

FOR MORE INFORMATION

Dawkins, R. *The Ancestor's Tale*. Boston: Houghton Mifflin, 2004. A masterful overview of life's history as revealed by systematics.

Gould, S. J. "What Is a Species?" *Discover*, December 1992. Discusses the difficulties of distinguishing separate species.

Mann, C., and Plummer, M. *Noah's Choice: The Future of Endangered Species*. New York: Knopf, 1995. A thought-provoking look at the hard choices we must make with regard to protecting biodiversity. Which species will we choose to preserve? What price are we willing to pay?

Margulis, L., and Sagan, D. *What Is Life?* London: Weidenfeld & Nicolson, 1995. A lavishly illustrated survey of life's diversity. Also includes an account of life's history and a meditation on the question posed by the title.

May, R. M. "How Many Species Inhabit the Earth?" *Scientific American*, October 1992. Although no one knows the precise answer to this question, an effective estimate is crucial to our effort to manage our biological resources.

19

The Diversity of Prokaryotes and Viruses

Workers prepare to decontaminate the Hart Office Building in Washington, D.C., after it was attacked with a biological weapon.

CASE STUDY AGENTS OF DEATH

IN THE FALL OF 2001, a long-standing fear became a terrible reality when residents of the United States were attacked with a biological weapon. The weapon, which killed five people and caused six others to become gravely ill, was simply a culture of bacteria. The bacteria had been placed in envelopes and mailed to the Hart Office Building in Washington, D.C., and to media company offices, where they were unknowingly inhaled by victims who opened the harmless-looking envelopes. The attack, though comparatively small, nonetheless dramatically illustrated the possibility and potential destructive power of a larger attack.

The bacterium used in the attack was *Bacillus anthracis*, which causes the disease anthrax. Anthrax normally infects domestic animals such as goats and sheep, but can also infect humans. The bacterium is a dangerous, often deadly infectious agent with properties that make it especially attractive to developers of biological weapons. Anthrax bacteria are easily isolated from infected animals, are cheap and easy to culture in large quantities, and, once produced, can be dried into a powder that remains viable for years. The powder is easily "weaponized" by packing it into a missile warhead or other delivery device, and a small volume of bacteria can infect a very large number of people. Areas contaminated with anthrax bacteria are very difficult to decontaminate.

Since the attack, it has become apparent that much of our ability to defend against such biological attacks depends on our understanding of the disease-causing microbes (as single-celled organisms are collectively known) that can be used as biological weapons. Scientific investigation of microbes can provide the knowledge required to detect an attack, destroy dangerous microbes in the environment, and prevent and treat infections. Fortunately, biologists already know quite a bit about microorganisms. In this chapter, we'll explore some of that knowledge.

19.1 WHICH ORGANISMS MAKE UP THE PROKARYOTIC DOMAINS— BACTERIA AND ARCHAEA?

Earth's first organisms were prokaryotes, single-celled microbes that lacked organelles such as the nucleus, chloroplasts, and mitochondria. (See Chapter 4 for a comparison of prokaryotic and eukaryotic cells.) For the first 1.5 billion years or more of life's history, all life was prokaryotic. Even today, prokaryotes are extraordinarily abundant. A drop of seawater contains hundreds of thousands of prokaryotic organisms, and a spoonful of soil contains billions. The average human body is home to trillions of prokaryotes, which live on the skin, in the mouth, and in the stomach and intestines. In terms of abundance, prokaryotes are Earth's predominant form of life.

Bacteria and Archaea Are Fundamentally Different

Two of life's three domains, Bacteria and Archaea, consist entirely of prokaryotes. Bacteria and archaea are superficially similar in appearance under the microscope, but have striking differences in their structural and biochemical features that reveal the ancient evolutionary separation between them. For example, the cell walls of bacterial cells contain molecules of the polymer *peptidoglycan*, which

helps strengthen the cell wall. Peptidoglycan is unique to bacteria, and the cell walls of archaea do not contain it. Bacteria and archaea also differ in the structure and composition of plasma membranes, ribosomes, and RNA polymerases and in the mechanics of basic processes such as transcription and translation.

Classification of Prokaryotes Within Each Domain Is Difficult

The sharp biochemical differences between archaea and bacteria make distinguishing the two domains a straightforward matter, but classification within each domain poses challenges. Prokaryotes are tiny and structurally simple and simply do not exhibit the huge array of anatomical and developmental differences that can be used to infer the evolutionary history of plants, animals, and other eukaryotes. Consequently, prokaryotes have been classified on the basis of such features as shape, means of locomotion, pigments, nutrient requirements, the appearance of colonies (groups of individuals that descended from a single cell), and staining properties. For example, the **Gram stain**, a staining technique, distinguishes two types of cell-wall construction in bacteria. Depending on the results of the stain, these bacteria are classified as either *gram-positive* or *gram-negative*.

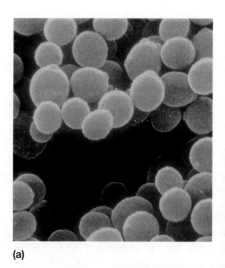

(a)

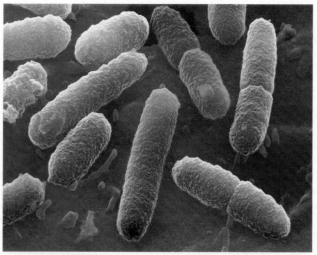

(b)

FIGURE 19-1 Three common prokaryote shapes
(a) Spherical bacteria of the genus *Micrococcus*, (b) rod-shaped bacteria of the genus *Escherichia*, and (c) corkscrew-shaped bacteria of the genus *Borrelia*.

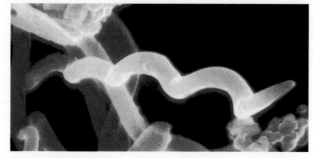

(c)

In recent years our understanding of the evolutionary history of the prokaryotic domains has been greatly expanded by comparisons of DNA and RNA nucleotide sequences. On the basis of this new information, some biologists now classify bacteria into 13 to 15 kingdoms and archaea into about 3 kingdoms. Prokaryote classification, however, is a rapidly changing field, and consensus on kingdom-level classification has thus far proved elusive. With new DNA sequence data being generated at a furious pace, and with new and distinctive types of bacteria and archaea being discovered and described on a regular basis, the revision of prokaryote classification schemes will likely continue for some time to come.

Prokaryotes Differ in Size and Shape

Both bacteria and archaea are normally very small, ranging from about 0.2 to 10 micrometers in diameter. (In comparison, the diameters of eukaryotic cells range from about 10 to 100 micrometers.) About 250,000 average-sized bacteria or archaea could congregate on the period at the end of this sentence, though a few species of bacteria are larger. The largest known bacterium, *Thiomargarita namibiensis*, is as much as 700 micrometers in diameter, making it visible to the naked eye.

The cell walls that surround prokaryotic cells give characteristic shapes to different types of bacteria and archaea. The most common shapes are spherical, rodlike, and corkscrew-shaped (**FIG. 19-1**).

19.2 HOW DO PROKARYOTES SURVIVE AND REPRODUCE?

The abundance of prokaryotes is due in large part to adaptations that allow members of the two prokaryotic domains to inhabit and exploit a wide range of environments. In this section, we discuss some of the traits that help prokaryotes survive and thrive.

Some Prokaryotes Are Mobile

Many bacteria and archaea adhere to a surface or drift passively in liquid surroundings, but some can move about. Many of these mobile prokaryotes have **flagella** (singular, *flagellum*). Prokaryote flagella may appear singly at one end of a cell, in pairs (one at each end of the cell), as a tuft at one end of the cell (**FIG. 19-2a**), or scattered over the entire cell surface. Flagella can rotate rapidly, propelling the organism through its liquid environment. By using flagella to move, prokaryotes can disperse into new habitats, migrate toward nutrients, and leave unfavorable environments.

The structure of prokaryote flagella is different from and simpler than the structure of eukaryotic flagella (see page 67 for a description of the eukaryotic flagellum). In bacterial flagella, a unique wheel-like structure embedded

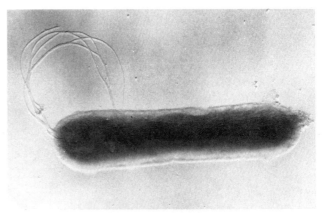

(a)

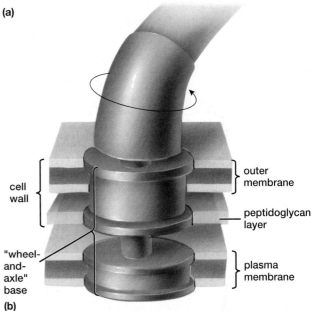

(b)

cell wall

outer membrane

peptidoglycan layer

"wheel-and-axle" base

plasma membrane

FIGURE 19-2 The prokaryote flagellum
(a) A flagellated archaean of the genus *Aquifex* uses its flagella to move toward favorable environments. **(b)** In bacteria, a unique "wheel-and-axle" arrangement anchors the flagellum within the cell wall and plasma membrane, enabling the flagellum to rotate rapidly.

in the bacterial membrane and cell wall allows the flagellum to rotate (**FIG. 19-2b**). Archaeal flagella are thinner than bacterial flagella and are constructed of different proteins. The structure of the archaeal flagellum, however, is not yet as well understood as that of the bacterial flagellum.

Many Bacteria Form Films on Surfaces

The cell walls of some bacterial species are surrounded by sticky layers of protective slime, composed of polysaccharide or protein, which protects the bacteria and helps them adhere to surfaces. In many cases, slime-secreting bacteria of one or more species aggregate in colonies to form communities known as *biofilms*. One familiar biofilm is dental plaque, which is formed by the bacteria that inhabit the

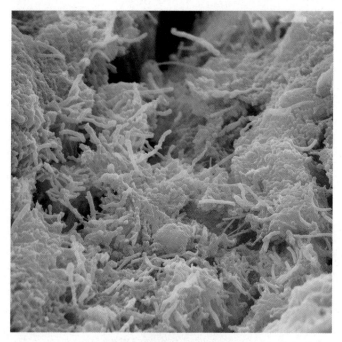

FIGURE 19-3 The cause of tooth decay
Bacteria in the human mouth form a slimy biofilm that helps them cling to tooth enamel and protects them from threats in the environment. In this micrograph, individual bacteria (colored green or yellow) are visible, embedded in the brown biofilm. The bacteria-laden biofilm can cause tooth decay.

mouth (**FIG. 19-3**). The protection afforded by biofilms helps defend the embedded bacteria against a variety of attacks, including those launched by antibiotics and disinfectants. As a result, biofilms formed by bacteria harmful to humans can be very difficult to eradicate. The persistence of biofilms is unfortunate, because the surfaces on which biofilms form include contact lenses, surgical sutures, and medical equipment such as catheters. In addition, many infections of the human body take the form of biofilms, including those responsible for tooth decay, gum disease, and ear infections.

Protective Endospores Allow Some Bacteria to Withstand Adverse Conditions

When environmental conditions become inhospitable, many rod-shaped bacteria form protective structures called **endospores**. An endospore, which forms inside a bacterium, contains genetic material and a few enzymes encased within a thick protective coat (**FIG. 19-4**). Metabolic activity ceases until the spore encounters favorable conditions, at which time metabolism resumes and the spore develops into an active bacterium.

Endospores are resistant to even extreme environmental conditions. Some can withstand boiling for an hour or more. Endospores are also able to survive for extraordinarily long periods. In the most extreme example of such

longevity, scientists recently discovered bacterial spores that had been sealed inside rock for 250 million years. After being carefully extracted from their rocky "tomb," the spores were incubated in test tubes. Amazingly, live bacteria developed from the ancient spores, which were older than the oldest dinosaur fossils.

Endospores are one of the main reasons that the bacterial disease anthrax has become an agent of biological terrorism. The bacterium that causes anthrax forms endospores, which provide the means by which terrorists (or governments) can disperse the bacteria. The spores can be stored indefinitely and can survive the harsh conditions they might encounter while traveling to their destination, including the stress of a missile launch and high-altitude travel. When they reach their target, the spores can survive dispersal into the atmosphere, remaining viable until inhaled by a potential victim.

Prokaryotes Are Specialized for Specific Habitats

Prokaryotes occupy virtually every habitat, including those where extreme conditions keep out other forms of life. For example, some bacteria thrive in near-boiling environments, such as the hot springs of Yellowstone National Park (**FIG. 19-5**). Many archaea live in even hotter environments, including springs where the water actually boils, or deep-ocean vents, where superheated water is spewed through cracks in Earth's crust at temperatures of up to 230°F (110°C). Prokaryotes can also survive at extremely high pressures, such as are found 1.7 miles (2.8 kilometers) below Earth's surface, where scientists recently discovered a new bacterial species. Bacteria and archaea are also found in very cold environments, such as Antarctic sea ice.

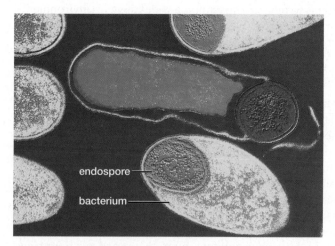

endospore
bacterium

FIGURE 19-4 Spores protect some bacteria
Resistant endospores have formed inside bacteria of the genus *Clostridium*, which causes the potentially fatal food poisoning called botulism. **QUESTION** What might explain the observation that most species of endospore-forming bacteria live in soil?

FIGURE 19-5 **Some prokaryotes thrive in extreme conditions** Hot springs harbor bacteria and archaea that are both heat and mineral tolerant. Several species of cyanobacteria paint these hot springs in Yellowstone National Park with vivid colors; each is confined to a specific area determined by temperature range. **QUESTION** Some of the enzymes that have important uses in molecular biology procedures are extracted from prokaryotes that live in hot springs. Can you guess why?

Even extreme chemical conditions fail to impede invasion by prokaryotes. Thriving colonies of bacteria and archaea live in the Dead Sea, where a salt concentration seven times that of the oceans precludes all other life, and in waters that are as acidic as vinegar or as alkaline as household ammonia. Of course, rich bacterial communities also reside in a full range of more moderate habitats, including in and on the healthy human body. But an animal need not be healthy to harbor bacteria. Recently, a colony of bacteria was found dormant within the intestinal contents of a mammoth that had lain in a peat bog for 11,000 years.

No single species of prokaryote, however, is as versatile as these examples may suggest. In fact, most prokaryotes are specialists. One species of archaea that inhabits deep-sea vents, for example, grows optimally at 223°F (106°C) and stops growing altogether at temperatures below 194°F (90°C). Clearly, this species could not survive in a less-extreme habitat. Bacteria that live on the human body are also specialized; different species colonize the skin, the mouth, the respiratory tract, the large intestine, and the urogenital tract.

Prokaryotes Exhibit Diverse Metabolisms

Prokaryotes are able to colonize diverse habitats partly because they have evolved diverse methods of acquiring energy and nutrients from the environment. For example, unlike eukaryotes, many prokaryotes are **anaerobes**; their metabolisms do not require oxygen. Their ability to inhabit oxygen-free environments allows prokaryotes to exploit habitats that are off-limits to eukaryotes. Some anaerobes, such as many of the archaea found in hot springs and the

bacterium that causes tetanus, are actually poisoned by oxygen. Others are opportunists, engaging in anaerobic respiration when oxygen is lacking and switching to aerobic respiration (a more efficient process) when oxygen becomes available. Many prokaryotes, of course, are strictly aerobic, and require oxygen at all times.

Whether aerobic or anaerobic, different prokaryote species can extract energy from an amazing array of substances. Prokaryotes subsist not only on the sugars, carbohydrates, fats, and proteins that we normally think of as foods, but also on compounds that are inedible or even poisonous for humans, including petroleum, methane (the main component of natural gas), and solvents such as benzene and toluene. Prokaryotes can even metabolize inorganic molecules, including hydrogen, sulfur, ammonia, iron, and nitrite. The process of metabolizing inorganic molecules sometimes yields by-products that are useful to other organisms. For example, certain bacteria release sulfates or nitrates, crucial plant nutrients, into the soil.

Some species of bacteria, such as the *cyanobacteria* (**FIG. 19-6**), use photosynthesis to capture energy directly from sunlight. Like green plants, photosynthetic bacteria possess chlorophyll. Most species produce oxygen as a by-product of photosynthesis, but some, known as the sulfur bacteria, use hydrogen sulfide (H_2S) instead of water (H_2O) in photosynthesis, releasing sulfur instead of oxygen. No photosynthetic archaea are known.

Prokaryotes Reproduce by Binary Fission

Most prokaryotes reproduce asexually by a simple form of cell division called binary fission (see Chapter 11), which produces genetically identical copies of the original cell

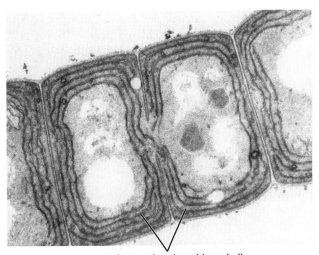

membranes bearing chlorophyll

FIGURE 19-6 **Cyanobacteria** Electron micrograph of a section through a cyanobacterial filament. Chlorophyll is located on the membranes visible within the cells.

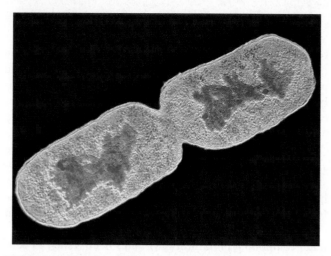

FIGURE 19-7 Reproduction in prokaryotes
Prokaryotic cells reproduce by binary fission. In this color-enhanced electron micrograph, an *Escherichia coli*, a normal component of the human intestine, is dividing. Red areas are genetic material. **QUESTION** What is the main advantage of binary fission, compared to sexual reproduction?

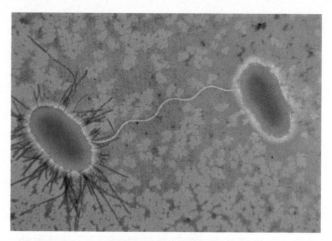

FIGURE 19-8 Conjugation: Prokaryotic "mating"
During conjugation, one prokaryote acts as a donor, transferring DNA to the recipient. In this photo, two *Escherichia coli* are connected by a long sex pilus. The sex pilus will retract, drawing the recipient bacterium (at right) to the donor bacterium. The donor bacterium is bristling with nonsex pili that help it attach to surfaces.

(FIG. 19-7). Under ideal conditions, a prokaryotic cell can divide about once every 20 minutes, potentially giving rise to sextillions (10^{21}) of offspring in a single day. This rapid reproduction allows prokaryotes to exploit temporary habitats, such as a mud puddle or warm pudding. Rapid reproduction also allows bacterial populations to evolve quickly. Recall that many mutations, the source of genetic variability, are the result of mistakes in DNA replication during cell division (see Chapter 10). Thus, the rapid, repeated cell division of prokaryotes provides ample opportunity for new mutations to arise and also allows mutations that enhance survival to spread quickly.

Prokaryotes May Exchange Genetic Material Without Reproducing

Although prokaryote reproduction is generally asexual and does not involve genetic recombination, some bacteria and archaea nonetheless exchange genetic material. In these species, DNA is transferred from a donor to a recipient in a process called **conjugation**. The cell membranes of two conjugating prokaryotes fuse temporarily to form a cytoplasmic bridge across which DNA travels. In bacteria, donor cells may use specialized extensions called *sex pili* that attach to a recipient cell, drawing it closer to allow conjugation (**FIG. 19-8**). Conjugation produces new genetic combinations that may allow the resulting bacteria to survive under a greater variety of conditions. In some cases, genetic material may be exchanged even between individuals of different species.

DNA transferred during bacterial conjugation is contained within a structure called a **plasmid**, a small, circular DNA molecule that is separate from the single bacterial chromosome. Plasmids may carry genes for antibiotic resistance or even alleles of genes also found on the main bacterial chromosome. Researchers in molecular genetics have made extensive use of bacterial plasmids, as described in Chapter 13.

19.3 HOW DO PROKARYOTES AFFECT HUMANS AND OTHER EUKARYOTES?

Although they are largely invisible to us, prokaryotes play a crucial role in life on Earth. Plants and animals (including humans) are utterly dependent on prokaryotes. Prokaryotes help plants and animals obtain vital nutrients and help break down and recycle wastes and dead organisms. We could not survive without prokaryotes, but their impact on us is not always beneficial. Some of humanity's most deadly diseases stem from microbes.

Prokaryotes Play Important Roles in Animal Nutrition

Many eukaryotic organisms depend on close associations with prokaryotes. For example, most animals that eat leaves, including cattle, rabbits, koalas, and deer, can't themselves digest cellulose, the principal component of plant cell walls. Instead, these animals depend on certain bacteria that have the unusual ability to break down cellulose. Some of these bacteria live in the animals' digestive tracts, where they help liberate nutrients from plant tissue that the animals are unable to break down themselves. Without the bacteria, leaf-eating animals could not survive.

Prokaryotes also have important impacts on human nutrition. Many foods, including cheese, yogurt, and sauerkraut, are produced by the action of bacteria. Bacteria also inhabit your intestines. These bacteria feed on undigested food and synthesize such nutrients as vitamin K and vitamin B_{12}, which the human body absorbs.

Prokaryotes Capture the Nitrogen Needed by Plants

Humans could not live without plants, and plants are entirely dependent on bacteria. In particular, plants are unable to capture nitrogen from that element's most abundant reservoir, the atmosphere. Plants need nitrogen

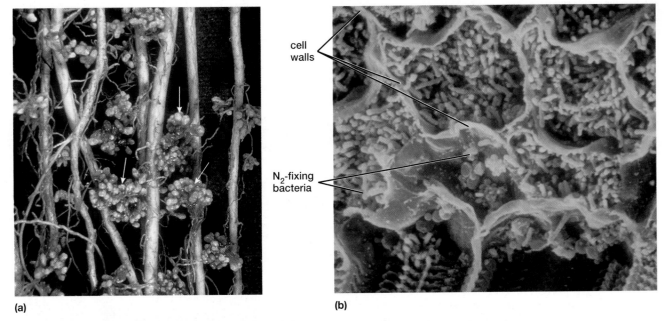

FIGURE 19-9 Nitrogen-fixing bacteria in root nodules
(a) Special chambers called nodules on the roots of a legume (alfalfa) provide a protected and constant environment for nitrogen-fixing bacteria. **(b)** This scanning electron micrograph shows the nitrogen-fixing bacteria inside cells within the nodules. **QUESTION** If all of Earth's nitrogen-fixing prokaryotes were to die suddenly, what would happen to the concentration of nitrogen gas in the atmosphere?

to grow. To acquire it, they depend on **nitrogen-fixing bacteria**, which live both in soil and in specialized nodules, small, rounded lumps on the roots of certain plants (legumes, which include alfalfa, soybeans, lupines, and clover; **FIG. 19-9**). The nitrogen-fixing bacteria capture nitrogen gas (N_2) from air trapped in the soil and combine it with hydrogen to produce ammonium (NH_4^+), a nitrogen-containing nutrient that plants can use directly.

Prokaryotes Are Nature's Recyclers

Prokaryotes play a crucial role in recycling waste. Most prokaryotes obtain energy by breaking down complex organic (carbon-containing) molecules. Such prokaryotes find a plentiful source of organic molecules in the waste products and dead bodies of plants and animals. By consuming and thereby decomposing these wastes, prokaryotes ensure that wastes do not accumulate in the environment. In addition, decomposition by prokaryotes releases the nutrients contained in wastes. Once released, the nutrients become available for reuse by living organisms.

Prokaryotes perform their recycling service wherever organic matter is found. They are important decomposers in lakes and rivers, in the oceans, and in the soil and groundwater of forests, grasslands, deserts, and other terrestrial environments. The recycling of nutrients by prokaryotes and other decomposers provides the basis for continued life on Earth.

Prokaryotes Can Clean Up Pollution

Many of the pollutants that are produced as by-products of human activity are organic compounds. As such, these pollutants can potentially serve as food for archaea and bacteria. Many of them are, in fact, consumed; the range of

compounds consumed by prokaryotes is staggering. Nearly anything that human beings can synthesize can be broken down by some prokaryote, including detergents, many toxic pesticides, and harmful industrial chemicals such as benzene and toluene.

Even oil can be broken down by prokaryotes. Soon after the tanker *Exxon Valdez* dumped 11 million gallons of crude oil into Alaska's Prince William Sound in 1989, researchers sprayed oil-soaked beaches with a fertilizer that encouraged the growth of natural populations of oil-eating bacteria. Within 15 days, the oil deposits on these beaches were noticeably reduced in comparison with unsprayed areas.

The practice of manipulating conditions to stimulate breakdown of pollutants by living organisms is known as *bioremediation.* Improved methods of bioremediation could dramatically increase our ability to clean up toxic waste sites and polluted groundwater. A great deal of current research is therefore aimed at identifying prokaryote species that are especially effective for bioremediation and at discovering practical methods for manipulating these organisms to improve their effectiveness.

Some Bacteria Pose a Threat to Human Health

Despite the benefits some bacteria provide, the feeding habits of certain bacteria threaten our health and well-being. These **pathogenic** (disease-producing) bacteria synthesize toxic substances that cause disease symptoms. (So far, no pathogenic archaea have been identified.)

Some Anaerobic Bacteria Produce Dangerous Poisons

Some bacteria produce toxins that attack the nervous system. Examples of such pathogens include *Clostridium tetani*, which causes tetanus, and *Clostridium botulinum*,

Although the prospect of biological weapons is chilling, you are far more likely to encounter harmful microorganisms from a more mundane source: your food. The nutrients that you consume during meals and snacks can also provide sustenance for a wide variety of disease-causing bacteria and protists. Some of these invisible diners may accompany your lunch to your digestive tract and take up residence there, causing unpleasant symptoms. The Centers for Disease Control estimates that U.S. residents experience an astonishing 76 million cases of food-borne illness each year, resulting in 325,000 hospitalizations and 5200 deaths.

The most frequent culprits in food-borne diseases are bacteria. Species in the genera *Escherichia*, *Salmonella*, *Listeria*, *Streptococcus*, and *Campylobacter* are responsible for an especially large number of illnesses, with *Campylobacter* currently claiming the largest number of victims.

How can you protect yourself from the bacteria and protists that share our food supply? It's easy: clean, cook, and chill. Cleaning helps prevent the spread of pathogens. Wash your hands before preparing food, and wash all utensils and cutting boards after preparing each item. Thorough cooking is the best way to ensure that any bacteria and protists present in food are killed. Meats, in particular, must be thoroughly cooked; do *not* eat meat that is still pink inside (**FIG. E19-1**). Fish should be cooked until it is opaque and flakes easily with a fork; cook eggs until both white and yolk are firm. Finally, keep food cold. Pathogens multiply most rapidly at temperatures between 40°F and 140°F. So get your groceries home from the store and into the refrigerator or freezer as quickly as possible. Don't leave cooked leftovers unrefrigerated for more than two hours. Thaw frozen foods in the refrigerator, not at room temperature. A little bit of attention to food safety can save you from unwelcome guests in your food.

FIGURE E19-1 Rare beef is a haven for dangerous bacteria

which causes botulism (a sometimes lethal food poisoning). Both of these bacterial species are anaerobes that survive as spores until introduced into a favorable, oxygen-free environment. For example, a deep puncture wound may allow tetanus bacteria to penetrate a human body and reach a place where they will be protected from contact with oxygen. As they multiply, the bacteria release their paralyzing poison into the bloodstream. For botulism bacteria, a sealed container of canned food that has been improperly sterilized may provide a haven. Thriving on the nutrients in the can, these anaerobes produce a toxin so potent that a single gram could kill 15 million people. Perhaps inevitably, this potent poison has caught the attention of biological weapon designers, who are presumed to have added it to their arsenals.

Humans Battle Bacterial Diseases Old and New

Bacterial diseases have had a significant impact on human history. Perhaps the most infamous example is bubonic plague, or "Black Death," which killed 100 million people during the mid-fourteenth century. In many parts of the world, one-third or more of the population died. Plague is caused by the highly infectious bacteria *Yersinia pestis*, which is spread by fleas that feed on infected rats and then move to human hosts. Although bubonic plague has not reemerged as a large-scale epidemic, about 2000 to 3000 people worldwide are still diagnosed with the disease each year.

Some bacterial pathogens seem to emerge suddenly. Lyme disease, for example, was unknown until 1975. This disease, named after the town of Old Lyme, Connecticut, where it was first described, is caused by the spiral-shaped bacterium *Borrelia burgdorferi*. The bacterium is carried by deer ticks, which transmit it to the humans they bite. At first, the symptoms resemble flu, with chills, fever, and body aches. If untreated, weeks or months later the victim may experience rashes, bouts of arthritis, and in some cases abnormalities of the heart and nervous system. Both physicians and the general public are becoming more familiar with the disease, so more victims are receiving treatment before serious symptoms develop.

Perhaps the most frustrating pathogens are those that come back to haunt us long after we believed that we had them under control. Tuberculosis, a bacterial disease once almost vanquished in developed countries, is again on the rise in the United States and elsewhere. Two sexually transmitted bacterial diseases, gonorrhea and syphilis, have reached epidemic proportions around the globe. Cholera, a water-transmitted bacterial disease that flourishes when raw sewage contaminates drinking water or fishing areas, is under control in developed countries but remains a major killer in poorer parts of the world.

Some Common Bacterial Species Can Be Harmful

Some pathogenic bacteria are so widespread and common that we cannot expect to ever be totally free of their damaging effects. For example, different species of the abundant streptococcus bacterium produce several diseases. One streptococcus causes strep throat. Another, *Streptococcus*

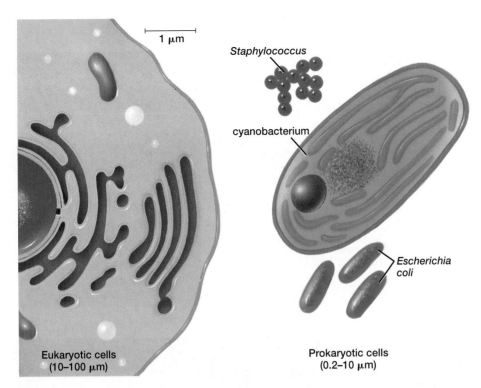

1 μm

Staphylococcus

cyanobacterium

Escherichia coli

Eukaryotic cells
(10–100 μm)

Prokaryotic cells
(0.2–10 μm)

Viruses
(0.05–0.2 μm)

pneumoniae, causes pneumonia by stimulating an allergic reaction that clogs the lungs with fluid. Yet another streptococcus has gained fame as the "flesh-eating bacterium." A small percentage of people who become infected with this one experience severe symptoms, described luridly in tabloid newspapers with such headlines as "Killer Bug Ate My Face." About 800 Americans each year are victims of necrotizing fasciitis (as the "flesh-eating" infection is more properly known), and about 15% of these victims die. The streptococci enter through broken skin and produce toxins that either destroy flesh directly or stimulate an overwhelming and misdirected attack by the immune system against the body's own cells. A limb can be destroyed in hours, and in some cases only amputation can halt the rapid tissue destruction. In other cases, these rare strep infections sweep through the body, causing death within a matter of days.

One of the most common bacterial inhabitants of the human digestive system, *Escherichia coli*, is also capable of doing harm. Different populations of *E. coli* may differ genetically, and some genetic differences can transform this normally benign species into a pathogen. One particularly notorious strain, known as O157:H7, infects about 70,000 Americans each year, about 60 of whom die from its effects. Most O157:H7 infections result from consumption of contaminated beef. About a third of the cattle in the United States carry O157:H7 in their intestinal tracts, and the bacteria can be transmitted to humans when a slaughterhouse inadvertently grinds some gut contents into hamburger. Once in a human digestive system, O157:H7 bacteria attach firmly to the wall of the intestine and begin to release a toxin that causes intestinal bleeding and that spreads to and damages other organs as well. The best defense against O157:H7 is to cook all meat thoroughly. (For more tips on protecting yourself from food-borne bacteria, see "Links to Life: Unwelcome Dinner Guests.")

Most Bacteria Are Harmless

Although some bacteria assault the human body, most of the bacteria with which we share our bodies are harmless, and many are beneficial. For example, the normal bacterial community in the vagina creates an environment that is hostile to infections by parasites such as yeasts. The bacteria that harmlessly inhabit our intestines are an important source of vitamin K. As the late physician, researcher, and author Lewis Thomas so aptly put it, "Pathogenicity is, in a sense, a highly skilled trade, and only a tiny minority of all the numberless tons of microbes on the Earth has ever been involved in it; most bacteria are busy with their own business, browsing and recycling the rest of life."

19.4 WHAT ARE VIRUSES, VIROIDS, AND PRIONS?

The particles known as **viruses** are generally found in close association with living organisms, but most biologists do not consider them to be alive. Viruses lack the traits that characterize life. For example, they are not cells—nor are they composed of cells. Further, they cannot on their own accomplish the basic tasks that living cells perform. Viruses have no ribosomes on which to make proteins, no cytoplasm, no ability to synthesize organic molecules, and no capacity to extract and use the energy stored in such molecules. They possess no membranes of their own and cannot grow or reproduce on their own. The simplicity of viruses seems to place them outside the realm of living things.

A Virus Consists of a Molecule of DNA or RNA Surrounded by a Protein Coat

Viruses are tiny; they are much smaller than even the smallest prokaryotic cell (**FIG. 19-10**). Virus particles are so

FIGURE 19-11 Viruses come in a variety of shapes
Viral shape is determined by the nature of the virus's protein coat.

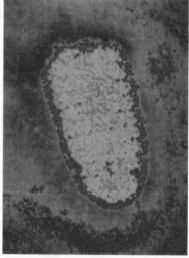

rabies virus

bacterophage

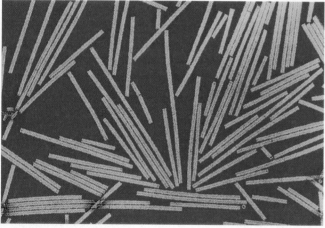

tobacco mosaic viruses

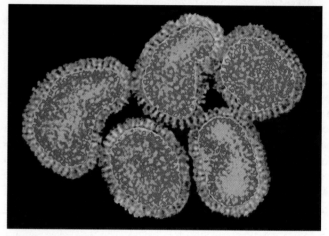

influenza viruses

small (0.05–0.2 micrometer in diameter) that they can be seen only under the enormous magnification of an electron microscope. Under such magnification, one can see that viruses assume a great variety of shapes (**FIG. 19-11**).

Viruses consist of two major parts: a molecule of hereditary material and a coat of protein surrounding the molecule. The hereditary molecule may be either DNA or RNA and may be single-stranded or double-stranded, linear or circular. The protein coat may be surrounded by an envelope formed from the plasma membrane of the host cell (**FIG. 19-12**).

Viruses Are Parasites

Viruses are parasites of living cells. (Parasites live in or on host organisms, harming their hosts in the process). A virus can reproduce only inside a **host** cell—the cell that a virus or other infectious agent infects. Viral reproduction begins when a virus penetrates a host cell. After the virus enters the host cell, the viral genetic material takes command. The hijacked host cell is forced to use the instructions encoded in the viral genes to produce the components of new viruses. The pieces are rapidly assembled, and an army of new viruses bursts forth to invade and conquer neighboring cells (see "A Closer Look: Viruses—How the Nonliving Replicate").

Viruses Are Host Specific

Each type of virus is specialized to attack a specific host cell. As far as we know, no organism is immune to all viruses. Even bacteria fall victim to viral invaders; viruses that infect bacteria are called **bacteriophages** (**FIG. 19-13**). Bacteriophages may soon become important in treating diseases caused by bacteria, as many disease-causing bacteria have become increasingly resistant to antibiotics. Treatments based on bacteriophages could also take advantage of the viruses' specificity, attacking only the targeted bacteria and not the many other harmless or beneficial bacteria in the body.

In multicellular organisms such as plants and animals, different viruses specialize in attacking particular cell types. Viruses responsible for the common cold, for example, attack the membranes of the respiratory tract. Those causing measles infect the skin, and the rabies virus attacks nerve cells. One type of herpes virus specializes in the mucous membranes of the mouth and lips, causing cold sores; a second type produces similar sores on or near the genitals. Herpes viruses take up permanent residence in the body, erupting periodically (typically during times of stress) as infectious sores. The devastating disease AIDS

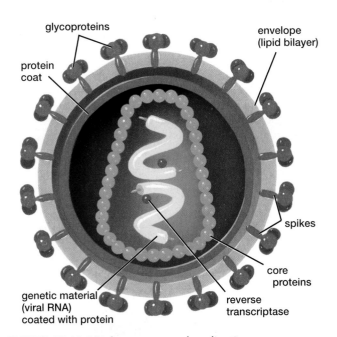

FIGURE 19-12 Viral structure and replication
A cross section of the virus that causes AIDS. Inside, genetic material is surrounded by a protein coat and molecules of reverse transcriptase, an enzyme that catalyzes the transcription of DNA from the viral RNA template after the virus enters the host cell. This virus is among those that also have an outer envelope that is formed from the host cell's plasma membrane. Spikes made of glycoprotein (protein and carbohydrate) project from the envelope and help the virus attach to its host cell. **QUESTION** Why are viruses unable to replicate outside of a host cell?

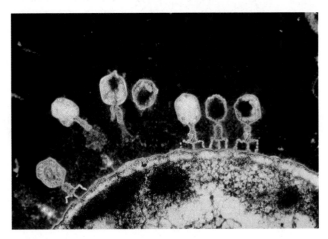

FIGURE 19-13 Some viruses infect bacteria
In this electron micrograph, bacteriophages are seen attacking a bacterium. They have injected their genetic material inside, leaving their protein coats clinging to the bacterial cell wall. The black objects inside the bacterium are newly forming viruses. **QUESTION** In biotechnology, viruses are often used to transfer genes from the cells of one species to the cells of another. Which properties of viruses make them useful for this purpose?

(acquired immune deficiency syndrome), which cripples the body's immune system, is caused by a virus that attacks a specific type of white blood cell that controls the body's immune response. Viruses have also been linked to some types of cancer, such as T-cell leukemia (a cancer of the white blood cells), liver cancer, and cervical cancer.

Viral Infections Are Difficult to Treat

Because viruses are closely tied to the cellular machinery of their hosts, the illnesses they cause are difficult to treat. The antibiotics that are often effective against bacterial infections are useless against viruses, and antiviral agents may destroy host cells as well as viruses. Despite the difficulty of attacking viruses as they "hide" within cells, a number of antiviral drugs have been developed. Many of these drugs destroy or block the function of enzymes that the targeted virus requires for replication.

Unfortunately, the benefits of most antiviral drugs are limited because viruses quickly evolve resistance to the drugs. Mutation rates are very high in viruses, in part because viruses lack mechanisms for correcting errors that occur during replication of DNA or RNA. It is thus almost inevitable that when a population of viruses is under attack by an antiviral drug, a mutation will arise that confers resistance to the drug. The resistant viruses prosper and replicate in great numbers, eventually spreading to new human hosts. Ultimately, resistant viruses predominate, and a formerly helpful antiviral drug is rendered ineffective.

The difficulty of treating viral infections makes the possibility of virus-based biological weapons troubling. Of particular concern is the smallpox virus. Smallpox has been eradicated as a naturally occurring disease, and the only known cultures of the virus are held at two well-guarded government labs, one in Russia and one in the United States; nevertheless, other samples may exist in unknown locations. Because of this possibility, plans to destroy the remaining stocks of the virus have been indefinitely deferred so that the stored viruses can be used in research to develop a more effective smallpox vaccine. Another potential threat is the virus that causes Ebola hemorrhagic fever, a serious disease of African origin that kills more than 90% of its victims. Ebola is especially worrisome, both as an emerging infectious disease and as a possible biological weapon, because there is currently no effective treatment and no vaccine to prevent infections.

Some Infectious Agents Are Even Simpler than Viruses

Viroids are infectious particles that lack a protein coat and consist of nothing more than short, circular strands of RNA. Despite their simplicity, viroids are able to enter the nucleus of a host cell and direct the synthesis of new viroids. About a dozen crop diseases, including cucumber pale fruit disease, avocado sunblotch, and potato spindle tuber disease, are caused by viroids.

Prions are even more puzzling than viroids. In the 1950s, physicians studying the Fore, a primitive tribe in New Guinea, were puzzled to observe numerous cases of a fatal

Viruses—How the Nonliving Replicate

Viruses multiply, or replicate, using their own genetic material, which—depending on the virus—consists of single-stranded or double-stranded RNA or DNA. This material serves as a template (or blueprint) for the viral proteins and genetic material required to make new viruses. Viral enzymes may participate in replication as well, but the overall process depends on the biochemical machinery that the host cell uses to make its own proteins.

Viral replication follows a general sequence:

1. *Penetration.* Viruses may be engulfed by their host cell (endocytosis). Some viruses have surface proteins that bind to receptors on the host cell's plasma membrane and stimulate endocytosis. Other viruses are coated with an envelope that can fuse with the host's membrane. The viral genetic material is then released into the cytoplasm.

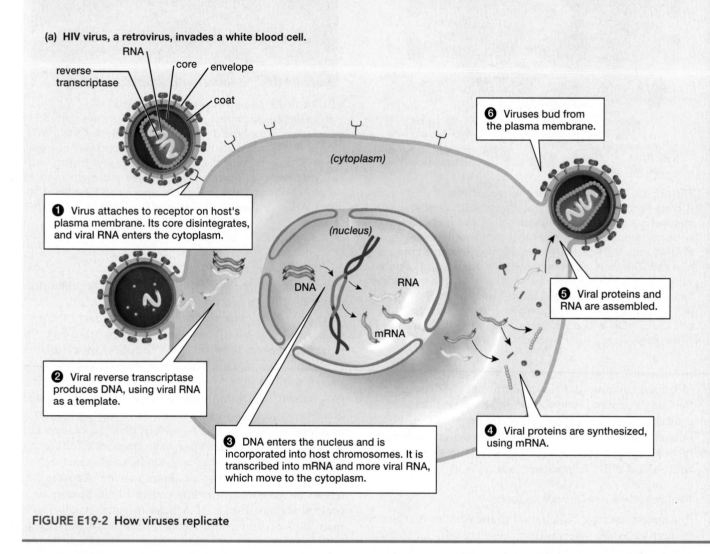

(a) **HIV virus, a retrovirus, invades a white blood cell.**

RNA

reverse transcriptase

core envelope

coat

6 Viruses bud from the plasma membrane.

(cytoplasm)

1 Virus attaches to receptor on host's plasma membrane. Its core disintegrates, and viral RNA enters the cytoplasm.

(nucleus)

DNA

RNA

mRNA

5 Viral proteins and RNA are assembled.

2 Viral reverse transcriptase produces DNA, using viral RNA as a template.

3 DNA enters the nucleus and is incorporated into host chromosomes. It is transcribed into mRNA and more viral RNA, which move to the cytoplasm.

4 Viral proteins are synthesized, using mRNA.

FIGURE E19-2 How viruses replicate

degenerative disease of the nervous system, which the Fore called *kuru*. The symptoms of kuru—loss of coordination, dementia, and ultimately death—were similar to those of the rare but more widespread *Creutzfeldt-Jakob disease* in humans and of *scrapie* and *bovine spongiform encephalopathy*, diseases of domestic livestock (see "Puzzling Proteins" and "Puzzling Proteins Revisited" in Chapter 3). Each of these diseases typically results in brain tissue that is spongy—riddled with holes. The researchers in New Guinea eventually determined that kuru was transmitted by ritual cannibalism; members of the Fore tribe honored their dead by consuming their brains. This prac-

tice has since stopped, and kuru has virtually disappeared. Clearly, kuru was caused by an infectious agent transmitted by infected brain tissue—but what was that agent?

In 1982, Nobel Prize–winning neurologist Stanley Prusiner published evidence that scrapie (and, by extension, kuru, Creutzfeldt-Jakob disease, and a number of other, similar afflictions) is caused by an infectious agent that consists only of protein. This idea seemed preposterous at the time, because most scientists believed that infectious agents must contain genetic material such as DNA or RNA in order to replicate. But Prusiner and his colleagues were able to isolate the infectious agent from scrapie-in-

2. *Replication*. The viral genetic material is copied many times.

3. *Transcription*. Viral genetic material is used as a blueprint to make messenger RNA (mRNA).

4. *Protein synthesis*. In the host cytoplasm, viral mRNA is used to synthesize viral proteins.

5. *Viral assembly*. The viral genetic material and enzymes are surrounded by their protein coat.

6. *Release*. Viruses emerge from the host cell by "budding" from the cell membrane or by bursting the cell.

Here, two types of viral life cycles are depicted. **FIGURE E19-2a** shows the *human immunodeficiency virus (HIV)*, the *retrovirus* that causes AIDS. Retroviruses use single-stranded RNA as a template to make double-stranded DNA by using a viral enzyme called *reverse transcriptase*. Many other retroviruses exist, and several cause cancers or tumors. **FIGURE E19-2b** shows the *herpes virus*, which contains double-stranded DNA that is transcribed into mRNA.

(b) **Herpes virus, a double-stranded DNA virus, invades a skin cell.**

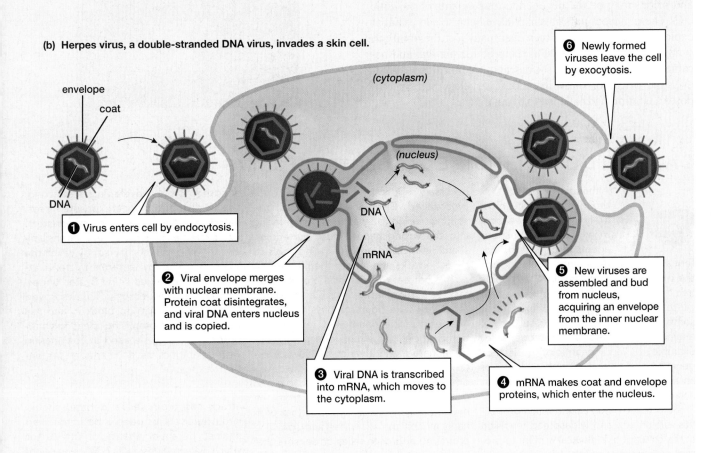

❶ Virus enters cell by endocytosis.

❷ Viral envelope merges with nuclear membrane. Protein coat disintegrates, and viral DNA enters nucleus and is copied.

❸ Viral DNA is transcribed into mRNA, which moves to the cytoplasm.

❹ mRNA makes coat and envelope proteins, which enter the nucleus.

❺ New viruses are assembled and bud from nucleus, acquiring an envelope from the inner nuclear membrane.

❻ Newly formed viruses leave the cell by exocytosis.

fected hamsters and demonstrate that it contained no nucleic acids. The researchers called these infectious protein particles *prions* (**FIG. 19-14**).

How can a protein replicate itself and be infectious? Not all researchers are convinced that this is possible. However, recent findings have sketched the outline of a possible mechanism for prion replication. It turns out that prions consist of a single protein produced by normal nerve cells. Some copies of this normal protein molecule, for reasons still poorly understood, become folded into the wrong shape and are thus transformed into infectious prions. Once present, prions can apparently induce other, normal copies of the protein molecule to become transformed into prions. Eventually, the concentration of prions in nerve tissue may get high enough to cause cell damage and degeneration. Why would a slight alteration to a normally benign protein turn it into a dangerous cell-killer? No one knows.

Another peculiarity of prion-caused diseases is that they can be inherited as well as transmitted by infection. Recent research has shown that certain small mutations in the gene that codes for "normal" prion protein increase the likelihood that the protein will fold into its abnormal form. If one of these mutations is genetically passed on to offspring, the tendency to develop a prion disease may be inherited.

FIGURE 19-14 Prions: Puzzling proteins
A section from the brain of a cow infected with bovine spongiform encephalopathy contains fibrous clusters of prion proteins.

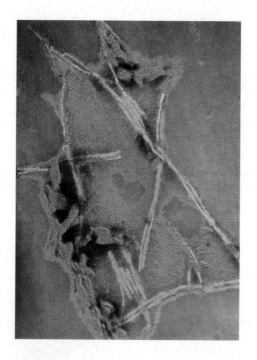

No One Is Certain How These Infectious Particles Originated

The origin of viruses, viroids, and prions is obscure. Some scientists believe that the huge variety of mechanisms for self-replication among these particles reflects their status as evolutionary remnants of the very early history of life, before evolution settled on the more familiar large, double-stranded DNA molecules. Another possibility is that viruses, viroids, and prions may be the degenerate descendants of parasitic cells. These ancient parasites may have been so successful at exploiting their hosts that they eventually lost the ability to synthesize all of the molecules required for survival and became dependent on the host's biochemical machinery. Whatever the origin of these infectious particles, their success poses a continuing challenge to living things.

CASE STUDY REVISITED   AGENTS OF DEATH

Although anthrax is thought to be the most likely biological weapon, many other infectious agents also have the potential to become weapons. These include the viruses that cause smallpox and Ebola hemorrhagic fever, and the bacteria that cause plague. Evidence also exists that some countries are trying to use genetic engineering to "improve" pathogens: for example, by adding antibiotic-resistance genes to plague bacteria so that victims of an attack would be more difficult to treat and more likely to die.

Before 2001, humankind relied on politics, diplomacy, and widespread revulsion at the concept of biological warfare to protect us from its terrifying destructive potential. Now, however, it is painfully clear that humanity also includes people willing to unleash biological weapons. Unfortunately, little expertise is required to culture pathogenic bacteria or viruses, and the necessary supplies and equipment are easily acquired. Given the difficulty of preventing biological weapons from falling into the wrong hands, much current research is focused on developing tools to detect attacks and render them harmless.

It is not easy to detect a biological attack, as pathogens are invisible and symptoms may take hours or days to appear after an attack. Nonetheless, rapid detection is crucial if an effective response is to be mounted, and a variety of new detection technologies are being rapidly developed. Detectors must be able to distinguish released pathogens from the multitude of harmless microbes that ordinarily inhabit the air, water, and soil. One promising approach relies on sensors that incorporate living human immune cells that have been genetically altered to glow when receptor molecules in their cell membranes bind to a particular pathogen.

Once an attack is detected, the main task is to care for those who have been exposed. Thus, developing fast-acting, easily distributed postexposure treatments is a top priority for researchers. For example, biologists have intensively investigated the mechanism by which the toxin released by anthrax bacteria attacks and damages cells. Better understanding of this process has improved investigators' ability to block it, and has yielded several promising ideas for antidotes that could be used in conjunction with antibiotics as a treatment for anthrax exposure.

Consider This The threat of biological attack has prompted a debate: should large numbers of people be immunized against potential agents of attack for which vaccinations exist? Mass vaccinations are costly and would inevitably cause some deaths due to occasional adverse reactions. Is the increased protection and peace of mind that would come with vaccinations worth the price?

CHAPTER REVIEW

SUMMARY OF KEY CONCEPTS

19.1   Which Organisms Make Up the Prokaryotic Domains—Bacteria and Archaea?

Members of the domains Bacteria and Archaea—the bacteria and archaea—are unicellular and prokaryotic. Archaea and bacteria are not closely related and differ in several fundamental features, including cell wall composition, ribosomal RNA sequence, and membrane lipid structure. A cell wall determines the characteristic shapes of prokaryotes: round, rodlike, or spiral.

19.2 How Do Prokaryotes Survive and Reproduce?

Certain types of bacteria can move about using flagella; others form spores that disperse widely and withstand inhospitable environmental conditions. Bacteria and archaea have colonized nearly every habitat on Earth, including hot, acidic, very salty, and anaerobic environments.

Prokaryotes obtain energy in a variety of ways. Some, including the cyanobacteria, rely on photosynthesis. Others are chemosynthetic, breaking down inorganic molecules to obtain energy. Heterotrophic forms are capable of consuming a wide variety of organic compounds. Many are anaerobic, able to obtain energy from fermentation when oxygen is not available. Prokaryotes reproduce by binary fission and may exchange genetic material by conjugation.

Web Tutorial 19.1 Bacterial Conjugation

19.3 How Do Prokaryotes Affect Humans and Other Eukaryotes?

Some bacteria are pathogenic, causing disorders such as pneumonia, tetanus, botulism, and the sexually transmitted diseases gonorrhea and syphilis. Most bacteria, however, are harmless to humans and play important roles in natural ecosystems. Some live in the digestive tracts of ruminants and break down cellulose. Nitrogen-fixing bacteria enrich the soil and aid in plant growth. Many others live off the dead bodies and wastes of other organisms, liberating nutrients for reuse.

19.4 What Are Viruses, Viroids, and Prions?

Viruses are parasites consisting of a protein coat that surrounds genetic material. They are noncellular and unable to move, grow, or reproduce outside a living cell. They invade cells of a specific host and use the host cell's energy, enzymes, and ribosomes to produce more virus particles, which are liberated when the cell ruptures. Many viruses are pathogenic to humans, including those causing colds and flu, herpes, AIDS, and certain forms of cancer.

Viroids are short strands of RNA that can invade a host cell's nucleus and direct the synthesis of new viroids. To date, viroids are known to cause only certain diseases of plants.

Prions have been implicated in diseases of the nervous system, such as kuru, Creutzfeldt-Jakob disease, and scrapie. Prions are unique in that they lack genetic material. They are composed solely of mutated prion protein, which may act as an enzyme, catalyzing the formation of more prions from normal prion protein.

Web Tutorial 19.2 Retrovirus Replication

Web Tutorial 19.3 Herpes Virus Replication

KEY TERMS

anaerobe *page 375*
bacteriophage *page 380*
conjugation *page 376*
endospore *page 374*

flagellum *page 373*
Gram stain *page 372*
host *page 380*

nitrogen-fixing bacterium *page 377*
pathogenic *page 377*
plasmid *page 376*

prion *page 381*
viroid *page 381*
virus *page 379*

THINKING THROUGH THE CONCEPTS

1. Describe some of the ways in which bacteria obtain energy and nutrients.

2. What are nitrogen-fixing bacteria, and what role do they play in ecosystems?

3. What is an endospore? What is its function?

4. What is conjugation? What role do plasmids play in conjugation?

5. Why are prokaryotes especially useful in bioremediation?

6. Describe the structure of a typical virus. How do viruses replicate?

APPLYING THE CONCEPTS

1. In some developing countries, antibiotics can be purchased without a prescription. Why do you think this is done? What biological consequences would you predict?

2. Before the discovery of prions, many (perhaps most) biologists would have agreed with the statement, "It is a fact that no infectious organism or particle can exist that lacks nucleic acid (such as DNA or RNA)." What lessons do prions teach us about nature, science, and scientific inquiry? You may wish to review Chapter 1 to help answer this question.

3. Argue for and against the statement, "Viruses are alive."

FOR MORE INFORMATION

Costerton, J., and Stewart, P. "Battling Biofilms." *Scientific American*, July 2001. How biofilms form and how to fight them.

Madigan, M., and Marrs, B. "Extremophiles." *Scientific American*, April 1997. Prokaryotes that prosper under extreme conditions, and potential industrial uses of the enzymes that they use to do so.

Prusiner, S. "The Prion Diseases." *Scientific American*, January 1995. A description of prions and the research that led to their discovery, from the viewpoint of the most influential scientist in the field.

Prusiner, S. "Detecting Mad Cow Disease." *Scientific American*, July 2004. Overview of the public health perspective on bovine spongiform encephalopathy and of emerging methods of testing cows for the disease.

Villarreal, L. "Are Viruses Alive?" *Scientific American*, December 2004. An overview of what we know about viruses and their impact on life.

Young, J., and Collier, R. J. "Attacking Anthrax." *Scientific American*, March 2002. A summary of recent research that could help develop new techniques for detecting and treating anthrax.

20

The Diversity of Protists

The photosynthetic protist *Caulerpa taxifolia* is
an unwanted invader in temperate seas.

CASE STUDY GREEN MONSTER

IN CALIFORNIA, IT IS A CRIME to possess, transport, or sell *Caulerpa*. Is *Caulerpa* an illicit drug or some kind of weapon? No, it is merely a small green seaweed. Why, then, do California's lawmakers want to ban it from their state?

The story of *Caulerpa's* rise to public enemy status begins in the early 1980s at the Wilhelmina Zoo in Stuttgart, Germany. There, the keepers of a saltwater aquarium found that the tropical seaweed *Caulerpa taxifolia* was an attractive companion and background for the tropical fish on display. Even better, years of captive breeding at the zoo had yielded a strain of the seaweed that was well suited to life in an aquarium. The new strain was very hardy and could survive in waters considerably cooler than the tropical wa-

ters in which wild *Caulerpa* is found. The aquarium strain was tough but attractive, and the zoo staff was happy to send cuttings to other institutions that wished to use it in aquarium displays.

One institution that received a cutting was the Oceanographic Museum of Monaco, which occupies a stately building that stands directly on the shore of the Mediterranean Sea. In 1984, a visiting marine biologist discovered a small patch of *Caulerpa* growing in the waters just below the museum. Presumably, someone cleaning an aquarium had dumped the water into the sea and thereby inadvertently introduced *Caulerpa* to the Mediterranean.

By 1989, the *Caulerpa* patch had grown to cover a few acres. It grew as a continuous mat that seemed to exclude

most of other organisms that normally inhabit the Mediterranean sea bottom. The local herbivores, such as sea urchins and fish, did not feed on *Caulerpa*.

It soon became apparent that *Caulerpa* spreads rapidly, is not controlled by predation, and displaces native species. By the mid-1990s, biologists were alarmed to find *Caulerpa* all along the Mediterranean coast from Spain to Italy. Today it grows in extensive beds throughout the Mediterranean and covers an ever-expanding area on the sea floor.

Despite the threat it poses to ecosystems, *Caulerpa* is a fascinating creature. We will return to *Caulerpa* and its biology after an overview of the protists, a group that includes green seaweeds such as *Caulerpa* along with a host of organisms.

20.1 WHAT ARE PROTISTS?

Two of life's domains, Bacteria and Archaea, contain only prokaryotes. The third domain, Eukarya, includes all eukaryotic organisms. The most conspicuous Eukarya are members of the kingdoms Plantae, Fungi, and Animalia, which we will discuss in Chapters 21 through 24. The remaining eukaryotes constitute a diverse collection of evolutionary lineages collectively known as **protists** (Table 20-1). The term "protist" does not describe a true evolutionary unit united by shared features, but is a term of convenience that means "any eukaryote that is not a plant, animal, or fungus." About 60,000 protist species have been described.

Table 20-1 The Major Groups of Protists

Group	Subgroup	Locomotion	Nutrition	Representative features	Representative Genus
Excavates	Diplomonads	Swim with flagella	Heterotrophic	Lack mitochondria; inhabit soil or water or are parasitic	*Giardia* (intestinal parasite of mammals)
	Parabasalids	Swim with flagella	Heterotrophic	Lack mitochondria; parasites or commensal symbionts	*Trichomonas* (causes the sexually transmitted disease trichomoniasis)
Euglenozoans	Euglenids	Swim with one flagellum	Autotrophic; photosynthetic	Have an eyespot; all freshwater	*Euglena* (common pond dweller)
	Kinetoplastids	Swim with flagella	Heterotrophic	Inhabit soil or water or are parasitic	*Trypanosoma* (causes African sleeping sickness)
Stramenopiles (Chromists)	Water molds	Swim with flagella (gametes)	Heterotrophic	Filamentous	*Plasmopara* (causes downy mildew)
	Diatoms	Glide along surfaces	Autotrophic; photosynthetic	Have silica shells; most marine	*Navicula* (glides toward light)
	Brown algae	Nonmotile	Autotrophic; photosynthetic	Seaweeds of temperate oceans	*Macrocystis* (forms kelp forests)
Alveolates	Dinoflagellates	Swim with two flagella	Autotrophic; photosynthetic	Many bioluminescent; often have cellulose	*Gonyaulax* (causes red tide)
	Apicomplexans	Nonmotile	Heterotrophic	All parasitic; form infectious spores	*Plasmodium* (causes malaria)
	Ciliates	Swim with cilia	Heterotrophic	Most complex single cells	*Paramecium* (fast-moving pond dweller)
Cercozoans	Foraminifera	Extend thin pseudopods	Heterotrophic	Have calcium carbonate shells	*Globigerina*
Radiolarians		Extend thin pseudopods	Heterotrophic	Have silica shells	*Actinomma*
Amoebozoans	Lobose amoebas	Extend thick pseudopods	Heterotrophic	Have no shells	*Amoeba* (common pond dweller)
	Acellular slime molds	Slug-like mass oozes over surfaces	Heterotrophic	Form multinucleate plasmodium	*Physarum* (forms a large, bright orange mass)
	Cellular slime molds	Amoeboid cells extend pseudopods; slug-like mass crawls over surfaces	Heterotrophic	Form pseudoplasmodium with individual amoeboid cells	*Dictyostelium* (often used in laboratory studies)
Red algae		Nonmotile	Autotrophic; photosynthetic	Some deposit calcium carbonate; most marine	*Porphyra* (used as food in Japan)
Green algae		Swim with flagella (some species)	Autotrophic; photosynthetic	Closest relatives of land plants	*Ulva* (sea lettuce)

Most Protists Are Single-Celled

Most protists are single-celled, and they are invisible to us as we go about our daily lives. If we could somehow shrink to their microscopic scale, we might be more impressed with their spectacular and beautiful forms, their varied and active lifestyles, their astonishingly diverse modes of reproduction, and the structural and physiological innovations that are possible within the limits of a single cell. In reality, however, their small size makes them challenging to observe. A microscope and a good supply of patience are required to appreciate the majesty of protists.

Although most protists are single-celled, some are visible to the naked eye, and a few are genuinely large. Some larger protists are aggregations or colonies of single-celled individuals, while others are multicellular organisms.

Protists Use Diverse Modes of Nutrition

Three major modes of nutrition are represented among protists. Protists may ingest their food, absorb nutrients from their surroundings, or capture solar energy directly by photosynthesis.

Protists that ingest their food are generally predators. Predatory single-celled protists may have flexible cell membranes that can change shape to surround and engulf food such as bacteria. Protists that feed in this manner typically use fingerlike extensions called **pseudopods** (**FIG. 20-1**) to engulf prey. Other predatory protists use cilia to create tiny currents that sweep food particles into mouthlike openings in the cell. Whatever the means by which food is ingested, once it is inside the protist cell, it is typically contained in a membrane-surrounded *food vacuole* for digestion.

Protists that absorb nutrients directly from the surrounding environment may be free-living or may live inside the bodies of other organisms. The free-living types live in soil and other environments that contain dead organic matter, where they act as decomposers. Most absorptive feeders, however, live inside other organisms. In most cases, these protists are parasites whose feeding activity harms the host species.

Photosynthetic protists are abundant in oceans, lakes, and ponds. Most float free in the water column, but some live in close association with other organisms, such as corals or clams. These associations appear to be mutually beneficial; some of the solar energy captured by the photosynthetic protists is used by the host organism, which provides shelter and protection for the protists.

Protist photosynthesis takes place in organelles called *chloroplasts*. As described in Chapter 17, chloroplasts are the descendants of ancient photosynthetic bacteria that took up residence inside a larger cell in a process known as *endosymbiosis*. In addition to the original instance of endosymbiosis that created the first protist chloroplast, there have been several different later occurrences of *secondary endosymbiosis* in which a non-photosynthetic protist engulfed a photosynthetic, chloroplast-containing protist. Ultimately, most components of the engulfed species disappeared, leaving only a chloroplast surrounded by four membranes: two from the original, bacteria-derived chloroplast, one from the engulfed protist, and one from the food vacuole that originally contained the engulfed protist. Multiple occurrences of secondary endosymbiosis account for the presence of photosynthetic species in a number of different, unrelated protist groups.

Past classifications of protists grouped species according to their mode of nutrition, but improved understanding of the evolutionary history of protists has revealed that the old categories did not accurately reflect phylogeny. Nonetheless, biologists still use terminology that refers to groups of protists that share particular characteristics but are not necessarily related. For example, photosynthetic protists are collectively known as **algae** (singular, alga), and single-celled, non-photosynthetic protists are collectively known as **protozoa** (singular, protozoan).

Protists Use Diverse Modes of Reproduction

In prokaryotes, reproduction is strictly asexual; an individual divides to yield two individuals that are genetically identical to the parent cell. Most protists also reproduce asexually, creating new individuals by mitotic cell division (**FIG. 20-2a**). Many protists, however, are also capable of sexual reproduction, in which two individuals contribute genetic material to an offspring that is genetically different from either parent. The presence of sexual reproduction in protists but not in prokaryotes suggests that sex first arose in eukaryotes, some time after the evolutionary split between Eukarya and the prokaryotic domains.

In many protist species that are capable of sexual reproduction, most reproduction is nonetheless asexual. Sexual reproduction occurs only infrequently, at a particular time of year or under certain circumstances, such as a crowded environment or a shortage of food. The details of sexual reproduction and the resulting life cycles vary tremendously among different types of protists. However, protist

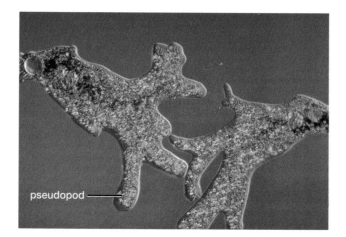

FIGURE 20-1 Pseudopods
Some single-celled protists can extend projections that are used to engulf food or move the organism about.

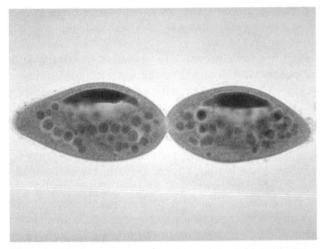

(a)

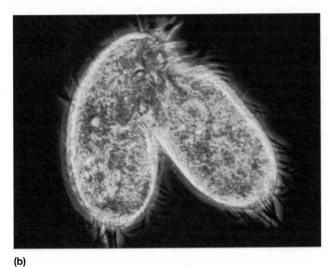

(b)

FIGURE 20-2 Protistan reproduction and gene exchange
(a) *Paramecium*, a ciliate, reproduces asexually by cell division.
(b) *Euplotes*, a ciliate, exchanges genetic material across a cytoplasmic bridge. QUESTION What do biologists mean when they say that sex and reproduction are uncoupled in most protists?

reproduction never includes the formation and development of an embryo, as occurs during reproduction in plants and animals. Nonreproductive processes that combine the genetic material of different individuals are also common among protists (**FIG. 20-2b**).

Protists Have Significant Effects on Humans

Although most of us do not notice protists as we go about our day-to-day business, they have important impacts, both positive and negative, on human lives. The primary positive impact actually benefits all living things and stems from the ecological role of photosynthetic marine protists. Just as plants do on land, algae in the oceans capture solar energy and make it available to the other organisms in the ecosystem. Thus, the marine ecosystems on which humans depend for food in turn depend on algae. Further, in the process of

using photosynthesis to capture energy, the algae release oxygen gas that helps replenish the atmosphere.

On the negative side of the ledger, many human diseases are caused by parasitic protists. The diseases caused by protists include some of humanity's most prevalent ailments and some of its deadliest afflictions. Protists also cause a number of plant diseases, some of which attack crops that are important to humans. In addition to causing diseases, some marine protists release toxins that can accumulate to harmful levels in coastal areas.

The following sections include information about the particular protists responsible for these helpful and harmful effects.

20.2 WHAT ARE THE MAJOR GROUPS OF PROTISTS?

Genetic comparisons are helping systematists gain a better understanding of the evolutionary history of protist groups. Because systematists strive to devise classification systems that reflect evolutionary history, the new information has fostered a revision of protist classification. Some protist species that had been previously grouped together on the basis of physical similarity actually belong to independent evolutionary lineages that diverged from one another very early in the history of eukaryotes. Conversely, some protist groups bearing little physical resemblance to one another have been revealed to share a common ancestor, and thus have been classified together in new kingdoms. The process of revising protist classification, however, is far from complete. Thus, our understanding of the eukaryotic tree of life is still "under construction"; many of the branches are in place, but others await new information that will allow systematists to place them alongside their closest evolutionary relatives.

In the following sections, we'll explore a brief sampling of protist diversity.

Excavates Lack Mitochondria

Excavates are named for a feeding groove that gives the appearance of having been "excavated" from the surface of the cell. Excavates also lack mitochondria. It is likely that the excavates' ancestors did possess mitochondria but that the organelles were lost early in the evolutionary history of the group. The two largest groups of excavates are the diplomonads and the parabasalids.

Diplomonads Have Two Nuclei

The single cells of **diplomonads** have two nuclei and move about by means of multiple flagella. A parasitic diplomonad, *Giardia*, is an increasing problem in the United States, particularly to hikers who drink from apparently pure mountain streams. *Cysts* (tough structures that enclose the organism during one phase of its life cycle) of this diplomonad are released in the feces of infected humans, dogs, or other animals; a single gram of feces may contain 300 million cysts. Once outside the animal's body, the cysts may enter freshwater streams and even community reservoirs.

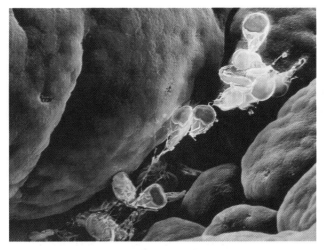

FIGURE 20-3 Giardia: The curse of campers
A diplomonad (genus *Giardia*) that may infect drinking water, causing gastrointestinal disorders, is shown here in a human small intestine.

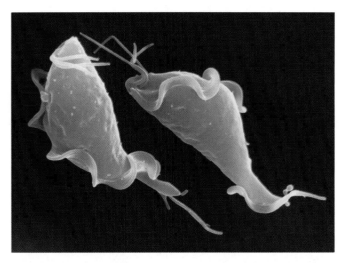

FIGURE 20-4 Trichomonas causes a sexually transmitted disease
The parabasalid *Trichomonas vaginalis* infects the urinary and reproductive tracts of men and women. Women, however, are far more likely to experience unpleasant symptoms.

If a mammal drinks infected water, the cysts develop into the adult form (**FIG. 20-3**) in the small intestine of their mammalian host. In humans, infections can cause severe diarrhea, dehydration, nausea, vomiting, and cramps. Fortunately, these infections can be cured with drugs, and deaths from *Giardia* infections are uncommon.

Parabasalids Include Mutualists and Parasites

All known **parabasalids** live inside animals. For example, this group includes several species that inhabit the digestive systems of some species of wood-eating termites. The termites cannot themselves digest wood, but the parabasalids can. Thus, the insect and the protist are in a mutually beneficial relationship. The termite delivers food to the parabasalids in its gut; as the parabasalids digest the food, some of the energy and nutrients released become available for use by the termite.

In other cases, the host animal does not benefit from a parabasalid's presence but is instead harmed. For example, in humans the parabasalid *Trichomonas vaginalis* causes the sexually transmitted disease trichomoniasis (**FIG. 20-4**). *Trichomonas* inhabits the mucus layers of the urinary and reproductive tracts and uses its flagella to move along them. When conditions are favorable, a *Trichomonas* population can grow rapidly. Infected females may experience unpleasant symptoms, including vaginal itching and discharge. Infected males do not usually display symptoms, but they can transmit the infection to sexual partners.

Euglenozoans Have Distinctive Mitochondria

In most **euglenozoans,** the folds of the inner membrane of the cell's mitochondria have a distinctive shape that appears under the microscope as a stack of discs. Two major groups of euglenozoans are the euglenids and the kinetoplastids.

Euglenids Lack a Rigid Covering and Swim by Means of Flagella

Euglenids are single-celled protists that live mostly in fresh water and are named after the group's best-known representative, *Euglena* (**FIG. 20-5**), a complex single cell that moves about by whipping its flagellum through water. Many euglenids are photosynthetic, but some species instead absorb or engulf food. Euglenids lack a rigid outer covering, so some can move by wriggling as well as by whipping their flagella. Some euglenids also possess simple light-sensing organelles consisting of a photoreceptor, called an *eyespot*, and an adjacent patch of pigment. The pigment shades the

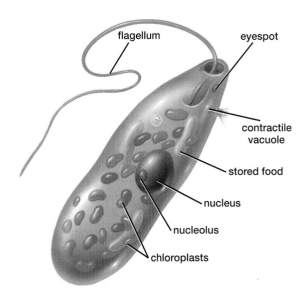

FIGURE 20-5 Euglena, a representative euglenid
Euglena's elaborate single cell is packed with green chloroplasts, which will disappear if the protist is kept in darkness.

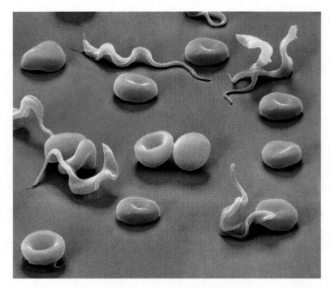

FIGURE 20-6 A disease-causing kinetoplastid
This photomicrograph shows human blood that is heavily infested with the corkscrew-shaped, parasitic kinetoplastid *Trypanosoma*, which causes African sleeping sickness. Note that the *Trypanosoma* are larger than red blood cells.

photoreceptor only when light strikes from certain directions, enabling the organism to determine the direction of the light source. Using information from the photoreceptor, the flagellum propels the protist toward light levels appropriate for photosynthesis.

Some Kinetoplastids Cause Human Diseases

The DNA in the mitochondria of **kinetoplastids** is arranged in distinctive structures called kinetoplasts. Most kinetoplastids possess at least one flagellum, which may propel the organism, sense the environment, or ensnare food. Some kinetoplastids are free-living, inhabiting soil and water; others live inside other organisms in a relationship that may be mutually beneficial or parasitic. A dangerous parasitic kinetoplastid in the genus *Trypanosoma* is responsible for African sleeping sickness, a potentially fatal disease (**FIG. 20-6**). Like many parasites, this organism has a complex life cycle, part of which is spent in the tsetse fly. While feeding on the blood of a mammal, the fly transmits the trypanosome to the mammal. The parasite then develops in the new host (which may be a human) and enters the bloodstream. It may then be ingested by another tsetse fly that bites the host, thus beginning a new cycle of infection.

Stramenopiles Include Photosynthetic and Non-Photosynthetic Organisms

The **stramenopiles** (also known as *chromists*) form a group whose shared ancestry was discovered through genetic comparison. The group is designated as a kingdom by some systematists. All members of the group have fine, hairlike projections on their flagella (though in many stramenopiles, flagella are present only at certain stages of the life cycle). Despite their shared evolutionary history, however, stra-

menopiles display a wide range of forms. Some are photosynthetic and some are not; most are single-celled, but some are multicellular. Three major stramenopile groups are the water molds, the diatoms, and the brown algae.

Water Molds Have Had Important Impacts on Humans

The **water molds**, or *oomycetes*, form a small group of protists, many of which are shaped as long filaments that aggregate to form cottony tufts. These tufts are superficially similar to structures produced by some fungi, but this resemblance is due to convergent evolution (see Chapter 14), not shared ancestry. Many water molds are decomposers that live in water and damp soil. Some species have profound economic impacts on humans. For example, a water mold causes a disease of grapes, known as *downy mildew* (**FIG. 20-7**). Its inadvertent introduction into France from the United States in the late 1870s nearly destroyed the French wine industry. Another oomycete has destroyed millions of avocado trees in California; still another is responsible for *late blight*, a devastating disease of potatoes. When accidentally introduced into Ireland about 1845, this protist destroyed nearly the entire potato crop, causing the devastating potato famine during which as many as 1 million people in Ireland starved and many more emigrated to the United States.

Diatoms Encase Themselves Within Glassy Walls

The **diatoms**, photosynthetic stramenopiles found in both fresh and salt water, produce protective shells of *silica* (glass), some of exceptional beauty (**FIG. 20-8**). These shells consist of top and bottom halves that fit together like a pillbox or petri dish. Accumulations of diatoms' glassy walls over millions of years have produced fossil deposits of "diatomaceous earth" that may be hundreds of meters thick. This slightly abrasive substance is widely used in products such as toothpaste and metal polish.

FIGURE 20-7 A parasitic water mold
Downy mildew, a plant disease caused by the water mold *Plasmopara*, nearly destroyed the French wine industry in the 1870s. QUESTION Although water molds are stramenopiles, they look and function very much like fungi. What is the cause of this similarity?

Diatoms form part of the **phytoplankton**, the single-celled photosynthesizers that float passively in the upper layers of Earth's lakes and oceans. Phytoplankton play an immensely important ecological role. Marine phytoplankton account for nearly 70% of all the photosynthetic activity on Earth, absorbing carbon dioxide, recharging the atmosphere with oxygen, and supporting the complex web of aquatic life. Diatoms, as key components of the phytoplankton, are so important to marine food webs that they have been called the "pastures of the sea."

Brown Algae Dominate in Cool Coastal Waters

Though most photosynthetic protists, such as diatoms, are single-celled, some form multicellular aggregations that are commonly known as *seaweeds*. Although some seaweeds seem to resemble plants, they are not closely related to plants and lack many of the distinctive features of the plant kingdom. For example, none of the seaweeds have roots or shoots, and none form embryos during reproduction.

The chromists include one group of seaweeds, the brown algae. The brown algae are named for the brownish-yellow pigments that (in combination with green chlorophyll) increase the seaweed's light-gathering ability and produce a brown to olive-green color.

Almost all brown algae are marine. The group includes the dominant seaweed species that dwell along rocky shores in the temperate (cooler) oceans of the world, including the eastern and western coasts of the United States. Brown algae live in habitats ranging from nearshore, where they cling to rocks that are exposed at low tide, to far offshore. Several species use gas-filled floats to support their bodies (**FIG. 20-9a**). Some of the giant kelp found along the Pacific coast reach heights of 325 feet (100 meters) and may grow more than 6 inches (15 centimeters) in a single day. With their dense growth and towering height (**FIG. 20-9b**), kelp form undersea forests that provide food, shelter, and breeding areas for marine animals.

Alveolates Include Parasites, Predators, and Phytoplankton

The **alveolates** are single-celled organisms that have distinctive, small cavities beneath the surface of their cells. Like the stramenopiles, the alveolates form a distinct lineage that may eventually be given kingdom status. Also like the stramenopiles, the evolutionary link among the

(a)

(b)

FIGURE 20-9 Brown algae, a multicellular protist
(a) *Fucus*, a genus found near shores, is shown here exposed at low tide. Notice the gas-filled floats, which provide buoyancy in water. (b) The giant kelp *Macrocystis* forms underwater forests off southern California.

FIGURE 20-8 Some representative diatoms
This photomicrograph illustrates the intricate, microscopic beauty and variety of the glassy walls of diatoms.

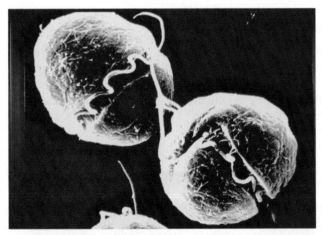

FIGURE 20-10 Dinoflagellates
Two dinoflagellates covered with protective cellulose armor. Visible on each is a flagellum in a groove that encircles the body.

FIGURE 20-11 A red tide
The explosive reproductive rate of certain dinoflagellates under the right environmental conditions can produce concentrations so great that their microscopic bodies dye the seawater red or brown.

alveolates was long obscured by the variety of structures and ways of life among the group members, but was revealed by molecular comparisons. Some alveolates are photosynthetic, some are parasitic, and some are predatory. The major alveolate groups are the dinoflagellates, apicomplexans, and ciliates.

Dinoflagellates Swim by Means of Two Whiplike Flagella

Though most **dinoflagellates** are photosynthetic, there are also some non-photosynthetic species. Dinoflagellates are named for the motion created by their two whiplike flagella ("dino" is Greek for "whirlpool"). One flagellum encircles the cell, and the second projects behind it. Some dinoflagellates are enclosed only by a cell membrane; others have cellulose walls that resemble armor plates (**FIG. 20-10**). Although some species live in fresh water, dinoflagellates are especially abundant in the ocean, where they are an important component of the phytoplankton and a food source for larger organisms. Many dinoflagellates are bioluminescent, producing a brilliant blue-green light when disturbed. Specialized dinoflagellates live within the tissues of corals, some clams, and even other protists, where they provide their hosts with nutrients from photosynthesis and remove carbon dioxide. Reef-building corals live only in the shallow, well-lit waters in which their embedded dinoflagellates can survive.

Warm water that is rich in nutrients may bring on a dinoflagellate population explosion. Dinoflagellates can become so numerous that the water is dyed red by the color of their bodies, causing a "red tide" (**FIG. 20-11**). During red tides, fish die by the thousands, suffocated by clogged gills or by the oxygen depletion that results from the decay of billions of dinoflagellates. One type of dinoflagellate, *Pfisteria*, even eats fish directly, first secreting chemicals that dissolve the victim's flesh. But dinoflagellate explosions can benefit oysters, mussels, and clams, which have a feast, filtering millions of the protists from

the water and consuming them. In the process, however, their bodies accumulate concentrations of a nerve poison produced by the dinoflagellates. Humans who eat these mollusks may be stricken with potentially lethal paralytic shellfish poisoning.

Apicomplexans Are Parasitic and Have No Means of Locomotion

All **apicomplexans** (sometimes known as *sporozoans*) are parasitic, living inside the bodies and sometimes inside the individual cells of their hosts. They form infectious spores, resistant structures transmitted from one host to another through food, water, or the bite of an infected insect. As adults, apicomplexans have no means of locomotion. Many have complex life cycles, a common feature of parasites. A well-known example is the malarial parasite *Plasmodium* (**FIG. 20-12**). Parts of its life cycle are spent in the stomach, and later the salivary glands, of the female *Anopheles* mosquito. When the mosquito bites a human, it passes the *Plasmodium* to the unfortunate victim. The apicomplexan develops in the victim's liver, then enters the blood, where it reproduces rapidly in red blood cells. The release of large quantities of spores through the rupture of the blood cells causes the recurrent fever of malaria. Uninfected mosquitoes may acquire the parasite by feeding on the blood of a malaria victim, spreading the parasite when they bite another person.

Although the drug chloroquine kills the malarial parasite, drug-resistant populations of *Plasmodium* are, unfortunately, spreading rapidly throughout Africa, where the

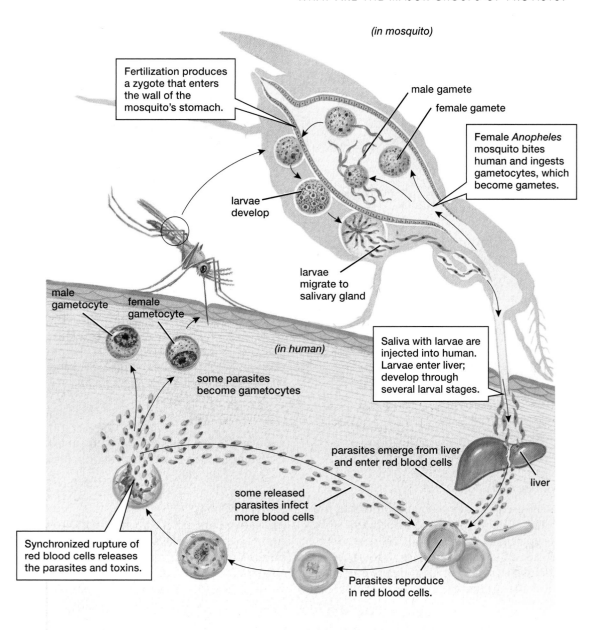

(in mosquito)

Fertilization produces a zygote that enters the wall of the mosquito's stomach.

male gamete

female gamete

Female *Anopheles* mosquito bites human and ingests gametocytes, which become gametes.

larvae develop

larvae migrate to salivary gland

male gametocyte

female gametocyte

(in human)

some parasites become gametocytes

Saliva with larvae are injected into human. Larvae enter liver; develop through several larval stages.

parasites emerge from liver and enter red blood cells

liver

some released parasites infect more blood cells

Synchronized rupture of red blood cells releases the parasites and toxins.

Parasites reproduce in red blood cells.

FIGURE 20-12 The life cycle of the malaria parasite

disease is prevalent. Programs to eradicate mosquitoes have failed because the mosquitoes rapidly evolve resistance to pesticides.

Ciliates Are the Most Complex of the Alveolates

Ciliates, which inhabit fresh and salt water, represent the peak of unicellular complexity. They possess many specialized organelles, including **cilia** (singular, cilium), the short hairlike outgrowths after which they are named. The cilia may cover the cell or may be localized. In the well-known freshwater genus *Paramecium*, rows of cilia cover the organism's entire body surface (**FIG. 20-13**). Their coordinated beating propels the cell through the water at a rate of one millimeter per second—a protistan speed record. Although only a single cell, *Paramecium* responds to its environment

as if it had a well-developed nervous system. Confronted with a noxious chemical or a physical barrier, the cell immediately backs up by reversing the beating of its cilia and then proceeds in a new direction. Some ciliates, such as *Didinium*, are accomplished predators (**FIG. 20-14**).

Cercozoans Have Thin Pseudopods and Elaborate Shells

Protists in a number of different groups possess flexible plasma membranes that they can extend in any direction to form pseudopods for use in locomotion and for engulfing food. The pseudopods of **cercozoans** are thin and threadlike. In most species in these groups, the pseudopods extend through hard shells. The largest group within the cercozoans is the foramineferans.

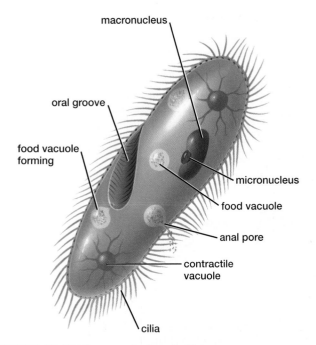

FIGURE 20-13 The complexity of ciliates
The ciliate *Paramecium* illustrates some important ciliate organelles. The oral groove acts as a mouth, food vacuoles—miniature digestive systems—form at its apex, and waste is expelled by exocytosis through an anal pore. Contractile vacuoles regulate water balance.

FIGURE 20-14 A microscopic predator
In this scanning electron micrograph, the predatory ciliate *Didinium* attacks a *Paramecium*. Notice that the cilia of *Didinium* are confined to two bands, whereas *Paramecium* has cilia over its entire body. Ultimately, the predator will engulf and consume its prey. This microscopic drama could take place on a pinpoint with room to spare.

Fossil Foramineferan Shells Form Chalk

The **foraminiferans** are primarily marine protists that produce beautiful shells. Their shells are constructed mostly of calcium carbonate (chalk; **FIG. 20-15a**). These elaborate shells are pierced by myriad openings through which pseudopods extend. The chalky shells of dead foraminiferans, sinking to the ocean bottom and accumulating over

millions of years, have resulted in immense deposits of limestone such as those that form the famous White Cliffs of Dover, England.

Radiolarians Have Glassy Shells

Radiolarians are not members of the cercozoan group, but form a separate lineage that is thought to be closely related to cercozoans. Like foraminiferans, radiolarians have thin pseudopods that extend through hard shells. The shells of radiolarians, however, are made of glass-like silica

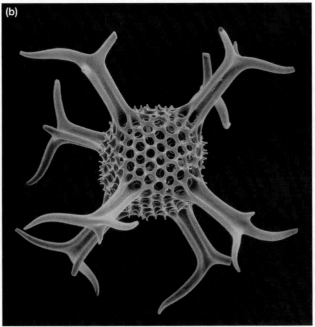

FIGURE 20-15 Foraminiferans and radiolarians
(a) The chalky shells of foraminiferans show numerous interior chambers. (b) The delicate, glassy shell of a radiolarian. Pseudopods, which sense the environment and capture food, extend out through the openings in the shell.

(FIG. 20-15b). In some areas of the ocean, radiolarian shells raining down over vast stretches of time have accumulated to form thick layers of "radiolarian ooze."

Amoebozoans Inhabit Aquatic and Terrestrial Environments

Amoebozoans move by extending finger-shaped pseudopods, which may also be used for feeding. They generally do not have shells. The major groups of amoebozoans are the amoebas and the slime molds.

Amoebas Have Thick Pseudopods and No Shells

Amoebas, sometimes known as *lobose amoebas* to distinguish them from other protists that have pseudopods, are common in freshwater lakes and ponds (FIG. 20-16). Many amoebas are predators that stalk and engulf prey, but some species are parasites. One parasitic form causes amoebic dysentery, a disease that is prevalent in warm climates. The dysentery-causing amoeba multiplies in the intestinal wall, triggering severe diarrhea.

Slime Molds Are Decomposers That Inhabit the Forest Floor

The physical form of *slime molds* seems to blur the boundary between a colony of different individuals and a single, multicellular individual. The life cycle of the slime mold consists of two phases: a mobile feeding stage and a stationary reproductive stage called a *fruiting body*. There are two main types of slime mold: acellular and cellular.

Acellular Slime Molds Form a Multinucleate Mass of Cytoplasm Called a Plasmodium

The **acellular slime molds,** also known as *plasmodial* slime molds, consist of a mass of cytoplasm that may spread thinly over an area of several square meters. Although the mass contains thousands of diploid nuclei, the nuclei are not confined in separate cells surrounded by plasma

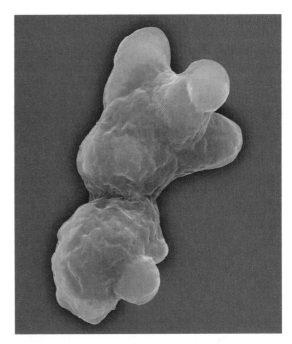

FIGURE 20-16 The amoeba
Lobose amoebas are active predators that move through water to engulf food with thick, blunt pseudopods.

membranes, as in most multicellular organisms. This structure, called a **plasmodium**, explains why these protists are described as "acellular" (without cells). The plasmodium oozes through decaying leaves and rotting logs, engulfing food such as bacteria and particles of organic material. The mass may be bright yellow or orange—a large plasmodium can be rather startling (FIG. 20-17a). Dry conditions or starvation stimulate the plasmodium to form a fruiting body, on which haploid spores are produced (FIG. 20-17b). The spores are dispersed and germinate under favorable conditions, eventually giving rise to a new plasmodium.

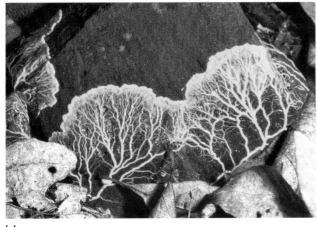
(a)

(b)

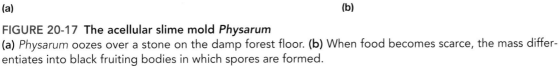

FIGURE 20-17 The acellular slime mold *Physarum*
(a) *Physarum* oozes over a stone on the damp forest floor. (b) When food becomes scarce, the mass differentiates into black fruiting bodies in which spores are formed.

Cellular Slime Molds Live as Independent Cells but Aggregate into a Pseudoplasmodium When Food Is Scarce

The **cellular slime molds** live in soil as independent haploid cells that move and feed by producing pseudopods. In the best-studied genus, *Dictyostelium*, individual cells release a chemical signal when food becomes scarce. This signal attracts nearby cells into a dense aggregation that forms a slug-like mass called a **pseudoplasmodium** ("false plasmodium") because, unlike a true plasmodium, it actually consists of individual cells (**FIG. 20-18**). The pseudoplasmodium then behaves like a multicellular organism. After crawling toward a source of light, the cells in the aggregation take on specific roles, forming a fruiting body. Haploid spores formed within the fruiting body are dispersed by wind and germinate directly into new single-celled individuals.

Red Algae Live Primarily in Clear Tropical Oceans

The red algae are multicellular, photosynthetic seaweeds (**FIG. 20-19**). These protists range in color from bright red to nearly black, and they derive their color from red pigments that mask their green chlorophyll. Red algae are found almost exclusively in marine environments. They dominate in deep, clear tropical waters, where their red pigments absorb the deeply penetrating blue-green light and transfer this light energy to chlorophyll, where it is used in photosynthesis.

Some species of red algae deposit calcium carbonate, which forms limestone, in their tissues and contribute to the formation of reefs. Other species are harvested for food in Asia. Red algae also contain certain gelatinous

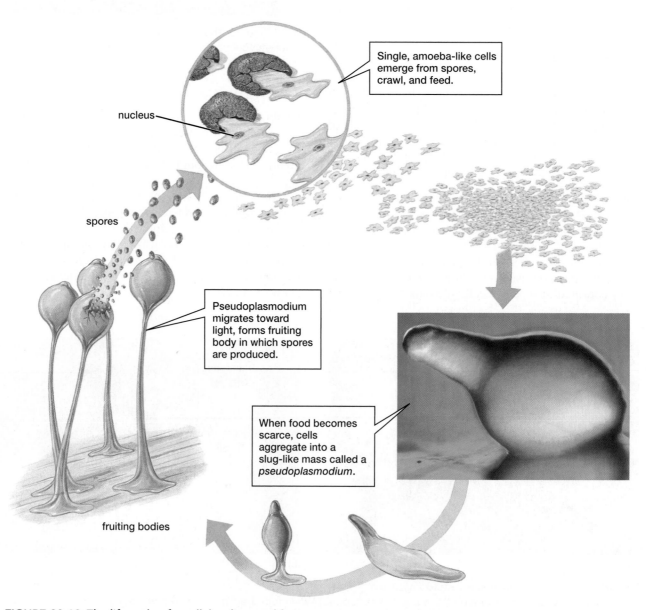

Single, amoeba-like cells emerge from spores, crawl, and feed.

nucleus

spores

Pseudoplasmodium migrates toward light, forms fruiting body in which spores are produced.

When food becomes scarce, cells aggregate into a slug-like mass called a *pseudoplasmodium*.

fruiting bodies

FIGURE 20-18 The life cycle of a cellular slime mold

FIGURE 20-19 Red algae
Red coralline algae from the Pacific Ocean off California. Coralline algae, which deposit calcium carbonate within their bodies, also contribute to coral reefs in tropical waters.

substances with commercial uses, including carrageenan (used as a stabilizing agent in products such as paints, cosmetics, and ice cream) and agar (a substrate for growing bacterial colonies in laboratories). However, the major importance of these and other algae lies in their photosynthetic ability; the energy they capture helps support non-photosynthetic organisms in marine ecosystems.

Green Algae Live Mostly in Ponds and Lakes

The green algae, a large and diverse group of photosynthetic protists, include both multicellular and unicellular species. Most species live in freshwater ponds and lakes, but some live in the seas. Some green algae, such as *Spirogyra*, form thin filaments from long chains of cells (**FIG. 20-20a**). Other species of green algae form colonies containing clusters of cells that are somewhat interdependent and constitute a structure intermediate between unicellular and multicellular forms. These colonies range from a few cells to a few thousand cells, as in species of *Volvox*. Most green algae are small, but some marine species are large. For example, the green alga *Ulva*, or sea lettuce, is similar in size to the leaves of its namesake (**FIG. 20-20b**).

The green algae are of special interest because, unlike other groups that contain multicellular, photosynthetic protists, green algae are closely related to plants. Plants share a common ancestor with some types of green algae, and many researchers believe that the very earliest plants were similar to today's multicellular green algae.

(a)

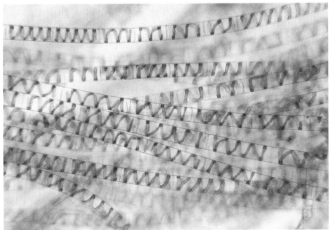

(b)

FIGURE 20-20 Green algae
(a) *Spirogyra* is a filamentous green alga composed of strands only one cell thick. (b) *Ulva* is a multicellular green alga that assumes a leaflike shape.

EVOLUTIONARY CONNECTIONS

Our Unicellular Ancestors

Some of today's microbes are probably quite similar to the ancient species that ultimately gave rise to the complex multicellular organisms that are now the most conspicuous inhabitants of Earth. For example, the external appearance of many modern prokaryotes is basically indistinguishable from that of 3.5-billion-year-old fossilized cells. Similarly, the metabolism of today's anaerobic, heat-loving archaea is probably similar to the methods of energy acquisition used by Earth's earliest inhabitants, long before any oxygen entered the atmosphere. Likewise, modern purple sulfur bacteria and cyanobacteria are probably not too different from the first photosynthetic organisms that appeared more than 2 billion years ago.

Life might still consist solely of prokaryotic, single-celled organisms if protists, with their radical eukaryotic design, had not appeared nearly 2 billion years ago. As you learned in the discussion of the endosymbiont theory in Chapter 17, eukaryotic cells may have originated when one prokaryote, perhaps a bacterium capable of aerobic respiration, took up residence inside a partner, forming the first "mitochondrion." A separate but equally crucial merger may have occurred when a photosynthetic bacterium (probably resembling a cyanobacterium) took up residence within a non-photosynthetic partner and became the first "chloroplast." The foundations of multicellularity were laid with the eukaryotic cell, whose intricacy allowed specialization of entire cells for specific functions within a multicellular aggregation. Thus, primitive protists, some absorbing nutrients from the environment, some photosynthesizing, and others consuming their food in chunks, almost certainly followed divergent evolutionary paths that led to the three multicellular kingdoms—the fungi, plants, and animals—that are the subjects of the following four chapters.

CASE STUDY REVISITED   GREEN MONSTER

Caulerpa taxifolia, the invasive seaweed that threatens to overrun the Mediterranean, is a green alga. This species and other members of its genus have very unusual bodies. Outwardly, they appear plantlike, with rootlike structures that attach to the seafloor and other structures that look like stems and leaves and rise to a height of several inches. Despite its seeming similarity to a plant, however, a *Caulerpa* body consists of a single, extremely large cell. The entire body is surrounded by a single, continuous cell membrane. The interior consists of cytoplasm that contains numerous cell nuclei but is not subdivided. That a single cell can take such a complex shape is extraordinary.

A potential problem with *Caulerpa's* single-celled organization might arise when its body is damaged, perhaps by wave action or when a predator takes a bite out of it. When the cell membrane is breached, there is nothing to prevent all of the cytoplasm from leaking out, an event that would be fatal. But *Caulerpa* has evolved a defense against this potential calamity. Shortly after the cell membrane breaks, it is quickly filled with a "wound plug" that closes the gap. After the plug is established, the cell begins to grow and regenerates any lost portion of the body.

This ability to regenerate is a key component of the ability of aquarium-strain *Caulerpa taxifolia* to spread rapidly in new environments. If part of a *Caulerpa* body breaks off and drifts to a new location, it can regenerate a whole new body. The regenerated individual becomes the founder of a new, quickly growing colony.

And these quickly growing colonies might appear anywhere in the world. Authorities in many countries worry that the aquarium strain of *Caulerpa* could invade their coastal waters, unwittingly transported by ships from the Mediterranean or released by careless aquarists. In fact, invasive *Caulerpa* is no longer restricted to the Mediterranean. It has been found in two locations on the California coast and in at least eight bodies of water in Australia. Local authorities in both countries have attempted to control the invading algae, but it is impossible to say if their efforts will be successful. *Caulerpa taxifolia* is a resourceful foe.

Consider This Is it important to stop the spread of *Caulerpa*? Governments invest substantial resources to combat introduced species and prevent their populations from increasing and dispersing. Why might this be a wise use of funds? Can you think of some arguments against spending time and money for this purpose?

CHAPTER REVIEW

SUMMARY OF KEY CONCEPTS

20.1 What Are Protists?

"Protist" is a term of convenience that refers to any eukaryote that is not a plant, animal, or fungus. Most protists are single, highly complex eukaryotic cells, but some form colonies and some, such as seaweeds, are multicellular. Protists exhibit diverse modes of nutrition, reproduction, and locomotion. Photosynthetic protists form much of the phytoplankton, which plays a key ecological role. Some protists cause human diseases; others are crop pests.

20.2 What Are the Major Groups of Protists?

Protist groups include excavates (diplomonads and parabasalids), eugleonozoans (euglenids and kinetoplastids), stramenopiles (water molds, diatoms, and brown algae), alveolates (dinoflagellates, apicomplexans, and ciliates), cercozoans (which include foraminiferans), amoebozoans (amoebas and slime molds), red algae, and green algae (the closest relatives of plants).

Web Tutorial 20.1 The Life Cycle of the Malaria Parasite

KEY TERMS

acellular slime mold
page 397
algae *page 389*
alveolate *page 393*
amoeba *page 397*
amoebozoan *page 397*
apicomplexan *page 394*
cellular slime mold *page 398*

cercozoan *page 395*
cilia *page 395*
ciliate *page 395*
diatom *page 392*
dinoflagellate *page 394*
diplomonad *page 390*
euglenid *page 391*
euglenozoan *page 391*

excavate *page 390*
foraminiferan *page 396*
kinetoplastid *page 392*
parabasalid *page 391*
phytoplankton *page 393*
plasmodium *page 397*
protist *page 388*
protozoa *page 389*

pseudoplasmodium
page 398
pseudopod *page 389*
radiolarian *page 396*
stramenopile *page 392*
water mold *page 392*

THINKING THROUGH THE CONCEPTS

1. List the major differences between prokaryotes and protists.

2. What is secondary endosymbiosis?

3. What is the importance of dinoflagellates in marine ecosystems? What can happen when they reproduce rapidly?

4. What is the major ecological role played by single-celled algae?

5. Which protist group consists entirely of parasitic forms?

6. Which protist groups include seaweeds?

7. Which protist groups include species that use pseudopods?

APPLYING THE CONCEPTS

1. Recent research shows that ocean water off southern California has become 2 to 3°F (1 to 1.5°C) warmer over the past four decades, possibly due to the greenhouse effect. This warming has indirectly led to a depletion of nutrients in the water and thus a decline in photosynthetic protists such as diatoms. What effects is this warming likely to have on life in the oceans?

2. The internal structure of many protists is much more complex than that of cells of multicellular organisms. Does this mean that the protist is engaged in more-complex activities than the multicellular organism is? If not, why are protistan cells more complicated?

3. Why would the lives of multicellular animals be impossible if prokaryotic and protistan organisms did not exist?

FOR MORE INFORMATION

Amato, I. "Plankton Planet." *Discover*, August 2004. A brief overview of the organisms that compose the phytoplankton. Includes beautiful photos.

Jacobs, W. "Caulerpa." *Scientific American*, December 1994. Description of the distinctive structure and physiology of *Caulerpa*, written by a scientist who has studied these protists for decades.

Raloff, J. "Taming Toxins." *Science News*, November 30, 2002. Describes a possible new strategy for combating red tides and other blooms of toxic dinoflagellates.

The Diversity of Plants

The huge, foul-smelling flower of the stinking corpse lily
is a treat for visitors to Asian rain forests.

CASE STUDY   QUEEN OF THE PARASITES

THE FLOWER OF THE STINKING CORPSE LILY makes a strong impression. For one thing, it's huge; a single flower may be three feet across. It also has a rather strange appearance, consisting largely of fleshy lobes that are almost fungus-like. But the thing that makes a stinking corpse lily almost impossible to ignore is its aroma, which has been described as "a penetrating smell more repulsive than any buffalo carcass in an advanced stage of decomposition." Though utterly revolting to humans, the smell is attractive to blowflies and other insects that normally feed on and lay their eggs in decaying flesh. When such insects visit a male stinking corpse lily, they may carry away pollen that can fertilize a nearby female flower.

Close examination of a stinking corpse lily reveals that it has no visible leaves, roots, or stems. In fact, it is a parasite, and its body is completely embedded in the tissue of its host, a vine of the genus *Tetrastigma*. Without leaves, the stinking corpse lily cannot produce any food of its own, but instead draws all of its nutrition from its host. The parasite becomes visible outside the body of its host only when one of its cabbage-shaped flower buds pushes through the surface of the host's stem and its gigantic, stinking flower opens for a week or so before shriveling and falling off. If a male and a female flower happen to be open simultaneously and close together, the female flower may be fertilized and produce

seeds. A seed that is dispersed in animal droppings and happens to land on a *Tetrastigma* stem may germinate and penetrate a new host.

When you think of plants, you might first think of their most obvious feature: green leaves that capture solar energy by photosynthesis. It may seem odd, then, that this chapter about plants begins with a peculiar plant that does not photosynthesize. Oddities such as the stinking corpse lily, however, serve as reminders that evolution does not always follow a predictable pathway, and that even an adaptation as seemingly valuable as the ability to live on sunlight can be discarded.

21.1 WHAT ARE THE KEY FEATURES OF PLANTS?

Plants are the most conspicuous living things in almost every landscape on Earth. Unless you are in a polar region, a harsh desert, or a densely populated urban area, you live surrounded by plants. The plants that dominate Earth's forests, grasslands, parks, lawns, orchards, and farm fields are such familiar parts of the backdrop to our daily lives that we tend to take them for granted. But if we take some time to look more closely at our green companions, we might gain a greater appreciation for the adaptations that have made them so successful and for the properties that make them essential to our own survival.

What distinguishes members of the plant kingdom from other organisms? Perhaps the most noticeable feature of plants is their green color. The color comes from the presence of the pigment chlorophyll in many plant tissues. Chlorophyll plays a crucial role in photosynthesis, the process by which plants use energy from sunlight to convert water and carbon dioxide to sugar. Chlorophyll and photosynthesis, however, are not unique to plants; they are also present in many types of protists and prokaryotes. Instead, the key distinctive feature of plants is their reproductive cycle, which features *alternation of generations*.

Plants Have Alternating Multicellular Haploid and Diploid Generations

The plant life cycle is characterized by **alternation of generations** (FIG. 21-1), in which separate diploid and haploid generations alternate with one another. (Recall that a diploid organism has two sets of chromosomes; a haploid organism, one set.) In the diploid generation, a plant body consists of diploid cells and is known as the **sporophyte**. Certain cells of sporophytes undergo meiosis to produce haploid reproductive cells called *spores*. The haploid spores develop into multicellular, haploid plants called **gametophytes**.

A gametophyte ultimately produces male and female haploid *gametes* by mitosis. Gametes, likes spores, are reproductive cells but, unlike spores, an individual gamete by itself cannot develop into a new individual. Instead, two gametes of opposite sexes must meet and fuse to form a new individual. In plants, gametes produced by gametophytes fuse to form a diploid **zygote**, which develops into a diploid sporophyte, and the cycle begins again.

Plants Have Multicellular, Dependent Embryos

In plants, zygotes develop into multicellular embryos that are retained within and receive nutrients from the tissues of the parent plant. That is, a plant embryo is attached to and dependent on its parent as it grows and develops. Such multicellular, dependent embryos are not found among photosynthetic protists; they distinguish plants from their nearest relatives among the algae.

Plants Play a Crucial Ecological Role

Plants provide food, directly or indirectly, for all of the animals, fungi, and non-photosynthetic microbes on land. Plants use photosynthesis to capture solar energy, and they convert part of the captured energy to leaves, shoots, seeds, and fruits that are eaten by other organisms. Many of these consumers of plant tissue are themselves eaten by still other organisms. Plants are the main providers of energy and

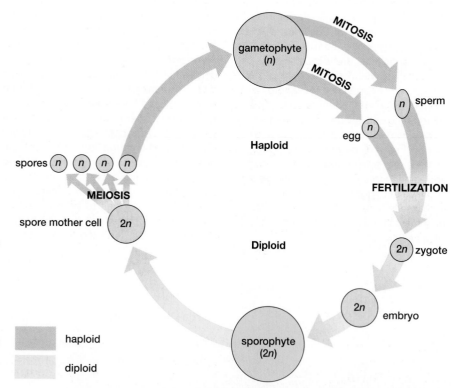

FIGURE 21-1 Alternation of generations in plants
As shown in this generalized depiction of a plant life cycle, a diploid sporophyte generation produces haploid spores through meiosis. The spores develop into a haploid gametophyte generation that produces haploid gametes by mitosis. The fusion of these gametes results in a diploid zygote that develops into the sporophyte plant.

nutrients to terrestrial ecosystems, and all life on land depends on plants' ability to manufacture food from sunlight.

In addition to their role as food suppliers, plants make other essential contributions to their fellow organisms. For example, plants produce oxygen gas as a by-product of photosynthesis, and by doing so they continually replenish oxygen in the atmosphere. Without plants' contribution, atmospheric oxygen would be rapidly depleted by the oxygen-consuming respiration of Earth's multitude of organisms.

Plants also help create and maintain soil. When a plant dies, its stems, leaves, and roots become food for fungi, prokaryotes, and other decomposers. Decomposition breaks the plant tissue into tiny particles of organic matter that become part of the soil. Organic matter improves the ability of soil to hold water and nutrients, thereby making the soil more fertile and better able to support the growth of living plants. The roots of those living plants help hold the soil together and keep it in place. Soils from which vegetation has been removed are susceptible to erosion by wind and water.

Plants Provide Humans with Necessities and Luxuries

All inhabitants of terrestrial ecosystems depend on plants' contributions to those ecosystems, but human reliance on plants is especially pronounced. It would be difficult to exaggerate the degree to which human populations depend on plants. Neither our explosive population growth nor our rapid technological advance would have been possible without plants.

Plants Provide Shelter, Fuel, and Medicine

Plants are the source of the wood that is used to construct housing for a large portion of Earth's human population. For much of human history, wood was also the main fuel for warming dwellings and for cooking. Wood is still the most important fuel in many parts of the world. Coal, another important fuel, is composed of the remains of ancient plants that have been transformed by geological processes.

Plants have also supplied many of the medicines on which modern health care depends. Important drugs that were originally found in and extracted from plants include aspirin, the heart medication digitalin, the cancer treatments Taxol® and vinblastine, the malaria drug quinine, the painkillers codeine and morphine, and many more.

In addition to harvesting useful material from wild plants, humans have domesticated a host of useful plant species. Through generations of selective breeding, humans have modified the seeds, stems, roots, flowers, and fruits of favored plant species in order to provide us with food and fiber. It is difficult to imagine life without corn, rice, potatoes, apples, tomatoes, cooking oil, cotton, and the myriad other staples that domestic plants provide.

Plants Provide Pleasure

Despite the obvious contributions of plants to human well-being, our relationship with plants seems to be based on something more profound than their ability to help us meet our material needs. Though we appreciate the practical value of wheat and wood, our most emotionally powerful connections with plants are purely sensual. Many of life's pleasures come to us courtesy of our plant partners. We delight in the beauty and fragrance of flowers, and present them to others as symbols of our most sublime and inexpressible emotions. Quite a few of us spend hours of our leisure time tending gardens and lawns, for no reward other than the pleasure and satisfaction we derive from observing the fruits of our labor. In our homes, we reserve space not only for members of our families, but also for our houseplant companions. We feel compelled to line our streets with trees, and we seek refuge from the stress of daily life in parks with abundant plant life. Our mornings are enhanced by the aroma of coffee or tea, and our evenings by a nice glass of wine. Clearly, plants help fill our desires as well as our needs.

21.2 WHAT IS THE EVOLUTIONARY ORIGIN OF PLANTS?

The ancestors of plants were photosynthetic protists, probably similar to today's algae. Like modern algae, the organisms that gave rise to plants presumably lacked true roots, stems, leaves, and complex reproductive structures such as flowers or cones. All of these features appeared later in the evolutionary history of plants (**FIG. 21-2**).

Green Algae Gave Rise to Plants

Of today's different groups of algae, green algae are probably most similar to ancestral plants. This supposition stems from the close phylogenetic relationship between the two groups. DNA comparisons have shown that green algae are plants' closest living relatives, and the hypothesis that plants evolved from green algal ancestors is supported by other evidence as well. For example, green algae and plants use the same type of chlorophyll and accessory pigments in photosynthesis. In addition, both plants and green algae store food as starch and have cell walls made of cellulose. In contrast, the photosynthetic pigments, food-storage molecules, and cell walls of other photosynthetic protists, such as the red algae and the brown algae, differ from those of plants.

The Ancestors of Plants Lived in Fresh Water

Most green algae live in fresh water, which suggests that the early evolutionary history of plants took place in freshwater habitats. In contrast to the nearly constant environmental conditions of the ocean, freshwater habitats are highly variable. Water temperature can fluctuate seasonally or even daily. Changing rates of rainfall and evaporation can cause frequent changes in the concentration of chemicals in the water and can even cause an aquatic habitat to dry up periodically. Ancient freshwater green algae must have evolved characteristics that enabled them to withstand temperature extremes and periods of dryness. These adaptations to the difficulties of freshwater life laid the foundation for the descendants of early algae to evolve the traits that made life on land possible.

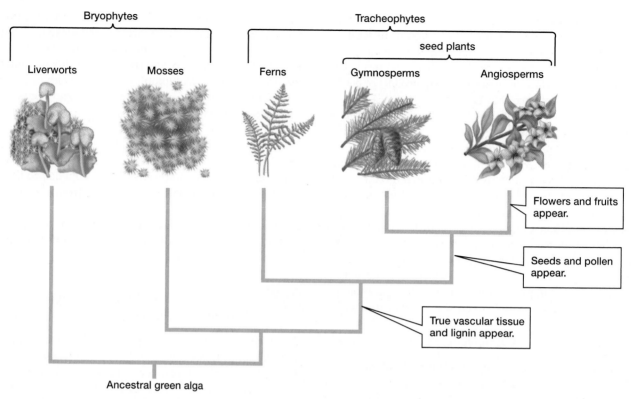

FIGURE 21-2 Evolutionary tree of some major plant groups

21.3 HOW HAVE PLANTS ADAPTED TO LIFE ON LAND?

Most plants live on land. Life on land has many advantages for plants, including access to sunlight unimpeded by water that might block its rays and access to nutrients contained in surface rocks. These advantages, however, come at a cost. On land, the supportive buoyancy of water is missing, the body is not bathed in a nutrient solution, and the air tends to dry things out. In addition, gametes (sex cells) and zygotes (fertilized sex cells) cannot be carried by water currents or propelled by flagella, as they are in many water-dwelling organisms. As a result, life on land has favored the evolution in plants of structures that support the body and conserve water, vessels that transport water and nutrients to all parts of the plant, and processes that disperse gametes and zygotes by methods that are independent of water.

Plant Bodies Resist Gravity and Drying

Some of the key adaptations to life on land arose early in plant evolution, and they are now common to virtually all land plants. They include

- Roots or rootlike structures that anchor the plant and/or absorb water and nutrients from the soil
- A waxy **cuticle** that covers the surfaces of leaves and stems and that limits the evaporation of water
- Pores called **stomata** (singular, stoma) in the leaves and stems that open to allow gas exchange but close when

water is scarce, reducing the amount of water lost to evaporation

Other key adaptations occurred somewhat later in the transition to terrestrial life, and are now widespread but not universal among plants (most nonvascular plants, a group described later, lack them):

- Conducting vessels that transport water and minerals upward from the roots and that move photosynthetic products from the leaves to the rest of the plant body
- The stiffening substance **lignin**, which is a rigid polymer that impregnates the conducting vessels and supports the plant body, helping the plant expose maximum surface area to sunlight

Plant Embryos Are Protected, and Plant Sex Cells May Disperse Without Water

All plants protect developing embryos within parental tissues, but the most widespread groups of plants are characterized by especially well-protected and well-provisioned embryos and by waterless dispersal of sex cells. The key adaptations of these plant groups are *pollen*, *seeds*, and, in the flowering plants, *flowers* and *fruits*. Early seed plants produced dry, microscopic pollen grains that allowed wind, instead of water, to carry the male gametes. Seeds provided protection and nourishment for developing embryos and the potential for more effective dispersal. Later came the evolution of flowers, which enticed animal pollinators that delivered pollen more precisely than did wind. Fruits also attracted animal foragers, which consumed the fruit and dispersed its indigestible seeds in their feces.

21.4 WHAT ARE THE MAJOR GROUPS OF PLANTS?

Two major groups of land plants arose from ancient algal ancestors (Table 21-1). One group, the **bryophytes** (also called *nonvascular plants*), requires a moist environment to reproduce and thus straddles the boundary between aquatic and terrestrial life, much like the amphibians of the animal kingdom. The other group, the **vascular** plants (also called *tracheophytes*), has been able to colonize drier habitats.

Bryophytes Lack Conducting Structures

Bryophytes retain some characteristics of their algal ancestors. They lack true roots, leaves, and stems. They do possess rootlike anchoring structures called *rhizoids* that bring water and nutrients into the plant body, but bryophytes are nonvascular; they lack well-developed structures for conducting water and nutrients. They must instead rely on slow diffusion or poorly developed conducting tissues to distribute water and other nutrients. As a result, their body size is limited. Size is also limited by the absence of any stiffening agent in their bodies. Without such material, they cannot grow upward very far. Most bryophytes are less than 1 inch (2.5 centimeters) tall.

Bryophytes Include the Hornworts, Liverworts, and Mosses

The bryophytes include three phyla: hornworts, liverworts, and mosses. Hornworts and liverworts are named for their shapes. Hornwort sporophytes generally have a spiky shape that appears hornlike to some observers (**FIG. 21-3a**). The gametophytes of certain liverwort species have a lobed form reminiscent of the shape of a liver (**FIG. 21-3b**). Hornworts and liverworts are most abundant in areas where moisture is plentiful, such as in moist forests and near the banks of streams and ponds.

Mosses are the most diverse and abundant of the bryophytes (**FIG. 21-3c**). Like hornworts and liverworts, mosses are most likely to be found in moist habitats. Some mosses, however, have a waterproof covering that retains moisture, preventing water loss. In addition, many of these mosses are also able to survive the loss of much of the water in their bodies; they dehydrate and become dormant during dry periods but absorb water and resume growth when moisture returns. Such mosses can survive in deserts, on bare rock, and in far northern and southern latitudes where humidity is low and liquid water is scarce for much of the year.

Mosses of the genus *Sphagnum* are especially widespread, living in moist habitats in northern regions around the world. In many of these wet northern habitats, *Sphagnum* is the most abundant plant, forming extensive mats (**FIG 21-3d**). Because decomposition is slow in cold climates and because *Sphagnum* contains compounds that inhibit bacteria, dead *Sphagnum* may decay only very slowly. As a result, partially decayed moss tissue can accumulate in deposits that can, over thousands of years, become hundreds of feet thick. These deposits are known as peat. Peat has long been harvested for use as fuel, a practice that continues today in some northern areas. Now, however, peat is more often harvested for use in horticulture. Dried peat can absorb many times its own weight in water, making it useful as a soil conditioner and as a packing material for transporting live plants.

Table 21-1 Features of the Major Plant Groups

Group	Subgroup	Relationship of Sporophyte and Gametophyte	Transfer of Reproductive Cells	Early Embryonic Development	Dispersal	Water and Nutrient Transport Structures
Bryophytes		Gametophyte dominant—sporophyte develops from zygote	Motile sperm swims to stationary egg retained on gametophyte	Occurs within archegonium of gametophyte	Haploid spores carried by wind	Absent
Vascular Plants	Ferns	Sporophyte dominant—develops from zygote retained on gametophyte	Motile sperm swims to stationary egg retained on gametophyte	Occurs within archegonium of gametophyte	Haploid spores carried by wind	Present
	Conifers	Sporophyte dominant—microscopic gametophyte develops within sporophyte	Wind-dispersed pollen carries sperm to stationary egg in cone	Occurs within a protective seed containing a food supply	Seeds containing diploid sporophyte embryo dispersed by wind or animals	Present
	Flowering plants	Sporophyte dominant—microscopic gametophyte develops within sporophyte	Pollen, dispersed by wind or animals, carries sperm to stationary egg within flower	Occurs within a protective seed containing a food supply; seed encased in fruit	Fruit, carrying seeds, dispersed by animals, wind, or water	Present

FIGURE 21-3 Bryophytes
The plants shown here are less than $\frac{1}{2}$ inch (about 1 centimeter) in height. **(a)** The hornlike sporophytes of hornworts grow upward from archegonia that are embedded in the gametophyte body. **(b)** Liverworts grow in moist, shaded areas. This female plant bears umbrella-like archegonia, which hold the eggs. Sperm must swim up the stalks through a film of water to fertilize the eggs. **(c)** Moss plants, showing the stalks that carry spore-bearing capsules. **(d)** Mats of *Sphagnum* moss cover moist bogs in northern regions. QUESTION Why are all bryophytes short?

The Reproductive Structures of Bryophytes Are Protected

Among the bryophytes' adaptations to terrestrial existence are their enclosed reproductive structures, which prevent the gametes from drying out. There are two types of structure: **archegonia** (singular, archegonium), in which eggs develop, and **antheridia** (singular, antheridium), where sperm are formed (**FIG. 21-4**). In some bryophyte species, both archegonia and antheridia are located on the same plant; in other species, each individual plant is either male or female.

In all bryophytes, the sperm must swim to the egg, which emits a chemical attractant, through a film of water. (Bryophytes that live in drier areas must time their reproduction to coincide with rains.) The fertilized egg is retained in the archegonium, where the embryo grows and matures into a small diploid sporophyte that remains attached to the parent gametophyte plant. At maturity, the sporophyte produces haploid spores by meiosis within a capsule. When the capsule is opened, spores are released

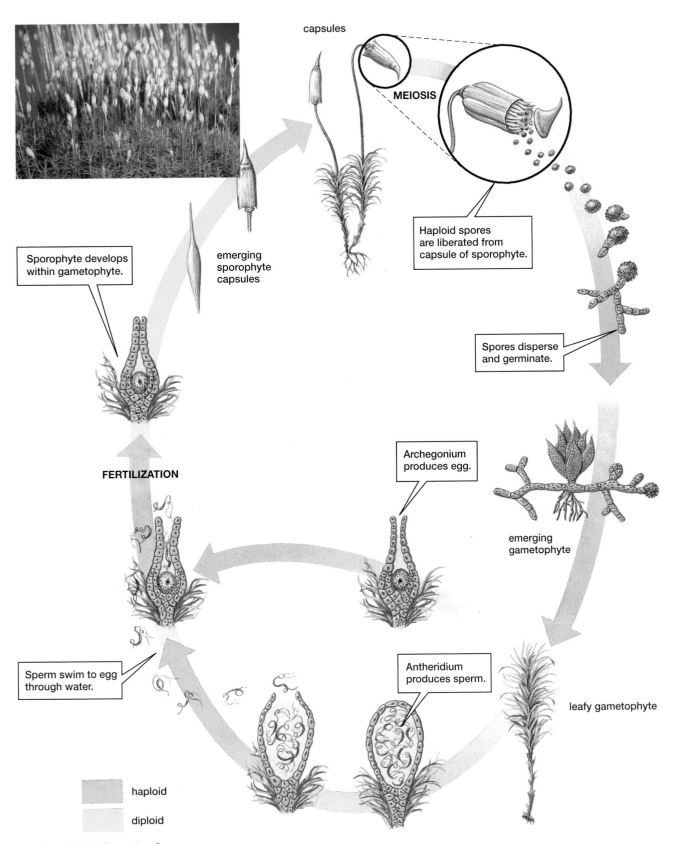

capsules

MEIOSIS

Haploid spores
are liberated from
capsule of sporophyte.

Spores disperse
and germinate.

Sporophyte develops
within gametophyte.

emerging
sporophyte
capsules

Archegonium
produces egg.

emerging
gametophyte

FERTILIZATION

Antheridium
produces sperm.

Sperm swim to egg
through water.

leafy gametophyte

haploid

diploid

FIGURE 21-4 Life cycle of a moss
The leafy green gametophyte (lower right) is the haploid generation that produces sperm and eggs. The sperm must swim through a film of water to the egg. The zygote develops into a stalked, diploid sporophyte that emerges from the gametophyte plant. The sporophyte is topped by a brown capsule in which haploid spores are produced by meiosis. These are dispersed and germinate, producing another green gametophyte generation. **(Inset)** Moss plants. The short, leafy green plants are haploid gametophytes; the reddish brown stalks are diploid sporophytes.

409

and dispersed by the wind. If a spore lands in a suitable environment, it may develop into another haploid gametophyte plant.

Vascular Plants Have Conducting Vessels That Also Provide Support

Vascular plants are distinguished by specialized groups of conducting cells called **vessels**. The vessels are impregnated with the stiffening substance lignin and serve both supportive and conducting functions. Vessels allow vascular plants to grow taller than nonvascular plants, both because of the extra support provided by lignin and because the conducting cells allow water and nutrients absorbed by the roots to move to the upper portions of the plant. Another difference between vascular plants and bryophytes is that in vascular plants, the diploid sporophyte is the larger, more conspicuous generation; in nonvascular plants, the haploid gametophyte is more evident.

The vascular plants can be divided into two groups: the seedless vascular plants and the seed plants.

The Seedless Vascular Plants Include the Club Mosses, Horsetails, and Ferns

Like the bryophytes, seedless vascular plants have swimming sperm and require water for reproduction. As their name implies, they do not produce seeds but rather propagate by spores. The present-day seedless vascular plants—the club mosses, horsetails, and ferns—are much smaller than their ancestors, which dominated the landscape in the Carboniferous period (350–290 million years ago). The bodies of ancient seedless vascular plants—transformed by heat, pressure, and time—are burned today as coal. The seedless vascular plants were once dominant, but today the more versatile seed plants have become dominant.

Club Mosses and Horsetails Are Small and Inconspicuous

The club mosses are now limited to representatives a few inches in height (**FIG. 21-5a**). Their leaves are small and scalelike, resembling the leaflike structures of mosses. Club mosses of the genus *Lycopodium*, commonly known as ground pine, form a beautiful ground cover in some temperate coniferous and deciduous forests.

Modern horsetails belong to a single genus, *Equisetum*, that contains only 15 species, most less than 3 feet tall (**FIG. 21-5b**). The bushy branches of some species lend them the common name horsetails; the leaves are reduced to tiny scales on the branches. They are also called "scouring rushes" because early European settlers of North America used them to scour pots and floors. All species of *Equisetum* deposit large amounts of silica (glass) in their outer layer of cells, giving them an abrasive texture.

Ferns Are Broad-Leaved and More Diverse

The ferns, with 12,000 species, are the most diverse of the seedless vascular plants (**FIG. 21-5c**). In the tropics, tree ferns still reach heights reminiscent of their ancestors from the Carboniferous period (**FIG. 21-5d**). Ferns are the only seedless vascular plants that have broad leaves.

In ferns, haploid spores are produced in structures called *sporangia* that form on special leaves of the sporophyte (**FIG. 21-6**). The spores are dispersed by the wind and give rise to tiny, haploid gametophyte plants, which produce sperm and eggs. The gametophyte generation retains two traits that are reminiscent of the bryophytes. First, the small gametophytes lack conducting vessels. Second, as in bryophytes, the sperm must swim through water to reach the egg.

The Seed Plants Dominate the Land, Aided by Two Important Adaptations: Pollen and Seeds

The seed plants are distinguished from bryophytes and seedless vascular plants by their production of pollen and seeds. **Pollen** grains are tiny structures that carry sperm-producing cells. Pollen grains are dispersed by wind or by animal pollinators such as bees. In this way, sperm move through the air to fertilize egg cells. This airborne transport means that the distribution of seed plants is not limited by the need for water through which sperm can swim to the egg. Seed plants are fully adapted to life on dry land.

Analogous to the eggs of birds and reptiles, **seeds** consist of an embryonic plant, a supply of food for the embryo, and a protective outer coat (**FIG. 21-7**). The *seed coat* maintains the embryo in a state of suspended animation or dormancy until conditions are proper for growth. The stored food helps sustain the emerging plant until it develops roots and leaves and can make its own food by photosynthesis. Some seeds possess elaborate adaptations that allow them to be dispersed by wind, water, and animals.

In seed plants, gametophytes (which produce the sex cells) are greatly reduced in size. The female gametophyte is a small group of haploid cells that produces the egg. The male gametophyte is the pollen grain.

Seed plants are grouped into two general types: *gymnosperms*, which lack flowers, and *angiosperms*, the flowering plants.

Gymnosperms Are Nonflowering Seed Plants

Gymnosperms evolved earlier than the flowering plants. Early gymnosperms coexisted with the forests of seedless vascular plants that dominated the Carboniferous period. During the subsequent Permian period (290–248 million years ago), however, gymnosperms became the predominant plant group and remained so until the rise of the flowering plants more than 100 million years later. Despite their success, most of these early gymnosperms are now extinct. Today, four phyla of gymnosperms survive: the ginkgos, the cycads, the gnetophytes, and the conifers.

Only One Ginkgo Species Survives

Ginkgos probably have a long evolutionary history. They were widespread during the Jurassic period, which began 208 million years ago. Today, however, they are represented

(a)

(c)

(b)

(d)

FIGURE 21-5 Some seedless vascular plants
Seedless vascular plants are found in moist woodland habitats. **(a)** The club mosses (sometimes called ground pines) grow in temperate forests. This specimen is releasing spores. **(b)** The giant horsetail extends long, narrow branches in a series of rosettes. Its leaves are insignificant scales. At right is a cone-shaped spore-forming structure. **(c)** The leaves of this deer fern are emerging from coiled fiddleheads. **(d)** Although most fern species are small, some, such as this tree fern, retain the large size that was common among ferns of the Carboniferous period. QUESTION In each of these photos, is the pictured structure a sporophyte or a gametophyte?

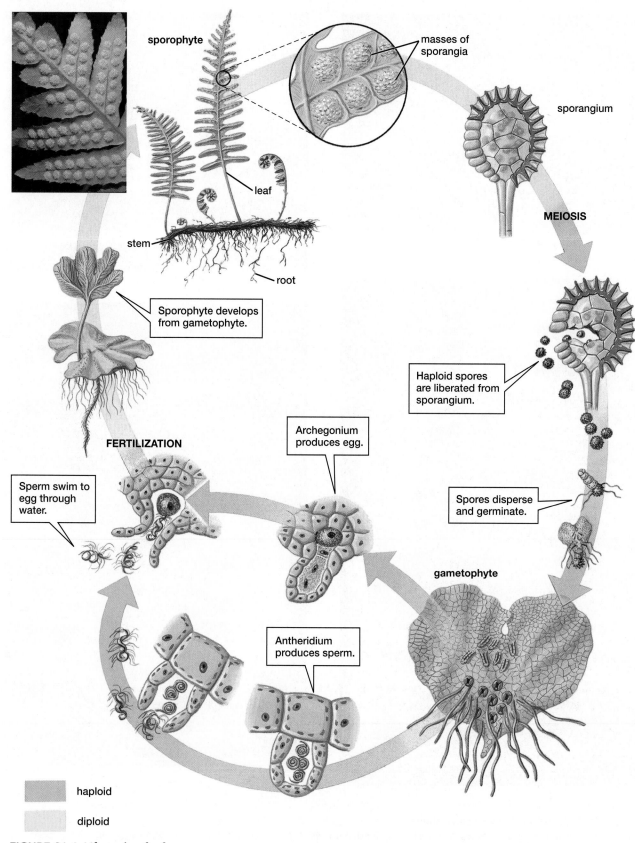

sporophyte

masses of
sporangia

sporangium

leaf

stem

root

MEIOSIS

Sporophyte develops
from gametophyte.

Haploid spores
are liberated from
sporangium.

Archegonium
produces egg.

FERTILIZATION

Spores disperse
and germinate.

Sperm swim to
egg through
water.

Antheridium
produces sperm.

gametophyte

haploid

diploid

FIGURE 21-6 Life cycle of a fern
The dominant plant body (upper left) is the diploid sporophyte. Haploid spores, formed in sporangia located on the underside of certain leaves, are dispersed by the wind to germinate on the moist forest floor into
inconspicuous haploid gametophyte plants. On the lower surface of these small, sheetlike gametophytes,
male antheridia and female archegonia produce sperm and eggs. The sperm must swim to the egg, which
remains in the archegonium. The zygote develops into the large sporophyte plant. **(Inset)** Underside of a
fern leaf, showing clusters of sporangia.

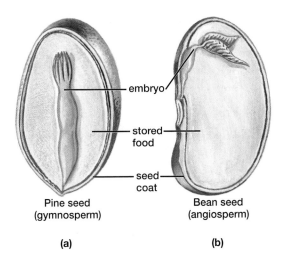

FIGURE 21-7 Seeds

Seeds from **(a)** a gymnosperm and **(b)** an angiosperm. Both consist of an embryonic plant and stored food confined within a seed coat. Seeds exhibit diverse adaptations for dispersal, including **(c)** the dandelion's tiny, tufted seeds that float in the air and **(d)** the massive, armored seeds (protected inside the fruit) of the coconut palm, which can survive prolonged immersion in seawater as they traverse oceans. QUESTION Can you think of some adaptations that help protect seeds from destruction by animal consumption?

by the single species *Ginkgo biloba*, the maidenhair tree. Ginkgo trees are either male or female; female trees bear foul-smelling, fleshy seeds the size of cherries (**FIG. 21-8a**). Ginkgos have been maintained by cultivation, particularly in Asia; if not for this cultivation, they might be extinct today. Because they are more resistant to pollution than are most other trees, ginkgos (normally, the male trees) have been extensively planted in U.S. cities. Recently, the leaves of the ginkgo have gained attention as a herbal remedy that purportedly improves memory.

Cycads Are Restricted to Warm Climates

Like ginkgos, cycads were diverse and abundant in the Jurassic but have since dwindled. Today there are approximately 160 species, most of which dwell in tropical or subtropical climates. Cycads have large, finely divided leaves and bear a superficial resemblance to palms or large ferns (**FIG. 21-8b**). Most cycads are about 3 feet (1 meter) in height, although some species can reach 65 feet (20 meters). Cycads grow slowly and live for a long time; one Australian specimen is estimated to be 5000 years old.

The tissues of cycads contain potent toxins. Despite the presence of these toxins, people in some parts of the world use cycad seeds, stems, and roots for food. Careful preparation and processing removes the toxins before the plants

are consumed. Nonetheless, cycad toxins are the suspected cause of neurological problems that occur with some frequency in populations that use cycads for food. Cycad toxins can also harm grazing livestock.

About half of all cycad species are classified as threatened or endangered. The main threats to cycads are habitat destruction, competition from introduced species, and harvesting for the horticultural trade. A large specimen of a rare cycad can sell for thousands of dollars. Because cycads grow slowly, recovery of endangered populations is uncertain.

Gnetophytes Include the Odd Welwitschia

The gnetophytes include about 70 species of shrubs, vines, and small trees. Leaves of gnetophyte species in the genus *Ephedra* contain alkaloid compounds that act in humans as stimulants and appetite suppressants. For this reason, *Ephedra* is widely used as an energy booster and weight-loss aid. However, following reports of sudden deaths of *Ephedra* users and publication of several studies linking *Ephedra* consumption to increased risk of heart problems, the U.S. Food and Drug Administration banned the sale of products containing *Ephedra*.

The gnetophyte *Welwitschia mirabilis* is among the most distinctive of plants (**FIG. 21-8c**). Found only in the extremely dry deserts of southwest Africa, *Welwitschia* has a

FIGURE 21-8 Gymnosperms
(a) This ginkgo, or maidenhair tree, is female and bears fleshy seeds the size of large cherries. **(b)** A cycad. Common in the age of dinosaurs, these are now limited to about 160 species. Like ginkgos, cycads have separate sexes. **(c)** The leaves of the gnetophyte *Welwitschia* may be hundreds of years old. **(d)** The needle-shaped leaves of conifers are protected by a waxy surface layer.

deep taproot that can extend as far as 100 feet (30 meters) down into the soil. Above the surface, the plant has a fibrous stem. Two (and only two) leaves grow from the stem. The leaves are never shed and remain on the plant for its entire life, which can be very long. The oldest *Welwitschia* are more than 2000 years old, and a typical lifespan is about 1000 years. The strap-like leaves continue to grow for that entire period, spreading over the ground. The older portions of the leaves, whipped by the wind for centuries, may shred or split, giving the plant its characteristic gnarled and well-worn appearance.

Conifers Are Adapted to Cool Climates

Though the other gymnosperm phyla are drastically reduced from their former prominence, the **conifers** still dominate large areas of our planet. Conifers, whose 500 species include pines, firs, spruce, hemlocks, and cypresses, are most abundant in the cold latitudes of the far north and at high elevations where conditions are dry. Not only is

rainfall limited in these areas, but water in the soil remains frozen and unavailable during the long winters.

Conifers are adapted to these dry, cold conditions in three ways. First, most conifers retain green leaves throughout the year, enabling these plants to continue photosynthesizing and growing slowly during times when most other plants become dormant. For this reason, conifers are often called evergreens. Second, conifer leaves are actually thin needles covered with a thick, waterproof surface that minimizes evaporation (**FIG. 21-8d**). Finally, conifers produce an "antifreeze" in their sap that enables them to continue transporting nutrients in below-freezing temperatures. This substance gives them their fragrant piney scent.

Conifer Seeds Develop in Cones

Reproduction is similar in all conifers, so let's examine the reproductive cycle of a pine tree (**FIG. 21-9**). The tree itself is the diploid sporophyte, and it develops both male and female cones. The male cones are relatively small

414

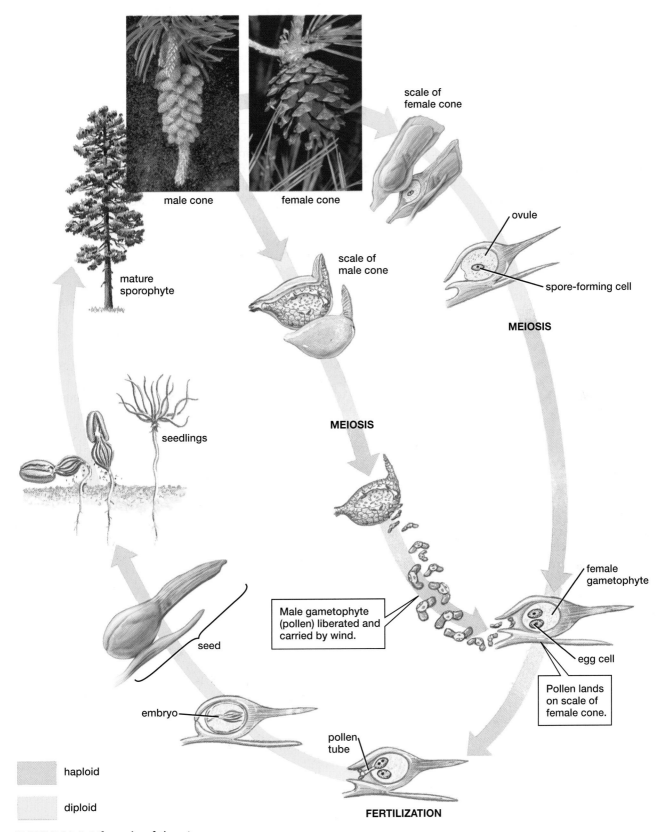

scale of
female cone

ovule

spore-forming cell

MEIOSIS

scale of
male cone

male cone

female cone

mature
sporophyte

MEIOSIS

seedlings

female
gametophyte

Male gametophyte
(pollen) liberated and
carried by wind.

egg cell

Pollen lands
on scale of
female cone.

seed

embryo

pollen
tube

haploid

diploid

FERTILIZATION

FIGURE 21-9 Life cycle of the pine
The pine tree is the sporophyte generation (upper left) and bears both male and female cones. Haploid female gametophytes develop within the scales of female cones and produce egg cells. Male cones produce pollen, the male gametophytes. A pollen grain, dispersed by the wind, may land on the scale of a female cone. It then grows a pollen tube that penetrates the female gametophyte and conducts sperm to the egg. The fertilized egg develops into an embryonic plant enclosed in a seed. The seed is eventually released from the cone, germinates, and grows into a sporophyte tree.

(normally $\frac{3}{4}$ of an inch—about 2 centimeters—or less), delicate structures that release clouds of pollen during the reproductive season and then disintegrate. These clouds of pollen are immense; inevitably, some pollen grains land by chance on a female cone.

Each female cone consists of a series of woody scales arranged in a spiral around a central axis. At the base of each scale are two **ovules** (immature seeds), within which diploid spore cells form and undergo meiosis to produce haploid female gametophytes. These gametophytes then develop and produce egg cells. If a pollen grain from a male cone lands nearby, it sends out a pollen tube that slowly burrows into the female gametophyte. After nearly 14 months, the tube finally reaches the egg cell and releases the sperm that fertilize it. The fertilized egg becomes enclosed in a seed as it develops into a tiny embryonic plant. The seed is liberated when the cone matures and its scales separate.

Angiosperms Are Flowering Seed Plants

Modern flowering plants, or **angiosperms**, have dominated Earth for more than 100 million years. The group is incredibly diverse, with more than 230,000 species. Angiosperms range in size from the diminutive duckweed (**FIG. 21-10a**) to the towering eucalyptus tree (**FIG. 21-10b**). From desert cactus to tropical orchids to grasses to parasitic stinking corpse lilies, angiosperms rule over the plant kingdom.

Flowers Attract Pollinators

Three major adaptations have contributed to the enormous success of angiosperms: flowers, fruits, and broad leaves. **Flowers**, the structures in which both male and female gametes are formed, may have evolved when gymnosperm ancestors formed an association with animals (most likely insects) that carried their pollen from plant to plant. According to this scenario, the relationship between these ancient gymnosperms and their animal pollinators was so beneficial that natural selection favored the evolution of showy flowers that advertised the presence of pollen to insects and other animals (**FIG. 21-10b, e**). The animals benefited by eating some of the protein-rich pollen, whereas the plant benefited from the animals' unwitting transportation of pollen from plant to plant. With this animal assistance, many flowering plants no longer needed to produce prodigious quantities of pollen and send it flying on the fickle winds to ensure fertilization. But there are also many wind-pollinated angiosperms (**FIG. 21-10c, d**).

In the angiosperm life cycle (**FIG. 21-11**), flowers develop on the dominant sporophyte plant. Male gametophytes (pollen) are formed inside a structure called the *anther*; the female gametophyte develops from an ovule within a part of the flower called the *ovary*. The egg, in turn, develops within the female gametophyte. Fertilization occurs when the pollen forms a tube through the *stigma*, a sticky pollen-catching structure of the flower, and bores into the ovule. There, the zygote develops into an embryo enclosed in a seed formed from the ovule.

Fruits Encourage Seed Dispersal

The ovary surrounding the seed of an angiosperm matures into a **fruit**, the second adaptation that has contributed to the success of angiosperms. Just as flowers encourage animals to transport pollen, so, too, many fruits entice animals to disperse seeds. If an animal eats a fruit, many of the enclosed seeds may pass through the animal's digestive tract unharmed, perhaps to fall at a suitable location for germination. Not all fruits, however, depend on edibility for dispersal. Dog owners are well aware, for example, that some fruits (called burs) disperse by clinging to animal fur. Other fruits, such as those of maples, form wings that carry the seed through the air. The variety of dispersal mechanisms made possible by fruits has helped the angiosperms invade nearly all possible terrestrial habitats.

Broad Leaves Capture More Sunlight

The third feature that gives angiosperms an advantage in warmer, wetter climates is broad leaves. When water is plentiful, as it is during the warm growing season of temperate and tropical climates, broad leaves provide an advantage by collecting more sunlight for photosynthesis. In regions with seasonal variation in growing conditions, many trees and shrubs drop their leaves during periods when water is in short supply, because being leafless reduces evaporative water loss. In temperate climates, such periods occur during the fall and winter, at which time most temperate angiosperm trees and shrubs drop their leaves. In the tropics and subtropics, most angiosperms are evergreen, but species that inhabit certain tropical climates where periods of drought are common may drop their leaves to conserve water during the dry season.

The advantages of broad leaves are offset by some evolutionary costs. In particular, broad, tender leaves are much more appealing to herbivores than are the tough, waxy needles of conifers. As a result, angiosperms have developed a range of defenses against mammalian and insect herbivores. These adaptations include physical defenses such as thorns, spines, and resins that toughen the leaves. But the evolutionary struggle for survival has also led to a host of chemical defenses—compounds that make plant tissue poisonous or distasteful to potential predators. Many of the compounds responsible for chemical defense have properties that humans have exploited for medicinal and culinary uses. Medicines such as aspirin and codeine, stimulants such as nicotine and caffeine, and spicy flavors such as mustard and peppermint are all derived from angiosperm plants.

More Recently Evolved Plants Have Smaller Gametophytes

The evolutionary history of plants has been marked by a tendency for the sporophyte generation to become increasingly prominent, and for the longevity and size of the gametophyte generation to shrink (see Table 21-1). Thus, the earliest plants are believed to have been similar to today's nonvascular plants, which have a sporophyte that is

FIGURE 21-10 Angiosperms
(a) The smallest angiosperm is the duckweed, found floating on ponds. These specimens are about $\frac{1}{8}$ inch (3 millimeters) in diameter. (b) The largest angiosperms are eucalyptus trees, which can reach 325 feet (100 meters) in height. Both (c) grasses and many trees, such as (d) this birch, in which flowers are shown as buds (green) and blossoms (brown), have inconspicuous flowers and rely on wind for pollination. More conspicuous flowers, such as those on (e) this butterfly weed and on a eucalyptus tree (b, **inset**), entice insects and other animals that carry pollen between individual plants. EXERCISE List the advantages and disadvantages of wind pollination. Do the same for pollination by animals. Why do both types of pollination persist among the angiosperms?

417

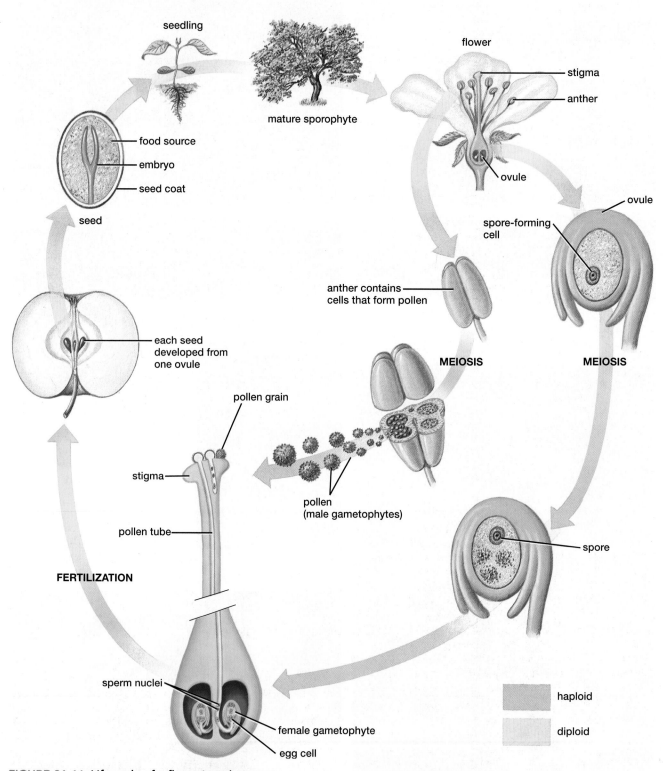

seedling

mature sporophyte

flower

stigma

anther

ovule

food source

embryo

seed coat

seed

spore-forming cell

ovule

anther contains cells that form pollen

MEIOSIS

MEIOSIS

each seed developed from one ovule

pollen grain

stigma

pollen tube

pollen (male gametophytes)

spore

FERTILIZATION

sperm nuclei

female gametophyte

egg cell

haploid

diploid

FIGURE 21-11 Life cycle of a flowering plant
The dominant plant body (upper right) is the diploid sporophyte, whose flowers normally produce both male and female gametophytes. Male gametophytes (pollen grains) are produced within anthers. The female gametophyte develops from a spore within the ovule, and contains one egg cell. A pollen grain that lands on a stigma grows a pollen tube that burrows down to the ovule and into the female gametophyte. There it releases its sperm, one of which fuses with the egg to form a zygote. The ovule gives rise to the seed, which contains the developing embryo and its food source. The seed is dispersed, germinates, and develops into a mature sporophyte.

smaller than the gametophyte and remains attached to it. In contrast, plants that originated somewhat later, such as ferns and the other seedless vascular plants, feature a life cycle in which the sporophyte is dominant, and the gametophyte is a much smaller, independent plant. Finally, in the most recently evolved group of plants, the seed plants, gametophytes are microscopic and barely recognizable as an alternate generation. These tiny gametophytes, however, still produce the eggs and sperm that unite to form the zygote that develops into the diploid sporophyte.

CASE STUDY REVISITED QUEEN OF THE PARASITES

 The approximately 17 parasitic plant species of the genus *Rafflesia*, which includes the stinking corpse lily, are found in the moist forests of Southeast Asia, a habitat that is disappearing rapidly as forests are cleared for agriculture and development. The geographic range of the stinking corpse lily is limited to the dwindling forests of the Malaysian peninsula and the Indonesian islands of Borneo and Sumatra; the species is rare and endangered. The government of Indonesia has established some parks and reserves that help protect the stinking corpse lily, but—as is often the case in developing countries—a forest that is protected on paper may still be vulnerable in reality. Perhaps the best hope for the continued survival of the largest *Rafflesia* is the growing realization among the rural residents of Sumatra and Borneo that the spectacular, putrid-smelling flowers of the stinking corpse lily might lure interested tourists to their countries. Under an innovative conservation program that seeks to take advantage of this potential for ecotourism, people who live in the vicinity of the stinking corpse lily can become caretakers of the plants. These assigned caretakers watch over the plants and, in return, may charge a small fee to curious visitors. Local inhabitants have been given an economic incentive to protect this rare parasitic plant.

Consider This A parasitic lifestyle is unusual among plants, but it is not exactly rare. Fifteen different plant families contain parasitic species, and systematists estimate that parasitism has evolved at least nine different times over the evolutionary history of plants. Given the obvious benefits of photosynthesis, why has parasitism (which is often accompanied by loss of photosynthetic capability) evolved repeatedly in photosynthetic plants?

CHAPTER REVIEW

SUMMARY OF KEY CONCEPTS

21.1 What Are the Key Features of Plants?
The kingdom Plantae consists of eukaryotic, photosynthetic, multicellular organisms. Unlike their green algae relatives, plants have multicellular, dependent embryos and exhibit alternation of generations in which a haploid gametophyte generation alternates with a diploid sporophyte generation. Plants play a key ecological role, capturing energy for use by inhabitants of terrestrial ecosystems, replenishing atmospheric oxygen, and creating and stabilizing soils.

21.2 What Is the Evolutionary Origin of Plants?
Photosynthetic protists, probably green algae, gave rise to the first plants. Ancestral plants were probably similar to modern multicellular green algae, which have photosynthetic pigments, starch molecules, and cell wall components—including cellulose—similar to those of plants. The freshwater heritage of green algae may have endowed them with qualities that enabled their descendants to invade land.

21.3 How Have Plants Adapted to Life on Land?
Plants also have a number of key adaptations for a terrestrial existence: rootlike structures for anchorage and for absorption of water and nutrients; a waxy cuticle to slow the loss of water through evaporation; stomata that can open, allowing gas exchange, and that can also close, preventing water loss; conducting vessels to transport water and nutrients throughout the plant; and a stiffening substance, called lignin, to impregnate the vessels and support the plant body.

Plant reproductive structures suitable for life on land include a reduced male gametophyte (pollen) that allows wind to replace water in carrying sperm to eggs; seeds that nourish, protect, and help disperse developing embryos; flowers that attract animals, which carry pollen more precisely and efficiently than wind; and fruits that entice animals to disperse seeds.

Web Tutorial 21.1 Adaptations in Plant Evolution

21.4 What Are the Major Groups of Plants?
Two major groups of plants, bryophytes and vascular plants, arose from the ancient algal ancestors. Bryophytes, including the liverworts and mosses, are small, simple land plants that lack conducting vessels. Although some have adapted to dry areas, most live in moist habitats. Bryophyte reproduction requires water through which the sperm swim to the egg.

In vascular plants, a system of vessels—stiffened by lignin—conducts water and nutrients absorbed by the roots into the upper portions of the plant and supports the body as well. Owing to this support system, seedless vascular plants, including the club mosses, horsetails, and ferns, can grow larger than bryophytes. As in bryophytes, the sperm of seedless vascular plants must swim to the egg for sexual reproduction to occur, and the gametophyte lacks conducting vessels.

Vascular plants with seeds have two major additional adaptive features: pollen and seeds. Seed plants are often classified into two categories: gymnosperms and angiosperms. Gymnosperms include ginkgos, cycads, gnetophytes, and the highly successful conifers. These plants were the first fully terrestrial plants to evolve. Their success on dry land is partially due to the evolution of the male gametophyte into the pollen grain. Pollen protects and transports the male gamete, eliminating the need for the sperm to swim to the egg. The seed, a protective resting structure containing an embryo and a supply of food, is a second important adaptation contributing to the success of seed plants.

Angiosperms, the flowering plants, dominate much of the land today. In addition to pollen and seeds, angiosperms also produce flowers and fruits. The flower allows angiosperms to use animals as pollinators. In contrast to wind, animals can in some cases carry pollen farther and with greater accuracy and less waste. Fruits may attract animal consumers, which incidentally disperse the seeds in their feces.

There has been a general evolutionary trend toward reduction of the haploid gametophyte, which is dominant in bryophytes but microscopic in seed plants.

Web Tutorial 21.2 Life Cycle of a Fern

KEY TERMS

alternation of generations *page 404*
angiosperm *page 416*
antheridium *page 408*
archegonium *page 408*
bryophyte *page 407*

conifer *page 414*
cuticle *page 406*
flower *page 416*
fruit *page 416*
gametophyte *page 404*
gymnosperm *page 410*

lignin *page 406*
ovule *page 416*
pollen *page 410*
seed *page 410*
sporophyte *page 404*
stomata *page 406*

vascular *page 407*
vessel *page 410*
zygote *page 404*

THINKING THROUGH THE CONCEPTS

1. What is meant by "alternation of generations"? What two generations are involved? How does each reproduce?

2. Explain the evolutionary changes in plant reproduction that adapted plants to increasingly dry environments.

3. Describe evolutionary trends in the life cycles of plants. Emphasize the relative sizes of the gametophyte and sporophyte.

4. From which algal group did green plants probably arise? Explain the evidence that supports this hypothesis.

5. List the structural adaptations necessary for the invasion of dry land by plants. Which of these adaptations are possessed by bryophytes? By ferns? By gymnosperms and angiosperms?

6. The number of species of flowering plants is greater than the number of species in the rest of the plant kingdom. What feature(s) are responsible for the enormous success of angiosperms? Explain why.

7. List the adaptations of gymnosperms that have helped them become the dominant trees in dry, cold climates.

8. What is a pollen grain? What role has it played in helping plants colonize dry land?

9. The majority of all plants are seed plants. What is the advantage of a seed? How do plants that lack seeds meet the needs served by seeds?

APPLYING THE CONCEPTS

1. You are a geneticist working for a firm that specializes in plant biotechnology. Explain what *specific* parts (fruit, seeds, stems, roots, etc.) of the following plants you would try to alter by genetic engineering, what changes you would try to make, and why: (a) corn, (b) tomatoes, (c) wheat, and (d) avocados.

2. Prior to the development of synthetic drugs, more than 80% of all medicines were of plant origin. Even today, indigenous tribes in remote Amazonian rain forests can provide a plant product to treat virtually any ailment. Herbal medicine is also widely and successfully practiced in China. Most of these drugs are unknown to the Western world. But the forests from which much of this plant material is obtained are being converted to agriculture. We are in danger of losing many of these potential drugs before they can be discovered. What steps can you suggest to preserve these natural resources while also allowing nations to direct their own economic development?

3. Only a few hundred of the hundreds of thousands of species in the plant kingdom have been domesticated for human use. One example is the almond. The domestic almond is nutritious and harmless, but its wild precursor can cause cyanide poisoning. The oak makes potentially nutritious seeds (acorns) that contain very bitter-tasting tannins. If we could breed the tannin out of acorns, they might become a delicacy. Why do you suppose we have failed to domesticate oaks?

FOR MORE INFORMATION

Diamond, J. "How to Tame a Wild Plant." *Discover*, September 1994. Cultivated plants have ecological and genetic properties that make them well suited for agriculture.

Joyce, C. *Earthly Goods: Medicine-Hunting in the Rainforest.* Boston: Little, Brown, 1994. Science and adventure combine in this account of prospecting for new medicines and the people who do it.

Kaufman, P. B. *Plants—Their Biology and Importance.* New York: Harper & Row, 1989. Complete, readable coverage of all aspects of plant taxonomy, physiology, and evolution.

McClintock, J. "The Life, Death, and Life of a Tree." *Discover*, May 2002. The author describes the biology of California's majestic redwood trees and the threat they face from humanity's appetite for their wood.

Milot, V. "Blueprint for Conserving Plant Diversity." *BioScience*, June 1989. Points out the importance of preserving genetic diversity in endangered plant species.

Pollan, M. *The Botany of Desire.* New York: Random House, 2001. A literate look at the mutually beneficial relationship between humans and plants.

Russell, S. A. *Anatomy of a Rose: Exploring the Secret Life of Flowers.* New York, 2001. An elegant, beautifully written exploration of the biology and influence on humans of flowers.

22

The Diversity of Fungi

These honey mushrooms are part of the visible
portion of the largest organism on Earth.

AT A GLANCE

CASE STUDY HUMONGOUS FUNGUS

WHAT IS THE LARGEST ORGANISM on Earth? A reasonable guess might be the world's largest animal, the blue whale, which can be 100 feet long and weigh 300,000 pounds. But the blue whale is dwarfed by the General Sherman tree, a giant sequoia specimen that is 275 feet high and whose weight is estimated at 6200 *tons*. Even these two behemoths, however, are pipsqueaks compared to the real record-holder, the fungus *Armillaria ostoyae*, also known as the honey mushroom. The largest known *Armillaria* is a specimen in Oregon that spreads over 2200 acres (about 3.4 square miles) and probably weighs even more than the General Sherman tree. Despite its huge size, no one has

actually seen the monster fungus, since it is largely underground. Its only aboveground parts are brown mushrooms that sprout occasionally from the creature's gigantic body. Just beneath the surface, however, the fungus spreads through the soil by means of long, string-like structures called rhizomorphs. These rhizomorphs extend until they encounter the tree roots on which *Armillaria* subsists, causing "root rot" that weakens or kills trees. This "root rot" provides aboveground evidence of *Armillaria*'s existence; the giant Oregon specimen was first identified by examining aerial photos to find forested areas with many dead trees.

How can researchers be sure that the Oregon fungus is truly one single individual

and not many intertwined individuals? The strongest evidence is genetic. Researchers gathered *Armillaria* tissue samples from throughout the area thought to be inhabited by a single individual and compared DNA extracted from the samples. All were genetically identical, demonstrating that they came from the same individual.

It may seem strange that the world's largest organisms went unnoticed until very recently, but the lives of fungi typically take place outside of our view. Nonetheless, fungi play a fascinating role in human affairs. Read on to find out more about the inconspicuous but often influential members of kingdom Fungi.

22.1 WHAT ARE THE KEY FEATURES OF FUNGI?

When you think of a fungus, you probably picture a mushroom. Most fungi, however, do not produce mushrooms. And even in fungus species that do produce mushrooms, the mushrooms are just temporary reproductive structures extending from a main body that is typically concealed beneath the soil or inside a piece of decaying wood. So, to fully appreciate the kingdom Fungi, we must take our cue from *mycologists*—the scientists who study fungi—and look beyond the conspicuous structures we encounter on the forest floor, at the edges of our lawns, and as a pizza topping. A closer look at the fungi reveals a group of mostly multicellular organisms that play a key role in the web of life and whose way of life differs in fascinating ways from that of plants or animals.

Fungal Bodies Consist of Slender Threads

The body of almost all fungi is a **mycelium** (FIG. 22-1a), which is an interwoven mass of one-cell-thick, threadlike filaments called **hyphae** (singular, hypha; FIG. 22-1b, c). Depending on the species, hyphae either consist of single elongated cells with numerous nuclei or are subdivided—by partitions called **septa** (singular, septum)—into many cells, each containing from one to many nuclei. Pores in the septa allow cytoplasm to stream between cells, distributing nutrients. Like plant cells, fungal cells are surrounded by cell walls. Unlike plant cells, however, fungal cell walls are strengthened by *chitin*, the same substance found in the exoskeletons of arthropods.

Fungi cannot move. They compensate for this lack of mobility with hyphae that can grow rapidly in any direction within a suitable environment. In this way, the fungal mycelium can quickly infuse itself into aging bread or cheese, beneath the bark of decaying logs, or into the soil. Periodically, the hyphae grow together and differentiate into reproductive structures that project above the surface beneath which the mycelium grows. These structures, including mushrooms, puffballs, and the powdery molds on unrefrigerated food, represent only a fraction of the complete fungal body, but are typically the only part of the fungus that we can easily see.

Fungi Obtain Their Nutrients from Other Organisms

Like animals, fungi survive by breaking down nutrients stored in the bodies or wastes of other organisms. Some fungi digest the bodies of dead organisms. Others are parasitic, feeding on living organisms and causing disease. Others live in close, mutually beneficial relationships with other organisms that provide food. There are even a few predatory fungi, which attack tiny worms in soil.

Unlike animals, fungi do not ingest food. Instead, they secrete enzymes that digest complex molecules outside their bodies, breaking down the molecules into smaller subunits that can be absorbed. Fungal filaments can penetrate deeply into a source of nutrients and are only one cell thick, presenting an enormous surface area through which to secrete enzymes and absorb nutrients. This mode of securing nutrition serves fungi well. Almost every biological material can be consumed by at least one fungal species, so nutritional support for fungi is likely to be present in nearly every terrestrial habitat.

Fungi Propagate by Spores

Unlike plants and animals, fungi do not form embryos. Instead, fungi reproduce by means of **spores**—tiny, lightweight reproductive packages that are extraordinarily mobile, even though most lack a means for self-propulsion. Spores are distributed far and wide as hitchhikers on the outside of an-

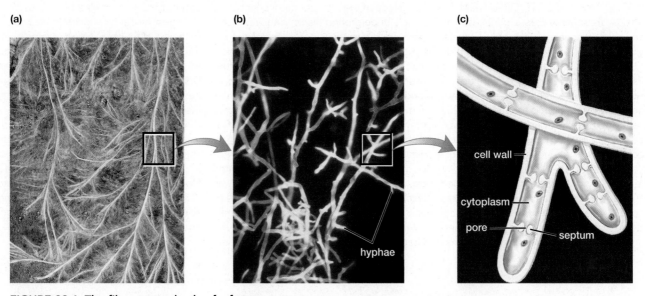

(a)   (b)   (c)

cell wall

cytoplasm

pore

septum

hyphae

FIGURE 22-1 The filamentous body of a fungus
(a) A fungal mycelium spreads over decaying vegetation. The mycelium is composed of (b) a tangle of microscopic hyphae, only one cell thick, portrayed in cross section (c) to show their internal organization. QUESTION Which features of a fungus's body structure are adaptations related to its method of acquiring nutrients?

imal bodies, as passengers inside the digestive systems of animals that have eaten them, or as airborne drifters, cast aloft by chance or shot into the atmosphere by elaborate reproductive structures (**FIG. 22-2**). Spores are often produced in great numbers (a single giant puffball may contain 5 trillion spores; see Fig. 22-9a). The fungal combination of prodigious reproductive capacity and highly mobile spores ensures that fungi are ubiquitous in terrestrial environments and accounts for the inevitable growth of fungi on every uneaten sandwich and container of leftovers.

Most Fungi Can Reproduce Both Sexually and Asexually

In general, fungi are capable of both asexual and sexual reproduction. For the most part, fungi reproduce asexually by default under stable conditions, with sexual reproduction occurring mainly under conditions of environmental change or stress. Asexual and sexual reproduction both ordinarily involve the production of spores within special fruiting bodies that project above the mycelium.

Asexual Reproduction Produces Haploid Spores by Mitosis

The bodies and spores of fungi are haploid (contain only a single copy of each chromosome). A haploid mycelium produces haploid asexual spores by mitosis. If an asexual spore is deposited in a favorable location, it will begin mitotic divisions and develop into a new mycelium. This sim-

FIGURE 22-2 Some fungi can eject spores
A ripe earthstar mushroom, struck by a drop of water, releases a cloud of spores that will be dispersed by air currents.

ple reproductive cycle results in the rapid production of genetically identical clones of the original mycelium.

Sexual Reproduction Produces Haploid Spores by Meiosis

Diploid structures form only during a brief period of the sexual portion of the fungal life cycle. Sexual reproduction begins when a filament of one mycelium comes into contact with a filament from a second mycelium that is of a different, but compatible, mating type (the different mating types of fungi are analogous to the different sexes of animals, except that there are often more than two mating types). If conditions are suitable, the two hyphae may fuse, so that nuclei from the two different hyphae share a common cell. This merger of hyphae is followed (immediately in some species, after some delay in others) by fusion of the two different haploid nuclei to form a diploid zygote. The zygote then undergoes meiosis to form haploid sexual spores. These spores are dispersed, germinate, and divide by mitosis to form new haploid mycelia. Unlike the cloned offspring of asexual spores, these sexually produced fungal bodies are genetically distinct from either parent.

22.2 WHAT ARE THE MAJOR GROUPS OF FUNGI?

Among the three kingdoms of multicellular eukaryotes, fungi and animals are more closely related to one another than either is to plants. That is, the common ancestor of fungi and animals lived more recently than did the common ancestor of plants, animals, and fungi (see Fig. 18-6). A person eating a salad of lettuce leaves topped by a sliced mushroom is more closely related to the mushroom than the mushroom is to the lettuce.

Fungi are very diverse. Nearly 100,000 species of fungi have been described, but this number represents only a fraction of the true diversity of these organisms. Many new species are discovered and described each year, and mycologists estimate that the number of undiscovered species of fungus is well over a million. Fungus species are classified into four phyla: Chytridiomycota (chytrids), Zygomycota (zygomycetes), Ascomycota (ascomycetes), and Basidiomycota (basidiomycetes) (**FIG. 22-3**, Table 22-1).

Chytrids Produce Swimming Spores

Unlike other types of fungi, most **chytrids** live in water. The chytrids (**FIG. 22-4**) are further distinguished from other fungi by their swimming spores, which require water for dispersal (even soil-dwelling chytrids require a film of water for reproduction). A chytrid spore propels itself through the water by means of a single flagellum located on one end of the spore. No other fungus group has flagella.

Research by fungal systematists suggests that the chytrids form an ancient group that predates and gave rise to the other groups of modern fungi. This conclusion is bolstered by the fossil record, which indicates that the oldest known fossil fungi are chytrids that were found in rocks

FIGURE 22-3 Evolutionary tree of the major groups of fungi

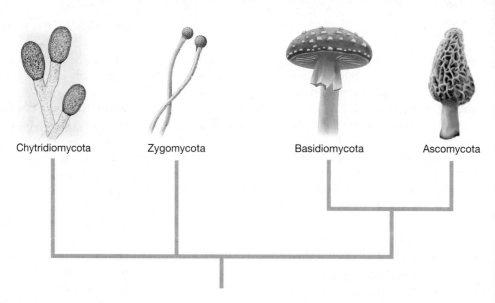

Chytridiomycota   Zygomycota   Basidiomycota   Ascomycota

more than 600 million years old. Ancestral fungi may well have been similar in habit to today's aquatic and marine chytrids, so fungi (like plants and animals) probably originated in a watery environment before colonizing land.

Most chytrid species feed on dead aquatic plants or other debris in watery environments, but some species are parasites of plants or animals. One such parasitic chytrid is believed to be a major cause of the current worldwide die-off of frogs, which threatens many species and has apparently already caused the extinction of several. No one yet understands exactly why this fungal disease emerged as a major cause of death in frogs. One hypothesis is that frog populations under stress from pollution and other environmental challenges might be more susceptible to infection by chytrids. (For more on the decline of frogs, see "Earth Watch: Frogs in Peril" in Chapter 24.)

Zygomycetes Can Reproduce by Forming Diploid Spores

The **zygomycetes** generally live in soil or on decaying plant or animal material. This group includes species belonging to the genus *Rhizopus,* which cause the familiar annoyances of soft fruit rot and black bread mold. The life cycle of the black bread mold, which reproduces both asexually and sexually, is depicted in **FIGURE 22-5**. Asexual reproduction in zygote fungi is initiated by the formation of haploid spores in black spore cases called **sporangia**. These spores disperse through the air and, if they land on a suitable substrate (such as a piece of bread), germinate to form new haploid hyphae.

If two hyphae of different mating types of zygote fungi come into contact, sexual reproduction may ensue. The

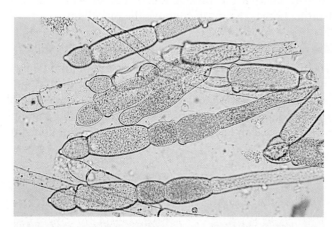

FIGURE 22-4 Chytrid filaments
These filaments of the chytrid fungus *Allomyces* are in the midst of sexual reproduction. The orange structures visible on many of the filaments will release male gametes; the clear structures will release female gametes. Chytrid gametes are flagellated, and these swimming reproductive structures aid dispersal of members of this mostly aquatic phylum.

Table 22-1 **The Phyla of Fungi**				
Common Name (Phylum)	**Reproductive Structures**	**Cellular Characteristics**	**Economic and Health Impacts**	**Representative Genera**
Chytrids (Chytridiomycota)	Produce haploid or diploid flagellated spores	Septa are absent	Contribute to decline of frog populations	*Batrachochytrium* (frog pathogen)
Zygomycetes (Zygomycota)	Produce diploid sexual zygospores	Septa are absent	Cause soft fruit rot and black bread mold	*Rhizopus* (causes black bread mold); *Pilobolus* (dung fungus)
Ascomycetes (Ascomycota)	Haploid sexual ascospores formed in saclike ascus	Septa are present	Cause molds on fruit; can damage textiles; cause Dutch elm disease and chestnut blight; include yeasts and morels	*Saccharomyces* (yeast); *Ophiostoma* (causes Dutch elm disease)
Basidiomycetes (Basidiomycota)	Sexual reproduction involves production of haploid basidiospores on club-shaped basidia	Septa are present	Cause smuts and rusts on crops; include some edible mushrooms	*Amanita* (poisonous mushroom); *Polyporus* (shelf fungus)

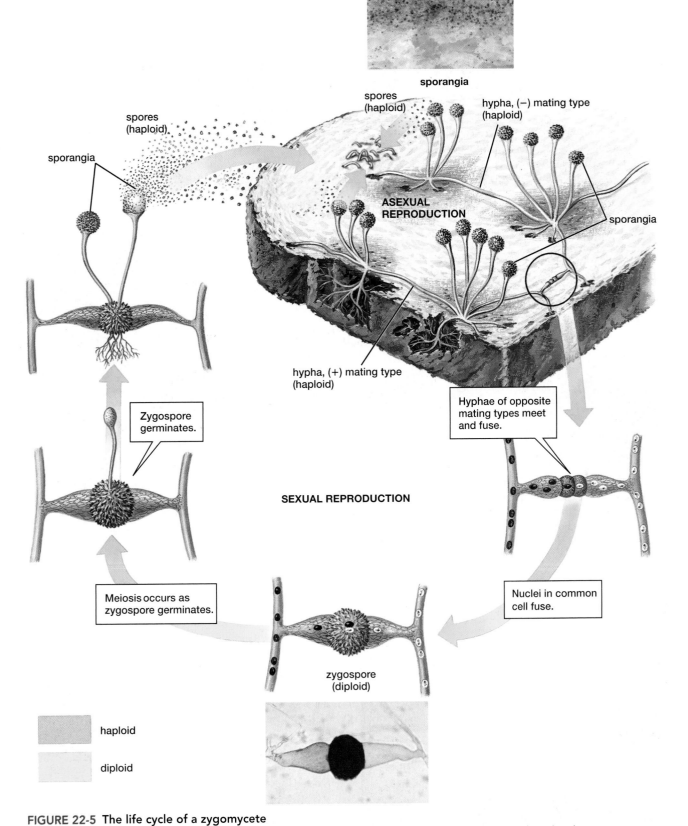

sporangia

spores (haploid)

spores (haploid)

hypha, (−) mating type (haploid)

sporangia

sporangia

ASEXUAL REPRODUCTION

hypha, (+) mating type (haploid)

Hyphae of opposite mating types meet and fuse.

Zygospore germinates.

SEXUAL REPRODUCTION

Nuclei in common cell fuse.

Meiosis occurs as zygospore germinates.

zygospore (diploid)

haploid

diploid

FIGURE 22-5 The life cycle of a zygomycete

Top: During asexual reproduction in the black bread mold (genus *Rhizopus*), haploid spores, produced within sporangia, disperse and germinate on food such as bread. Bottom: During sexual reproduction, hyphae of different mating types (designated + and − on the bread) contact one another and fuse, producing a diploid zygospore. The zygospore undergoes meiosis and germinates, producing sporangia that liberate haploid spores.

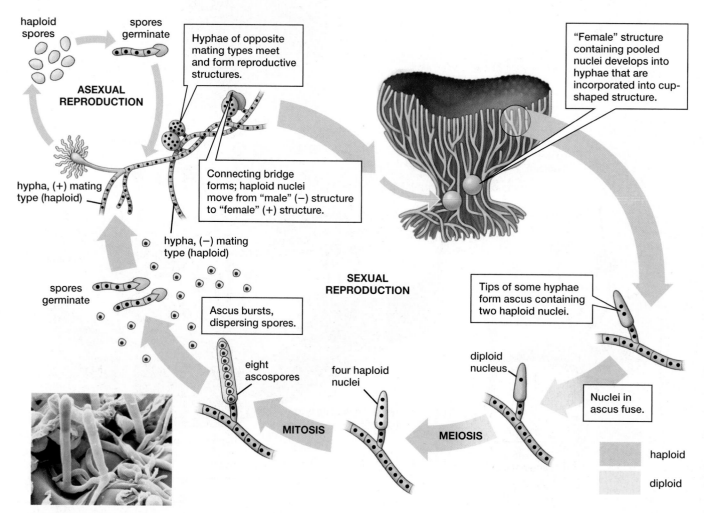

FIGURE 22-6 The life cycle of a typical ascomycete
Top: In ascomycete asexual reproduction, haploid hyphae give rise to stalked structures that produce haploid spores. Bottom: In sexual reproduction, haploid nuclei of different mating types fuse to form diploid zygotes that divide and give rise to haploid ascospores. The ascospores develop inside a structure called the the ascus; some asci rising from hyphae are shown in the photo.

(a)   (b)

FIGURE 22-7 Diverse ascomycetes
(a) The cup-shaped fruiting body of the scarlet cup fungus. (b) The morel, an edible delicacy. (Consult an expert before sampling any wild fungus—some are deadly!)

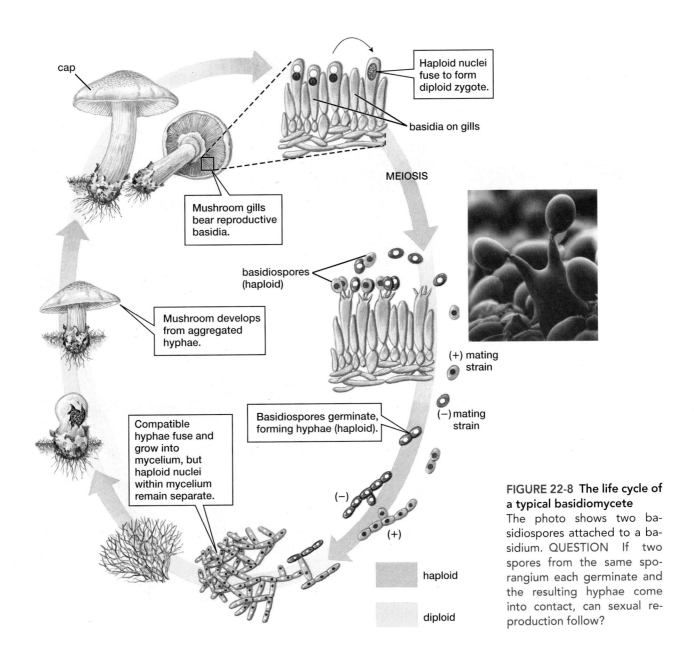

cap

Haploid nuclei fuse to form diploid zygote.

basidia on gills

MEIOSIS

Mushroom gills bear reproductive basidia.

basidiospores (haploid)

Mushroom develops from aggregated hyphae.

Basidiospores germinate, forming hyphae (haploid).

(+) mating strain

(−) mating strain

Compatible hyphae fuse and grow into mycelium, but haploid nuclei within mycelium remain separate.

(−)

(+)

haploid

diploid

FIGURE 22-8 The life cycle of a typical basidiomycete
The photo shows two basidiospores attached to a basidium. QUESTION If two spores from the same sporangium each germinate and the resulting hyphae come into contact, can sexual reproduction follow?

two hyphae "mate sexually," and their nuclei fuse to produce diploid **zygospores**: tough, resistant structures that give this group its name. Zygospores can remain dormant for long periods until environmental conditions are favorable for growth. Like asexually produced spores, zygospores disperse and germinate, but instead of producing new hyphae directly, they undergo meiosis. As a result, they form structures that bear haploid spores, which develop into new hyphae.

Ascomycetes Form Spores in a Saclike Case

The **ascomycetes**, or **sac fungi**, also reproduce both asexually and sexually (**FIG 22-6**). Asexual spores of sac fungi are produced at the tips of specialized hyphae. During sexual reproduction, spores are produced by a complex sequence of events that begins when hyphae of two different mating types fuse. This sequence culminates in the formation of **asci** (singular, ascus), saclike cases that contain several spores and that give this phylum its name.

Some ascomycetes live in decaying forest vegetation and form either beautiful cup-shaped reproductive structures (**FIG. 22-7a**) or corrugated, mushroom-like fruiting bodies called *morels* (**FIG. 22-7b**). This phylum also includes many of the colorful molds that attack stored food and destroy fruit and grain crops and other plants, as well as yeasts (some of the few unicellular fungi) and the species that produces penicillin, the first antibiotic.

Basidiomycetes Produce Club-Shaped Reproductive Structures

Basidiomycetes are called the **club fungi** because they produce club-shaped reproductive structures. Members of this phylum typically reproduce sexually (**FIG. 22-8**); hyphae of different mating types fuse to form filaments in which each cell contains two nuclei, one from each parent. The nuclei do not fuse until the formation of specialized, club-shaped diploid cells called **basidia** (singular, basidium). Basidia give rise to haploid reproductive **basidiospores** by meiosis.

(a)

(b)

FIGURE 22-9 Diverse basidiomycetes
(a) The giant puffball *Lycopedon giganteum* may produce up to 5 trillion spores. (b) Shelf fungi, some the size of dessert plates, are conspicuous on trees. (c) The spores of stinkhorns are carried on the outside of a slimy cap that smells terrible to humans, but appeals to flies. The flies lay their eggs on the stinkhorn, and inadvertently disperse the spores that stick to their bodies. QUESTION Are the structures shown in these photos haploid or diploid?

(c)

The formation of basidia and basidiospores takes place in special fruiting bodies, which are familiar to most of us as mushrooms, puffballs, shelf fungi, and stinkhorns (**FIG. 22-9**). These reproductive structures are actually dense aggregations of hyphae that emerge under proper conditions from a massive underground mycelium. On the undersides of mushrooms are leaflike gills on which basidia are produced. Basidiospores are released by the billions from the gills of mushrooms or through openings in the tops of puffballs and are dispersed by wind and water.

Falling on fertile ground, a mushroom basidiospore may germinate and form haploid hyphae. These hyphae grow outward from the original spore in a roughly circular pattern as the older hyphae in the center die. The subterranean body periodically sends up numerous mushrooms, which emerge in a ringlike pattern called a fairy ring (**FIG. 22-10**). The diameter of the fairy ring reveals the approximate age of the fungus—the wider the diameter, the older the underlying fungus. Some fairy rings are estimated to be 700 years old, and basidiomycete mycelia can be even older than that. For example, the researchers who discovered the gigantic *Armillaria* in Oregon estimate that it took at least 2400 years to grow to its current size.

22.3 HOW DO FUNGI INTERACT WITH OTHER SPECIES?

Many fungi live in direct contact with another species for a prolonged period. Such intimate, long-term relationships are known as *symbiosis*. In many cases, the fungal member of a symbiotic relationship is parasitic and harms its host. But some symbiotic relationships are mutually beneficial.

Lichens Are Formed by Fungi That Live with Photosynthetic Algae or Bacteria

Lichens are symbiotic associations between fungi and single-celled green algae or cyanobacteria (**FIG. 22-11**). Lichens are sometimes described as "fungi that have learned to garden," because the fungal member of the partnership "tends" the photosynthetic algal or bacterial partner by providing shelter and protection from harsh conditions. In this protected environment, the photosynthetic member of the partnership uses solar energy to manufacture simple sugars, producing food for itself but also some excess food that is consumed by the fungus. In fact, the fungus often consumes the lion's share of the photosynthetic product (up to 90% in some species), leading some researchers to conclude that the symbiotic relationship in lichens is really much more one-sided than it is

FIGURE 22-10 A mushroom fairy ring
Mushrooms emerge in a fairy ring from an underground fungal mycelium, growing outward from a central point where a single spore germinated, perhaps centuries ago.

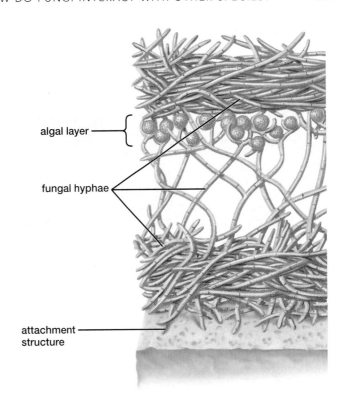

FIGURE 22-11 The lichen: A symbiotic partnership
Most lichens have a layered structure bounded on the top and bottom by an outer layer formed from fungal hyphae. The fungal hyphae emerge from the lower layer, forming attachments that anchor the lichen to a surface, such as a rock or a tree. An algal layer in which the alga and fungus grow in close association lies beneath the upper layer of hyphae.

usually portrayed. This view was bolstered by the discovery that, in lichens that include algal symbionts, fungal hyphae actually penetrate the cell walls of the algae, as do the hyphae of fungi that parasitize plants.

Thousands of different fungal species (mostly ascomycetes) form lichens (**FIG. 22-12**), combining with one of a much smaller number of algal or bacterial species. Together, these organisms form a unit so tough and self-sufficient that lichens are among the first living things to colonize newly formed volcanic islands. Brightly colored lichens also invade other inhospitable habitats ranging from deserts to the Arctic, and they can even grow on bare rock. Understandably, lichens in extreme environments grow very slowly; arctic colonies, for example, expand as

(a)

(b)

FIGURE 22-12 Diverse lichens
(a) A colorful encrusting lichen, growing on dry rock, illustrates the tough independence of this symbiotic combination of fungus and algae. (b) A leafy lichen grows from a dead tree branch.

FIGURE 22-13 Mycorrhizae enhance plant growth
Hyphae of mycorrhizae entwining about the root of an aspen tree. Plants grow significantly better in a symbiotic association with these fungi, which help make nutrients and water available to the roots.

slowly as 1 to 2 inches per 1000 years. Despite their slow growth, lichens can persist for long periods of time; some arctic lichens are more than 4000 years old.

Mycorrhizae Are Fungi Associated with Plant Roots

Mycorrhizae (singular, mycorrhiza) are important symbiotic associations between fungi and plant roots. More than 5000 species of mycorrhizal fungi (including representatives of all the major groups of fungi) can grow in intimate association with about 80% of plants that have roots, including most trees. The hyphae of mycorrhizal fungi surround the plant root and invade the root cells (**FIG. 22-13**).

Mycorrhizae Help Feed Plants

The association between plants and mycorrhizae benefits both the fungi and their plant partners. The mycorrhizal fungi receive energy-rich sugar molecules that are produced photosynthetically by plants and passed from their roots to the fungi. In return, the fungi digest and absorb minerals and organic nutrients from the soil, passing some of them directly into the root cells. Experiments have shown that phosphorus and nitrogen, key nutrients that are crucial for plant growth, are among the molecules that mycorrhizae move from soil to roots. Mycorrhizal fungi also absorb water and pass it to the plant—an advantage for plants in dry, sandy soils.

The partnership between mycorrhizae and plants makes a crucial contribution to the health of Earth's plants. Plants that are deprived of mycorrhizal fungi tend to be smaller and less vigorous than are plants with mycorrhizal partners. Thus, the presence of mycorrhizae increases the overall productivity of Earth's plant communities and thereby increases their ability to support the animals and other organisms that depend on them.

Mycorrhizae May Have Helped Plants Invade Land

Some scientists believe that mycorrhizal associations may have been important in the invasion of land by plants more

than 400 million years ago. Such a relationship between an aquatic fungus and a green alga (ancestral to terrestrial plants) could have helped the alga acquire the water and mineral nutrients it needed to survive out of water.

The fossil record is consistent with the hypothesis that mycorrhizae played a role in plants' colonization of land. The oldest fossils of terrestrial fungi are about 460 million years old, roughly the same age as the oldest fossils of terrestrial plants. This finding suggests that fungi and plants came ashore at the same time, perhaps together. In addition, plant fossils formed not long after the invasion of land exhibit distinctive root structures similar to those that form today in response to the presence of mycorrhizae. These fossils show that fully developed mycorrhizae were present very early in the evolution of land plants, suggesting that a simpler plant-fungus association might have been present even earlier.

Endophytes Are Fungi That Live Inside Plant Stems and Leaves

The intimate association between fungi and plants is not limited to root mycorrhizae. Fungi have also been found living inside the aboveground tissues of virtually every plant species that has been tested for their presence. Some of these *endophytes* (organisms that live inside other organisms) are parasites that cause plant diseases, but many, perhaps most, are beneficial to the host plant. The best-studied examples of beneficial fungal endophytes are the ascomycete species that live inside the leaf cells of many species of grass. These fungi produce substances that are distasteful or toxic to insects and grazing mammals and thus help protect the grass plants from those predators.

The antipredator protection provided by the fungal endophytes is sufficiently effective that agricultural scientists are working hard to discover a way to grow grasses free of the endophytes. Horses, cows, and other agriculturally important grazers tend to avoid eating grasses that contain the endophytes. When the only available food is endophyte-containing grass, animals that eat it experience poor health and slow growth.

Some Fungi Are Important Recyclers

As mycorrhizae and endophytes, some fungi play a major role in the growth and preservation of plant tissue. Other fungi, however, play a similarly major role in its destruction. Alone among organisms, fungi can digest both lignin and cellulose, the molecules that make up wood. When a tree or other woody plant dies, only fungi are capable of decomposing its remains.

Fungi are Earth's undertakers, consuming not only dead wood but the dead of all kingdoms. The fungi that are *saprophytes* (feeding on dead organisms) return the dead tissues' component substances to the ecosystems from which they came. The extracellular digestive activities of saprophytic fungi liberate nutrients that can be used by plants. If fungi and bacteria were suddenly to disappear, the consequences would be disastrous. Nutrients would remain

FIGURE 22-14 Corn smut
This basidiomycete pathogen destroys millions of dollars' worth of corn each year. Even a pest like corn smut has its admirers, though. In Mexico this fungus is known as *huitlacoche* and is considered to be a great delicacy.

locked in the bodies of dead plants and animals, the recycling of nutrients would grind to a halt, soil fertility would rapidly decline, and waste and organic debris would accumulate. In short, ecosystems would collapse.

22.4 HOW DO FUNGI AFFECT HUMANS?

The average person gives little thought to fungi, except perhaps for an occasional, momentary appreciation for the mushrooms on a pizza. Nonetheless, fungi affect our lives in more ways than you might imagine.

Fungi Attack Plants That Are Important to People

Fungi cause the majority of plant diseases, and some of the plants that they infect are important to humans. For example, fungal pathogens have a devastating effect on the world's food supply. Especially damaging are the basidiomycete plant pests descriptively called *rusts* and *smuts*, which cause billions of dollars' worth of damage to grain crops annually (**FIG. 22-14**). Fungal diseases also affect the appearance of our landscape. The American elm and the American chestnut, two tree species that were once prominent in many of America's parks, yards, and forests, were destroyed on a massive scale by the ascomycetes that cause Dutch elm disease and chestnut blight. Today, few people can recall the graceful forms of large elms and chestnuts, which are now almost entirely absent from the landscape.

Fungi continue to attack plant tissues long after they have been harvested for human use. To the dismay of homeowners, a host of different fungal species attack wood, causing it to rot. Some ascomycete molds secrete the enzymes cellulase and protease, which can cause significant damage to cotton and wool textiles, especially in warm, humid climates where molds flourish.

The fungal impact on agriculture and forestry is not entirely negative, however. Fungal parasites that attack insects and other arthropod pests can be an important ally in pest control (**FIG. 22-15a**). Farmers who wish to reduce their dependence on toxic and expensive chemical pesticides are increasingly turning to biological methods of pest control, including the application of "fungal pesticides." Fungal pathogens are currently used to control termites, rice weevils, tent caterpillars, aphids, citrus mites, and other pests. In

(a)

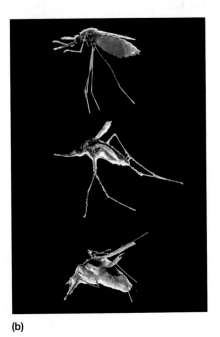

(b)

FIGURE 22-15 Helpful fungal parasites
Fungal pathogens can be helpful to humans. For example, fungi such as **(a)** the *Cordyceps* species that has killed this grasshopper are used by farmers to control insect pests. **(b)** Some fungi might one day be used to help protect humans from disease. A malaria-carrying mosquito infected by *Beauveria* is transformed from a healthy animal (top) to a fungus-encrusted corpse in less than two weeks.

addition, biologists have discovered that certain fungi attack and kill the mosquito species that transmit malaria (**FIG. 22-15b**). Plans are under way to enlist these fungi in the fight against one of the world's deadliest diseases.

Fungi Cause Human Diseases

The kingdom Fungi includes parasitic species that attack humans directly. Some of the most familiar fungal diseases are those caused by ascomycetes that attack the skin, resulting in athlete's foot, jock itch, and ringworm. These diseases, though unpleasant, are not life threatening and can usually be treated with antifungal ointments. Prompt treatment can also usually control another common fungal disease, vaginal infections caused by the yeast *Candida albicans* (**FIG. 22-16**). Fungi can also infect the lungs if victims inhale spores of disease-causing fungal species such as those that cause valley fever and histoplasmosis. Like other fungal infections, these diseases can, if promptly diagnosed, be controlled with antifungal drugs. If untreated, however, they can develop into serious, systemic infections. Singer Bob Dylan, for instance, became gravely ill with histoplasmosis when a fungus infected the pericardial membrane surrounding his heart.

Fungi Can Produce Toxins

In addition to their role as agents of infectious disease, some fungi produce toxins that are dangerous to humans. Of particular concern are toxins produced by fungi that grow on grains and other foodstuffs that have been stored in too-moist conditions. For example, molds of the genus *Aspergillus* produce highly toxic, carcinogenic compounds known as aflatoxins. Some foods, such as peanuts, seem especially susceptible to attack by *Aspergillus*. Since aflatoxins were discovered in the 1960s, food growers and processors have developed methods for reducing the growth of *Aspergillus* in stored crops, so aflatoxins have been largely eliminated from the nation's peanut butter supply.

One infamous toxin-producing fungus is the ascomycete *Claviceps purpurea*, which infects rye plants and causes a disease known as ergot. This fungus produces several toxins, which can affect humans if infected rye is ground into flour and consumed. This happened frequently in northern Europe in the Middle Ages, with devastating effects. At that time, ergot poisoning was typically fatal, and victims experienced terrible symptoms before dying. One ergot toxin is a vasoconstrictor, which constricts blood vessels and reduces blood flow. The effect can be so extreme that gangrene develops and limbs actually shrivel and fall off. Other ergot toxins cause symptoms that include a burning sensation, vomiting, convulsive twitching, and vivid hallucinations. Today, new agricultural techniques have effectively eliminated ergot poisoning, but the hallucinogenic drug LSD, which is derived from a component of the ergot toxins, still remains as a legacy of this disease.

Many Antibiotics Are Derived from Fungi

Fungi have also had positive impacts on human health. The modern era of lifesaving antibiotic medicines was ushered in by the discovery of penicillin, which is produced by an ascomycete mold (**FIG. 22-17**; see Fig. 1-5). Penicillin is still used, along with other fungi-derived antibiotics such as oleandomycin and cephalosporin, to combat bacterial diseases. Other important drugs are also derived from fungi, including cyclosporin, which is used to suppress the immune response during organ transplants so that the body is less likely to reject the transplanted organs.

FIGURE 22-17 Penicillium
Penicillium growing on an orange. Reproductive structures, which coat the fruit's surface, are visible, while hyphae beneath draw nourishment from inside. The antibiotic penicillin was first isolated from this fungus. QUESTION Why do some fungi produce antibiotic chemicals?

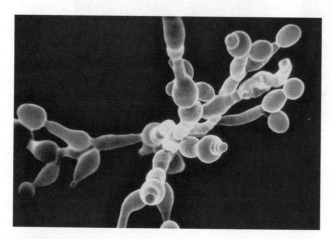

FIGURE 22-16 The unusual yeast
Yeasts are unusual, normally nonfilamentous ascomycetes that reproduce most commonly by budding. The yeast shown here is *Candida*, a common cause of vaginal infections.

EARTH WATCH   The Case of the Disappearing Mushrooms

Mycologists (scientists who study fungi) and gourmet cooks may seem to have little in common, but recently, they have been united by a common concern: mushrooms are rapidly declining in numbers, average size, and species diversity. Although the problem is most easily recognized in Europe, where people have been gathering wild mushrooms for centuries, American mycologists are also alarmed; the same decline may be occurring here as well. Why are mushrooms disappearing? Overhunting of edible mushrooms is not the culprit, because poisonous forms are equally affected. The loss is evident in all types of mature forests, so changing forest management practices could not be the cause. The most likely cause is air pollution, because the loss of mushrooms is greatest where the air contains the highest levels of ozone, sulfur, and nitrogen.

Although mycologists have not yet determined exactly how air pollution harms mushrooms, the evidence is clear. In Holland, for example, the average number of fungal species per 1000 square meters has dropped from 37 down to 12 over the past several decades. Twenty out of 60 fungal species surveyed in England are declining. Concern is intensified by the fact that the mushrooms most affected are those whose hyphae form mycorrhizal associations with tree roots. Trees with diminished mycorrhizae may have less resistance to periodic droughts or extreme cold spells. Because air pollution is also harming forests directly, the additional loss of mycorrhizae could be devastating.

Fungi Make Important Contributions to Gastronomy

Fungi make important contributions to human nutrition. The most obvious components of this contribution are the fungi that we consume directly: wild and cultivated basidiomycete mushrooms and ascomycetes, such as morels and the rare and prized truffle (see "Evolutionary Connections: Fungal Ingenuity—Pigs, Shotguns, and Nooses"). The role of fungi in cuisine, however, also has less-visible manifestations. For example, some of the world's most famous cheeses, including Roquefort, Camembert, Stilton, and Gorgonzola, gain their distinctive flavors from ascomycete molds that grow on them as they ripen. Perhaps the most important and pervasive fungal contributors to our food supply, however, are the single-celled ascomycetes (a few species are basidiomycetes) known as *yeasts*.

Wine and Beer Are Made Using Yeasts

The discovery that yeasts could be harnessed to enliven our culinary experience is surely a key event in human history. Among the many foods and beverages that depend on yeasts for their production are bread, wine, and beer, which are consumed so widely that it is difficult to imagine a world without them. All derive their special qualities from fermentation by yeasts. Fermentation occurs when yeasts extract energy from sugar and, as by-products of the metabolic process, emit carbon dioxide and ethyl alcohol.

As yeasts consume the fruit sugars in grape juice, the sugars are replaced by alcohol, and wine is the result. Eventually, the increasing concentration of alcohol kills the yeasts, ending fermentation. If the yeasts die before all available grape sugar is consumed, the wine will be sweet; if the sugar is exhausted, the wine will be dry.

Beer is brewed from grain (usually barley), but yeasts cannot effectively consume the carbohydrates that compose grain kernels. For the yeasts to do their work, the barley grains must have sprouted (recall that grains are actually seeds). Germination converts the kernels' carbohydrates to sugar, so the sprouted barley provides an excellent food source for the yeasts. As with wine, fermentation converts sugars to alcohol, but beer brewers capture the carbon dioxide by-product as well, giving the beer its characteristic bubbly carbonation.

Yeasts Make Bread Rise

In bread making, carbon dioxide is the crucial fermentation product. The yeasts added to bread dough do produce alcohol as well as carbon dioxide, but the alcohol evaporates during baking. In contrast, the carbon dioxide is trapped in the dough, where it forms the bubbles that give bread its light, airy texture (and saves us from a life of eating sandwiches made of crackers). So the next time you're enjoying a slice of French bread with Camembert cheese and a nice glass of chardonnay, or a slice of pizza and a cold bottle of your favorite brew, you might want to quietly give thanks to the yeasts. Our diets would certainly be a lot duller without the help we get from fungal partners.

EVOLUTIONARY CONNECTIONS

Fungal Ingenuity—Pigs, Shotguns, and Nooses

Natural selection, operating over millions of years on the diverse forms of fungi, has produced some remarkable adaptations by which fungi disperse their spores and obtain nutrients.

The Rare, Delicious Truffle

Although many fungi are prized as food, none are as avidly sought as the truffle (FIG. 22-18). The finest Italian truffles may sell for as much as $1500 per pound, and especially large specimens can fetch even higher prices. A 2.2-pound white Italian truffle recently sold at auction for $53,800.

Truffles are the spore-containing structures of an ascomycete that forms a mycorrhizal association with the roots of oak trees. Although it develops underground, the

LINKS TO LIFE   Collect Carefully

In the early 1980s, doctors at a California hospital noticed a curious trend. Over a period of a few months, the number of patients admitted for treatment of poisoning had risen dramatically, and many of those admitted had died. What caused this sudden outbreak of poisoning? Further investigation revealed that, in almost all cases, the victims were recent immigrants from Laos or Cambodia. Struggling to adjust to their new country, they had been thrilled to find that California's forests contained mushrooms that looked just like the ones they had collected for food back in Asia. Unfortunately, the similarity was superficial; the mushrooms were, in fact, poisonous species. The immigrants' nostalgic pursuit of "comfort food" had tragic consequences.

In general, immigrants from countries where mushroom collecting is common have proved to be especially susceptible to poisoning by toxic mushrooms. But they are not the only victims. Each year, a number of small children, inexperienced collectors, and unlucky guests at gourmet dinners make unexpected trips to the hospital after eating poisonous wild mushrooms.

It can be fun and rewarding to collect wild mushrooms, which offer some of the richest and most complex flavors a human can experience. But if you decide to go collecting, be careful, because some of the deadliest poisons known to humankind are found in mushrooms. Especially noted for their poisons are certain species in the genus *Amanita*, which have evocative common names such as death cap and destroying angel (**FIG. E22-1**). These names are apt, because even a single bite of one of these mushrooms can

be lethal. Damage from *Amanita* toxins is most severe in the liver, where the toxins tend to accumulate. Often, a victim of *Amanita* poisoning can be saved only by undergoing a liver transplant. So be sure to protect your health by inviting an expert to join your mushroom-hunting expeditions.

FIGURE E22-1 The destroying angel
Mushrooms produced by the basidiomycete *Amanita virosa* can be deadly.

FIGURE 22-18 The truffle
Truffles, rare ascomycetes (each about the size of a small apple), are a gastronomic delicacy.

truffle has evolved an effective mechanism to entice animals to dig it up. As its spores ripen, the truffle releases an odor that animals, especially pigs, find appealing. When a pig follows the smell to the truffle, digs it up, and devours it, millions of spores are scattered to the winds. Traditional-

ly, truffle collectors used muzzled pigs to hunt their quarry; a good truffle pig can smell an underground truffle from 150 feet (about 50 meters) away. Today, dogs are the most common assistants to truffle hunters.

The Shotgun Approach to Spore Dispersal

If you get close enough to a pile of horse manure to scrutinize it, you may observe the beautiful, delicate reproductive structures of the zygomycete *Pilobolus* (**FIG. 22-19**). Despite their daintiness, these are actually fungal shotguns. The clear bulbs, capped with sticky black spore cases, extend from hyphae that penetrate the dung. As the bulbs mature, the sugar concentration inside them increases and water is drawn in by osmosis. Meanwhile, the bulb begins to weaken just below its cap. Suddenly, like an overinflated balloon, the bulb bursts and blows its spore-carrying top up to 3 feet (1 meter) away.

The airborne spores may well land in some leaves of grass; *Pilobolus* bends toward the light as it grows, thereby increasing the chances that its spores will be directed toward open pasture. Spores that adhere to grass remain there until consumed by a grazing herbivore, perhaps a horse. Later (likely some distance away), the horse will deposit a fresh

FIGURE 22-19 An explosive zygomycete
The delicate, translucent reproductive structures of the zygomycete *Pilobolus* literally blow their tops when ripe, dispersing the black caps with their payload of spores.

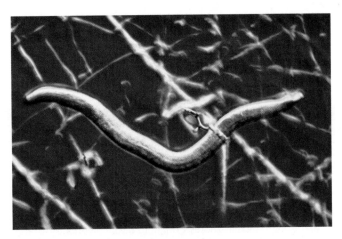

FIGURE 22-20 Nemesis of nematodes
Arthrobotrys, the nematode (roundworm) strangler, traps its prey in a noose-like modified hypha that swells when the inside of the loop is contacted.

pile of manure containing *Pilobolus* spores that have passed unharmed through its digestive tract. The spores germinate and growing hyphae penetrate the manure (a rich source of nutrients), ultimately sending up new projectiles to continue this ingenious cycle.

The Nematode Nemesis

Microscopic *nematodes* (roundworms) abound in rich soil, and fungi have evolved several fascinating forms of nematode-nabbing hyphae that allow them to exploit this rich source of protein. Some fungi produce sticky pods that ad-here to passing nematodes and penetrate the worm body with hyphae, which then begin digesting the worm from within. One species shoots a microscopic, harpoon-like spore into passing nematodes; the spore develops into a new fungus inside the worm. The fungal strangler *Arthrobotrys* produces nooses formed from three hyphal cells. When a nematode wanders into the noose, its contact with the inner parts of the noose stimulates the noose cells to swell with water (**FIG. 22-20**). In a fraction of a second, the hole constricts, trapping the worm. Fungal hyphae then penetrate and feast on their prey.

CASE STUDY REVISITED HUMONGOUS FUNGUS

Why do *Armillaria* fungi grow so large? Their size is due in part to their ability to form rhizomorphs, which consist of hyphae bundled together inside a protective rind. The enclosed hyphae carry nutrients to the rhizomorphs, allowing them to extend long distances through nutrient-poor areas to reach new sources of food. The *Armillaria* fungus can thus grow beyond the boundaries of a particular food-rich area.

Another factor that may contribute to the gigantic size of the Oregon *Armillaria* is the climate in which it was found. In this dry region, fungal fruiting bodies form only rarely, so the colossal *Armillaria* rarely produces spores. In the absence of spores that might grow into new individuals, the existing individual faces little competition for resources, and is free to grow and fill an increasingly large area.

The discovery of the Oregon specimen is merely the latest chapter in a long-running, good-natured "fungus war" that began in 1992 with the discovery of the first humongous fungus, a 37-acre *Armillaria gallica* growing in Michigan. Since that initial landmark discovery, research groups in Michigan, Washington, and Oregon have engaged in a friendly competition to find the largest fungus. Will the current record holder ever be topped? Stay tuned.

Consider This Because the entire Oregon *Armillaria* grew from a single spore, all of its cells are genetically identical. Its parts, however, are not all physiologically dependent on one another, so it is unlikely that any substances are transported through the entire 3.4-square-mile mycelium. And there is no continuous skin or bark or membrane that covers the entire mycelium and separates it from the environment as a unit. Is the fungus's genetic unity sufficient evidence for it to be considered a single individual, or is greater physiological integration required? Do you think that the claim of "world's largest organism" is valid?

CHAPTER REVIEW

SUMMARY OF KEY CONCEPTS

22.1   What Are the Key Features of Fungi?

Fungal bodies generally consist of filamentous hyphae, which are either multicellular or multinucleated and form large, intertwined networks called *mycelia*. Fungal nuclei are generally haploid. A cell wall of chitin surrounds fungal cells.

All fungi are heterotrophic, secreting digestive enzymes outside their bodies and absorbing the liberated nutrients.

Fungal reproduction is varied and complex. Asexual reproduction can occur either through fragmentation of the mycelium or through asexual spore formation. Sexual spores form after compatible haploid nuclei fuse to form a diploid zygote, which undergoes meiosis to form haploid sexual spores. Both asexual and sexual spores produce haploid mycelia through mitosis.

Web Tutorial 22.1 The Structure and Reproduction of Fungi

22.2   What Are the Major Groups of Fungi?

The major phyla of fungi and their characteristics are summarized in Table 22-1.

Web Tutorial 22.2 Classification of Fungi

22.3   How Do Fungi Interact with Other Species?

A lichen is a symbiotic association between a fungus and algae or cyanobacteria. This self-sufficient combination can colonize bare rock. Mycorrhizae are associations between fungi and the roots of most vascular plants. The fungus derives photosynthetic nutrients from the plant roots and, in return, carries water and nutrients into the root from the surrounding soil. Endophytes are fungi that grow inside the leaves or stems of plants and that may help protect the plants that harbor them. Saprophytic fungi are extremely important decomposers in ecosystems. Their filamentous bodies penetrate rich soil and decaying organic material, liberating nutrients through extracellular digestion.

22.4   How Do Fungi Affect Humans?

The majority of plant diseases are caused by parasitic fungi. Some parasitic fungi can help control insect crop pests. Others can cause human diseases, including ringworm, athlete's foot, and common vaginal infections. Some fungi produce toxins that can harm humans. Nonetheless, fungi add variety to the human food supply, and fermentation by fungi helps make wine, beer, and bread.

KEY TERMS

asci   *page 429*
ascomycete   *page 429*
basidia   *page 429*
basidiomycete   *page 429*
basidiospore   *page 429*

chytrid   *page 425*
club fungus   *page 429*
hyphae   *page 424*
lichen   *page 430*

mycelium   *page 424*
mycorrhizae   *page 432*
sac fungus   *page 429*
septa   *page 424*

sporangium   *page 426*
spore   *page 424*
zygospore   *page 429*
zygomycete   *page 426*

THINKING THROUGH THE CONCEPTS

1. Describe the structure of the fungal body. How do fungal cells differ from most plant and animal cells?

2. What portion of the fungal body is represented by mushrooms, puffballs, and similar structures? Why are these structures elevated above the ground?

3. What two plant diseases, caused by parasitic fungi, have had an enormous impact on forests in the United States? In which division are these fungi found?

4. List some fungi that attack crops. To which phyla do they belong?

5. Describe asexual reproduction in fungi.

6. What is the major structural ingredient in fungal cell walls?

7. List the phyla of fungi, describe the feature that gives each its name, and give one example of each.

8. Describe how a fairy ring of mushrooms is produced. Why is the diameter related to its age?

9. Describe two symbiotic associations between fungi and organisms from other kingdoms. In each case, explain how each partner in these associations is affected.

APPLYING THE CONCEPTS

1. Dutch elm disease in the United States is caused by an *exotic*—that is, an organism (in this case, a fungus) introduced from another part of the world. What damage has this introduction done? What other fungal pests fall into this category? Why are parasitic fungi particularly likely to be transported out of their natural habitat? What can governments do to limit this importation?

2. The discovery of penicillin revolutionized the treatment of bacterial diseases. However, penicillin is now rarely prescribed. Why is this? *Hint:* Refer to Chapter 15.

3. The discovery of penicillin was the result of a chance observation by an observant microbiologist, Alexander Fleming. How would you search systematically for new antibiotics produced by fungi? Where would you look for these fungi?

4. Fossil evidence indicates that mycorrhizal associations between fungi and plant roots existed in the late Paleozoic era, when the invasion of land by plants began. This evidence suggests an important link between mycorrhizae and the successful invasion of land by plants. Why might mycorrhizae have been important fungi in the colonization of terrestrial habitats by plants?

5. General biology texts in the 1960s included fungi in the plant kingdom. Why do biologists no longer consider fungi as legitimate members of the plant kingdom?

6. What ecological consequences would occur if humans, using a new and deadly fungicide, destroyed all fungi on Earth?

FOR MORE INFORMATION

Barron, G. "Jekyll-Hyde Mushrooms." *Natural History*, March 1992. Fungi include a variety of forms; some absorb decaying plant material, and others prey on microscopic worms.

Dix, N. J., and Webster, J. *Fungal Ecology*. London: Chapman & Hall, 1995. A rather technical but comprehensive and readable account of fungal diversity that focuses on the roles of fungi in different ecological communities.

Hudler, G. W. *Magical Mushrooms, Mischievous Molds*. Princeton, NJ: Princeton University Press, 1998. Engaging treatment of the fungi, focusing on their importance in human affairs.

Kiester, E. "Prophets of Gloom." *Discover*, November 1991. Lichens can be used as indicators of air quality and provide evidence of deteriorating environmental conditions.

Money, N. P. *Carpet Monsters and Killer Spores: A Natural History of Toxic Mold*. New York: Oxford University Press, 2004. An engaging account of the biology, toxicology, and legal status of some of the fungi that feed on human habitations.

Schaechter, E. *In the Company of Mushrooms*. Cambridge: Harvard University Press, 1997. An accessible account of the world of mushrooms, written in a warm, personal style.

Vogel, S. "Taming the Wild Morel." *Discover*, May 1988. Describes the research that has allowed these rare delicacies to be cultivated in the laboratory.

Animal Diversity I: Invertebrates

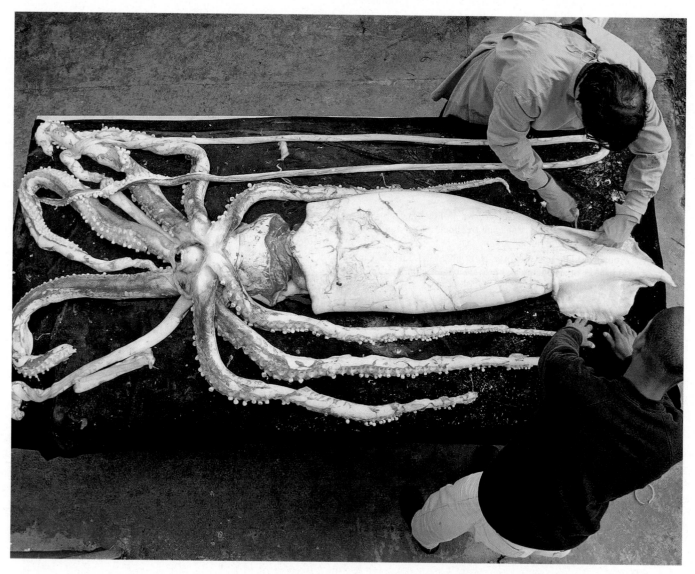

Although the giant squid is Earth's largest invertebrate animal, a single, brief video is our only observation of a living giant squid in its natural habitat.

CASE STUDY THE SEARCH FOR A SEA MONSTER

EVERYONE LOVES A GOOD MYSTERY, and a mystery involving a gigantic, fearsome predator is even better. Consider *Architeuthis*, the giant squid. It is the world's largest invertebrate animal, reaching lengths of 60 feet (18 meters) or more. Each of its huge eyes, the largest in the animal kingdom, can be as large as a human head. The squid's ten tentacles, two of which are longer than the others, are covered with powerful suckers. The suckers contain sharp, clawlike hooks to better grasp prey, which is then pulled toward the squid's mouth, where a heavily muscled beak tears the food apart. The giant squid is one of the most imposing organisms on

Earth, yet we know almost nothing of its habits and lifestyle. What does it eat? Does it swim with its head elevated or pointed downward? How does it mate? Does it live alone or in groups? Even these basic questions about the behavior of the giant squid remain unanswered, because a single brief video recording is our only observation of an adult giant squid alive in its natural habitat.

Our limited scientific knowledge of the giant squid comes almost entirely from specimens that have been found dead or dying, washed up on beaches, caught in fishermen's nets, or contained in the stomachs of sperm whales—which con-

sume vast quantities of squid, including the occasional giant squid. More than 200 such specimens have been reported over the past century, and written accounts of squid corpses go back to the sixteenth century. Live squids, however, remain elusive because they inhabit deep ocean waters, beyond the reach of human divers.

Clyde Roper, a biologist at the Smithsonian Institution, has devoted much of his professional life to a quest to view and study live giant squids. We'll explore some of Dr. Roper's undersea search methods at the close of this chapter, following a survey of the extraordinary diversity of invertebrate animal life.

23.1 WHAT ARE THE KEY FEATURES OF ANIMALS?

It is difficult to devise a concise definition of the term "animal." No single feature fully characterizes animals, so the group is defined by a list of characteristics. None of these characteristics is unique to animals, but together they distinguish animals from members of other kingdoms:

- Animals are multicellular.
- Animals obtain their energy by consuming the bodies of other organisms.
- Animals typically reproduce sexually. Although animal species exhibit a tremendous diversity of reproductive styles, most are capable of sexual reproduction.
- Animal cells lack a cell wall.
- Animals are motile (able to move about) during some stage of their lives. Even the stationary sponges have a free-swimming *larval* stage (a juvenile form).

- Most animals are able to respond rapidly to external stimuli due to the activity of nerve cells, muscle tissue, or both.

23.2 WHICH ANATOMICAL FEATURES MARK BRANCH POINTS ON THE ANIMAL EVOLUTIONARY TREE?

By the Cambrian period, which began 544 million years ago, most of the animal phyla that currently populate Earth were already present. Unfortunately, the Precambrian fossil record is very sparse, and does not reveal the sequence in which the animal phyla arose. Therefore, animal systematists have looked to features of animal anatomy, embryological development, and DNA sequences for clues about the evolutionary history of animals. These investigations have shown that certain features mark major branching points on the animal evolutionary tree and represent milestones in the evolution of the different body plans of

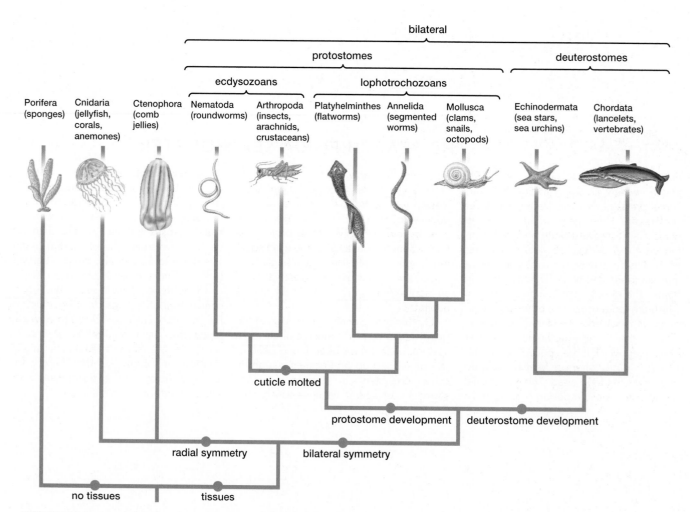

FIGURE 23-1 An evolutionary tree of some major animal phyla

modern animals (**FIG. 23-1**). In the following sections, we will explore these evolutionary milestones and their legacies in the bodies of modern animals.

Lack of Tissues Separates Sponges from All Other Animals

One of the earliest major innovations in animal evolution was the appearance of **tissues**—groups of similar cells integrated into a functional unit, such as a muscle. Today, the bodies of almost all animals include tissues; the only animals that have retained the ancestral lack of tissues are the sponges. In sponges, individual cells may have specialized functions, but they act more or less independently and are not organized into true tissues. This unique feature of sponges suggests that the split between sponges and the evolutionary branch leading to all other animal phyla must have occurred very early in the history of animals. An ancient common ancestor without tissues gave rise to both the sponges and the remaining tissue-containing phyla.

Animals with Tissues Exhibit Either Radial or Bilateral Symmetry

The evolutionary advent of tissues coincided with the first appearance of body symmetry; all animals with true tissues also have symmetrical bodies. An animal is said to be symmetrical if it can be bisected along at least one plane such that the resulting halves are mirror images of one another. Unlike the asymmetrical sponges, any symmetrical animal has an upper, or *dorsal*, surface and a lower, or *ventral*, surface.

The symmetrical, tissue-bearing animals can be divided into two groups, one containing animals that exhibit **radial symmetry** (**FIG. 23-2a**) and one with animals that exhibit **bilateral symmetry** (**FIG. 23-2b**). In radial symmetry, any plane through a central axis divides the object into roughly equal halves. In contrast, a bilaterally symmetrical animal can be divided into roughly mirror-image halves only along one particular plane through the central axis.

The difference between radially and bilaterally symmetrical animals reflects another major branching point in the animal evolutionary tree. This split separated the ancestors of the radially symmetrical cnidarians (jellyfish, anemones, and corals) and ctenophores (comb jellies) from the ancestors of the remaining animal phyla, all of which are bilaterally symmetrical.

Radially Symmetrical Animals Have Two Embryonic Tissue Layers; Bilaterally Symmetrical Animals Have Three

The distinction between radial and bilateral symmetry in animals is closely tied to a corresponding difference in the number of tissue layers, called *germ layers*, that arise during embryonic development. Embryos of animals with radial symmetry have two germ layers: an inner layer of **endoderm** (which gives rise to the tissues that line most hollow organs) and an outer layer of **ectoderm** (which gives rise to the tissues that cover the body and line its inner cavities and nerve tissues). Embryos of bilaterally symmetrical animals add a third germ layer. Between the endoderm and ectoderm is a layer of **mesoderm** (which forms muscle and, when present, the circulatory and skeletal systems).

The parallel evolution of symmetry type and number of germ layers helps us make sense of the potentially puzzling case of the echinoderms (starfish, sea cucumbers, and sea urchins). Adult echinoderms are radially symmetrical, yet our evolutionary tree places them squarely within the bilaterally symmetrical group. It turns out that echinoderms have three germ layers, as well as several other characteristics (some described later) that unite them with the bilaterally symmetrical animals. So, the immediate ancestors of echinoderms must have been bilaterally symmetrical, and the group subsequently evolved radial symmetry (a case of convergent evolution). Even now, larval echinoderms retain bilateral symmetry.

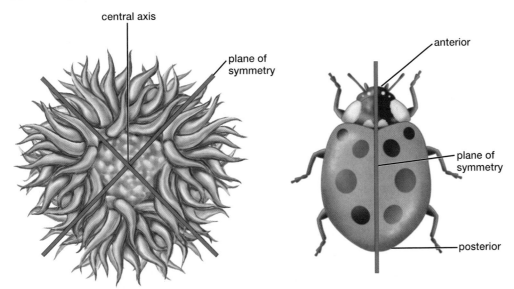

(a) Radial symmetry

central axis

plane of symmetry

(b) Bilateral symmetry

anterior

plane of symmetry

posterior

FIGURE 23-2 Body symmetry and cephalization
(a) Animals with radial symmetry lack a well-defined head. Any plane that passes through the central axis divides the body into mirror-image halves. **(b)** Animals with bilateral symmetry have an anterior head end and a posterior tail end. The body can be split into two mirror-image halves only along a particular plane that runs down the midline.

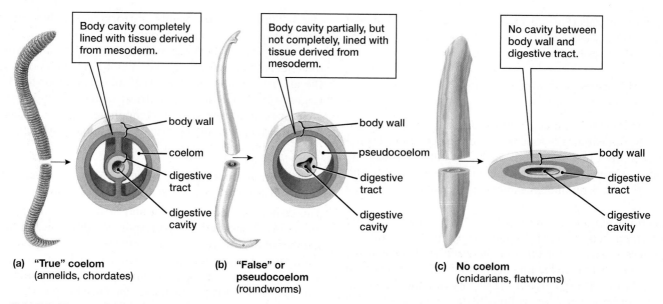

FIGURE 23-3 Body cavities
(a) Annelids have a true coelom. **(b)** Roundworms are pseudocoelomates. **(c)** Flatworms have no cavity between the body wall and digestive tract. (Tissues shown in blue are derived from ectoderm, those in red from mesoderm, and those in yellow from endoderm.)

Bilateral Animals Have Heads

Radially symmetrical animals tend either to be *sessile* (fixed to one spot, like sea anemones) or to drift around on currents (like jellyfish). Such animals may encounter food or threats from any direction, so a body that essentially "faces" all directions at once is advantageous. In contrast, most bilaterally symmetrical animals are *motile* (move under their own power in any given direction). Resources such as food are most likely to be encountered by the part of the animal that is closest to the direction of movement. The evolution of bilateral symmetry was therefore accompanied by **cephalization**, the concentration of sensory organs and a brain in a defined head region. Cephalization produces an *anterior* (head) end, where sensory cells, sensory organs, clusters of nerve cells, and organs for ingesting food are concentrated. The other end of a cephalized animal is designated *posterior* and may feature a tail (see Fig. 23-2b).

Most Bilateral Animals Have Body Cavities

The members of many bilateral animal phyla have a fluid-filled cavity between the digestive tube (or gut, where food is digested and absorbed) and the outer body wall. In an animal with a body cavity, the gut and body wall are separated by a space, creating a "tube-within-a-tube" body plan. Body cavities are absent in radially symmetrical animals, so it is likely that this feature arose sometime after the split between radially and bilaterally symmetrical animals.

A body cavity can serve a variety of functions. In the earthworm it acts as a kind of skeleton, providing support for the body and a framework against which muscles can act. In other animals, internal organs are suspended within the fluid-filled cavity, which serves as a protective buffer between them and the outside world.

Body Cavity Structure Varies Among Phyla

The most widespread type of body cavity is a **coelom**, a fluid-filled cavity that is completely lined with a thin layer of tissue that develops from mesoderm (**FIG. 23-3a**). Phyla whose members have a coelom are called *coelomates*. The annelids (segmented worms), arthropods (insects, spiders, crustaceans), mollusks (clams and snails), echinoderms, and chordates (which include humans) are coelomate phyla.

Members of some phyla have a body cavity that is *not* completely surrounded by mesoderm-derived tissue. This type of cavity is known as a **pseudocoelom**, and phyla whose members have one are collectively known as *pseudocoelomates* (**FIG. 23-3b**). The roundworms (nematodes) are the largest pseudocoelomate group.

Some phyla of bilateral animals have no body cavity at all and are known as *acoelomates*. For example, flatworms have no cavity between their gut and body wall; instead, the space is filled with solid tissue (**FIG. 23-3c**).

Simpler Body Cavities Evolved from Coelomate Body Plans

Because acoelomate and pseudocoelomate body plans appear to be more "primitive" than a coelomate body plan, the acoelomate and pseudocoelomate phyla were once presumed to represent a distinct lineage that diverged early in animal evolutionary history, before the origin of the coelom. Now, however, animal systematists recognize that the various acoelomate and pseudocoelomate phyla are not all closely related to one another, but instead form branches at various points on the animal evolutionary tree (see Fig. 23-1). Thus, acoelomate and pseudocoelomate body plans are not evolutionary precursors of the coelom, but instead are modifications of it.

(a)
(b)
(c)

FIGURE 23-4 The diversity of sponges
Sponges come in a wide variety of sizes, shapes, and colors. Some, such as **(a)** this fire sponge, grow in a free-form pattern over undersea rocks. **(b)** Tiny appendages attach this tubular sponge to rocks, whereas **(c)** this reef sponge with flared tubular openings attaches to a coral reef. QUESTION Sponges are often described as the most "primitive" of animals. How can such a primitive organism have become so diverse and abundant?

Bilateral Organisms Develop in One of Two Ways

Among the bilateral animal phyla, embryological development follows a variety of pathways. These diverse developmental pathways, however, can be grouped into two categories, known as **protostome** and **deuterostome** development. In protostome development, the body cavity forms within the space between the body wall and the digestive cavity. In deuterostome development, the body cavity forms as an outgrowth of the digestive cavity. The two types of development also differ in the pattern of cell division immediately after fertilization and in the method by which the mouth and anus are formed. Protostomes and deuterostomes represent distinct evolutionary branches within the bilateral animals. Annelids, arthropods, and mollusks exhibit protostome development; echinoderms and chordates are deuterostomes.

Protostomes Include Two Distinct Evolutionary Lines

The protostome animal phyla fall into two groups, which correspond to two different lineages that diverged early in the evolutionary history of protostomes. One group, the *ecdysozoans*, includes phyla such as the arthropods and roundworms, whose members have bodies covered by an outer layer that is periodically shed. The other group is known as the *lophotrochozoans* and includes phyla whose members have a special feeding structure called a lophophore, as well as phyla whose members pass through a particular type of developmental stage called a trochophore larva. The mollusks, annelids, and flatworms are examples of lophotrochozoan phyla.

23.3 WHAT ARE THE MAJOR ANIMAL PHYLA?

It's easy to overlook the differences among the multitude of small, boneless animals in the world. Even Carolus Linnaeus, the originator of modern biological classification,

recognized only two phyla of animals without backbones (insects and worms). Today, however, biologists recognize about 27 phyla of animals, some of which are summarized in Table 23-1.

For convenience, biologists often place animals in one of two major categories: **vertebrates**, those with a backbone (or vertebral column), and **invertebrates**, those lacking a backbone. The vertebrates, which we will discuss in Chapter 24, are perhaps the most conspicuous animals from a human viewpoint, but less than 3% of all known animal species on Earth are vertebrates. The vast majority of animals are invertebrates.

The earliest animals probably originated from colonies of protists whose members had become specialized to perform distinct roles within the colony. In our survey of the invertebrate animals, we begin with the sponges, whose body plan most closely resembles the probable ancestral protozoan colonies.

Sponges Have a Simple Body Plan

Sponges (phylum Porifera) are found in most marine and aquatic environments. Most of Earth's 5000 or so sponge species live in saltwater, where they inhabit ocean waters warm and cold, deep and shallow. In addition, some sponges live in freshwater habitats such as lakes and rivers. Adult sponges live attached to rocks or other underwater surfaces. They do not generally move, though researchers have demonstrated that some species, at least when captive in aquaria, are able to move about (very slowly—a few millimeters per day). Sponges come in a variety of shapes and sizes. Some species have a well-defined shape, but others grow free-form over underwater rocks (**FIG. 23-4**). The largest sponges can grow to more than 3 feet (1 meter) in height.

Sponges lack true tissues and organs. In some ways, a sponge resembles a colony of single-celled organisms. The colony-like properties of sponges were revealed in

Table 23-1 Comparison of the Major Animal Phyla

Common name (Phylum)		Sponges (Porifera)	Hydra, Anemones, Jellyfish (Cnidaria)	Flatworms (Platyhelminthes)
Body Plan	Level of organization	Cellular—lack tissues and organs	Tissue—lack organs	Organ system
	Germ layers	Absent	Two	Three
	Symmetry	Absent	Radial	Bilateral
	Cephalization	Absent	Absent	Present
	Body cavity	Absent	Absent	Absent
	Segmentation	Absent	Absent	Absent
Internal Systems	Digestive system	Intracellular	Gastrovascular cavity; some intracellular	Gastrovascular cavity
	Circulatory system	Absent	Absent	Absent
	Respiratory system	Absent	Absent	Absent
	Excretory system (fluid regulation)	Absent	Absent	Canals with ciliated cells
	Nervous system	Absent	Nerve net	Head ganglia with longitudinal nerve cords
	Reproduction	Sexual; asexual (budding)	Sexual; asexual (budding)	Sexual (some hermaphroditic); asexual (body splits)
	Support	Endoskeleton of spicules	Hydrostatic skeleton	Hydrostatic skeleton
	Number of known species	5000	9000	20,000

an experiment performed by embryologist H. V. Wilson in 1907. Wilson mashed a sponge through a piece of silk, thereby breaking it apart into single cells and cell clusters. He then placed these tiny bits of sponge into seawater and waited for three weeks. By the end of the experiment, the cells had reaggregated into a functional sponge, demonstrating that individual sponge cells had been able to survive and function independently.

All sponges have a similar body plan. The body is perforated by numerous tiny pores, through which water enters, and by fewer, large openings, through which it is expelled. Within the sponge, water travels through canals. As water passes through the sponge, oxygen is extracted, microorganisms are filtered out and taken into individual cells where they are digested, and wastes are released (**FIG. 23-5**).

Sponges have three major cell types, each with a specialized role. Flattened *epithelial cells* cover their outer body surfaces. Some epithelial cells are modified into *pore cells*, which surround pores, controlling their size and regulating the flow of water. The pores close when harmful sub-

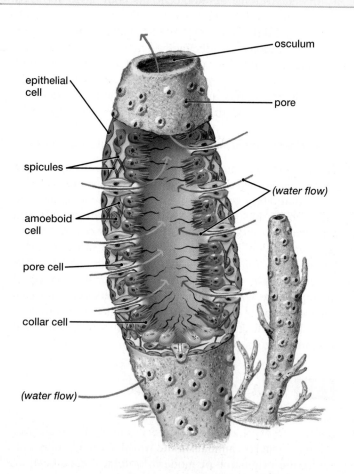

FIGURE 23-5 The body plan of sponges
Water enters through numerous tiny pores in the sponge body and exits through oscula. Microscopic food particles are filtered from the water.

Segmented Worms (Annelida)	Snails, Clams, Squid (Mollusca)	Insects, Arachnids, Crustaceans (Arthropoda)	Roundworms (Nematoda)	Sea Stars, Sea Urchins (Echinodermata)
Organ system	Organ system	Organ system	Organ system	Organ system
Three	Three	Three	Three	Three
Bilateral	Bilateral	Bilateral	Bilateral	Bilateral larvae, radial adults
Present	Present	Present	Present	Absent
Coelom	Coelom	Coelom	Pseudocoel	Coelom
Present	Absent	Present	Absent	Absent
Separate mouth and anus	Separate mouth and anus	Separate mouth and anus	Separate mouth and anus	Separate mouth and anus (normally)
Closed	Open	Open	Absent	Absent
Absent	Gills, lungs	Tracheae, gills, or book lungs	Absent	Tube feet, skin gills, respiratory tree
Nephridia	Nephridia	Excretory glands resembling nephridia	Excretory gland	Absent
Head ganglia with paired ventral cords; ganglia in each segment	Well-developed brain in some cephalopods; several paired ganglia, most in the head; nerve network in body wall	Head ganglia with paired ventral nerve cords; ganglia in segments, some fused	Head ganglia with dorsal and ventral nerve cords	Head ganglia absent; nerve ring and radial nerves; nerve network in skin
Sexual (some hermaphroditic)	Sexual (some hermaphroditic)	Normally sexual	Sexual (some hermaphroditic)	Sexual (some hermaphroditic); asexual by regeneration (rare)
Hydrostatic skeleton	Hydrostatic skeleton	Exoskeleton	Hydrostatic skeleton	Endoskeleton of plates beneath outer skin
9000	50,000	1,000,000	12,000	6500

stances are present. *Collar cells* maintain a flow of water through the sponge by beating flagella that extend into the inner canal. The collars that surround these flagella act as fine sieves, filtering out microorganisms that are then ingested by the cell. Some of the food is passed to the *amoeboid cells*. These cells roam freely between the epithelial and collar cells, digesting and distributing nutrients, producing reproductive cells, and secreting small skeletal projections called *spicules*. Spicules may be composed of calcium carbonate (chalk), silica (glass), or protein and form an internal skeleton that provides support for the body (see Fig. 23-5). Natural household sponges, which are now largely replaced by factory-made cellulose imitations, are actually sponge skeletons.

Sponges may reproduce asexually by **budding**, in which the adult produces miniature versions of itself that drop off and assume an independent existence. Alternatively, they may reproduce sexually through the fusion of sperm and eggs. Fertilized eggs develop inside the adult into active larvae that escape through the openings in the sponge body. Water currents disperse the larvae to new areas, where they settle and develop into adult sponges.

Because sponges remain in one spot and have no protective shell, they are vulnerable to predators such as fish, turtles, and sea slugs. Many sponges, however, have evolved chemical defenses against predation. The bodies of these sponges contain chemicals that are toxic or distasteful to potential predators. Fortunately for people, a number of these chemicals have proved to be valuable medicines for humans. For example, the drug spongistatin, a compound first isolated from a sponge, is an emerging treatment for the fungal infections that frequently sicken AIDS patients. Other medicines derived from sponges include the very promising new anticancer drugs discodermolide and halichondrin. The discovery of these and other medicines has raised hopes that, as more species are screened by researchers, sponges will become a major source of new drugs.

Cnidarians Are Well-Armed Predators

Like sponges, the roughly 9000 known species of *cnidarians* (phylum Cnidaria)—jellyfish, sea anemones, corals, and hydrozoans—are confined to watery habitats, and most species are marine. Most cnidarians are small, from a few millimeters to a few inches in diameter, but the largest jellyfish can be eight feet across and have tentacles 150 feet long. All cnidarians are carnivorous predators.

The cells of cnidarians are organized into distinct tissues, including contractile tissue that acts like muscle. The nerve cells are organized into tissue called a *nerve net*, which branches through the body and controls the contractile tissue to bring about movement and feeding

FIGURE 23-6 Cnidarian diversity
(a) A red-spotted anemone spreads its tentacles to capture prey. **(b)** A small medusa. **(c)** A close-up of coral reveals bright yellow polyps in various stages of tentacle extension. At the lower right, areas where the coral has died expose the calcium carbonate skeleton that supports the polyps and forms the reef. A strikingly patterned crab (an arthropod) sits atop the coral, holding tiny white anemones in its claws. Their stinging tentacles help protect the crab. **(d)** A sea wasp, a cnidarian whose stinging cells contain one of the most toxic of all known venoms. QUESTION In each of these photos, is the pictured organism a polyp or a medusa?

behavior. However, most cnidarians lack true organs and have no brain.

Cnidarians come in a bewildering and beautiful variety of forms (**FIG. 23-6**), all of which are actually variations on two basic body plans: the *polyp* (**FIG. 23-7a**) and the *medusa* (**FIG. 23-7b**). The generally tubular polyp is adapted to a life spent quietly attached to rocks. The polyp has tentacles, extensions that reach upward to grasp and immobilize prey. The bell-shaped body of the medusa ("jellyfish") floats in the water and is carried by currents, trailing its tentacles like multiple fishing lines.

Many cnidarian species have life cycles that include both polyp and medusa stages, though some species live only as polyps and others only as medusae. Both polyps and medusae develop from just two germ layers—the interior endoderm and the exterior ectoderm; between those layers is a jellylike substance. Polyps and medusae are radially

symmetrical, with body parts arranged in a circle around the mouth and digestive cavity (see Fig. 23-2a). This arrangement of parts is well suited to these sessile or free-floating animals because it enables them to respond to prey or threats from any direction.

Cnidarian tentacles are armed with *cnidocytes*, cells containing structures that, when stimulated by contact, explosively inject poisonous or sticky filaments into prey (**FIG. 23-8**). These stinging cells, found only in cnidarians, are used to capture prey. Cnidarians do not hunt actively but instead wait for victims to blunder, by chance, into the grasp of their enveloping tentacles. Stung and firmly grasped, the prey is forced through an expansible mouth into a digestive sac, the *gastrovascular cavity*. Digestive enzymes secreted into this cavity break down some of the food, and further digestion occurs within the cells lining the cavity. Because the gastrovascular cavity has only a single

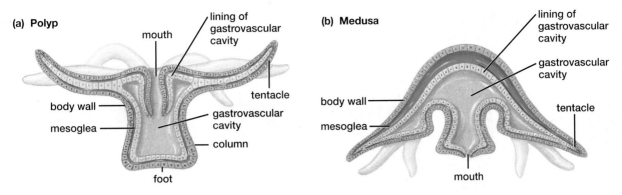

FIGURE 23-7 Polyp and medusa
(a) The polyp form is seen in hydra (see Fig. 23-8), sea anemones (Fig. 23-6a), and the individual polyps within a coral (Fig. 23-6c). **(b)** The medusa form, seen in the jellyfish (Fig. 23-6b), resembles an inverted polyp. (Tissues shown in blue are derived from ectoderm, those in yellow from endoderm.)

opening, undigested material is expelled through the mouth when digestion is completed. Although this two-way traffic prevents continuous feeding, it is adequate to support the low energy demands of these animals.

Cnidarians may reproduce sexually or asexually. Reproductive cycles vary considerably among different types of cnidarians, but one pattern is fairly common in species with both polyp and medusa stages. In such species, polyps typically reproduce by asexual budding that gives rise to new polyps. Under certain conditions, however, budding may instead give rise to medusae. After a medusa grows to maturity, it may release gametes (sperm or eggs) into the water. If a sperm and egg should meet, they may fuse to form a zygote that develops into a free-swimming, ciliated larva. The larva may eventually settle on a hard surface, where it develops into a polyp.

The venom of some cnidarians can cause painful stings in humans unfortunate enough to come into contact with them, and the stings of a few jellyfish species can even be life threatening. The most deadly of these species is the sea wasp, *Chironex fleckeri*, which is found in the waters off northern Australia and Southeast Asia. The amount of venom in a single sea wasp could kill up to 60 people, and the victim of a serious sting may die within minutes of being stung.

One group of cnidarians, the corals, is of particular ecological importance (see Fig. 23-6c). Coral polyps form colonies, and each member of the colony secretes a hard skeleton of calcium carbonate. The skeletons persist long after the organisms die, serving as a base to which other individuals may attach themselves. The cycle continues until, after thousands of years, massive coral reefs are formed.

Coral reefs are found in both cold and warm oceans. Cold-water reefs form in deep waters and, though widely distributed, have only recently attracted the attention of researchers and are not yet well studied. The more familiar warm-water coral reefs are restricted to clear, shallow waters in the tropics. Here, coral reefs form undersea habitats that are the basis of an ecosystem of stunning diversity and unparalleled beauty.

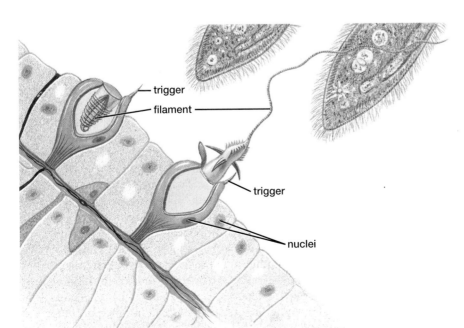

FIGURE 23-8 Cnidarian weaponry: the cnidocyte
At the slightest touch to the trigger of a special structure in their cnidocytes, cnidarians, such as this hydra, violently expel a poisoned filament.

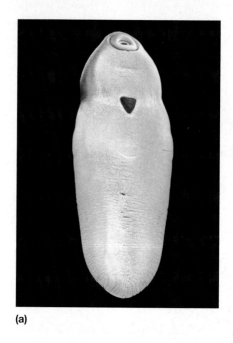

(a)

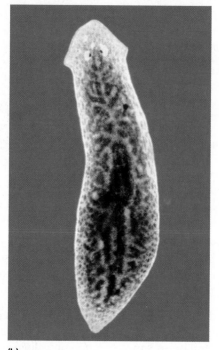

(b)

(c)

FIGURE 23-9 Flatworm diversity
(a) This fluke is an example of a parasitic flatworm. **(b)** Eyespots are clearly visible in the head of this free-living, freshwater flatworm. **(c)** Many of the flatworms that inhabit tropical coral reefs are brightly colored.

Flatworms Have Organs but Lack Respiratory and Circulatory Systems

The *flatworms* (phylum Platyhelminthes) are aptly named; their width is much greater than their height, giving them a ribbonlike profile. Many of the approximately 20,000 flatworm species are parasites (**FIG. 23-9a**). (**Parasites** are organisms that live in or on the body of another organism, called a *host*, which is harmed as a result of the relationship.) Nonparasitic, free-living flatworms inhabit aquatic, marine, and moist terrestrial habitats. They tend to be small and inconspicuous (**FIG. 23-9b**), but some are brightly colored, spectacularly patterned residents of tropical coral reefs (**FIG. 23-9c**).

Unlike cnidarians, flatworms have well-developed organs, in which tissues are grouped into functional units. For example, most free-living flatworms have sense organs, including eyespots (see Fig. 23-9b) that detect light and dark, and cells that respond to chemical and tactile stimuli. To process information, a flatworm has clusters of nerve cells called **ganglia** (singular, ganglion) in its head, forming a simple brain. Paired neural structures called **nerve cords** conduct nervous signals to and from the ganglia.

Despite the presence of some organs, flatworms lack respiratory and circulatory systems. In the absence of a respiratory system, gas exchange is accomplished by direct diffusion between body cells and the environment. This mode of respiration is possible because the small size and flat shape of flatworm bodies ensure that no body cell is very far from the surrounding environment. Without a circulatory system, nutrients move directly from the digestive tract to body cells. The digestive cavity has a branching structure that reaches all parts of the body and allows di-

gested nutrients to diffuse into nearby cells. The digestive cavity has only one opening to the environment, so undigested waste must pass out through the same opening that also serves as a mouth.

Flatworms are bilaterally symmetrical, rather than radially symmetrical (see Fig. 23-2). This body plan and its accompanying cephalization foster active movement. Cephalized, bilaterally symmetrical animals possess an anterior end, which is the first part of a moving animal to encounter the environment ahead. Consequently, sense organs are concentrated in the anterior portion of the body, thus enhancing the animal's ability to respond appropriately to any stimuli it encounters (for example, eating food items and retreating from obstacles).

Flatworms can reproduce both sexually and asexually. Free-living forms may reproduce by cinching themselves around the middle until they separate into two halves, each of which regenerates its missing parts. All forms can reproduce sexually; most are **hermaphroditic**—that is, they possess both male and female sexual organs. This trait is a great advantage to parasitic forms because it allows each worm to reproduce through self-fertilization, even when it is the only individual present in its host.

Some parasitic flatworms can infect humans. For example, tapeworms can infect people who eat improperly cooked beef, pork, or fish that has been infected by the worms. Worm larvae form encapsulated resting structures, called *cysts*, in the muscles of these animals. The cysts hatch in the human digestive tract, where the young tapeworms attach themselves to the lining of the intestine. There they may grow to a length of more than 20 feet (7 meters), absorbing digested nutrients directly through their outer surface and eventually releasing packets of eggs that are shed in the host's feces. If pigs, cows, or fish eat food contaminated with infected human feces, the eggs hatch in the animal's digestive tract, releasing larvae that burrow into its muscles and form cysts, thereby continuing the infective cycle (**FIG. 23-10**).

Another group of parasitic flatworms is the *flukes*. Of these, the most devastating are liver flukes (common in Asia) and blood flukes, such as those of the genus *Schistosoma*, which cause the disease schistosomiasis. Like most parasites, flukes have a complex life cycle that includes an intermedi-

ate host (a snail, in the case of *Schistosoma*). Prevalent in Africa and parts of South America, schistosomiasis affects an estimated 200 million people worldwide. Its symptoms include diarrhea, anemia, and possible brain damage.

Annelids Are Composed of Identical Segments

Charles Darwin, arguably the greatest of all biologists, devoted substantial time to the study of earthworms. In fact, he wrote an entire book about them. Darwin was impressed by

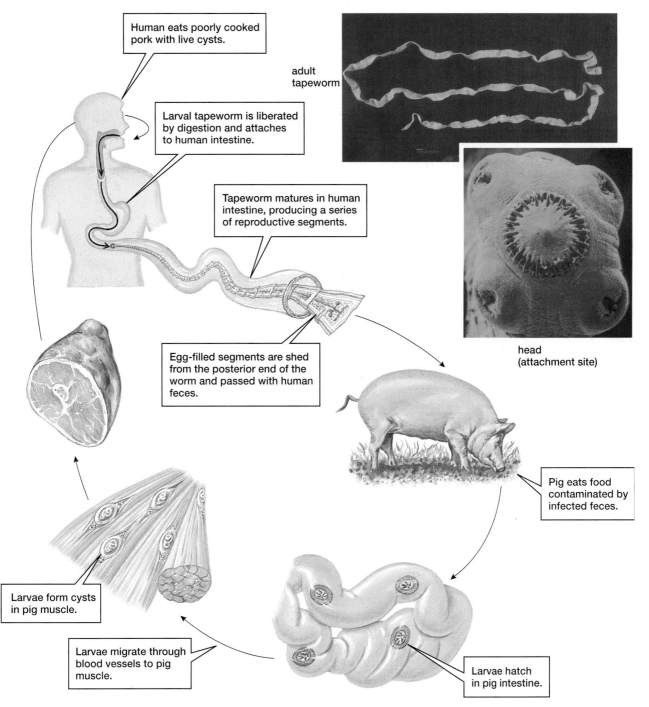

Human eats poorly cooked pork with live cysts.

Larval tapeworm is liberated by digestion and attaches to human intestine.

Tapeworm matures in human intestine, producing a series of reproductive segments.

Egg-filled segments are shed from the posterior end of the worm and passed with human feces.

Larvae form cysts in pig muscle.

Larvae migrate through blood vessels to pig muscle.

adult tapeworm

head (attachment site)

Pig eats food contaminated by infected feces.

Larvae hatch in pig intestine.

FIGURE 23-10 The life cycle of the human pork tapeworm
Each reproductive unit, or proglottid, is a self-contained reproductive factory that includes both male and female sex organs. QUESTION Why have tapeworms evolved a long, flat shape?

FIGURE 23-11 An annelid, the earthworm
This diagram shows an enlargement of segments, many of which are repeating similar units separated by partitions. QUESTION What advantage does a digestive system with two openings have relative to digestive systems with only a single opening (like that of the flatworms)?

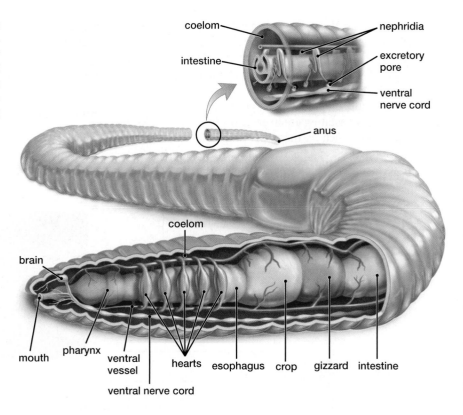

earthworms' role in improving soil fertility. More than a million earthworms may live in an acre of land, beneath the surface of which the worms tunnel through soil, consuming and excreting soil particles and organic matter. These actions help ensure that air and water can move easily through the soil and that organic matter is continually mixed into it, creating conditions that are favorable for plant growth. In Darwin's view, the activity of the earthworms has had such a significant impact on agriculture that "it may be doubted whether there are many other animals which have played so important a part in the history of the world."

Earthworms are examples of *annelids* (phylum Annelida), the segmented worms. A prominent feature of annelids is the division of the body into a series of repeating segments. Externally, these segments appear as ringlike depressions on the surface. Internally, most of the segments contain identical copies of nerves, excretory structures, and muscles. **Segmentation** is advantageous for locomotion, because the body compartments are each controlled by separate muscles and collectively are capable of more complex movement than could be achieved with only one set of muscles to control the whole body.

Another feature that distinguishes annelids from flatworms is a fluid-filled true coelom between the body wall and the digestive tract (see Fig. 23-3a). The incompressible fluid in the coelom of many annelids is confined by the partitions between the segments and serves as a **hydrostatic skeleton**, a rigid framework against which muscles can act. The hydrostatic skeleton makes possible such feats as burrowing through soil.

Annelids have well-developed organ systems. For example, annelids have a **closed circulatory system** that dis-

tributes gases and nutrients throughout the body. In closed circulatory systems (including yours), blood remains confined to the heart and blood vessels. In the earthworm, for example, blood with oxygen-carrying hemoglobin is pumped through well-developed vessels by five pairs of "hearts" (**FIG. 23-11**). These hearts are actually short segments of specialized blood vessels that contract rhythmically. The blood is filtered and wastes are removed by excretory organs called *nephridia* (singular, nephridium), which are found in many of the segments. Nephridia resemble the individual tubules of the vertebrate kidney. The annelid nervous system consists of a simple ganglionic brain in the head and a series of repeating paired segmental ganglia joined by a pair of ventral nerve cords that pass along the length of the body. The annelid digestive system includes a tubular gut that runs from the mouth to the anus. This kind of digestive tract, with two openings and a one-way digestive path, is much more efficient than the single-opening digestive systems of cnidarians and flatworms. Digestion in annelids occurs in a series of compartments, each specialized for a different phase of food processing (see Fig. 23-11).

Sexual reproduction is common among annelids. Some species are hermaphroditic; others have separate sexes. Fertilization may be external or internal. External fertilization, in which sperm and eggs are released into the surrounding environment, is found mainly in species that live in water. In internal fertilization, two individuals copulate and sperm are transferred directly from one to another. In hermaphroditic species, sperm transfer may be mutual, with each individual transferring sperm to the other. In addition, some annelids can reproduce asexually, typically by

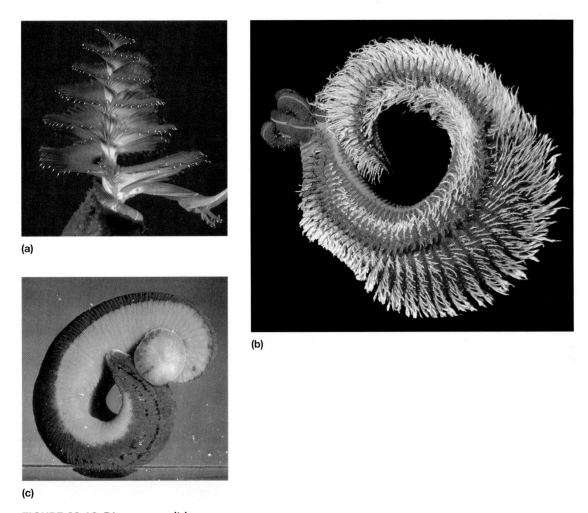

FIGURE 23-12 Diverse annelids
(a) A polychaete annelid projects brightly spiraling gills from a tube attached to rock. When the gills retract, the tube is covered by the trap door visible on the lower right. **(b)** This tiny polychaete (seen here through a microscope) lives among seaside rocks near the tide line. **(c)** This leech, a freshwater annelid, shows numerous segments. The sucker encircles its mouth, allowing it to attach to its prey. QUESTION Why does pouring salt on a leech harm it?

fragmentation in which the body breaks into two pieces, each of which regenerates the missing part.

The 9000 species in the phylum Annelida fall into three main subgroups: the *oligochaetes*, the *polychaetes*, and the *leeches*. The oligochaetes include the familiar earthworm and its relatives. Polychaetes live primarily in the ocean. Some polychaetes have paired fleshy paddles on most of their segments, used in locomotion. Others live in tubes from which they project feathery gills that both exchange gases and sift the water for microscopic food (**FIG. 23-12a, b**). Leeches (**FIG. 23-12c**) live in freshwater or moist terrestrial habitats and are either carnivorous or parasitic. Carnivorous leeches prey on smaller invertebrates; parasitic leeches suck the blood of larger animals. One parasitic leech species, the medicinal leech, has become a tool of modern medicine and was recently approved for use as a "medical device" by the U.S. Food and Drug Administration (see "Links to Life: Physicians' Assistants").

Most Mollusks Have Shells

If you have ever enjoyed a bowl of clam chowder, a plate of oysters on the half shell, or some sautéed scallops, you are indebted to the *mollusks* (phylum Mollusca). The mollusks include species with a range of lifestyles, from passive forms that go through their adult lives in one spot, filtering microorganisms from water, to active, voracious predators of the ocean depths. Mollusks also include the largest invertebrate animals and the most intelligent ones. Mollusks are very diverse; in number of known species, the 50,000 mollusks rank second (although a distant second) only to the arthropods. Except for some snails and slugs, mollusks are water dwellers.

Most mollusks protect their bodies with hard shells of calcium carbonate. Others, however, lack shells and escape predation by moving swiftly or by having an unappealing taste. Mollusks have a *mantle*, an extension of the body wall that forms a chamber for the gills and, in shelled

Although invertebrate animals cause or transmit many human illnesses, some have been recruited to make a positive contribution to human health. Consider leeches, for example. For more than 2000 years, healers enlisted these parasitic annelids for treatment of almost every human illness or injury. For much of human medical history, treatment with leeches was based on the hope that the creatures would suck out the "tainted" blood that was believed to be the primary cause of disease. As the actual causes of disease were discovered, however, medical use of leeches declined. By the beginning of the twentieth century, leeches no longer had a place in the toolkit of modern medicine and had become a symbol of the ignorance of an earlier age. Today, however, medical use of leeches is making a surprising comeback.

Currently, leeches are used to treat a surgical complication known as venous insufficiency. This complication is especially common in reconstructive surgery, such as reattaching a severed finger or repairing a disfigured face. In such cases, surgeons are often unable to reconnect all of the veins that would normally carry blood away from tissues. Eventually, new veins will grow, but in the meantime blood may accumulate in the repaired tissue. Unless the excess blood is removed, it will coagulate, causing clots that can deprive the tissue of the oxygen and nutrients it needs to live. Fortunately, leeches can help. Applied to the affected area, the leeches get right to work, making a small, painless incision and sucking blood into their stomachs. To aid them in their blood-removal task, the leeches' saliva contains a mixture of chemicals that dilate blood vessels and prevent blood from clotting. Although the chemical brew in the saliva is an adaptation that helps leeches consume blood more efficiently, it also helps the patient by promoting blood flow in the damaged tissue. In this way, leeches provide a painless, effective treatment for venous insufficiency and have thereby regained their status as medical assistants to humanity.

Another invertebrate animal that has been recruited for medical duty is the blowfly or, more precisely, blowfly larvae, commonly known as maggots (**FIG. E23-1**). Blowfly maggots have proved to be effective at ridding wounds and ulcers of dead and dying tissue. If such tissue is not removed, it can interfere with healing or lead to infection. Traditionally, dead tissue in wounds is removed by a physician wielding a scalpel, but maggots offer an increasingly common alternative treatment. In this treatment, a bandage containing day-old, sterile maggots is applied to the wound. The maggots consume dead or dying tissue, secreting digestive enzymes that do not harm healthy skin or bone. After a few days, the maggots have grown to the size of rice kernels and are removed. The treatment is repeated until the wound is clean.

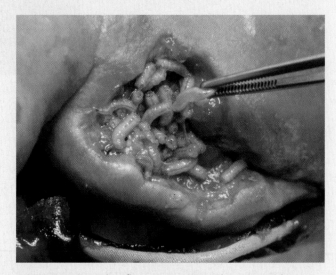

FIGURE E23-1 **Blowfly maggots can clean wounds.**

---

species, secretes the shell. Mollusks also have well-developed circulatory systems with a feature not seen in annelids: the **hemocoel**, or blood cavity. Blood empties into the hemocoel, where it bathes the internal organs directly. This arrangement, known as an **open circulatory system**, is also present in most arthropods. The nervous system, like that of annelids, consists of ganglia connected by nerves, but many more of the ganglia are concentrated in the brain. Reproduction is sexual; some species have separate sexes, and others are hermaphroditic. Although mollusks are enormously diverse, a simplified diagram of the body plan of a mollusk is shown in **FIGURE 23-13**.

Among the many classes of mollusks, we will discuss three in more detail: gastropods, bivalves, and cephalopods.

Gastropods Are One-Footed Crawlers

Snails and slugs—collectively known as *gastropods*—crawl on a muscular *foot*, and many are protected by shells that vary widely in form and color (**FIG. 23-14a**). Not all gastropods are shelled, however. Sea slugs, for example, lack

shells, but their brilliant colors warn potential predators that they are poisonous or foul tasting (**FIG. 23-14b**).

Gastropods feed with a *radula*, a flexible ribbon of tissue studded with spines that is used to scrape algae from rocks or to grasp larger plants or prey (see Fig. 23-13). Most snails use gills, typically enclosed in a cavity beneath the shell, for respiration. Gases can also diffuse readily through the skin of most gastropods; sea slugs rely on this mode of gas exchange. The few gastropod species that live in terrestrial habitats (including the destructive garden snails and slugs) use a simple lung for breathing.

Bivalves Are Filter Feeders

Included among the *bivalves* are scallops, oysters, mussels, and clams (**FIG. 23-15**). Members of the class lend exotic variety to the human diet and are important members of the nearshore marine community. Bivalves possess two shells connected by a flexible hinge. A strong muscle clamps the shells closed in response to danger; this muscle is what is served when you order scallops in a restaurant.

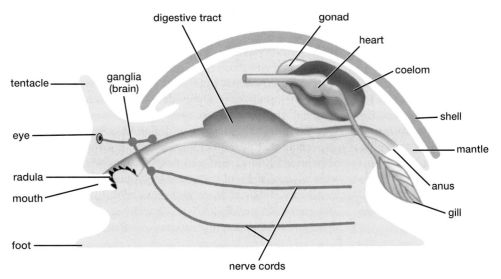

FIGURE 23-13 A generalized mollusk
The general body plan of a mollusk, showing the mantle, foot, gills, shell, radula, and other features that are seen in most (but not all) mollusk species.

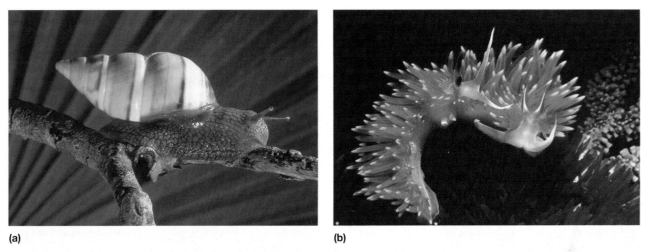

(a)

(b)

FIGURE 23-14 The diversity of gastropod mollusks
(a) A Florida tree snail displays a brightly striped shell and eyes at the tip of stalks that retract instantly if touched. (b) Spanish shawl sea slugs prepare to mate. The brilliant colors of many sea slugs warn potential predators that they are distasteful.

(a)

(b)

FIGURE 23-15 The diversity of bivalve mollusks
(a) This swimming scallop parts its hinged shells. The upper shell is covered with an encrusting sponge. (b) Mussels attach to rocks in dense aggregations exposed at low tide. White barnacles are attached to the mussel shells and surrounding rock.

Clams use a muscular foot for burrowing in sand or mud. In mussels, which live attached to rocks, the foot is smaller and helps secrete threads that anchor the animal to the rocks. Scallops lack a foot and move by a sort of jet propulsion achieved by flapping their shells together. Bivalves are filter feeders, using their gills as both respiratory and feeding structures. Water circulates over the gills, which are covered with a thin layer of mucus that traps microscopic food particles. Food is conveyed to the mouth by the beating of cilia on the gills. Probably because they are filter feeders and do not move extensively, bivalves "lost their heads" as they evolved.

Cephalopods Are Marine Predators

The *cephalopods* include octopuses, nautiluses, cuttlefish, and squids (**FIG. 23-16**). The largest invertebrate, the giant squid, belongs to this group. All cephalopods are predatory carnivores, and all are marine. In these mollusks, the foot has evolved into tentacles with well-developed chemosensory abilities and suction disks for detecting and grasping prey. Prey grasped by the tentacles may be immobilized by a paralyzing venom in the saliva before being torn apart by beak-like jaws.

Cephalopods move rapidly by jet propulsion, which is generated by the forceful expulsion of water from the mantle cavity. Octopuses may also travel along the seafloor by using their tentacles like multiple undulating legs. The rapid movements and generally active lifestyles of cephalopods are made possible in part by their closed circulatory systems. Cephalopods are the only mollusks with closed circulation, which transports oxygen and nutrients more efficiently than open circulatory systems do.

Cephalopods have highly developed brains and sensory systems. The cephalopod eye rivals our own in complexity and exceeds it in efficiency of design. The cephalopod brain, especially that of the octopus, is (for an invertebrate brain) exceptionally large and complex. It is enclosed in a skull-like case of cartilage and endows the octopus with highly developed capabilities to learn and remember. In the laboratory, octopuses can rapidly learn to associate certain symbols with food and to open a screw-cap jar to obtain food.

Arthropods Are the Dominant Animals on Earth

In terms of both number of individuals and number of species, no other animal phylum comes close to the *arthropods* (phylum Arthropoda), which include insects, arachnids, myriapods, and crustaceans. About 1 million arthropod species have been discovered, and scientists estimate that millions more remain undescribed.

All arthropods have an **exoskeleton**, an external skeleton that encloses the arthropod body like a suit of armor. Secreted by the *epidermis* (the outer layer of skin), the exoskeleton is composed chiefly of protein and a polysaccharide called *chitin*. The external skeleton protects against predators and is responsible for arthropods' greatly increased agility relative to their wormlike ancestors. The exoskeleton is thin and flexible in places, allowing movement

(a)

(b)

(c)

FIGURE 23-16 The diversity of cephalopod mollusks **(a)** An octopus can crawl rapidly by using its eight suckered tentacles, and it can alter its color and skin texture to blend with its surroundings. In emergencies, this mollusk can jet backward by vigorously contracting its mantle. Octopuses and squid can emit clouds of dark purple ink to confuse pursuing predators. **(b)** The squid moves entirely by contracting its mantle to generate jet propulsion, which pushes the animal backward through the water. **(c)** The chambered nautilus secretes a shell with internal, gas-filled chambers that provide buoyancy. Note the well-developed eyes and the tentacles used to capture prey.

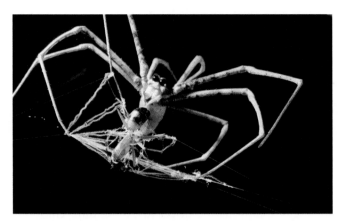

FIGURE 23-17 The exoskeleton allows precise movements
A net-throwing spider begins to wrap a captured insect in silk. Such dexterous manipulations are made possible by the exoskeleton and jointed appendages characteristic of arthropods.

FIGURE 23-18 The exoskeleton must be molted periodically
A newly emerged praying mantis (a predatory insect) hangs beside its outgrown exoskeleton (left).

of the paired, jointed appendages. By providing stiff but flexible appendages and rigid attachment sites for muscles, the exoskeleton makes possible the flight of the bumblebee and the intricate, delicate manipulations of the spider as it weaves its web (**FIG. 23-17**). The exoskeleton also contributed enormously to the arthropod invasion of the land (arthropods were the earliest terrestrial animals) by providing a watertight covering for delicate, moist tissues such as those used for gas exchange. (Arthropods were the earliest terrestrial animals; see Chapter 17.)

Like a suit of armor, the arthropod exoskeleton poses some problems. First, because it cannot expand as the animal grows, the exoskeleton periodically must be shed, or **molted**, and replaced with a larger one (**FIG. 23-18**). Molting uses energy and leaves the animal temporarily vulnerable until the new skeleton hardens. ("Soft-shelled" crabs are simply regular "hard-shelled" crabs that are caught just after molting.) The exoskeleton is also heavy, and its weight increases exponentially as the animal grows. It is no coincidence that the largest arthropods are crustaceans (crabs and lobsters), whose watery habitat supports much of their weight.

Arthropods are segmented, but their segments tend to be few and specialized for different functions such as sensing the environment, feeding, and movement (**FIG. 23-19**). For example, in insects, sensory and feeding structures are concentrated on the front segment, known as the *head*, and digestive structures are largely confined to the *abdomen*, which is the rear segment. Between the head and the abdomen is the *thorax*, the segment to which structures used in locomotion, such as wings and walking legs, are attached.

Efficient gas exchange is required to supply adequate oxygen to the muscles that allow the rapid flight, swimming, or running displayed by many arthropods. In aquatic forms such as crustaceans, gas exchange is accomplished by gills. In terrestrial arthropods, gas exchange is performed either by lungs (in arachnids) or by *tracheae* (sin-

gular, trachea), a network of narrow, branching respiratory tubes that open to the surrounding environment and that penetrate all parts of the body. Most arthropods have open circulatory systems, like those of mollusks, in which blood directly bathes the organs in a hemocoel.

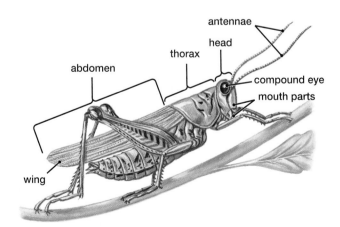

FIGURE 23-19 Segments are fused and specialized in insects
Insects, such as this grasshopper, show fusion and specialization of body segments into a distinct head, thorax, and abdomen. Segments are visible on the abdomen beneath the wings.

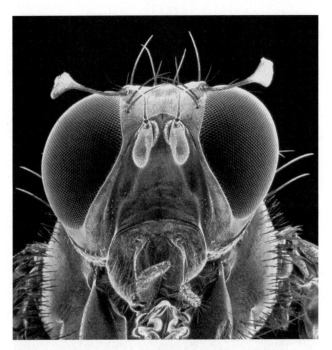

FIGURE 23-20 Arthropods possess compound eyes
This scanning electron micrograph shows the compound eye of a fruit fly. Compound eyes consist of an array of similar light-gathering and sensing elements whose orientation gives the arthropod a wide field of view. Insects have reasonably good image-forming ability and good color discrimination.

Most arthropods possess a well-developed sensory system, including **compound eyes**, which have multiple light detectors (**FIG. 23-20**), and acute chemical and tactile senses. The arthropod nervous system is similar in plan to that of annelids, but more complex. It consists of a brain, composed of fused ganglia, and a series of additional ganglia along the length of the body that are linked by a ventral nerve cord. The capacity for finely coordinated movement, combined with sophisticated sensory abilities and a well-developed nervous system, has allowed the evolution of complex behavior.

Insects Are the Only Flying Invertebrates

The number of described *insect* species is about 850,000, roughly three times the total number of known species in all other classes of animals combined (**FIG. 23-21**). Insects have a single pair of antennae and three pairs of legs, normally supplemented by two pairs of wings. Insects' capacity for flight distinguishes them from all other invertebrates and has contributed to their enormous success (see Fig. 23-21c). As anyone who has pursued a fly can testify, flight helps in escaping from predators. It also allows the insect to find widely dispersed food. Swarms of locusts (see Fig. 23-21d) can travel 200 miles a day in search of food; researchers tracked one swarm on a journey that totaled almost 3000 miles. Flight requires rapid and efficient gas exchange, which insects accomplish by means of tracheae.

During their development, insects undergo **metamorphosis**, a radical change from a juvenile body form to an adult body form. In insects with complete metamorphosis, the immature stage, called a **larva** (plural, larvae), is worm shaped (for example, the maggot of a housefly or the caterpillar of a moth or butterfly; see Fig. 23-21e). The larva hatches from an egg, grows by eating voraciously and shedding its exoskeleton several times, and then forms a nonfeeding stage called a **pupa**. Encased in an outer covering, the pupa undergoes a radical change in body form, emerging in its adult winged form. The adults mate and lay eggs, continuing the cycle. Metamorphosis may include a change in diet as well as in shape, thereby eliminating competition for food between adults and juveniles and in some cases allowing the insect to exploit different foods when they are most available. For instance, a caterpillar that feeds on new green shoots in springtime metamorphoses into a butterfly that drinks nectar from the summer's blooming flowers. Some insects undergo a more gradual metamorphosis (called incomplete metamorphosis), hatching as young that bear some resemblance to the adult and then gradually acquiring more-adult features as they grow and molt.

Biologists classify the amazing diversity of insects into several dozen orders. We describe three of the largest orders here.

Butterflies and Moths
This is perhaps the most conspicuous and best-studied group of insects. The brightly colored, often iridescent wing patterns of many butterfly and moth species arise from pigments and light-refracting structures borne in the scales that cover the wings of all members of this group. (You may have noticed the scales as a powdery substance that rubs off onto your hand when you handle a butterfly or moth.) Butterflies fly mainly during the day, moths at night (though there are exceptions to this general rule, such as the hummingbird-like hawk moths that are often seen feeding by day in flower gardens). The evolution of butterflies and moths has been closely tied to the evolution of flowering plants. Butterflies and moths at all stages of life feed almost exclusively on flowering plants. Many species of flowering plants in turn depend on butterflies and moths for pollination.

Bees, Ants, and Wasps
These insects are known to many people by their painful stings. Many species in this group are equipped with a barbed stinger that extends from the abdomen and can be used to inject venom into the victim of a sting. The venom may be extremely toxic but, fortunately for human stinging victims, each insect carries only a tiny amount. Nonetheless, the amount is often sufficient to cause considerable pain. Only females have stingers, which are used by many stinging species to help defend a nest from attack by a potential predator. Defense, however, is not the only use for stingers. Many wasps, for example, have a parasitic form of reproduction: a wasp lays an egg inside the body of another species, typically a caterpillar, which becomes food for the wasp larva after it hatches. Before laying its egg, the wasp may sting the caterpillar, paralyzing it.

(a) (b)

(c) (d) (e)

FIGURE 23-21 The diversity of insects
(a) The rose aphid sucks sugar-rich juice from plants. (b) A mating pair of Hercules beetles. The large "horns" are found only on the male. (c) A June beetle displays its two pairs of wings as it comes in for a landing. The outer wings protect the abdomen and the inner wings, which are relatively thin and fragile. (d) Insects such as this locust can cause devastation of food crops and natural vegetation. (e) Caterpillars are larval forms of moths or butterflies. This caterpillar larva of the Australian fruit-sucking moth displays large eyespot patterns that may frighten potential predators, who mistake them for eyes of a large animal.

The social behavior of some ant and bee species is extraordinarily intricate. Such species form huge colonies with complex organization in which individuals specialize in particular tasks such as foraging, defense, reproduction, and rearing larvae. The organization and division of labor in these insect societies may require levels of communication and learning comparable to those present in vertebrates. The remarkable tasks accomplished by social insects include the manufacture and storage of food (honey) by honeybees and "farming" by ant species that cultivate fungi in underground chambers or "milk" aphids by inducing them to secrete a nutritious liquid.

Beetles

The largest insect order is the beetles, which account for about a third of all known insect species. As might be expected in such a large group, beetles exhibit a huge variety of shapes, sizes, and lifestyles. All beetles, however, have a hard, protective, exoskeletal structure that covers their wings. Many destructive agricultural pests are beetles, such as the Colorado potato beetle, the grain weevil, and the

Japanese beetle. However, other beetles, such as lady beetles, are predators that are used to control insect pests.

Among the many fascinating adaptations of beetles, one of the most impressive is found in the bombardier beetle. This species defends itself against ants and other enemies by emitting a toxic spray from a nozzle-like structure at the end of its abdomen. The beetle is able to precisely aim the spray, which emerges with explosive force and at temperatures higher than 200°F (93°C). The beetle is able to safely carry this weapon because it is not permanently present in the animal's body. Instead, it is quickly manufactured when needed by mixing two substances, normally carried in separate glands. Each substance is harmless by itself, but when the two are mixed, they form a boiling, caustic liquid.

Most Arachnids Are Predatory Meat Eaters

The *arachnids* include spiders, mites, ticks, and scorpions (**FIG. 23-22**). All members of the class Arachnida lack antennae and have eight walking legs, and most are carnivorous. Many subsist on a liquid diet of blood or predigested

FIGURE 23-22 The diversity of arachnids
(a) The tarantula is one of the largest spiders but is relatively harmless. **(b)** Scorpions, found in warm climates such as the deserts of the southwestern United States, paralyze their prey with venom from a stinger at the tip of the abdomen. A few species can harm humans. **(c)** Ticks before (left) and after feeding on blood. The uninflated exoskeleton is flexible and folded, allowing the animal to become grotesquely bloated while feeding.

prey. For example, spiders, the most numerous arachnids, first immobilize their prey with a paralyzing venom. They then inject digestive enzymes into the helpless victim (typically an insect) and suck in the resulting soup. Arachnids breathe by using either tracheae, lungs, or both.

In contrast to the compound eyes of insects and crustaceans, arachnids have simple eyes, each with a single lens. Most spiders have eight eyes placed in such a way as to give them a panoramic view of predators and prey. The eyes are sensitive to movement, and in some spider species—especially those that hunt actively and have no webs—the eyes can probably form images. Most spider perception, however, is not through their eyes but through sensory hairs. All spiders are hairy, and many of their hairs have sensory functions. Some of a spider's hairs are touch sensitive and help the animal perceive prey, mates, and surroundings. Other hairs are sensitive to chemicals and function as organs of smell and taste. Hairs also respond to vibrations in the air, ground, or web, allowing spiders to detect nearby movement by predators, prey, or other spiders.

Among the distinctive features of spiders is their production of protein threads known as silk. Spiders manufacture silk in special glands in their abdomens and use it to perform a variety of functions, such as building webs that capture prey, wrapping up and immobilizing cap-

tured prey, constructing protective shelters for themselves, making cocoons to surround their eggs, and making "draglines" that connect a spider to a web or other surface and support its weight if it drops from its perch. Each of these functions requires silk with different properties, and most spiders make several kinds of silk. Spider silk is an amazingly light, strong, and elastic fiber. Dragline silk can be as strong as or even stronger than a steel wire of the same size, yet it is as elastic as rubber. Human engineers have long sought to develop a fiber with this combination of strength and elasticity. Despite careful study of the structure of spider silk, no comparable man-made substance has been successfully manufactured. Some researchers have applied the techniques of biotechnology to the problem, inserting spider genes that code for silk proteins into mammal or bacterial cells that can be maintained in the lab. The hope is that such cells can be induced to produce silk.

Myriapods Have Many Legs

The *myriapods* include the centipedes and millipedes, whose most prominent feature is an abundance of legs (**FIG. 23-23**). Most millipede species have between 100 and 300 legs. Centipedes are not quite as leggy; a typical species has around 70 legs, but many species have fewer. Both centipedes and millipedes have one pair of antennae. The legs

FIGURE 23-23 **The diversity of myriapods**
(a) Centipedes and (b) millipedes are common nocturnal arthropods. Each segment of a centipede's body holds one pair of legs, while each millipede segment has two pairs.

and antennae of centipedes are longer and more delicate than those of millipedes. Myriapods have very simple eyes that detect light and dark but do not form images. In some species, the number of eyes can be high—up to 200. Myriapods respire by means of tracheae.

Myriapods inhabit terrestrial environments exclusively, living mostly in the soil or leaf litter or under logs and rocks. Centipedes are generally carnivorous, capturing prey (mostly other arthropods) with their frontmost legs, which are modified into sharp claws that inject poison into prey. Bites from large centipedes can be painful to humans. In contrast, most millipedes are not predators but instead feed on decaying vegetation and other debris. When attacked, many millipedes defend themselves by secreting a foul-smelling, distasteful liquid.

Most Crustaceans Are Aquatic

The *crustaceans*, including crabs, crayfish, lobster, shrimp, and barnacles, make up the only class of arthropods whose members live primarily in the water (**FIG. 23-24**). Crustaceans range in size from microscopic maxillopods that live in the spaces between grains of sand to the largest of all arthropods, the Japanese spider crab, with legs spanning nearly 12 feet (4 meters). Crustaceans have two pairs of

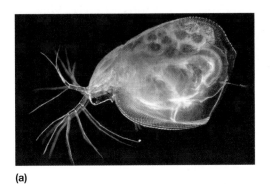
(a)

(b)

(c)

(d)

FIGURE 23-24 **The diversity of crustaceans**
(a) The microscopic waterflea is common in freshwater ponds. Notice the eggs developing within the body. (b) The sowbug, found in dark, moist places such as under rocks, leaves, and decaying logs, is one of the few crustaceans to invade the land successfully. (c) The hermit crab protects its soft abdomen by inhabiting an abandoned snail shell. (d) The gooseneck barnacle uses a tough, flexible stalk to anchor itself to rocks, boats, or even animals such as whales. Other types of barnacles attach with shells that resemble miniature volcanoes (see Fig. 23-15b). Early naturalists thought barnacles were mollusks until they observed barnacles' jointed legs (seen here extending into the water).

sensory antennae, but the rest of their appendages are highly variable in form and number, depending on the habitat and lifestyle of the species. Most crustaceans have compound eyes similar to those of insects, and nearly all respire by means of gills.

Roundworms Are Abundant and Mostly Tiny

Although you may be blissfully unaware of their presence, *roundworms* (phylum Nematoda) are nearly everywhere. Roundworms, also called *nematodes*, have colonized nearly every habitat on Earth, and they play an important role in breaking down organic matter. They are extremely numerous; a single rotting apple may contain 100,000 roundworms. Billions thrive in each acre of topsoil. In addition, almost every plant and animal species hosts several parasitic nematode species.

In addition to being abundant and ubiquitous, roundworms are diverse. Although only about 12,000 roundworm species have been named, there may be as many as 500,000. Most, such as the one shown in **FIGURE 23-25**, are microscopic, but some parasitic forms reach a meter in length.

Nematodes have a rather simple body plan, featuring a tubular gut and a fluid-filled pseudocoelom that surrounds the organs and forms a hydrostatic skeleton. A tough, flexible, nonliving cuticle encloses and protects the thin, elongated body and is periodically molted. The molting of roundworms reveals that they share a common evolutionary heritage with arthropods and other ecdysozoan phyla. Sensory organs in the roundworm head transmit information to a simple "brain," composed of a nerve ring.

Like flatworms, nematodes lack circulatory and respiratory systems. Because most nematodes are extremely thin and have low energy requirements, diffusion suffices for gas exchange and distribution of nutrients. Most nematodes reproduce sexually, and the sexes are separate; the male (normally smaller than the female) fertilizes the female by placing sperm inside her body.

During your life, you may become host to one of the 50 species of roundworms that infect humans. Most such worms are relatively harmless, but there are important exceptions. For example, hookworm larvae (found in soil) can bore into human feet, enter the bloodstream, and travel to the intestine, where they cause continuous bleeding. Another dangerous nematode parasite, *Trichinella*, causes the disease trichinosis. *Trichinella* worms can infect people who eat improperly cooked infected pork, which can contain up to 15,000 larval cysts per gram (**FIG. 23-26a**). The cysts hatch in the human digestive tract and invade blood vessels and muscles, causing bleeding and muscle damage.

Parasitic nematodes can also endanger domestic animals. Dogs, for example, are susceptible to heartworm, which is transmitted by mosquitoes (**FIG. 23-26b**). In the southern United States, and increasingly in other parts of the country, heartworm poses a severe threat to the health of unprotected pets.

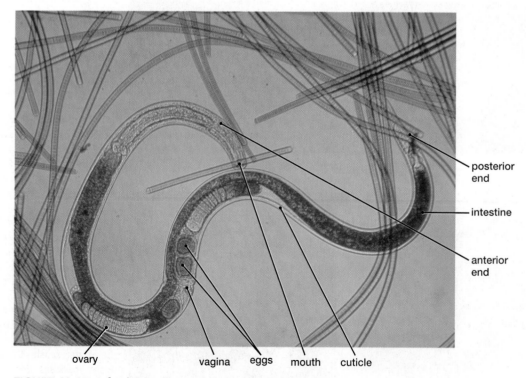

ovary   vagina   eggs   mouth   cuticle

posterior end

intestine

anterior end

FIGURE 23-25 A freshwater nematode
Eggs can be seen inside this female freshwater nematode, which feeds on algae.

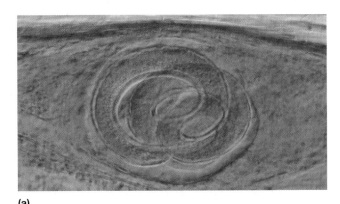

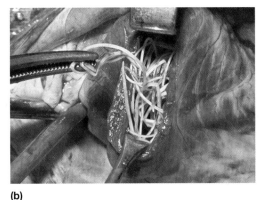

(a)   (b)

FIGURE 23-26 Some parasitic nematodes
(a) Encysted larva of the *Trichinella* worm in the muscle tissue of a pig, where it may live for up to 20 years.
(b) Adult heartworms in the heart of a dog. The juveniles are released into the bloodstream, where they may be ingested by mosquitoes and passed to another dog by the bite of an infected mosquito.

Echinoderms Have a Calcium Carbonate Skeleton

Echinoderms (phylum Echinodermata) are found only in marine environments, and their common names tend to evoke their saltwater habitats: sand dollars, sea urchins, sea stars (or starfish), sea cucumbers, and sea lilies (**FIG. 23-27**). The name "echinoderm" (Greek, "hedgehog skin") stems from the bumps or spines that extend from the skin of most echinoderms. These spines are especially well developed in sea urchins and much reduced in sea stars and sea cucumbers. Echinoderm bumps and spines are actually extensions of an **endoskeleton** (internal skeleton) composed of plates of calcium carbonate that lie beneath the outer skin.

Echinoderms exhibit deuterostome development and are linked by common ancestry with the other deuterostome phyla, including the chordates (described later). Deuterostomes form a group of branches on the larger evolutionary tree of bilaterally symmetrical animals, but in echinoderms bilateral symmetry is expressed only in embryos and free-swimming larvae. An adult echinoderm, in contrast, is radially symmetrical and lacks a head. This absence of cephalization is consistent with the sluggish or sessile existence of echinoderms. Most echinoderms move only very slowly as they feed on algae or small particles sifted from sand or water. Some echinoderms are slow-motion predators. Sea stars, for example, slowly pursue even slower-moving prey, such as bivalve mollusks.

Echinoderms move on numerous tiny *tube feet*, delicate cylindrical projections that extend from the lower surface of the body and terminate in a suction cup. Tube feet are part of a unique echinoderm feature, the *water-vascular system*, which functions in locomotion, respiration, and food capture (**FIG. 23-28**). Seawater enters through an opening (the *sieve plate*) on the animal's upper surface and is conducted through a circular central canal, from which branch a number of radial canals. These canals conduct water to the tube feet, each of which is controlled by a muscular squeeze bulb. Contraction of the bulb forces water into the tube foot, causing it to extend. The suction cup may be pressed against the seafloor or a food object, to which it adheres tightly until pressure is released.

Echinoderms have a relatively simple nervous system with no distinct brain. Movements are loosely coordinated by a system consisting of a nerve ring that encircles the

(a)   (b)   (c)

FIGURE 23-27 The diversity of echinoderms
(a) A sea cucumber feeds on debris in the sand. (b) The sea urchin's spines are actually projections of the internal skeleton. (c) The sea star has reduced spines and typically has five arms.

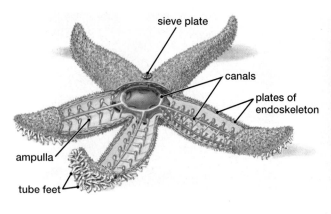

(a)

(b)

FIGURE 23-28 The water-vascular system of echinoderms
(a) Changing pressure inside the seawater-filled water-vascular system extends or retracts the tube feet. **(b)** The sea star often feeds on mollusks such as this mussel. A feeding sea star attaches numerous tube feet to the mussel's shells, exerting a relentless pull. Then, the sea star turns the delicate tissue of its stomach inside out, extending it through its centrally located ventral mouth. The stomach can fit through an opening in the bivalve shells which measures less than 1 millimeter. Once insinuated between the shells, the stomach tissue secretes digestive enzymes that weaken the mollusk, causing it to open further. Partially digested food is transported to the upper portion of the stomach, where digestion is completed.

esophagus, radial nerves to the rest of the body, and a nerve network through the epidermis. In sea stars, simple receptors for light and chemicals are concentrated on the arm tips, and sensory cells are scattered over the skin. In some brittle star species, light receptors are associated with tiny lenses, smaller than the width of a human hair, that gather light and focus it on receptors. The optical quality of these "microlenses" is excellent, far superior to that of any human-created lens of comparable size.

Echinoderms lack a circulatory system, although movement of the fluid in their well-developed coelom serves this function. Gas exchange occurs through the tube feet and, in some forms, through numerous tiny "skin gills" that project through the epidermis. Most species have separate sexes and reproduce by shedding sperm and eggs into the water, where fertilization occurs.

Many echinoderms are able to regenerate lost body parts, and these regenerative powers are especially potent in sea stars. In fact, a single arm of a sea star is capable of developing into a whole animal, provided that part of the central body is attached to it. Before this ability was widely appreciated, mussel fishermen often tried to rid mussel beds of predatory sea stars by hacking them into pieces and throwing the pieces back. Needless to say, the strategy backfired.

The Chordates Include the Vertebrates

The phylum Chordata, which contains the vertebrate animals, also includes a few groups of invertebrates, such as the sea squirts and the lancelets. We will discuss these invertebrate chordates and their vertebrate relatives in Chapter 24.

CASE STUDY REVISITED THE SEARCH FOR A SEA MONSTER

Clyde Roper's search for the giant squid has led him to organize three major expeditions. The first of these searched waters near the Azores Islands in the Atlantic Ocean. Because sperm whales are known to prey upon giant squid, Roper believed that the whales might lead him to the squids. To test this idea, he and his team affixed video cameras to sperm whales, thus allowing the scientists to see what the whales were seeing. These "crittercams" revealed a great deal of new information about sperm whale behavior but, alas, no footage of giant squids.

The next Roper-led expedition took place in the Kaikoura Canyon, an area of very deep water (3300 feet, or 1000 meters) off the coast of New Zealand. The scientists chose this spot because deep-sea fishing boats had recently captured several giant squid in the vicinity. Cameras were again deployed on sperm whales; but this time the mobile cameras were supplemented by a stationary, baited camera and an unmanned, remote-controlled submarine. Again, however, a large investment of time, money, and equipment yielded no squid sightings.

A few years later, Roper assembled a team of scientists for a return to Kaikoura Canyon. This time, the group was able to use Deep Rover, a one-person submarine that could carry an observer to depths of 2200 feet. The scientists used Deep

Rover to explore the canyon, following sperm whales in hopes that the huge mammals would lead them to giant squid. Unfortunately, the team again failed to find a squid.

Although Roper has pursued his search for the giant squid with extraordinary persistence, he is not alone in seeking a glimpse of the creature. Other research teams have also been on the lookout for giant squid, and it was one of these groups that finally secured the first (and so far only) visual record of living giant squid. Working off the coast of Japan, the researchers placed a video camera on a long, baited fishing line. Long hours of dragging the line through the water at a depth of 3000 feet were eventually rewarded with images of a giant squid that attacked the bait (**FIG. 23-29**).

Consider This Steve O'Shea, another scientist interested in the giant squid, captured a few juvenile giant squid in 2002. The tiny animals, only a few millimeters long, survived in captivity for just a few hours, but their identity as giant squid was confirmed by comparing their DNA to that of preserved adult specimens. O'Shea believes that with more research and experience, he could learn to raise the young animals to adulthood. Given that research funds are limited, which approach is better? Would we learn more from viewing wild adult squid in the depths of the ocean, or by capturing tiny juveniles from surface waters and figuring out how to raise them in the lab?

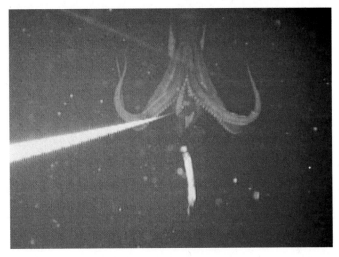

FIGURE 23-29 **A giant squid approaches a baited line**

CHAPTER REVIEW

SUMMARY OF KEY CONCEPTS

23.1   What Are the Key Features of Animals?

Animals are multicellular, sexually reproducing, heterotrophic organisms. Most can perceive and react rapidly to environmental stimuli and are motile at some stage in their lives. Their cells lack a cell wall.

23.2   Which Anatomical Features Mark Branch Points on the Animal Evolutionary Tree?

The earliest animals had no tissues, a feature retained by modern sponges. All other modern animals have tissues. Animals with tissues can be divided into radially symmetrical and bilaterally symmetrical groups. During embryonic development, radially symmetrical animals have two germ layers; bilaterally symmetrical animals have three. Bilaterally symmetrical animals also tend to have sense organs and clusters of neurons concentrated in the head, a process called *cephalization*. Bilateral phyla can be divided into two main groups, one of which undergoes protostome development, the other of which undergoes deuterostome development. Protostome phyla can in turn be divided into ecdysozoans and lophotrochozoans. Some phyla of bilaterally symmetrical animals lack body cavities, but most have either pseudocoeloms or true coeloms.

Web Tutorial 23.1 The Architecture of Animals

23.3   What Are the Major Animal Phyla?

The bodies of sponges (phylum Porifera) are typically free-form in shape and are sessile. Sponges have relatively few types of cells. Despite the division of labor among the cell types, there is little coordination of activity. Sponges lack the muscles and nerves required for coordinated movement, and digestion occurs exclusively within the individual cells.

The hydra, anemones, and jellyfish (phylum Cnidaria) have tissues. A simple network of nerve cells directs the activity of contractile cells, allowing loosely coordinated movements. Digestion is extracellular, occurring in a central gastrovascular cavity with a single opening. Cnidarians exhibit radial symmetry, an adaptation to both the free-floating lifestyle of the medusa and the sedentary existence of the polyp.

Flatworms (phylum Platyhelminthes) have a distinct head with sensory organs and a simple brain. A system of canals forming a network through the body aids in excretion. They lack a body cavity.

The segmented worms (phylum Annelida) are the most complex of the worms, with a well-developed closed circulatory system and excretory organs that resemble the basic unit of the vertebrate kidney. The segmented worms have a compartmentalized digestive system, like that of vertebrates, which processes food in a sequence. Annelids also have a true coelom, a fluid-filled space between the body wall and the internal organs.

The snails, clams, and squid (phylum Mollusca) lack a skeleton; some forms protect the soft, moist, muscular body with a single shell (many gastropods and a few cephalopods) or a pair of hinged shells (the bivalves). The lack of a waterproof external covering limits this phylum to aquatic and moist terrestrial habitats. Although the body plan of gastropods and bivalves limits the complexity of their behavior, the cephalopod's tentacles are capable of precisely controlled movements. The octopus has the most complex brain and the best-developed learning capacity of any invertebrate.

Arthropods, the insects, arachnids, millipedes and centipedes, and crustaceans (phylum Arthropoda), are the most diverse and abundant organisms on Earth. They have invaded nearly every available terrestrial and aquatic habitat. Jointed appendages and well-developed nervous systems make possible complex, finely coordinated behavior. The exoskeleton (which conserves water and provides support) and specialized respiratory structures (which remain moist and protected) enable the insects and arachnids to inhabit dry land. The diversification of insects has been enhanced by their ability to fly. Crustaceans, which include the largest arthropods, are restricted to moist, usually aquatic habitats and respire by using gills.

The pseudocoelomate roundworms (phylum Nematoda) possess a separate mouth and anus and a cuticle layer that is molted.

The sea stars, sea urchins, and sea cucumbers (phylum Echinodermata) are an exclusively marine group. Like other complex invertebrates and chordates, echinoderm larvae are bilaterally symmetrical; however, the adults show radial symmetry. This, in addition to a primitive nervous system that lacks a definite brain, adapts them to a relatively sedentary existence. Echinoderm bodies are supported by a nonliving internal skeleton that sends projections through the skin. The water-vascular system, which functions in locomotion, feeding, and respiration, is a unique echinoderm feature.

The phylum Chordata includes two invertebrate groups, the lancelets and tunicates, as well as the vertebrates.

KEY TERMS

bilateral symmetry   *page 443*
budding   *page 447*
cephalization   *page 444*
closed circulatory system   *page 452*
coelom   *page 444*
compound eye   *page 458*
deuterostome   *page 445*

ectoderm   *page 443*
endoderm   *page 443*
endoskeleton   *page 463*
exoskeleton   *page 456*
ganglion   *page 450*
hemocoel   *page 454*
hermaphroditic   *page 450*
hydrostatic skeleton   *page 452*

invertebrate   *page 445*
larva   *page 458*
mesoderm   *page 443*
metamorphosis   *page 458*
molt   *page 457*
nerve cord   *page 450*
open circulatory system   *page 454*

parasite   *page 450*
protostome   *page 445*
pseudocoelom   *page 444*
pupa   *page 458*
radial symmetry   *page 443*
segmentation   *page 452*
tissue   *page 443*
vertebrate   *page 445*

THINKING THROUGH THE CONCEPTS

1. List the distinguishing characteristics of each of the phyla discussed in this chapter, and give an example of each.

2. Briefly describe each of the following adaptations, and explain its adaptive significance: bilateral symmetry, cephalization, closed circulatory system, coelom, radial symmetry, segmentation.

3. Describe and compare respiratory systems in the three major arthropod classes.

4. Describe the advantages and disadvantages of the arthropod exoskeleton.

5. State in which of the three major mollusk classes each of the following characteristics is found:

    a. two hinged shells

    b. a radula

    c. tentacles

    d. some sessile members

    e. the best-developed brains

    f. numerous eyes

6. Give three functions of the water-vascular system of echinoderms.

7. To what lifestyle is radial symmetry an adaptation? Bilateral symmetry?

APPLYING THE CONCEPTS

1. The class Insecta is the largest taxon of animals on Earth. Its greatest diversity is in the Tropics, where habitat destruction and species extinction are occurring at an alarming rate. What biological, economic, and ethical arguments can you advance to persuade people and governments to preserve this biological diversity?

2. Discuss at least three ways in which the ability to fly has contributed to the success and diversity of insects.

3. Discuss and defend the attributes you would use to define biological success among animals. Are humans a biological success by these standards? Why?

FOR MORE INFORMATION

Adis, J., Zompro, O., Moombolah-Goagoses, E., and Marais, E. "Gladiators: A New Order of Insects." *Scientific American*, November 2002. An unusual insect, discovered fossilized in amber, is found to be a member of a previously unknown order. Later, living representatives of the new group were discovered in Africa.

Brusca, R. C., and Brusca, G. J. *Invertebrates*. Sunderland, MA: Sinauer, 1990. A thorough survey of the invertebrate animals in textbook format but readable and filled with beautiful and informative drawings.

Chadwick, D. H. "Planet of the Beetles." *National Geographic*, March 1998. The beauty and diversity of beetles, which comprise one-third of the world's insects, are described in text and photographs.

Conniff, R. "Stung." *Discover*, June 2003. The function, evolution, and diversity of stinging by ants, bees, and wasps.

Conover, A. "Foreign Worm Alert." *Smithsonian*, August 2000. Escaped night crawlers, annelids imported for use as fishing bait, threaten North American ecosystems.

Hamner, W. "A Killer Down Under." *National Geographic*, August 1994. Among the most poisonous animals in the world is the box jellyfish, which lives off the coast of northern Australia.

Kunzig, R. "At Home with the Jellies." *Discover*, September 1997. An account of the biologists who study jellyfish, and some of their findings. Includes excellent photos.

Morell, V. "Life on a Grain of Sand." *Discover*, April 1995. The sand beneath shallow waters is home to an incredible range of microscopic creatures.

Scigliano, E. "Through the Eye of an Octopus." *Discover*, October 2003. How intelligent are cephalopods? How do researchers try to find out?

Stix, G. "A Toxin Against Pain." *Scientific American*, April 2005. Fish-killing venom produced by a predatory snail contains substances that may prove to be valuable medicines.

24 Animal Diversity II: Vertebrates

Would you be shocked to learn that dinosaurs still walked the Earth? The discovery of modern coelacanth fishes was no less surprising.

CASE STUDY FISH STORY

MARJORIE COURTNEY-LATIMER received a phone call on December 22, 1938, that would lead to one of the most spectacular discoveries in biological history. The call was from a local fisherman whom Courtney-Latimer, the curator of a small museum in South Africa, had asked to collect some fish specimens for the museum. His boat had returned from its most recent voyage and was waiting at the town dock. Dutifully, Courtney-Latimer went to the boat and began sorting through the fish that were strewn across the deck. Later, she wrote, "I noticed a blue fin sticking up from beneath the pile. I uncovered the specimen, and, behold, there appeared the most beautiful fish I had ever seen." In addition to its beauty,

the fish had some odd features, including fins that were stumpy and lobed, unlike the fins of any other living species.

Courtney-Latimer did not recognize the strange fish, but she knew it was unusual. She tried to find a place to refrigerate it, but in her small town she was unable to find a cold storage facility willing to store a fish. In the end, she was able to save only the skin. Undaunted, she made some drawings of the fish and used them to attempt an identification. To her amazement, the creature did not resemble any species known to inhabit the waters off South Africa, but did seem similar to members of a family of fishes known as coelacanths. The only problem with this assessment was that coelacanths

were known only from fossils. The earliest coelacanth fossils were found in 400-million-year-old rocks and, as far as anyone knew, the group had been extinct for 80 million years!

Perplexed, Courtney-Latimer sent her drawings to J. L. B. Smith, a fish expert at Rhodes University. Smith was astounded when he saw the sketch, later writing that "a bomb seemed to burst in my brain." Although bitterly disappointed that the specimen's bones and internal organs had been lost, Smith arranged to view the preserved skin. Ultimately, he confirmed the astonishing news that coelacanths still swam Earth's waters.

24.1 WHAT ARE THE KEY FEATURES OF CHORDATES?

In both number of species and number of individuals, Earth's animals are overwhelmingly boneless invertebrates. Nonetheless, when we think of animals we tend to think of **vertebrates**—fish, reptiles, amphibians, birds, and mammals. Our bias toward vertebrates arises in part because, compared to invertebrates, they are generally larger and more conspicuous; a person is simply more likely to notice a crow or a squirrel than a flatworm or a clam. But our affinity for vertebrates also stems from their similarity to us. We are, after all, vertebrates ourselves.

All Chordates Share Four Distinctive Structures

Humans are members of the phylum Chordata (**FIG. 24-1**), which we share not only with birds and apes but also with the tunicates (sea squirts) and small fishlike creatures called lancelets. What characteristics do we share with these creatures that seem so different from us? All chordates have deuterostome development (which is also characteristic of echinoderms; see Chapter 23) and are further united by four features that all possess at some stage of their lives: a dorsal, hollow nerve chord; a notochord; pharyngeal gill slits; and a post-anal tail.

Dorsal, Hollow Nerve Cord

The **nerve cord** of chordates is hollow and lies above the digestive tract, running lengthwise along the dorsal (upper) portion of the body. In contrast, the nerve cords of other animals are solid and lie in a ventral position, below the digestive tract (see Figs. 23-11 and 23-13). During embryonic development in chordates, the nerve cord develops a thickening at its anterior end that becomes a brain.

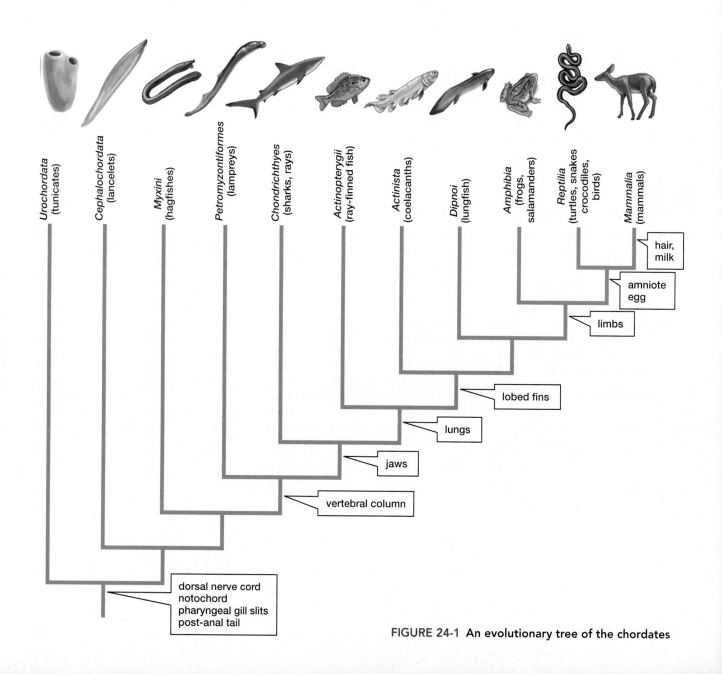

FIGURE 24-1 An evolutionary tree of the chordates

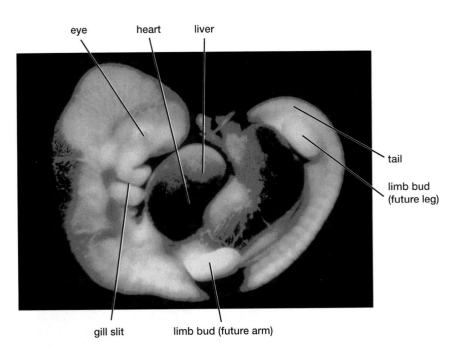

eye   heart   liver

tail

limb bud
(future leg)

gill slit   limb bud (future arm)

FIGURE 24-2 Chordate features in the human embryo
This 5-week-old human embryo is about 1 centimeter long and clearly shows a tail and external gill slits (more properly called grooves, since they do not penetrate the body wall). Although the tail will disappear completely, the gill grooves contribute to the formation of the lower jaw.

Notochord

The **notochord** is a stiff but flexible rod, located between the digestive tract and the nerve cord, that extends along the length of the body. It provides support for the body and an attachment site for muscles. In many chordates, the notochord is present only during early stages of development and disappears as a skeleton develops.

Pharyngeal Gill Slits

Pharyngeal gill slits are located in the pharynx (the cavity behind the mouth). They may form functional openings for gills (organs for gas exchange in water) or may appear only as grooves during an early stage of development.

Post-Anal Tail

The posterior portion of a chordate body extends past the anus to form a **post-anal tail**. Other animals lack this kind of tail, because their digestive tracts extend the full length of the body.

This list of distinctive chordate structures may seem puzzling because, although humans are chordates, at first glance we seem to lack every feature except the nerve cord. But evolutionary relationships are sometimes seen most clearly during early stages of development, and it is during our embryonic life that we develop, and lose, our notochord, our gill slits, and our tails (**FIG. 24-2**). Humans share these chordate features with all other vertebrates and with two invertebrate chordate groups, the lancelets and the tunicates.

The Invertebrate Chordates Live in the Seas

The invertebrate chordates lack the backbone that is the defining feature of vertebrates. These chordates comprise two groups, the lancelets and the tunicates. The small (2 inches, or about 5 centimeters, long), fishlike lancelet spends most of its time half-buried in the sandy sea bottom, filtering tiny food particles from the water. As can be seen

in **FIGURE 24-3a**, all four chordate features are present in the adult lancelet.

The tunicates form a larger group of marine invertebrate chordates that includes the sea squirts. It is difficult to imagine a less likely relative of humans than the immobile, filter-feeding, vaselike sea squirt (**FIG. 24-3b**). Its ability to move is limited to forceful contractions of its saclike body, which can send a jet of seawater into the face of anyone who plucks it from its undersea home; hence the name sea squirt. Although adult sea squirts are immobile, their larvae swim actively and possess the four chordate features (see Fig. 24-3b).

Vertebrates Have a Backbone

In vertebrates, the embryonic notochord is normally replaced during development by a backbone, or vertebral column. The **vertebral column** is composed of bone or **cartilage**, a tissue that resembles bone but is less brittle and more flexible. This column supports the body, offers attachment sites for muscles, and protects the delicate nerve cord and brain. It is also part of a living internal skeleton that can grow and repair itself. Because this internal skeleton provides support without the armor-like weight of the arthropod exoskeleton, it has allowed vertebrates to achieve great size and mobility.

Vertebrates show other adaptations that have contributed to their successful invasion of most habitats. One such adaptation is paired appendages. These first appeared as fins in fish and served as stabilizers for swimming. Over millions of years, some fins were modified by natural selection into legs that allowed animals to crawl onto dry land, and later into wings that allowed some to take to the air. Another adaptation that has contributed to the success of vertebrates is an increase in the size and complexity of their brains and sensory structures, which allow vertebrates to perceive their environment in detail and to respond to it in a great variety of ways.

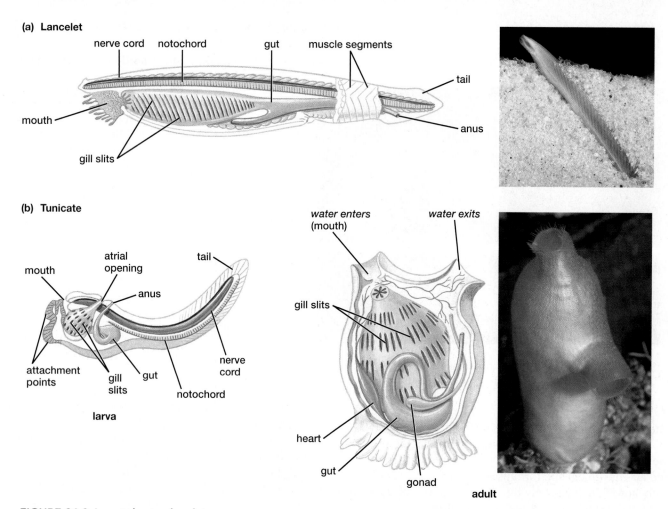

(a) Lancelet

nerve cord   notochord   gut   muscle segments

tail

mouth

gill slits

anus

(b) Tunicate

mouth

atrial opening

tail

anus

attachment points

gill slits

gut

nerve cord

notochord

larva

water enters (mouth)

water exits

gill slits

heart

gut

gonad

adult

FIGURE 24-3 Invertebrate chordates
(a) A lancelet, a fishlike invertebrate chordate. The adult organism exhibits all the diagnostic features of chordates. **(b)** The sea squirt larva (left) also exhibits all the chordate features. The adult sea squirt (a type of tunicate, middle) has lost its tail and notochord and has assumed a sedentary life, as shown in the photo (right).

24.2 WHAT ARE THE MAJOR GROUPS OF VERTEBRATES?

The evolutionary ancestor of vertebrates was probably an organism similar to today's lancelets. The earliest known vertebrates, whose fossils were found in 530-million-year-old rocks, resembled lancelets but had brains, skulls, and eyes. Today, vertebrates include lampreys, cartilaginous fishes, bony fishes, amphibians, reptiles, birds, and mammals.

Some Vertebrates Lack Jaws

The mouths of the earliest vertebrates did not include jaws. The early history of vertebrates was characterized by an array of strange, now-extinct jawless fishes, many of which were protected by bony armor plates. Today, two groups of jawless fishes survive: the hagfishes (class Myxini) and the lampreys (class Petromyzontiformes). Although both hagfishes and lampreys have eel-shaped bodies and smooth, unscaled skin, the two groups represent distinct, early branches of the chordate evolutionary tree. The branch leading to modern hagfishes is the more ancient of the two.

Hagfishes Are Slimy Residents of the Ocean Floor

The hagfish body is stiffened by a notochord, but its "skeleton" is limited to a few small cartilage elements, one of which forms a rudimentary braincase. Because hagfishes lack skeletal elements that surround and protect the nerve cord, most systematists do not consider them to be vertebrates. Instead, hagfishes represent the chordate group that is most closely related to the vertebrates.

Hagfishes are exclusively marine (**FIG. 24-4a**). They live near the ocean floor, often burrowing in the mud, and feed primarily on worms. They will, however, eagerly attack dead and dying fish, using pincerlike teeth to burrow into their prey's body and consume the soft internal organs. Hagfishes are regarded with great disgust by some fishermen because they secrete massive quantities of slime as a defense against predators. Despite their well-deserved reputation as "slimeballs of the sea," hagfishes are avidly pursued by many commercial fishermen, because the leather industry in some parts of the world provides a market for hagfish skin. Most leather items that purport to be "eel skin" are in fact made from tanned hagfish skin.

(a)

(b)

FIGURE 24-4 Jawless fishes
(a) Hagfishes live in communal burrows in mud, feeding on worms. (b) Some lampreys are parasitic, attaching to fish (such as this carp) with sucker-like mouths lined with rasping teeth (**inset**).

(a)

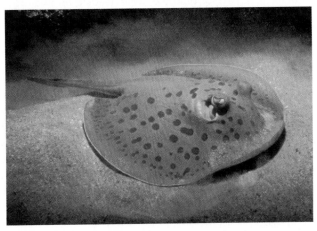
(b)

FIGURE 24-5 Cartilaginous fishes
(a) A sand tiger shark displaying several rows of teeth. As outer teeth are lost, they are replaced by the new ones behind them. Both sharks and rays lack a swim bladder and tend to sink toward the bottom when they stop swimming. (b) The tropical blue-spotted stingray swims by graceful undulations of lateral extensions of the body.

Some Lampreys Parasitize Fish

A lamprey is recognizable by the large, rounded sucker that surrounds its mouth and by the single nostril on the top of its head. The nerve cord of a lamprey is protected by segments of cartilage, so lampreys are considered to be true vertebrates. They live in both fresh and salt waters, but the marine forms must return to fresh water to spawn.

Some lamprey species are parasitic. A parasitic lamprey uses its tooth-lined mouth to attach itself to a larger fish (**FIG. 24-4b**). Using rasping teeth on its tongue, the lamprey excavates a hole in the host's body wall, through which it sucks blood and body fluids. Beginning in the 1920s, lampreys spread into the Great Lakes. There, in the absence of effective predators, they have multiplied prodigiously and greatly reduced commercial fish populations, including the lake trout. Vigorous measures to control the lamprey population have allowed some recovery of the other fish populations of the Great Lakes.

Jawed Fishes Rule Earth's Waters

About 425 million years ago, jawless fishes that were ancestors of the lampreys and hagfishes gave rise to a group of fish that possessed an important new structure: jaws. Jaws allowed fish to grasp, tear, or crush their food, permitting them to exploit a much wider range of food sources than could jawless fish. Although the earliest forms of jawed fishes have been extinct for 230 million years, they gave rise to the groups of jawed fishes that survive today: the cartilaginous fishes, the ray-finned fishes, and the lobe-finned fishes.

Cartilaginous Fishes Are Marine Predators

The class Chondrichthyes includes 625 marine species, among them the sharks, skates, and rays (**FIG. 24-5**). These cartilaginous fishes are graceful predators that lack any

473

(a)

(b)

(c)

FIGURE 24-6 The diversity of ray-finned fishes
Ray-finned fishes have colonized nearly every aquatic habitat. **(a)** This female deep-sea anglerfish attracts prey with a living lure that projects just above her mouth. The fish is ghostly white; at the 6000-foot depth where anglers live, no light penetrates and thus colors are unnecessary. Male deep-sea anglerfish are extremely small and remain attached to the female as permanent parasites, always available to fertilize her eggs. Two parasitic males can be seen attached to this female. **(b)** This tropical green moray eel lives in rocky crevices. A small fish (a banded cleaner goby) on its lower jaw eats parasites that cling to the moray's skin. **(c)** The tropical seahorse may anchor itself with its prehensile tail (adapted for grasping) while feeding on small crustaceans. QUESTION With regard to water regulation (maintaining the proper amount of water in the body), how does the challenge faced by a freshwater fish differ from that faced by a saltwater fish?

bone in their skeleton, which is formed entirely of cartilage. The body is protected by a leathery skin roughened by tiny scales. Members of this group respire using gills. Although some must swim to circulate water through their gills, most can pump water across their gills. Like all fishes, the cartilaginous fishes have two-chambered hearts. Some cartilaginous fishes are very large. A whale shark, for example, can grow to more than 45 feet (15 meters) in length, and a manta ray may be more than 20 feet (7 meters) wide and weigh more than 3000 pounds.

Although some sharks feed by filtering plankton (tiny animals and algae) from the water, most sharks are fearsome predators of larger prey such as other fishes, marine mammals, sea turtles, crabs, or squid. Many sharks attack their prey with strong jaws that contain several rows of razor-sharp teeth; the back rows move forward as the front teeth are lost to age and use (see Fig. 24-5a).

Most sharks avoid humans, but large sharks of some species can be dangerous to swimmers and divers. However, shark attacks on people are rare. A U.S. resident is 30 times more likely to die from a lightning strike than from a shark attack, and a beachgoer is far more likely to drown than to be bitten by a shark. Nonetheless, shark attacks do occur. In the United States during 2004, for example, there were 30 documented attacks, two of them fatal.

Skates and rays are mostly bottom dwellers with flattened bodies, wing-shaped fins, and thin tails (see Fig. 24-5b). Most skates and rays eat invertebrates. Some species defend themselves with a spine near their tail that can inflict dangerous wounds, and others produce a powerful electrical shock that can stun their prey.

Ray-Finned Fishes Are the Most Diverse Vertebrates

Just as our size bias makes us overlook the most diverse invertebrate groups, our habitat bias makes us overlook the most diverse vertebrates. The most diverse and abundant vertebrates are not the birds or the predominantly terrestrial mammals. The vertebrate diversity crown belongs instead to the lords of the oceans and freshwater, the ray-finned fishes (class Actinopterygii). About 24,000 species have been identified, and scientists estimate that perhaps twice this number exist, including many in deep waters and remote areas. Ray-finned fish are found in nearly every watery habitat, both freshwater and marine.

Ray-finned fish are distinguished by the structure of their fins, which are formed by webs of skin supported by bony spines. In addition, ray-finned fishes have skeletons made of bone, a trait they share with the lobe-finned fishes and limbed vertebrates discussed later in this chapter. The skin of ray-finned fishes is covered with interlocking scales that provide protection while allowing for flexibility. Most ray-finned fish have a swim bladder, a sort of internal balloon that allows a fish to float effortlessly at any level. The swim bladder evolved from lungs, which were present (along with gills) in the ancestors of modern ray-finned fishes.

The ray-finned fishes include not only a large number of species but also a huge variety of different forms and lifestyles (**FIG. 24-6**). These range from snakelike eels to flattened flounders; from sluggish bottom feeders that probe the seafloor to speedy, streamlined predators that range in open water; from brightly colored reef dwellers to translucent, luminescent deep-sea dwellers; from the massive 3000-pound mola to the tiny stout infantfish, which weighs in at about 0.00003 ounces (1 milligram).

Ray-finned fishes are an extremely important source of food for humans; we collectively consume a prodigious number of fish. Unfortunately, however, our appetite for ray-finned fish, combined with increasingly effective high-tech methods for finding and catching them, has had a devastating impact on fish populations. Biologists report that

(a)

(b)

FIGURE 24-7 Lungfish are lobe-finned fish
Among the fishes, **(a)** lungfish are the group most closely related to land-dwelling vertebrates. **(b)** Lungfish may wait out dry periods sealed in burrows in the mud.

populations of almost all economically important ray-finned fish species have declined drastically. Large, predatory fish such as tuna and cod have been especially hard-hit; today's populations of these species contain less than 10% of the number present when commercial fishing began. If overfishing continues, fish stocks are likely to collapse. The solution to this problem—catch fewer fish—is simple in concept but very difficult to implement in practice, due to economic and political factors.

Lobe-Finned Fishes Include the Closest Living Relatives of Tetrapods

Although almost all fish with bony skeletons belong to the ray-finned group, some bony fishes are members of a different group, the lobe-finned fishes. Lobe-finned fishes have fleshy fins that contain rod-shaped bones surrounded by a thick layer of muscle. Living fish with this feature actually constitute two distinct lineages that have been evolving separately for hundreds of millions of years. One lineage includes the coelacanths (Actinista), discussed at length in this chapter's Case Study (see the chapter-opening photo). The other lineage includes the lungfish (Dipnoi), of which only six species have survived to modern times (**FIG. 24-7a**). These survivors are the closest living relatives of the tetrapods, which, instead of fins, have limbs that can support their weight on land and digits (fingers or toes) on the ends of those limbs.

Lungfish, which are found in freshwater habitats in Africa, South America, and Australia, have both gills and lungs. They tend to live in stagnant waters that may be low in oxygen, and their lungs allow them to supplement their supply of oxygen by breathing air directly. Lungfish of several species are able to survive even if the pool they inhabit dries up completely. These fish burrow into mud and seal themselves in a mucus-lined chamber (**FIG 24-7b**). There, they breathe through their lungs as their metabolic rate declines drastically. When the rains return and the pool refills, the lungfish leave their burrows and resume their normal underwater way of life.

In addition to the coelacanths and lungfish, several other lineages of lobefins arose early in the evolutionary history of jawed fish. Some groups of early lobe-finned fishes developed modified fleshy fins that, in an emergency, could be used as legs, allowing the fish to drag itself from a drying puddle to a deeper pool. We know from fossils that at least one species even evolved actual limbs, although the function of limbs in a water-dwelling organism is not well understood. One group of such ancestors ultimately gave rise to the vertebrates that made the first tentative invasion of the land: the amphibians.

Amphibians Live a Double Life

The 4800 species of amphibians (class Amphibia) straddle the boundary between aquatic and terrestrial existence (**FIG. 24-8**). The limbs of amphibians show varying degrees of adaptation to movement on land, from the belly-dragging crawl of salamanders to the long leaps of frogs and toads. A three-chambered heart (in contrast to the two-chambered heart of fishes) circulates blood more efficiently, and lungs replace gills in most adult forms. Amphibian lungs, however, are poorly developed and must be supplemented by the skin, which serves as an additional respiratory organ. This respiratory function requires that the skin remain moist, a constraint that greatly restricts the range of amphibian habitats on land.

Amphibians are also tied to moist habitats by their breeding behavior, which requires water. Their fertilization is external and takes place in water, where the sperm can swim to the eggs. The eggs must remain moist, as they are protected only by a jellylike coating that leaves them vulnerable to water loss by evaporation. Different amphibian species keep their eggs moist in different ways, but many species simply lay their eggs in water. In some amphibian species, fertilized eggs develop into aquatic larvae such as the tadpoles of some frogs and toads. These aquatic larvae undergo a dramatic transformation into

Frogs and toads have lived in Earth's ponds and swamps for nearly 150 million years, somehow surviving the Cretaceous catastrophe that extinguished the dinosaurs and so many other species about 65 million years ago. Their evolutionary longevity, however, doesn't protect them from the environmental changes wrought by human activities. Over the past two decades, herpetologists (biologists who study reptiles and amphibians) from around the world have documented an alarming decline in amphibian populations. Thousands of species of frogs, toads, and salamanders are dramatically decreasing in number, and many have apparently gone extinct.

This is a worldwide phenomenon; population crashes have been reported from every part of the globe. Yosemite toads and yellow-legged frogs are disappearing from the mountains of California. Tiger salamanders have been nearly wiped out in the Colorado Rockies. Leopard frogs, eagerly chased by children, are becoming rare in the United States. Logging destroys the habitats of amphibians from the Pacific Northwest to the tropics (**FIG. E24-1**), but even amphibians in protected areas are dying. In the Monteverde Cloud Forest Preserve in Costa Rica, the golden toad was common in the early 1980s but has not been seen since 1989. The gastric brooding frog of Australia fascinated biologists by swallowing its eggs, brooding them in its stomach, and later regurgitating fully formed offspring. This species was abundant and seemed safe in a national park. Suddenly, in 1980, the gastric brooding frog disappeared and hasn't been seen since.

The causes of the worldwide decline in amphibian diversity are not fully understood, but researchers have discovered that frogs and toads in many places are succumbing to infection by a pathogenic fungus. The fungus has been found in the skin of dead and dying frogs in widespread locations, including Australia, Central America, and the west-

ern United States. In those places, discovery of the fungus has coincided with massive frog and toad die-offs, and most herpetologists agree that the fungus is causing the deaths.

It seems unlikely, however, that the fungus alone is responsible for the worldwide decline of amphibians. For one thing, die-offs have occurred in many places where the fungus has not been found. In addition, many herpetologists believe that the fungal epidemic would not have arisen if the frogs and toads had not first been weakened by other stresses. So, if the fungus is not doing all of the damage on its own, what are the other possible causes of amphibian decline? All of the most likely causes stem from human modification of the biosphere—the portion of Earth that sustains life.

Habitat destruction, especially the draining of wetlands that are especially hospitable to amphibian life, is one major cause of the decline. Amphibians are also vulnerable to toxic substances in the environment. For example, researchers found that frogs exposed to trace amounts of atrazine, a widely used herbicide that is found in virtually all fresh water in the United States, suffered severe damage to their reproductive tissues. The unique biology of amphibians makes them especially susceptible to poisons in the environment. Amphibian bodies at all stages of life are protected only by a thin, permeable skin that pollutants can easily penetrate. To make matters worse, the double life of many amphibians exposes their permeable skin to a wide range of aquatic and terrestrial habitats and to a correspondingly wide range of environmental toxins.

Amphibian eggs can also be damaged by ultraviolet (UV) light, according to research by Andrew Blaustein, an ecologist at Oregon State University. Blaustein demonstrated that the eggs of some species of frogs in the Pacific Northwest are sensitive to damage from UV light and that the most sen-

(a)

(b)

(c)

FIGURE 24-8 "Amphibian" means "double life"
The double life of amphibians is illustrated by the bullfrog's transition from **(a)** a completely aquatic larval tadpole to **(b)** an adult leading a semiterrestrial life. **(c)** The red salamander is restricted to moist habitats in the eastern United States. Salamanders hatch in a form that closely resembles the adult. QUESTION What advantages might amphibians gain from their "double life"?

sitive species are experiencing the most drastic declines. Unfortunately, many parts of Earth are subject to increasingly intense UV radiation levels, because atmospheric pollutants have caused a thinning of the protective ozone layer.

Another disturbing trend among frogs and toads is the increasing incidence of grotesquely deformed individuals. Researchers at the Environmental Protection Agency recently demonstrated that developing frogs exposed to current natural levels of UV light grow deformed limbs much more frequently than do frogs protected from UV radiation. Other researchers have shown that deformities are more common in frogs exposed to low concentrations of commonly used pesticides. In addition, a growing body of evidence suggests that some deformities—especially the most common one, extra limbs—are caused by parasitic infections during embryonic development. Many frogs with extra legs are infested with a parasitic flatworm, and researchers have shown that tadpoles experimentally infected with the flatworm developed into deformed adults. Why have the parasites, which have long coexisted with frogs, suddenly begun causing so many deformities? A likely explanation is that exposure to UV radiation, pesticides, and herbicides weakens frogs' immune response. A compromised immune system leaves a developing tadpole more vulnerable to parasitic infection.

Many scientists believe that the troubles of amphibians signal an overall deterioration of Earth's ability to support life. According to this line of reasoning, the highly sensitive amphibians are providing an early warning of environmental degradation that will eventually affect more-resistant organisms as well. Equally worrisome is the observation that amphibians are not just sensitive indicators of the health of the biosphere but also crucial components of many ecosystems. They may keep insect populations in check, in turn serving as food for larger carnivores. Their decline will further disrupt the balance of these delicate communities.

Margaret Stewart, an ecologist at the State University of New York, Albany, aptly summarized the problem: "There's a famous saying among ecologists and environmentalists: 'Everything is related to everything else.' . . . You can't wipe out one large component of the system and not see dramatic changes in other parts of the system."

FIGURE E24-1 Amphibians in danger
The corroboree toad, shown here with its eggs, is rapidly declining in its native Australia. Tadpoles are developing within the eggs. The thin water-permeable and gas-permeable skin of the adult and the jellylike coating around the eggs make them vulnerable to both air and water pollutants.

(a)   (b)   (c)

FIGURE 24-9 The diversity of reptiles
(a) This milk snake has a color pattern very similar to that of the poisonous coral snake, which potential predators avoid. This mimicry helps the harmless milk snake elude predation. (b) The outward appearance of the American alligator, found in swampy areas of the South, is almost identical to that of 150-million-year-old fossil alligators. (c) The tortoises of the Galápagos Islands, Ecuador, may live to be more than 100 years old.

semiterrestrial adults, a metamorphosis that gives the amphibians their name, which means "double life." Their double life and their thin, permeable skin have made amphibians particularly vulnerable to pollutants and to environmental degradation, as described in "Earth Watch: Frogs in Peril."

Reptiles and Birds Are Adapted for Life on Land

The reptiles include lizards, snakes, alligators, crocodiles, turtles (**FIG. 24-9**), and birds. Reptiles evolved from an amphibian ancestor about 250 million years ago. Early reptiles, the dinosaurs, ruled the land for nearly 150 million years.

Reptiles Haves Scales and Shelled Eggs

Some reptiles, particularly desert dwellers such as tortoises and lizards, are completely independent of their aquatic origins. They achieved this independence through a series of adaptations, three of which are outstanding: (1) Reptiles evolved a tough, scaly skin that resists water loss and protects the body. (2) Reptiles evolved internal fertilization, in which the male deposits sperm within the female's body. (3) Reptiles evolved a shelled **amniote egg**, which can be buried in sand or dirt, far from water and hungry predators. The shell prevents the egg from drying out on land. An internal membrane, the **amnion**, encloses the embryo in the watery environment that all developing animals require (**FIG. 24-10**).

In addition to these features, reptiles have more-efficient lungs than do earlier vertebrates and do not use their skin as a respiratory organ. The three-chambered heart became modified, allowing better separation of oxygenated and deoxygenated blood, and the limbs and skeleton evolved features that provided better support and more efficient movement on land.

Lizards and Snakes Share a Common Evolutionary Heritage

Lizards and snakes together form a distinct lineage containing about 6800 species. The common ancestor of snakes and lizards had limbs, which are retained by most lizards but have been lost in snakes. The limbed ancestry of snakes is revealed by remnants of hindlimb bones that are present in some snake species.

Most lizards are small predators that eat insects or other small invertebrates, but a few lizard species are quite large. The Komodo dragon, for example, can reach 10 feet in length and weigh over 200 pounds. These giant lizards live in Indonesia and have powerful jaws and inch-long teeth that enable them to prey on large animals including

deer, goats, and pigs. The Komodo dragon, however, does not rely on its teeth alone to kill its prey. Its mouth is home to at least 50 different species of bacteria, many of which are harmful to animals. If an animal bitten by a Komodo dragon is not immediately killed, it is very likely to develop an infection that will kill it within a few days. The lizard simply waits patiently until its wounded prey dies. Why is the Komodo dragon itself not harmed by the deadly bacteria in its mouth? The lizard's blood contains antimicrobial compounds that apparently protect it from infection.

Most snakes are active, predatory carnivores and have a variety of adaptations that help them acquire food. For example, many snakes have special sense organs that help track prey by sensing small temperature differences between a prey's body and its surroundings. Some snake species immobilize prey with venom that is delivered through hollow teeth. Snakes also have a distinctive jaw joint that allows the jaws to distend so that the snake can swallow prey much larger than its head.

Alligators and Crocodiles Are Adapted for Life in Water

Crocodilians, as the 21 species of alligators and crocodiles are collectively known, are found in coastal and inland waters of the warmer regions of Earth. They are well adapted to an aquatic lifestyle, with eyes and nostrils located high on their heads so that they are able to remain submerged for long periods with only the uppermost portion of the head above the water's surface. Crocodilians have strong jaws and conical teeth that they use to crush and kill the fish, birds, mammals, turtles, and amphibians that they eat.

Parental care is extensive in crocodilians, which bury their eggs in mud nests. Parents guard the nest until the young hatch and then carry their newly hatched young in their mouths, moving them to safety in the water. Young crocodilians may remain with their mother for several years.

Turtles Have Protective Shells

The 240 species of turtles occupy a variety of habitats, including deserts, streams and ponds, and the ocean. This variety of habitats has fostered a variety of adaptations; but all turtles are protected by a hard, boxlike shell that is fused to the vertebrae, ribs, and collarbone. Turtles have no teeth but have instead evolved a horny beak. The beak is used to eat a variety of foods; some turtles are carnivores, some are herbivores, and some are scavengers. The largest turtle, the leatherback, is an ocean dweller that can grow to 6 feet (2 meters) or more in length and feeds largely on jellyfish. Leatherbacks and other marine turtles must return to land to breed and often undertake extraordinary long-distance migrations to reach the beaches on which they bury their eggs.

Birds Are Feathered Reptiles

One very distinctive group of reptiles is the birds (**FIG. 24-11**). Although the 9600 species of birds have traditionally been classified as a group separate from reptiles, biologists have shown that birds are really a subset of an evolutionary group that includes both birds and the groups that

FIGURE 24-10 The amniote egg
An anole lizard struggles free of its egg. The amniote egg encapsulates the developing embryo in a fluid-filled membrane (the amnion), ensuring that development occurs in a watery environment, even if the egg is far from water.

(a)

(b)

(c)

FIGURE 24-11 The diversity of birds
(a) The delicate hummingbird beats its wings about 60 times per second and weighs about 0.15 ounce (4 grams).
(b) This young frigate bird, a fish-eater from the Galápagos Islands, has nearly outgrown its nest. (c) The ostrich, the largest of all birds, weighs more than 300 pounds (136 kilograms); its eggs weigh more than 3 pounds (1500 grams). QUESTION Although the ancestor of all birds could fly, many bird species—such as the ostrich—cannot. Why do you suppose flightlessness has evolved repeatedly among birds?

have been traditionally designated as reptiles (see page 361 in Chapter 18 for a more complete explanation). The first birds appear in the fossil record roughly 150 million years ago (**FIG. 24-12**) and are distinguished from other reptiles by feathers, which are essentially a highly specialized version of reptilian body scales. Modern birds retain scales on their legs—evidence of the ancestry they share with the rest of the reptiles.

Bird anatomy and physiology are dominated by adaptations that help them to fly. In particular, birds are exceptionally light for their size. Hollow bones reduce the weight of the bird skeleton to a fraction of that of other vertebrates, and many bones present in other reptiles have been lost in the course of evolution or fused with other bones. Reproductive organs shrink considerably during nonbreeding periods, and female birds possess only a single ovary, further minimizing weight. The shelled egg that contributed to the reptiles' success on land frees the mother bird from carrying her developing offspring internally. Feathers form lightweight extensions to the wings and the tail for the lift and control required for flight, and they provide lightweight protection and insulation for the body. The nervous system of birds accommodates the special demands of flight with extraordinary coordination and balance, combined with acute eyesight.

Birds are also able to maintain body temperatures high enough to allow their muscles and metabolic processes to operate at peak efficiency, supplying the power to fly regardless of the outside temperature. This physiological ability to maintain an internal temperature that is usually higher than that of the surrounding environment is characteristic of both birds and mammals, which are therefore sometimes described as warm blooded or endothermic. In contrast, the body temperature of invertebrates, fish, amphibians, and reptiles varies with the temperature of their environment, though these animals may exert some control of their body temperature by their behavior (such as basking in the sun or seeking shade).

Warm-blooded animals such as birds have a high metabolic rate, which increases their demand for energy and requires efficient oxygenation of tissues. Therefore, birds must eat frequently and possess circulatory and respiratory adaptations that help meet the need for efficiency. A bird's heart has four chambers, thus preventing the mixing of oxygenated and deoxygenated blood (alligators and crocodiles also have four-chambered hearts). The respiratory system of birds is supplemented by air sacs that provide a continuous supply of oxygenated air to the lungs, even while the bird exhales.

FIGURE 24-12 Archaeopteryx, the earliest-known bird
Archaeopteryx is preserved in 150-million-year-old limestone. Feathers, a feature unique to birds, are clearly visible; but the reptilian ancestry of birds is also apparent: like a modern reptile (but unlike a modern bird), *Archaeopteryx* had teeth, a tail, and claws.

Mammals Provide Milk to Their Offspring

One branch of the reptile evolutionary tree gave rise to a group that evolved hair and diverged to form the mammals (class Mammalia). The mammals first appeared approximately 250 million years ago but did not diversify and become prominent on land until after the dinosaurs went extinct roughly 65 million years ago. In most mammals, fur protects and insulates the warm body. Like birds, alligators, and crocodiles, mammals have four-chambered hearts that increase the amount of oxygen delivered to the tissues. Legs designed for running rather than crawling make many mammals fast and agile.

Mammals are named for the milk-producing **mammary glands** used by all female members of this class to suckle their young. In addition to these unique glands, the mammalian body has sweat, scent, and sebaceous (oil-producing) glands, none of which are found in other vertebrates. The mammalian nervous system has contributed significantly to the success of the mammals by making possible behavioral adaptation to changing and varied environments. The brain is more highly developed than in any other vertebrate group, giving mammals unparalleled curiosity and learning ability. Their highly developed brain allows mammals to alter their behavior on the basis of experience and helps them survive in a changing environment. Relatively long periods of parental care after birth allow some mammals to learn extensively under parental guidance. Humans and other primates are good examples. In fact, the large brains of humans have been the major factor leading to human domination of Earth.

The 4600 species of mammals include three main evolutionary lineages: monotremes, marsupials, and placental mammals.

Monotremes Are Egg-Laying Mammals

Unlike other mammals, **monotremes** lay eggs rather than giving birth to live young. This group includes only three species: the platypus and two species of spiny anteaters, also known as echidnas (**FIG. 24-13**). Monotremes are found only in Australia (the platypus and short-beaked echidna) and New Guinea (the long-beaked echidna).

Echidnas are terrestrial and eat insects or earthworms that they dig out of the ground. Platypuses forage for food in the water, diving below the surface to capture small vertebrate and invertebrate animals. Platypus bodies are well adapted for this aquatic lifestyle, with a streamlined shape, webbed feet, a broad tail, and a fleshy, duck-like bill that is used to probe for food.

Monotreme eggs have leathery shells and are incubated for 10–12 days by the mother. Echidnas have a special pouch for incubating eggs, but platypus eggs are held for incubation between the mother's tail and belly. Newly hatched monotremes are tiny and helpless and feed on milk secreted by the mother. Monotremes, however, lack nipples. Milk from the mammary glands oozes through ducts on the mother's abdomen and soaks the fur around the ducts. The young then suck the milk from the fur.

Marsupial Diversity Reaches Its Peak in Australia

In all mammals except the monotremes, embryos develop in the uterus, a muscular organ in the female reproductive

(a)

(b)

FIGURE 24-13 Monotremes
(a) Monotremes, such as this platypus, lay leathery eggs resembling those of reptiles. Platypuses live in burrows that they dig on the banks of rivers, lakes, or streams. (b) The short limbs and heavy claws of spiny anteaters (also known as echidnas) help them unearth insects and earthworms to eat. The stiff spines that cover a spiny anteater's body are modified hairs.

(a)   (b)   (c)

FIGURE 24-14 Marsupials
(a) Marsupials, such as the wallaby, give birth to extremely immature young who immediately grasp a nipple and develop within the mother's protective pouch (inset). (b) The wombat is a burrowing marsupial whose pouch opens toward the rear of its body to prevent dirt and debris from entering the pouch during tunnel digging. One of the wombat's predators is (c) the Tasmanian devil, the largest carnivorous marsupial.

tract. The lining of the uterus combines with membranes derived from the embryo to form the **placenta**, a structure that allows gases, nutrients, and wastes to be exchanged between the circulatory systems of the mother and embryo.

In **marsupials**, embryos develop in the uterus for only a short period. Marsupial young are born at a very immature stage of development. Immediately after birth, they crawl to a nipple, firmly grasp it, and, nourished by milk, complete their development. In most, but not all, marsupial species, this postbirth development takes place in a protective pouch.

Only one marsupial species, the Virginia opossum, is native to North America. The majority of the 275 species of marsupials are found in Australia, where marsupials such as kangaroos have come to be seen as emblematic of the island continent. Kangaroos are the largest and most conspicuous of Australia's marsupials; the largest species, the red kangaroo, may be 7 feet tall and can make 30-foot leaps when moving at top speed. Though kangaroos are perhaps the most familiar marsupials, the group encompasses species with a range of sizes, shapes, and lifestyles, including koalas, wombats, and the Tasmanian devil (**FIG. 24-14**).

Placental Mammals Inhabit Land, Air, and Sea

Most mammal species are **placental** mammals, so named because their placentas are far more complex than those of marsupials. Compared to marsupials, placental mammals retain their young in the uterus for a much longer period, so that offspring complete their embryonic development before being born.

The placental mammals have evolved a remarkable diversity of form. The bat, mole, impala, whale, seal, monkey, and cheetah exemplify the radiation of mammals into nearly all habitats, with bodies finely adapted to their varied lifestyles (**FIG. 24-15**). The largest groups of placental mammals, in terms of number of species, are the bats and the rodents.

Rodents account for almost 40% of all mammal species. Most rodent species are rats or mice, but the group also includes squirrels, hamsters, guinea pigs, porcupines, beavers, woodchucks, chipmunks, and voles. The largest rodent, the capybara, is found in South America and can weigh up to 110 pounds (50 kilograms). Capybara meat is widely consumed in South America, mostly from animals harvested by hunters but increasingly from animals raised on ranches.

About 20% of mammal species are bats, the only mammals to have evolved wings and powered flight. Bats are nocturnal and spend the daylight hours roosting in caves, rock crevices, trees, or human houses. Most bat species have evolved adaptations for feeding on a particular kind of food. Some bats eat fruit; others feed on nectar from night-blooming flowers. Most bats are predators, including species that hunt frogs, fish, or even other bats. A few species (vampire bats) subsist entirely on blood that they lap up from incisions they make in the skin of sleeping mammals or birds. Most predatory bats, however, feed on flying insects, which they detect by echolocation. To echolocate, a bat emits short pulses of high-pitched sound (too high for humans to hear). The sounds bounce off objects in the surrounding environment to produce echoes, which the bat hears and uses to identify and locate insect prey.

(a)

(b)

(c)

(d)

FIGURE 24-15 The diversity of placental mammals
(a) A humpback whale gives its offspring a boost. **(b)** A bat, the only type of mammal capable of true flight, navigates at night by using a kind of sonar. Large ears help the animal detect echoes as its high-pitched cries bounce off nearby objects. **(c)** Mammals are named after the mammary glands with which females nurse their young, as illustrated by this mother cheetah. **(d)** The male orangutan can weigh up to 75 kilograms (165 pounds). These gentle, intelligent apes occupy swamp forests in limited areas of the tropics but are endangered by hunting and habitat destruction.

EVOLUTIONARY CONNECTIONS

Are Humans a Biological Success?

Physically, human beings are fairly unimpressive biological specimens. For such large animals, we are not very strong or very fast, and we lack natural weapons such as fangs or claws. It is the human brain, with its tremendously developed cerebral cortex, that truly sets us apart from other animals. Our brains give rise to our minds, which, in bursts of solitary brilliance and in the collective pursuit of common goals, have created wonders. No other animal could sculpt the graceful columns of the Parthenon, much less reflect on the beauty of this ancient Greek temple. We alone can eradicate smallpox and polio, domesticate other life-forms, penetrate space with rockets, and fly to the stars in our imaginations.

And yet, are we, as it appears at first glance, the most successful of all living things? The duration of human existence is a mere instant in the 3.5-billion-year span of life on Earth. But during the last 300 years, the human population has increased from 0.5 billion to 6 billion and now grows by 1 million people every 4 days. Is this a measure of our success? In expanding our range over the globe, we have driven at least 300 other species to extinction. Within your lifetime, the rapid destruction of tropical rain forests and other diverse habitats may wipe out millions of species of plants, invertebrates, and vertebrates, most of which we

LINKS TO LIFE   Do Animals Belong in Laboratories?

Vertebrate animals are the subjects of much laboratory research, in part because biologists, like most people, tend to be more interested in vertebrates than in other types of organisms. Much of the use of vertebrates in research, however, stems from their similarity to humans. Researchers often hope to answer questions about human biology by using information gained from experiments on rats, mice, dogs, monkeys, and other vertebrates (**FIG. E24-2**). Many such experiments would be considered unethical if performed on humans. For example, it is impermissible to intentionally expose humans to disease-causing organisms, to inject humans with untested drugs, to perform experimental surgery on a healthy human, or to intentionally kill a human for research purposes. Nonetheless, such manipulations are routinely performed on laboratory animals.

Some observers and activists argue that nonhuman animals are entitled to protection against suffering inflicted by experimental research. In this view, humans have no right to subject members of other species to treatment that would be unethical if applied to humans, and the pursuit of experiments that cause animals to suffer cannot be justified. Many scientists, however, object strenuously to the assertion that research on animals is unethical, arguing that the advance of scientific knowledge, including that leading to lifesaving treatments for humans, requires research on vertebrate animals.

Where do you stand? Is experimental research on animals always unethical? Or is it acceptable in some types of research but unacceptable in others? Or are you satisfied with the current system, in which scientists are largely free to use animals in research, but the animals are protected by regulations intended to limit their suffering?

FIGURE E24-2 Rats are among the vertebrates most commonly used in research

will never know. Many of our activities have altered the environment in ways that are hostile to life, including our own. Acid from power plants and automobiles rains down on the land, threatening our forests and lakes—and eroding the marble of the Parthenon. Deserts spread as land is stripped by overgrazing and the demand for firewood. Our aggressive tendencies, spurred by pressures of expanding wants and needs, their scope magnified by our technological prowess, have given us the capacity to destroy ourselves and most other life-forms as well.

The human mind is the source of our most pressing problems—and our greatest hope for solving them. Will we devote our mental powers to reducing our impact, controlling our numbers, and preserving the ecosystems that sustain us and other life? Are we a phenomenal biological success—or a brilliant catastrophe? Perhaps the next few centuries will tell.

CASE STUDY REVISITED   FISH STORY

After Marjorie Courtney-Latimer's discovery of the coelacanth, J. L. B. Smith dedicated himself to searching for more coelacanth specimens in the waters off South Africa. He didn't find one until 1952, when fishermen from the Comoro Islands, having seen leaflets that offered a reward for a coelacanth, contacted Smith with the news that they had one in their possession. Smith immediately booked a flight to the Comoros, and reportedly wept for joy upon holding the 88-pound coelacanth awaiting him.

In the years since Smith's trip, about 200 additional coelacanths have been caught by fishermen, mostly in waters around the Comoros but also around nearby Madagascar and off the coasts of Mozambique and South Africa. Scientists thought that the fish's range was restricted to this relatively small area in the western Indian Ocean, and it was therefore some-

thing of a shock when a few specimens were discovered in Indonesia, more than 6000 miles away. DNA tests showed that these Indonesian coelacanths were members of a second species.

Although the coelacanth specimens have revealed a great deal about the creatures' anatomy, their habits and behavior remain comparatively mysterious. Observations from research submarines suggest that coelacanths spend much of their time in caves and beneath rocky overhangs at depths of 300 to 1200 feet. Radio tracking suggests that they may venture into open water at night, presumably to forage for food. Almost all individuals observed (or caught) have been at least 3 feet long, which suggests that young fish must travel to sites away from the main adult populations to mature, though no such location has been discovered.

The known populations of coelacanths are small, consisting of a few hundred indi-

viduals, and appear to be declining. Part of this decline is due to fishing, though coelacanths are mostly caught accidentally by fishermen searching for more commercially desirable species. Conservation efforts in South Africa and the Comoros thus focus largely on introducing fishing methods that will reduce the chances of accidentally snaring a coelacanth.

Consider This   Many accounts of coelacanths refer to them as "living fossils," a term that is also applied to alligators, gingko trees, horseshoe crabs, and other species whose modern appearance matches that of ancient fossils. The implication of the living fossil designation is that these organisms have evolved very little over a very long period. Do you think this implication is accurate? Is it accurate to say that "living fossils" have evolved more slowly or undergone less evolutionary change than other species?

CHAPTER REVIEW

SUMMARY OF KEY CONCEPTS

24.1   What Are the Key Features of Chordates?

The phylum Chordata includes two invertebrate groups, the lancelets and tunicates, as well as the familiar vertebrates. All chordates possess a notochord; a dorsal, hollow nerve cord; pharyngeal gill slits; and a post-anal tail at some stage in their development. The vertebrates are a subphylum of chordates and have a backbone, which is part of their living endoskeleton.

Web Tutorial 24.1   Chordates

24.2   What Are the Major Groups of Vertebrates?

Hagfishes are jawless, eel-shaped chordates that lack a true backbone and, therefore, are not true vertebrates. Lampreys are jawless vertebrates; the best-known lamprey species are parasites of fish.

All amphibians have legs, and most have simple lungs for breathing in air rather than in water. Most are confined to relatively damp terrestrial habitats by their need to keep their

skin moist, their use of external fertilization, and their eggs and larvae, which develop in water.

Reptiles, with well-developed lungs, dry skin covered with relatively waterproof scales, internal fertilization, and an amniote egg with its own water supply, are well adapted to the driest terrestrial habitats.

Birds are also fully terrestrial and have additional adaptations that allow the muscles to respond rapidly regardless of the temperature of the environment, such as an elevated body temperature. The bird body is molded for flight, with feathers, hollow bones, efficient circulatory and respiratory systems, and well-developed eyes.

Mammals have insulating hair and give birth to live young that are nourished with milk. The mammalian nervous system is the most complex in the animal kingdom, providing mammals with enhanced learning ability that helps them adapt to changing environments.

KEY TERMS

amnion *page 478*	**marsupial** *page 480*	**pharyngeal gill slit** *page 471*	**vertebral column** *page 471*
amniote egg *page 478*	**monotreme** *page 480*	**placenta** *page 481*	**vertebrate** *page 470*
cartilage *page 471*	**nerve cord** *page 470*	**placental** *page 481*	
mammary gland *page 480*	**notochord** *page 471*	**post-anal tail** *page 471*	

THINKING THROUGH THE CONCEPTS

1. Briefly describe each of the following adaptations, and explain the adaptive significance of each: vertebral column, jaws, limbs, amniote egg, feathers, placenta.

2. List the vertebrate groups that have each of the following:
   a. a skeleton of cartilage
   b. a two-chambered heart
   c. an amniote egg
   d. warm-bloodedness
   e. a four-chambered heart
   f. a placenta
   g. lungs supplemented by air sacs

3. List four distinguishing features of chordates.

4. Describe the ways in which amphibians are adapted to life on land. In what ways are amphibians still restricted to a watery or moist environment?

5. List the adaptations that distinguish reptiles from amphibians and help reptiles adapt to life in dry terrestrial environments.

6. List the adaptations of birds that contribute to their ability to fly.

7. How do mammals differ from birds, and what adaptations do they share?

8. How has the mammalian nervous system contributed to the success of mammals?

APPLYING THE CONCEPTS

1. Are hagfishes vertebrates or invertebrates? On which characteristics did you base your answer? Is it important to be able to place them in one category or the other? Why?

2. Is the decline of amphibian populations of concern to humans? What about the increase in frog deformities? Why is it important to understand the causes of these phenomena?

3. Discuss and defend the attributes you would use to define biological success among animals. Are humans a biological success by these standards? Why?

FOR MORE INFORMATION

Attenborough, David. *The Life of Birds*. Princeton, NJ: Princeton University Press, 1998. A thoughtful survey of birds and their adaptations, with beautiful photographs.

Attenborough, David. *The Life of Mammals*. Princeton, NJ: Princeton University Press, 2002. An overview of mammals and how they live, with beautiful photographs.

Blaustein, A. R. "Amphibians in a Bad Light." *Natural History*, October 1994. Recent declines in amphibian population size and overall diversity are linked to possible harm from ultraviolet light that is penetrating a depleted ozone layer.

Blaustein, A., and Johnson, P. T. J. "Explaining Frog Deformities." *Scientific American*, February 2003. Dramatic increases in the occurrence of deformed frogs are caused by a parasite epidemic exacerbated by environmental degradation.

Duellman, W. E. "Reproductive Strategies of Frogs." *Scientific American*, July 1992. Free-living tadpoles are only one way in which these amphibians progress from egg to adult.

Pauly, D., and Watson, R. "Counting the Last Fish." *Scientific American*, July 2003. A summary of the evidence that fish stocks are in catastrophic decline and a discussion of what to do about it.

Perkins, S. "The Latest Pisces of an Evolutionary Puzzle." *Science News*, May 5, 2001. A summary of recent research on coelacanths in their natural habitat.

Raloff, J. "Empty Nets." *Science News*, June 4, 2005. An update on the threat that overfishing by humans poses to populations of cartilaginous and bony fishes.

Behavior and Ecology

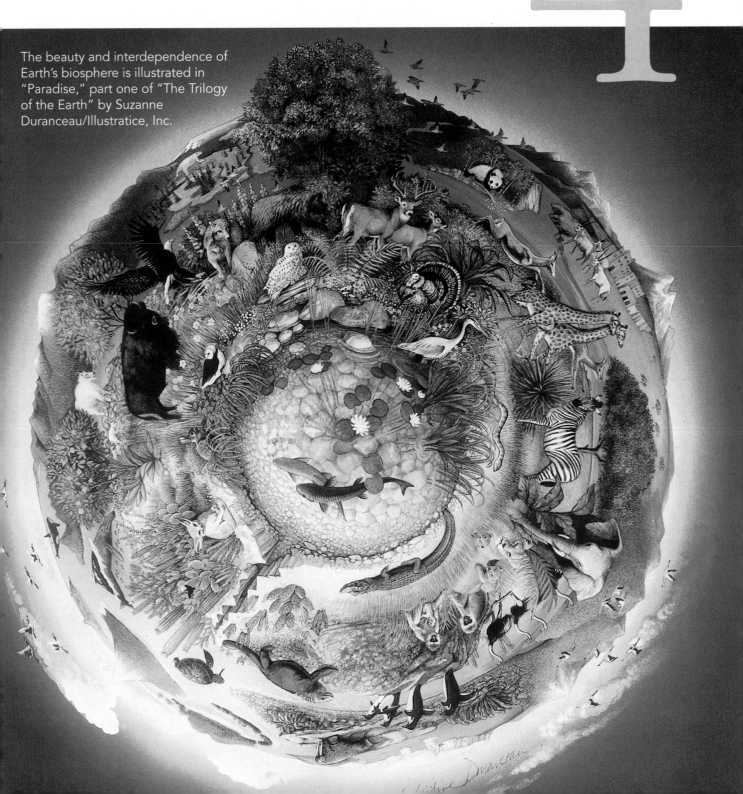

The beauty and interdependence of Earth's biosphere is illustrated in "Paradise," part one of "The Trilogy of the Earth" by Suzanne Duranceau/Illustratice, Inc.

25

Animal Behavior

Both this male scorpionfly and this male human are exceptionally
attractive to females of their species. The secret to their sex appeal
may be that both have highly symmetrical bodies.

CASE STUDY   SEX AND SYMMETRY

WHAT MAKES A MAN SEXY? According to a growing body of research, it's his symmetry. Female sexual preference for symmetrical males was first documented in insects. For example, biologist Randy Thornhill found that symmetry accurately predicts the mating success of male Japanese scorpionflies. In Thornhill's experiments and observations, the most successful males were those whose left and right wings were equal or nearly equal in length. Males with one wing longer than the other were less likely to copulate; the greater the difference between the two wings, the lower the likelihood of success.

Thornhill's work with scorpionflies led him to wonder if the effects of male sym-metry also extend to humans. To test the hypothesis that female humans find symmetrical males more attractive, Thornhill and colleagues began by measuring symmetry in some young adult males. Each man's degree of symmetry was assessed by measurements of his ear length and the width of his foot, ankle, hand, wrist, elbow, and ear. From these measurements, the researchers derived an index that summarized the degree to which the size of these features differed between the right and left sides of the body.

The researchers next gathered a panel of female observers who were unaware of the nature of the study and showed them photos of the faces of the measured males. As predicted by the researchers' hypothesis, the panel judged the most symmetrical men to be most attractive. A survey of the male subjects revealed that the more symmetrical men also tended to begin having sex earlier in life and to have had a larger number of sexual partners. Apparently, a man's sexual activity and attractiveness to women are correlated with his body symmetry.

Why might females prefer symmetrical males? Consider this question as you read about animal behavior.

25.1 HOW DO INNATE AND LEARNED BEHAVIORS DIFFER?

Behavior is any observable activity of a living animal. For example, a moth flies toward a bright light, a honeybee flies toward a cup of sugar-water, and a housefly flies toward a piece of rotting meat. Bluebirds sing, wolves howl, and frogs croak. Mountain goats butt heads in ritual combat; chimpanzees groom one another; ants attack a termite that approaches an anthill. Humans smoke cigarettes, play tennis, and plant gardens. Even the most casual observer sees many examples of animal behavior each day, and a careful observer encounters a virtually limitless number of fascinating behaviors.

Innate Behaviors Can Be Performed Without Prior Experience

Innate behaviors are performed in reasonably complete form even the first time an animal of the right age and motivational state encounters a particular stimulus. (The proper motivational state for feeding, for example, would be hunger.) Scientists can demonstrate that a behavior is innate by depriving an animal of the opportunity to learn it. For example, red squirrels, which in the wild bury nuts in the fall for retrieval during the winter, can be raised from birth in a bare cage on a liquid diet, providing them with no experience of nuts, digging, or burying. Nonetheless, such a squirrel will, when presented with nuts for the first time, carry one to the corner of its cage, and then make covering and patting motions with its forefeet. Nut burying is therefore an innate behavior.

Innate behaviors can also be recognized by their occurrence immediately after birth, before any opportunity for learning presents itself. The cuckoo, for example, lays its eggs in the nest of another bird species, to be raised by the unwitting adoptive parent. Soon after a cuckoo egg hatches, the cuckoo chick performs the innate behavior of shoving the nest owner's eggs (or baby birds) out of the nest, eliminating its competitors for food (**FIG. 25-1**).

Learned Behaviors Are Modified by Experience

Natural selection may favor innate behaviors in many circumstances. For instance, it is clearly to the advantage of a herring gull chick to peck at its parent's bill as soon as possible after hatching, because pecking stimulates the parent to feed the chick. But in other circumstances, rigidly fixed behavior patterns may be less useful. For example, a male red-winged blackbird presented with a stuffed female blackbird will often attempt to copulate with the stuffed bird, a behavior that obviously will produce no offspring. In many situations, a degree of behavioral flexibility is advantageous.

The capacity to make changes in behavior on the basis of experience is called **learning**. This deceptively simple definition encompasses a vast array of phenomena. A toad learns to avoid distasteful insects; a baby shrew learns which adult is its mother; a human learns to speak a language; a sparrow learns to use the stars for navigation.

(a)

(b)

FIGURE 25-1 Innate behavior
(a) The cuckoo chick, just hours after it hatches and before its eyes have opened, evicts the eggs of its foster parents from the nest. **(b)** The parents, responding to the stimulus of the cuckoo chick's wide-gaping mouth, feed the chick, unaware that it is not related to them. QUESTION The cuckoo chick benefits from its innate behavior, but the foster parent harms itself with its innate response to the cuckoo chick's begging. Why hasn't natural selection eliminated this disadvantageous innate behavior?

Each of the many examples of animal learning represents the outcome of a unique evolutionary history, so learning is as diverse as animals themselves. Nonetheless, it can be useful to categorize types of learning, as long as we keep in mind that the categories are only rough guides; many examples of learning will not fit neatly into any category.

Habituation Is a Decline in Response to a Repeated Stimulus

A common form of simple learning is **habituation**, defined as a decline in response to a repeated stimulus. The ability to habituate prevents an animal from wasting its energy and attention on irrelevant stimuli. This form of learning is displayed by even the simplest animals. For example, a sea anemone will retract its tentacles when touched but gradually stops retracting if touching is continued (**FIG. 25-2**).

The ability to habituate is clearly adaptive. If a sea anemone withdrew every time it was brushed by a strand of waving seaweed, the animal would waste a great deal of energy, and its retracted posture would prevent it from snaring food. Humans habituate to many stimuli; city dwellers habituate to nighttime traffic sounds, as country dwellers do to choruses of crickets and tree frogs. Each may initially find the other's habitat unbearably noisy at first, but each habituates after a time.

Conditioning Is a Learned Association Between a Stimulus and a Response

A more complex form of learning is **trial-and-error learning**, in which animals acquire new and appropriate responses to stimuli through experience. Many animals are faced with naturally occurring rewards and punishments and can learn to modify their responses to them. For example, a hungry toad that captures a bee quickly learns to avoid future encounters with bees (**FIG. 25-3**). After only one experience with a stung tongue, a toad modifies its response to flying insects, ignoring bees (and even other insects that resemble them).

Trial-and-error learning is an important factor in the behavioral development of many animal species and often occurs during play and exploratory behavior (see "Evolutionary Connections: Why Do Animals Play?"). This type of learning also plays a key role in human behavior—allowing, for example, a child to learn which foods taste good or bad, that a stove can be hot, and not to pull a cat's tail.

Some interesting properties of trial-and-error learning have been revealed by a laboratory technique known as **operant conditioning**. During operant conditioning, an animal learns to perform a behavior (such as pushing a lever or pecking a button) to receive a reward or to avoid punishment. This technique is most closely associated with the American comparative psychologist B. F. Skinner, who designed the "Skinner box," in which an animal is isolated and allowed to train itself. The box might contain a lever that, when pressed, ejects a food pellet. If the animal accidentally bumps the lever, a food reward appears. After a few such occurrences, the animal learns the connection between pressing the lever and receiving food and begins to press the lever repeatedly.

Operant conditioning has been used to train animals to perform tasks far more complex than pressing a lever, but perhaps the most interesting revelation fostered by the technique is that species differ in their propensity for learning particular associations. In particular, species seem to be predisposed to learn behaviors that are relevant to their own needs. For example, if a rat is given a distinctively flavored food that has been mixed with a substance that makes the rat sick, the animal learns to avoid eating that food in the future. In contrast, it is very difficult to train a rat to rear up on its hind legs in response to a particular

Touched for the first time, anemone withdraws.

After many touches, the anemone habituates and no longer responds.

FIGURE 25-2 Habituation in a sea anemone

(a) A naive toad is presented with a bee.

(b) While trying to eat the bee, the toad is stung painfully on the tongue.

(c) Presented with a harmless robber fly, which resembles a bee, the toad cringes.

(d) The toad is presented with a dragonfly.

(e) The toad immediately eats the dragonfly, demonstrating that the learned aversion is specific to bees and insects resembling bees.

FIGURE 25-3 Trial-and-error learning in a toad

sound or visual cue. The difference can be explained by asking which learning task is more likely to benefit a wild Norway rat (the species from which lab rats are descended). Clearly, an ability to avoid foods that induce illness is beneficial to an animal like the Norway rat, which is well known for eating a tremendous variety of foods. However, the animal gains no obvious benefit from learning to stand up in response to a noise. In general, the specific learning abilities of each species have evolved to support its particular mode of life.

Insight Is Problem Solving Without Trial and Error

In certain situations, animals seem able to solve problems suddenly, without the benefit of prior experience. This kind of sudden problem solving is sometimes called **insight learning**, because it seems at least superficially similar to the process by which humans mentally manipulate concepts to arrive at a solution. We cannot, of course, know for sure if nonhuman animals experience similar mental states when they solve problems.

In 1917, the animal behaviorist Wolfgang Kohler showed that a hungry chimpanzee, without any training, could stack boxes to reach a banana suspended from the ceiling. This type of mental problem solving was once believed to be limited to very intelligent types of animals such as primates,

but similar abilities may also be present in species that we tend to view as less intelligent. For example, R. Epstein and colleagues performed an experiment that showed that pigeons may be capable of insight learning. In the experiment, pigeons (whose wings had been clipped to prevent flight) were first trained to perform two unrelated tasks in return for food rewards. The tasks were to push a small box around the cage and to peck at a small plastic banana. Later, the trained birds were presented with a novel situation: a plastic banana that hung above their reach in a cage that also contained a small box. Many of the pigeons pushed the box to a position beneath the plastic banana and climbed atop the box to peck the faux fruit. Apparently, a pigeon trained to execute the necessary physical movements can also solve the suspended banana problem.

There Is No Sharp Distinction Between Innate and Learned Behaviors

Although the terms "innate" and "learned" can be useful tools to help us describe and understand behaviors, these words also have the potential to lull us into an oversimplified view of animal behavior. In practice, no behavior is unambiguously innate or unequivocally learned; all behaviors are an intimate mixture of the two.

Seemingly Innate Behavior Can Be Modified By Experience

Behaviors that seem to be performed correctly on the first attempt without prior experience can later be modified by experience. For example, a newly hatched herring gull chick is able to peck at a red spot on its parent's beak (**FIG. 25-4**), an innate behavior that causes the parent to regurgitate food for the chick to eat. Biologist Niko Tinbergen studied this pecking behavior and found that the pecking response of very young chicks was triggered by the long, thin shape and red color of the parent's bill. In fact, when Tinbergen offered newly hatched chicks a thin, red rod with white stripes painted on it, they pecked at it more often than at a real beak. Within a few days, however, the chicks learned enough about the appearance of their parents that they began pecking more frequently at models more closely resembling the parents. After one week, the young gulls recognized their parents' appearance enough to prefer models of their own species to models of a closely related species. Eventually, the young birds learned to beg only from their own parents.

Habituation (a decline in response to a repeated stimulus) can also fine-tune an organism's innate responses to environmental stimuli. For example, young birds crouch down when a hawk flies over but ignore harmless birds such as geese. Early observers hypothesized that only the very specific shape of predatory birds provoked crouching. Using an ingenious model (**FIG. 25-5**), Niko Tinbergen and Konrad Lorenz (two of the founding fathers of **ethology**, the study of animal behavior) tested this hypothesis. When moved in one direction, the model resembled a goose and was ignored by the chicks. When its movement was reversed, however, the model resembled a hawk and elicited crouching behavior. Further research revealed that newborn chicks instinctively crouch when *any* object moves over their heads. Over time, their response habituates to things that soar by harmlessly and frequently, such as leaves, songbirds, and geese. Predators are much less com-

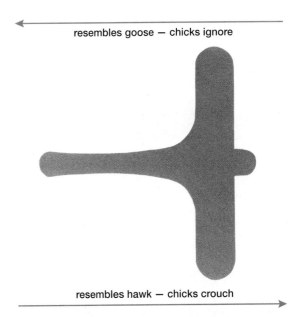

resembles goose — chicks ignore

resembles hawk — chicks crouch

FIGURE 25-5 Habituation modifies innate responses
Konrad Lorenz and his student Niko Tinbergen used this model to investigate the response of chicks to the shape of objects flying overhead. The chicks' response depended on which direction the model moved. Moving toward the right, the model resembles a predatory hawk; but when it moves left, it resembles a harmless goose.

mon, and the novel shape of a hawk continues to elicit instinctive crouching. Thus, learning modifies the innate response, making it more advantageous.

Learning May Be Governed by Innate Constraints

Learning always occurs within boundaries that help increase the chances that only the appropriate behavior is acquired. For example, even though young robins hear the singing of sparrows, warblers, finches, and other bird species that share nesting areas with robins, the young birds do not imitate the songs of these other species. Instead, young robins learn only the songs of adult robins. The robin's ability to learn songs is limited to those of its own species, and the songs of other species are excluded from the learning process.

The innate constraints on learning are perhaps most strikingly illustrated by **imprinting**, a special form of learning in which an animal's nervous system is rigidly programmed to learn a certain thing only at a certain period of development. This causes a strong association to be formed during a particular stage, called a sensitive period, in the animal's life. During this stage, the animal is primed to learn specific information, which is then incorporated into behaviors that are not easily altered by further experience.

Imprinting is best known in birds such as geese, ducks, and chickens. These birds learn to follow the animal or object that they most frequently encounter during an early sensitive period. In nature, a mother bird is likely to be nearby during the sensitive period, so her offspring imprint on her. In the laboratory, however, these birds may

FIGURE 25-4 Innate behaviors can be modified by experience
A herring gull chick pecks at the red spot on its mother's bill, causing her to regurgitate food.

FIGURE 25-6 Konrad Lorenz and imprinting
Konrad Lorenz, known as the "father of ethology," is followed by goslings that imprinted on him shortly after they hatched. They follow him as they would their mother.

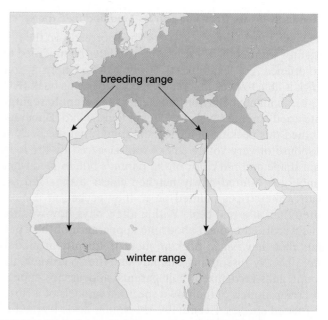

FIGURE 25-7 Genes influence migratory behavior
Blackcap warblers from western Europe begin their fall migration by flying in a southwesterly direction, but those from eastern Europe fly to the southeast when they begin migrating. If members of the two populations are crossbred in captivity, the hybrid offspring orient in a direction intermediate between the migratory directions of the parents: due south. QUESTION If young blackcap warblers from a wild population in western Europe were transported to eastern Europe and reared to adulthood in a normal environment, in which direction would you expect them to orient?

imprint on a toy train or other moving object (**FIG. 25-6**). If given a choice, however, they select an adult of their own species.

All Behavior Arises Out of Interaction Between Genes and Environment

Many early ethologists saw innate behaviors as rigidly controlled by genetic factors and viewed learned behaviors as determined exclusively by an animal's environment. Today, however, ethologists realize that just as no behavior is wholly innate or wholly learned, no behavior can be caused strictly by genes or strictly by the environment. Instead, all behavior develops out of an interaction between genes and the environment. The relative contributions of heredity and learning vary among animal species and among behaviors within an individual.

The precise nature of the link among genes, environments, and behaviors is not well understood in most cases. The chain of events between the transcription of genes and the performance of a behavior may be so complex that we will never decipher it fully. Nonetheless, a great deal of evidence demonstrates the existence of both genetic and environmental components in the development of behaviors. For example, consider bird migration. Even though it is well known that migratory birds must learn by experience how to navigate with celestial cues, this learning is not the only factor involved.

Bird Migration Behavior Has an Inherited Component

At the close of the summer, many birds disappear from their breeding habitats and head for their winter territory, which may be hundreds or even thousands of miles to the south. Many of these migrating birds are traveling for the first time, because they were hatched only a few months earlier. Amazingly, these naive birds depart at the proper time, head in the proper direction, and locate the proper wintering location, even though they often do not simply follow more-experienced birds (which typically depart a few weeks in advance of the first-year birds). Somehow, these young birds execute a very difficult task the first time they try it. Thus, it seems that birds must be born with the ability to migrate; it must be "in their genes." Indeed, birds hatched and raised in isolation indoors still orient in the proper migratory direction when autumn comes, apparently without the need for any learning or experience.

The conclusion that birds must have a genetically controlled ability to migrate in the right direction has been further supported by hybridization experiments with blackcap warblers. This species breeds in Europe and migrates to Africa, but populations from different areas travel by different routes. Blackcaps from western Europe travel in a southwesterly direction to reach Africa, whereas birds from eastern Europe travel to the southeast (**FIG. 25-7**). If birds from the two populations are crossbred in captivity, however, the hybrid offspring exhibit migratory orientation

FIGURE 25-8 Active visual signals
The wolf signals aggression by lowering its head, ruffling the fur on its neck and along its back, facing its opponent with a direct stare, and exposing its fangs. These signals can vary in intensity, communicating different levels of aggression.

FIGURE 25-9 A passive visual signal
The female mandrill's colorfully swollen buttocks serve as a passive visual signal that she is fertile and ready to mate.

due south, which is intermediate between the orientation of the two parents. This result suggests that parental genes—of which offspring inherit a mixture—influence migratory direction.

25.2 HOW DO ANIMALS COMMUNICATE?

Animals frequently make information available for sharing. The sounds uttered, movements made, and chemicals emitted by animals can reveal their location, level of aggression, readiness to mate, and so on. If this information evokes a response from other individuals, and if that response tends to benefit the sender and the receiver, then a communication channel can form. **Communication** is defined as the production of a signal by one organism that causes another organism to change its behavior in a way beneficial to one or both.

Although animals of different species may communicate (picture a cat, its tail erect and bushy, hissing at a dog), most animals communicate only with members of their own species. Potential mates must communicate, as must parents and offspring. Communication is also often used to help resolve the conflicts that arise when members of a species compete directly with one another for food, space, and mates.

The ways in which animals communicate are astonishingly diverse and use all of the senses. In the following sections, we will look at communication by visual displays, sound, chemicals, and touch.

Visual Communication Is Most Effective over Short Distances

Animals with well-developed eyes, from insects to mammals, use visual signals to communicate. Visual signals can be *active*, in which a specific movement (such as baring

fangs) or posture (such as lowering the head) conveys a message (**FIG. 25-8**). Alternatively, visual signals may be *passive*, in which case the size, shape, or color of the animal conveys important information, commonly about its sex and reproductive state. For example, when female mandrills become sexually receptive, they develop a large, brightly colored swelling on their buttocks (**FIG. 25-9**). Active and passive signals can be combined, as illustrated by the lizard in **FIGURE 25-10**.

FIGURE 25-10 Active and passive visual signals combined
A male South American *Anolis* lizard raises his head high in the air (an active visual signal), revealing a brilliantly colored throat pouch (a passive visual signal) that warns others to keep their distance.

Like all forms of communication, visual signals have both advantages and disadvantages. On the plus side, they are instantaneous, and active signals can be rapidly changed to convey a variety of messages in a short period. Visual communication is quiet and unlikely to alert distant predators, although the signaler does make itself conspicuous to those nearby. On the negative side, visual signals are generally ineffective in dense vegetation or in darkness, although female fireflies do communicate with potential mates at night by using species-specific patterns of flashes. Finally, visual signals are limited to close-range communication.

Communication by Sound Is Effective over Longer Distances

The use of sound overcomes many of the shortcomings of visual displays. Like visual displays, sound signals reach receivers almost instantaneously. But unlike visual signals, sound can be transmitted through darkness, dense forests, and murky water. Sound signals can also be effective over longer distances than can visual signals. For example, the low, rumbling calls of African elephants can be heard by elephants several miles away, and the songs of humpback whales are audible for hundreds of miles. Likewise, the howls of a wolf pack carry for miles on a still night. Even the small kangaroo rat produces a sound (by striking the desert floor with its hind feet) that is audible 150 feet (45 meters) away.

Sound signals are similar to visual displays in that they can be varied to convey rapidly changing messages. (Think of speech and the emotional nuances conveyed by the human voice during a conversation.) Changes in motivation can be signaled by a change in the loudness or pitch of the sound. An individual can convey different messages by variations in the pattern, volume, and pitch of the sound produced. In a study of vervet monkeys in Kenya in the 1960s, ethologist Thomas Struhsaker found that the monkeys produced different calls in response to threats from each of their major predators: snakes, leopards, and eagles. In 1980, other researchers reported that the response of other vervet monkeys to each of these calls is appropriate to the particular predator. For example, the "bark" that warns of a leopard or other four-legged carnivore causes monkeys on the ground to take to trees and prompts those in trees to climb higher. The "rraup" call, signaling an eagle or other hunting bird, causes monkeys on the ground to look upward and take cover; monkeys already in trees will drop to the shelter of lower, denser branches. The "chutter" call that indicates the presence of a snake causes the monkeys to stand up and search the ground for the predator.

The use of sound is by no means limited to birds and mammals. Male crickets produce species-specific songs that attract female crickets of the same species. The female mosquito's annoying whine as she prepares to bite alerts nearby males that she may soon have the blood meal necessary for laying eggs. Male water striders vibrate their legs, sending species-specific patterns of vibrations through the water, attracting mates and repelling other

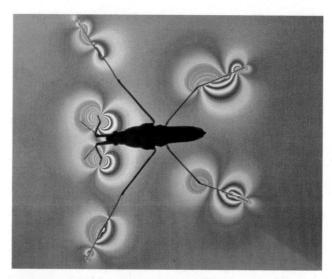

FIGURE 25-11 Communication by vibration
The light-footed water strider relies on the surface tension of water to support its weight. By vibrating its legs, the water strider sends signals that radiate out over the surface of the water. These vibrations advertise the strider's species and sex to others nearby.

males (**FIG. 25-11**). From these rather simple signals to the complexities of human language, sound is one of the most important forms of communication.

Chemical Messages Persist Longer but Are Hard to Vary

Chemical substances that are produced by individuals and that influence the behavior of other members of the species are called **pheromones**. Pheromones can carry messages over long distances and, unlike sound, take very little energy to produce. Pheromones may not even be detected by other species, whereas predators might be attracted to visual or sound signals. Like a signpost, a pheromone persists over time and can convey a message after the animal has departed. Wolf packs, hunting over areas as large as 385 square miles (1000 square kilometers), warn other packs of their presence by marking the boundaries of their travels with urine that contains pheromones. As anyone who has walked a dog can attest, the domesticated dog reveals its wolf ancestry by staking out its neighborhood with urine that carries a chemical message, "I live in this area."

Such communication requires animals to synthesize a different chemical for each message. As a result, chemical signaling systems communicate fewer different messages than do sight- or sound-based systems. In addition, pheromone signals cannot easily convey rapidly changing messages. Nonetheless, chemicals effectively convey a few simple but critical messages.

Many pheromones cause an immediate change in the behavior of the animal that detects them. For example, foraging termites that discover food lay a trail of pheromones from the food to the nest, and other termites follow the trail

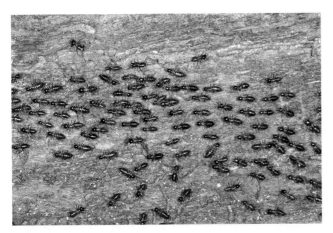

FIGURE 25-12 Communication by chemical messages
A trail of pheromones, secreted by termites from their own colony, orients foraging termites toward a source of food.

(FIG. 25-12). Pheromones can also stimulate physiological changes in the animal that detects them. For example, the queen honeybee produces a pheromone called *queen substance*, which prevents other females in the hive from becoming sexually mature. Similarly, mature males of some mouse species produce urine containing a pheromone that influences female reproductive physiology. The pheromone stimulates newly mature females to become fertile and sexually receptive. It will also cause a female mouse that is newly pregnant by another male to abort her litter and become sexually receptive to the new male.

Humans have harnessed the power of pheromones to combat insect pests. The sex attractant pheromones of some agricultural pests, such as the Japanese beetle and the gypsy moth, have been successfully synthesized. These synthetic pheromones can be used to disrupt mating or to lure these insects into traps. Controlling pests with pheromones has major environmental advantages over conventional pesticides, which kill beneficial as well as harmful insects and foster the evolution of pesticide-resistant insects. In contrast, each pheromone is specific to a single species and does not promote the spread of resistance, because insects resistant to the attraction of their own pheromones do not reproduce successfully.

Communication by Touch Helps Establish Social Bonds

Communication by physical contact often serves to establish and maintain social bonds among group members. This function is especially apparent in humans and other primates, which have many gestures—including kissing, nuzzling, patting, petting, and grooming—with important social functions (**FIG. 25-13a**). Touch may even be essential to human well-being. For example, research has shown that when the limbs of premature human infants were stroked and moved for 45 minutes daily, the infants were more active, responsive, and emotionally stable and gained weight more rapidly than did premature infants who received standard hospital treatment.

Communication by touch is not limited to primates, however. In many other mammal species, close physical contact helps cement the bond between parent and offspring. Species in which sexual activity is preceded or accompanied by physical contact can be found throughout the animal kingdom (**FIG. 25-13b**).

25.3 HOW DO ANIMALS COMPETE FOR RESOURCES?

The Darwinian contest to survive and reproduce stems from the scarcity of resources relative to the reproductive potential of populations. The resulting competition underlies many of the most frequent types of interactions between animals.

Aggressive Behavior Helps Secure Resources

One of the most obvious manifestations of competition for resources such as food, space, or mates is **aggression**, or antagonistic behavior, between members of the same species.

FIGURE 25-13 Communication by touch
(a) An adult olive baboon grooms a juvenile. Grooming both reinforces social relationships and removes debris and parasites from the fur. **(b)** Touch is also important in sexual communication. These land snails engage in courtship behavior that will culminate in mating.

(a)   (b)

(a)

(b)

FIGURE 25-14 Aggressive displays
(a) Threat display of the male baboon. Despite the potentially lethal fangs so prominently displayed, aggressive encounters between baboons rarely cause injury. **(b)** The aggressive display of many male fish, such as these sarcastic fringeheads, includes elevating the fins and flaring the gill covers, thus making the body appear larger.

Although the expression "survival of the fittest" evokes images of the strongest animal emerging triumphantly from among the dead bodies of its competitors, in reality most aggressive encounters between members of the same species end without physical damage to the participants. Natural selection has favored the evolution of symbolic displays or rituals for resolving conflicts. During fighting, even the victorious animal can be injured and might not survive to pass on its genes. Aggressive *displays,* in contrast, allow the competitors to assess each other and acknowledge a winner on the basis of size, strength, and motivation rather than wounds inflicted.

During aggressive displays, animals may exhibit weapons such as claws and fangs (**FIG. 25-14a**), and they often do things to make themselves appear larger (**FIG. 25-14b**). Competitors often stand upright and erect their fur, feathers, ears, or fins (see Fig. 25-8). The displays are typically accompanied by intimidating sounds (growls, croaks, roars, chirps) whose loudness can help decide the winner. Fighting tends to be a last resort when displays fail to resolve a dispute.

In addition to aggressive visual and vocal displays, many animal species engage in ritualized combat. Deadly weapons may clash harmlessly (**FIG. 25-15**) or may not be used at all. In many cases, these encounters involve shoving rather than slashing. The ritual thus allows contestants to assess the strength and the motivation of their rivals, and the loser slinks away in a submissive posture that minimizes the size of its body.

Dominance Hierarchies Help Manage Aggressive Interactions

Aggressive interactions use a lot of energy, can cause injury, and can disrupt other important tasks such as finding food, watching for predators, or raising young. Thus, there are advantages to resolving conflicts with minimal aggression. In a **dominance hierarchy**, each animal establishes a

FIGURE 25-15 Displays of strength
Ritualized combat of fiddler crabs. Oversized claws, which could severely injure another animal, grasp harmlessly. Eventually one crab, sensing greater vigor in his opponent, retreats unharmed.

FIGURE 25-16 A dominance hierarchy
The dominance hierarchy of the male bighorn sheep is signaled by the size of the horns; these rams increase in status from right to left. The backward-curving horns, clearly not designed to inflict injury, are used in ritualized combat.

FIGURE 25-17 Jane Goodall observing chimps

rank that determines its access to resources. Although aggressive encounters occur frequently while the dominance hierarchy is being established, once each animal learns its place in the hierarchy, disputes are infrequent, and the dominant individuals obtain most access to the resources needed for reproduction, including food, space, and mates. For example, domestic chickens, after squabbling, sort themselves into a reasonably stable "pecking order." Thereafter, all birds in the group defer to the dominant bird, all but the dominant bird give way to the second most dominant, and so on. In wolf packs, one member of each sex is the dominant, or "alpha," individual to whom all others of that sex are subordinate. Among male bighorn sheep, dominance is reflected in horn size (**FIG. 25-16**).

Perhaps the most thoroughly studied dominance hierarchy is that of chimpanzees. Ethologist Jane Goodall (**FIG. 25-17**) has devoted more than 30 years to meticulously observing chimpanzee behavior in the field at Gombe National Park in Tanzania, describing and documenting the animals' complex social organization. Chimps live in groups, and dominance hierarchies among males are a key aspect of their social life. Many males devote a significant amount of time to maintaining their position in the hierarchy, largely by way of an aggressive *charging display* in which a male rushes forward, throws rocks, leaps up to shake vegetation, and otherwise seeks to intimidate rival males. The advantages that accrue to dominant males, however, are not clear. According to Goodall, low-ranking males can still secure access to food and successful copulations—although not quite as easily as can higher-ranking males. In Goodall's view, little evolutionary advantage accrues to a dominant male, and the function of chimpanzee dominance hierarchies remains unexplained.

Animals May Defend Territories That Contain Resources

In many animal species, competition for resources takes the form of **territoriality**, the defense of an area where important resources are located. The defended area may include places to mate, raise young, feed, or store food. Territorial animals generally restrict most or all of their activities to the defended area and advertise their presence there. Territories may be defended by males, females, a mated pair, or entire social groups (as in the defense of a nest by social insects). However, territorial behavior is most commonly seen in adult males, and territories are normally defended against members of the same species, who compete most directly for the resources being protected.

Territories are as diverse as the animals defending them. For example, a territory can be a tree where a woodpecker stores acorns (**FIG. 25-18**), small depressions in the

FIGURE 25-18 A feeding territory
Acorn woodpeckers live in communal groups that excavate acorn-sized holes in dead trees and then stuff the holes with green acorns for dining during the lean winter months. The group defends the trees vigorously against other groups of acorn woodpeckers and against acorn-eating birds of other species, such as jays.

sand used as nesting sites by cichlid fish, a hole in the sand that is home to a crab, or an area of forest providing food for a squirrel.

Territoriality Reduces Aggression

Acquiring and defending a territory requires considerable time and energy, yet territoriality is seen in animals as diverse as worms, arthropods, fish, birds, and mammals. The fact that organisms as distantly related as worms and humans independently evolved similar behavior suggests that territoriality provides some important advantages. Although the particular benefits depend on the species and the type of territory it defends, we can make some broad generalizations. First, as with dominance hierarchies, once a territory is established through aggressive interactions, relative peace prevails if boundaries are recognized and respected. The saying "good fences make good neighbors" also applies to nonhuman territories. One reason is that an animal is highly motivated to defend its territory and will often defeat even larger, stronger animals that attempt to invade it. Conversely, an animal outside its territory is much less secure and more easily defeated. This principle was demonstrated by Niko Tinbergen in an experiment using stickleback fish (**FIG. 25-19**).

Competition for Mates May Be Based on Territories

For males of many species, successful territorial defense has a direct impact on reproductive success. In these species, males defend territories, and females are attracted to high-quality territories, which might have features such as large size, abundant food, and secure nesting areas. Males who successfully defend the best territories have the greatest chance of mating and passing on their genes. For example, as experiments have shown, male stickleback fish that defend large territories are more successful in attract-

ing mates than are males that defend small territories. Females that select males with the best territories increase their own reproductive success and pass their genetic traits (typically including their mate-selection preferences) to their offspring.

Animals Advertise Their Occupancy

Territories are advertised through sight, sound, and smell. If a territory is small enough, its owner's mere presence, reinforced by aggressive displays toward intruders, can provide sufficient defense. A mammal that has a territory but cannot always be present may use pheromones to scent-mark the boundaries of its territory. Male rabbits use pheromones secreted by glands in their chin and by anal glands to mark their territories. Hamsters rub the areas around their dens with secretions from special glands in their flanks.

Vocal displays are a common form of territorial advertisement. Male sea lions defend a strip of beach by swimming up and down in front of it, calling continuously. Male crickets produce a specific pattern of chirps to warn other males away from their burrows. Birdsong is a striking example of territorial defense. The husky trill of the male seaside sparrow is part of an aggressive display, warning other males to steer clear of his territory (**FIG. 25-20**). In fact, male sparrows that are unable to sing are unable to defend territories. The importance of singing to seaside sparrows' territorial defense was elegantly demonstrated by ornithologist M. Victoria McDonald, who captured territorial males and performed an operation that left them temporarily unable to sing but still able to utter the other, shorter and quieter signals in their vocal repertoires. The songless males were unable to defend territories or attract mates, but regained their lost territories when they recovered their singing ability.

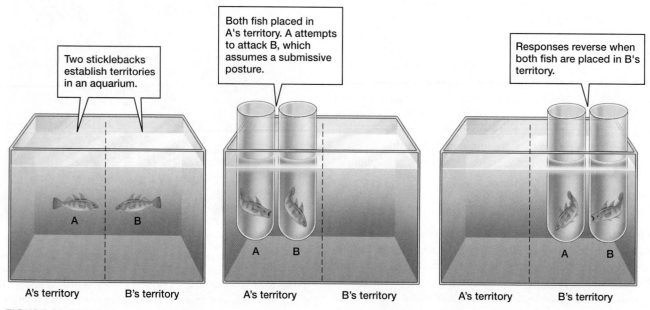

Two sticklebacks establish territories in an aquarium.

Both fish placed in A's territory. A attempts to attack B, which assumes a submissive posture.

Responses reverse when both fish are placed in B's territory.

A's territory   B's territory   A's territory   B's territory   A's territory   B's territory

FIGURE 25-19 Territory ownership and aggression
Niko Tinbergen's experiment demonstrating the effect of territory ownership on aggressive motivation.

FIGURE 25-20 Defense of a territory by song
A male seaside sparrow announces ownership of his territory.

25.4 HOW DO ANIMALS FIND MATES?

In many sexually reproducing animal species, mating involves copulation or other close contact between males and females. Before animals can successfully mate, however, they must identify one another as members of the same species, as members of the opposite sex, and as being sexually receptive. In many species, finding an appropriate potential partner is only the first step. Often the male must demonstrate his quality before the female will accept him as a mate. The need to fulfill all of these requirements has resulted in the evolution of a diverse and fascinating array of courtship behaviors.

Signals Encode Sex, Species, and Individual Quality

Individuals that waste energy and gametes by mating with members of the wrong sex or wrong species are at a disadvantage in the contest to reproduce. Thus, natural selection favors behaviors by which animals communicate their sex and species to potential mates.

Many Mating Signals Are Acoustic

Animals often use sounds to advertise their sex and species. Consider the raucous nighttime chorus of male tree frogs, each singing a species-specific song. Male grasshoppers and crickets also advertise their sex and species by their calls, as does the female mosquito with her high-pitched whine.

Signals that advertise sex and species may also be used by potential mates in comparisons among rival suitors. For example, the male bellbird uses its deafening song to defend large territories and attract females from great distances. A female flies from one territory to another, alighting near each male in his tree. The male, beak gaping, leans directly over the flinching female and utters an ear-splitting note. The female apparently endures this noise to compare the songs of the various males, perhaps choosing the loudest as a mate.

Visual Mating Signals Are Also Common

Many species use visual displays for courting. The firefly, for example, flashes a message that identifies its sex and species. Male fence lizards bob their heads in a species-specific rhythm, and females distinguish and prefer the rhythm of their own species. The elaborate construction projects of the male gardener bowerbird and the scarlet throat of the male frigate bird serve as flashy advertisements of sex, species, and male quality (**FIG. 25-21**). Sending these extravagant signals must be risky, as they make it much easier for predators to locate the sender. For males, the added risk is an evolutionary necessity, because females won't mate with males that lack the appropriate signal. Females, in contrast, typically do

FIGURE 25-21 Sexual displays
(a) During courtship, a male gardener bowerbird builds a bower out of twigs and decorates it with colorful items that he gathers. (b) A male frigate bird inflates his scarlet throat pouch to attract passing females. QUESTION The male bowerbird provides no protection, feeding, or other care to his mate or offspring. Why, then, do females carefully compare the bowers of different males before choosing a mate?

(b)

(a)

FIGURE 25-22 Sexual dimorphism in guppies
As in many animal species, the male guppy (left) is brighter and more colorful than the female.

not need to attract males or assume the risk associated with a conspicuous signal, so in many species females are drab in comparison to males (**FIG. 25-22**).

The intertwined functions of sex recognition and species recognition, advertisement of individual quality, and synchronization of reproductive behavior commonly require a complex series of signals, both active and passive, by both sexes. Such signals are beautifully illustrated by the complex underwater "ballet" executed by the male and female three-spined stickleback fish (**FIG. 25-23**).

Chemical Signals Can Bring Mates Together

Pheromones can also play an important role in reproductive behavior. A sexually receptive female silk moth, for example, sits quietly and releases a chemical message so powerful that it can be detected by males up to 3 miles (5 kilometers) away. The exquisitely sensitive and selective receptors on the antennae of the male silk moth respond to just a few molecules of the substance, allowing him to travel upwind along a concentration gradient to find the female (**FIG. 25-24a**).

Water is an excellent medium for dispersing chemical signals, and fish commonly use a combination of pheromones and elaborate courtship movements to ensure the synchronous release of gametes. Mammals, with their highly developed sense of smell, often rely on pheromones released by the female during her fertile periods to attract males (**FIG. 25-24b**).

25.5 WHAT KINDS OF SOCIETIES DO ANIMALS FORM?

Sociality is a widespread feature of animal life. Most animals interact at least a little with other members of their species. Many spend the bulk of their lives in the company of others, and a few species have developed complex, highly structured societies. Social interaction can be cooperative or competitive but is typically a mixture of the two.

Group Living Has Advantages and Disadvantages

Living in a group has both costs and benefits. A species will not evolve social behavior unless the benefits of doing so outweigh the costs.

Here are some costs that social animals may encounter:

- Increased competition within the group for limited resources
- Increased risk of infection from contagious diseases
- Increased risk that offspring will be killed by other members of the group
- Increased risk of being spotted by predators

Social animals may realize these benefits:

- Increased abilities to detect, repel, and confuse predators
- Increased hunting efficiency or increased ability to spot localized food resources
- Advantages resulting from the potential for division of labor within the group
- Increased likelihood of finding mates

Sociality Varies Among Species

The degree to which animals of the same species cooperate varies from one species to the next. Some types of animals, such as the mountain lion, are basically solitary; interactions between adults consist of brief aggressive encounters and mating. Other types of animals cooperate on the basis of changing needs. For example, the coyote is solitary when food is abundant but hunts in packs in the winter when food becomes scarce.

Loose social groups, such as pods of dolphins, schools of fish, flocks of birds, and herds of musk oxen (**FIG. 25-25**), can provide benefits. For example, the characteristic spacing of fish in schools or the V-pattern of geese in flight provides a hydrodynamic or aerodynamic advantage for each individual in the group, reducing the energy required for swimming or flying. Some biologists hypothesize that herds of antelope or schools of fish confuse predators; their myriad bodies make it difficult for the predator to focus on and pursue a single individual.

A Few Species Form Complex Societies

At the other end of the social spectrum are a few highly integrated cooperative societies, primarily among insects and mammals. As you read the following section, you may notice that some cooperative societies are based on behavior that seems to sacrifice the individual for the good of the group. There are many examples: Young, mature Florida scrub jays may remain at their parents' nest and help them raise subsequent broods instead of breeding; worker ants often die in defense of their nest; Belding ground squirrels may sacrifice their own lives to warn the rest of their group of an approaching predator. These behaviors are examples of **altruism**—behavior that decreases the reproductive success of one individual to benefit another.

(a) A male, inconspicuously colored, leaves the school of males and females to establish a breeding territory.

(b) As his belly takes on the red color of the breeding male, he displays aggressively at other red-bellied males, exposing his red underside.

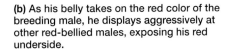

(c) Having established a territory, the male begins nest construction by digging a shallow pit that he will fill with bits of algae cemented together by a sticky secretion from his kidneys.

(d) After he tunnels through the nest to make a hole, his back begins to take on the blue courting color that makes him attractive to females.

(e) An egg-carrying female displays her enlarged belly to him by assuming a head-up posture. Her swollen belly and his courting colors are passive visual displays.

(f) Using a zigzag dance, he leads her to the nest.

(g) After she enters, he stimulates her to release eggs by prodding at the base of her tail.

(h) He enters the nest as she leaves and deposits sperm, which fertilize the eggs.

**FIGURE 25-23
Courtship of the three-spined stickleback**

Forming Groups with Relatives Fosters the Evolution of Altruism

How could altruistic behavior evolve? When individuals perform self-sacrificing deeds, why aren't the alleles that contribute to this behavior eliminated from the gene pool? One possibility is that other members of the group are close relatives of the altruistic individual. Because close relatives share alleles, the altruistic individual may promote the survival of its own alleles through behaviors that maximize the survival of its close relatives. This concept is called **kin selection**. Kin selection helps explain the self-sacrificing behaviors that contribute to the success of cooperative societies. Cooperative behavior is illustrated in the following sections, which describe two examples of complex societies, one in an insect species and one in a mammal species.

(a)

(b)

FIGURE 25-24 Pheromone detectors
(a) Male moths find females not by sight but by following airborne pheromones released by females. These odors are sensed by receptors on the male's huge antennae, whose enormous surface area maximizes the chances of detecting the female scent. **(b)** When dogs meet, they typically sniff each other near the base of the tail. Scent glands there broadcast information about the bearer's sex and interest in mating. QUESTION Female dogs use a pheromone to signal readiness to mate, but female mandrills (see Fig. 25-9) signal mating readiness with a visual signal. What differences would you predict between the two species' methods of searching for food?

Honeybees Live Together in Rigidly Structured Societies

Perhaps the most puzzling of all animal societies are those of the bees, ants, and termites. Scientists have long struggled to explain the evolution of a social structure in which most individuals never breed, but instead labor intensively to feed and protect the offspring of a different individual. Whatever its evolutionary explanation, the intricate organization of a social insect colony makes a compelling story. In these communities, the individual is a mere cog in an intricate, smoothly running machine and could not survive by itself.

Individual social insects are born into one of several castes within the society. These castes are groups of similar individuals that perform a specific function. Honeybees emerge from their larval stage into one of three major preordained roles. One role is that of *queen*. Only one queen

is tolerated in a hive at any time. Her functions are to produce eggs (up to 1000 per day for a lifetime of 5 to 10 years) and regulate the lives of the workers. Male bees, called *drones*, serve merely as mates for the queen. Soon after the queen hatches, drones lured by her sex pheromones swarm around her, and she mates with as many as 15 of them. This relatively brief "orgy" supplies her with sperm that will last a lifetime, enough to fertilize more than 3 million eggs. Their sexual chore accomplished, the drones become superfluous and are eventually driven out of the hive or killed.

The hive is run by the third class of bees, sterile female *workers*. A worker's tasks are determined by her age and

FIGURE 25-25 Cooperation in loosely organized social groups
A herd of musk oxen functions as a unit when threatened by predators such as wolves. Males form a circle, horns pointed outward, around the females and young.

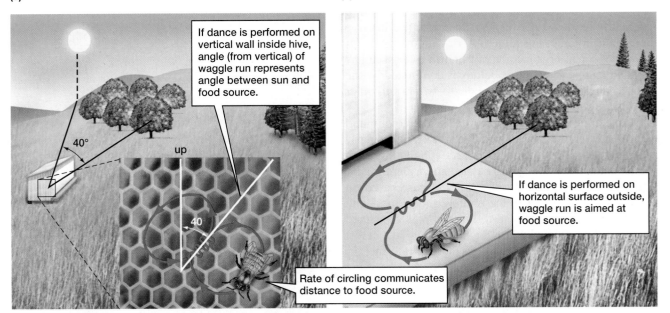

FIGURE 25-26 Bee language: the waggle dance
A forager, returning from a rich source of nectar, performs a waggle dance that communicates the distance and direction of the food source as other foragers crowd around her, touching her with their antennae. The bee moves in a straight line while shaking her abdomen back and forth ("waggling") and buzzing her wings. She repeats this over and over in the same location, circling back in alternating directions.

by conditions in the colony. A newly emerged worker starts life as a waitress, carrying food such as honey and pollen to the queen, to other workers, and to developing larvae. As she matures, special glands begin wax production, and she becomes a builder, constructing perfectly hexagonal cells of wax in which the queen deposits her eggs and the larvae develop. She also takes shifts as a maid, cleaning the hive and removing the dead, and as a guard, protecting the hive against intruders. Her final role in life is that of a forager, gathering pollen and nectar, food for the hive. She spends nearly half of her two-month life in this role. Acting as a forager scout, she seeks new and rich sources of nectar and, if she finds one, returns to the hive and communicates its location to other foragers. She communicates by means of the **waggle dance**, an elegant form of symbolic expression (**FIG. 25-26**).

Pheromones play a major role in regulating the lives of social insects. Honeybee drones are drawn irresistibly to the queen's sex pheromone (*queen substance*), which she releases during her mating flights. Back at the hive, she uses the same substance to maintain her position as the only fertile female. The queen substance is licked off her body and passed among all the workers, rendering them sterile. The queen's presence and health are signaled by her continuing production of queen substance; a decrease in production alters the behavior of the workers. Almost immediately they begin building extra-large "royal cells." The workers feed the larvae that develop in these cells a special glandular secretion known as "royal jelly." This unique food alters the development of the growing larvae so that, instead of a worker, a new queen emerges from the royal cell. The old queen then leaves the hive, taking a swarm of workers with her to establish residence elsewhere. If more than one new queen emerges, a battle to the death ensues. The victorious queen takes over the hive.

Naked Mole Rats Form a Complex Vertebrate Society

The nervous systems of vertebrates are far more complex than those of insects, and we might therefore expect vertebrate societies to be proportionately more complex. With the exception of human society, however, they are not. Perhaps the most unusual society among nonhuman mammals is that of the naked mole rat (**FIG. 25-27**). These nearly

FIGURE 25-27 A naked mole rat queen rests atop a group of workers

blind, nearly hairless relatives of guinea pigs live in large underground colonies in southern Africa and have an ant-like form of social organization not known to exist in any other mammalian society. The colony is dominated by the queen, a single reproducing female to whom all other members are subordinate.

The queen is the largest individual in the colony and maintains her status by aggressive behavior, particularly shoving. She prods and shoves workers, stimulating them to become more active. As in honeybee hives, there is a division of labor among the workers, in this case based on size. Small, young rats clean the tunnels, gather food, and tunnel. Tunnelers line up head to tail and pass excavated dirt along the completed tunnel to an opening. Just below the opening, a larger mole rat flings the dirt into the air, adding it to a cone-shaped mound. Biologists observing this behavior from the surface dubbed it "volcanoing." In addition to volcanoing, large mole rats defend the colony against predators and members of other colonies.

If another female begins to become fertile, the queen apparently senses changes in the estrogen levels of the subordinate female's urine. The queen then selectively shoves the would-be breeder, causing stress that prevents her rival from ovulating. Although all adult males are fertile, large males are more likely to mate with the queen than are small ones. When the queen dies, a few of the females gain weight and begin shoving one another. The aggression may escalate until a rival is killed. Ultimately, a single female becomes dominant. Her body lengthens, and she assumes the queenship and begins to breed. Litters averaging 14 pups are produced about four times a year. During the first month, the queen nurses her pups, and the workers feed the queen. Then the workers begin feeding the pups solid food.

25.6 CAN BIOLOGY EXPLAIN HUMAN BEHAVIOR?

The behaviors of humans, like those of all other animals, have an evolutionary history. Thus, the techniques and concepts of ethology can help us understand and explain human behavior. Human ethology, however, will remain a less-rigorous science than animal ethology. We cannot treat people as laboratory animals, devising experiments that control and manipulate the factors that influence their attitudes and actions. In addition, some observers argue that human culture has been freed from the constraints of its evolutionary past for so long that we cannot explain our behavior in terms of biological evolution. Nevertheless, many scientists have taken an ethological, evolutionary approach to human behavior, and their work has had a major impact on our view of ourselves.

The Behavior of Newborn Infants Has a Large Innate Component

Because newborn infants have not had time to learn, we can assume that much of their behavior is innate. The rhythmic movement of an infant's head in search of its

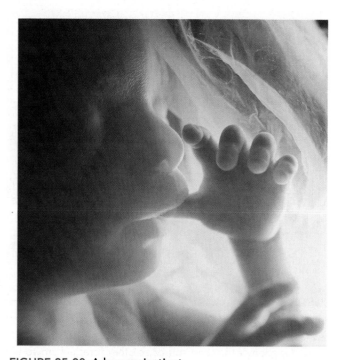

FIGURE 25-28 A human instinct
Thumb sucking is a difficult habit to discourage in young children, because sucking on appropriately sized objects is an instinctive, food-seeking behavior. This fetus sucks its thumb at about 4-1/2 months of development.

mother's breast is an innate behavior that is expressed in the first days after birth. Sucking, which can be observed even in a human fetus, is also innate (**FIG. 25-28**). Other behaviors seen in newborns and even premature infants include grasping with the hands and feet and making walking movements when the body is held upright and supported.

Another example is smiling, which can occur soon after birth. Initially, smiling can be induced by almost any object looming over the newborn. This initial indiscriminate response, however, is soon modified by experience. Infants up to two months old will smile in response to a stimulus consisting of two dark, eye-sized spots on a light background, which at that stage of development is a more potent stimulus for smiling than is an accurate representation of a human face. But as the child's development continues, learning and further development of the nervous system interact to limit the response to more-correct representations of a face.

Newborns in their first three days of life can be conditioned to produce certain rhythms of sucking when their mother's voice is used as reinforcement. In experiments, infants preferred their own mothers' voices to other female voices, as indicated by their responses (**FIG. 25-29**). The infant's ability to learn his or her mother's voice and respond positively to it within days of birth has strong parallels to imprinting and may help initiate bonding with the mother.

Young Humans Acquire Language Easily

One of the most important insights from studies of animal learning is that animals tend to have an inborn predilection for specific types of learning that are important to their species' mode of life. In humans, one such inborn predilection seems to be for the acquisition of language. Young children are able to acquire language rapidly and nearly effortlessly; they typically acquire a vocabulary of 28,000 words before the age of eight years. Research suggests that we are born with a brain that is already primed for this early facility with language. For example, a human fetus begins responding to sounds during the third trimester of pregnancy, and researchers have demonstrated that infants are able to distinguish among consonant sounds by six weeks after birth. In this demonstration, an infant responded to various consonant sounds by sucking on a pacifier that contained a force transducer to record the sucking rate. When one sound (such as "ba") was presented repeatedly, the infant became habituated and decreased her sucking rate. But when a new sound (such as "pa") was presented, sucking rate increased, revealing that the infant perceived the new sound as different.

Behaviors Shared by Diverse Cultures May Be Innate

Another way to study the innate bases of human behavior is to compare simple acts performed by people from diverse cultures. This comparative approach, pioneered by ethologist Irenaus Eibl-Eibesfeldt, has revealed several gestures that seem to form a universal, and therefore prob-

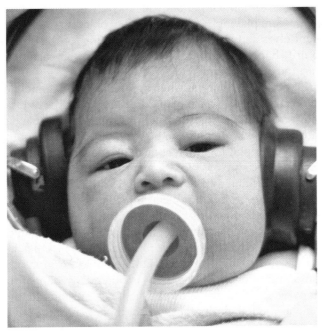

FIGURE 25-29 Newborns prefer their mother's voices
Using a nipple connected to a computer that plays audio tapes, researcher William Fifer demonstrated that newborns can be conditioned to suck at specific rates for the privilege of listening to their own mothers' voices through headphones. For example, if the infant sucks faster than normal, her mother's voice is played; if she sucks more slowly, another woman's voice is played. Researchers found that infants easily learned and were willing to work hard at this task just to listen to their own mothers' voices, presumably because they had become used to her voice in the womb.

ably innate, human signaling system. Such gestures include facial expressions for pleasure, rage, and disdain and greeting movements such as an upraised hand or the "eye flash" (in which the eyes are widely opened and the eyebrows rapidly elevated). The evolution of these "hardwired" gestures presumably depended on the advantages that accrued to both senders and receivers from sharing information about the emotional state and intentions of the sender. A species-wide method of communication was perhaps especially important before the advent of language and later remained useful during encounters between people who shared no common language.

Certain complex social behaviors are widespread among diverse cultures. For example, the incest taboo (avoidance of mating with close relatives) seems to be universal across human cultures (and even across many species of nonhuman primates). It seems unlikely, however, that a shared belief could be encoded in our genes. Some biologists have suggested that the taboo is instead a cultural expression of an evolved, adaptive behavior. According to this hypothesis, close contact among family members early in life suppresses sexual desire, an effect that arose because of the negative consequences of inbreeding (such as a higher incidence of genetic diseases). The hypothesis does not require us to assume an innate social belief, but rather proposes that we inherit a learning program that causes us to undergo a kind of imprinting early in life.

Humans May Respond to Pheromones

Although the main channels of human communication are through the eyes and ears, humans also seem to respond to chemical messages. The possible existence of human pheromones was hinted at in the early 1970s, when biologist Martha McClintock found that the menstrual cycles of roommates and close friends tended to become synchronized. McClintock suggested that the synchrony resulted from some chemical signal between the women, but almost 30 years passed before she and her colleagues uncovered more conclusive evidence that a pheromone was at work.

In 1998, McClintock's research group asked nine female volunteers to wear cotton pads in their armpits for eight hours each day during their menstrual cycles. The pads were then disinfected with alcohol and swabbed above the upper lips of another set of 20 female subjects (who reported that they could detect no odors other than alcohol on the pads). The subjects were exposed to the pads in this way each day for two months, with half the group sniffing secretions from women in the early (preovulation) part of the menstrual cycle while the other half was exposed to secretions from later in the cycle (postovulation). Women exposed to early-cycle secretions had shorter-than-usual menstrual cycles, and women exposed to late-cycle secretions had delayed menstruation. It appears that women release different pheromones, with different effects on receivers, at different points in the menstrual cycle.

Although McClintock's experiment offers strong evidence for the existence of human pheromones, little else is known about chemical communication in humans. The actual molecules that caused the effects documented by Mc-Clintock remain unknown, as does their function—what benefit would a woman gain from influencing the menstrual cycles of other women? Receptors for chemical messages have not yet been found in humans, and we don't know if the "menstrual pheromones" are the first known example of an important communication system or merely an isolated case of a vestigial ability. Despite the hopeful advertisements for "sex attraction pheromones" on late-night television, chemical communication in humans is a scientific mystery awaiting a solution.

Studies of Twins Reveal Genetic Components of Behavior

Twins present an opportunity to examine the hypothesis that differences in human behavior are related to genetic differences. If a particular behavior is heavily influenced by genetic factors, we would expect to find similar expression of that behavior in *identical twins* (which arise from a single fertilized egg and have identical genes) but not in *fraternal twins* (which arise from two individual eggs and are no more similar genetically than are other siblings). Data from twin studies, and from other within-family investigations, have tended to confirm the heritability of many human behavioral traits. These studies have documented a significant genetic component for traits such as activity level, alcoholism, sociability, anxiety, intelligence, dominance, and even political attitudes. On the basis of tests designed to measure many aspects of personality, identical twins are about two times more similar in personality than are fraternal twins.

The most fascinating twin findings come from observations of identical twins separated soon after birth, reared in different environments, and reunited for the first time as adults. Identical twins reared apart have been found to be as similar in personality as those reared together, indicating that the differences in their environments had little influence on their personality development. They have been found to share nearly identical taste in jewelry, clothing, humor, food, and names for children and pets. In some cases these separated twins share personal idiosyncrasies such as giggling, nail biting, drinking patterns, hypochondria, and mild phobias.

Biological Investigation of Human Behavior Is Controversial

The field of human behavioral genetics is controversial, especially among nonscientists, because it challenges the long-held belief that environment is the most important determinant of human behavior. As discussed earlier in this chapter, we now recognize that all behavior has some genetic basis and that complex behavior in nonhuman animals typically combines elements of both innate and learned behaviors. Thus, it seems likely that our own behavior is influenced by both our evolutionary history and our cultural heritage. The debate over the relative importance of heredity and environment in determining human behav-

ior continues and is unlikely ever to be fully resolved. Human ethology is not yet recognized as a rigorous science, and it will always be hampered because we can neither view ourselves with detached objectivity nor experiment with people as if they were laboratory rats. Despite these limitations, there is much to be learned about the interaction of learning and innate tendencies in humans.

EVOLUTIONARY CONNECTIONS

Why Do Animals Play?

To the delight of thousands of visitors—and the puzzlement of behavioral biologists—Pigface, a giant 50-year-old African softshell turtle, would spend hours each day batting a ball around his aquatic home in the National Zoo in Washington, D.C. Play has always been somewhat of a mystery. It has been observed in many birds and in most mammals, but until zookeepers tossed Pigface a ball some years ago, play had never been seen in animals as evolutionarily ancient as turtles.

Animals at play are fascinating. Pygmy hippopotamuses push one another, shake and toss their heads, splash in the water, and pirouette on their hind legs. Otters delight in elaborate acrobatics. Bottlenose dolphins balance fish on their snouts, throw objects, and carry them in their mouths while swimming. Baby vampire bats chase, wrestle, and slap each other with their wings. Even octopuses have been seen playing a game: pushing objects away from themselves and into a current, then waiting for the objects to drift back, only to push them back into the current to start the cycle again.

Play can be solitary, as when a single animal manipulates an object—think of a cat with a ball of yarn, or the dolphin with its fish, or a macaque monkey making and playing with a snowball. Or, play can be social. Often, young of the same species play together, but parents may join them (**FIG. 25-30a**). Social play typically includes chasing, fleeing, wrestling, kicking, and gentle biting (**FIG. 25-30b, c**).

Play seems to lack any clear immediate function, and it is abandoned in favor of feeding, courtship, and escaping from danger. Young animals play more frequently than do adults. Play typically borrows movements from other behaviors (attacking, fleeing, stalking, and so on) and uses considerable energy. Also, play is potentially dangerous. Many young humans and other animals are injured, and some are killed, during play. In addition, play can distract an animal from the presence of danger while making it conspicuous to predators. So, why do animals play?

The most logical conclusion is that play must have survival value and that natural selection has favored those individuals who engage in playful activities. One of the best explanations for the survival value of play is the "practice theory." It suggests that play allows young animals to gain experience in behaviors that they will use as adults. By performing these acts repeatedly in play, the animal practices skills that will later be important in hunting, fleeing, or social interactions.

(b)

(c)

(a)

FIGURE 25-30 Young animals at play

Recent research supports and extends that proposal. Play is most intense early in life when the brain develops and crucial neural connections form. John Byers, a zoologist at the University of Idaho, has observed that species with large brains tend to be more playful than species with small brains. Because larger brains are generally linked to greater learning ability, this relationship supports the idea that adult skills are learned during juvenile play. Watch children roughhousing or playing tag, and you will see how play fosters strength and coordination and develops skills that might have helped our hunting ancestors survive. Quiet play with other children, using dolls, blocks, and other toys, helps children prepare to interact socially, nurture their own children, and deal with the physical world.

Although Shakespeare tells us "play needs no excuse," there is good evidence that the tendency to play has evolved as an adaptive behavior in animals capable of learning. Play is quite literally serious fun.

CASE STUDY REVISITED   SEX AND SYMMETRY

In the experiment described at the beginning of this chapter, women found males with the most symmetrical bodies to be the most attractive. But how did the women know which males were most symmetrical? After all, the researchers' measurement of male symmetry was based on small differences in the size of body parts that the female judges did not even see during the test.

Perhaps male body symmetry is reflected in facial symmetry, and females prefer symmetrical faces. To test this hypothesis, a group of researchers used computers to alter photos of male faces, either increasing or decreasing their symmetry (**FIG. 25-31**). Then female observers rated each face for attractiveness. The observers had a strong preference for more symmetrical faces.

Some evidence suggests that women need not even look at men to determine symmetry. In one study, researchers measured the body symmetry of 80 men and then issued a clean T-shirt to each one. Each subject wore his shirt to bed for two consecutive nights. A panel of 82 women sniffed the shirts and rated their scents for

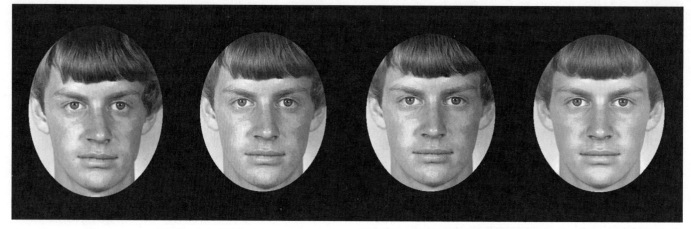

FIGURE 25-31 Faces of varying symmetry
Researchers used sophisticated software to modify facial symmetry. From left: a face modified to be less symmetrical; the original, unmodified face; a face modified to be more symmetrical; a perfectly symmetrical face.

"pleasantness" and "sexiness." Which shirts had the sexiest, most pleasant scents? The ones worn by the most symmetrical men. The researchers concluded that women can identify symmetrical men by their scent.

Why would females prefer to mate with symmetrical males? The most likely explanation is that symmetry indicates good physical condition. Disruptions of normal embryological development can cause bodies to be asymmetrical, so a highly symmetrical body indicates healthy, normal development. Females that mate with individuals whose health and vitality are announced by their symmetrical bodies might have offspring that are similarly healthy and vital.

Consider This Is our perception of human beauty determined by cultural standards, or is it part of our biological makeup, the product of our evolutionary heritage? What evidence would persuade you that beauty is a biological phenomenon, or that it is a cultural one?

CHAPTER REVIEW

SUMMARY OF KEY CONCEPTS

25.1 How Do Innate and Learned Behaviors Differ?

Although all animal behavior is influenced by both genetic and environmental factors, biologists can distinguish between behaviors whose development is not highly dependent on external factors and behaviors that require more extensive environmental stimuli in order to develop. Behaviors in the first category are sometimes designated as innate and can be performed properly the first time an animal encounters the appropriate stimulus. Behavior that changes in response to input from an animal's social and physical environment is said to be learned. Learning can modify innate behavior to make it more appropriate.

Although the distinction between innate and learned behavior is conceptually useful, the distinction is not sharp in naturally occurring behaviors. Learning allows animals to modify innate responses so that they occur only with appropriate stimuli. Imprinting, a form of learning with innate constraints, is possible only at a certain time in an animal's development.

Web Tutorial 25.1 Observing Behavior: Homing in Digger Wasps

25.2 How Do Animals Communicate?

Communication allows animals of the same species to interact effectively in their quest for mates, food, shelter, and other resources. Animals communicate through visual signals, sound, chemicals (pheromones), and touch. Visual communication is quiet and can convey rapidly changing information. Visual signals are active (body movements) or passive (body shape and color). Sound communication can also convey rapidly changing information, and it is effective where vision is impossible. Pheromones can be detected after the sender has departed, conveying a message over time. Physical contact reinforces social bonds and is a part of mating.

25.3 How Do Animals Compete for Resources?

Although many competitive interactions are resolved through aggression, serious injuries are rare. Most aggressive encounters are settled by means of displays that communicate the motivation, size, and strength of the combatants.

Some species establish dominance hierarchies that minimize aggression. On the basis of initial aggressive encounters, each animal acquires a status in which it defers to more dominant individuals and dominates subordinates. When resources are limited, dominant animals obtain the largest share and are most likely to reproduce.

Territoriality, a behavior in which animals defend areas where important resources are located also minimizes aggressive encounters. In general, territorial boundaries are respected, and the best-adapted individuals defend the richest territories and produce the most offspring.

25.4 How Do Animals Find Mates?

Successful reproduction requires that animals recognize the species, sex, and sexual receptivity of potential mates. In many species, animals also assess the quality of potential mates. These requirements have contributed to the evolution of sexual displays that use all forms of communication.

25.5 What Kinds of Societies Do Animals Form?

Social living has both advantages and disadvantages, and species vary in the degree to which their members cooperate. Some species form cooperative societies. The most rigid and highly organized are those of the social insects such as the honeybee, in which the members follow rigidly defined roles throughout life. These roles are maintained through both genetic programming and the influence of certain pheromones. Naked mole rats exhibit the most complex and rigid vertebrate social interactions, resembling those of social insects.

Web Tutorial 25.2 Communication in Honeybees

25.6 Can Biology Explain Human Behavior?

The degree to which human behavior is genetically influenced is highly controversial. Because we cannot freely experiment on humans, and because learning plays a major role in nearly all human behavior, investigators must rely on studies of newborn infants and comparative cultural studies, correlations between certain behaviors and physiology (which suggest a role for pheromones), and studies of identical and fraternal twins. Evidence is mounting that our genetic heritage plays a role in personality, intelligence, simple universal gestures, our responses to certain stimuli, and our tendency to learn specific things such as language at particular stages of development.

KEY TERMS

aggression *page 497*
altruism *page 502*
behavior *page 490*
communication *page 495*
dominance hierarchy *page 498*

ethology *page 493*
habituation *page 491*
imprinting *page 493*
innate *page 490*
insight learning *page 492*
kin selection *page 503*

learning *page 490*
operant conditioning *page 491*
pheromone *page 496*
territoriality *page 499*

trial-and-error learning *page 491*
waggle dance *page 505*

THINKING THROUGH THE CONCEPTS

1. Explain why neither "innate" nor "learned" adequately describes the behavior of any given organism.

2. Explain why animals play.

3. List four senses through which animals communicate, and give one example of each form of communication. After each, present both advantages and disadvantages of that form of communication.

4. A bird will ignore a squirrel in its territory but will act aggressively toward a member of its own species. Explain why.

5. Why are most aggressive encounters among members of the same species relatively harmless?

6. Discuss advantages and disadvantages of group living.

7. In what ways do naked mole rat societies resemble those of the honeybee?

APPLYING THE CONCEPTS

1. Male mosquitoes orient toward the high-pitched whine of the female, and female mosquitoes, the only sex that sucks blood, are attracted to the warmth, humidity, and carbon dioxide exuded by their prey. Using this information, design a mosquito trap or killer that exploits a mosquito's innate behaviors. Then, design one for moths.

2. You raise honeybees but are new at the job. Trying to increase honey production, you introduce several queens into the hive. What is the likely outcome? What different things could you do to increase production?

3. Describe and give an example of a dominance hierarchy. What role does it play in social behavior? Give a human parallel, and describe its role in human society. Are the two roles similar? Why or why not? Repeat this exercise for territorial behavior in humans and in another animal.

4. You are manager of an airport. Planes are being endangered by large numbers of flying birds, which can be sucked into and disable the engines. Without harming the birds, what might you do to discourage them from nesting and flying near the airport and its planes?

FOR MORE INFORMATION

de Waal, F. "The End of Nature versus Nurture." *Scientific American*, December 1999. An eminent ethologist makes the case for taking a nuanced biological approach to understanding human behavior.

Lorenz, K. *King Solomon's Ring: New Light on Animal Ways.* New York: Thomas Y. Crowell, 1952. Beautifully written and filled with interesting anecdotes; provides important insights into early ethology.

Pinker, S. *The Language Instinct.* New York: William Morrow, 1994. An entertaining account of linguists' current understanding of how we develop the ability to use language and how that ability may have evolved.

Sherman, P., and Alcock, J. *Exploring Animal Behavior.* Sunderland, MA: Sinauer, 1998. A collection of articles in which behavioral biologists describe their research and their conclusions about the mechanisms, function, and evolution of behavior.

Weiner, J. *Time, Love, Memory: A Great Biologist and His Quest for the Origins of Behavior.* New York: Knopf, 2000. A wonderful science writer chronicles the study of the genetics of behavior, especially the work of a great pioneer of the field, Seymour Benzer.

Wheelwright, J. "Study the Clones First." *Discover*, August 2004. An account of how studying twins can help uncover the genetic component of human behavior.

26 Population Growth and Regulation

A bemused-looking statue sits on the desolate landscape of Easter Island. If they could speak, would the Easter Island statues tell of a population that outgrew the environment's capacity to sustain it?

CASE STUDY THE MYSTERY OF EASTER ISLAND

WHY DO CIVILIZATIONS VANISH? Among those who have pondered this question were the first Europeans to reach Easter Island. When they arrived in the 1700s, these voyagers were mystified by the enormous stone statues that dominate the island's barren landscape. The island's few inhabitants had no record or memory of the statues' creators and did not possess the technology that would have been needed to transport and erect such huge and heavy structures. Moving such heavy objects over the 6 miles from the nearest stone quarry and then maneuvering them into an upright position would have required long ropes and strong timbers. Easter Island, however, was devoid of anything that could

have furnished wood or fibers for rope. There were no trees at all on the island, and none of the few shrubs there grew higher than 10 feet.

An important clue to the mystery of Easter Island was revealed by scientists who studied pollen grains that they found in layers of ancient sediments. Because the age of each sediment layer can be determined, and because each plant species can be identified by the unique appearance of its pollen, pollen analysis can show how vegetation has changed over time.

The pollen record of Easter Island showed that before the arrival of humans, the island supported a diverse forest, in-

cluding toromiro trees that make excellent firewood, hauhau trees that could supply fiber for rope, and Easter Island palm trees with long, straight trunks that would have made good rollers for moving statues. But by A.D. 1400, about a thousand years after the arrival of humans, almost all of Easter Island's trees were gone. Some scientists hypothesize that a thousand years of clearing land for agriculture and cutting trees for firewood and construction materials had destroyed the forest.

Apparently, the culture responsible for the statues had disappeared along with the forest. Could there have been a connection between the two disappearances?

This chapter begins our study of **ecology** (derived from the Greek word "oikos," meaning "a place to live"). Ecology refers to the study of interrelationships between living things and their nonliving environment. The environment consists of an **abiotic** (nonliving) component—including soil, water, and weather—and a **biotic** (living) component, which includes all forms of life. An **ecosystem** includes everything, both living and nonliving, within a defined area, such as the island described in this chapter's Case Study. Within the ecosystem, all interacting populations of organisms form the **community**. Easter Island once hosted a flourishing community of several species of now-extinct trees, as well as shrubs and grasses, insects, microorganisms, and many species of birds.

What keeps natural populations from overpopulating their habitat and starving? What happens when different organisms compete for the same type of food, for space, or for other resources? Why has the human population continued to expand while other populations have fluctuated, remained stable, or declined? Look for answers to these questions in this chapter as we explore how populations grow and how population growth is controlled. In the rest of this unit we progress to *communities* and the interactions within them, describe the natural laws that govern how ecosystems function, and explore the diversity of ecosystems that make up the *biosphere* which encompasses all life on Earth. We conclude by examining human impacts on the biosphere and our attempts to conserve biodiversity.

26.1 HOW DOES POPULATION SIZE CHANGE?

A **population** consists of all the members of a particular species that live within an ecosystem. On Easter Island, for example, the Easter Island palm trees, the hauhau trees, and the toromiru trees each constituted a different population.

Studies of undisturbed ecosystems show that some populations tend to remain relatively stable in size over time, others fluctuate in a roughly cyclical pattern, and still others vary sporadically in response to complex environmental variables. But, in contrast to most nonhuman species, the global human population has exhibited steady growth for centuries. Here we examine how and why populations grow and then examine the forces that control this growth.

Three factors determine whether and how much the size of a population changes: births, deaths, and migration in or out. Organisms join a population through birth or **immigration** (migration in) and leave it through death or **emigration** (migration out). A population remains stable if, on the average, as many individuals leave as join. Population growth occurs when the number of births plus immigrants exceeds the number of deaths plus emigrants. Populations decline when the reverse occurs. A simple equation for the change in population size within a given time period is as follows:

$$(\textbf{births} - \textbf{deaths}) + (\textbf{immigrants} - \textbf{emigrants})$$
$$= \textbf{change in population size}$$

In many natural populations, organisms moving in and out contribute relatively little to population change, making birth and death rates the primary factors that influence population growth. For simplicity, then, we will omit immigration and emigration from future calculations of population change.

The size of any population results from the interaction between two major opposing factors that determine birth and death rates: *biotic potential* and *environmental resistance*. **Biotic potential** is the maximum rate at which the population could increase, assuming ideal conditions that allow a maximum birth rate and minimum death rate. **Environmental resistance** refers to the curbs on population growth that are set by the living and nonliving environment. Environmental resistance limits population growth and a population's ultimate size by increasing deaths and decreasing births. Examples of environmental resistance include such interactions among species as *predation* and *parasitism*. It includes *competition* that occurs both within a species and among different species that use the same resources. Environmental resistance also encompasses the always-limited availability of nutrients, energy, and space, as well as short-term natural events such as storms, fires, freezing weather, floods, and droughts.

For long-lived organisms in nature, the interaction between biotic potential and environmental resistance usually results in a balance between the size of a population and the resources available to support it. To understand how populations grow and how their size is regulated, let's examine each of these forces in more detail.

Biotic Potential Can Produce Exponential Growth

Population Growth Is a Function of the Birth Rate, the Death Rate, and Population Size

A change in population size over time is a function of birth rate, death rate, and the number of individuals in the original population. The **birth rate** (b) and the **death rate** (d) are often expressed as the number of births (or deaths) per individual during a specific unit of time, such as a month or a year.

The **growth rate** (r) of a population is a measure of the change in population size per individual per unit of time. This value is determined by subtracting the death rate (d) from the birth rate (b):

$$r \quad = \quad b \quad - \quad d$$
$$(\textbf{growth rate}) \ = \ (\textbf{birth rate}) \ - \ (\textbf{death rate})$$

If the death rate exceeds the birth rate, the growth rate will be negative and the population will decline. To calculate the annual growth rate of a population of 1000, in which 150 births and 50 deaths occur each year, we can use this simple equation:

$$r \ = \quad 0.15 \quad - \quad 0.05 \ = \ 0.1 \text{ or } 10\% \text{ per year}$$
(pop. growth rate)   (birth rate)   (death rate)

To determine the number of individuals added to a population in a given time period (G), multiply the growth rate (r) by the original population size (N):

$$G \quad = \quad r \quad \times \quad N$$

| (population growth per unit time) | (growth rate) | (population size) |

In this example, population growth (rN) equals $0.1 \times 1000 = 100$ in the first year. If this growth rate is constant, then the following year the population size (N) starts at 1100, and 110 (rN) individuals join it. During the third year, 121 individuals are added, and so on.

This pattern of continuously accelerating increase in population size is called **exponential growth**. During exponential growth, a population (over a given time period) grows by a fixed percentage of its size at the beginning of that time period. Thus, an increasing number of individuals are added to the population during each succeeding time period, causing population size to grow at an ever-accelerating pace. The graph of exponential population growth that is often called a **J-curve**, after its shape. Populations approximate exponential growth whenever births consistently exceed deaths. This occurs if, on the average, each individual produces more than one surviving offspring during its lifetime. Although the number of offspring an individual produces each year varies from millions (for an oyster) to one or fewer (for a human), each organism—whether working alone or as part of a sexually reproducing pair—has the potential to replace itself many times over during its lifetime. This high biotic potential evolved because it helps ensure that, in a world filled with forces of environmental resistance, some offspring survive to bear their own young. Several factors influence biotic potential, including the following:

- The age at which the organism first reproduces
- The frequency at which reproduction occurs
- The average number of offspring produced each time the organism reproduces
- The length of the organism's reproductive life span
- The death rate of individuals under ideal conditions

We will use examples in which these factors differ to illustrate the concept of exponential growth. The bacterium *Staphylococcus* (FIG. 26-1a) normally resides harmlessly in and on the human body, where its population growth is restricted by environmental resistance. But in an ideal culture medium—such as warm custard—where *Staphylococcus* may accidentally be introduced, each bacterial cell might divide every 20 minutes. The population would double three times each hour (in this case, threatening the consumer with food poisoning). The larger the population grows, the more cells there are to divide. The reproductive potential of bacteria is so great that, hypothetically, the offspring of a single bacterium could coat Earth in a layer 7 feet deep within 48 hours!

In contrast, the golden eagle is a relatively long-lived, rather slowly reproducing species (FIG. 26-1b). Let's assume that the golden eagle can live 30 years and reaches sexual maturity at age 4 years, and that each pair of eagles produces two offspring annually for the remaining 26 years (red line). Figure 26-1 compares the potential population growth of eagles to that of bacteria, assuming no deaths occur in either population during the time graphed. Although the time scale differs tremendously between our slowly and rapidly reproducing examples, notice that the shapes of the curves are virtually identical: both populations exhibit the J-curve characteristic of exponential growth. Figure 26-1b also shows what happens if eagle reproduction begins at age 6 years (green line) instead of 4. Exponential growth still occurs, but the time required to reach a particular size increases considerably. This result has important implications for the human population: delayed reproduction significantly slows population growth. If each woman bears three children in her early teens, the population will grow much faster than if each woman bears five children but has her first after age 30.

So far we have looked only at population growth based on maximum birth rates. Even under ideal conditions, however, deaths are inevitable, and biotic potential takes minimum death rates into account. **FIGURE 26-2** compares three hypothetical bacterial populations experiencing different death rates. Notice that the shapes of the three curves are the same. As long as births exceed deaths, the population eventually approaches infinite size. As death rates increase, however, it takes longer to reach a given population size.

26.2 HOW IS POPULATION GROWTH REGULATED?

Exponential Growth Occurs Only Under Special Conditions

In 1859 Charles Darwin wrote, "There is no exception to the rule that every organic being naturally increases at so high a rate that, if not destroyed, the Earth would soon be covered by the progeny of a single pair." But in nature, exponential growth occurs only under special circumstances and only for a limited time.

Populations That Undergo Boom-and-Bust Cycles Show Exponential Growth

Exponential growth occurs in populations that undergo regular cycles in which rapid population growth is followed by a sudden, massive die-off. These **boom-and-bust cycles** occur in a variety of organisms for complex and varied reasons. Many short-lived, rapidly reproducing species—from algae to insects—have seasonal population cycles that are linked to predictable changes in rainfall, temperature, or nutrient availability (FIG. 26-3). In temperate climates, insect populations grow rapidly during

(a)

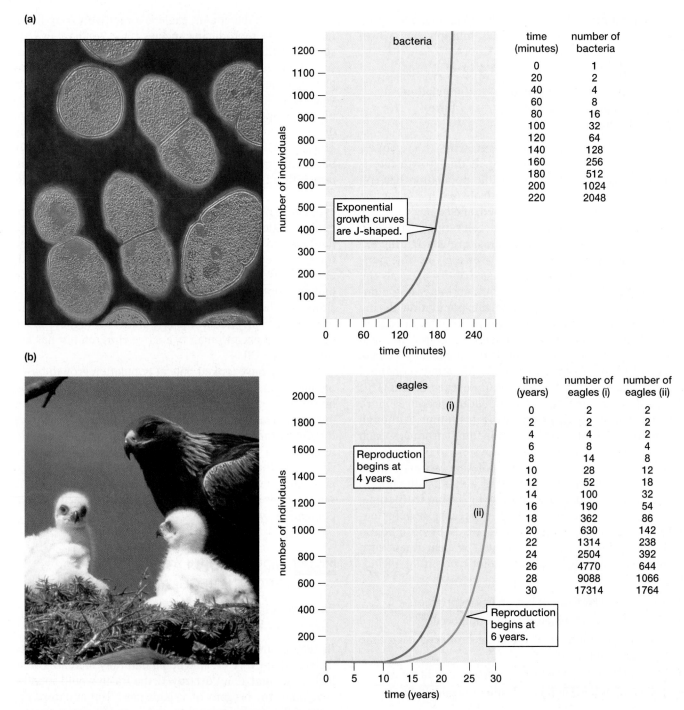

time (minutes)	number of bacteria
0	1
20	2
40	4
60	8
80	16
100	32
120	64
140	128
160	256
180	512
200	1024
220	2048

Exponential growth curves are J-shaped.

(b)

time (years)	number of eagles (i)	number of eagles (ii)
0	2	2
2	2	2
4	4	2
6	8	4
8	14	8
10	28	12
12	52	18
14	100	32
16	190	54
18	362	86
20	630	142
22	1314	238
24	2504	392
26	4770	644
28	9088	1066
30	17314	1764

Reproduction begins at 4 years.

Reproduction begins at 6 years.

FIGURE 26-1 Exponential growth curves are J-shaped
All such curves share a similar J shape; the major difference is the time scale. **(a)** Growth of a population of bacteria, starting with a single individual and with a doubling time of 20 minutes. **(b)** Growth of a population of eagles, starting with a single pair of hatchlings, for ages at first reproduction of 4 years (red line) and 6 years (green line). Notice in the table that after 26 years, the population of eagles that began reproducing at age 4 is nearly seven times that of the eagles that began reproducing at age 6.

the spring and summer, then crash with the killing hard frosts of winter. More-complex factors produce roughly four-year cycles for small rodents such as voles and lemmings, and much longer population cycles in hares, muskrats, and grouse.

Lemming populations, for instance, may grow until the rodents overgraze their fragile arctic tundra ecosystem. Lack of food, increasing populations of predators, and social stress caused by overpopulation may all contribute to a sudden high mortality. Many deaths occur as waves of

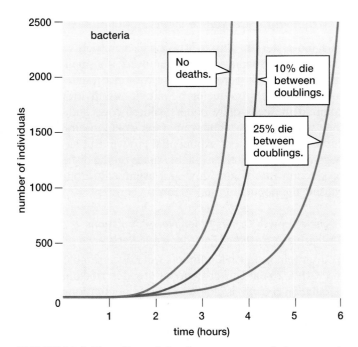

FIGURE 26-2 The effect of death rates on population growth
The graphs assume that a bacterial population doubles every 20 minutes. Notice that the population in which a quarter of the bacteria die every 20 minutes reaches 2500 only 2 hours and 20 minutes later than one in which no deaths occur. QUESTION What would the death rate need to be for these populations to stabilize?

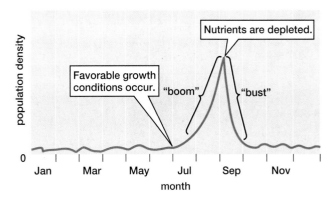

FIGURE 26-3 A boom-and-bust population cycle
Population density of cyanobacteria (blue-green algae) in an annual boom-and-bust cycle in a lake. Algae survive at a low level through the fall, winter, and spring. Early in July, conditions become favorable for growth, and exponential growth occurs through August. Nutrients soon become depleted, and the population "goes bust."

lemmings emigrate from regions of high population density. During these dramatic mass movements, lemmings are easy targets for predators. Many drown as they encounter bodies of water and try to swim across. The reduced lemming population eventually contributes to a decline in predator numbers (see "Scientific Inquiry: Cycles in Predator and Prey Populations") and a recovery of the plant community on which the lemmings normally feed. These responses, in turn, set the stage for the next round of exponential growth in the lemming population (**FIG. 26-4**).

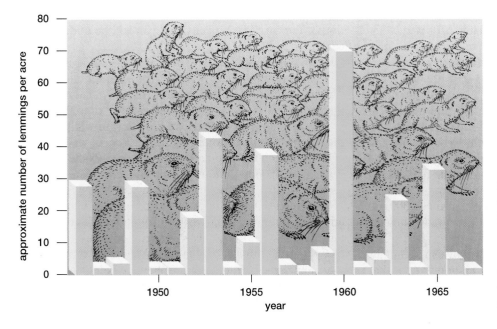

FIGURE 26-4 Lemming population cycles follow a boom-and-bust pattern
Lemming population density follows roughly a four-year cycle (data from Point Barrow, Alaska). QUESTION What factors might make the data in this graph somewhat erratic and irregular?

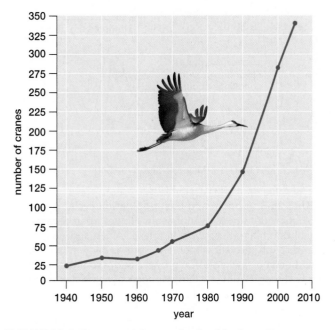

FIGURE 26-5 Exponential growth of wild whooping cranes Hunting and habitat destruction had reduced the world's whooping crane population to about 20 before they were protected in 1940. By 2005, their wild population had grown to 340 individuals. Notice the J-curve characteristic of exponential growth.

Exponential Growth Occurs When Organisms Invade New, Favorable Habitats

In populations that do not show boom-and-bust cycles, exponential growth may occur temporarily under special circumstances—for example, if food supply is increased, or if other population-controlling factors, such as predators (or human hunters) are eliminated. For example the whooping crane population has grown exponentially since they were protected from hunting and human disturbance in 1940 (**FIG. 26-5**). Exponential growth can also occur when individuals invade a new habitat with favorable conditions and little competition. For instance, if a farm field is plowed and then abandoned, it provides an ideal habitat for weedy annual plants and perennial grasses, whose populations may initially increase exponentially. Invasive species also show explosive population growth. **Invasive species** are organisms with a high biotic potential that are introduced (deliberately or accidentally) into ecosystems where they did not evolve and where they encounter little environmental resistance. For example, people introduced cane toads into Australia in 1935 to control beetles that were destroying sugar cane. The cane toads encountered few predators, and the females lay 8000 to 35,000 eggs at a time. Spreading outward from their release point, they now inhabit an area of about 300,000 square miles and are migrating rapidly into new habitats, threatening native species both by eating them and displacing them. Their population continues to grow exponentially.

As you will learn in the next section, all populations that exhibit exponential growth must eventually either stabilize or drop suddenly and dramatically, an event described as a "population crash."

Environmental Resistance Limits Population Growth

Imagine a sterile culture dish in which nutrients are constantly replenished and wastes removed. If a small number of living skin cells were added, they would attach to the bottom and begin reproducing by mitotic cell division. If you counted the cells daily under a microscope and graphed their numbers, for a while your graph would resemble the J-curve characteristic of exponential growth. But as the cells began to occupy all the available space on the dish, their reproduction rate would slow and eventually drop to zero, causing the population size to remain constant.

Logistic Growth Occurs When New Populations Stabilize as a Result of Environmental Resistance

Your graph of skin cell numbers will now resemble the one in **FIGURE 26-6a**. This growth pattern, described as **logistic population growth**, is characteristic of populations that increase up to the maximum number that their environment can sustain, and then stabilize.

The curve that results when logistic growth is plotted is sometimes called an **S-curve** after its general shape. The mathematical formula that produces a logistic growth curve consists of the formula for exponential growth ($G = rN$) multiplied by a factor that imposes limits on this growth. For real populations, these limits are imposed by the environment. The *logistic formula* includes a variable (K) that is described as the *carrying capacity* of the ecosystem. **Carrying capacity (K)** is the maximum population size that can be sustained by an ecosystem for an extended time without damage to the ecosystem. The equation for the S-curve for logistic population growth is

$$G = rN\,[(K - N)/K]$$

To understand this new multiplier $[(K - N)/K]$, let's start with $(K - N)$. When we subtract the current population (N) from carrying capacity (K), we get the number of individuals that can still be added to the current population. Now if we divide this new number by K, we get the fraction of the carrying capacity that can still be added to the current population before it stops growing ($G = 0$). As you can see, when N is very small, $(K - N)/K$ is close to 1, and the equation resembles that for exponential growth. This produces the initial portion of the S-curve, which resembles a J-curve. As N increases over time, however, $K - N$ will approach zero. The growth rate will slow, and the steeply rising portion of the initial J-curve will begin to level off. When the population size (N) equals the carrying capacity (K), population growth will cease ($G = 0$) as seen on the final portion of the S-curve (Fig. 26-6a).

Although the math of the logistic equation will not permit this, in nature, an increase in N above K can be sustained for a short time. This is dangerous, however, because a population above K is living at the expense of resources that cannot regenerate as fast as they are being depleted. A small overshoot above K is likely to be followed by a decrease in N until the resources recover and the original K is restored. But this may not have happened on Easter Island.

SCIENTIFIC INQUIRY   Cycles in Predator and Prey Populations

If we assume that a certain prey species is eaten exclusively by a particular predator, it seems logical that both populations might show cyclic changes, with changes in the predator population size lagging behind changes in the prey population size. For example, a large hare population would provide abundant food for lynx and their offspring, which would then survive in large numbers. The increased lynx population would eat more hares, reducing the hare population. With fewer prey, fewer lynx would survive and reproduce, so the lynx population would decline a short time later.

Does this out-of-phase population cycle of predators and prey actually occur in nature? A classic example of such a cycle was demonstrated by using the ingenious method of counting all the pelts of northern Canada lynx and snowshoe hares purchased from trappers by the Hudson Bay Company between 1845 and 1935. The availability of pelts (which was assumed to reflect population size) showed dramatic, closely linked population cycles of these predators and their prey (**FIG. E26-1**). However, many uncontrolled variables could have influenced the relationship between hares and lynx. For example, hare populations sometimes fluctuate even without lynx present, possibly because, in the absence of predators, the hare population overshoots the ability of the environment to support it, reducing the food supply. Further, lynx do not feed exclusively on hares but can eat a variety of small mammals. Density-independent environmental variables, such as exceptionally severe winters, also could have adversely affected both populations and produced similar cycles. Recently, researchers tested the hare–predator relationship more rigorously on 1-kilometer square test plots in northern Canada. The hare population increased by a factor of two when more food was supplied, by a factor of three when predators were excluded, and by a factor of 11 with added food and no predators. This suggest that food availability and predation both contribute to natural boom and bust cycles in hares.

To test the predator-prey cycle hypothesis in an even more controlled manner, investigators performed laboratory

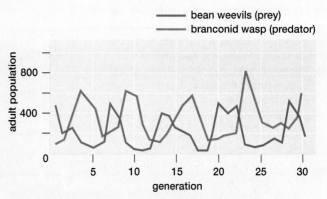

FIGURE E26-2 **Experimental predator-prey cycles** Out-of-phase fluctuations in laboratory populations of the bean weevil and its braconid wasp predator.

studies on populations of small predators and their prey. The study illustrated in **FIGURE E26-2** involved tiny braconid wasp predators and their bean weevil prey. Abundant food was supplied to the weevils, no other food was available to the wasps, and other variables were carefully controlled. As predicted, the two populations showed regular cycles, with the predator population rising and falling slightly later than the prey population. The wasps lay their eggs on weevil larvae, which provide food for the newly hatched wasps. A large weevil population ensures a high survival rate for wasp offspring, increasing the predator population. Then, under intense predation pressure, the weevil population plummets, reducing the food available and therefore the population size of the next generation of wasps. Reduced predation then allows the weevil population to increase rapidly, and so on.

It is highly unlikely that such a straightforward example will ever be found in nature, but this type of predator-prey interaction clearly can contribute to the fluctuations observed in many natural populations.

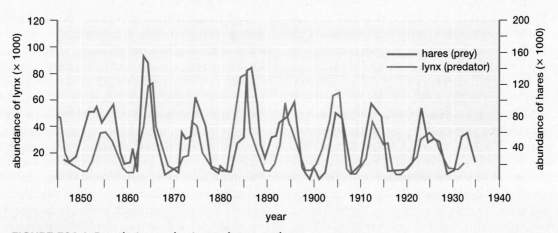

FIGURE E26-1 **Population cycles in predators and prey** Populations of snowshoe hares and their lynx predators, graphed on the basis of the number of pelts received by the Hudson Bay Company.

FIGURE 26-6 The S-curve of logistic population growth
(a) During logistic growth, the population will remain small for a time, then will expand increasingly rapidly for a time. Then the growth rate slows and growth eventually ceases at or near the carrying capacity (*K*). The result is a curve shaped like a "lazy s." **(b)** In nature, populations can overshoot carrying capacity (*K*), but only for a limited time. Three possible results are illustrated.

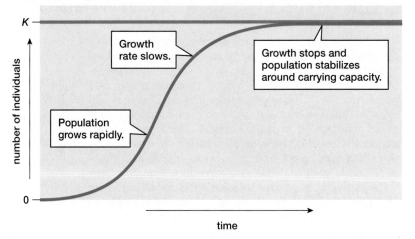

(a) A logistic growth curve stabilizes at *K*.

Growth rate slows.

Growth stops and population stabilizes around carrying capacity.

Population grows rapidly.

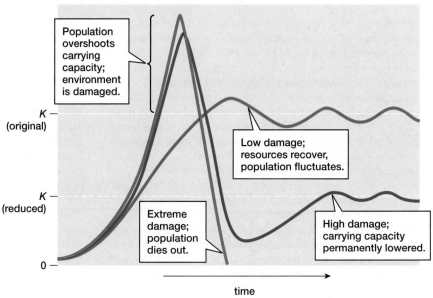

(b) Consequences of exceeding *K*.

Population overshoots carrying capacity; environment is damaged.

Low damage; resources recover, population fluctuates.

Extreme damage; population dies out.

High damage; carrying capacity permanently lowered.

If a population far exceeds the carrying capacity of its environment, consequences are more severe, since excess demands decimate essential resources. This can permanently and severely reduce *K*, causing the population to decline to a fraction of its former size or disappear entirely (**FIG. 26-6b**). For example, overgrazing by cattle on some dry western grasslands has reduced the grass cover and encouraged the growth of sage, which cattle will not eat. Once established, sage replaces edible grasses and reduces the land's carrying capacity for cattle. Reindeer introduced onto an island with no large predators may rapidly increase in number before their population plummets and remains low, as shown in **FIGURE 26-7**. The barren landscape that replaced the lush forests of Easter Island is a dramatic example of what may happen when overpopulation eliminates critical resources (such as trees), permanently and dramatically reducing the capacity of the island to sustain people, and driving many to its natural populations to extinction. Islands are particularly vulnerable to such drastic events, partly because their

populations cannot emigrate. For the expanding human population, however, all of Earth is an island.

Logistic population growth can occur in nature when a species moves into a new habitat, as ecologist John Connell documented for barnacles he counted as they colonized bare rock along a rocky ocean shoreline (**FIG. 26-8**). Initially, new settlers may find ideal conditions that allow their population to grow almost exponentially. As population density increases, however, individuals begin to compete, particularly for space, energy, and nutrients. These forms of environmental resistance can reduce the reproductive rate and the average life span, as has been demonstrated for laboratory populations of fruit flies (**FIG. 26-9**). The death rate of offspring may also increase. As environmental resistance increases, population growth slows and eventually stops. In nature, conditions are never completely stable, so both *K* and the population size will vary somewhat from year to year.

In nature, environmental resistance ultimately maintains populations at or below the carrying capacity of their

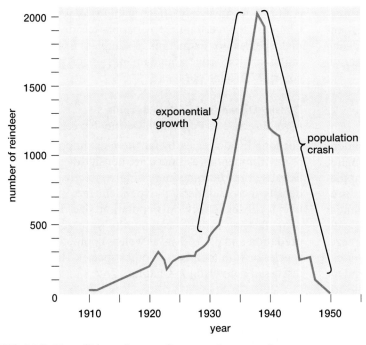

FIGURE 26-7 The effects of exceeding carrying capacity
Exceeding carrying capacity can damage an ecosystem, reducing its ability to support the population. In 1911, 25 reindeer were introduced onto one of the Pribilof Islands (St. Paul) in the Bering Sea off Alaska. Food was plentiful, and the reindeer encountered no predators on the island. The herd grew exponentially (note the initial J shape) until it reached 2000 reindeer in 1926. At this point, the small island was seriously overgrazed, food was scarce, and the population declined dramatically. By 1950, only eight reindeer remained.

environment. Environmental resistance can be classified into two broad categories. **Density-independent** factors limit population size regardless of the population density (number of individuals per given area). **Density-dependent** factors increase in effectiveness as the population density

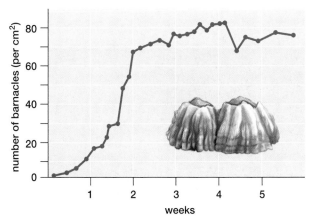

FIGURE 26-8 A logistic curve in nature
Barnacles are crustaceans whose larvae are carried in ocean currents to rocky seashores where they settle and then attach permanently to rock and grow into the shelled adult form. On a bare rock, the number of settling larvae produce a logistic growth curve as competition for space limits their population density. *Source*: Based on data from J. H. Connell, *Ecological Monographs* 31(1), 1961: 61–104.

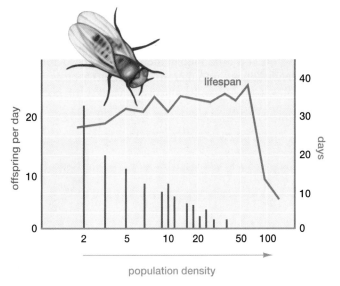

FIGURE 26-9 Density-dependent environmental resistance
In response to crowding, laboratory populations of fruit flies show a decrease in both reproductive rate and life span. In this graph, population density (horizontal axis) increases from left to right. Notice that the number of offspring produced per day decreases as population density increases. The life span remains relatively constant until population density reaches a critical level, causing life span to drop off dramatically. *Source*: Based on data from R. Pearl, J. R. Miner, and S. L. Parker, *American Naturalist* 61 (1927): 289–318.

increases. Note that nutrients, energy, and space, the primary determinants of carrying capacity, are all density-dependent regulators of population size. In the following sections, we will look more closely at how population growth is controlled.

Density-Independent Factors Limit Populations Regardless of Density

Perhaps the most important natural density-independent factors are climate and weather. Hurricanes, droughts, floods, and fire can have profound effects on local populations, particularly those of small, short-lived species, regardless of population density. Many insects and annual plant populations are limited in size by the number of individuals that can be produced before the first hard freeze. Such populations typically do not reach the carrying capacity of their environment, because density-independent factors intervene first. Weather is largely responsible for the boom-and-bust population cycles described earlier. Weather can also cause significant variations within natural populations from year to year.

Organisms with a life span of more than a year have evolved various mechanisms that compensate for seasonal changes, circumventing these density-independent population control factors. Many mammals, for example, develop thick coats and store fat for the winter; some also hibernate. Other animals, including many birds, migrate long distances to find food and a hospitable climate. Many trees and bushes survive the rigors of winter by entering a period of dormancy, dropping their leaves and drastically slowing their metabolic activities.

Human activities can also limit the growth of natural populations in density-independent ways. Pesticides and pollutants can cause drastic declines in natural populations. Before it was banned in the 1970s, the pesticide DDT drastically reduced populations of predatory birds, including eagles, ospreys, and pelicans; a variety of pollutants continue to adversely affect wildlife, as you will learn in Chapter 28. Overhunting by people has driven entire animal species to extinction. Examples from the United States are the once-abundant passenger pigeon and the colorful Carolina parakeet.

Density-Dependent Factors Become More Effective as Population Density Increases

For long-lived species, by far the most important elements of environmental resistance are density-dependent factors. Because they become increasingly effective as population density increases, density-dependent factors exert a negative feedback effect on population size. Density-dependent factors include community interactions, such as predation and parasitism, as well as competition within the species or with members of other species. These factors are discussed below and in Chapter 27.

Predators Often Switch Prey Based on Its Abundance

Both *predation* and *parasitism* involve one organism feeding on another and harming it in the process. Although the distinction is not always clear-cut, predation typically occurs when one organism, the **predator**, kills another, its **prey**, in order to eat it. Parasitism occurs when one organism, the **parasite**, lives on another, its **host** (usually a much larger organism) and feeds on the host's body without killing it—at least not immediately. While predators must kill their prey to feed, parasites benefit by having their hosts remain alive.

Predation includes wolves working together to kill an elk (**FIG. 26-10**) and a Venus flytrap plant digesting a fly. Predation becomes increasingly important as prey populations multiply, because most predators eat a variety of prey, depending on what is most abundant and easiest to find. Coyotes might eat more field mice when the mouse population is high but switch to eating more ground squirrels as the mouse population declines. In this way, preda-

FIGURE 26-10 Predators help control prey populations
A pack of grey wolves has brought down an elk that may have been weakened by age or parasites.

tors often exert density-dependent population control over more than one prey population.

Predator populations often grow as their prey becomes more abundant. For instance, predators such as the arctic fox and snowy owl, which feed heavily on lemmings, regulate the number of offspring they produce according to the abundance of lemmings. Snowy owls produce up to 13 chicks when lemmings are abundant but may not reproduce at all in years when lemmings are scarce. In some cases, an increase in predators might cause a crash of the prey population, which in turn may result in a decline in the predator population. This pattern can result in out-of-phase **population cycles** of both predators and prey (see "Scientific Inquiry: Cycles in Predator and Prey Populations").

Predators may maintain their prey at well below carrying capacity. A dramatic example of this phenomenon is the prickly pear cactus, which was introduced into Australia from Latin America. Lacking natural predators, its population grew exponentially and the plant spread uncontrollably, destroying millions of acres of valuable pasture and range land. Finally, in the 1920s, a cactus moth (a predator of the prickly pear) was imported from Argentina and released to feed on the cacti. Within a few years, the cacti were virtually eliminated. Today the moth continues to maintain its prey cacti at very low population densities, well below the carrying capacity of the ecosystem.

Some predators contribute to the health of prey populations by culling those that are genetically weak or poorly adapted. If a prey population has exceeded the carrying capacity of its environment, some individuals may be weakened by lack of food or unable to find adequate shelter. In such cases, predation may maintain prey populations near a density that can be sustained by the resources of the ecosystem.

Parasites Spread More Rapidly Among Dense Populations

In contrast to a predator, a parasite feeds on a larger organism, its host, usually harming it but not killing it directly. Examples include any organism that can cause disease, including bacteria, fungi, intestinal worms, ticks, and protists such as the malaria parasite. Insects that feed on plants—such as the gypsy moth that feeds on trees—are also parasites. Most parasites cannot travel long distances, so they spread more readily among hosts in dense populations. For example, plant diseases and pest insects spread readily through acres of densely planted crops, and childhood diseases spread rapidly through schools and daycare centers. Parasites influence population size by weakening their hosts and making them more susceptible to death from other causes, such as harsh weather. Organisms weakened by parasites are also less able to fight off other infections, escape predators, or reproduce.

Parasites, like predators, more often contribute to the death of less-fit individuals, producing a balance in which the prey population is regulated but not eliminated. This balance can be destroyed if parasites (or predators) are introduced into regions where local prey species have had no opportunity to evolve defenses against them. The smallpox virus, inadvertently carried by traveling Europeans, caused heavy losses of life among native inhabitants of the continental United States, Hawaii, South America, and Australia. The chestnut blight fungus, introduced from Asia, has almost eliminated wild chestnut trees from U.S. forests. Introduced rats and mongooses have exterminated several of Hawaii's native bird populations.

Competition for Resources Helps Control Populations

The resources that determine carrying capacity (space, energy, and nutrients) may be inadequate to support all the organisms that need them. **Competition**, defined as the interaction among individuals who attempt to use the same limited resource, limits population size in a density-dependent manner. There are two major forms of competition: **interspecific competition** (competition among individuals of different species), and **intraspecific competition** (competition among individuals of the same species). Because the needs of members of the same species for water and nutrients, shelter, breeding sites, light, and other resources are almost identical, intraspecific competition is more intense than interspecific competition.

Organisms have evolved several ways to deal with intraspecific competition. Some organisms, including most plants and many insects, engage in **scramble competition**, a kind of free-for-all with resources as the prize. For example, gypsy moth females each lay a mass of up to 1000 eggs on tree trunks in eastern North America. As the eggs hatch, armies of caterpillars crawl up the tree (**FIG. 26-11**).

FIGURE 26-11 Scramble competition
Gypsy moths gather on tree trunks to lay egg masses that each produce many hundreds of caterpillars (inset).

Huge outbreaks of this invasive species can completely strip large trees of their leaves in a few days. Under these conditions, competition for food may be so great that most of the caterpillars die before they can metamorphose into egg-laying moths. Plant seeds may also germinate in dense aggregations. As they grow, those that germinated first begin to shade the smaller ones, while their more extensive root systems absorb most of the water. Seedlings that emerge later wither and die.

Many animals (and even a few plants) have evolved **contest competition**, which involves social or chemical interactions used to limit access to important resources. Territorial species—such as wolves, many fish, rabbits, and songbirds—defend an area that contains important resources such as food or places to raise young. When the population exceeds the available resources, only the best-adapted individuals are able to defend territories that supply adequate food and shelter. Those without territories may not reproduce (reducing the future population), or they may fail to obtain adequate food or shelter and become easy prey for predators.

As population densities increase and competition becomes more intense, some animals react by emigrating. Large numbers leave their homes to colonize new areas, and many, sometimes most, die in the quest. For example, the mass movements of lemmings seem to occur in response to overcrowding. Emigrating swarms of locusts plague the African continent, stripping all vegetation in their path (**FIG. 26-12**).

Density-Independent and Density-Dependent Factors Interact to Regulate Population Size

The size of a population at any given time is the result of complex interactions between density-independent and density-dependent forms of environmental resistance.

For example, a stand of pines weakened by drought (a density-independent factor) may more readily fall victim to the pine bark beetle (a density-dependent parasite). Likewise, a caribou weakened by hunger (density dependent) and attacked by parasites (density dependent) is more likely to be killed by an exceptionally cold winter (a density-independent factor). The demands of the growing human population are reducing the carrying capacities of many ecosystems for their animal and plant populations, radically reducing their population size. Bulldozing grasslands and their prairie dog towns to build shopping malls, or cutting rain forests for agriculture, reduces populations in a density-independent fashion. However, the end result is a reduced carrying capacity of the environment, which in turn exerts density-dependent limits on future population sizes.

26.3 HOW ARE POPULATIONS DISTRIBUTED IN SPACE AND TIME?

Populations Exhibit Different Spatial Distributions

The spatial pattern in which individuals are dispersed within a given area is that population's *distribution*. Distribution may vary with time, changing with the breeding season, for example. Ecologists recognize three major types of spatial distribution: *clumped*, *uniform*, and *random* (**FIG. 26-13**).

There are many populations whose members live in groups and who exhibit a **clumped distribution** (Fig. 26-13a). Examples include family or social groupings, such as elephant herds, wolf packs, prides of lions, flocks of birds, and schools of fish. What are the advantages of clumping? Flocks provide many eyes that can search for localized food, such as a tree full of fruit or a lake with fish. Schooling fish and flocks of birds may confuse predators with their sheer numbers. Packs of predators can cooperate to hunt more effectively. Some species form temporary groups for mating and caring for their young. Other plant or animal populations cluster not for social reasons, but because resources are localized. Cottonwood trees, for example, cluster along streams and rivers in grasslands.

Organisms with a **uniform distribution** maintain a relatively constant distance between individuals. This type of distribution is most common among animals that defend territories and exhibit territorial behaviors designed to protect scarce resources. Male Galápagos iguanas establish regularly spaced breeding territories. For animals that remain together to raise their young, spacing often refers to pairs rather than to individuals. Shorebirds are also often found in evenly spaced nests, just out of reach of one another. Other territorial species, such as the tawny owl, mate for life and continuously occupy well-defined, relatively uniformly spaced territories. Some plants, such as sage, release into the soil around them chemicals that inhibit germination of other plants and thus space themselves relatively evenly (Fig. 26-13a). A uniform distribution helps ensure adequate resources for each individual.

FIGURE 26-12 Emigration
In response to overcrowding and lack of food, locusts emigrate in swarms, devouring all the vegetation in their path. **QUESTION** What benefits does mass emigration give to animals such as locusts or lemmings? Can you see any parallels in human emigration?

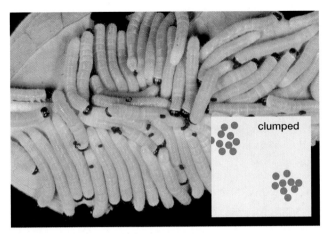

(a)

(b)

(c)

FIGURE 26-13 Population distributions
(a) Clumped: a gathering of caterpillars. (b) Uniform: creosote bushes in the desert. (c) Random: trees and other plants in a rain forest.

Organisms with a **random distribution** are relatively rare. Such individuals do not form social groups. The resources they need are more or less equally available throughout the area they inhabit, and those resources are not scarce enough to require territorial spacing. Trees and other plants in rain forests come close to being randomly distributed (Fig. 26-13c). There are probably no vertebrate species that maintain a random distribution throughout the year, because they must breed—a behavior that makes social interaction inevitable.

Survivorship in Populations Follows Three Basic Patterns

Different populations show dramatic differences in death rates or (more optimistically) survivorship at different ages. Some produce large numbers of offspring, most of which usually die before reaching reproductive age. Others produce few offspring, which are each given far more resources and often survive to reproduce. To determine the pattern of survivorship, researchers may create a *life table* (**FIG. 26-14a**). **Life tables** track groups of organisms born at the same time, following them throughout their lives, and recording how many continue to survive in each succeeding year (or other unit of time). If these numbers are graphed, they reveal the **survivorship curves** characteristic of the species in question in the particular environment where the data were collected. Three types of survivorship curve—described as "late loss," "constant loss," and "early loss," according to the part of the life cycle during which most deaths occur—are shown in **FIGURE 26-14b**. Survivorship curves reflect the number of offspring produced and the amount of parental care and protection that the offspring receive.

Late-loss populations produce convex survivorship curves. Such populations have relatively low juvenile death rates, and many or most individuals survive to old age. Late-loss survivorship curves are characteristic of humans and other large and long-lived animals such as elephants and mountain sheep. These species produce relatively few offspring, which are protected by their parents during early life.

Populations with *constant-loss* survivorship curves have a fairly constant death rate; their survivorship graphs appear as a relatively straight line. In these populations, individuals have an equal chance of dying at any time during their life span. This phenomenon is seen in some birds such as gulls and the American robin, and in laboratory populations of organisms that reproduce asexually, such as hydra and bacteria.

Early-loss survivorship produces a concave curve and is characteristic of organisms that produce large numbers of offspring. These offspring receive little parental care; they are largely left to compete on their own. Many exhibit scramble competition for resources early in life. The death rate is very high among the young, but those that reach adulthood have a good chance of surviving to old age. Most invertebrates, most plants, and many fish exhibit such early-loss survivorship curves. Even some mammalian

(a) Number of survivors by age, out of 100,000 born alive: United States, 2002

Age	Total	Male	Female
0	100,000	100,000	100,000
10	99,105	99,014	99,199
20	98,672	98,436	98,922
30	97,740	97,091	98,424
40	96,419	95,381	97,500
50	93,563	91,809	95,364
60	87,711	84,637	90,826
70	75,335	70,087	80,556
80	52,178	44,370	59,621
90	20,052	13,925	25,411
100	2,095	1,005	2,954

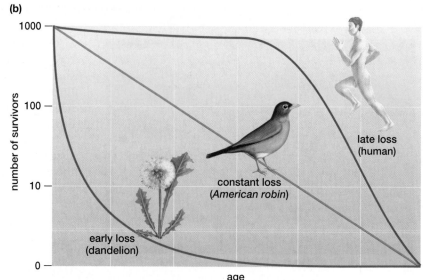

(b)

FIGURE 26-14 **Life tables and survivorship curves**
(a) A life table of U.S. residents in 2002. Graphing this data produces the human survivorship curve shown in (b). **(b)** Three types of survivorship curve are shown. Because the life spans differ, the percentages of survivors (rather than ages) are used. (*Source:* National Vital Statistics Reports, Vol. 53, No. 6, November 10, 2004)

populations have early-loss survivorship curves; in some populations of black-tailed deer, 75% of the population dies within the first 10% of its life span.

26.4 HOW IS THE HUMAN POPULATION CHANGING?

Demographers Track Changes in the Human Population

Demography is the study of the changing human population. Using complex life tables, *demographers* measure human populations in different countries and world regions, track population changes, and make comparisons between developing and developed countries. They examine birth and death rates by race, sex, education levels, and socioeconomic status both within and between countries. Demographers not only track past and current trends, they attempt to explain these changes, evaluate their impacts, and make predictions for the future. The data gathered by demographers are useful for formulating policies in areas such as public health, housing, education, employment, immigration, and environmental protection.

The Human Population Continues to Grow Rapidly

Compare the graph of human population growth in **FIGURE 26-15** with the exponential growth curves in Figure 26-1. The time spans are different, but each has the J-shaped curve characteristic of exponential growth. It took more than 1 million years for the human population to reach 1 billion. In the inset of Figure 26-15, note the decreasing times to add billions of people; an estimated 6% of all the humans who have ever lived on Earth are alive today. But also notice that we have been adding billions at a relatively constant rate since the 1970s. This suggests that, although the human population is growing rapidly, it

may no longer be growing exponentially. Are humans starting to enter the final bend of the S-shaped growth curve shown in Figure 26-6 that will eventually lead to a stable population? Time will tell. However, Earth's human population (currently over 6.5 billion) now grows by about 75–80 million yearly; more than 203,000 people are added every day, and about 1.5 million every week! Why hasn't environmental resistance put an end to our continued growth? What is the carrying capacity of Earth for humans? We explore this question further in "Earth Watch: Have We Exceeded Earth's Carrying Capacity?"

Like all populations, humans have encountered environmental resistance; but unlike nonhuman populations, we have responded to environmental resistance by devising ways to overcome it. As a result, the human population has grown for an unprecedented time span. To accommodate our growing numbers, we have altered the face of the globe. Human population growth has been spurred by a series of "revolutions," each of which circumvented environmental resistance and increased Earth's carrying capacity for people.

Technological Advances Have Increased Earth's Capacity to Support People

Primitive people produced a *technical and cultural revolution* when they discovered fire, invented tools and weapons, built shelters, and designed protective clothing. Tools and weapons allowed them to hunt more effectively and increased their food supply, while shelter and clothing increased the habitable areas of the globe.

Domesticated crops and animals supplanted hunting and gathering by about 8000 B.C. This *agricultural revolution* provided people with a larger, more dependable food supply, further increasing Earth's carrying capacity for humans. An increased food supply resulted in a longer life

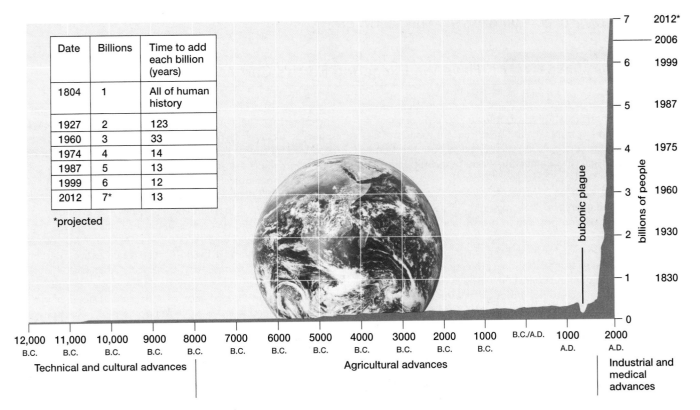

Date	Billions	Time to add each billion (years)
1804	1	All of human history
1927	2	123
1960	3	33
1974	4	14
1987	5	13
1999	6	12
2012	7*	13

*projected

Technical and cultural advances

Agricultural advances

Industrial and medical advances

FIGURE 26-15 Human population growth
The human population from the Stone Age to the present has shown continued exponential growth as various advances overcame environmental resistance. Note the dip in the fourteenth century caused by the bubonic plague. Note also the time intervals over which additional billions were added. (Inset) Earth is an island of life in a sea of emptiness; its space and resources are limited. QUESTION The human population continues to grow rapidly, but evidence suggests we have already exceeded Earth's carrying capacity at current levels of technology. What do you think this curve will look like when we reach the year 2500? 3000? Explain.

span and more childbearing years, but a high death rate from disease still restrained population growth.

Human population growth continued slowly for thousands of years until the *industrial–medical revolution* began in England in the mid-eighteenth century, spreading throughout Europe and North America in the nineteenth century. Medical advances dramatically decreased the death rate by reducing environmental resistance caused by disease. These advances included the discovery of bacteria and their role in infection, which led to the control of bacterial diseases through improved sanitation and the use of antibiotics. Viruses were also discovered, leading to the development of vaccines for diseases such as smallpox.

Today, the countries of the world are often described as either *developed* or *developing*. People in developed countries (including North America, Europe, Australia, New Zealand, and Japan) benefit from a relatively high standard of living, with access to modern technology and medical care (including contraception). Income is relatively high; education and employment opportunities are available to both sexes; and death rates from infectious diseases are relatively low. Yet fewer than 20% of the world's inhabitants live in developed countries. Most of the people in developing countries (Central and South America, much of Asia, and Africa) lack many of these advantages.

The Demographic Transition Helps Stabilize Populations

In developed countries, the industrial–medical revolution resulted in an initial rise in population due to a decrease in death rates, which was followed by a decline in birth rates, resulting in a relatively stable population. This changing population dynamic, in which a population experiences rapid growth and then returns to a stable (although much larger) size, is called a **demographic transition** (FIG. 26-16).

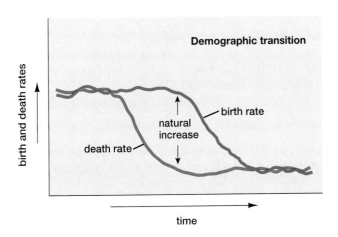

FIGURE 26-16 The demographic transition

527

EARTH WATCH   Have We Exceeded Earth's Carrying Capacity?

In Côte d'Ivoire (Ivory Coast), a small country in West Africa, the government is waging a battle to protect some of its rapidly dwindling tropical rain forest from thousands of illegal hunters, farmers, and loggers. Officials burn the homes of the squatters, who immediately return and rebuild. One illegal resident is Sep Djekoule. "I have 10 children and we must eat," he explains. "The forest is where I can provide for my family, and everybody has that right." His words illustrate the conflict between population growth and environmental protection, between the "right" to have many children and the ability to provide for them using Earth's finite resources. The middle-level United Nations projection is that by the year 2050, the human population will have reached 9 billion and will still be increasing, although at a much lower rate. How many people can Earth support?

Ecologists agree that the concept of carrying capacity is a nebulous one for people, because people use technology to overcome environmental resistance by increasing food supplies, curing disease, and prolonging life. Further, since the time people lived in caves, our steadily increasing expectations for comfort, mobility, and convenience have reduced Earth's capacity to sustain us. We can and should use technology to reduce our destructive impact, for example by improving agricultural practices, conserving energy and water, reducing pollutants, and recycling far more paper, plastic, and metal. However, our ability to reproduce far exceeds our ability to increase Earth's capacity to support us.

A large group of scientists from around the world is engaged in an ongoing project to assess humanity's impact on global ecosystems. They are comparing the resource demands of the global human population with the ability of the world's ecosystems to supply these resources, which include agricultur-al land, fish and other wild foods, wood, space, and energy. The researchers estimate the amount of biologically productive space or *biocapacity* needed to absorb the carbon dioxide produced by energy use and to supply the resource demands of an average person at current levels of technology. They call this area an **ecological footprint**. Their most recent estimate (based on 2002 data) is that Earth then had 4.5 acres available for each of its 6.2 billion human inhabitants. The average ecological footprint, however, was 5.4 acres. This suggests that even in 2002, when Earth supported over 300 million fewer people than today's 6.5+ billion, humanity's collective footprint exceeded Earth's biocapacity by about 20%. Disturbingly, these estimates are intended to be conservative; they do not take into account the depletion of underground stores of fresh water or the need to leave significant portions of the biosphere untouched to provide habitat for wild species.

A population that exceeds carrying capacity damages the ecosystem and reduces its ability to support that population. In the following paragraphs, we provide evidence that humanity is depleting Earth's resource base and reducing Earth's capacity to sustain us.

Each year, overgrazing and deforestation decrease the productivity of land, especially in developing countries. In a world where the United Nations estimates that over 850 million people lack adequate food, a significant portion of the world's agricultural land suffers from erosion, reducing its fertility that supports both crops and grazing livestock (**FIG. E26-3**). The quest for food drives people to deforest and attempt to farm land that is poorly suited for agriculture. The demand for wood also causes large areas to be deforested annually, allowing runoff of much-needed fresh water, erosion of precious topsoil, pollution of rivers, and an overall reduction in the ability of the land to support future crops or forests. The

---

The decline in birth rates that ends the demographic transition is a result of many factors, including better education, increased access to contraceptives, a shift of populations to cities (where having children provides fewer advantages than in agricultural areas), and more women working outside the home. In most developed countries, the demographic transition has occurred and populations have more or less stabilized. Populations stabilize when the adults of reproductive age have just enough children to replace themselves, a situation called **replacement-level fertility (RLF)**. Because not all children survive to maturity, RLF is slightly higher than 2 (2.1).

Population Growth Is Unevenly Distributed

In developing countries, such as most of Central and South America, Asia (excluding China and Japan), and Africa (excepting those African countries devastated by the AIDS epidemic), medical advances have decreased death rates and increased the life span, but birth rates remain relatively high. These countries are in different stages of the demographic transition. Although China is a developing country, its leaders have recognized the negative impacts of continued growth and instituted social reforms that have brought the birth rate down below RLF. In other developing nations, children may be the only support for elderly parents, they may contribute significantly to the labor force (particularly on farms, but sometimes in factories as well), and they can also be a source of social prestige. In some countries, religious beliefs promote large families and perpetuate high birth rates. Many women who would like to limit their family size lack access to contraceptives. In Nigeria, the most populous country in Africa, only 8% of women use modern contraceptive methods, and the average woman bears 6 children. Nigeria is suffering from loss of forests and wildlife, soil erosion, and water pollution. With 43% of its more than 134 million people under age 15, continued population growth is a certainty.

Population growth is highest in the countries that can least afford it. This creates a type of positive feedback; as more people share the same limited resources, poverty increases. Poverty diverts children away from schools and into activities that help support their families. The lack of both education and access to contraceptives then con-

demand for wood, food, and recently, *biofuels* (crops such as soybeans grown for fuel) drive the destruction of tens of millions of acres of rain forest annually, and the extinction of species on an unprecedented scale (see Chapter 30). Worldwide, the amount of cropland per person has declined by over half in the past 50 years.

In many developing countries, including India and China (each home to over 1 billion people), clean fresh water is in short supply. In these countries, underground water supplies, called *aquifers*, are rapidly being depleted to irrigate cropland. Because irrigated land supplies about 40% of human food crops, future water shortages could rapidly lead to food shortages.

The total world fish harvest peaked in the late 1980s and has been gradually declining since then, despite increased investments in fishing equipment, improved technology for finding fish, and increased harvests of smaller and less-desirable fish species. Almost 70% of commercial ocean fish populations have been fully exploited or overfished, and many formerly abundant fish populations, such as cod harvested off New England, Canada, and in the North Sea, have declined dramatically because of overfishing.

Our present population, at its present level of technology, is clearly "overgrazing" the world ecosystem. As more than 5.2 billion people in less-developed countries strive to increase their standard of living, the damage to Earth's ecosystems accelerates. In estimating how many people Earth can—or should—support, it is important to recognize that people desire far more from life than just remaining alive. The standard of living in developed countries is already an unattainable luxury for most of Earth's inhabitants.

Inevitably, the human population will stop growing. Either we must voluntarily reduce our birth rates, or various forces of

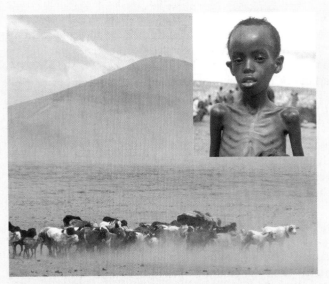

FIGURE E26-3 Deforestation can lead to the loss of productive land
Human activities, including overgrazing, deforestation, and poor agricultural practices, reduce the productivity of the land. (Inset) An expanding human population, coupled with a loss of productive land, can lead to tragedy.

environmental resistance, including disease and starvation, will dramatically increase human death rates. The choice is ours. Hope for the future lies in recognizing the signs of "human overgrazing" and acting to reduce our population before we decimate our biodiversity and irrevocably damage the biosphere.

tributes to continued high birth rates. Of the more than 6.5 billion people on Earth today, over 5.2 billion reside in developing countries. Encouragingly, birth rates in some developing countries have begun to decline and approach RLF, because their governments are taking steps to encourage smaller families and increase access to contraceptives. But the prospects for population stabilization—*zero population growth*—in the near future are nil. For the year 2050, the United Nations predicts that the population will be over 9 billion and growing (although much more slowly than at present), with 7.8 billion people living in the developing nations (**FIG. 26-17**).

The Current Age Structure of a Population Predicts Its Future Growth

Data gathered by demographers allow the **age structure** of human populations to be determined. Age-structure diagrams show age groups on the vertical axis and the numbers (or percentages) of individuals in each age group on the horizontal axis, with males and females shown on opposite sides. Age-structure diagrams all rise to a peak that

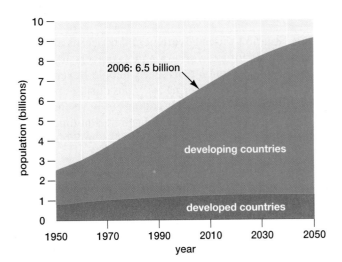

FIGURE 26-17 Mid-level UN population projections

reflects the maximum human life span, but the shape of the rest of the diagram reveals whether the population is expanding, stable, or shrinking. If the reproductive-age adults (about 15 to 44 years) are having more children (between 0 and 14 years) than are needed to replace themselves, the population is above RLF and is expanding; its age structure will resemble a pyramid (**FIG. 26-18a**). If the adults of reproductive age have just the number of children needed to replace themselves, the population is at RLF. A population that has been at RLF for many years will have an age-structure diagram with relatively straight sides (**FIG. 26-18b**). In shrinking populations, the reproducing adults are having fewer children than required to replace themselves, causing the age structure to narrow at the base (**FIG. 26-18c**).

FIGURE 26-19 shows the average age structures for the populations of developed and developing countries for the year 2006 with projections for 2025 and 2050. Even if rapidly growing countries were to achieve RLF immediately, their population increase would continue for decades, because today's children create a momentum for future growth as they enter their reproductive years and begin their own families—even if they have only two children each. This fuels China's population growth of 0.6% annually, even with a fertility rate at RLF. Less than 20% of individuals in a stable population are in the prereproductive (1–14 years) age group. In Mexico, this age group makes up 31% of the population, and in many African nations, children comprise well over 40% of the population.

Fertility in Europe Is Below Replacement Level

FIGURE 26-20 illustrates growth rates for various world regions. In Europe, the average annual change in population is −0.1%, with an average fertility rate of 1.4, substantially below RLF, as many women delay or forgo having children for various reasons relating both to family economics and lifestyle.

This situation raises governmental concerns about the availability of future workers and taxpayers to support the increasing percentage of elderly people. Several European countries are offering or considering incentives (such as large tax breaks) for couples to have children at an earlier age, which shortens the generation time and increases population. Japan's nearly 128 million people (42% of the entire U.S. population) inhabit an area about the size of Montana. Yet despite this tremendous crowding, their government is concerned about Japan's low fertility rate (1.3) and provides a variety of subsidies that encourage larger families.

Although a reduced and eventually stable population will ultimately offer tremendous benefits for both people and the biosphere that sustains them, current economic structures in countries throughout the world are based on growing populations. The difficult adjustments required by stabilizing or declining populations lead governments to adopt policies that encourage more childbearing and continued growth.

(a) Population pyramids for Mexico

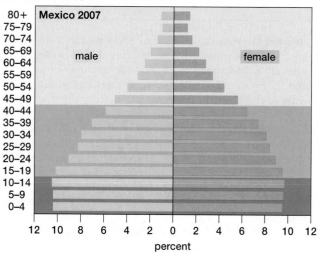

(b) Population pyramids for Sweden

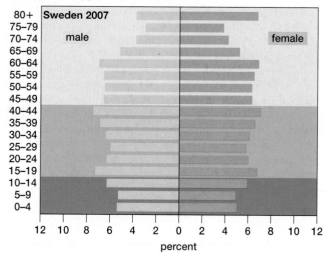

(c) Population pyramids for Italy

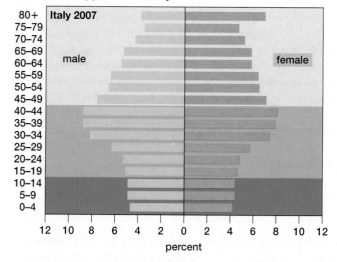

FIGURE 26-18 Age structure diagrams
(a) Mexico is growing quite rapidly. **(b)** Sweden has a stable population. **(c)** Italy's population is shrinking. (*Source:* Data provided by the U.S. Census Bureau; http://www.census.gov).

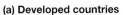

(a) Developed countries

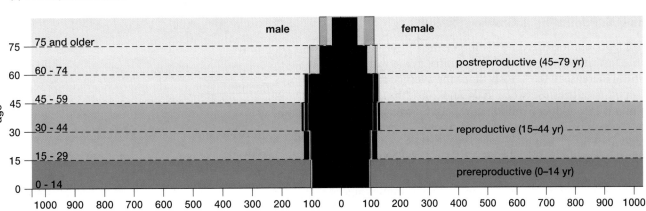

(b) Developing countries

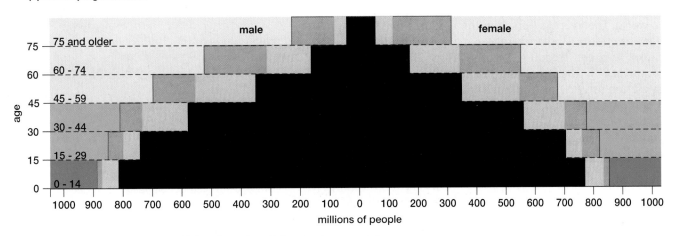

FIGURE 26-19 Age structures of developed and developing countries
Notice that the predicted excess number of children over parents in developing countries is smaller in 2025 and in 2050, as these populations approach RLF. But as huge numbers of young people enter childbearing years, growth will continue. (*Source:* Data provided by the U.S. Census Bureau; http://www.census.gov). QUESTION How does a fertility rate above RLF create a positive feedback effect in population growth?

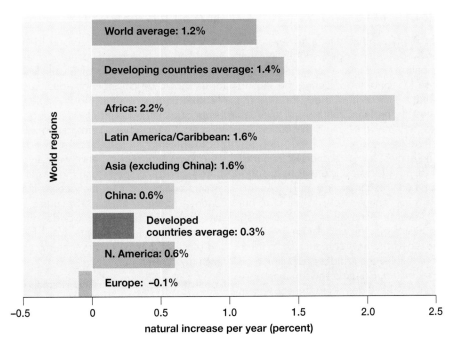

FIGURE 26-20 Population change by world regions
Growth rates shown are due to natural increase (births − deaths) expressed as the percentage increase per year for various regions of the world. These figures do not include immigration or emigration. (*Source:* Data from the Population Reference Bureau, World Population Data Sheet, 2005; www.prb.org/pdf05/05WorldDataSheet_Eng.pdf). QUESTION Why are there such big population differences between developed and developing countries?

The U.S. Population Is Growing Rapidly

With a population of over 300 million, the U.S. (**FIG. 26-21**) is the fastest-growing developed country in the world. The natural increase of 0.6% is six times the average rate of developed countries. Between 2004 and 2005, the U.S. grew by about 1%, adding 3 million new people (over 8000 each day). The fertility rate is currently about 2.0, just slightly below RLF (2.1). But immigration to the U.S. adds about 1 million people legally and an estimated 500,000 illegally each year, accounting for about half of the population increase. The average fertility rate of these immigrants is above RLF, compounding their impact on population growth. This situation assures continued U.S. population growth for the indefinite future.

The rapid growth of the U.S. population has major environmental implications both for this country and for the world. The average person in the U.S. uses five times as much energy as the average person worldwide (see "Links to Life: Treading Lightly—How Big Is Your 'Footprint'?"). The 3 million people added to the U.S. population use 2.5 times as much energy as do the nearly 18 million people that expanded India's population during the same year. The inexorable spread of housing, commercial establishments, and energy-extraction enterprises degrades or destroys natural habitats, reducing the carrying capacity of a variety of ecosystems for other forms of life.

When and how will human numbers ever stabilize? How many people can Earth support? There are no certain answers, but in "Earth Watch: Have We Exceeded Earth's Carrying Capacity?" we explore them in more detail.

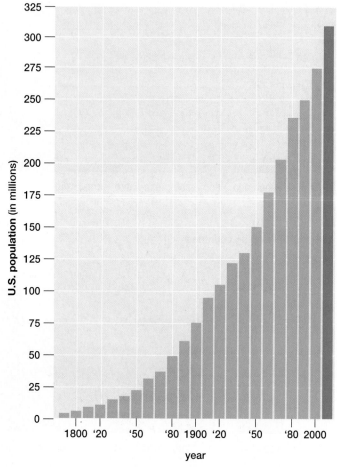

FIGURE 26-21 U.S. population growth
Since 1790, U.S. population growth has produced a J-shaped curve similar to that seen for exponential growth. Each bar reflects the date below it. QUESTION At what stage of the S-curve is the U.S. population? What factors do you think will cause it to stabilize, and when?

CASE STUDY REVISITED   THE MYSTERY OF EASTER ISLAND

The pre-history of Easter Island is shrouded in mystery. Did people deforest the island, or did rats that stowed away on their canoes wreak ecological havoc, consuming both tree seeds and native wildlife? Fossils reveal that Easter Island was once home to parrots, owls, herons, and numerous other bird species. At least 25 species of seabirds, including albatrosses, boobies, and frigate birds, once nested on the island. Today, Easter Island contains no native land birds or mammals. No animal larger than an insect is a permanent resident there.

When the forest and its animals disappeared, the island's human population probably lacked adequate food. With no wood for canoes or boats, there was no way to leave the island and fish in the open ocean. Ominously, garbage heaps from the post-forest period contain a few examples of gnawed human bones, suggesting that cannibalism may have emerged in response to lack of food. During the second half of the 19th century, the island's history was lost as its inhabitants were taken as slaves and ravaged by diseases carried from foreign lands.

The earliest human settlers of Easter Island found a forested isle that held abundant natural resources and a diverse array of plant and animals species useful to humans. Over time, however, some scientists hypothesize that the population grew until it exceeded the environment's capacity to support it. Like the reindeer population on St. Paul Island, Easter Island's human population may have damaged the ecosystem on which it depended. This may have caused the population to decline dramatically, and its society to disintegrate.

What can we learn from Easter Island? According to biologist and author Jared Diamond, "the meaning of Easter Island for us should be chillingly obvious. Easter Island is Earth writ small. Today, again, a rising population confronts shrinking resources. We too have no emigration valve, because all human societies are linked by international transport, and we can no more escape into space than the Easter Islanders could flee into the ocean. If we continue to follow our present course, we shall have exhausted the world's major fisheries, tropical rain forests, fossil fuels, and much of our soil by the time my sons reach my current age."

Consider This Easter Island is small (66 square miles) and it is 1200 miles from the nearest habitable island. Why was it particularly susceptible to a population crash?

LINKS TO LIFE   Treading Lightly—How Big Is Your "Footprint"?

You now know that an "ecological footprint" measures a person's environmental impact. While animal populations tend to have the minimal footprints needed to sustain health and reproduce, human ecological footprints differ tremendously among different countries and among individuals within those countries. What determines the size of a person's footprint? If you search the Internet for "ecological footprint," you will find Web sites that further describe this concept, compare different countries, and allow you to calculate the size of your own footprint. You'll find that your energy use, the type of home you live in, and even the food you eat all influence your footprint.

Individuals in the U.S., on average, have larger footprints than do people in any other country on Earth. U.S. residents' footprints average 24 acres per person, compared to the global average of 5.4 acres per person—and the Earth's biocapacity, estimated at 4.5 acres per person. If all 6.3 billion people on Earth lived as extravagantly as the average U.S. citizen, we would need 5.4 Earths to supply their demands.

People in the Netherlands and Canada also enjoy a comfortable standard of living, with footprints of 14 and 17, respectively. Nonetheless, we would need 3.8 Earths to support our current global population at the average Canadian living standard.

But, you might ask, what is wrong with eating meat or imported fruits, driving a car, or living in your own house with a big yard? In fact, there is nothing inherently wrong with any of these options; they have become environmentally harmful only because of humanity's ongoing failure to limit its population. Individuals must recognize that the decision to bear more than two children will result in more footprints treading planet Earth and fewer resources to go around. For example, if 1 billion people inhabited Earth, each could live in reasonable luxury without damaging the planet. Fewer footprints would also allow us to set aside enough untouched land for the continued survival and well-being of the millions of irreplaceable species that provide Earth's wealth of biodiversity.

CHAPTER REVIEW

SUMMARY OF KEY CONCEPTS

26.1 How Does Population Size Change?
Individuals join populations through birth or immigration and leave through death or emigration. The ultimate size of a stable population results from interactions among biotic potential (the maximum possible growth rate) and environmental resistance (which limits population growth).

All organisms have the biotic potential to more than replace themselves over their lifetime, resulting in population growth. Populations tend to grow exponentially, with increasing numbers of individuals added during each successive time period. Populations cannot exhibit exponential growth for long; they either stabilize or undergo periodic boom-and-bust cycles as a result of environmental resistance.

Web Tutorial 26.1 Population Growth and Regulation

26.2 How Is Population Growth Regulated?
Environmental resistance restrains population growth by increasing the death rate or decreasing the birth rate. The maximum size at which a population may be sustained indefinitely by an ecosystem is called the carrying capacity, K, determined by limited resources such as space, nutrients, and light. Environmental resistance generally maintains populations at or below the carrying capacity. In nature, populations can overshoot K temporarily by depleting their resource base. Depending on the amount of damage to critical resources, this leads to (1) the population oscillating around K; (2) the population crashing, then stabilizing at a reduced K; (3) the population being eliminated from the area.

Population growth is restrained by density-independent forms of environmental resistance (weather, climate) and density-dependent forms of resistance (competition, predation, parasitism).

Web Tutorial 26.2 Human Population Growth

26.3 How Are Populations Distributed in Space and Time?
Populations can be classified into three major types of distribution: clumped, uniform, and random. Clumped distribution may occur for social reasons or around limited resources. Uniform distribution is normally the result of territorial spacing. Random distribution is rare, occurring only when individuals do not interact socially and when resources are abundant and evenly distributed.

Populations show specific survivorship curves that describe the likelihood of survival at any given age. Late-loss (convex) curves are characteristic of long-lived species with few offspring, which receive parental care. Species with constant-loss curves have an equal chance of dying at any age. Early-loss (concave) curves are typical of organisms that produce numerous offspring, most of which die before reaching maturity.

26.4 How Is the Human Population Changing?
The human population has exhibited exponential growth for an unprecedented time, the result of a combination of high birth rates and technological, agricultural, industrial, and medical advances that have overcome several types of environmental resistance and increased Earth's carrying capacity for

humans. Age-structure diagrams depict numbers of males and females in various age groups in different countries. Expanding populations have pyramidal age structures; stable populations show rather straight-sided age structures; and shrinking populations are illustrated by age structures that are constricted at the base.

Most of Earth's people live in developing countries with growing populations. Although birth rates have declined considerably in many places, momentum from previous high birth rates assures continued substantial population growth. The United States is the fastest-growing developed country, owing both to higher birth rates and higher immigration rates. Recently, scientists have estimated the amount of biologically productive space needed to supply the demands of an average person at current levels of technology. This "ecological footprint" provides evidence that the demands of Earth's 6.5 billion people exceed the sustainably available resources. Declines in several critical resources suggest that we are damaging our world ecosystem, decreasing its future ability to sustain us. As the U.S. population continues to surge, and as people in less-developed countries strive to increase their standard of living, the damage will accelerate. Unlike other animals, people can make conscious decisions to reverse damaging trends.

KEY TERMS

abiotic   *page 514*
age structure   *page 529*
biotic   *page 514*
biotic potential   *page 514*
birth rate   *page 514*
boom-and-bust cycle
   page 515
carrying capacity (K)
   page 519
clumped distribution
   page 524
community   *page 514*
competition   *page 523*
contest competition
   page 524

death rate   *page 514*
demographic transition
   page 527
demography   *page 526*
density-dependent   *page 520*
density-independent
   page 520
ecological footprint
   page 528
ecology   *page 514*
ecosystem   *page 514*
emigration   *page 514*
environmental resistance
   page 514

exponential growth   *page 515*
growth rate   *page 514*
host   *page 522*
immigration   *page 514*
interspecific competition
   page 523
intraspecific competition
   page 523
invasive species   *page 519*
J-curve   *page 515*
life table   *page 525*
logistic population growth
   page 519

parasite   *page 522*
population   *page 514*
population cycle   *page 523*
predator   *page 522*
prey   *page 522*
random distribution   *page 525*
replacement-level fertility
   (RLF)   *page 528*
scramble competition
   page 523
S-curve   *page 519*
survivorship curve   *page 525*
uniform distribution   *page 524*

THINKING THROUGH THE CONCEPTS

1. Define *biotic potential* and *environmental resistance*.

2. Draw the growth curve of a population before it encounters significant environmental resistance. What is the name of this type of growth, and what is its distinguishing characteristic?

3. Distinguish between density-independent and density-dependent forms of environmental resistance.

4. What is logistic population growth? What is *K*?

5. Describe three different possible consequences of exceeding carrying capacity. Sketch these scenarios on a graph. Explain your answer.

6. List three density-dependent forms of environmental resistance, and explain why each is density dependent.

7. Distinguish between populations showing concave and convex survivorship curves.

8. Draw the general shape of age-structure diagrams characteristic of (a) expanding, (b) stable, and (c) shrinking populations. Label all the axes. Explain why you can predict near-term future growth by the current age structure of populations.

9. Given that the U.S. birth rate is currently around replacement-level fertility, why is our population growing?

10. Discuss some of the reasons that making the transition from a growing to a stable population can be economically difficult.

APPLYING THE CONCEPTS

1. Explain natural selection in terms of biotic potential and environmental resistance.

2. The United States has a long history of accepting large numbers of immigrants. Discuss the pros and cons of allowing high levels of legal immigration. Should illegal immigration be made more difficult? What are the implications of immigration for population stabilization?

3. What factors encourage rapid population growth in developing countries? What will it take to change this growth?

4. Contrast age structures in rapidly growing versus stable human populations. Why would a rapidly growing population continue to grow even if families all immediately started having only two children? For how long would the population increase?

5. Why is the concept of carrying capacity difficult to apply to human populations?

6. Search the Internet for "ecological footprint," and calculate your footprint using a questionnaire found on one of the resulting Web sites. For five of your daily activities, explain how and why each contributes to your ecological footprint.

FOR MORE INFORMATION

Cohen, J. "Human Population Grows Up." *Scientific American*, September, 2005. Massive changes are in store as the human population swells.

Korpimaki, E., and Krebs, C. J. "Predation and Population Cycles of Small Mammals." *BioScience*, November 1996. A review of recent studies designed to evaluate cycles of predators and their prey.

Myers, N. "Biotic Holocaust." *International Wildlife*, March–April 1999. Human activities are causing extinction of species unprecedented since the disappearance of the dinosaurs. How can we reverse this trend?

Pauly, D., and Watson, R. "Counting the Last Fish." *Scientific American*, July 2003. Overfishing is causing the collapse of fisheries worldwide.

Potts, M. "The Unmet Need for Family Planning." *Scientific American*, January 2000. Reducing population growth and increasing the quality of life requires increased access to contraceptives in developing countries.

Wackernagel, M., et al. "Tracking the Ecological Overshoot of the Human Economy." *Proceedings of the National Academy of Sciences* 99, July 2002. A groundbreaking and conservative assessment of the human ecological footprint suggests that we have already surpassed Earth's ability to sustain our population at current living standards.

Wilson, E. O. "The Bottleneck." *Scientific American*, February 2002. Expanding human population combined with shrinking resources creates a bottleneck for humanity. This fascinating article by a biologist awarded both the National Medal of Science and a Pulitzer Prize compares and contrasts environmental and economic viewpoints.

Community
Interactions

Workers blast jets of hot water at zebra mussels coating the interior of a
Michigan water-treatment plant. (Inset) Zebra mussels smother a crayfish.

CASE STUDY INVASION OF THE ZEBRA MUSSELS

IN 1989 RESIDENTS OF MONROE, MICHIGAN, a town on the Lake Erie shore, suddenly found themselves without water. Their schools, industries, and businesses were closed for two days while workers labored to fix the problem: Zebra mussels had clogged their water-treatment plant. The town's problem was not unique; at another treatment plant on Lake Erie, zebra mussel populations reached 600,000 per square yard (see the opening photo). Where did they come from?

Sometime in 1985 or 1986, a trading vessel bringing cargo from Europe discharged fresh water into Lake St. Clair, located between Lake Huron and Lake Erie at the border between Ontario and Michigan. The water, used for ballast during the ship's transatlantic voyage, contained stowaways—millions of zebra mussel larvae. Although these mollusks are native to the Caspian and Black Seas (large inland seas between Europe and Asia), they found ideal conditions in North America. Spreading throughout the Great Lakes and the Mississippi and Ohio River drainage systems, they have now reached as far south as New Orleans and as far west as Oklahoma.

Microscopic mussel larvae can be carried in currents for hundreds of miles. Using sticky threads, they attach themselves to nearly any underwater surface, including piers, pipes, machinery, underwater debris, boat hulls, and even sand and silt. Since they can survive out of water for days, mussels clinging to small boats may be portaged to other lakes and rivers, where they quickly move in. An adult female can produce 100,000 eggs each year, and the mussel menace has proven unstoppable. The mussels cover and suffocate other forms of shellfish, threatening many rare varieties with extinction. Think about the zebra mussel as you read about the community interactions that characterize healthy ecosystems. Why have these invaders been so enormously successful? Will anything control them?

27.1 WHY ARE COMMUNITY INTERACTIONS IMPORTANT?

An ecological **community** consists of all the interacting populations within an ecosystem; in other words, a community is the *biotic*, or living, component of an ecosystem. In the previous chapter, you learned that community interactions such as predation, parasitism, and competition help limit the size of populations. A community's interacting web of life tends to maintain a balance between resources and the numbers of individuals consuming them. When populations interact and influence each other's ability to survive and reproduce, they serve as agents of natural selection. For example, in killing prey that are easiest to catch, *predators* leave behind individuals with better defenses against predation. These individuals leave the most offspring, and over time their inherited characteristics increase within the prey population. Thus, as community interactions limit population size, they simultaneously shape the bodies and behaviors of the interacting populations. The process by which two interacting species act as agents of natural selection on one another over evolutionary time is called **coevolution**.

The most important community interactions are competition, predation, parasitism, and mutualism. If we assume that each of these interactions involves two species, the types of interactions can be characterized according to whether each of the two species is harmed or benefits, as shown in Table 27-1. These interactions have shaped both the bodies and the behavior of organisms.

27.2 WHAT IS THE RELATIONSHIP BETWEEN THE ECOLOGICAL NICHE AND COMPETITION?

The Ecological Niche Defines the Place and Role of Each Species in Its Ecosystem

The concept of the *ecological niche* is important to our understanding of how competition within and between species selects for adaptations in body form and behavior. Although the word *niche* may call to mind a small "cubbyhole," in ecology it means much more. Each species occupies a unique **ecological niche** that encompasses all aspects

of its way of life, including its physical home or *habitat*. The primary habitat of a white-tailed deer, for example, is the eastern deciduous forest. In addition, the niche includes all the physical environmental factors necessary for survival and reproduction, such as nesting or denning sites, the range of temperatures under which the organism can survive, the amount of moisture it requires, the pH of the water or soil it may inhabit, the type of nutrients it requires, and the degree of shade it can tolerate. The ecological niche encompasses the entire "role" that a given species performs within an ecosystem, including what the species eats (or whether it derives energy from photosynthesis) and the other species with which it competes. Although different species share many aspects of their niche with others, no two species ever occupy exactly the same ecological niche, as explained in the following sections.

Competition Occurs Whenever Two Organisms Attempt to Use the Same, Limited Resources

Competition is an interaction that may occur between individuals or species as they attempt to use the same, limited resources, particularly energy, nutrients, or space. **Interspecific competition** describes competitive interactions between different species, which may both use similar food or breeding sites or which may compete for light. Interspecific competition is detrimental to both of the

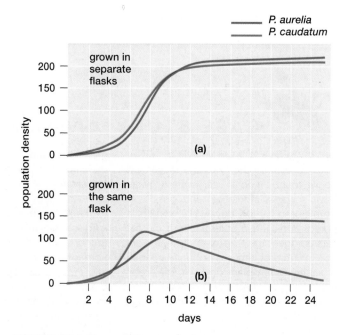

FIGURE 27-1 Competitive exclusion
(a) Raised separately with a constant food supply, both *Paramecium aurelia* and *P. caudatum* show the S-curve typical of a population that initially grows rapidly and then stabilizes. **(b)** Raised together and forced to occupy the same niche, *P. aurelia* consistently outcompetes *P. caudatum* and causes that population to die off. (Modified from G. F. Gause, *The Struggle for Existence*. Baltimore: Williams & Wilkins, 1934.) QUESTION Explain how competitive exclusion could contribute to the threat posed by an invasive species.

Table 27-1 Interactions Among Organisms		
Type of Interaction	Effect on Organism A	Effect on Organism B
Competition between A and B	Harms	Harms
Predation by A on B	Benefits	Harms
Symbiosis		
Parasitism by A on B	Benefits	Harms
Commensalism of A with B	Benefits	No effect
Mutualism between A and B	Benefits	Benefits

species involved because it reduces their access to resources that are in limited supply. The degree of interspecific competition depends on how similar the requirements of the two species are. In other words, the more the ecological niches of two species overlap, the greater the amount of competition between them.

Adaptations Reduce the Overlap of Ecological Niches Among Coexisting Species

Just as no two organisms can occupy exactly the same physical space at the same time, no two species can inhabit exactly the same ecological niche simultaneously and continuously. This important concept, often called the **competitive exclusion principle**, was formulated in 1934 by Russian biologist G. F. Gause. This principle leads to the hypothesis that if two species with the same niche are placed together and forced to compete for limited resources, inevitably one will outcompete the other, and the less well-adapted of the two will die out. Gause used two species of the protist *Paramecium (P. aurelia* and *P. caudatum)* to demonstrate this principle. In laboratory flasks, both species thrived on bacteria and fed in the same parts of the flask. Grown separately, each population flourished (**FIG. 27-1a**). But when Gause placed the two species together in a flask, one (*P. aurelia*) always eliminated, or "competitive-

ly excluded," the other (**FIG. 27-1b**). Gause then repeated the experiment, replacing *P. caudatum* with a different species, *P. bursaria*, which tended to feed in a different part of the flask. In this case, the two species of *Paramecium* were able to coexist indefinitely because they occupied slightly different niches. Invasive species such as zebra mussels have niches that overlap significantly with those of native species, such as freshwater clams, which they are able to outcompete. For more on invasive species, see "Earth Watch: Invasive Species Disrupt Community Interactions."

Ecologist R. MacArthur tested Gause's laboratory findings under natural conditions by investigating five species of North American warbler. These birds all hunt for insects and nest in the same type of spruce tree. Although the niches of these birds appear to overlap considerably, MacArthur found that each species concentrates its search in specific areas of the tree, employs different hunting tactics, and nests at slightly different times. By *partitioning*, or dividing, the resources provided by the spruce trees they share, the warblers minimize the overlap of their niches and reduce competition among the different species (**FIG. 27-2**).

As MacArthur discovered, when two species with similar requirements coexist, each typically occupies a smaller niche than either would by itself. This phenomenon, called **resource partitioning**, is an evolutionary adaptation that reduces the harmful effects of interspecific competition.

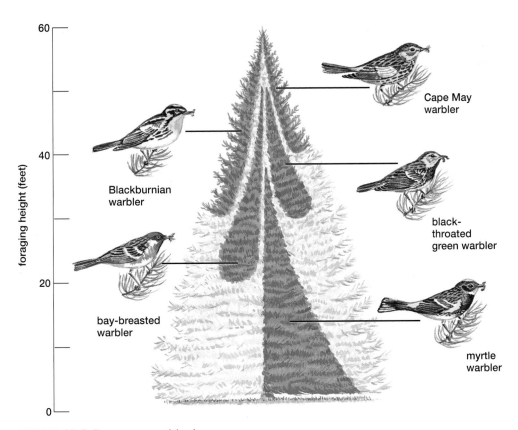

FIGURE 27-2 Resource partitioning
Each of these five insect-eating species of North American warblers searches for food in different regions within spruce trees. They reduce competition by occupying very similar, but not identical, niches.

Resource partitioning is the outcome of the coevolution of species with extensive—but not total—niche overlap. Because natural selection favors individuals with fewer competitors, over evolutionary time the competing species develop physical and behavioral adaptations that minimize their competitive interactions. A dramatic example of resource partitioning was discovered by Charles Darwin among finches of the Galápagos Islands. The finches that shared the same island had evolved different bill sizes and shapes and different feeding behaviors that reduced the competition among them (described in Chapter 16).

Interspecific Competition May Reduce the Population Size and Distribution of Each Species

Although natural selection leads to a reduction of niche overlap between different species, those with similar niches still directly compete for limited resources. This interspecific competition may restrict the size and distribution of competing populations.

A classic study of the effects of interspecific competition was performed by ecologist J. Connell, using barnacles (crustaceans that attach permanently to rocks and other surfaces; gray barnacles coat the rocks in Figure 27-14a). Barnacles of the genus *Chthamalus* share the rocky shores of Scotland with another genus, *Balanus*, and their niches overlap considerably. Both live in the **intertidal zone**, an area of the shore that is alternately covered and exposed by the tides. Connell found that *Chthamalus* dominates the upper shore and *Balanus* dominates the lower. When he scraped off *Balanus*, the *Chthamalus* population increased, spreading downward into the area its competitor had once inhabited. Where the habitat is appropriate for both genera, *Balanus* conquers because it is larger and grows faster. But *Chthamalus* tolerates drier conditions, so on the upper shore, where only high tides submerge the barnacles, it has

a competitive advantage. As this example illustrates, interspecific competition can limit both the size and the distribution of the competing populations.

Competition Within a Species Is a Major Factor Controlling Population Size

Individuals of the same species have essentially identical requirements for resources and thus occupy exactly the same ecological niche. For this reason, **intraspecific competition**—or competition among individuals of the same species—is the most intense form of competition. As explained in Chapter 26, intraspecific competition exerts strong density-dependent environmental resistance, limiting population size. The evolutionary result of interspecific competition is that individuals who are better adapted to obtain scarce resources are more likely to reproduce successfully, passing these traits to their offspring.

27.3 WHAT ARE THE RESULTS OF INTERACTIONS BETWEEN PREDATORS AND THEIR PREY?

Predators kill and eat other organisms. Ecologists sometimes include **herbivores** (animals that eat plants) in this general category because they can have a major influence on the size and distribution of plant populations. Here we will define predation in its broadest sense, to include the grass-eating pika (**FIG. 27-3a**), the zebra mussel that filters microscopic algae from water, the goby fish eating a zebra mussel, and the bat homing in on a moth (**FIG. 27-3b**). Most predators are either larger than their prey or hunt collectively, as wolves do when bringing down an elk (see Fig. 26-10). Predators are generally less abundant than their prey; you will learn why in the next chapter.

(a)   (b)

FIGURE 27-3 Forms of predation
(a) A pika, whose preferred food is grass, is a small relative of the rabbit and lives in the Rocky Mountains. The tough stems of grass have evolved under predation pressure by herbivores. **(b)** A long-eared bat uses a sophisticated echolocation system to hunt moths, which in turn have evolved special sound detectors and behaviors to avoid capture. QUESTION Describe some other examples of coevolution of predators and prey.

Invasive species are species introduced into an ecosystem where they did not evolve, and they are harmful to human health, the environment, or the economy of a region. Invasive species often spread widely because they find few forms of environmental resistance, such as predators or parasites, in their new environment. Their unchecked population growth may seriously damage the ecosystem as they displace, outcompete, and prey on native species. Not all nonnative species become pests, but those that do often have high reproductive rates, effective means of moving into new habitats, and the ability to thrive under a relatively wide range of environmental conditions. Invasive plants may spread by runners as well as seeds, and some can form new plants from plant fragments. Invasive animals are usually not "picky eaters." By evading the checks and balances imposed by millennia of coevolution, invasive species are wreaking havoc on natural ecosystems throughout the world.

Both starlings and English sparrows have spread dramatically since their deliberate introduction into the eastern U.S. in the 1890s. Their success has harmed some native songbirds, such as bluebirds, with which they compete for nesting sites. Red fire ants from South America were accidentally introduced into Alabama on shiploads of lumber in the 1930s and have since spread throughout the southern U.S. Fire ants kill native ants, birds, and young reptiles. Their mounds can ruin farm fields, and their fiery stings and aggressive temperament can make backyards uninhabitable. The Asian long-horned beetle, which arrived around 1996 in wooden pallets and boxes shipped from China, is now devouring hardwood trees in the eastern and midwestern U.S.

Invasive plants also threaten natural communities. In the 1920s, the Japanese vine kudzu was planted extensively in the southern U.S. to control erosion. Today kudzu is a major pest, overgrowing and killing trees and underbrush and occasionally engulfing small houses (**FIG. E27-1a**). The water hyacinth, introduced from South America as an ornamental plant, now clogs about 2 million acres of lakes and waterways in the southern U.S., slowing boat traffic and displacing natural vegetation (**FIG. E27-1b**). Purple loosestrife, introduced as an ornamental plant in the early 1800s, aggressively invades wetlands, where it displaces native plants and reduces both food and habitat for native animals (**FIG. E27-1c**).

A microscopic invader, the West Nile virus, was first recognized in the U.S. in 1999, when crows began dying in large numbers in New York City. The virus replicates in birds, which are bitten by mosquitoes, who then bite and infect more birds, people, and some other mammals, including horses. The birds, horses, and people in this country lack the immunity that comes with long association with the virus, and so they are more vulnerable than populations in Africa and the Middle East where the virus is common.

Ecologists estimate that the thousands of invasive species in the U.S. are responsible for reducing populations of about 400 native species to the point where they are considered threatened or endangered with extinction. Recently, wildlife officials have made cautious attempts to reestablish these checks and balances by importing predators or parasites to attack selected invasive species. This type of control is fraught with danger, however, because introducing additional nonnative predators or parasites into an ecosystem can have unpredicted and possibly disastrous consequences for native species. For example, in 1958 a large predatory Florida snail, the rosy wolf snail, was imported into Hawaii to feed on another invasive pest, the giant African snail. The rosy wolf snail has now become a major menace, threatening several species of native Hawaiian snails with extinction.

Despite the risks of importing these biological control organisms, there often seems to be little alternative, because poisons kill native and nonnative organisms indiscriminately. Learning from past disasters, biologists now carefully screen proposed *biocontrols* to make sure they are specific for the intended invasive species. For example, a small fly from South America whose larvae feed selectively on fire ants is now being released in the southern U.S. Scientists are researching other possible imported insects to feed on invasive foreign weeds—such as kudzu and purple loosestrife—ideally without attacking native plants.

(a) kudzu

(b) water hyacinth

(c) purple loosestrife

FIGURE E27-1 Invasive species
(a) The Japanese vine kudzu will rapidly cover entire trees and houses. **(b)** Water hyacinths, originally from South America, today clog waterways in the southern United States. **(c)** Purple loosestrife displaces native plants and reduces food and habitat for native animals in wetlands.

(a) sand dab (fish)

(b) nightjar (bird)

FIGURE 27-4 Camouflaged by blending in
(a) The sand dab is a flat, bottom-dwelling ocean fish with a mottled color that closely resembles the sand on which it rests. **(b)** This nightjar bird on its nest in Central America is barely visible among the surrounding leaf litter.

(a) moth

(b) leafy sea dragon

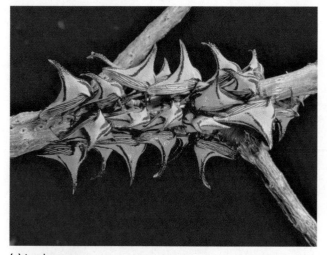

(c) treehoppers

(d) cactus

FIGURE 27-5 Camouflaged by resembling specific objects
(a) A moth whose color and shape resemble a bird dropping sits motionless on a leaf. **(b)** The leafy sea dragon (an Australian "seahorse" fish) has evolved extensions of its body that mimic the algae in which it normally hides. **(c)** Florida treehopper insects avoid detection by resembling thorns on a branch. **(d)** This cactus of the American Southwest is appropriately called the "living rock cactus." QUESTION How could such camouflage evolve?

Predator-Prey Interactions Shape Evolutionary Adaptations

To survive, predators must feed and prey must avoid becoming food. Therefore, predator and prey populations exert intense environmental pressure on one another, resulting in coevolution. As prey become more difficult to catch, predators must become more adept at hunting. Coevolution has endowed the mountain lion with tearing teeth and claws and has given the hunted fawn dappled spots and the behavior of lying still as it awaits its mother. It has produced the keen eyesight of the hawk, and the earthy camouflage coloration of its small mammal prey. Evolution under predation pressure has also produced the poisons and bright colors of the poison arrow frog and the coral snake (see Figs. 27-7 and 27-9b). In the following sections, we examine a few of the evolutionary results of predator-prey interactions. In "Earth Watch: Invasive Species Disrupt Community Interactions," you learned about what happens when natural checks and balances are circumvented by transporting organisms into ecological communities in which they did not evolve.

Some Predators and Prey Have Evolved Counteracting Behaviors

Bat and moth adaptations (see Fig. 27-3) provide an excellent example of how both physical structures and behaviors are molded by coevolution. Most bats are nighttime hunters that navigate and locate prey by echolocation. They emit extremely high-frequency and high-intensity pulses of sound and, by analyzing the returning echoes, create an "image" of their surroundings. Under selection pressure from this specialized prey-location system, certain moths (a favored prey of bats) have evolved simple ears that are particularly sensitive to the frequencies used by echolocating bats. When they hear a bat, these moths take evasive action, flying erratically or dropping to the ground. The bats, in turn, have evolved the ability to counter this defense by switching the frequency of their sound pulses away from the moth's sensitivity range. Some moths interfere with the bats' echolocation by producing their own high-frequency clicks. In response, when hunting a clicking moth, a bat may turn off its own sound pulses temporarily and zero in on the moth by following the moth's clicks.

Camouflage Conceals Both Predators and Their Prey

An old saying in detective novels is that the best hiding place may be right out in plain sight. Both predators and prey have evolved colors, patterns, and shapes that resemble their surroundings. Such disguises, called **camouflage**, render animals inconspicuous even when they are in plain sight (**FIG. 27-4**).

Some animals closely resemble specific objects such as leaves, twigs, seaweed, thorns, or even bird droppings (**FIG. 27-5a–c**). Camouflaged animals tend to remain motionless rather than to flee their predators; a fleeing "bird dropping" would be quite conspicuous! Whereas many camouflaged animals resemble plants, a few types of plants have evolved to resemble rocks, which their herbivorous predators ignore (**FIG. 27-5d**).

Predators that ambush are also aided by camouflage. For example, a spotted cheetah becomes inconspicuous in the grass as it watches for grazing mammals. The frogfish closely resembles the algae-covered rocks and sponges on which it sits motionless, dangling a small lure from its upper lip (**FIG. 27-6**). Small fish notice only the lure and are engulfed as they approach it.

(a) cheetah (b) frogfish

FIGURE 27-6 Camouflage assists predators
(a) As it waits for prey, a cheetah blends into the background of the grass. (b) Combining camouflage and aggressive mimicry, a frogfish waits in ambush, its camouflaged body matching the sponge-encrusted rock on which it rests. Above its mouth dangles a lure that closely resembles a small fish. The lure attracts small predators, who will find themselves prey.

FIGURE 27-7 Warning coloration
The South American poison arrow frog advertises its poisonous skin with bright and contrasting color patterns.

Bright Colors Often Warn of Danger

Some animals have evolved very differently, exhibiting bright **warning coloration** (FIG. 27-7; see also Fig. 27-8 and Fig. 27-9). These animals are usually distasteful, and many are poisonous—such as the yellow jacket with its bright yellow and black stripes. Poisoning a predator is small consolation for an organism that has already been eaten; thus, the bright colors declare, "Eat me at your own risk!" One unpleasant experience is enough to teach predators to avoid such conspicuous prey.

Some Prey Organisms Gain Protection Through Mimicry

Mimicry refers to a situation in which one species has evolved to resemble another organism. By sharing similar warning-color patterns, several poisonous species may all benefit. Mimicry among different distasteful species is called *Müllerian mimicry*. For example, toxic monarch butterflies have wing patterns strikingly similar to those of equally distasteful (and nearly indistinguishable) viceroy butterflies (FIG. 27-8). Birds that become ill from consuming one will also avoid the other. A toad that is stung while attempting to eat a bee is likely to avoid not only bees, but other black and yellow striped insects—such as venomous hornets and yellow jackets—without ever tasting one. A common color pattern results in faster learning by predators—and less predation on all similarly colored species.

Once warning coloration evolved, there arose a selective advantage for harmless animals to resemble poisonous ones, an adaptation called *Batesian mimicry*. Thus the harmless hoverfly avoids predation by resembling a bee (FIG. 27-9a), and the nonpoisonous scarlet king snake is protected by brilliant warning coloration closely resembling that of the deadly coral snake (FIG. 27-9b).

Certain prey species use another form of mimicry: **startle coloration.** Several insects and even some vertebrates (such as the false-eyed frog) have evolved patterns of color that closely resemble the eyes of a much larger, and possibly dangerous, animal (FIG. 27-10). If a predator gets close, the prey suddenly flashes its eyespots, startling the predator and allowing the prey to escape. A sophisticated variation on the theme of prey mimicking dangerous animals is seen in snowberry flies, which are hunted by territorial jumping spiders. When a fly spots an approaching spider, it spreads its wings, moving them back and forth in a jerky dance. Seeing this display, the spider is likely to flee the fly. Why? Researchers have observed that the markings on the fly's

monarch butterfly (distasteful)

viceroy butterfly (distasteful)

FIGURE 27-8 Müllerian mimicry
Nearly identical warning coloration protects both the distasteful monarch (left) and the equally distasteful viceroy butterfly (right).

(a) bee (poisonous)

haverfly (non-poisonous)

(b) coral snake (venomous)

scarlet king snake (non-venomous)

FIGURE 27-9 Batesian mimicry
(a) A bee, which is capable of stinging (left), is mimicked by the stingless hoverfly (right). **(b)** The warning coloration of the poisonous coral snake (left) is mimicked by the harmless scarlet king snake (right).

(a) false-eyed frog

(b) peacock moth

(c) swallowtail butterfly caterpillar

FIGURE 27-10 Startle coloration
(a) When threatened, the false-eyed frog raises its rump, which resembles the eyes of a larger predator.
(b) The peacock moth from Trinidad is well camouflaged, but should a predator approach, it opens its
wings to reveal spots resembling large eyes. **(c)** Predators of this caterpillar larva of the swallowtail butterfly are deterred by its resemblance to a snake. The caterpillar's head is the "snake's" nose.

jumping spider (predator)   snowberry fly (prey)

FIGURE 27-11 A prey mimics its predator
In response to the approach of a jumping spider (left), the snowberry fly spreads its wings, revealing a pattern that resembles spider legs (right). The fly enhances the effect by performing a jerky, side-to-side dance that resembles the leg-waving display of a jumping spider defending its territory.

wings look like the legs of a jumping spider, and the fly's jerky movements mimic the behavior of a jumping spider driving another spider from its territory (**FIG. 27-11**). Thus, natural selection has finely tuned both the behavior and the appearance of the fly to avoid predation by jumping spiders.

Predators May Use Mimicry to Attract Prey

Some predators have evolved **aggressive mimicry**, a "wolf-in-sheep's-clothing" approach, in which they entice their prey to come close by resembling something attractive to the prey. For example, by using a rhythm of flashes that is unique to each species, female fireflies attract males to mate. But in one species, the females sometimes mimic the flashing pattern of a different species, attracting males that they kill and eat. The frogfish (see Fig. 27-6b) is not only camouflaged, but exhibits aggressive mimicry by dangling, just above its mouth, a wriggling lure that resembles a small fish. A hungry fish attracted to the lure is quickly swallowed.

Predators and Prey May Engage in Chemical Warfare

Both predators and prey use a variety of toxic chemicals for attack and defense. The venom of spiders and poisonous snakes, such as the coral snake (see Fig. 27-9), serves both to paralyze prey and to deter predators. Many plants also produce defensive toxins. For example, lupine plants, whose flowers grace both gardens and mountain meadows, produce chemicals called *alkaloids*, which deter attack by the blue butterfly, whose larvae feed on the lupine's buds. In fact, different individuals of the same species of lupine produce different forms of alkaloids, thus making it more difficult for the butterflies to evolve resistance to the toxin.

Certain mollusks (including squid, octopus, and some sea slugs) emit clouds of ink when attacked. These colorful chemical "smoke screens" confuse predators and mask the prey's escape. A dramatic example of chemical defense is seen in the bombardier beetle. In response to the bite of an ant, the beetle releases secretions from special glands into an abdominal chamber. There, enzymes catalyze an explosive chemical reaction that shoots a toxic, boiling-hot spray onto the attacker (**FIG. 27-12a**).

Plants and Herbivores Have Coevolutionary Adaptations

Although we have classified them as predators, herbivores (animals that eat plants) don't fit neatly into our categories. A grazing horse or cow uproots and kills some grass, but most of the time acts more like a lawn mower, cropping but not killing the plants. Regardless of how we categorize them, herbivores exert strong selective pressure on plants to avoid being eaten. Plants have evolved a variety of chemical adaptations that deter their herbivorous "predators." Many, such as the milkweed, synthesize toxic and distasteful chemicals. As plants evolved toxic chemicals for defense, certain insects evolved increasingly efficient ways to detoxify or even use the chemicals. The result is that nearly every toxic plant is eaten by at least one type of insect. For example, monarch butterflies lay their eggs on milkweed; when their larvae hatch, they consume the toxic plant (**FIG. 27-12b**). The caterpillars not only tolerate the milkweed poison, but store it in their tissues as a defense against their own predators. The stored toxin is retained in the metamorphosed monarch butterfly (see Fig. 27-8).

Grasses have evolved tough silicon (glassy) substances in their blades that make them difficult to chew, selecting for grazing animals with longer, harder teeth. Over an evolutionary time scale, as grasses evolved tougher blades that discouraged predation, horses evolved longer teeth with thicker enamel coatings that resist wear and abrasion from the tough grasses.

(a) bombardier beetle

FIGURE 27-12 Chemical warfare
(a) The bombardier beetle sprays a hot toxic brew in response to a leg pinch. **(b)** A monarch caterpillar feeds on milkweed that contains a powerful toxin. QUESTION Why is the caterpillar colored with bright stripes?

(b) monarch caterpillar

27.4 WHAT IS SYMBIOSIS?

Symbiosis, which literally means "living together," is the close interaction between organisms of different species for an extended time. Considered in its broadest sense, symbiosis includes *parasitism*, *mutualism*, and *commensalism*. Although one species always benefits in these symbiotic relationships, the second species may be unaffected, harmed, or helped. **Commensalism** is a relationship in which one species benefits while the other is relatively unaffected. Barnacles that attach themselves to the skin of a whale, for example, get a free ride through nutrient-rich waters without harming the whale. In parasitism and mutualism, the participants act on each other as strong agents of natural selection, as discussed in the following sections.

Parasitism Harms, but Does Not Immediately Kill, the Host

In parasitism, one organism benefits by feeding on another. **Parasites** live in or on their prey, which are called *hosts*, usually harming or weakening them but not immediately killing them. Although it is sometimes difficult to distinguish clearly between a predator and a parasite, parasites are generally much smaller and more numerous than their hosts. Familiar parasites include tapeworms, fleas, and the many types of disease-causing protozoa, bacteria, and viruses. Many parasites, particularly worms and protozoa (such as the malarial parasite), have complex life cycles involving two or more hosts. There are few parasitic vertebrates; the lamprey eel, which attaches itself to a host fish and sucks its blood, is one example.

The variety of infectious bacteria and viruses and the precision of the immune system that counters their attacks are evidence of the powerful forces of coevolution between parasitic microorganisms and their hosts. Consider the malaria parasite, which has provided strong environmental pressure for humans to carry a defective hemoglobin gene that causes sickle-cell anemia: the parasite can't survive in the affected red blood cells. In certain areas of Africa where malaria is common, up to 20% of the human population carries the sickle-cell gene.

Another example is *Trypanosoma*, a parasitic protozoan that causes both human sleeping sickness and a disease called *nagana* in cattle. African antelope, which coevolved with this parasite, are relatively unaffected by it. Cattle, which are not native to Africa, suffer but survive infection if they have been bred in an infested area for many generations. Newly imported cattle, however, generally die from nagana if not treated.

In Mutualistic Interactions, Both Species Benefit

When two species interact in a way that benefits both, the relationship is called **mutualism**. If you see colored patches on rocks, they are probably lichens, a mutualistic association

(a) lichen

(b) clownfish

FIGURE 27-13 Mutualism
(a) This brightly colored lichen growing on bare rock is a mutualistic relationship between an alga and a fungus. **(b)** The clownfish snuggles unharmed among the stinging tentacles of the anemone. Note the bright "warning" color of the clownfish. Although the fish itself is defenseless, the coloration may warn potential predators of the threat posed by the anemone.

of an alga and a fungus (**FIG. 27-13a**). The fungus provides support and protection while deriving food from the photosynthetic alga, whose bright colors are actually light-trapping pigments. Mutualistic associations also occur in the digestive tracts of cows and termites, where protists and bacteria find food and shelter while helping their hosts extract nutrients, as well as in our own intestines, where bacteria synthesize certain vitamins. The nitrogen-fixing bacteria inhabiting special chambers on the roots of legume plants are another important example. These bacteria obtain food and shelter from the plant and in return trap nitrogen in a form the plant can use. Some mutualistic

partners have coevolved to the extent that neither can survive alone. An example is the ant–acacia mutualism described in "Scientific Inquiry: Ants and Acacias—An Advantageous Association."

Mutualistic relationships involving vertebrates are rare and typically less intimate and extended. The clownfish, which is coated with a layer of protective mucus, takes shelter among the venomous tentacles of certain species of anemones. The anemone provides the fish with protection from predators. In return, the clownfish drives away other fish that eat anemones, cleans dirt and debris from its host, and may bring bits of food to its anemone (**FIG. 27-13b**).

(a)

(b)

FIGURE 27-14 Keystone species
(a) The starfish *Pisaster* is a keystone species along the rocky coast of the Pacific Northwest. **(b)** The elephant is a keystone species on the African savanna.

SCIENTIFIC INQUIRY   Ants and Acacias—An Advantageous Association

Daniel Janzen, a doctoral student at the University of Pennsylvania, was walking down a road in Veracruz, Mexico, when he saw a flying beetle alight on a thorny tree, only to be driven off by an ant. Looking more closely, he saw that the tree, a bull's-horn acacia, was covered with ants. A large ant colony of the genus *Pseudomyrmex* made its home inside the enlarged thorns of the plant; the ants easily excavated the thorns' soft pulpy interiors to provide shelter (**FIG. E27-2**).

To determine whether the ants were important to the tree, Janzen began stripping the thorns by hand until he found and removed the thorn that housed the ant queen, thus destroying the colony. Later, he turned to more efficient but dangerous methods, using an insecticide to eliminate all the ants on a large stand of acacias. Though the acacias were unharmed by the poison, Janzen became ill from it, and all the ants were killed. Within a year of spraying the insecticide, Janzen found nearly all the acacia trees dead, consumed by insects and other herbivores and shaded out by competing plants. The ground surrounding the trees, which the ants normally kept trimmed, was overgrown with vegetation. The trees were apparently dependent on their resident ants for survival.

Wondering if the ants could survive without the tree, Janzen painstakingly peeled ant-inhabited thorns off 100 acacia trees, suffering multiple stings in the process. Janzen housed each ant colony in a jar supplied with local non-acacia vegetation and insects for food, but all the ant colonies starved. Carefully examining the acacia, he found swollen structures filled with sweet syrup at the base of the leaves and protein-rich capsules on the leaf tips (Fig. E27-2, inset). Together, these materials provide a balanced diet for the ants.

Janzen's experiments strongly suggest that these species of ant and acacia have an obligatory mutualistic relationship: Neither can survive without the other. Of course, further observations are required to support this hypothesis. The fact that the ants starved in Janzen's jars did not rule out their possible survival elsewhere; but, in fact, this species of ant has never been found living independently. Similarly, the bull's-horn acacia has never been found without a resident ant colony. Thus, a chance observation followed by careful research led to the discovery of an important mutualistic association.

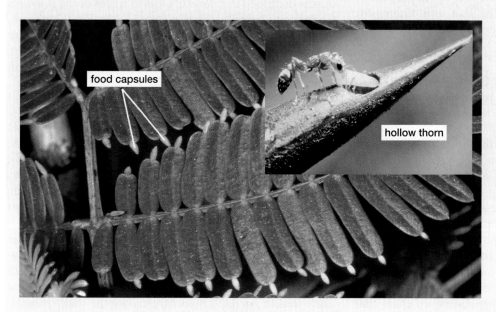

food capsules

hollow thorn

FIGURE E27-2 A mutualistic relationship
Yellow, protein-rich capsules produced at the tips of acacia leaves provide food for the resident ants. (Inset) A hole in the enlarged thorn of the bull's-horn acacia provides shelter for members of the ant colony. The ant entering the thorn is carrying a food capsule. As the ant colony grows, it invades more thorns.

27.5 HOW DO KEYSTONE SPECIES INFLUENCE COMMUNITY STRUCTURE?

In some communities, a particular key species, called a **keystone species**, plays a major role in determining community structure—a role that is out of proportion to its abundance in the community. Removal of the keystone species dramatically alters the community. For example, in 1969 Robert Paine, an ecologist at the University of Washington, removed predatory starfish *Pisaster* (**FIG. 27-14a**) from sections of Washington's rocky intertidal coast. Mussels (two-shelled or "bivalve" mollusks that are a favored

prey of *Pisaster*) became so abundant that they outcompeted other invertebrates and algae that normally coexist in intertidal communities. Another marine invertebrate, the lobster, may be a keystone species off the east coast of Canada. Overfishing of the lobster allowed its prey, sea urchins, to increase in numbers. The population explosion of sea urchins nearly eliminated certain types of algae on which the urchins feed, leaving large expanses of bare rock where a diverse community once existed. The sea otter appears to be a keystone species along the coast of western Alaska. Starting around 1990, observers noted an alarming decline in otter numbers, resulting in an increase in their sea urchin prey. This led to overgrazing by sea urchins on the kelp forests that provide critical undersea habitat for a variety of marine species. What killed the sea otters? Killer whales, which formerly fed primarily on seals and sea lions, were seen increasingly dining on sea otters, as their favored prey disappeared. Scientists hypothesize that the seal and sea lion decline, in turn, is a result of overfishing by humans in the North Pacific, depleting the food supply of these fish-eaters. In the African savanna, the African elephant is a keystone predator. By grazing on small trees and bushes, elephants prevent the encroachment of forests and help maintain the grassland community (**FIG. 27-14b**). In Chapter 30, you will learn of another keystone species: the wolf.

It is difficult to identify keystone species; doing so properly requires that the species be selectively removed and the community studied for several years before and after its removal. However, many ecological studies performed since the concept was introduced provide evidence that keystone species are important in a wide variety of communities. Why is it important to study keystone species? As human activities infringe on natural ecosystems, it becomes increasingly urgent to understand community interactions and to preserve species that are crucial to maintaining the natural community.

27.6 SUCCESSION: HOW DO COMMUNITY INTERACTIONS CAUSE CHANGE OVER TIME?

In a mature terrestrial ecosystem, the populations that make up the community interact with one another and with their nonliving environment in intricate ways. But this tangled web of life did not spring fully formed from bare rock or naked soil; rather, it emerged in stages over a long period, by a process called succession. **Succession** is a structural change in a community and its nonliving environment over time. It is a kind of "community relay" in which assemblages of plants and animals replace one another in a sequence that is somewhat predictable.

Succession is preceded by a **disturbance**, an event that disrupts the ecosystem either by altering its community, its abiotic structure, or both. In the case of primary succession, the disturbance may be a glacier scouring the landscape down to bare rock, or it may be a volcano covering an ecosystem with new rock or producing an entirely new island (**FIG. 27-15a**). In secondary succession, the disturbance is far more limited. For example, beavers, landslides, or people may dam streams, causing marshes, ponds, or lakes to form. A landslide or avalanche may strip a swath of trees from a mountainside. Fire is another common disturbance. Each disturbance sets the stage for succession. Volcanic eruptions, as in the case of Mount St. Helens, leave behind a nutrient-rich environment that encourages rapid invasion of new life (**FIG. 27-15b**). Forest fires, while destroying an existing community, also release nutrients and create conditions that favor rapid succession (**FIG. 27-15c**).

The precise changes that occur during succession are as diverse as the environments in which succession occurs, but we can recognize certain general stages. In each case, succession is begun by a few hardy plant invaders called **pioneers**. The pioneers may alter the ecosystem in ways that favor competing plants, which gradually displace them. If allowed to continue, succession progresses to a diverse and relatively stable **climax community**. Alternatively, recurring disturbances maintain many communities in **subclimax** stages. Our discussion of succession will focus on plant communities, which dominate the landscape and provide both food and habitat for animals.

There Are Two Major Forms of Succession: Primary and Secondary

Succession takes two major forms: primary and secondary. During **primary succession**, a community gradually colonizes bare rock, sand, or a clear glacial pool where there is no trace of a previous community. This building of a community "from scratch" typically requires thousands or even tens of thousands of years. During **secondary succession**, a new community develops after an existing ecosystem is disturbed in a way that leaves traces of the previous community behind, such as soil and seeds. For this reason, secondary succession happens much more rapidly than does primary succession; it may take only a few centuries. In the following examples, we examine these processes in more detail.

Primary Succession Can Begin on Bare Rock

FIGURE 27-16 illustrates primary succession on Isle Royale, Michigan, an island in Lake Superior. Bare rock, such as that exposed by a retreating glacier, liberates nutrients such as minerals by *weathering*. In weathering, cracks form as the rock alternately freezes and thaws, contracting and expanding. Chemical action such as acid rain further breaks down the surface.

Weathered rock provides a place for lichens, a pioneer species, to attach where there are no competitors and plenty of sunlight. Lichens can photosynthesize, and they obtain minerals by dissolving some of the rock with an acid they secrete. As the lichens spread over the rock, drought-resistant, sun-loving mosses begin growing in the cracks. Fortified by nutrients liberated by the lichens, the moss forms a dense mat that traps dust, tiny rock particles, and bits of organic debris. It may cover and kill the lichen that made its growth possible. As some mosses die each year, their bodies add to a growing nutrient base, and the living moss mat acts like a sponge, trapping moisture. Within the moss, seeds of larger plants, such as bluebell and yarrow, germinate. When these plants die, their bodies contribute to a growing layer of soil. As woody shrubs such as blueberry and juniper take advantage of the newly formed soil, the moss and remaining lichens may be shaded out and buried by decaying leaves and vegetation. Eventually, trees such as jack pine, blue spruce, and aspen take root in the deeper crevices, and the sun-loving shrubs are shaded out. Within the forest, shade-tolerant seedlings of taller or

(a) Mt. Kilauea, Hawaii

(b) Mt. St. Helens, Washington State

(c) Yellowstone, Wyoming

FIGURE 27-15 Succession in progress
Pairs of photographs illustrate primary and secondary succession. **(a)** Primary succession. Left: The Hawaiian volcano Mount Kilauea has erupted repeatedly since 1983, sending rivers of lava over the surrounding countryside. Right: A pioneer fern takes root in a crack in hardened lava. **(b)** Secondary succession. Left: On May 18, 1980, the explosion of Mount St. Helens in Washington State devastated the surrounding pine forest ecosystem. Right: Twenty years later, life abounds on the once-barren landscape. Because traces of the former ecosystem remained, this is an example of secondary succession. **(c)** Secondary succession. Left: In the summer of 1988, extensive fires swept through the forests of Yellowstone National Park in Wyoming. Right: Trees and flowering plants are thriving in the sunlight, and wildlife populations are rebounding as secondary succession occurs. **QUESTION** People have suppressed fires for decades. What are the implications of fire suppression for forest ecosystems and succession?

lichens and moss on bare rock   bluebell, yarrow   blueberry, juniper   jack pine, black spruce, aspen   balsam fir, paper birch, white spruce, climax forest

time (years)

0 ——————————————————————————————→ 1000

FIGURE 27-16 Primary succession
Primary succession as it occurs on bare rock in upper Michigan.

faster-growing trees, such as balsam fir, paper birch, and white spruce, thrive. In time they tower over and replace the original trees, which are intolerant of shade. After a thousand years or more, a tall climax forest thrives on what was once bare rock.

An Abandoned Farm Will Undergo Secondary Succession

FIGURE 27-17 illustrates secondary succession on an abandoned farm in the southeastern United States. The pioneer species—fast-growing annual weeds such as crabgrass, ragweed, and sorrel—root in the rich soil already present and thrive in direct sunlight. They generally produce large numbers of easily dispersed seeds that help them colonize open spaces, but they don't compete well against longer-lived (perennial) species that gradually grow larger and shade out the pioneers. After a few years, perennial plants such as asters, goldenrod, and broom sedge grass invade, as do woody shrubs such as blackberry. These multiply rapidly and dominate for the next few decades. Eventually, they are replaced by pines and fast-growing deciduous trees, such as tulip poplar and sweet gum, which sprout from windblown seeds. These trees become prominent after about 25 years, and a pine forest dominates the field for the rest of the first century. Meanwhile, shade-resistant, slow-growing hardwoods such as oak and hickory take root beneath the pines. After the first century, they begin

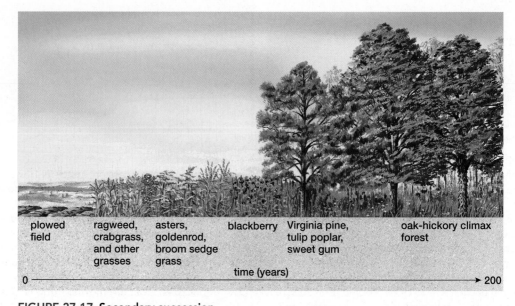

plowed field   ragweed, crabgrass, and other grasses   asters, goldenrod, broom sedge grass   blackberry   Virginia pine, tulip poplar, sweet gum   oak-hickory climax forest

time (years)

0 ——————————————————————————————→ 200

FIGURE 27-17 Secondary succession
Secondary succession as it might occur on a plowed, abandoned farm field in the southeastern United States.

to tower over and shade the pines, which eventually die from lack of sun. A relatively stable climax forest dominated by oak and hickory is present by the end of the second century.

Succession Also Occurs in Ponds and Lakes

In freshwater ponds or lakes, succession occurs not only through changes within the pond or lake but also through an influx of nutrients from outside the ecosystem. Sediments and nutrients carried in by runoff from the surrounding land have a particularly large impact on small freshwater lakes, ponds, and bogs, which gradually undergo succession to dry land (FIG. 27-18). In forests, meadows may be produced by lakes undergoing succession. As the lake fills in from the edges, grasses colonize the newly exposed soil. As the lake shrinks and the meadow grows, trees will encroach around the meadow's edges. If you return to a forest lake 20 years later, it probably will be a bit smaller.

Succession Culminates in the Climax Community

Succession ends with a relatively stable climax community, which perpetuates itself if it is not disturbed by external forces (such as fire, parasites, introduced species, or human activities). The populations within a climax community have ecological niches that allow them to coexist without replacing one another. In general, climax communities have more species and more types of community interactions than do early stages of succession. The plant species that dominate climax communities generally live longer and tend to be larger than pioneer species; this trend is particularly evident in ecosystems where forest is the climax community.

In your travels, you have undoubtedly noticed that the type of climax community varies dramatically from one area to the next. For example, if you drive through Colorado, you will see a shortgrass prairie climax community on the eastern plains (in those rare areas where it has not been replaced by farms), pine-spruce forests in the mountains, tundra on the mountain summits, and sagebrush-dominated climax community in the western valleys. The exact nature of the climax community is determined by numerous geological and climatic variables, including temperature, rainfall, elevation, latitude, type of rock (which determines the type of nutrients available), and exposure to sun and wind. Natural events such as hurricanes, avalanches, and fires started by lightning may destroy sections of climax forest, reinitiating secondary succession and producing a patchwork of various successional stages within an ecosystem.

In many forests throughout the United States, rangers now allow fires set by lightning to run their course, recognizing that this natural process is important for the maintenance of the entire ecosystem. Fires liberate nutrients used by plants and kill some (but usually not all) of the trees they engulf. Sunlight can then reach the forest floor, encouraging the growth of *subclimax plants*, which belong to a successional stage earlier than the climax stage. The combination of climax and subclimax regions within the ecosystem provides habitats for a larger number of species.

Human activities may dramatically alter the climax vegetation. Large stretches of grasslands in the western U.S., for example, are now dominated by sagebrush due to overgrazing. The grass that usually outcompetes sagebrush is selectively eaten by cattle, allowing the sagebrush to prosper.

Some Ecosystems Are Maintained in a Subclimax Stage

Some ecosystems are not allowed to reach the climax stage but are maintained in a subclimax stage by frequent disturbance. The tallgrass prairie that once covered northern Missouri and Illinois is a subclimax stage of an ecosystem whose climax community is deciduous forest. The prairie was maintained by periodic fires, some set by lightning and others deliberately set by Native Americans to increase grazing land for buffalo. Forest now encroaches, and limited prairie preserves are maintained by carefully managed burning.

Agriculture also depends on the artificial maintenance of carefully selected subclimax communities. Grains are specialized grasses characteristic of the early stages of succession, and much energy goes into preventing competitors (weeds and shrubs) from taking over. The suburban

(a)   (b)

FIGURE 27-18 Succession in a small freshwater pond
In small ponds, succession is speeded by an influx of materials from the surroundings. **(a)** In this small pond, dissolved minerals carried by runoff from the surroundings support aquatic plants, whose seeds or spores were carried in by the winds or by birds and other animals. **(b)** Over time, the decaying bodies of aquatic plants build up soil that provides anchorage for more terrestrial plants. Finally, the pond is entirely converted to dry land.

lawn is a painstakingly maintained subclimax ecosystem. Mowing (a disturbance) destroys woody invaders, and some suburbanites also use herbicides to selectively kill pioneers such as crabgrass and dandelions.

The study of succession is the study of variations in communities over time. The climax communities that form during succession are strongly influenced by climate and geography—the distribution of ecosystems in space. Deserts, grasslands, and deciduous forests are climax communities formed over broad geographical regions with similar environmental conditions. These extensive areas of characteristic plant communities are called **biomes**. Although the communities within the various biomes differ radically in the types of populations they support, communities worldwide are structured according to general rules. These principles of ecosystem structure, as well as some of the great biomes of the world, are described in the following chapters.

EVOLUTIONARY CONNECTIONS

Is Camouflage Splitting a Species?

The insect known as the walking stick is aptly named; its elongated, camouflaged body blends in beautifully with the plants on which it feeds and hides from predatory birds and lizards. In the Santa Ynez Mountains of California, a single species of walking stick (*Timena cristinae*) exhibits two distinct, genetically determined color patterns: green with a white stripe and solid green. Researcher Cristina Sandoval found that the striped form most often hides in and prefers to feed on chamise bushes, where it almost disappears among the needlelike leaves (**FIG. 27-19**, top). In contrast, she found the solid-colored form feeding mostly on wild blue lilac (**FIG. 27-19**, bottom), camouflaged among the lilac's solid green leaves.

Birds and lizards voraciously eat both colors of walking sticks. Therefore, striped forms that prefer striped leaves will be better camouflaged, allowing more to survive to reproduce and pass their plant preference on to their offspring. Solid green walking sticks that prefer solid leaves will have a similar survival advantage. Sandoval and colleagues from Simon Frazer University in Canada brought both insect forms into the laboratory and allowed them to mate. They observed that walking sticks from chamise plants preferred to mate with others from chamise plants, and those from wild lilacs preferred others from wild lilacs, indicating that natural selection had favored behavioral (as

FIGURE 27-19 Color variants of "walking sticks" prefer different plants
(top) The striped form of walking stick is well hidden among the needlelike leaves of its preferred food, the chamise bush. **(bottom)** The solid-colored version of the same species blends well with the leaves of the wild lilac, which it prefers. This photo shows a mating pair. In the lab, the insects preferred to mate with others of the same color pattern.

well as color) differences that accompanied the food preferences. This selective mating assures that the offspring will resemble the parents' host plant. Although the two color forms are still capable of interbreeding, the scientists hypothesize that they are observing the early stages of the splitting of a single species into two different species. Inherited traits that cause differently colored insects to resemble and prefer dining on different plant species create a type of *ecological isolation* (described in Chapter 16) in which the two color forms are unlikely to encounter one another, and they are unlikely to mate if they do. This sets the stage for further divergence of both physical and behavioral traits as the two forms encounter different selection pressures based on their preference for different plants.

CASE STUDY REVISITED INVASION OF THE ZEBRA MUSSELS

About five years after the zebra mussel arrived, scientists were pleased to see a native sponge growing on top of zebra mussels. Both sponges and mussels feed by filtering water and removing microscopic algae, so these species compete with one another for food. In some study areas, the number of mussels has declined, partly as a result of being smothered by sponges and partly from being eaten by another invasive species: the round goby. In 1990 a biologist at the University of Michigan found a round goby in the St. Clair River. The goby probably arrived the same way the mussels did, and it also originated in southeastern Europe. Recognizing one of its favorite prey, the five-inch-long predator immediately began feasting on small zebra mussels and expanding its range into habitats already invaded by mussels; gobies are now found in all five Great Lakes.

Is this a serendipitous solution to the mussel problem? Unfortunately, it is not. The gobies ignore the largest zebra mussels, so these continue to spawn. Further, the gobies are not picky eaters. In addition to mussels, they will eat the eggs and young of any other fish in their habitat, including native smallmouth bass, walleye, perch, and sculpins. Researchers are now investigating ways to halt the goby's spread toward the Mississippi River. Meanwhile, zebra mussels are invading new waterways.

Consider This Although the round goby was introduced accidentally, some nonnative predators have been imported to control invasive pests, and some officials have even proposed importing nonnative predators to control native pest species, such as grasshoppers. Discuss the implications of importing such "biological controls" for ecological communities and for native species. Describe the types of studies that should be conducted before any new predator is imported.

CHAPTER REVIEW

SUMMARY OF KEY CONCEPTS

27.1 Why Are Community Interactions Important?

Community interactions influence population size, and the interacting populations within communities act on one another as agents of natural selection. Thus, community interactions also shape the bodies and behaviors of members of the interacting populations.

27.2 What Is the Relationship Between the Ecological Niche and Competition?

The ecological niche defines all aspects of a species' habitat and interactions with its living and nonliving environments. Each species occupies a unique ecological niche. Interspecific competition occurs when the niches of two populations within a community overlap. When two species with the same niche are forced (under laboratory conditions) to occupy the same ecological niche, one species always outcompetes the other. Species within natural communities have evolved in ways that avoid excessive niche overlap, with behavioral and physical adaptations that allow resource partitioning. Interspecific competition limits both the size and the distribution of competing populations. Intraspecific competition is most intense because individuals of the same species occupy the same ecological niche. Competition of both types exerts density-dependent controls on population growth.

Web Tutorial 27.1 Competitive Exclusion and Resource Partitioning

27.3 What Are the Results of Interactions Between Predators and Their Prey?

Predators eat other organisms and are generally larger and less abundant than their prey. Predators and prey act as strong agents of selection on one another. Prey animals have evolved a variety of protective colorations that render them either inconspicuous (camouflage) or startling (startle coloration) to their predators. Some prey are poisonous and exhibit warning coloration by which they are readily recognized and avoided. The situation in which some animals have evolved to resemble others is called mimicry. Both predators and prey have evolved a variety of toxic chemicals for attack and defense. Plants that are preyed on have evolved elaborate defenses, ranging from poisons to thorns to overall toughness. These defenses, in turn, have selected for predators that can detoxify poisons, ignore thorns, and grind down tough tissues.

27.4 What Is Symbiosis?

Symbiotic relationships involve two species that interact closely over an extended time, and include parasitic, commensal, and mutualistic associations. In parasitism, the parasite feeds on a larger, less abundant host, usually harming it but not killing it immediately. In commensalism, one species benefits, typically by finding food more easily in the presence of the other species, which is not affected by the association. Mutualism benefits both symbiotic species.

27.5 How Do Keystone Species Influence Community Structure?

Keystone species have a greater influence on community structure than can be predicted by their numbers. For example, if the African elephant were driven to extinction, the African grasslands it now inhabits might revert to forests.

Web Tutorial 27.2 The Importance of Keystone Species

27.6 Succession: How Do Community Interactions Cause Change over Time?

Succession is a progressive change over time in the types of populations that comprise a community. Primary succession, which may take thousands of years, occurs where no remnant of a previous community existed (such as on rock scraped bare by a glacier or cooled from molten lava, on a sand dune, or in a newly formed glacial lake). Secondary succession occurs much more rapidly because it builds on the remains of a disrupted community, such as an abandoned field or the aftermath of a fire. Secondary succession on land is initiated by fast-growing, readily dispersing pioneer plants, which are eventually re-placed by longer-lived, generally larger and more shade-toler-ant species. Uninterrupted succession ends with a climax com-munity, which tends to be self-perpetuating unless acted on by outside forces, such as fire or human activities. Some ecosys-tems, including tallgrass prairie and farm fields, are maintained in relatively early stages of succession by periodic disruptions.

Web Tutorial 27.3 Primary Succession

KEY TERMS

aggressive mimicry   *page 546*
biome   *page 554*
camouflage   *page 543*
climax community   *page 550*
coevolution   *page 538*
commensalism   *page 547*
community   *page 538*
competition   *page 538*
competitive exclusion principle   *page 539*

disturbance   *page 550*
ecological niche   *page 538*
herbivore   *page 540*
interspecific competition   *page 538*
intertidal zone   *page 540*
intraspecific competition   *page 540*
invasive species   *page 541*

keystone species   *page 549*
mimicry   *page 544*
mutualism   *page 547*
parasite   *page 547*
pioneer   *page 550*
primary succession   *page 550*
resource partitioning   *page 539*

secondary succession   *page 550*
startle coloration   *page 544*
subclimax   *page 550*
succession   *page 550*
symbiosis   *page 547*
warning coloration   *page 544*

THINKING THROUGH THE CONCEPTS

1. Define an ecological *community*, and list three important types of community interactions.

2. Describe four very different ways in which specific plants and animals protect themselves from being eaten. In each, de-scribe an adaptation that might evolve in predators of these species that would overcome their defenses.

3. List two important types of symbiosis; define and provide an example of each.

4. Which type of succession would occur on a clear-cut (a region in which all the trees have been removed by logging) in a na-tional forest, and why?

5. List two subclimax and two climax communities. How do they differ?

6. Define *succession*, and explain why it occurs.

APPLYING THE CONCEPTS

1. Herbivorous animals that eat seeds are considered by some ecologists to be predators of plants, and herbivorous animals that eat leaves are considered to be parasites of plants. Discuss the validity of this classification scheme.

2. An ecologist visiting an island finds two very closely related species of birds, one of which has a slightly larger bill than the other. Interpret this finding with respect to the competitive exclusion principle and the ecological niche, and explain both concepts.

3. Think about the case of the camouflaged frogfish and its prey. As the frogfish sits camouflaged on the ocean floor, wiggling its lure, a small fish approaches the lure and is eaten, while a very large predatory fish fails to notice the frogfish. Describe all the possible types of community interactions and adaptations that these organisms exhibit. Remember that predators can also be prey and that community interactions are complex!

4. Design an experiment to determine whether the kangaroo is a keystone species in the Australian outback.

5. Why is it difficult to study succession? Suggest some ways you would approach this challenge for a few different ecosystems.

FOR MORE INFORMATION

Amos, W. H. "Hawaii's Volcanic Cradle of Life." *National Geographic*, July 1990. A naturalist explores succession on lava flows.

Enserink, M. "Biological Invaders Sweep In"; Kaiser, J. "Stemming the Tide of Invading Species"; and Malkoff, D. "Fighting Fire with Fire." *Science*, September 17, 1999. A series of articles covering the problems posed by invasive species.

Freindel, S. "If All the Trees Fall in the Forest ... " *Discover*, December 2002. The imported fungus responsible for chestnut blight killed 3.5 billion chestnuts in the 1920s. Now, a new invasive species of fungus threatens a variety of native trees, including live oaks and redwoods.

Gutin, J. C. "Purple Passion." *Discover*, August 1999. The invasive plant called purple loosestrife can grow 10 feet tall. Introduced to the U.S. East Coast 200 years ago, it is now rapidly spreading westward, threatening native species.

Harder, B. "Stemming the Tide." *Science News*, April 13, 2002. How can ships' ballast water be prevented from spreading invasive species such as the zebra mussel?

Power, M., et al. "Challenges in the Quest for Keystones." *Bioscience*, September 1996. A comprehensive review of the importance of keystone species and the challenges of studying them.

Stewart, D. "Good Bugs Gone Bad." *National Wildlife*, August/September 2005. A biocontrol run amuck, the Asian ladybug both eats and displaces native ladybugs.

Withgott, J. "California Tries to Rub Out the Monster of the Lagoon." *Science*, March 22, 2002. An invasive tropical alga now blankets coastal areas in the Mediterranean and Australia, while California desperately tries to keep it from invading the U.S. West Coast.

28

How Do Ecosystems Work?

A grizzly bear intercepts a salmon on its spawning journey as it struggles up a waterfall in an attempt to reach the same streambed where it hatched years earlier.

CASE STUDY WHEN THE SALMON RETURN

SOCKEYE SALMON of the Pacific Northwest have a remarkable life cycle. Hatching in shallow depressions in the gravel bed of a swiftly flowing stream, they follow the stream's path into ever-larger rivers that finally enter the ocean. Emerging into estuaries—wetlands where fresh water and salt water mix—the salmon's remarkable physiology allows the fish to adapt to the change to salt water before they reach the sea. The small percentage of young salmon that evade predators grows to adulthood, feeding on crustaceans and smaller fish.

Years later, their bodies undergo another transformation. As they reach sexual maturity, a compelling instinctive drive—still poorly understood despite decades of research—lures them back to fresh water, but not any stream or river will do. The salmon swim along the coast (probably navigating by sensing Earth's magnetic field) until the unique scent of their home stream entices them to swim inland. Battling swift currents, leaping up small waterfalls, undulating over shallow sandbars, and evading human traps, they carry their precious cargo of

sperm and eggs back home to renew the cycle of life. The fishes' journey back to their birthplace is remarkable in another way. Nutrients almost always flow downstream, carried from the land into the ocean; the salmon, filled with muscle and fat acquired from feeding in the ocean, not only battle against the flow of the current in their upstream journey; they also reverse the usual movement of nutrients. What awaits the salmon at their journey's end? How does their journey affect the web of life upstream?

28.1   WHAT ARE THE PATHWAYS OF ENERGY AND NUTRIENTS?

The activities of life, from the migration of the salmon to active transport of molecules through a cell membrane, are powered by the *energy* of sunlight. The molecules of life are constructed of chemical building blocks that are obtained as *nutrients* from the environment. Solar energy continuously bombards Earth, is used and transformed in the chemical reactions that power life, and is ultimately converted to heat energy that radiates back into space. Chemi-cal nutrients, in contrast, remain on Earth. While they may change in form and distribution, and may even be transported among different ecosystems, nutrients are constantly recycled. Thus, two basic laws underlie ecosystem function. First, energy moves through the communities within ecosystems in a continuous one-way flow, needing constant replenishment from an outside source, the sun. Second, nutrients constantly cycle and recycle within and among ecosystems (**FIG. 28-1**). These laws shape the complex interactions among populations within ecosystems, and between communities and their abiotic environment.

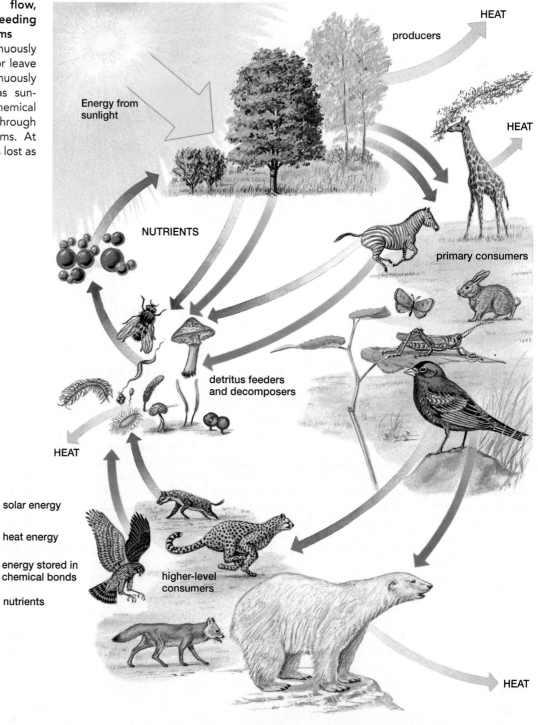

FIGURE 28-1 Energy flow, nutrient cycling, and feeding relationships in ecosystems Nutrients, which are continuously recycled, neither enter nor leave the cycle. Energy, continuously supplied to producers as sunlight, is captured in chemical bonds and transferred through various levels of organisms. At each level, some energy is lost as heat.

HEAT

producers

Energy from sunlight

HEAT

NUTRIENTS

primary consumers

detritus feeders and decomposers

HEAT

solar energy

heat energy

energy stored in chemical bonds

nutrients

higher-level consumers

HEAT

28.2 HOW DOES ENERGY FLOW THROUGH COMMUNITIES?

Energy Enters Communities Through Photosynthesis

Ninety-three million miles away, the sun fuses hydrogen molecules into helium molecules, releasing tremendous quantities of energy. A tiny fraction of this energy reaches Earth in the form of electromagnetic waves, including heat, light, and ultraviolet energy. Of the energy that reaches Earth, much is reflected by the atmosphere, clouds, and Earth's surface. Still more is absorbed as heat by Earth and its atmosphere, leaving only about 1% to power all life. Of this 1%, which reaches Earth's surface as light, green plants and other photosynthetic organisms capture 3% or less. The teeming life on this planet is thus supported by less than 0.03% of the energy reaching Earth from the sun.

During photosynthesis (see Chapter 7), pigments such as chlorophyll absorb specific wavelengths of sunlight. This solar energy is then used in reactions that store energy in chemical bonds, producing sugar and other high-energy molecules (**FIG. 28-2**). Photosynthetic organisms, from mighty oak trees to single-celled diatoms in the ocean, are called **autotrophs** (Greek, "self-feeders") or **producers**, because they produce food for themselves using nonliving nutrients and sunlight. In doing so, they directly or indirectly produce food for nearly all other forms of life as well. Organisms that cannot photosynthesize, called **heterotrophs** (Greek, "other-feeders") or **consumers**, must acquire energy and many of their nutrients prepackaged in the molecules that comprise the bodies of other organisms.

The quantity of life that a particular ecosystem can support is determined by the energy captured by the producers in that ecosystem. The energy that photosynthetic organisms store and make available to other members of the community over a given period is called **net primary productivity**. Net primary productivity can be measured in units of energy (calories) stored by autotrophs in a specified unit of area (such as square yards, acres, or hectares) during a specified time (often a year). Primary productivity can also be measured as the **biomass**, or dry weight of

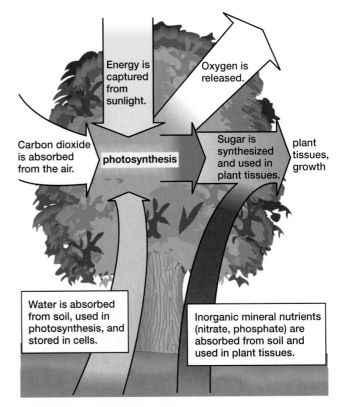

FIGURE 28-2 Primary productivity
Photosynthetic organisms, which capture solar energy and acquire inorganic nutrients from their environment, ultimately provide essentially all of the energy and most of the nutrients for organisms in higher trophic levels.

organic material stored in producers that is added to the ecosystem per unit area during a specified time. The productivity of an ecosystem is influenced by many environmental variables, including the amount of nutrients available to the producers, the amount of sunlight reaching them, the availability of water, and the temperature. In the desert, for example, lack of water limits productivity; in the open ocean, light is a limiting factor in deep water, and lack of nutrients limit productivity in surface water. When resources are abundant, as in estuaries and tropical rain forests, productivity is high. Some average productivity measurements for a variety of ecosystems are shown in **FIGURE 28-3**.

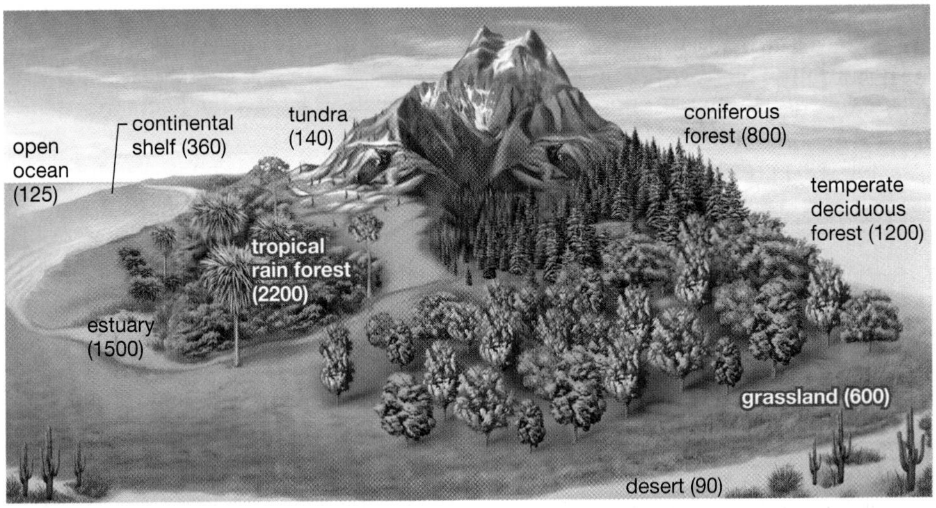

FIGURE 28-3 Ecosystem productivities compared
The average net primary productivity, in grams of organic material per square meter per year, of some terrestrial and aquatic ecosystems is illustrated. Notice the enormous differences. QUESTION What factors contribute to these differences in productivity?

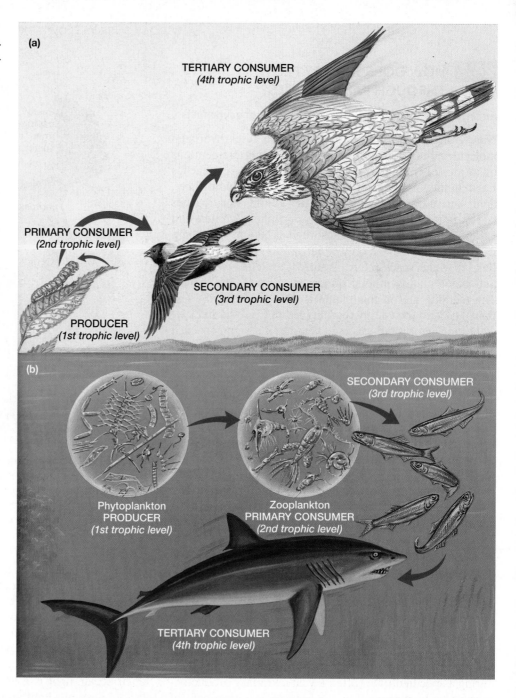

FIGURE 28-4 Food chains
(a) A simple terrestrial food chain.
(b) A simple marine food chain.

(a)

TERTIARY CONSUMER
(4th trophic level)

PRIMARY CONSUMER
(2nd trophic level)

SECONDARY CONSUMER
(3rd trophic level)

PRODUCER
(1st trophic level)

(b)

SECONDARY CONSUMER
(3rd trophic level)

Phytoplankton
PRODUCER
(1st trophic level)

Zooplankton
PRIMARY CONSUMER
(2nd trophic level)

TERTIARY CONSUMER
(4th trophic level)

Energy Is Passed from One Trophic Level to Another

Energy flows through communities from photosynthetic producers through several levels of consumers. Each category of organisms is called a **trophic level** (literally, "feeding level"). Producers—from redwood trees to cyanobacteria—form the first trophic level, obtaining their energy directly from sunlight (see Fig. 28-1). Consumers occupy several trophic levels. Some consumers feed directly and exclusively on producers, the most abundant living energy source in any ecosystem. These **herbivores** (meaning "plant eaters"), ranging from grasshoppers to giraffes, are also called **primary consumers**; they form the second trophic level. **Carnivores** (meaning "meat eaters")—such as the spider, eagle, and wolf—are predators that feed primarily on primary consumers. Carnivores, also called **secondary consumers**, form

the third trophic level. Some carnivores occasionally eat other carnivores; when doing so, they occupy the fourth trophic level, **tertiary consumers.**

Food Chains and Food Webs Describe the Feeding Relationships Within Communities

To illustrate who feeds on whom in a community, it is common to identify a representative of each trophic level that eats a representative of the level below it. This linear feeding relationship is called a **food chain.** As illustrated in **FIGURE 28-4**, different ecosystems have radically different food chains.

Natural communities, however, rarely contain well-defined groups of primary, secondary, and tertiary consumers. A **food web** shows many interconnecting food chains and more accurately describes the actual feeding relationships within a given community (**FIG. 28-5**). Some

FIGURE 28-5 A simplified grassland food web

animals, such as raccoons, bears, rats, and humans, are **omnivores** (Latin, "eating all")—that is, at different times they act as primary, secondary, and occasionally tertiary (third-level) consumers. Many carnivores will eat either herbivores or other carnivores, thus acting as secondary or tertiary consumers, respectively. An owl, for instance, is a secondary consumer when it eats a mouse, which feeds on plants; but it is a tertiary consumer when it eats a shrew, which feeds on insects. A shrew that eats a carnivorous insect is a tertiary consumer, and the owl that fed on the shrew is then a quaternary (fourth-level) consumer. When it is digesting a spider, a carnivorous plant, such as the sundew, can "tangle the web" hopelessly by serving simultaneously as a photosynthetic producer and a secondary consumer.

Detritus Feeders and Decomposers
Release Nutrients for Reuse

Among the most important strands in the food web are *detritus feeders* and *decomposers*. **Detritus feeders** are an army of mostly small and often unnoticed animals that live on the refuse of life: molted exoskeletons, fallen leaves, wastes, and dead bodies (*detritus* means "debris"). The network of detritus feeders is extremely complex and includes earthworms, mites, protists, centipedes, some insects, a land-dwelling crustacean called a pill bug or roly-poly, nematode worms, and even a few large vertebrates such as vultures. They consume dead organic matter, extract some of the energy stored within it, and excrete it in a further decomposed state. Their excretory products serve as food for other detritus feeders and for decomposers. **Decomposers** are primarily fungi and bacteria that digest food outside their bodies by secreting digestive enzymes into the environment. The black or gray fuzz you may notice on tomatoes and bread crusts left too long in your refrigerator are fungal decomposers hard at work. They absorb the nutrients and energy-rich compounds that they need, releasing those that remain.

Through the activities of detritus feeders and decomposers, the bodies and wastes of living organisms are reduced to simple molecules—such as carbon dioxide, water, minerals, and organic molecules—that return to the atmosphere, soil, and water. By liberating nutrients for reuse, detritus feeders and decomposers form a vital link in the nutrient cycles of ecosystems. In some ecosystems, such as deciduous forests, more energy passes through the detritus feeders and decomposers than through the primary, secondary, or tertiary consumers.

What would happen if detritus feeders and decomposers disappeared? This portion of the food web, although inconspicuous, is absolutely essential to life on Earth. Without it, communities would gradually be smothered by accumulated wastes and dead bodies. The nutrients stored in these bodies would be unavailable to enrich the soil. The quality of the soil would become poorer and poorer until plant life could no longer be sustained. With plants eliminated, energy would cease to enter the community; the higher trophic levels, including humans, would disappear as well.

Energy Transfer Through Trophic Levels Is Inefficient

As discussed in Chapter 6, a basic law of thermodynamics is that energy use is never completely efficient. For example, as your car burns gasoline, about 75% of the energy released is lost as heat. This is also true in living systems. For example, splitting the chemical bonds of adenosine triphosphate (ATP) to cause muscular contraction produces heat as a by-product; this is why walking briskly on a cold day will warm you. Small amounts of waste heat are produced by all the biochemical reactions that keep cells alive. Compost piles can achieve internal temperatures of over 130°F (54.4°C) as a result of the heat liberated by decomposer microorganisms.

Energy transfer from one trophic level to the next is also quite inefficient. When a caterpillar (a primary consumer) eats the leaves of a tomato plant (a producer), only some of the solar energy originally trapped by the plant is available to the insect. Some energy was used by the plant for growth and maintenance, and more was lost as heat during these processes. Some energy was converted into the chemical bonds of molecules such as cellulose, which the caterpillar cannot digest. Therefore, only a fraction of the energy captured by the first trophic level is available to organisms in the second trophic level. The energy consumed by the caterpillar is in turn partially used to power crawling and the gnashing of mouthparts. Some is used to construct the indigestible exoskeleton, and much is given off as heat. All this energy is unavailable to the songbird in the third trophic level when it eats the caterpillar. The bird loses energy as body heat, uses more in flight, and converts a considerable amount into indigestible feathers, beak, and bone. All this energy will be unavailable to the hawk that catches it. A simplified model of energy flow through the trophic levels in a deciduous forest ecosystem is illustrated in **FIGURE 28-6**.

Energy Pyramids Illustrate Energy Transfer
Between Trophic Levels

Studies of a variety of communities indicate that the net transfer of energy between trophic levels is roughly 10% efficient, although transfer among levels within different communities varies significantly. This means that, in general, the energy stored in primary consumers (herbivores) is only about 10% of the energy stored in the bodies of producers. In turn, the bodies of secondary consumers possess roughly 10% of the energy stored in primary consumers. In other words, for every 100 calories of solar energy captured by grass, only about 10 calories are converted into herbivores, and only 1 calorie is converted into carnivores. This inefficient energy transfer between trophic levels is called the "10% law." An **energy pyramid**, which shows maximum energy at the base and steadily diminishing

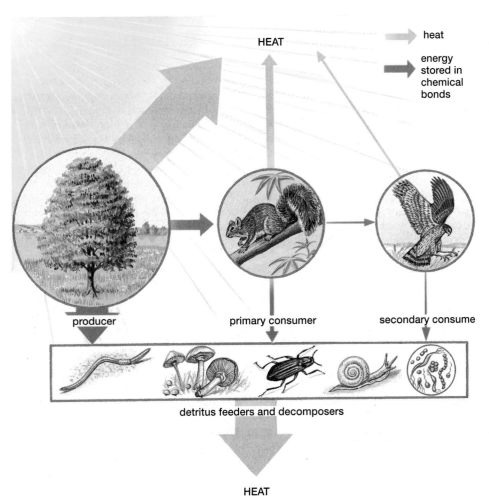

FIGURE 28-6 Energy transfer and loss
The width of the arrows is roughly proportional to the quantity of energy transferred between trophic levels as chemical energy or lost as heat in a forest community. QUESTION Why is so much energy lost as heat? Explain this effect in terms of the second law of thermodynamics (introduced in Chapter 6), and relate it to the energy pyramid in Figure 28-7.

heat

energy stored in chemical bonds

HEAT

producer

primary consumer

secondary consume

detritus feeders and decomposers

HEAT

amounts at higher levels, illustrates the energy relationships between trophic levels graphically (**FIG. 28-7**). Ecologists sometimes use biomass as a measure of the energy stored at each trophic level. Because the dry weight of organisms' bodies at each trophic level is roughly proportional to the amount of energy stored in the organisms at that level, a *biomass pyramid* for a given community often has the same general shape as its energy pyramid.

What does this mean for community structure? If you wander through an undisturbed ecosystem, you will notice that the predominant organisms are plants. Plants have the most energy available to them, because they trap it directly from sunlight. The most abundant animals will be those that feed on plants, and carnivores will be relatively rare. The inefficiency of energy transfer also has important implications for human food production. The lower the trophic level we utilize, the more food energy is available to us; in other words, far more people can be fed on grain than on meat.

An unfortunate side effect of the inefficiency of energy transfer, coupled with human production and release of toxic chemicals, is that certain persistent toxic chemicals become concentrated in the bodies of carnivores, including people. This effect is described in "Earth Watch: Food Chains Magnify Toxic Substances."

tertiary consumer
(1 calorie)

secondary consumer
(10 calories)

primary consumer
(100 calories)

producer
(1000 calories)

FIGURE 28-7 An energy pyramid for a prairie ecosystem
The width of each rectangle is proportional to the energy stored at that trophic level. A biomass pyramid for this ecosystem would look quite similar.

EARTH WATCH   Food Chains Magnify Toxic Substances

In the 1940s, the properties of the new insecticide DDT seemed close to miraculous. In the Tropics, DDT saved millions of lives by killing the mosquitoes that spread malaria. Increased crop yields resulting from DDT's destruction of insect pests saved millions more from starvation. But DDT was entering food chains and unraveling the complex web of life. In the mid-1950s, the World Health Organization sprayed DDT on the island of Borneo to control malaria. A caterpillar that fed on the thatched roofs of houses was relatively unaffected, but a predatory wasp that fed on the caterpillar was destroyed. Eaten by the burgeoning caterpillar population, thatched roofs collapsed. Gecko lizards that ate poisoned insects accumulated high concentrations of DDT in their bodies. Both they, and the village cats that ate the geckos, died of DDT poisoning. With the cats eliminated, the rat population exploded. Villages were threatened with an outbreak of plague, carried by the uncontrolled rats. The outbreak was avoided by airlifting new cats to the villages.

In the U.S., wildlife biologists during the 1950s and 1960s witnessed an alarming decline in populations of several predatory birds, especially fish-eaters such as bald eagles, cormorants, ospreys, and brown pelicans. The decline pushed some, including the brown pelican and the bald eagle, close to extinction (all have shown significant recovery since the pesticide was banned in the U.S. in 1973). The aquatic ecosystems supporting these birds had been sprayed with relatively low amounts of DDT to control insects. In the tissues of the top predators, scientists found concentrations of DDT up to 1 million times greater than in the water where their fish prey lived. The birds were victims of **biological magnification**, the process by which toxic substances accumulate in increasingly high concentrations in animals occupying higher trophic levels.

DDT and other substances that undergo biological magnification share two properties that make them dangerous. First, decomposer organisms cannot readily break them down into harmless substances—that is, they are not **biodegradable**. Second, they tend to be stored in the body, particularly in fat, accumulating over the years in the bodies of long-lived animals. Exposure to high levels of pesticides and other persistent pollutants has been linked to some types of cancer, infertility, heart disease, suppressed immune function, and neurological damage in children.

Today, mercury contamination is a particular cause for concern. It is an extremely potent neurotoxin that bioaccumulates in muscle as well as fat. Its level in predatory fish consumed by people has become high enough that the U.S. Food and Drug Administration has advised women of childbearing age and young children not to eat swordfish and shark, and to limit their consumption of albacore tuna, because these long-lived ocean predators have accumulated enough mercury to pose a possible health hazard. In the U.S., coal-fired power plants are the largest single source of mercury contamination; but atmospheric mercury can be wafted thousands of miles and be deposited in what should be pristine environments, such as the Arctic. About half the mercury deposited on U.S. soil and water comes from overseas. Researchers have found neurological damage, including lowered IQ, corresponding to elevated mercury levels in mother's hair samples in two different island populations that consume high levels of marine fish and mammals. Inuit natives living north of the Arctic Circle have high levels of mercury and other bioaccumulating pollutants from consuming large quantities of predatory marine mammals and fish.

A class of chemicals called *endocrine disruptors*—including some widely used pesticides, plasticizers (which make plastic flexible), and flame retardants—have become widespread in the environment. Like DDT, they accumulate in fat and either mimic or interfere with the actions of animal hormones. There is compelling evidence that these chemicals are interfering with the reproduction and development of fish (including salmon), fish-eating birds such as cormorants (**FIG. E28-1**), frogs, salamanders, alligators, and many other animals. Endocrine disruptors are also suspected of causing reduced sperm counts in people.

To reduce human health hazards as well as loss of wildlife, we must understand the properties of pollutants and the workings of food webs. When we eat tuna or swordfish, for example, we act as tertiary or even quaternary consumers, and so we are vulnerable to bioaccumulating substances. In addition, the long human life span provides more time for substances stored in our bodies to accumulate to toxic levels.

FIGURE E28-1 The price of pollution
Deformities such as the twisted beak of this double-crested cormorant from Lake Michigan have been linked to bioaccumulating chemicals. Abnormalities of the reproductive and immune systems are also common in many types of organisms exposed to these pollutants. Predatory animals are especially vulnerable owing to biological magnification.

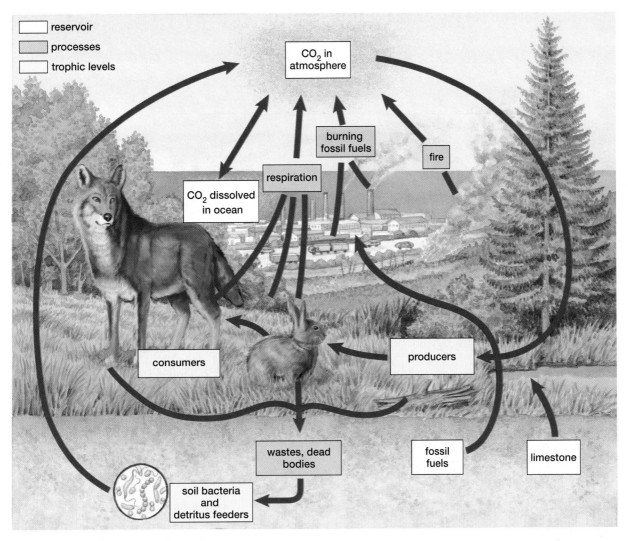

Legend:
- □ reservoir
- ▨ processes
- □ trophic levels

CO₂ in atmosphere

burning fossil fuels

fire

respiration

CO₂ dissolved in ocean

consumers

producers

wastes, dead bodies

fossil fuels

limestone

soil bacteria and detritus feeders

FIGURE 28-8 The carbon cycle

28.3 HOW DO NUTRIENTS MOVE WITHIN AND AMONG ECOSYSTEMS?

In contrast to the energy of sunlight, nutrients do not flow down onto Earth in a steady stream from above. Essentially the same pool of nutrients has been supporting life for more than 3 billion years. *Nutrients* are elements and small molecules that form the chemical building blocks of life. Some, called *macronutrients*, are required by organisms in large quantities. These include water, carbon, hydrogen, oxygen, nitrogen, phosphorus, sulfur, and calcium. *Micronutrients*, including zinc, molybdenum, iron, selenium, and iodine, are required only in trace quantities. **Nutrient cycles**, also called **biogeochemical cycles**, describe the pathways these substances follow as they move from communities to nonliving portions of ecosystems and back again to communities.

The ultimate sources and storage sites of nutrients are called **reservoirs**. The major reservoirs are generally in the nonliving, or abiotic, environment. For example, carbon has several major reservoirs: it is stored as carbon dioxide gas in the atmosphere, in dissolved form in oceans, in rock as limestone, and as fossil fuels underground. In the following sections, we describe the cycles of carbon, nitrogen, phosphorus, and water.

Carbon Cycles Through the Atmosphere, Oceans, and Communities

Chains of carbon atoms form the framework of all organic molecules, the building blocks of life. Carbon enters the living community through capture of carbon dioxide (CO_2) during photosynthesis by producers. On land, producers acquire CO_2 from the atmosphere, where it represents a mere 0.036% of all atmospheric gases. Aquatic producers in the ocean, such as seaweeds and diatoms, find abundant CO_2 for photosynthesis dissolved in the water; in fact, far more CO_2 is stored in the oceans than in the atmosphere. Producers return some CO_2 to the atmosphere and ocean during cellular respiration and incorporate the rest into their bodies. Primary consumers, such as cows, shrimp, or tomato hornworms, eat the producers and acquire the carbon stored in their tissues. These herbivores also release some carbon through respiration and store the rest, which is sometimes consumed by organisms in higher trophic levels. All living things eventually die, and their bodies are broken down by detritus feeders and decomposers. Cellular respiration by these organisms returns CO_2 to the atmosphere and oceans. Carbon dioxide passes freely between these two great reservoirs (**FIG. 28-8**).

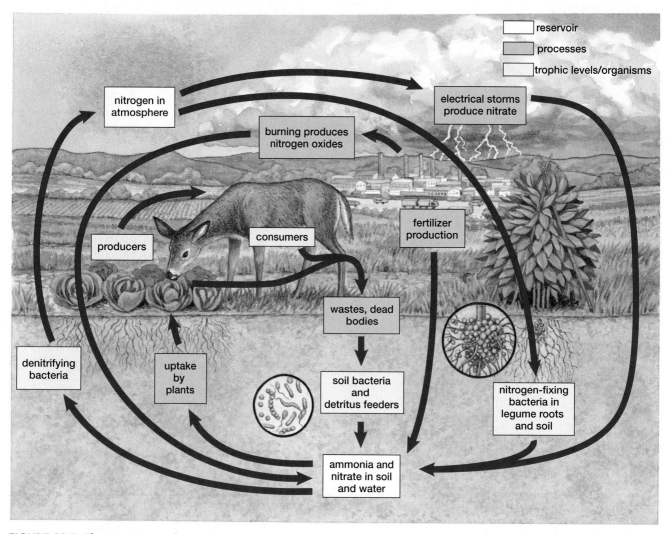

FIGURE 28-9 The nitrogen cycle
QUESTION What incentives caused people to capture nitrogen from the air and pump it into the nitrogen cycle? What are some consequences of human augmentation of the nitrogen cycle?

Some carbon cycles much more slowly. For example, mollusks and microscopic marine organisms extract CO_2 dissolved in water and combine it with calcium to form calcium carbonate ($CaCO_3$) from which they construct their shells. After death, the shells of these organisms collect in undersea deposits, are buried, and may eventually be converted to limestone. Geological events may expose the limestone, which dissolves gradually as water runs over it, making the carbon available to living organisms once again.

Another long-term portion of the carbon cycle is the production of fossil fuels. **Fossil fuels** form from the remains of ancient forms of life. The carbon in the organic molecules of these prehistoric organisms was transformed by high temperatures and pressures over millions of years into coal, oil, and natural gas. The energy of prehistoric sunlight is also trapped in these fossil fuels, captured by ancient autotrophs and then passed upward through various trophic levels before being sequestered in the high-energy hydrocarbons that we burn today. When people burn fossil fuels to use this stored energy, CO_2 is released into the atmosphere. In addition to the burning of fossil

fuels, human activities such as cutting and burning Earth's great forests (where much carbon is stored) increase the amount of CO_2 in the atmosphere, as we will discuss later in this chapter.

The Major Reservoir for Nitrogen Is the Atmosphere

The atmosphere contains about 78% nitrogen gas (N_2) and is thus the major reservoir for this important nutrient. Nitrogen is a crucial component of proteins, many vitamins, and the nucleic acids DNA and RNA. Interestingly, neither plants nor animals can extract this gas from the atmosphere. Instead, plants must be supplied with nitrate (NO_3^-) or ammonia (NH_3). But how is atmospheric nitrogen converted to these molecules? Ammonia is synthesized by certain bacteria that live in water and soil. Some have entered a symbiotic association with plants called *legumes* (including alfalfa, soybeans, clover, and peas), upon which the bacteria live in special swellings on the roots. Legumes are extensively planted on farms, where they fertilize the soil. Decomposer bacteria can also produce ammonia from the amino acids and urea found in dead bodies and wastes. Still other bacteria convert ammonia to nitrate.

Nitrogen is also combined with oxygen by nonbiological processes: electrical storms and the combustion of forests and fossil fuels produce nitrogen oxides. Synthetic fertilizers often contain ammonia, nitrate, or both. Plants incorporate the nitrogen from ammonia and nitrate into amino acids, proteins, nucleic acids, and vitamins. These nitrogen-containing molecules from the plant are eventually consumed by primary consumers, detritus feeders, or decomposers. As it is passed through the food web, some of the nitrogen is released in wastes and dead bodies, which decomposer bacteria in soil or water convert back to nitrate and ammonia. This form of nitrogen is available to plants; nitrates and ammonia in soil and water constitute a second reservoir. The nitrogen cycle is completed by a continuous return of nitrogen to the atmosphere by *denitrifying bacteria*. These residents of wet soil, swamps, and estuaries break down nitrate, releasing nitrogen gas back to the atmosphere (**FIG. 28-9**).

Nitrogen compounds produced by people now dominate the nitrogen cycle, creating serious environmental concerns. When they enter ecosystems, excess nitrogen compounds may change the composition of plant communities by overfertilizing them, or they may destroy forests and freshwater communities by acidifying the environment, as discussed later in this chapter.

The Phosphorus Cycle Has No Atmospheric Component

Phosphorus is a crucial component of biological molecules, including energy transfer molecules (ATP and NADP), nucleic acids, and the phospholipids of cell membranes. It is also a major component of vertebrate teeth and bones. In contrast to the cycles of carbon and nitrogen, the phosphorus cycle has no atmospheric component. The major storage site for phosphorus in ecosystems is rock, where it is bound to oxygen in the form of phosphate. As phosphate-rich rocks are exposed and eroded, rainwater dissolves the phosphate. Dissolved phosphate is readily absorbed through the roots of plants and by other autotrophs, such as photosynthetic protists and cyanobacteria, which incorporate it into biological molecules. From these producers, phosphorus is passed through food webs (**FIG. 28-10**). At each level, excess phosphate is excreted. Ultimately, detritus feeders and decomposers return the phosphorus that remains in dead bodies back to the soil and water in the form of phosphate. It may then be reabsorbed by autotrophs, or it may become bound to sediment and eventually reincorporated into rock.

Some of the phosphate dissolved in fresh water is carried to the oceans. Although much of this phosphate ends up in marine sediments, some is absorbed by marine producers and eventually incorporated into the bodies of invertebrates and fish. Some of these, in turn, are consumed

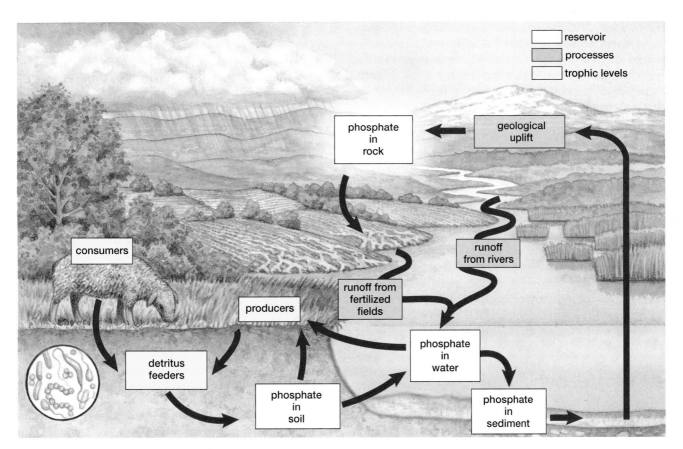

FIGURE 28-10 The phosphorus cycle

FIGURE 28-11 The hydrologic cycle

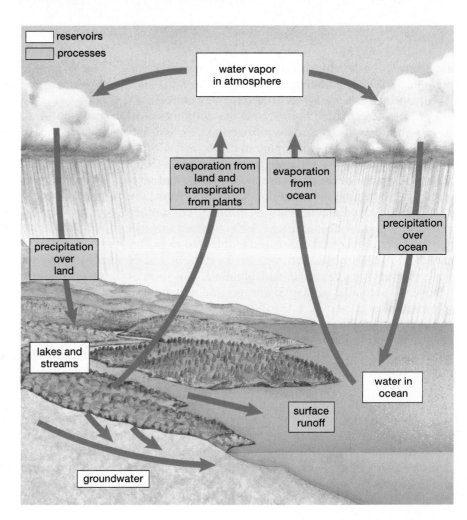

by seabirds, which excrete large quantities of phosphorus back onto the land. At one time, the guano (droppings) deposited by seabirds along the western coast of South America was mined; it provided a major source of the world's phosphorus. Phosphate-rich rock is also mined, and the phosphate is incorporated into fertilizer. Soil that erodes from fertilized fields carries large quantities of phosphates into lakes, streams, and the ocean, where it stimulates the growth of producers. In lakes, phosphorus-rich runoff from land can stimulate such a rich growth of algae and bacteria that the natural community interactions of the lake are disrupted.

Most Water Remains Chemically Unchanged During the Hydrologic Cycle

The water cycle, or **hydrologic cycle** (**FIG. 28-11**), differs from most other nutrient cycles in that most water remains in the form of water throughout the cycle and is not used in the synthesis of new molecules. The major reservoir of water is the ocean, which covers about three-quarters of Earth's surface and contains more than 97% of Earth's water. Another 2% is trapped in ice, leaving only 1% as liquid fresh water. The hydrologic cycle is driven by solar energy, which evaporates water, and by gravity, which draws it back to Earth in the form of precipitation (rain, snow, sleet, and dew). Most evaporation occurs from the oceans, and much water returns directly to them as precipitation. Water falling on land takes various paths. Some is evaporated from the soil, lakes, and streams. A portion runs off the land back to the oceans, and a small amount enters underground reservoirs. Because the bodies of living things are roughly 70% water, some of the water in the hydrologic cycle enters the living communities of ecosystems. It is absorbed by the roots of plants; much of this water is evaporated back to the atmosphere from plants' leaves. A small amount is combined with carbon dioxide during photosynthesis to produce high-energy molecules. Eventually these molecules are broken down during cellular respiration, releasing water back to the environment. Consumers get water from their food or by drinking.

Lack of Access to Water for Irrigation and Drinking Is a Growing Human Problem

As the human population has grown, fresh water has become scarce in many regions of the world. Additionally, contaminated, untreated drinking water is a major prob-

lem in developing countries, where over 1 billion people drink it. Impure water spreads diseases that kill millions of children each year. In both Africa and India, where water contamination poses significant threats, people are starting to use sunlight to kill disease-causing organisms. They place water in plastic bottles and shake them to increase the oxygen levels in the water. Then they put the bottles in a sunny spot, allowing the combination of oxygen, warmth, and ultraviolet (UV) light to create free radicals that kill bacteria. With no technology other than plastic bottles, these people are generating safe drinking water.

Currently, about 10% of the world's food is grown on cropland irrigated with water drawn from aquifers, which are natural underground reservoirs. Unfortunately, in many areas of the world—including China, India, Northern Africa, and the midwestern United States—this groundwater is being "mined" for agriculture; that is, it is removed faster than it is replenished. Parts of the High Plains aquifer, which extends from the Texas Panhandle north to South Dakota, have been depleted by about 50%. In India, two-thirds of crops are grown using underground water for irrigation, draining aquifers far faster than they are being replenished. One promising solution is to devise ways of trapping the heavy monsoon rains, whose water usually pours into rivers and eventually into the ocean. People of a village in India have found that by digging a series of holding ponds, they can capture rainwater that would formerly run off. Their system allows the water to

FIGURE 28-12 A natural substance out of place
This bald eagle was killed by an oil spill off the coast of Alaska.

FIGURE 28-13 Acid deposition is corrosive
This limestone statue at Rheims Cathedral in France is being dissolved by acid deposition.

percolate down into the soil and helps replenish the underground water supplies. During the dry season, the people can then tap these supplies for irrigation.

28.4 WHAT CAUSES "ACID RAIN"?

Many of the environmental problems that plague modern society have resulted from human interference in ecosystem function. Primitive peoples were sustained solely by the energy flowing from the sun, and they produced wastes that were readily taken back into the nutrient cycles. But as the population grew and technology increased, humans began to act more and more independently of these natural processes. The Industrial Revolution, which began in earnest in the mid-nineteenth century, resulted in a tremendous increase in our reliance on energy from fossil fuels (rather than from sunlight) for heat, light, transportation, industry, and agriculture. In mining and transporting these fuels, we have exposed ecosystems to a variety of substances that are foreign and often toxic to them (**FIG. 28-12**). In the following sections, we describe two environmental problems of global proportion that are primarily a direct result of human reliance on fossil fuels: acid deposition and global warming.

Overloading the Sulfur and Nitrogen Cycles Causes Acid Deposition

Although volcanoes, hot springs, and decomposer organisms all release sulfur dioxide, human industrial activities, primarily burning fossil fuels in which sulfur is trapped, account for about 75% of the sulfur dioxide emissions worldwide. This is far more than natural ecosystems can absorb and recycle. The nitrogen cycle is also being overwhelmed. Although natural processes—such as the activity of nitrogen-fixing bacteria and decomposer organisms, fires, and lightning—produce nitrogen oxides and ammonia, about 60% of the nitrogen that is available to Earth's ecosystems now results from human activities. Burning of fossil fuels combines atmospheric nitrogen with oxygen, producing most of the emissions of nitrogen oxides. On farms, ammonia and nitrate are often supplied by chemical fertilizers produced by using the energy in fossil fuels to convert atmospheric nitrogen into compounds that plants can use.

Excess production of nitrogen oxides and sulfur dioxide was identified in the late 1960s as the cause of a growing environmental threat: *acid rain*, more accurately called **acid deposition**. When combined with water vapor in the atmosphere, nitrogen oxides and sulfur dioxide are converted to nitric acid and sulfuric acid, respectively. Days later, and often hundreds of miles from the source, these acids fall to Earth with rainwater, eating away at statues and buildings (**FIG. 28-13**), damaging trees and crops, and rendering lakes lifeless. Sulfuric acid may form particles that visibly cloud the air, even under dry conditions. In the U.S., the Northeast, Mid-Atlantic, Upper Midwest, and West regions, as well as the state of Florida, are the most vulnerable, because the rocks and soils that predominate there are less able to buffer acids.

Acid Deposition Damages Life in Lakes and Forests

In the Adirondack Mountains, acid rain has made about 25% of all the lakes and ponds too acidic to support fish. But by the time the fish die, much of the food web that sustains them has been destroyed. Clams, snails, crayfish, and insect larvae die first, then amphibians, and finally fish. The result is a crystal-clear lake—beautiful but dead. The impact is not limited to aquatic organisms. Acid rain also interferes with the growth and yield of many farm crops by leeching out essential nutrients such as calcium and potassium and killing decomposer microorganisms, thus preventing the return of nutrients to the soil. Plants, poisoned and deprived of nutrients, become weak and vulnerable to infection and insect attack. High in the Green Mountains of Vermont, scientists have witnessed the death of about half of the red spruce and beech trees and one-third of the sugar maples since 1965. The snow, rain, and heavy fog that commonly cloak these eastern mountaintops are highly acidic. At a monitoring station atop Mount Mitchell in North Carolina, the pH of fog has been recorded at 2.9—more acidic than vinegar (FIG. 28-14).

Acid deposition increases the exposure of organisms to toxic metals, including aluminum, mercury, lead, and cadmium, which are far more soluble in acidified water than in water of neutral pH. Aluminum dissolved from rock may inhibit plant growth and kill fish. The tap water in some households has been found to be dangerously contaminated with lead dissolved by acidic water from lead solder in old pipes. Fish in acidified water often have dangerous levels of mercury in their bodies, because mercury is subject to *biological magnification* as it is passes through trophic levels (see "Earth Watch: Food Chains Magnify Toxic Substances").

The Clean Air Act Has Significantly Reduced Sulfur, but Not Nitrogen, Emissions

In the U.S., amendments to the Clean Air Act in 1990 resulted in substantial reductions in emissions of both sulfur dioxide and nitrogen oxides from power plants. Overall sulfur emissions have decreased considerably throughout the U.S., improving air quality and rain acidity in some regions. Nitrogen oxide and ammonia release is not as strictly limited under the Clean Air Act. Although emissions of nitrogen oxides have dropped in some regions, atmospheric nitrogen compounds have shown a small overall increase, primarily because more gasoline is being burned by automobiles. Release of ammonia (NH_3) mostly from livestock and fertilizers, has increased by about 19% in the U.S. since 1985.

Unfortunately, damaged ecosystems recover slowly. A recent survey of Adirondack lakes found hopeful signs that about 60% of them are becoming less acid, although full recovery is still decades away. Some southeastern soils have become saturated with acid-releasing substances, and in these areas freshwater acid levels are increasing. High-elevation forests remain at risk throughout the U.S. Many scientists believe that considerable additional reductions in emissions, with far stricter controls on nitrogen emissions, will be needed to prevent further deterioration and allow the recovery of damaged ecosystems.

28.5   WHAT CAUSES GLOBAL WARMING?

Interfering with the Carbon Cycle Contributes to Global Warming

Between 345 million and 280 million years ago, huge quantities of carbon were diverted from the carbon cycle when, under the warm, wet conditions of the Carboniferous period, the bodies of prehistoric organisms were buried in sediments, escaping decomposition. Over time, heat and pressure converted their bodies, with their stored energy derived from the sun, into fossil fuels such as coal, oil, and natural gas. Without human intervention, this carbon would have remained underground. Beginning with the Industrial Revolution, however, we have increasingly relied on the energy stored in these fuels. One researcher estimates that a typical gas tank holds the transformed remains of 1000 tons of prehistoric life, largely microscopic phytoplankton. As we burn fossil fuels in our power plants, factories, and cars, we harvest the energy of prehistoric sunlight and release CO_2 into the atmosphere. Since 1850, the CO_2 content of the atmosphere has increased from 280 parts per million (ppm) to 381 ppm, or almost 36%. According to recent analysis of gas bubbles trapped in ancient Antarctic ice, the atmospheric CO_2 content is now

FIGURE 28-14 Acid deposition can destroy forests Acid rain and fog have destroyed this forest atop Mount Mitchell in North Carolina.

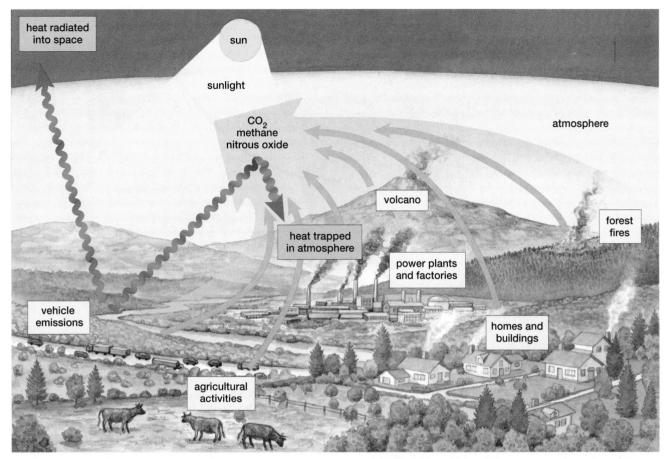

FIGURE 28-15 Increases in greenhouse gas emissions contribute to global warming
Incoming sunlight warms Earth's surface and is radiated back to the atmosphere. Greenhouse gases, released by natural processes but tremendously increased by human activities, absorb some of this heat, trapping it in the atmosphere.

about 27% higher than at any time during the past 650,000 years, and CO_2 is increasing at an unprecedented rate of about 1.5 parts per million yearly. Burning fossil fuels accounts for 80–85% of the CO_2 added to the atmosphere each year.

A second source of added atmospheric CO_2 is **deforestation**, which destroys tens of millions of forested acres annually and accounts for 15–20% of CO_2 emissions. Deforestation is occurring principally in the Tropics, where rain forests are rapidly being converted to marginal agricultural land. The carbon stored in the massive trees in these forests returns to the atmosphere (primarily through burning) after they are cut.

Collectively, human activities release almost 7 billion tons of carbon (in the form of CO_2) into the atmosphere each year. About half of this carbon is absorbed into the oceans, plants, and soil, while the remaining 3.5 billion tons remains in the atmosphere, fueling global warming.

Greenhouse Gases Trap Heat in the Atmosphere

Atmospheric CO_2 acts something like the glass in a greenhouse; it allows solar energy to enter, then absorbs and holds that energy once it has been converted to heat (FIG. 28-15). Several other **greenhouse gases** share this property, including nitrous oxide (N_2O) and methane (CH_4) which are both released by agricultural activities, landfills, wastewater treatment, coal mining, and burning fossil fuels. The **greenhouse effect**, which is the ability of greenhouse gases to trap the sun's energy in a planet's atmosphere as heat, is a natural process. By keeping our atmosphere relatively warm, the greenhouse effect allows life on Earth as we know it. However, there is overwhelming consensus among atmospheric scientists that human activities have amplified the natural greenhouse effect, producing a phenomenon called **global warming**.

Historical temperature records have revealed a global temperature increase paralleling the rise in atmospheric CO_2 (FIG. 28-16). Nineteen of the twenty hottest years on record have occurred since 1980, and the six hottest years happened between 1998 and 2005, which set an all-time record.

If greenhouse gas emissions are not curtailed, the Intergovernmental Panel on Climate Change (IPCC) predicts that average global temperatures will rise from the current

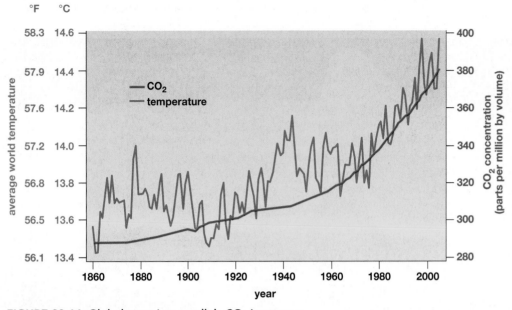

FIGURE 28-16 **Global warming parallels CO_2 increases**
The CO_2 concentration of the atmosphere (blue line) has increased steadily since 1860. Average global temperatures (red line) have also increased, roughly paralleling the increasing atmospheric CO_2.

average of about 58°F (14.4°C) to between 61.5°C (16°C) and 66°C (19°C) by the year 2100 (**FIG. 28-17**).

Seemingly small overall temperature changes can have enormous impacts. For example, average air temperatures during the peak of the last Ice Age (20,000 years ago) were only about 5°C lower than at present. The predicted rapid temperature increase is of particular concern because it is likely to exceed the rate at which natural selection can allow most organisms to adapt. The temperature change will not be distributed evenly worldwide; temperatures in both the arctic and parts of the U.S., for example, are predicted to increase considerably faster than the global average.

Global Warming Will Have Severe Consequences

As geochemist James White at the University of Colorado quipped, "If the Earth had an operating manual, the chapter on climate might begin with the caveat that the system has been adjusted at the factory for optimum comfort, so don't touch the dials." Earth has begun to experience the consequences of global warming, and all indications are that they will be severe and, in some regions, catastrophic.

A Meltdown Is Occurring

Throughout the world, ice is melting (see "Earth Watch: Poles in Peril"). Worldwide, glaciers are retreating and disappearing (**FIG. 28-18**). In Glacier National Park, where 150 glaciers once graced the mountainsides, only 35 remain, and scientists estimate that these may all disappear within the next 30 years. Greenland's ice sheet is melting at twice the rate of a decade ago, releasing 53 cubic miles (221 cubic kilometers) of water into the Atlantic annually. As polar ice caps and glaciers melt and ocean waters expand in response to atmospheric warming, sea levels will rise, threatening coastal cities and flooding coastal wetlands. Alaskan permafrost is melting, dumping mud into rivers, destroying salmon spawning grounds, and releasing CO_2 into the atmosphere as trapped organic matter decomposes. In Siberia, a region of frozen peat the size of France and Germany combined is melting, creating giant bogs that could release billions of tons of methane (a far more powerful heat-trapping gas than CO_2) into the atmosphere. Permafrost melting provides an example of positive feedback, in which an outcome of global warming, in this case the release of additional greenhouse gases, accelerates the warming process.

More Extremes in Weather Are Predicted

Many scientists believe that global warming is already affecting our weather. Recent studies have documented that during the past 35 years, both the intensity and duration of hurricanes have increased by 50%, doubling the number in

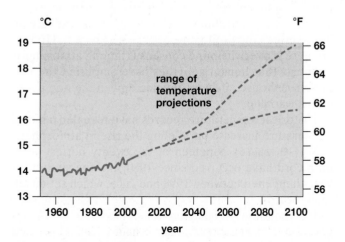

FIGURE 28-17 **Projected range of temperature increases**

FIGURE 28-18 Glaciers are melting
Photos taken from the same vantage point in 1904 (top) and 2004 (bottom) document the retreat of Carroll Glacier in Glacier Bay, Alaska.

drought and overly dense forests resulting from fire suppression in the past, have swept through large areas of the western U.S. and Alaska, releasing still more carbon dioxide into the atmosphere. As the world warms, tree distributions will change, based on their tolerance for heat. For example, sugar maples may disappear from northeastern U.S. forests, while some southeastern forests could be replaced by grasslands. Coral reefs, already stressed by human activities, are likely to suffer further damage from warmer waters, which drive out the symbiotic algae that provide them with energy. Corals are further threatened because, as the oceans absorb more CO_2, their waters are becoming more acidic, making it more difficult for corals to form their limestone skeletons.

Reports of changes keep coming in from throughout the world. The growing season in Europe has increased by more than 10 days over the past 28 years. Mexican Jays in southern Arizona are nesting 10 days earlier than they did in 1971. Many species of butterflies and birds have shifted their ranges northward. In the United Kingdom and the northeastern U.S., spring flowers are blooming earlier. While individual cases could be due to other factors, the cumulative weight of data from diverse sources worldwide provides strong evidence that warming-related biological changes have begun. Global warming is also predicted to increase the range of tropical disease-carrying organisms, such as malaria-transmitting mosquitoes, with negative consequences for human health.

the highest categories of wind speed and destruction (Categories 4 and 5), such as Hurricane Katrina, which devastated New Orleans in 2005. As the world warms, experts predict that droughts will last longer and be more severe, while other regions experience flooding. Scientists at the National Center for Atmospheric Research report that since the 1970s, the area of Earth impacted by severe drought has doubled from about 15% to about 30% as a result of increased temperatures and local decreases in rainfall. Agricultural disruption resulting from such extremes in weather could be disastrous for nations that are barely able to feed themselves.

Wildlife Is Affected

Biologists worldwide are documenting changes in plant and animal wildlife related to warming. The impact of global warming on forests could be profound. Fires, fuelled by

How Are People Responding to the Threat?

Under the landmark Kyoto Treaty, negotiated in 1997 and implemented in 2005, thirty-five industrialized countries have pledged to reduce their collective emissions of greenhouse gases to levels 5.2% below 1990 levels. The treaty exempts developing countries (where most of the

EARTH WATCH　　Poles in Peril

On Earth's opposite poles, Arctic and Antarctic ice is melting. The Antarctic Peninsula is uniquely vulnerable to global warming because its average year-round temperature hovers close to the freezing point of water. Over the past 50 years, the temperature around the peninsula has increased by about 4.5°F (about 2.5°C), far faster than the global average. Since 1995, over 2000 square miles of the ice shelf off the Antarctic Peninsula have disintegrated; based on ice core samples, scientists believe that these shelves had persisted for thousands of years. The loss of floating ice shelves has far-reaching consequences. The sea ice creates conditions that favor abundant growth of phytoplankton and algae. These primary producers provide food for larval krill, shrimplike crustaceans that are a keystone species in the Antarctic food web. Krill comprise a major portion of the diet of seals, penguins, and several species of whales. But over the past 30 years, krill populations in the southwest Atlantic have plummeted by about 80%. Researcher Angus Atkinson of the British Antarctic Survey hypothesizes that the decline is linked to the loss of sea ice. One likely scenario is that as the ice shelves shrink, algae that grows on the underside of the ice dwindles, and the krill that rely on it starve. Researchers are concerned that the impact of the krill loss may reverberate up the food chain, starving whales and seals and perhaps penguins as well. Adélie penguins spend their winters on the Antarctic ice shelves, feeding on krill. Although most Antarctic penguin populations remain healthy, researcher William Fraser, who has been studying Antarctic penguins for 30 years, reports that the Adélie penguin population in the western Antarctic Peninsula has lost about 10,000 breeding pairs since 1975.

At the far end of Earth, arctic temperatures have risen almost twice as rapidly as the world average, causing a 20–30% decrease in late-summer arctic sea ice during the past 30 years. Larger changes are projected for the next century, including temperature increases of about 7°F–14°F (4°C–8°C). In a disturbing example of positive feedback, melting ice will accelerate warming, because ice reflects 80–90% of the solar energy that hits it, but the ocean water exposed when ice disappears absorbs most solar energy, converting it to heat.

Arctic sea ice is critical for polar bears and for ringed seals, their major food source. Complete loss of sea ice, which some scientists predict within the next century, would mean almost certain extinction for polar bears in the wild. In Canada's Hudson Bay, sea ice is breaking up 3 weeks earlier than it did 30 years ago, depriving the bears of a prime opportunity to hunt ringed seals on the ice (**FIG. E28-2**). As a result, Hudson Bay polar bears now start their summers with 15% less weight (150 pounds for an adult male). Leaner females are producing fewer cubs with a lower survival rate, and the local bear population has declined by 22% since 1987. Hungry polar bears are increasingly invading northern Canadian and Alaskan towns, where they are sometimes shot. Polar bears are powerful swimmers; but as the ice floes have retreated, they have now been seen swimming 60 miles offshore, a far greater distance than usual for them. Several bears were spotted floating dead after a storm and are believed to have drowned, being too far out at sea to swim to safety.

The Arctic National Wildlife Refuge is the site of the largest number of onshore polar bear dens in Alaska. In late fall, polar bears congregate along the refuge shoreline. More bears are gathering there as ice retreats farther from the shoreline. Yet there is continuing political pressure in the U.S. to open the refuge to oil drilling. Ironically, polar bears are threatened not only with climate change, but with drilling for yet more oil to fuel the country's voracious appetite for fossil fuels, which will contribute to future global warming.

FIGURE E28-2　Polar bears on thin ice
The loss of arctic sea ice threatens the survival of polar bears.

world's population resides), whose emissions per person are extremely low, and whose attempts to increase living standards cannot currently be implemented without increases in greenhouse emissions. Although 159 countries have ratified (agreed to implement) the treaty, the U.S.—the world's largest generator of greenhouse gases—has refused (as of this writing) to do so. Encouragingly, several U.S. states (including California) and many city mayors have pledged to adopt Kyoto-type standards independently. Although worldwide efforts are essential, our individual choices, collectively, can also have a big impact, as described in "Links to Life: Making a Difference."

LINKS TO LIFE   Making a Difference

With less than 5% of the world's population, the U.S. is responsible for about 25% of world's greenhouse gases. The total greenhouse gas emissions produced by the U.S. amount to about 6 tons (5 metric tonnes) of carbon per person each year—more than any other country on Earth.

Can one person's actions make a difference? Jonathan Foley of the University of Wisconsin thinks so. He is on the cutting edge of climatological research, having led a team that developed one of the first computer models of global climate change to consider the impact of biological systems and human land use (such as converting forests to cropland) on climate. In 1998 Jon and his wife, Andrea, recognizing that greenhouse gas emissions and the resulting climate change can be significantly impacted by individual decisions and choices, made a choice of their own: to cut their family's energy use and carbon dioxide emissions in half. The Foleys and their young daughter lived in a five-bedroom house 30 miles from their work; Jon and Andrea each used a separate car to commute about 60 miles daily. First, they moved to a smaller house much closer to work. A visitor to the Foleys' new home—warm and cozy in winter and cool in summer—would never realize how little energy it consumes. Cracks have been sealed and the attic insulated. Every appliance has been chosen for energy efficiency. Compact fluorescent bulbs, using 75% less energy than incandescents, shed light throughout. Decorative ceiling fans reduce the need for summer air conditioning. Solar collectors supply over two-thirds of the family's water heating needs, while low-emittance window glass lets sunlight in while reducing heat loss in winter. The Foleys can now ride bicycles or take the bus to work, but they also enjoy their Toyota Prius hybrid gas/electric car—which gets nearly 50 mpg in city driving. Have they reached their goals? Within two years of their resolution, the Foleys, who now have two daughters, cut their energy use by roughly 65%. Foley says:

Cutting your greenhouse gas emissions doesn't have to be a "sacrifice" at all. We have cut our emissions more than 50%, and we now have lower energy bills, a more comfortable house, more time to spend with our family, and a higher quality of life. Americans have a lot to gain by cutting fossil fuel use: reduced greenhouse gas emissions, improved air quality in our cities, less dependence on foreign oil supplies, and so on. This is a win-win scenario, so why not go for it?

Recently, innovative programs throughout the world (such as Carbonfund.org) are providing additional ways for individuals to "go for it." *Carbon offset* initiatives allow people to help compensate for the carbon they release by investing in projects that encourage energy efficiency, renewable energy use, and reforestation. For example, if your car gets 30 mpg and you drive 12,000 miles/yr, your car will release about 3.5 tons of CO_2 (or about 1 ton of carbon). Carbonfund.org allows you to select projects where your investment will reduce CO_2 emissions for about $5 per ton. This and many other carbon offset initiatives (search "Carbon offsetting" at http://www.ecobusinesslinks.com) provide a great way to augment personal lifestyle choices and further reduce your impact. Can you make a difference? The answer is an unequivocal "Yes!"

CASE STUDY REVISITED   WHEN THE SALMON RETURN

Researchers investigating the sockeye salmon's return to an Alaskan stream witness an awesome sight. Hundreds of brilliant red bodies writhe in water so shallow it barely covers them. A female beats her tail, excavating a shallow depression in the gravel where she releases her coral-colored eggs; meanwhile, a male showers them with sperm. But after their long and strenuous migration, these adult salmon are dying. Their flesh is tattered, their muscles wasted, and the final act of reproduction saps the last of their energy. Soon the stream is clogged with dying, dead, and decomposing bodies— an abundance of nutrients unimaginable at any other time of the year. Eagles, grizzly bears, and gulls gather to gorge themselves on the fleeting bounty. Flies breed in the carcasses, feeding spiders, birds,

and trout. The breeding cycles of local mink populations have evolved around the event; females lactate just when the salmon provide them with abundant food. Isotope studies reveal that up to one-fourth of the nitrogen incorporated into the leaves of trees and shrubs near these streams comes from the bodies of salmon. Historically, researchers estimate that 500 million pounds of salmon migrated upstream in the U.S. Pacific Northwest each year, contributing hundreds of thousands of pounds of nitrogen and phosphorus to the Columbia River watershed alone. Now, due to factors including overfishing, river damming, diversion of water for irrigation, runoff from agriculture, and pollution of the estuaries (where several salmon species spend a significant part of their life cycle), migratory salmon populations in the region have

declined by over 90% in the past century. The web of life that relied on the mighty annual upstream flow of nutrients has been disrupted.

Consider This Some salmon populations have been so thoroughly depleted that they qualify for protection under the Endangered Species Act. Some people argue that because these salmon are also raised commercially in hatcheries and artificial ponds, they should not be afforded this protection. Meanwhile, researchers studying chinook salmon raised in hatcheries noted a 25% decline in the average size of eggs of hatchery-reared fish over just four generations. These eggs produce smaller juvenile fish. Based on this information, explain why ecologists and conservationists are arguing for protection of wild salmon.

CHAPTER REVIEW

SUMMARY OF KEY CONCEPTS

28.1 What Are the Pathways of Energy and Nutrients?

Ecosystems are sustained by a continuous flow of energy from sunlight and a constant recycling of nutrients.

Web Tutorial 28.1 Energy Flow and Food Webs

28.2 How Does Energy Flow Through Communities?

Energy enters the biotic portion of ecosystems when it is harnessed by autotrophs during photosynthesis. Net primary productivity is the amount of energy that autotrophs store in a given unit of area over a given period of time.

Trophic levels describe feeding relationships in ecosystems. Autotrophs are the producers, the lowest trophic level. Herbivores occupy the second level as primary consumers. Carnivores act as secondary consumers when they prey on herbivores and as tertiary or higher-level consumers when they eat other carnivores. Omnivores, which consume both plants and other animals, occupy multiple trophic levels.

Feeding relationships in which each trophic level is represented by one organism are called *food chains*. In natural ecosystems, feeding relationships are far more complex and are described as food webs. Detritus feeders and decomposers, which digest dead bodies and wastes, use and release the energy stored in these substances and liberate nutrients for recycling. In general, only about 10% of the energy captured by organisms at one trophic level is converted to the bodies of organisms in the next higher level. The higher the trophic level, the less energy is available to sustain it. As a result, plants are more abundant than herbivores, and herbivores are more common than carnivores. The storage of energy at each trophic level is illustrated graphically as an energy pyramid. The energy pyramid leads to biological magnification, the process by which toxic substances accumulate in increasingly high concentrations in progressively higher trophic levels.

28.3 How Do Nutrients Move Within and Among Ecosystems?

A nutrient cycle depicts the movement of a particular nutrient from its reservoir (usually in the abiotic, or nonliving, portion of the ecosystem) through the biotic, or living, portion of the ecosystem and back to its reservoir, where it is again available to producers. Carbon reservoirs include the oceans, atmosphere, and fossil fuels. Carbon enters producers through photosynthesis. From autotrophs it is passed through the food web and released to the atmosphere as CO_2 during cellular respiration.

The major reservoir of nitrogen is the atmosphere. Nitrogen gas is converted by bacteria and human industrial processes into ammonia and nitrate, which can be used by plants.

Nitrogen passes from producers to consumers and is returned to the environment through excretion and the activities of detritus feeders and decomposers.

The reservoir of phosphorus is in rocks as phosphate, which dissolves in rainwater. Phosphate is absorbed by photosynthetic organisms, then passed through food webs. Some is excreted, and the rest is returned to the soil and water by decomposers. Some is carried to the oceans, where it is deposited in marine sediments. Humans mine phosphate-rich rock to produce fertilizer.

The major reservoir of water is the oceans. Solar energy evaporates water, which returns to Earth as precipitation. Water flows into lakes and underground reservoirs and in rivers, which flow to the oceans. Water is absorbed directly by plants and animals and is also passed through food webs. A small amount is combined with CO_2 during photosynthesis to form high-energy molecules.

Web Tutorial 28.2 The Carbon Cycle and Global Warming

Web Tutorial 28.3 The Nitrogen Cycle

Web Tutorial 28.4 The Hydrologic Cycle

28.4 What Causes "Acid Rain"?

Environmental disruption occurs when human activities interfere with the natural functioning of ecosystems. Human industrial processes release toxic substances and produce more nutrients than nutrient cycles can efficiently process. Through massive consumption of fossil fuels, we have overloaded the natural cycles of carbon, sulfur, and nitrogen. Fossil fuel combustion releases sulfur dioxide and nitrogen oxides. In the atmosphere, these substances are converted to sulfuric acid and nitric acid, which fall to Earth as acid deposition, including acid rain. Acidification of freshwater ecosystems has substantially reduced their ability to support life, particularly in the eastern U.S. At high elevations, acid deposition has significantly damaged many eastern forests and threatens other forests throughout the U.S.

28.5 What Causes Global Warming?

Burning fossil fuels has substantially increased atmospheric carbon dioxide, a greenhouse gas. This increase is correlated with increased global temperatures, leading nearly all atmospheric scientists to conclude that global warming is due to human industrial activities. Global warming is causing ancient ice to melt and is influencing the distribution and seasonal activities of wildlife. Scientists believe global warming is beginning to have a major impact on precipitation and weather patterns, with unpredictable results.

KEY TERMS

acid deposition *page 571*
autotroph *page 561*
biodegradable *page 566*
biogeochemical cycle
  page 567
biological magnification
  page 566
biomass *page 561*
carnivore *page 562*

consumer *page 561*
decomposer *page 564*
deforestation *page 573*
detritus feeder *page 564*
energy pyramid *page 564*
food chain *page 562*
food web *page 562*
fossil fuel *page 568*
global warming *page 573*

greenhouse effect *page 573*
greenhouse gas *page 573*
herbivore *page 562*
heterotroph *page 561*
hydrologic cycle *page 570*
net primary productivity
  page 561
nutrient cycle *page 567*
omnivore *page 564*

primary consumer *page 562*
producer *page 561*
reservoir *page 567*
secondary consumer
  page 562
tertiary consumer *page 562*
trophic level *page 562*

THINKING THROUGH THE CONCEPTS

1. What makes the flow of energy through ecosystems fundamentally different from the flow of nutrients?

2. What is an autotroph? What trophic level does it occupy, and what is its importance in ecosystems?

3. Define *primary productivity*. Would you predict higher productivity in a farm pond or an alpine lake? Defend your answer.

4. List the first three trophic levels. Among the consumers, which are most abundant? Why would you predict that there will be a greater biomass of plants than herbivores in any ecosystem? Relate your answer to the "10% law."

5. How do food chains and food webs differ? Which is the more accurate representation of actual feeding relationships in ecosystems?

6. Define *detritus feeders* and *decomposers*, and explain their importance in ecosystems.

7. Trace the movement of carbon from its reservoir through the biotic community and back to the reservoir. How have human activities altered the carbon cycle, and what are the implications for future climate?

8. Explain how nitrogen gets from air into a plant body.

9. Trace a phosphorus molecule from a phosphate-rich rock into the DNA of a carnivore. What makes the phosphorus cycle fundamentally different from the carbon and nitrogen cycles?

10. Trace the movement of a water molecule from the ocean, through a plant body, and back to the ocean, describing all the intermediate stages and processes.

APPLYING THE CONCEPTS

1. What could your college or university do to reduce its contribution to acid rain and global warming? Be specific and, if possible, offer practical alternatives to current practices.

2. Define and give an example of *biological magnification*. What qualities are present in materials that undergo biological magnification? In which trophic level are the problems worst, and why?

3. Discuss the contribution of population growth to (a) acid rain and (b) global warming.

4. Describe what would happen to a population of deer if all predators were removed and hunting banned. Include effects on vegetation as well as on the deer population itself. Relate your answer to carrying capacity as discussed in Chapter 26.

FOR MORE INFORMATION

Gorman, C. "Global Warming: How It Affects Your Health." *Time*, April 3, 2006. The warming trend may cause more deaths from weather extremes and a possible spread of malaria-carrying mosquitoes.

Kluger, J. "The Turning Point." *Time*, April 3, 2006. Polar ice caps are melting, droughts are increasing, wildlife is vanishing, and the effects of global warming may create positive feedback cycles that further increase the problem.

Krajick, K. "Long-Term Data Show Lingering Effects from Acid Rain." *Science*, April 13, 2001. Acid rain's damaging effects linger, and current control levels are inadequate to restore ecosystem health.

Milius, S. "Decades of Dinner." *Science News*, May 7, 2005. The body of a whale on the seafloor forms the basis of an underwater community.

Moore, K. D., and Moore, J. W. "The Gift of Salmon." *Discover*, May 2003. Salmon migrating upstream to spawn and die reverse the usual travel of nutrients and help replenish nutrients carried downstream during the rest of the year.

Pearce, F. "The Parched Planet." *New Scientist*, February 2006. Drought combined with unsustainable mining of groundwater threatens food production, particularly in developing countries.

Walsh, B. "The Impacts of Asia's Giants." *Time*, April 3, 2006. How India and China develop will have profound impacts on the future of the planet.

Wright, K. "Our Preferred Poison." *Discover*, March 2005. Mercury bioaccumulation threatens animals in the top trophic levels, including people.

29 Earth's Diverse Ecosystems

Kahindi Samson snares a butterfly. Insets: **(top)** a dark blue pansy butterfly.
(bottom) Pupae are identified and sorted for shipment.

CASE STUDY   WINGS OF HOPE

TO HELP SUPPORT AND FEED his five younger brothers and sisters, Kahindi Samson began sneaking into the Arabuko-Sokoke forest when he was 12 years old, snaring its endangered antelope and cutting the old-growth trees that provide homes for the rare Sokoke Scops owl. This precious Kenyan forest is protected by the government as the largest remnant of coastal forest in East Africa and a final refuge for endangered birds and mammals that have been displaced by the growing human population. But to the farmers on surrounding land, the forest was the enemy, home to elephants and baboons that emerged at night to eat their crops. Most wanted to see the forest felled.

Ian Gordon, a butterfly ecologist, watched the poaching and illegal tree-cutting with alarm; Arabuko-Sokoke forest is home to 250 species of butterflies. Unwilling to stand by helplessly, Gordon founded Project Kipepeo ("kipepeo" means "butterfly" in Swahili). His mission was to convince skeptical local farmers to grow butterflies instead of crops. Today, Samson enters the forest with a license and a butterfly net. In a cage outside his home, he places his catch of pregnant female butterflies. As their eggs hatch, Samson fattens the caterpillars on leaves he collects from the forest. Within a month, the caterpillars are ready to form pupae and be shipped to the United

States and Europe. There they will hatch amidst lush tropical vegetation in butterfly gardens, where they will delight visitors who have never witnessed the splendor of rain-forest butterflies. Samson is one of about 650 local butterfly workers who now rely on the Arabuko-Sokoke forest for their livelihood and make a far better living than before. "We used to wish the forest would go away," says butterfly farmer Priscilla Kiti, "but these days we earn our living mostly in butterfly farming, so if they cut the forest, things are going to be very difficult."

29.1 WHAT FACTORS INFLUENCE EARTH'S CLIMATE?

The distribution of life, particularly on land, is dramatically affected by both weather and climate. **Weather** refers to short-term fluctuations in temperature, humidity, cloud cover, wind, and precipitation in a region over periods of hours or days. **Climate**, in contrast, refers to patterns of weather that prevail over years or centuries in a particular region. The amount of sunlight and water and the range of temperatures determine the climate of a given region. Whereas weather affects individual organisms, climate influences and limits the overall distribution of entire species.

Both Climate and Weather Are Driven by the Sun

Both climate and weather are driven by a great thermonuclear engine: the sun. Solar energy reaches Earth in a range of wavelengths—from short, high-energy ultraviolet (UV) rays, through visible light, to the longer infrared wavelengths that produce heat. The solar energy that reaches Earth drives the wind, the ocean currents, and the global water cycle. Before it reaches Earth's surface, however, sunlight is modified by the atmosphere. A layer relatively rich in ozone (O_3) is found in the middle atmosphere. This *ozone layer* absorbs much of the sun's high-energy UV radiation, which can damage biological molecules (see "Earth Watch: The Ozone Hole—A Puncture in Our Protective Shield"). Dust, water vapor, and clouds scatter light, reflecting some of the energy back into space. Carbon dioxide, water vapor, methane, and other *greenhouse gases* selectively absorb energy at infrared wavelengths, trapping heat in the atmosphere. Human activities have significantly increased levels of greenhouse gases, as described in Chapter 28.

Only about half the solar energy reaching the atmosphere actually strikes Earth's surface. Of this, a small fraction is immediately reflected back into space; another small fraction is captured by photosynthetic plants and microorganisms and used to power photosynthesis, and the rest is absorbed as heat. Eventually, nearly all the incoming solar energy is radiated back into space, either as light or as infrared radiation (heat). The solar energy temporarily stored as heat by the atmosphere and Earth's surface maintains Earth's relative warmth.

Earth's Physical Features Also Influence Climate

Many physical factors influence climate. Among the most important are Earth's curvature and its tilted axis as it rotates around the sun. These factors cause uneven heating of the surface and seasonal changes in the directness of sunlight north and south of the equator. Uneven heating, in conjunction with Earth's rotation, generates air and ocean currents, which in turn are modified by irregularly shaped landmasses.

Earth's Curvature and Tilt Influence the Angle at Which Sunlight Strikes

The amount of sunlight that strikes a given area of Earth's surface has a major effect on average yearly temperatures. At the equator, sunlight hits Earth's surface nearly at a right angle, making the weather there consistently warm. Farther north or south, the sun's rays strike Earth's surface at a greater slant. This angle spreads the same amount of sunlight over a larger area, producing lower overall temperatures (**FIG. 29-1**).

Latitude is a measure of the distance north or south from the equator, expressed in degrees. The equator is defined as 0° latitude, and the poles are at 90° north and south latitudes. Earth is tilted on its axis, so as it makes its

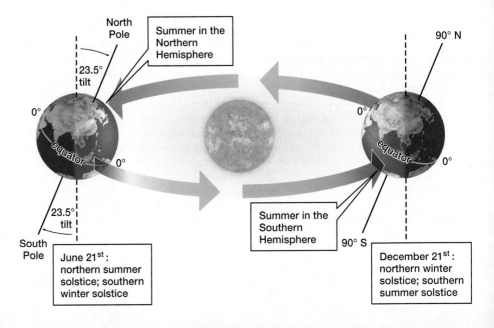

FIGURE 29-1 Earth's curvature and tilt produce seasons and climate Temperatures are highest and most uniform at the equator and lowest and most variable at the poles. Equatorial sunlight is more perpendicular and hits this band year-round, while polar sunlight is seasonal, and its angle spreads it over a much larger land surface. The tilt of Earth on its axis causes seasonal variations in the range and directness of sunlight. QUESTION Describe the seasons and day length if Earth were not tilted on its axis. Would there still be a temperature gradient from the equator to the poles?

North Pole

23.5° tilt

0°

0°

equator

23.5° tilt

South Pole

Summer in the Northern Hemisphere

June 21st: northern summer solstice; southern winter solstice

Summer in the Southern Hemisphere

90° N

0°

0°

equator

90° S

December 21st: northern winter solstice; southern summer solstice

year-long journey around the sun, higher latitudes experience regular changes in the angle of sunlight, resulting in pronounced seasons. When the Northern Hemisphere tilts toward the sun, it receives more direct sunlight and experiences summer. During the Northern Hemisphere's winter, the Southern Hemisphere is tilted toward the sun (see Fig. 29-1). Throughout the year, sunlight continues to hit the equator directly, so this region remains warm and experiences little seasonal variation.

Air Currents Produce Broad Climatic Regions

Air currents are generated by Earth's rotation and by differences in temperature between different air masses. Warm air is less dense than cold air, so as the sun's direct rays fall on the equator, heated air rises there. The warm air near the equator is also laden with water evaporated by solar heat (**FIG. 29-2a**). As the water-saturated air rises, it cools somewhat. Cool air cannot retain as much moisture, so water condenses from the rising air and falls as rain. The direct rays of the sun and the rainfall produced when warm, moist air rises and is cooled create a band around the equator called the *Tropics*. This region is both the warmest and the wettest on Earth. The cooler dry air then flows north and south from the equator. Near 30° N and

30° S, the air has cooled enough to sink. As it sinks, the air is again warmed by heat radiated from Earth. By the time it reaches the surface, it is both warm and very dry. Not surprisingly, the major deserts of the world are found at these latitudes (Fig. 29-2a, b). This air then flows back toward the equator. Farther north and south, this general circulation pattern is repeated, dropping moisture at around 60° N and 60° S and creating extremely dry conditions at the North and South Poles.

Notice in Figure 29-2a that the air currents appear to be deflected toward the right (relative to their direction of travel) in the Northern Hemisphere, and to the left in the Southern Hemisphere. This is because Earth is rotating from East to West beneath the air masses. Movement of Earth's surface relative to its atmosphere causes earthbound observers—and ecosystems—to experience prevailing winds whose direction depends on the circulation pattern of the air above them. In the U.S., prevailing *westerlies* occur (Fig. 29-2a; coming generally from the south, and deflected to the right), while Mexico experiences prevailing *northeasterly trade winds* (Fig. 29-2a; from the north, deflected right). This wind deflection is an example of the *Coriolis effect*, which refers to the effects of Earth's rotation on large masses of air and water that flow freely relative to Earth's surface.

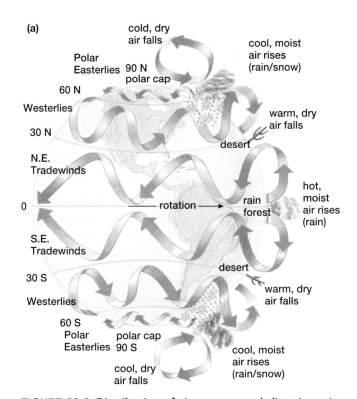

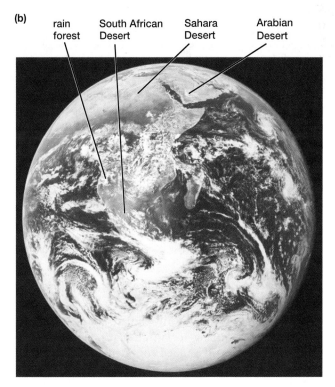

FIGURE 29-2 Distribution of air currents and climatic regions
(a) Rainfall is determined primarily by the distribution of temperatures and by Earth's rotation. The interaction of these two factors creates air currents that rise and fall predictably with latitude, producing broad climatic regions. (b) Some of these regions are visible in this photograph of the African continent taken from *Apollo 11*. Along the equator are heavy clouds that drop moisture on the central African rain forests. Note the lack of clouds over the Sahara and Arabian Deserts near 30° N and the South African Desert near 30° S.

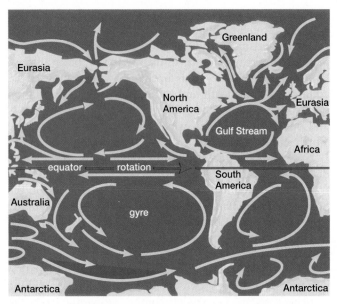

FIGURE 29-3 Ocean circulation patterns are called gyres
Gyres travel clockwise in the Northern Hemisphere and counterclockwise in the Southern Hemisphere. These currents tend to distribute warmth from the equator to northern and southern coastal areas.

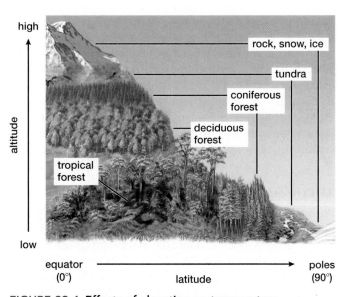

FIGURE 29-4 Effects of elevation on temperature
In terms of temperature, climbing a mountain in the Northern Hemisphere is like going north; in both cases, increasingly cool temperatures produce a similar series of biomes.

Ocean Currents Moderate Nearshore Climates

Ocean currents are driven by Earth's rotation, by winds, and by direct heating of water by the sun. Continents interrupt the currents, breaking them into roughly circular patterns called **gyres**. The Coriolis effect causes gyres to rotate clockwise in the Northern Hemisphere and counterclockwise in the Southern Hemisphere (**FIG. 29-3**). Because water both heats and cools more slowly than does land or air, ocean currents tend to moderate temperature extremes. Coastal areas, then, generally have less-variable climates than do areas near the center of continents. For example, the Gulf Stream—part of the Atlantic Ocean gyre (Fig. 29-3)—carries warm water from equatorial regions north along the eastern coast of North America, creating a warmer, moister climate than is found farther inland. It then carries the still-warm water farther north and east, warming the western coast of Europe before returning south.

Continents and Mountains Complicate Weather and Climate

If Earth's surface were uniform, climate zones would occur in bands corresponding to latitude (see Fig. 29-2a). The presence of irregularly shaped continents (which heat and cool relatively quickly) amidst oceans (which heat and cool more slowly) alters the flow of wind and water and contributes to the irregular distribution of ecosystems.

Variations in elevation within continents further complicate the situation. As elevation increases, the atmosphere becomes thinner and retains less heat. The temperature drops approximately 3.5°F (2°C) for every 1000-foot (305-

meter) elevation gain. This characteristic explains why snowcapped mountains are found even in the Tropics (**FIG. 29-4**). Mountains also modify rainfall patterns. When water-laden air is forced to rise as it meets a mountain, it cools. Cooling reduces the air's ability to retain water, and the water condenses as rain or snow on the windward (near) side of the mountain. The cool, dry air is warmed again as it travels down the far side of the mountain and absorbs water from the land, creating a local dry area called a **rain shadow**. For example, the Sierra Nevada range of the western U.S. wrings moisture from westerly winds blowing off the Pacific Ocean, producing the Mojave Desert in the mountains' rain shadow to the east (**FIG. 29-5**).

El Niño Periodically Disrupts the Usual Ocean-Atmosphere Interactions

The tropical western Pacific ocean typically harbors a huge pool of warm water that is pushed westward by the Northeast Trade Winds (see Fig. 29-2a). Water evaporating from this warm mass falls as rain over countries bordering the western Pacific, such as Indonesia and Australia. To replace the westward-moving water, cooler water from the ocean depths, rich in nutrients, wells up along the western coast of South America, bringing Peru a rich harvest of fish. But for unknown reasons, at intervals of around 3 to 7 years, the trade winds die down, and an **El Niño** (literally "boy child") occurs. Named by Peruvian fishermen in reference to the baby Jesus, El Niño brings rain to Peru during the normally arid December, and the diminishing trade winds allow warm water to spread back eastward across the Pacific until it reaches the western coast of South America. Meanwhile, drought plagues Indonesia, eastern Australia, and southern Africa. An El Niño of immense strength disrupted weather

FIGURE 29-5 **Mountains create rain shadows**

Water is carried from ocean by prevailing winds.

Water is released as air rises and cools.

Dry air sinks, warms and absorbs water from the land.

moist climate

dry climate in rain shadow

worldwide during 1997–1998, causing disastrous floods in both Peru and east Africa, and droughts and fires in Indonesia. In the U.S., Texas sizzled in withering heat and drought, while Florida was deluged. As weather patterns return to normal after an El Niño, the interaction between ocean and atmosphere may overshoot, reversing the pattern and causing a **La Niña** ("girl child"). Researchers are working to create computer models that will allow them to predict these events and their worldwide impacts well before they occur—a task of staggering complexity. Meanwhile, scientists debate the impact of global warming; some predict that El Niños will become more frequent and severe.

29.2 WHAT CONDITIONS DOES LIFE REQUIRE?

From the lichens on bare rock to the thermophilic (Greek for "heat-loving") algae in the hot springs of Yellowstone National Park to the bacteria thriving under the pressure-cooker conditions of a deep-sea vent, Earth teems with life. These diverse and extreme habitats all share the ability to provide, to varying degrees, four fundamental resources that are required for life:

- Nutrients from which to construct living tissue
- Energy to power metabolic activities
- Liquid water to serve as a medium in which metabolic activities occur
- Appropriate temperatures in which to carry out these processes

As we will see in the following sections, these resources are unevenly distributed over Earth's surface. Their availability limits the types of organisms that can exist within the various terrestrial and aquatic ecosystems on Earth.

Ecosystems are extraordinarily diverse, yet clear patterns exist. The community that is characteristic of each ecosystem is dominated by organisms specifically adapted to particular environmental conditions. Variations in temperature and in the availability of light, water, and nutrients shape the adaptations of organisms that inhabit an ecosystem. The desert community, for example, is dominated by plants adapted to heat and drought. The cacti of the Mojave

Desert in the American Southwest are strikingly similar to the euphorbia of the deserts of Africa and of the nearby Canary Islands, although these plants are only distantly related genetically. Their reduced leaves and thick, water-storing stems are adaptations for dry climates (**FIG. 29-6**). Likewise, the plants of the arctic tundra and those of the alpine tundra on high mountaintops show growth patterns clearly recognizable as adaptations to a cold, dry, windy climate.

29.3 HOW IS LIFE ON LAND DISTRIBUTED?

Terrestrial organisms are restricted in their distribution largely by the availability of water and by temperature. Terrestrial ecosystems receive plenty of light, even on an overcast day, and most soils provide adequate nutrients. Water, however, is limited and very unevenly distributed,

(a)   (b)

FIGURE 29-6 **Environmental demands mold physical characteristics**
Evolution in response to similar environments has molded the bodies of **(a)** American cacti and **(b)** Canary Islands euphorbia into nearly identical shapes, although they are in different families. QUESTION Describe the similar selection pressures that operate on these two different families of plants.

EARTH WATCH   The Ozone Hole—A Puncture in Our Protective Shield

A fraction of the radiant energy produced by the sun—*ultraviolet* (UV) radiation—is so energetic that it can damage biological molecules. In small quantities, UV radiation helps human skin produce vitamin D and causes tanning in fair-skinned people. But in larger doses, UV causes sunburn and premature aging of skin, skin cancer, and cataracts (a condition in which the lens of the eye becomes cloudy).

Fortunately, most UV radiation is filtered out by ozone in the stratosphere, a layer of atmosphere extending from 10 to 50 kilometers (6 to 30 miles) above Earth. In pure form, *ozone* (O_3) is an explosive, highly poisonous gas. In the stratosphere, the normal concentration of ozone is about 0.1 part per million (ppm), compared with 0.02 ppm in the lower atmosphere. This ozone-enriched layer is called the **ozone layer**. Ultraviolet light striking ozone and oxygen gas causes reactions that break down as well as regenerate ozone. In the process, the UV radiation is converted to heat, and the overall level of ozone remains reasonably constant—at least it did until humans intervened.

In 1985 British atmospheric scientists published the startling news that springtime levels of stratospheric ozone over Antarctica had declined by more than 40% since 1977. In the *ozone hole* over Antarctica, ozone now dips to about one-third of its pre-depletion levels (**FIG. E29-1**). Although ozone layer depletion is most severe over Antarctica, the ozone layer is somewhat reduced over most of the world, including nearly all of the continental U.S. Satellite data reveal that, since the early 1970s, UV radiation has risen by nearly 7% per decade in the Northern Hemisphere and by almost 10% per decade in the Southern Hemisphere. Epidemiological studies indicate that for every 1% increase in lifetime exposure to UV radiation, the lifetime risk of skin cancer also increases by about 1%. But human health effects are only one cause for concern. Photosynthesis by phytoplankton, the producers for marine ecosystems, is reduced under the ozone hole above Antarctica. Some types of trees and farm crops are also harmed by increased UV radiation.

The ozone hole is caused by human production and release of chlorofluorocarbons (CFCs). Developed in 1928, these gases were widely used as coolants in refrigerators and air conditioners, as aerosol spray propellants, in the production of foam plastic, and as cleansers for electronic parts. CFCs are very stable and were considered safe. Their stability, however, has proved to be a major problem because they remain chemically unchanged as they gradually rise into the stratosphere. There, under intense bombardment by UV light, the CFCs break down, releasing chlorine atoms. Chlorine catalyzes the breakdown of ozone to oxygen gas (O_2) while remaining unchanged itself. Clouds composed of ice particles over the Antarctic regions provide a surface on which the reaction can occur.

Fortunately, people have taken the first steps toward "plugging" the ozone hole. In a series of treaties beginning in 1987, industrialized nations throughout the world agreed to phase out ozone-depleting chemicals, eliminating CFCs by 1996. Ground-level global atmospheric chlorine levels (an indicator of CFC use) peaked in 1994, and by 1999, scientists had detected chlorine reductions in the stratosphere. In 2005, the National Oceanic and Atmospheric Association reported that ozone concentrations had leveled off between 1996 and 2002. Because CFCs persist for 50 to 100 years and take a decade or more to ascend into the stratosphere, current release of CFCs by developing countries—adding to the millions of tons already released by developed countries—means that significant recovery could take 40 years. In a spirit of continued cooperation, developed countries have recently pledged to help developing countries devise alternatives to CFCs, and China has now pledged to stop producing CFCs by 2007.

FIGURE E29-1 Satellite image of the Antarctic ozone hole The ozone hole recorded in September, 2006, is shown in blue and purple on this NASA satellite image. At 11.4 million square miles (29.5 million square kilometers) it has tied the previous record set in 2000. (*Image courtesy of NASA.*)

both in place and in time. Terrestrial organisms must be adapted to obtain water when it is available and to conserve it when it is scarce.

Like water, temperatures favorable to life are also unevenly distributed in place and time. At the South Pole, even in summer, the average temperature is usually well below freezing; not surprisingly, life is scarce there. Places such as central Alaska have favorable temperatures for plant growth only during the summer, whereas the Tropics have a uniformly warm, moist climate in which life abounds.

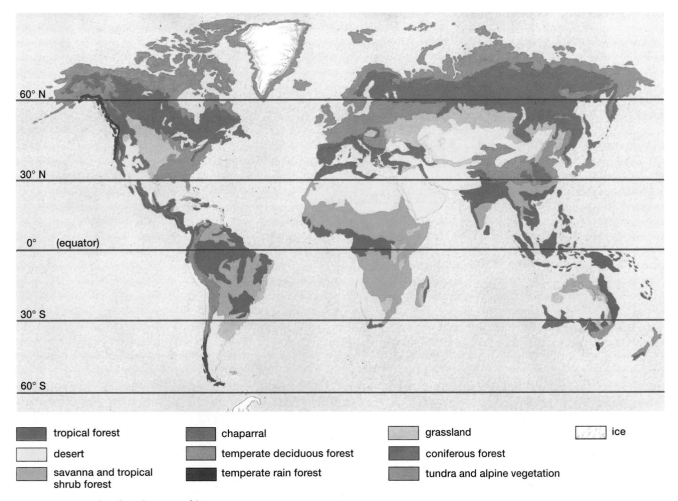

■ tropical forest	■ chaparral	■ grassland	▢ ice
▢ desert	■ temperate deciduous forest	■ coniferous forest	
■ savanna and tropical shrub forest	■ temperate rain forest	■ tundra and alpine vegetation	

FIGURE 29-7 The distribution of biomes
Although mountain ranges and the sheer size of continents complicate the pattern of biomes, note the overall consistencies. Tundras and coniferous forests are in the northernmost parts of the Northern Hemisphere, whereas the deserts of Mexico, the Sahara, Saudi Arabia, South Africa, and Australia are located about 20° to 30° N and S.

Terrestrial Biomes Support Characteristic Plant Communities

Terrestrial communities are dominated and defined by their plant life. Because plants can't escape from drought, sun, or winter weather, they tend to be extremely well adapted to the climate of a particular region. Large land areas with similar environmental conditions and characteristic plant communities are called **biomes** (FIG. 29-7), which are generally named after the major type of vegetation found there. The predominant vegetation of each biome is determined by the complex interplay of rainfall and temperature (FIG. 29-8). These factors determine the soil moisture available for plant metabolic activities and to replace water lost to evaporation from their leaves. In addition to the overall average rainfall and temperature, the way that these factors vary seasonally determines which plants can grow in a given

region. Arctic tundra plants, for example, must be adapted to marshy conditions in the early summer but cold and extremely dry conditions for much of the rest of the year, when water is solidly frozen and unavailable. In the following sections we discuss the major biomes, beginning at the equator and working our way poleward. We also discuss some of the impacts of human activities on these biomes. You will learn more about human impacts on Earth in Chapter 30.

Tropical Rain Forests

Near the equator, the temperature averages between 77°F and 86°F (25°C and 30°C) with little variation, and rainfall ranges from 100 to 160 inches (250 to 400 centimeters) annually. These evenly warm, evenly moist conditions combine to create the most diverse biome on Earth, the **tropical rain forest**, dominated by huge broadleaf evergreen trees

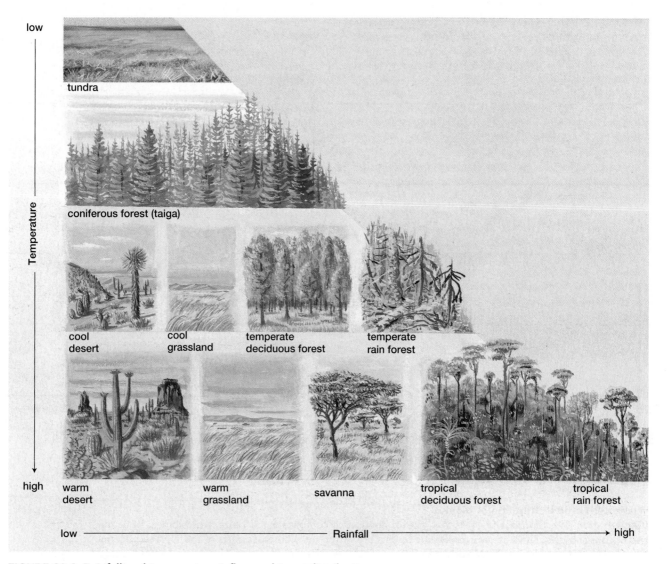

FIGURE 29-8 Rainfall and temperature influence biome distribution
Together, rainfall and temperature determine the soil moisture available to support plants.

(FIG. 29-9). Extensive rain forests are found in Central and South America, Africa, and Southeast Asia.

Biodiversity refers to the total number of species within an ecosystem and the resulting complexity of interactions among them; in short, it defines the biological "richness" of an ecosystem. Rain forests have the highest biodiversity of any ecosystem on Earth. Although rain forests cover only 6% of Earth's total land area, ecologists estimate that they are home to between 5 million and 8 million species, representing half to two-thirds of the world's total. For example, a recent survey of a 2.5-acre site in the upper Amazon basin revealed 283 different species of trees, most of which were represented by a single individual. In a 3-square-mile (about 5-square-kilometer) tract of rain forest in Peru, scientists counted more than 1300 butterfly species and 600 bird species. For comparison, the entire U.S. is home to only 400 butterfly and 700 bird species.

Tropical rain forests typically have several layers of vegetation. The tallest trees reach 150 feet (50 meters) and tower above the rest of the forest. Below is a fairly continuous canopy of treetops at about 90 to 120 feet (30 to 40 meters). Another layer of shorter trees typically stands below the canopy. Huge woody vines, often over 300 feet (over 100 meters) long, grow up the trees, reaching the sunlight far above. These layers capture most of the sunlight. Many of the plants that live in the dim green light that filters through to the forest floor have enormous leaves to trap the little available energy.

Because edible plant material close to the ground in tropical rain forests is relatively scarce, much of the animal life—including a huge variety and number of birds, monkeys, and insects—is arboreal, or tree dwelling. Competition for the nutrients that do reach the ground is intense among both animals and plants. Even such unlikely sources of food as the droppings of monkeys are in great demand. For example, dung beetles feed and lay their eggs on monkey droppings. When ecologists attempted to collect droppings of the South American howler monkeys to

FIGURE 29-9 The tropical rain-forest biome
(a) Towering trees draped with vines reach for the light in the dense tropical rain forest. Amid their branches dwells the most diverse assortment of animals on Earth, including (b) a fruit-eating toucan, (c) a golden-eyed leaf frog, (d) a tree-dwelling orchid, and (e) a howler monkey. QUESTION How does a biome with such poor soil support the highest plant productivity and the greatest animal diversity on Earth?

find out what the monkeys had been eating, they found themselves in a race with the beetles, hundreds of which would arrive within minutes after a dropping hit the ground.

Almost as soon as bacteria or fungi release any nutrients from dead plants or animals into the soil, rain-forest trees and vines absorb the nutrients. Virtually all the nutrients in a rain forest are already tied up in the vegetation, so the soil is relatively infertile and thin.

Human Impact
Because of infertile soil and heavy rains, agriculture is risky and destructive in rain forests. If the trees are carried away for lumber, few nutrients remain to support crops. If the nutrients are released to the soil by burning the natural vegetation, the heavy year-round rainfall quickly dissolves and erodes them away, leaving the soil depleted after only a few seasons of cultivation. The exposed soil, which is rich in iron and aluminum, then takes on an impenetrable, bricklike

FIGURE 29-10 Amazon rain forest cleared by burning
The burned area will be converted to ranching or agriculture, but both are doomed to failure due to poor soil quality. Spreading fires and the smoke they produce also threaten adjacent uncut forests and their myriad inhabitants.

quality as it bakes in the tropical sun. As a result, secondary succession on cleared rain-forest land is slow; even small forest cuttings take about 70 years to regenerate. Despite their unsuitability for agriculture, rain forests are being felled for lumber or burned down for ranching or farming at an alarming rate (FIG. 29-10). The demand for biofuels (fuels produced from biomass, including palm and soybean oil) is driving rapid destruction of rain forests to grow these crops. Estimates of rain-forest destruction range up to 65,000 square miles (42 million acres, or about 170,000 square kilometers) per year, or about 1.3 acres each second. In recent years Brazil alone has lost about 10,000 square miles (6000 square kilometers) annually. For comparison, the state of Connecticut occupies about 5000 square miles.

The rain forest left standing in cutover areas is often in fragments too small to allow reproduction of trees and to provide adequate foraging and breeding habitat for larger animals. Fires in clear-cut areas also spread along the ground into adjacent forest, killing young trees and disrupting community interactions. In West Africa, forest burning is causing acid rain, which damages the remaining trees. Further, much of the rainfall in a rain forest comes from water transpired through the forest's leaves. As large tracts of rain forests disappear, the region becomes drier, more stressed, and more susceptible to fires. Researchers estimate that about 20% of the carbon released into the atmosphere during the past decade came from cutting and burning tropical rain forests, exacerbating global warming.

At least 40% of the world's rain forests are now gone. The impact of these losses of irreplaceable biodiversity is incalculable. Although disastrous losses continue, some areas have been set aside as protected preserves, and some reforestation efforts are under way. Local residents are becoming more involved in conservation efforts, as illustrated in both our Case Study and "Links to Life: Crave Chocolate and Save Rain Forests?" In Brazil, a union of

people who harvest natural rubber from tree sap is fighting to preserve large tracts of land for rubber production and harvesting fruits and nuts. These are important steps toward *sustainable use*: deriving benefits from an ecosystem in a manner that can be sustained indefinitely.

Tropical Deciduous Forests

Slightly farther from the equator, the rainfall is not nearly as constant, and there are pronounced wet and dry seasons. In these areas, which include much of India as well as parts of Southeast Asia, South America, and Central America, **tropical deciduous forests** grow. During the dry season, the trees cannot get enough water from the soil to compensate for evaporation from their leaves. As a result, the plants have adapted to the dry season by shedding their leaves, thereby minimizing water loss. If the rains fail to return on schedule, the trees delay the formation of new leaves until the drought passes.

Savanna

Along the edges of the tropical deciduous forest, the trees gradually become more widely spaced, with grasses growing between them. Eventually, grasses become the dominant vegetation, with only scattered trees and thorny scrub forests here and there; this biome is the **savanna** (FIG. 29-11). Savanna grasslands typically have a rainy season during which virtually all of the year's precipitation falls—12 inches (30 centimeters) or less. When the dry season arrives, it comes with a vengeance. Rain might not fall for months, and the soil becomes hard, dry, and dusty. Grasses are well adapted to this type of climate, growing very rapidly during the rainy season and dying back to drought-resistant roots during the arid times. Only a few specialized trees, such as the thorny acacia or the water-storing baobab, can survive the devastating savanna dry seasons. In areas where the dry season becomes even more pronounced, virtually no trees

LINKS TO LIFE Crave Chocolate and Save Rain Forests?

Cacao trees—the beans of which are used to produce chocolate—are native to the rain forests of Central and South America, but due to tremendous worldwide demand for chocolate (people in the U.S. alone consume about 3 billion pounds each year), cacao trees are now cultivated in equatorial countries throughout the world. In an attempt to increase production, vast stretches of rain forest have been cleared and cacao trees planted in large plantations. The cleared rain forest quickly loses its fertility and eventually can be used only for pasture. Lacking the diverse environment in which cacao trees evolved, pollination rates are low, and about one-third of the cocoa crop is lost each year to pests. Dousing the plantations with herbicides, fungicides, and pesticides reduces biodiversity and creates toxic runoff.

In 2000 a group of chocolate companies, recognizing the need to protect both cacao trees and their rain-forest habi-

tat, established the World Cocoa Foundation (WCF), whose mission includes producing "quality cocoa in a sustainable, environmentally friendly manner." This includes using natural controls for pests wherever possible. Several South American countries have recognized the value of cultivating cacao under the conditions in which it evolved—beneath a dense rain-forest canopy. This environment provides outstanding habitat for a variety of rain-forest species. For example, Brazilian researchers hope that the golden lion tamarin, a recently discovered and critically endangered primate, will be saved by extending its habitat into Brazilian rain forest that is also used by cacao farmers. The Atlantic rain forest of Brazil has been reduced to less than 8% of its original size, but some officials hope that, thanks to careful cultivation of cacao, portions of the rain forest might be restored. Thanks to the World Cocoa Foundation, you can feel even better about satisfying your chocolate cravings.

FIGURE 29-11 **The African savanna**
(a) Elephants roam beneath a rainbow. (b) A red-billed oxpecker looks up at a sleeping white rhino. Oxpeckers feed on parasites that live on rhino skin. (c) Large herds of grazing animals, such as zebras, can still be seen on African preserves. The herds of herbivores provide food for the greatest assortment of large carnivores on Earth. (d) Cheetahs feast on their prey (both rhinos and cheetahs are endangered).

FIGURE 29-12 Poaching threatens African wildlife
Rhinoceros horns, believed by some to have aphrodisiac properties, fetch staggering prices and encourage poaching. The black rhino is now nearly extinct.

can grow, and the savanna imperceptibly grades into tropical grassland.

The African savanna has probably the most diverse and impressive array of large mammals on Earth. These mammals include numerous herbivores such as antelope, wildebeest, water buffalo, elephants, and giraffes and carnivores such as the lion, leopard, hyena, and wild dog.

Human Impact

Africa's rapidly expanding human population threatens the wildlife of the savanna. Poaching has driven the black rhinoceros to the brink of extinction (**FIG. 29-12**) and endangers the African elephant, a keystone species in this ecosystem. The abundant grasses that make the savanna a suitable habitat for so much wildlife also make it suitable for grazing domestic cattle. Fences increasingly disrupt the migration of the great herds of wild herbivores in search of food and water. Ecologists have discovered that the native herbivores are much more efficient at converting grass into meat than are cattle. Perhaps the future African savanna may support herds of domesticated antelope and other large native grazers in place of cattle.

Deserts

Even drought-resistant grasses need at least 10 to 20 inches (25 to 50 centimeters) of rain a year, depending on its seasonal distribution and the average temperature. When less than 10 inches of rain fall, **desert** biomes result. Although we tend to think of them as hot, deserts are defined by their lack of rain rather than by their temperatures. In the Gobi Desert of Asia, for example, temperatures average below freezing for half the year, while summer temperatures average 105°F to 110°F (29°C to 43°C). Desert biomes are found on every continent, typically around 20° to 30° N and S latitude as well as in the rain shadows of major mountain ranges.

As with all biomes, deserts include a variety of environments. At one extreme are parts of the Sahara desert in Africa and the Atacama desert in Chile, where it almost never rains and no vegetation grows (**FIG. 29-13a**). More commonly, deserts are characterized by widely spaced vegetation and large areas of bare ground (**FIG. 29-13b**). In many cases, the perennial plants are bushes or cacti with large, shallow root systems. The shallow roots quickly soak up soil moisture after the infrequent desert storms. The rest of the plant is typically covered with a waterproof, waxy coating to prevent evaporation of precious water. The thick stems of cacti and other succulents store water. The spines of cacti are leaves modified for protection and water conservation, presenting almost no surface area for evaporation. In many deserts, all the rain falls in just a few storms, and specialized annual wildflowers take advantage of the brief period of moisture to race through germination, growth, flowering, and seed production in a month or less (**FIG. 29-14**).

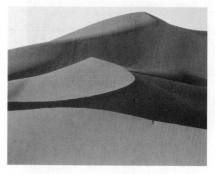
(a)

(b)

(c)

FIGURE 29-13 The desert biome
(a) Under the most extreme conditions of heat and drought, deserts can be almost devoid of life, such as these sand dunes of the Sahara Desert in Africa. **(b)** Throughout much of Utah and Nevada, the Great Basin Desert presents a monotonous landscape of widely spaced shrubs, such as sagebrush and greasewood. These shrubs often secrete a growth inhibitor from their roots, preventing germination of nearby plants and thus reducing competition for water. **(c)** The kangaroo rat is an elusive inhabitant of North American deserts.

FIGURE 29-15 Desertification
A human population above carrying capacity has reduced the ability of many arid (dry) regions, such as the Sahel area of Africa, to support life; this process is called desertification.

FIGURE 29-14 The Sonoran Desert
After a relatively wet spring, this Arizona desert is carpeted with wildflowers. Through much of the year—sometimes for several years—annual wildflower seeds lie dormant, waiting for adequate spring rains to fall. QUESTION How might desert plant seeds "determine" if rainfall is adequate for germination?

The animals of the deserts, like the plants, are specially adapted to survive heat and drought. Few animals are seen during hot summer days. Many, including the burrowing owl, kangaroo rat, desert toad, desert tortoise, and sidewinder rattlesnake, take refuge from the heat in underground burrows that are relatively cool and moist. The desert jackrabbit shelters in the shade of rocks and bushes; its enormous ears and long legs are adaptations that radiate heat. Reptiles such as desert snakes, turtles, and lizards adjust their activity depending on the temperature. In summer, they may be active only around dawn and dusk. Nocturnal animals that take advantage of cool night temperatures include jackrabbits, bats, burrowing owls, and kangaroo rats (see Fig. 29-13c). Many of the smaller desert animals survive without ever drinking; they acquire the water they need from their food and from water produced as a by-product of cellular respiration in their tissues. Larger animals, such as desert bighorn sheep, are dependent on permanent water holes during the driest times of the year.

Human Impact

Desert ecosystems are fragile. Ecologists studying the soil of the Mojave Desert in southern California recently found tread marks left by tanks in 1940 when General Patton trained tank crews in preparation for entry into World War II. The desert soil is stabilized and enriched by micro-

scopic cyanobacteria whose filaments intertwine among sand grains. The tanks, and now off-road vehicles that careen about the desert for recreation, destroy this crucial network. This damage allows the soil to erode and reduces nutrients available to the desert's slow-growing plants. Ecologists estimate that desert soil may require hundreds of years to fully recover from heavy vehicle use.

Human activities are contributing to desertification, the process by which relatively dry, drought-prone regions are converted to desert as a result of human activities. The UN estimates that desertification is affecting about one-third of Earth's land. Its main cause is inappropriate use of the land, including overharvesting bushes and trees for wood, overgrazing of grasses by livestock, and depletion of surface water and groundwater to grow crops. Loss of vegetation, which humidifies the air and stabilizes soil, allows soil to erode and droughts to intensify, decreasing the land's productivity. Desertification of land is a consequence of the human population exceeding its carrying capacity in a fragile ecosystem. For example, the Sahel—dry savanna south of the Sahara Desert—has been significantly overgrazed and degraded by a growing human population (**FIG. 29-15**). Encouragingly, in Niger, West Africa, "Project Eden" is helping to reduce desertification by providing farmers with perennial fruit trees that grow under arid conditions, stabilize soils, and provide food. Gambia's president has also initiated programs that support reforestation of the Sahel.

Chaparral

In many coastal regions that border on deserts, such as southern California and much of the Mediterranean, we find a unique type of vegetation called **chaparral**. The annual rainfall in these regions is up to 30 inches, nearly all of which falls during cool, wet winters that alternate with hot,

FIGURE 29-16 The chaparral biome
Limited to coastal regions and maintained by fires set by lightning, this hardy biome is characterized by drought-resistant shrubs and small trees. Some of the shrubs in this chaparral near San Francisco, California, turn brilliant red in the fall.

FIGURE 29-17 Tallgrass prairie
In the central U.S., moisture-bearing winds out of the Gulf of Mexico produce summer rains, allowing a lush growth of tall grasses and abundant wildflowers. Periodic fires, now carefully managed, prevent the encroachment of forest. QUESTION Why is tallgrass prairie one of the most endangered biomes in the world?

dry summers. The proximity of the sea provides a slightly longer rainy season in the winter and frequent fogs during the spring and fall. Chaparral plants consist mainly of drought-resistant shrubs and small trees. Their leaves are usually small, and are often coated with tiny hairs or protective layers that reduce evaporation during the dry summer months. These hardy shrubs are also able to withstand the frequent summer fires started by lightning (FIG. 29-16).

Grasslands

In the temperate regions of North America, deserts occur in the rain shadows east of the mountain ranges, such as the Sierra Nevada and Rocky Mountains. Eastward, as the rainfall gradually increases, the land supports more and more grasses, giving rise to the prairies of the Midwest. Most **grassland**, or **prairie**, biomes are located in the centers of continents, such as North America and Eurasia, and receive 10 to 30 inches (25 to 75 centimeters) of rain annually. In general, they have a continuous cover of grass and virtually no trees except along the rivers. From the tallgrass prairies of Iowa, Missouri, and Illinois where the rainfall is relatively high (FIG. 29-17), to the drier shortgrass prairies of eastern Colorado, Wyoming, and Montana (FIG. 29-18), the North American grassland once stretched across almost half the continent.

Water and fire are the crucial factors in the competition between grasses and trees. The hot, dry summers and frequent droughts of the shortgrass prairies can be tolerated by grass but are fatal to trees. In the more eastern tallgrass prairies, forests are the climax ecosystems; historically, however, the trees were destroyed by frequent fires, caused by lightning or deliberately set by Native Americans to maintain grazing land for bison. Although fire kills trees, the root systems of grass usually survive. The grasslands of North America once supported huge herds of bison—as many as 60 million in the early nineteenth century. Pronghorn antelope can still be seen in some prairies of the western U.S., where bobcats and coyotes are the major large predators (a prairie food web is illustrated in Fig. 28-5). Grasses growing and decomposing for thousands of years produced what may be the most fertile soil in the world. An acre of natural tallgrass prairie in the U.S. supports 200–400 different native plants.

Human Impact

The development of plows that could break through the dense grass turf set the stage for converting the midwestern U.S. prairies into the "breadbasket" of North America, so named because enormous quantities of grain are cultivated in its fertile soil. The tallgrass prairie, now one of the most endangered ecosystems worldwide, has nearly all been converted to agricultural land. Only about 1% remains, in tiny protected remnants maintained by periodic controlled burning.

On the dry western shortgrass prairie, cattle have replaced the bison and pronghorn antelope. Overgrazing can destroy grasses, which may be replaced by woody sagebrush (FIG. 29-19). Several states of the western and midwestern U.S., recognizing the importance of these biomes to wildlife,

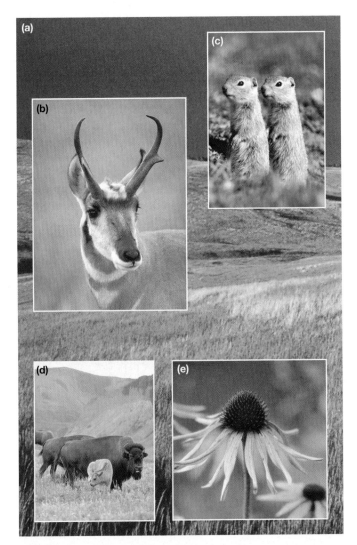

FIGURE 29-18 Shortgrass prairie
The lands east of the Rocky Mountains receive relatively little rainfall, and **(a)** shortgrass prairie results, characterized by low-growing bunchgrasses such as buffalo grass and grama grass. **(b)** Pronghorn antelope, **(c)** prairie dogs, and **(d)** protected bison herds occupy this biome, in which **(e)** wildflowers such as this coneflower abound.

FIGURE 29-19 Sagebrush desert or shortgrass prairie?
Biomes are influenced by human activities as well as by temperature, rainfall, and soil. The shortgrass prairie field on the right has been overgrazed by cattle, causing the grasses to be replaced by sagebrush.

are protecting expanses of shortgrass prairie and restoring tallgrass prairie. Cattle ranchers increasingly recognize that limiting grazing on fragile lands keeps them productive.

Temperate Deciduous Forests

At their eastern edge, the North American grasslands merge into the **temperate deciduous forest** biome, also found in Western Europe and East Asia (**FIG. 29-20**). More precipitation occurs there than in the grasslands (30 to 60 inches, or 75 to 150 centimeters), particularly during the summer. The soil retains enough moisture for trees to grow, and the resulting forest shades out grasses. In contrast to tropical forests, the temperate deciduous forest biome has cold winters, usually with at least several hard frosts and long periods of below-freezing weather. Winter in this biome has an effect on the trees similar to that of the dry season in the tropical deciduous forests: During periods of subfreezing temperatures, liquid water is not available. To

FIGURE 29-20 The temperate deciduous forest biome
(a) In temperate deciduous forests of the eastern U.S., **(b)** the white-tailed deer is the largest herbivore, and **(c)** birds such as the blue jay are abundant. **(d)** In spring, a profusion of woodland wildflowers (such as these hepaticas) blooms briefly before the trees produce leaves.

reduce evaporation when water is in short supply, the trees drop their leaves in the fall. They produce leaves again in the spring, when liquid water becomes available. During the brief time in spring when the ground has thawed but the trees have not yet blocked off all the sunlight, abundant wildflowers grace the forest floor.

Insects and other arthropods are numerous and conspicuous in deciduous forests. The decaying leaf litter on the forest floor also provides food and habitat for bacteria, earthworms, fungi, and small plants. Many arthropods feed on these or on each other. A variety of vertebrates—including mice, shrews, squirrels, raccoons, deer, bears, and many species of birds—dwell in the deciduous forests.

Human Impact

Large predatory mammals such as black bears, wolves, bobcats, and mountain lions were formerly abundant, but hunting and habitat loss has severely reduced their numbers and effectively eliminated wolves from deciduous forests. In many deciduous forests, deer are plentiful due to lack of natural predators. Clearing for lumber, agriculture, and housing has dramatically reduced deciduous forests in the U.S. from their original extent, and virgin deciduous forests are now almost nonexistent. Over the past 50 years, however, Forest Service data show that forest cover in the U.S. (both evergreen and deciduous) has increased as a result of regrowth of forests on abandoned farms, paper recycling that decreases demand for wood pulp, more efficient lumber milling and tree farming techniques, and the use of alternative building materials.

Temperate Rain Forests

On the U.S. Pacific coast, from the lowlands of the Olympic Peninsula in Washington State to southeast Alaska, lies a **temperate rain forest** biome (**FIG. 29-21**). Temperate rain forests, which are relatively rare, are also located along the southeastern coast of Australia and the southwestern coasts of New Zealand and Chile. As in the tropical rain forest, there is no shortage of liquid water year-round. This abundance of water is due to two factors. First, there is a tremendous amount of rain. The Hoh River rain forest in Olympic National Park receives more than 160 inches (400 centimeters) of rain annually, more than 24 inches (60 centimeters) in the month of December alone. Second, the moderating influence of the Pacific Ocean prevents severe frost from occurring along the coast, so the ground seldom freezes and liquid water remains available.

The abundance of water means that the trees have no need to shed their leaves in the fall, and almost all the trees are evergreens. In contrast to the broadleaf evergreen trees of the Tropics, temperate rain forests are dominated by conifers. The ground and typically the trunks of the trees are covered with mosses and ferns. As in tropical rain forests, so little light reaches the forest floor that tree seedlings usually cannot become established. Whenever one of the forest giants falls, however, it opens up a patch of light, and new seedlings quickly sprout, commonly right atop the fallen log. This event produces a "nurse log" (see Fig. 29-21b).

Taiga

North of the grasslands and temperate forests, the **taiga**, also called the **northern coniferous forest** (**FIG. 29-22**), stretches from east to west across all of North America and Eurasia, including parts of the northern U.S. and much of southern Canada. Conditions in the taiga are much harsher than in temperate deciduous forests, with long, cold winters and short growing seasons. This severely limits the trees' ability to photosynthesize and acquire both energy and nutrients. As a result, the taiga is populated almost entirely by evergreen coniferous trees with narrow, waxy needles that

FIGURE 29-21 The temperate rain-forest biome
(a) The Hoh River temperate rain forest in Olympic National Park. Coniferous trees do not block the light as effectively as do broadleaf trees, so ferns, mosses, and wildflowers grow in the pale green light of the forest floor. **(b)** The dead feeds the living, as new trees grow on this "nurse log," and **(c)** flowering foxglove and **(d)** fungi find ideal conditions amid the moist, decaying vegetation.

FIGURE 29-22 **The taiga (northern coniferous forest) biome**
(a) The small needles and pyramidal shape of conifers allow them to shed heavy snows. Taiga predators include **(b)** the Canada lynx and **(c)** the great horned owl.

FIGURE 29-23 **Clear-cutting**
Clear-cutting, as seen in this Oregon forest, is relatively simple and cheap, but its environmental costs are high. Erosion diminishes the fertility of the soil, slowing new growth. Further, the dense stands of same-age trees that typically regrow are more vulnerable to attack by parasites than a natural stand of trees of various ages would be.

Tundra

The last biome encountered before we reach the polar ice caps is the arctic **tundra**, a vast treeless region bordering the Arctic Ocean (**FIG. 29-24**). Conditions in the tundra are severe. Winter temperatures in the arctic tundra often reach −40°F (−55°C) or below, winds howl at 30 to 60

remain on the trees year-round. The waxy coating and small surface area of the needles reduce water loss by evaporation during the cold months. The tree conserves energy by not regenerating all its leaves each spring, and it is instantly ready to take advantage of good growing conditions when spring arrives. Because of the harsh climate in the taiga, the diversity of life there is much lower than in many other biomes. Vast stretches of central Alaska, for example, are covered by a somber forest that consists almost exclusively of black spruce and an occasional birch. Large mammals such as the wood bison, grizzly bear, moose, and wolf still roam the taiga, as do smaller animals such as the wolverine, fox, snowshoe hare, and deer. Wolf populations are found in the taiga of Canada, and down into Idaho, Michigan, Wisconsin, Minnesota, and Montana (where they have been reintroduced into Yellowstone National Park).

Human Impact

The taiga is a major source of lumber. Clear-cutting, the removal of all the trees in a given area for use in papermaking and construction, has destroyed huge expanses of forest in both Canada and the U.S. Pacific Northwest (**FIG. 29-23**). However, owing to the remoteness of the northernmost taiga and the severity of its climate, a greater percentage of the taiga remains in undisturbed condition than any other North American biome except the tundra.

FIGURE 29-24 **The tundra biome**
(a) Life on the tundra is adapted to cold. **(b)** Plants such as dwarf willows and perennial wildflowers (such as this dwarf clover) grow low to the ground, escaping the chilling tundra wind. Tundra animals, such as **(c)** caribou and **(d)** arctic foxes, can regulate blood flow in their legs, keeping them just warm enough to prevent frostbite while preserving precious body heat for the brain and vital organs.

miles (50 to 100 kilometers) per hour, and precipitation averages 10 inches (25 centimeters) or less each year, making this region a "freezing desert." Even during the summer, the temperatures can drop to freezing, and the growing season may last only a few weeks. Somewhat less cold but similar conditions produce alpine tundra on mountaintops above the altitude where trees can grow.

The cold climate of the arctic tundra results in **permafrost**, a permanently frozen layer of soil typically no more than about 1.5 feet below the surface. As a result, when summer thaws come, the water from melted snow and ice cannot soak into the ground, and the tundra becomes a huge marsh. Trees cannot survive in the tundra, partly because permafrost severely limits the depths to which their roots can penetrate.

Nevertheless, the tundra supports a surprising abundance and variety of life. The ground is carpeted with small perennial flowers, dwarf willows only a few inches tall, and a large lichen called "reindeer moss," a favorite food of caribou. The standing water provides a superb habitat for mosquitoes. These and other insects provide food for numerous birds, most of which migrate long distances to nest and raise their young during the brief summer feast. The tundra vegetation supports lemmings, which are eaten by wolves, snowy owls, arctic foxes, and even grizzly bears.

Human Impact

The tundra is among the most fragile of all the biomes because of its short growing season. A willow 4 inches (10 centimeters) high may have a trunk 3 inches (7 centimeters) in diameter and be 50 years old. Human activities in the tundra leave scars that persist for centuries. Fortunately for the tundra inhabitants, the impact of civilization is localized around oil-drilling sites, pipelines, mines, and military bases.

Rainfall and Temperature Limit the Plant Life of a Biome

Terrestrial biomes are greatly influenced by both temperature and rainfall, whose effects interact. Temperature strongly influences the effectiveness of rainfall in providing soil moisture for plants and standing water for animals to drink. The hotter it is, the more rapidly water evaporates, both from the ground and from plants. As a result of this interaction of temperature with rainfall (and to a lesser extent, the distribution of rain throughout the year), areas that receive almost exactly the same rainfall can have startlingly different vegetation, all the way from desert to taiga. Now that you are familiar with biomes, let's travel from southern Arizona to central Alaska, visiting ecosystems that each receive about 12 inches (28 centimeters) of rain annually.

The Sonoran Desert near Tucson, Arizona (see Fig. 29-14), has an average annual temperature of 68°F (20°C) and receives about 12 inches of rain each year. The landscape is dominated by giant saguaro cactus and low-growing, drought-resistant bushes. Going about 900 miles (1500 kilometers) north will bring you to eastern Montana, where rainfall is about the same; however, you will be amid short-grass prairie (see Fig. 29-18) largely because the average

temperature is much lower, about 45°F (7°C). Much farther north, central Alaska receives about the same annual rainfall, yet it is covered with taiga forest (see Fig. 29-22). As a result of the low average annual temperature (about 25°F, or −4°C), permafrost underlies much of the ground there. During the summer thaw, the taiga earns its Russian name "swamp forest," although its rainfall is about the same as that of the Sonoran Desert.

29.4 HOW IS LIFE IN WATER DISTRIBUTED?

Freshwater Ecosystems Include Lakes, Streams, and Rivers

Although they are as diverse as terrestrial ecosystems, aquatic ecosystems share three general features. First, because water is slower to heat and cool than air, temperatures in aquatic ecosystems are more moderate than those in terrestrial ecosystems. Second, water absorbs light; even in very clear water, below 650 feet (200 meters) little light is left to power photosynthesis. Suspended *sediment* (nonliving particles carried by moving water) or microorganisms greatly reduce light penetration. Finally, nutrients in aquatic ecosystems tend to be concentrated near the bottom sediments, so where nutrients are highest, the light levels are lowest. Of the four requirements for life, aquatic ecosystems provide abundant water and appropriate temperatures. Thus, the availability of energy and nutrients largely determines the quantity of life and the distribution of life in aquatic ecosystems.

Freshwater Lakes Have Distinct Regions of Life

Freshwater lakes form when large natural depressions fill with water from sources including groundwater seepage, streams, or runoff from rain or melting snow. Lakebeds have diverse origins. Many were excavated by glaciers as they scoured the landscape millennia ago; others formed when landslides or debris deposited by slow-flowing rivers dammed up water behind them. A few, such as Crater Lake in Oregon, occupy remnant cones of extinct volcanoes.

Although lakes vary tremendously in size, depth, and nutrient content, moderate to large lakes in temperate climates share some common features, including distinct zones of life. The distribution, quantity, and type of life in lakes depends largely on access to light, nutrients, and in some cases a place for attachment (the bottom). Although small lakes, called *ponds*, often have ample light and nutrients all the way to the bottom, larger lakes have *life zones* that correspond to specific depths (**FIG. 29-25**).

Near the shore is a shallow **littoral zone**, where plants find abundant light, anchorage, and nutrients from the bottom sediments. Littoral zone communities are the most diverse, including plants such as cattails, bullrushes, and water lilies near the shore, with submerged plants and algae flourishing in deeper littoral waters. The greatest diversity of animal life is also found in the littoral zone; vertebrates include frogs, fish (such as pike, bluegill, and perch), and aquatic snakes and turtles. Littoral invertebrates include

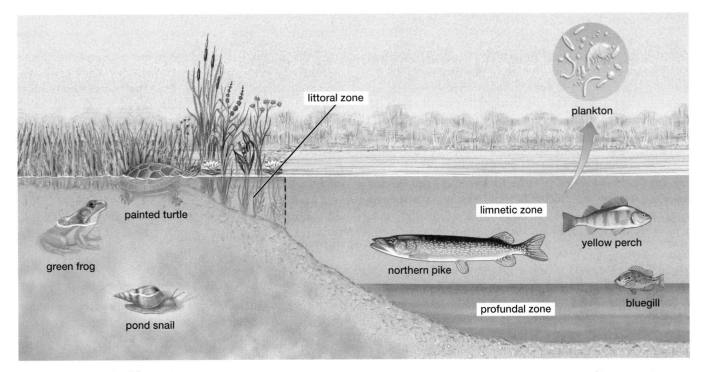

FIGURE 29-25 Lake life zones
There are three life zones in a "typical" large lake: a nearshore littoral zone with rooted plants, an open-water limnetic zone, and a deep, dark profundal zone. Fish swim among all the zones.

crustaceans (such as crayfish), insect larvae, snails, and flatworms. Littoral waters are also home to small, unattached organisms collectively called **plankton**. There are two forms of plankton: **phytoplankton** (Greek, "drifting plants"), which includes photosynthetic protists and bacteria, as well as algae that may form masses of microscopic filaments; and **zooplankton** (Greek, "drifting animals"), which includes non-photosynthetic protists and tiny crustaceans.

As the water increases in depth, plants are unable to anchor to the bottom and still photosynthesize. This open-water region is divided into an upper *limnetic zone* and a lower *profundal zone* (see Fig. 29-25). In the **limnetic zone**, enough light penetrates to support photosynthesis, and plankton, fish, and floating plants such as duckweed predominate. Below lies the **profundal zone**, where light is too weak to support photosynthesis. This area is nourished both by organic matter that drifts down from the littoral and limnetic zones and by incoming sediment. It is inhabited primarily by decomposers and detritus feeders, such as bacteria, snails and insect larvae, and by fish that swim freely among the life zones.

Freshwater Lakes Are Classified According to Their Nutrient Content

Freshwater lakes are sometimes classified according to their nutrient levels, as *eutrophic* (Greek, "well fed") or *oligotrophic* (Greek, "poorly fed"). As you might predict, many fall in between and can be described as *mesotrophic* (Greek, "middle fed"). Here we describe the two extremes.

Oligotrophic lakes are very low in nutrients and support relatively little life. Many were formed by glaciers that scraped depressions in bare rock and are fed by mountain streams. Because there is little sediment or microscopic life

to cloud the water, oligotrophic lakes are clear, and light penetrates deeply. Fish such as trout, which require well-oxygenated water, often thrive in oligotrophic lakes because there is little organic material to decay and deplete the oxygen.

Eutrophic lakes receive relatively large inputs of sediments, organic material, and inorganic nutrients (such as phosphorus) from their surroundings, allowing them to support dense plant communities (**FIG. 29-26**). They are murky from suspended sediment and dense phytoplankton

FIGURE 29-26 A eutrophic lake
Rich in dissolved nutrients due to runoff from the land, often provided by human activities such as farming, eutrophic lakes support dense growths of algae, phytoplankton, and both floating and rooted plants.

599

populations, so the lighted limnetic zone is shallower. The dead bodies of the producers and other forms of life fall into the profundal zone, where they are used as food by decomposer organisms. The metabolic activities of these decomposers use up oxygen, so that the profundal zone of eutrophic lakes is often very low in oxygen.

Although large lakes may persist for millions of years, over time, as nutrient-rich sediment accumulates, oligotrophic lakes gradually tend to become eutrophic, a process called *eutrophication*. This same process—operating over geologic time for large lakes—may eventually cause them to undergo succession to dry land (see Chapter 27).

Human Impact

Human activities can greatly accelerate the process of eutrophication, because nutrients are carried into lakes from farms, feedlots, sewage, and even fertilized suburban lawns. Over-enriched lakes become clogged with microorganisms whose dead bodies are attacked by bacteria that deplete the water of oxygen. Normal community interactions are disrupted as organisms in higher trophic levels are smothered. Lake Erie was suffering severe eutrophication due to phosphate detergents and runoff from fertilized farm fields, which together nourish dense growths of phytoplankton. Agreements between the U.S. and Canada have greatly improved Lake Erie's water quality, and have helped prevent eutrophication of the larger Great Lakes. Invasive species remain a problem in large lakes, with the Great Lakes hosting over 150 different nonnative species, including the zebra mussel.

Acid rain (see Chapter 28) poses a very different threat, particularly to small freshwater lakes and ponds. In the Adirondack Mountains of New York State, roughly 25% of the lakes have been rendered nearly lifeless by acid rain. Because power plants now emit less sulfur dioxide, many of these lakes are showing signs of improvement.

Streams Collect Surface Water and Funnel It into Rivers

Streams often begin in mountains where runoff from rain and melting snow cascades over impervious rock—the *source* region shown in **FIGURE 29-27**. Little sediment reaches the streams at this point, phytoplankton is sparse, and the water is clear and cold. Algae adhere to rocks in the streambed, where insect larvae find food and shelter. Turbulence keeps mountain streams well oxygenated, providing a good home for trout that feed on insect larvae.

At lower elevations, in the *transition* region, small side streams or *tributaries* merge, forming wider, slower-moving streams and small rivers. The water warms slightly, and more sediment is carried in, providing nutrients that allow aquatic plants, algae, and phytoplankton to proliferate. Fish such as black bass, bluegill, and channel catfish (which require less oxygen than trout do) are found here.

As the land becomes still lower and flatter, the river warms, widens, and slows, meandering back and forth. Side streams carry in nutrient-rich sediment and deposit it the riverbed. The water becomes murky with sediment and dense growths of phytoplankton. Decomposer bacteria deplete the oxygen in deeper water, but carp and catfish can still thrive where oxygen is relatively low. When precipitation or snowmelt is high, the river may flood surrounding flat land, or *floodplain*, depositing a rich layer of sediment over the adjoining terrestrial ecosystem.

Rivers drain into lakes or into other rivers that usually lead ultimately to oceans. As the river approaches sea level, the land flattens, flow rates diminish, and sediment is deposited. This interrupts the river's flow, breaking it into small channels amidst the rich sediment. Saltwater from the ocean mixes with the incoming fresh water, creating **estuaries**, wetlands that support enormous biological productivity and diversity.

Most Wetlands in the U.S. Are Freshwater Habitats

Most wetlands in the U.S., variously called marshes, swamps, bogs, or fens, are freshwater ecosystems. Some are isolated, while others occur near lakes or within river floodplains. Wetlands act as giant sponges, absorbing water when it is abundant, and helping to recharge underground supplies. They provide breeding, sheltering, and feeding sites for freshwater fish and many species of birds and mammals. The Everglades of Southern Florida are among the largest wetlands in the world.

Human Impact

Rivers have been channelized (deepened and straightened) to facilitate boat traffic, prevent flooding, and allow farming along their banks. This change has caused increased erosion, as the rivers flow more rapidly, and loss of the nutrients that flooding formerly provided to the nearby floodplain. China is currently building Three Gorges Dam on the Yangtze River to provide water for farming. The dam is trapping nutrient-laden water that no longer enriches the East China Sea, where it formerly fostered an extensive growth of phytoplankton. The phytoplankton, in turn, supported one of the world's largest fisheries, which is now threatened by the world's largest dam. In the U.S., northwestern Pacific and northeastern Atlantic salmon populations, which must spawn in clean, free-flowing rivers, have been greatly reduced by hydroelectric dams, water diversion for agriculture, erosion from logging operations, overfishing, and acidification of water from acid rain. On both coasts of the U.S., state, federal, and local groups are working to restore the rivers and streams that support rich wildlife communities, including endangered salmon populations.

Logging, filling, and draining freshwater wetlands for housing, commercial uses, and agriculture has reduced the extent of freshwater wetlands by about half in the U.S. In addition to eliminating wildlife habitat, this loss contributes to the severity of floods. Fortunately, local, state, and federal agencies have enacted laws and forged partnerships to protect existing wetlands and restore some that have been degraded. As a result, the rate of wetland loss has diminished in the U.S. One of the largest ecosystem restorations ever attempted, the Comprehensive Everglades Restoration Plan, is currently under way (see Chapter 30).

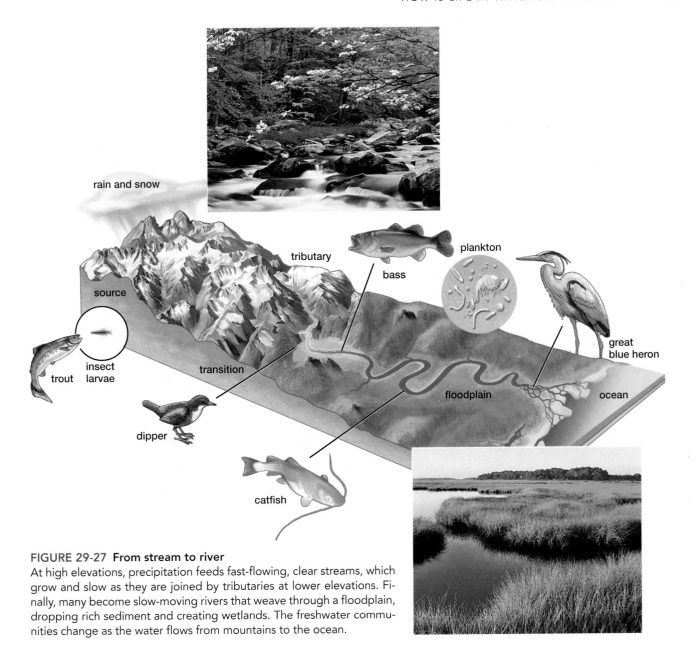

FIGURE 29-27 From stream to river
At high elevations, precipitation feeds fast-flowing, clear streams, which grow and slow as they are joined by tributaries at lower elevations. Finally, many become slow-moving rivers that weave through a floodplain, dropping rich sediment and creating wetlands. The freshwater communities change as the water flows from mountains to the ocean.

Marine Ecosystems Cover Much of Earth

In the oceans, the upper layer of water, where the light is strong enough to support photosynthesis (to a depth of about 650 feet, or 200 meters), is called the **photic zone**. Below the photic zone lies the **aphotic zone**, where nearly all the energy needed to support life is extracted from the excrement and bodies of organisms that sink or swim into this deep region (**FIG. 29-28**).

As in lakes, most of the nutrients in the oceans are near the bottom, where there is not enough light for photosynthesis. Photic zone nutrients are constantly incorporated into the bodies of organisms and carried into the ocean depths when the creatures die. These are replenished from two major sources: runoff from the land and **upwelling** from the ocean depths. Upwelling occurs around Antarctica and along western coastlines—as in California, Peru, and West

Africa—where prevailing winds displace surface water, causing it to be replaced by cold, nutrient-laden water from below. Not surprisingly, the major concentrations of life in the oceans are found where abundant light is combined with a source of nutrients, which occurs most commonly in regions of upwelling and in shallow coastal waters.

Coastal Waters Support the Most Abundant Marine Life
Intertidal and Nearshore Zones
The most abundant life in the oceans is found in a narrow strip surrounding Earth's landmasses, where the water is shallow and a steady flow of nutrients washes off the land. Coastal waters include the **intertidal zone**, a region alternately covered and uncovered with the rising and falling tides; and the **nearshore zone**, relatively shallow but constantly submerged areas including bays and

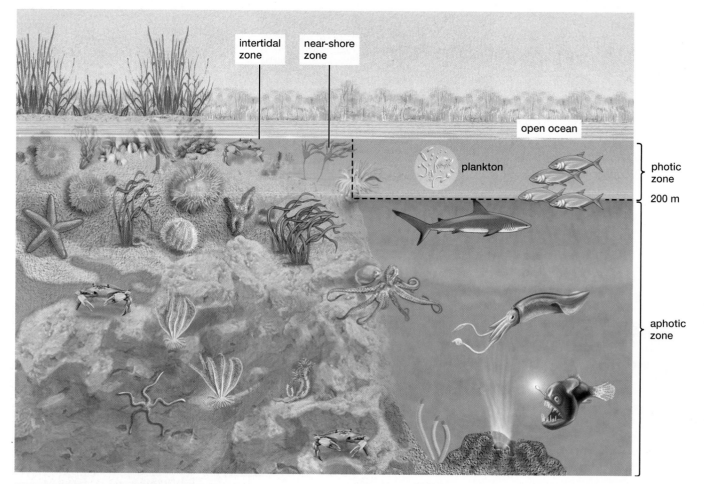

FIGURE 29-28 Ocean life zones
Photosynthesis can occur only in the upper photic zone, which includes the intertidal and nearshore zones and the upper waters of the open ocean. Life that remains in the aphotic zone relies on energy-rich material that drifts down from the photic zone or, in the unique case of hydrothermal vent communities, on energy stored in hydrogen sulfide.

coastal wetlands (**FIG. 29-29**). Coastal wetlands include salt marshes, the gradually sloping coastal areas that are protected from waves, and estuaries, wetlands formed where rivers meet the oceans. Both contain accumulations of rich sediment. The nearshore zone is the only part of the ocean where large salt-tolerant plants or seaweeds can grow, anchored to the bottom. In addition, the abundance of nutrients and sunlight in this zone promotes the growth of a veritable soup of photosynthetic phytoplankton. Associated with these plants and protists are animals from nearly every phylum: annelid worms, sea anemones, sea jellies, sea urchins, sea stars, mussels, snails, fish, and sea otters, to name just a few. A large number and variety of organisms live permanently in coastal waters, but many that spend most of their lives in the open ocean come into the coastal waters to reproduce. Bays, salt marshes, and estuaries in particular are the breeding grounds for a wide variety of organisms, such as crabs, shrimp, and an array of fish, including most of our commercially important species. Off the coast of California, great undersea forests of kelp provide food

and shelter for a rich assemblage of fish and invertebrates that in turn provide food for otters and seals (see Fig. 29-29d). The productivity of freshwater and saltwater wetlands is second only to that of rain forests.

Human Impact

As the population increases in coastal states, and resources such as oil become increasingly scarce, the conflict between the preservation of coastal wetlands as wildlife habitat and the development of these areas for housing, harbors, marinas, and energy extraction will become increasingly intense. Estuaries are also threatened by runoff from farming operations, which provide a glut of nutrients that foster excessive growth of producers, whose decaying bodies deplete the water of oxygen, killing fish and invertebrates. Extensive loss of coastal wetlands, which serve as wave buffer zones, increased the damage done to New Orleans, Louisiana, by Hurricane Katrina in 2005. Fortunately, conservation efforts have slowed the loss of wetlands in the U.S., and some, such as the Florida Everglades, are being restored.

FIGURE 29-29 Nearshore ecosystems
(a) A salt marsh in the eastern U.S. Expanses of shallow water fringed by marsh grass (*Spartina*) provide excellent habitat and breeding grounds for many marine organisms and shorebirds. (b) Although the shifting sands present a challenge to life, grasses stabilize them, and animals such as (inset) this *Emerita* crab burrow in the sandy intertidal zone. (c) A rocky intertidal shore in Oregon, where animals and algae grip the rock against the pounding waves and resist drying during low tide. (Inset) Colorful sea stars cling to the rocks surrounded by algae (*Fucus*). (d) Towering kelp sway through the clear water off southern California, providing the basis for a diverse community of invertebrates, fishes, and (inset) an occasional sea otter. QUESTION Nearshore ecosystems have the highest productivity in the ocean. What factors explain this? Which of the ecosystems pictured here would you expect to have the highest productivity, and why?

Coral Reefs

In warm tropical waters with just the right combination of bottom depth, wave action, and nutrients, specialized algae and corals build reefs from their own calcium carbonate skeletons. **Coral reefs** are most abundant in tropical waters of the Pacific and Indian Oceans, the Caribbean, and the Gulf of Mexico as far north as southern Florida, where the maximum water temperatures range between 72°F and 82°F (22°C and 28°C).

Reef-building corals, which are related to anemones, harbor photosynthetic unicellular algae within their tissues in a mutualistic relationship. The algae represent up to half the weight of the coral polyp and give the corals their diverse, brilliant colors (**FIG. 29-30**). These corals

thrive within the photic zone at depths of less than 130 feet (40 meters), where light penetrates the clear water and provides energy for photosynthesis. The algae benefit from the high nitrogen, phosphorus, and carbon dioxide levels in the coral tissues. In return, algae provide food for the coral and help produce calcium carbonate, which forms the coral skeleton. The skeletons of corals accumulate over thousands of years, providing an anchorage, shelter, and food for the most diverse community of algae, invertebrates, and fish in the oceans (see Fig. 29-30). Coral reefs might be considered the "ocean's rain forests," since they are home to more than 90,000 known species, with probably ten times that number yet to be identified. The Great Barrier Reef in Australia supports

FIGURE 29-30 Coral reefs
(a) Coral reefs, composed of the bodies of corals and algae, provide habitat for an extremely diverse community of extravagantly colored animals. **(b)** Many fish, including this blue tang, feed on coral (note the bright yellow corals in the background). A vast array of invertebrates such as **(c)** this sponge and **(d)** blue-ringed octopus live among the corals of Australia's Great Barrier Reef. This tiny octopus (6 inches, fully extended) is one of the world's most venomous creatures. QUESTION Why does "bleaching" threaten the life of a coral reef? What causes bleaching?

more than 200 species of coral alone, and a single reef may harbor 3000 identified species of fish, invertebrates, and algae.

Human Impact

Anything that diminishes the water's clarity harms the coral's photosynthetic algal partners and hinders coral growth. Runoff from farming, agriculture, logging, and construction carries silt and an excess of nutrients that promote eutrophication that reduce both sunlight and oxygen. Silt has ruined several reefs near Honolulu, Hawaii. In the Philippines, rain-forest logging has dramatically increased erosion, destroying reefs as well as rain forests.

In many tropical countries, mollusks, turtles, fish, crustaceans, and the corals themselves are being harvested from reefs faster than they can reproduce, primarily for sale to tourists, shell enthusiasts, and aquarium owners in developed countries. Removing predatory fish and invertebrates from reefs may disrupt the ecological balance of the community, allowing an explosion in populations of coral-eating sea urchins or sea stars.

Complex interactions involving both human disturbances and global warming are hastening the spread of diseases among corals. *Coral bleaching* occurs when waters become too warm, causing the corals to expel their colorful symbiotic algae and making them appear bleached white. In 2002, Australia's Great Barrier Reef suffered bleaching in almost 60% of its coral. The algae will return if the water cools; but without their algal partners, corals will gradually starve. Florida's coral reefs suffer from bleaching, infections, and sediments that cloud the water and promote growth of harmful algae. More than a million divers and snorkelers visit the reefs annually, sometimes damaging the coral.

But there is also some good news. A fishing ban in reserves in the Florida Keys has begun to restore several important species. Many coral reefs are also protected within the world's largest marine sanctuaries, Australia's Great Barrier Reef Marine Park and the Northwestern Hawaiian Islands Marine National Monument. Collectively, about 20,000 species thrive in these two hotspots of biodiversity.

The Open Ocean

Beyond the coastal regions lie vast areas of the ocean in which the bottom is too deep to allow plants to anchor and still receive enough light to grow. Most life in the open ocean (**FIG. 29-31**) is limited to the upper photic zone, where life-forms are **pelagic**—that is, free-swimming or floating—for their entire lives. The food web of the open ocean is dependent on phytoplankton consisting of microscopic photosynthetic protists, mainly diatoms and dinoflagellates (Fig. 29-31d). These organisms are consumed by zooplankton, such as tiny crustaceans that are relatives of crabs and lobsters (Fig. 29-31e). Zooplankton in turn serve as food for larger invertebrates, small fish, and even marine mammals such as the humpback whale (see Fig. 29-31b).

To remain afloat in the photic zone, where sunlight and food are abundant, many members of the planktonic community have buoyant oil droplets in their cells or spines that slow their rate of sinking (see **FIG. 29-31d**). Most fish have gas-filled swim bladders that regulate their buoyancy. Some animals swim to stay in the photic zone. Many small crustaceans migrate to the surface at night to feed, then sink into the dark depths during daylight, thus avoiding visual predators such as fish. The amount of pelagic life varies tremendously from place to place. The blue clarity of tropical waters is due to a lack of nutrients, which limits the concentration of plankton in the water. Nutrient-rich waters that support a large plankton community are greenish and relatively murky.

Human Impact

Two major threats to the open ocean are pollution and overfishing. Oceangoing vessels dump millions of plastic containers overboard daily, and plastic refuse blows off the land. The plastic looks like food to unsuspecting sea turtles, gulls, porpoises, seals, and whales, many of which die after trying to consume it. Oil contaminates the open ocean from oil-tanker spills, runoff from improper disposal on land, leakage from offshore oil wells, and natural seepage. The Mississippi River carries nutrient-laden runoff from nitrogen-fertilized farms into the Gulf of Mexico, where it fuels an excessive growth of phytoplankton. The plankton die, sink to the seafloor, and provide a feast for decomposer bacteria that deplete the deep waters of oxygen. During the warmer months of the year, this creates a 7000-square-mile bottom *dead zone* where the marine community is nearly eliminated. The dead zone threatens both the local ecological community and the fishing industry that relies on a healthy ecosystem. Similar dead zones are developing in coastal waters throughout the world.

Increased demand for fish, fueled by growing human populations using more efficient fishing techniques, has caused fish to be harvested in unsustainable numbers (see Chapter 30). The once abundant cod populations of eastern Canada have nearly vanished, despite more than a decade of severe fishing restrictions; the New England cod fishery may be destined to follow. Populations of haddock, swordfish, tuna, and many other types of seafood have also declined dramatically because of overfishing. Dredging for fish, scallops, and crabs has not only depleted many of these populations, but can seriously damage seafloor ecosystems, harming many other species as well. Shark populations have declined precipitously, and many are now endangered due to overfishing. These slow-growing predators form a keystone species in ocean food webs. Because many types of sharks do not breed until they are a decade or more old and produce small numbers of offspring, their populations are slow to recover.

Marine reserves are increasingly being established throughout the world, causing substantial improvements in the diversity, number, and size of marine animals within these areas. Nearby areas benefit because the reserves act as

FIGURE 29-31 The open ocean
The open ocean supports abundant life in the photic zone, including marine mammals such as **(a)** porpoises and **(b)** humpback whales and **(c)** fish such as the blue jack. **(d)** The photosynthetic phytoplankton are the producers on which most other marine life ultimately depends. Phytoplankton are eaten by **(e)** zooplankton, represented by this tiny crustacean. Spines help keep planktonic organisms afloat within the photic zone.

FIGURE 29-32 Denizens of the deep
(a) A viperfish, whose huge jaws and sharp teeth allow it to grasp and swallow its prey whole, entices victims with a luminescent lure. **(b)** A deep ocean squid. **(c)** The skeleton of a whale provides an undersea nutrient bonanza. **(d)** A "zombie worm" can insert its rootlike lower body deep into the bones of the decomposing whale carcass.

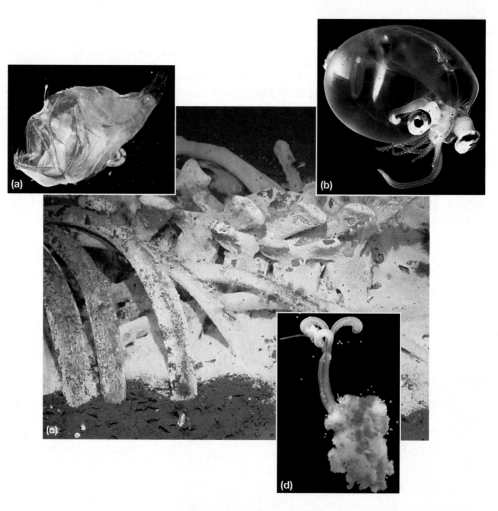

nurseries that restore populations outside the reserve. Many countries have established quotas on sharks and other fish whose populations are threatened, and dredging has recently been banned along the western coast of the U.S. to protect marine communities. Fish farming, or *aquaculture*, can help meet the demand for some types of seafood, including shrimp and salmon, but the "farms" must be carefully designed and managed to prevent farmed species from escaping and to avoid damaging local ecosystems.

Unique Communities Cover the Ocean Floor

Deep Ocean Communities

Below the photic zone, the only available energy in most regions comes from the excrement and dead bodies that drift down from above. Nevertheless, life within the aphotic zone is found in amazing quantity and variety, including fishes of bizarre shapes, squid, corals, worms, sea cucumbers, sea stars, and mollusks (**FIG. 29-32**). Here, many animals generate their own light using complex metabolic pathways that release light energy, a phenomenon known as bioluminescence. Some fish maintain colonies of bioluminescent bacteria in special visible chambers on their bodies. Bioluminescence may help bottom dwellers to see (many have large eyes), attract prey (Fig. 29-32a), or attract mates. Little is known of the behavior and ecology of these amazing and exotic creatures, which never survive being brought to the surface.

Recently, entire communities—including species new to science—have been found feeding on bodies of whales, each of which brings an average of 40 tons of food onto the ocean floor (Fig. 29-32c). First, fish and crabs strip the carcass of muscle and blubber and then dense masses of worms, clams, and snails swarm over the massive skeleton, extracting fats stored in the bones. The bodies of bone-eating "zombie worms," first described in 2005, consist mostly of rootlike structures that tunnel into the bone, a feeding strategy never before observed (Fig. 29-32d). Anaerobic bacteria then continue bone breakdown, and an extremely diverse community of clams, worms, mussels, and crustaceans moves in to feed on the bacteria during this stage of decomposition, which can last for many decades.

Hydrothermal Vent Communities

In 1977, geologists exploring the Galápagos Rift (an area of the Pacific floor where the plates that form Earth's crust are separating) found vents they called "black smokers" emitting superheated water that was blackened with sulfides and other minerals. Surrounding these vents was a rich **hydrothermal vent community** of pink fish, blind white crabs, enormous mussels, white clams, sea anemones, giant tube worms, and a snail sporting iron-laden armor plates (**FIG. 29-33**). Hundreds of new species have been found in these specialized habitats, which have now been discovered in many deep-sea regions where tectonic plates are separating, allowing material from Earth's interior to spew out.

In this unique ecosystem, sulfur bacteria serve as the primary producers. They harvest energy from an unlikely source that is deadly to most other forms of life—hydrogen sulfide discharged from cracks in Earth's crust. This process, called *chemosynthesis*, replaces photosynthesis in the vent communities, which flourish more than a mile below the ocean surface. Many vent animals consume the microorganisms directly, while others, such as the giant tube worm

FIGURE 29-33 Hydrothermal vent communities
"Black smokers" spew superheated water rich in minerals that provide both energy and nutrients to the vent community. Giant red tube worms may reach 9 feet (nearly 3 meters) in length and live up to 250 years. Parts of these worms are colored red by a hydrogen sulfide–trapping form of hemoglobin. (**inset**) The foot of this vent community snail is protected by scales coated with iron sulfide.

(which lacks a digestive tract), harbor the bacteria within their bodies and derive all their energy from them. The worm, which can reach a length of 9 feet, derives its red color from a unique form of hemoglobin that transports hydrogen sulfide to the symbiotic bacteria. These tube worms hold the record for invertebrate longevity—an astonishing 250 years.

The bacteria and archaea that inhabit the vent communities hold the record for survival at high temperatures. Some can survive water temperatures of 248°F (106°C); water at this depth can reach temperatures higher than boiling, because of the tremendous pressure. Scientists are investigating how the enzymes and other proteins of these heat-loving microbes can continue to function at temperatures that would destroy the proteins in our bodies, and they are looking for ways to put these amazing proteins to commercial use.

CASE STUDY REVISITED   WINGS OF HOPE

The Arabuko-Sokoke forest remains under siege from squatters who want to clear land and establish homes within its confines. But where the farmers gather their butterflies, the forest suffers far less poaching; the farmers now report poachers rather than joining them. Over several years of monitoring, project manager Ian Gordon sees no evidence that butterfly populations are being reduced. With full stomachs and money to buy a few minor luxuries, the people can now afford to support the philosophy of one village elder, who states, "The forest is here, we found the forest here, and we have to leave it here for our children's generation."

Consider This Most conservationists agree that "fencing and fining" is not the way to preserve habitat; local residents must actively support and participate in its preservation. Design or research other projects that fit the model of sustainable use of rain forests or other endangered ecosystems, such as coral reefs.

CHAPTER REVIEW

SUMMARY OF KEY CONCEPTS

29.1 What Factors Influence Earth's Climate?

The availability of sunlight, water, and appropriate temperatures determines the climate of a given region. Sunlight maintains Earth's temperature. Equal amounts of solar energy are spread over a smaller surface at the equator than farther north and south, making the equator relatively warm, while higher latitudes have lower overall temperatures. Earth's tilt on its axis causes dramatic seasonal variation at northern and southern latitudes.

Rising warm air and sinking cool air in regular patterns from north to south produce areas of low and high moisture. These patterns are modified by the topography of continents and by ocean currents.

Web Tutorial 29.1 Tropical Atmospheric Circulation and Global Climate

29.2 What Conditions Does Life Require?

The requirements for life on Earth include nutrients, energy, liquid water, and a reasonable temperature. The differences in the form and abundance of living things in various locations on Earth are largely attributable to differences in the interplay of these four factors.

29.3 How Is Life on Land Distributed?

On land, the crucial limiting factors are temperature and liquid water. On continents, large regions with similar climates will have similar vegetation, determined by the interaction of temperature and rainfall or the availability of water. These regions are called biomes.

Tropical forest biomes, located near the equator, are warm and wet, dominated by huge broadleaf evergreen trees. Most nutrients are tied up in vegetation, and most animal life is arboreal. Rain forests, home to at least 50% of all species, are rapidly being cut for agriculture, although the soil is extremely poor.

The African savanna is an extensive grassland with pronounced wet and dry seasons. It is home to the world's most diverse and extensive herds of large mammals.

Most deserts, which receive less than 10 inches of rain annually, are located between 20° and 30° N and S latitude and in the rain shadows of mountain ranges. In deserts, plants are widely spaced and have adaptations to conserve water. Animals have both behavioral and physiological mechanisms to conserve water and avoid excessive heat.

Chaparral exists in desertlike conditions that are moderated by their proximity to a coastline, allowing small trees and bushes to thrive. Grasslands, concentrated in the centers of continents, have a continuous grass cover and few trees. They produce the world's richest soils and have largely been converted to agriculture.

Temperate deciduous forests, whose broadleaf trees drop their leaves in winter to conserve moisture, dominate the eastern half of the U.S. and are also found in Western Europe and East Asia. Higher precipitation occurs there than in the grasslands. Wet temperate rain forests, dominated by evergreens, are found on the northern Pacific coast of the U.S. The taiga, or northern coniferous forest, covers much of the northern U.S., southern Canada, and northern Eurasia. It is dominated by conifers whose small, waxy needles are adapted for water conservation and year-round photosynthesis.

The tundra is a frozen desert where permafrost prevents the growth of trees and bushes remain stunted. Nonetheless, diverse arrays of animal life and perennial plants flourish in this fragile biome, which is found on mountain peaks and in the Arctic.

29.4 How Is Life in Water Distributed?

Energy and nutrients are the major limiting factors in the distribution and abundance of life in aquatic ecosystems. Nutrients are found in bottom sediments and are washed in from surrounding land, concentrating them near shore and in deep water.

In freshwater lakes, the nearshore littoral zone receives both sunlight and nutrients and supports the most diverse community. The limnetic zone is the lighted region of open water where photosynthesis can occur. In the deep profundal zone of large lakes, light is inadequate for photosynthesis and the community is dominated by heterotrophic organisms. Oligotrophic lakes are clear, low in nutrients, and support sparse communities. Eutrophic lakes are rich in nutrients and support dense communities. During succession, lakes tend to go from an oligotrophic to a eutrophic condition.

Streams begin at a source region, often in mountains, where water is provided by rain and snow. Source water is generally clear, high in oxygen, and low in nutrients. Streams join at lower elevations, carrying sediment from land and supporting a larger community in this transition region, where rivers form. On their way to lakes or oceans, rivers enter relatively flat floodplains where they deposit nutrients, take a meandering path, and spill over the land when precipitation is high.

Most life in the oceans is found in shallow water, where sunlight can penetrate, and is concentrated near the continents and in areas of upwelling, where nutrients are most plentiful. Coastal waters, consisting of the intertidal zone and the nearshore zone, contain the most abundant life. Producers include aquatic plants anchored to the bottom and photosynthetic protists called phytoplankton. Coral reefs are confined to warm, shallow seas. These calcium carbonate reefs form a complex habitat supporting the most diverse undersea ecosystem, now threatened by silt, overfishing, and global warming.

In the open ocean, most life is found in the photic zone, where light supports phytoplankton. In the lower aphotic zone, life is supported by nutrients that drift down from the photic zone. Many ocean fisheries have been overexploited.

The deep ocean is dark, many species are bioluminescent, and all are adapted to the tremendous water pressure. Whale carcasses provide a nutrient bonanza that supports a succession of unique communities over a span of many decades. Specialized vent communities, supported by chemosynthetic bacteria, thrive at great depths in the superheated waters where Earth's crustal plates are separating.

KEY TERMS

aphotic zone   *page 601*
biodiversity   *page 588*
biome   *page 587*
chaparral   *page 593*
climate   *page 582*
coral reef   *page 603*
desert   *page 592*
El Niño   *page 584*
estuary   *page 600*
eutrophic lake   *page 599*
grassland   *page 594*

gyre   *page 584*
hydrothermal vent community   *page 606*
intertidal zone   *page 601*
La Niña   *page 585*
limnetic zone   *page 599*
littoral zone   *page 598*
nearshore zone   *page 601*
northern coniferous forest   *page 596*
oligotrophic lake   *page 599*

ozone layer   *page 586*
pelagic   *page 605*
permafrost   *page 598*
photic zone   *page 601*
phytoplankton   *page 599*
plankton   *page 599*
prairie   *page 594*
profundal zone   *page 599*
rain shadow   *page 584*
savanna   *page 590*
taiga   *page 596*

temperate deciduous forest   *page 595*
temperate rain forest   *page 596*
tropical deciduous forest   *page 590*
tropical rain forest   *page 587*
tundra   *page 597*
upwelling   *page 601*
weather   *page 582*
zooplankton   *page 599*

THINKING THROUGH THE CONCEPTS

1. Explain how air currents contribute to the formation of the Tropics and the large deserts.

2. What are large, roughly circular ocean currents called? What effect do they have on climate, and where is that effect strongest?

3. What are the four major requirements for life? Which two are most often limiting in terrestrial ecosystems? In ocean ecosystems?

4. Explain why traveling up a mountain takes you through biomes similar to those you would encounter by traveling north for a long distance.

5. Where are the nutrients of the tropical forest biome concentrated? Why is life in the tropical rain forest concentrated high above the ground?

6. Explain two undesirable effects of agriculture in the tropical rain-forest biome.

7. List some adaptations of (a) desert plants and (b) desert animals to heat and drought.

8. What human activities damage deserts? What is desertification?

9. How are trees of the taiga adapted to a lack of water and a short growing season?

10. How do deciduous and coniferous biomes differ?

11. What single environmental factor best explains why there is shortgrass prairie in Colorado, tallgrass prairie in Illinois, and deciduous forest in Ohio?

12. Where are the world's largest populations of large herbivores and carnivores found?

13. Where is life in the oceans most abundant, and why?

14. Why is the diversity of life so high in coral reefs? What human impacts threaten them?

15. Distinguish among the limnetic, littoral, and profundal zones of lakes in terms of their location and the communities they support.

16. Distinguish between oligotrophic and eutrophic lakes. Describe (a) a natural scenario and (b) a human-created scenario under which an oligotrophic lake might be converted to a eutrophic lake.

17. Compare the source, transition, and floodplain zones of streams and rivers.

18. Distinguish between the photic and aphotic zones. How do organisms in the photic zone obtain nutrients? How are nutrients obtained in the aphotic zone?

19. What unusual primary producer forms the basis for hydrothermal vent communities?

APPLYING THE CONCEPTS

1. In which terrestrial biome is your college or university located? Discuss similarities and differences between your location and the general description of that biome in the text. In the city or town where your campus is located, how has human domination modified community interactions?

2. During the 1960s and 1970s, many parts of the U.S. and Canada banned the use of detergents containing phosphates. Until that time, almost all laundry detergents and many soaps and shampoos had high concentrations of phosphates. What environmental concern do you think prompted these bans, and what ecosystem has benefited most from the bans?

3. In developing countries, where CFCs are still manufactured, there is a growing trade in illegally sold CFCs, which will reduce the speed of recovery of the ozone layer. What steps would you suggest developed and developing countries might take to reduce the impetus for illegal use of CFCs?

4. Global warming is expected to make most areas warmer, but it is also expected to change rainfall in much less predictable ways. Why is it especially important to be able to predict rainfall changes in tropical areas?

5. More northerly forests are far better able to regenerate after logging than are tropical rain forests. Explain why this is true.

FOR MORE INFORMATION

Burroughs, D. "On the Wings of Hope." *International Wildlife*, July–August 2000. Kenyan butterflies are saving a unique forest and its people; this article is the basis for our Case Study.

Falkowski, P. G. "The Ocean's Invisible Forest." *Scientific American*, August 2002. Describes the productivity of phytoplankton and their importance in capturing carbon dioxide; speculates about their impact on global warming.

Milius, S. "Decades of Dinner." *Science News*, May 7, 2005. A whale carcass supports a unique and changing ocean floor community over a period of decades.

Myers, A. "Will the Class of 2003 Save the Cod?" *Blue Planet*, winter/spring 2006. The cod population of New England is in serious trouble. Hope lies in reducing catch so those hatched in recent years can make it to reproductive age.

Pauly, D., and Watson, R. "Counting the Last Fish." *Scientific American*, July 2003. Large predatory fish stocks have been decimated worldwide by overfishing, and because fisheries are increasingly exploiting smaller fish that feed lower on the food chain, they are endangering these populations as well.

Pearce, F. "Forests Paying the Price for Biofuels." *New Scientist*, November 22, 2005. Biofuels are not "green" when rain forests are cut down and replaced with soybean and palm plantations to generate these fuels.

Raloff, J. "Clipping the Fin Trade." *Science News*, October 12, 2002. Documents the overfishing of sharks, usually for their fins alone, and explains how difficult it is for shark populations to recover.

Schrope, M. "The Undiscovered Oceans." *New Scientist*, November 12, 2005. A largely unexplored world of exotic and diverse creatures covers much of Earth.

Stolzenburg, W. "Understanding the Underdog." *Nature Conservancy*, Fall 2004. Prairie dogs, decimated by farming, ranching, and advancing civilization over most of their former range, serve as a keystone species of the American grasslands.

30

Conserving Earth's Biodiversity

An artist's rendering of the ivory-billed woodpecker.

AT A GLANCE

CASE STUDY BACK FROM EXTINCTION

"YOU NEVER KNOW when you get up in the morning what earth-shaking event might take place and change your life forever," writes Tim Gallagher, an ornithologist at Cornell University. For Gallagher, a life-changing series of events began when he read an Internet posting by a kayaker, describing a large bird he spotted in a remote bayou of Arkansas. The description matched that of an ivory-billed woodpecker, a bird that almost everyone believed to be extinct. Gallagher, however, had never given up hope that, somewhere, the ivory-billed woodpecker survived. Since the late 1980s when some of the birds were observed in Cuba, there had been no reliable reports of ivory-billed woodpeckers anywhere in the world.

Ivory-bills were never abundant. A crucial part of their diet consists of large beetle grubs excavated from the wood of recently dead but still-standing trees in old-growth hardwood forests. These forests once covered extensive regions of the southeastern United States, and ivory-billed woodpeckers were seen as far north as North Carolina, south throughout Florida and Louisiana, and as far west as eastern Texas. But during the past century, logging eliminated most of these magnificent forest habitats, along with the ivory-bill's most important food. The last reliable sighting in the U.S. was in 1944, when a female ivory-bill was spotted in the cutover remnants of an old-growth hardwood forest in Louisiana. This ancient forest was the largest remaining primeval forest in the south and home to the last well-documented population of ivory-billed woodpeckers in the U.S. It was logged in spite of public outcry and offers from the Audubon Society and the State of Louisiana to purchase the land. Over the next 60 years, hope waned. Then, in 2005, bird lovers and ornithologists were astonished and thrilled when the journal *Science* published an article written by John Fitzpatrick, Tim Gallagher, and other ornithologists describing the rediscovery of the ivory-billed woodpecker. A male bird was spotted repeatedly in the Cache River National Wildlife Refuge in Arkansas..

Hailed as a "miracle," the sighting crowned 20 years of conservation efforts by The Nature Conservancy and its partners to protect and restore 120,000 acres of an area called the Big Woods. This region of wetlands, rivers, and hardwood forest is located on the floodplain of the Mississippi River. Since the sightings, The Nature Conservancy has joined with the Cornell Laboratory of Ornithology to raise millions of dollars to conserve an additional 200,000 acres of forest and rivers in this region over the next decade. The ivory-bill would be extinct today were it not for successful efforts to preserve a remnant of its former habitat.

Will the ivory-billed woodpecker escape extinction? Is there more than one bird left? Will birdwatchers, eager for a sighting, invade its last remaining refuge and inadvertently "love the bird to death?"

Conservation biologists seek to apply the principles of biology—particularly from ecology, genetics, and evolutionary biology—to improve the well-being and maintain the diversity of life on Earth. Conservation biologists also work closely with policymakers, lawyers, geographers, economists, ethicists, and historians, because conservation is necessarily a social enterprise. The goal of **conservation biology** is to preserve the diversity of living organisms, both for its own sake and for the benefits that biological diversity provides for people.

30.1 WHAT IS BIODIVERSITY, AND WHY SHOULD WE CARE ABOUT IT?

Biodiversity is simply the variety of life: the amazing diversity of living organisms, their genes, the ecosystems of which they are a part, and the interactions among them. Conservation biology seeks to preserve the diversity of species and the genetic diversity within a species, as well as preserving entire ecosystems and the complex community interactions within them.

Conservation biology must operate at the species, population, and community level. Each species is unique and irreplaceable. Although extinctions occur naturally over evolutionary time, conservation biology seeks to prevent extinctions caused by human activities. Within each species, genetic diversity produces slightly different adaptations among individuals that allow the species to thrive in a range of environments, and to evolve in response to changing conditions. Reasonably large populations must be maintained to preserve adequate genetic diversity within the species (see Chapter 15). Finally, the intricate network of community interactions is crucial to maintaining properly functioning ecosystems that, in turn, sustain human health, well-being, and ultimately, survival.

Ecosystem Services: Practical Uses for Biodiversity

Many of us work in offices and live in cities, much of our food comes in packages from a supermarket, and we may spend weeks without glimpsing an ecosystem in its natural state. Why should we care about preserving ecosystems and the communities they support? Many would say that ecosystems are worth preserving for their own sake. A more immediate, practical reason is simple self-interest: these ecosystems, both directly and indirectly, support us (**FIG. 30-1**).

In recent decades, scientists, economists, and policymakers have come to recognize that nature provides free but usually unrecognized benefits. These **ecosystem services** are the processes through which natural ecosystems and their living communities sustain and fulfill human life. Ecosystem services include purifying air and water, replenishing oxygen, pollinating plants and dispersing their seeds, providing wildlife habitat, decomposing wastes, controlling erosion and flooding, controlling pests, and providing recreational opportunities. These services are literally priceless because they sustain humanity. But because they are provided free and their economic value is difficult to

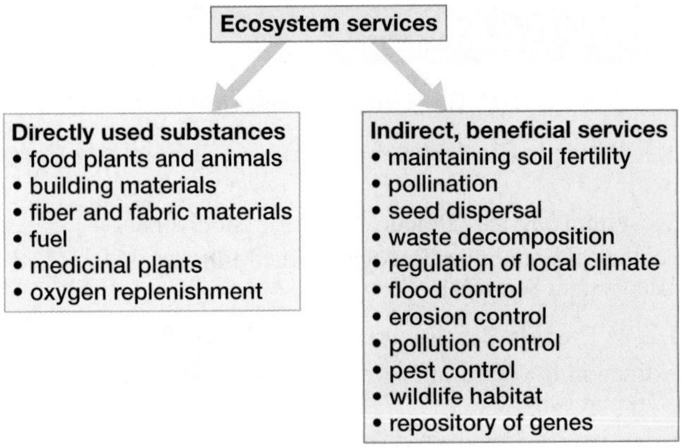

Ecosystem services

Directly used substances	Indirect, beneficial services
• food plants and animals • building materials • fiber and fabric materials • fuel • medicinal plants • oxygen replenishment	• maintaining soil fertility • pollination • seed dispersal • waste decomposition • regulation of local climate • flood control • erosion control • pollution control • pest control • wildlife habitat • repository of genes

FIGURE 30-1 Ecosystem services

measure, ecosystem services are almost always ignored. When land is converted to housing, for example, there is usually no incentive for the developer to preserve ecosystems and their services, but there is considerable economic incentive to destroy them. People have almost never even attempted to weigh the true costs against the economic benefits of altering the environment.

In 2005, the *Millennium Ecosystem Assessment* was released. This report, the result of four years of effort by over 1300 scientists in 95 countries and drawing on the best available information on the world's ecosystems, concluded that 60% of all Earth's ecosystem services were being degraded or used in an unsustainable manner by people. These findings underscore the need to preserve Earth's remaining natural ecosystems, and to work to restore those that have been damaged.

People Use Some Ecosystem Goods Directly

Healthy ecosystems provide a variety of resources directly to people. Nearly everyone can purchase wild-caught fish and other sea life that can thrive only in a healthy marine environment. Hunting for food and sport is important to the economy of many rural areas. In Africa, most types of wild animals are harvested for food, and they provide an important source of inexpensive protein for a growing and often poorly nourished population (see "Earth Watch: Tangled Troubles—Logging, Fishing, and Bushmeat"). In many less-developed countries, rural residents rely on wood from local forests for heat and cooking. Rain forests provide valuable hardwoods such as teak for consumers worldwide. Traditional medicines, used by about 80% of the world's people, are derived primarily from plants. About 25% of prescription medications contain active ingredients that are now—or were originally—extracted from plants. The antiviral drug Tamiflu is based on a chemical extracted from the seedpods of Chinese star anise (**FIG. 30-2**). Cancer researchers are excited

FIGURE 30-2 Star anise seedpods

about a compound isolated from a South American rain-forest plant (*Forsteronia refracta*) that inhibits the growth of cancerous, but not noncancerous, breast cells grown in the laboratory.

Ecosystem Services Also Benefit People Indirectly

The indirect services provided by healthy, diverse ecosystems are far-reaching and make a much greater contribution to human welfare than do goods harvested directly from nature. Here we describe just a few important examples.

Soil Formation

It can take hundreds of years to build up a single inch of soil. The rich soils of the U.S. Midwest accumulated under natural grasslands over thousands of years. Farming has converted these grasslands into one of the most productive agricultural regions in the world.

Soil, with its diverse community of decomposer and detritivore organisms (bacteria, fungi, worms, many insects, and others), plays a major role in breaking down wastes and recycling nutrients. People rely on soils to decompose waste products from industry, sewage, agriculture, and forestry. Thus, soil serves some of the same functions as a water purification plant. Soil communities are also crucial to nearly every nutrient cycle. For example, nitrogen-fixing bacteria in soil convert atmospheric nitrogen into a form that plants can use.

Erosion and Flood Control

Plants block wind that blows away loose soil. Their roots stabilize the soil and increase its ability to hold water, reducing both soil erosion and flooding. The massive flooding along the Missouri River in the U.S. in 1993 resulted, in part, from conversion of the natural riverside forests, marshes, and grasslands to farmland. This greatly increased the runoff and accompanying soil erosion in the wake of heavy rains (**FIG. 30-3a**).

Wetland ecosystems (marshes), in addition to their immense value as wildlife habitat, act like enormous sponges that absorb storm water. They also cushion the impact of waves that batter the coastline. The catastrophic flooding of New Orleans during Hurricane Katrina in August 2005 was a grim reminder of the value of coastal wetlands and the consequences of destroying them. In its natural state, the Mississippi River's silt-laden waters replenished the wetlands with sediment and fortified a series of outlying islands that served as a natural barrier to the force of incoming storms. Now, dredged, walled, and diverted, the waters of the Mississippi no longer sustain these natural ecosystems; southern Louisiana has lost 1000 square miles of wetlands in the past 50 years. Huge levees (built at enormous cost to replace ecosystem services that were formerly provided free) temporarily substituted for some of the protection afforded by wetlands; then Katrina struck, breeching the levees and flooding 80% of the city (**FIG. 30-3b**).

Climate Regulation

By providing shade, reducing temperature, and serving as windbreaks that reduce evaporation, plant communities have a major impact on local climates. Forests dramatically influence the water cycle, returning water to the atmosphere through transpiration (evaporation from leaves). In the Amazon rain forest, one-third to one-half of the rain consists of water transpired by leaves. Extensive clear-cutting of rain forests can cause the local climate to become hotter and drier, making it harder for the ecosystem to regenerate and damaging nearby forests as well.

Trees also affect global climate. They absorb carbon dioxide from the atmosphere, storing the carbon in their trunks, roots, and branches. As much as 20% of the carbon dioxide produced by human activities results from deforestation; as the trees are burned or decomposed, they release CO_2, which contributes to global warming.

(a)

(b)

FIGURE 30-3 Loss of flood control services
(a) Conversion of natural ecosystems to agriculture contributed to flooding of the Missouri River after unusually heavy rains in 1993. **(b)** New Orleans in the wake of Hurricane Katrina in 2005.

Genetic Resources

Crop plants, such as corn, wheat, and apples, have wild ancestors that humans have selectively bred for centuries to produce modern domestic crops. According to the United Nations Food and Agriculture Organization, 75% of human food is supplied by only 12 food crops. Many more wild plants might be developed into food sources that are more nutritious or better suited to a variety of growing conditions. Researchers have identified genes in wild plants that might be transferred into crops to increase productivity and provide greater resistance to disease, drought, and salt accumulation in irrigated soil. For example, some wild relatives of wheat have considerable salt tolerance, and researchers are working to transfer the genes that confer the ability to thrive in salty water from these wild plants into domestic wheat. The climates and soils of many developing countries are not well suited for food plants grown successfully in developed countries. Because scientists have just begun to explore the genetic treasure house of biodiversity, it promises to become an increasingly important resource in the future—but only if it is preserved.

Recreation

Many, perhaps most, people experience great pleasure in "returning to nature." Each year in the U.S., about 350 million visitors flock to protected public lands such as national parks and wildlife refuges. Small towns in Arkansas near the Big Woods preserve hope that their economies will be revitalized by income from tourists attracted by the discovery of the ivory-billed woodpecker. In many rural areas, the local economy depends on money spent by visitors who come to hike, camp, hunt, fish, or photograph nature.

Ecotourism, in which people travel to observe unique biological communities, is a rapidly growing industry worldwide. Examples of ecotourism destinations include tropical coral reefs and rain forests, the Galápagos Islands, the African savanna, and even Antarctica (**FIG. 30-4**).

Ecological Economics Recognizes the Monetary Value of Ecosystem Services

The relatively new discipline of *ecological economics* attempts to place values on ecosystem services and to assess the trade-offs that occur when natural ecosystems are damaged to make way for human profit-making activities. Consider a proposal to drain a wetland to irrigate a crop. If the loss of benefits from the wetland (neutralizing pollutants, controlling floods, providing breeding grounds for fish, birds, and many other animals) were factored into the decision, people might decide that the wetland is more valuable than the crop. In "Earth Watch: Restoring the Everglades," we describe a massive and costly project to undo human manipulation of the largest wetland ecosystem in the U.S.

One way to assess the economic value of ecosystem services is by estimating the cost of disasters that would have been prevented or reduced by natural ecosystems—if they had been left undisturbed. For example, the 1993 flooding along the Missouri River (see Fig. 30-3a) caused an estimated $12 billion in damage, much of which could have been avoided by appropriate land use decisions in past decades. These losses were dwarfed by the estimated $100 billion required to recover from the flood damage to New Orleans caused by Hurricane Katrina, which almost certainly would have been greatly reduced if people had not disrupted the flow of the Mississippi (see Fig. 30-3b).

Because the profits from damaging ecosystems go to individuals but the costs are borne by society as a whole, governing bodies must become far more involved in planning to make ecological economics work. The cost of catastrophic flooding, borne by U.S. taxpayers, reflects our failure to understand the value of ecosystem services and to plan development in a way that sustains them. Will we apply these lessons in rebuilding New Orleans? Time will tell.

(a)

(b)

(c)

FIGURE 30-4 Ecotourism
Carefully managed ecotourism represents a sustainable use of natural ecosystems, generating revenue without damaging the environment. (a) Investigating a coral reef reef in Figi. Visitors are warned never to touch the coral so it won't be damaged. (b) A "photographic safari" in Africa. (c) Ecotourists visit Antarctica.

FIGURE 30-5 Recently discovered large mammals (a) The secretive African kipunji was discovered in rapidly disappearing African tropical rain forest. (b) The shy Australian snubnose dolphin is named for its short snout.

(a)   (b)

An excellent example of government planning to preserve ecosystem services comes from New York City, which obtains much of its water from the nearby Catskill Mountains. The mountain forests, meadows, and soil purify water and provide New York City with almost half of its drinking water, once ranked among the cleanest in the nation. In 1997, realizing that its water was being polluted by sewage and agricultural runoff as the Catskills were developed, city officials calculated that it would cost $6 to $8 billion to build a water filtration plant, plus an additional $300 million annually to run it. Recognizing that the same service was provided free by the Catskill Mountains, government officials decided to invest in protecting them. A large fund now supports projects within the Catskills to reduce agricultural runoff, upgrade sewer systems, and purchase land to halt development and maintain functioning ecosystems and the water-purifying services they provide.

30.2 IS EARTH'S BIODIVERSITY DIMINISHING?

Extinction Is a Natural Process, but Rates Have Risen Dramatically

Scientists hypothesize that, in the absence of cataclysmic events, extinctions occur naturally at a very low rate, called the *background extinction rate*. In contrast, the fossil record provides evidence of five previous **mass extinctions**, during which many forms of life were eradicated in a relatively short time. The most recent happened roughly 65 million years ago, abruptly ending the age of dinosaurs. The causes of mass extinctions are uncertain, but sudden changes in the environment (such as would be caused by enormous meteor impacts or extremely rapid climate changes) are the most likely explanations.

The majority of biologists believe that human activities are now causing a sixth mass extinction, one rivaling these prehistoric events. A minority believe that the extinctions caused by people will not substantially alter the overall diversity of life or the way most natural communities function. Some are optimistic that we will find ways to preserve most of the remaining biodiversity.

Because our knowledge of biodiversity is limited, it is difficult to accurately measure extinction rates. Extinctions of birds and mammals are best documented, although these represent only about 0.1% of the world's total species. Since the 1500s, we have lost about 2% of all mammal species and 1.3% of all bird species. At background extinction rates, one species of bird might go extinct every 400 years; but in the past 400 years, at least 132 bird species (and likely far more) have been driven to extinction—almost entirely by human activities. The World Conservation Union (IUCN)* recently estimated that the current extinction rate is between 100 and 1000 times the background rate predicted in the absence of people. Although even the most conservative figure is alarming, the range underscores the uncertainty of such estimates. Because scientists have identified only a fraction Earth's total biodiversity, we have certainly lost many undescribed species. For example, a new species of dolphin, the Australian snubnose, and a new genus of monkey, the African kipunji, were discovered in 2005 (FIG. 30-5). Only about 1000 individuals of each species remain, and both are threatened by human activities, so they could easily have become extinct before ever being discovered, as, undoubtedly, have many other species.

*The World Conservation Union, often referred to as IUCN (International Union for the Conservation of Nature and Natural Resources), is the world's largest conservation network. It includes 111 government agencies, over 800 nongovernmental conservation-related organizations, and about 10,000 scientists and other experts from 181 different countries.

EARTH WATCH   Restoring the Everglades

In 1948, the U.S. Congress authorized the Central and Southern Florida Project, creating an extensive series of canals, levees, and other structures to control flooding, irrigate farms, and provide drinking water for new Florida developments in the massive marshland that dominated central and southern Florida. The project also transformed the meandering, 103-mile-long Kissimmee River into a straightened, 56-mile-long channel, eliminating most of its surrounding wetlands (**FIG. E30-1**). As the Everglades and other wetlands in southern Florida diminished, so did the wildlife that depended on them. The natural water-purifying functions of the wetlands were also lost, compounding the pollution problem as new farms and cities sprang up. Native plants, birds, fish, and other species declined, while invasive

species flourished. During the next 50 years, people became aware that a serious mistake had been made. The huge diversity of species and richness of community interactions that had made the Everglades a unique ecosystem was rapidly being lost.

With half of the Everglades' original area converted to agriculture, housing, and other forms of development, Florida and the U.S. government launched the Comprehensive Everglades Restoration Plan. Approved in 2000, the 30-year plan is intended to restore 18,000 square miles (over 11,500,000 acres) of wetlands at a projected cost of $7.8 billion. One of the largest ecosystem restorations ever attempted, the plan will remove 240 miles of canals and levees, reestablish natural river flow, restore wetlands, and recycle

(a) Pre-Channelized Kissimmee River

(b) Channelized River

FIGURE E30-1 Florida's Kissimmee River

Increasing Numbers of Species Are Threatened with Extinction

The IUCN has established a "Red List" that classifies at-risk species. Species may be described as **critically endangered**, **endangered**, or **vulnerable**, depending on how likely they are to become extinct in the near future. The ivory-billed woodpecker, for example, is critically endangered. Species that fall into any of the three categories listed above are described as **threatened**. In 2004, the Red List showed 15,589 threatened species, including 12% of all birds, 23% of mammals (**FIG. 30-6**), 32% of amphibians,

FIGURE 30-6 IUCN classification of mammals
Of Earth's 4776 known species of mammals, about 23% are threatened; an additional 12% are "near threatened," which means they are close to being classified as "vulnerable." The "least concern category" (not threatened or near threatened) applies to only slightly more than half of the world's mammals.

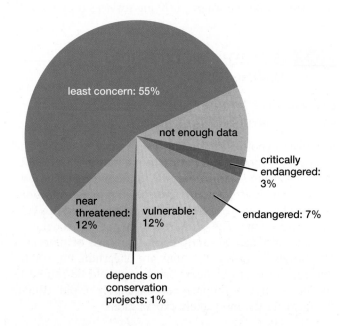

least concern: 55%

not enough data

critically endangered: 3%

endangered: 7%

vulnerable: 12%

near threatened: 12%

depends on conservation projects: 1%

some waste water (**FIG. E30-2**). As a result of restoration efforts, Florida now has over 41,000 acres of reconstructed wetlands, some of which are being used as giant water treatment areas. Restoration of the Kissimmee River will eventually restore 47 miles of river. Bird populations have already rebounded along restored portions of the river, and water quality has improved.

A 30-year, $8 billion program to undo human destruction of an ecosystem provides evidence that we are becoming aware of the economic and intrinsic values of natural communities.

FIGURE E30-2 Restoring the Everglades
(a) Birds such as this snowy egret are rebounding along restored segments of the Kissimmee River. **(b)** The area of Florida impacted by the Comprehensive Everglades Restoration Plan.

and 42% of turtles and tortoises. There are 1272 threatened species in the U.S. alone. Many scientists fear that a large number of currently endangered species are on their way to extinction. Why is this happening?

30.3 WHAT ARE THE MAJOR THREATS TO BIODIVERSITY?

Two major interrelated factors underlie the worldwide decline in biodiversity: (1) the increasingly large fraction of the Earth's resources used to support human lives and lifestyles and (2) the direct impacts of human activities, such as habitat destruction and pollution, on the rest of life on Earth.

Humanity Is Depleting Earth's "Ecological Capital"

The human **ecological footprint** (see Chapter 26) estimates the area of Earth's surface required to produce the resources we use and absorb the wastes we generate, expressed in acres of average productivity. A complementary concept, **biocapacity**, estimates the sustainable resources and waste-absorbing capacity actually available on Earth. While related to the concept of *carrying capacity* explained in Chapter 26, both the footprint and biocapacity calculations are subject to change as new technologies influence the way people use resources. Although the data are incomplete, scientists use the best estimates available, based primarily on statistics provided by international agencies such as the United Nations (UN). The calculations are intended to be conservative and to avoid overstating human impacts, and they do not include any land set aside to protect biodiversity.

How does humanity's footprint compare with Earth's biocapacity? In 2002, the biocapacity available for each of Earth's 6.2 billion people was 4.5 acres, but the average human footprint was 5.4 acres (24 acres for U.S. residents). This finding suggests that humanity had exceeded the capacity of Earth to support it on a continuing basis by over

EARTH WATCH	Tangled Troubles—Logging, Fishing, and Bushmeat

The "bushmeat" trade in Africa is a prime example of how threats to biodiversity interact and amplify one another. Historically, rural Africans have supplemented their diet by hunting a variety of animals, collectively called bushmeat. Traditional subsistence hunting by small tribes using primitive weapons did not pose a serious threat to animals. Now, as logging roads penetrate deep into rain forests, hunters follow, using shotguns and snares to kill any animal large enough to eat. Communities that spring up along logging roads develop a culture of hunting and selling bushmeat, and they become dependent on this new and profitable industry. Logging trucks are sometimes used to carry the meat to urban markets. The World Conservation Society estimates that the bushmeat harvest in equatorial Africa exceeds a million tons annually. Because many of the hunted animals play an important role in dispersing tree seeds, loss of these animals reduces the ability of the logged forest to regenerate.

The threat to African wildlife is also compounded by commercial overfishing off the West African coast. A 2004 study in Ghana documented a significant link between declining fish catches, increased poaching in Ghana's nature reserves, and increased sales of bushmeat in coastal villages. This strongly suggests that bushmeat is now substituting for protein traditionally supplied by fishing.

Because bushmeat hunters pay no attention to the sex, age, size, or rarity of the animal, many threatened species are declining rapidly. For example, despite estimates that only about 2000 to 3000 endangered pygmy hippos remain in the wild, pygmy hippo meat has been found in bushmeat markets. African elephants and rhinoceros also find their way into bushmeat markets.

The profitability of bushmeat has helped overcome traditional African taboos against eating primates. Although one-third of all primates (apes, monkeys, lemurs, and others) are threatened with extinction, in some bushmeat markets 15% of the meat comes from primates. In Cameroon, Africa, endangered gorillas are a favored target of poachers because of their large size. Even endangered chimpanzees and bonobos, our closest relatives, end up in cooking pots (**FIG. E30-3**). The impact of hunting on primates is difficult to assess because many are butchered on the spot and eaten or sold as pieces of unidentifiable meat. Experts now believe that bushmeat hunting is an even greater threat than habitat loss for Africa's great apes and that the combined threats of hunting and habitat loss make extinction in the wild a very real possibility for some of these magnificent and intelligent species.

Recognizing the threats to wildlife, several central African countries are working to reduce illegal timber harvesting and wildlife poaching. Together they have established a set of protected areas in the African rain forest of the Congo River Basin. Although enormous logging enterprises continue along the borders of these reserves, protecting them is a crucial step toward preserving some of Africa's rich natural heritage.

FIGURE E30-3 Bushmeat
Primates are threatened by bushmeat hunters using powerful rifles.

20% (**FIG. 30-7**). Since then, the human population has grown by over 250 million, while Earth's total biocapacity has not increased. Running such an "ecological deficit" is possible only on a temporary basis. Imagine a bank account that must support you for the rest of your life. If you preserve the capital and live on the interest, the account will sustain you indefinitely. But if you withdraw the capital to support an extravagant lifestyle or a growing family, you will soon run out of money. By degrading Earth's ecosystems, humanity is drawing down Earth's "ecological capital." As the population grows and less-developed countries such as India and China (each with a population over 1 billion) raise their living standards, the strain on Earth's resources will only increase.

As explained in the following sections, human activities are damaging the ability of ecosystems throughout the world to continue to sustain human and other forms of life.

Human Activities Threaten Biodiversity in Several Important Ways

Habitat destruction, overexploitation, harmful interactions with invasive species, pollution, and global warming pose the greatest hazards to natural populations.

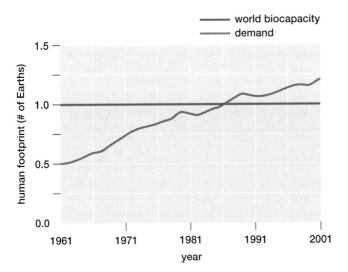

FIGURE 30-7 Human demand exceeds Earth's estimated biocapacity
The estimated ratio of demand versus biocapacity from 1961 to 2002, with biocapacity fixed at 1. Humanity's ecological footprint has increased steadily over the past 40 years. According to these estimates, in 1961, we were using about half of Earth's biocapacity. Now it would require over 1.2 Earths to support us all, at current rates of consumption, in a sustainable manner. (Modified from "Humanity's Footprint 1961–2002," provided by the Global Footprint Network.)

Imperiled species usually face multiple threats simultaneously, as highlighted in "Earth Watch: Tangled Troubles—Logging, Fishing, and Bushmeat." For example, the major decline in frog populations worldwide results from a combination of habitat destruction, invasive species, pollution, and a virulent fungal infection that many experts believe is linked to global warming (see "Earth Watch: Frogs in Peril" in Chapter 24). Coral reefs, home to about one-third of marine fish species, suffer from a combination of overharvesting, pollution (including silt eroded from nearby land that has been logged), and global warming.

Habitat Destruction Is the Most Serious Threat to Biodiversity

Since people began to farm around 11,000 years ago, Earth has lost about half of its total forest cover. More alarmingly, approximately half of all tropical rain forests have been cut down in just the last 50 years. In addition to providing wood for export, rain-forest land is being extensively converted to agriculture to supply world demand for beef, coffee, soybeans, palm oil, sugarcane, and other crops (**FIG. 30-8a, b**).

The IUCN has identified habitat destruction as the leading threat to biodiversity worldwide as rivers are dammed, wetlands are drained, and grasslands and forests are converted to agriculture, roads, housing, and industry. Loss of habitat impacts over 85% of all threatened mammals, birds, and amphibians. Reptiles such as sea turtles also suffer. In Florida, seawalls built to protect coastal developments contribute to beach erosion and block female sea turtles as they seek nesting sites. In "Earth Watch: Saving Sea Turtles," you will learn about a successful and innovative turtle conservation program in South America.

A serious threat to wildlife is **habitat fragmentation**, in which natural ecosystems are split into small pieces

(a)

(b)

FIGURE 30-8 Habitat destruction
The loss of habitat due to human activities is the greatest single threat to biodiversity worldwide. **(a)** Clear-cutting a rain forest. **(b)** This image of soybean plantations created within Bolivian rain forest was photographed by astronauts from the International Space Station in 2001.

FIGURE 30-9 Habitat fragmentation
Fields isolate forest patches in Paraguay.

surrounded by regions devoted to human activities (**FIG. 30-9**). Some species of U.S. songbirds, such as the oven-bird and Acadian flycatcher, may need up to 600 acres of continuous forest to find food, mates, and breeding sites. The ivory-billed woodpecker requires even larger tracts of undisturbed forest. Big cats are also threatened by habitat fragmentation. Cougars in the mountains near Los Angeles and endangered Florida panthers are often killed while attempting to cross highways that cut through their ranges. In the 1970s, India established a series of forest reserves intended to protect the endangered Bengal tiger. Originally interconnected by forests, the reserves have now become islands in a sea of development, forcing the estimated 5000 remaining tigers into about 160 isolated patches of woodland.

Habitat fragmentation may result in populations that are too small to survive. To be functional, a preserve must support a **minimum viable population (MVP)**. This is the smallest isolated population that can persist in spite of natural events, including inbreeding, disease, fires, and floods. The MVP for any species is influenced by many factors including the quality of the environment, the species' average life span, its fertility, and how many young usually reach maturity. Some wildlife experts believe that a minimum viable population of Bengal tigers must include at least 50 females—more than are found in many of India's tiger reserves.

Overexploitation Threatens Many Species

Overexploitation refers to hunting or harvesting natural populations at a rate that exceeds their ability to replenish their numbers. Overexploitation has increased as growing demand is coupled with technological advances that greatly increase our efficiency at harvesting wild animals and plants. The IUCN estimates that overexploitation impacts about 30% of threatened mammals and birds.

Overfishing is the single greatest threat to marine life, causing dramatic declines of many species, including cod, sharks, red snapper, and swordfish. The population of western Atlantic bluefin tuna, a high-priced delicacy in Japan, has dropped by about 97% since 1960. Enormous fishing nets intentionally trap huge numbers of commercial fish, but each year they unintentionally catch hundreds of thousands of marine mammals—including whales, porpoises, and dolphins—and have endangered 10 species of dolphins. Most species of marine turtles are endangered due to overharvesting of the adults and their eggs for food (**FIG. 30-10**).

Rapidly growing populations in less-developed countries increase the demand for animal products, as hunger and poverty drive people to harvest all that can be sold or eaten, legally or illegally, without regard to its rarity. As Callum Rankine of the World Wildlife Fund explains, "It's extremely difficult to get people to live sustainably. Often they are just concerned with trying to live." Compounding the problem, rich consumers fuel the demand for endangered animals by paying high prices for illegal products such as elephant-tusk ivory, rhinoceros horn, and exotic rain-forest birds. The demand for wood in developed countries encourages unsustainable logging and tree poaching in rain forests; in fact, less than 1% of rain-forest wood is harvested sustainably.

Invasive Species Displace Native Wildlife and Disrupt Community Interactions

Humans have transported a multitude of species around the world—cattle to the Americas and redwood trees to England, for example. In many cases, these introduced species cause no great harm. Sometimes, however, nonnative species become *invasive*: they increase in number at

FIGURE 30-10 Overexploitation
Illegally harvested green sea turtle eggs are sold in a market in Borneo.

Earth's largest turtles are in trouble, with six of the seven sea turtle species endangered or critically endangered. Sea turtles don't begin to breed until they are 30 to 50 years old. Then they must swim up to 1800 miles to reach their breeding grounds, probably the same beaches on which they hatched. Dragging themselves ashore, females excavate a hole in the sand, deposit their eggs, and return to sea (**FIG. E30-4a**). Baby turtles emerge after about two months and begin the difficult journey to adulthood. Seabirds and crabs attack them as they make their short journey back to the ocean (**FIG. E30-4b**). Once there, the turtles are a tasty target for a variety of fish. Although relatively few reach breeding age, under natural conditions, enough would survive to maintain the turtle population. Unfortunately, the turtles' nesting beaches attract poachers who find nesting females and their eggs easy prey. Turtle meat and eggs are a delicacy for many people, their shells make beautiful jewelry, and their skin makes fancy leather. Adults are also caught, both deliberately and accidentally, in fish lines and nets. Tourists are attracted to the same beaches as turtles are, and may frighten nesting females. Because females require a specific type of sand to lay eggs, augmenting beach sand for tourists (often done in Florida) can prevent them from laying eggs. Bright lights from beachfront developments disorient hatchlings as they attempt to navigate toward the sea.

Since 1980, the conservation organization TAMAR has reduced these threats for the five species of sea turtles that nest along the Brazilian coast, becoming a model for integrated conservation efforts everywhere (**FIG. E30-4c**). TAMAR founders realized that fishermen and local villagers must participate or the project would fail. Now, most of their employees are fishermen. Having formerly killed sea turtles, they now free turtles caught in nets and patrol the beaches during nesting season. TAMAR biologists tag females and trace their travels. The fishermen fend off the (now rare) turtle poachers, identify nests in risky locations, and relocate the eggs to better beach sites or to a nearby hatchery. Hatchlings are counted; each year TAMAR helps about 350,000 of them to reach the sea.

TAMAR has been successful because, rather than simply banning turtle hunting, the project organizers have engaged local communities as partners in turtle protection. Money flows into the local economies as ecotourists flock to see baby turtles, visit turtle museums, buy souvenirs made by local residents, and learn about the program. TAMAR sponsors communal gardens, daycare centers, and environmental education activities. The organization has also created artificial floating islands that attract fish for fishermen, so they don't feel the need to kill sea turtles. Recognizing that the economic benefits derived from preserving turtles far outweigh the money that can be made from hunting them, local residents eagerly participate in turtle conservation. The success of TAMAR not only underscores the need for community support for the sustainable use of any natural resource, but highlights how successful such efforts can be.

(a)

(b)

FIGURE E30-4
Endangered sea turtles
(a) A female green turtle scoops sand with powerful flippers, creating a cavity where she will bury about 100 eggs. (b) After incubating in the sand for about two months, the eggs hatch. Here a hatchling heads for the sea, where (if it survives) it will spend 25 to 50 years before reaching sexual maturity. (c) TAMAR's 22 stations help turtles along the entire Brazilian coast (blue).

(c)

the expense of native species, competing for food or habitat or preying on them directly (see Chapter 27). Introduced species often make native species more vulnerable to extinction from other causes, such as disease or habitat destruction. Approximately 7000 invasive species have become established in the U.S., and nearly half of threatened U.S. species suffer from competition with or predation by invasive species.

Island species are particularly vulnerable. Island populations are small, many island species are unique, and they have nowhere to go if conditions change. For example, 99% of Hawaii's 414 imperiled plants and 98% of its 42 threatened bird species are endangered by invasive species. Mongooses (**FIG. 30-11a**) were deliberately imported in the 1800s to control accidentally introduced rats. Now both mongooses and rats pose major threats to Hawaii's native ground-nesting birds. Wild pigs and goats, released by early Polynesian settlers to provide food, have decimated native Hawaiian plants. On the island of Guam, invasive brown tree snakes have wiped out most of the island's native bird species, driving some to extinction.

Lakes are also particularly vulnerable. The Great Lakes of the northern U.S. now host at least 87 invasive species, including the zebra mussel and lamprey eel. Lake Victoria in Africa, one of the world's largest lakes, is home to at least 300 different species of cichlid fish. Enormous Nile perch (**FIG. 30-11b**), introduced to Lake Victoria in the 1950s to provide human food, threaten about 200 of these species with extinction.

Pollution Is a Multifaceted Threat to Biodiversity

Pollution takes many forms. Pollutants include synthetic chemicals such as plasticizers, flame retardants, and pesticides that enter the air, soil, and water and then accumulate to toxic levels in animal tissues. Some of these are endocrine disruptors that interfere with normal development or reproduction. Serious threats also result from naturally occurring substances released in unnaturally high quantities. Some, such as the mercury, lead, and arsenic released by mining and manufacturing, are directly toxic to both people and wildlife. But nutrients in excessive amounts also become pollutants. For example, burning fossil fuels releases oxidized nitrogen and sulfur that disrupt the natural biogeochemical cycles for these plant nutrients, causing acid rain that threatens forests and lakes (see Chapter 28).

Global Warming Is an Emerging Threat to Biodiversity

The use of fossil fuels, coupled with deforestation, has substantially increased atmospheric carbon dioxide levels. As predicted by climatologists, this increase has been accompanied by an overall increase in global temperatures. In response to global warming, species are shifting their ranges further toward the poles, plants and animals are beginning springtime activities earlier in the year, and glaciers, ice shelves, and ice caps are melting (see Chapter 28). Some meteorologists hypothesize that global warming is also causing more extremes in weather, such as heat waves, droughts, floods, and stronger hurricanes and storms.

The rapid pace of human-induced climate change challenges the ability of species to adapt through natural selection. Recently, conservation biologist Chris Thomas, working with 18 other scientists in 6 biodiversity-rich regions throughout the world, concluded that global warming is now as great a threat to biodiversity as is direct habitat destruction. They estimated that about 1 million species will be in danger of extinction by 2050 as a result of global warming and the numerous disruptions caused by this change in Earth's climate.

(a)

(b)

FIGURE 30-11 Invasive species
(a) The mongoose, imported from India to prey on rats, threatens native Hawaiian ground-nesting birds.
(b) The Nile perch, introduced into Lake Victoria for fishermen, has proven to be a disaster for native fish.

30.4 HOW CAN CONSERVATION BIOLOGY HELP TO PRESERVE BIODIVERSITY?

The Foundations of Conservation Biology

Four important goals of conservation biology are as follows:

- Understand the impact of human activities on species, populations, communities, and ecosystems.
- Preserve and restore natural communities.
- Reverse the escalating loss of Earth's biodiversity caused by human activities.
- Foster sustainable use of Earth's resources.

Conservation biology incorporates the philosophical and ethical principles that biodiversity has intrinsic value and that other forms of life have a right to exist, independent of their value to people. Consequently, we should try to prevent human-caused extinctions of individual species. Another important principle is that the elaborate relationships among organisms that have evolved over millennia should be preserved within their natural environments.

Conservation biology also has the very practical goal of sustaining human well-being by understanding and protecting the natural environment. It recognizes that people, as well as other life-forms, evolved within this environment and rely on the services it provides.

Conservation Biology Is an Integrated Science

Conservation biology applies knowledge from diverse disciplines to conserve species and foster the survival of healthy, self-sustaining, and genetically diverse populations within natural communities.

Within the broad field of biology, conservation efforts enlist the help of ecologists, wildlife managers, geneticists, botanists, and zoologists. But effective conservation depends on expertise and support from people outside the field of biology as well. These include government leaders at all levels, who establish environmental policy and laws; environmental lawyers, who help enforce laws protecting species and their habitats; and ecological economists, who help place a value on ecosystem services. In addition, social scientists provide insight into the ways that people in different cultural groups use their environments. Educators help students understand how ecosystems function, how they support human life, and how people can either disrupt or preserve them. Conservation organizations identify areas of concern, provide educational materials, and organize grassroots support by individuals. Finally, individual choices and actions ultimately determine whether conservation efforts succeed.

Conserving Wild Ecosystems

Each conservation effort is unique because each wild area and each threatened species faces different challenges to survival. Core reserves, connected by wildlife corridors, can be very successful ways to conserve natural ecosystems and their diverse communities, including species that are threatened.

Core Reserves Preserve All Levels of Biodiversity

Core reserves are natural areas protected from most human uses except very low-impact recreation. These reserves encompass enough space to preserve ecosystems with all their biodiversity. "Earth Watch: Restoring a Keystone Predator" explains how reintroducing wolves to Yellowstone National Park, a core reserve in Wyoming, is enhancing other populations and restoring many community interactions. Because natural storms, fires, and floods are important to maintain ecosystems, core reserves should be large enough to sustain these events without loss of species.

To establish effective core reserves, conservationists must know the *minimum critical areas* required to sustain minimum viable populations of the species that require the most space. Minimum critical areas vary significantly among species, but for a given species, they also depend on the availability of food, water, and shelter. The minimum critical area required to maintain a minimum viable population of cougars in southern California, for example, is estimated at about 800 square miles.

Corridors Connect Critical Animal Habitats

In today's crowded world, an individual core reserve is seldom large enough to maintain biodiversity and complex community interactions by itself. **Wildlife corridors**, which are strips of protected land linking core reserves, allow animals to move freely and safely between habitats that would otherwise be isolated (**FIG. 30-12**). Corridors effectively increase the size of smaller reserves by connecting them. Both core reserves and corridors, ideally, are surrounded by buffer zones supporting human activities that

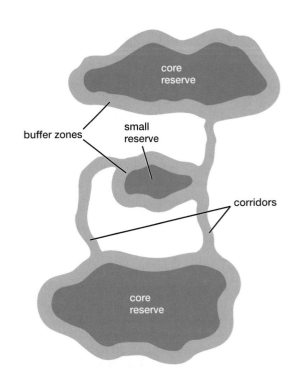

FIGURE 30-12 Corridors connect reserves

EARTH WATCH   Restoring a Keystone Predator

A predator is described as a *keystone predator* when its hunting activities have major effects on the community structure of an ecosystem. Research in Yellowstone National Park in the western U.S. (**FIG. E30-5a**) is documenting the complex interrelationships within this natural community and the critical role played by a keystone predator—the wolf. Considered a threat to elk and bison herds, wolves were deliberately exterminated from Yellowstone in 1928. Based on tree-ring data and aerial photographs, researchers have determined that this event marked the beginning of the end for regeneration of aspen trees (**FIG. E30-5b**). Aspen groves, which shelter a diverse community of plants and birds, have declined by over 95% since the park was established in 1872. New research suggests that elk, the major prey of wolves, eat nearly all the young aspen, as well as young willow and cottonwood trees.

From 1995 to 1996, after years of planning, study, and public comment, the U.S. Fish and Wildlife Service captured 21 gray wolves in Canada and released them into Yellowstone National Park (**FIG. E30-5c**). The wolves are thriving and now number about 250. Many ecologists are convinced that reintroduction of this keystone predator has had far-reaching and favorable impacts on the Yellowstone ecosystem.

A recent study by ecologists William Ripple and Eric Larsen suggests that wolf predation not only controls elk numbers but also changes elk behavior. With wolves nearby, elk tend to avoid those streamside aspen, willow, and cottonwood groves where they are less able to spot or escape attacking wolves. Now, with elk steering clear of them, these plant communities are regenerating, providing more habitat for songbirds and better stream conditions for trout. As their favorite trees have rebounded, beaver have returned and built dams on the streams, creating marshlands that provide habitat for mink, muskrat, otter, ducks, and rare boreal toads. Succulent plants in beaver marshes are favored food for grizzly bears as they emerge from hibernation. Grizzlies also feed on elk carcasses left by wolves, as do bald and golden eagles. In a further twist of the tangled web of community interactions, wolves compete with and kill coyotes, which eat rodents. With rodent populations on the rise, red foxes that feed on rodents are proliferating, and biologists are anticipating a rebound in other small predators such as weasels and wolverines.

Some ecologists, noting the complexity of species interactions, correctly point out that the "wolf effect" needs more time and study before it is conclusively demonstrated. But the evidence so far suggests that, as the Wolf Restoration Project leader Douglas Smith puts it, "Wolves are to Yellowstone what water is to the Everglades."

FIGURE E30-5 Impact of a keystone predator (a) The location of Yellowstone National Park. **(b)** Remnants of a once-thriving aspen grove in Yellowstone National Park attest to the lack of aspen regeneration since the early 1900s. (Photo courtesy of Dr. William Ripple) **(c)** Wolves now roam Yellowstone, delighting visitors and exerting far-reaching and positive effects on the natural community.

are compatible with wildlife. Buffer zones prevent high-impact uses such as clear-cutting, mining, freeways, and housing from impacting wildlife in the core region.

In Costa Rica, government and conservation organizations provide tax incentives, called *conservation easements*, to property owners in critical habitats who protect their land from development. This program has led many citizens to participate in creating wildlife corridors linking protected areas. Low-impact ecotourism just outside these corridors brings much-needed income to the local communities.

A wildlife corridor can be as narrow as an underpass beneath a highway. For example, in densely populated southern California, plans for a development of over 1000 new houses near San Diego were abandoned and the freeway exits closed after wildlife biologists tracked a cougar that was using the Coal Canyon underpass to move between suitable habitats. Now an official wildlife corridor, the underpass and its surroundings are being restored to a more natural state, encouraging cougars and other wild animals to cross safely beneath the freeway (FIG. 30-13).

In the northern Rocky Mountains, a coalition of conservation groups and scientists has proposed a series of wildlife corridors linking existing core reserves, such as Yellowstone National Park, with nearby ecosystems. These interconnected habitats would sustain populations of grizzly bears, elk, and mountain lions.

30.5 WHY IS SUSTAINABILITY THE KEY TO CONSERVATION?

Sustainable Living and Sustainable Development Promote Long-Term Ecological and Human Well-Being

Natural ecosystems share certain "operating principles" that are frequently violated by nonsustainable human development. Four important features required for sustainability are

- Diverse communities with a richness of community interactions
- Relatively stable populations that remain within the carrying capacity of the environment
- Recycling and efficient use of raw materials
- Reliance on renewable sources of energy

Respect for nature's operating principles is central to sustainability. In the landmark document *Caring for the Earth*, the IUCN states that **sustainable development** "meets the needs of the present without compromising the ability of future generations to meet their own needs." It explains that

Humanity must take no more from nature than nature can replenish. This in turn means adopting lifestyles and development paths that respect and work within nature's limits. It can be done without rejecting the many benefits that modern technology has brought, provided that technology also works within those limits.

Commercial fishing is a prime example of technology working outside nature's limits. Using sonar, enormous fishing nets, and trawls that may scrape entire communities from the ocean floor, commercial fisherman have harvested far more than can be replenished, endangering both commercial and noncommercial species in the process. Sustainable fishing requires that we preserve spawning grounds, limit fish catches, and improve technology to avoid unintended damage.

Unfortunately, in modern human society, "sustainable development" is almost an oxymoron, because "development" so often means replacing natural ecosystems with human infrastructure such as housing and retail developments. Traditionally, many economists and businesspeople have insisted that without continued growth, humanity cannot prosper. People in "developed" countries have, indeed, achieved economic growth and a high quality of life.

(a)

(b)

FIGURE 30-13 Wildlife corridors
(a) National Park Service wildlife biologists identify and track cougars using ear tags and GPS radio collars, such as worn by this tranquilized animal. (b) Asphalt has been removed and traffic barred from the underpass at Coal Canyon beneath the Riverside Freeway near San Diego to allow cougars to move safely between habitats on either side.

But they have done this not only by exploiting, in an unsustainable manner, the direct and indirect services provided free by ecosystems but also by using large quantities of nonrenewable energy.

Now, however, evidence from all parts of the world shows that such human growth activities are unraveling the complex tapestry of natural communities and undermining Earth's ability to support life. As individuals and governments recognize the need to change, there are increasing numbers of projects intended to meet human needs sustainably. We describe a few of these in the following sections.

Biosphere Reserves Provide Models for Conservation and Sustainable Development

A world network of **Biosphere Reserves** has been designated under the "Man and the Biosphere" program of the UN. The goal of Biosphere Reserves is to maintain biodiversity and evaluate techniques for sustainable human development while preserving local cultural values. Biosphere Reserves consist of three regions. A central *core reserve*, while protected, allows research and sometimes tourism and some traditional sustainable cultural uses. A surrounding *buffer zone* permits low-impact human activities and development. Outside the buffer zone is a *transition area* that is flexible in size and use, supporting settlements, tourism, fishing, and agriculture, all (ideally) operated sustainably (**FIG. 30-14**). The first Biosphere Reserve was designated in the late 1970s, and there are now over 480 sites worldwide.

Biosphere Reserves are entirely voluntary and managed by the countries and regional areas where they are located. This has greatly reduced opposition to them, but as a result of their voluntary nature, few completely adhere to the ideal Biosphere Reserve model. In the U.S., most of the 47 biosphere core reserves are national parks and national forests. Much of the land in buffer and tran-sition zones may be privately owned, and some of the landowners may be unaware of its designation. Often, funding is inadequate to compensate them for restricting development and for promoting and coordinating sustainable development, particularly in the transition zones.

The Chihuahuan Desert Biosphere Reserve is an innovative regional reserve established in 1977 and consisting of three separate reserves within the Chihuahuan Desert. Collectively, these areas meet the minimum UN criteria (**FIG. 30-15**). Big Bend National Park in Texas serves as the "protected core reserve" area, supporting research and tourism but no private development. The Jornada portion of the reserve, located in New Mexico, is considered the "buffer zone." Here, researchers investigate sustainable rangeland management in dry ecosystems. In Mexico, the Mapimi reserve serves as the "transition area." Over 70,000 people live in this reserve, and scientists are working with them to shift to more sustainable farming practices and conserve desert species such as the endangered Bolson tortoise, the largest land reptile in North America. Scientists now hope to reintroduce this species to Big Bend National Park and to establish wildlife corridors linking these widely separated reserves.

Biosphere Reserves have met the same obstacles encountered by most efforts to preserve biodiversity and change the way people use natural resources, but they are gradually succeeding. The concept offers an elegant model for conservation in the context of sustainable development and provides a framework for both present and future efforts.

(a)

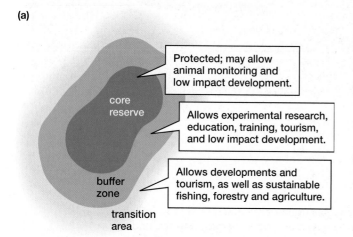

Protected; may allow animal monitoring and low impact development.

Allows experimental research, education, training, tourism, and low impact development.

Allows developments and tourism, as well as sustainable fishing, forestry and agriculture.

core reserve

buffer zone

transition area

FIGURE 30-14 The concept of a biosphere reserve

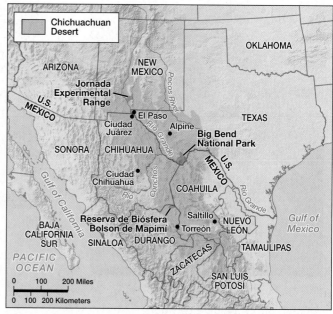

FIGURE 30-15 A unique regional biosphere reserve This regional reserve within the Chihuahuan Desert (shown in brown) consists of three smaller reserves (green) in the U.S. and Mexico.

Sustainable Agriculture Helps Preserve Natural Communities

The greatest loss of habitat occurs as people convert natural ecosystems to monoculture farming, in which large expanses of land are devoted to single crops. In the midwestern U.S., the original natural grasslands have been almost entirely converted to agriculture. Farming is necessary to feed humanity, and farmers face pressure to produce large amounts of food at the lowest possible cost. In some cases, this has led to unsustainable approaches to farming that interfere with ecosystem services. For example, allowing fields to lie fallow (unplanted) after harvesting allows rich soil to erode. A widely used herbicide that kills weeds in fields has been found to be a potent endocrine disruptor, and insecticides often indiscriminately kill both pests and their natural predators. In many regions throughout the world, irrigation practices are depleting underground water supplies faster than they can be replaced by natural processes.

Fortunately, farmers are increasingly recognizing that sustainable agriculture ultimately saves money while preserving the land (Table 30-1). The **no-till** cropping technique, which leaves the residue of harvested crops in the fields to form mulch for the next year's crops, represents one component of sustainable agriculture. In the U.S., it is now used on about 20% of croplands, saving farmers over 300 million gallons of fuel annually compared with conventional farming methods. Since 1980, this practice has helped reduce soil erosion by about 30%.

Most no-till farmers use herbicides to kill the cover crop and weeds (**FIG. 30-16**) and other pesticides to control fungal and insect outbreaks. Many organic farmers use no-till methods, and they also avoid using synthetic herbicides, insecticides, or fertilizers. Organic farming relies on natural predators to control pests and on soil microorganisms to degrade animal and crop wastes, thereby recycling their nutrients. Diverse crops reduce outbreaks of pests and diseases that attack a single type of plant, and no-till agriculture protects the soil.

In contrast to "factory farms," which may devote hundreds or even thousands of acres to a single crop, organic farms tend to be small. Because the loss of ecosystem services is not factored into the costs of unsustainable farming practices, food produced unsustainably tends to be cheaper, at least in the short term. Thus, consumers who wish to support sustainable agriculture must be prepared to pay a bit more for food produced this way. Although only about 0.5% of cropland in the U.S. is devoted to organic farming, consumer demand is driving steady growth of this sustainable approach. Many projects, such as the University of California's Sustainable Agriculture Research and Education Program, support research and educate farmers and the general public about the advantages of sustainable agriculture, how to practice it, and how to support it. In "Earth Watch: Preserving Biodiversity with Shade-Grown Coffee," you'll learn about one form of sustainable tropical agriculture.

The Future Is in Your Hands

Human Population Growth Is Unsustainable

The root causes of environmental degradation are simple: too many people using too many resources and generating too much waste. The solutions, unfortunately, are complex. Sustainable development includes progress toward a good quality of life, including adequate food and clothing, clean air and water, good health care and working conditions, educational and career opportunities, and access to unspoiled natural environments. Most of Earth's people live in less-developed countries and lack at least some of these

(a)

(b)

FIGURE 30-16 No-till crops
(a) A cover crop of wheat is killed with an herbicide before the grain is harvested. Cotton seedlings thrive amid the dead wheat, which anchors soil and reduces evaporation. **(b)** Later in the season, the same field shows a healthy cotton crop mulched by the dead wheat.

Table 30-1 **Agricultural Practices Affect Sustainability**		
	Unsustainable Agriculture	**Sustainable Agriculture**
Soil erosion	Allows soil to erode far faster than it can be replenished, because the remains of crops are plowed under, leaving the soil exposed until new corps grow.	Erosion is greatly reduced by no-till agriculture. Wind erosion is reduced by planting strips of trees as windbreaks around fields.
Pest control	Uses large amounts of pesticides to control crop pests.	Trees and shrubs near fields provide habitat for insect-eating birds and predatory insects. Reducing insecticide use helps to protect birds and insect predators.
Fertilizer use	Uses large amounts of synthetic fertilizer.	No-till agriculture retains nutrient-rich soil. Animal wastes are used as fertilizer. Legumes that replenish soil nitrogen (such as soybeans and alfalfa) are alternated with crops that deplete soil nitrogen (such as corn and wheat).
Water quality	Runoff from bare soil contaminates water with pesticides and fertilizers. Excessive amounts of animal wastes drain from feedlots.	Animal wastes are used to fertilize fields. Plant cover left by no-till agriculture reduces nutrient runoff.
Irrigation	May excessively irrigate crops, using groundwater pumped from natural underground storage at a rate faster than the water is replenished by rain or snow.	Modern irrigation technology reduces evaporation and delivers water only when and where it is needed. No-till agriculture reduces evaporation.
Crop diversity	Relies on a small number of high-profit crops, which encourages outbreaks of insects or plant diseases and leads to reliance on large quantities of pesticides.	Alternating crops and planting a wider variety of crops reduces the likelihood of major outbreaks of insects and diseases.
Fossil fuel use	Uses large amounts of nonrenewable fossil fuels to run farm equipment, produce fertilizer, and apply fertilizers and pesticides.	No-till agriculture reduces the need for plowing and fertilizing.

basic amenities (**FIG. 30-17**). The United Nations estimates that over 850 million people (nearly one out of seven) lack adequate food.

Discussions of sustainability and preservation of biodiversity often skirt an inescapable fact. As stated in the IUCN document *Who Will Care for the Earth?* a central issue is "how to bring human populations into balance with the natural ecosystems that sustain them." Through technological advances and by depleting Earth's ecological capital, we have achieved a population that far exceeds this balance. Yet we currently add 75 to 80 million people to Earth every year. This growth is incompatible with a sustainable increase in the quality of life for the 6.5 billion already here—and with saving what is left of Earth's biodiversity for those who are to come.

Changes in Lifestyle and Use of Appropriate Technologies Are Also Essential

The only truly effective way to conserve life on Earth is to make fundamental changes in the way we interact with our natural environment, collectively and individually. Without a sustainable approach, there can be no long-term improvement in the quality of human life.

In addition to making responsible reproductive choices, we can reduce our consumption of energy and nonrenewable fossil fuels by conserving and using energy-saving technologies. In the absence of as yet unproven technologies such as nuclear fusion, sustainable living must ultimately rely on renewable energy sources (solar, wind, and wave energy, for example). We can emulate natural ecosystems by recycling nonrenewable resources. Our choices as consumers can provide markets for foods and durable goods that are produced sustainably. These are all changes

FIGURE 30-17 Poverty
Billions of people lack the resources required for a good quality of life. This housing in Cuidad Juarez, Mexico, is just across the border from El Paso, Texas (both are within the Chihuahuan Desert shown in Figure 30-15).

that we can, and must, make. Humans clearly have the ability to destroy nature, but we also have the ability and a moral responsibility to protect it.

This chapter has provided some examples of human activities moving in the right direction. Look around your campus and community—what is being done sustainably? What isn't? What would it take to make the necessary changes? In "Links to Life: What Can Individuals Do?" we suggest some ways that individuals can live more sustainably and help protect life on Earth.

EARTH WATCH   Preserving Biodiversity with Shade-Grown Coffee

Viewed from a short distance away, "rustic" Mexican coffee plantations are indistinguishable from tropical rain forest. Coffee is one of the few major crops that can be grown in shade (cacao, the source of chocolate, is another). Rustic plantations often consist of nearly intact rain forest with dozens of tree species forming layers of canopy up to 60 feet high (**FIG. E30-6**). The trees protect the soil from erosion, trap water, and humidify the air, creating cool shade that reduces the growth of weeds. The trees also provide a home for over 150 different species of birds. The birds feed on the diverse community of insects that inhabit both the trees and the moist soil created by natural decomposers that break down the falling leaves. The waist-high coffee bushes will continue to produce beans in this shady environment for about 30 years.

Coffee can also be grown in full sunlight, which increases the yield. Beginning in the 1970s, there was a major forest-clearing effort to make way for coffee monocultures. In Columbia, nearly 70% of the coffee plantations are now grown in full sunlight. Lacking the nutrients provided by forest decomposition, these farms require large amounts of expensive fertilizers. The sunny, fertilized soil and the lack of natural predators promote the growth of weeds and insect pests, which must be controlled by herbicides and insecticides. The number of bird species is reduced by about 95% in this artificial environment, as birds are poisoned by pesticides and deprived of both their habitat and their insect prey. As coffee plantations and other activities have reduced rain-forest cover over the past 30 years, there has also been a decline in populations of both native birds and those that breed in North America but spend the winter in Central and South America's tropical forests. These include wood thrushes, fly-catchers, scarlet tanagers, vireos, warblers, and redstarts.

The U.S. consumes over one-third of the world's coffee; two-thirds of this is grown in Latin America and the Caribbean. Fortunately, most coffee grown in Mexico, and more than half of that grown in Costa Rica, remains in shady plantations. The shading canopy trees may also provide a source of food or income from citrus fruits, bananas, guavas, or lumber. Coffee im-porters and consumers are gradually awakening to the importance of these traditional plantations in maintaining biodiversity, particularly of migratory birds. The Smithsonian Migratory Bird Center certifies coffee plantations as Bird Friendly™ if they meet high standards for forest diversity and canopy cover. The Rainforest Alliance certifies coffee grown sustainably with the Rainforest Alliance Certified seal of approval. The higher price of these coffees reflects the value of the ecosystem services they preserve, but discriminating coffee drinkers also find the added flavor of shade-grown coffee well worth the cost.

FIGURE E30-6 Rustic coffee plantations preserve biodiversity
Domingo Silva practices sustainable agriculture in his shaded coffee plantation in Oaxaca, Mexico. The farm, which has been in his family for four generations, produces a variety of fruits as well as coffee beans. (Inset) Coffee beans are one of the few crops that thrive in shade.

CASE STUDY REVISITED   BACK FROM EXTINCTION

The scarcity of old-growth trees and the beetle grubs they harbor means that a breeding pair of ivory-billed woodpeckers requires up to 6 square miles of uncut forest under ideal conditions. But in the disrupted forest that remains, a pair might require 20 to 30 square miles to obtain adequate resources. No one knows whether there is even a single successful breeding pair in the Big Woods, but researchers will be avidly, if carefully, searching. Fearful of publicity and the possibility of a deluge of eager birdwatchers, the ornithologists who first spotted an ivory-bill kept their find a secret for over a year, surreptitiously leading expeditions and gathering evidence for their publication. As news of their discovery hit the presses, the U.S. Fish and Wildlife Service restricted access to about 5000 acres near the sightings, allowing only a few researchers inside. However, much of the rest of the Big Woods remains open to the public. While some birders, lured by the possibility of a once-in-a-lifetime sighting, have been drawn irresistibly to the area, many care so deeply about the bird's survival that they are willing to leave it alone. Describing the ivory-billed woodpecker as a "symbol of wilderness," one avid birder explained, "I don't really need to see it or photograph it. I just need to know it is there."

Consider This What makes people care so much about a bird that they will probably never see? Why do others care so little that they, like the owners of the logging company that destroyed the ivory-billed woodpeckers' last known habitat, are willing to see it become extinct?

LINKS TO LIFE   What Can Individuals Do?

"There are no passengers on spaceship earth. We are all crew."

—Marshall McLuhan

Sustainable living is, ultimately, an ethic that must permeate all levels of human society, beginning with individuals. The adage "Reduce, Reuse, and Recycle" provides excellent advice to minimize your impact on Earth's life-support systems. Of these "three Rs," reducing consumption is the most important. Here are some ways to make a difference:

CONSERVE ENERGY

- **Heating and cooling:** Don't heat your house to over 68°F in winter or air condition it below 78°F in summer. Turn down the heat or cooling while you're away. When you purchase or remodel a home, consider energy-efficient features such as passive solar heating, good insulation, an attic fan, double-glazed windows (with "low-e" coating to reduce heat transfer), and good weather stripping. Plant deciduous trees on the south side of your home for shade in summer and sun in winter. If possible, purchase renewable energy from your energy provider.

- **Hot water:** Take shorter showers and switch to low-flow shower heads. Wash only full loads in your washing machine and dishwasher; use cold water to wash clothes; don't prewash your dishes. Insulate and turn down the temperature on your water heater.

- **Appliances:** Compare Energy Star ratings when you choose a major appliance. Don't use your dryer in the summer—put up a clothesline. Turn off unused lights and appliances. Replace incandescent light bulbs with fluorescent or LED bulbs wherever possible.

- **Transportation:** Choose the most fuel-efficient car that meets your needs, and use it efficiently by combining errands. Use public transit, carpool, walk, bicycle, or telecommute when possible.

CONSERVE MATERIALS

- **Recycle:** Look into recycling options in your community, and recycle everything that is accepted. Explore composting (there are excellent sites on the Internet). Fruit and vegetable waste, leaves, and lawn clippings can protect and fertilize your plants. Support and encourage campus and community recycling efforts.

- **Buy recycled material:** Purchase recycled paper products. Decking and carpet is now made from recycled plastic bottles.

- **Reuse:** Reuse anything possible, such as manila envelopes, file folders, and both sides of paper. Refill your water bottle, and use refill cups when possible. Reuse your grocery bags. Give away—rather than throwing away—serviceable clothing, toys, and furniture. Make rags out of old clothes and use them instead of disposable cleaning materials.

- **Conserve water:** If you live in a dry area, plant drought-resistant vegetation around your home to reduce water usage.

SUPPORT SUSTAINABLE PRACTICES

- **Food choices:** Buy locally grown and organically grown produce that does not require long-distance shipping. Look for shade-grown coffee with the "Bird-Friendly™" or "Rainforest Alliance Certified" seal of approval, and request it at your local Starbucks. Reduce meat, particularly beef, consumption. Obtain "The Fish List" from http://www.thefishlist.org for ocean-friendly choices when you purchase seafood.

- **Limit or avoid the use of harmful chemicals:** Harsh cleaners, insecticides, and herbicides contaminate water and soil.

MAGNIFY YOUR EFFORTS

- **Support organized conservation efforts:** Join conservation groups and donate money for conservation efforts. Find these on the Internet, and sign up for e-mail alerts that educate you about environmental legislation and make it easy for you to contact your government representatives and express your views. Check out http://www.energyaction.net/main/ and join the Campus Climate Challenge.

- **Volunteer:** Join grassroots efforts to change the world—this is where it all begins. Volunteer for local campus and community projects that improve the environment.

- **Make your vote count:** Investigate candidates' stands and voting records on conservation issues, and consider this information when making your choices.

- **Educate:** Through your words and actions, share your concern for sustainability with your family, friends, and community. Write letters to the editor of your school or local newspaper, to local businesses, and to elected officials. Look for ways your campus could conserve energy, recruit other concerned students, and lobby for change.

- **Reduce population growth:** Consider the consequences of the enormous and expanding human population when you plan your family. Adoption, for example, allows people to have large families while simultaneously contributing to the welfare of humanity and the environment.

CHAPTER REVIEW

SUMMARY OF KEY CONCEPTS

30.1 What Is Biodiversity, and Why Should We Care About It?

Biodiversity includes genetic diversity, species diversity, and diverse community interactions. It is a source of goods such as food, fuel, building materials, and medicines. Biodiversity provides ecosystem services such as forming soil, purifying water, controlling floods, moderating climate, and providing genetic reserves and recreational opportunities. The emerging discipline of ecological economics attempts to measure the contribution of ecosystem goods and services to the economy and estimates the costs of losing them to unsustainable development.

30.2 Is Earth's Biodiversity Diminishing?

Natural communities have a low background extinction rate. Many biologists believe that human activities are currently causing a mass extinction, increasing extinction rates by a factor of 100 to 1000. About 15,600 plants and animals are now threatened with extinction.

30.3 What Are the Major Threats to Biodiversity?

Human use of natural resources has exceeded Earth's ability to replenish what we are taking from it. By exceeding Earth's biocapacity, we damage its ability to sustain future life. Major threats to biodiversity include habitat destruction and fragmentation as ecosystems are converted to human uses; overexploitation as wild animals and plants are harvested beyond their ability to regenerate; pollution, including global warming; and the introduction of invasive species.

Web Tutorial 30.1 Habitat Destruction and Fragmentation

30.4 How Can Conservation Biology Help to Preserve Biodiversity?

Conservation biology seeks to identify the diversity of life, explore the impact of human activities on natural ecosystems, and apply this knowledge to conserve species and foster the survival of healthy, self-sustaining communities. It is based on the premise that biodiversity has intrinsic value. Conservation biology integrates knowledge from many areas of science and requires the efforts of government leaders, environmental lawyers, conservation organizations, and most importantly, individuals. Conservation efforts include establishing wildlife reserves connected by wildlife corridors, with the goal of preserving functional communities and self-sustaining populations.

30.5 Why Is Sustainability the Key to Conservation?

Sustainable development meets present needs without compromising the future. It requires that people maintain biodiversity, recycle raw materials, and rely on renewable resources. Biosphere reserves promote conservation and sustainable development. A shift to sustainable farming is crucial to conserving soil and water, reducing pollution and energy use, and preserving biodiversity.

Human population growth is unsustainable and is driving consumption of resources beyond nature's ability to replenish them. We must bring our population into line with Earth's ability to support us, leaving room and resources for all forms of life. Individuals must make responsible reproductive choices and reduce resource consumption so that it does not exceed what Earth can replenish.

KEY TERMS

biocapacity   *page 617*
biodiversity   *page 612*
Biosphere Reserves   *page 626*
conservation biology
    page 612
core reserves   *page 623*

critically endangered species
    page 616
ecological footprint
    page 617
ecosystem services   *page 612*
endangered species
    page 616

habitat fragmentation
    page 619
mass extinction   *page 615*
minimum viable population
    (MVP)   *page 620*
no-till   *page 627*

overexploitation   *page 620*
sustainable development
    page 625
threatened species   *page 616*
vulnerable species   *page 616*
wildlife corridors   *page 623*

THINKING THROUGH THE CONCEPTS

1. Define conservation biology. What are some of the disciplines it draws on, and how does each discipline contribute to it?

2. What are the three different levels of biodiversity, and why is each one important?

3. What is ecological economics? Why is it important?

4. List the types of goods and services that natural ecosystems provide.

5. What four specific threats to biodiversity are described in this chapter? Provide an example of each.

6. Why is the TAMAR turtle project a good model for conservation and sustainable development?

7. How can purchasing rustic-grown coffee imported from Latin America help preserve wildlife in the U.S.?

8. What is "bushmeat"? What types of animals are particularly endangered by it? What types of development promote it? What factors drive this development?

9. What types of evidence support the hypothesis that the wolf is a keystone species in Yellowstone National Park?

APPLYING THE CONCEPTS

1. What are the ethical foundations of conservation biology? Do you agree with them? Why or why not?

2. List some reasons that the ecological footprints of U.S. residents are by far the largest in the world. Looking at your own life, how could you reduce the size of your footprint? How does the ecological footprint of U.S. residents extend into the tropics?

3. Search for and describe some examples of habitat destruction, pollution, and invasive species in the region around your home or campus. Predict how each of these might affect specific local populations of native animals and plants.

4. Identify a dense suburban development near your home or school. Redesign it in any ways necessary to make it into a sustainable development (this would make a good group project).

5. What economic arguments would conventional farmers be likely to raise against switching to organic farming techniques and other sustainable agricultural methods? What would be the advantages to farmers? How does this affect consumers?

6. Some U.S. government officials are encouraging the use of biofuels (gasoline supplemented with palm or soybean oil, or ethanol) to reduce reliance on oil imports. Discuss the use of biofuels from as many angles as possible.

FOR MORE INFORMATION

Daly, H. E. "Economics in a Full World." *Scientific American*, September 2005. We must think in new ways to develop a sustainable economy.

Fitzpatrick, J. W., and coauthors. "Ivory-billed Woodpecker (*Campephilus principalis*) Persists in Continental North America." *Science*, June 3, 2005. The original research article reporting the rediscovery of the ivory-billed woodpecker.

Graham-Rowe, D., and Holmes, B. "Goodbye Cruel World." *New Scientist*, November 20, 2004. Loss of species is correlated with human population growth and bushmeat harvesting.

Graham-Rowe, D., and Holmes, B. "The World Can't Go on Living Beyond Its Means." *New Scientist*, April 2, 2005. According to the 2005 Millennium Ecosystem Assessment, about 60% of Earth's ecosystem services are being degraded.

Groom, M. J., Meffe, G. K., and Carroll, C. R. *Principles of Conservation Biology*, 3rd ed. Sinauer Associates, 2006. An introductory text providing broad and diverse coverage of this rapidly developing discipline.

Levy, S. "A Top Dog Takes Over." *National Wildlife*, August/September 2004. Explores the far-reaching impact of wolf reintroduction into Yellowstone National Park.

Lovins, A. "More Profit with Less Carbon." *Scientific American*, September, 2005. Saving energy will save consumers money while increasing profits for businesses and reducing global warming.

Milius, S. "Comeback Bird." *Science News*, June 11, 2005. The amazing story of the rediscovery of the ivory-billed woodpecker.

Musser. G. "The Climax of Humanity." *Scientific American*, September, 2005. Our choices during the next few decades could lead to environmental sustainability or environmental collapse.

Pimm, S. L. "Sustaining the Variety of Life." *Scientific American*, September 2005. Saving biodiversity on a budget.

Tangley, L. "Out of Sync." *National Wildlife*, April/May 2005. Global warming is disrupting the life cycles of many animals and plants.

Animal Anatomy and Physiology

The animal body is an exquisite expression of the elegance with which evolution has linked form to function. All the animal body systems work in concert to maintain life.

31

Homeostasis and the Organization of the Animal Body

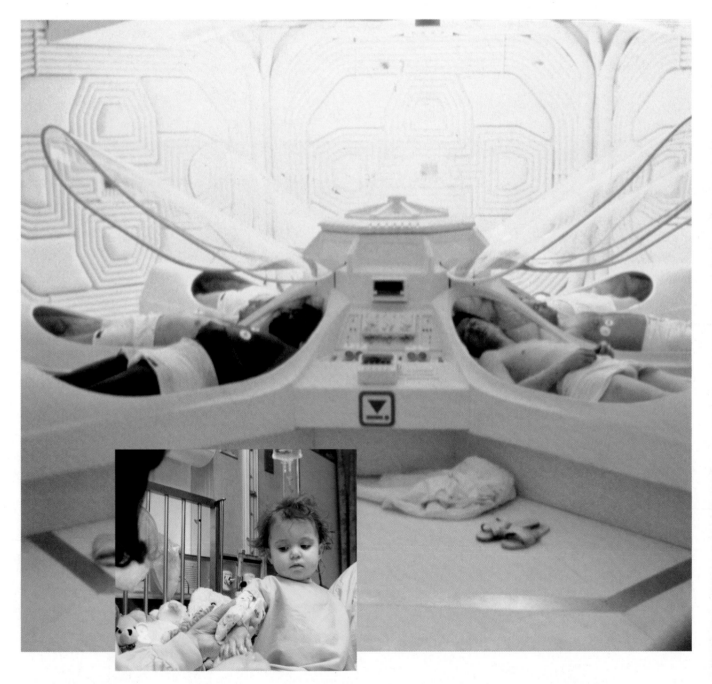

In this movie scene, space travelers prepare to emerge from capsules where they have existed in suspended animation during the long flight. **(Inset)** Caregivers tend to Erika Nordby's frostbitten fingers at a hospital in Alberta, Canada.

CASE STUDY   LIFE ON HOLD?

AFTER DECADES IN SPACE, the ship's sensors detect an Earth-type planet. Electrical signals activate pods in which young travelers rest, icy and immobile. As the pods gradually thaw, heart monitors record slow, then increasing activity, chests begin to rise and fall, and eyelids flutter. These travelers through space—and time—look around in momentary confusion until they recall the stasis pod closing over them as they began their journey. They smile and arise a bit stiffly to start their new life.

This image of suspended animation is captivating and, for now, a complete fantasy. But closer to home, placing life on hold has urgent practical implications. Each year, while thousands of people die waiting for organ transplants, hundreds of donated organs must be discarded because they cannot survive long enough to reach a matched recipient. Tens of thousands of people die from lack of oxygen caused by blood loss, heart attacks, and

strokes before they can be treated. Could many of these victims—as well as donated organs—be saved by placing them in a state of suspended animation in which their cells require almost no oxygen while they are transported?

Biologists have long been fascinated by the ability of hibernating mammals to seemingly suspend life. For example, the body temperatures of hibernating ground squirrels can approach freezing while their heart rates, breathing rates, and oxygen use all drop by about 97%. Can people's lives also be put on hold by the cold? Some accidents suggest they can. In sub-zero temperatures through the predawn darkness of Alberta, Canada, a frantic mother followed tiny footprints to find her 13-month-old toddler rigid and face down in the snow. Erika Nordby was not breathing, her heart was still, and her body temperature was only 61°F (16°C; normal is about 98.6°F, or 37°C). Para-

medics performed CPR, and doctors wrapped her in a heated blanket. Two hours after she was discovered, Erika's heart began beating on its own. A day later, she was alert and hungrily sucking milk from a bottle (see the chapter opener inset). When a backcountry skier fell through the ice of a Norwegian lake, it took her companions 80 minutes to bring her to the surface. Her body temperature was 57°F, she was not breathing, and her heart had stopped. After 9 hours of resuscitation, Anna Bagenholm's heart began to beat on its own; she eventually made an excellent recovery.

What happened to Erika and Anna's bodies when they lost the fight to stay warm? Why, having been clinically dead for some time, are they now alive and healthy? Could cold-induced suspended animation hold the key to saving donated organs as well as people deprived of oxygen by trauma or stroke?

31.1 HOMEOSTASIS: HOW DO ANIMALS REGULATE THEIR INTERNAL ENVIRONMENT?

Whether you dive into a swimming pool, hike in the desert, or swim in the ocean, your cells remain isolated from outside conditions. They are bathed in extracellular fluid containing a specific, complex mixture of dissolved substances that must be maintained regardless of conditions on the outside. Many animals have evolved elaborate physiological mechanisms that allow them to maintain precise internal conditions despite lifetimes spent in harsh environments. For example, desert dwellers such as the kangaroo rat have kidneys that allow them to conserve water, while freshwater animals such as the trout or frog must excrete copious amounts of water. Ocean-dwelling fish secrete excess salt from their gills. Because the cells of the animal body cannot survive if the internal environment deviates from a narrow range of acceptable states, cells devote a large portion of their energy to actions that keep the cellular environment stable.

This "internal constancy" was first recognized by French physiologist Claude Bernard in the mid-nineteenth century. Later, in the 1920s, Walter B. Cannon coined the term **homeostasis** to describe the processes by which an organism maintains its internal environment within the narrow range of conditions necessary for optimal cell functioning. Although the word *homeostasis* (meaning "to stay the same") implies a static, unchanging state, the internal environment in fact seethes with activity as the body continuously adjusts to internal and external changes.

The Internal Environment Is Maintained in a State of Dynamic Constancy

The internal state of an animal body can be described as a *dynamic constancy*. Many physical and chemical changes do occur (the dynamic aspect), but the net result of all this activity is that physical and chemical parameters are kept within the range that cells require to function (the constant aspect). Examples of conditions within the fluid surrounding cells that are regulated by homeostatic mechanisms include

- Temperature
- Water and salt levels
- Glucose levels
- pH
- Oxygen and carbon dioxide levels

Why are cells so particular about their surroundings? As you will learn in later chapters of this unit, appropriate levels of various types of salts are required for such life-sustaining processes as neuronal activity and muscle contraction. Cellular requirements for energy and reliance on the complex three-dimensional structure of proteins to regulate metabolic activities also constrain their environment.

Under normal conditions, animal cells are constantly generating and using large quantities of ATP to sustain life processes (see Chapter 8). The reactions that produce ATP (glycolysis followed by cellular respiration) require a continuous supply of high-energy molecules (primarily glucose) and a continuous supply of oxygen to complete the series of reactions that generate most of the cell's ATP. Thus, energy production helps explain the importance of both glucose and oxygen levels.

Each of the many reactions required to generate ATP is catalyzed by a specific protein whose ability to function is critically dependent on its three-dimensional structure, maintained, in part, by hydrogen bonds. These crucial but vulnerable bonds can be disrupted by an environment that is too hot, too salty, or too acidic or basic (see Chapter 3). The need to maintain these bonds and the protein function that depends on them helps explain the requirement for a narrow range of temperature, pH, and salt.

Because increasing temperatures speed up chemical reactions, higher temperatures increase the demand for ATP, as well as the rate at which cells generate ATP, thus increasing cellular demands for both oxygen and glucose. When Anna fell through the ice, her supply of oxygen was cut off. Fortunately, the very low temperature of the water also dramatically reduced her cells' oxygen requirements, keeping them alive.

Animals May Be Categorized by How They Regulate Body Temperature

You are probably familiar with descriptions of mammals and birds as "warm-blooded" and reptiles, amphibians, fish and invertebrates as "cold-blooded." In fact, the bodies of desert pupfish may reach over 100°F (37.8°C) as their desert pools heat in the summer sun, while hummingbirds may allow their bodies to cool down to 55°F (12.8°C) during the night to save energy (**FIG. 31-1a, b**). To avoid confusion, scientists often classify animals according to their major source of body warmth. Animals are **endotherms** (Greek, "inside heat") if they produce most of their heat by metabolic reactions; birds and mammals are all endotherms. A few fish—tuna and some large sharks—as well as some butterflies and bees also can warm their bodies considerably by using metabolic heat. Animals are **ectotherms** (Greek, "outside heat") if they derive most of their heat from the environment, for example by basking in the sun (**FIG. 31-1c**). Reptiles, amphibians, and most fish and invertebrates are ectotherms. In general, endotherms have higher metabolic rates than ectotherms do, allowing them to keep their bodies at consistently warm temperatures. Because their bodies generate so much heat, endotherms, including people, can be endangered by conditions in which they cannot escape the heat or cool themselves (see "Links to Life: Heat or Humidity?"). Ectotherms generally have lower and more variable body temperatures than endotherms do, because they are more dependent on environmental heat. However, through behavior or by occupying a very constant environment, ectotherm body tem-

FIGURE 31-1 Warm-blooded or cold-blooded?
(a) Because "cold-blooded" fish such as this desert pupfish may be quite warm, and (b) "warm-blooded" animals like this hummingbird can become quite cool, scientists prefer to classify animals as endothermic or ectothermic depending on the source of body warmth. (c) This lizard basking in the sun illustrates a behavioral mechanism that reptiles, which are ectotherms, use to regulate body temperature.

peratures may also remain quite stable. For example, the pupfish mentioned earlier can tolerate water temperatures ranging from 36°F to 113°F (2.2°C–45°C), but can breed only within a narrow range of temperatures. During breeding season, a pupfish can regulate its temperature quite precisely by swimming to different areas of its pond or hot spring as the temperature changes. In the deep ocean, the temperature is so constant (around 37.5°F, or 3°C) that ectothermic deep-sea fish experience almost no variation in body temperature.

Because warmer temperatures increase the rate of metabolic reactions, there are both costs and benefits to keeping warm. If their bodies become cool (as they do at night), butterflies and bees cannot fly; and lizards are too

sluggish to hunt or escape predators effectively. These ectotherms, and many others, warm themselves during the day so they can carry on normal behavior. Bees shiver and butterflies beat their wings to generate metabolic heat while lizards seek a warm, sunlit stone. Then, as they rest in sheltered places during the night, their bodies cool and conserve energy. Hummingbirds' daytime body temperature of about 105°F (41°C) contributes to their ability to beat their wings at the incredible rate of 80 times per second, which in turn allows them to hover precisely as they suck nectar from flowers (Fig. 31-1b). However, if these endotherms were to try to maintain this high body temperature through a very cool night, they would exhaust their energy reserves and starve. So when nights are cool, a

LINKS TO LIFE   Heat or Humidity?

It's not the heat, it's the humidity! Actually, the discomfort you feel on a hot, muggy day results from both these factors. Meteorologists have developed a formula, the heat index, that generates an "apparent temperature" by taking humidity into account. For example, a temperature of 90°F (32.2°C) has a heat index of 100°F (37.8°C) at 60% humidity, but it feels like only 85°F (29.4°C) at 10% humidity. Air temperature close to human body temperature prevents the body from radiating the excess heat it generates, while high humidity undermines your body's ability to cool itself by evaporating sweat. Heat and humidity together disrupt your body's attempts to maintain the narrow range of temperature that promotes homeostasis.

In extreme cases, excessive heat can lead to a deadly condition called *hyperthermia* or heat stroke. Although anyone

can succumb, the elderly and the very young are most susceptible because their ability to regulate body temperature is less efficient. During heat stroke, the body's homeostatic mechanisms are overcome, and body temperature rises to 106°F (41.1°C) or higher, often accompanied by dehydration. Sweating ceases and the perception of thirst may be lost. Dozens of young children in the United States die every summer from being left in closed cars, where temperatures can reach lethal levels in as little as 15 minutes.

As you gripe about the heat and humidity this summer, keep in mind that your discomfort has evolved as a warning system that homeostasis is threatened. So run through a sprinkler, seek shade or air conditioning, or drink a tall, cool glass of water!

hummingbird can shift its temperature downward and maintain a body temperature as much as 50°F (10°C) cooler than during the daytime.

Feedback Systems Regulate Internal Conditions

An organism's internal environment is maintained by mechanisms collectively known as *feedback systems. Negative feedback systems* counteract the effects of changes in the internal environment and are responsible for maintaining homeostasis. *Positive feedback systems* also occur in organisms. These drive rapid, self-limiting changes such as those occurring when a mother gives birth.

Negative Feedback Reverses the Effects of Changes

The most important mechanism governing homeostasis is **negative feedback**, in which a change in the environment causes responses that "feed back" and counteract the change. The overall result of negative feedback is to return the system to its original condition by counteracting the initial change (**FIG. 31-2**).

A familiar example of negative feedback is illustrated by your home thermostat. In a thermostat, a stimulus (temperature dropping below a *set point,* the thermostat setting) is detected by a thermometer, which signals a control device that switches on a heater. The heater restores the temperature to the set point, and the heater is switched off. Continuously repeated on-off cycles keep your home's temperature near the set point. Note that the rise and fall of temperature around the set point produces a dynamic equilibrium similar to that in animal bodies, rather than an absolutely constant temperature. The thermostat's negative feedback mechanism requires a *control center* with a set point, a *sensor* (the thermometer), and an *effector* (the furnace), which accomplishes the change.

Negative Feedback Maintains Body Temperature

How do people and other endothermic animals maintain their internal temperature despite extreme fluctuations in the temperature around them, such as those experienced by Erika in the snow on a Canadian winter night? The set point in the temperature control system, which varies by only about 1°F in healthy humans, is located in a control center in the *hypothalamus*, a region deep in the brain that controls many homeostatic responses. Nerve endings in the hypothalamus, abdomen, skin, and large veins act as temperature sensors and transmit this information to the hypothalamus. When body temperature drops, the hypothalamus activates various effector mechanisms that tend to raise body temperature. When normal body temperature is restored, sensors signal the hypothalamus to switch off these temperature control mechanisms. For example, Erika undoubtedly began shivering soon after she walked out the door into the snow. Shivering uses rapid, reflexive contractions of skeletal muscles to burn stored fuel and generate heat (Fig. 31-2b). The blood vessels supplying nonvital areas of her body (such as her face, hands, feet, and skin) became constricted, reducing heat loss and diverting warm blood to vital inner regions (brain, heart, and other internal organs). Then, her hypothalamus initiated a series of chemical signals that raised her metabolic rate, generating more heat to maintain vital functions. But with limited energy reserves and a small body with a large surface area, the child's temperature dropped quickly. Below about 85°F (29°C), the activity of her hypothalamus was depressed to the point where it no longer attempted to control her body temperature.

Negative feedback mechanisms abound in physiological systems. In the chapters that follow, you will find many examples of homeostatic control that operate by negative

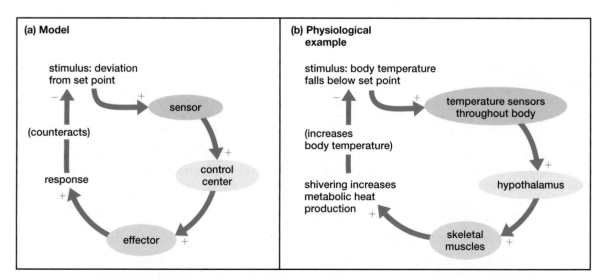

FIGURE 31-2 Negative feedback maintains homeostasis
Negative feedback maintains a set point by using a sensor to detect a deviation from the set point (stimulus). The receptor signals a control center, which activates an effector mechanism to counteract the stimulus.
QUESTION What would happen if a cold, shivering mammal ingested a poison that destroyed all of its body's nerve endings that detect heat?

feedback, including the systems regulating blood oxygen content, water balance, blood-sugar levels, hormone levels, and other components of the internal environment.

Positive Feedback Drives Certain Short-Term Physiological Events

When you first think about it, positive feedback is a rather frightening concept, because a change in a **positive feedback** system produces a response that actually intensifies the original change. Positive feedback, as you can imagine, tends to create accelerating reactions. For example, in nuclear fission, each particle that is split from an atom triggers the splitting of another atom, the pieces of which trigger the fission of other atoms, and so on. Uncontrolled, the energy released by this chain reaction is responsible for the devastation produced by an atomic bomb. A familiar biological example of positive feedback occurs when members of a population average more offspring than needed to replace themselves, causing the population to grow exponentially. Ecologist Paul Ehrlich coined the apt expression "population bomb" to describe the positive feedback of unchecked population growth. Negative feedback ultimately terminates processes driven by positive feedback. The atomic bomb eventually runs out of atoms to split, and growing populations run out of food or space.

Positive Feedback Produces Labor and Childbirth

In physiological systems, events governed by positive feedback are self-limiting and relatively rare. One important instance occurs during childbirth. The early contractions of labor begin to force the baby's head against the cervix at the base of the uterus, causing the cervix to dilate (open). Stretch-receptor neurons in the cervix react to this expansion by signaling the hypothalamus, which responds by triggering the release of a hormone (oxytocin) that stimulates more and stronger uterine contractions. Stronger contractions create further pressure on the cervix, which in turn causes release of more hormone. The birth of the baby relieves pressure on the cervix, providing negative feedback that halts this positive feedback cycle.

The Body's Internal Systems Act in Concert

The systems of the animal body are all "team players" that work together in a coordinated manner to maintain a relatively constant internal environment. Numerous mechanisms are constantly at work, responding to various stimuli that continuously change as the animal's activities and external environment change.

Fortunately, evolution has ensured that the various systems work together. The systems that take substances into the body act in concert with those responsible for transporting substances within the body and those that remove substances from the body. Each cell is indirectly connected to all of the others by an elaborate network of blood vessels and nerves that can carry molecules and messages to the appropriate locations. To maintain homeostasis, chemical signals act only on appropriate target cells that are specialized to receive and respond to specific signals. There are often many links in this chain of communication. Exposure to cold, for example, causes the hypothalamus to release chemical signals that travel in the bloodstream to a nearby gland (the pituitary). The pituitary gland then releases a hormone that triggers the thyroid gland to release a different hormone. This hormone acts to increase the body's metabolic rate, generating more heat. The hypothalamus stops releasing the chemical signal when its blood temperature receptors indicate that body temperature has been restored to normal. Thus, using a variety of routes and mechanisms, messages are carried from sensors to effectors and back again, allowing negative feedback mechanisms to maintain homeostasis.

31.2 HOW IS THE ANIMAL BODY ORGANIZED?

From simple, free-living cells, evolutionary change has produced astonishingly complex systems consisting of trillions of specialized cells that accomplish hundreds of functions simultaneously. The parts fit together with a degree of precision and integration about which human engineers can only dream. This complexity is based on a simple organizational hierarchy:

Cells → Tissues → Organs → Organ systems

An example of this hierarchy is illustrated in **FIGURE 31-3**. You learned in Chapter 1 that cells are the building blocks of all life. The animal body incorporates cells into **tissues**; each tissue is composed of dozens to billions of structurally similar cells that act in concert to perform a particular function. Tissues are the building blocks of **organs**, the discrete structures that perform complex functions. Examples of organs include the stomach, small intestine, kidneys, and urinary bladder. Organs, in turn, are organized into **organ systems**, groups of organs that function in a coordinated manner. For example, the digestive system is an organ system made up of the stomach, small intestine, large intestine, and other organs that work together to allow us to digest food and absorb nutrients from it. Major organ systems of vertebrates are illustrated in Table 31-1.

Animal Tissues Are Composed of Similar Cells That Perform a Specific Function

A tissue is composed of cells that are similar in structure and perform a specialized function. Tissue may also include extracellular components produced by these cells, as in the case of cartilage and bone. Here we present a brief overview of the four major categories of animal tissue and the major cell types that comprise these tissues: epithelial tissue, connective tissue, muscle tissue, and nerve tissue.

Epithelial Tissue Covers the Body, Lines Its Cavities, and Forms Glands

Epithelial tissue consists of one or more layers of densely packed epithelial cells underlain by an extracellular layer, the *basal lamina*, composed of collagen and other fibrous

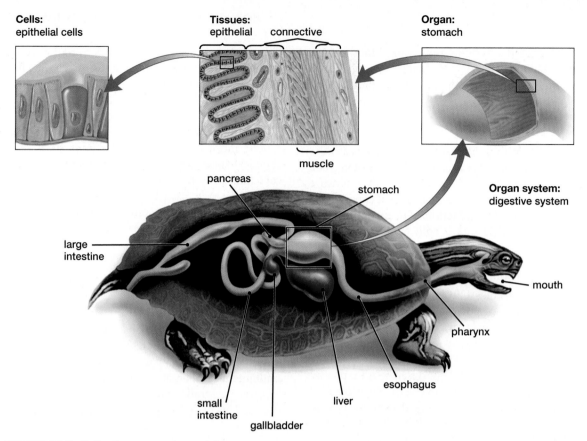

FIGURE 31-3 Cells, tissues, organs, and organ systems
The animal body is composed of cells, which make up tissues, which combine to form organs that work in harmony as organ systems.

proteins (**FIG. 31-4**). There are a variety of epithelial tissue types, each adapted to a particular function. Epithelial tissues cover the entire outside of the body and line all its inner cavities, including the digestive and respiratory tracts, heart and blood vessels, reproductive organs, and the organs of the urinary system.

Epithelial tissue is often classified by cell layers: *simple epithelium* is only one cell thick and lines the respiratory tract, much of the digestive tract, and the circulatory system. *Stratified epithelium*, which is several cells thick and can withstand considerable wear and tear, is found in the mouth and skin. Within each of these categories we find additional specializations. For example, the simple epithelium that lines the lungs consists of a single layer of thin, flattened cells whose shape makes them ideal for allowing rapid diffusion of gases (Fig. 31-4a). The simple epithelium that lines the trachea is actually quite complex, consisting of both short and elongated cells bearing cilia, interspersed with glandular *goblet cells* that secrete mucus (Fig. 31-4b). The mucus traps inhaled debris, and the cilia sweep it up out of the trachea. The stratified epithelium of the skin is covered with many layers of dead cells. These and the underlying epithelial cells are heavily laced with a strong, elastic protein (*keratin*) that makes skin flexible, water-resistant, and relatively tough (Fig. 31-4c).

An important property of epithelial cells is that they are continuously lost and replaced by mitotic cell division. For example, consider the abuse suffered by the epithelium that lines your mouth. Scalded by coffee and scraped by

corn chips, it would be destroyed within a few days if it did not replace itself continuously. The stomach lining, abraded by food and attacked by acids and protein-digesting enzymes, is completely replaced every 2 to 3 days. The skin's epithelial tissue, the epidermis (Fig. 31-4c), is renewed about twice a month.

Glands are cells or groups of cells specialized to secrete (release) large quantities of substances outside the cell. All glands are composed of specialized epithelial cells. Some glands are single cells, such as the mucus-secreting goblet cells of the tracheal epithelium (see Fig. 31-4b). Most glands are multicellular, including those that secrete saliva, sweat, milk, or hormones. Glands are classified into two broad categories: *exocrine glands* and *endocrine glands*. **Exocrine glands** secrete non-hormonal substances into a body cavity or onto the body surface, often through a narrow tube or *duct*. Examples of exocrine glands are sweat glands and *sebaceous* (oil-secreting) glands; both are found in the skin and are derived from skin epithelium (look ahead to Fig. 31-11). *Salivary glands* release saliva into the mouth; other exocrine glands release digestive enzymes into the stomach and small intestine. **Endocrine glands** lack ducts; they secrete *hormones* that diffuse into nearby capillaries. **Hormones** are chemicals produced in small quantities and transported in the bloodstream to regulate the activity of (often-distant) cells. Endocrine glands and their hormones are covered in detail in Chapter 37.

(a) Lining of lungs (simple)

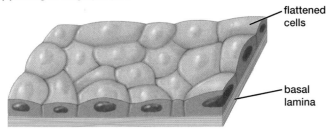

flattened cells

basal lamina

(b) Lining of trachea (simple)

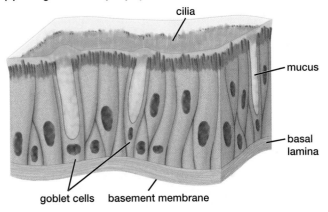

cilia

mucus

basal lamina

goblet cells   basement membrane

(c) Skin epidermis (stratified)

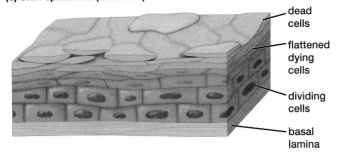

dead cells

flattened dying cells

dividing cells

basal lamina

FIGURE 31-4 Examples of epithelial tissue
(a) Thin, flattened cells in a single layer form the epithelial tissue that lines lungs, where exchange of gases by diffusion is important. **(b)** Elongated epithelial cells bearing cilia and interspersed with mucus-secreting goblet cells line the trachea. **(c)** The epidermis of the skin consists of multilayered, stratified epithelial tissue covered by a protective layer of dead cells. All epithelial tissue includes a thin layer of fibrous protein, called the basal lamina, beneath the epithelial cells.

Connective Tissues Have Diverse Structures and Functions

Connective tissues serve mainly to support and bind other tissues. Most connective tissues consist of cells embedded in a *matrix* of extracellular substances. The noncellular matrix of connective tissue commonly includes fluid and various types of flexible protein fibers, the most abundant of which is **collagen**. These proteins are typically secreted by the connective tissue cells. Connective tissues can be placed into three main categories.

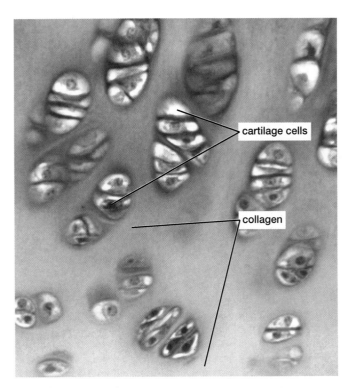

cartilage cells

collagen

FIGURE 31-5 Cartilage
The cells of the cartilage are stained dark purple and surrounded by clear spaces. The homogeneous material stained pale purple is the matrix of collagen secreted by the cartilage cells.

Loose Connective Tissue

This is the most abundant form, consisting of a thick fluid containing scattered cells that secrete protein fibers and collagen protein. This flexible tissue connects, supports, and surrounds other tissue types, and it forms an internal framework for organs such as the liver. Loose connective tissue combines with epithelial tissue to form *membranes*. For example, the *skin* (stratified epithelial tissue underlain by loose connective tissue) is a membrane that encloses the entire outer body surface, and *mucous membranes* (simple ciliated epithelial tissue underlain by loose connective tissue) line the inner cavities of the digestive, reproductive, respiratory, and urinary systems.

Fibrous Connective Tissue

This group includes **tendons** (which connect muscles to bones) and **ligaments** (which connect bones to bones). Fibrous connective tissue contains collagen fibers, which are densely packed in an orderly parallel arrangement—a design that gives tendons and ligaments their flexibility and tremendous strength.

Specialized Connective Tissues

This diverse group includes *cartilage, bone, fat, blood,* and *lymph*. **Cartilage** is flexible and resilient, consisting of widely spaced cells surrounded by a thick, nonliving matrix. This matrix is composed of collagen secreted by the cartilage cells (**FIG. 31-5**). Cartilage covers the ends of bones at joints, provides the supporting framework for the respiratory passages, supports the ear and nose, and forms shock-absorbing pads between the vertebrae.

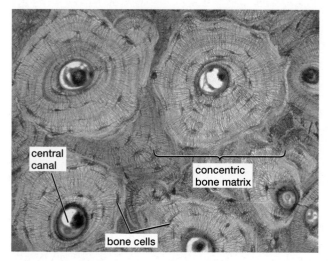

FIGURE 31-6 Bone
Concentric circles of bone (a specialized connective tissue), deposited around a central canal that contains a blood vessel, are clearly visible in this micrograph. Individual bone cells appear as dark spots trapped in small chambers within the hard matrix that the cells themselves deposit.

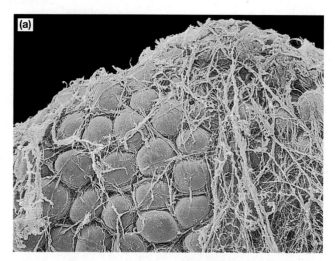

FIGURE 31-7 Adipose tissue
(a) Adipose tissue, shown here from a human abdomen, is specialized connective tissue made up almost exclusively of fat cells. A droplet of oil occupies most of the volume of the cell. Yellow fibrous connective tissue covers the cells. (b) A hooded seal at four days of age has doubled her birth weight by drinking mother's milk that consists of 61% fat, derived from the mother's own stores of blubber. At 100 pounds, she is almost too fat to move. The fat will feed and insulate the pup as the ice floes break up and she dives into icy water to learn to hunt and feed on her own. QUESTION Why are young mammals often fatter than adults?

Bone (**FIG. 31-6**) resembles cartilage, but its matrix is hardened by deposits of calcium phosphate. Bone forms in concentric circles around a central canal, which contains a blood vessel. (We will discuss cartilage and bone in depth in Chapter 39.) Fat cells, collectively called **adipose tissue** (**FIG. 31-7a**), are modified for long-term energy storage. Adipose tissue is especially important in the physiology of animals adapted to cold environments because it not only stores energy but also serves as insulation (**FIG. 31-7b**). Although they are liquids, **blood** and **lymph** are considered specialized forms of connective tissues because they are composed largely of extracellular fluids in which individual cells are suspended. The cellular portion of blood (**FIG. 31-8**) consists of red blood cells (which transport oxygen), white blood cells (which fight infection), and cell fragments called *platelets* (which aid in blood clotting). These are all suspended in extracellular fluid called *plasma*. Lymph consists largely of fluid that has leaked out of blood capillaries (the smallest of the blood vessels) and is carried back to the circulatory system within lymph vessels. You will learn more about blood and lymph in Chapter 32.

Muscle Tissue Has the Ability to Contract

The long, thin cells of muscle tissue contract (shorten) when stimulated and then relax passively. There are three types of muscle tissue: skeletal, cardiac, and smooth. **Skeletal muscle** (**FIG. 31-9**) is generally under voluntary, or conscious, control. As its name implies, its main function is to move the skeleton, as occurs when you walk or turn the pages of this text. **Cardiac muscle** is located only in the heart. Unlike skeletal muscle, it is spontaneously active and involuntary (not under conscious control). Cardiac muscle cells are interconnected by gap junctions, through which electrical signals spread rapidly through the heart, stimulating the cardiac muscle cells to contract in a coordinated fashion. **Smooth muscle**, so named because it lacks the orderly arrangement of proteins that causes stripes seen in cardiac and skeletal muscles, is embedded in the walls of the digestive tract, uterus, bladder, and large blood vessels. Smooth muscle

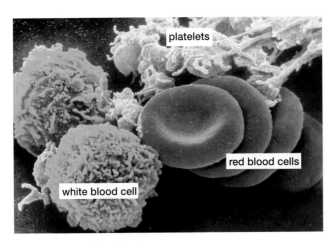

FIGURE 31-8 Blood cells
Blood contains three types of cellular components, shown in this color-enhanced scanning electron micrograph. The cells are suspended in plasma, which is also a component of this specialized connective tissue. QUESTION Why is blood considered a form of connective tissue?

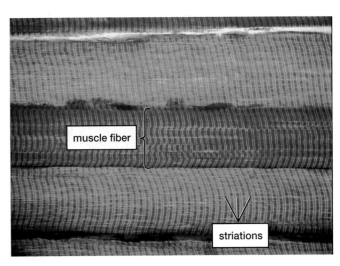

FIGURE 31-9 Muscle tissue consists of contractile cells called muscle fibers
A regular arrangement of fibrous proteins inside the muscle cells (called muscle fibers) of skeletal muscle gives this muscle tissue stripes or "striations" when viewed under a microscope.

produces slow, sustained contractions that are involuntary. Muscles and muscle contraction are covered in Chapter 39.

Nerve Tissue Is Specialized to Transmit Electrical Signals

You owe your ability to sense and respond to the world to **nerve tissue**, which makes up the brain, the spinal cord, and the nerves that travel from them to all parts of the body. Nerve tissue is composed of two types of cells: nerve cells, also called **neurons**, and glial cells. Neurons are specialized to generate electrical signals and to conduct these signals to other neurons, muscles, or glands (**FIG. 31-10**). Glial cells surround, support, electrically insulate, and protect neurons. Glial cells also regulate the composition of the interstitial fluid in the nervous system, allowing neurons to function optimally. We discuss nerve tissue in Chapter 38.

Organs Include Two or More Interacting Tissue Types

Organs are formed from at least two tissue types that function together. If an organ is hollow, such as the bladder or blood vessels, its interior is lined with a membrane consisting of epithelial tissue underlain by connective tissue. Different organs have different types and proportions of connective, glandular, muscular, and nervous tissues. Most organs function as part of organ systems that are discussed in other chapters of this unit. In the following section, we describe skin (which is not part of an organ system) as a representative organ that includes all four of the tissue types.

The Skin Illustrates the Properties of Organs

The structure of the skin is, in a general sense, representative of many organs. An outer layer of epithelial tissue is underlain by connective tissue that contains a blood supply, a nerve supply, muscle (in some cases), and glandular structures derived from the epithelium. Although we take our skin for granted, it is so important as a barrier against infection and water loss that large-scale destruction of skin, such as by extensive burns, can prove fatal. The skin also helps maintain homeostasis by dissipating or maintaining body heat, and serving as a barrier to disease organisms.

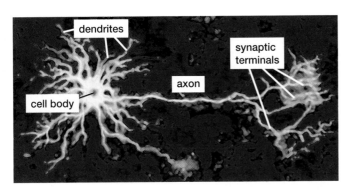

FIGURE 31-10 Nerve tissue
Nerve cells are specialized to receive and transmit signals. This micrograph shows a human neuron that stimulates muscle cells to contract. The dendrites are specialized to receive signals from other neurons, the axon carries the neuron's output signal to the muscle, and the synaptic terminals transmit the signal to the muscle cells.

FIGURE 31-11 Skin is an organ
Mammalian skin, a representative organ, shown in cross section. The skin is a membrane containing embedded glands, muscles, and nerve cells. QUESTION Why is skin considered to be an organ but blood is considered to be a tissue?

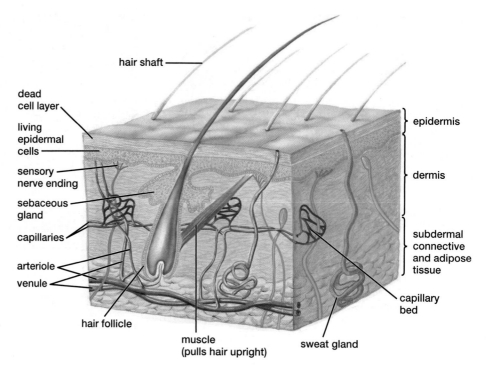

The **epidermis**, or outer layer of the skin, is a specialized multilayered (stratified) epithelial tissue (**FIG. 31-11**). It is covered by a protective layer of dead cells produced by the underlying living epidermal cells. These dead cells are packed with the protein keratin, which helps make the skin elastic, tough, and relatively waterproof.

Immediately beneath the epidermis lies a layer of loose connective tissue, the **dermis**. The loosely packed cells of the dermis are permeated by *arterioles* (small arteries). Arterioles feed blood pumped from the heart into a dense meshwork of capillaries that nourish both the dermal and epidermal tissue and empty into a network of *venules* (small veins) in the dermis. Loss of heat through the skin is precisely regulated by neurons controlling the degree of dilation (expansion) of the arterioles. When cooling is required, the arterioles dilate and flood the capillary beds with blood, thus releasing excess heat. When heat conservation is required, the arterioles supplying the skin capillaries are constricted; this is why doctors found that some of the skin on Erika's hands had frozen, producing "frostbite." Lymph vessels collect and carry off extracellular fluid within the dermis. Various sensory nerve endings responsive to temperature, touch, pressure, vibration, and pain are scattered throughout the dermis and epidermis and provide information to the nervous system.

The dermis is also packed with glands derived from epithelial tissue. Glands called **hair follicles** produce hair from protein-containing secretions. Sweat glands produce watery secretions that cool the skin and excrete substances such as salts and urea. *Sebaceous glands* secrete an oily substance (*sebum*) that lubricates the epithelium.

In addition to the epithelial, connective, and nerve tissues already mentioned, skin often contains muscle tissue. Tiny muscles attached to the hair follicles can cause the hairs of the skin to "stand on end" in response to signals from motor neurons. Most mammals are able to increase the thickness of their insulating fur in cold weather by erecting individual hairs, but this reaction is useless for retaining heat in people, who merely experience "goose bumps" when these tiny muscles contract.

Organ Systems Consist of Two or More Interacting Organs

Organ systems consist of two or more individual organs (in some cases, located in different regions of the body) that work together, performing a common function. An example is the digestive system, in which the mouth, esophagus, stomach, intestines, and other organs (such as the liver and pancreas) that supply digestive secretions all function together to convert food into nutrient molecules (see Fig. 31-3). The major organ systems of the vertebrate body and their representative organs and functions are described in Table 31-1.

Table 31-1 Major Vertebrate Organ Systems

Organ System	Major Structures	Physiological Role	Organ System	Major Structures	Physiological Role
Circulatory system	Heart, blood vessels, blood	Transports nutrients, gases, hormones, metabolic wastes; also assists in temperature control	Endocrine system	A variety of hormone-secreting glands and organs, including the hypothalamus, pituitary, thyroid, pancreas, adrenals, ovaries, and testes	Controls physiological processes, typically in conjunction with the nervous system
Lymphatic/ immune system	Lymph, lymph nodes and vessels, white blood cells	Carries fat and excess fluids to blood; destroys invading microbes	Nervous system	Brain, spinal cord, peripheral nerves	Controls physiological processes in conjunction with the endocrine system; senses the environment, directs behavior
Digestive system	Mouth, esophagus, stomach, small and large intestines, glands producing digestive secretions	Supplies the body with nutrients that provide energy and materials for growth and maintenance	Muscular system	Skeletal muscle	Moves the skeleton
				Smooth muscle	Controls movement of substances through hollow organs (digestive tract, large blood vessels)
				Cardiac muscle	Initiates and implements heart contractions
Urinary system	Kidneys, ureters, bladder, urethra	Maintains homeostatic conditions within bloodstream; filters out cellular wastes, certain toxins, and excess water and nutrients	Skeletal system	Bones, cartilage, tendons, ligaments	Provides support for the body, attachment sites for muscles, and protection for internal organs
Respiratory system	Nose, pharynx, trachea, lungs (mammals, birds, reptiles, amphibians), gills (fish and some amphibians)	Provides a large area for gas exchange between the blood and the environment; allows oxygen acquisition and carbon dioxide elimination	Reproductive system	Male: testes, seminal vesicles, prostate gland, penis	Male: produces sperm, inseminates female
				Female: ovaries, oviducts, uterus, vagina, mammary glands	Female: produces egg cells, nurtures developing offspring

CASE STUDY REVISITED   LIFE ON HOLD?

To stay alive, animals require a constant supply of ATP, which in turn requires a constant supply of oxygen. Cold slows all biochemical reactions; both Erika's and Anna's metabolic rates (the rate at which cells produce and use energy), and thus their need for oxygen, were radically reduced by their low body temperatures. While the brain suffers permanent damage after about five minutes without oxygen at normal body temperatures, their brains survived unharmed for far longer in the extreme cold. Doctors routinely take advantage of this fact during open heart surgery, lowering the patient's body temperature from its normal 98.6°F (37°C) to about 65°F (18°C) for about 45 minutes while the heart is stopped. Animal studies suggest that even

deeper cooling (such as experienced by Erika and Anna) may extend the time that a person could safely remain in a state of suspended animation.

At Massachusetts General Hospital, trauma surgeon Hasan Alam works with pigs, completely draining their blood and replacing it with a frigid nutrient solution. In this state, with their bodies cooled to 50°F (10°C), the pigs have no heartbeat, they don't breathe, and their brains are electrically silent. Yet as warm blood is pumped back into the animals even after two and one-half hours in this condition, the heart resumes beating and the animal revives. Behavioral tests suggest that the pigs have experienced no lasting damage from their experience. In the near future, clinical trials using this process on trauma victims who face almost certain death

from blood loss may begin. By replacing a victim's remaining blood with a near-freezing nutrient solution, doctors hope to gain precious time to operate and repair the damage before reanimating the patient with warmed blood.

Consider This Patients who are near death from blood loss are unable to give informed consent to an experimental procedure. For this reason, Dr. Alam is working to inform the entire community served by his hospital about the experimental blood replacement proposal, and he is encouraging anyone who does not wish to participate to wear a bracelet stating this choice. Describe some advantages and the potential problems with this approach to obtaining "volunteers" for these clinical trials.

CHAPTER REVIEW

SUMMARY OF KEY CONCEPTS

31.1 Homeostasis: How Do Animals Regulate Their Internal Environment?

Homeostasis refers to the dynamic equilibrium within the animal body by which physiological conditions including temperature, salt, oxygen, glucose, pH, and water levels are maintained within a range in which proteins can function and energy can be made available. Animals differ in temperature regulation. Ectotherms derive most of their body warmth from the environment and tend to tolerate larger extremes of body temperature. Endotherms derive most of their heat from metabolic activities and tend to regulate their body temperature within a narrower range.

Homeostatic conditions are maintained through negative feedback, in which a change triggers a response that counteracts the change and restores conditions to a set point. There are a few instances of positive feedback, in which a change initiates events that intensify the change (such as uterine contractions leading to childbirth), but these situations are all self-limiting through negative feedback. Within the animal body, multiple feedback mechanisms work in concert.

Web Tutorial 31.1 Homeostasis

31.2 How Is the Animal Body Organized?

The animal body is composed of organ systems consisting of two or more organs. Organs, in turn, are made up of tissues. A

tissue is a group of cells and extracellular material that form a structural and functional unit; it is specialized for a specific task. Animal tissues include epithelial, connective, muscle, and nerve tissue.

Epithelial tissue forms membranous coverings over internal and external body surfaces and also gives rise to glands. Connective tissue usually contains considerable extracellular material, called matrix, and includes dermal tissue, bone, cartilage, tendons, ligaments, fat, and blood. Muscle tissue is specialized to produce movement by contracting. There are three types of muscle tissue: skeletal, cardiac, and smooth. Nerve tissue, including neurons and glial cells, is specialized to generate and conduct electrical signals.

Organs include at least two tissue types that function together. Mammalian skin is a representative organ. The epidermis, an epithelial tissue, covers and protects the dermis beneath it. The dermis contains blood and lymph vessels, sweat and sebaceous glands, and tiny muscles that erect the hairs. Animal organ systems include the digestive, urinary, immune, respiratory, circulatory/lymphatic, nervous, muscular, skeletal, endocrine, and reproductive systems, summarized in Table 31-1.

KEY TERMS

THINKING THROUGH THE CONCEPTS

1. Define and compare *ectotherms* and *endotherms*. Provide an example of each. Are "cold-blooded" and "warm-blooded" accurate ways to describe them? Explain.

2. Define *homeostasis*, and explain how negative feedback helps to maintain it. Explain one example of homeostasis in the human body.

3. Explain positive feedback, and provide one physiological example. Explain why this type of feedback is relatively rare in physiological processes.

4. Explain what goes on in your body to restore temperature homeostasis when you become overheated by exercising on a hot, humid day.

5. Describe the structure and functions of epithelial tissue.

6. What property distinguishes connective tissue from all other tissue types? List three general types of connective tissue, and briefly describe the function of each type.

7. Describe the skin, a representative organ. Include the various tissues that compose it, and briefly describe the role of each tissue.

APPLYING THE CONCEPTS

1. Why does life on land present particular difficulties in maintaining homeostasis?

2. In response to cold, explain how behavior and unconscious mechanisms work together to maintain thermal homeostasis.

3. Third-degree burns are usually painless. Skin regenerates only from the edges of these wounds. Second-degree burns regenerate from cells located at the burn edges, in hair follicles, and in sweat glands. First-degree burns are painful but heal rapidly from undamaged epidermal cells. Using this information, draw the depth of first-, second-, and third-degree burns on Figure 31-11.

4. Imagine you are a healthcare professional teaching a prenatal class for fathers. So that a layperson could understand the feedback relationships involved in initiating labor, design a real-world analogy to sensors, electrical currents, motors, and so on.

FOR MORE INFORMATION

Bruemmer, F. "Five Days with Fat Hoods." *International Wildlife*, January–February 1999. The rapid growth and prodigious fat-storing ability of the hooded seal adapts it to maintaining homeostasis under the extreme conditions of the far north.

Nuland, S. *The Wisdom of the Body*. New York: Alfred A. Knopf, 1997. Human physiology as seen through a surgeon's eyes. A firsthand account of the beauty and power of the body's mechanisms for maintaining homeostasis.

Roth, M. B., and Nystul, T. "Buying Time in Suspended Animation." *Scientific American*, June 2005. This engagingly written article explains how animal studies suggest that suspended animation may in the future help to buy time for donated organs or trauma victims.

Storey, K. B., and Storey, J. M. "Frozen and Alive." *Scientific American*, December 1990. Some animals have special adaptations that allow them to withstand freezing.

Trivedi, Bijal. "Life on Hold." *New Scientist*, January 21, 2006. Reviews the evidence from animal studies that suggests that through cold or chemical means, human organs or damaged human bodies might gain precious time by being placed in suspended animation.

Circulation

At a memorial service, Darryl Kile's teammates watch him wind up for a pitch.
(Inset) A plaque in this coronary artery (an artery supplying the heart) has
stimulated the formation of a blood clot that is totally blocking the vessel
and preventing blood from reaching a portion of the heart muscle.
This blockage will cause a heart attack.

CASE STUDY SUDDEN DEATH

ON JUNE 22, 2002, players for the St. Louis Cardinals prepared for their upcoming game against the Chicago Cubs. As game time approached they were first puzzled, then concerned, by the unexplained absence of their pitcher, Darryl Kile. Their concern turned to shock and grief when 33-year-old Kile was found dead in his hotel room, apparently having died in his sleep. One of the country's top pitchers and noted for his exceptional curveball, Kile was an athlete in his prime. But an autopsy revealed that two of his three coronary arteries (supplying blood to the heart muscle) were 80% to 90% blocked by atherosclerosis, in which arteries are narrowed by fatty deposits called plaque (see the inset in this chapter's opening photo). Kile's heart was also enlarged, a result of its heroic efforts to force blood through the partially blocked arteries. Some inherited genetic traits favor plaque buildup, and can cause life-threatening levels of plaque to accumulate at a much younger age than in people without these risk factors. The fact that Kile's father died of a heart attack at the age of 44 suggests that Darryl Kile may have been in this high-risk group.

Atherosclerosis often begins in childhood. Millions of children in the United States have elevated blood cholesterol levels, exacerbated by high-fat diets and lack of exercise; but decades usually pass before the disease is recognized. How does the heart work? How it is threatened by atherosclerosis and hypertension? What treatments might have benefited Darryl Kile had he known about his condition?

32.1   WHAT ARE THE MAJOR FEATURES AND FUNCTIONS OF CIRCULATORY SYSTEMS?

Billions of years ago, the first cells were nurtured by the sea in which they evolved. The sea brought them nutrients, which diffused into the cells and washed away any wastes that diffused out. Today, microorganisms and some simple multicellular animals still rely almost exclusively on diffusion to exchange gases with the environment. Sponges, for example, circulate seawater through pores in their bodies, bringing the environment close to each cell. As larger, more complex animals evolved, individual cells became increasingly distant from the outside world. However, the constant demands of a cell require short diffusion distances, so that adequate nutrients reach the cell and the cell isn't poisoned by its own wastes. With the evolution of the circulatory system, a sort of "internal sea" was created, serving the same purpose as the sea did for the first cells. This internal sea transports food and oxygen close to each cell and carries away wastes produced by the cells.

All circulatory systems have three major parts:

- A fluid, **blood**, that serves as a medium of transport
- A system of channels, or **blood vessels**, that conduct the blood throughout the body
- A pump, the **heart**, which keeps the blood circulating

Animals Have Two Types of Circulatory Systems

Animals have one of two major types of circulatory systems: open and closed. **Open circulatory systems** are present in some invertebrates, including arthropods—such as crustaceans, spiders, and insects—and mollusks, such as snails and clams. An animal with an open circulatory system has one or more simple "hearts," a network of blood vessels, and a large open space within the body called a **hemocoel** (**FIG. 32-1a**). Within the hemocoel (which may occupy 20% to 40% of the body volume), tissues and internal organs are directly bathed in blood. In insects the heart is a modified portion of the dorsal blood vessel consisting of a series of contracting chambers. When the chambers contract, valves in the heart are pressed shut, forcing the blood out through vessels into the hemocoel. When the heart chambers relax, blood is drawn back into them from the hemocoel.

Closed circulatory systems are present in some invertebrates, including the earthworm (**FIG. 32-1b**) and very active mollusks (such as squid and octopuses). Closed circulatory systems are also a characteristic of all vertebrates, including humans. In closed circulatory systems, the blood (whose volume is only 5% to 10% of the body volume) is confined to the heart and a continuous series of blood vessels. Closed circulatory systems allow more rapid blood flow, more efficient transport of wastes and nutrients, and higher blood pressure than is possible in open

FIGURE 32-1 Open and closed circulatory systems (a) (top) In open circulatory systems of arthropods, a heart pumps blood through vessels into the hemocoel, where blood directly bathes the other organs. (bottom) The grasshopper provides a good example of an open circulatory system (insect blood lacks hemoglobin and is almost clear or very pale green). **(b)** (top) In a closed circulatory system, blood remains confined to the heart(s) and the blood vessels. (bottom) In the earthworm, five contractile vessels serve as hearts that pump blood through major ventral and dorsal vessels, from which smaller interconnecting vessels branch. Earthworm blood, like ours, contains red hemoglobin.

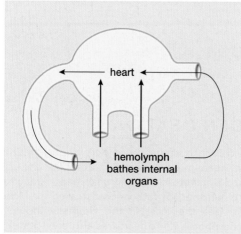
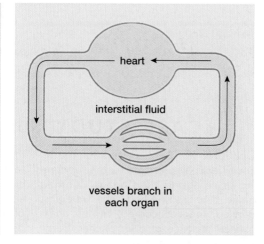
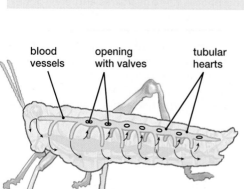
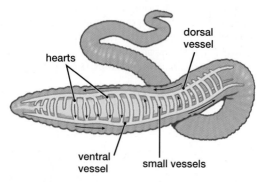

(a) Open circulatory system

(b) Closed circulatory system

systems. In the earthworm, five contractile vessels serve as hearts, pumping blood through major vessels from which smaller vessels branch.

The Vertebrate Circulatory System Has Diverse Functions

The circulatory system supports all the other organ systems in the body. The circulatory systems of humans and other vertebrates perform the following functions:

- Transport oxygen from the lungs or gills to the tissues, and transport carbon dioxide from the tissues to the lungs or gills
- Distribute nutrients from the digestive system to all body cells
- Transport waste products and toxic substances to the liver (where many of them are detoxified) and kidneys for excretion
- Distribute hormones from the glands and organs that produce them to the tissues upon which they act
- Regulate body temperature, which is achieved partly by adjustments in blood flow
- Prevent blood loss by means of the clotting mechanism
- Protect the body from bacteria and viruses by circulating antibodies and white blood cells

In the following sections we examine the three parts of the circulatory system: the heart, the blood, and the vessels, with an emphasis on the human system. Finally, we describe the lymphatic system, which works closely with the circulatory system.

32.2 HOW DOES THE VERTEBRATE HEART WORK?

Increasingly Complex and Efficient Hearts Have Arisen During Vertebrate Evolution

No circulatory system can operate without a dependable pump. Blood must be moved through the body continuously throughout an animal's life; the vertebrate heart consists of muscular chambers capable of strong contractions. Chambers called **atria** (singular, **atrium**) collect blood. Atrial contractions send blood into the **ventricles**, chambers whose contractions circulate blood through the body. During the course of vertebrate evolution, the heart has become increasingly complex, with more separation between oxygenated blood (which has picked up oxygen from the lungs or gills) and deoxygenated blood (which, in passing through body tissues, has lost oxygen).

The hearts of fishes, the first vertebrates to evolve, consist of two contractile chambers: a single atrium that empties into a single ventricle (**FIG. 32-2a**). Blood pumped from the ventricle passes first through the gill *capillaries*, thin-walled vessels where blood picks up oxygen and gives off carbon dioxide. The blood then travels to the rest of the body, delivering oxygen to the tissues and picking up carbon dioxide in the body capillaries.

Over evolutionary time, as fish gave rise to amphibians and amphibians to reptiles, a three-chambered heart evolved, consisting of two atria and one ventricle (**FIG. 32-2b**). In the three-chambered hearts of amphibians and most reptiles, deoxygenated blood from the body is delivered into the right atrium, while blood from the lungs travels into the left atrium. Both atria empty into the single ventricle. Although some mixing does occur, the deoxygenated blood

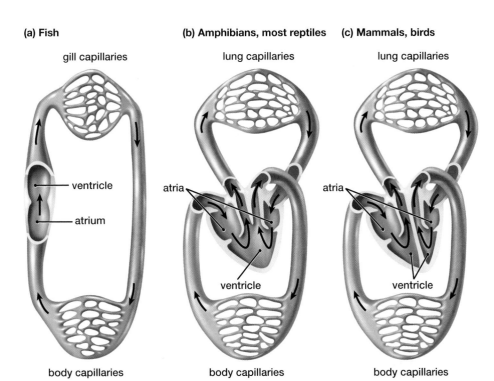

(a) Fish

gill capillaries

ventricle

atrium

body capillaries

(b) Amphibians, most reptiles

lung capillaries

atria

ventricle

body capillaries

(c) Mammals, birds

lung capillaries

atria

ventricle

body capillaries

FIGURE 32-2 Evolution of the vertebrate heart (a) The earliest vertebrate heart is illustrated by the two-chambered heart of fishes. (b) Amphibians and most reptiles have hearts with two atria, from which blood empties into a single ventricle. Many reptiles have a partial wall down the middle of the ventricle. (c) The hearts of birds and mammals are two separate pumps that prevent mixing of oxygenated and deoxygenated blood. Oxygenated blood is depicted as bright red and deoxygenated blood as bright blue.

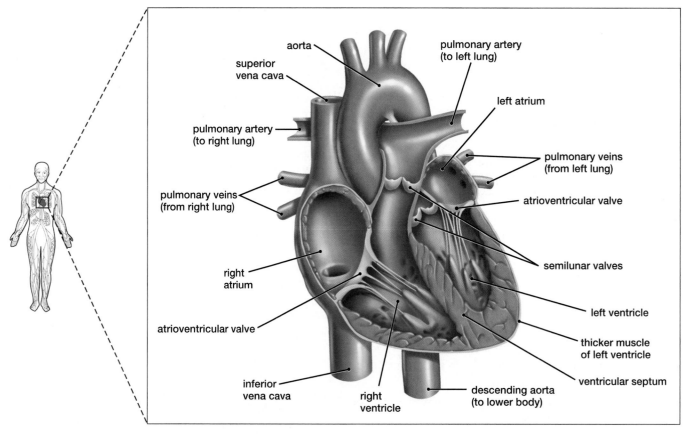

FIGURE 32-3 The human heart and its valves and vessels
The heart is drawn facing you, so right and left appear reversed. The muscular wall of the left ventricle is thickened because it must pump blood throughout the body. One-way semilunar valves separate the aorta from the left ventricle, and the pulmonary artery from the right ventricle. Atrioventricular valves separate each atrium from its corresponding ventricle.

tends to remain in the right portion of the ventricle and is pumped into vessels that enter the lungs, while most of the oxygenated blood remains in the left portion of the ventricle and is pumped to the rest of the body. This separation is enhanced in reptiles by a partial wall between the right and left portions of the ventricle. The four-chambered hearts of birds and mammals (FIG. 32-2c) have separate right and left ventricles that isolate oxygenated from deoxygenated blood.

The Vertebrate Heart Consists of Muscular Chambers That Form Two Separate Pumps

Veins Carry Blood to the Heart, and Arteries Carry Blood Away

The hearts of birds and mammals, including humans, can be considered as two separate pumps, each with two chambers. In each pump, an atrium receives and briefly stores the blood before passing it to a ventricle that propels it through the body (FIG. 32-3). One pump, consisting of the right atrium and right ventricle, deals with deoxygenated blood. The right atrium receives oxygen-depleted blood from the body through two large **veins**, which are vessels that carry blood toward the heart: the *superior vena cava* and the *inferior vena cava*. After being filled with blood, the right atrium contracts, forcing the blood into the right ventricle. Contraction of the right ventricle sends the oxygen-depleted blood to the lungs via pulmonary **arteries**,

which are vessels that carry blood away from the heart. The other pump, consisting of the left atrium and ventricle, deals with oxygenated blood. Oxygen-rich blood from the lungs enters the left atrium through pulmonary veins and is then squeezed into the left ventricle. Strong contractions of the left ventricle, the heart's most muscular chamber, send the oxygenated blood coursing out through a major artery, the *aorta*, to the rest of the body.

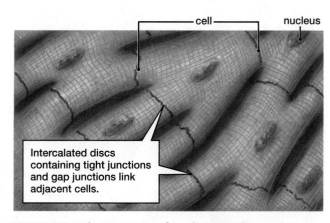

FIGURE 32-4 The structure of cardiac muscle
Cardiac muscle cells are branched, and intercalated disks link adjacent cells. QUESTION If a muscle is repeatedly exercised, it increases in size. Why is the resting heart rate of a well-conditioned athlete slower than the heart rate of a less-active person?

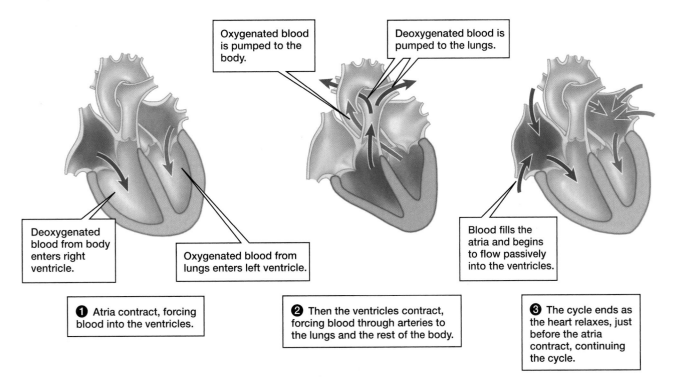

Oxygenated blood is pumped to the body.

Deoxygenated blood is pumped to the lungs.

Deoxygenated blood from body enters right ventricle.

Oxygenated blood from lungs enters left ventricle.

Blood fills the atria and begins to flow passively into the ventricles.

❶ Atria contract, forcing blood into the ventricles.

❷ Then the ventricles contract, forcing blood through arteries to the lungs and the rest of the body.

❸ The cycle ends as the heart relaxes, just before the atria contract, continuing the cycle.

FIGURE 32-5 The cardiac cycle

Cardiac Muscle Is Present Only in the Heart

Each **cardiac muscle** cell is small, branched, and contains strands of protein that give it a striped appearance (**FIG. 32-4**). Cardiac muscle cells are linked to one another by *intercalated disc*s that appear as bands between the cells. Intercalated discs contain both tight junctions (*desmosomes*) and gap junctions (pores connecting adjacent cells). Desmosomes prevent the strong heart contractions from pulling the muscle cells apart. Gap junctions allow the electrical signal that triggers contractions to spread directly and rapidly from one muscle cell to adjacent ones. This causes the interconnected regions of heart muscle to contract almost synchronously.

The Coordinated Contractions of Atria and Ventricles Produce the Cardiac Cycle

The human heart beats about 100,000 times each day. During each beat, the two atria first contract in synchrony, emptying their contents into the ventricles. A fraction of a second later, the two ventricles contract simultaneously, forcing blood into arteries that exit the heart. Both atria and ventricles then relax briefly before this **cardiac cycle** repeats (**FIG. 32-5**). At a typical resting heart rate, the cardiac cycle lasts just under 1 second. The cardiac cycle is involved in blood pressure measurement (**FIG. 32-6**); *systolic*

cuff

Stethoscope detects pulse sounds.

Cuff is inflated, putting pressure on the artery.

FIGURE 32-6 Measuring blood pressure
First, the cuff is inflated until its pressure closes off the arm's main artery; this pressure is then gradually reduced. When rhythmic blood sounds are first heard through the stethoscope, the contractions of the left ventricle have overcome the pressure in the cuff and blood is flowing. This pressure is recorded as the upper (and higher) reading: systolic pressure. Next, cuff pressure is further reduced until no pulse sounds are audible, indicating that blood is flowing continuously through the artery. In other words, the pressure between ventricular contractions is adequate to overcome the cuff pressure. This is the lower reading: the diastolic pressure. The numbers are in millimeters of mercury, a standard measure of pressure also used in barometers. EXERCISE Sketch a graph that shows how blood pressure inside an artery changes during the cardiac cycle. On the graph, label the points that correspond to the systolic and diastolic blood pressure measured by the blood pressure cuff.

Cardiovascular disease (CVD; disorders of the heart and blood vessels) is the leading cause of death in the U.S. According to the American Heart Association, CVD claims a life every 35 seconds, over 900,000 deaths annually in the U.S.—and no wonder. The heart must contract vigorously more than 2.5 billion times during an average lifetime without once stopping to rest, forcing blood through vessels whose total length would encircle the globe twice. Since these vessels may become constricted, weakened, or clogged, the cardiovascular system is a prime candidate for malfunction.

ATHEROSCLEROSIS OBSTRUCTS BLOOD VESSELS

Atherosclerosis (from the Greek *athero*, meaning "gruel" or "paste," and *scleros*, "hard") causes the walls of the large arteries to thicken and lose their elasticity. This is caused by deposits called **plaques** within the wall of the artery. A major risk factor for plaque formation is high blood levels of *LDL cholesterol,* which consists of cholesterol molecules bound to carrier molecules made up of lipids and proteins (called *low-density lipoprotein: LDL*). Plaque formation seems to be initiated by minor damage to the endothelium lining the artery, which may be caused by high blood pressure, toxins from cigarette smoke, or other factors. The damaged endothelium attracts white blood cells, which burrow beneath it and ingest large quantities of cholesterol and other lipids. The bloated bodies of these macrophages contribute to a growing fatty core of plaque. Meanwhile, smooth muscle cells from below the endothelium migrate into the core, absorb more fat and cholesterol, and contribute to the plaque. They also produce proteins that form a fibrous cap that replaces the damaged endothelium and covers the fatty core (**FIG. E32-1**).

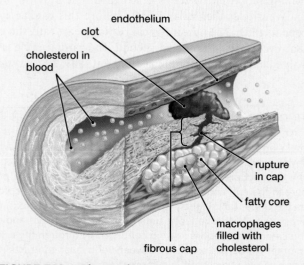

FIGURE E32-1 Plaque clogs an artery
When the fibrous cap ruptures, a blood clot forms, obstructing the artery.

The fibrous cap may rupture, exposing the blood to clot-promoting factors within the plaque. The clot further obstructs the artery and may completely block it (see chapter opener photo, inset). Or the clot may be swept away in the flowing blood until it lodges in and blocks a narrower part of the artery. Arterial clots are responsible for the most serious consequences of atherosclerosis: *heart attacks* and *strokes.*

A **heart attack** occurs when one of the arteries supplying the heart muscle is blocked. Deprived of nutrients and oxygen, the heart muscle supplied by the blocked artery rapidly and painfully dies. Although heart attacks are the major cause of death from atherosclerosis, this disease causes plaques and clots to form in arteries throughout the body. Clots obstructing arteries in the brain are the major causes of **stroke**, which is sometimes called a *brain attack.* Strokes can have serious consequences because brain cells have very high oxygen demands and can die within minutes if deprived of blood. Depending on the extent and location of brain damage, strokes can cause a variety of neurological problems including partial paralysis; difficulties in remembering, communicating, or learning; mood swings; or personality changes.

TREATMENT OF ATHEROSCLEROSIS

Atherosclerosis is promoted by high blood pressure, cigarette smoking, obesity, diabetes, lack of exercise, genetic predisposition, and high LDL-cholesterol levels in blood. Traditional treatment of atherosclerosis includes changes in diet and lifestyle, and if this fails, drugs may be prescribed to lower cholesterol. If a person has had a heart attack or suffers from **angina**, which is chest pain caused by insufficient blood flow to the heart, he or she may be a candidate for surgery to widen or bypass the obstructed artery.

Angioplasty refers to techniques that widen obstructed coronary arteries (**FIG. E32-2**). All of these procedures involve threading a thin, flexible tube through an artery in the upper leg or arm and guiding it into the clogged artery. In balloon angioplasty the tube is tipped with a small balloon that is inflated, compressing the plaque and allowing blood to flow more freely (**FIG. E32-2b**). Alternatively, the balloon may be equipped with small, rotating blades that shear off the plaque as it is compressed. Another approach is a high-speed, diamond-coated drill that grinds away the plaque in microscopic pieces that are carried away in the blood (**FIG. E32-2c**). After doctors remove the plaque, they often insert a wire mesh tube or *stent* into the artery to help keep it open (**FIG. E32-2d**).

Coronary bypass surgery bypasses obstructed coronary arteries with segments of vein (usually obtained from the patient's leg; **FIG. E32-3**) or artery (often from the patient's forearm). But removing a vein is a painful, time-consuming operation in itself, and thousands of patients who might benefit from bypass surgery do not have vessels suitable for grafting. In an effort to help these patients, researchers are working to develop artificial vessels. Collagen protein from pigs or cows, molded into tubes, can either be placed in a nutrient broth with blood vessel cells or grafted directly into

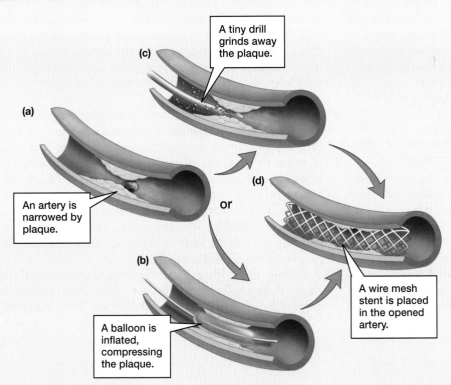

FIGURE E32-2 Angioplasty unclogs arteries
(a) A narrowed artery may be opened by (b) inflating a tiny balloon inside or (c) drilling out the plaque. Following angioplasty, (d) a metal mesh stent is often inserted to maintain the opening.

(c) A tiny drill grinds away the plaque.

(a) An artery is narrowed by plaque.

or

(d)

(b) A balloon is inflated, compressing the plaque.

A wire mesh stent is placed in the opened artery.

the vessels of experimental animals. Living blood vessel cells will invade and cover these tubes to form at least temporarily functional vessels. A new process uses the patient's own cells, grown on a plastic mold in the laboratory. Although this procedure shows promise, it will require years of development and testing before it is used in people.

If a heart attack occurs, rapid treatment can minimize the damage and significantly increase the victim's chances of survival. Blood clots in coronary arteries or in the brain can be dissolved by injecting a "clot-busting" protein (*tissue plasminogen activator, tPA*) that activates an enzyme (plasmin) that breaks down fibrin, the protein holding blood clots together. This drug is effective only if it is administered within a few hours after the blockage occurs, so many patients miss this opportunity for a better recovery. Recently, a more potent clot-busting agent has been tested in people, with encouraging results. This protein is called *DSPA*, which stands for *Desmodus salivary plasminogen activator*. It was first discovered in the saliva of vampire bats (of the genus *Desmodus*), who secrete the substance so that clotting doesn't interfere with their blood meal. This unique protein not only works better, but it can be administered up to nine hours after the clot forms, so it may eventually help many more people than tPA can. Although heart disease is still the leading cause of death in the U.S., steady progress in treatment has significantly reduced the rate of early deaths and disability from atherosclerosis.

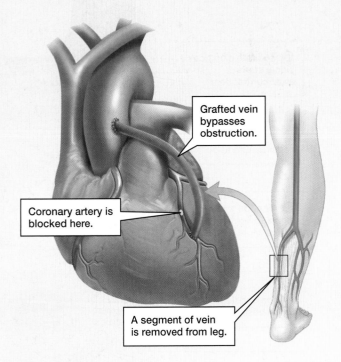

Grafted vein bypasses obstruction.

Coronary artery is blocked here.

A segment of vein is removed from leg.

FIGURE E32-3 Coronary bypass surgery

pressure (the higher of the two readings) is measured during ventricular contraction, and *diastolic pressure* is measured between contractions. High blood pressure, also called **hypertension**, is caused by the constriction of arterioles, causing resistance to blood flow and strain on the heart. In most cases, the cause is unknown. An approximate borderline reading for high blood pressure is 140/90. High blood pressure also contributes to "hardening of the arteries" as described in "Health Watch: Repairing Broken Hearts."

Valves Maintain the Direction of Blood Flow, and Electrical Impulses Coordinate the Sequence of Contractions

When the ventricles contract, blood must be directed out through the arteries and not back up into the atria. Then, once blood has entered the arteries, it must be prevented from flowing back as the heart relaxes. The directionality of blood flow is maintained by one-way valves (see Figs. 32-3 and 32-5). Pressure in one direction opens them easily, but reverse pressure forces them closed. **Atrioventricular valves** allow blood to flow from the atria into the ventricles (but not the reverse), and **semilunar valves**

(from Latin, meaning "half-moon" valves) allow blood to enter the pulmonary artery and the aorta when the ventricles contract (and prevent it from returning as the ventricles relax). You may notice that no valves separate the left and right atria from the pulmonary veins and the vena cavas, respectively. In fact, when the atria contract, some blood does flow backward into these veins; but since the ventricles fill adequately despite this backflow, no valves have evolved here.

The contraction of the heart is initiated and coordinated by a **pacemaker**, a cluster of specialized heart muscle cells that produce spontaneous electrical signals at a regular rate. These electrical signals are transmitted among the heart muscle cells and stimulate them to contract. The heart's primary pacemaker is the **sinoatrial (SA) node**, located in the upper wall of the right atrium (**FIG. 32-7**). The gap junctions linking adjacent cardiac cells allow the electrical signals to pass freely and rapidly from cells near the pacemaker to adjoining atrial cells.

During the cardiac cycle, the atria contract first and empty their contents into the ventricles, then refill while the ventricles contract. Thus, there must be a delay be-

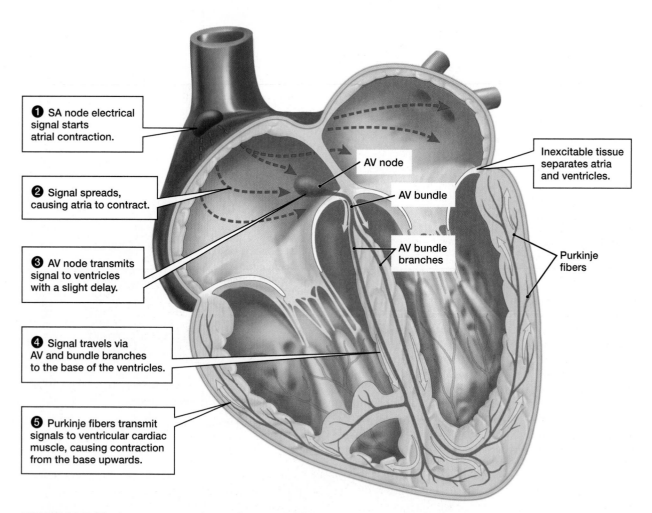

❶ SA node electrical signal starts atrial contraction.

❷ Signal spreads, causing atria to contract.

❸ AV node transmits signal to ventricles with a slight delay.

❹ Signal travels via AV and bundle branches to the base of the ventricles.

❺ Purkinje fibers transmit signals to ventricular cardiac muscle, causing contraction from the base upwards.

AV node

AV bundle

AV bundle branches

Inexcitable tissue separates atria and ventricles.

Purkinje fibers

FIGURE 32-7 The heart's pacemaker and its connections

tween the atrial and the ventricular contractions. How is this accomplished? Starting at the SA node, an electrical impulse creates a wave of contraction that sweeps through the muscles of the right and left atria, which contract in smooth synchrony. The signal then reaches a barrier of electrically inexcitable tissue between the atria and the ventricles. Here, the excitation is channeled through the **atrioventricular (AV) node**, a small mass of specialized muscle cells located in the floor of the right atrium (see Fig. 32-7). The impulse is conducted slowly at the AV node, briefly postponing the ventricular contraction. This delay gives the atria time to complete the transfer of blood into the ventricles before ventricular contraction begins. From the AV node, the signal to contract spreads along specialized tracts of rapidly-conducting muscle fibers—starting with the *AV bundle*, which sends *AV bundle branches* to the lower portion of both ventricles. Here, the tracts branch further to form **Purkinje fibers** that carry the electrical signal to contract upward within the ventricular walls (see Fig. 32-7). The impulse passes rapidly from these fibers into and then through the interconnected cardiac muscle fibers, causing the ventricles to contract simultaneously from the base upward, forcing blood up into the pulmonary artery and aorta.

A variety of disorders can interfere with the complex series of events that produce the normal cardiac cycle. When the pacemaker fails, or if other areas of the heart become more excitable and usurp the pacemaker's role, uncoordinated, irregular, weak contractions called *fibrillation* may occur. Fibrillation of the ventricles is soon fatal because blood cannot be pumped by the quivering muscle. A defibrillating machine applies a jolt of electricity to the heart, synchronizing the contraction of the ventricular muscle cells and sometimes allowing the pacemaker to resume its normal coordinating function.

The Nervous System and Hormones Influence Heart Rate

The heart rate is finely tuned to the body's activity level, whether you are running to class or basking in the sun. On its own, the SA node pacemaker would maintain a steady rhythm of about 100 beats per minute. However, nerve impulses and hormones significantly alter heart rate. In a resting individual, activity of the parasympathetic nervous system, which regulates body systems during periods of rest (see Chapter 38), slows heart rate to about 70 beats per minute (this resting rate is typically lower in athletes). When exercise or stress creates a demand for greater blood flow to the muscles, the sympathetic nervous system, which prepares the body for emergency action, accelerates the heart rate. Simultaneously, the hormone epinephrine (also known as adrenaline) increases the heart rate while mobilizing the body to respond to threatening or exciting events. For example, when astronauts landed on the moon, their heart rates were more than 170 beats per minute, even though they were sitting in their spacecraft.

32.3 WHAT IS BLOOD?

Blood, which has been called the "river of life," transports dissolved nutrients, gases, hormones, and wastes throughout the body. Blood has two major components: (1) the fluid called **plasma** and (2) the cellular components—*red blood cells, white blood cells*, and *platelets*—that are suspended in the plasma (**FIG. 32-8**). The cellular components of blood typically account for 40% to 45% of its volume; the other 55% to 60% is plasma. The average human has 5 to 6 liters of blood, constituting about 8% of total body weight. Blood components are summarized in Table 32-1.

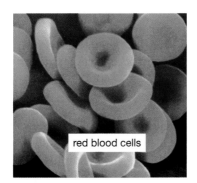

(a) Erythrocytes

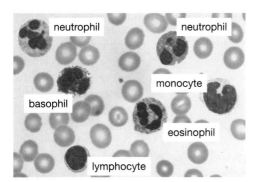

(b) White blood cells

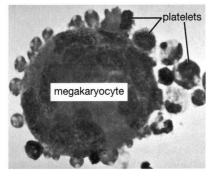

(c) Megakaryocyte forming platelets

FIGURE 32-8 Types of blood cells
(a) This false-color scanning electron micrograph clearly shows the biconcave disk shape of red blood cells. **(b)** This composite photo of stained white blood cells shows the five different types. **(c)** Platelets are membrane-enclosed pieces of cytoplasm, here shown budding off of a single megakaryocyte. QUESTION Why does dietary iron deficiency cause anemia (too few red blood cells to provide adequate oxygen)?

Table 32-1 Composition of Blood

Plasma Components (55% of blood)	Functions
Water	Dissolves other components
	Makes blood fluid
Proteins	
Albumin	Maintains blood osmolarity
Prothrombin	Gives rise to thrombin, which promotes clotting
Fibrinogen	Gives rise to fibrin, which promotes clotting
Globulins	Fight infection; transport substances; promote clotting
Salts (sodium, potassium, calcium, magnesium, bicarbonate, chloride)	Maintain osmolarity; maintain pH; allow neuronal activity; allow muscle contraction

Cellular Components (45% of blood)	Functions
Red blood cells (5,000,000 per mm^3)	Transport oxygen and some carbon dioxide
White blood cells (5,000–10,000 per mm^3)	Fight infection and disease
Platelets (250,000 per mm^3)	Important in blood clotting

Substances Transported in Blood

Metabolic wastes: carbon dioxide, urea, ammonia

Oxygen: to allow energy production using cellular respiration

Nutrients: glucose, amino acids, lipids, vitamins

Hormones: influence growth, development, metabolic activities

Plasma Is Primarily Water in Which Proteins, Salts, Nutrients, and Wastes Are Dissolved

Plasma, which is straw-colored, is about 90% water. Dissolved in the plasma are a variety of proteins and salts. The blood also serves as a transport system for hormones, nutrients, gases, and wastes (see Table 32-1). Plasma proteins are the most abundant of these dissolved substances. The three major plasma proteins are *albumins*, which help maintain the blood's osmotic pressure (which controls the flow of water across plasma membranes); *globulins*, which transport nutrients and play a role in the immune system; and *fibrinogen*, which is important in blood clotting and is discussed later in this chapter.

Red Blood Cells Carry Oxygen from the Lungs to the Tissues

The most abundant cells in the blood are those that carry oxygen; they are called red blood cells or **erythrocytes**. Each cubic millimeter of blood (a small droplet) contains about 5 million erythrocytes, which make up about 99% of all blood cells and typically constitute about 40% of the total blood volume in females and about 45% in males. A red blood cell resembles a ball of clay squeezed between a thumb and forefinger (see Fig. 32-8a). This shape, which results when

the cell loses its nucleus during development, provides a larger surface area than would a spherical cell of the same volume, and it increases the cell's ability to absorb and release oxygen through its plasma membrane.

The red color of erythrocytes is caused by the large, iron-containing protein **hemoglobin** (**FIG. 32-9**; described in Chapter 3), which makes up about one-third of the weight of each red blood cell. One hemoglobin molecule can bind and carry four molecules of oxygen (one on each heme group, shown as green discs in Fig. 32-9). Hemoglobin permits blood to hold far more oxygen than would be possible if the oxygen were just dissolved in the plasma. Hemoglobin takes on a bright cherry-red color when it binds oxygen, and it becomes a deeper red when oxygen is released. Since

FIGURE 32-9 Hemoglobin

deoxygenated blood is found in veins, which appear bluish when viewed through the skin, the color conventions in most diagrams depict arteries as red and veins as blue. Hemoglobin binds loosely to oxygen, picking it up in the capillaries of the lungs—where oxygen concentration is high—and releasing it in other tissues of the body where oxygen concentration is lower. After releasing its oxygen, some of the hemoglobin picks up carbon dioxide from the tissues for transport back to the lungs. The role of blood in gas exchange is discussed further in Chapter 33.

An "artificial blood" containing purified hemoglobin extracted from human red blood cells is being widely tested in trauma centers on victims of massive blood loss. The product can be stored far longer than fresh blood, does not transmit disease, and can be transfused immediately because it is compatible with any blood type. This can make a life-or-death difference to a victim of massive bleeding.

Red Blood Cells Have a Relatively Short Life Span

Red blood cells, or erythrocytes, are formed in the bone marrow, the soft interior portion of bones found in the chest, upper arms, upper legs, and hips. During their development, mammalian erythrocytes lose their nuclei. Without a nucleus, the cell cannot divide or synthesize new enzymes and other cellular components coded by genetic material. Thus, erythrocytes are short-lived, with an average life span of about four months. Every second, more than 2 million red blood cells die and are replaced by new ones from the bone marrow. Dead or damaged red blood cells are removed from circulation, primarily in the liver and spleen, and broken down to release their iron. Blood carries the salvaged iron back to the bone marrow, where it is used to make more hemoglobin and packaged into new red blood cells. Although this recycling process is efficient, small amounts of iron are excreted daily and must be replenished by the diet. Bleeding from injury or menstruation also depletes iron stores.

Negative Feedback Regulates Red Blood Cell Numbers

The number of red blood cells in the blood determines how much oxygen it can carry; this number is maintained by a negative feedback system that involves the hormone **erythropoietin**. Erythropoietin is produced by the kidneys and released into the blood in response to oxygen deficiency. This lack of oxygen may be caused by a loss of blood, insufficient production of hemoglobin, high altitude (where less oxygen is available), or lung disease that interferes with gas exchange in the lungs. Erythropoietin stimulates the rapid production of new red blood cells by the bone marrow. When adequate oxygen levels are restored, erythropoietin production declines, and the rate of red blood cell production returns to normal (**FIG. 32-10**).

White Blood Cells Help Defend the Body Against Disease

White blood cells, or **leukocytes**, come in five types: *neutrophils*, *eosinophils*, *basophils*, *lymphocytes*, and *monocytes*. Together, these make up less than 1% of all blood cells (see Fig. 32-8b). Like all blood cells, they are produced in bone marrow, and they all function in various ways to protect the body against disease, using the circulatory system to travel to the site of invasion. Some, such as monocytes, travel to wounds where bacteria have gained entry, then ooze out through narrow openings in capillary walls. After leaving the capillaries, monocytes change into amoeba-like cells called **macrophages** (literally, "big eaters") that engulf foreign particles, including bacteria and cancer cells (**FIG. 32-11**). These macrophages typically die in the process, and their dead bodies accumulate and contribute to a white substance, called *pus*, often seen at infection sites. The **lymphocytes**, described in Chapter 36, produce antibodies that help provide immunity against

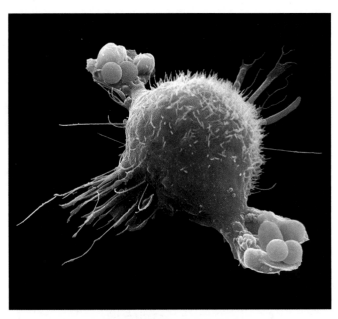

FIGURE 32-11 A white blood cell attacks bacteria These bacteria (small green spheres) are *Escherichia coli*. Some forms of these intestinal bacteria can cause disease if they enter the bloodstream.

disease. Cells that give rise to lymphocytes migrate from the bone marrow through the bloodstream to tissues of the lymphatic system, such as the thymus, spleen, and lymph nodes, described later in this chapter.

Platelets Are Cell Fragments That Aid in Blood Clotting

Platelets, which are crucial to blood clotting, are pieces of large cells called *megakaryocytes*. Megakaryocytes remain in the bone marrow, where they pinch off membrane-enclosed chunks of their cytoplasm to form platelets (see Fig. 32-8c). The platelets then enter the blood and play a central role in blood clotting. Like red blood cells, platelets lack a nucleus, but their life span is even shorter—about 10 to 12 days.

Blood clotting is a complex process that keeps animals from bleeding to death, not only from trauma but also from normal wear and tear on the body. Clotting (**FIG. 32-12a**) begins when blood comes in contact with injured tissue, for example, a break in the blood vessel wall. The ruptured surface exposes collagen protein that causes platelets to adhere and partially block the opening. Both the adhering platelets and the ruptured cells release a variety of substances, initiating complex sequences of reactions among circulating plasma proteins. One important outcome of these chemical reactions is the production of the enzyme **thrombin** from its inactive form, *prothrombin*. Thrombin then catalyzes the conversion of the plasma protein **fibrinogen** into insoluble strands of protein called **fibrin**. Fibrin proteins adhere to one another, forming a fibrous

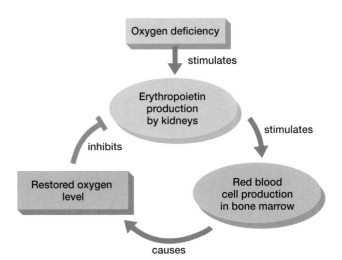

FIGURE 32-10 Red blood cell regulation by negative feedback
QUESTION Some endurance athletes cheat by "blood doping": injecting large doses of erythropoietin. Why does this provide a competitive advantage?

(a) Steps in clot formation

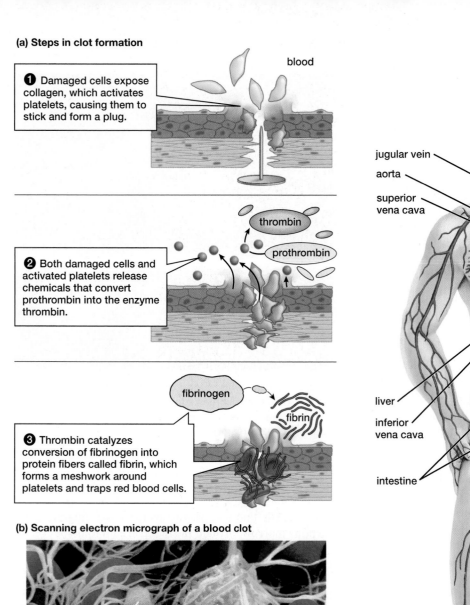

❶ Damaged cells expose collagen, which activates platelets, causing them to stick and form a plug.

blood

❷ Both damaged cells and activated platelets release chemicals that convert prothrombin into the enzyme thrombin.

thrombin

prothrombin

❸ Thrombin catalyzes conversion of fibrinogen into protein fibers called fibrin, which forms a meshwork around platelets and traps red blood cells.

fibrinogen

fibrin

(b) Scanning electron micrograph of a blood clot

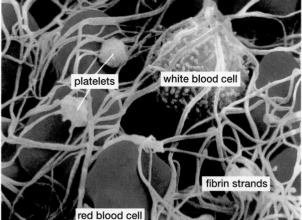

platelets

white blood cell

fibrin strands

red blood cell

FIGURE 32-12 Blood clotting
(a) Injured tissue and adhering platelets cause a series of biochemical reactions among blood proteins that lead to clot formation. A simplified sequence is shown here. **(b)** Threadlike fibrin proteins produce a tangled, sticky mass that traps red blood cells and eventually forms a clot.

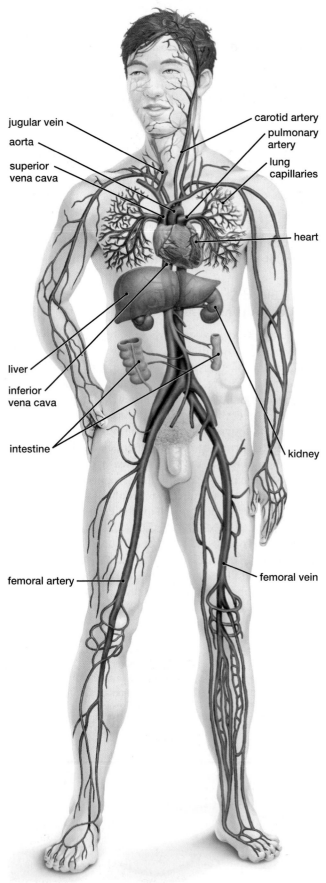

jugular vein

carotid artery

aorta

pulmonary artery

superior vena cava

lung capillaries

heart

liver

inferior vena cava

intestine

kidney

femoral artery

femoral vein

FIGURE 32-13 The human circulatory system
Most veins (right) carry deoxygenated blood to the heart, and most arteries (left) conduct oxygenated blood away from the heart. The pulmonary veins (which carry oxygenated blood) and arteries (which carry deoxygenated blood) are exceptions. All organs receive blood from arteries, send it back through veins, and are nourished by microscopic capillaries (lung capillaries are greatly enlarged).

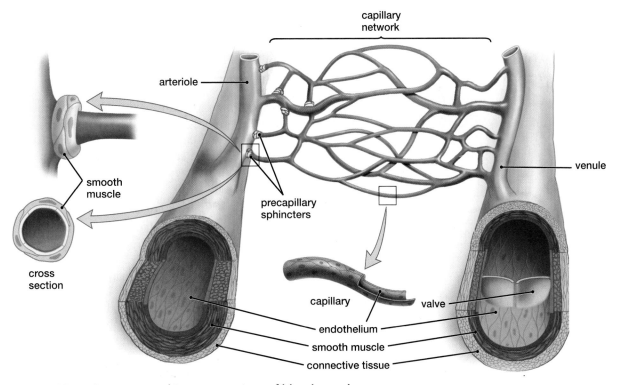

FIGURE 32-14 Structures and interconnections of blood vessels
Arteries and arterioles are more muscular than veins and venules. Oxygenated blood moves from arteries to arterioles to capillaries. Capillaries, which have walls only one cell thick, empty deoxygenated blood into venules, which empty into veins. Precapillary sphincters regulate the movement of blood from arterioles into capillaries.

network around the aggregated platelets. This protein web traps red blood cells and more platelets (**FIG. 32-12b**), increasing the density of the clot. Platelets adhering to the fibrous mass send out sticky projections that grip one another. Within about half an hour, the platelets contract, pulling the fibrin web tighter and forcing liquid out. This action creates a denser, stronger clot (on the skin it is called a *scab*) and also constricts the wound, pulling the damaged surfaces closer together in a way that promotes healing.

Despite blood's clotting ability, each year in the U.S. tens of thousands of people bleed to death from gunshots and other forms of trauma. Researchers have developed bandages impregnated with large quantities of thrombin and fibrinogen that stimulate rapid clotting. Other scientists have genetically engineered cows and pigs that secrete human fibrinogen in their milk. These developments hold promise for better control of bleeding and more rapid wound healing for future victims.

32.4 WHAT ARE THE TYPES AND FUNCTIONS OF BLOOD VESSELS?

The blood flows in well-defined channels called *blood vessels*. Some of the major blood vessels of the human circulatory system are diagrammed in **FIGURE 32-13**. As it leaves

the heart, blood travels from arteries to *arterioles* to capillaries to *venules* to veins, which return it to the heart. The structure of these vessels is shown in **FIGURE 32-14**.

Arteries and Arterioles Are Thick-Walled Vessels That Carry Blood Away from the Heart

Arteries carry blood away from the heart. These vessels have thick walls embedded with smooth muscle and elastic connective tissue (see Fig. 32-14). With each surge of blood from the ventricles, the arteries expand slightly, like thick-walled balloons. As their resilient walls recoil between heartbeats, the arteries actually help pump the blood and keep it flowing steadily through the smaller vessels. Arteries branch into smaller-diameter vessels called **arterioles**, which play a major role in determining how blood is distributed within the body, as described later.

Capillaries Are Microscopic Vessels That Allow the Blood and Body Cells to Exchange Nutrients and Wastes

The entire circulatory system is an elaborate device that allows each cell of the body to exchange nutrients and wastes by diffusion. Arterioles conduct blood to **capillaries**, which are so microscopically small that red blood cells

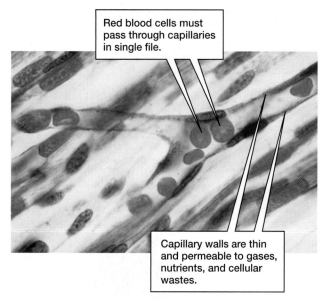

Red blood cells must pass through capillaries in single file.

Capillary walls are thin and permeable to gases, nutrients, and cellular wastes.

FIGURE 32-15 Red blood cells travel single file through a capillary
QUESTION Why does oxygen move out of capillaries in body tissues while carbon dioxide moves in (instead of the other way around)?

must pass through them single file (**FIG. 32-15**). Capillaries are so numerous that most body cells are no more than 100 micrometers (about as thick as four pages of this book) from a capillary. The human body contains about 50,000 miles (80,600 kilometers) of capillaries, enough to encircle the globe twice. Both blood pressure and the rate of blood flow drops very quickly as blood moves through this narrow, lengthy capillary network, allowing more time for diffusion to occur.

With walls only one cell endothelial cell thick (see Fig. 32-4), capillaries are well adapted to their role of exchanging materials between the blood and the fluid bathing body cells. The high pressure within the capillaries that branch from arterioles causes fluid to leak continuously from the blood plasma into the spaces surrounding the capillaries. The resulting substance, called **interstitial fluid**, consists primarily of water containing dissolved nutrients, hormones, gases, cellular wastes, some proteins, and white blood cells. This medium acts as an intermediary between body cells and capillary blood, providing the cells with nutrients and accepting their wastes and other secretions.

Substances take various routes across the capillary walls. Gases, water, and lipid-soluble hormones and fatty acids can diffuse directly through capillary cell membranes. Small, water-soluble nutrients such as salts, glucose, and amino acids travel into the interstitial fluid through narrow spaces between adjacent capillary cells. White blood cells can also ooze through these crevices. Large proteins may be ferried across the endothelial cell membranes in vesicles. As a result of the filtering action of the capillary walls, the composition of interstitial fluid differs from that of blood. While concentrations of ions and glucose are very similar, interstitial fluid lacks red blood cells and platelets, and has far less protein than blood plasma.

Pressure within the capillaries drops as blood travels toward the venules, and the high osmotic pressure of the blood that remains inside the capillaries (due to the presence of albumins and other large proteins) draws water back into the vessels by osmosis as blood approaches the venous end of the capillaries. As water moves into the capillaries and the blood within becomes more dilute, dissolved substances in the interstitial fluid tend to diffuse back into the capillaries as well. Thus, much of the interstitial fluid (about 85%) is restored to the bloodstream through the capillary walls on the venous side of the capillary network. As you will learn later in this chapter, the lymphatic system restores the remaining fluid (look ahead to Fig. 32-17).

Veins and Venules Carry Blood Back to the Heart

After picking up carbon dioxide and other cellular wastes from cells, capillary blood drains into larger vessels called **venules**, which empty into still larger veins (see Fig. 32-14). Veins provide a low-resistance pathway that conducts blood back toward the heart. The walls of veins are thinner, less muscular, and more expandable than those of arteries, although both contain a layer of smooth muscle. Because blood pressure in the veins is low, contractions of skeletal muscle during exercise and breathing help return blood to the heart by squeezing the veins and forcing blood through them.

When veins are compressed, why isn't blood forced away from the heart as well as toward it? Veins have one-way valves that allow blood to flow only toward the heart (**FIG. 32-16**). When you sit or stand for a long time, the lack of activity allows blood to accumulate in the veins of the lower legs, so you may find your feet swollen after a long airplane flight. Long periods of inactivity can contribute to the formation of varicose veins, in which the valves become stretched and weakened and the veins become permanently swollen.

If blood pressure should fall—for instance, after extensive bleeding—veins can help restore it. In such cases, the sympathetic nervous system (which prepares the body for emergency action) automatically stimulates contraction of the smooth muscles in the vein walls. This action decreases the internal volume of the veins and raises blood pressure, speeding up the return of blood to the heart.

Arterioles Control the Distribution of Blood Flow

Arterioles carry blood to capillaries, and their muscular walls are influenced by nerves, hormones, and chemicals produced by nearby tissues. Therefore, arterioles contract and relax in response to the needs of the tissues and organs they supply. In a suspense novel, you might read: "Her face paled as she gazed at the bloodstained floor." Skin becomes pale when the arterioles supplying the skin capillaries with blood constrict because the nervous system stimulates the smooth muscle in the arteriole walls to contract. This contraction raises blood pressure overall; but selective constriction redirects blood to the heart and muscles, where it may be needed for vigorous action, and away from the skin, where it is less essential.

On a hot summer day, however, you become flushed as arterioles in the skin expand and bring more blood to the skin capillaries. This enables the body to dissipate excess heat to the outside, thus maintaining a relatively constant internal

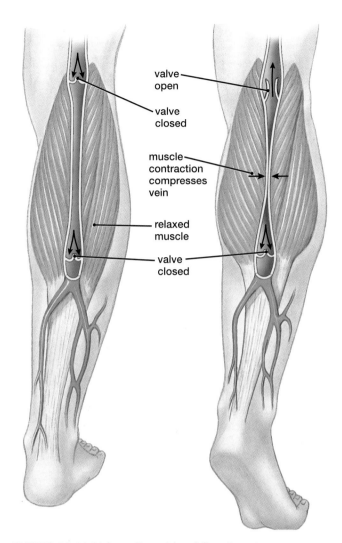

FIGURE 32-16 Valves direct blood flow in veins
Veins and venules have one-way valves that keep blood flowing in the proper direction. When the vein is compressed by nearby muscles, these valves allow blood to flow toward the heart, but clamp shut to prevent backflow.

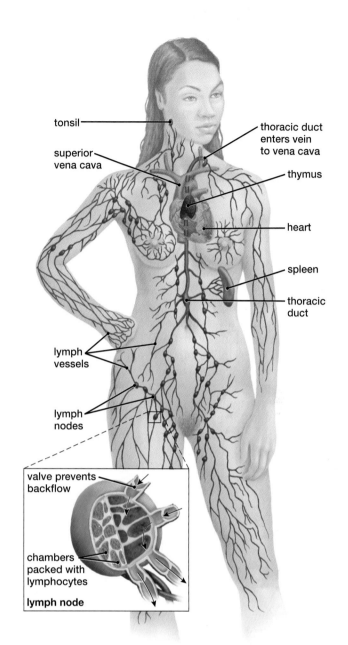

FIGURE 32-17 The human lymphatic system
Lymph vessels, lymph nodes, and two auxiliary lymph organs, the thymus and spleen, are illustrated. Lymph is returned to the circulatory system through the thoracic duct. (**Inset**) A cross section of a lymph node, filled with channels lined with white blood cells that remove foreign matter from the lymph.

temperature. In extremely cold weather, fingers and toes can become frostbitten because the arterioles that supply blood to the extremities constrict. Blood is shunted to vital organs, such as the heart and brain, which cannot function properly if their temperature drops. By minimizing blood flow to heat-radiating extremities, the body conserves heat.

The flow of blood in capillaries is regulated by tiny rings of smooth muscle called **precapillary sphincters,** which surround the junctions between arterioles and capillaries (see Fig. 32-14). These open and close in response to local chemical changes that signal the needs of nearby tissues. For example, the accumulation of carbon dioxide, lactic acid, or other cellular wastes signals the need for increased blood flow to the tissues. These signals cause the precapillary sphincters and the muscles in the walls of nearby arterioles to relax, allowing more blood to flow through the capillaries.

What happens when blood vessels rupture, are narrowed by deposits of cholesterol, or are blocked by clots, damming the "river of life"? We explore these questions in "Health Watch: Repairing Broken Hearts."

32.5 HOW DOES THE LYMPHATIC SYSTEM WORK WITH THE CIRCULATORY SYSTEM?

The **lymphatic system** consists of a network of lymph capillaries and larger vessels that empty into the circulatory system, numerous small *lymph nodes*, patches of lymphocyte-rich connective tissue (including the *tonsils*), and two additional organs: the *thymus* and the *spleen* (**FIG. 32-17**). Although not strictly part of the circulatory system, the lymphatic system is closely associated with it. The lymphatic system serves to:

- Return excess interstitial fluid to the bloodstream
- Transport fats from the small intestine to the bloodstream
- Defend the body by exposing bacteria and viruses to white blood cells

Lymphatic Vessels Resemble the Veins and Capillaries of the Circulatory System

Like blood capillaries, *lymph capillaries* form a complex network of microscopically narrow, thin-walled vessels into which substances move readily. Lymph capillary walls are composed of cells with openings between them that act as one-way valves. These openings allow relatively large particles, along with fluid, to be carried into the lymph capillary. Unlike blood capillaries, which form a continuous connected network, lymph capillaries "dead-end" in the body's tissues (**FIG. 32-18**). Materials collected by the lymph capillaries flow into larger lymph vessels. Lymph is pumped by rhythmic contractions of smooth muscles in the walls of these vessels. Further impetus for

lymph flow comes from the contraction of nearby muscles, such as those used in breathing and walking. As in blood veins, the direction of flow is regulated by one-way valves (**FIG. 32-19**).

The Lymphatic System Returns Fluids to the Blood

As described earlier, dissolved substances are exchanged between the capillaries and body cells by means of interstitial fluid that is pressure-filtered from blood plasma through capillary walls and bathes nearly all the body's cells. In an average person, about 3 or 4 more liters of fluid leave the blood capillaries than are reabsorbed by them each day. One function of the lymphatic system is to return this excess fluid and its dissolved molecules to the blood. As interstitial fluid accumulates, its pressure forces the fluid through openings between the cells in the lymph capillaries (see Fig. 32-18). The lymphatic system transports this fluid, now called **lymph**, back to the circulatory system. The importance of the lymphatic system in returning fluid to the bloodstream is illustrated by the condition known as *elephantiasis* (**FIG. 32-20**). This disfiguring disorder is caused by a parasitic roundworm that colonizes lymphatic vessels, scarring them and preventing them from draining off excess fluid.

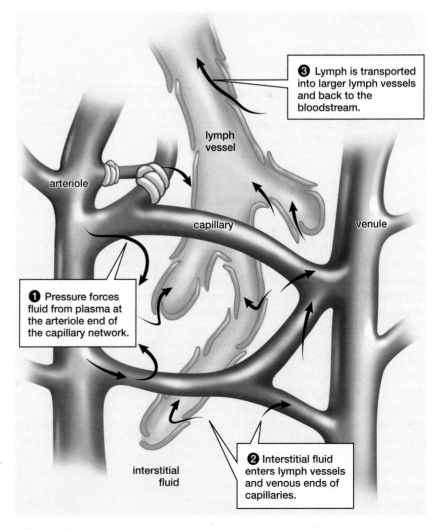

❸ Lymph is transported into larger lymph vessels and back to the bloodstream.

lymph vessel

arteriole

capillary

venule

❶ Pressure forces fluid from plasma at the arteriole end of the capillary network.

interstitial fluid

❷ Interstitial fluid enters lymph vessels and venous ends of capillaries.

FIGURE 32-18 Lymph capillary structure
Lymph capillaries end blindly in the body tissues, where pressure from the accumulation of interstitial fluid forces the fluid into the lymph capillaries as well as into the venous side of the capillary network.

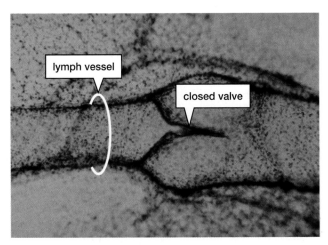

FIGURE 32-19 A valve in a lymph vessel
Like blood-carrying veins, lymph vessels have internal one-way valves that direct the flow of lymph toward the large veins into which they empty.

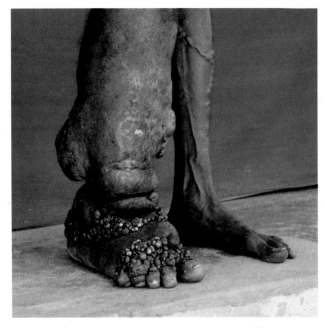

FIGURE 32-20 Elephantiasis results from blocked lymphatic vessels
When a parasitic worm scars lymph vessels, preventing fluid from returning to the bloodstream, the affected area can become massively swollen.

The Lymphatic System Transports Fats from the Small Intestine to the Blood

After a fatty meal, fat-transporting particles may make up 1% of the lymphatic fluid. How does this happen? As you will learn in Chapter 34, the small intestine is richly supplied with lymph capillaries. After absorbing digested fats, intestinal cells release fat-transporting particles into the interstitial fluid. These particles are too large to diffuse into blood capillaries but can easily move through the openings between lymph capillary cells. Once in the lymph, they are dumped into veins that merge into the superior vena cava, a large vein that carries blood into the heart.

The Lymphatic System Helps Defend the Body Against Disease

In addition to its other roles, the lymphatic system helps to defend the body against foreign invaders, such as bacteria and viruses. In the linings of the respiratory, digestive, and urinary tracts are patches of connective tissue containing large numbers of lymphocytes. The largest of these patches are the **tonsils**, located in the cavity behind the mouth. The lymph vessels are interrupted periodically by kidney-bean-shaped structures about 1 inch (2.5 centimeters) long called **lymph nodes** (see Fig. 32-17). Lymph is filtered through narrow spaces within the lymph nodes, which contain masses of macrophages and lymphocytes. Some infections cause lymph nodes to swell as they accumulate white blood cells, bacteria, debris from dead cells, and fluid.

The thymus and the spleen are organs of the lymphatic system (see Fig. 32-17). The **thymus** is located beneath the breastbone slightly above the heart. It is particularly active in infants and young children but decreases in size and importance in early adulthood. Some types of immature lymphocytes produced in bone marrow travel in the bloodstream to the thymus, where they mature. The **spleen** is located on the left side of the abdominal cavity, between the stomach and diaphragm. Just as the lymph nodes filter lymph, the spleen filters blood, exposing it to macrophages and lymphocytes that destroy foreign particles and aged red blood cells.

CASE STUDY REVISITED SUDDEN DEATH

Darryl Kile may have been entirely unaware of the insidious progress of atherosclerosis that threatened and suddenly claimed his life. Although extremely narrowed coronary arteries can cause chest pain (*angina*) and warn of impending trouble, less than 20% of heart attacks are caused by the plaque buildup itself. Most occur because the fibrous cap that contains the plaque within the artery wall breaks open and stimulates the formation of a blood clot within the artery (see inset in the chapter opener photo). Such clots can break off suddenly and be carried to a narrower part of the artery, where they completely block blood flow. In the U.S., about 3000 young people between the ages of 15 and 34 die of heart attacks each year. In some individuals, genetic factors contribute to exceptionally high levels of LDL cholesterol, hypertension, and atherosclerosis. But with lifestyle changes and the help of drugs that lower blood pressure, reduce LDL cholesterol, and decrease inflammation, even people with special risk factors may lead long, active lives and avoid the tragedy that befell Darryl Kile—and his family, friends, and teammates.

Consider This Do risk factors such as high cholesterol (total and LDL), hypertension, or angina run in your family? Has anyone in your immediate family died prematurely of heart disease? It is never too early to find out! Even in the absence of a family history, a complete physical exam that includes a blood lipid evaluation and an electrocardiogram can help you evaluate your own risks and take action to help safeguard your cardiovascular health.

CHAPTER REVIEW

SUMMARY OF KEY CONCEPTS

32.1   What Are the Major Features and Functions of Circulatory Systems?

Circulatory systems transport blood rich in dissolved nutrients and oxygen close to each cell, where nutrients can be released and wastes absorbed. All circulatory systems have three major parts: blood, a fluid; vessels, a system of channels to conduct the blood; and one or more hearts that pump the blood. Invertebrates have open or closed circulatory systems. In open systems, blood is pumped by the heart into a hemocoel, where the blood directly bathes internal organs. Nearly all vertebrates have closed systems, in which the blood is confined to the heart and blood vessels.

Web Tutorial 32.1 Circulatory Systems

32.2   How Does the Vertebrate Heart Work?

Vertebrate circulatory systems transport gases, hormones, and wastes; distribute nutrients; help regulate body temperature; and defend the body against disease.

The vertebrate heart evolved from two chambers in fishes, to three in amphibians and most reptiles, to four in birds and mammals. In the four-chambered heart, blood is pumped separately to the lungs and through the body, maintaining complete separation of oxygenated and deoxygenated blood. Deoxygenated blood is collected from the body in the right atrium and passed to the right ventricle, which pumps it to the lungs. Oxygenated blood from the lungs enters the left atrium, passes to the left ventricle, and is pumped to the rest of the body.

The cardiac cycle consists of two stages: (1) atrial contraction, followed by (2) ventricular contraction. The direction of blood flow is maintained by valves within the heart. The contractions of the heart are initiated and coordinated by the sinoatrial node, the heart's pacemaker. Heart rate can be modified by the nervous system and by hormones such as epinephrine.

Web Tutorial 32.2 Two-Chambered Hearts

Web Tutorial 32.3 Three-Chambered Hearts

Web Tutorial 32.4 The Human Cardiovascular System

Web Tutorial 32.5 Blood Pressure

32.3   What Is Blood?

Blood is composed of both fluid and cellular materials. The fluid plasma consists of water that contains proteins, hormones, nutrients, gases, and wastes. Red blood cells, or erythrocytes, are packed with iron-containing protein called hemoglobin, which carries oxygen. Their numbers are regulated by the hormone erythropoietin. There are five types of white blood cells, or leukocytes, that fight infection. Platelets, which are fragments of megakaryocytes, are important for blood clotting.

32.4   What Are the Types and Functions of Blood Vessels?

Blood leaving the heart travels (in sequence) through arteries, arterioles, capillaries, venules, veins, and then back to the heart. Each vessel is specialized for its role. Elastic, muscular arteries help pump the blood. The thin-walled capillaries allow the exchange of materials between the body cells and the blood. Veins provide a path of low resistance back to the heart, with one-way valves that maintain the direction of blood flow. The distribution of blood is regulated by the constriction and dilation of arterioles under the influence of the sympathetic nervous system and local factors such as the amount of carbon dioxide in the tissues. Local factors also regulate precapillary sphincters, which control blood flow to the capillaries.

32.5   How Does the Lymphatic System Work with the Circulatory System?

The human lymphatic system consists of lymphatic vessels, tonsils, lymph nodes, and the thymus and spleen. The lymphatic system removes excess interstitial fluid that leaks through blood capillary walls. It transports fats to the bloodstream from the small intestine and fights infection by filtering the lymph through lymph nodes, where white blood cells ingest foreign invaders, such as viruses and bacteria. The thymus, which is most active in young children, contains lymphocytes that function in immunity. The spleen filters blood past macrophages and lymphocytes, which remove bacteria and damaged blood cells.

KEY TERMS

angina *page 654*
arteriole *page 661*
artery *page 652*
atherosclerosis *page 654*
atrioventricular (AV) node *page 657*
atrioventricular valve *page 656*
atrium *page 651*
blood *page 650*
blood clotting *page 659*
blood vessel *page 650*
capillary *page 661*

cardiac cycle *page 653*
cardiac muscle *page 653*
closed circulatory system *page 650*
erythrocyte *page 658*
erythropoietin *page 659*
fibrin *page 659*
fibrinogen *page 659*
heart *page 650*
heart attack *page 654*
hemocoel *page 650*
hemoglobin *page 658*
hypertension *page 656*

interstitial fluid *page 662*
leukocyte *page 659*
lymph *page 664*
lymphatic system *page 663*
lymph node *page 665*
lymphocyte *page 659*
macrophage *page 659*
open circulatory system *page 650*
pacemaker *page 656*
plaque *page 654*
plasma *page 657*
platelet *page 659*

precapillary sphincter *page 663*
Purkinje fibers *page 657*
semilunar valve *page 656*
sinoatrial (SA) node *page 656*
spleen *page 665*
stroke *page 654*
thrombin *page 659*
thymus *page 665*
tonsil *page 665*
vein *page 652*
ventricle *page 651*
venule *page 662*

THINKING THROUGH THE CONCEPTS

1. Trace the flow of blood through the circulatory system, starting and ending with the right atrium.

2. List three types of blood cells and describe their principal functions.

3. What are five functions of the vertebrate circulatory system?

4. In what way do veins and lymph vessels resemble one another? Describe how fluid is transported in each of these vessels.

5. Describe three important functions of the lymphatic system.

6. Distinguish among plasma, interstitial fluid, and lymph.

7. Describe veins, capillaries, and arteries, noting their similarities and differences.

8. Trace the evolution of the vertebrate heart from two chambers to four chambers.

9. Explain in detail what causes the vertebrate heart to beat.

10. Describe the cardiac cycle, and relate the contractions of the atria and ventricles to the two readings taken during the measurement of blood pressure.

11. Describe how the number of red blood cells is regulated by a negative feedback system.

12. Describe the formation of an atherosclerotic plaque. What are the risks associated with atherosclerosis?

APPLYING THE CONCEPTS

1. Discuss the steps you can take now and in the future to reduce your risks of developing heart disease.

2. Considering the prevalence of cardiovascular disease and the high, increasing costs of treating it, certain treatments may not be available to all who might need them. What factors would you take into account if you had to ration cardiovascular procedures, such as heart transplants or bypass surgery?

3. Joe, an overweight 45-year-old executive of a major corporation who works 60-hour weeks, has been feeling chest pain when he tries to play basketball with his son on weekends. What treatments or lifestyle changes might Joe's physician recommend? If Joe's angina does not respond and becomes more severe, what treatment options might Joe's physician use? Explain how each option works.

FOR MORE INFORMATION

Aldhous, P. "Print Me a Heart and a Set of Arteries." *New Scientist*, April 15, 2006. Using a device that resembles an inkjet printer, bioengineers are "printing" cells in thin layers onto a supporting gel, where they may begin behaving like a real organ.

Ditlea, S. "The Trials of an Artificial Heart." *Scientific American*, July 2002. Describes the workings and clinical trials of artificial hearts and other bridging devices, with a discussion of ethical considerations.

Fox, C. "Can Stem Cells Save Dying Hearts?" *Discover*, September 2005. Human trials of injecting stem cells from the patient's own bone marrow into diseased hearts have begun.

Gibbons, R., and associates. "Waiting for Organ Transplantation." *Science*, January 14, 2000. Statistics about the need, availability, and waiting time for organ transplantation suggest that new sources of organs are needed.

Jain, R. K., and Carmeliet, P. F. "Vessels of Death or Life." *Scientific American*, December 2001. By learning to manipulate angiogenesis (the formation of blood vessels), researchers may find ways to thwart tumors or bring more blood to the heart.

Libby, A. "Atherosclerosis: The New View." *Scientific American*, May 2002. Beautifully illustrated description of plaque formation with an emphasis on the role of inflammation in heart disease.

Martindale, D. "Reactive Reasoning." *Scientific American*, April 2005. A compound in the blood called C-reactive protein may be central to the inflammation that triggers atherosclerotic plaque formation.

Wang, L. "Blood Relatives." *Science News*, March 31, 2001. Describes a variety of approaches and problems in making artificial blood, and the most likely prospects for the near future.

33 Respiration

These students say they will quit smoking later.
Only about one in three will succeed.

AT A GLANCE

 CASE STUDY LIVES UP IN SMOKE

COLLEGE STUDENTS CLUSTER just outside the building doors, lighting up between classes. Coughs punctuate their conversation. Like most adult smokers, many of these students picked up the habit in high school. "I started as a social smoker in high school," says a junior at the University of Illinois quoted in the university's student newspaper. "The next thing you know, I'm addicted. But I like it. I like smoking and I don't care [about the harmful effects]." This university, where about 30% of the students have used tobacco during the past month, is seeing an increase in student smoking. According to John Kirkwood, the American Lung Association president, every day in the United States, more than 4000 kids under age 18 light up their first cigarette. For many, this is the beginning of a lifelong struggle with addiction. According to the Centers for Disease Control, about 22% of U.S. high school students use tobacco, although fewer than 10% are frequent smokers.

Statistically, young smokers will suffer more frequent and severe respiratory infections, and smoking may retard the growth of their lungs. Most of these students are aware of the dangers but say they will stop "when the time comes." Are these new smokers likely to quit? What are their chances of dying from smoking? How does smoking interfere with the respiratory system from the bronchi to the bloodstream?

33.1 WHY EXCHANGE GASES?

Late again! Sprinting up two flights of stairs to your classroom, you feel your calves "burning." Remembering what you learned in Chapter 8, you think, "Ah ha! That's lactic acid building up—my muscle cells are fermenting glucose because they can't get enough oxygen for cellular respiration." As you slip into your seat, quietly panting and feeling your heart pounding, the discomfort eases. Less exertion coupled with rapid breathing ensures that adequate oxygen (O_2) is now available; the lactic acid is being reconverted to pyruvate and then broken down into carbon dioxide (CO_2) and water, while providing additional energy.

You are experiencing firsthand the relationship between cellular respiration and the act of breathing, also called *respiration*. Each cell in your body (and in all organisms) must continuously expend energy to maintain itself. When you call on your muscles to carry you upstairs quickly, the demands are extreme. As cellular respiration converts the energy in nutrients (such as sugar) into ATP that can be used by cells, the process requires a steady supply of O_2 and generates CO_2 as a waste product. The rapid beating of your heart as you relax after your sprint upstairs reminds you that the circulatory system works in close harmony with the respiratory system. It extracts O_2 from air in your lungs, carries it within diffusing distance of each cell, and then picks up CO_2 for release in the lungs.

How does breathing help support cellular respiration? What does the inside of a lung look like, and how is it adapted for gas exchange? Why aren't our lungs outside our bodies, where they would be directly exposed to the air? How do aquatic animals respire? In this chapter, we will explore the specialized structures of respiratory systems and how they work.

33.2 WHAT ARE SOME EVOLUTIONARY ADAPTATIONS FOR GAS EXCHANGE?

Gas exchange in all organisms ultimately relies on diffusion. Cellular respiration depletes O_2 and increases CO_2 levels, creating a concentration gradient that favors the diffusion of CO_2 out of cells and the diffusion of O_2 into them. Although animal respiratory systems are amazingly diverse, they all meet three requirements that facilitate diffusion:

- Respiratory surfaces must remain moist, because gases must be dissolved in water when they diffuse into or out of cells.
- Respiratory surfaces must be very thin to facilitate diffusion of gases through them.
- The respiratory system must have a large surface area in contact with the environment to allow adequate gas exchange.

(a) Flatworm

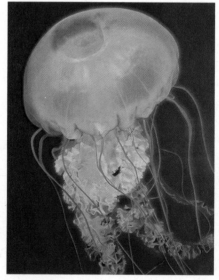

(b) Sea jelly

(c) Sponge

FIGURE 33-1 Some animals lack specialized respiratory structures
Most animals that lack a respiratory system have low metabolic demands and a large, moist body surface. **(a)** The flattened body of this marine flatworm exchanges gases with the water. **(b)** Cells in the bell-shaped body of a sea jelly have a low metabolic rate, and seawater flowing in and out of the bell during swimming allows adequate gas exchange. **(c)** Flagellated cells draw currents of water through numerous openings in the body of the sponge. These currents carry microscopic food particles and allow the cells to exchange gases with the water.

In the following sections we examine some diverse respiratory systems in animals, each shaped by the environment in which it evolved.

Some Animals in Moist Environments Lack Specialized Respiratory Structures

For some animals that live in moist environments, the outside of the body, covered by a thin, gas-permeable skin, provides an adequate surface area for the diffusion of gases. If the body is extremely small and elongated, as in microscopic roundworms, gases need to diffuse only a short distance to reach all the cells. Alternatively, an animal's body may be thin and flattened, as in flatworms, where most cells are close to the moist skin through which gases diffuse (**FIG. 33-1a**).

If energy demands are sufficiently low, the relatively slow rate of gas exchange by diffusion may suffice even for a larger, thicker-bodied organism. For example, sea jellies can be quite large, but cells that are far from the surface are relatively inert and require little O_2 (**FIG. 33-1b**).

Another adaptation for gas exchange involves bringing the watery environment close to each cell. Sponges, for example, circulate seawater through channels within their bodies (**FIG. 33-1c**).

Some animals combine a large skin surface—through which diffusion occurs—with a well-developed circulatory system. For example, in the earthworm, gases diffuse through the moist skin and are distributed throughout the body by an efficient circulatory system (see Fig. 32-1b). Blood in skin capillaries rapidly carries off O_2 that has diffused through the skin, maintaining a concentration gradient that favors the inward diffusion of O_2. The worm's elongated shape ensures a large skin surface relative to its internal volume. This system is quite effective, since a worm's sluggish metabolism demands relatively little O_2. To remain effective as a gas-exchange organ, the skin must stay moist; a dry earthworm will suffocate.

Respiratory Systems Facilitate Gas Exchange by Diffusion

Most animals have evolved specialized respiratory systems that interface closely with their circulatory systems to exchange gases between the cells and the environment. The transfer of gases from the environment to the blood, and then to the cells and back again, usually occurs in stages that alternate bulk flow with diffusion. During **bulk flow**, fluids or gases move in bulk through relatively large spaces, from areas of higher pressure to areas of lower pressure. Bulk flow contrasts with diffusion, in which molecules move individually from areas of higher concentration to areas of lower concentration (see Chapter 5). For animals with well-developed respiratory systems—from insects to people—gas exchange occurs in the following stages, as illustrated for mammals in **FIGURE 33-2**.

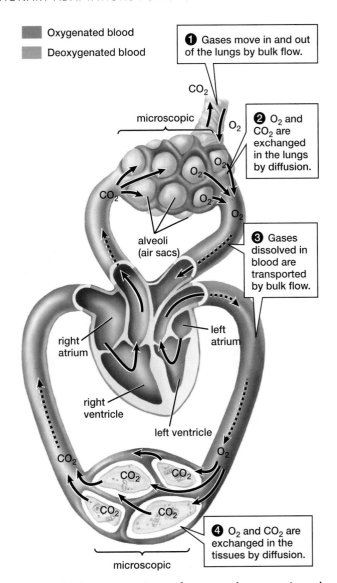

Oxygenated blood
Deoxygenated blood

1 Gases move in and out of the lungs by bulk flow.

CO_2

microscopic

O_2

2 O_2 and CO_2 are exchanged in the lungs by diffusion.

O_2 CO_2 O_2 O_2 O_2 O_2

CO_2

alveoli (air sacs)

3 Gases dissolved in blood are transported by bulk flow.

right atrium

left atrium

right ventricle

left ventricle

O_2

CO_2

CO_2 CO_2

CO_2 CO_2

4 O_2 and CO_2 are exchanged in the tissues by diffusion.

microscopic

FIGURE 33-2 An overview of gas exchange using the mammal as a model

1. Air or water, relatively high in O_2 and low in CO_2, is moved past a respiratory surface by bulk flow; this is usually facilitated by muscular movements.

2. O_2 and CO_2 are exchanged through the respiratory surface by diffusion; O_2 diffuses into the capillaries of the circulatory system and CO_2 diffuses out, both along their concentration gradients.

3. Gases are transported between the respiratory system and the tissues by the bulk flow of blood as it is pumped throughout the body by the heart.

4. Gases are exchanged between the tissues and the circulatory system by diffusion. At the tissues, O_2 diffuses out of the capillaries and into nearby tissues and CO_2 diffuses into capillaries from tissues along their concentration gradients.

FIGURE 33-3 External gills on a mollusk

Gills Facilitate Gas Exchange in Aquatic Environments

Gills are the respiratory structures of many aquatic animals. The simplest type of gill, found in certain amphibians (see Fig. 33-5a) and nudibranch (literally, "naked gill") mollusks (**FIG. 33-3**), consists of many thin projections of the body surface into the surrounding water. In general, gills are elaborately branched or folded to increase their surface area. In some animals, gill size is determined by the availability of O_2 in the surrounding water. For example, salamanders living in stagnant water (which has little opportunity to mix with air) have larger gills than do those living in well-aerated water. Gills have a dense profusion of capillaries just beneath their delicate outer membranes. These capillaries bring blood close to the surface, where gas exchange occurs.

The fish's body protects its delicate gill membranes beneath a bony flap, the *operculum*. Fish create a continuous current over their gills by pumping water into their mouths and ejecting it through the operculum just behind the gills (see Fig. E33-2 in "A Closer Look"). Fish can increase the flow of water by swimming with their mouths open; some fast swimmers, such as mackerel, tuna, and a few (but not all) sharks, rely heavily on swimming to ventilate their gills. Fish face a challenge in extracting O_2 from water. While O_2 makes up about 21% of air, it comprises far less than 1% of water. Because water is about 800 times as dense as air, pumping water over gills uses up far more energy than does breathing air. Fish have evolved a very efficient method called *countercurrent exchange* for exchanging gases with water. Within the gill, water and blood flow in opposite directions, maintaining relatively constant concentration gradients, as described in "A Closer Look: Gills and Gases—Countercurrent Exchange."

Terrestrial Animals Have Internal Respiratory Structures

Gills are useless in air because they collapse and dry out, but respiratory surfaces need to remain moist. Therefore, as animals made the transition from water to dry land over evolutionary time, it was necessary for them to evolve respiratory structures whose thin surface membranes were protected, supported, and covered with a film of water. Natural selection has produced a variety of these structures, including tracheae in insects and lungs in vertebrates.

Insects Respire Using Tracheae

Insects use a system of elaborately branching internal tubes called **tracheae**, which convey air throughout the body (**FIG. 33-4a**). Reinforced with chitin (which also supports the insect's external skeleton), tracheae penetrate the body tissues (**FIG. 33-4b**) and branch into microscopic channels called *tracheoles* that allow gas exchange through their fluid-filled endings (**FIG. 33-4c**). Each cell is close to a tracheole, minimizing diffusion distances. Air enters and leaves the tracheae through a series of openings called **spiracles**, located along each side of the body. Some large insects use muscular pumping movements of the abdomen to enhance air movement through the tracheae.

Most Terrestrial Vertebrates Respire Using Lungs

Lungs are chambers containing moist respiratory surfaces that are protected within the body, where water loss is minimized and the body wall provides support. The first vertebrate lung probably appeared in a freshwater fish and consisted of an outpocketing of the digestive tract. Gas exchange in this simple lung helped the fish survive in stagnant water, where O_2 is scarce. Amphibians, which straddle the boundary between aquatic and terrestrial life, may use gills in the larval stage and lungs in their more terrestrial adult form. For example, the purely aquatic tadpole exchanges its gills for lungs as it develops into a more terrestrial frog (**FIG. 33-5a, b**). Frogs and salamanders also use their moist skin as a supplemental respiratory surface.

The scales of reptiles (**FIG. 33-5c**) reduce the loss of water through the skin and allow reptiles to survive in dry environments. But scales also reduce the diffusion of gases through the skin, so the lungs of reptiles are better developed than those of amphibians.

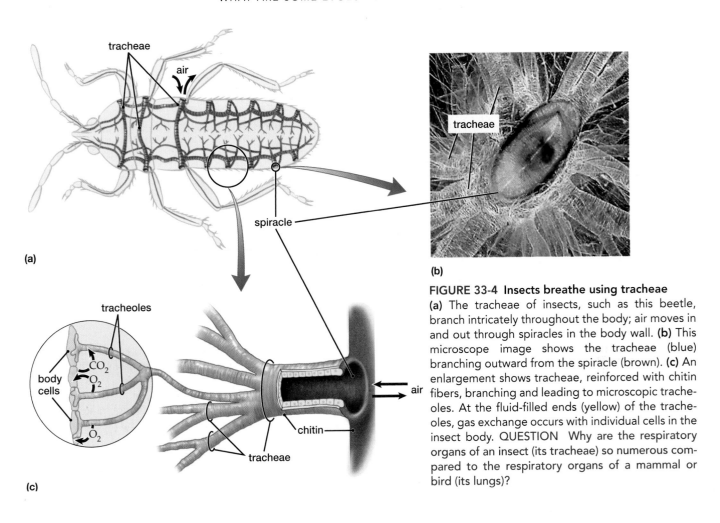

FIGURE 33-4 Insects breathe using tracheae
(a) The tracheae of insects, such as this beetle, branch intricately throughout the body; air moves in and out through spiracles in the body wall. (b) This microscope image shows the tracheae (blue) branching outward from the spiracle (brown). (c) An enlargement shows tracheae, reinforced with chitin fibers, branching and leading to microscopic tracheoles. At the fluid-filled ends (yellow) of the tracheoles, gas exchange occurs with individual cells in the insect body. QUESTION Why are the respiratory organs of an insect (its tracheae) so numerous compared to the respiratory organs of a mammal or bird (its lungs)?

FIGURE 33-5 Amphibians and reptiles have different respiratory adaptations
(a) The bullfrog, an amphibian, begins life as a fully aquatic tadpole with feathery external gills that will later become enclosed in a protective chamber. (b) During metamorphosis into an air-breathing adult, the frog's gills are lost and replaced by simple saclike lungs. In both tadpole and adult, gas exchange also occurs by diffusion through the skin, which must be kept moist to function as a respiratory surface. (c) Reptiles such as this snake, which are adapted to life on land, are covered with scales that restrict gas exchange through the skin. To compensate for this loss, reptilian lungs are more efficient than are those of amphibians. QUESTION How do the respiratory adaptations of amphibians influence the range of habitats in which amphibians are found?

During **countercurrent exchange**, fluids that differ in some property—such as temperature or concentration of a dissolved substance—flow past one another in opposite directions. During this *countercurrent flow*, they transfer some heat or solute from the fluid with the larger amount to the fluid with the smaller amount. Countercurrent exchange is used to reduce heat loss from vertebrate limbs, in the kidney to reabsorb water, and in gills to maximize gas exchange. When fluids flow in opposite directions, this maintains a concentration gradient of the dissolved substance from one fluid to the other over a relatively large distance, promoting maximum diffusion. **FIGURE E33-1a** shows *concurrent exchange*, in which two fluids flow in the same direction. Although the starting gradient is larger, the two fluids eventually acquire equal concentrations of the solute (red color), eliminating the diffusion gradient. In contrast, countercurrent exchange (**FIG. E33-1b**) maintains a gradient and results in far greater transfer of the solute.

Fish gills use countercurrent exchange as an efficient mechanism of transferring O_2 from the water flowing over them into their gill capillaries (**FIG. E33-2a**). Fish gills consist of a series of filaments (**FIG. E33-2b**). Each filament is served by an incoming blood vessel carrying deoxygenated blood and an outgoing vessel containing oxygenated blood, and each filament is covered with thin folds of tissue called *lamellae* (singular, lamella). Blood flows in capillaries across each lamella from the incoming to the outgoing vessel, while water flows through the filaments and across the lamellae in the opposite direction (**FIG. E33-2c**). During its transit across the lamella, capillary blood picks up O_2 (and releases CO_2) into the countercurrent of water. As illustrated

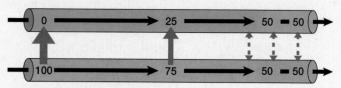

(a) Concurrent exchange

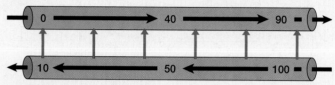

(b) Countercurrent exchange

FIGURE E33-1 Concurrent vs. countercurrent exchange

in Figures E33-1 and E33-2c, water and blood maintain a roughly equal concentration gradient of gases as they pass one another. Thus, although the most highly oxygenated water encounters the most highly oxygenated blood, this arrangement maintains a steady concentration gradient that favors diffusion of O_2 from the water into the blood (and carbon dioxide from the blood into the water) all the way across each lamella. Countercurrent exchange is so efficient that some fish can extract 85% of the O_2 from water flowing over their gills.

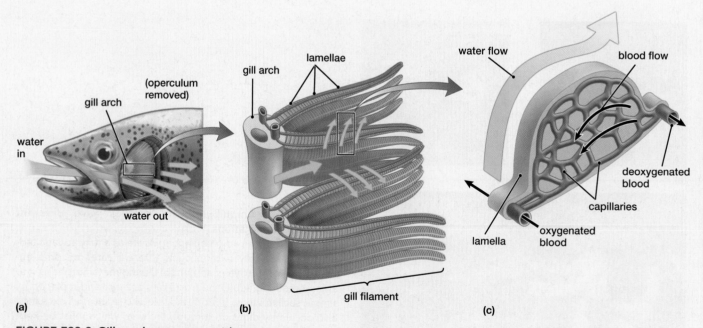

FIGURE E33-2 Gills exchange gases with water
(a) Fish pump water in through their mouths and out over their gills. **(b)** Water flows past a dense array of paired filaments attached to bony gill arches. **(c)** Lamellae protrude from each gill filament. Water flows over the lamellae in the opposite direction from the blood flowing through them in capillary beds.

(a) Bird respiratory system

(b) Parabronchi in lung tissue

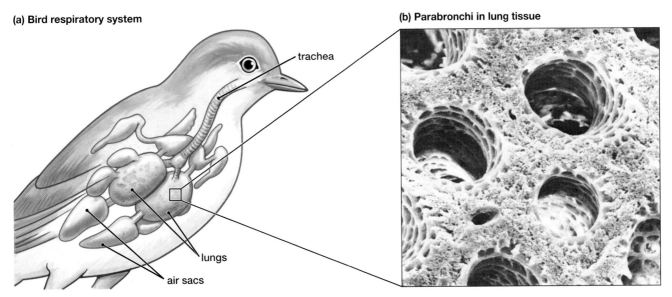

trachea

lungs

air sacs

FIGURE 33-6 The bird respiratory system is extremely efficient
(a) In addition to their lungs, birds have air sacs that allow more efficient gas exchange. **(b)** In birds, tubular gas-exchange organs called *parabronchi* allow air to flow entirely through the lungs as it passes to and from the air sacs. Air flows through open-ended channels; the porous region between the channels is filled with capillaries and air spaces where gas exchange occurs.

All birds and mammals breathe using lungs. The bird lung has adaptations that allow extremely efficient gas exchange (**FIG. 33-6**) to support the enormous energy demands of flight—sometimes at thousands of feet up, where O_2 is limited. In contrast to other vertebrate lungs, which are saclike, bird lungs are relatively rigid and are filled with hollow, thin-walled tubes called *parabronchi*, which allow air to pass through them in both directions (Fig. 33-6b). Birds also differ from other vertebrates in using from seven to nine flexible air sacs as bellows to pump air in and out. As a bird breathes in, it draws air through its lungs, where O_2 is extracted. The same breath pulls air into air sacs, some of which are located beyond the lungs (Fig. 33-6a). As the bird breathes out, oxygenated air from the air sacs is forced back through the lungs, allowing the bird to extract O_2 even as it exhales.

33.3 HOW DOES THE HUMAN RESPIRATORY SYSTEM WORK?

The respiratory system in humans and other lung-breathing vertebrates can be divided into two parts: the **conducting portion** and the **gas-exchange portion**. The conducting portion consists of a series of passageways that carry air into and out of the gas-exchange portion, where gases are exchanged with the blood in tiny sacs in the lungs.

The Conducting Portion of the Respiratory System Carries Air to the Lungs

The conducting portion brings air to the lungs; it also contains the apparatus that makes speaking possible. Air enters through the nose or the mouth, passes through the nasal cavity or oral cavity into a common chamber, the **pharynx**, and then travels through the **larynx,** or "voice box" (**FIG. 33-7**). The opening to the larynx is guarded by the *epiglottis*, a flap of tissue supported by cartilage. During normal breathing, the epiglottis is tilted upward, as shown in Figure 33-7, allowing air to flow freely into the larynx. During swallowing, the epiglottis tilts downward and covers the larynx, directing substances into the esophagus instead. If an individual attempts to inhale and swallow at the same time, this reflex may fail and food can become lodged in the larynx, blocking air from entering the lungs. What should you do if you see this happen? Use the *Heimlich maneuver* (**FIG. 33-8**), which is easy to perform and has saved countless lives.

Within the larynx are the **vocal cords**, bands of elastic tissue controlled by muscles. Muscular contractions can cause the vocal cords to partially obstruct the opening within the larynx. Exhaled air causes them to vibrate, producing the tones of speech or song. Stretching the cords changes the pitch of the tones, which can be articulated into words by movements of the tongue and lips.

Inhaled air travels past the larynx into the **trachea**, a flexible tube whose walls are reinforced with semicircular bands of stiff cartilage. Within the chest, the trachea splits into two large branches called **bronchi** (singular, bronchus), one leading to each lung. Inside the lung, each bronchus branches repeatedly into ever-smaller tubes called **bronchioles**. Bronchioles lead finally to the microscopic **alveoli** (singular, alveolus), the tiny air sacs where gas exchange occurs (see Fig. 33-7). During its passage

(a) Human respiratory system

(b) Alveoli with capillaries

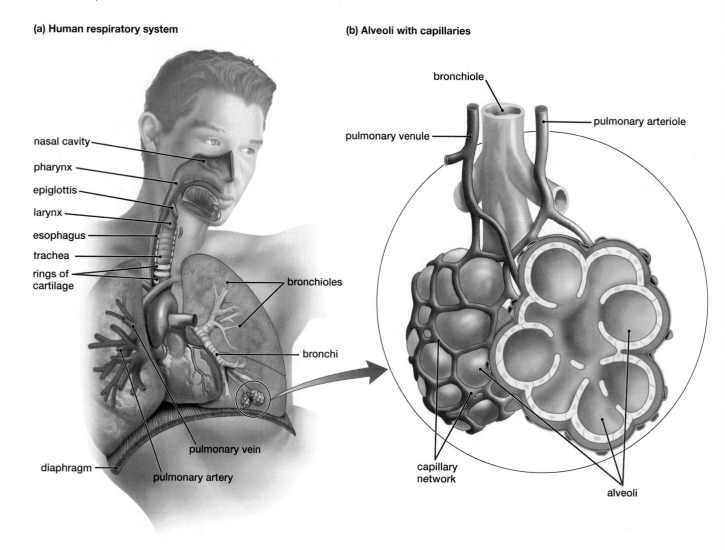

FIGURE 33-7 The human respiratory system
(a) Air enters the nasal cavity or mouth and passes through the pharynx and the larynx into the trachea. The epiglottis prevents food from going down the trachea. The trachea splits into two bronchi, which lead into the two lungs. Each bronchus branches into numerous bronchioles that conduct air into microscopic alveoli, enmeshed in capillaries, where gas exchange occurs. The pulmonary artery carries deoxygenated blood (blue) to the lungs; the pulmonary vein carries oxygenated blood (red) back to the heart. **(b)** Close-up of alveoli (interiors shown in cutaway section) and their surrounding capillaries.

through the conducting system, air is warmed and moistened. Much of the dust and bacteria it carries is trapped in mucus secreted by cells that line the respiratory passages. The mucus, with its trapped debris, is continuously swept upward toward the pharynx by cilia that line the bronchioles, bronchi, and trachea. Upon reaching the pharynx, the mucus is coughed up or swallowed. Smoking interferes with this cleansing process by paralyzing the cilia (see "Health Watch: Smoking—A Life and Breath Decision"). *Asthma* occurs when smooth muscle tissue in the bronchioles becomes hyperexcitable and mucus production increases, usually because of an allergy to an airborne substance, such as pollen. The bronchioles spasm and re-

duce the diameter of the airway. The asthma victim has particular difficulty breathing out through the constricted bronchioles, since the reduced air volume caused by exhalation allows the mucus-clogged passageways to collapse more readily.

Gas Exchange Occurs in the Alveoli

The lung provides an enormous moist surface for gas exchange. The dense network of bronchioles conducts air to tiny structures, the alveoli, which cluster about the end of each bronchiole like a bunch of grapes. In a typical adult, the two lungs combined have approximately 300 million

After carrying O_2 through all the tissues of the body, blood is pumped to the lungs by the heart. The incoming blood surrounding the alveoli is low in O_2 (because the cells have used it up) and high in CO_2 (which is released by the cells; see Fig. 33-2). Carbon dioxide diffuses out of the blood, where its concentration is high, into the air in the alveoli, where its concentration is lower (see Fig. 33-9). Carbon dioxide concentration in the blood is especially high after heavy exertion—for example, if you sprinted up the stairs on your way to class. As CO_2 diffuses into the air in the alveoli, O_2 diffuses from the air, where its concentration is high, into the blood, where its concentration is low. Blood from the lungs, now oxygenated and purged of CO_2, returns to the heart, which pumps it back through the body tissues. In the tissues, O_2 diffuses into the cells because the concentration of O_2 is lower in the cells than in the blood.

Carbon Dioxide and Oxygen Are Transported Using Different Mechanisms

Carbon dioxide (CO_2) is constantly being produced by cellular respiration in all cells. Capillary blood picks up CO_2 and carries it to the lungs, where it diffuses out and is exhaled. In

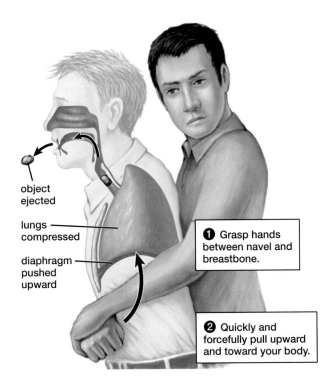

FIGURE 33-8 The Heimlich maneuver can save lives If a person is choking and unable to breathe, use the Heimlich maneuver to push upward on the victim's diaphragm, forcing air out of the lungs and often dislodging the object. Repeat if necessary.

object ejected

lungs compressed

diaphragm pushed upward

❶ Grasp hands between navel and breastbone.

❷ Quickly and forcefully pull upward and toward your body.

alveoli. These microscopic (0.2 millimeter in diameter) chambers give magnified lung tissue the appearance of a pink sponge. Alveoli provide an extensive surface area for diffusion, totaling about 1500 square feet (143 square meters; about 80 times the skin surface area of a human adult). A network of capillaries covers about 85% of the alveolar surface (see Fig. 33-7b). The walls of the alveoli, which consist of a single layer of epithelial cells, form the innermost portion of the *respiratory membrane*. The respiratory membrane through which gases diffuse consists of the epithelium of the alveolus and the endothelial cells that form the wall of each capillary. These two layers are held together by collagen fibers. Because the alveolar walls and the adjacent capillary walls are each only one cell thick, gases must diffuse only a short distance to move between the air and the blood (**FIG. 33-9**). The alveoli are lined with a thin layer of fluid containing an oily secretion (called a *surfactant*) that reduces surface tension and prevents the alveoli from collapsing during exhalation. The lung disease *emphysema*, which is most often caused by cigarette smoking, causes the alveoli to rupture, severely reducing the area available for gas exchange (emphysema is discussed in greater detail in "Health Watch: Smoking— A Life and Breath Decision").

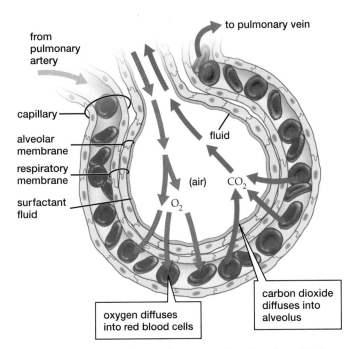

from pulmonary artery

to pulmonary vein

capillary

alveolar membrane

respiratory membrane

surfactant fluid

fluid

(air)

O_2

CO_2

oxygen diffuses into red blood cells

carbon dioxide diffuses into alveolus

FIGURE 33-9 Gas exchange between alveoli and capillaries The alveoli and capillary walls are only one cell thick and are very close to one another, with cells coated in a thin layer of fluid. This allows gases to dissolve and diffuse easily between the lungs and circulatory system.

HEALTH WATCH   Smoking—A Life and Breath Decision

About 440,000 people in the United States die of smoking-related diseases each year; these include lung cancer, emphysema, chronic bronchitis, heart disease, stroke, and other forms of cancer.

Tobacco smoke has a dramatic impact on the human respiratory tract. As smoke is inhaled, toxic substances such as nicotine and sulfur dioxide paralyze the cilia that line the respiratory tract; a single cigarette can inactivate them for a full hour. Because these ciliary sweepers remove inhaled particles, smoking inhibits them just when they are most needed. The visible portion of cigarette smoke consists of billions of microscopic carbon particles. Adhering to these particles are about 200 different toxic substances, of which more than a dozen are known or probable *carcinogens* (cancer-causing substances). With the cilia incapacitated, the particles stick to the walls of the respiratory tract and enter the lungs.

Cigarette smoke also impairs the white blood cells that defend the respiratory tract by engulfing foreign particles and bacteria. Consequently, still more bacteria, dust, and smoke particles enter the lungs. In response to the irritation of cigarette smoke, the respiratory tract produces more mucus, another method of trapping foreign particles. But without the cilia sweeping it along, the mucus builds up and can obstruct the airways, producing the familiar "smoker's cough." Microscopic smoke particles accumulate in the alveoli over the years until the lungs of a heavy smoker are literally blackened (compare the normal lung in **FIG. E33-3a** to the diseased lung in **FIG. E33-3b**). The longer the delicate tissues of the lungs are exposed to the carcinogens on the trapped particles, the greater the chance that cancer will develop (**FIG. E33-3c**).

Some smokers will develop *chronic bronchitis*, a persistent lung infection characterized by coughing, swelling of the lining of the respiratory tract, an increase in mucus production, and a decrease in the number and activity of cilia. The result is a decrease in air flow to the alveoli. *Emphysema* (Fig. E33-3b) occurs when toxic substances in cigarette smoke cause the body to produce substances that lead to brittle and ruptured alveoli. The loss of the alveoli, where gas exchange occurs, leads to oxygen deprivation of all body tissues. In an individual with emphysema, breathing becomes labored and grows increasingly difficult, sometimes leading to death.

Carbon monoxide, present in high levels in cigarette smoke, binds tenaciously to red blood cells in place of oxygen, reducing the blood's oxygen-carrying capacity and increasing the workload on the heart. Chronic bronchitis and emphysema compound this problem. Smoking also promotes *atherosclerosis*, or thickening of the arterial walls by fatty deposits that can lead to heart attacks (see Chapter 32). As a result, smokers are 70% more likely than nonsmokers to die of heart disease. Smokers' wounds and broken bones take longer to heal, and the skin of smokers often develops premature wrinkles. The carbon monoxide in cigarette smoke may also contribute to the reproductive problems experienced by women who smoke, because it deprives the developing fetus of oxygen. These complications include an increased incidence of infertility, miscarriage, and lower-birth-weight babies—and later, to more learning and behavioral problems in children.

"Passive smoking," or breathing secondhand smoke, poses real health hazards for both children and adults. Researchers have concluded that children whose parents smoke are more likely to contract bronchitis, pneumonia, ear infections, coughs, and colds and to have decreased lung capacity. Children who grow up with smokers are more likely to develop asthma and allergies; for children with asthma, secondhand smoke increases the number and severity of asthma attacks. Among adults, studies have concluded that nonsmoking spouses of smokers face a 30% higher risk of both heart attack and lung cancer than do spouses of nonsmokers. A recent study links even relatively infrequent exposure to secondhand smoke with an increased risk of

---

the blood, CO_2 is transported in three different ways (**FIG. 33-10a**): a little is dissolved in the plasma, more is loosely bound to **hemoglobin** (an iron-containing protein in red blood cells), and most of it combines with water to form bicarbonate ion (HCO_3^-) in the following reaction:

$$CO_2 + H_2O \xrightarrow{\text{(carbonic anhydrase)}} H^+ + HCO_3^-$$

This reaction occurs in the red blood cells, which contain the enzyme *carbonic anhydrase*. Most HCO_3^- then diffuses into the plasma, where it helps maintain the proper pH of the blood (around pH 7.4; the remaining H^+ remains in the red blood cell bound to hemoglobin). This reaction reverses when the capillaries flow past the alveoli, where CO_2 is low (**FIG. 33-10b**). The production of bicarbonate ions and the binding of CO_2 to hemoglobin both reduce the concentration of dissolved CO_2 in the blood and increase the gradient for CO_2 to diffuse from cells into the blood.

As capillary blood, rich in CO_2 from the tissues, flows past the alveoli of the lungs (or membranes of other gas-exchange organs), it releases CO_2 and picks up O_2. Nearly all the O_2 carried by the blood is bound to hemoglobin molecules, which can each carry up to four O_2 molecules (Fig. 33-10b). By removing oxygen from solution in the plasma, hemoglobin maintains a concentration gradient that favors the diffusion of oxygen from the air into the blood. This allows mammalian blood to carry about 70 times more oxygen than if the oxygen were simply dissolved in the plasma. As hemoglobin binds oxygen, the protein undergoes a slight change in shape which alters its color. Oxygenated blood is a bright cherry-red; deoxygenated blood is dark maroon-red and appears bluish through the skin.

Carbon monoxide (CO) is a toxic gas produced by combustion, such as that occurring in engines, furnaces, and cigarettes when the fuel is not completely burned to form carbon dioxide. CO can be deadly because it binds to

atherosclerosis. Government agencies report that second-hand smoke is responsible for an estimated 3000 lung-cancer deaths and a minimum of 35,000 deaths from heart disease in nonsmokers in the U.S. each year.

For smokers who quit, however, healing begins immediately; and the chances of heart attack, lung cancer, and numerous other smoking-related illnesses gradually diminish (see "Links to Life: Quitters Are Winners").

FIGURE E33-3 Smoking damages the lungs
(a) A normal lung. (b) Lung of a smoker who died of emphysema is both blackened and collapsed. (c) A lung cancer is visible as a pale mass; the lung tissue surrounding it is blackened by trapped smoke particles. Smoking causes about 87% of all lung cancers—it is the leading cause of cancer deaths in the U.S.

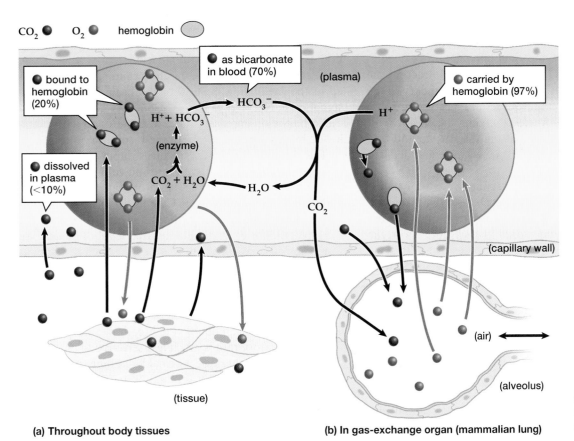

FIGURE 33-10 The mechanism of gas exchange

(a) Throughout body tissues

(b) In gas-exchange organ (mammalian lung)

LINKS TO LIFE   Quitters Are Winners

Have you or someone you know quit smoking? The American Lung Association has some encouraging news—a timeline of improvements experienced by former smokers as their bodies begin to recover, starting from the last cigarette puffed. If you stop smoking, after 20 minutes your blood pressure and pulse rate drop. After 8 hours the level of carbon monoxide in your blood drops, while blood O_2 increases to normal levels. After 24 hours your chance of a heart attack decreases. After 48 hours you regain more ability to smell and taste your food. From 2 weeks to 3 months later you are able to exercise more easily as circulation and respiratory function improve. From 1 to 9 months later you cough less, and you have less sinus congestion and more energy. After 1 year you have about half the risk of coronary heart disease as does a smoker. After 5 years your risk of a stroke begins to drop. After 10 years your risk of lung cancer may be half that of a continuing smoker, and your risk of cancers of the pancreas, kidney, bladder, esophagus, throat, and mouth also decline. After 15 years your risk of death from all smoking-related causes is nearly as low as that of people who have never smoked. (Modified from "What Are the Benefits of Quitting Smoking?" published online by the American Lung Association at http://www.lungusa.org)

hemoglobin more than 200 times as tightly as oxygen. People can die from breathing air with as little as 0.1% CO. Hemoglobin bound to CO is bright red (like oxygenated hemoglobin), but it is incapable of transporting O_2 because its O_2 binding sites are occupied. While most victims of asphyxiation have bluish lips and nail beds because their hemoglobin is deoxygenated, the lips and nail beds of victims of carbon monoxide poisoning (such as could occur from breathing car exhaust in a closed space) are brighter red than normal.

Air Is Inhaled Actively and Exhaled Passively

Breathing occurs in two stages: (1) **inhalation**, when air is actively drawn into the lungs, and (2) **exhalation**, when it is passively expelled from the lungs. Inhalation is accomplished by enlarging the chest cavity by contracting the **diaphragm**, a muscle that forms the lower boundary of the chest cavity. At rest, this thin muscle domes upward, but contraction pulls it downward, expanding the chest cavity. The rib muscles also contract, lifting the ribs up and outward (**FIG. 33-11a**). When the chest cavity is expanded, the lungs inflate within it, because an airless space with a layer of fluid holds them tightly against the inner wall of the chest. As the lungs expand with the chest cavity during inhalation, their increased volume draws in air. A puncture wound to the chest is dangerous in part because it can allow air to penetrate between the chest wall and the lungs, breaking the seal and preventing the lungs from inflating when the chest cavity expands.

Although air can be forcibly exhaled, at rest exhalation occurs automatically when the muscles that cause inhalation are relaxed. As it relaxes, the diaphragm domes upward; at the same time, the ribs fall down and inward, decreasing the size of the chest cavity and forcing air out of the lungs (**FIG. 33-11b**). Additional air can be forced out by contracting the abdominal muscles. After exhalation, the lungs still contain air, which helps prevent the thin alveoli from collapsing and fills the spaces within the conducting portion of the respiratory system. A typical breath moves about a pint (500 milliliters) of "new" air into the respiratory system. Of this, only about 1.5 cups (350 milliliters) reach the alveoli where gas exchange occurs. During exercise, deeper breathing can move several times more air than occurs at rest.

Breathing Rate Is Controlled by the Respiratory Center of the Brain

Imagine having to think about every breath. Fortunately, breathing occurs rhythmically and automatically without conscious thought. But, unlike the heart muscle, the muscles used in breathing are not self-activating; each contraction is stimulated by impulses from nerve cells. These impulses originate in the **respiratory center**, which is located in the medulla, a portion of the brain just above the spinal cord. Nerve cells in the respiratory center generate cyclic bursts of impulses that cause the contraction (followed by passive relaxation) of the respiratory muscles.

The respiratory center receives input from several sources and adjusts the breath rate and volume to meet the body's changing needs. The respiratory rate is regulated to maintain a constant level of CO_2 in the blood, as monitored by CO_2 receptors in the medulla (just above the spinal cord). For example, an elevated level of blood CO_2 caused by an increase in cellular activity, such as happens in your muscle cells when you run upstairs, signals a need for more O_2. The receptors then stimulate the respiratory center, causing you to breathe faster and more deeply. These receptors are extremely sensitive; an increase in CO_2 of only 0.3% can double the breathing rate.

The respiratory rate is much less sensitive to changes in O_2 concentration because normal breathing supplies an overabundance of O_2. But if blood O_2 levels fall drastically, receptors in the aorta and carotid arteries stimulate the respiratory center.

(a) Inhalation   **(b) Exhalation**

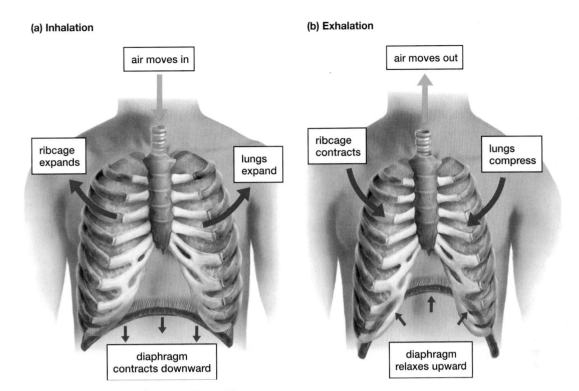

FIGURE 33-11 The mechanics of breathing
(a) During inhalation, rhythmic nerve impulses from the brain stimulate the diaphragm to contract (pulling it downward) and the muscles surrounding the ribs to contract (moving them up and outward). The result is an increase in the size of the chest cavity, causing air to rush in. **(b)** Relaxation of these muscles (exhalation) allows the diaphragm to dome upward and the rib cage to collapse, forcing air out of the lungs. QUESTION Imagine that a woman living at sea level travels to a high mountaintop; there, she inhales with a muscular contraction of exactly the same strength as had been her habit while at rest in her sea-level home. Would the resulting inhalation contain a volume of air that was larger, smaller, or the same as at sea level? Why?

CASE STUDY REVISITED LIVES UP IN SMOKE

Mark Twain once said, "Quitting smoking is easy, I've done it a thousand times." Nicotine is a powerfully addictive drug, as likely to lead to addiction as cocaine or heroin is. Most smokers use cigarettes daily, averaging about 14. Researchers have found that, like cocaine and heroin, nicotine activates the brain's reward center. The brain adapts by becoming less sensitive, requiring larger quantities of nicotine to experience the same rewarding effect, and causing the reward center to feel understimulated when nicotine is withdrawn. Withdrawal symptoms can include nicotine craving, depression, anxiety, irritability, difficulty concentrating,

headaches, and disturbed sleep. So it's no wonder that for some people, the only way to quit smoking is not to start smoking at all. Although at least 70% of smokers would like to quit, only about 2.5% of smokers are successful each year. By age 60, about 33% will have succeeded, most after having made 2–3 attempts. Of those who continue to smoke, about one-third will die from a smoking-related cause. This junior at the University of Illinois expresses a common attitude among young smokers: "I know it's bad for me and it could kill me. I'm 21 years old; I'm not dumb to it. My grandpa died of cancer from smoking all his life." A nonsmoking student there explains: "People are going to do it no matter what. You can show

them pictures of [cancerous] lungs or rotting teeth but it's not going to affect people all the same way.... We have the right to choose how we live our lives."

Consider This Direct medical costs for treating lung cancer are approximately $5 billion annually in the U.S. Nearly 90% of these cancers are preventable by avoiding smoking. Most of this cost is borne by insurance. Rising insurance rates are predicted to increase the percentage of uninsured workers and decrease wages as businesses struggle to contain costs. Discuss the responsibilities that go along with the "right to choose" to smoke.

CHAPTER REVIEW

SUMMARY OF KEY CONCEPTS

33.1 Why Exchange Gases?

The respiratory system supports cellular respiration. Oxygen-rich air is inhaled and supplies O_2 to the blood, which carries it to cells throughout the body. Blood also picks up CO_2 (a product of cellular respiration) from body cells and transports it to the lungs, where it is released into the atmosphere.

33.2 What Are Some Evolutionary Adaptations for Gas Exchange?

The exchange of O_2 and CO_2 between the body and the environment by diffusion across a moist surface is required by nearly all forms of life. In watery environments, animals with very small or flattened bodies may rely exclusively on diffusion through the body surface. Animals with low metabolic demands and/or well-developed circulatory systems may also lack specialized respiratory structures. Larger, more active animals have evolved specialized respiratory systems. Animals in aquatic environments often possess gills, such as those of fish and many amphibians. On land, respiratory surfaces must be protected and moistened internally. This has selected for the evolution of tracheae in insects and lungs in terrestrial vertebrates.

The transfer of gases between respiratory systems and tissues requires both bulk flow and diffusion. Air or water moves by bulk flow past the respiratory surface, and gases in blood are also carried by bulk flow. Gases move by diffusion across membranes between the respiratory system and the capillaries and between the capillaries and the tissues.

33.3 How Does the Human Respiratory System Work?

The human respiratory system consists of a conducting portion consisting of the nose and mouth, pharynx, larynx, trachea, bronchi, and bronchioles, and a gas-exchange portion, composed of alveoli (microscopic air sacs). Blood, within a dense capillary network surrounding the alveoli, releases CO_2 and absorbs O_2 from the air.

Most of the O_2 in the blood is bound to hemoglobin within red blood cells. By removing O_2 from solution, hemoglobin maintains a favorable concentration gradient that allows O_2 to readily diffuse from the air into the blood. Hemoglobin then transports the O_2 to the body tissues, where it diffuses out along its concentration gradient. Carbon dioxide diffuses into the blood from the tissues and is transported in three ways: as bicarbonate; bound to hemoglobin; or dissolved in blood plasma.

Breathing involves actively drawing air into the lungs by contracting the diaphragm and the rib muscles, which expands the chest cavity. Relaxing these muscles reduces the volume of the chest cavity, expelling the air. Respiration is controlled by nerve impulses that originate in the medulla's respiratory center. The respiratory center is influenced by receptors, such as those in the medulla that monitor CO_2 levels in the blood.

Web Tutorial 33.1 The Human Respiratory System

KEY TERMS

alveoli *page 675*
bronchi *page 675*
bronchiole *page 675*
bulk flow *page 671*
conducting portion *page 675*
countercurrent exchange
  page 674

diaphragm *page 680*
exhalation *page 680*
gas-exchange portion
  page 675
gill *page 672*
hemoglobin *page 678*

inhalation *page 680*
larynx *page 675*
lung *page 672*
pharynx *page 675*
respiratory center *page 680*
spiracle *page 672*

trachea (in birds/mammals)
  page 675
tracheae (in insects)
  page 672
vocal cords *page 675*

THINKING THROUGH THE CONCEPTS

1. Describe countercurrent exchange. What are the benefits of this process? How does it work in fish gills?

2. Trace the route taken by air in the vertebrate respiratory system, listing the structures through which it flows and the point at which gas exchange occurs.

3. Explain some characteristics of animals in moist environments that may supplement respiratory systems or make them unnecessary.

4. How are human respiratory movements initiated? How are they modified, and why are these controls adaptive?

5. What events occur during human inhalation? Exhalation? Which of these is always an active process?

6. Trace the pathway of an O_2 molecule in the human body, starting with the nose and ending with a body cell.

7. Describe the effects of smoking on the human respiratory system.

8. Explain how bulk flow and diffusion interact to promote gas exchange between air and blood and between blood and tissues.

9. Compare CO_2 and O_2 transport in the blood. Include the source and destination of each.

10. Explain how the structure and arrangement of alveoli make them well suited for their role in gas exchange.

APPLYING THE CONCEPTS

1. Heart-lung transplants are performed in some cases when both have been damaged, for example by cigarette smoking, but donors are scarce. Based on your knowledge of the respiratory and circulatory systems and of lifestyle factors that might damage them, what criteria would you use in selecting a recipient for such a transplant?

2. Nicotine is a drug in tobacco that is responsible for several of the effects that smokers crave. Discuss the advantages and disadvantages of low-nicotine cigarettes.

3. Discuss why a brief exposure to CO, carbon monoxide, is much more dangerous than a brief exposure to CO_2.

4. Review insect and mammalian respiratory systems and circulatory systems (see Chapter 32). In what way does the respiratory system of an insect serve some of the functions of the circulatory system of mammals?

5. Mary, a strong-willed 3-year-old, threatens to hold her breath until she dies if she doesn't get her way. Can she carry out her threat? Explain.

FOR MORE INFORMATION

Platt, C. "Here, Breathe This Liquid." *Discover*, October 2001. Perfusing ice-cold, O_2-enriched perfluorocarbons into the lungs might increase the survival chances of patients whose hearts have stopped. After the heart is restarted, the chilled liquid would rapidly cool the body and reduce the brain's requirements and the damaging chemical reactions that occur in O_2-starved brain tissue.

Rist, C. "The Physics of … Singing." *Discover* August, 1999. The author describes how the vocal cords turn air flow into musical notes.

Seppa, N. "Secondary Smoke Carries High Price." *Science News*, January 17, 1998. Research suggests that smoking causes atherosclerosis, that the damage continues after a smoker quits, and that exposure to secondhand smoke significantly increases the buildup of plaque in the carotid arteries of nonsmokers.

Wheelwright, J. "Toxic Inheritance." *Discover*, March, 2006. There is evidence that pollutants inhaled by a mother may alter her DNA in ways that can affect her children.

34 Nutrition and Digestion

Recovered anorexic Carré Otis now sees a healthy reflection in her mirror. Anorexics look in the mirror and see an overweight person, no matter how skeletal their bodies.

CASE STUDY DIETING TO DEATH?

FORMER SUPERMODEL CARRÉ OTIS explains, "The sacrifices I made were life threatening. I had entered a world that seemed to support a 'whatever it takes' mentality to maintain abnormal thinness." For many models, performers, and others in the public eye, meeting expectations for thinness is a continuing battle that can lead to disaster. Celebrity victims include Jane Fonda, the late Princess Diana, and Elton John. At 5 feet 10 inches, Carré once weighed only 100 pounds. Now maintaining a healthy weight—at the expense of her modeling career—Carré has become a spokesperson for the National Eating Disorders Association. She hopes to help others avoid the damage her body suffered. "It was common for the young girls I worked with to have a heart attack; if an eating disorder is not treated, it can be a fatal disease."

Natural selection has provided animals with strong drives to eat when nutrients are needed (and even when not, if good food is available), but in some people, these natural impulses can go terribly awry. In recent decades we have seen an increase in both overeating and *eating disorders*, ailments characterized by a severe disruption of normal eating behavior.

Eating disorders include two particularly debilitating illnesses, *anorexia nervosa* and *bulimia nervosa*. People with bulimia—who may be anorexic, but sometimes maintain a normal weight—engage in binge eating, which means they consume enormous amounts of food in a short time. They follow their binges with self-induced vomiting or overdosing with laxatives to purge the food from their bodies. Bulimics may also exercise excessively to burn off the calories they have consumed. Their repeated vomiting damages the digestive tract and upsets the normal balance of salts in the blood, which can lead to heart disorders.

People with anorexia experience an intense fear of gaining weight, and they achieve extreme weight loss—often weighing over 30% less than their healthy weight—by eating very little food and sometimes exercising compulsively. Although skeletal in appearance, anorexics still see themselves as fat. About half of all anorexics also develop bulimia. The consequences are disastrous. Anorexics become emaciated, losing both fat and muscle. In the process they often disrupt their digestive, cardiac, endocrine, and reproductive functions. People with anorexia who are between ages 18 and 24 are 12 times more likely to die than others in their age group.

More than 90% of diagnosed eating disorders occur in females; the incidence of anorexia is between 0.5% and 1%, and for bulimia it is 1–2% of women in the United States. How can eating disorders damage the digestive system? What causes them, and how are they treated? Look for answers in "Case Study Revisited."

34.1   WHAT NUTRIENTS DO ANIMALS NEED?

All foods, from broccoli to a chocolate milkshake, contain nutrients that you need to survive. **Nutrients** fall into six major categories: lipids, carbohydrates, proteins, minerals, vitamins, and water. These substances provide the body with its basic needs, including energy and the raw materials to synthesize the molecules of life—enzymes, structural proteins, genetic material, energy carriers, the calcium-based components of bone, and the lipid-based components of all cell membranes, to name just a few.

Energy Is Derived from Nutrients and Measured in Calories

Cells rely on a continuous supply of energy to maintain their incredible complexity and wide range of activities. Deprived of this energy, cells begin to die within seconds. Two types of nutrients provide most of the energy in the animal diet: carbohydrates and fats (in the U.S., an average person also gets about 15% of his or her energy from protein). These molecules are broken down by digestion, and their subunits are metabolized during cellular respiration, releasing energy that is captured in adenosine triphosphate (ATP; see Chapter 3).

The energy in nutrients is measured in calories. A **calorie** is the amount of energy required to raise the temperature of 1 gram of water by 1 degree Celsius. The calorie content of foods is measured in units of 1000 calories (*kilocalories*), also known as **Calories** (with a capital *C*). The average human body at rest burns roughly 70 Calories per hour, but this value is influenced by body size, muscle mass, age, sex, and genetic factors. Exercise significantly boosts caloric requirements; well-trained athletes can temporarily raise their calorie consumption from a resting rate of about 1 Calorie per minute to nearly 20 Calories per minute during vigorous exercise (Table 34-1).

Lipids Include Fats, Phospholipids, and Cholesterol

Although they are sometimes viewed as the enemy in our overweight society, fats and other lipids are essential nu-

trients. Lipids are a diverse group of molecules including *triglycerides* (fats), *phospholipids*, and *cholesterol* (see Chapter 3). Triglycerides are used primarily as a source of energy. Phospholipids are important components of all cellular membranes, and cholesterol is used in the synthesis of cellular membranes, sex hormones, and bile (which aids in fat breakdown). Some animal species can synthesize all the lipid "building blocks" necessary to make the specialized lipids they need. Others must acquire some of these, called **essential fatty acids**, from their food. For example, humans are unable to synthesize linoleic acid (required for the synthesis of certain phospholipids); we need to obtain this essential fatty acid from our diets.

Animals Store Energy as Fat

When an animal's diet provides more energy than it expends through metabolic activities, most of the excess carbohydrate, fat, or protein is converted to fat for storage. About 3600 Calories are stored in each pound of fat. Fats have two major advantages as energy-storage molecules. First, they are the most concentrated energy source, containing more than twice the energy per unit weight of either carbohydrates or proteins (about 9 Calories per gram for fats compared with about 4 per gram for proteins and carbohydrates). Second, lipids are *hydrophobic*—that is, they repel water, rather than dissolving in it. Fat deposits, therefore, do not cause any extra accumulation of water in the body. For both these reasons, fats store more calories with less weight than do other molecules. Minimizing weight allows an animal to move faster (important for escaping predators and hunting prey) and to use less energy when it moves (important when food supplies are limited). Since people evolved under the same food constraints as other animals, we have a strong tendency to eat when food is available—often in excess of our needs, because we may require the energy later. Some modern societies now have access to almost unlimited high-calorie food. In this environment, our natural tendency to overeat can become a liability, and we need to exert considerable willpower to avoid becoming obese (see "Health Watch: Dying for a Cheeseburger").

Table 34-1 Approximate Energy Consumed by a 150-Pound Person for Different Activities

Activity	Calories/hr	500 Calories Cheeseburger	340 Calories Ice cream cone	70 Calories Apple	40 Calories Broccoli, 1 cup
			Time to "Work Off"		
Running (6 mph)	700	43 min	26 min	6 min	3 min
Cross-country skiing (moderate)	560	54 min	32 min	7.5 min	4 min
Roller skating	490	1 hr 1 min	37 min	8.6 min	5 min
Bicycling (11 mph)	420	1 hr 11 min	43 min	10 min	6 min
Walking (3 mph)	250	2 hr	1 hr 12 min	17 min	10 min
Frisbee® playing	210	2 hr 23 min	1 hr 26 min	20 min	11 min
Studying	100	5 hr	3 hr	42 min	24 min

Patrick Deuel's doctor gave him an ultimatum: be hospitalized and lose weight or die. But a wall had to be removed to get Patrick out of his bedroom, where at 1072 pounds he was trapped by his very size. After almost a year of supervised diet and exercise, Patrick had lost 421 pounds and was ready for *gastric bypass* surgery. His doctor sealed off most of his stomach, leaving only a tiny pouch that was attached midway along his small intestine (**FIG. E34-1a**). His stomach can now hold only morsels of food, and even these bypass part of the small intestine, where calorie-rich nutrients are absorbed.

Obesity is a growing epidemic in the U.S. The percentage of overweight adults has more than doubled since 1980, and among children and adolescents, it has more than tripled. A simple way to estimate body fat is to calculate **body mass index** (BMI), which is done by multiplying your weight (in kilograms) by your height squared (in centimeters). The BMI calculation applies to people with average amounts of muscle and assumes that weight in excess of this is fat, so it won't be accurate for heavily muscled people, such as bodybuilders. Many Internet sites provide tables or do this calculation for you: search for "BMI." A BMI between 20 and 24 is considered healthy. A recent study by the Centers for Disease Control concluded that 66% of all U.S. adults are overweight (BMI ≥ 25) and 32% are obese (BMI ≥ 34).

GENETICALLY FAT?

Are some people just "born to be fat?" Patrick's doctor described him as "obese" when he was only three months old. Researchers have found at least 134 different genes that influence weight in some way. People differ genetically in how their bodies respond to exercise, how full they feel after eating, how much fat they store, and how many calories their bodies consume at rest. Although our genes haven't changed appreciably in the past 20 years, the number of overweight people in the U.S. has increased dramatically. Despite genetic differences, everyone who is overweight has eaten more than his or her body needs, succumbing to a drive that helped our early ancestors survive famine. The National Institutes of Health (NIH) advises people to eat frequent small meals, and choose fruits, vegetables, whole grains, and low-fat meat and dairy products. In response to consumer demand, several fast-food chains now include healthy options on their menus. The NIH also recommends that people exercise at moderate intensity (walking, for example) for at least 30 minutes on most days. To lose weight or maintain a weight loss, some people may require additional exercise. A lot of work? Perhaps, but well worth it. Overweight individuals have a higher risk of heart disease and strokes, diabetes, cancer, sleep apnea (breathing problems during sleep), osteoarthritis, liver disease, and gallstones.

SURGICALLY SLIM?

Some people have unusual difficulty controlling their weight, and doctors may recommend surgery for those whose weight poses a serious health risk. Stuart Logan, for example, began packing on the pounds in first grade. At age 16, weighing 585 pounds, he opted for *gastric band* surgery, recently approved in the U.S. As shown in **FIG. E34-1b**, this process limits food intake by placing an adjustable band around the upper portion of the stomach. Three months later (having reducing his food intake by about half), Logan had lost 41 pounds and could finally fit in the chair at his barbershop. While some studies show that the gastric band produces less-dramatic results than gastric bypass does, it is a simpler surgery, it is reversible, and it has fewer complications than gastric bypass, which sometimes leads to nutrient deficiencies because it bypasses part of the small intestine. Meanwhile, Patrick has had a second surgery to remove an 80-pound flap of abdominal skin and tissue that used to house much of the 600 pounds he has now lost. Now under 400 pounds, he rejoices at being able to see his knees and walk again.

(a) Gastric bypass surgery

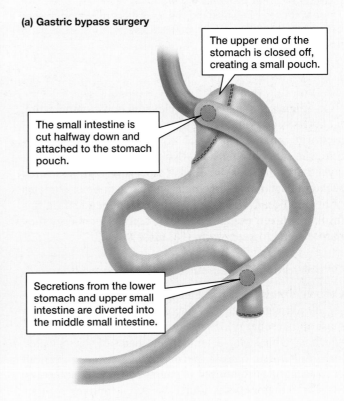

The upper end of the stomach is closed off, creating a small pouch.

The small intestine is cut halfway down and attached to the stomach pouch.

Secretions from the lower stomach and upper small intestine are diverted into the middle small intestine.

(b) Gastric band surgery

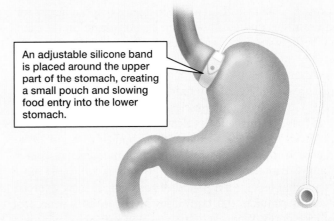

An adjustable silicone band is placed around the upper part of the stomach, creating a small pouch and slowing food entry into the lower stomach.

FIGURE E34-1 Weight-loss surgeries

FIGURE 34-1 Fat provides insulation
These walruses can withstand the icy waters of the polar seas because a thick layer of fat beneath the skin insulates them from the cold.

In animals that maintain an elevated body temperature, fat deposits do double duty by providing insulation as well as storing energy. Fat, which conducts heat at only one-third the rate of other body tissues, is typically stored in a layer beneath the skin. Birds (especially flightless birds such as penguins) and mammals who live in polar climates or in cold ocean waters are particularly dependent on this insulating layer, which reduces the amount of energy they must expend to keep warm (**FIG. 34-1**).

Carbohydrates Are a Source of Quick Energy

Carbohydrates (described in Chapter 3) include *monosaccharide* sugars (such as glucose, from which cells derive most of their energy), *disaccharide* sugars (such as sucrose, which is "table sugar"), and longer chains of sugars called *polysaccharides. Starch, glycogen, and cellulose* are all polysaccharides composed of chains of glucose molecules. Starch is a major energy source for people and many other animals, and the principal energy-storage material of plants. Glycogen is used by animals for short-term energy storage. Cellulose, the major structural component of plant cell walls, is the most abundant carbohydrate on the planet; but few species of animals are able to digest it, as we describe later.

Animals, including humans, store the carbohydrate **glycogen** (a large, highly branched chain of glucose molecules) in the liver and muscles. Athletes sometimes "carbo-load" by eating meals rich in carbohydrates such as potatoes and pasta to allow their bodies to build up as much glycogen as possible. Although humans can potentially accumulate hundreds of pounds of fat, most store somewhat less than a pound of glycogen. During exercise such as running, the body uses glycogen as a source of quick energy. When activity is prolonged, as in running a marathon, glycogen can be used up; the runner may "hit the wall" and experience extreme fatigue. When this occurs, the runner must rely mostly on fatty acids from stored body fat for energy. Because the metabolic transformations are more complex, it takes about twice as long to extract energy from fatty acids as from glycogen. A glycogen-depleted runner, even one with plenty of fat, may need to slow her pace dramatically to compensate for this reduction in available energy. For this reason, marathon runners often drink sugary solutions during the race.

Amino Acids Form the Building Blocks of Protein

In the digestive tract, protein from food is broken down into its amino acid subunits, which can be used to synthesize new proteins. These perform many different roles in the body, acting as enzymes, receptors on cell membranes, oxygen transport molecules (hemoglobin), structural proteins (hair and nails), antibodies, and muscle proteins. Excess amino acids can serve as an energy source or be converted into fat for storage.

Humans can synthesize 11 of the 20 different amino acids used in proteins. The nine amino acids that we cannot synthesize are called **essential amino acids**, and must be obtained from protein-rich foods such as meat, milk, eggs, corn, beans, and soybeans. Because many plant proteins are deficient in some of the essential amino acids, vegetarians must consume a variety of plants (for example, legumes, grains, and corn) whose proteins collectively provide all nine of the essential amino acids. Protein deficiency can cause a debilitating condition called *kwashiorkor* (**FIG. 34-2a**), which is most often encountered in poverty-stricken countries.

Minerals Are Elements Required by the Body

Minerals are elements that play many crucial roles in animal nutrition (Table 34-2). Since no organism can manufacture minerals, they must be obtained through the diet, either from food or dissolved in drinking water. Minerals such as calcium, magnesium, and phosphorus are major constituents of bones and teeth. Sodium, calcium, and potassium are essential for muscle contraction and the conduction of nerve impulses. Iron is a central component of each hemoglobin molecule in blood, and iodine is found in hormones produced by the thyroid gland. We also require trace amounts of several other minerals, including zinc and magnesium (both required for the function of some enzymes), copper (needed for hemoglobin synthesis), and chromium (used in the metabolism of sugar).

Vitamins Play Many Roles in Metabolism

"Take your vitamins!" is a familiar refrain in many households with children. But why are vitamins so important? **Vitamins** are a diverse group of organic compounds that animals require in small amounts for normal cell function, growth, and development. Many vitamins are required for the proper functioning of enzymes that control metabolic reactions throughout the body. In general, the body cannot synthesize vitamins (or cannot do so in adequate amounts), so they must be obtained from food. Our modern diet is

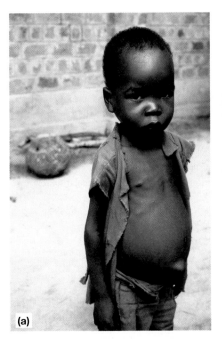
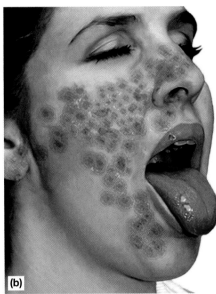
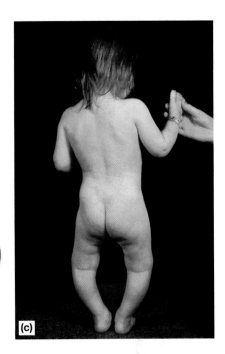

(a)   (b)   (c)

FIGURE 34-2 Symptoms of protein and vitamin deficiency
(a) Kwashiorkor is caused by protein deficiency. Low levels of blood albumin protein decrease the blood's osmotic strength and allow fluids to leak from blood capillaries into the abdominal area. Muscles are also wasted due to lack of protein. **(b)** Pellagra, characterized by scaly, reddish-brown skin lesions and a reddened, swollen tongue, is a symptom of niacin deficiency. **(c)** Rickets, which causes bone deformation, is a result of vitamin D deficiency.

Table 34-2 Minerals, Sources, and Functions for Humans

Mineral	Dietary Sources	Major Functions in Body	Deficiency Symptoms
Calcium	Milk, cheese, green vegetables, legumes	Bone and tooth formation Blood clotting Nerve impulse transmission	Stunted growth Rickets, osteoporosis Convulsions
Phosphorus	Milk, cheese, meat, poultry, grains	Bone and tooth formation Acid-base balance	Weakness Demineralization of bone Loss of calcium
Potassium	Meats, milk, fruits	Acid-base balance Body water balance Nerve function	Muscular weakness Paralysis
Chlorine	Table salt	Formation of gastric juice Acid-base balance	Muscle cramps Apathy Reduced appetite
Sodium	Table salt	Acid-base balance Body water balance Nerve function	Muscle cramps Apathy Reduced appetite
Magnesium	Whole grains, green leafy vegetables	Activation of enzymes in protein synthesis	Growth failure Behavioral disturbances Weakness, spasms
Iron	Eggs, meats, legumes, whole grains, green vegetables	Constituent of hemoglobin and enzymes involved in energy metabolism	Iron-deficiency anemia (weakness, reduced resistance to infection)
Fluorine	Fluoridated water, tea, seafood	Maintenance of teeth and probably bone structure	High frequency of tooth decay
Zinc	Widely distributed in foods	Constituent of enzymes involved in digestion	Growth failure Small sex glands
Iodine	Sea fish and shellfish, dairy products, many vegetables, iodized salt	Constituent of thyroid hormones	Goiter
Chromium	Fruits, vegetables, whole grains	Metabolism of sugar and fats	Reduced glucose tolerance Elevated insulin in blood

Table 34-3 Vitamins, Sources, and Functions for Humans

Vitamin	Dietary Sources	Functions in Body	Deficiency Symptoms
Water soluble			
B complex			
Vitamin B_1 (thiamin)	Milk, meat, bread	Coenzyme in metabolic reactions	Beriberi (muscle weakness, peripheral nerve changes, edema, heart failure)
Vitamin B_2 (riboflavin)	Widely distributed in foods	Constituent of coenzymes in energy metabolism	Reddened lips, cracks at corner of mouth, lesions of eye
Niacin	Liver, lean meats, grains, legumes	Constituent of two coenzymes in energy metabolism	Pellagra (skin and gastrointestinal lesions; nervous, mental disorders)
Vitamin B_6 (pyridoxine)	Meats, vegetables, whole-grain cereals	Coenzyme in amino acid metabolism	Irritability, convulsions, muscular twitching, dermatitis, kidney stones
Pantothenic acid	Milk, meat	Constituent of coenzyme A, with a role in energy metabolism	Fatigue, sleep disturbances, impaired coordination
Folic acid	Legumes, green vegetables, whole wheat	Coenzyme involved in nucleic and amino acid metabolism	Anemia, gastrointestinal disturbances, diarrhea, retarded growth, birth defects
Vitamin B_{12}	Meats, eggs, dairy products	Coenzyme in nucleic acid metabolism	Pernicious anemia, neurological disorders
Biotin	Legumes, vegetables, meats	Coenzymes required for fat synthesis, amino acid metabolism, and glycogen formation	Fatigue, depression, nausea, dermatitis, muscular pains
Choline	Egg yolk, liver, grains, legumes	Constituent of phospholipids, precursor of the neurotransmitter acetylcholine	None reported in humans
Vitamin C (ascorbic acid)	Citrus fruits, tomatoes, green peppers	Maintenance of cartilage, bone, and dentin (hard tissue of teeth); collagen synthesis	Scurvy (degeneration of skin, teeth, gums, blood vessels; epithelial hemorrhages)
Fat soluble			
Vitamin A (retinol)	Beta-carotene in green, yellow, and red vegetables. Retinol added to dairy products	Constituent of visual pigment. Maintenance of epithelial tissues	Night blindness, permanent blindness
Vitamin D	Cod liver oil, eggs, dairy products	Promotes bone growth and mineralization. Increases calcium absorption	Rickets (bone deformities) in children; skeletal deterioration
Vitamin E (tocopherol)	Seeds, green leafy vegetables, margarines, shortenings	Antioxidant, may reduce cellular damage from free radicals	Possibly anemia
Vitamin K	Green leafy vegetables. Product of intestinal bacteria	Important in blood clotting	Bleeding, internal hemorrhages

now so different from the natural diet on which we evolved that people may suffer vitamin deficiencies even with adequate food. For example, our skin can manufacture vitamin D when it is exposed to sunlight, but most of us spend so much time indoors that we do not synthesize enough and must augment our natural supply with fortified foods or supplements. The vitamins considered essential in human nutrition are listed in Table 34-3.

Some vitamins, such as C and E, also function as *antioxidants*. As our cells generate and use energy, damaging molecules called *free radicals* are produced. These molecules react with and can damage DNA, in some cases causing cancer. Free radicals can also promote atherosclerosis; over a lifetime, they contribute to the deterioration of physiological functioning associated with aging. In cells grown in the laboratory, antioxidants combine with free radicals to limit their damaging effects, and these vitamins may have similar effects in the body.

Water-Soluble Vitamins

Human vitamins are often grouped into two categories: water soluble and fat soluble. Water-soluble vitamins include vitamin C and the nine compounds that make up the B-vitamin complex. Since these substances dissolve in the blood plasma and are filtered out by the kidneys, they are not stored in appreciable amounts. Therefore, the body's supply of these vitamins must be constantly replenished by diet. Most water-soluble vitamins work in conjunction with enzymes to promote chemical reactions that supply energy or synthesize biological molecules. Because each vitamin participates in several metabolic processes, a deficiency of a single vitamin can have wide-ranging effects (see Table 34-3). For example, deficiency of the B vitamin niacin causes the cracked, scaly skin of pellagra (**FIG. 34-2b**), as well as some nervous disorders. Folic acid, another B vitamin, is required to synthesize thymine, a component of DNA; folic acid deficiency impairs cell division throughout the body. As you might predict, it is particularly important for pregnant women to get enough folic acid to supply the rapidly growing fetus. Folic acid deficiency can also lead to a reduction in red blood cells and anemia. For folic acid to function properly, trace amounts of vitamin B_{12} are required. In the human diet, vitamin B_{12} can be obtained only from eating animal protein, so strict vegetarians require supplements of this vitamin.

Fat-Soluble Vitamins

The fat-soluble vitamins A, D, E, and K have a variety of functions (see Table 34-3). Vitamin K, for example, helps regulate blood clotting. Vitamin A deficiency can lead to reduced night vision because it is used to produce the light-capturing molecule in the retina of the eye. Vitamin D is required for normal bone formation; a deficiency can lead to rickets (see **FIG. 34-2c**). U.S. researchers have recently discovered that many adult women, particularly those with dark skin who cannot synthesize as much vitamin D in the presence of sunlight, have inadequate levels of this vitamin. Children born to vitamin D–deficient mothers are at particular risk for rickets, the prevalence of which is increasing in the U.S. Fat-soluble vitamins can be stored in body fat and may accumulate in the body over time. For this reason, high doses of certain fat-soluble vitamins (vitamin A, for example) are toxic.

The Human Body Is About Two-Thirds Water

Although a typical person can survive for weeks without food, death occurs in a few days without water. Most metabolic reactions occur in a watery solution, and water directly participates in hydrolysis reactions that break down nutrients—such as protein, carbohydrates, and fats—into simpler molecules (see Chapter 3). Water is the principal component of saliva, blood, lymph, interstitial fluid, and the cytosol within each cell. By sweating, people rely on evaporation of water to cool themselves. Water is necessary to eliminate metabolic wastes in urine, as described in Chapter 35. The average daily adult water requirement is about 10 cups (2500 milliliters), but this need can increase dramatically with exercise, heat, or low humidity. There is considerable water in the solid food of a typical diet, enough for about half of the usual daily requirement. You may recall that cellular respiration also generates water (Chapter 8); this makes up about 10% of the average needs, while the rest is obtained by drinking fluids. Excess water is excreted in urine.

Nutritional Guidelines Help People Obtain a Balanced Diet

Most people in the U.S. are fortunate to live amid an abundance of food. But the overwhelming variety of foods available in a typical U.S. supermarket and the easy availability of "fast food" can contribute to obesity and poor nutrition. To help people make informed choices, the U.S. government has recently placed nutritional guidelines called "My Pyramid" on an interactive Web site. The Web site, which provides 12 individualized sets of nutritional recommendations, is a major departure from the familiar food pyramid. Check it out by entering "mypyramid" into an internet search engine.

Still more nutritional information can be found on the labels of commercially packaged foods, which provide complete information about calorie, fiber, fat, sugar, and vitamin content (**FIG. 34-3**). Some fast-food chains also provide fliers that list nutritional information about their products.

NUTRITION FACTS

Serving Size		1 Cup (55g)
Servings Per Package		8
Amount Per Serving		
Calories 210		Calories from Fat 0
		% Daily Value*
Total Fat 0g		0%
Saturated Fat 0g		0%
Cholesterol 0mg		0%
Sodium 20mg		1%
Total Carbohydrate 46g		15%
Dietary Fiber 6g		24%
Sugars 12g		
Protein 6g		
Vitamin A*	•	Vitamin C 2%
Calcium 4%	•	Iron 18%
Thiamin 38%	•	

*Percent Daily Values are based on a 2,000 calorie diet. Your Daily Values may be higher or lower depending on your calorie needs:

	Calories	2,000	2,500
Total Fat	Less than	65g	80g
Saturated Fat	Less than	20g	25g
Cholesterol	Less than	300mg	300mg
Sodium	Less than	2,400mg	2,400mg
Total Carbohydrate		300g	375g
Dietary Fiber		25g	30g

Calories per gram:
Fat 9 • Carbohydrate 4 • Protein 4

FIGURE 34-3 Complete food labels
The U.S. government requires complete nutritional labeling of foods, as illustrated by this sample. The weight (in grams) of various nutrients—such as fat, cholesterol, and sodium—is shown as a percentage of the recommended daily allowance, assuming a 2000-Calorie diet. At the bottom of the label, the total recommended number of grams of these nutrients is listed for a 2000- and a 2500-Calorie diet.

34.2 HOW IS DIGESTION ACCOMPLISHED?

An Overview of Digestion

After a meal, you may hear your stomach gurgling and churning; these noises are generated during one of the several phases of digestion. **Digestion** is the process that physically grinds up and chemically breaks down food. The **digestive systems** of animals take in food and then digest its complex molecules into simpler molecules that can be absorbed. Material that cannot be broken down or used is then expelled from the body.

Animals eat the bodies of other organisms, but these bodies may resist becoming food. The plant body, for example, supports each cell with a wall of indigestible cellulose. Animal bodies may be covered with equally indigestible fur, scales, or feathers. In addition, the complex lipids, proteins, and carbohydrates in food do not occur in a form that can be used directly. These nutrients must be broken down before they can be absorbed and distributed to the cells of the animal that has consumed them, where they recombine in unique ways. Different types of animals acquire nutrients with various types of digestive tracts, each finely tuned to meet the challenges of a unique diet and lifestyle. Amid this diversity, however, all digestive systems must accomplish certain tasks:

1. *Ingestion.* The food is brought into the digestive tract through an opening, usually called a **mouth**.

2. *Mechanical breakdown.* The food is physically broken down into smaller pieces by gizzards or teeth, as well as by the churning action of the digestive tract. The particles produced have a greater surface area, allowing digestive enzymes to attack them more effectively.

3. *Chemical breakdown.* Particles of food are exposed to enzymes and other digestive fluids that break down large molecules into smaller subunits.

4. *Absorption.* The small subunits are transported out of the digestive cavity and into cells.

5. *Elimination.* Indigestible materials are expelled from the body.

In the following sections, we explore a few of the diverse mechanisms by which animal digestive systems accomplish these functions. From a simplified perspective, animals are machines for converting food into more animals, allowing them to renew their bodies and to reproduce. Natural selection has produced behaviors and digestive adaptations that allow animals to acquire nutrients from the varied environments they inhabit, and to take advantage of almost every conceivable food source.

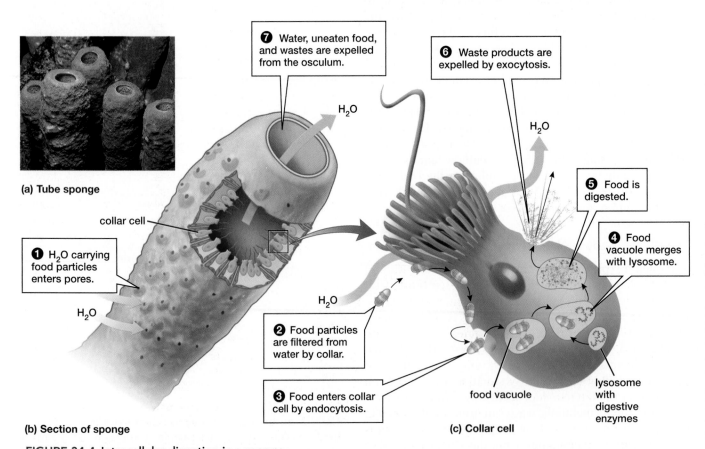

FIGURE 34-4 Intracellular digestion in a sponge
(a) Tube sponges photographed in the Virgin Islands. **(b)** The anatomy of a simple sponge showing the direction of water flow and the location of the collar cells. **(c)** Here, a single collar cell is enlarged to show the intracellular digestion of single-celled organisms, which are filtered from the water, trapped on the outside of the collar, engulfed, and digested.

In Sponges, Digestion Occurs Within Single Cells

Sponges are the only animals that rely exclusively on individual cells to digest their food using **intracellular digestion**. As you might suspect, this limits their food to microscopic organisms or particles. Sponges attach permanently to rocks, circulating seawater through pores in their bodies. As shown in **FIGURE 34-4**, specialized *collar cells* inside the sponge filter microscopic organisms from the water and ingest them using the process of phagocytosis ("cell eating"; see Chapter 5). Once ingested by a cell, the food is enclosed in a **food vacuole**, a space surrounded by a membrane that serves as a temporary stomach. The vacuole fuses with **lysosomes**, membrane-enclosed packets of digestive enzymes within the cell. Food is broken down within the vacuole into smaller molecules that can be absorbed into the cell cytoplasm. Undigested remnants remain in the vacuole, which eventually expels its contents back into the seawater using the process of exocytosis.

A Sac with One Opening Forms the Simplest Digestive System

Larger, more complex organisms evolved a chamber within the body where chunks of food are broken down by enzymes that act outside the cells, by a process called **extracellular digestion**. One of the simplest of these chambers is found in cnidarians, such as sea anemones, hydra, and jellyfish. These animals possess a digestive sac called a **gastrovascular cavity**, which has a single opening through which food is ingested and wastes are ejected (**FIG. 34-5**). The animal's stinging tentacles capture food (such as the tiny "water flea"—a crustacean) and escort it through the mouth into the gastrovascular cavity. Gland cells lining the cavity secrete enzymes that begin digestion of the prey. Nutritive cells lining the cavity then absorb the nutrients and engulf partly digested food particles by phagocytosis. Further digestion is intracellular, within food vacuoles in the nutritive cells. Undigested wastes are expelled back through the mouth, and so only one meal can be processed at a time.

Digestion in a Tube Allows Animals to Feed More Frequently

A saclike digestive system is unsuitable for active animals that must eat frequently, or for animals whose food offers so little nutrition that they must feed continually. The needs of these animals are met by a digestive system consisting of a one-way tube divided into a series of compartments with an opening at each end. A tubular digestive tract allows an animal to eat frequently, because outgoing wastes do not interfere with incoming food. Most animals, including all vertebrates and invertebrates such as earthworms, mollusks, arthropods, and echinoderms, have digestive systems that are basically tubes that begin with a mouth and end with an anus. Specialized regions within the tube process the food in an orderly sequence, first

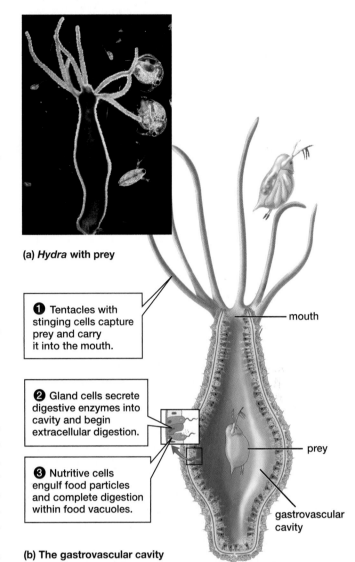

(a) *Hydra* with prey

❶ Tentacles with stinging cells capture prey and carry it into the mouth.

❷ Gland cells secrete digestive enzymes into cavity and begin extracellular digestion.

❸ Nutritive cells engulf food particles and complete digestion within food vacuoles.

mouth

prey

gastrovascular cavity

(b) The gastrovascular cavity

FIGURE 34-5 Digestion in a sac
(a) A hydra has captured and ingested a tiny crustacean. **(b)** After the processes described, undigested waste is then expelled back through the mouth.

physically grinding it up, then enzymatically breaking it down using extracellular digestion, and finally absorbing the nutrients into the body.

Digestive Specializations

Specialized tubular digestive tracts allow different types of animals to eat a wide range of foods and to extract the maximum amount of nutrients from them. **Carnivores**, such as wolves, cats, seals, and predatory birds, eat other animals. **Herbivores** eat only plants. These animals include seed-eating birds, grazing animals such as deer, camels, and cows, and many rodents such as mice. Animals such as people, bears, and raccoons are called **omnivores** because they are adapted to digest both animal and plant sources of food.

Special Adaptations Allow Ruminants to Digest Cellulose

The cellulose surrounding each plant cell is potentially one of the most abundant food energy sources on Earth; nevertheless, if people were restricted to a cow's diet of grass, we

FIGURE 34-6 The ruminant digestive system
Arrows trace the path of food through the digestive tract. QUESTION In addition to the ability to digest cellulose, what other nutritional benefits might ruminants gain by having microorganisms in their guts?

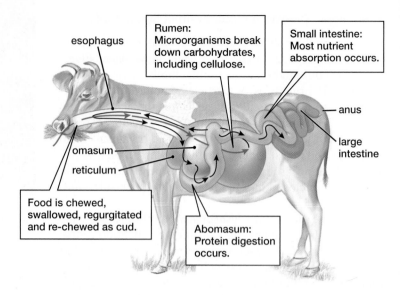

would soon starve. Although cellulose, like starch, consists of long chains of glucose molecules, because of the way bonds link the glucose molecules, it resists the attack of animal digestive enzymes. **Ruminant** animals—cows, sheep, goats, camels, and hippos, to name a few—have evolved elaborate digestive systems housing microorganisms that can break down cellulose. *Rumination,* or "cud chewing," is the process of regurgitating food and rechewing it, and it is one of several adaptations that enable these animals to digest tough plant material. Ruminant stomachs consist of several chambers (**FIG. 34-6**). The first chamber is the *rumen,* a large fermentation vat; a cow rumen can hold nearly 40 gallons (about 150 liters). This chamber is home to many species of bacteria and other types of microorganisms. In addition to digesting plant sugars and starches, these microorganisms produce **cellulase**, an enzyme that breaks down cellulose into its component sugars. After being processed in the rumen, the plant material enters the *reticulum* and is formed into masses called *cud.* The cud is regurgitated, chewed, and swallowed back into the rumen. The extra chewing exposes more of the cellulose and cell contents to the rumen's microorganisms, which digest it further. Gradually, the partially digested plant material and microorganisms are released into the remaining chambers, traveling through the narrow *omasum* and then into the larger *abomasum,* where protein digestion occurs. Here the cow digests not only plant proteins but also the microorganisms from its rumen. The cow then absorbs most of the products of digestion through the walls of its small intestine.

Small Intestine Length Is Correlated with Diet

Since most digestion and absorption of nutrients occurs in the small intestine, a longer small intestine provides herbivores with more opportunity to extract nutrients from plants, whose cell walls are difficult to break down. In general, carnivores, whose diets are mainly protein, have much shorter small intestines than herbivores do, since proteins are relatively easy to digest and protein digestion begins in

the stomach. This is strikingly illustrated during frog development. The juvenile tadpole is an algae-eating herbivore with a long intestine. When it metamorphoses into a carnivorous (usually insect-eating) adult frog, the intestine shortens to about one-third of its previous length.

Teeth Accommodate Different Diets

Teeth are adapted to diet. The varied, omnivorous diet of humans has selected for flat incisors for biting, pointed canines for tearing, premolars for grinding, and molars for crushing and chewing (**FIG. 34-7a**). If you have a dog, look carefully in its mouth. Carnivores have small incisors but greatly enlarged canines for stabbing and tearing flesh. They have a reduced set of molars and premolars with specialized sharp edges for shearing through tendon and bone (**FIG. 34-7b**). Herbivores such as horses have reduced canines, and their incisors are adapted for snipping leaves. They also have wide, flat premolars and molars that can grind up tough, cellulose-containing plants (**FIG. 34-7c**). Many herbivores have teeth that grow continuously throughout their lives, compensating for the wear.

Birds Have Gizzards That Grind Food

Birds lack teeth and swallow their food whole, after which it passes through a muscular, tubular esophagus. In seed-eating birds, the food may be stored and softened by water in a large, expandable crop. The food then passes gradually into a two-part stomach (**FIG. 34-8**). The first part secretes protein-digesting enzymes, while the second part is modified into a grinding *gizzard*—a thick-walled, muscular chamber. Many bird species swallow small, sharp-edged stones that lodge in the gizzard and act like teeth, crushing and grinding the food under pressure from the gizzard's muscular contractions. In carnivorous birds, such as owls, the gizzard is smaller, and bones, hair, and feathers of prey are trapped in the gizzard and regurgitated. The small intestine of birds receives digestive secretions from the pancreas and liver, and much digestion occurs here. Nutrients are also absorbed through the small intestine. The large in-

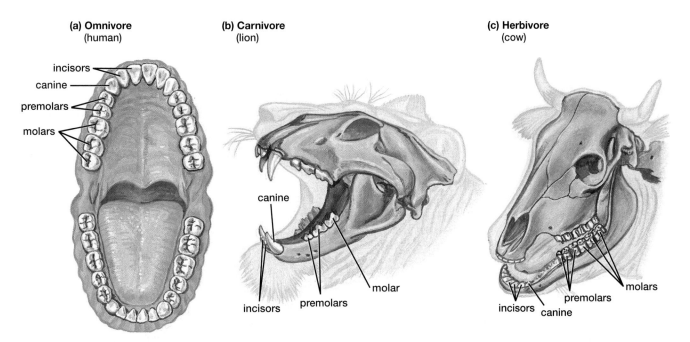

(a) Omnivore
(human)

incisors
canine
premolars
molars

(b) Carnivore
(lion)

canine

incisors   premolars   molar

(c) Herbivore
(cow)

molars

incisors   premolars
canine

FIGURE 34-7 Teeth evolved to suit different diets

testine is extremely short and empties into the *cloaca*, a multipurpose chamber that serves the urinary, reproductive, and digestive systems.

34.3 HOW DO HUMANS DIGEST FOOD?

The human digestive system (**FIG. 34-9**), which is adapted for processing the wide variety of foods in our omnivorous diet, provides a good example of the mammalian digestive system. Food travels in a continuous tube from mouth to anus; along the way it is subjected to a precisely orchestrated succession of digestive operations. By the time its passage is complete, the food has been chopped, mashed, mixed, churned, and bathed in a series of powerful chemicals. Nearly everything of nutritional value has been extracted and absorbed, and the residue is ejected. This sequential breakdown of food requires coordinated action from the variety of structures that make up the digestive system.

Mechanical and Chemical Breakdown of Food Begins in the Mouth

You take a bite, your mouth "waters," and you begin chewing. This begins both the mechanical and the chemical breakdown of food. In people and other mammals, the mechanical work is done mostly by teeth. *Incisors* at the front of the mouth snip off pieces of food, the pointed *canine* teeth beside them are useful for tearing the pieces apart, and the *premolars* and *molars* at the back of the mouth have flat surfaces for grinding food to a paste (see Fig. 34-7). In adult humans, 32 teeth of varying shapes and sizes cut and grind food into small pieces.

While the teeth pulverize the food, the first phase of chemical digestion occurs as three pairs of salivary glands

pour out saliva in response to the smell, feel, taste, and (if you're hungry) even the thought of food. Altogether, your salivary glands will produce about 1 to 1.5 quarts (about 1.0–1.5 liters) of saliva daily. Saliva contains the digestive enzyme **amylase**, which begins the breakdown of starches into sugar (Table 34-4). Saliva has other functions as well. It contains a bacteria-killing enzyme and antibodies that help guard against infection. Saliva also lubricates the food to facilitate swallowing and dissolves some food molecules

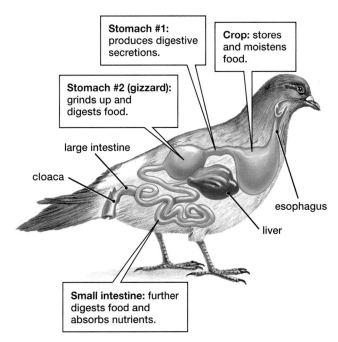

Stomach #1: produces digestive secretions.

Crop: stores and moistens food.

Stomach #2 (gizzard): grinds up and digests food.

large intestine

cloaca

esophagus

liver

Small intestine: further digests food and absorbs nutrients.

FIGURE 34-8 Bird digestive adaptations
QUESTION How can birds grind their food without teeth?

FIGURE 34-9 The human digestive tract

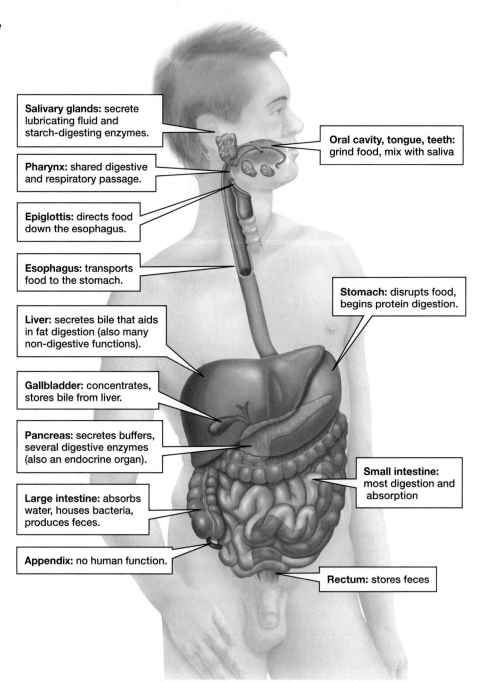

Salivary glands: secrete lubricating fluid and starch-digesting enzymes.

Pharynx: shared digestive and respiratory passage.

Epiglottis: directs food down the esophagus.

Esophagus: transports food to the stomach.

Liver: secretes bile that aids in fat digestion (also many non-digestive functions).

Gallbladder: concentrates, stores bile from liver.

Pancreas: secretes buffers, several digestive enzymes (also an endocrine organ).

Large intestine: absorbs water, houses bacteria, produces feces.

Appendix: no human function.

Oral cavity, tongue, teeth: grind food, mix with saliva

Stomach: disrupts food, begins protein digestion.

Small intestine: most digestion and absorption

Rectum: stores feces

Table 34-4 Digestive Secretions and Their Sources

Site of Digestion	Secretion	Source of Secretion	Role in Digestion
Mouth	Salivary amylase	Salivary glands	Breaks down starch into disaccharides
	Mucus, water	Salivary glands	Lubricates, dissolves food
Stomach	Hydrochloric acid	Cells lining stomach	Allows pepsin to work, kills some bacteria, aids in mineral absorption
	Pepsin	Cells lining stomach	Breaks down proteins into large peptides
	Mucus	Cells lining stomach	Protects stomach from digesting itself
Small intestine	Sodium bicarbonate	Pancreas	Neutralizes acidic chyme from stomach
	Pancreatic amylase	Pancreas	Breaks down starch into disaccharides
	Protease	Pancreas	Breaks down proteins into large peptides
	Lipase	Pancreas	Breaks down lipids into fatty acids and glycerol
	Bile	Liver	Emulsifies lipids
	Peptidases	Small intestine	Split small peptides into amino acids
	Disaccharidases	Small intestine	Split disaccharides into monosaccharides
	Mucus	Small intestine	Protects intestine from digestive secretions

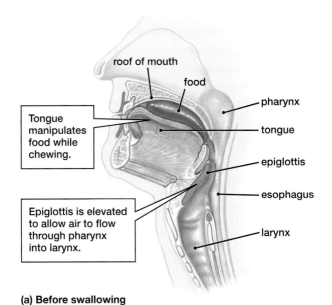

Tongue manipulates food while chewing.

Epiglottis is elevated to allow air to flow through pharynx into larynx.

roof of mouth
food
pharynx
tongue
epiglottis
esophagus
larynx

(a) Before swallowing

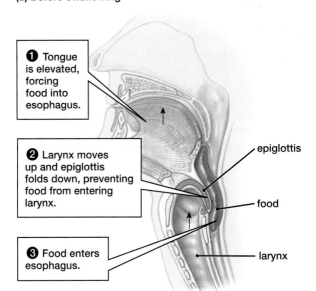

❶ Tongue is elevated, forcing food into esophagus.

❷ Larynx moves up and epiglottis folds down, preventing food from entering larynx.

❸ Food enters esophagus.

epiglottis
food
larynx

(b) During swallowing

FIGURE 34-10 The challenge of swallowing
(a) Swallowing is complicated by the fact that both the esophagus (part of the digestive system) and the larynx (part of the respiratory system) open into the pharynx. **(b)** During swallowing, the larynx moves upward beneath a small flap of cartilage, the epiglottis. The epiglottis folds down over the larynx, sealing off the opening to the respiratory system and directing food down the esophagus instead.

such as acids and sugars, carrying them to *taste buds* on the tongue. The taste buds bear sensory receptors that help identify the type and quality of the food.

With the help of the muscular tongue, the food is manipulated into a mass and pressed backward into the **pharynx**, a muscular cavity connecting the mouth with the esophagus (**FIG. 34-10a**). The pharynx also connects the nose and mouth with the larynx that leads to the trachea, which conducts air to the lungs. This arrangement occasionally causes

problems, as anyone who has ever choked on a piece of food can attest. Normally, however, the swallowing reflex (triggered by food entering the pharynx) elevates the larynx so it meets the **epiglottis**, a flap of tissue that blocks off the respiratory passages, directing food into the esophagus (**FIG. 34-10b**).

The Esophagus Conducts Food to the Stomach

Swallowing forces food into the esophagus, a muscular tube that propels food from the mouth to the stomach. Mucus secreted by cells that line the esophagus helps protect it from abrasion and also lubricates the food during its passage. Muscles surrounding the esophagus produce a wave of contraction that begins just above the swallowed mass and progresses down the esophagus, forcing the food toward the stomach. This muscular action, called **peristalsis**, occurs throughout the digestive tract (**FIG. 34-11**), conducting food products through the esophagus, stomach, intestines, and finally out through the anus. Peristalsis is so effective that a person can actually swallow when upside down.

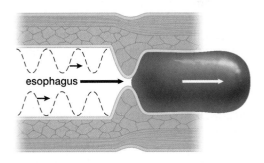

esophagus

FIGURE 34-11 Peristalsis

The **stomach** in humans is an expansible muscular sac. Its comfortable capacity is about a quart (1 liter) in an adult, but this varies with body size. Food is retained in the stomach by two rings of circular muscle (*sphincters*). The sphincter at the top, called the *lower esophageal sphincter* (**FIG. 34-12**), keeps food and acid stomach secretions from sloshing up into the esophagus while the stomach churns. If the lower esophageal sphincter weakens, *acid reflux* can occur as stomach acids enter the esophagus and attack its unprotected lining. A second sphincter, the *pyloric sphincter*, separates the lower portion of the stomach from the upper small intestine. This muscle regulates the passage of food into the small intestine, as described later.

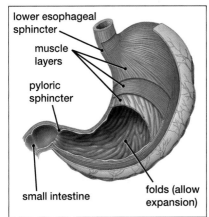

lower esophageal sphincter
muscle layers
pyloric sphincter
small intestine
folds (allow expansion)

FIGURE 34-12 The stomach

The stomach has three major functions. First, it stores food and releases it gradually into the small intestine at a rate suitable for proper digestion and absorption. Folds in the stomach wall (see Fig. 34-12) allow it to expand so we can eat rather large, infrequent meals. Carnivores carry this ability to an extreme. A lion, for instance, may consume about 40 pounds (18 kilograms) of meat at one meal, and then spend the next few days quietly digesting it. A second function of the stomach is to break down food mechanically; its muscular walls produce a variety of churning contractions that disrupt large pieces of food.

The third function of the stomach is the chemical breakdown of food. The stomach lining is packed with cell clusters called *gastric glands* that include several types of cells. Gastric glands secrete pepsinogen, hydrochloric acid (HCl), and mucus. The stomach is where protein digestion begins. *Pepsinogen* is the inactive form of the protein-digesting enzyme *pepsin*. Pepsin is a **protease**, an enzyme that helps break proteins into shorter chains of amino acids called *peptides* (see Table 34-4). Pepsin is secreted in the form of pepsinogen to prevent it from digesting the very cells that produce it. Hydrochloric acid, which gives the fluid in the stomach a very acidic pH of 1 to 3, converts the pepsinogen into pepsin, which functions best in this acidic environment. Stomach acid also promotes absorption of calcium and iron, and it kills many bacteria that are present on food.

As you may have noticed, the stomach produces all the ingredients necessary to digest itself. Indeed, this is what happens when a person develops ulcers (see "Health Watch: Ulcers: Digesting the Digestive Tract"). However, mucus released by gastric gland cells coats the stomach walls, providing a protective barrier to self-digestion. The protection is not perfect, however, because the cells lining the stomach are digested to some extent and must be replaced every few days. When the protective barriers are breached, an ulcer can occur.

Food in the stomach is gradually converted to a thick, acidic liquid called **chyme**, which consists of partially digested food and digestive secretions. Peristaltic waves (about three per minute) then propel the chyme toward the small intestine, forcing about one teaspoon of chyme through the pyloric sphincter with each wave. Depending on the size of the meal and the type of food eaten, it takes 2 to 6 hours to empty the stomach completely after a meal of solid food. Churning movements of an empty stomach are felt as "hunger pangs."

Only a few substances, including some drugs and alcohol, can enter the bloodstream through the stomach wall. Because food in the stomach slows alcohol absorption, the advice "never drink on an empty stomach" is based on sound physiological principles.

Most Digestion Occurs in the Small Intestine

The **small intestine** of a human adult is about 1 inch in diameter and is the longest portion of the digestive tract, about 10 feet long in a living person (reports of 20 feet or more are based on measurements from cadavers in which all muscle tone is lost). The functions of the small intestine are to digest food into small molecules and to absorb these molecules into the bloodstream. Digestion within the small intestine is accomplished with the aid of enzymes and other digestive secretions from three sources: the liver, the pancreas, and the cells of the small intestine itself (**FIG. 34-13**). The small intestine completes carbohydrate digestion that began in the mouth and protein digestion that began in the stomach. In addition, all fat digestion occurs in the small intestine.

The Liver and Gallbladder Provide Bile, Which Helps Break Down Fats

The **liver** is perhaps the most versatile organ in the body. The liver stores fats and carbohydrates for energy, regulates blood glucose levels, synthesizes blood proteins, stores iron and certain vitamins, converts toxic ammonia (released when amino acids are metabolized) into **urea**, and detoxifies harmful substances we ingest such as nicotine and alcohol. The role of the liver in digestion is to produce *bile*, a liquid stored and concentrated in the **gallbladder** and released into the small intestine through a tube called the *bile duct* (see Figs. 34-9 and 34-13).

Bile is a complex mixture composed of **bile salts**, water, other salts, and cholesterol. Bile salts are synthesized in the liver from cholesterol and amino acids. Although they assist in fat breakdown, bile salts are not enzymes. They have a hydrophilic end that is attracted to water, and a hydrophobic end that interacts with fats. As a result, bile salts disperse fats into microscopic particles in the watery chyme, much like a dish detergent disperses fat from a fry pan. The fat particles are easily attacked by **lipases**, lipid-digesting enzymes produced mostly by the pancreas.

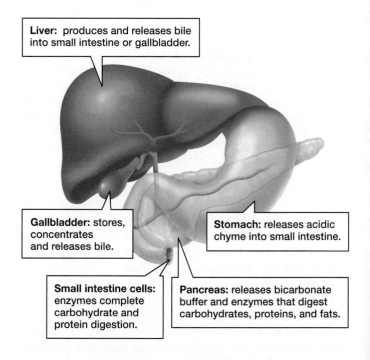

Liver: produces and releases bile into small intestine or gallbladder.

Gallbladder: stores, concentrates and releases bile.

Stomach: releases acidic chyme into small intestine.

Small intestine cells: enzymes complete carbohydrate and protein digestion.

Pancreas: releases bicarbonate buffer and enzymes that digest carbohydrates, proteins, and fats.

FIGURE 34-13 Digestion in the small intestine

HEALTH WATCH   Ulcers: Digesting the Digestive Tract

Ulcers occur when localized areas of the tissue layers that line the stomach or upper portion of the small intestine become eroded. Ulcer victims can experience burning pain in the stomach area as well as vomiting and nausea; in severe cases, blood may appear in the feces due to bleeding at the site of tissue destruction (**FIG. E34-2**). Doctors formerly believed that ulcers were caused mainly by overproduction of acid (thought to be stress-related), and they treated their patients with antacids and stress-reduction programs. Now, however, the Centers for Disease Control report that the bacterium *Helicobacter pylori* causes about 80% to 90% of all ulcers and that appropriate antibiotics (used in conjunction with acid-reducing medications) can cure most ulcers. What caused this change in our understanding of ulcers? In 1983 J. R. Warren, an Australian pathologist, noticed that samples of inflamed stomach tissue were consistently infected with a spiral-shaped bacterium. He worked with Barry Marshall, a trainee in internal medicine, to isolate and culture the bacterium, later named *Helicobacter pylori*. The researchers proposed that *H. pylori* causes the inflammation that may lead to ulcers, but the medical community was skeptical. How could bacteria survive, much less flourish, in the acidic, protein-digesting environment of the stomach? To prove their point, Marshall and another volunteer swallowed a batch of the bacteria and later provided samples of their own *H. pylori*–infected stomach tissue. More research and epidemiological studies supported Warren and Marshall's hypothesis, and these researchers were awarded the Nobel Prize for Physiology or Medicine in 2005.

Scientists now know that *Helicobacter pylori* colonize the protective layer of mucus coating the stomach wall and upper intesitinal walls. In the process, these bacteria weaken the mucus layer and increase stomach acid production, making the stomach and upper intestinal lining more susceptible to attack by acid and protein-digesting enzymes. The body's immune response to the infection also contributes to the destruction of tissue. Interestingly, although about half of the world's population harbors *H. pylori*, most infected people do not have ulcers or other obvious symptoms of infection.

Some ulcers are caused by other factors, including prolonged use of painkillers such as aspirin or ibuprofen, which interfere with the mechanisms that protect the stomach and intestinal cells from acids and digestive enzymes. Other factors that may aggravate ulcers and slow their healing include smoking, caffeine, and possibly alcohol.

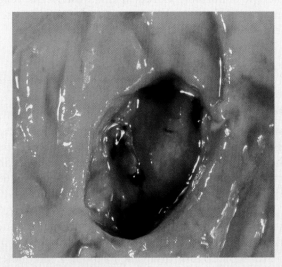

FIGURE E34-2 An ulcer
This ulcer was photographed through a fiber-optic viewing device called an endoscope.

The Pancreas Supplies Several Digestive Secretions to the Small Intestine

The **pancreas** lies in the loop between the stomach and small intestine (see Fig. 34-10). It consists of two major types of cells. One type produces hormones involved in blood-sugar regulation (described in Chapter 37), and the other produces a digestive secretion called **pancreatic juice**, which is released into the small intestine. Pancreatic juice neutralizes the acidic chyme and digests carbohydrates, lipids, and proteins. About 1 quart (1 liter) of pancreatic juice is released into the small intestine each day. This secretion contains water, sodium bicarbonate, and several digestive enzymes (see Table 34-4). Sodium bicarbonate (the active ingredient in baking soda) neutralizes the acidic chyme in the small intestine, producing a slightly basic pH. In contrast to the stomach's digestive enzymes, which require an acidic pH, pancreatic digestive enzymes require a more basic (alkaline) pH to function properly.

The pancreatic digestive enzymes break down three major types of nutrients (see Table 34-4). Pancreatic amylase breaks down carbohydrates, lipases attack lipids, and several proteases break down proteins and peptides.

The Digestive Process Is Completed by Cells of the Intestinal Wall

The wall of the small intestine is studded with cells that are specialized to complete the digestive process and absorb the small molecules that result. These cells have various enzymes on their external membranes, which form the lining of the small intestine. The enzymes include *peptidases*, which complete the breakdown of peptides into amino acids, and *disaccharidases*, which break down disaccharides into monosaccharides (see Chapter 3). For example, the disaccharidase known as *lactase* breaks down lactose (milk sugar) into the monosaccharides glucose and galactose. Because these enzymes are actually embedded in the membranes of the cells that line the small intestine, this final phase of digestion occurs *as* the nutrients are being absorbed into the epithelial cells. Like the stomach, the small intestine is protected from digesting itself by large amounts of mucous secretions from specialized cells in its lining.

You may know someone who has **lactose intolerance** and who may experience bloating, cramping, and diarrhea after consuming milk. In fact, many mammals—including most of the world's human population—make large amounts of lactase as infants, when milk is their primary food, but produce very little as adults. Consider that milk is not available to adult mammals, and people only began having access to milk after they domesticated other mammals. People of northern European and western European descent are rather unusual in that most continue to secrete lactase through adulthood.

Most Absorption Occurs in the Small Intestine

The Intestinal Lining Provides a Huge Surface Area for Absorption

The small intestine is not only the principal site of chemical digestion but also the major site of nutrient **absorption** into the blood. The small intestine has numerous folds and projections, giving it an internal surface area that is 600 times that of a smooth tube of the same length (**FIG. 34-14a**). Minute, fingerlike projections called **villi** (from Latin meaning "hairs"; singular, villus) cover the entire folded surface of the intestinal wall (**FIG. 34-14b**). Villi, which are about 1/25th of an inch (about 1 millimeter) long, make the intestinal lining appear velvety to the naked eye. Villi move gently back and forth in the chyme

as it passes through the intestine, increasing their exposure to the molecules to be digested and absorbed. The epithelial cells that coat each villus bear fringes of microscopic projections called **microvilli** ("tiny hairs"; **FIG. 34-14d**). Taken together, these specializations of the lining of the small intestine give it a surface area of about 2700 square feet (about 250 square meters), almost the size of a tennis court.

Unsynchronized contractions of the circular muscles of the intestine, called **segmentation movements**, slosh the chyme back and forth, bringing nutrients into contact with the absorptive surface of the small intestine. When absorption is complete, coordinated peristaltic waves conduct the leftovers into the *large intestine*.

Nutrients Are Transported Through the Intestinal Wall in Several Ways

Nutrients absorbed by the small intestine include water, monosaccharides, amino acids and short peptides, fatty acids produced by lipid digestion, vitamins, and minerals. The mechanisms by which this absorption occurs are varied; many nutrients move using carrier proteins, either by active transport or facilitated diffusion (described in Chapter 5).

Each villus of the small intestine is provided with a rich supply of blood capillaries and a single lymph capillary, called a **lacteal**, to carry off the absorbed nutrients and distribute them throughout the body (**FIG. 34-14c**). Most of the nutrients pass through the cells lining the small intestine and enter the bloodstream through the capillaries in each villus, but glycerol and fatty acids (produced when lipase attacks fat) take a different route. After diffusing into intestinal cells, these fat subunits are assembled into particles called *chylomicrons*, consisting of triglycerides, choles-

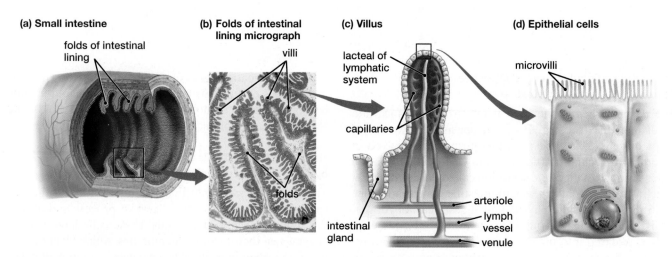

(a) Small intestine
folds of intestinal lining

(b) Folds of intestinal lining micrograph
villi
folds

(c) Villus
lacteal of lymphatic system
capillaries
intestinal gland
arteriole
lymph vessel
venule

(d) Epithelial cells
microvilli

FIGURE 34-14 The structure of the small intestine
(a) Visible folds in the intestinal lining are carpeted with tiny projections called villi **(b)** seen in this micrograph projecting from the folded membrane. **(c)** Each villus contains a network of capillaries and a central lymph vessel called a lacteal. Most digested nutrients enter the capillaries, but fats enter the lacteal. **(d)** The plasma membranes of the epithelial cells covering each villus bear microvilli. QUESTION What might the anatomy of the digestive system be like if the folds, villi, and microvilli of the small intestine had not evolved?

FIGURE 34-15 Nerves and hormones influence the digestive system

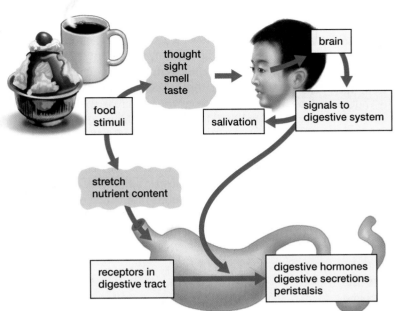

terol, and protein, which are then released inside the villus. The chylomicrons, too large to enter capillaries, enter the lacteal, whose walls are more porous. From the lacteal they are carried within the lymphatic system, which eventually empties into a large vein near the heart, as described in Chapter 32.

Water Is Absorbed and Feces Are Formed in the Large Intestine

The **large intestine** in an adult human is about 5 feet long and about 3 inches in diameter, so it is both wider and shorter than the small intestine. The large intestine has two parts: for most of its length it is called the **colon**, but its final 6-inch compartment is called the **rectum**. Into the large intestine flow the leftovers of digestion—the indigestible cell walls from fruits and vegetables, small amounts of unabsorbed protein and fat, and some remaining nutrients, including water. These support a flourishing population of intestinal bacteria (although, among mammals, only ruminants harbor intestinal microorganisms

that can digest cellulose). The bacteria of the large intestine earn their keep by synthesizing vitamin B_{12}, thiamin, riboflavin, and vitamin K. A typical human diet would be deficient in vitamin K without these helpful bacteria. Cells lining the large intestine absorb these vitamins as well as leftover water and salts.

After absorption is complete, any remaining material is compacted into semisolid **feces**. Feces consist of water, indigestible wastes, some leftover nutrients, some breakdown products of red blood cells, and the dead bodies of bacteria (bacteria account for about one-third of the dry weight of feces). The feces are transported by peristaltic movements until they reach the rectum. Expansion of this chamber stimulates the urge to defecate. The anal opening is controlled by two sphincter muscles, an inner one that is involuntary and an outer muscle that can be consciously controlled. Although defecation is a reflex (as any new parent can attest), it comes under voluntary control in early childhood.

Digestion Is Controlled by the Nervous System and Hormones

Imagine that a waiter places a chef salad in front of you, and you hungrily begin to eat. Without conscious control, your body coordinates a complex series of events that converts the salad into nutrients circulating in your blood. Not surprisingly, the secretions and muscular activity of the digestive tract are coordinated by both nerves and hormones (**FIG. 34-15** and Table 34-5).

Table 34-5 Some Important Digestive Hormones

Hormone	Site of Production	Stimulus for Production	Effect
gastrin	stomach	Peptides and amino acids in stomach	Stimulates acid secretion by cells in stomach
secretin	small intestine	Acid in small intestine	Stimulates bicarbonate production by pancreas; increases bile output by liver
cholecystokinin	small intestine	Amino acids, fatty acids in small intestine	Stimulates secretion of pancreatic enzymes and release of bile by gallbladder
gastric inhibitory peptide	small intestine	Fatty acids and sugars in small intestine	Inhibits stomach movements and release of stomach acid

Food Triggers Nervous System Responses

The sight, smell, taste, and sometimes the thought of food generate signals from the brain that act on salivary glands and many other parts of the digestive tract, preparing it to digest and absorb food. For example, these nerve impulses cause the stomach to begin secreting acid and protective mucus. As food enters and moves through the system, its bulk stimulates local nervous reflexes that cause peristalsis and segmentation movements.

Hormones Help Regulate Digestive Activity

Four major hormones are secreted by the digestive system. These all enter the bloodstream and circulate through the body, acting on specific receptors within the digestive tract. Like most hormones, they are regulated by negative feedback. For example, nutrients in chyme such as amino acids and peptides from protein digestion stimulate cells in the stomach lining to release the hormone **gastrin** into the bloodstream. Gastrin travels back to the stomach cells and stimulates further acid secretion, which promotes protein digestion. When the pH of the stomach reaches a low level (high acidity), this inhibits gastrin secretion, which in turn inhibits further acid production (**FIG. 34-16**).

Gastrin also stimulates muscular activity of the stomach, which helps it break down food and send chyme into the small intestine.

Three hormones are released by cells of the upper small intestine in response to chyme. Together, these help control both the chemical environment inside the small intestine and the rate at which the chyme enters, promoting optimal digestion and absorption of nutrients. The hormones **secretin** and **cholecystokinin** stimulate release of digestive fluids into the small intestine: bicarbonate and digestive enzymes from the pancreas and bile from the liver and gallbladder. **Gastric inhibitory peptide** is produced in response to the presence of fatty acids and sugars in chyme. This hormone stimulates the pancreas to release the hormone insulin, which helps body cells absorb sugar, into the bloodstream. Gastric inhibitory peptide (as its name suggests) also inhibits both acid production and peristalsis in the stomach. As a result, it slows the rate at which chyme is pumped into the small intestine, providing additional time for digestion and absorption to occur.

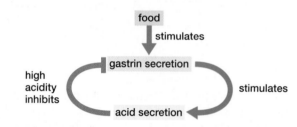

FIGURE 34-16 Negative feedback controls stomach acidity

CASE STUDY REVISITED DIETING TO DEATH?

Eating disorders, as you might predict, can cause severe malnutrition, and in the most extreme cases, death from starvation. Imbalances in blood levels of various salts can interfere with heart muscle contractions, causing death from cardiac arrest. Eating disorders also play havoc with the digestive system; most damage occurs because of frequent vomiting and use of laxatives. The protective enamel of teeth is eroded by repeated exposure to the strong acid in the vomited stomach contents. Stomach acid eats away at tissues of the throat, gums, and esophagus as well. The explosive pressure of vomiting can cause small tears or (in extreme cases) rupture of the esophagus—a true medical emergency. Repeated vomiting also damages the stomach lining. Overuse of laxatives causes the large intestine to become reliant on them, producing constipation if they are withdrawn.

The causes of eating disorders are numerous and poorly understood. Genes

apparently play a role; people with a close relative who has an eating disorder are about five times as likely to develop such a disorder themselves. Mental problems (such as anxiety and depression) and personality traits (such as perfectionism, low self-esteem, and a high need for acceptance and achievement) seem to predispose people to eating disorders. Many cases begin during adolescence, when bodies and brains are undergoing rapid changes. As TV, magazines, and Internet ads bombard susceptible young people with the message that being skinny is a route to acceptance, beauty, and riches, their attempts to meet these impossible standards sometimes spiral out of control into eating disorders.

Unfortunately, these disorders are difficult to treat. Victims are usually given nutritional therapy that may include hospitalization to help them recover from malnutrition. Psychotherapy is usually necessary, and antidepressant drugs are

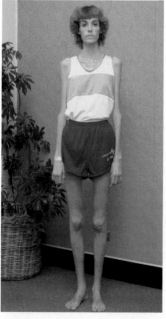

FIGURE 34-17 A 17 year old anorexia victim

helpful in some cases. Because many victims hide or deny their problems, and because treatment is expensive, the majority of sufferers are inadequately treated, so that only about half of anorexia victims report complete recovery. For Carré Otis, the wake-up call came when, at age 30, she required surgery to repair her heart, damaged by years of malnutrition. "I needed to make a change, or clearly my body would not hold up. At that moment, I finally realized how out of control I was, and knew that I was not ready to die. I was ready to embark on the road to recovery."

Consider This Media that glamorize slimness have been blamed for the prevalence of eating disorders. Why do you think extreme thinness is made so appealing? Are there appropriate measures that a free society can take to reverse or limit this message?

CHAPTER REVIEW

SUMMARY OF KEY CONCEPTS

34.1 What Nutrients Do Animals Need?
Each type of animal has specific nutritional requirements. These requirements include molecules that can be broken down to liberate energy, such as lipids, carbohydrates, and proteins; chemical building blocks used to construct complex molecules, such as amino acids that can be linked together to form proteins; and minerals and vitamins that facilitate the diverse chemical reactions of metabolism.

Web Tutorial 34.1 The Digestion and Absorption of Food

34.2 How Is Digestion Accomplished?
Digestive systems must accomplish five tasks: ingestion, mechanical and chemical breakdown of food, absorption, and elimination of wastes. Digestive systems convert the complex molecules of the animal and plant bodies that have been eaten into simpler molecules that can be used. Animal digestion at its simplest is intracellular, as occurs within the individual cells of a sponge. Extracellular digestion, used by more complex animals, occurs within a body cavity. The simplest form is a saclike gastrovascular cavity in organisms such as *Hydra*. Still more complex animals use a tubular compartment with specialized chambers in which food is processed in a well-defined sequence.

34.3 How Do Humans Digest Food?
In humans, digestion begins in the mouth, where food is physically broken down by chewing and chemical digestion is initi-

ated by saliva. Food is then conducted to the stomach by peristaltic waves of the esophagus. In the acidic environment of the stomach, food is churned into smaller particles, and protein digestion begins. Gradually, the liquefied food, now called chyme, is released into the small intestine. There, it is neutralized by sodium bicarbonate from the pancreas. Secretions from the pancreas and liver and from the cells of the intestine itself complete the breakdown of proteins, fats, and carbohydrates. In the small intestine, the simple molecular products of digestion are absorbed into the bloodstream for distribution to the body cells. The large intestine absorbs the remaining water and converts indigestible material to feces, which are temporarily stored in the rectum and eliminated through the anus.

Digestion is regulated by the nervous system and hormones. The smell and taste of food and the action of chewing trigger the secretion of saliva in the mouth and the production of gastrin by the stomach. Gastrin stimulates stomach-acid production. As chyme enters the small intestine, three additional hormones are produced by intestinal cells: secretin, which causes sodium bicarbonate production to neutralize the acidic chyme; cholecystokinin, which stimulates bile release and causes the pancreas to secrete digestive enzymes into the small intestine; and gastric inhibitory peptide, which inhibits acid production and peristalsis by the stomach. This inhibition slows the movement of food into the intestine.

KEY TERMS

absorption *page 700*
amylase *page 695*
bile *page 698*
bile salt *page 698*
body mass index *page 687*
calorie *page 686*
Calorie *page 686*
carnivore *page 693*
cellulase *page 694*
cholecystokinin *page 702*
chyme *page 698*
colon *page 701*
digestion *page 692*
digestive system *page 692*
epiglottis *page 697*

essential amino acid *page 688*
essential fatty acid *page 686*
extracellular digestion *page 693*
feces *page 701*
food vacuole *page 693*
gallbladder *page 698*
gastric inhibitory peptide *page 702*
gastrin *page 702*
gastrovascular cavity *page 693*
glycogen *page 688*
herbivore *page 693*

intracellular digestion *page 693*
lacteal *page 700*
lactose intolerance *page 700*
large intestine *page 701*
lipase *page 698*
liver *page 698*
lysosome *page 693*
microvillus *page 700*
mineral *page 688*
mouth *page 692*
nutrient *page 686*
omnivore *page 693*
pancreas *page 699*
pancreatic juice *page 699*

peristalsis *page 697*
pharynx *page 697*
protease *page 698*
rectum *page 701*
ruminant *page 694*
secretin *page 702*
segmentation movement *page 700*
small intestine *page 698*
stomach *page 697*
urea *page 698*
villi *page 700*
vitamin *page 688*

THINKING THROUGH THE CONCEPTS

1. List four general types of nutrients, and describe the role of each in nutrition.

2. Describe two different types of digestive tract specializations, including their function and relationship to the animal's diet.

3. List and describe the function of the three principal secretions of the stomach.

4. Why is the stomach both muscular and expandable?

5. List the digestive substances secreted into the small intestine, and describe the origin and function of each.

6. Name and describe the muscular movements that usher food through the human digestive tract.

7. Vitamin C is an essential vitamin for humans but not for dogs. Certain amino acids are essential for humans but not for plants. Explain.

8. Name four structural or functional adaptations of the human small intestine that ensure good digestion and absorption.

9. Describe protein digestion in the stomach and small intestine.

APPLYING THE CONCEPTS

1. The food label on a soup can shows that the product contains 10 grams of protein, 4 grams of carbohydrate, and 3 grams of fat. How many Calories are in this soup?

2. Small birds have high metabolic rates, efficient digestive tracts, and high-calorie diets. Some birds consume an amount of food equivalent to 34% of their body weight every day. Why do you think birds rarely consume leaves or grass?

3. Control of the human digestive tract involves several feedback loops and messages that coordinate activity in one chamber with those taking place in subsequent chambers. List the coordinating events you discovered in this chapter, in order, beginning with tasting, chewing, and swallowing a piece of meat and ending with residue that enters the large intestine. What turns on and what shuts off each process?

4. Symbiotic protozoa in the digestive tracts of termites produce cellulase used by their hosts. In return, termites provide protozoa with food and shelter. Imagine that the human species is gradually invaded, over many generations, by symbiotic proto-

zoa capable of producing cellulase. What evolutionary adaptive changes in body structure and function might occur simultaneously?

5. Trace a ham and cheese with lettuce sandwich through the human digestive system, discussing what happens to each part of the sandwich as it passes through each region of the digestive tract.

6. One of the common remedies for constipation (difficulty eliminating feces) is a laxative solution that contains magnesium salts. In the large intestine, magnesium salts are absorbed very slowly by the intestinal wall, remaining in the intestinal tract for long periods of time. Thus, the salts affect water movement in the large intestine. Based on this information, explain the laxative action of magnesium salts.

FOR MORE INFORMATION

Blaser, M. J. "An Endangered Species in the Stomach." *Scientific American*, February 2005. Describes the benefits and problems associated with the presence of ulcer-causing bacteria (*Helicobacter pylori)* in the stomach.

Pennisi, E. "The Dynamic Gut." *Science*, March 25, 2005. The digestive systems of several vertebrates change to accommodate different feeding regimens.

Pennisi, E. "A Mouth Full of Microbes." *Science*, March 25, 2005. Researchers are learning more about the bacterial community of the human mouth.

Pollen, M. "Power Steer." *New York Times Magazine*, March 31, 2002. Excellent description of how the cattle industry has caused major problems—including the development of antibiotic-resistant bacteria—by ignoring the specialized ruminant digestive tract.

Raloff, J. "Still Hungry?" *Science News*, April 2, 2005. Research sheds light on "hunger hormones."

Raloff, J. "Vitamin Boost." *Science News*, October 9, 2004. New benefits of vitamin D and dangers of deficiencies are being discovered.

Trivedi, B. "Slimming for Slackers." *New Scientist*, October 1, 2005. The type of bacteria living in your intestine may determine how many calories you extract from your food.

Vogel, S. "Why We Get Fat." *Discover*, April 1999. Americans are getting fatter for many reasons—environmental, genetic, and behavioral.

CHAPTER

35

The Urinary System

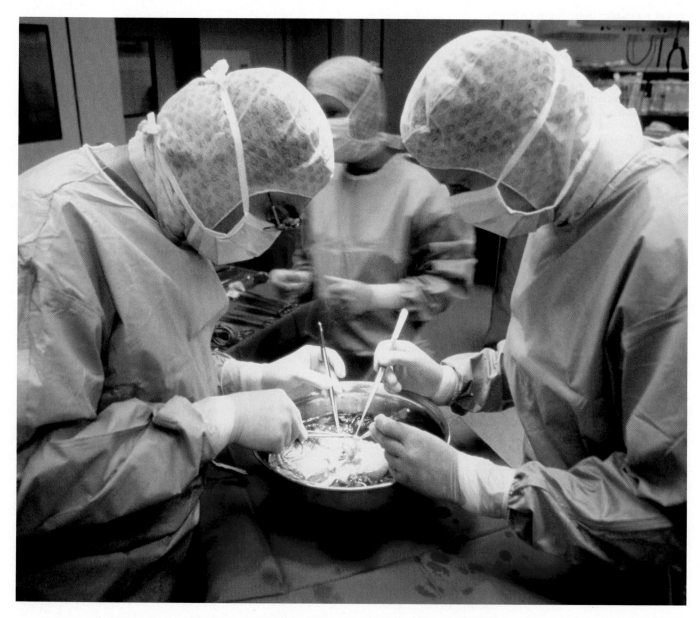

Surgeons prepare a donated kidney for transplant.

CASE STUDY A PERFECT MATCH

KAY BURT IS A SURVIVOR. Her story began in 1966—at the dawn of kidney transplantation—and now involves three generations of her family. Born with congenitally small kidneys, Kay was dying by the time she was 14 years old. The primitive dialysis machines of the time could not keep her alive, and her weight dropped to only 57 pounds. When her doctor suggested a radical procedure—kidney transplantation—her father, who was a compatible donor for Kay, unhesitatingly offered one of his. At that time, kidney transplantation was in its early stages of development; Kay's was only the fifth kidney transplant in the entire state of Texas. Her surgery was a grueling 12-hour ordeal. When the transplanted kidney began to function normally in Kay's body, it seemed like a miracle. Kay's doctor warned her never to become pregnant, fearing that the developing baby could damage the transplanted organ or its delicate connections. Despite this advice, five years later Kay gave birth to a healthy daughter, Cherry. After three decades of good health, Kay's life was threatened again in 1998. Just two weeks after her beloved father's death, his kidney—which had functioned in her body for 32 years—also failed, forcing Kay to go back on dialysis.

Each year in the United States, tens of thousands of people lose kidney function. About 16,000 will receive kidney transplants, about 67,000 patients currently await kidney transplants, and about 3500 die while awaiting a transplant each year. Over 300,000 people who have lost kidney function are kept alive by hemodialysis (often simply called "dialysis"). What is hemodialysis, and how does it work? What other "medical miracles" might help people with urinary system problems in the future? How have Kay and her family dealt with her new setback?

35.1 WHAT ARE THE BASIC FUNCTIONS OF URINARY SYSTEMS?

In this chapter we explore the workings of urinary systems and discover that they do far more than just produce *urine*. The **urinary system** serves many crucial functions relating to **homeostasis** (see Chapter 31), helping maintain the chemical composition of the blood and extracellular fluid within the narrow bounds required for cellular metabolism. One critical element in homeostasis is water balance, which is crucial to maintaining the proper concentration or **osmolarity** of solutes (dissolved substances) in cells and in their extracellular environment. In the body, dissolved substances include ions such as sodium (Na^+), chloride (Cl^-), and calcium (Ca^{++}) as well as urea, sugars, amino acids, and proteins—to name a few. Solutions with higher osmolarity have more dissolved particles than do those with lower osmolarity.

As with many physiological systems, the urinary system is a master of multitasking. An important function of the urinary system is **excretion**, a general term referring to the elimination of wastes or excess substances from the body. Excretion also occurs through the respiratory system (carbon dioxide) and the digestive system (undigested material). The urinary system eliminates many waste products of cellular metabolism—for example, urea produced by amino acid breakdown, as well as excess water, excesses of certain vitamins, and some drugs.

Whether we are looking at flatworms, fishes, or people, urinary systems (often called *excretory systems*) perform similar functions using the same basic sequence of processes:

1. Either blood or interstitial fluid is filtered, removing water and small dissolved molecules.
2. Nutrients are selectively reabsorbed from the filtrate.
3. Excess water, excess nutrients, and dissolved wastes are excreted from the body.

It is important to recognize that urinary systems must be in intimate association with the **interstitial fluid**—the watery substance that bathes all cells—in order to remove wastes. In the simplest animals, this fluid is filtered directly by the excretory system. In animals with a circulatory system, the excretory system filters blood as it passes through the circulatory system, as explained in the examples that follow.

35.2 WHAT ARE SOME EXAMPLES OF INVERTEBRATE EXCRETORY SYSTEMS?

Protonephridia Filter Interstitial Fluid in Flatworms

The first specialized excretory structures to arise during animal evolution probably resembled **protonephridia** (literally, "before the nephridium"). Protonephridia consist of tubules that carry fluid and wastes to an excretory pore that empties them to the outside. Hollow cells containing beating cilia (called *flame cells* because the beating cilia resemble a flickering flame) produce a current that draws water and dissolved wastes into the tubules and directs it out through the excretory pores. Freshwater flatworms, which use a protonephridial system, must excrete large amounts of water that enters their bodies by osmosis. Flatworms lack circulatory systems, and their two protonephridia branch throughout the body, where they collect excess water and some wastes that are emptied out through numerous excretory pores (**FIG. 35-1a**). Flatworms also have a large body surface through which most cellular wastes diffuse out.

Malpighian Tubules Filter the Blood of Insects

Insects have an open circulatory system, in which blood fills a body cavity (the *hemocoel*) and bathes tissues directly. Insect excretory systems consist of **Malpighian tubules**, small tubes that extend outward from the intestine and end within the blood of the hemocoel. Wastes and salts

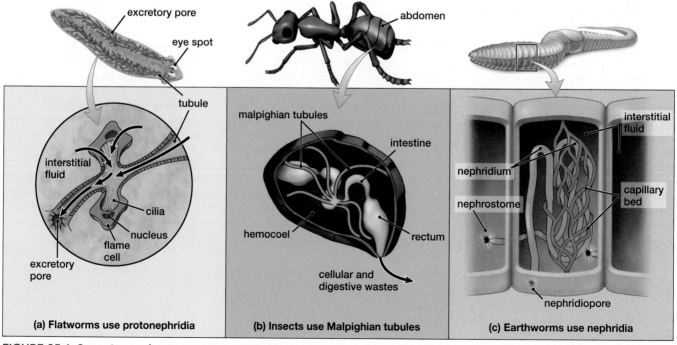

(a) Flatworms use protonephridia   (b) Insects use Malpighian tubules   (c) Earthworms use nephridia

FIGURE 35-1 Some invertebrate excretory systems

move from the surrounding blood into the tubules by diffusion and by active transport, and water follows by osmosis (FIG.35-1b). In the intestine and rectum, important salts are secreted back into the blood by active transport, and water follows by osmosis. Insects can produce very concentrated urine, which is excreted along with feces.

Nephridia Filter Coelomic Fluid in Earthworms

Earthworms, mollusks, and several other invertebrates have simple structures called **nephridia** (singular, nephridium). In the earthworm, interstitial fluid fills the body cavity (*coelom*) that surrounds the internal organs, collecting both wastes and nutrients from the blood and tissues. From the coelom, the fluid is collected by cilia into a funnel-shaped **nephrostome** and swept along a narrow, twisted tubule surrounded by capillaries (FIG. 35-1c). Salts and other nutrients are reabsorbed back into the blood within the capillaries, leaving water and wastes behind. The resulting urine is then excreted through the **nephridiopore**, an opening in the body wall. Each of the earthworm's many segments contains a pair of nephridia. As you study the kidney tubules of vertebrates, notice how they resemble nephridia.

35.3 WHAT ARE THE FUNCTIONS OF VERTEBRATE URINARY SYSTEMS?

Vertebrate excretory systems must filter wastes out of the blood while retaining nutrients and adequate water.

Vertebrate Kidneys Filter the Blood

In the **kidneys**, water and all its dissolved molecules (excluding proteins) are first filtered out of the blood. The kidneys then reabsorb water and nutrients back into the blood, leaving behind toxic substances and cellular wastes, as well as excess vitamins, salts, hormones, and water. Some additional wastes are actively transported into the remaining fluid, which becomes **urine**. The rest of the urinary system channels and stores the urine until it is expelled from the body. The vertebrate urinary system helps maintain homeostasis in several ways:

- Regulates blood levels of ions such as sodium, potassium, chloride, and calcium
- Regulates the water content of the blood
- Maintains proper pH of the blood
- Retains important nutrients such as glucose and amino acids in the blood
- Secretes hormones, such as *erythropoietin* which stimulates red blood cell production
- Eliminates cellular waste products such as *urea*

Excretion of Nitrogenous Wastes Is Adapted to the Environment

One important function of excretory systems is to eliminate cellular wastes that result from protein digestion. These are called *nitrogenous wastes* because they contain nitrogen de-

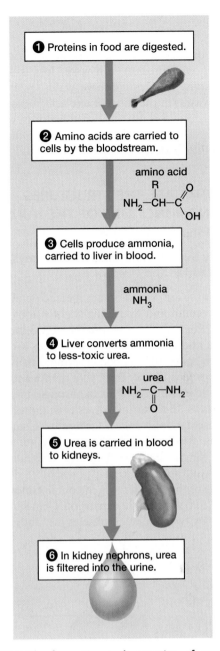

FIGURE 35-2 The formation and excretion of urea
QUESTION In some animals, ammonia is not converted to urea but is carried by the blood until it is excreted. In what kinds of environments would you expect to find such animals?

rived from the amino acids of proteins. Animals excrete nitrogenous wastes as *ammonia, urea,* or *uric acid.*

When amino acids are taken into cells, some are used directly to synthesize new proteins. Others have their amino ($-NH_2$) groups removed and are then used as a source of energy or in synthesizing other molecules. The amino groups are released as **ammonia** (NH_3), which is toxic. Fish excrete ammonia, which diffuses readily from the capillaries in their gills into the surrounding water (see Fig. 35-8a). In terrestrial vertebrates, ammonia travels in the blood to the liver, where energy-requiring reactions convert the ammonia to **urea**, a far less toxic substance (FIG. 35-2). Urea is filtered from the blood by the kidneys and excreted in urine.

Because urea is water soluble, some water must be excreted along with urea, even if water loss is undesirable. Birds and some reptiles conserve water by excreting nitrogenous wastes in the form of **uric acid**, a more complex molecule. Uric acid is not very soluble; birds, reptiles, and insects excrete it as a paste along with the feces so they lose only a little water.

35.4 WHAT ARE THE STRUCTURES AND FUNCTIONS OF THE HUMAN URINARY SYSTEM?

The Urinary System Consists of the Kidneys, Ureters, Bladder, and Urethra

Human kidneys are paired organs located on either side of the spinal column and extending slightly above the waist (**FIG. 35-3**). Each kidney is approximately 5 inches long, 3 inches wide, and 1 inch thick and resembles a kidney bean in both shape and color. Blood carrying dissolved cellular wastes enters each kidney through a **renal artery**. After the blood has been filtered, it exits through the **renal vein** (**FIGS. 35-3** and **35-4**). Urine leaves each kidney through a narrow, muscular tube called the **ureter**. Using rhythmic contractions, the ureters transport urine to the *urinary bladder*, or simply **bladder**, a hollow, muscular chamber that collects and stores urine.

The walls of the bladder, which contain smooth muscle, are capable of considerable expansion. Urine is retained in the bladder by two sphincter muscles just above the open-

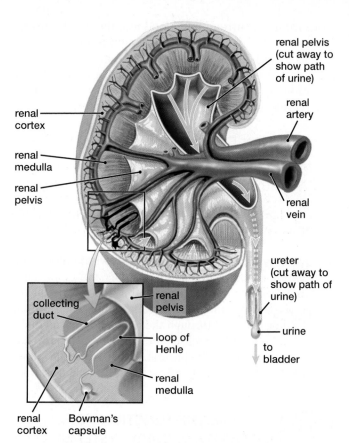

FIGURE 35-4 Cross section of a kidney
A nephron is drawn considerably larger than normal (and further enlarged in the inset) to show its location in the kidney. Collecting ducts empty the urine into channels that lead to the renal pelvis, a chamber that funnels urine into the ureter and out of the kidney.

ing through which the bladder empties into the urethra. When the bladder is stretched by stored urine, receptors in its walls respond and trigger reflexive contractions. The sphincter nearest the bladder, the *internal sphincter*, opens during this reflex. However, the lower *external sphincter* is under voluntary control, allowing the brain to suppress the reflex unless the bladder becomes overly full. The average adult bladder can hold, if necessary, about a pint (500 milliliters) of urine, but the desire to urinate is triggered by considerably smaller amounts. Urine exits the body via the **urethra**, a single narrow tube about 1.5 inches long in the female and about 8 inches long in the male (because it extends the length of the penis).

Urine Is Formed in the Nephrons of the Kidneys

Each kidney contains a solid outer layer consisting of the **renal cortex** which overlies the **renal medulla**. In this outer layer, urine is formed. Beneath the cortex and medulla lies the **renal pelvis**, a subdivided chamber that collects urine and funnels it into the ureter (see Fig. 35-4). Under the microscope, the outer layer of the kidney is revealed to contain a densely packed array of tiny individual filters, or **nephrons**, which are richly supplied with blood vessels. More than 1 million nephrons are packed into the outer layer of each human kidney. Each nephron has three major parts: the **glomerulus**, a dense knot of

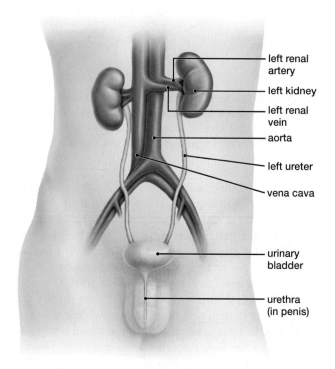

FIGURE 35-3 The human urinary system
A simplified diagram of the human urinary system and its blood supply.

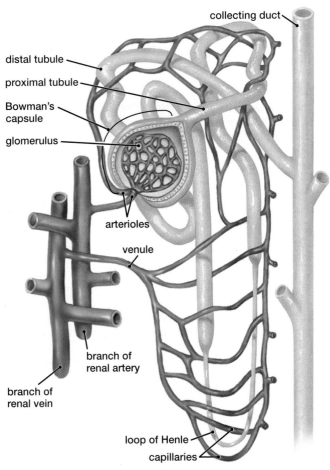

collecting duct

distal tubule

proximal tubule

Bowman's capsule

glomerulus

arterioles

venule

branch of renal artery

branch of renal vein

loop of Henle

capillaries

FIGURE 35-5 An individual nephron and its blood supply

capillaries where fluid is filtered out of the blood through the porous capillary walls; **Bowman's capsule**, a cuplike structure surrounding the glomerulus; and the long, twisted **tubule** (from the Latin, "little tube"). The tubule is further subdivided into the *proximal tubule*, the *loop of Henle*, and the *distal tubule*. **Collecting ducts** are larger tubes that collect fluid from many nephrons and conduct it from the cortex through the medulla and into the renal pelvis (**FIGS. 35-5** and **35-6**). The processes by which these structures produce urine and help maintain homeostasis are summarized in the following sections and described in more detail in "A Closer Look: The Nephron and Urine Formation."

Blood Is Filtered Through Capillaries in the Glomerulus

Nearly one-quarter of the volume of blood pumped by each heartbeat travels through the kidneys, which receive more than one quart (about one liter) every minute. Each nephron receives blood from an arteriole that branches from the renal artery. Within Bowman's capsule, the arteriole branches further into microscopic capillaries that intertwine in a mass called the glomerulus (see Fig. 35-6). The walls of the glomerular capillaries are extremely permeable to water and small dissolved molecules but are impermeable to blood cells, fat droplets, and most large proteins, such as albumin. Beyond the glomerulus, the capillaries reunite to form an arteriole whose diameter is smaller than that of the incoming arteriole (this differs from the usual situation, in which blood flows from capillaries into venules). The difference in diameter between the incoming and outgoing arteriole creates pressure within the glomerulus that drives water, carrying many dissolved substances, through the

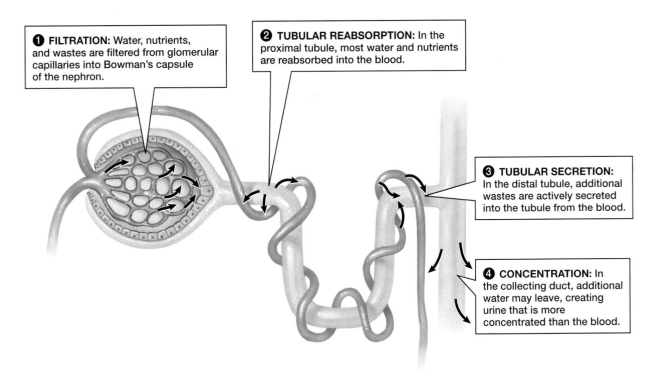

❶ FILTRATION: Water, nutrients, and wastes are filtered from glomerular capillaries into Bowman's capsule of the nephron.

❷ TUBULAR REABSORPTION: In the proximal tubule, most water and nutrients are reabsorbed into the blood.

❸ TUBULAR SECRETION: In the distal tubule, additional wastes are actively secreted into the tubule from the blood.

❹ CONCENTRATION: In the collecting duct, additional water may leave, creating urine that is more concentrated than the blood.

FIGURE 35-6 Urine formation in the nephron

A CLOSER LOOK   The Nephron and Urine Formation

The complex structure of the nephron is finely adapted to its function. In **FIGURE E35-1**, the nephron is presented in diagram form to illustrate the processes occurring in each part. The circled numbers in the illustration refer to the following descriptions.

① *Filtration.* Water and dissolved substances are forced out of the glomerular capillaries into Bowman's capsule and then funneled into the proximal tubule.

② *Tubular reabsorption.* In the proximal tubule, most nutrients remaining in the filtrate are actively pumped out through the walls of the tubule to be reabsorbed into the blood. These include salts, amino acids, sugars, and vitamins. The proximal tubule is highly permeable to water, so water accompanies the nutrients, moving out of the tubule and back into the blood by osmosis.

③ The loop of Henle (unique to birds and mammals) is essential for the concentration of urine. It maintains a salt concentration gradient in the extracellular fluid that surrounds it, with the highest concentration at the bottom of the loop. The descending portion of the loop of Henle is very permeable to water but not to salt or other dissolved substances. As the filtrate passes through the descending portion, water leaves by osmosis as the concentration of the surrounding fluid increases.

④ The thin portion of the ascending loop of Henle is relatively impermeable to water and urea but is permeable to salt, which moves out of the filtrate by diffusion. Why? Although the osmolarities inside and outside the tubule are about equal, at this portion of the loop, the urea level is higher outside and the salt level is higher inside. Thus, the concentration gradient favors the diffusion of salt outward. Because water cannot follow it, the filtrate now becomes less concentrated than its surroundings.

⑤ The thick portion of the ascending loop of Henle is also impermeable to water and urea. Here, salt is actively pumped out of the filtrate, leaving water and wastes behind.

⑥ The watery filtrate, low in salt but retaining wastes such as urea, now arrives at the distal portion of the tubule, lower in osmolarity (more dilute) than when it entered the loop. Now, more salt is pumped out, and because the distal tubule is water-permeable, water follows by osmosis. Tubular secretion is especially active in the distal tubule, where substances such as K^+, H^+, NH_3, and some drugs and toxins are actively pumped into the tubule from the extracellular fluid.

⑦ By the time the filtrate reaches the collecting duct, very little salt is left, and about 99% of its water has been reabsorbed into the bloodstream. Because it has lost both salt and water, the fluid now has about the same osmolarity as when it entered the loop of Henle. The collecting duct conducts the urine through the increasing concentration gradient in the extracellular fluid created by the loop of Henle. The collecting duct is very permeable to water when antidiuretic hormone (ADH) is present, allowing water to move out by osmosis as the concentration of the external fluid increases. If ADH is absent, the collecting duct remains impermeable to water, and the urine remains dilute and watery.

⑧ The lower portion of the collecting duct is also permeable to urea. Therefore, as the filtrate moves farther down the collecting duct, some urea diffuses out, increasing the osmolarity of the surrounding fluid. If ADH is present, water moves out as well. In the most extreme case, the osmolarity of urine in the collecting duct can reach equilibrium with the high osmolarity of the external fluid; in people, this is about four times the osmolarity of the blood.

---

porous capillary walls. This process is called **filtration**, and the resulting fluid is called **filtrate** (see Fig. 35-6, step ①, and Fig. E35-1, step ①). The watery filtrate, resembling blood plasma without its proteins, is collected in Bowman's capsule and then travels through the nephron.

After exiting the glomerulus, the blood is more concentrated, containing about 20% less fluid. Beyond the glomerulus, the outgoing arteriole branches into highly porous capillaries that entwine around the length of the tubule (see Figs. 35-5 and 35-6). Blood in these surrounding capillaries is low in both nutrients and water, but these vital molecules are gradually reacquired from the filtrate. The capillaries then conduct the blood, with most of its water and nutrients restored, into venules that transport it away from the kidneys. The processes by which water and nutrients move from the filtrate back into the blood are described in the following sections.

Diseases that prevent adequate filtration of blood can cause kidney failure, resulting in the need for dialysis (see "Health Watch: When the Kidneys Collapse") or, ideally, a kidney transplant (as Kay Burt was fortunate to receive). If filtration is impaired, blood volume may increase (causing high blood pressure), the salt and pH balance is disrupted, and toxic wastes are retained in the blood.

The Filtrate Is Converted to Urine in the Tubule of the Nephron

The filtrate collected in Bowman's capsule contains a mixture of wastes and essential nutrients, including large amounts of water. The tubule of the nephron will restore the nutrients and most of the water to the blood, while retaining wastes and excess water for elimination; in this way, the balance of water and nutrients required for homeostasis is maintained. This is accomplished by two processes: *tubular reabsorption* and *tubular secretion*.

Tubular reabsorption occurs primarily in the proximal tubule. During tubular reabsorption, tubule cells use active transport to pump nutrients such as salts, amino acids,

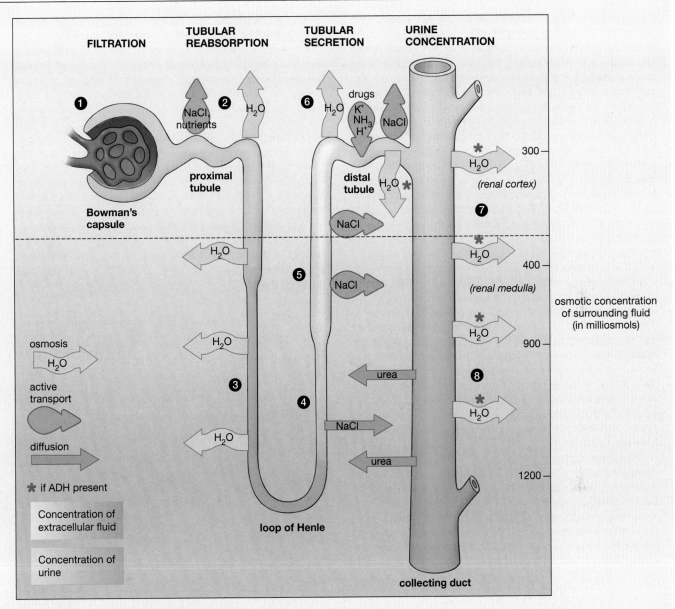

FIGURE E35-1 Details of urine formation
A single nephron, showing the movement of materials through different regions. The concentration of dissolved substances in the filtrate inside the nephron increases from top to bottom in this diagram. Outside the nephron, darker shades of blue represent higher concentrations of salt and urea in the surrounding fluid. The dashed line marks the boundary between the renal cortex and the renal medulla.

and glucose from the filtrate, out of the tubule, and into the interstitial fluid. From the interstitial fluid, the nutrients diffuse into the surrounding capillaries. Water follows the nutrients passively, moving out of the tubule and into the capillaries by osmosis (see Fig. 35-6, step ②, and Fig. E35-1, step ②). As it passes through the length of the tubule, about 99% of the water is reabsorbed.

Tubular secretion occurs primarily in the distal tubule. During tubular secretion, tubule cells use active transport to secrete wastes and excess substances into the distal tubule (see Fig. 35-6, step ③, and Fig. E35-1, step ⑥). These substances are pumped out of the interstitial fluid (which got them by diffusion from the nearby blood capillaries). Wastes secreted into the tubule for excretion include excess hydrogen and potassium ions, ammonia, and many drugs. These are added to the filtrate, which becomes urine as it leaves the distal tubule and enters the collecting duct. Unfortunately, in victims of kidney failure, these wastes

HEALTH WATCH   When the Kidneys Collapse

If the kidneys fail, death usually occurs within two weeks. Each year in the United States, about 90,000 people die because of urinary system disease. The most common causes of kidney failure are diabetes and high blood pressure, both of which damage the glomerular capillaries, but kidneys are also vulnerable to infection and overdoses of some painkilling medicines. Kidney failure is typically treated by hemodialysis.

First used in 1945, hemodialysis operates on a simple principle: substances will diffuse from areas of higher concentration to areas of lower concentration across an artificial permeable membrane. This process is called *dialysis*; therefore, the filtration of blood according to this principle is called **hemodialysis**. During hemodialysis, the patient's blood is diverted from the body and pumped through narrow tubes made of a special cellophane membrane suspended in dialyzing fluid. Like glomerular capillaries, the membrane has pores too small to permit the passage of blood cells and large proteins but large enough to pass small molecules such as water, sugar, salts, amino acids, and urea. Dialyzing fluid has normal blood levels of salts and nutrients and no waste products; so molecules whose concentrations are higher than normal in the patient's blood (such as the waste product urea) diffuse into the dialyzing fluid, which is continuously replenished to maintain the gradient. The patient usually remains attached to the dialysis machine for 4 to 6 hours, three times a week (**FIG. E35-2**). People relying on dialysis can survive for many years, but their diet and fluid intake must be carefully regulated. Even with dialysis, these patients' blood composition fluctuates, and toxic substances reach higher-than-normal levels between sessions.

Peritoneal dialysis is a technique in which dialyzing fluid is pumped through a tube implanted directly into the abdominal cavity. The abdominal cavity is lined with a natural membrane called the *peritoneum*. Waste products from blood circulating in capillaries within the peritoneum gradually diffuse into the dialysis fluid, which is then drained out through the tube. Patients can perform this procedure at home, replacing the dialysis fluid about four times daily, or can link their implanted tube to a machine that circulates the fluid through the abdominal cavity during the night.

What does the future hold for victims of urinary system diseases? At the University of Michigan, Dr. David Humes has developed a prototype "bioartificial kidney." Humes's device circulates blood through a cartridge filled with tiny tubes lined with living proximal tubule kidney cells, which are grown from cells extracted from donated human kidneys. Clinical trials using the device on intensive-care patients whose kidneys were failing demonstrated that the bioartificial kidney—used in conjunction with a hemodialysis machine—improved kidney function in patients near death from kidney failure. In the future, Humes hopes to use this technology in an implanted bioartificial kidney. Researchers are also continuing work on *xenotransplantation*—a process that would allow people to receive kidneys from animals such as pigs whose cells have been genetically modified to prevent the recipient's immune system from attacking and rejecting them.

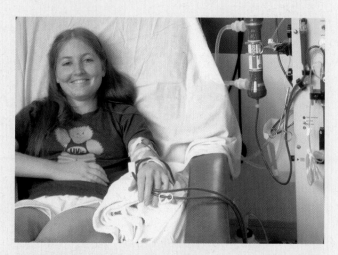

FIGURE E35-2 A patient on dialysis

can accumulate and reach toxic levels in the blood, a condition that will be fatal if not treated by dialysis or a kidney transplant.

The Loop of Henle Allows Urine to Become Concentrated

The kidneys of mammals and birds are able to produce urine that has a higher concentration of dissolved materials than their blood does. Urine can become concentrated because there is a concentration gradient of dissolved salts and urea in the interstitial fluid of the kidney medulla that surrounds the collecting ducts. This gradient is produced by the **loop of Henle**, the portion of the tubule that separates the proximal and distal regions. The loop of Henle is located within the medulla of the kidney. Loops of Henle can differ in length; the longer the loop, the larger the concentration gradient that it can produce because there is a greater length of tubule to transport salt out into the surrounding fluid (see Fig. E35-1, steps ④ and ⑤). The most concentrated fluid (with the greatest amount of dissolved substances and the least amount of water) surrounds the bottom of the loop within the medulla. As urine travels through the medulla within the collecting ducts en route to the renal pelvis (see Fig. 35-4, inset), it passes through this concentration gradient. Along the way, additional water may leave the filtrate by osmosis through the walls of the collecting duct (to be carried off by the surrounding capillaries), leaving more concentrated wastes in the urine within the collecting duct. As it moves through the collecting duct, the urine can become about as concentrated as the surrounding fluid. In humans, this is up to four times the osmolarity of the blood. Because the rest of the excretory system does not allow water to enter or urea to leave, the urine remains concentrated.

It is important to produce concentrated urine when water is scarce and to produce dilute, watery urine when there is excess water in the blood. Whether or not the urine becomes concentrated depends on the permeability of the collecting duct to water. This permeability is controlled by the amount of *antidiuretic hormone* in the blood, as we will describe shortly.

35.5 HOW DO MAMMALIAN KIDNEYS HELP MAINTAIN HOMEOSTASIS?

Every drop of blood in your body passes through a kidney about 350 times a day; in this way, the kidney is able to fine-tune the composition of the blood and thereby help maintain homeostasis. The importance of this task is illustrated by the fact that kidney failure quickly results in death.

The Kidneys Regulate the Water Content of the Blood

One important function of the kidney is to regulate the water content of the blood. Human kidneys filter out about half a cup of fluid from the blood each minute. Without reabsorption of water, a person would produce about 50 gallons of urine daily (and would need to drink continuously). Water reabsorption occurs passively by osmosis as the filtrate travels through the tubule and the collecting duct.

The amount of water reabsorbed into the blood is controlled by a negative feedback mechanism (see Chapter 31) involving the amount of **antidiuretic hormone** (**ADH**; also called *vasopressin*) circulating in the blood. Antidiuretic hormone is produced the secretory cells of the hypothalamus and released into the blood *via* the posterior pituitary gland (see Chapter 37). Release of ADH is triggered when receptor cells of the hypothalamus detect an increase in blood osmolarity, and when receptors in the heart detect a decrease in blood pressure (both are symptoms of reduced water in the blood). Antidiuretic hormone increases the permeability of both the distal tubule and the collecting duct to water, thereby allowing more water to be reabsorbed from the urine. In response to ADH binding to receptors on their plasma membranes, cells of the distal tubule and collecting duct insert *aquaporin* (Latin, "water pore") proteins into their membranes, increasing their permeability to water (aquaporins are described in Chapter 5).

Imagine yourself lost in the desert, staggering in the searing sunlight, perspiring copiously and losing water with every breath. As your blood volume drops, its osmolarity rises, triggering ADH release from your pituitary. This increases water reabsorption, and you produce concentrated urine (**FIG. 35-7**). Finally, you stumble upon an oasis with a clear spring and drink until you can hold no more. This causes your blood volume to rise and its os-

molarity to fall below normal levels, triggering a decrease in your ADH output. Reduced ADH makes your distal tubules and collecting ducts less permeable to water, so relatively little water is reabsorbed after the urine leaves the loop of Henle. Your bladder now begins to fill with urine that is more dilute than your blood. In extreme cases, urine flow can exceed one quart (about one liter) per hour. As the water content of your blood is reduced to normal, its increased osmolarity and decreased volume will stimulate some ADH release, maintaining homeostasis by keeping the blood's water content within narrow limits.

Kidneys Release Hormones That Help Regulate Blood Pressure and Oxygen Levels

When blood pressure falls, the kidneys release the protein **renin** into the bloodstream. Renin acts as an enzyme, catalyzing the formation of a second hormone, **angiotensin**, from a protein that circulates in the blood. Angiotensin, in turn, causes arterioles to constrict, elevating blood pressure. Constriction of the arterioles that carry blood to the kidneys also reduces the rate of blood filtration, causing less water to be removed from the blood. Water retention causes an increase in blood volume and, consequently, an increase in blood pressure.

In response to low blood oxygen levels, the kidneys release a second hormone, **erythropoietin** (see Chapter 32).

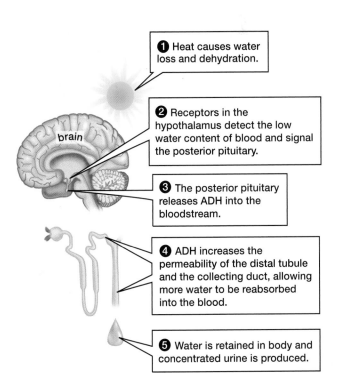

1 Heat causes water loss and dehydration.

brain

2 Receptors in the hypothalamus detect the low water content of blood and signal the posterior pituitary.

3 The posterior pituitary releases ADH into the bloodstream.

4 ADH increases the permeability of the distal tubule and the collecting duct, allowing more water to be reabsorbed into the blood.

5 Water is retained in body and concentrated urine is produced.

FIGURE 35-7 Dehydration stimulates ADH release and water retention

QUESTION Alcohol inhibits ADH release. Predict how alcohol consumption affects the body's water balance.

Erythropoietin travels in the blood and stimulates the bone marrow to produce more oxygen-transporting red blood cells. Kidney failure almost always causes anemia because the kidneys do not produce enough erythropoietin to stimulate adequate red blood cell production. Physicians may now give human erythropoietin made by genetic engineering techniques to patients who have anemia caused by kidney failure.

Kidneys Monitor and Regulate Dissolved Substances in the Blood

As the kidney filters the blood, it monitors and regulates blood composition and adjusts tubular secretion and reabsorption rates, maintaining a constant internal environment. Substances the kidney regulates, in addition to water, include nutrients such as glucose, amino acids, vitamins, urea, and various ions including sodium, potassium, chloride, and sulfate. The kidney maintains a constant blood pH by regulating the amount of hydrogen and sodium bicarbonate ions that are secreted into the urine. This remarkable organ also eliminates potentially harmful substances, including some drugs, food additives, pesticides, and toxic substances such as nicotine from cigarette smoke.

Vertebrate Kidneys Are Adapted to Diverse Environments

Freshwater and Saltwater Environments Pose Special Challenges

Animals that are constantly immersed in a solution that has a lower or a higher osmolarity than their body fluids (*hypotonic*) have evolved special mechanisms to maintain homeostasis of water and salt within their bodies. This process is called **osmoregulation**. Freshwater fish, for example, live in a hypotonic environment; their bodies maintain concentrations of dissolved substances about 4–6 times the osmolarity of their freshwater surroundings. As these fish circulate water over their gills to exchange gases, some water continuously leaks into their bodies by osmosis, while salts diffuse out. To counteract loss of salt, freshwater fish have active transport proteins in their gills that use energy to pump salt into their bodies from their dilute surroundings, against its concentration gradient. To compensate for entry of excess water, freshwater fish never drink, and their kidneys can excrete large quantities of extremely dilute urine. Their food also provides salts, which are reabsorbed by their kidneys using active transport (**FIG. 35-8a**).

Saltwater fishes, living in a *hypertonic* environment, face the opposite problem. Because they live in water with a solute concentration two to three times that of their body fluids, water is constantly leaving their tissues by osmosis, and salt is constantly diffusing in. Most saltwater fish drink continuously to restore their lost water, but drinking seawater brings in large amounts of salt. The excess salt is excreted by active transport proteins in their gills. Fish

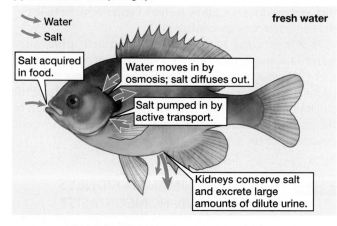

(a) Freshwater fish (bluegill)

→ Water
→ Salt

fresh water

Salt acquired in food.

Water moves in by osmosis; salt diffuses out.

Salt pumped in by active transport.

Kidneys conserve salt and excrete large amounts of dilute urine.

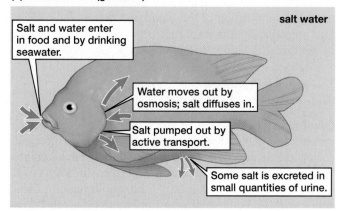

(b) Saltwater fish (garibaldi)

salt water

Salt and water enter in food and by drinking seawater.

Water moves out by osmosis; salt diffuses in.

Salt pumped out by active transport.

Some salt is excreted in small quantities of urine.

FIGURE 35-8 Osmoregulation in fishes

kidney tubules lack loops of Henle, and so fish cannot produce urine that is more concentrated than their blood. To conserve water, they excrete only small quantities of urine containing salts that their gills cannot eliminate (**FIG. 35-8b**). The continuous requirement for active transport in freshwater and saltwater fish means that in both environments, the fish use substantial amounts of energy in osmoregulation.

The class of marine fishes that includes sharks and rays takes a different approach to osmoregulation. These fishes store large quantities of urea in their tissues—enough to kill most other vertebrates. The stored urea gives their tissues approximately the same osmolarity as the surrounding seawater, so they avoid losing water by osmosis.

Loops of Henle Determine the Urine-Concentrating Ability of Mammals

Kidney structure is adapted to the availability of water in animals' natural habitats. Mammals that must conserve water produce urine that is more concentrated than their blood. Longer loops of Henle allow a higher concentration of salt and urea to be produced in the interstitial fluid of

LINKS TO LIFE   Too Much to Drink?

People who drink alcohol have observed that the volume of urine they produce afterward seems to exceed the volume of liquids they drank. Why? One of alcohol's many effects is to inhibit ADH release, and without ADH, urine remains very dilute and watery. Because alcohol causes you to excrete more urine than necessary to restore water balance, ironically, having "too much to drink" can actually dehydrate you.

Meanwhile, many people may be getting more water than they need. You may have heard advice such as "Always drink before you're thirsty," or "Drink 64 ounces of water each day in addition to other fluids." To the delight of the bottled water industry, many people are taking this serious-

ly. For healthy people on normal diets, drinking before you are thirsty is completely unnecessary—evolution has fine-tuned our desire for fluids to match our needs. Carried to an extreme, drinking too much fluid is dangerous. If the body accumulates water faster than the kidneys can eliminate it, crucial salts (particularly sodium) in the interstitial fluid become diluted and excess water moves into cells by osmosis. In the brain, excess water causes swelling that can be fatal. A study of runners in the 2002 Boston Marathon concluded that 13% of them probably drank dangerous quantities of fluid in a mistaken attempt to stay hydrated; and tragically, one 28-year-old participant died as a result. The bottom line: everything in moderation.

---

the kidney medulla, thus allowing urine to become more concentrated as it travels through the collecting ducts. As you might predict, animals living in very dry climates have the longest loops of Henle (as do marine mammals, such as whales), whereas those living in watery environments have relatively short loops.

The beaver, for example, has only short-looped nephrons and is unable to concentrate its urine to more than twice its blood osmolarity. Human kidneys have a mixture of long- and short-looped nephrons and can concentrate urine to about four times the osmolarity of blood. The masters of urine concentration are desert rodents such as kangaroo rats, which can produce urine with 14 times the osmolarity of their blood (**FIG. 35-9**). As you might predict, kangaroo rats have only very long-looped nephrons. With their extraordinary ability to conserve water, kangaroo rats do not need to drink; they rely entirely on water derived from their food and from metabolic reactions that produce water.

FIGURE 35-9 A well-adapted desert dweller
The desert kangaroo rat of the southwestern United States seldom drinks, partly because its long loops of Henle allow it to produce very concentrated urine.

CASE STUDY REVISITED A PERFECT MATCH

When Kay Burt's daughter, Cherry, realized her mother might be dependent on hemodialysis for the rest of her life, she immediately offered to donate a kidney. Kay's remarkable luck continued; her daughter, whom doctors had ordered her not to conceive, was a good match. Kay's second transplant operation, 32 years after her first, was another success; she made history by receiving transplanted kidneys from both her father and her daughter.

Kidneys are now routinely transplanted from both living and recently deceased

donors. Fortunately, the number of living donors is on the rise, and studies indicate that transplants from unrelated but matching donors have a high success rate. (An ideal match between donor and recipient is based on having the same key, genetically determined blood and tissue proteins.) Recipients of living-donor kidneys have better long-term survival rates. Some transplant centers are now using endoscopic surgery to remove kidneys from living donors. This technique, in which the surgeon views the operation through a tiny camera inserted into the body, allows the kidney to be extracted

through an incision about 2.5 inches long—far less traumatic than the 9-inch incision of traditional methods. Endoscopic surgery dramatically reduces pain, hospitalization time, and overall recovery time for the donor.

Consider This You read in your local paper that a family is looking for a kidney donor for their child. You have never met the family. Would you be willing to undergo compatibility testing, and if you were found to be a compatible donor, consider donating a kidney?

CHAPTER REVIEW

SUMMARY OF KEY CONCEPTS

35.1   What Are the Basic Functions of Urinary Systems?

All urinary (excretory) systems perform similar functions using the same basic processes. First, the blood or other fluid that bathes cells is filtered, removing water and small dissolved molecules; second, nutrients are selectively reabsorbed from the filtrate; and third, any remaining water and dissolved wastes are excreted from the body.

35.2   What Are Some Examples of Invertebrate Excretory Systems?

The flatworm's simple excretory system consists of a network of tubules that branch through the body, collecting wastes and excess water for excretion through excretory pores. Insects use Malpighian tubules that filter blood in the hemocoel of their open circulatory systems, releasing concentrated urine into the intestine for elimination. Earthworms and mollusks use nephridia, which resemble vertebrate nephrons, to filter the interstitial fluid that fills the body cavity. Wastes and excess water are released through excretory pores.

35.3   What Are the Functions of Vertebrate Urinary Systems?

Urinary systems eliminate cellular wastes and toxic substances while retaining vital nutrients in the blood. Urinary systems play a crucial role in homeostasis, regulating the water content, the ion content, and the pH of the blood. They also secrete hormones such as erythropoietin, which stimulates red blood cell production.

35.4   What Are the Structures and Functions of the Human Urinary System?

The human urinary system consists of kidneys, ureters, bladder, and urethra. Kidneys produce urine, which is conducted by the ureters to the bladder, a storage organ. Distension of the muscular bladder wall triggers urination, during which urine passes out of the body through the urethra.

Each kidney consists of more than 1 million individual nephrons in an outer layer, which consists of the renal medulla and the overlying renal cortex. Urine formed in the nephrons enters collecting ducts that travel through the medulla and empty into the renal pelvis, from which the urine is funneled into the ureter.

Each nephron is served by an arteriole branching from the renal artery. The arteriole further branches into a mass of porous-walled capillaries called the glomerulus. There, water and dissolved substances are filtered from the blood by pressure. The filtrate is collected in the cup-shaped Bowman's capsule and conducted along the tubular portion of the nephron. During tubular reabsorption, nutrients are actively pumped out of the filtrate through the walls of the tubule. Nutrients then enter capillaries surrounding the tubule, and water follows by osmosis. Some wastes remain in the filtrate; others are actively pumped into the tubule by tubular secretion. The tubule forms the loop of Henle, which creates a salt concentration gradient surrounding it. After completing its passage through the tubule, the filtrate enters the collecting duct, which passes through the concentration gradient in the kidney medulla. During this passage (if ADH is present) water leaves the urine, which can become four times as concentrated as the blood. In the absence of ADH, the urine remains dilute.

Web Tutorial 35.1 The Mammalian Kidney

35.5   How Do Mammalian Kidneys Help Maintain Homeostasis?

The kidneys are important organs of homeostasis. The water content of the blood is regulated by antidiuretic hormone (ADH), produced in the hypothalamus and released into the blood by the posterior pituitary gland. Dehydration stimulates the release of ADH, which increases water absorption into the blood through the distal tubule and the collecting duct. The kidneys also control blood pH, remove toxic substances, and regulate ions such as sodium, chloride, potassium, and sulfate. The kidneys excrete excess glucose, vitamins, and amino acids.

The kidneys also secrete hormones. Renin, released in response to low blood pressure, catalyzes the formation of angiotensin, which constricts arterioles and elevates blood pressure. Erythropoietin, released when the oxygen content of the blood is reduced, stimulates the bone marrow to produce red blood cells.

Mammalian kidneys are adapted to the animal's environment. Animals such as beavers that live where water is abundant tend to have short loops of Henle and produce dilute urine, while desert animals have very long loops of Henle and can produce very concentrated urine.

KEY TERMS

ammonia   *page 709*
angiotensin   *page 715*
antidiuretic hormone (ADH)   *page 715*
bladder   *page 710*
Bowman's capsule   *page 711*
collecting duct   *page 711*
erythropoietin   *page 715*
excretion   *page 708*
filtrate   *page 712*
filtration   *page 712*

glomerulus   *page 710*
hemodialysis   *page 714*
homeostasis   *page 708*
interstitial fluid   *page 708*
kidney   *page 709*
loop of Henle   *page 714*
Malpighian tubules   *page 708*
nephridiopore   *page 709*
nephridium   *page 709*
nephron   *page 710*

nephrostome   *page 709*
osmolarity   *page 708*
osmoregulation   *page 716*
protonephridium   *page 708*
renal artery   *page 710*
renal cortex   *page 710*
renal medulla   *page 710*
renal pelvis   *page 710*
renal vein   *page 710*
renin   *page 715*

tubular reabsorption   *page 712*
tubular secretion   *page 713*
tubule   *page 711*
urea   *page 709*
ureter   *page 710*
urethra   *page 710*
uric acid   *page 710*
urinary system   *page 708*
urine   *page 709*

THINKING THROUGH THE CONCEPTS

1. Explain the two major functions of urinary systems.

2. Trace a urea molecule through the mammalian body, starting with an ammonia molecule in the bloodstream and ending outside the body.

3. What is the function of the loop of Henle? The collecting duct? Antidiuretic hormone?

4. Describe and compare the processes of filtration, tubular reabsorption, and tubular secretion.

5. Describe the role of the kidneys as organs of homeostasis.

6. Compare and contrast the excretory systems of humans, earthworms, and flatworms. In what general ways are they similar? How do they differ?

APPLYING THE CONCEPTS

1. Discuss the differences in function of the two major capillary beds in the kidneys: the glomerular capillaries and those surrounding the tubules.

2. Explain why a whale would have nephrons with long loops of Henle. What would you predict about the loops of Henle in a seal? What about a river otter?

3. Some "quick weight loss" diets are high in protein and very low in carbohydrates, and they require a person to drink more water than usual. Explain why the extra water is important.

4. In his poem "The Rime of the Ancient Mariner," Samuel Taylor Coleridge wrote: "Water, water, every where, nor any drop to drink." Seawater has more than four times the osmolarity of blood. Why can't a person keep from dying of thirst by drinking seawater?

FOR MORE INFORMATION

Graham-Rowe, D. "Army Rations Rehydrated with Urine." *New Scientist*, July 21, 2004. A ration with a special built-in filter will allow soldiers to rehydrate the food with muddy water or even urine in an emergency.

Khamsi, R. "Bio-engineered Bladders Successful in Patients." *New Scientist*, April 4, 2006. A patient's own cells, grown on a biodegradable mold, can be used to replace a poorly functioning bladder.

Seppa, N. "Homegrown Defender." *New Scientist*, June 10, 2006. Scientists have discovered a peptide that defends the urinary tract against infections.

36

Defenses Against Disease

A sneeze propels thousands of microscopic droplets out of the nose and mouth at a rate of about 200 miles per hour! As many as 100,000 bacteria may be released during a single sneeze; viruses are also spread in this manner.

CASE STUDY FIGHTING THE FLU

HAVE YOU EVER BEEN sitting in class when the person next to you lets loose with an explosive, full-bodied sneeze? As you draw in a breath to whisper "Gesundheit," you unwittingly inhale thousands of microscopic droplets laden with "germs," including common cold and flu viruses and bacteria from the flourishing populations that grow in everyone's nose and mouth. The symptoms of a cold and of some types of flu—sneezing, coughing, and a runny nose—are ideally suited to spread these viruses. In fact, natural selection favors viruses that cause symptoms that help them spread to other victims.

After class, your sneezing classmate admits she's been feeling achy and feverish. Remembering your own experiences with similar symptoms, you advise her to take some aspirin and go home to bed. You realize that your classmate's sneeze may have exposed you to the same virus that caused her illness. Remembering being miserable with the flu last year, you fervently hope you are now immune. Should you get a flu shot? Might she have that bird flu you've been hearing so much about—and what are its symptoms? What is the likelihood that someone in your class might have it? Is there a vaccine for it? Was your suggestion to take aspirin good advice? How will your body cope with all the microbes you have unknowingly inhaled?

36.1 WHAT ARE THE BASIC MECHANISMS OF DEFENSE AGAINST DISEASE?

Our environment teems with microscopic organisms or **microbes**, including viruses, bacteria, fungi, and protists. Most do no harm, but some are **parasites**—organisms living on or within the bodies of others and doing harm in the process. Disease-causing microbes are described as **pathogens**, and new pathogens are appearing with disturbing regularity. Since the early 1980s, several viruses have emerged as threats to people—HIV, Ebola virus, hantavirus, and (more recently) West Nile virus and SARS (severe acute respiratory syndrome). We have witnessed outbreaks of relatively rare but deadly strains of the common intestinal bacterium *Escherichia coli* and the respiratory bacterium *Streptococcus pyogenes* ("flesh-eating bacteria"). Meanwhile, hundreds of "cold virus" strains and constantly mutating strains of influenza (flu) virus continue to threaten us.

Given the prevalence and diversity of disease-causing organisms, you might wonder, "Why don't we get sick more often?" Over evolutionary time, animals and their parasites have engaged in a constantly escalating battle. As animals evolve more sophisticated defense systems, parasites in turn evolve more effective tactics for penetrating those defenses. This "evolutionary arms race" has honed our defenses into a stunningly complex system designed to resist parasitic invasion and attack.

Vertebrate Animals Have Three Major Lines of Defense

Vertebrates (animals with backbones) have evolved three major forms of protection against disease: *nonspecific external barriers, nonspecific internal defenses,* and *specific internal defenses,* also called the *immune response* (**FIG. 36-1**). First, the body defends itself with nonspecific external barriers that prevent most disease-causing microbes from entering. These barriers are primarily anatomical structures such as skin, hair, and cilia and secretions such as tears, saliva, and mucus. Such barriers cover the body and line body cavities that are continuous with the outside; for example, the surfaces of the respiratory, digestive, and urogenital tracts. Second, if the external barriers are breached, a variety of nonspecific internal defenses are called into action. Some white blood cells engulf foreign particles or destroy infected cells. Chemicals released by damaged body cells and proteins released by white blood cells are responsible for inflammation and *fever* (in animals that regulate body temperature), which can also deter infection. Both the nonspecific barriers and the nonspecific internal defenses operate regardless of the exact nature of the invader, repelling, killing, or neutralizing the threat. The third and final line of defense consists of specific internal defenses or the **immune response**, in which immune cells selectively destroy the particular toxin or microbe and then "remember" the invader, allowing a faster response if it reappears in the future.

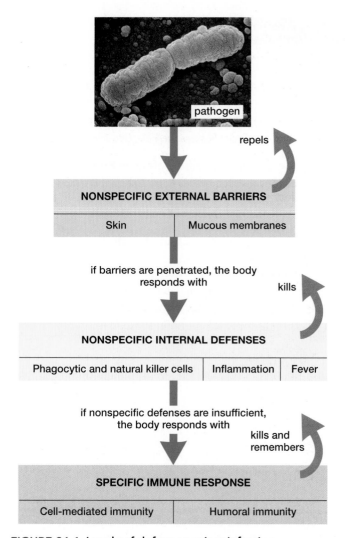

FIGURE 36-1 **Levels of defense against infection**

Invertebrate Animals Possess the First Two Lines of Defense

Invertebrates were the first animals to evolve, and they far outnumber vertebrates; but researchers know remarkably little about their defensive systems. Scientists have found that the relatively simple invertebrate system relies entirely on the two nonspecific lines of defense just described. Invertebrates exhibit an enormous range of physical barriers, ranging from external skeletons to a variety of slimy secretions. Internally, invertebrates have white blood cells that attack pathogens and secrete proteins to neutralize these invaders or the toxins they release. Interestingly, researchers have found very similar proteins in the blood of vertebrates (including people) and the invertebrate horseshoe crab (*Limulus*; FIG. 36-2). In these diverse organ-

FIGURE 36-2 **A horseshoe crab**

isms, the proteins help ward off infections. This finding suggests that genes coding for these molecules have been passed down through the evolutionary lineage of animals for hundreds of millions of years and so are likely to be extremely important in fighting infections.

The specific internal defenses of the immune system arose with the vertebrates. Although some texts include some nonspecific defenses as part of the "immune system" and discuss "invertebrate immune systems," here we define the term more narrowly. Although invertebrates have many, and very effective, defenses against disease, they lack defenses that are specific for particular diseases and so, by our definition, they lack an immune system.

36.2 HOW DO NONSPECIFIC DEFENSES FUNCTION?

Skin and Mucous Membranes Form External Barriers to Invasion

An ideal defense strategy prevents invaders from entering the body in the first place. In animal bodies, this first line of defense consists of the two surfaces with direct exposure to the environment: the *skin* and the *mucous membranes* of the digestive, respiratory, and urogenital tracts.

The Skin and Its Secretions Block Entry and Provide an Inhospitable Environment for Microbial Growth

Any virus and bacteria-laden droplets from your neighbor's sneeze that land on your skin will encounter an outer surface of dry, dead cells. Most microbes that come in contact with the skin do not obtain the water and nutrients they need to survive. The few bacteria and fungi that manage to gain a foothold on skin are usually shed before they can do harm, because skin cells are constantly sloughed off and replaced. Secretions from sweat glands, sebaceous (oil) glands, and—in the external ear canal—wax-secreting glands all contain natural antibiotics that inhibit the growth of bacteria foreign to the body. These multiple defenses make the unbroken skin an extremely effective barrier against microbial invasion.

Antimicrobial Secretions, Mucus, and Ciliary Action Defend the Mucous Membranes Against Microbes

The warm, moist mucous membranes surrounding the eyes and lining the digestive, respiratory, and urogenital tracts are far more hospitable to microbes than the skin is; but they also possess effective defense mechanisms. The mucus and tears secreted by these membranes contain antibacterial enzymes—called *lysozymes*—that destroy bacterial cell walls. Mucus also physically traps microbes and other foreign particles that enter the body through the nose or mouth (**FIG. 36-3**). Cilia on the membranes lining the respiratory tract sweep the mucus, and the particles it has trapped, up into the nose and mouth where it is coughed or sneezed out of the body, or swallowed. If microbes are swallowed, they enter the stomach, where they encounter

extreme acidity and protein-digesting enzymes, both of which are often lethal to them. Further along in the digestive tract, the intestine is inhabited by harmless, resident bacteria that secrete substances that destroy invading foreign bacteria or fungi. In the urinary tract, the slight acidity of urine inhibits bacterial growth. In females, acidic secretions and mucus help protect the vagina. Despite these defenses, many disease-causing organisms enter the body through the mucous membranes; for example, many viruses enter your lungs when you inhale the fallout from your classmate's sneeze. Some of these will probably penetrate the membranes. What happens next?

Nonspecific Internal Defenses Combat Disease

Invading parasites that penetrate the skin or mucous membranes encounter an array of internal defenses, some of which are nonspecific—that is, they do not target specific invaders. Nonspecific internal defenses fall into three major categories. First, the body has a standing army of white blood cells, or **leukocytes**, some of which are specialized to attack and destroy foreign cells. Second, an injury, with its combination of tissue damage and massive invasion of microbes, provokes an *inflammatory response* that brings leukocytes to the scene. Third, if an infection occurs, the body may produce a **fever**, an elevated temperature that both slows down microbial reproduction and enhances the body's own fighting abilities.

Phagocytic Cells and Natural Killer Cells Destroy Invading Microbes

The body contains several types of leukocytes called **phagocytes** that ingest foreign invaders and cellular debris by phagocytosis. Two important types of leukocytes are **macrophages** and **neutrophils**. Both travel within the bloodstream, ooze through capillary walls, and patrol the

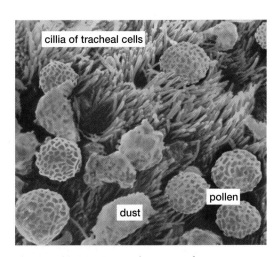

FIGURE 36-3 The protective function of mucus
The ciliated, mucus-coated epithelium lining the respiratory tract traps microbes and foreign debris.

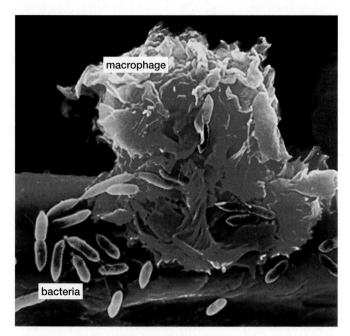

FIGURE 36-4 An attacking macrophage
This macrophage (purple) is busily engulfing bacteria (green).

body's tissues. Macrophages (literally, "big eaters") and neutrophils both consume bacteria and foreign substances that penetrate the body's delicate mucous membranes (**FIG. 36-4**). Macrophages also act as housekeepers, scavenging dead and dying cells that are produced as the body continuously renews its tissues and suffers various traumas. If they were not engulfed, the leaking contents of these dead cells could damage healthy neighboring cells. As described later, macrophages also function in the immune response by exposing proteins from the pathogens they have ingested on their plasma membranes so that immune system cells can identify and respond to these invaders.

Natural killer cells are another type of leukocyte. In general, they attack the body's own cells that have become cancerous or have been invaded by viruses, recognizing these cells by abnormal molecules on their surfaces. Virus-infected cells acquire viral proteins, and cancerous cells bear atypical molecules on their plasma membranes. Rather than engulfing their victims (as neutrophils and macrophages do), natural killer cells release proteins that bore holes in the membrane of the infected or cancerous cell and then secrete enzymes through these holes. Shot full of holes and then chewed on by enzymes, the attacked cell soon dies.

The Inflammatory Response Attracts Phagocytes to Infected or Injured Tissue

Whether you catch the flu from your classmate or accidentally get a splinter in your skin (**FIG. 36-5**), you will experience inflammation—an important component of the body's nonspecific internal defenses. The **inflammatory response** causes injured tissues to become warm, red, swollen, and painful (inflammation literally means "to set

on fire"). This defense mechanism has several functions: it attracts phagocytic cells to the area, promotes blood clotting, and causes pain that stimulates protective behaviors.

As shown in Figure 36-5, the inflammatory response can be initiated by cells that are damaged by pathogens or by trauma. These cells release substances that cause connective tissue cells called **mast cells** to release **histamine** (and other chemicals) into the wounded area. Histamine is a chemical that makes capillary walls leaky and relaxes the smooth muscle surrounding arterioles, increasing blood flow. Extra blood flowing through the leaky capillaries drives fluid from the blood through capillary walls into the region around the wound. Other chemical messengers released by wounded cells and mast cells, and some substances produced by the microbes themselves, attract phagocytic macrophages and neutrophils. These squeeze out through the leaky capillary walls and ingest bacteria, dirt, and cellular debris caused by the injury (see Figures 36-4 and 36-5). In the case of a dirty wound, *pus*—a thick whitish mixture of dead bacteria, tissue debris, and living and dead white blood cells—may accumulate. Macrophages release **cytokines**, protein messenger molecules that allow cells to communicate with one another in responding to pathogens and foreign substances. These cytokines supplement histamine in making the capillaries leaky, attracting more leukocytes to the area, and helping them to penetrate through the capillary walls.

The swollen, painful tissues in the throat of a flu victim and the fluid secretions that lead to sneezing, a runny nose, and coughing are a direct result of the inflammatory response. In addition, leaky capillaries and injured cells release chemicals that lead to blood clotting, which blocks damaged blood vessels (see Chapter 32). Clotting seals off the wound from the outside world and prevents more microbes from entering. Finally, sensations of pain, activated by swelling and by chemicals released by the injured tissue, alert the injured person (or other animal) to protect the area from further damage.

Fever Combats Large-Scale Infections

If invaders breach other defenses and mount a large-scale infection, they may cause fever, as occurs when you contract the flu. Although high fevers can be dangerous, and even moderate ones can be very unpleasant, fever is actually part of the body's nonspecific response to infection. The onset of fever is controlled by the hypothalamus, the part of the brain housing the temperature-sensing nerve cells that are the body's thermostat. In humans, this thermostat is set at about 98.6°F (37°C). Certain macrophages, as they respond to the infection, release a cytokine called *endogenous pyrogen* (literally, "self-produced fire-maker"). This cytokine travels in the bloodstream to the hypothalamus and raises the thermostat's set point, triggering responses that increase body temperature. These responses include increased fat metabolism, constriction of blood vessels supplying the skin, and heat-producing and conserving behaviors such as shivering and bundling under blankets. Cytokines also cause other cells to reduce the concentration of iron in the blood.

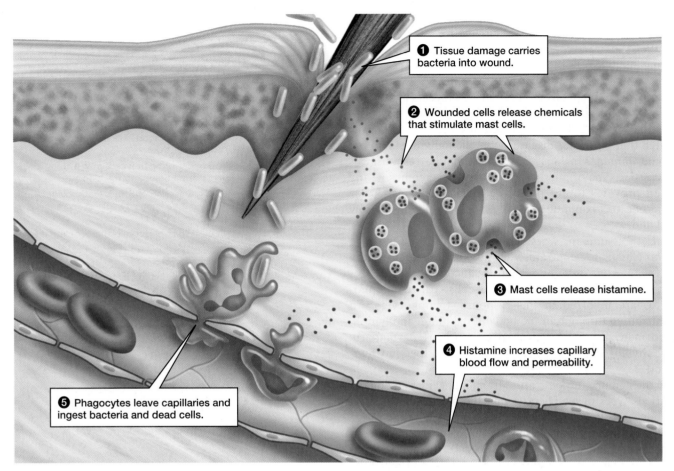

1 Tissue damage carries bacteria into wound.

2 Wounded cells release chemicals that stimulate mast cells.

3 Mast cells release histamine.

4 Histamine increases capillary blood flow and permeability.

5 Phagocytes leave capillaries and ingest bacteria and dead cells.

FIGURE 36-5 The inflammatory response
QUESTION Why can white blood cells, but not red blood cells, leave a capillary?

Fever both enhances the body's normal defenses and harms invading microbes. In the case of bacterial infection, fever increases the activity of the white blood cells that engulf bacteria. Immune cells (discussed later) multiply more rapidly. At the same time, feverish body temperatures force many bacteria to reproduce more slowly and to require more iron, so the elevated temperature and reduced iron in the blood work together to help keep infectious bacteria in check.

Fever also stimulates cells attacked by viruses to produce more of the antiviral cytokine *interferon*, which then travels to other cells and increases their resistance to viral attack. Fever is a defense mechanism, as illustrated by an experiment in which people with viral infections were treated either with aspirin (which reduces fever) or with a *placebo* (an inactive substance that looks like the real drug and serves as a control). The people given aspirin had far more viruses in their noses and throats, and they sneezed and coughed out far more viruses than did those given the placebo. This finding supports the hypothesis that fever helps control viral infections. Because fevers are a sign of infection—and because high fevers are dangerous—seek advice from a health professional if you get one.

36.3 WHAT ARE THE KEY CHARACTERISTICS OF THE IMMUNE RESPONSE?

External barriers, phagocytic cells, natural killer cells, the inflammatory response, and fever are all *nonspecific* defenses, which prevent or overcome microbial invasions of the body. Unfortunately, these nonspecific defenses are not impregnable. When they fail to do the job, the body mounts a highly specific and coordinated immune response directed against the particular organism that has successfully colonized the body. This response involves an army of specialized white blood cells that secrete an array of different chemicals and communicate in complex ways. The immune response is made possible by the **immune system**, which consists of cells and organs scattered throughout the body. The immune system includes macrophages and specialized leukocytes (white blood cells) called **lymphocytes** that participate in the immune response. It encompasses various proteins such as cytokines and antibodies secreted into the blood by immune cells, as well as the *complement proteins* (described later) which are secreted primarily by the liver. The immune system also includes the bone marrow where the immune cells are produced and structures where these cells reside, differentiate, proliferate, and attack invaders. These structures include the thymus (see Chapter 33), spleen, lymph nodes, and lymphatic vessels (see Chapter 32). Although the circulatory system is not an "official" part of the immune system, it is important in carrying immune cells throughout the body.

The essential features of the immune response to infection were recognized more than 2000 years ago by the Greek historian Thucydides. He observed that, occasionally, a person would contract a disease, recover, and never catch that particular disease again; the individual had become *immune*. In most cases, the immune response attacks one type of microbe or its toxin, overcomes it, and then provides future protection against that particular microbe or toxin, but not others, unless they are closely related.

Let's assume the worst happens and the flu virus from your classmate penetrates your nonspecific defenses and activates your immune response. Macrophages and specialized leukocytes (white blood cells) called lymphocytes are involved in this response. Lymphocytes are distributed throughout the body in the blood and lymph, and many are clustered in specific organs, particularly the thymus, lymph nodes, and spleen (described in Chapter 32).

The immune response arises from interactions among the various types of lymphocytes and the molecules that they produce. Table 36-1 provides a brief overview of the major cell types that participate in both nonspecific defenses and specific immune responses. The key cellular players in the immune response are lymphocytes called **B cells** and **T cells**. Like all blood cells, B cells and T cells arise from stem cells that reside in bone marrow. These stem cells produce lymphocyte precursor cells that migrate from the marrow into the blood. Some lymphocyte precursors travel in the blood to the thymus, where they complete their development into T (for *thymus*) cells. In contrast, B cells differentiate in the bone marrow. Although they play quite different roles, immune responses produced by both B cells and T cells include the same three fundamental steps. First, the cells of the immune system recognize the invader; second, they launch an attack; and third, they retain a memory of the invader that allows them to ward off future infections.

Cells of the Immune System Recognize the Invader

Foreign Invaders Exhibit Characteristic Antigens

Immune cells recognize foreign molecules that are specific to the invading microbe or toxin; these molecules serve as **antigens** (short for "*anti*body response *gen*erating"). In general, only large, complex molecules such as proteins, polysaccharides, and glycoproteins can act as antigens. Antigens are located on the surfaces of invading microbes or cancer cells, and viral antigens become incorporated into the plasma membranes of infected body cells. Viral or bacterial antigens are also exposed on the plasma membranes of the macrophages that engulf them. Other types of antigen may be dissolved in the blood or interstitial fluid; for example, snake venom and toxins released by bacteria are dissolved antigens.

Table 36-1 The Body's Cellular Armory

Mast Cells		Connective tissue cells that release histamine; important in the inflammatory response
Neutrophils		White blood cells that engulf invading microbes
Macrophages		White blood cells that engulf invading microbes and present antigens
Natural killer cells		White blood cells that destroy infected or cancerous cells
B cells		White blood cells that produce antibodies
Plasma cells		Offspring of B cells that secrete antibodies into the bloodstream
Memory B cells		Offspring of B cells that provide future immunity against invasion by the same antigen
T cells		White blood cells that regulate the immune response or kill certain types of cells
Cytotoxic T cells		Offspring of T cells that destroy foreign eukaryotic cells, infected body cells, or cancerous body cells
Helper T cells		Offspring of T cells that stimulate immune responses by both B cells and cytotoxic T cells
Memory T cells		Offspring of T cells that provide future immunity against invasion by the same antigen

Antibodies and T Cell Receptors Recognize and Bind to Foreign Antigens

Cells of the immune system generate two types of proteins that recognize, bind to, and then help destroy specific antigens. **Antibody** proteins are produced by B cells and their offspring (plasma cells and memory B cells), and **T-cell receptor** proteins are produced by T cells.

Antibodies are Y-shaped proteins composed of two pairs of peptide chains: one pair of identical large (heavy) chains and one pair of identical small (light) chains (**FIG. 36-6**). Both heavy and light chains consist of a **constant region**, which is similar in all antibodies of the same type, and a **variable region**, which differs among individual antibodies. The combination of light and heavy chains results in antibodies with two functional parts: the two "arms" of the Y and the "stem" of the Y. The variable regions (located at the arm tips) form binding sites whose shapes and electrical charges make them specific for only one or a few types of antigen molecules.

There are five classes of antibodies, distinguished by differences in the stem region. Antibodies on B cells activate the B cell when they bind an antigen, causing it to divide and produce antibody-secreting plasma cells (see Fig. 36-9). The secreted antibodies bind to antigens and help dispose of them in various ways. Some make the antigen more attractive to phagocytes or link pathogens together into a clump that is more readily recognized and phagocytosed (**FIG. 36-7a**). Some antibodies interfere with structures that allow bacteria to adhere to their host cells or block sites on viruses that

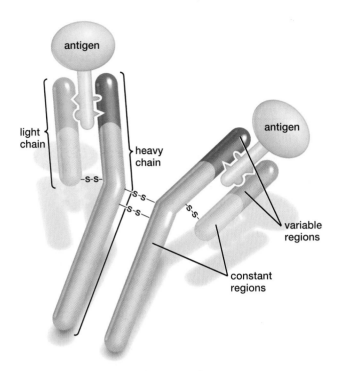

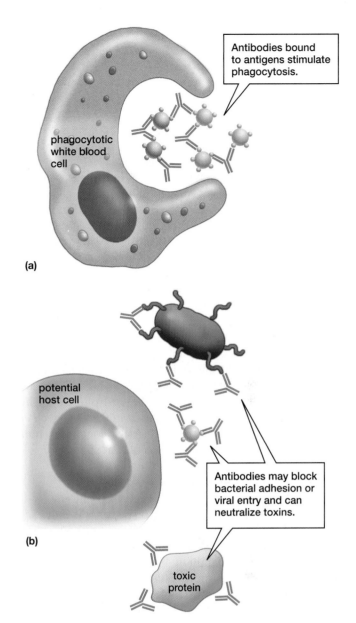

FIGURE 36-6 Antibody structure
Antibodies are proteins composed of two pairs of peptide chains (light chains and heavy chains) arranged in a Y shape. Constant regions on both chains form the stem of the Y; variable regions on the two chains form a specific binding site at the end of each arm of the Y. Different antibodies have different variable regions, forming unique binding sites. QUESTION Why do antibody molecules have both constant and variable regions?

FIGURE 36-7 Some functions of antibodies

promote viral entry into their host cells. Other antibodies bind bacterial toxins and render them harmless (**FIG. 36-7b**). Antibodies may also activate blood proteins of the *complement system* (described later), which can destroy pathogens. One class of antibodies can cross the placenta and provide immunity to a developing child, and another class is secreted in breast milk. Both of these help the newborn baby, whose immune system is not fully developed, to resist the assaults of the outside world.

T cell receptors exhibit both differences from and similarities to antibodies. Unlike many antibodies, T cell receptors remain attached to the surfaces of T cells. They consist of two peptide chains of about equal size. Like antibody arms, the ends of the two chains form highly specific binding sites for an antigen. T cell receptors trigger a response in the T cell only when they encounter antigen molecules either borne on the membranes of cancerous or infected cells or presented on the membranes of macrophages that have ingested the invading microbes. Unlike antibodies, the T cell receptor proteins themselves do not directly inactivate pathogens. Instead, they alter the activity of the T cell to which they are attached, as described later.

Cells of the Immune System Can Recognize and Produce Specific Responses to Millions of Types of Molecules

During your lifetime, your body will be challenged by a multitude of invaders. Your classmates may sneeze cold and flu viruses into the air you breathe. Your food may

host large bacterial populations or toxins they have released. You may drink contaminated water, or a tick carrying Lyme disease bacteria might bite you. There is no escaping these pervasive and persistent assaults. Fortunately, the immune system recognizes and responds to virtually all of the millions of potentially harmful antigens that an animal encounters, because immune cells produce millions of different antibodies and T cell receptors, each specific for a different antigen.

The mechanisms by which cells of the immune system recognize antigens are both complex and fascinating. Antibodies and T cell receptors are, after all, proteins, and proteins are encoded by genes. It would seem that, to recognize millions of different antigens, each person would have to possess millions of different genes for antibodies and T cell receptors. But there are only about 21,000 genes in the entire human genome. This means that a relatively small number of genes must code for millions of antibodies and T cell receptors. How can this occur?

Antibodies Are Pieced Together

There are no genes for entire antibody molecules. The key to producing huge numbers of antibodies from a far smaller number of genes is to assemble the antibodies from pieces combined in random ways. The genome of people and other vertebrates includes genes that code for many different versions of the antibody regions that can be pieced together in many millions of combinations. As each individual B cell develops, it retains only a few randomly selected genes (including both light and heavy chains with both constant and variable regions); the rest are cut out from the DNA as the B cell matures, as illustrated in **FIGURE 36-8**. Therefore, each B cell produces only a single type of antibody, specified by the chance recombination of variable-region and constant-region genes. This antibody is different from the antibodies produced by all other B cells (except its own daughter cells). The random selection of antibody-region genes from a large pool of choices yields a vast number of possible unique combinations. It might help to think of antibody gene formation in terms of dealing from two decks, each containing hundreds of cards. Each B cell is dealt a "hand" of two variable-region genes, one randomly chosen from the "light-chain variable-region deck" and one from the "heavy-chain variable-region deck." With each "deck" containing hundreds of genes, there will be millions of combinations.

The Immune System Does Not Design Antibodies or T Cell Receptors Expressly to Bind Invading Antigens

The end result of this recombination of genes is that each individual B cell produces its own unique antibodies. The human body contains a "standing army" of perhaps 100 million different antibodies (and even more T cell receptors), so antigens almost always encounter antibodies or T cell receptors that will bind them. It is important to recognize that the immune cells do not "design" antibodies and T cell receptors to fit invading antigens in the way that a tailor might design custom clothes for a customer. Instead, the immune system randomly synthesizes millions of different antibodies and receptors. Like clothes in a department store, the array of antibodies and receptors is simply there, waiting. In our clothing analogy, given enough racks of clothing from which to choose, each of us will find something suitable that fits. The binding of antigen to antibody triggers changes in the B cells. These changes usually lead to the destruction of microbes bearing that antigen. We examine these changes in more detail later.

The Immune System Distinguishes "Self" from "Nonself"

Why doesn't your immune system attack and destroy your own body cells? The surface of every cell in the body bears large proteins and polysaccharides, just as microbes do. Some of these proteins, collectively called the **major histocompatibility complex (MHC)**, are unique to each individual (except identical twins, who have the same genes and hence the same MHC proteins). So why don't these "self" antigens arouse an individual's own immune system? The key seems to be the continuous presence of the body's own antigens while the immune cells mature. Some newly differentiating immune cells do indeed produce antibodies or T cell receptors that can bind the body's own proteins and polysaccharides, treating them as antigens. However, these misguided, immature immune cells are quickly destroyed. By destroying these delinquent cells before they can mature, the immune system distinguishes "self" from "nonself," retaining

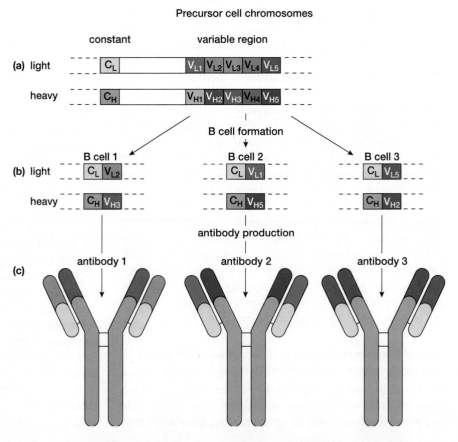

FIGURE 36-8 Recombination produces antibody genes
(a) Each precursor cell of the immune system contains at least one gene for the constant regions (C) of the light (C_L) and heavy (C_H) chains of antibodies and many genes for the variable regions (V_{L1-5} and V_{H1-5}). **(b)** During the development of each B cell, these genes are recombined, moving one of the variable-region genes next to a constant-region gene. For each light and each heavy chain, each B cell generates a "recombined antibody gene" that differs from the recombined antibody genes generated by any other B cell. **(c)** The antibodies synthesized by each B cell in (b).

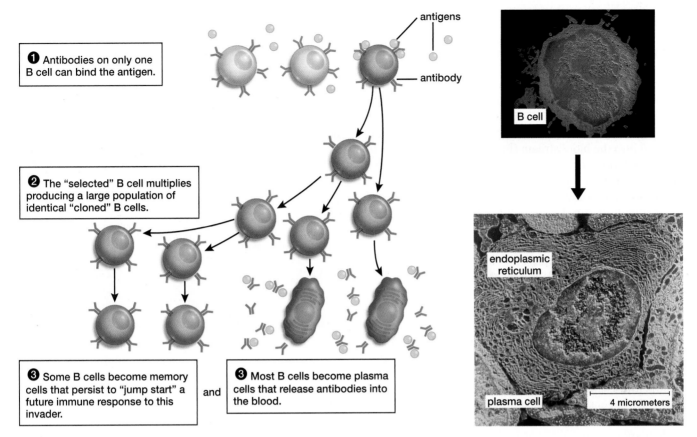

FIGURE 36-9 Clonal selection among B cells and plasma cell production
(**left**) Antigens "select" a B cell by binding to its specific antibody, causing it to divide and then differentiate into plasma cells and memory B cells. (**right**) B cell before (top) and after (bottom) conversion into a plasma cell. The plasma cell is much larger and is packed with rough endoplasmic reticulum that synthesizes antibodies to be released into the plasma.

In the figure:

❶ Antibodies on only one B cell can bind the antigen.

antigens
antibody
B cell

❷ The "selected" B cell multiplies producing a large population of identical "cloned" B cells.

❸ Some B cells become memory cells that persist to "jump start" a future immune response to this invader.

and

❸ Most B cells become plasma cells that release antibodies into the blood.

endoplasmic reticulum
plasma cell
4 micrometers

only those immune cells that do not attack the body's own molecules.

Because an individual's MHC proteins are unique, they act as foreign antigens in other people's bodies. This is why organ transplants are sometimes rejected. Physicians must find a donor whose MHC proteins are as similar as possible to those of the recipient, but even with this tissue matching, transplant patients must also take drugs that suppress their immune systems. Without these drugs, the recipient's immune system attacks the foreign MHC proteins on the donor's cells, destroying the transplanted tissue.

Cells of the Immune System Launch an Attack

Once a body has been invaded, say by a flu virus, the immune cells launch two types of attack: humoral immunity and cell-mediated immunity. **Humoral immunity** is provided by B cells and the antibodies that they secrete into the bloodstream, which attack pathogens before they can enter body cells. **Cell-mediated immunity** is produced by a type of T cell called *cytotoxic T cells*, which attack cancerous or infected cells, killing both the cell and any pathogens inside it. These two types of immune responses require considerable communication among the cell types. Both humoral and cell-mediated immunity are stimulated by **helper T cells**. The T cell receptors of helper T cells bind

microbial antigens that are exposed on the plasma membranes of macrophages that have engulfed the invaders. Upon binding an antigen, the helper T cells proliferate and release a variety of cytokines, which stimulate cell division and development among both B cells and cytotoxic T cells that are specific for that antigen (look ahead to Fig. 36-11). Very little immune response, either cell-mediated or humoral, can occur without the chemical boost provided by helper T cells. That is why AIDS, which destroys helper T cells, is such a deadly disease. Although humoral and cell-mediated immunity are not completely independent, we consider them separately in the following sections for ease of understanding.

Humoral Immunity Is Produced by Antibodies Dissolved in Blood

Each B cell bears a specific type of antibody on its surface. When an infection occurs, the antibodies borne by a few B cells are able to bind to antigens on the invader. Antigen–antibody binding causes these B cells to divide rapidly. This process is called **clonal selection** because the resulting population of cells is composed of "clones" (genetically identical to the parent B cells) that have been "selected" to multiply by the presence of particular invading antigens (**FIG. 36-9**). The daughter cells differentiate into two cell types: **memory B cells** and **plasma cells**. Memory cells do not release antibodies, but do play an important role in future immunity to the particular disease that stimulated their production (described later). Plasma cells become enlarged and packed with rough endoplasmic reticulum, in which huge quantities of specific antibody proteins are synthesized (Fig. 36-9). These antibodies are released into the bloodstream (hence the name "humoral" immunity; to the ancient Greeks, blood was one of the four "humors," or body fluids).

Because antibodies circulate in the bloodstream, humoral immunity can defend only against pathogens in the blood or interstitial fluid. Bacteria, bacterial toxins, and some fungi and protists are vulnerable to the humoral immune response. Pathogenic microbes such as viruses are safe from antibody attack as long as they are inside a body cell. Antibodies can attack viruses either as they first enter the body, or after they have replicated in a cell, ruptured it, and been released into the body fluids.

Humoral immunity destroys foreign chemicals or microbes in several ways, depending on the class of the antibody. As noted earlier, circulating antibodies can inactivate pathogens by binding them or causing them to clump together. Pathogens bound by antibodies are more easily engulfed by macrophages. One class of antibodies, upon binding an antigen (particularly a bacterial antigen), initiates a series of reactions among a group of blood proteins called the **complement system**. The antibody-coated bacteria are killed when these reactions produce pore-forming proteins that punch holes in the bacterial membrane. Some complement proteins help phagocytic white blood cells to engulf the invaders.

To fight the viral infection you contract from your sneezing classmate, you will also need the help of cell-mediated immune responses. Both humoral and cell-mediated forms of immune response can take two weeks to become fully active against an unfamiliar infectious organism.

Cell-Mediated Immunity Is Produced by T Cells

How can cells of the immune system recognize body cells that are cancerous or harboring viruses? Why are transplanted organs sometimes rejected? These events occur because of cell-mediated immunity. Three types of T cells contribute to cell-mediated immunity: *helper T cells*, *cytotoxic T cells*, and *memory T cells*.

Recall that helper T cells respond to antigens on the surfaces of macrophages, proliferate, and produce cytokines that stimulate T cell division and differentiation. Some of these T cells develop into **cytotoxic T cells.** T cell receptors on cytotoxic T cell membranes bind to antigens on the surface of infected or cancerous cells (**FIG. 36-10**). The cytotoxic T cells then release proteins (similar to those produced by natural killer cells) that insert large pores in the membranes of the target cells and enzymes that enter through these holes and cause the abnormal cells to break down. After the infection is over, some helper T cells develop into **memory T cells.** Like memory B cells, these help

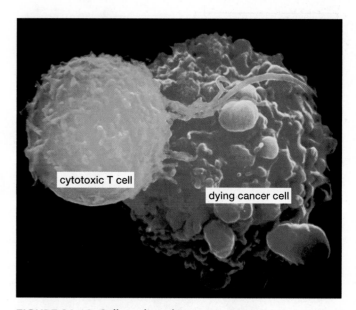

FIGURE 36-10 Cell-mediated immunity in action
A cytotoxic T cell attacks a cancer cell (red) in this scanning electron micrograph. QUESTION Cancer consists of the body's own cells, so why does the immune system attack cancer cells?

protect the body against future infection. **FIGURE 36-11** compares the humoral immune response with the cell-mediated immune response.

Cells of the Immune System Remember Past Victories

After recovering from a disease, you will remain immune to that particular strain of microbe for many years, perhaps a lifetime. Retaining immunity is the function of memory cells. Plasma cells and cytotoxic T cells directly fight disease-causing organisms, but they generally live only a few days. Both B cells (which produce plasma cells) and cytotoxic T cells, however, leave behind smaller populations of memory cells that may survive for many years. If the body is invaded by microbes to which the body has already mounted an immune response, the appropriate memory cells recognize the invaders immediately. These memory cells multiply rapidly and generate huge populations of plasma cells (from memory B cells) and cytotoxic T cells (from memory T cells). Because memory cells are primed to respond, the body will be able to fend off a second attack before it gains a foothold (**FIG. 36-12**). Why, then, do you repeatedly suffer from colds and flu? The problem, as described in "Health Watch: Fighting Influenza—Is a Bird Flu Pandemic Imminent?" is that each year's flu is caused by a different virus. Unfortunately, there are also so many cold viruses that you will encounter different versions of these throughout your life as well.

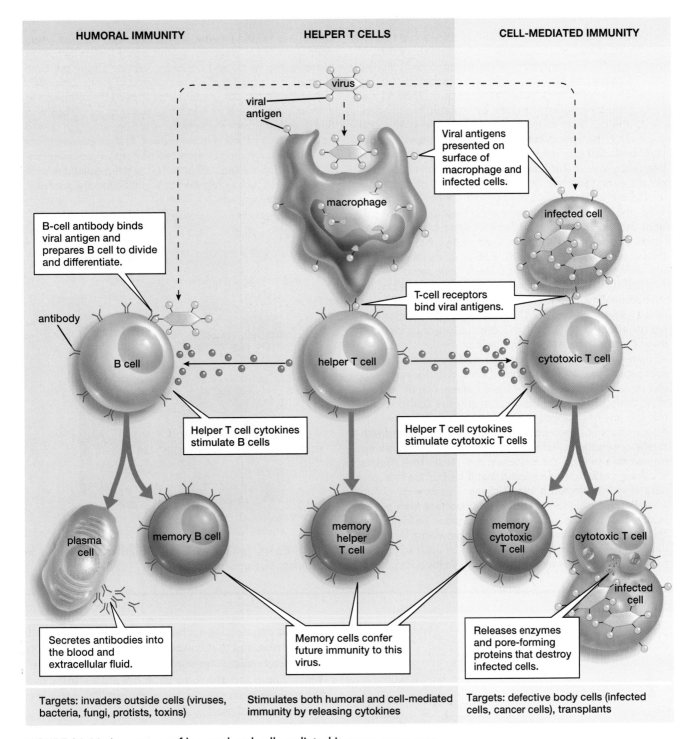

HUMORAL IMMUNITY

HELPER T CELLS

CELL-MEDIATED IMMUNITY

virus

viral antigen

Viral antigens presented on surface of macrophage and infected cells.

macrophage

infected cell

B-cell antibody binds viral antigen and prepares B cell to divide and differentiate.

antibody

T-cell receptors bind viral antigens.

B cell

helper T cell

cytotoxic T cell

Helper T cell cytokines stimulate B cells

Helper T cell cytokines stimulate cytotoxic T cells

plasma cell

memory B cell

memory helper T cell

memory cytotoxic T cell

cytotoxic T cell

infected cell

Secretes antibodies into the blood and extracellular fluid.

Memory cells confer future immunity to this virus.

Releases enzymes and pore-forming proteins that destroy infected cells.

Targets: invaders outside cells (viruses, bacteria, fungi, protists, toxins)

Stimulates both humoral and cell-mediated immunity by releasing cytokines

Targets: defective body cells (infected cells, cancer cells), transplants

FIGURE 36-11 A summary of humoral and cell-mediated immune responses

36.4 HOW DOES MEDICAL CARE AUGMENT THE IMMUNE RESPONSE?

Vaccinations Stimulate the Development of Memory Cells

You now know that memory cells survive and are primed to fight off a repeat attack from a specific pathogen after you've had the disease. Wouldn't it be great if you could create this state of "immunological preparedness" without having to suffer the disease first? *Vaccines* accomplish this feat by priming the body with antigens from the disease organism. These shortcuts to immunity may be created in several ways:

• Disease organisms are killed or weakened so they are unable to cause disease but retain their surface antigen molecules. Most flu vaccines are made this way.

• Specific antigenic molecules are purified from the disease organism.

Smallpox has been feared by humankind since ancient times. This highly contagious disease formerly killed about 30% of its victims and left many more disfigured with pitted skin. The pits are caused by blisters that fill with pus before forming scabs. Documents from India dating back to 1100 BC describe protecting against smallpox by exposing healthy people to the pus from blisters of smallpox victims. Although some died, many recipients of this earliest form of vaccination developed only mild symptoms and were able to resist later exposure to the disease. By the eighteenth century, people in England had observed that victims of cowpox (a related but far less serious disease) did not get smallpox. In 1796, the English surgeon and experimental biologist Edward Jenner obtained fluid from cowpox blisters on the hands of a dairymaid and inoculated an 8-year-old boy with this bacteria-laden material (**FIG. E36-1**). A few months later, in a daring and risky experiment, Jenner inoculated the boy with material from a smallpox lesion. Fortunately, the boy remained healthy. After repeating these results, Jenner published his findings in 1798. This early form of vaccination was rapidly adopted in Europe and eventually worldwide, dramatically reducing deaths from smallpox.

Somewhat surprisingly, nearly a century passed before vaccination was applied to other infectious diseases. This was done in the late 1800s by Louis Pasteur, one of history's most distinguished scientists and among the first to recognize the role of microbes in causing disease. As is often the case in scientific experiments, chance events, combined with careful observations, led to Pasteur's early work on vaccines. He was able to grow the bacterium that causes fowl cholera in culture medium, and he found that it caused the fatal disease when he injected it into chickens. The story goes that when Pasteur injected the chickens with bacteria from an old culture that had been left over the summer holidays, the chickens became ill but survived. Upon growing a fresh culture and needing chickens to inoculate, he used some that had survived from his earlier experiment. To his surprise, the chickens remained healthy. He hypothesized that the weakened bacteria could protect against later infection by healthy bacteria, and he coined the term "vaccine" (from the Latin "vacca," meaning "cow") in recognition of Jenner's pioneering work with cowpox. Pasteur later applied the technique to anthrax in sheep and then to rabies, saving a young boy who had been bitten repeatedly by a rabid dog. Although he had no knowledge of the immune system, Pasteur's experiments marked the beginning of the science of immunology. As Pasteur himself remarked, "Chance favors the prepared mind." This insight is just as relevant to scientific inquiry now as it was over a century ago.

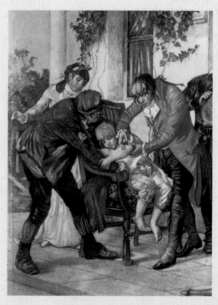

FIGURE E36-1 Jenner exposes a child to cowpox as a vaccine against smallpox

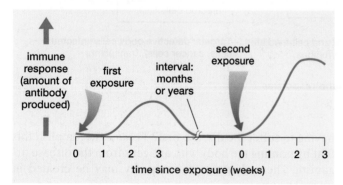

FIGURE 36-12 Memory cells enhance the immune response The immune system responds sluggishly to the first exposure to a disease-causing organism as B and T cells are selected and multiply. A second exposure activates memory cells formed during the first response, making the second response both faster and larger.

- Genetic engineering techniques are used to insert the genetic material that codes for a specific antigen from a pathogenic microorganism into harmless bacteria or yeast. These are grown in large quantities and express the antigen from the disease organism. The antigen can be purified and used as a vaccine.

- A related pathogen that causes no (or very mild) symptoms can be used to stimulate an immune response to a much more deadly pathogen with which it shares antigenic molecules. This technique was employed by Edward Jenner to create the first systematic vaccinations against smallpox, using viruses that caused far less deadly cowpox, as described in "Scientific Inquiry: The Discovery of Vaccines."

In each of these cases, a later exposure to the antigen will cause memory cells—bearing antibodies specific for the disease—to spring into action, just as if you had contracted the disease. Proliferating far more rapidly than the first time around, the immune cells destroy the disease organisms before they can produce symptoms (see Fig. 36-12).

Today, many diseases—including polio, diphtheria, typhoid fever, mumps, and measles—can be controlled through **vaccination**. Smallpox, the deadliest of these, has been completely eradicated since 1980 thanks to a massive vaccination program sponsored by the World Health Organization. There are now only two places where smallpox viruses are legally kept in cold storage, against the remote possibility that "bioterrorists" possess stocks of smallpox virus and more vaccines might be needed in the future.

Antibiotics Slow Down Microbial Reproduction

Any infection starts a race between the invading microbes and the immune response. If the initial infection is massive or if the microbes produce particularly toxic by-products, the nonspecific defenses may be overcome and the immune response may come too late. Further, some pathogens have evolved ingenious defenses that help them evade the immune system. For example, the malarial parasite undergoes three stages of its complex life cycle within humans, but each stage exposes different antigens for too short a time for the immune system to stage an effective attack.

Antibiotics are drugs that retard the growth and multiplication of many pathogens, including bacteria, fungi, and protists (but not viruses). Although antibiotics usually don't destroy every microbe, they give the body's internal defenses enough time to finish the job. One problem with antibiotics, however, is that they are potent agents of natural selection (see Chapter 15). The occasional mutant microbe that is resistant to an antibiotic will flourish and pass on the gene(s) for resistance to its offspring. The result is that resistant mutants proliferate, while susceptible microbes die off. Consequently, many antibiotics have now become ineffective for various diseases, and medical researchers work steadily to develop new ones.

Until recently, little could be done to help people suffering from viral infections except to treat the symptoms and hope that the immune system would triumph. Now, drugs are available that target viruses at different stages of their infectious cycle, which includes attachment to a host cell, replication of viral parts using the host cell's machinery, assembly of the virus within the host cell, and release of many viruses into the interstitial fluid, which allows them to infect new cells. Antiviral drugs are not prescribed for most viruses, but they are used to treat HIV, severe herpes virus infections, and in some cases, the flu virus (see "Health Watch: Fighting Influenza—Is a Bird Flu Pandemic Imminent?").

36.5 WHAT HAPPENS WHEN THE IMMUNE SYSTEM MALFUNCTIONS?

Allergies Are Misdirected Immune Responses

As your classmate sneezes nearby, you entertain a fleeting hope that she is merely allergic to something and not coming down with the flu. More than 35 million people in the United States suffer from **allergies**, immune reactions to harmless substances. Allergic individuals react to antigens of specific harmless substances as if the materials were pathogens, and their responses are directed toward ridding their bodies of the substances. Recent research indicates that people can inherit a tendency for their immune systems to overreact this way. Common allergies include those to pollen, dust, mold spores, bee or wasp venoms, and a specific protein in cat saliva that is the most common cause of cat allergies. Allergy symptoms flare when these allergens trigger an inflammatory response.

An allergic reaction begins when an antigen (such as pollen) enters the body and is recognized by a B cell carrying an antibody to some molecule on the pollen's surface. This B cell proliferates, producing plasma cells that pour out "allergy antibodies" against the pollen antigens. The antibodies attach to mast cells, priming them to respond to later exposures to the antigen by releasing histamine (**FIG. 36-13**). Histamine causes inflammation (see Fig. 36-5) and, in the respiratory tract, increased mucus secretion. Because

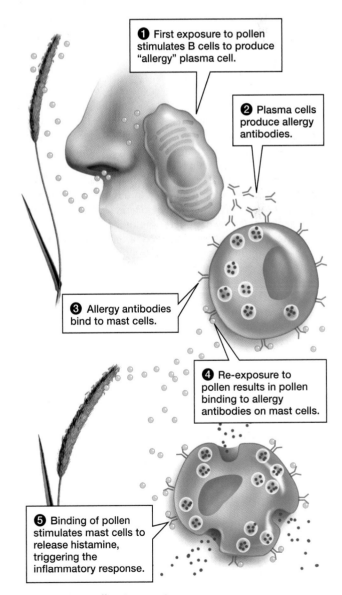

❶ First exposure to pollen stimulates B cells to produce "allergy" plasma cell.

❷ Plasma cells produce allergy antibodies.

❸ Allergy antibodies bind to mast cells.

❹ Re-exposure to pollen results in pollen binding to allergy antibodies on mast cells.

❺ Binding of pollen stimulates mast cells to release histamine, triggering the inflammatory response.

FIGURE 36-13 Allergic reactions

Every winter, a wave of *influenza*, or flu, sweeps across the world. Hundreds of thousands of the elderly, the newborn, and those already suffering from illness perish while millions more suffer the respiratory distress, fever, and muscle aches of severe cases. Occasionally, devastating flu varieties appear, causing worldwide epidemics called *pandemics*. During the great flu pandemic of 1918, a total of 20 million people worldwide died in one winter. In 1968, the Hong Kong flu infected 50 million people in the U.S., causing 70,000 deaths in just 6 weeks.

FLU VIRUSES

Flu is caused by viruses that invade the cells of the respiratory tract, turning each cell into a factory for manufacturing new viruses. The outer surface of a flu virus is studded with spike-like proteins, some of which are recognized by the immune system as antigens. Most people survive the flu because their immune systems eventually inactivate the viruses or kill virus-infected body cells. Why, then, don't our memory T and B cells make us immune to new flu outbreaks?

The answer lies in a flu virus's amazing ability to change. Flu virus genes consist of RNA, which mutates rapidly. Mutations in key surface proteins can prevent the immune system from fully recognizing and attacking the virus. Since your memory cells then provide only partial protection, you may get the flu year after year.

DEADLY NEW STRAINS

On rare occasions, dramatically new and deadly flu viruses appear and cause pandemics. These viruses carry entirely new surface proteins that the human immune system has never before encountered. To respond, the immune system must start from scratch, selecting entirely new lines of B cells and T cells to attack the intruder. In the meantime, the virus multiplies so rapidly that many individuals die or become so weakened that they contract some other fatal disease.

Where do the genes that encode these new flu antigens come from? Birds, especially migratory waterfowl such as ducks, may carry viruses strikingly similar to human flu viruses, although they show no symptoms. Both human and bird viruses can infect pigs, so both viruses will sometimes meet within the same pig cell. On rare occasions, viruses produced within this doubly infected pig cell will acquire RNA from both human and bird viruses (**FIG. E36-2**). From the

human virus, they pick up the genes needed to infect human cells; from the bird virus, they obtain surface antigen proteins that the human immune system is unprepared for. The hybrid viruses can now move easily from pigs to humans. The pandemics of 1957 and 1968 both arose from viruses that combined traits from bird and human influenza. Although the "mixing vessel" host for these particular strains has not been positively identified, the pig is a likely candidate. Recently, in a remarkable example of medical detective work, scientists recovered flu viruses that caused the 1918 pandemic from the body of a young woman buried in the permanently frozen soil of the far north. After sequencing the genes, the researchers concluded that the avian virus had not recombined with human flu but had acquired mutations that made it substantially different from other bird viruses and able to infect people.

A new strain of bird flu, called H5N1, has the potential to become the world's next pandemic. It has moved from wild birds into domesticated poultry in Asia, where hundreds of millions of chickens have been killed in an attempt to stop its spread. Like the 1918 flu virus, H5N1 has developed the ability to spread from birds (in this case, chickens) to people, where it has a high mortality rate. As of 2006, both poultry-to-person and person-to-person transmission is rare, but further mutations could change that. Ominously, the H5N1 virus has now been found in pigs. Scientists are concerned that either pigs or people simultaneously infected with both H5N1 and human influenza could become mixing vessels from which a much more infectious version of bird flu may arise. As a result, the World Health Organization (WHO) has recommended that the world prepare for another pandemic.

SWATTING THE FLU BUG

Vaccinations are the best defense against influenza, but how can we vaccinate against a constantly mutating virus? Each year, the WHO collects samples of influenza virus from 112 centers in 83 countries. Researchers analyze the different strains and identify the three that appear most likely to spread widely. These viruses are grown in fertilized chicken eggs, and viral proteins are then isolated and injected into people to stimulate an immune response that protects against subsequent infections by these strains. Flu shots are often quite effective, especially if the correct strains have been selected. People who are in particular danger from flu

---

airborne substances such as pollen grains typically enter the nose and throat, allergic reactions create the runny nose, sneezing, and congestion typical of "hay fever."

If you suffer excessively from allergies, particularly to pollen or bee or wasp venom, you might be a candidate for allergy shots, also called *allergy vaccinations*, that gradually "train" the immune system to ignore specific allergens. The treatment usually consists of a series of injections of tiny but increasing amounts of the purified allergen. Eventually, allergic symptoms subside, and the beneficial effects may last several years. Although the exact mechanism is not known, allergy vaccinations decrease production of antibodies to allergens, thereby diminishing the inflammatory responses they cause.

An Autoimmune Disease Is an Immune Response Against the Body's Own Molecules

Fortunately, our immune systems rarely mistake our own cells for invaders. Occasionally, however, something goes awry, and "anti-self" antibodies are produced. The reason for this is not understood; one hypothesis is that pathogens may occasionally bear antigens that resemble proteins on an individual's body cells. After an infection with such a pathogen, the antibodies may then mistakenly attack some of the individual's own cells, mistaking them for the pathogen. The result is an **autoimmune disease**, in which the body mounts an immune response against a particular cell type that it should recognize as "self." Some types of ane-

may be given *neuraminidase inhibitors* (such as Tamiflu™). The viral enzyme neuraminidase allows newly formed viruses to escape from their host cells. Administered immediately after flu symptoms begin, the drugs interrupt the viral life cycle and hasten recovery.

Unfortunately, however, the world is poorly prepared for a pandemic. Existing vaccines and previous immunity would be useless. Efforts to develop a vaccine against H5N1 are under way, but the rapid mutation rate of this virus could undermine a vaccine's usefulness. The antiquated method of growing viruses in chicken eggs is time-consuming and vac-cine-production facilities are limited; these factors hinder the rapid production of huge numbers of doses that would be needed to avert a pandemic. Antiviral drugs such as Tamiflu™ can be effective; but already, one strain of H5N1 seems to have become resistant to that drug. Public health officials are hoping to stockpile both antiviral drugs and vaccines against bird flu and make them available as soon as an outbreak is detected, in an effort to halt a potential pandemic before it spreads. Exactly when another pandemic will occur, and how effective the world's response will be, remains an open question.

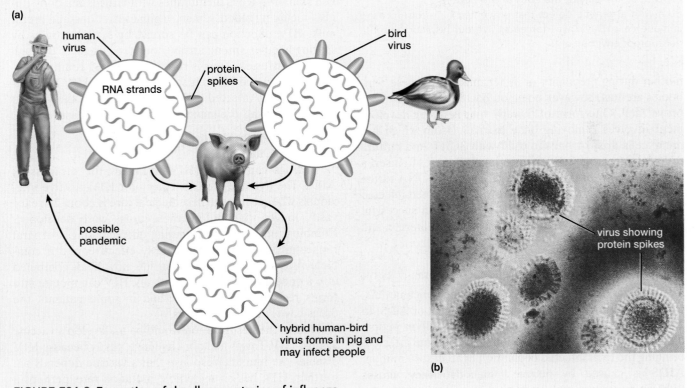

FIGURE E36-2 Formation of deadly new strains of influenza
(a) Rare recombination of genes from bird and human flu viruses in pigs (or people) can result in deadly new strains of flu. **(b)** Transmission electron micrograph of flu viruses. The protein spikes attach to cells in the human respiratory system, helping the virus gain entry into the cells. If these spike proteins are derived from the bird virus, they may be totally new to the human immune system.

mia, for example, are caused by antibodies that destroy an individual's red blood cells. Many cases of juvenile-onset diabetes occur because the immune system attacks the insulin-secreting cells of the pancreas. Multiple sclerosis occurs when immune cells launch a misdirected attack against the insulating fatty sheath that coats parts of neurons in the brain and spinal cord. Rheumatoid arthritis results when the immune system attacks the cartilage of the joints. Unfortunately, there is currently no known cure for autoimmune diseases. Some types of therapy can alleviate the symptoms—for instance, administering insulin to diabetics or blood transfusions to people with anemia. Frequently, the victims are given drugs that suppress the immune system. Immune suppression, however, also reduces immune responses to the everyday assaults of disease microbes, so this therapy has major drawbacks.

An Immune Deficiency Disease Disables the Immune System

David, the "bubble boy," lived all of his 12-year life in a germ-proof "bubble," isolated from direct contact with every unsterilized object, including other people (**FIG. 36-14**). Occasionally, a child like David is born with **severe combined immune deficiency (SCID)**, a disorder in which few or no immune cells are formed due to a defective gene. A child with SCID may survive the first few months of postnatal life, protected by antibodies acquired from the

735

FIGURE 36-14 David, the "boy in the bubble"
Born with a genetic defect that prevented him from forming immune cells, David was forced to live out his short life in a germ-proof environment.

mother during pregnancy or in her milk. Once these antibodies are lost, however, common bacterial infections can prove fatal. One form of therapy that is under development involves removing bone marrow (from which immune cells arise) from the child with SCID. In a culture dish, doctors use genetic engineering techniques to insert a functional copy of the defective gene into the DNA of the marrow cells. The modified marrow cells are then injected into the bloodstream of the SCID victim. In successful cases, the cells return to the bone marrow, proliferate, and stimulate the formation of healthy immune cells.

AIDS Is a Devastating Immune Deficiency Disease

The most common and widespread immune deficiency disease is **acquired immune deficiency syndrome**, or **AIDS**. In 2005, the UN estimated that more than 3 million people died of AIDS and nearly 5 million more became infected, bringing the total infected population to at least 40 million. AIDS is caused by **human immunodeficiency viruses (HIV)**. These viruses undermine the immune system by infecting and destroying the helper T cells, which stimulate both the cell-mediated and humoral immune responses. AIDS does not kill people directly, but AIDS victims become increasingly susceptible to other diseases as their helper T cell populations decline. Although AIDS was first recognized in 1981, genetic studies show that the virus is almost certainly derived from viruses that have been infecting chimpanzees in Africa for thousands of years. Researchers believe that this ancestral virus mutated and acquired the ability to infect people between the mid-1940s and the early 1950s.

The Human Immunodeficiency Virus Infects and Destroys Helper T Cells

How does HIV wreak havoc on the human immune system? Like most viruses, it enters its host cell (in this case a helper T cell) and hijacks the cell's metabolic machinery, forcing it to make more HIV particles that then emerge, taking an outer coating of T cell membrane with them (**FIG. 36-15**). Early in the infection, as the immune system fights the virus, the victim may develop a fever, rash, muscle

aches, headaches, and enlarged lymph nodes. After several months, the rate of viral replication slows. Enough helper T cells remain that infected individuals are able to resist disease, and they generally feel quite well. In some cases, this condition persists for several years. If untreated, however, helper T cell levels eventually diminish, severely weakening the immune response, and at this point the person is considered to have AIDS. As HIV levels skyrocket, they kill more helper T cells, and the person becomes easy prey for other infections. The life expectancy for untreated AIDS victims is about 1 to 2 years.

Human Immunodeficiency Virus Is Transmitted by Body Fluids

HIV cannot survive for very long outside the body. The virus can be transmitted only by the direct contact of broken skin or mucous membranes with virus-laden body fluids—including blood, semen, vaginal secretions, and breast milk. HIV infection can be spread by sexual activity, by sharing needles among intravenous drug users, or through blood transfusions (this is rare in developed countries because all donated blood is screened for anti-HIV antibodies). A woman infected with HIV can transmit the virus to her child during pregnancy, childbirth, or breast feeding.

There Are Partially Effective Treatments, but No Cures, for AIDS

New drugs can disable HIV and slow the progress of AIDS. The average life expectancy for HIV-positive individuals who receive the best medical care is about 24 years, vastly longer than before these drugs were developed. Combinations of drugs targeting different stages of viral replication have been particularly effective, and a complete AIDS treatment regimen has now been combined into a once-a-day pill. Unfortunately, HIV can mutate into forms resistant to the drugs; and in some patients, the drugs have severe side effects.

Clearly, the best solution would be to develop a vaccine against HIV. This is a major challenge, partly because HIV disables the immune response that a vaccine depends on. Further, HIV has an incredible mutation rate, perhaps a thousand times faster than that of flu viruses. Single infected individuals may harbor different strains of HIV in blood and in semen because of mutations that occurred within their bodies after they were first infected. Despite billions of dollars invested in research and in animal and human trials, no HIV vaccine has yet proven effective.

Cancer Can Evade or Overwhelm the Immune Response

Cancer is one of the most dreaded words in the English language, and with good reason. Cancer is second only to heart disease as the leading cause of mortality in the U.S. A sobering 40% of U.S. citizens will eventually contract some form of cancer. Despite decades of intensive research, a satisfactory cure for cancer remains elusive. Why can't we cure or prevent cancer?

Unlike most other diseases, cancer is not a straightforward invasion of the body by a foreign organism. Although some cancers are triggered by viruses, cancer is essentially a failure of the mechanisms controlling the growth of the body's own cells—a disease in which the body destroys it-

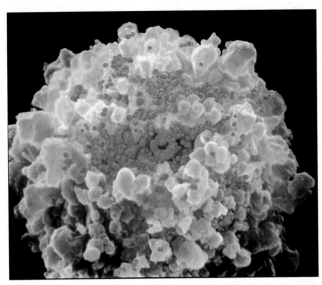

(a)

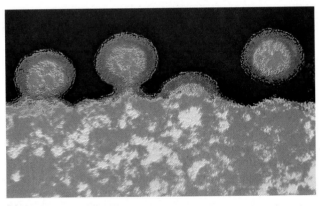

(b)

FIGURE 36-15 HIV causes AIDS
(a) The red specks in this scanning electron micrograph are HIVs that have just emerged from the large, green helper T cell that was infected earlier. **(b)** In this higher-magnification transmission electron micrograph, HIVs are seen emerging from the helper T cell and acquiring an outer envelope of its plasma membrane (green) in the process. This will help them infect new cells.

self. The mechanisms used in regulation of cell division, and the genetic mutations that cause cancer, are discussed in Chapter 11.

Cancer is a disease characterized by the unchecked growth of malignant tumor cells. A **tumor** is a population of cells that has escaped normal regulatory processes and grows at an abnormal rate. The cells in a *benign tumor* remain confined to one area, but the cells of a *malignant tumor* grow uncontrollably and spread to other areas of the body. As a malignant tumor grows, it uses increasing amounts of the body's energy and nutrient supplies and literally squeezes out vital organs nearby.

The Immune System Defends Against Most Cancerous Cells

Cancer cells form in our bodies every day. It is almost impossible to avoid some *carcinogens* (cancer-causing agents), such as gamma rays from the sun, radioactivity from the rocks beneath our feet, and naturally produced carcinogens in our food. Fortunately, natural killer cells and cytotoxic T cells screen the body for cancer cells and destroy nearly all of them before they have a chance to proliferate and spread. Cancer cells are, of course, "self" cells (the body's own cells), and the immune response ideally does not respond to "self." How are these cancer cells weeded out? It is likely that the very processes that cause cells to become cancerous also cause new and slightly different proteins to appear on their surfaces. Cytotoxic T cells encounter these new proteins, recognize them as non-self antigens, and destroy the cancer cells (see Fig. 36-10). However, some cancer cells may evade detection because they do not bear antigens that allow the immune system to recognize them as foreign. Some types of cancers, such as leukemia, suppress the immune system; and others simply grow so fast that the immune response can't keep up. If the tumor grows and spreads, the individual's health depends on medical treatment.

Medical Treatments for Cancer Depend on Selectively Killing Cancerous Cells

The three main forms of cancer treatment are surgery, radiation, and chemotherapy. Surgically removing the tumor is the first step in treating many cancers, but it can be difficult to remove every bit of cancerous tissue. Tumors can be bombarded with radiation, which can destroy even microscopic clusters of cancer cells by disrupting their DNA, preventing their cell division and growth. Unfortunately, neither surgery nor radiation is effective against cancer that has spread throughout the body.

Chemotherapy, or drug treatment, is commonly used to supplement surgery and/or radiation or to treat cancers that cannot be treated any other way. Chemotherapy drugs attack the machinery of cell division, and so they are somewhat selective for cancer cells, which divide much more frequently than do normal cells. Unfortunately, other cells of the body divide too, and chemotherapy inevitably also kills some healthy cells. Damage to dividing cells in patients' hair follicles and intestinal lining by chemotherapy produces its well-known side effects of hair loss, nausea, and vomiting.

A tremendous amount of research has been devoted to the search for cancer treatments that are effective and have few unpleasant side effects. Developing a "cancer vaccine" is a high priority. Other approaches include (1) developing therapies that stimulate the immune system to attack tumors and (2) developing antibodies or T cell receptors that recognize tumor cells and could be used to deliver drugs or radioactive particles directly to tumor cells without affecting healthy cells. Research and clinical trials continue, and cancer patients are beginning to benefit from some of these innovative new treatments.

Without constant surveillance by the body's defenses, it is unlikely that any of us would survive more than a few years. However, we can reduce our own chances of developing cancer by avoiding cigarette smoke and by eating a diet high in fruits and vegetables. Other behaviors that lower the risk of cancer include avoiding excessive alcohol consumption, using sunscreen, and getting regular exercise.

CASE STUDY REVISITED FIGHTING THE FLU

To reassure you that you have not contracted bird flu, your doctor explains that although the death rate from bird flu is high, very few people worldwide have ever contracted this type of flu. Its symptoms are similar to those for other flus, but deaths are caused by pneumonia-like symptoms that sometimes develop. Your doctor gives you a regular flu shot—unfortunately, too late. It takes a few weeks for the flu shot to activate your specific immune system to recognize and remember the virus. A few days after intercepting your classmate's sneeze, your head aches, your throat feels raw, and you develop chills leading to a fever. Not only that, your muscles ache and your skin is hypersensitive. What's going on? Most of the unpleasant symptoms of flu and other infections are caused by the body's nonspecific defenses. As macrophages ingest the virus, they release cytokines such as the pyrogen that causes fever. Cytokines also sensitize pain endings throughout the body; your back aches, old injuries throb, and your skin protests as you change clothes. Cytokines also make you tired. Why do our bodies produce these unpleasant symptoms? Biologists hypothesize that feeling miserable sent our ancestors to the safety of their caves—and sends us to bed—to conserve energy for the battles our bodies are waging against the microscopic invaders.

When you finally recover from the flu, you'll be safe from reinfection this year—until next year's mutated form shows up. If flu shots are available and safe for you, consider getting one early in the flu season.

Consider This When an infection such as a cold or flu makes you feel lousy, how many of your symptoms are caused by your immune system's efforts to fight the infection? Many over-the-counter drugs are intended to suppress symptoms (fever, cough, runny nose, aches, etc.). Argue for and against taking medicines that suppress the symptoms of illness.

CHAPTER REVIEW

SUMMARY OF KEY CONCEPTS

36.1 What Are the Basic Mechanisms of Defense Against Disease?

First, nonspecific external barriers, including anatomical structures and secretions such as skin, hair, cilia, and mucus, prevent disease-causing organisms from easily entering the body. Second, nonspecific internal defenses consisting of a variety of white blood cells not only destroy microbes, toxins, and cancerous and infected body cells but also scavenge dead and dying cells. Finally, the specific immune response selectively destroys the particular toxin or microbe and "remembers" the invader, allowing a faster response if it reappears in the future.

Web Tutorial 36.1 The Inflammatory Response

36.2 How Do Nonspecific Defenses Function?

The skin and its secretions physically block the entry of microbes into the body and inhibit their growth. The mucous membranes of the respiratory and digestive tracts secrete antibiotic substances and mucus that traps microbes. If microbes do enter the body, they are engulfed by white blood cells. Natural killer cells secrete proteins that kill infected or cancerous cells. Injuries stimulate the inflammatory response, in which chemicals are released that attract phagocytic white blood cells, increase blood flow, and make capillaries leaky. Later, blood clots wall off the injury site. Fever is caused by endogenous pyrogens, chemicals released by white blood cells in response to infection. High temperatures inhibit bacterial growth and accelerate the immune response.

Web Tutorial 36.2 Humoral Versus Cell-Mediated Immunity

36.3 What Are the Key Characteristics of the Immune Response?

The immune response involves two types of immune cells or lymphocytes: B cells and T cells. B cells give rise to plasma cells, which secrete antibodies into the bloodstream, causing humoral immunity. Cytotoxic T cells destroy some microbes, cancer cells, and virus-infected cells on contact, causing cell-mediated immunity. Helper T cells stimulate both the humoral and cell-mediated immune responses. Immune responses have three steps: (1) recognition, (2) attack, and (3) memory.

First, antibodies (made by B cells or their offspring, plasma cells) and T cell receptors (on T cells) recognize foreign antigens and trigger the immune response. Antibodies are Y-shaped proteins composed of a constant region and a variable region. Antigens are molecules that generate an antibody response. Antibodies bind and help destroy antigens. Each B cell synthesizes only one type of antibody, unique to that particular cell and its progeny. The diversity of antibodies arises from gene shuffling during B cell development. Each antibody has specific sites that bind only one or a few types of antigen. Normally, only foreign antigens are recognized by the B cells.

Second, antibodies attack the invaders. Antigens from an invader bind to and activate only those B and T cells with the complementary antibodies or T cell receptors. In humoral immunity, B cells with the proper antibodies, stimulated by the presence of particular antigens, divide to produce plasma cells that synthesize massive quantities of the antibody. The circulating antibodies destroy antigens and antigen-bearing microbes. In cell-mediated immunity, T cells with the proper receptors bind antigens and divide rapidly. Cytotoxic T cells bind to antigens on microbes, infected cells, or cancer cells and kill the cells. Helper T cells chemically stimulate both the B-cell and cytotoxic T-cell responses.

Finally, some progeny cells of both B and T cells are long-lived memory cells. If the same antigen reappears in the bloodstream, these memory cells are immediately activated, dividing rapidly and causing an immune response that is much faster and more effective than the original response.

Web Tutorial 36.3 Memory B Cells and the Immune Response

36.4 How Does Medical Care Augment the Immune Response?

Antibiotics kill microbes or slow down their reproduction, allowing the body's defenses more time to respond and exterminate the invaders. Vaccinations are injections of antigens from disease organisms, in some cases the weakened or dead microbes themselves. An immune response is evoked by the antigens, providing memory and a rapid response should a real infection occur later.

36.5 What Happens When the Immune System Malfunctions?

Allergies are immune responses to normally harmless foreign substances. B cells treat these as antigens and produce "allergy antibodies" that bind to mast cells. When exposed to the antigen, the mast cells release histamine, causing a local inflammatory response. Autoimmune diseases arise when the immune system mistakes the body's own cells for foreign invaders and destroys them. Immune deficiency diseases occur when the immune system cannot respond strongly enough to ward off usually minor diseases.

Infection with human immunodeficiency viruses (HIV) nearly always leads to AIDS (acquired immune deficiency syndrome). These viruses invade and destroy helper T cells. Without helper T cells to stimulate the immune responses of B cells and cytotoxic T cells, an individual with AIDS is extremely susceptible to a wide assortment of infections, which are eventually fatal. New antiviral drug combinations have vastly increased the life expectancy and quality of life for HIV-infected victims with access to good medical care.

Cancer is a population of the body's cells that multiplies without control. Cancerous cells may be recognized as "different" by the immune system and destroyed by natural killer cells and cytotoxic T cells. A few evolve the capacity to evade the immune system; some attack immune cells; and others multiply too fast for the immune system to keep up. In these cases, cancer develops.

Web Tutorial 36.4 HIV: The AIDS Virus

KEY TERMS

acquired immune deficiency syndrome (AIDS) *page 736*
allergy *page 733*
antibody *page 726*
antigen *page 726*
autoimmune disease *page 734*
B cell *page 726*
cancer *page 737*
cell-mediated immunity *page 729*
clonal selection *page 729*

complement system *page 730*
constant region *page 726*
cytokine *page 724*
cytotoxic T cell *page 730*
fever *page 723*
helper T cell *page 729*
histamine *page 724*
human immunodeficiency virus (HIV) *page 736*
humoral immunity *page 729*
immune response *page 722*

immune system *page 725*
inflammatory response *page 724*
leukocyte *page 723*
lymphocyte *page 726*
macrophage *page 723*
major histocompatibility complex (MHC) *page 728*
mast cell *page 724*
memory B cell *page 729*
memory T cell *page 730*
microbe *page 722*

natural killer cell *page 724*
neutrophil *page 723*
parasite *page 722*
pathogen *page 722*
phagocyte *page 723*
plasma cell *page 729*
severe combined immune deficiency (SCID) *page 735*
T cell *page 726*
T-cell receptor *page 726*
tumor *page 737*
vaccination *page 733*
variable region *page 726*

THINKING THROUGH THE CONCEPTS

1. List the human body's three lines of defense against invading microbes. Which are nonspecific (that is, act against all types of invaders), and which are specific (act only against a particular type of invader)? Explain your answers.

2. How do natural killer cells and cytotoxic T cells destroy their targets?

3. Describe humoral immunity and cell-mediated immunity. Include in your answer the types of immune cells involved in each, the location of antibodies and receptors that bind foreign antigens, and the mechanisms by which invading cells are destroyed.

4. How does the immune system construct so many different antibodies?

5. How does the body distinguish "self" from "nonself"?

6. Diagram the structure of an antibody. What parts bind to antigens? Why does each antibody bind only to a specific antigen?

7. What are memory cells? How do they contribute to long-lasting immunity to specific diseases?

8. What is a vaccination? How does it confer immunity to a disease?

9. How does the inflammatory response help the body resist disease? What allergy symptoms does it cause?

10. Distinguish between autoimmune diseases and immune deficiency diseases, and give one example of each.

11. Describe the causes and eventual outcome of AIDS. How do AIDS treatments work? How is HIV spread?

12. Why is cancer sometimes fatal?

APPLYING THE CONCEPTS

1. Why is it essential that antibodies and T-cell receptors bind only relatively large molecules (such as proteins) and not relatively small molecules (such as amino acids)?

2. The essay "Health Watch: Is a Bird Flu Pandemic Imminent?" states that the flu virus is different each year. If that is true, what good does it do to get a "flu shot" each winter?

3. Argue for and against the experiment that Jenner performed. Would he have been able to do this in modern times?

4. Organ transplant patients typically receive the drug cyclosporine. This drug inhibits the production of a cytokine that stimulates helper T cells to proliferate. How does cyclosporine prevent rejection of transplanted organs? Some patients who received successful transplants many years ago are now developing various kinds of cancers. Propose a hypothesis to explain this phenomenon.

FOR MORE INFORMATION

Hamilton, G. "Filthy Friends." *Scientific American*, April 16, 2005. Does keeping our environment ultraclean alter the immune system in ways that predispose people to allergies and asthma?

MacKenzie, D. "The Bird Flu Threat." *New Scientist*, January 7, 2006. Could bird flu give rise to the next pandemic?

Taubenberger, J. K., Reid, A. H., and Fanning, T. G. "Capturing a Killer Flu Virus" *Scientific American*, January 2005. Describes the detective work leading to the resurrection of the deadliest flu in recorded history.

37

Chemical Control of the Animal Body: The Endocrine System

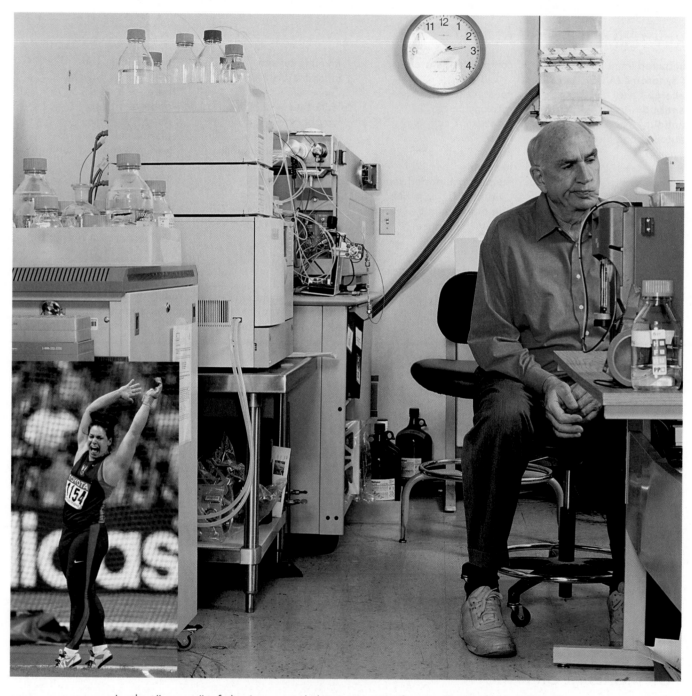

In the "game" of drug use and detection, Professor Catlin finds himself in competition with some of the world's greatest athletes. He is frustrated by this ongoing battle and the tainting of sports by cheaters. (Inset) Melissa Price celebrates a winning hammer throw.

AT A GLANCE

CASE STUDY LOSING ON ARTIFICIAL HORMONES

IT WAS SUMMER OF 2003 when officials at the United States Anti-Doping Agency received an anonymous tip. The informant, now identified as a well-known track and field coach, claimed that an undetectable steroid was being used by professional athletes. As proof, the informant mailed the agency a used syringe, still containing traces of the substance. Officials asked Dr. Don Catlin and his team of scientists at the UCLA Olympic Analytic Laboratory to identify it.

In an unmarked, nondescript building, Catlin works with a team of about 40. Filled with millions of dollars of high-tech equipment, this world-renowned lab can test for more than 200 substances banned by athletic associations. Of these, probably the most notorious are anabolic steroids. *Anabolic steroid* is a term used to describe any of dozens of performance-enhancing drugs whose chemical composition resembles that of the male sex hormone testosterone. Bodybuilders often take them, and some Olympic and professional athletes have been caught illegally exploiting them. The chemical signature they leave in the body can be detected in urine for months after a person stops using them.

With a team of chemists, Professor Catlin set about analyzing the tiny amount of substance that could be rinsed from the syringe to determine its chemical formula. Based on its structure, Catlin dubbed it tetrahydrogestrinone, or THG. The molecule was new to science, proving that creating new, undetectable versions of synthetic testosterone is a big enough business to engage the efforts of skilled chemists, and suggesting that more versions may be in the works.

The U.S. Food and Drug Administration (FDA) quickly banned the new substance. Because officials keep urine samples collected at important athletic competitions in cold storage for years, they retrieved samples for retesting. The results were disturbing. Although THG was probably manufactured by only one California lab, the substance had tainted professional football and baseball as well as track and field competitions. For example, hammer thrower Melissa Price (inset) tested positive for THG after winning the 2003 U.S. Championship. Although she denies using THG, she and several other athletes have been stripped of their titles and banned from competition for two years based on results of the urine tests devised by Catlin.

As you read this chapter, notice how many different effects the same hormone can have throughout the body. What are the effects of taking anabolic steroids? Are there health risks associated with their use?

37.1 HOW DO ANIMAL CELLS COMMUNICATE?

In all multicellular organisms, individual cells must remain in continuous communication with one another (Table 37-1). In some specialized tissues, such as heart muscle, *gap junctions* directly link the insides of cells, allowing the flow of ions and electrical signals. More commonly, cells release chemical signaling molecules sometimes called "messenger molecules" that affect other cells, either nearby or distant. Like conversation at a crowded party, this communication is directed to specific "target" cells (the people you are talking to) and not to others (nearby guests holding their own conversations). To ensure that the chemical message reaches the appropriate targets, cells have **receptors**—specialized protein molecules that bind only to specific chemical messengers. Receptors may be located either on the plasma membrane or inside target cells. Upon binding to its receptor, the chemical triggers some type of change within the target cell.

There are three general classes of messenger molecules, each using a different distribution system: *local hormones* diffuse through the interstitial fluid to cells in the immediate vicinity; *endocrine hormones* are released into the blood, where they are distributed to both nearby and distant cells; and *neurotransmitters* are released across a very narrow gap (the *synaptic cleft*) between a specialized region of a neuron and its target (see Table 37-1). This chapter deals with endocrine hormones and local hormones called *prostaglandins*. You will learn more about neurotransmitters in Chapter 38.

37.2 WHAT ARE THE CHARACTERISTICS OF ANIMAL HORMONES?

Local Hormones Diffuse to Nearby Target Cells

Most cells secrete **local hormones** into their immediate vicinity; for example, the cytokines described in Chapter 36 allow immune cells to communicate. **Prostaglandins**, modified fatty acids synthesized from membrane phospholipids, are another type of local hormone (look ahead to Table 37-2). Unlike most other hormones, which are synthesized by a limited number of specialized cells, prostaglandins are produced by cells throughout the body. Research on this diverse and potent family of compounds is still in its infancy; several prostaglandins are known, and a great many more undoubtedly await discovery. Researchers have found a prostaglandin that causes constriction of arteries in the umbilical cord at birth, so bleeding is halted. Another works in conjunction with oxytocin during labor, stimulating uterine contractions. Some prostaglandins contribute to inflammation (such as occurs in arthritic joints) and stimulate pain receptors. Drugs such as aspirin and ibuprofen provide relief from these symptoms by blocking enzymes that lead to prostaglandin synthesis. The use of local hormones such as prostaglandins to communicate with nearby cells is called *paracrine communication* ("para," appropriately, means "beside"), whereas *endocrine communication* ("endo" means "inside") uses chemicals that travel within the bloodstream, often over considerable distances.

Hormones of the Endocrine System Are Transported by the Circulatory System

Endocrine hormones are chemical messages produced by specialized cells; often they are released in response to some stimulus from inside or outside the body. There are three classes of vertebrate endocrine hormones (Table 37-2): **peptide hormones**, made from chains of amino acids; **amino acid derived hormones**, which are synthesized from one or two amino acids; and **steroid hormones**, which resemble cholesterol, from which most steroid hormones are synthesized.

Endocrine hormones are carried by the circulatory system and influence target cells bearing specific receptors for them. The changes induced by hormonal messages may

Table 37-1 How Cells Communicate

Communication		Chemical Messengers	Mechanism of Transmission	Examples
Direct		Ions, small molecules	Direct movement through gap junctions linking cytosol of adjacent cells	Ions flowing between cardiac muscle cells
Paracrine		Local hormones	Diffusion through interstitial fluid to nearby cells bearing receptors	Prostaglandins
Endocrine		Hormones	Carried in the bloodstream to near or distant cells bearing receptors	Insulin
Synaptic		Neurotransmitters	Diffusion from a neuron across a narrow space (synaptic cleft) to a cell bearing receptors	Acetylcholine

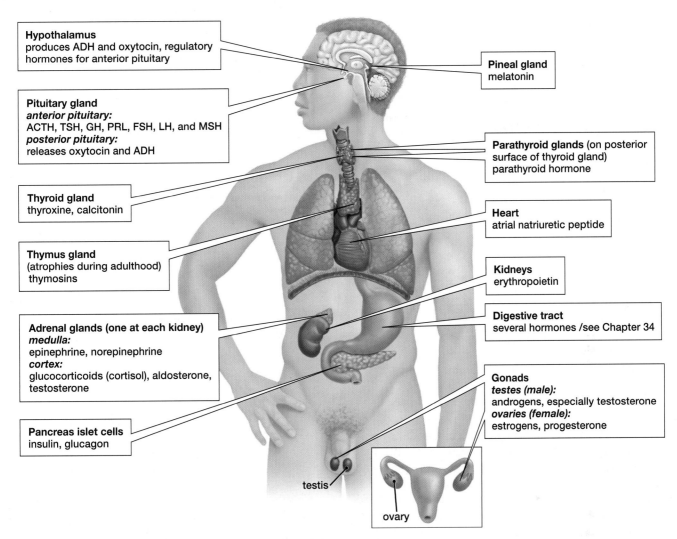

Hypothalamus
produces ADH and oxytocin, regulatory hormones for anterior pituitary

Pineal gland
melatonin

Pituitary gland
anterior pituitary:
ACTH, TSH, GH, PRL, FSH, LH, and MSH
posterior pituitary:
releases oxytocin and ADH

Parathyroid glands (on posterior surface of thyroid gland)
parathyroid hormone

Thyroid gland
thyroxine, calcitonin

Heart
atrial natriuretic peptide

Thymus gland
(atrophies during adulthood)
thymosins

Kidneys
erythropoietin

Adrenal glands (one at each kidney)
medulla:
epinephrine, norepinephrine
cortex:
glucocorticoids (cortisol), aldosterone, testosterone

Digestive tract
several hormones /see Chapter 34

Gonads
testes (male):
androgens, especially testosterone
ovaries (female):
estrogens, progesterone

Pancreas islet cells
insulin, glucagon

testis

ovary

FIGURE 37-1 Major mammalian endocrine glands and their secretions

be prolonged and irreversible, such as those occurring at puberty or during the metamorphosis of a tadpole into a frog or a caterpillar into a butterfly. More typically, the induced changes are transient and reversible, helping to control and regulate the interrelated physiological systems that compose the animal body. Regulation of the body requires communication; in animal bodies, hormones provide much of that communication. In fact, the **endocrine system**—consisting of hormones and the various cells that secrete them—can be viewed as the "postal system of physiology," moving information and instructions between cells that may be some distance apart. The major mammalian endocrine glands are illustrated using the human body in **FIGURE 37-1.**

Hormones Bind Specific Receptors on Target Cells

Because nearly all cells have a blood supply, a hormone released into the bloodstream will reach nearly every cell of the body. However, a given hormone acts only on **target cells** bearing receptors for that particular hormone molecule, thus exerting precise control; cells that lack the appropriate receptor will not respond to the hormonal message (**FIG. 37-2**). In addition, a given hormone may have several different effects, depending on the nature of the receptor on the target cell it contacts. Receptors for hormones are found in two general locations on target cells: on the plasma membrane and inside the cell, within the cytosol or the nucleus.

Many peptide and amino acid derived hormones are soluble in water but not in lipids; therefore, these hormones cannot diffuse through the phospholipid bilayer of the plasma membrane. Instead, most of these hormones bind to receptors on the target cell's plasma membrane (**FIG. 37-3**). These *membrane receptors* are large proteins that span the plasma membrane, so a hormone binding to the external portion of the receptor can cause a shape change in the portion of the receptor protein protruding into the cell. This physical transformation triggers reactions (starting with a molecule called a *G protein* associated with the receptor)

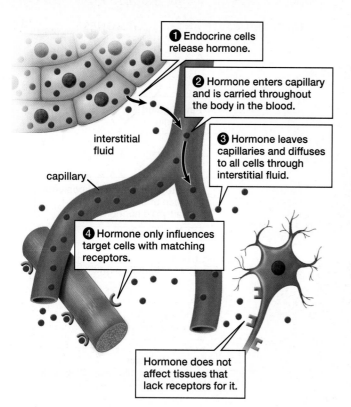

1 Endocrine cells release hormone.

2 Hormone enters capillary and is carried throughout the body in the blood.

3 Hormone leaves capillaries and diffuses to all cells through interstitial fluid.

interstitial fluid

capillary

4 Hormone only influences target cells with matching receptors.

Hormone does not affect tissues that lack receptors for it.

FIGURE 37-2 A hormone reaches its target

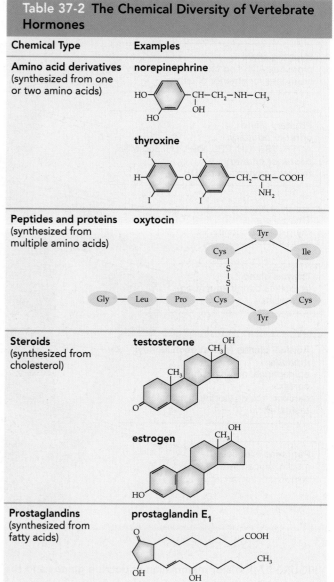

Table 37-2 The Chemical Diversity of Vertebrate Hormones

Chemical Type	Examples
Amino acid derivatives (synthesized from one or two amino acids)	norepinephrine, thyroxine
Peptides and proteins (synthesized from multiple amino acids)	oxytocin
Steroids (synthesized from cholesterol)	testosterone, estrogen
Prostaglandins (synthesized from fatty acids)	prostaglandin E₁

that generate a **second messenger** molecule inside the cell. The second messenger transfers the signal from the first messenger (the hormone) to other molecules within the cell, often starting a series of biochemical reactions (see Fig. 37-3). Although a variety of molecules can act as second messengers, in many cases hormones binding to receptors cause ATP to be converted to **cyclic AMP** (*cAMP*), a nucleotide that regulates many cellular activities (see Chapter 3). Cyclic AMP, in its role of second messenger, initiates a chain of reactions inside the cell. Each reaction in the chain involves an increasing number of molecules, amplifying the original signal. The end result varies with the target cell; channels may be opened in the plasma membrane, or substances may be synthesized or secreted. For example, the hormone *epinephrine* (also called *adrenaline*) binds to membrane receptors on heart muscle, triggers cAMP formation, and starts a chain of molecular events that causes stronger contraction in the cardiac muscle. This is one of several ways in which epinephrine helps your body prepare for emergency situations, as described later in this chapter.

Steroid hormones, in contrast, are lipid soluble and so can diffuse through cell membranes and bind to receptors inside the cell—either in the cytosol or in the nucleus. Having latched onto a steroid hormone, these receptors act by regulating the activity of genes. Some steroid receptors are in the nucleus; others wait in the cytosol, where they bind the hormone and transport it into the nucleus. In the nucleus, the receptor–hormone complex attaches to DNA and stimulates particular genes to transcribe messenger RNA. The messenger RNA then travels into the cytosol and directs the synthesis of a protein (**FIG. 37-4**). In hens, for example, the steroid protein estrogen promotes tran-

scription of the albumin gene, causing the synthesis of albumin (egg-white protein) that is packaged into the egg to feed the developing chick. Steroid hormones may take minutes or even days to exert their full effects. Researchers have also found receptors for steroid hormones on the plasma membrane; these receptors make steroids extremely versatile in their signaling mechanisms.

Hormone Release Is Regulated by Feedback Mechanisms

If a hormone is to be useful for physiological control, there must be a way to turn its message both on and off. In animals, the switching mechanism usually involves negative feedback: secretion of a hormone stimulates a response in target cells; the response then inhibits further secretion of

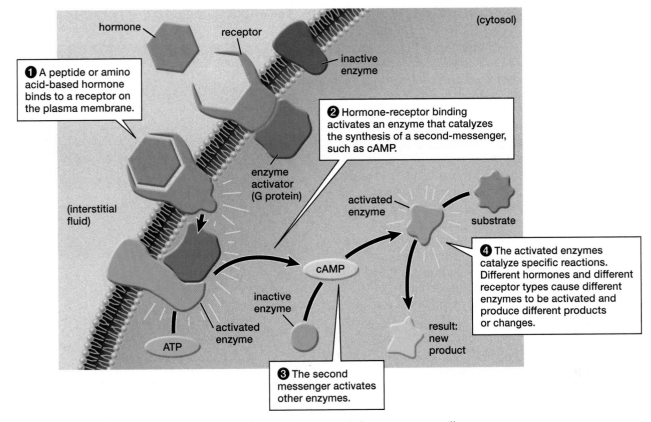

FIGURE 37-3 How peptide or amino acid derived hormones influence target cells

① A peptide or amino acid-based hormone binds to a receptor on the plasma membrane.

② Hormone-receptor binding activates an enzyme that catalyzes the synthesis of a second-messenger, such as cAMP.

③ The second messenger activates other enzymes.

④ The activated enzymes catalyze specific reactions. Different hormones and different receptor types cause different enzymes to be activated and produce different products or changes.

the hormone. Most hormones exert powerful effects on the body that could be harmful if prolonged; therefore, control of hormone release by negative feedback is especially important. Suppose you have jogged a few miles on a hot, sunny day and have lost a pint of water through perspiration. In response, your pituitary gland releases *antidiuretic*

hormone (ADH), which causes your kidneys to reabsorb more water and to produce concentrated urine (see Chapter 35). However, if you arrive home and drink a quart of Gatorade®, you will more than replace the water you lost. Continued retention of excess water could raise blood pressure and strain the circulatory system. Negative feedback

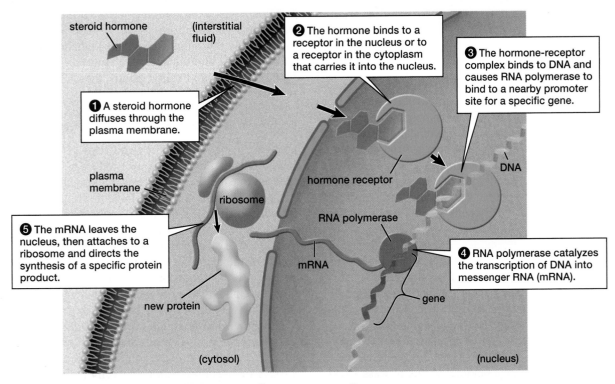

① A steroid hormone diffuses through the plasma membrane.

② The hormone binds to a receptor in the nucleus or to a receptor in the cytoplasm that carries it into the nucleus.

③ The hormone-receptor complex binds to DNA and causes RNA polymerase to bind to a nearby promoter site for a specific gene.

④ RNA polymerase catalyzes the transcription of DNA into messenger RNA (mRNA).

⑤ The mRNA leaves the nucleus, then attaches to a ribosome and directs the synthesis of a specific protein product.

FIGURE 37-4 How many steroid hormones influence target cells

ensures that ADH secretion is turned off when the water content of the blood returns to normal, allowing the kidneys to begin eliminating the excess. Look for more examples of negative feedback as you read through this chapter.

In a few cases, positive feedback stimulates hormone release, at least for a short time. For example, as described in Chapter 31, contractions of the uterus early in childbirth stretch the cervix and cause the posterior pituitary to release the hormone **oxytocin**. Oxytocin stimulates stronger contractions of the uterus, which cause more oxytocin to be released, creating a positive feedback cycle. Simultaneously, oxytocin causes uterine cells to release prostaglandins, which further enhance uterine contractions—another example of positive feedback. But positive feedback systems elicit negative feedback that limits their duration. In this case, the uterine contractions cause the eventual birth of the baby, eliminating the stretching of the cervix and halting the positive feedback cycle that sustained and enhanced the uterine contractions.

Vertebrate and Invertebrate Endocrine Hormones Show Striking Similarities

Although invertebrates make up about 95% of all species on Earth, their hormones are not as well known as those of vertebrates. One reason is that invertebrate life cycles, and therefore the hormones that regulate them, are far more diverse than those of vertebrates. However, researchers have discovered that invertebrates have both peptide and steroid hormones that utilize the basic signaling mechanisms described earlier. For example, scientists have found estrogen and testosterone in snails, and these hormones appear to regulate sexual differentiation as the animals develop, as they do in people. This makes snails a target for endocrine-disrupting pollutants, as described in "Earth Watch: Endocrine Deception."

Insect Molting Is Controlled by a Steroid Hormone

Insect hormones have been relatively well studied because of their potential use in controlling insect pests. Insects are supported by an external skeleton composed of nonliving, rigid cuticle that they must molt periodically in order to grow. Molting is controlled by the steroid hormone **ecdysone**, often called *molting hormone*. Like many vertebrate steroid hormones, ecdysone acts on receptors located within the nucleus and affects gene transcription. As old cuticle becomes tight, sensory cells stimulate release of a hormone that in turn stimulates ecdysone secretion. As occurs with vertebrate hormones, ecdysone affects cells throughout the insect's body. It initiates a complex process in which the epithelial cells detach from the old cuticle and secrete a soft new cuticle beneath it. The insect then expands its body by pumping itself full of air. This splits open the old cuticle and stretches out the new one to accommodate some future growth. As the insect emerges, it leaves an insect-shaped cuticle behind (**FIG. 37-5**). Researchers have exploited their understanding of this process to devise pesticides that are selective for insects and far less

FIGURE 37-5 Insect molting
A pale cicada emerges from its shed cuticle.

toxic to vertebrates than are many of the poisons traditionally sprayed on crops. These new insecticides permanently bind to ecdysone receptors, stimulating them and causing larval insects to molt prematurely and die.

37.3 WHAT ARE THE STRUCTURES AND HORMONES OF THE MAMMALIAN ENDOCRINE SYSTEM?

Endocrinologists don't fully understand how animal hormones work. New hormones and new roles for known hormones are discovered nearly every year. Key functions of the major endocrine glands and endocrine organs, however, have been known for many years. Here we will focus on the endocrine functions of the hypothalamus–pituitary complex, the thyroid and parathyroid glands, the pancreas, the sex organs, and the adrenal glands (see Fig. 37-1). Table 37-3 lists these and other glands, their major hormones, and their principal functions.

Mammals Have Both Exocrine and Endocrine Glands

There are two basic types of glands: exocrine glands and endocrine glands. **Exocrine glands** produce secretions that are released outside the body ("exo" means "out of" in Greek) or into the digestive tract (a hollow tube continuous with the outside world). Exocrine gland secretions are released through tubes or openings called **ducts**. The exocrine glands include the sweat glands and oil-producing (sebaceous) glands of the skin, the tear-producing (lacrimal) glands of the eye, and the milk-producing (mammary) glands; also included are glands that produce digestive secretions, such as the salivary glands and some cells of the pancreas.

Endocrine glands, sometimes called *ductless glands*, release their hormones within the body ("endo" means "inside of"). An endocrine gland generally consists of clusters of hormone-producing cells embedded within a network of capillaries. The cells secrete their hormones

Table 37-3 Major Mammalian Endocrine Glands and Hormones

Endocrine Gland	Hormone	Type of Chemical	Principal Function
Hypothalamus (to anterior pituitary)	Releasing and inhibiting hormones	Peptides	At least nine hormones; releasing hormones stimulate release of hormones from anterior pituitary; inhibiting hormones inhibit release of hormones from anterior pituitary
Anterior pituitary	Follicle-stimulating hormone (FSH)	Peptide	*Females:* stimulates growth of egg follicles, estrogen secretion, perhaps ovulation *Males:* stimulates spermatogenesis
	Luteinizing hormone (LH)	Peptide	*Females:* stimulates ovulation, growth of corpus luteum, and secretion of estrogen and progesterone. *Males:* stimulates secretion of testosterone
	Thyroid-stimulating hormone (TSH)	Peptide	Stimulates thyroid to release thyroxine
	Adrenocorticotropic hormone (ACTH)	Peptide	Stimulates adrenal cortex to release hormones, especially glucocorticoids, such as cortisol
	Growth hormone (GH)	Peptide	Stimulates growth, protein synthesis, and fat metabolism; inhibits sugar metabolism
	Prolactin (PRL)	Peptide	Stimulates milk synthesis in and secretion from mammary glands
	Melanocyte-stimulating hormone (MSH)	Peptide	Promotes synthesis of brown skin pigment, melanin
Hypothalamus (via posterior pituitary)	Antidiuretic hormone (ADH)	Peptide	Promotes reabsorption of water from kidneys; constricts arterioles
	Oxytocin	Peptide	*Females:* stimulates contraction of uterine muscles during childbirth, milk ejection, and maternal behaviors *Males:* may facilitate ejaculation of sperm
Thyroid	Thyroxine	Amino acid derivative	Increases metabolic rate of most body cells; increases body temperature; regulates growth and development
Parathyroid	Parathyroid hormone	Peptide	Increases blood calcium by stimulating calcium release from bones, absorption by intestines, and reabsorption by the kidneys
Pancreas	Insulin	Peptide	Decreases blood glucose levels by increasing uptake of glucose into cells and converting glucose to glycogen, especially in liver; regulates fat metabolism
	Glucagon	Peptide	Converts glycogen to glucose, raising blood glucose levels
Ovaries[a]	Estrogen	Steroid	Causes development of female secondary sexual characteristics and maturation of eggs; promotes growth of uterine lining
	Progesterone	Steroid	Stimulates development of uterine lining and formation of placenta
Testes[a]	Testosterone	Steroid	Stimulates development of genitalia and male secondary sexual characteristics; stimulates spermatogenesis
Adrenal medulla	Epinephrine (adrenaline) and norepinephrine (noradrenaline)	Amino acid derivatives	Increase levels of sugar and fatty acids in blood; increase metabolic rate; increase rate and force of contractions of the heart; constrict some blood vessels
Adrenal cortex	Glucocorticoids (cortisol)	Steroid	Increase blood sugar; regulate sugar, lipid, and fat metabolism; anti-inflammatory effects
	Aldosterone	Steroid	Increases reabsorption of salt in kidney
	Testosterone	Steroid	Causes masculinization of body features, growth
Other Sources of Hormones			
Pineal gland	Melatonin	Amino acid derivative	Regulates seasonal reproductive cycles and sleep–wake cycles; may regulate onset of puberty
Thymus	Thymosin	Peptide	Stimulates maturation of cells of immune system
Kidney	Renin	Peptide	Acts on blood proteins to produce hormone (angiotensin) that regulates blood pressure
	Erythropoietin	Peptide	Stimulates red blood cell synthesis in bone marrow
Heart	Atrial natriuretic peptide (ANP)	Peptide	Increases salt and water excretion by kidneys; lowers blood pressure
Digestive tract[b]	Secretin, gastrin, cholecystokinin, and others	Peptides	Control secretion of mucus, enzymes, and salts in digestive tract; regulate peristalsis
Fat cells	Leptin	Peptide	Regulates appetite; stimulates immune function; promotes blood vessel growth; required for onset of puberty

[a]See Chapters 40 and 41.

[b]See Chapter 34.

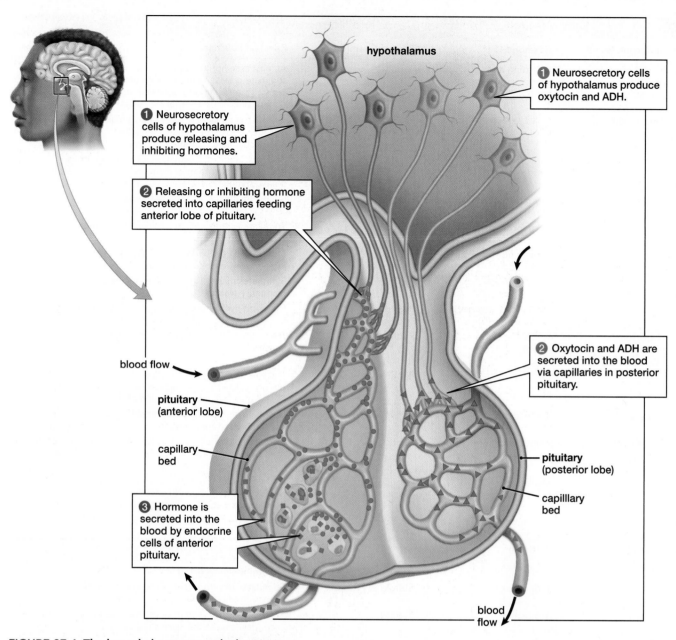

hypothalamus

❶ Neurosecretory cells of hypothalamus produce releasing and inhibiting hormones.

❶ Neurosecretory cells of hypothalamus produce oxytocin and ADH.

❷ Releasing or inhibiting hormone secreted into capillaries feeding anterior lobe of pituitary.

❷ Oxytocin and ADH are secreted into the blood via capillaries in posterior pituitary.

blood flow

pituitary (anterior lobe)

capillary bed

pituitary (posterior lobe)

capilllary bed

❸ Hormone is secreted into the blood by endocrine cells of anterior pituitary.

blood flow

FIGURE 37-6 The hypothalamus controls the pituitary
Neurosecretory cells of the hypothalamus control hormone release in the anterior lobe of the pituitary by producing releasing or inhibiting hormones (left). These cells secrete their hormones into a capillary network that carries them to the anterior pituitary. There, each hormone stimulates endocrine cells with appropriate receptors, while leaving other types unaffected. The posterior lobe of the pituitary (right) is an extension of the hypothalamus. Neurosecretory cells in the hypothalamus have endings on a capillary bed in the posterior lobe of the pituitary, where the cells release oxytocin or antidiuretic hormone (ADH). QUESTION What advantage is gained by having nerve cells in the hypothalamus participate in controlling the release of pituitary hormones?

into the interstitial fluid surrounding the capillaries (see Fig. 37-2). The hormones then enter the capillaries by diffusion and are carried throughout the body in the bloodstream. We focus on endocrine glands in the rest of this chapter.

The Hypothalamus Controls Secretions of the Pituitary Gland

If the endocrine system is the body's postal service, then the hypothalamus is the main post office. Together, these structures coordinate the action of many key hormonal messaging systems. The **hypothalamus** is a part of the brain that contains clusters of specialized nerve cells called *neurosecretory cells*. **Neurosecretory cells** synthesize peptide hormones, store them, and release them when stimulated. The **pituitary gland** is a pea-sized gland that dangles from the hypothalamus by a stalk. The pituitary consists of two distinct parts: the **anterior pituitary** and the **posterior pituitary** (FIG. 37-6). The hypothalamus controls the release of hormones from both parts. The anterior pituitary

(a)

(b)

FIGURE 37-7 Growth hormone effects on the body (a) Too little growth hormone, or a lack of functioning receptors for it, creates one type of dwarfism. In the mid-1800s, "Tom Thumb" (Charles Stratton), whose adult height was 33 inches (84 cm), was a hit with the Barnum and Bailey Circus. He is shown here with P. T. Barnum. (b) Too much growth hormone causes gigantism, as illustrated by Robert Wadlow, who reached a height of 8 feet 11 inches (2.72 meters). Wadlow is shown here with his two younger brothers who did not suffer from this disorder. QUESTION Why is gigantism usually more difficult to treat than dwarfism?

is a true endocrine gland, composed of several types of hormone-secreting cells enmeshed in a network of capillaries. The posterior pituitary, however, consists mainly of a capillary bed and the endings of neurosecretory cells whose cell bodies are in the hypothalamus.

Hypothalamic Hormones Control the Anterior Pituitary

Neurosecretory cells of the hypothalamus produce at least nine peptide hormones that regulate the release of hormones from the anterior pituitary. These peptides are called **releasing hormones** or **inhibiting hormones**, depending on whether they stimulate or inhibit the release of a particular pituitary hormone. Releasing and inhibiting hormones are synthesized in nerve cells in the hypothalamus, secreted into a capillary bed in the stalk connecting the hypothalamus to the pituitary, and travel a short distance through blood vessels to a second capillary bed that surrounds the endocrine cells of the anterior pituitary. There, the releasers and inhibitors diffuse out of the capillaries and influence pituitary hormone secretion.

Because the releasing and inhibiting hormones are secreted very close to the anterior pituitary, they are produced only in tiny amounts. Not surprisingly, they were extremely difficult to isolate and study. Andrew Schally and Roger Guillemin, U.S. endocrinologists who shared the Nobel Prize in Medicine in 1977 for characterizing several of these hormones, used the brains of millions of sheep and pigs (obtained from slaughterhouses) to extract enough releasing hormone for chemical analysis.

The Anterior Pituitary Produces and Releases a Variety of Hormones

The anterior pituitary produces several peptide hormones. Four of these regulate hormone production in other endocrine glands. **Follicle-stimulating hormone (FSH)** and **luteinizing hormone (LH)** stimulate the production of sperm and testosterone in males and the production of eggs, estrogen, and progesterone in females. We will discuss the roles of FSH and LH further in Chapter 40. **Thyroid-stimulating hormone (TSH)** stimulates the thyroid gland to release its hormones, and **adrenocorticotropic hormone (ACTH;** "hormone that stimulates the adrenal cortex") causes the release of the hormone *cortisol* from the adrenal cortex, described later in this chapter.

The remaining hormones of the anterior pituitary do not act on other endocrine glands. **Prolactin**, in conjunction with other hormones, stimulates the development of mammary glands (milk-producing exocrine glands within the breasts) during pregnancy. **Melanocyte-stimulating hormone (MSH)** causes the synthesis of the skin pigment melanin, which determines skin color. **Growth hormone** regulates the body's growth by acting on nearly all the body's cells—increasing protein synthesis, fat utilization, and the storage of carbohydrates. As a vertebrate animal matures, growth hormone stimulates bone growth, which influences the ultimate size of the adult. Much of the normal variation in human height is due to differences in the secretion of growth hormone from the anterior pituitary. Too little growth hormone—or defective receptors for it—causes some cases of *dwarfism*; too much can cause *gigantism* (**FIG. 37-7**). In adulthood, bones lose their ability to lengthen; but growth hormone continues to be secreted throughout life, helping to regulate protein, fat, and sugar metabolism.

A major advance in the treatment of pituitary dwarfism occurred when molecular biologists successfully inserted the gene for human growth hormone into bacteria, which then churned out large quantities of the substance. Previously, the main commercial source of growth hormone was human cadavers, from which tiny amounts were extracted at great cost. Thanks to the new, cheaper source children with underactive pituitary glands, who would previously have been extremely short, can now achieve normal height. This is why there are almost no modern photographs of people with this specific type of dwarfism.

The Posterior Pituitary Releases Hormones Produced by Cells in the Hypothalamus

The posterior pituitary contains the endings of two types of neurosecretory cells whose cell bodies are located in the hypothalamus. These neurosecretory cell endings are enmeshed in a capillary bed into which they release hormones to be carried into the bloodstream (see Fig. 37-6). Two peptide hormones are synthesized in the hypothalamus and released from the posterior pituitary: *antidiuretic hormone (ADH)* and oxytocin.

Antidiuretic hormone (literally, "hormone that prevents urination") helps prevent dehydration. As you learned in Chapter 35, ADH increases the permeability to water of the collecting ducts of nephrons in the kidney, causing more water to be reabsorbed from the urine and retained in the body. Interestingly, alcohol inhibits the release of ADH, greatly increasing urination, so alcoholic drinks can actually be dehydrating.

Oxytocin causes contractions of the muscles of the uterus during childbirth. It also triggers the "milk letdown reflex" in nursing mothers by causing muscle tissue within the mammary glands of the breasts to contract in response to stimulation by the suckling infant. This contraction reflex ejects milk from the saclike milk glands into the nipples (**FIG. 37-8**).

Recent studies using laboratory animals indicate that oxytocin also has behavioral effects. In rats, for example, oxytocin injections cause virgin females to exhibit maternal behavior

such as building a nest, licking other rats' pups, and retrieving pups that have strayed. Oxytocin may also have a role in male reproductive behavior, enhancing movement of sperm through the reproductive tract and causing male rats to ejaculate more readily.

The Thyroid and Parathyroid Glands Influence Metabolism and Calcium Levels

Lying at the front of the neck, nestled just below the larynx (**FIG. 37-9a**), the **thyroid gland** produces two hormones: *thyroxine* and *calcitonin*. Calcitonin is a peptide important in regulating calcium blood levels in several types of mammals, but it seems to play only a minor role in humans. It is sometimes administered as a drug to reduce osteoporosis. **Thyroxine**, or thyroid hormone, is an iodine-containing amino acid derivative. Because it cannot diffuse across membranes, it is carried into cells by membrane transporter proteins. Thyroxine works by binding to nuclear receptors that regulate gene activity. It influences most of the cells in the body, elevating their metabolic rate and stimulating the

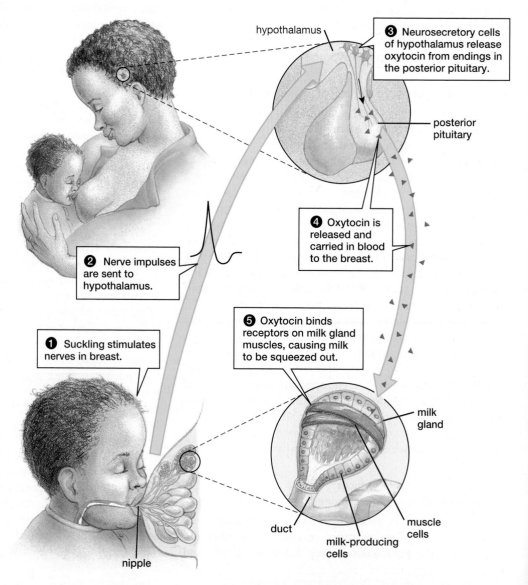

FIGURE 37-8 Hormones and breast-feeding
The control of milk letdown by oxytocin is regulated by feedback between a baby and its mother. This cycle continues until the infant is full and stops suckling. With the nipple no longer being stimulated, oxytocin release stops, the duct muscles relax, and milk flow ceases.

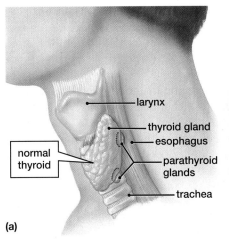

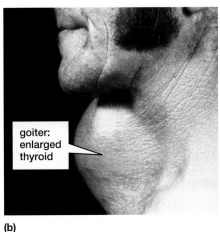

(a)

(b)

FIGURE 37-9 The thyroid and parathyroid glands
(a) The thyroid, bearing tiny parathyroid glands, wraps around the front of the larynx in the neck. (b) Goiter, a condition in which the thyroid gland becomes greatly enlarged, is caused by an iodine-deficient diet.

synthesis of enzymes that break down glucose and provide energy. In adults, levels of thyroxine determine the overall metabolic rate—that is, the resting rate of cellular metabolism. Normal levels of thyroxine are required for mental alertness. Low thyroxine makes people feel mentally and physically sluggish; they may lose appetite but still gain weight, and they become less tolerant of cold (the body generates less of its own heat when its metabolic rate is low). Excess thyroxine leads to restlessness and irritability, increased appetite, and intolerance to heat.

In juvenile animals, including humans, thyroxine helps regulate growth by stimulating both metabolic rate and nervous system development. Undersecretion of thyroid hormone early in life can cause *cretinism,* a condition characterized by retardation of both mental and physical development. Fortunately, early diagnosis and thyroxine supplementation can reverse this condition. Conversely, oversecretion of thyroxine in developing vertebrates can trigger precocious development. In 1912, in one of the first demonstrations of hormone action, a physiologist discovered that thyroxine can induce early metamorphosis in tadpoles (see "Evolutionary Connections: The Evolution of Hormones").

Levels of thyroxine in the bloodstream are fine-tuned by negative feedback loops. Thyroxine release is stimulated by thyroid-stimulating hormone (TSH) from the anterior pituitary, which in turn is stimulated by a releasing hormone from the hypothalamus. The amount of TSH released from the pituitary is regulated by negative feedback. Adequate levels of thyroxine circulating in the bloodstream inhibit the secretion of both the releasing hormone (from the hypothalamus) and TSH (from the anterior pituitary), thus inhibiting further release of thyroxine from the thyroid gland (**FIG. 37-10**).

An iodine-deficient diet can reduce the production of thyroxine and trigger a feedback mechanism that attempts

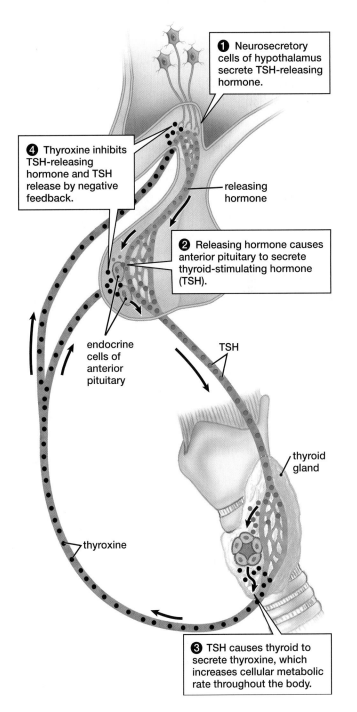

❶ Neurosecretory cells of hypothalamus secrete TSH-releasing hormone.

❹ Thyroxine inhibits TSH-releasing hormone and TSH release by negative feedback.

releasing hormone

❷ Releasing hormone causes anterior pituitary to secrete thyroid-stimulating hormone (TSH).

endocrine cells of anterior pituitary

TSH

thyroid gland

thyroxine

❸ TSH causes thyroid to secrete thyroxine, which increases cellular metabolic rate throughout the body.

FIGURE 37-10 Negative feedback in thyroid gland function
QUESTION A common test of thyroid gland function is to measure the amount of thyroid-stimulating hormone circulating in the blood. What would you conclude if such a test found an abnormally high level of TSH?

to restore normal hormone levels by dramatically increasing the number of thyroxine-producing cells. The thyroid gland becomes enlarged and may bulge from the neck, a condition called **goiter** (see Fig. 37-9b). Goiter was once common in some regions of the U.S. where iodine is low in soil and water, but widespread use of iodized salt has now all but eliminated this condition in developed countries.

The four small disks of the **parathyroid glands** are embedded in the back of the thyroid gland (see Fig. 37-9a). The parathyroids secrete **parathyroid hormone (PTH)**, which controls the concentration of calcium in the blood and interstitial fluid. Calcium is essential for many processes, including nerve and muscle function, so the calcium concentration in body fluids must be kept within narrow limits. The bones serve as a "bank" into which calcium can be deposited or withdrawn as necessary. If blood calcium levels drop, parathyroid hormone causes the bones to release some calcium. It also causes the kidneys to reabsorb more calcium as urine is formed. The increased blood calcium then inhibits further release of parathyroid hormone in a negative feedback loop (**FIG. 37-11**).

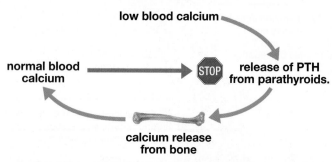

FIGURE 37-11 **Negative feedback regulates blood calcium levels**

The Pancreas Is Both an Exocrine and an Endocrine Gland

The **pancreas** is a gland that produces both exocrine and endocrine secretions. The exocrine portion synthesizes digestive secretions that are released into the *pancreatic duct* and flow into the small intestine (see Chapter 34). The endocrine portion consists of clusters of cells called **islet cells**, which produce peptide hormones. One type of islet cell produces the hormone **insulin**; another type produces the hormone **glucagon**.

Insulin and glucagon work in opposition to regulate carbohydrate and fat metabolism: insulin reduces the blood glucose level; glucagon increases it (**FIG. 37-12**). Together, the two hormones help keep the blood glucose level nearly constant. When blood glucose rises (for example, after you've eaten), insulin is released. Insulin causes body cells to take up glucose and either metabolize it for energy or convert it to fat or *glycogen* (a polysaccharide made of long chains of glucose molecules) for storage. When blood glucose levels drop (for example, after you've skipped breakfast or run a 10-kilometer race), glucagon is released.

Glucagon activates a liver enzyme that breaks down glycogen (which is stored primarily in the liver), releasing glucose into the blood. Glucagon also promotes lipid breakdown, which releases fatty acids that can be metabolized for energy.

Lack of insulin production or the failure of target cells to respond to insulin results in **diabetes mellitus**. There are several causes of diabetes; but in all cases, blood glucose levels are high and fluctuate with food intake. For reasons that are not yet fully understood, diabetes causes a wide range of circulatory problems that can result in high blood pressure, atherosclerosis, and increased levels of LDL (bad) cholesterol. Diabetes can indirectly cause heart attacks, blindness, and kidney failure. Scientists have inserted the human insulin gene into bacterial and other cells that can be grown in laboratories in large quantities, making human insulin readily available. A new treatment is described in "Links to Life: Closer to a Cure for Diabetes."

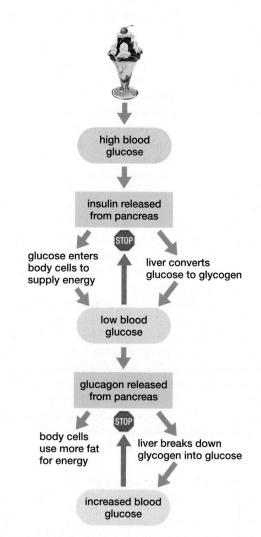

FIGURE 37-12 **The pancreas controls blood glucose levels**
Islet cells of the pancreas produce insulin and glucagon, which cooperate in a two-part negative feedback loop that controls blood glucose concentrations. QUESTION How would blood glucose be affected in a person who was born with a mutation that prevented glucagon receptors from binding to glucagon?

The Sex Organs Secrete Steroid Hormones

The sex organs do far more than produce sperm or eggs. The **testes** in males and the **ovaries** in females are also important endocrine organs (see Fig. 37-1). The testes secrete several steroid hormones, collectively called **androgens**. The most important of these is **testosterone**. The ovaries secrete two types of steroid hormones: **estrogen** and **progesterone**. The roles of the sex hormones in sperm and egg production, the menstrual cycle, pregnancy, and development are discussed in Chapters 40 and 41.

The sex hormones also play a key role in *puberty,* the phase of life during which the reproductive systems of both sexes become mature and functional. Puberty is accompanied by behavioral changes that make it such an interesting time for teenagers and their parents. Puberty begins when, for reasons not fully understood, the hypothalamus starts to secrete increasing amounts of releasing hormones, which in turn stimulate the anterior pituitary to secrete more luteinizing hormone (LH) and follicle-stimulating hormone (FSH) into the bloodstream. Both LH and FSH stimulate target cells in the testes or ovaries to produce higher levels of sex hormones. The resulting increase in circulating sex hormones ultimately affects tissues throughout the body that bear the appropriate receptors. Both sexes develop pubic and underarm hair. Testosterone, secreted by the testes of males, stimulates the development of male secondary sexual attributes, including body and facial hair, muscle growth, and a larger larynx ("voice box"), which lowers the voice. Testosterone also promotes sperm cell production. Estrogen from the ovaries in females stimulates growth of mammary glands and maturation of the female reproductive system, including production of mature egg cells. Progesterone, secreted by the ovaries during pregnancy, prepares the reproductive tract to receive and nourish the fertilized egg. Although there is a surge of sex hormone production at puberty, sex hormones are present from the fetal stage onward. They influence development in both sexes, and they continue to affect both behavior and brain function throughout life.

Over the past few decades, research on a wide variety of animals as well as a few studies of human populations has revealed that common environmental pollutants from agricultural and industrial activities can disrupt hormone systems. This is particularly true for the functioning of sex hormones, as described in "Earth Watch: Endocrine Deception."

The Adrenal Glands Have Two Parts That Secrete Different Hormones

Imagine how your body feels when you are startled, afraid, or angry. These physical reactions are the result of hormones produced by the adrenal glands, which act in conjunction with the part of your nervous system that prepares you to cope with emergencies. Like the pituitary gland and pancreas, the **adrenal glands** (Latin for "on the kidney") are two glands in one: the *adrenal medulla* and the *adrenal cortex* (**FIG. 37-13**). The **adrenal medulla** is located in the center

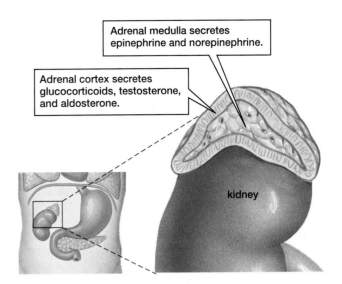

Adrenal medulla secretes epinephrine and norepinephrine.

Adrenal cortex secretes glucocorticoids, testosterone, and aldosterone.

kidney

FIGURE 37-13 The adrenal glands
An adrenal gland sits atop each kidney. The cortex consists of endocrine cells that secrete steroid hormones. The inner medulla, derived from nervous tissue during development, secretes epinephrine and norepinephrine.

of each gland (*medulla* means "marrow" in Latin). It consists of secretory cells derived during development from nervous tissue, and its hormone secretion is controlled directly by the nervous system. The adrenal medulla produces two hormones—**epinephrine** and, in much smaller quantities, **norepinephrine** (also called *adrenaline* and *noradrenaline,* respectively)—in response to stress. These hormones, which are amino acid derivatives, prepare the body for emergency action. They increase the heart and respiratory rates, cause blood glucose levels to rise, and direct blood flow away from the digestive tract toward the brain and muscles. They also cause the air passages of the lungs to expand, allowing more efficient exchange of gases. For this reason, substances that mimic epinephrine are often administered to asthmatics, whose airways become constricted during an asthma attack. The adrenal medulla is activated by the sympathetic nervous system, which prepares the body to respond to emergencies, as described in Chapter 38.

The outer layer of the adrenal gland forms the **adrenal cortex** ("cortex" is Latin for "bark"), which secretes three types of steroid hormones collectively called **glucocorticoids**. Glucocorticoid release is stimulated by ACTH from the anterior pituitary in response to a releasing hormone from the hypothalamus. Hormone levels are controlled by negative feedback; circulating glucocorticoids inhibit the release of both the hypothalamic releasing hormone and ACTH.

The glucocorticoid **cortisol** is secreted in the largest quantity. It is released when the body is stressed by trauma, infection, exposure to temperature extremes, emotional turmoil, or final exams. The effects of cortisol help the body cope with short-term stressors, raising blood glucose levels by stimulating glucose production and promoting more breakdown of fats for energy production. Cortisol also inhibits the immune system and inflammation—if you

EARTH WATCH   Endocrine Deception

Human activities have introduced a vast number and amount of foreign substances into the environment, and these substances are now found in our water, air, and food. All of us are exposed to them daily. Some have disrupted the reproduction of exposed wildlife. These compounds are extremely diverse in chemical structure and originate from a variety of sources, including pesticides (DDT, methoxychlor), plastics (bisphenol A, phthalates), detergents (nonylphenol ethoxylate), and industrial processes (PCBs). These **endocrine disrupters** have been studied using laboratory animals, cell cultures, and wild animals that have been exposed in their natural habitats. Some cause a reduction in thyroid hormone; others either mimic or block reproductive hormones, depending on the site of action or the species.

Scientists have identified a wide variety of harmful effects caused by endocrine disrupters, including feminization in males, masculinization in females, reproductive cancers, malformed sex organs, altered blood hormone levels, and reduced fertility. After a chemical plant near Lake Apopka in Florida released large quantities of several known estrogen disruptors into the water, wildlife biologists noted an alarming decline in the alligator population of the lake. Many eggs were not hatching. Males had high estrogen, low testosterone, smaller penises, and abnormal testes. Females typically had exceptionally high estrogen levels and abnormal ovaries. In another study, researchers discovered that male freshwater fish downstream from sewage outlets in both the U.S. and England produce an egg-yolk protein normally found only in females. Researchers suspect that this feminization of males is caused by human estrogen (both natural and synthetic estrogen from birth control pills) excreted in female urine. There is increasing concern over two new forms of birth control: contraceptive patches and vaginal rings, which retain high levels of synthetic estrogen after use and may be flushed down toilets or discarded into landfills, where the estrogen may be released into waterways or leach into groundwater.

Some of the most devastating endocrine disrupters, including DDT and PCBs, have been banned in developed countries (although they remain in our air, water, or soil), but many others are still widely used and persist in the environment. As more chemicals are investigated, new chemicals found to be endocrine disrupters are discovered regularly. Atrazine, the most widely used weed killer in the U.S., causes a tenfold decrease in testosterone as well as sexual abnormalities in frogs at concentrations that are common in streams and are well below the maximum levels permitted by the U.S. Environmental Protection Agency. Atrazine and other endocrine disruptors may be contributing to a worldwide decline in amphibians and the recent extinction of several species.

The effects of endocrine disruptors are likely to extend to people. Researchers are now investigating a possible link between herbicides, which are widespread in streams and groundwater in the Midwest, and reduced human semen quality. Recently, PBDEs (polybrominated diphenyl ethers), used as flame retardants in manufactured products including computers, plastics, carpet, and furniture, have leached into air, water, and the human food supply; and they can be found in human breast milk. Although levels in humans are still very low, the effects on people are unknown; animal research has revealed toxicity similar to that of PCBs, which cause nervous system damage and birth defects in both animals and humans. Both PCBs and PBDEs are believed to disrupt thyroid function.

Although high levels of endocrine disrupters are known to be harmful, no one knows what effects long-term, low-level exposure to these substances (alone and in various combinations) will have on human and other animal populations, particularly during vulnerable early developmental stages. How many of the thousands of industrial chemicals we use act as endocrine disrupters? What are their mechanisms of action? What levels of exposure do various human and wildlife populations experience? Is there a threshold of exposure for toxic effects to occur? Does exposure of people or wildlife to the "cocktail" of different endocrine disrupters now present in the environment produce more dramatic effects than the chemicals do separately? Answers to these questions will allow us to formulate appropriate controls their use. Unfortunately, these are difficult and complex questions; as we seek answers, more and more of these chemicals are entering the environment.

---

find yourself getting sick around exam time, cortisol secretion may be a contributing factor. Suppressing the immune response, which is quite energy demanding and makes people and other animals feel lethargic, can help an organism cope with more immediate and possibly life-threatening situations. Hydrocortisone (synthetic cortisol) creams are sold to treat rashes and bites, evidence of cortisol's effectiveness in suppressing the inflammation that causes these itchy swellings.

You may have noticed that many different hormones are involved in glucose metabolism: thyroxine, insulin, glucagon, epinephrine, and the glucocorticoids. Why? The reason can probably be traced to a metabolic requirement of the brain. Although most body cells can produce energy from fats and proteins as well as from carbohydrates, brain cells can burn only glucose. Thus, blood glucose levels cannot be allowed to fall too low, or brain cells rapidly starve, leading to unconsciousness and possibly death.

The adrenal cortex also secretes the hormone **aldosterone**, which regulates the sodium content of the blood. Sodium ions, derived from salt in the diet, are the most abundant positive ions in blood and interstitial fluid. The sodium ion gradient across plasma membranes (high outside, low inside) is a factor in many cellular events, including the production of electrical signals by nerve cells. If blood sodium falls, the adrenal cortex releases aldosterone, which causes the kidneys and sweat glands to retain sodium; this enables salt and other sources of dietary sodium to raise blood sodium levels, shutting off further aldosterone secretion (an example of negative feedback).

In both women and men, the adrenal cortex also produces the male sex hormone testosterone, although normally in much smaller amounts than produced by the testes. Tumors of the adrenal cortex can lead to excessive testosterone release, causing masculinization of women. Many of the "bearded ladies" who once appeared in circus sideshows probably had this condition.

Hormones Are Produced by the Pineal Gland, Thymus, Kidneys, Heart, Digestive Tract, and Fat Cells

The **pineal gland** is located between the two hemispheres of the brain, just above and behind the hypothalamus (see Fig. 37-1). Named for its resemblance to a pine cone, the pineal gland is smaller than a pea. In 1646 philosopher René Descartes described it as "the seat of the rational soul." Since then, scientists have learned a bit more about this organ, but many of its functions are still poorly understood.

The pineal gland produces the hormone **melatonin**, synthesized from an amino acid. Melatonin is secreted in a daily rhythm, which in mammals is regulated by the eyes. In some vertebrates, such as the frog, the pineal itself contains photoreceptive cells; the skull above it is thin, so the pineal can detect sunlight and thus day length. By responding to day lengths characteristic of different seasons, the pineal appears to regulate the seasonal reproductive cycles of many mammals. Despite years of research, the function of melatonin and the pineal gland in humans is still not well understood; but darkness increases melatonin production, and bright light inhibits it. One hypothesis is that the pineal gland and melatonin secretion influence sleep–wake cycles; melatonin is available as a sleeping aid. Overproduction of melatonin may contribute to a form of depression called *seasonal affective disorder (SAD)* that some people experience during the short days of winter. Sitting in front of a bank of bright lights in the morning helps alleviate SAD symptoms for many victims.

The **thymus** is located in the chest cavity behind the sternum (breastbone; see Fig. 37-1). In addition to harboring white blood cells, the thymus produces the hormone **thymosin**, which stimulates the development of specialized white blood cells (T cells) that play an important role in the immune system (see Chapter 36). The thymus is extremely large in infants but, under the influence of sex hormones, begins decreasing in size after puberty.

The kidneys, which play a central role in maintaining body fluid homeostasis, are important endocrine organs as well. When the oxygen content of the blood drops, the kidneys produce the hormone **erythropoietin**, which increases red blood cell production (see Chapter 32). The kidneys also produce an enzyme called **renin**, in response to low blood pressure, such as that caused by bleeding. Renin is an enzyme that catalyzes the production of the hormone

angiotensin from proteins in the blood. Angiotensin raises blood pressure by constricting arterioles. It also stimulates the release of aldosterone by the adrenal cortex, causing the kidneys to retain sodium in the blood, increasing its osmolarity. Higher osmolarity attracts and retains water, increasing blood volume.

The heart seems an unlikely endocrine organ, but in 1981 a substance extracted from atrial tissue of the heart and injected into rats was found to cause an increase in the output of salt and water by the kidneys. The substance was **atrial natriuretic peptide (ANP)**, which is released by atrial cells when blood volume becomes excessive and stretches the heart muscle. Atrial natriuretic peptide reduces blood volume by decreasing the release of both ADH and aldosterone, allowing the kidneys to excrete more salt and water.

The stomach and small intestine produce a variety of peptide hormones that help regulate digestion. These hormones include **gastrin**, **secretin**, and **cholecystokinin**, discussed in Chapter 34.

Can fat be an endocrine organ? In 1995 researchers described the peptide hormone **leptin** (derived from a word meaning "slender"), which is released by adipose (fat) cells. Mice missing the gene for leptin became obese (**FIG. 37-14**), and leptin injections caused them to lose weight. The researchers hypothesized that adipose tissue, by releasing leptin, tells the body how much fat it has stored and therefore how much to eat. Unfortunately, trials of leptin as a human weight-loss aid have not been encouraging. Many obese people have high levels of leptin but seem to be relatively insensitive to it. However, researchers are discovering surprising new functions for leptin and are finding leptin receptors in unexpected places, such as blood vessels and white blood cells. Leptin appears to stimulate the growth of new capillaries and to speed wound healing. It also stimulates the immune system, and its presence seems to be required for the onset of puberty.

FIGURE 37-14 Leptin helps regulate body fat
The mouse on the left has been genetically engineered to lack the gene for the hormone leptin.

LINKS TO LIFE   Closer to a Cure for Diabetes

In the most common form of diabetes, called *type 2*, the pancreas either produces some, but not enough, insulin or the victim's cells lose their ability to respond to it (insulin resistance). Although the causes are poorly understood, there is a clear correlation between insulin-resistant type 2 diabetes and obesity. Obesity rates have doubled since 1980, and the incidence of type 2 diabetes has doubled since 1970. This form of diabetes generally occurs in people over 40; but more and more children are being diagnosed, paralleling the increase in childhood obesity.

While type 2 diabetes can often be controlled by improvements in diet and lifestyle, type 1 diabetes occurs when a person's immune system attacks and kills the insulin-producing islet cells of the pancreas. This form of diabetes often strikes early in life, and its victims rarely survive for a full life span. Their lives, filled with multiple daily blood tests and insulin injections, are never normal. For the 1 million people in the U.S. who suffer from this form of diabetes, islet cell transplantation offers a ray of hope. A research team led by James Shapiro at the University of Alberta, Canada, removed pancreases from recently deceased donors and infused extracted islet cells into a vein feeding the livers of diabetes victims. Some of the cells took up residence and began secreting insulin. Five years later, most of these people still require insulin, but at lower dosages, and their blood-sugar levels are more stable. Unfortunately, they must take immunosuppressant drugs to prevent tissue rejection.

Only about 3000 donor pancreases become available each year in the U.S., and most successful islet transplants require two donor pancreases because the fragile islet cells are damaged by cold-storage and transport time. Recently, doctors performed the first islet cell transplant from a living donor; a mother donated part of her pancreas to her adult daughter. The transplanted cells began making insulin immediately, giving doctors hope that healthier cells from living donors may have a better outcome than those from cadavers. Time will tell.

Research continues to expand our understanding of the multiple effects of hormones and the wide variety of organs and cells that produce them. In time, this understanding will lead to a wealth of new medical treatments. It should also increase our respect for these substances and our recognition that any hormone a person takes may influence physiological systems throughout the body.

EVOLUTIONARY CONNECTIONS

The Evolution of Hormones

Vertebrate endocrine systems were once considered unique to our phylum, but over the past decades physiologists have discovered that hormones are evolutionarily ancient. Insulin, for example, is found not only in vertebrates but also in protists, fungi, and bacteria, although the function of insulin in most of these organisms is not known. Protists also manufacture ACTH, even though (as single cells) they have no adrenal glands to stimulate. Yeasts have receptors for estrogen but certainly no ovaries. Thyroid hormones have been found in invertebrates such as worms, insects, and mollusks. Even among vertebrates, the effects of chemically identical hormones, secreted by the same glands, can vary dramatically from organism to organism. Let's look briefly at the diverse effects of thyroxine.

Some fish species undergo radical physiological changes during their lifetimes. A salmon, for example, begins life in fresh water, migrates to the ocean, and returns to fresh water to spawn. In the stream where the salmon hatched, fresh water tends to enter the fish's tissues by osmosis; in salt water, the fish tends to lose water. The salmon's migrations, therefore, require complete revamping of salt and water control. In salmon, one function of thyroxine is to produce the metabolic changes necessary to go back and forth from life in streams to life in the ocean.

In amphibians, thyroxine has the dramatic effect of triggering metamorphosis. In 1912, in one of the first demonstrations of the action of any hormone, tadpoles were fed minced horse thyroid. As a result, the tadpoles metamorphosed prematurely into miniature adult frogs (**FIG. 37-15**). In high mountain lakes in Mexico, where the water is deficient in the iodine needed to synthesize thyroxine, natural selection has produced one species of salamander that can reproduce while still in its juvenile form.

Thyroxine regulates the seasonal molting of most vertebrates. From snakes to birds to the family dog, surges of thyroxine stimulate the shedding of skin, feathers, or hair. In humans (who don't regularly migrate, metamorphose, or molt), thyroxine regulates growth and metabolism.

The diversity of life on Earth rests on a conservative foundation: a relative handful of chemicals coordinate activities within single cells and among groups of cells. Life's diversity originated in part as natural selection favored adaptive changes in the systems used to deliver the chemicals as well as the body's responses to them. Early in their evolution, animals developed a complement to hormonal communication that provides faster, more precise delivery of chemical messages: the nervous system, described in the next chapter.

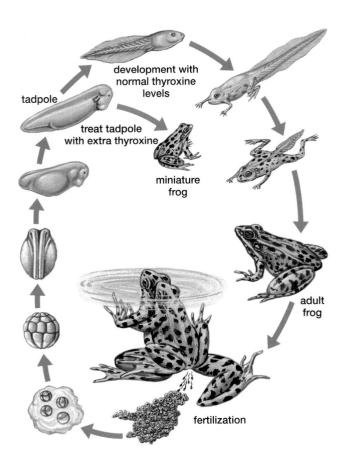

FIGURE 37-15 Thyroxine controls metamorphosis in amphibians
The life cycle of the frog begins with fertilization of the eggs (bottom). The fertilized eggs develop into an aquatic, fishlike tadpole, which grows and ultimately metamorphoses into an adult frog. Metamorphosis is triggered by a surge of thyroxine from the tadpole's thyroid gland. If injected with extra thyroxine, a young tadpole will metamorphose ahead of schedule into a miniature adult frog. QUESTION What would happen if tadpoles were given a drug that blocked the production of thyroxine?

CASE STUDY REVISITED LOSING ON ARTIFICIAL HORMONES

Athletes willing to cheat are drawn to anabolic steroids because, like natural testosterone, they increase muscle mass. Hormones have evolved to be present in miniscule amounts at proper times in the normally developing and functioning body. Taking relatively large dosages can mean trouble because, as with many hormones, anabolic steroids exert their effects throughout the body.

In males, artificially increased levels of anabolic steroids create a negative feedback effect that can reduce the natural production of testosterone, reduce testicle size and sperm count, and cause breast development. In females, anabolic steroids can interfere with menstrual periods and increase facial hair. For both sexes, steroids cause acne and may depress the immune system. Mood swings and sudden aggressiveness are sufficiently common to have produced the slang term "roid rage." Anabolic steroids have been linked to increases in blood pressure and decreases in the good (HDL) form of cholesterol; both are risk factors for heart attacks and strokes. Based on a large poll, the Centers for Disease Control (CDC) concluded that about 6% of U.S. high school students had used anabolic steroids at some time. Because anabolic steroids can cause bone growth to shut down prematurely, young steroid abusers may never reach their full potential height.

Professional baseball and football players have tarnished their sports by taking THG. The 2006 Tour de France was marred when nine cyclists were banned just before the race began, based on indications that they had taken performance-enhancing drugs. Some Olympians have been stripped of their medals. To Catlin, who runs the Olympic Analytic Laboratory, desecration of the Olympics is especially tragic. "The notion of the Olympic Games to me is the cleanest, purest kind of event ever," he says. "People in every country in the world can compete and the best man or woman crosses the finish line first. What could be worse than to think they are tainted?"

Consider This Given what you know about steroids and other hormones, do you think that high school and college athletes should be routinely tested, just as Olympic athletes are? Explain your answer.

CHAPTER REVIEW

SUMMARY OF KEY CONCEPTS

37.1 How Do Animal Cells Communicate?

Within multicellular organisms, communication among cells occurs through gap junctions directly linking cells, by diffusion of chemicals to nearby cells (local hormones and neurotransmitters), and by transport of chemicals within the bloodstream (endocrine hormones). Extracellular chemical messengers act selectively on target cells that bear specific receptors for the chemical.

37.2   What Are the Characteristics of Animal Hormones?

Most cells release local hormones, such as prostaglandins, to communicate with nearby cells. The endocrine system is a collection of glands and organs that release endocrine hormones, which are transported in the bloodstream to other parts of the body; there, they affect the activity of specific target cells bearing receptors for the hormones. Hormones are synthesized either from amino acids (amino acid derived hormones and peptides) or from lipids (steroid hormones).

Most hormones act on their target cells in one of two ways: Peptide hormones and amino acid derived hormones bind to receptors on the surfaces of target cells and activate intracellular second messengers, such as cyclic AMP, which then alter the cell's metabolism. Steroid hormones can bind to surface receptors or diffuse through the plasma membranes of their target cells and bind with receptors in the cytosol or nucleus. The hormone–receptor complex promotes the transcription of specific genes within the nucleus. Thyroid hormones are ferried across the plasma membrane and enter the nucleus, where they bind to receptors associated with the chromosomes and influence gene transcription.

Hormone action is commonly regulated through negative feedback, a process in which a hormone causes changes that inhibit further secretion of that hormone.

37.3   What Are the Structures and Hormones of the Mammalian Endocrine System?

Many hormones are produced by endocrine glands, which are clusters of cells embedded within a network of capillaries. Hormones are secreted into the extracellular fluid and diffuse into capillaries. The major endocrine glands of the human body are the hypothalamus–pituitary complex, the thyroid and parathyroid glands, the pancreas, the sex organs, and the adrenal glands. The hormones released by these glands and their actions are summarized in Table 37-3. Other structures that produce hormones include the pineal gland, thymus, kidneys, heart, stomach and small intestine, and fat cells.

Web Tutorial 37.1 Hypothalamic Control of the Pituitary

Web Tutorial 37.2 How Hormones Influence Target Cells

KEY TERMS

adrenal cortex   *page 753*
adrenal gland   *page 753*
adrenal medulla   *page 753*
adrenocorticotropic
  hormone (ACTH)   *page 749*
aldosterone   *page 754*
amino acid derived hormone
  page 742
androgen   *page 753*
angiotensin   *page 755*
anterior pituitary   *page 748*
antidiuretic hormone (ADH)
  page 750
atrial natriuretic peptide
  (ANP)   *page 755*
cholecystokinin   *page 755*
cortisol   *page 753*
cyclic AMP   *page 744*
diabetes mellitus   *page 752*
duct   *page 746*

ecdysone   *page 746*
endocrine disrupter
  page 754
endocrine gland   *page 746*
endocrine hormone
  page 742
endocrine system   *page 743*
epinephrine   *page 753*
erythropoietin   *page 755*
estrogen   *page 753*
exocrine gland   *page 746*
follicle-stimulating hormone
  (FSH)   *page 749*
gastrin   *page 755*
glucagon   *page 752*
glucocorticoid   *page 753*
goiter   *page 752*
growth hormone   *page 749*
hypothalamus   *page 748*
inhibiting hormone   *page 749*

insulin   *page 752*
islet cell   *page 752*
leptin   *page 755*
local hormone   *page 742*
luteinizing hormone (LH)
  page 749
melanocyte-stimulating
  hormone (MSH)   *page 749*
melatonin   *page 755*
neurosecretory cell   *page 748*
norepinephrine   *page 753*
ovary   *page 753*
oxytocin   *page 746*
pancreas   *page 752*
parathyroid gland   *page 752*
parathyroid hormone
  page 752
peptide hormone   *page 742*
pineal gland   *page 755*
pituitary gland   *page 748*

posterior pituitary   *page 748*
progesterone   *page 753*
prolactin   *page 749*
prostaglandin   *page 742*
receptor   *page 742*
releasing hormone   *page 749*
renin   *page 755*
second messenger   *page 744*
secretin   *page 755*
steroid hormone   *page 742*
target cell   *page 743*
testis   *page 753*
testosterone   *page 753*
thymosin   *page 755*
thymus   *page 755*
thyroid gland   *page 750*
thyroid-stimulating hormone
  (TSH)   *page 749*
thyroxine   *page 750*

THINKING THROUGH THE CONCEPTS

1. What are the three types of molecules used as endocrine hormones in vertebrates? Give an example of each.

2. What is the difference between an endocrine gland and an exocrine gland? Which type releases hormones?

3. When peptide hormones attach to target cell receptors, what cellular events follow? How do steroid hormones behave?

4. Diagram the process of negative feedback, and give an example of it in the control of hormone action.

5. What are the major endocrine glands in the human body, and where are they located?

6. Describe the structure of the hypothalamus–pituitary complex. Which pituitary hormones are neurosecretory? What are their functions?

7. Describe how releasing hormones regulate the secretion of hormones by cells of the anterior pituitary. Name the hormones of the anterior pituitary, and give one function of each.

8. Describe how the hormones of the pancreas act together to regulate the concentration of glucose in the blood.

9. Compare the adrenal cortex and adrenal medulla by answering the following questions: Where are they located within the adrenal gland? Which hormones do they produce? Which organs do their hormones target?

APPLYING THE CONCEPTS

1. A student decides to do a science project on the effect of the thyroid gland on frog metamorphosis. She sets up three aquaria with tadpoles. She adds thyroxine to the water of one, the drug thiouracil to a second, and nothing to the third. Thiouracil reacts with thyroxine in tadpoles to produce an ineffective compound. Assuming that the student uses appropriate physiological concentrations, predict what will happen.

2. If you had severe type 1 diabetes, would you consider an islet cell transplant? What pros and cons can you think of?

3. Suggest a hypothesis about the endocrine system to explain why many birds lay their eggs in the spring and why poultry farmers who produce eggs keep lights on at night.

4. Some parents who are interested in college sports scholarships for their children are asking physicians to prescribe growth hormone treatments, even though their children are in the normal height range. What biological and ethical dilemmas does this raise for the parents, children, physicians, coaches, and college scholarship boards?

5. Argue for and against banning or restricting the use of common endocrine disrupters, such as plasticizers and certain pesticides. What compromises can you suggest?

FOR MORE INFORMATION

Ashley, S. "Doping by Design." *Scientific American*, February 2004. New types of "designer steroids" pose problems for detection.

Christensen, D. "Transplanted Hopes." *Science News*, September 2, 2000. The author describes successful pancreatic islet cell transplants into diabetic patients.

Raloff, J. "Common Pollutants Undermine Masculinity." *Science News*, April 3, 1999. Several recent research articles report feminizing effects of estrogen disrupters in laboratory animals.

Schubert, C. "Burned by Flame Retardants?" *Science News*, October 13, 2001. Flame retardants from TVs, computers, drapes, and sofas are building up in our bodies.

Vogel, G. "A Race to the Starting Line." *Science*, July 30, 2004. This article describes the competition between athletes who take drugs and chemists who devise ways to detect them.

CHAPTER

38

The Nervous System and the Senses

Love: "A fire sparkling in lovers' eyes . . . a madness most discreet," or just the right mixture of chemicals deep within lovers' brains? (Inset) Prairie voles provide insights into the neurochemical basis of emotional love.

AT A GLANCE

 # CASE STUDY   HOW DO I LOVE THEE?

"But soft! What light through yonder
  window breaks?
It is the east, and Juliet is the sun."
            —*Romeo and Juliet*, Act II, scene II

IN SHAKESPEARE'S *ROMEO AND JULIET*, two teenagers fall in love the first time they meet. A few hours later, as Romeo watches Juliet gazing out of her window, he sees her as the sun that lights up his life. Shakespeare's *Romeo and Juliet* is one of the finest expressions of the power of romantic love, for which the lovers defy their families, risk their fortunes and their futures, and finally give up their lives.

Romance is, of course, not the only manifestation of love. A mother's love for her child is just as strong. People have also defined their lives and willingly died for love of God and love of country. Just what *is* love? Are all these types of love distinct, or are they related? What happens in the brain when two lovers meet, or a mother cradles her infant?

No one knows for sure—not in people, anyway. Perhaps surprisingly, neuroscientists know a lot about love—or at least monogamy, pair bonding, and sex—in a small rodent called the prairie vole. If Juliet had been a prairie vole, her first encounter with Romeo would have released a flood of oxytocin—the same hormone that causes uterine contractions during childbirth. The oxytocin would have bound to receptors in a tiny part of her brain called the nucleus accumbens, causing nerve cells there to release another chemical, called dopamine. Because of the dopamine, she would have felt wonderful and, what's more, she would have linked that euphoric feeling to Romeo. In a "Romeo vole," some of the molecules and brain regions would have differed, but the end result would have been similar: a torrent of dopamine, giving him the ultimate high, and he would have "known" that he could attain that feeling again only with Juliet. So, the two voles would mate—for about 24 hours straight—and bond for life. They would build a nest, live together, and raise their young together.

How do humans and other animals perceive their world, whether the sun's warmth or the faces and scents of their lovers? How do they evaluate what they perceive, and become calm or excited, fearful or smitten? Finally, how do they respond with appropriate behaviors such as resting, eating, or mating? Although most perceptions and behaviors are not completely understood, the answers to these questions are to be found in the senses and the nervous system. In this chapter, we will first investigate how nerve cells work. Then we will explore how the billions of cells in the brain communicate with one another and how clusters of cells control specific sensations and actions. Finally, we will see that specialized cells called receptors respond to stimuli in the environment—whether the external environment of the world around us or the internal environment within our own bodies—and relay that information to the brain, where it is interpreted and sometimes acted upon. And perhaps—in those few pounds of cells nestled in the skull, talking with one another through complex mixtures of chemicals, some broadcast widely, some squirted directly from one cell to another—we may even find the nature of love.

38.1   WHAT ARE THE STRUCTURES AND FUNCTIONS OF NEURONS?

An individual cell of the nervous system is called a nerve cell, or **neuron**. A neuron must perform four functions:

1. Receive information from the internal or external environment or from other neurons.
2. Process this information, often along with information from other sources, and produce an electrical signal.
3. Conduct the electrical signal, sometimes for a considerable distance, to a junction where it meets another cell.
4. Communicate with other cells, including other neurons, or the cells in muscles or glands.

Although neurons vary enormously in structure, a "typical" vertebrate neuron has four distinct parts that carry out these four functions: the *dendrites,* the *cell body,* the *axon,* and the *synaptic terminals* (**FIG. 38-1**).

Dendrites, branched tendrils protruding from the nerve cell body, respond to stimuli from other neurons or from the environment. Their many branches provide a large surface area for receiving signals. In neurons of the brain and spinal cord, dendrites usually respond to the chemicals, called **neurotransmitters**, that are released by other neurons. These dendrites have receptor proteins in their membranes that bind specific neurotransmitters and, as a result, produce electrical signals. Dendrites of *sensory neurons* have membrane adaptations that allow them to produce electrical signals in response to specific stimuli from the external environment, such as pressure, odor, or light, or the internal environment, such as body temperature, blood pH, or the position of a joint.

Electrical signals travel down the dendrites and converge on the neuron's **cell body**. The cell body performs two major functions. First, it contains the organelles that are usually found in most cells, such as the nucleus, endoplasmic reticulum and Golgi complex, and performs typical cellular activities such as synthesizing complex molecules and coordinating the cell's metabolism. The cell body also "adds up," or *integrates*, the various electrical signals it receives from the dendrites. As we will see shortly, some of these signals are positive and some are negative. If their sum is sufficiently positive, the neuron will produce a large, rapid, specialized electrical signal called an **action potential**.

In a typical neuron, a long, thin strand called an **axon** extends outward from the cell body and conducts the action potential from the cell body to the synaptic terminal at the axon's end, where it contacts another cell. Single axons may stretch from your spinal cord to your toes, a distance of about a meter (about 3 feet), making neurons the longest cells in the body. Axons are normally bundled together into **nerves**, much like wires are bundled in an electrical cable. In vertebrates, axons bundled into nerves emerge from the brain and spinal cord and extend to all regions of the body.

The site where a neuron communicates with another cell is called a **synapse**. A typical synapse consists of (1) the **synaptic terminal**, which is a swelling at the end of an axon of the "sending" neuron; (2) a dendrite or cell body of a

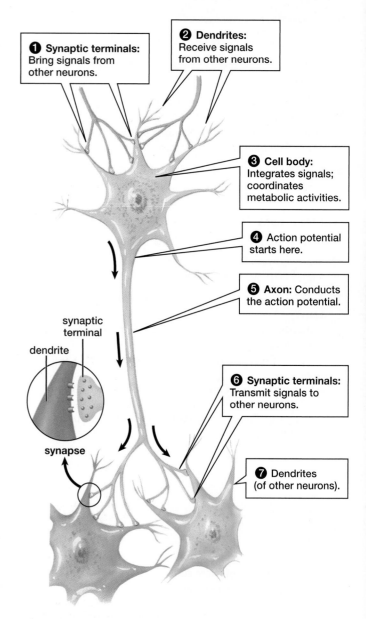

❶ **Synaptic terminals:** Bring signals from other neurons.

❷ **Dendrites:** Receive signals from other neurons.

❸ **Cell body:** Integrates signals; coordinates metabolic activities.

❹ **Action potential starts here.**

❺ **Axon:** Conducts the action potential.

❻ **Synaptic terminals:** Transmit signals to other neurons.

❼ **Dendrites** (of other neurons).

synaptic terminal

dendrite

synapse

FIGURE 38-1 A nerve cell, showing its specialized parts and their functions

"receiving" neuron (or sometimes the "receiving" part of a muscle or gland cell), and (3) a small gap that separates the two cells (Fig. 38-1; see also Fig. 38-3 later in this chapter). Most synaptic terminals contain a neurotransmitter that is released in response to an action potential reaching the terminal. At a synapse, the output of the first cell becomes the input to the second cell.

38.2   HOW IS NEURAL ACTIVITY PRODUCED AND TRANSMITTED?

Neurons Produce Electrical Voltages Across Their Membranes

In the 1930s, biologists developed ways to record electrical events inside individual neurons. They found that unstimu-

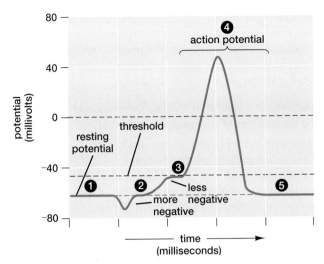

FIGURE 38-2 Electrical events during an action potential
(1) A neuron maintains a voltage across its plasma membrane called the resting potential, about –60 mV with respect to the outside. (2) Stimulation from the environment or other cells may make the neuron more negative (downward deflection) or less negative (upward deflection). (3) If the potential becomes about 10 to 20 mV less negative, the neuron reaches threshold, and (4) produces a brief positive potential called an action potential. (5) After a millisecond or two, the voltage across the neuron's plasma membrane returns to the resting potential.

lated, inactive neurons maintain a constant electrical voltage difference, or *potential*, across their plasma membranes, similar to the voltage across the poles of a battery. This voltage, called the **resting potential**, is always negative inside the cell and ranges from –40 to –90 millivolts (mV; thousandths of a volt).

If the neuron is stimulated, either naturally or by an experimenter using an electrical current, the potential inside the neuron can be made either more or less negative (**FIG. 38-2**). If the potential is made sufficiently less negative, it reaches a level called **threshold**, and an action potential is triggered. During an action potential, the neuron's voltage rapidly rises to about +50 mV inside the cell. Action potentials last a few milliseconds (thousandths of a second) before the cell's negative resting potential is restored. The plasma membranes of axons are specialized to conduct action potentials from a neuron's cell body to the axon's synaptic terminals. Unlike electrical voltages in metal wires, which get smaller with distance, action potentials are conducted from cell body to axon terminal, as far as a meter in a human or almost 20 meters in a blue whale, with no change in voltage. In "A Closer Look: Ions and Electrical Signaling in Neurons," we examine resting and action potentials more thoroughly.

How fast an action potential travels varies tremendously among axons. In general, the thicker the axon, the faster the action potential moves. A much more effective way of speeding conduction is to cover the axon with a fatty insulation called **myelin** (**FIG. 38-3**). Myelin is formed by spe-

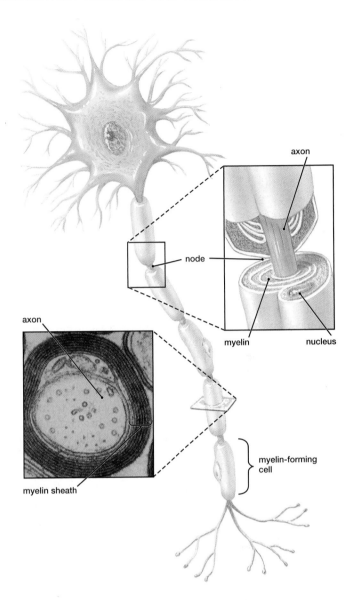

FIGURE 38-3 A myelinated axon
Many vertebrate axons are covered in a series of membrane wrappings of myelin, formed from windings of membrane of specialized non-neuronal cells. Action potentials occur only at the short segments of naked axons, called nodes, between each myelin wrapping.

cialized cells that flatten and wrap themselves around the axon, covering the axon with multiple wrappings of specially insulating plasma membrane, with scarcely any cytoplasm between the wraps. A given myelin-producing cell covers about 0.2 to 2 mm of axon, leaving naked *nodes* between the myelin wrappings. Instead of traveling continuously but fairly slowly down the axon (typically about 1 to 2 meters per second), action potentials in myelinated axons "jump" rapidly from node to node, traveling at a rate of 3 to 100 meters per second. Axons with both large diameters and myelin coatings conduct action potentials the fastest.

Neurons Communicate at Synapses

An action potential can be thought of as a packet of information moving down an axon (step ① in **FIG. 38-4**). Once it reaches the synaptic terminal, this information must be transmitted to another cell, such as another neuron, a cell in a muscle, or a gland. For simplicity, we will discuss only the transmission of information from one neuron to another neuron.

In ordinary English, the word "transmit" means "to send something," and that is exactly what happens when one nerve cell communicates with another. A synapse is the junction where the synaptic terminal of one neuron meets the dendrite of another neuron. However, the two neurons do *not* actually touch one another: a tiny gap separates the first, or **presynaptic neuron**, from the second, or **postsynaptic neuron**. Therefore, the presynaptic neuron must "send something" across this gap in order to communicate with the postsynaptic neuron. What it sends is a chemical appropriately called a neurotransmitter. How is the neurotransmitter "sent" across the gap, and how does the postsynaptic neuron respond to it?

The synaptic terminal at the end of an axon contains dozens of vesicles, each full of neurotransmitter molecules. When an action potential reaches the synaptic terminal, the inside of the terminal becomes positively charged (step ② in Fig. 38-4). This charge causes some of the vesicles to release neurotransmitter into the gap between the cells (step ③ in Fig. 38-4). This process is actually a specialized case of exocytosis (see Chapter 4). The outer surface of the plasma membrane of the postsynaptic neuron, just across the gap, is packed with receptor proteins. The neurotransmitter molecules rapidly diffuse across the gap and bind to these receptors (step ④ in Fig. 38-4).

Synapses Produce Excitatory or Inhibitory Postsynaptic Potentials

When a receptor binds a neurotransmitter, it causes a brief change in the resting potential of the postsynaptic neuron, called a **postsynaptic potential** or **PSP** (step ⑤ in Fig. 38-4). If the postsynaptic neuron becomes more negative, its resting potential moves farther away from threshold, reducing its likelihood of firing an action potential. This is called an **inhibitory postsynaptic potential (IPSP)**. If the postsynaptic neuron becomes less negative, then its resting potential will move closer to threshold, and it will be more likely to fire an action potential. Consequently, this is called an **excitatory postsynaptic potential (EPSP)**. The mechanisms by which neurotransmitters binding to receptors cause PSPs are explained in "A Closer Look: Ions and Electrical Signaling in Neurons."

Neurotransmitter Action Is Usually Brief

Consider what would happen if a presynaptic neuron started stimulating a postsynaptic cell, and never stopped. You might, for example, contract your biceps muscle, flex your arm, and have it stay flexed forever! Not surprisingly, the nervous system has several ways of terminating neuro-

transmitter action (step ⑥ in Fig. 38-4). Some neurotransmitters—notably acetylcholine, the transmitter that stimulates skeletal muscle cells—are rapidly broken down by enzymes in the synapse. Many neurotransmitters are transported back into the presynaptic neuron.

The Activity of a Neuron Is Largely Determined by the Summation of Postsynaptic Potentials

Most postsynaptic potentials are small, rapidly fading signals, but they travel far enough to reach the cell body. There, they determine whether an action potential will be produced. How? The dendrites and cell body of a single neuron often receive EPSPs and IPSPs from the synaptic terminals of thousands of presynaptic neurons. All the PSPs that reach the postsynaptic cell body at about the same time are "added up," a process called *integration* or *summation*. If the excitatory and inhibitory potentials, when added together, raise the electrical potential inside the neuron above threshold, the postsynaptic cell will produce an action potential.

The Nervous System Uses Many Neurotransmitters

Over the past few decades, researchers have found that the brain is a teeming cauldron, its neurons synthesizing and responding to a vast array of chemicals, including many hormones that were once thought unique to the endocrine system. We have already seen that oxytocin, more generally known for causing uterine contraction during childbirth, stimulates crucial actions in the brain that contribute to pair bonding in voles and perhaps to love in humans. Other hormones that control various functions of the digestive tract are also synthesized in the brain, where they influence appetite. At least 50 neurotransmitters have been identified, and the list is growing. Table 38-1 lists a few well-known neurotransmitters and some of their functions. "Health Watch: Drugs, Diseases, and Neurotransmitters" further explores the role of neurotransmitters in addiction and in neurological diseases.

38.3 HOW ARE NERVOUS SYSTEMS ORGANIZED?

The individual neuron uses a language of action potentials, yet somehow this basic language allows even simple animals to perform a variety of complex behaviors. One key to the versatility of the nervous system is the presence of networks of neurons that range from dozens to millions of cells. As in computers, small, simple elements can perform amazing feats when connected properly.

Information Processing in the Nervous System Requires Four Basic Operations

At a minimum, a nervous system must be able to perform four operations:

1. Determine the type of a stimulus.
2. Determine and signal the intensity of a stimulus.
3. Integrate information from many sources.
4. Initiate and direct appropriate responses.

Let's examine each of these operations.

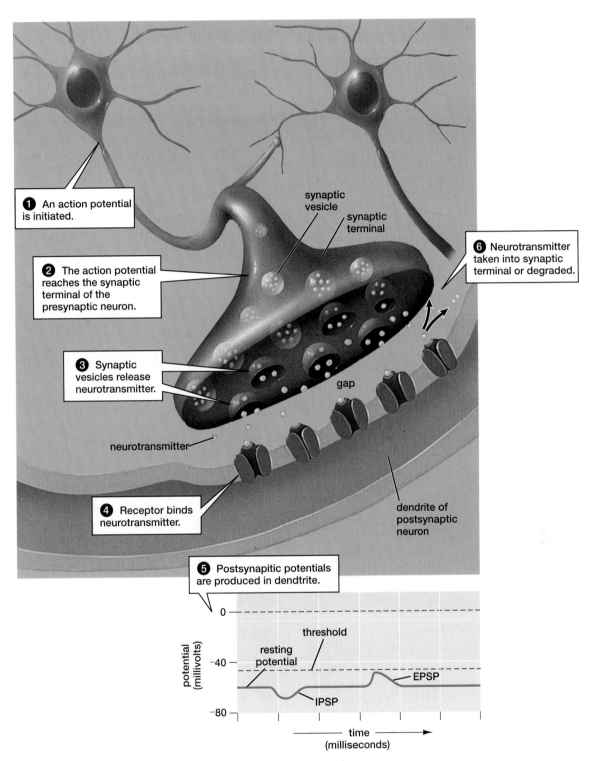

1 An action potential is initiated.

synaptic vesicle

synaptic terminal

6 Neurotransmitter taken into synaptic terminal or degraded.

2 The action potential reaches the synaptic terminal of the presynaptic neuron.

3 Synaptic vesicles release neurotransmitter.

gap

neurotransmitter

4 Receptor binds neurotransmitter.

dendrite of postsynaptic neuron

5 Postsynapitic potentials are produced in dendtrite.

0

threshold

resting potential

potential (millivolts)

−40

EPSP

IPSP

−80

time (milliseconds)

FIGURE 38-4 Structure and operation of the synapse

A synaptic terminal contains many neurotransmitter-filled vesicles. When an action potential enters the synaptic terminal, the vesicles release their neurotransmitter into the space between the neurons. The neurotransmitter diffuses rapidly across the gap and binds to receptors on the postsynaptic cell. In many cases, transmitter binding to receptors causes a change in the resting potential of the postsynaptic cell, called a postsynaptic potential (PSP). QUESTION Imagine an experiment in which the neurons pictured here are bathed in a solution containing a nerve poison. The presynaptic neuron is stimulated and produces an action potential, but this does *not* result in a PSP in the postsynaptic neuron. When the experimenter adds some neurotransmitter to the synapse, the postsynaptic neuron still produces no PSP. How does the poison act to disrupt nerve function?

A CLOSER LOOK   Ions and Electrical Signaling in Neurons

POTASSIUM PERMEABILITY PRODUCES THE RESTING POTENTIAL

The resting potential is based on a balance between chemical and electrical gradients, maintained by active transport and a membrane that is selectively permeable to specific ions. The ions of the cytoplasm consist mainly of positively charged potassium ions (K^+) and large, negatively charged organic molecules such as ATP and proteins, which cannot leave the cell. Outside the cell, the extracellular fluid contains mostly positively charged sodium ions (Na^+) and negatively charged chloride ions (Cl^-). These concentration differences are maintained by an active transport protein in the plasma membrane called the *sodium-potassium pump*, which simultaneously pumps K^+ into and Na^+ out of the cell.

In an unstimulated neuron, only K^+ can cross the plasma membrane, by traveling through specific membrane proteins called *potassium channels* (shown in yellow). Although *sodium channels* (shown in purple) are also present, they are closed. Because the K^+ concentration is higher inside the cell than outside, K^+ diffuses out, leaving the negatively charged organic ions behind (**FIG. E38-1**):

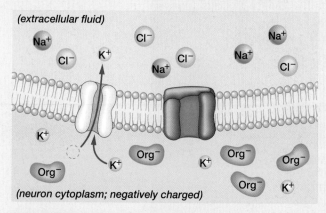

(extracellular fluid)

(neuron cytoplasm; negatively charged)

FIGURE E38-1 The resting potential

As the inside of the cell becomes more and more negatively charged, positive K^+ ions are attracted back into the cell. Eventually, the negative voltage inside the cell becomes large enough so that the rate of K^+ diffusing out is exactly counterbalanced by the rate of K^+ being pulled back in by electrical attraction. This negative voltage is the neuron's resting potential.

CHANGES IN PERMEABILITY TO SODIUM AND POTASSIUM PRODUCE THE ACTION POTENTIAL

Action potentials occur when the resting potential is changed, becoming less negative and reaching a voltage called the threshold (usually about 10 to 20 mV less negative than the resting potential). At threshold, Na^+ channels (purple) open, allowing a rapid influx of Na^+ (step ① in **FIG. E38-2**):

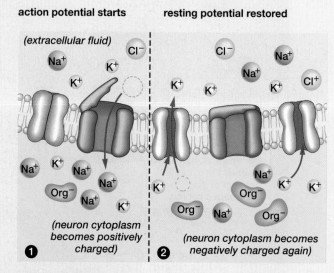

action potential starts   resting potential restored

(extracellular fluid)

(neuron cytoplasm becomes positively charged) ❶

(neuron cytoplasm becomes negatively charged again) ❷

FIGURE E38-2 The action potential

Soon after the Na^+ channels open, they spontaneously close again, and a different type of K^+ channel (orange) is opened by the positive charge inside the neuron, allowing K^+ to flow out of the cell and restore the negative resting potential (step ② in **FIG. E38-2**).

ACTION POTENTIALS ARE CONDUCTED DOWN AXONS WITHOUT CHANGING AMPLITUDE

An action potential usually starts where the axon emerges from a neuron's cell body. It then resembles a fast-moving wave of positive charge that travels, undiminished in size, along the axon to the synaptic terminal. The positive charge carried into the axon by Na^+ causes Na^+ channels farther along the axon to open. More Na^+ can then flow in, generating a new action potential (**FIG. E38-3**):

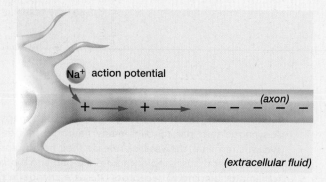

Na^+ action potential

(axon)

(extracellular fluid)

FIGURE E38-3 Na^+ entry during an action potential begins the movement of positive charge down an axon

As the wave of positive charges passes a given point along the axon, the resting potential is restored as K⁺ flows out (**FIG. E38-4**):

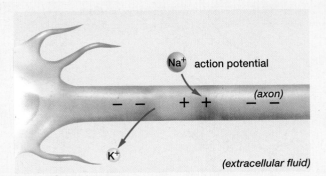

FIGURE E38-4 K⁺ leaving the axon restores the resting potential behind the advancing action potential

Only a tiny fraction of the total potassium and sodium in and around each neuron is exchanged during each action potential, so the concentration gradients of K⁺ and Na⁺ do not change appreciably.

Action potentials are all-or-none phenomena. That is, if the neuron does not reach threshold, there will be no action potential; but if threshold is reached, a full-sized action potential will occur and travel the entire length of the axon.

NEUROTRANSMITTERS MAY OPEN ION CHANNELS

When an action potential reaches a presynaptic terminal, it stimulates the release of neurotransmitter molecules. These diffuse across the synaptic gap and bind to receptor proteins on the postsynaptic cell. In many cases, the receptor proteins are linked to ion channels, and neurotransmitter binding opens the channels (**FIG. E38-5**).

If the channels are permeable to Na⁺ (Fig. E38-5, top), then Na⁺ ions diffuse into the cells down their concentration gradient, and make the cell less negative. If the postsynaptic neuron becomes sufficiently less negative, it may reach threshold and produce an action potential. Because they

"excite" the cell, possibly producing an action potential, such voltage changes in the postsynaptic cell are called **excitatory postsynaptic potentials (EPSPs)**. If the channels are permeable to K⁺ (Fig. E38-5, bottom), then K⁺ ions diffuse out of the cell, making it more negative. Making the cell more negative tends to inhibit the production of action potentials in the postsynaptic cell, so this voltage change is called an **inhibitory postsynaptic potential (IPSP)**.

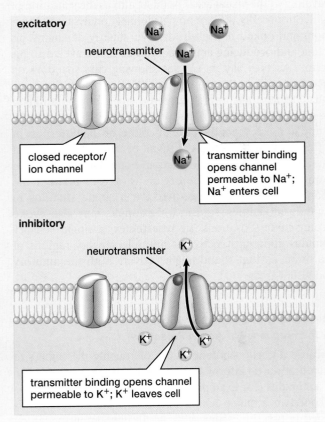

FIGURE E38-5 Neurotransmitter binding to receptor proteins opens ion channels

Table 38-1 Some Important Neurotransmitters

Neurotransmitter	Location in Nervous System	Some Functions
Acetylcholine	Motor neuron-to-muscle synapse; autonomic nervous system, brain	Activates skeletal muscles; activates target organs of parasympathetic nervous system
Dopamine	Midbrain	Important in control of movement; reward
Norepinephrine (noradrenaline)	Sympathetic nervous system	Activates target organs of sympathetic nervous system
Serotonin	Midbrain, pons, and medulla	Influences mood, sleep
Glutamate	Brain and spinal cord	Major excitatory neurotransmitter in CNS
Glycine	Spinal cord	Major inhibitory neurotransmitter in spinal cord
GABA (gamma amino butyric acid)	Throughout brain	Major inhibitory neurotransmitter in brain
Endorphins	Brain and spinal cord	Influence mood, reduce pain sensations
Nitric oxide	Brain	Important in forming memories

The Type of Stimulus Is Distinguished by Connectivity from the Senses to the Brain

If action potentials are the packets of information of almost all neurons, and if all action potentials are basically the same, then how can the brain of a pond snail or a person determine *what* a stimulus is—light, sound, odor—or *how strong* a stimulus is?

All nervous systems interpret *what* a stimulus is by monitoring *which neurons* are firing action potentials. For example, your brain interprets action potentials that occur in the axons of your optic nerves (originating in the eye and traveling to the visual areas of the brain) as the sensation of light. Supposedly, a German physiologist once sat in a dark room and poked himself in the eye, slightly damaging his retina and producing action potentials that traveled to his brain. (As they say in TV ads showing cars speeding on racetracks, do *not* try this yourself!) The result? He "saw stars" because his brain interpreted *any* action potential in his optic nerve as light. Thus, you can easily distinguish the sound of music from the taste of coffee, or the bitterness of coffee from the sweetness of sugar, because these different stimuli normally result in action potentials in different axons that end up in different areas of your brain. Some people, called synesthetes, perceive a single stimulus in multiple modalities—music, for example, may also evoke strong color sensations. In synesthetes, a single sensory stimulus apparently activates multiple sensory regions of the brain, so that sound might activate both the auditory and visual areas.

The Intensity of a Stimulus Is Coded by the Frequency of Action Potentials

Because all action potentials are of roughly the same size and duration, no information about the strength, or **intensity**, of a stimulus (for example, the loudness of a sound) can be encoded in a single action potential. Instead, intensity is coded in two other ways (**FIG. 38-5**). First, intensity can be signaled by the frequency of action potentials in a single neuron. The more intense the stimulus, the faster the neuron produces action potentials, or *fires*. Second, most nervous systems have many neurons that can respond to the same input. Stronger stimuli tend to excite more of these neurons, whereas weaker stimuli excite fewer. Thus, intensity can also be signaled by the number of similar neurons that fire at the same time. For example, the loud wail of a police siren activates many of your auditory neurons and causes them to fire action potentials very rapidly.

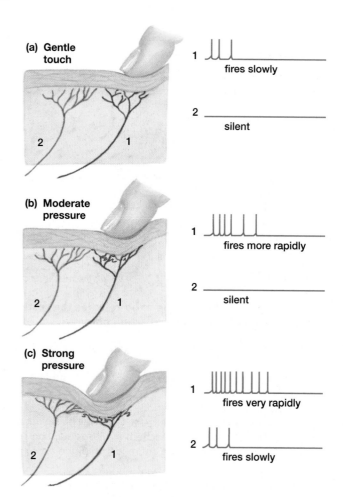

FIGURE 38-5 Signaling stimulus intensity
The intensity of a stimulus is signaled by the rate at which individual sensory neurons produce action potentials and by the number of sensory neurons activated. In this example, increasing pressure on the skin first causes faster firing and then causes an adjacent receptor to be activated. QUESTION How do you think the skin areas that are especially sensitive to touch differ from less-sensitive areas?

signals to fewer neurons. For example, many sensory neurons may converge onto a smaller number of brain cells. Some of these brain cells act as "decision-making cells," adding up the postsynaptic potentials that result from the synaptic activity of the sensory neurons. Depending on their relative strengths (and other internal factors such as hormones or metabolic activity), they produce appropriate outputs.

The Nervous System Processes Information from Many Sources

Your brain is continuously bombarded by sensory stimuli that originate both inside and outside the body. The brain must evaluate these inputs, determine which ones are important, and decide how to respond. Nervous systems, like individual neurons, integrate information from many sources: a large number of neurons may all funnel their

The Nervous System Produces Appropriate Outputs to Muscles and Glands

Action potentials from the decision-making neurons may travel to other parts of the brain, to the spinal cord, or to the sympathetic and parasympathetic nervous systems (described later). Ultimately, the output of the nervous system will stimulate activity in the muscles or glands that actually produce observable behaviors.

HEALTH WATCH   Drugs, Diseases, and Neurotransmitters

Chances are, you know someone who is addicted. How can substances such as nicotine, alcohol, and cocaine so profoundly influence people's lives? A big part of the answer lies in the effects of these drugs on neurotransmitters and in how the nervous system adapts to those insidious effects.

Addictive drugs activate the brain's "reward circuitry," creating feelings of intense pleasure. Cocaine is a good example. Synapses in the brain that use the neurotransmitters dopamine, serotonin, or norepinephrine contribute to our energy level and our overall sense of well-being. Normally, the presynaptic neuron, after releasing one of these neurotransmitters, immediately starts pumping it back in, thus limiting its effects. Researchers have found that cocaine works by blocking this pump mechanism. The result? When a person takes cocaine, these neurotransmitters remain in their synapses much longer and reach higher concentrations than normal, so their effects are enhanced. The user feels euphoric and energetic. However, eventually the brain attempts to restore the normal state. During repeated cocaine use, the postsynaptic neurons decrease their number of receptors for these neurotransmitters. When fewer receptors are present, the high levels of neurotransmitter caused by cocaine are now *required*, just for the user to feel normal. If cocaine is withdrawn, the postsynaptic neurons are inadequately stimulated, and the user experiences an emotional "crash" that can be relieved only by more cocaine. Over time, more and more cocaine is needed just to feel OK, let alone to get high: the user has become an addict (**FIG. E38-6**).

Alcohol stimulates receptors for the neurotransmitter GABA (gamma amino butyric acid), enhancing inhibitory neuronal signals, and blocks receptors for glutamate, reducing excitatory signals. When a person drinks frequently, the brain compensates by decreasing GABA receptors and increasing glutamate receptors. Without alcohol, an alcoholic feels jittery and nervous—in short, overstimulated. In extreme cases, withdrawal from alcohol can cause convulsions. Nicotine and other components of cigarette smoke also interfere with normal synaptic transmission, producing a variety of addictive effects. To overcome addictions, drug users must undergo the misery caused by a nervous system that is deprived of a drug to which it has adapted. Although receptors eventually return to normal levels, for unknown reasons, drug cravings often recur periodically.

You may also know someone with Parkinson's or Alzheimer's disease. Both are caused by the death of specific neurons in the brain and the loss of their neurotransmitters, which normally communicate with other neurons. In Parkinson's disease, dopamine-releasing neurons in the midbrain die, interfering with the complex control system that underlies smooth movements. Parkinson's patients experience tremors and have difficulty initiating movement. In Alzheimer's disease, neurons in the temporal lobes that produce the neurotransmitter acetylcholine die in large numbers. Memory loss is a prominent symptom of Alzheimer's.

The neurotransmitter serotonin acts in the brain and spinal cord. Too little serotonin can cause depression. The antidepressant Prozac® selectively blocks the re-uptake of serotonin into presynaptic neurons, enhancing the neurotransmitter's effects. You may have heard of Ecstasy (MDMA), a relative of the stimulant drug amphetamine. This drug causes a temporary massive increase in serotonin in synapses. Users report feelings of pleasure, increased energy, heightened sensory awareness, and improved rapport with other people. Increasing evidence from both animal and human research suggests that Ecstasy users may incur long-term damage to serotonin-producing neurons, and they may suffer from deficits in learning and memory. New studies with primates suggest that MDMA may also damage dopamine-producing neurons; if this is also true of humans, MDMA users may be particularly vulnerable to Parkinson's disease later in life.

The analgesic (pain-relieving) effects of plant-derived opiates, such as morphine, opium, codeine, and heroin, have been recognized for centuries. Since the brain has receptors that bind these molecules, researchers reasoned that perhaps these plant opiates resemble (then) unknown substances produced by the brain for which these receptors evolved. The search for such substances was rewarded in 1975 with the discovery of *opioids* (opiate-like substances); *endorphins* are a group of opioids. Certain opioids suppress pain in times of extreme stress, such as on a battlefield or a football field. Opioids released during strenuous exercise may account for the well-known "runner's high."

FIGURE E38-6 Addiction
An addict experiences extreme physical and emotional distress without his drug, because his nervous system has adapted to it.

Neural Pathways Direct Behavior

Most behaviors are controlled by neuron-to-muscle pathways composed of four elements:

1. **Sensory neurons**, which respond to a stimulus, either internal or external to the body.
2. **Interneurons**, which receive signals from many sources, including sensory neurons, hormones, neurons that store memories, and many others. Based on this input, interneurons often activate motor neurons.
3. **Motor neurons**, which receive instructions from sensory neurons or interneurons and activate muscles or glands.
4. **Effectors**, usually muscles or glands that perform the response directed by the nervous system.

The Simplest Behavior Is the Reflex

The simplest type of behavior in animals is the **reflex**, a largely involuntary movement of a body part in response to a stimulus. Reflexes occur without involving conscious portions of the brain; many occur entirely within the spinal cord and peripheral neurons. Examples of human reflexes include the familiar knee-jerk and pain-withdrawal reflexes, both of which are produced by neurons in the spinal cord. The pain-withdrawal reflex, which moves a body part away from a painful stimulus such as a thumbtack, uses one of each of the three types of neurons and an effector (see Fig. 38-10 later in this chapter). Although reflexes like these do not require the brain, other pathways inform the brain of pricked fingers and may trigger other, more complex behaviors (cursing, for example!).

Nearly all animals are capable of much more subtle and varied behavior than can be accounted for by simple reflexes. In principle, these more complex behaviors can be organized by interconnected neural pathways, in which several types of sensory input (along with memories, hormones, and other factors) converge on a set of interneurons. By integrating the postsynaptic potentials from several sources, the interneurons can "decide" what to do and can stimulate motor neurons to direct the appropriate activity in muscles and glands.

Complex Nervous Systems Are Centralized

In the animal kingdom, there are really only two nervous system designs: a diffuse nervous system, such as that of cnidarians (*Hydra*, jellyfish, and their relatives; **FIG. 38-6a**), and a centralized nervous system, found to varying degrees in more complex organisms. Not surprisingly, nervous system design is highly correlated with the animal's lifestyle. Radially symmetrical cnidarians have no "front end," so there has been no evolutionary pressure to concentrate the senses in one place. A *Hydra* sits anchored to the seafloor, so prey or predators are equally likely to come from any direction. Cnidarian nervous systems are composed of a network of neurons, often called a **nerve net**, woven through the animal's tissues. Here and there we find a cluster of neurons, called a **ganglion** (plural, ganglia), but nothing resembling a real brain.

Almost all other animals are bilaterally symmetrical, with definite head and tail ends. Because the head is usually the first part of the body to encounter food, danger, and potential mates, it is advantageous to have sense organs concentrated there. Sizable ganglia evolved that integrate the information gathered by the senses and initiate appropriate action. Over evolutionary time, the major sense organs of animals with increasingly complex nervous systems became localized in the head, and the ganglia became centralized into a brain. This trend, called cephalization, is clearly seen in the invertebrates (**FIG. 38-6b, c**). Cephalization reaches its peak in the vertebrates, in which nearly all the cell bodies of the nervous system are localized in the brain and spinal cord.

38.4 WHAT IS THE STRUCTURE OF THE HUMAN NERVOUS SYSTEM?

The vertebrate nervous system, including that of humans, can be divided into two parts: central and peripheral. Each of these has further subdivisions (**FIG. 38-7**). The **central nervous system (CNS)** consists of a **brain** and a **spinal cord** that extends down the dorsal part of the torso. The **peripheral**

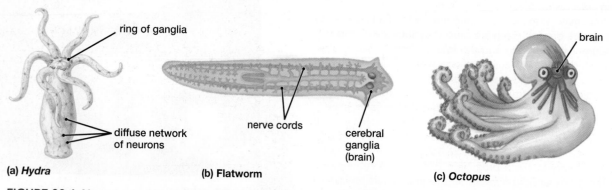

(a) *Hydra* — ring of ganglia / diffuse network of neurons

(b) Flatworm — nerve cords / cerebral ganglia (brain)

(c) *Octopus* — brain

FIGURE 38-6 Nervous system organization
(a) The diffuse nervous system of *Hydra* contains a few concentrations of neurons, particularly at the bases of tentacles, but no brain. Neural signals are conducted in virtually all directions throughout the body. **(b)** The flatworm has a nervous system that is less diffuse, with a cluster of ganglia in the head. **(c)** The genus *Octopus* has a large, complex brain and learning capabilities rivaling those of some mammals.

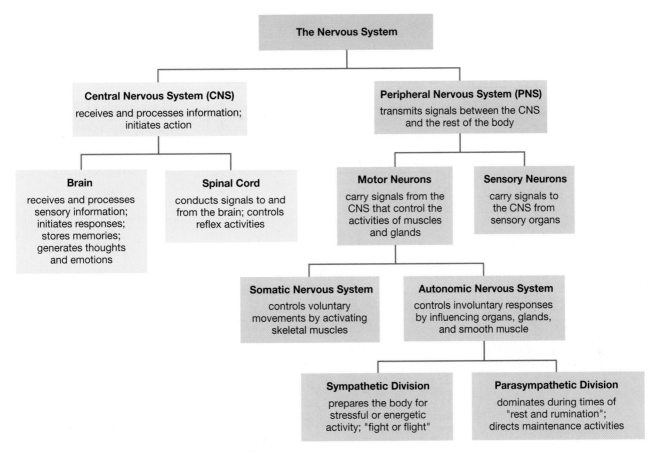

FIGURE 38-7 **Organization and functions of the vertebrate nervous system**

nervous system (PNS) consists of neurons that lie outside of the CNS (often in ganglia alongside the spinal cord, or in ganglia near target organs, such as ganglia in the head and neck that control the salivary glands) and the axons that connect these neurons with the CNS.

The Peripheral Nervous System Links the Central Nervous System to the Rest of the Body

The **peripheral nerves** allow the brain and spinal cord to communicate with the rest of the body, including the muscles, the sensory organs, and the organs of the digestive, respiratory, excretory, and circulatory systems. Within the peripheral nerves are axons of sensory neurons that bring sensory information *to* the central nervous system from all parts of the body. (You may think that these neuron fibers should be called dendrites, because they carry information toward the cell body. However, because they are long and conduct action potentials, neurobiologists call them axons.) Peripheral nerves also contain the axons of motor neurons that carry signals *from* the central nervous system to the organs and muscles.

The motor part of the peripheral nervous system can be subdivided into two parts: the **somatic nervous system** and the **autonomic nervous system**. Motor neurons of the somatic nervous system form synapses with skeletal muscles and control

voluntary movement. As you take notes, lift a coffee cup, or adjust your stereo, your somatic nervous system is in charge. The cell bodies of somatic motor neurons are located in the spinal cord. Their axons go directly to the muscles they control. (Muscles and their control are discussed in Chapter 39.)

Motor neurons of the autonomic nervous system control involuntary responses. They innervate the heart, smooth muscles, and glands. The autonomic nervous system is controlled primarily by the hypothalamus of the brain, described later in this chapter. It consists of two divisions: the **sympathetic division** and the **parasympathetic division** (FIG. 38-8). The two divisions of the autonomic nervous system innervate most of the same organs but usually produce opposite effects.

The sympathetic division releases the neurotransmitter norepinephrine (noradrenaline) onto its target organs, preparing the body for stressful or highly energetic activity, such as fighting, escaping, or taking an exam. During such "fight-or-flight" activities, the sympathetic nervous system curtails activity of the digestive tract, redirecting some of its blood supply to the muscles of the arms and legs. The heart rate accelerates. The pupils of the eyes open wider, admitting more light, and the air passages in the lungs expand, accommodating more air. These things may happen if you are suddenly called on in class to answer a question—especially if you don't know the answer!

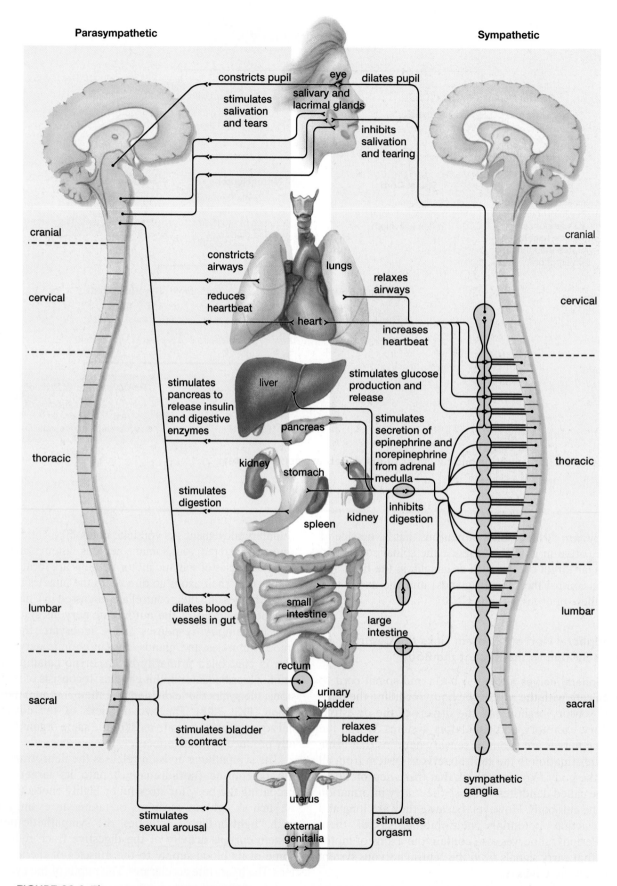

FIGURE 38-8 The autonomic nervous system
The autonomic nervous system has two divisions, sympathetic and parasympathetic, which supply nerves to many of the same organs but generally produce opposite effects. Activation of the autonomic nervous system is involuntarily commanded by signals from the hypothalamus.

The parasympathetic division, which releases acetylcholine onto its target organs, dominates during maintenance activities that can be carried on at leisure, often called "rest and rumination." Under its control, the digestive tract becomes active, the heart rate slows, and air passages in the lungs constrict.

The Central Nervous System Consists of the Spinal Cord and Brain

The spinal cord and brain make up the central nervous system. This portion of the nervous system receives and processes sensory information, generates thoughts, and directs responses. The CNS consists primarily of interneurons—probably about 100 billion of them!

The brain and spinal cord are physically protected in three ways. The first line of defense is a bony armor, consisting of the *skull,* which surrounds the brain, and the *vertebral column,* which protects the spinal cord. Beneath the bones lies a triple layer of connective tissue called the *meninges.* Between the layers of the meninges, the *cerebrospinal fluid,* a clear liquid resembling blood plasma, cushions the brain and spinal cord as it nourishes the cells of the CNS. The delicate cells of the brain are also protected from potentially damaging chemicals that reach the bloodstream because the walls of brain capillaries are far less permeable than capillaries in the rest of the body. This third line of defense is called the *blood–brain barrier.*

The Spinal Cord Is a Cable of Axons Protected by the Backbone

About as thick as your little finger, the spinal cord extends from the base of the brain to the lower back. It is protected by the bones of the vertebral column. Between the vertebrae, nerves carrying axons of sensory neurons and motor neurons arise from the dorsal and ventral portions of the spinal cord, respectively, and merge to form the peripheral nerves that innervate most of the body (**FIG. 38-9**). In the center of the spinal cord is a butterfly-shaped area of **gray matter**. Gray matter consists of the cell bodies of neurons that control voluntary muscles and the autonomic nervous system, plus interneurons that communicate with the brain and other parts of the spinal cord. The gray matter is surrounded by **white matter**, containing myelin-coated axons of neurons that extend up or down the spinal cord (the fatty myelin sheaths color these axon tracts white). These axons carry sensory signals from internal organs, muscles, and the skin up to the brain. Axons also extend downward from the brain, carrying signals that direct the motor portions of the peripheral nervous system.

If the spinal cord is severed, sensory input from below the cut cannot reach the brain, and motor output from the brain cannot reach motor neurons located below the cut. Therefore, body parts that are innervated by motor and sensory neurons located below the injury are paralyzed and feel numb, even though the motor and sensory neurons, the peripheral nerves, and the muscles remain intact.

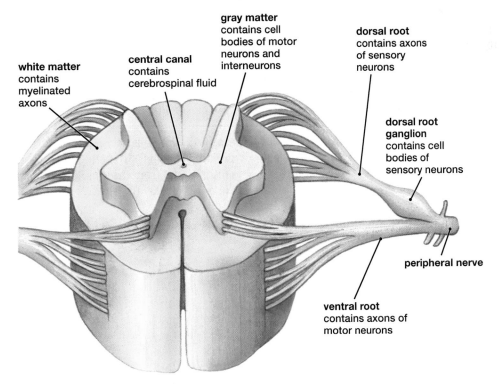

gray matter
contains cell bodies of motor neurons and interneurons

dorsal root
contains axons of sensory neurons

central canal
contains cerebrospinal fluid

white matter
contains myelinated axons

dorsal root ganglion
contains cell bodies of sensory neurons

peripheral nerve

ventral root
contains axons of motor neurons

FIGURE 38-9 The spinal cord
In cross section, the spinal cord has an outer region of myelinated axons (white matter) that travel to and from the brain and an inner, butterfly-shaped region of dendrites and the cell bodies of interneurons and motor neurons (gray matter). The cell bodies of the sensory neurons are outside the cord in the dorsal root ganglion.

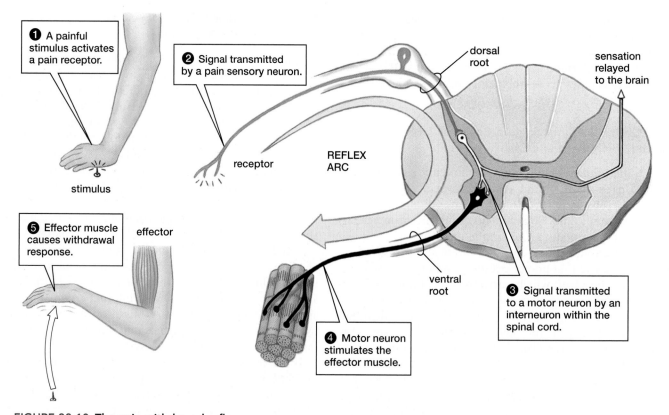

❶ A painful stimulus activates a pain receptor.

❷ Signal transmitted by a pain sensory neuron.

dorsal root

sensation relayed to the brain

receptor

REFLEX ARC

stimulus

❺ Effector muscle causes withdrawal response.

effector

ventral root

❸ Signal transmitted to a motor neuron by an interneuron within the spinal cord.

❹ Motor neuron stimulates the effector muscle.

FIGURE 38-10 The pain-withdrawal reflex
QUESTION Why does a paralyzed victim of a spinal cord injury, when pricked with a pin on a paralyzed part of his or her body, typically exhibit a normal pain-withdrawal reflex but feel no pain?

In addition to relaying neural signals between the brain and the rest of the body, the spinal cord contains the neural pathways for certain simple behaviors, such as reflexes. Let's examine the simple pain-withdrawal reflex, which involves neurons of both the central and the peripheral nervous systems (**FIG. 38-10**). The cell bodies of the sensory neurons from the skin (in this case, signaling pain) are located in **dorsal root ganglia**, clusters of neurons on spinal nerves just outside the spinal cord. Both interneuron and motor neuron cell bodies are found in the gray matter in the center of the spinal cord. Interneurons receive input from the sensory neurons and send their output to the motor neurons. (You can see why they are called "interneurons": they are literally "between other neurons.") Many spinal cord interneurons also have axons that extend up to the brain. Action potentials carried along these axons alert the brain to the painful event. The brain, in turn, sends action potentials down axons in the white matter to interneurons and motor neurons in the gray matter. These signals can modify spinal reflexes. With enough training or motivation, you can suppress the pain-withdrawal reflex. To rescue a child from a burning crib, for example, you could reach into the flames.

In addition to simple reflexes, the wiring for some fairly complex activities also resides within the spinal cord. All the neurons and interconnections needed for the basic movements of walking and running, for example, are contained in the spinal cord. The advantage of this semi-independent arrangement between brain and spinal cord is probably an increase in speed and coordination, because messages do not have to travel all the way up to the brain and back down again merely to swing one leg forward while walking. The brain's role in these "semi-automatic" behaviors is to initiate, guide, and modify the activity of spinal motor neurons, based on conscious decisions (where are you going; how fast should you walk?). To maintain balance, the brain also uses sensory input from the muscles to command motor neurons to adjust the way the muscles move.

The Brain Consists of Many Parts Specialized for Specific Functions

All vertebrate brains have the same general structure, with major modifications corresponding to lifestyle and intelligence. Vertebrate brains are divided into three parts: the **hindbrain, midbrain,** and **forebrain** (**FIG. 38-11a**). Scientists believe that in the earliest vertebrates, these three anatomical divisions were also functional divisions: the hindbrain governed automatic behaviors such as breathing and heart rate, the midbrain controlled vision, and the forebrain dealt largely with the sense of smell. In nonmammalian vertebrates, the three divisions remain prominent (**FIG. 38-11b, c**). However, in mammals—particularly in humans—the brain regions are significantly modified. Some have

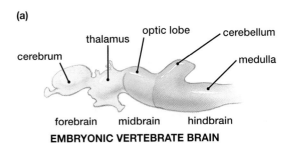

EMBRYONIC VERTEBRATE BRAIN

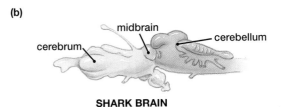

SHARK BRAIN

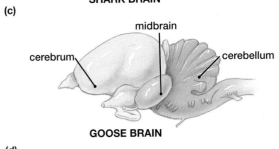

GOOSE BRAIN

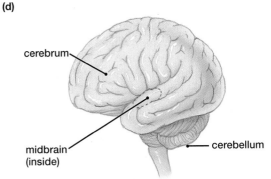

HUMAN BRAIN

FIGURE 38-11 A comparison of vertebrate brains
(a) The brains of modern vertebrate embryos, thought to be similar to the brains of the distant ancestors of today's vertebrates, consist of three distinct regions: the forebrain, midbrain, and hindbrain. (b) The brain of an adult shark maintains this basic organization. (c) In the goose, the midbrain is reduced, and the cerebrum and cerebellum are larger. (d) In mammals, especially humans, the cerebrum is very large compared to the other brain regions.

been reduced in size; others, especially the forebrain, are greatly enlarged (**FIG. 38-11d**). A section through the midline of the human brain reveals many of its structural features, as shown in **FIGURE 38-12**.

The Hindbrain Includes the Medulla, Pons, and Cerebellum

In humans, the hindbrain is comprised of the medulla, the pons, and the cerebellum (see Fig. 38-12). In both structure and function, the **medulla** is very much like an enlarged ex-

tension of the spinal cord. Like the spinal cord, the medulla has neuron cell bodies at its center, surrounded by a layer of myelin-covered axons. It controls several automatic functions, such as breathing, heart rate, blood pressure, and swallowing. Certain neurons in the **pons**, located above the medulla, appear to influence transitions between sleep and wakefulness and between stages of sleep. Other neurons influence the rate and pattern of breathing.

The **cerebellum** is crucial for coordinating movements of the body. It receives information both from command centers in the conscious areas of the brain that control movement and from position sensors in muscles and joints. By comparing information from these two sources, the cerebellum guides smooth, accurate motions and body position. The cerebellum is also involved in what is often called "motor learning." As you learn to write, throw a ball, or play the guitar, your forebrain directs your movements, often somewhat clumsily. After you have become proficient, your forebrain still "decides" what to do (for example, where to throw the ball), but your cerebellum becomes largely responsible for ensuring that the actions are carried out appropriately. Not surprisingly, the cerebellum is especially large in animals whose activities require coordinating fine movements or aerial maneuvers, such as bats and birds (see Fig. 38-11).

The Midbrain Contains the Reticular Formation

The midbrain is extremely reduced in humans (see Figs. 38-11 and 38-12). It contains an auditory relay center and a center that controls reflex movements of the eyes, as well as a portion of the **reticular formation**. The reticular formation consists of dozens of interconnected clusters of neurons that extend all the way from the central core of the medulla, passing through the midbrain into lower regions of the forebrain. These neurons receive input from virtually every sense, from every part of the body, and from many areas of the brain as well. The reticular formation plays a role in sleep and wakefulness, emotion, muscle tone, and some movements and reflexes. It filters sensory inputs before they reach the conscious regions of the brain, although the selectivity of the filtering seems to be set by higher brain centers, such as those that control conscious thought. Activities of the reticular formation allow you to read and concentrate in the presence of a variety of distracting stimuli, such as the music from your stereo and the smell of coffee. The fact that a mother wakens upon hearing the faint cry of her infant but sleeps through loud traffic noise outside her window testifies to the effectiveness of the reticular formation in screening inputs to the brain.

The Forebrain Includes the Thalamus, Limbic System, and Cerebral Cortex

The forebrain, also called the **cerebrum**, includes the *thalamus*, *limbic system*, and the *cerebral cortex*. In mammals, the cerebral cortex is much enlarged compared with that of fish, amphibians, and reptiles. This trend culminates in the human cerebral cortex (see Fig. 38-11).

FIGURE 38-12 The human brain
A section through the midline of the human brain reveals some of its major structures.

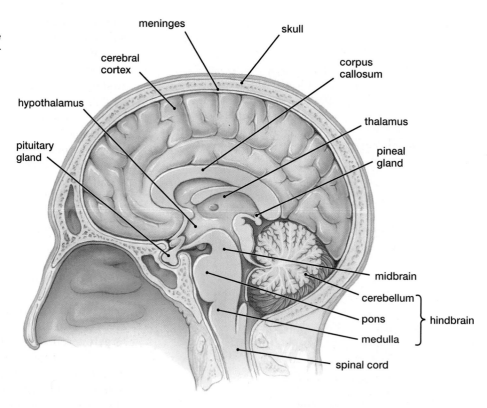

The **thalamus** is a complex relay station that channels sensory information from all parts of the body to the limbic system and cerebral cortex (see Fig. 38-12). Signals traveling from the cerebellum and limbic system to the cerebral cortex also pass through this busy thoroughfare. The **limbic system** is a diverse group of structures, including the *hypothalamus*, the *amygdala*, and the *hippocampus* as well as nearby regions of the cerebral cortex, located in an arc between the thalamus and cerebral cortex (**FIG. 38-13**). These structures work together to produce our most basic emotions, drives, and behaviors, including fear, rage, calm, hunger, thirst, pleasure, and sexual responses. Portions of the limbic system are also important in the formation of memories.

The **hypothalamus** (literally, "under the thalamus") contains many clusters of neurons. Some are neurosecretory cells that release hormones into the blood; others control the release of a variety of hormones from the pituitary gland (see Chapter 37). Other regions of the hypothalamus direct the activities of the autonomic nervous system. The hypothalamus, through its hormone production and neural connections, acts as a major coordinating center, maintaining homeostasis in a variety of ways: controlling body temperature, food intake, water balance, the menstrual cycle, and the sleep–wake cycle.

Clusters of neurons in the **amygdala** produce sensations of pleasure, fear, or sexual arousal when stimulated. Conscious humans whose amygdalas are electrically stimulated have reported feelings of rage or fear. Recent studies have revealed that damage to the amygdala early in life eliminates the ability both to feel fear and to recognize fearful facial expressions in other people.

The shape of the **hippocampus** as it curves around the thalamus inspired its name, which is derived from the Greek word meaning "seahorse." As with the amygdala and hypothalamus, behaviors that reflect a variety of emotions, including rage and sexual arousal, can be elicited by stimulating portions of the hippocampus. The hippocam-

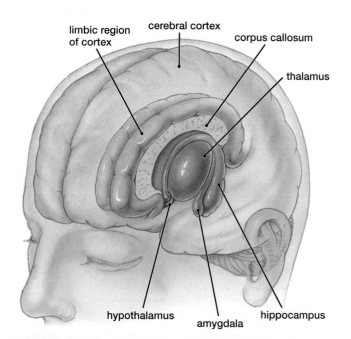

FIGURE 38-13 The limbic system and thalamus
The limbic system extends through several brain regions. The thalamus is a crucial relay center among the senses, the limbic system, and the cerebral cortex.

pus also plays an important role in the formation of long-term memory and is thus required for learning, discussed in more detail later in this chapter.

In humans, the largest part of the brain by far is the **cerebral cortex**, the outer layer of the forebrain. The cerebral cortex and underlying parts of the forebrain are divided into two halves, called **cerebral hemispheres**. These halves communicate with each other by means of a large band of axons, the **corpus callosum**. The cerebral cortex is the most sophisticated information-processing center known, but it is also the area of the brain about which scientists know the least. Billions of neurons are packed into this thin surface layer. The cortex is folded into **convolutions**, which greatly increase its area. In the cortex, cell bodies of neurons predominate, giving this outer layer—in preserved specimens—a gray color; hence the term "gray matter." These neurons receive sensory information, process it, create memories for future use, direct voluntary movements, and allow us to plan and think in ways we cannot yet understand.

The cerebral cortex is divided into four anatomical regions: the *frontal, parietal, occipital,* and *temporal lobes* (**FIG. 38-14**). Functionally, the cortex contains *primary sensory areas,* regions where signals originating in sensory organs such as the eyes and ears are received and converted into subjective impressions—for instance, light and sound. Nearby *association areas* interpret the sounds as speech or music, for example, and the visual stimuli as recognizable objects or words on this page. Association areas also link the stimuli with memories stored in the cortex and generate commands to produce speech. Research has revealed that the association areas of the brain do not always have the same function in the right and left hemispheres (see section 38.5).

Primary sensory areas in the parietal lobe interpret sensations of touch that originate in all parts of the body; these body parts are "mapped" in an orderly sequence (see Fig. 38-14). In an adjacent region of .the frontal lobe, *primary motor areas* command movements in corresponding areas of the body by stimulating the motor neurons in the spinal cord that innervate the muscles, allowing you to walk to class, throw a Frisbee®, or type a term paper. Like the primary sensory area, the primary motor area also has adjacent association areas, including the motor association area (also called the premotor area), which seems to be responsible for directing the motor area to produce movements. Behind the bones of the forehead lies another association area of the frontal lobe, which is important in complex reasoning functions such as decision making, predicting the consequences of actions, controlling aggression, and planning for the future.

Damage to the cortex from trauma, stroke, or a tumor results in specific deficits, such as problems with speech, difficulty reading, or the inability to sense or move specific parts of the body. Most brain cells of adults cannot be replaced, so if a brain region is destroyed, these deficits may be permanent. Fortunately, however, training can sometimes allow undamaged regions of the cortex to take control over and restore some of the lost functions.

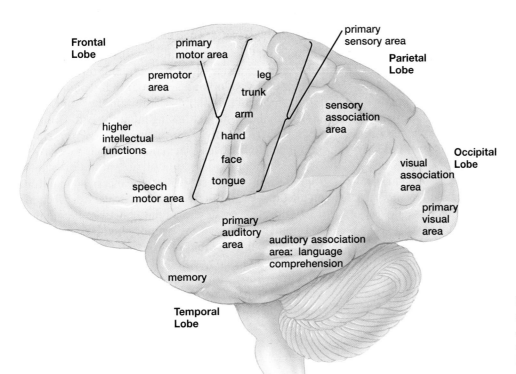

FIGURE 38-14 The cerebral cortex

A map of the human left cerebral cortex. A map of the right cerebral cortex would be similar, except that speech and language would not be as well developed.

38.5 HOW DOES THE BRAIN PRODUCE THE MIND?

Historically, people have always had difficulty imagining how a few pounds of soft material in the skull could produce the range of thoughts, emotions, and memories of the human mind. This "mind–brain problem" has occupied generations of philosophers and, more recently, neurobiologists. Beginning with observations of individuals with head injuries and progressing to sophisticated surgical, physiological, and biochemical experiments, the outlines of how the brain creates the mind are beginning to emerge. Here, we will highlight only a few of the most fascinating features.

The "Left Brain" and "Right Brain" Are Specialized for Different Functions

Although the cerebrum consists of two extremely similar-looking hemispheres, it has been known since the early 1900s that this symmetry does not extend to brain function. Much of what is known about the differences in hemisphere function comes from studies of accident or stroke victims with localized damage to one hemisphere, patients who have one hemisphere temporarily anesthetized, or those in whom the corpus callosum (which connects the two hemispheres) has been severed. This surgical procedure is still performed in certain cases of uncontrollable epilepsy, to prevent the spread of seizures from one hemisphere to the other. Because sensory input reaches both hemispheres, and each side guides appropriate responses based on its abilities, a person whose corpus callosum has been severed can function quite normally. As you read about hemispheric differences, remember that the brain performs as a coordinated unit. A person with an intact corpus callosum cannot use either side selectively; both hemispheres contribute in complex ways to our perceptions, thoughts, behavior, and abilities.

Beginning in the 1950s, Roger Sperry, a neurobiologist at the California Institute of Technology, studied people whose hemispheres had been surgically separated by cutting the corpus callosum. In his research, Sperry made use of the knowledge that axons within each optic tract (which are not severed by the surgery) follow a pathway that causes the left half of each visual field to be "seen" by the right cerebral hemisphere and the right half to be seen by the left hemisphere (FIG. 38-15). Sperry used an ingenious device that projected different images onto the left and right visual fields and thus sent different signals to each hemisphere.

When Sperry projected an image of a nude human onto just the left visual field, the subjects would blush and smile but would claim to have seen nothing. The same figure projected onto the right visual field was readily described verbally. These and later experiments have revealed that—in right-handed people—the left hemisphere is almost always dominant in speech, reading, writing, language comprehension, mathematical ability, and logical problem solving. The right side of the brain is superior to the left in

musical skills, artistic ability, recognizing faces, spatial visualization, and the ability to recognize and express emotions. For his pioneering work, Sperry shared the Nobel Prize for Physiology or Medicine in 1981.

Recent experiments, however, indicate that the left–right dichotomy is not as rigid as was once believed. An individual who has suffered a stroke that disrupted the blood supply to specific parts of the left hemisphere may lose the ability to speak, read, or write. Fortunately, these deficits can often be partially overcome through training, even though the left hemisphere itself has not recovered. This suggests that the right hemisphere has some latent language capabilities. Interestingly, females have a slightly larger corpus callosum than males do, suggesting a gender difference in the extent of interconnections between the two hemispheres. Imaging neural activity in the brains of normal subjects as they perform various mental tasks has provided further evidence of this difference. When subjects were asked to compare two word lists to find rhyming words, a specific region of the left cortex of male subjects became active, but in females, similar areas in *both* the left and right hemispheres were activated. (Further brain-imaging studies are described in "Scientific Inquiry: Neuroimaging—Peering into the 'Black Box.'").

The Mechanisms of Learning and Memory Are Subjects of Intensive Research

Although hypotheses abound about the cellular mechanisms of learning and memory, we are a long way from understanding these phenomena. However, we do know a fair amount about short-term "working" memory, long-term memory, and some of the brain sites involved in learning, memory storage, and recall in mammals, especially in humans.

Memory May Be Brief or Long Lasting

Experiments show that learning occurs in two phases: initial **working memory** followed by **long-term memory**. For example, if you look up a number in the phone book, you will probably remember the number only long enough to dial it—this is working memory. But if you call the number frequently, eventually you will remember the number more or less permanently—it is now stored in long-term memory.

Some working memory seems to be electrical in nature, involving the repeated activity of a particular neural circuit in the brain. As long as the circuit is active, the memory stays. In other cases, working memory involves temporary biochemical changes within neurons of a circuit, resulting in stronger synaptic connections between them.

In contrast, long-term memory seems to be structural—the result, perhaps, of persistent changes in the expression of certain genes. It may require the formation of new, long-lasting synaptic connections between specific neurons or the long-term strengthening of existing but weak synaptic connections (for example, increasing neurotransmitter release or increasing the number of receptors for the neurotransmitter). Converting working memory into long-term

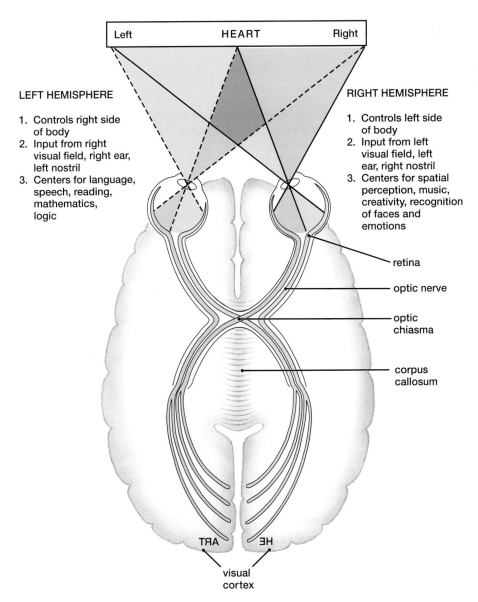

FIGURE 38-15 Specialization of the cerebral hemispheres
Each half of the retina of each eye "sees" the opposite visual field. The axons from the half-retinas that see the left visual field send information to the right hemisphere, and vice versa. Therefore, with a quick glance at the word "heart" (before you had a chance to move your eyes), the right hemisphere would perceive "he" and the left hemisphere would perceive "art." In addition to "seeing" different parts of the visual field, the two hemispheres typically control the opposite sides of the body and are specialized for a variety of functions.

memory seems to involve the hippocampus, which is believed to process new memories and transfer them to the cerebral cortex for permanent storage.

The Temporal Lobes Are Important for Memory

The mechanisms of learning, memory storage, and memory retrieval are subjects of extensive research. Intense electrical activity occurs in the hippocampus—deep within the brain's temporal lobe—during learning, providing strong evidence that the two are closely linked. Even more striking are the results of hippocampal damage. If both hippocampi (one in each hemisphere) are severely damaged, a person retains most of the memories formed before the damage occurred. However, the person usually cannot form new memories. A patient whose hippocampi and associated brain structures were surgically removed in 1953 (in an attempt to control seizures) can entertain himself by reading the same magazine over and over, and people whom he sees daily require reintroduction at each encounter. This example and others suggest that the hip-

pocampus is responsible for transferring information from working memory to long-term memory.

The temporal lobes of the cerebral hemispheres also seem to be important in the *retrieval*, or recall, of long-term memories. In a famous series of experiments in the 1940s, neurosurgeon Wilder Penfield electrically stimulated the temporal lobes of conscious patients undergoing brain surgery. The patients did not merely recall memories but felt that they were experiencing the past events right there in the operating room!

Insights About How the Brain Creates the Mind Come from Diverse Sources

Until about 100 years ago, the mind was a more appropriate subject for philosophers than for scientists, because tools for studying the brain did not yet exist. Through the first half of the twentieth century, the mind was treated by psychologists as a "black box" whose internal workings could be deduced only through the investigation of how

SCIENTIFIC INQUIRY   Neuroimaging—Peering into the "Black Box"

For most of human history, the brain has been a "black box"; its inputs and outputs were observable, but its internal workings were inherently unknowable. Now, however, new imaging techniques provide exciting insights into brain function. These include *PET* (positron emission tomography, described in Chapter 2) and *fMRI* (functional magnetic resonance imaging), which allow researchers to observe the brain in action.

Regions of the brain that are most active have higher energy demands; they use more glucose and attract a greater flow of oxygenated blood than do less-active areas. In PET scans, scientists inject the subject with a radioactive substance, such as a radioactive form of glucose, and then monitor levels of radioactivity that reflect differences in metabolic rate. These are translated by computer into colors on images of the brain. By monitoring radioactivity while a specific task is performed, scientists can identify which parts of the brain are most active during that task. In contrast, fMRI detects differences in the way oxygenated and deoxygenated blood respond to a powerful magnetic field applied by an enormous electromagnet that surrounds the body. Active brain regions can be distinguished with fMRI without using radioactivity and over much shorter time spans than required by PET.

Using fMRI or PET, researchers can observe changes as the brain performs a specific reasoning task or responds to an odor or a visual or auditory stimulus. Through brain scans, scientists have confirmed that different aspects of the processing of language occur in distinct areas of the cerebral cortex (**FIG. E38-7**). Using fMRI, other researchers analyzed the frontal lobe areas used in generating words in individuals who spoke two languages. In subjects who had grown up speaking two languages, the same region of the frontal lobe was in speaking each language. Subjects who had learned a second language later in life used different but adjacent frontal lobe areas for the two languages.

Functional MRI scans have also been used to determine the parts of the brain that are most active during various emotional states. For example, when a person is frightened, the amygdala lights up (**FIG. E38-8a**). When people in love view photos of their lovers, other areas in the brain are activated (**FIG. E38-8b**). Interestingly, most of these same areas are also activated by drugs of abuse, such as cocaine.

By the way, never let anyone tell you that you use only a small fraction of your brain! Although PET and fMRI images sometimes make it appear that only a small area of the brain is active, this is because activity of other regions is subtracted out during the imaging process, to show where brain activity has *changed* as a result of stimulation. The images shown here merely highlight areas where activity is even more intense than normal.

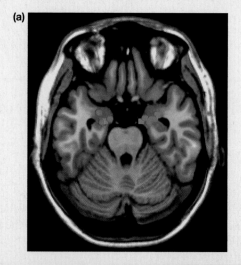

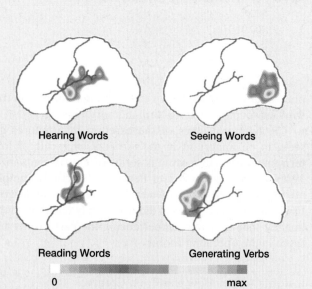

FIGURE E38-7 Localization of language tasks
Changes in glucose utilization, measured by PET, reveal different cortical regions involved in language-related tasks, based on research by Marcus Raichle of the Washington University School of Medicine in St. Louis. The scale ranges from white (lowest) to red (highest).

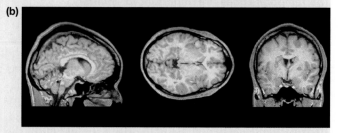

FIGURE E38-8 Localization of emotions
(a) A frightening experience activates the amygdala, a part of the forebrain that apparently produces emotions such as fear and rage. (b) Looking at photos of lovers activates multiple parts of the brain, including areas in the cerebral cortex (left) and the basal ganglia (right).

past and present experiences were interpreted and influenced behavior. New discoveries, however, are rapidly changing our views of the workings of the brain.

In recent decades, we have begun to understand the cellular bases of at least some psychological phenomena. Many forms of mental illness, such as schizophrenia, manic depression, and autism, once thought to be due to childhood trauma or inept parenting, are now known to result from neurotransmitter imbalances and/or structural abnormalities in the brain. Studies have also revealed strong heritability (and hence, a biological basis) for traits that were once considered entirely learned or acquired, such as alcoholism or a tendency toward shyness.

A striking illustration of how the physical structure of the brain is related to personality was unwittingly provided by Phineas Gage in 1848. A railroad construction foreman, Gage was setting an explosive charge when it triggered prematurely. The blast blew a 13-pound steel rod through his skull, damaging both of his frontal lobes (**FIG. 38-16**). Although Gage survived for many years after his accident, his personality changed radically. Before the accident, Gage was conscientious, industrious, and well liked. After his recovery, he became impetuous, profane, and incapable of working toward a goal. Subsequent research has implicated the frontal lobe in emotional expression, control of aggression, recognition of appropriate behavior, and the ability to work for delayed rewards.

Other sites of damage have revealed additional specializations. One patient with damage to a very small part of

his left frontal lobe was unable to name fruits and vegetables (although he could name everything else). Other victims of brain damage have developed a selective inability to recognize faces or, in a recent case, to recognize any object that is *not* a face, suggesting that the brain has a region specialized to recognize faces, separate from the regions that recognize objects in general. Researchers have recorded individual neurons in the temporal lobes of monkeys that are selective for particular orientations of faces, such as a profile or a head-on view. The association area of the posterior parietal lobe seems to integrate information from several different senses, enabling an individual to recognize his or her own body and its relationship to the environment. People with damage to one of the parietal lobes sometimes lose track of one side of their bodies, and of objects on that side as well. For example, they may ignore food on one side of their plates or become lost because they don't recognize corridors on one side of the building. One patient did not recognize his own leg, and tried repeatedly to throw it out of his hospital bed. With time, however, other areas of the brain usually compensate for this damage, restoring more normal perception.

In the past, much of our understanding of the human mind–brain connection came from the study of victims of brain damage such as that caused by a stroke, trauma, tumor, or surgery. Typically, the exact extent of the damage remained unknown until revealed by autopsy. New imaging techniques now permit insight into the functioning of normal, as well as diseased, brains (see "Scientific Inquiry: Neuroimaging—Peering into the 'Black Box'"). These and increasingly sophisticated techniques of the future will create ever-larger windows into the human brain and a clearer understanding of how the brain generates the human mind.

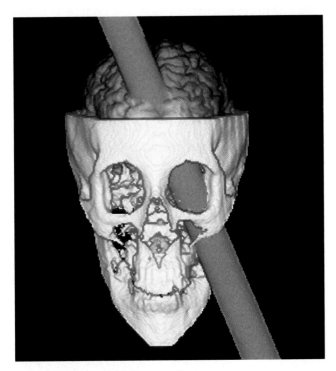

FIGURE 38-16 A revealing accident
Studies of the skull of Phineas Gage have enabled scientists to create this computer-generated reconstruction of the path taken by the steel rod that was blown through his head by an explosion.

38.6 HOW DO SENSORY RECEPTORS WORK?

The word *receptor* is used in several contexts in biology. In the most general sense, a **receptor** is a structure that changes when it is acted upon by a stimulus, causing a signal to be produced. A receptor may be a membrane protein that changes shape when it binds a specific hormone or neurotransmitter, as discussed in previous chapters. In neurobiology, a **sensory receptor** is an entire specialized cell (typically, a neuron) that produces an electrical signal in response to specific stimuli—that is, it translates environmental stimuli into the language of the nervous system. All sensory receptors produce electrical signals, but each receptor type is specialized to produce its signal only in response to a particular type of environmental stimulus (Table 38-2).

The stimulation of a sensory receptor causes an electrical signal called a **receptor potential**. The amplitude of a receptor potential varies with the intensity of the stimulus: the stronger the stimulus, the larger the receptor potential. In some sensory receptor neurons, the receptor potential may bring the cell above threshold and cause action potentials; a large receptor potential will bring neurons far

Table 38-2 Some Vertebrate Receptor Types

Type of Receptor	Sensory Cell Type	Stimulus	Location
Thermoreceptor	Free nerve ending	Heat, cold	Skin
Mechanoreceptor	Hair cell	Vibration, motion, gravity	Inner ear
	Specialized nerve endings and free nerve endings in skin (Pacinian corpuscle, Merkel's disc)	Vibration, pressure, touch	Skin
	Specialized nerve endings in muscles or joints (muscle spindle, Golgi tendon organ)	Stretch	Muscles, tendons
Photoreceptor	Rod, cone	Light	Retina of eye
Chemoreceptor	Olfactory receptor	Odor (airborne molecules)	Nasal cavity
	Taste receptor	Taste (waterborne molecules)	Tongue
Pain receptor	Free nerve ending	Chemicals released by tissue injury	Widespread in body

above threshold for a relatively long time, causing a high frequency of action potentials in its axon. In some very small sensory receptors, receptor potentials directly cause neurotransmitters to be released onto postsynaptic neurons, which in turn produce action potentials that travel to the central nervous system. In these cases, a large receptor potential causes the release of a large amount of transmitter onto the postsynaptic neuron, bringing its membrane potential far above threshold and producing a high frequency of action potentials. As we learned earlier, the frequency of action potentials is how the brain determines the intensity of a stimulus.

Some receptors, called *free nerve endings*, consist of branching dendrites of sensory neurons; other receptors have specialized structures that help them respond to a specific stimulus. Many sensory receptors are clustered into sensory organs, such as the eye, ear, skin, or tongue. Their electrical activity, after being processed by the brain, gives rise to the subjective perceptions of light, sound, touch, and taste that we describe as our "senses." In the following sections, we will explore some of the senses that determine the way we perceive the world.

38.7 HOW ARE MECHANICAL STIMULI DETECTED?

Mechanoreceptors are found throughout the body. They include receptors in the skin that respond to touch, vibration or pressure; stretch receptors in many internal organs; position sensors in the joints; and receptors in the inner ear that allow us to detect sound (see section 38.8 below).

The skin of humans and most other vertebrates is exquisitely sensitive to touch. Embedded in and directly beneath the skin are several distinct types of mechanoreceptor neurons, each with a dendrite that produces a receptor potential when its membrane is stretched or dented (**FIG. 38-17**). The dendrites of some touch receptors are free nerve endings that can produce sensations of itching or tickling, as well as touch. The endings of other receptors are enclosed in layers

of connective tissue—for example, the Pacinian corpuscle, which responds to rapid changes in pressure such as vibrations or a sharp poke. The density of mechanoreceptors in the skin varies tremendously over the surface of the body. Each square centimeter of fingertip has dozens of touch receptors, but on the back there may be less than one per square centimeter.

Mechanoreceptors in many hollow organs, such as the stomach, intestine, rectum, and urinary bladder, signal fullness by responding to stretch. Mechanoreceptors in the joints, often also responding to stretch, allow us to know whether our elbows are bent or straight, and how much, without having to look. Imagine how annoying it would be if you had to carefully watch your fork on its way to your mouth, to avoid stabbing yourself in the face!

38.8 HOW IS SOUND SENSED?

Sound is produced by a vibrating object—a drum, vocal cords, or the speakers of a stereo. These vibrations are usually transmitted through the air and intercepted by our ears, which are elaborate structures that detect the direction, pitch, and intensity of sound.

The Ear Converts Sound Waves into Electrical Signals

The ear of humans and most other vertebrates consists of three parts: the outer, middle, and inner ear (**FIG. 38-18a**). The **outer ear** consists of the **pinna** and the **auditory canal**. The pinna is the flap of skin-covered cartilage attached to the surface of the head. It collects sound waves and modifies them in ways that the brain uses to determine the location of the sound source. The air-filled auditory canal conducts the sound waves to the **middle ear**, consisting of the **tympanic membrane**, or *eardrum*; three tiny bones called the *hammer, anvil,* and *stirrup*; and the **auditory tube** (also called the *Eustachian tube*). The auditory tube connects the middle ear to the pharynx and equalizes the air pressure between the middle ear and the atmosphere. This tube may become swollen shut if you have a cold; if this

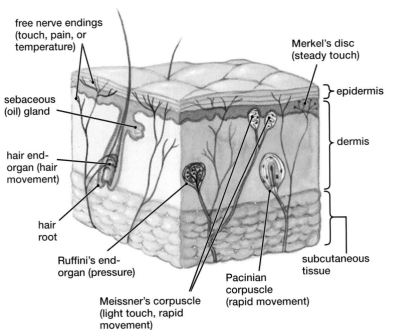

FIGURE 38-17 Receptors in the human skin
The diversity of receptors in the skin allows us to perceive touch, pressure, vibration, tickling, itching, pain, heat, and cold.

free nerve endings (touch, pain, or temperature)

sebaceous (oil) gland

hair end-organ (hair movement)

hair root

Ruffini's end-organ (pressure)

Meissner's corpuscle (light touch, rapid movement)

Merkel's disc (steady touch)

epidermis

dermis

subcutaneous tissue

Pacinian corpuscle (rapid movement)

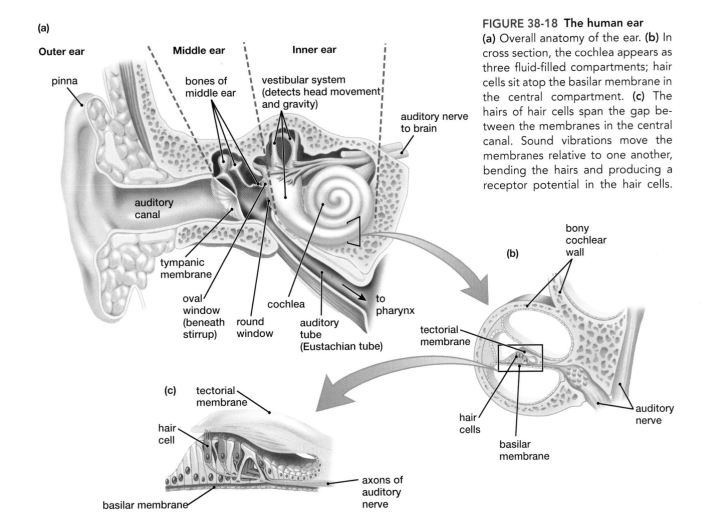

FIGURE 38-18 The human ear
(a) Overall anatomy of the ear. (b) In cross section, the cochlea appears as three fluid-filled compartments; hair cells sit atop the basilar membrane in the central compartment. (c) The hairs of hair cells span the gap between the membranes in the central canal. Sound vibrations move the membranes relative to one another, bending the hairs and producing a receptor potential in the hair cells.

(a)

Outer ear

Middle ear

Inner ear

pinna

bones of middle ear

vestibular system (detects head movement and gravity)

auditory nerve to brain

auditory canal

tympanic membrane

oval window (beneath stirrup)

round window

cochlea

auditory tube (Eustachian tube)

to pharynx

(b)

bony cochlear wall

tectorial membrane

hair cells

basilar membrane

auditory nerve

(c)

tectorial membrane

hair cell

basilar membrane

axons of auditory nerve

happens, air pressure changes (such as those experienced during aircraft takeoff and landing) can be painful.

Sound waves traveling down the auditory canal vibrate the tympanic membrane, which in turn vibrates the hammer, the anvil, and the stirrup. These bones transmit vibrations to the **inner ear**. The fluid-filled hollow bones of the inner ear form the spiral-shaped **cochlea** ("snail" in Latin) as well as structures of the *vestibular system* that detect head movement and the pull of gravity. The stirrup bone transmits vibrations to the fluid within the cochlea by vibrating a membrane in the cochlea called the *oval window*. The *round window* is a second membrane below the oval window; it allows fluid within the cochlea to shift back and forth as the stirrup bone vibrates the oval window.

Vibrations Are Converted into Electrical Signals in the Cochlea

The cochlea, in cross section, consists of three fluid-filled compartments. The central compartment houses the receptors and the supporting structures that activate them in response to sound vibrations. The floor of the central chamber is the **basilar membrane**, on top of which sit mechanoreceptors called **hair cells**. Hair cells have small cell bodies topped by hairlike projections that resemble stiff cilia. Some of these hairs are embedded in a gelatinous structure called the **tectorial membrane**, which protrudes into the central canal (**FIG. 38-18b, c**).

How do these structures allow the perception of sound? The oval window passes vibrations from the small bones of the middle ear to the fluid in the cochlea, which in turn vibrates the basilar membrane, causing it to move up and down. This movement bends the hairs of the hair cells, setting up a chain of events that culminates in receptor potentials. The receptor potentials cause the hair cells to release neurotransmitters onto neurons whose axons form the auditory nerve. Consequently, these axons produce action potentials that travel to auditory processing centers within the brain.

How do we perceive *loudness* (the magnitude of sound vibrations) and *pitch* (musical note; the frequency of sound vibrations)? Soft sounds cause small vibrations of the tympanic membrane, the bones of the middle ear, the oval window, and the basilar membrane. Therefore, the hairs bend only a little, so the hair cells produce small receptor potentials that cause the release of a tiny bit of neurotransmitter and result in a low frequency of action potentials in axons of the auditory nerve. Loud sounds cause large vibrations of all of these structures, which cause greater bending of the hairs and a larger receptor potential, producing high-frequency action potentials in the auditory nerve. Very loud sounds can damage the hair cells (**FIG. 38-19a**), resulting in hearing loss, a fate suffered by many rock musicians and their fans. In fact, many sounds in our everyday environment have the potential to damage hearing, especially if they are prolonged (**FIG. 38-19b**).

The structure of the basilar membrane allows the perception of pitch. Humans can detect sounds from about 30 vibrations per second (very low pitched) to about 20,000 vibrations per second (very high pitched). Like the strings of

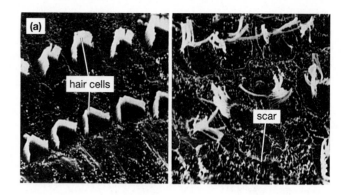

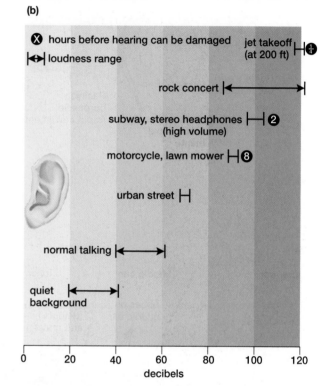

FIGURE 38-19 Loud sounds can damage hair cells
(a) Scanning electron micrographs show the effect of intense sound on the hair cells of the inner ear. The hairs of hair cells in a normal guinea pig; hairs emerge from each receptor in a V-shaped pattern (left). After 24-hour exposure to a sound level approached by loud rock music (2000 vibrations per second at 120 decibels), many of the hairs are damaged or missing, leaving "scars" (right). Hair cells in humans do not regenerate, so such hearing loss is permanent. [SEMs by Robert S. Preston, courtesy of Professor J. E. Hawkins, Kresge Hearing Research Institute, University of Michigan Medical School] **(b)** Sound levels of everyday noises and their potential to damage hearing. Sound intensity is measured in *decibels* on a logarithmic scale; a 10-decibel sound is 10 times as loud as a 1-decibel sound, and a 20-decibel sound is 100 times as loud. You feel pain at sound intensities above 120 decibels. [Source: Deafness Research Foundation, National Institute on Deafness and Other Communication Disorders.]

a harp, the basilar membrane is stiff and narrow at the end near the oval window but more flexible and wider near the tip of the cochlea. This progressive change in structure causes each successive portion of the membrane to resonate or vibrate in synchrony with a particular frequency of sound: high notes near the oval window and low notes near the tip of the cochlea. The brain interprets signals from hair cells near the oval window as high-pitched sound, whereas signals from hair cells located progressively closer to the tip of the cochlea are interpreted as increasingly lower in pitch.

38.9 HOW IS LIGHT SENSED?

Animal vision varies in its ability to provide sharp, accurate representations of the real world; in fact, several types of eyes have evolved independently. All forms of vision, however, use *photoreceptors*. These sensory cells contain receptor molecules called **photopigments** (because they are colored), which change shape when they absorb light. This shape change sets off a series of chemical reactions inside the photoreceptor cell that ultimately produces a receptor potential.

The Compound Eyes of Arthropods Produce a Mosaic Image

The arthropods (insects, spiders, and crustaceans) evolved **compound eyes**, which consist of a mosaic of many individual light-sensitive subunits called **ommatidia** (singular, **ommatidium**; **FIG. 38-20**). Each ommatidium functions as an on-off, bright-dim detector. Using a large number of ommatidia (up to 36,000 per eye in a dragonfly), most arthropods probably see a reasonably accurate—though grainy—image of the world. Compound eyes are excellent at detecting movement, an advantage in avoiding predators and in hunting. Many arthropods, such as bees and butterflies, also have good color perception.

The Mammalian Eye Collects and Focuses Light, Converting It into Electrical Signals

Mammalian eyes consist of two major parts. The **retina** is a multilayered sheet of photoreceptor cells and associated neurons. In response to light, the photoreceptors stimulate the neurons, ultimately producing action potentials in neurons whose axons form the optic nerve. The rest of the eye is a series of structures that transmit light, regulate the amount of light entering the eye, and focus the light on the photoreceptors (**FIG. 38-21**). As you read this, light reflected from the page first encounters the **cornea**, a transparent covering over the front of the eyeball that collects light waves and begins to focus them. Behind the cornea, light passes through a chamber filled with a watery fluid called **aqueous humor**, which provides nourishment for both the lens and cornea. The amount of light entering the eye is adjusted by the **iris**, a pigmented muscular tissue. The iris regulates the size of the **pupil**, the opening in its center. Light passing through the pupil encounters the **lens**, a structure that resembles a flattened sphere composed of transparent protein fibers. The lens is suspended behind the pupil by muscles that regulate its shape and allow fine focusing of the image. Behind the lens is a much larger chamber filled with **vitreous humor**, a clear jellylike substance that allows

(a) Compound eyes

(b) Ommatidia **Single ommatidium**

 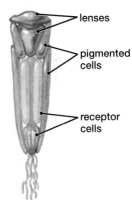

lenses

pigmented cells

receptor cells

FIGURE 38-20 Compound eyes
(a) Scanning electron micrograph of the head of a fruit fly, showing a compound eye on each side of the head. **(b)** Each eye is made up of numerous light-receptive ommatidia. Within each ommatidium are several receptor cells, capped by a lens. Pigmented cells surrounding each ommatidium prevent the passage of light to adjacent receptors.

light to pass freely while supporting and maintaining the shape of the eyeball.

After passing through the vitreous humor, light finally reaches the retina. Here, the light energy is converted into action potentials that are conducted to the brain. Behind the retina is the **choroid**, a darkly pigmented tissue. The choroid's rich blood supply helps nourish the cells of the retina. Its dark pigment absorbs stray light whose reflection inside the eyeball would interfere with clear vision. Surrounding the outer portion of the eyeball is the **sclera**, a tough connective tissue layer that is visible as the white of the eye and is continuous with the cornea.

Driving along a country road at night, have you ever been startled by apparently disembodied glowing eyes? In vertebrates that are most active at dusk (such as deer), the choroid may be modified to reflect light rather than to absorb it. By reflecting light that escaped the photoreceptors

(a) Anatomy of the human eye

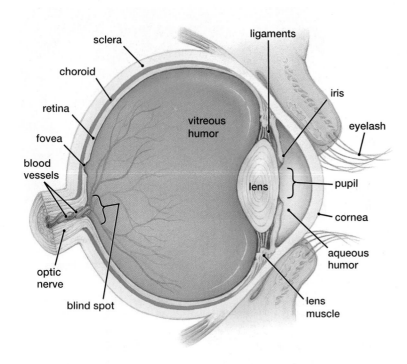

(b) Layers of the retina

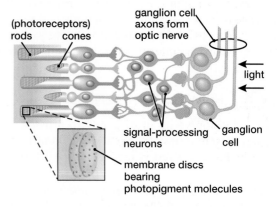

FIGURE 38-21 The human eye
(a) The anatomy of the human eye. **(b)** The human retina has rods and cones (photoreceptors), signal-processing neurons, and ganglion cells. Each rod and cone bears a long extension packed with membranes in which the light-sensitive molecules are embedded.

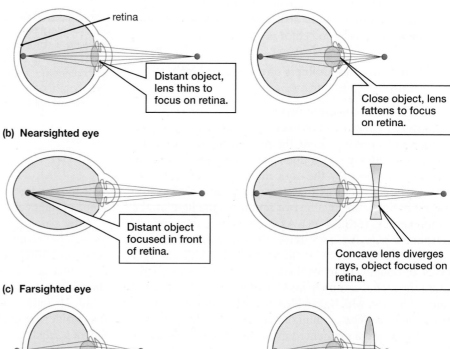

(a) Normal eye

retina

Distant object, lens thins to focus on retina.

Close object, lens fattens to focus on retina.

(b) Nearsighted eye

Distant object focused in front of retina.

Concave lens diverges rays, object focused on retina.

(c) Farsighted eye

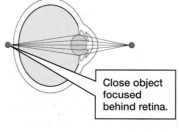

Close object focused behind retina.

Convex lens converges rays, object focused on retina.

FIGURE 38-22 Focusing in the human eye
QUESTION Today, many nearsighted and farsighted people choose to correct their vision problems with laser surgery on their corneas rather than with corrective lenses. For nearsightedness and farsightedness, how should corneas be reshaped to correct the problem?

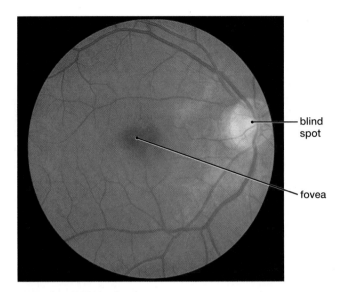

FIGURE 38-23 The human retina
A portion of the human retina, photographed through the cornea and lens of a living person. The blind spot and fovea are visible. Blood vessels supply oxygen and nutrients; notice that they are dense over the blind spot (where they won't interfere with vision) and scarce near the fovea.

during its initial passage, the choroid gives the receptors a second chance to detect it, maximizing the animal's ability to see in dim light. Reflective choroids give the eyes of these animals an eerie red or blue glow when bright light (such as that from car headlights) is reflected back through the wide-open pupil. Imagine how blindingly bright your headlights must be to a deer!

The Adjustable Lens Allows Focusing on Both Distant and Nearby Objects

The visual image is focused most sharply on a small area of the retina called the **fovea**. Although focusing begins at the cornea, whose rounded contour bends light rays, the lens is responsible for final, sharp focusing. The shape of the lens is adjusted by its encircling muscle. When viewed from the side, the lens is either rounded, to focus on nearby objects, or flattened, to focus on distant objects (**FIG. 38-22a**).

If your eyeball is too long or your cornea is too rounded, you will be **nearsighted**—the light from distant objects will focus in front of the retina. The eyes of **farsighted** people, with eyeballs that are too short or corneas that are too flat, focus light from nearby objects behind the retina. These conditions can be corrected by either contact lenses or glasses of the appropriate shape (**FIG. 38-22b, c**). Both nearsightedness and farsightedness can also be corrected with laser surgery that reshapes the cornea, producing a new corneal curvature that acts much like a corrective lens. As people age, the lens within the eye stiffens, causing farsightedness because the lens can no longer curve enough to focus on nearby objects. By their mid-forties, most people require glasses for close work such as reading.

Photoreceptors and Neurons in the Retina Capture Light, Process the Resulting Electrical Signals, and Ultimately Produce Action Potentials in the Optic Nerve

The vertebrate eye provides the sharpest vision in the animal kingdom, even though the complex, multilayered retina is, from an engineering perspective, "built backward." The photoreceptors, called **rods** and **cones** after their shapes, gather light at the rear of the retina (see Fig. 38-21b). Between the receptors and incoming light are several layers of neurons that process the signals from the photoreceptors. These neurons enhance our ability to detect edges, movement, dim light, and changes in light intensity. The retinal layer nearest the vitreous humor consists of **ganglion cells**, whose axons make up the **optic nerve**. The much-modified signal from the photoreceptors and intervening neurons is converted to action potentials in the ganglion cells. To reach the brain, ganglion cell axons pass back through the retina at a location called the **blind spot** (**FIG. 38-23**; see also Fig. 38-21a). This area lacks receptors, so objects focused there cannot be seen. You can locate your blind spot by closing your left eye and focusing steadily on the star below with your right eye (**FIG. 38-24**).

FIGURE 38-24 Finding your blind spot

Start with the book about a foot away and gradually move it closer. The circle will disappear when the image falls on your blind spot. During everyday life, your brain receives information from both eyes, and your eyes constantly flicker back and forth as well. Consequently, you do not usually experience a "hole" in your visual field.

Rods and Cones Differ in Distribution and Light Sensitivity

Photoreception in both rods and cones begins with the absorption of light by photopigment molecules that are embedded in the plasma membranes of the photoreceptors (see Fig. 38-21b). Light hitting a photopigment molecule changes its shape, setting off chemical reactions that ultimately produce a receptor potential in the photoreceptor.

Although cones are located throughout the retina, they are concentrated in the fovea, where the lens focuses images most sharply (see Fig. 38-21a and 38-22). The fovea looks like a dent near the center of the retina, because the layers of signal-processing neurons are spread apart, while still retaining their synaptic connections. This arrangement allows light to reach the cones of the fovea without having to pass through so many other layers of cells.

Human eyes have three varieties of cones, each containing a slightly different photopigment. Each type of photopigment is most strongly stimulated by a particular wavelength of light, corresponding roughly to red, green, or blue. The brain distinguishes color according to the relative intensity of stimulation of different cones. For example, the sensation of yellow is produced by fairly equal stimulation of red and green cones. About 4% to 8% of males have difficulty distinguishing red from green because they possess a defective gene for the red or green photopigment on the X chromosome (see Chapter 12).

Rods dominate in the peripheral portions of the retina. Rods, which are longer than cones and contain far more photopigment, are thus much more sensitive to light than cones are, so they are largely responsible for our vision in dim light. Unlike cones, rods do not distinguish colors. In moonlight, which is too dim to activate the cones, the world appears in shades of gray.

Not all animals have both rods and cones. Animals that are active almost entirely during the day (certain lizards, for example) may have all-cone retinas, whereas many nocturnal animals (such as rats and ferrets) and those dwelling in dimly lit habitats (such as deep-sea fishes) have mostly or entirely rods.

Binocular Vision Allows Depth Perception

The placement of vertebrate eyes on the head is determined by the lifestyle of the animal. Predators and omnivores usually have both eyes facing forward (**FIG. 38-25a**), but most herbivores have one eye on each side of the head (**FIG. 38-25b**). The forward-facing eyes of predators and omnivores have slightly different but extensively overlapping visual fields. This **binocular vision** allows depth perception, the accurate judgment of the distance of an object. These abilities are important to a cat about to pounce on a mouse or to a monkey leaping from branch to branch.

In contrast, the widely spaced eyes of herbivores have little overlap in their visual fields; some depth perception is sacrificed in favor of a nearly 360-degree field of view. This view allows these animals, who are frequently preyed on, to spot a predator approaching from any direction.

38.10   HOW ARE CHEMICALS SENSED?

Through chemical senses that use chemoreceptors (see Table 38-2), animals find food, avoid poisonous materials, locate homes, find mates, and maintain homeostasis. Chemoreceptors in certain large blood vessels and in the hypothalamus of the brain monitor levels of crucial molecules such as sugar, water, oxygen, and carbon dioxide in the blood; they also stimulate activities that maintain these levels within narrow limits. Terrestrial vertebrates have two separate senses for detecting chemicals outside the body: one for airborne molecules, called **smell** (*olfaction*), and one for chemicals dissolved in water or saliva, the sense of **taste** (*gustation*).

Olfactory Receptors Detect Airborne Chemicals

In humans and most other vertebrates, receptors for olfaction are nerve cells located in a patch of mucus-covered epithelial tissue in the upper portion of each nasal cavity (**FIG. 38-26**). The human olfactory epithelium is small compared with that of many other mammals (dogs, for example), whose sense of smell is hundreds of times more acute than ours. Olfactory receptors have hairlike dendrites that protrude into the nasal cavity and are embedded in the mucus. Odorous molecules in the air, such as those generated by coffee, diffuse into the mucus layer and bind to receptor proteins on the dendrites.

Humans produce about 500 different olfactory receptor proteins, but each olfactory receptor neuron expresses only one type. Each receptor protein is specialized to bind a particular type of molecule and stimulate the olfactory receptor to send a message to the brain. Many odors are complex mixtures of molecules that stimulate

(a)

(b)

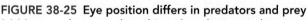

FIGURE 38-25 Eye position differs in predators and prey
(a) Most predators, such as this owl, and primates have eyes in front; both eyes can be focused on a target, providing binocular vision. **(b)** Most herbivorous prey animals, such as rabbits, have eyes at the sides, which allow them to scan for predators. QUESTION Why do some herbivorous or fruit-eating animals, such as monkeys and fruit bats, have both eyes in front?

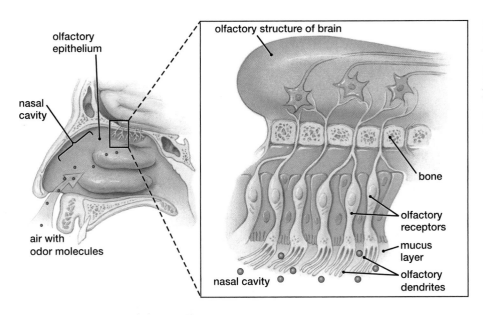

olfactory
epithelium

nasal
cavity

air with
odor molecules

olfactory structure of brain

bone

olfactory
receptors

mucus
layer

olfactory
dendrites

nasal cavity

FIGURE 38-26 Human olfactory receptors
The receptors for olfaction in humans are neurons bearing microscopic hairlike projections that protrude into the nasal cavity. The projections are embedded in a mucus layer in which odorous molecules dissolve before contacting the receptors.

several receptor proteins, so our perception of odors arises through the responses of multiple receptors.

Taste Receptors Detect Chemicals That Contact the Tongue

The human tongue bears about 10,000 **taste buds**, structures embedded in small bumps (called *papillae*) that cover the tongue's surface (**FIG. 38-27a**). Each taste bud consists of a cluster of 60 to 80 taste receptor cells surrounded by supporting cells in a small pit. The cells in the pit communicate with the mouth through a taste pore (**FIG. 38-27b**). Microvilli (thin membrane projections) of taste receptor cells protrude through the pore. Dissolved chemicals enter the pore and bind to receptor proteins on the microvilli, producing a receptor potential.

Four major types of taste receptors have long been known: sweet, sour, salty, and bitter. A fifth type, *umami* (a Japanese word loosely translated as "delicious"), has recently been identified. The umami receptor responds to glutamate, an amino acid that serves as a neurotransmitter; glutamate is part of MSG (monosodium glutamate), a seasoning that enhances the flavor of foods. Although it was formerly believed that taste buds for specific tastes were concentrated on specific areas of the tongue, recent research has shown they are more or less evenly distributed.

We perceive a great variety of tastes through two mechanisms. First, a particular substance may stimulate two or more receptor types to different degrees, making the substance taste "sweet and sour," for example. Second, and more important, a substance being tasted usually releases molecules into the air inside the mouth. These odorous molecules diffuse to the olfactory receptors, which contribute an odor component to the basic flavor. (Recall from Chapter 34 that the mouth and nasal passages are connected.)

To prove that what we call taste is really mostly smell, try holding your nose (and closing your eyes) while you eat

(a) The human tongue

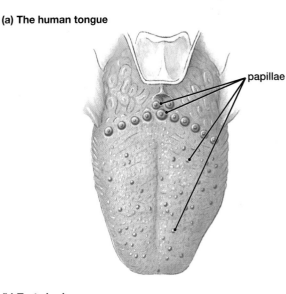

papillae

(b) Taste bud

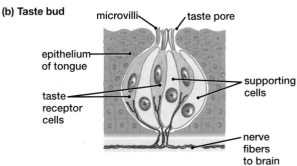

microvilli · taste pore

epithelium
of tongue

supporting
cells

taste
receptor
cells

nerve
fibers
to brain

FIGURE 38-27 Human taste receptors
(a) The human tongue is covered with papillae, bumps in which taste buds are embedded. Small papillae are located on the front two-thirds of the tongue; larger ones with more taste buds are far in the back. **(b)** Each taste bud consists of supporting cells surrounding 60 to 80 taste receptor cells, whose microvilli protrude into the taste pore. The microvilli bear protein receptors that bind tasty molecules that enter through the taste pore, producing a receptor potential.

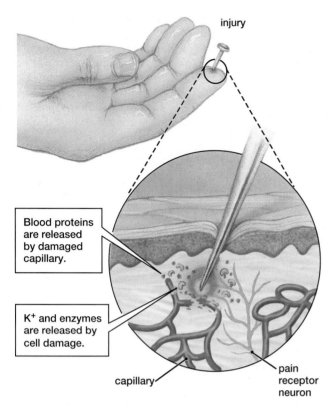

FIGURE 38-28 Pain perception
Pain perception is a specialized chemical sense. An injury damages both cells and blood vessels, releasing K$^+$, which activates pain receptor neurons. Damaged cells also release enzymes that convert certain blood proteins into bradykinin, which also stimulates pain-sensitive neurons.

different flavors of gourmet jelly beans. The flavors—from grape, cherry, and pear to buttered popcorn—will be indistinguishably sweet. Likewise, when you have a bad cold, notice how usually tasty foods seem bland and unappealing, and how your coffee tastes merely bitter.

Pain Is a Specialized Chemical Sense

Whether you burn, cut, or crush a fingertip, you will feel the same sensation: pain. Most pain is produced by tissue damage. Researchers have found that pain perception is actually a special kind of chemical sense (**FIG. 38-28**).

When cells and capillaries are damaged by a cut or a burn, their contents flow into the extracellular fluid. The cell contents include potassium ions, which stimulate **pain receptors**. Damaged cells also release enzymes that convert certain blood proteins into a chemical called bradykinin, which activates pain receptors. Each part of the body has a separate set of pain receptor neurons that provide input to particular brain cells. Hence, the brain can identify the location of the pain. Drugs that provide pain relief, such as morphine or Demerol®, block synapses in the pain pathways of the brain or spinal cord. As we mention in "Health Watch: Drugs, Diseases, and Neurotransmitters," the brain can modulate its perception of pain through its own narcotic-like endorphins.

EVOLUTIONARY CONNECTIONS

Uncommon Senses

We have reviewed the "common" senses of touch, sound, sight, smell, taste, and pain. But if this text focused on bats, it undoubtedly would include a large section on echolocation and almost no coverage of vision! Here, we'll discuss a few of the "uncommon" senses that have evolved in response to different environments.

Echolocation

Some animals that hunt in darkness or murky water have evolved a type of sonar called **echolocation**. Using echolocation, bats can navigate and hunt insect prey in total darkness. An echolocating bat emits pulses of noise at ultrasonic frequencies (higher than can be detected by the human ear) that bounce off nearby objects. The patterns of returning sound convey accurate information about the size, shape, surface texture, and location of objects in the environment. Little brown bats can detect wires only 1 millimeter thick from a distance of 2 meters (more than 6 feet). Several adaptations contribute to this remarkable sensitivity. The bat's enormous, elaborately folded outer ears collect the returning echoes and help the bat locate their source (**FIG. 38-29a**). As the bat emits its cry, muscles attached to the bones of the middle ear contract briefly, reducing the bones' vibrations and preventing the bat from being deafened by its own calls. The tympanic membrane and bones of the middle ear are exceptionally light and easily vibrated by the faint returning echoes.

Porpoises and dolphins produce ultrasonic clicks within their nasal passages and emit them through the front of the head (**FIG. 38-29b**). There, a large, oil-filled sac called the "melon" directs the sound forward in a broad beam (for navigation) or a narrow beam (for prey location). An echolocating porpoise can locate a pea-sized object on the floor of its tank and can distinguish among species of fish. Porpoises may also use the narrowly focused beam to stun fish with a blast of sound, making them easier to capture.

Detecting Electric Fields

Some fish, called weak electric fish, use electrical fields for **electrolocation** in much the same way that bats and porpoises use sound waves for echolocation. These fish produce high-frequency electrical signals from an electric organ just in front of their tails. They detect these signals via electroreceptor cells located along both sides of their bodies (**FIG. 38-30**). Objects near the fish distort the electric field that surrounds the fish. This distortion is detected by the electroreceptors, which send an altered pattern of action potentials to the brain. The fish uses this information to detect and localize nearby objects.

Detecting Magnetic Fields

Homing pigeons are famous for their ability to fly home after being released some distance away. Apparently, pigeons (and other birds that migrate long distances) can navigate either by the sun or, if the sun is hidden, by Earth's

(a)

(b)

FIGURE 38-29 Echolocation
(a) The enormous size and elaborate folds of a long-eared bat's external ears help it localize returning echoes. **(b)** The bottlenose porpoise focuses ultrasonic clicks by using the oil-filled sac in its forehead. QUESTION Why do echolocating porpoises lack the large external ears that are so helpful to echolocating bats?

magnetic field. They can accurately locate their home roost even under cloudy skies in terrain with few landmarks. However, when researchers strapped a small magnet to a pigeon's back, the bird could find its way home in sunny weather but lost its way under overcast skies. In cloudy weather, the experimental magnets confused the pigeon's internal magnetic compass. How do pigeons detect magnetic fields? Pigeons have deposits of magnetite (a magnetic iron compound) just beneath their skulls. These deposits may act as a built-in magnet that is used to tell direction.

Eels of eastern North America and western Europe swim out of streams and rivers into the Atlantic Ocean and migrate to the Sargasso Sea (near the West Indies) to spawn. Eels probably also use magnetic and electric fields for navigation. The currents of the Gulf Stream, flowing from the Gulf of Mexico along the East Coast of the United States and moving through Earth's magnetic field, generate an electric field. The electric field is roughly equivalent to that produced by a 1-volt battery with its poles more than 12 miles (7 km) apart. Researchers have discovered that eels can detect electric fields as weak as that of a 1-volt battery with poles *more than 3000 miles* (5000 kilometers) apart. For an eel, finding the Gulf Stream must be a piece of cake!

(a)

(b)

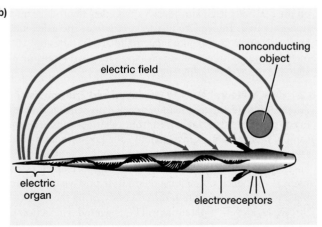

FIGURE 38-30 Electrolocation
(a) Weakly electric fish, such as the elephant nose fish, locate nearby objects by sensing distortions the objects produce in the fish's own electric field. **(b)** The electric field, which surrounds both sides of the body, is generated by electric organs near the tail and detected by electroreceptors along the sides of the body.

CASE STUDY REVISITED   HOW DO I LOVE THEE?

"... heaven is here,
Where Juliet lives."

—*Romeo and Juliet*, Act III, scene III

What happens in the human brain when we fall in love? Although people aren't just big prairie voles, you may be surprised—and perhaps disappointed—to find out that there are some striking similarities in both brain function and hormones during emotional encounters. For example, areas of the human brain that contain oxytocin and dopamine respond to pictures of faces of lovers and one's own children, but not to the faces of equally attractive and familiar people to whom the observer feels no emotional bond. Some of these areas are the same ones that are activated in prairie voles and that appear to be important in motivation and reward. In humans as in voles, oxytocin probably plays an important role in attraction and commitment. Oxytocin reduces stress and inhibits the amygdala, part of the brain involved in feeling fear. In a recent study, oxytocin was found to promote trust, even between complete strangers. Finally, oxytocin levels increase in women and men during sexual encounters.

What about the different kinds of love? Brain scans reveal similarities as well as differences between romantic and parental love. Some of the same brain areas are activated by photos of lovers and children; other areas are activated by one or the other, but not both. Still other brain areas, particularly those involved in critical decision making and social judgments, are turned off, at least while seeing the lover or child. The result is that lovers and children almost always seem to be better than they really are. This is probably especially important for parental love of newborns. Although often very cute, newborn infants can hardly reciprocate love the way adults can.

Will neurobiological explanations ruin the magic of love? Anthropologist Helen Fisher, who has probably studied the neurobiology of love more than anyone else, doesn't think so. "You can know every ingredient in a piece of chocolate cake, and ... it's [still] wonderful. In the same way, you can know all the ingredients of romantic love and still feel that passion."

Consider This In some ways, love strongly resembles addiction. For example, both activate "reward circuits" in the brain that release dopamine. Dr. Fisher notes that romantic love shows tolerance, withdrawal, and relapse. With both drugs and newfound love, small doses may spark interest, but soon small doses aren't nearly enough—you need more and more to satisfy your craving. If you don't get your love or drug "fix," you go into withdrawal—crying, depression, and desperately seeking a new "dose." Finally, there's relapse. Just as reformed drug addicts may develop sudden, urgent cravings if they encounter the drug again, people who have fallen "out of love" may still feel a surge of emotion if they see their former lovers again. But does love really resemble addiction to drugs? Or is it the other way around? It is extremely unlikely that humans have evolved reward centers and dopamine neurons so that we can become cocaine addicts. Rather, these circuits probably evolved to promote sex, pair bonding and infant care, and drugs like cocaine are addictive because they chemically activate these same circuits.

CHAPTER REVIEW

SUMMARY OF KEY CONCEPTS

38.1   What Are the Structures and Functions of Neurons?

Nervous systems are composed of individual cells called neurons. A neuron has four major specialized functions, which are reflected in its structure: Dendrites receive information from the environment or from other neurons. The cell body adds together electrical signals from the dendrites and from synapses on the cell body itself and "decides" whether to produce an action potential. The cell body also coordinates the cell's metabolic activities. The axon conducts the action potential to its output terminal: the synapse. Synaptic terminals transmit the signal to other nerve cells, glands, or muscles.

38.2   How Is Neural Activity Produced and Transmitted?

An unstimulated neuron maintains a negative resting potential inside the cell. Signals received from other neurons are small, rapidly fading changes in potential called postsynaptic potentials. Inhibitory and excitatory postsynaptic potentials (IPSPs and EPSPs) make the neuron less likely or more likely, respectively, to produce an action potential. If postsynaptic potentials, added together within the cell body, bring the neuron to threshold, an action potential will be triggered. The action potential is a wave of positive charge that travels, undiminished in magnitude, along the axon to the synaptic terminals.

A synapse, where two neurons communicate, consists of the synaptic terminal of the presynaptic neuron, a specialized region of the postsynaptic neuron, and the tiny gap between

them. Neurotransmitters from the presynaptic neuron, released in response to an action potential, bind to receptors in the postsynaptic cell's plasma membrane and produce either an EPSP or an IPSP.

There are many neurotransmitters. They are being intensively studied to explore their diverse roles in neurological disease, drug addiction, and all aspects of normal nervous system function.

Web Tutorial 38.1 Electrical Signals in Neurons

Web Tutorial 38.2 The Synapse

38.3   How Are Nervous Systems Organized?

Information processing in the nervous system requires four operations. The nervous system must (1) determine the type of stimulus, (2) signal the intensity of the stimulus, (3) integrate information from many sources, and (4) initiate and direct an appropriate response. The nervous system collects and processes sensory information from many sources.

Neural pathways normally have four elements: (1) sensory neurons, (2) interneurons, (3) motor neurons, and (4) effectors. Overall, nervous systems consist of numerous interconnected neural pathways, which may be either diffuse or centralized.

Web Tutorial 38.3 Reflex Arcs

38.4 What Is the Structure of the Human Nervous System?

The nervous system of humans and other vertebrates consists of the central nervous system and the peripheral nervous system. The peripheral nervous system is further subdivided into sensory and motor portions. The motor portion consists of the somatic nervous system (which controls voluntary movement) and the autonomic nervous system (which directs involuntary responses).

Within the central nervous system, the spinal cord contains neurons controlling voluntary muscles and the autonomic nervous system and neurons communicating with the brain and other parts of the spinal cord; axons leading to and from the brain; and neural pathways for reflexes and certain simple behaviors. The brain consists of three parts: the hindbrain, midbrain, and forebrain, each further subdivided into distinct regions.

The hindbrain in humans consists of the medulla and pons, which control involuntary functions (such as breathing), and the cerebellum, which coordinates complex motor activities (such as typing). In humans, the small midbrain contains much of the reticular formation, a filter and relay for sensory stimuli. The forebrain includes the thalamus, a sensory relay station that shuttles information to and from conscious centers in the forebrain; the limbic system, a diverse set of structures involved in emotion, learning, and the control of instinctive behaviors such as sex, feeding, and aggression; and the cerebral cortex, the center for information processing, memory, and initiation of voluntary actions. The cerebral cortex includes primary sensory and motor areas and association areas that analyze sensory information and plan movements.

38.5 How Does the Brain Produce the Mind?

The cerebral hemispheres are specialized. In general, the left hemisphere dominates speech, reading, writing, language comprehension, mathematical ability, and logical problem solving. The right hemisphere specializes in recognizing faces and spatial relationships, artistic and musical abilities, and recognition and expression of emotions.

Memory takes two forms: short-term memory is electrical or chemical, while long-term memory probably involves structural changes that increase the effectiveness or number of synapses. The hippocampus is an important site for the transfer of information from short-term into long-term memory. The temporal lobes are important for memory recognition of objects and faces, and understanding language.

Web Tutorial 38.4 The Human Brain

38.6 How Do Sensory Receptors Work?

Sensory receptors convert a stimulus from the internal or external environment to an electrical signal called a receptor poten-tial. Either directly or indirectly, receptor potentials ultimately result in action potentials in specific axons that connect to appropriate brain regions. Receptor cells are named after the stimulus to which they respond.

38.7 How Are Mechanical Stimuli Detected?

A variety of mechanoreceptors detect stimuli such as touch, vibration, pressure, stretch, or sound. In most cases, a mechanoreceptor produces a receptor potential in response to deformation or stretching of its plasma membrane.

38.8 How Is Sound Sensed?

In the vertebrate ear, air vibrates the tympanic membrane, which transmits vibrations to the bones of the middle ear and then to the oval window of the fluid-filled cochlea. Within the cochlea, vibrations bend the hairs of hair cells, which are receptors located between the basilar and tectorial membranes. This bending produces receptor potentials in the hair cells that cause action potentials in the axons of the auditory nerve, which leads to the brain.

Web Tutorial 38.5 The Human Ear

38.9 How Is Light Sensed?

In the vertebrate eye, light enters the cornea and passes through the pupil to the lens, which focuses an image on the fovea of the retina. Two types of photoreceptors, rods and cones, are located deep in the retina. They produce receptor potentials in response to light. These signals are processed through several layers of neurons in the retina and are translated into action potentials in the optic nerve, which leads to the brain. Rods are more abundant and more light-sensitive than cones, providing vision in dim light. Cones, which are concentrated in the fovea, provide color vision.

Web Tutorial 38.6 The Human Eye

38.10 How Are Chemicals Sensed?

Terrestrial vertebrates detect chemicals in the external environment either by smell (olfaction) or by taste. Each olfactory or taste receptor cell type responds to only one or a few specific types of molecules, allowing discrimination among tastes and odors. Olfactory neurons of vertebrates are located in a tissue that lines the nasal cavity. Taste receptors are located in clusters called taste buds on the tongue. Pain is a special type of chemical sense in which sensory neurons respond to chemicals released by damaged cells.

KEY TERMS

action potential *page 762*
amygdala *page 776*
aqueous humor *page 785*
auditory canal *page 782*
auditory tube *page 782*
autonomic nervous system *page 771*
axon *page 762*
basilar membrane *page 784*
binocular vision *page 788*
blind spot *page 787*

brain *page 770*
cell body *page 762*
central nervous system *page 770*
cerebellum *page 775*
cerebral cortex *page 777*
cerebral hemisphere *page 777*
cerebrum *page 775*
choroid *page 785*
cochlea *page 784*

compound eye *page 785*
cone *page 787*
convolution *page 777*
cornea *page 785*
corpus callosum *page 777*
dendrite *page 762*
dorsal root ganglion *page 774*
echolocation *page 790*
effector *page 770*
electrolocation *page 790*

excitatory postsynaptic potential (EPSP) *page 764*
farsighted *page 787*
forebrain *page 774*
fovea *page 787*
ganglion *page 770*
ganglion cell *page 787*
gray matter *page 773*
hair cell *page 784*
hindbrain *page 774*
hippocampus *page 776*

THINKING THROUGH THE CONCEPTS

1. List four major parts of a neuron, and explain the specialized function of each part.

2. Diagram a synapse. How are signals transmitted from one neuron to another at a synapse?

3. How does the brain perceive the intensity of a stimulus? The type of stimulus?

4. What are the four elements of a simple nervous pathway? Describe how these elements function in the human pain-withdrawal reflex.

5. Draw a cross section of the spinal cord. What types of neurons are located in the spinal cord? Explain why severing the cord paralyzes the body below the level where it is severed.

6. Describe the functions of the following parts of the human brain: medulla, cerebellum, reticular formation, thalamus, limbic system, and cerebrum.

7. What structure connects the two cerebral hemispheres? Describe the evidence that the two hemispheres are specialized for distinct intellectual functions.

8. Distinguish between long-term memory and working memory.

9. What are the names of the specific receptors used for taste, vision, hearing, smell, and touch?

10. Why are we apparently able to distinguish hundreds of different flavors if we have only five types of taste receptors? How are we able to distinguish so many different odors?

11. Describe the structure and function of the various parts of the human ear by tracing a sound wave from the air outside the ear to the receptor cells.

12. How does the structure of the inner ear allow for the perception of pitch? Of sound intensity?

13. Diagram the overall structure of the human eye. Label the cornea, iris, lens, sclera, retina, and choroid. Describe the function of each structure.

14. How does the lens change shape to allow focusing of distant objects? What defects make focusing on distant objects impossible, and what is this condition called? What type of lens can be used to correct it, and how does it do so?

15. List the similarities and differences between rods and cones.

16. Distinguish between taste and smell.

17. Describe how pain is signaled by tissue damage.

APPLYING THE CONCEPTS

1. In Parkinson's disease, which afflicts several million Americans, the cells that produce the neurotransmitter dopamine degenerate in a small part of the brain that is important in controlling movement. Some physicians have reported improvement after injecting cells taken from the same general brain region of an aborted fetus into appropriate parts of the brain of a Parkinson's patient. Discuss this type of surgery from as many viewpoints as possible: ethical, financial, practical, and so on. Based on your responses, is fetal transplant surgery the answer to curing Parkinson's disease?

2. If the axons of human spinal cord neurons were unmyelinated, would the spinal cord be larger or smaller? Would you move faster or slower? Explain your answer.

3. What is the adaptive value of reflexes? Why couldn't all behaviors be controlled by reflexes?

4. Explain the statement, "Your sensory perceptions are purely a creation of your brain." Discuss the implications for communicating with other humans, with other animals, and with intelligent life elsewhere in the universe.

FOR MORE INFORMATION

Anderson, A., and Middleton, L. "What Is This Thing Called Love?" *New Scientist*, April 26, 2006. A stimulating discussion of the natural history, biochemistry, and brain mechanisms of love.

Axel, R. "The Molecular Logic of Smell." *Scientific American*, October 1995. Describes research that uncovers some of the mechanisms by which the nose and brain decipher scents.

Beardsley, T. "The Machinery of Thought." *Scientific American*, August 1997. Using PET and functional MRI on monkeys and humans, researchers are learning more about where working memory resides.

Bower, B. "Creatures in the Brain." *Science News*, April 13, 1996. Using imaging techniques, scientists have discovered clues about how different regions of the brain are specialized for different concepts.

Damasio, A. R. "How the Brain Creates the Mind." *Scientific American*, December 1999. The author presents an intriguing hypothesis for how the sense of self emerges from the machinery of the brain.

Gazzaniga, M. S. "The Split Brain Revisited." *Scientific American*, July 1998. More insights into hemispheric specializations by one of the pioneers of human split-brain studies.

Goldsmith, T. H. "What Birds See." *Scientific American*, July 2006. A marvelously clear description of color vision in both birds and mammals.

Philips, H. "Just Can't Get Enough." *New Scientist*, August 26, 2006. Excessive cravings for sex, gambling, computer games, and drugs may all be based on activation of the dopamine reward system in the brain.

Raichle, M. E. "Visualizing the Mind." *Scientific American*, April 1994. Brain imaging techniques partially open the "black box" of the mind.

Smith, D. V., and Margolskee, R. F. "Making Sense of Taste." *Scientific American*, March 2001. Scientists are beginning to unravel the mechanisms by which taste receptors respond to various tastes.

Wuethrich, B. "Getting Stupid." *Discover*, March 2001. Heavy drinking may damage the brain, particularly in young people.

39

Action and Support: The Muscles and Skeleton

Astronaut Millie Hughes-Fulford floats in the weightless environment of the life sciences module in the space shuttle *Columbia* during a 1991 mission. *Columbia* was lost in a tragic explosion on February 1, 2003.

AT A GLANCE

 CASE STUDY HIDDEN HAZARDS OF SPACE TRAVEL

ASTRONAUTS CIRCLE EARTH while living and working together in one of humankind's most remarkable cooperative ventures—the International Space Station. Unfortunately, this once-in-a-lifetime experience does not come without risk. In addition to the obvious dangers of space flight, these astronauts face a more insidious threat: the loss of both muscle and bone due to weightlessness. Although being weightless seems like fun, our bodies are not adapted for it. Humans (and other terrestrial species) evolved under the relentless pull of gravity, which strengthens our bones and muscles on a daily basis. Astronauts routinely exercise to prevent muscle atrophy, but bone loss poses a more difficult problem. Research has shown that space travelers lose 0.5% to 2% of their total bone mass for every month they spend under weightless conditions. The loss is most pronounced in weight-bearing bones such as the lower portion of the vertebral column and the legs. Contrary to popular perception, bones are not merely dry scaffolding for the body; instead, they continuously change in response to the demands we place on them. In a process called "remodeling," bone thickens under stress and thins when stress is removed. How does remodeling happen? Does it change as we get older? What can be done to counteract bone loss in space?

39.1 AN INTRODUCTION TO THE MUSCULAR AND SKELETAL SYSTEMS

The system of muscles and skeleton that moves and supports the animal body is an engineering marvel. The flight of a bat, the pounce of a cat, the fluid grace of a ballet dancer, and the simple movements involved in walking to your classroom all depend on the same humble yet elegant mechanism. Muscle cells perform only one activity: they exert a force by contracting. Under the influence of natural selection, however, this simple unidirectional force has been applied to complex bony scaffolding that has been molded into structural elements such as wings, hands, and fins, with its action coordinated by the nervous system. The resulting capacity for movement gives animals the ability to search for food, seek out new habitats, flee from danger, and, on occasion, move in ways that we find awe-inspiring.

Muscles and skeletons also perform more mundane, yet crucial, functions. Pumping blood through the circulatory system, moving food through the digestive system, and breathing are some of the essential processes that depend on muscle contraction. The skeleton of terrestrial animals provides a framework against which muscles exert force to move the body (**FIG. 39-1**). Almost all animals depend on the support of a skeleton, either outside or inside the body, to maintain their shapes and protect internal organs. Without your skeleton, you'd be a formless, quivering mound of tissue.

39.2 HOW DO MUSCLES WORK?

All muscular work requires muscles to alternately contract and lengthen, but muscles are active only during contraction. The lengthening that follows contraction is passive, occurring when muscles relax and are stretched out by other forces. A relaxed muscle may be lengthened by contractions of opposing muscles, the weight of a limb, or forces such as pressure from food distending the muscular walls of the stomach.

Animals display a dazzling diversity of muscle function, adapted from a remarkable uniformity of muscle structure. Vertebrates have evolved three types of muscle: skeletal, cardiac, and smooth. All work on the same basic principles but differ in function, appearance, and

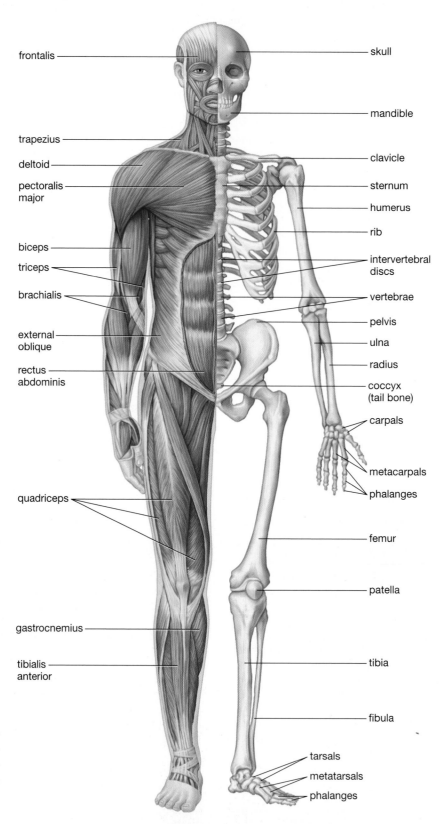

FIGURE 39-1 Muscular and skeletal systems These two systems work in harmony to allow coordinated movement. Here, some of the major muscles and bones are illustrated. The *axial skeleton*, forming the axis of the body, is in blue. The *appendicular skeleton*, which forms the appendages, is in bone color.

control (Table 39-1). Invertebrate muscles closely resemble vertebrate muscles, but show a tremendous range of adaptations to their diverse lifestyles. For example, bivalve mollusks (those with two shells, such as scallops and clams) possess a special type of smooth muscle that holds their shells tightly closed for hours using very little energy. In contrast to these sustained contractions, some flies have flight muscles that can contract at rate of 1000 times per second. In the following sections we describe vertebrate muscles, using the human muscular system as an example.

Skeletal muscle, so named because it moves the skeleton, appears striped when viewed through a microscope and is often referred to as *striated* (meaning "striped")

muscle. Most skeletal muscle is under voluntary, or conscious, control. It can produce contractions ranging from quick twitches (as in blinking) to powerful, sustained tension (as in carrying an armload of textbooks). **Cardiac muscle** is located only in the heart. It is *spontaneously active* and *involuntary* (that is, it initiates its own contractions and is not under conscious control), but it is also influenced by nerves and hormones. Like skeletal muscle, cardiac muscle is striated. **Smooth muscle**, as its name suggests, lacks the striations of skeletal and cardiac muscle. Smooth muscle lines the large blood vessels and most hollow organs, producing slow, sustained, and involuntary contractions. Muscle types are illustrated in Table 39-1.

Table 39-1 Location, Characteristics, and Functions of the Three Muscle Types

Property	Type of Muscle		
	Smooth	Cardiac	Skeletal
Muscle appearance	Nonstriated	Striated (striped)	Striated (striped)
Cell shape	Spindle	Branched	Spindle
Number of nuclei	One per cell	One per cell	Many per cell
Speed of contraction	Slow	Intermediate	Slow to rapid
Contraction stimuli	Spontaneous, stretch, nervous system, hormones	Spontaneous	Nervous system
Function	Controls movement of substances through hollow organs and tubes	Pumps blood	Moves the skeleton
Under voluntary control?	No	No	Yes

Note: Heart rate can be voluntarily changed after biofeedback training.

Cardiac muscle
- muscle fiber
- intercalated disks with gap junctions link adjacent cells
- nuclei

Skeletal muscle
- muscle fiber
- nuclei

Smooth muscle
- muscle fiber
- nucleus

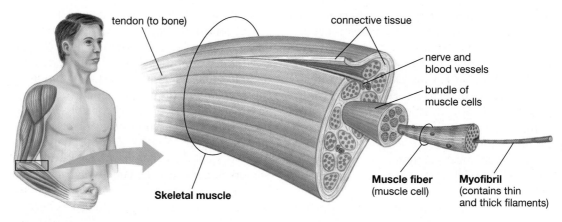

tendon (to bone)

connective tissue

nerve and
blood vessels

bundle of
muscle cells

Skeletal muscle

Muscle fiber
(muscle cell)

Myofibril
(contains thin
and thick filaments)

FIGURE 39-2 Skeletal muscle structure
A muscle is surrounded by connective tissue and attached to bones by tendons. It contains from a few to 1000 muscle cells called muscle fibers, often packaged into bundles within the muscle. Each fiber is packed with cylindrical subunits called myofibrils, which are composed of thick and thin filaments of protein.

The human body has about 650 skeletal muscles, which make up roughly 40% of the weight of an average person; some of these muscles are illustrated in Figure 39-1. The discussion that follows emphasizes skeletal muscle.

Skeletal Muscle Cell Structure and Function Are Closely Linked

In eukaryotic cells, movement of organelles within the cell, changes in cell shape, and cellular locomotion depend on interactions between microfilaments of the protein **actin** and strands of **myosin** protein. To produce this motion, actin and myosin slide past one another, changing the shape of the cell. This evolutionarily ancient mechanism also enables animal muscle cells to contract.

Skeletal muscles are attached to the skeleton by tough, fibrous cords of connective tissue called **tendons**. Tendons are composed of fibers of collagen protein, bundled together much like small wires within a larger cable. Each muscle is encased in connective tissue and includes bundles of muscle cells nourished by blood vessels and stimulated by nerves (**FIG. 39-2**).

Individual muscle cells, or **muscle fibers**, are among the largest cells in the human body. Ranging from 10 to 100 micrometers in diameter (a bit smaller than the period at the end of this sentence), each muscle fiber runs the entire length of the muscle, which can reach a foot in length (about 30 cm) in a human thigh. Skeletal muscle fibers are unusual in that they contain many nuclei located just beneath the cell's outer membrane; the largest fibers have several thousand nuclei.

Each muscle fiber consists of many **myofibrils**, contractile cylinders composed primarily of actin and myosin, extending from one end of the fiber to the other (Fig. 39-2). Each myofibril is surrounded by **sarcoplasmic reticulum**. Like the endoplasmic reticulum from which it is derived, sarcoplasmic reticulum consists of flattened, membrane-enclosed compartments (**FIG. 39-3a**). The fluid within the sarcoplasmic reticulum stores high concentrations of calcium ions, which play a key role in muscle contraction. Surrounding each muscle fiber is a plasma membrane, which tunnels deeply into the muscle fiber at intervals, producing membrane-lined channels called **T tubules** that are filled with extracellular fluid. T tubules form close connections with the sarcoplasmic reticulum, relaying signals that cause calcium release, which in turn allows muscle contraction, as described later.

Myofibrils are composed of subunits called **sarcomeres**, which are aligned end to end along the length of the myofibril, connected to one another by fibrous protein bands called **Z lines** (**FIG. 39-3b**). Each sarcomere contains a beautifully precise arrangement of actin and myosin filaments. Actin molecules (in association with two smaller accessory proteins, *troponin* and *tropomyosin*) form the **thin filaments**, each of which is anchored to a Z line at one end. Suspended between the thin filaments are **thick filaments**, composed of myosin proteins. Thick filaments can bind temporarily to thin filaments using a series of small projections called **myosin heads** (**FIG. 39-3c**). The regular arrangement of thick and thin filaments within each myofibril gives the muscle fiber its striated appearance.

Muscle Contraction Results from Thick and Thin Filaments Sliding Past One Another

The molecular structure and arrangement of thick and thin filaments allows them to both grip and slide past one another, shortening the sarcomeres and producing muscle contraction. The actin protein that makes up most of the thin filament is formed from a double chain of subunits, resembling a twisted double strand of pearls. Each subunit has a binding site for a myosin head. In a relaxed muscle cell, however, these binding sites on actin are covered by its accessory proteins, preventing the myosin heads from attaching (see Fig. 39-3c).

(a) Cross section of fiber

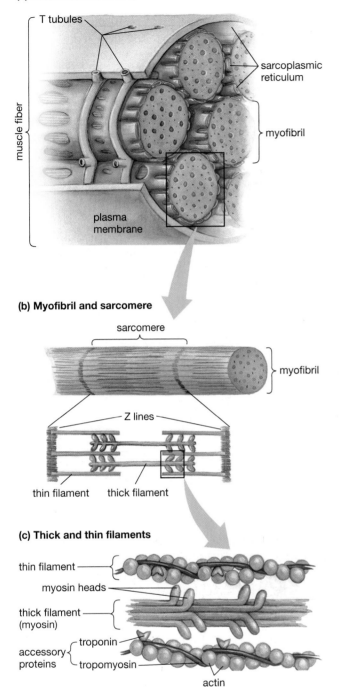

(b) Myofibril and sarcomere

(c) Thick and thin filaments

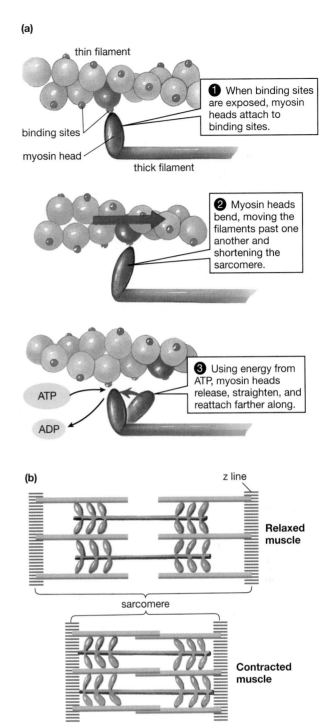

FIGURE 39-3 A skeletal muscle fiber
(a) Each muscle fiber is surrounded by plasma membrane that tunnels inside, forming T tubules. Sarcoplasmic reticulum surrounds each myofibril within the muscle cell. **(b)** Each myofibril consists of a series of subunits called sarcomeres, attached end to end by protein bands called Z lines. **(c)** Within each sarcomere are alternating thick and thin filaments, which are temporarily linked and pulled together by the myosin heads as the muscle contracts.

FIGURE 39-4 Muscle contraction
(a) Repeated cycles of myosin head attachment, bending, release, and reattachment result in muscle contraction. **(b)** Muscle contraction uses ATP and causes the thick and thin filaments to slide past one another toward the center of each sarcomere, shortening the muscle cell. QUESTION As the sarcomere shortens during muscle contraction, do the thick filaments shorten? Do the thin filaments?

When a muscle contracts, the accessory proteins of the thin filament move aside, exposing binding sites on the actin. As soon as the sites are exposed, myosin heads bind the actin, temporarily linking the thick and thin filaments. Using energy from the splitting of adenosine triphosphate (ATP), the myosin heads repeatedly bend, release, and reattach farther along the thin filament, much like a sailor pulling in an anchor line hand over hand (**FIG. 39-4a**). The activity of the myosin heads pulls the thin filaments past the thick filaments, shortening each sarcomere and contracting the muscle (**FIG. 39-4b**). Because the thick and thin filaments slide past one another during contraction, muscle contraction is described as using a *sliding filament mechanism*.

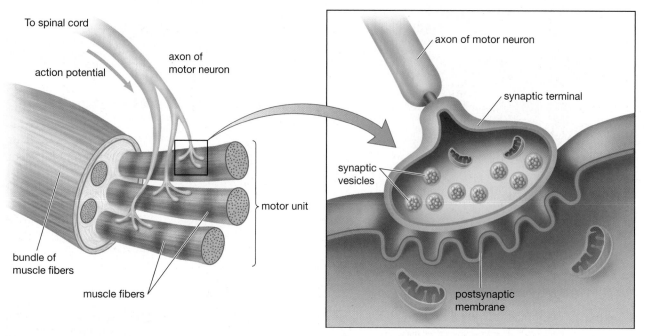

FIGURE 39-5 A motor unit and a neuromuscular junction
(a) A motor unit consists of a motor neuron and all the muscles fibers on which it synapses. These fibers contract together in response to an action potential in the motor neuron. **(b)** A neuromuscular junction in cross section. Action potentials in the motor neuron stimulate the muscle fiber membrane, which lies in folds beneath the terminal.

Skeletal Muscle Contraction Is Controlled by the Nervous System

Neurons called motor neurons activate skeletal muscles at specialized synapses called **neuromuscular junctions** (**FIG. 39-5b**), all of which use the neurotransmitter *acetylcholine* (described in Chapter 38). The nervous system can only excite (not inhibit) skeletal muscle, and every action potential in a motor neuron produces an action potential in a muscle fiber, causing all its sarcomeres to shorten and the fiber to contract. The nervous system controls the strength and degree of muscle contraction by controlling the number of muscle fibers stimulated and the frequency of action potentials in each fiber. Sustained rapid firing in the motor neurons that synapse on all the fibers of a particular muscle causes a sustained maximum contraction of that muscle, such as occurs when you carry an armful of books like this one.

Most motor neurons have many synaptic terminals, each synapsing on a different muscle fiber; therefore, a single action potential will cause the simultaneous contraction of a cluster of muscle cells. The group of fibers on which a single motor neuron forms synapses is called a **motor unit** (**FIG. 39-5a**). The number of muscle fibers in a motor unit varies from muscle to muscle. Muscles used for large-scale movements, such as the thigh muscles you use when climbing stairs or the muscles in your back that help maintain your posture, can each contain from hundreds up to 1000 muscle fibers. In muscles used for fine control of small body parts, such as those of the lips, eyes, and tongue, only a few muscle cells may be stimulated by each motor neuron. Therefore, the action potential fired by a single motor neuron can cause the contraction of a few muscle cells or many, depending on the size of the motor unit.

Muscle Contraction Depends on the Availability of Calcium Ions and ATP

An action potential in the muscle cell travels into the T tubules (see Fig. 39-3a) and opens calcium channels in the membrane of the sarcoplasmic reticulum, allowing calcium ions to flow out of the sarcoplasmic reticulum where they are stored and into the cytosol surrounding the thick and thin filaments. Once in the cytosol, calcium ions bind to the smaller accessory proteins (*troponin*) of the thin filament, causing them to change shape and pull the larger accessory proteins (*tropomyosin*) off the binding sites for myosin. As long as these binding sites are exposed and ATP is available, the myosin heads will repeatedly bind, bend, release, and reattach, contracting the muscle fiber. ATP powers the movement of the head and is necessary for the myosin to release from the actin.

As soon as the action potential is over, active transport proteins in the sarcoplasmic reticulum membrane pump the calcium ions back inside the sarcoplasmic reticulum. As the calcium ions leave the troponin, the accessory proteins return to a configuration that blocks the myosin binding sites, and the muscle fiber relaxes and can be stretched out.

You've probably heard of *rigor mortis*, in which muscles become rigid (without contracting) hours after death. This occurs because the dead muscle cells run out of ATP and can no longer pump calcium back into the sarcoplasmic reticulum. In the presence of calcium ions, the myosin heads bind the actin. Although no contraction occurs, the actin and myosin strands remain bound rigidly together because ATP is necessary for the strands to detach from one another. Rigor mortis passes many hours later as the muscle cells begin to decompose.

Muscles Require a Steady Supply of Energy to Perform Work

Contracting muscles require a continuous supply of ATP, but a skeletal muscle's ATP stores are used up after only a few seconds of high-intensity exercise. Skeletal muscles also stock a supply of creatine phosphate, an energy-storage molecule that quickly resynthesizes ATP from ADP, but this is also depleted rapidly. For prolonged or low-intensity exercise, muscle cells "burn" glucose and fatty acids using cellular respiration (see Chapter 8). The muscle obtains glucose as well as fatty acids from the blood, and gets additional glucose by breaking down glycogen (long chains of glucose molecules) stored in muscle. Cellular respiration requires a continuous supply of oxygen; in an exercising animal, this is limited by the cardiovascular system's ability to deliver oxygen to the muscles. Thus, a world-class marathoner may average only about 10 mph because the athlete's muscles are mostly powered by cellular respiration during this two-hour event. In contrast, a champion sprinter may sustain a speed of almost 22 mph per hour during a 200-meter dash that lasts about 20 seconds. How can sprinters run so fast? During short-term, high-intensity exertion, muscle cells generate ATP using glycolysis, which does not require oxygen. Glycolysis produces ATP from glucose very rapidly but very inefficiently and generates lactic acid as a by-product. The buildup of lactic acid in muscles contributes to the "burning" sensation when they are used at their maximum capacity. As the body rests after exercise and heavy breathing restores oxygen levels, most of the lactic acid is transported in the bloodstream to the liver. Here, lactic acid is resynthesized into glucose, some of which is returned to the muscles through the bloodstream and used to replenish glycogen stores.

Skeletal Muscle Plays an Important Role in Athletic Ability

Do bodybuilders have more muscle cells? Surprisingly, no. The total number of skeletal muscle fibers in an individual's body is established early in life, and although muscle fibers can be lost due to injury or normal aging processes, they are never replaced. They can, however, grow larger by adding more myofibrils. Repeated mechanical stresses on the muscle, such as occur during weightlifting, trigger the activation of genes that code for production of the actin and myosin proteins that make up myofibrils. Intensive weight training can triple the size of a muscle.

The old adage "use it or lose it" applies to both the muscular and skeletal systems, in space flight as in ordinary life. If you've ever broken a bone and had an arm or leg immobilized in a cast, you have probably seen firsthand that unused muscles lose mass over time. The weightlessness of space flight, or inactivity—such as might occur during prolonged bed rest—reduces the body's need to fight the forces of gravity and takes a toll on muscles. People who have spinal cord injuries that prevent signals from reaching their muscles can help prevent atrophy and eventual death of muscle fibers by using electrodes to stimulate motor neurons or the muscles themselves, causing them to contract without orders from the brain. Such devices, delivering appropriately timed stimulation, have allowed some patients whose legs are paralyzed to stand and even walk for short distances using a wheeled walker.

Genes Influence Muscle Composition

Next time you watch a track and field event, compare the legs of long-distance runners to those of sprinters. Although sprinters have larger calf muscles, marathoners have far greater endurance. Why? Skeletal muscle fibers come in two basic types, *fast-twitch* and *slow-twitch*. Slow-twitch fibers contract more slowly. They have abundant mitochondria (in which cellular respiration occurs) and a plentiful blood supply that provides oxygen. They also contain myoglobin, an oxygen-storing compound similar to the hemoglobin that carries oxygen in blood. Fast-twitch fibers, on the other hand, contract more quickly and powerfully. Because they have a smaller blood supply and fewer mitochondria, they are better adapted to use glycolysis, which does not require oxygen and supplies ATP much faster than does cellular respiration. While a typical adult has roughly even numbers of these two fiber types, champion sprinters have about 80% fast-twitch fibers, which allow bursts of amazing speed for a very short time. In contrast, world-class marathoners have about 80% slow-twitch fibers, which can contract for hours before exhausting the energy supplied by cellular respiration.

Are marathoners or sprinters born or made through practice? The proportions of fast-twitch and slow-twitch fibers vary greatly within the human population. Athletes with a high proportion of slow-twitch fibers are likely to excel in—and therefore gravitate to—endurance sports, while those with more fast-twitch fibers will find their niches in sports requiring bursts of speed. Researchers have found that intensive training can convert some slow-twitch fibers into fast-twitch fibers, but not the opposite. Weight training, which selectively bulks up existing fast-twitch fibers, changes the relative masses (but not numbers) of the different fiber types and can help train for sports that require bursts of activity.

Gene Therapy May Help Diseased Muscles but Create Athletic Dilemmas

Researchers have found many different genes coding for chemical messengers that influence muscles' ability to grow and repair themselves. When people and other animals experience prolonged inactivity or the weightlessness of outer space, muscles undergo *disuse atrophy* in which normal growth and repair mechanisms seem to be switched off by an unknown mechanism. During natural aging, muscle repair also becomes less efficient. To counteract this and help people with muscle-destroying diseases such as *muscular dystrophy*, scientists have turned to genetic engineering. They have introduced into mice certain synthetic genes that produce the muscle growth factor *IGF-1* (*insulin-like growth factor I*). The engineered mouse muscles were considerably larger and better able to repair damage and resist the weakening effects of aging than were those of normal mice. Researchers are now extending their studies to dogs with muscular dystrophy, hoping that a cure might be applicable to people.

Meanwhile, unscrupulous athletes and their trainers eagerly await this technology as a shortcut to medals, fame, and fortune, while the World Anti-Doping Agency (an international group that polices performance-enhancing cheating among athletes) has already asked scientists to come up with tests for illegal "gene doping."

Cardiac Muscle Powers the Heart

Cardiac muscle, like skeletal muscle, is striated due to its regular arrangement of sarcomeres with their alternating thick and thin filaments (see Table 39-1). The fibers of cardiac muscle are smaller than most skeletal muscle cells, and possess only a single nucleus. Unlike skeletal muscle, cardiac muscle fibers can initiate their own contractions. This ability is particularly well developed in the specialized cardiac muscle fibers of the heart's pacemaker (the *SA node*; see Chapter 32). Action potentials from the pacemaker spread rapidly through the gap junctions that interconnect cardiac muscle fibers. Gap junctions are concentrated in regions called *intercalated disks* (see Table 39-1) that also serve to attach adjacent cardiac muscle fibers to one another, preventing the forces of contraction from pulling them apart.

Smooth Muscle Produces Slow, Involuntary Contractions

Smooth muscle surrounds blood vessels and most hollow organs, including the uterus, bladder, and digestive tract. As its name suggests, smooth muscle is not striated, because it lacks the regular arrangement of sarcomeres that characterizes skeletal and cardiac muscles (see Table 39-1). Like cardiac muscle fibers, smooth muscle fibers are directly connected to one another by gap junctions, so their cells contract in synchrony, and each contains a single nucleus. Smooth muscle contraction is either slow and sustained (such as the constriction of arteries that elevates blood pressure during times of stress) or slow and wavelike (such as the peristaltic waves that move food through the digestive tract or expel a baby from the uterus during childbirth). Smooth muscle stretches easily and extensively, as can be observed in the bladder, the stomach, and the uterus. Smooth muscle contraction is involuntary and can be initiated by stretching, by hormones, by signals from the autonomic nervous system (see Chapter 38), or by some combination of these stimuli.

39.3   WHAT DOES THE SKELETON DO?

For most people, the word *skeleton* conjures up the image of human bones. But **skeletons**, broadly defined as supporting frameworks for the animal body, are diverse and need not be made of bone.

Three Types of Skeletons Are Found in Animals

Skeletons come in three radically different forms: *hydrostatic skeletons* (made of fluid), *exoskeletons* (on the outside of the animal), and *endoskeletons* (internal).

The **hydrostatic skeletons** of worms, mollusks (snails and their relatives), and cnidarians (sea jellies and their relatives; **FIG. 39-6**) are the simplest, consisting of a fluid-filled sac. Fluid, which cannot be compressed, provides excellent support. The anemone can expel most of its fluid skeleton, withdraw its tentacles, and contract down into a small lump. Because fluid is formless, to control the shape of their bodies, animals with hydrostatic skeletons rely on two layers of muscle—one circular, the other longitudinal—in the body wall. The wavelike movements of a burrowing earthworm, alternately extending to string-like thinness (as muscles encircling its body contract) and then fattening (as muscles along the length of its body contract), provide an excellent illustration of the flexibility of hydrostatic skeletons (look ahead to Fig. 39-8a).

Exoskeletons (literally, "outside skeletons") encase the bodies of arthropods (such as spiders, crustaceans, and insects). Exoskeletons vary tremendously in thickness and rigidity, from the thin, flexible covering of many insects and spiders to the armorlike covering of many crustaceans (**FIG. 39-7**). Exoskeletons are thin and flexible at the *joints* (movable junctures of adjacent body parts), allowing com-

FIGURE 39-6 The anemone supports its soft body with a hydrostatic skeleton

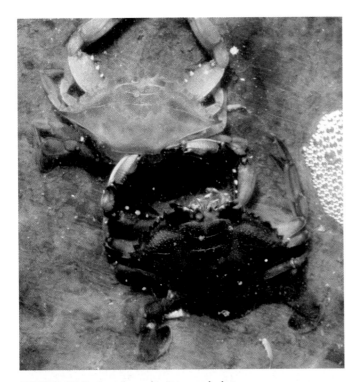

FIGURE 39-7 A crab molts its exoskeleton
Arthropods, such as this blue crab, must molt their exoskeletons in order to grow. They may molt 20 times during a six-year life span. Here a dark, newly molted crab has just crawled out of its old exoskeleton. QUESTION Why are thick, armor-like exoskeletons found mostly in water-dwelling animals, whereas land-dwelling insects and spiders tend to have thinner exoskeletons?

plex and skillful movements such as those of a web-spinning spider. Exoskeletons must be molted periodically to allow the animal to grow.

Endoskeletons, the internal skeletons of humans and other vertebrates, are found only in echinoderms (sea stars and their relatives) and chordates (animals with a notochord, most of which are vertebrates). Although we think of internal skeletons as "normal," this is actually the least common type of skeleton in the animal kingdom.

Regardless of the type of skeletons they possess, animals move by contractions of **antagonistic muscles**—muscles that work in opposition to one another, as illustrated in **FIGURE 39-8**.

The Vertebrate Skeleton Serves Many Functions

The bony endoskeleton of humans and most vertebrates serves a wide range of functions:

- The skeleton provides a rigid framework that supports the body and protects the internal organs. The brain and spinal cord are almost completely enclosed within the skull and vertebral column; the rib cage protects the lungs and the heart, while the pelvic girdle supports and partially protects abdominal organs.
- The vertebrate skeleton allows locomotion. Natural selection for efficient locomotion has produced the

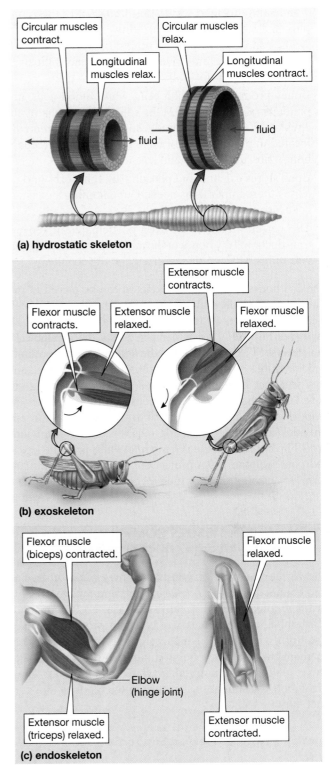

FIGURE 39-8 Antagonistic muscles work with diverse skeletons
(a) A hydrostatic skeleton, such as in this earthworm, is essentially a fluid-filled tube with muscular walls. Circular muscles form bands around the circumference of the tube and work in opposition to longitudinal muscles that run lengthwise. Because fluids are incompressible and can take any shape, if the circular muscles contract, the animal will become long and thin (left); if the longitudinal muscles contract, it will become short and fat (right). **(b)** Antagonistic flexor and extensor muscles attach to the inside of the arthropod exoskeleton and **(c)** to the outside of the vertebrate endoskeleton.

wonderfully formed wings, fins, limbs, and other complex skeletal structures that allow vertebrates to fly, swim, run, and leap. The bones of vertebrates are adapted for different types of locomotion, as illustrated in "Links to Life: Walk with a Dog."

- Bones produce red blood cells, white blood cells, and platelets (see Chapter 32). In adults, these cells of the circulatory system are produced by *red bone marrow*, located in porous areas of bone in the sternum (breastbone), ribs, upper arms and legs, and hips.

- Bone stores calcium and phosphorus. It absorbs and releases these minerals as needed, maintaining a constant concentration in the blood.

- The vertebrate skeleton even participates in sensory transduction. Tiny bones of the middle ear transmit sound vibrations between the eardrum and the cochlea.

The 206 bones of the human skeleton can be placed in two categories. The **axial skeleton** that forms the axis of the body includes the bones of the head, vertebral column, and rib cage (see Fig. 39-1). As its name suggests, the **appendicular skeleton** (see Fig. 39-1) includes the appendages or forelimbs and hindlimbs in terrestrial vertebrates (arms and hands; legs and feet in humans), as well as two supporting girdles. The *pectoral girdle*, which consists of the clavicle and scapula in people, links the arms to the axial skeleton and provides attachment sites for muscles of the trunk and arms. Hip bones form the *pelvic girdle*, which links the legs to the axial skeleton, helps protect the abdominal organs, and forms attachment sites for muscles of the trunk and legs.

39.4   WHICH TISSUES COMPRISE THE VERTEBRATE SKELETON?

The vertebrate skeleton consists of three types of specialized connective tissue: *cartilage*, *bone*, and *ligaments*. Both cartilage and bone are rigid tissues that consist of living cells embedded in a matrix of collagen protein. Collagen also forms the tough bands of connective tissue called **ligaments**, which join bones at joints, allowing the bones to move relative to one another while remaining attached. Ligaments are similar in structure to the tendons that connect muscles to bones.

Cartilage Provides Flexible Support and Connections

Cartilage plays many roles in the vertebrate skeleton. For example, during the development of the embryo, the skeleton forms first from cartilage that is later replaced by bone (**FIG. 39-9**). In the cartilaginous fishes (Class chondrichthyes, a group that includes sharks), the cartilage skeleton is never replaced by bone. Cartilage also covers the ends of bones at joints, supports the flexible portion of the nose and external ears, connects the ribs to the sternum (breastbone), and provides the framework for the larynx, trachea, and bronchi of the respiratory system. In addition, it forms tough, shock-absorbing pads that cush-

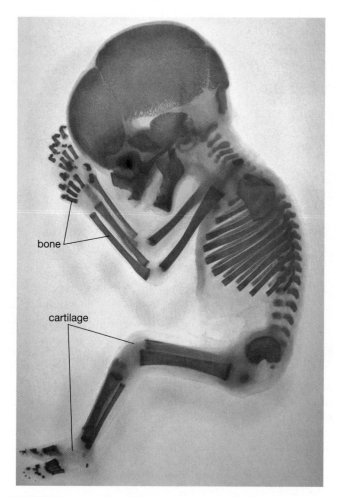

FIGURE 39-9 Bone replaces cartilage during development In this 16-week-old human fetus, bone is stained magenta. The clear areas at the wrists, knees, ankles, elbows, and breastbone show cartilage that will later be replaced by bone.

ion the knee joints and form the **intervertebral discs** between the vertebrae of the backbone (see Fig. 39-1).

The living cells of cartilage are called **chondrocytes**. These cells secrete a flexible, elastic, noncellular matrix of collagen that surrounds them and forms the bulk of the cartilage (**FIG. 39-10**, left). No blood vessels penetrate cartilage; to exchange wastes and nutrients, chondrocytes must rely on gradual diffusion of materials through the collagen matrix. As you might predict, cartilage cells have a very slow metabolic rate; and so damaged cartilage repairs itself very slowly, if at all.

Bone Provides a Strong, Rigid Framework for the Body

Bone is the most rigid form of connective tissue. Although bone resembles cartilage, the collagen fibers of bone are hardened by deposits of the mineral *calcium phosphate*. Bones, such as those supporting your arms and legs, consist of a hard outer shell of **compact bone**, with **spongy bone** in the interior (**FIG. 39-10**, middle). Compact bone is dense and strong and provides an attachment site for muscle. Spongy bone is lightweight, rich in blood vessels, and

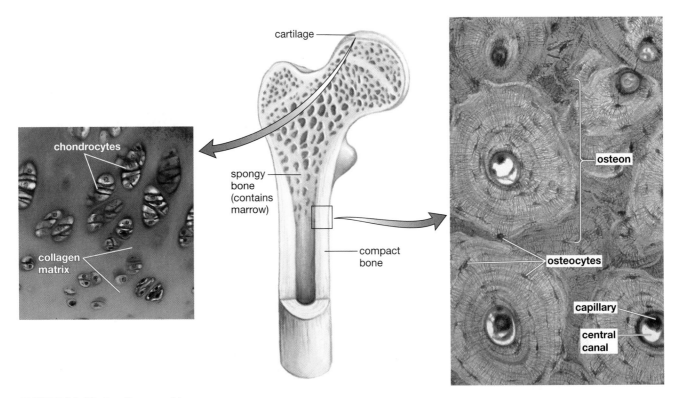

FIGURE 39-10 Cartilage and bone
(left) In cartilage, the chondrocytes, or cartilage cells, are embedded in an extracellular matrix of the protein collagen, which they secrete. **(middle)** A bone, such as in an arm or leg, consists of an outer layer of compact bone with spongy bone inside. For simplicity, blood vessels are not shown. **(right)** Osteons are clearly visible in this micrograph. Each includes a central canal containing a capillary that nourishes the osteocytes, which are embedded in the concentric rings of bone material.

highly porous. Bone marrow, where blood cells form, is found in cavities of spongy bone. In contrast to cartilage, bone is well supplied with blood capillaries.

There are three types of bone cells: **osteoblasts** (bone-forming cells), **osteocytes** (mature bone cells), and **osteoclasts** (bone-dissolving cells). Early in development, when bone replaces cartilage in the skeleton, osteoclasts invade and dissolve the cartilage; osteoblasts then replace it with bone. As bones grow, osteoblasts form a thin layer covering the outside of the bone. The osteoblasts secrete a hardened matrix of bone and gradually become entrapped within it. At this point the osteoblasts stop producing bone and mature into osteocytes (**FIG. 39-10**, right). Osteocytes are nourished by nearby capillaries and are connected to other osteocytes by thin cytoplasmic extensions through narrow channels in the bone. Although unable to produce more bone, osteocytes may secrete substances that control the continuous remodeling of bone.

Bone Remodeling Allows Skeletal Repair and Adaptation to Stresses

Each year, 5% to 10% of all the bone in your body dissolves and is replaced, through a process called *bone remodeling*. This process allows the skeleton to subtly alter its shape in response to the demands placed on it. Bones

that carry heavy loads or are subjected to extra stress become thicker, providing more strength and support. Archeologists excavating skeletons buried in volcanic ash at Pompeii could identify archers because the bones of their right and left arms differed considerably in thickness. A professional tennis player may have 30% more bone mass in his or her playing arm. In fact, normal stress is a major factor in maintaining bone strength. The bones of a limb immobilized in a cast rapidly lose significant amounts of calcium. People confined to wheelchairs initially lose 1% to 2% of bone mass per month, although the loss eventually stops. Bed rest and space flight also reduce normal stresses on the bone, leading to bone loss.

Bone remodeling is the result of the coordinated activity of the osteoclasts that dissolve bone and the osteoblasts that rebuild it. Osteoclasts cling to the bone surface, secreting acids and enzymes that dissolve the hard matrix and then tunnel into the bone, creating channels. These channels are invaded by capillaries and osteoblasts. The osteoblasts fill the channel with concentric deposits of new bone matrix, leaving only a small opening for the capillary. As a result of this process, hard bone is made up of tightly packed units called **osteons** (also called *Haversian systems*), each consisting of concentric layers of bone with embedded osteocytes. The concentric layers of bony material surround a *central canal* containing a capillary (Fig. 39-10, right). Osteoclasts

HEALTH WATCH   How Bones Heal

A sudden slip, an audible crack, and an oddly bent limb require a trip to the emergency room. There, the attending physician will coax the damaged bone back into its proper orientation and immobilize it with a cast or splint. The rest of the healing process is up to the body's own repair mechanisms. Over the next six weeks or so, the body orchestrates a systematic restoration of the bone's strength and integrity.

Healing begins when a large blood clot from vessels ruptured during the injury surrounds the break (**FIG. E39-1**, step ①). Phagocytic cells from the blood and osteoclasts from the damaged bone ingest and dissolve cellular debris and bone fragments. The fracture ruptures the *periosteum*, a

thin layer of connective tissue that surrounds the bone and is rich in capillaries and osteoblasts. The osteoblasts, in conjunction with cartilage-forming cells, secrete a *callus*, a porous mass of bone and cartilage that surrounds the break (**FIG. E39-1**, step ②). The callus replaces the original blood clot and holds the ends of the bones together while remodeling processes re-create the original shape of the bone. Once the callus is in place, osteoclasts, osteoblasts, and capillaries invade it. Nourished by the capillaries, osteoclasts break down cartilage while osteoblasts add new bone (**FIG. E39-1**, step ③). Finally, osteoclasts remove excess bone, restoring the bone's original shape, but often leaving a slight thickening (**FIG. E39-1**, step ④).

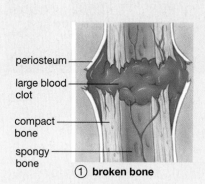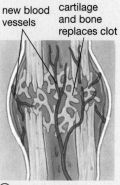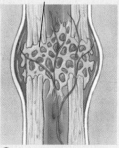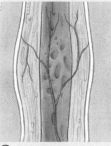

① broken bone   ② healing begins   ③ bony callus forms   ④ healed fracture

FIGURE E39-1 Stages of bone repair

and osteocytes also play a crucial role in the repair of bone fractures, as described in "Health Watch: How Bones Heal."

The continuous turnover of bone also allows the body to maintain constant levels of calcium in the blood; calcium from bones is retained in the blood if blood calcium levels drop, but is returned to bone if blood calcium is adequate or high. This process is regulated by the hormones calcitonin and parathyroid hormone, which cause bones to absorb calcium from the blood and release calcium into the blood, respectively (see Chapter 37). Early in life, the activity of osteoblasts outpaces that of osteoclasts, allowing bones to become larger and thicker as a child grows. In the aging body, however, the balance of power shifts to favor osteoclasts, and bones become more fragile as a result. (See "Health Watch: Osteoporosis—When Bones Become Brittle.")

39.5 HOW DOES THE BODY MOVE?

In addition to supporting the body, the skeleton facilitates movement by providing a framework that muscles can move. Movement of all skeletons is accomplished by the

action of pairs of antagonistic muscles: one muscle actively contracts, passively extending the other. Antagonistic muscles alter the configuration of the skeleton by causing movement around joints (in exoskeletons and vertebrate endoskeletons; see Fig. 39-8b, c) or by altering the shape of the internal fluid (in hydrostatic skeletons; see Fig. 39-8a).

Muscles Move the Skeleton Around Flexible Joints

In vertebrates, bones act as levers that can be moved by the attached skeletal muscles. Bones typically move around **joints**, the points at which two bones meet. Not all joints are movable; but in those that move, the portion of each bone that forms the joint is coated with a layer of cartilage, whose smooth, resilient surface allows the bone surfaces to slide past one another during movement (see Fig. 39-10, middle). On either side of a joint, skeletal muscles are attached to bones by tendons, and the bones themselves are joined by ligaments (**FIG. 39-11**).

Skeletal muscles are arranged in antagonistic pairs on opposite sides of many movable joints (see Figs. 39-8c and 39-11). When muscles pulling the bone in one direction contract, they move the bone around its joint, and

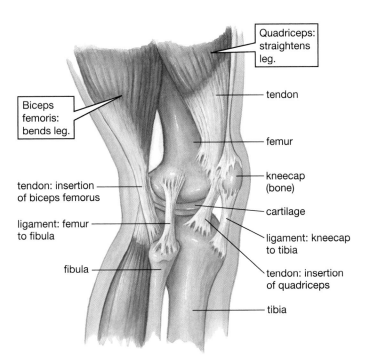

Quadriceps: straightens leg.

tendon

Biceps femoris: bends leg.

femur

kneecap (bone)

tendon: insertion of biceps femorus

cartilage

ligament: femur to fibula

ligament: kneecap to tibia

fibula

tendon: insertion of quadriceps

tibia

FIGURE 39-11 The human knee
The human knee showing antagonistic muscles (the biceps femoris and the quadriceps of the thigh), tendons, and ligaments. The complexity of this joint, coupled with the extreme stresses placed on it during activities such as jumping, running, or skiing, makes it very susceptible to injury.

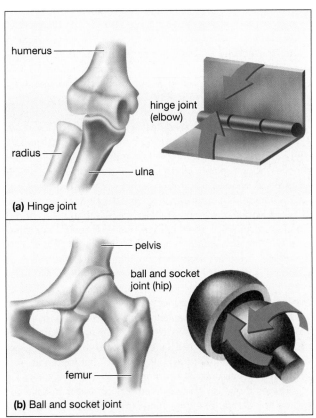

humerus

hinge joint (elbow)

radius

ulna

(a) Hinge joint

pelvis

ball and socket joint (hip)

femur

(b) Ball and socket joint

FIGURE 39-12 A ball-and-socket joint
(a) The human elbow is a good example of a hinge joint. **(b)** The human hip can rotate because it has a ball-and-socket joint. This consists of the rounded end of the femur (the ball) that fits into a cuplike depression (the socket) in the pelvic bone.

simultaneously stretch the relaxed opposing muscles (remember that muscles can only actively contract, not extend). **Hinge joints** are located in the elbows, knees, and fingers. Like a hinged door, these joints move in only two dimensions (**FIG. 39-12a**). Antagonistic muscle pairs, known as **flexor** and **extensor** muscles, lie in roughly the same plane as the joint (see Fig. 39-8c). The tendon at one end of each muscle, called the **origin**, is fixed to a bone that remains stationary, while the other end, the **insertion**, is attached to the bone on the far side of the joint, which is moved by the muscle. When the flexor muscle contracts, it bends the joint; when the extensor muscle contracts, it straightens the joint. In Figure 39-11, for example, contraction of the biceps femoris bends the leg at the knee, while contraction of the quadriceps straightens it. Alternate contractions of flexor and exten-

sor muscles cause the lower leg bones to swing back and forth at the knee joint.

Other joints, such as those of the hip and shoulder, are **ball-and-socket joints**, in which the round end of one bone fits into a hollow depression in another (**FIG. 39-12b**). Ball-and-socket joints allow movement in several directions; compare the wide-ranging swinging of your upper arm or upper leg with the limited bending of your knee or elbow. The range of motion in ball-and-socket joints is made possible by at least two pairs of muscles, oriented at right angles to each other, that provide flexibility of movement.

As a human grows and matures, bone density increases steadily, reaching a peak between ages 25 and 35. In middle age, the activity of osteoclasts starts to exceed that of osteoblasts, and bone density begins a slow, natural decline. Some loss is normal, but in people with **osteoporosis** (literally, "porous bones"; **FIG. E39-2a**), the loss is severe enough to weaken bones, making them vulnerable to fractures and deformities. In many cases, the vertebrae of individuals with osteoporosis compress, causing a hunchbacked appearance (**FIG. E39-2b**). In extreme cases, simple activities such as lifting a shopping bag, opening a window, or even sneezing can break a bone. Nearly one-third of women living to age 85 will fracture a hip weakened by osteoporosis.

The bones of women are less massive than those of men; and according to the National Osteoporosis Foundation, women are about four times as likely to suffer from osteoporosis. Another factor that differs between men and women is that, until menopause, women have high levels of the hormone estrogen, which stimulates osteoblasts and helps maintain bone density. After menopause, estrogen production drops dramatically, and women may lose 3% to 5% of their bone mass each year for several years. Nearly half of women over age 65 are estimated to have sufficiently low bone density that they are at risk for osteoporosis. Alcoholism and smoking also contribute to bone loss.

Bones thrive on moderate stress, but older people tend to be less active. Being inactive (or weightless, as astronauts have discovered) results in rapid loss of bone minerals. Even in elderly people, weight-bearing exercise such as walking or dancing can reduce bone loss and even sometimes increase bone mass.

Some women with osteoporosis, in consultation with their physicians, choose medical therapy to help maintain bone density. The hormone calcitonin, administered as a nasal spray, inhibits osteoclast activity and has beneficial effects on bone deposition. A drug that mimics estrogen's effects on bone is also available for use against osteoporosis. Several other drugs that intervene in various aspects of bone formation and breakdown are also under investigation. Someday astronauts in space, as well as women at risk for osteoporosis, may spend 10 to 20 minutes each day standing on a vibrating plate. Studies using several types of non-human animals reported impressive bone strengthening as a result of vibrations barely perceptible to human volunteers. Preliminary work with postmenopausal women suggests that this new therapy reduces bone loss. Researchers hypothesize that the minute stresses caused by the vibrations may activate bone formation much as do the everyday stresses imposed by muscle contractions.

Although medical intervention can partially reverse osteoporosis and slow its progression, there is currently no cure. Fortunately, much of the pain, incapacitation, and expense caused by fractures due to osteoporosis can be prevented. The best way to prevent osteoporosis is a combination of regular exercise and adequate calcium and vitamin D (which is essential for deposition of calcium in bone). Started early in life and continued, these measures help ensure that bone mass is as high as possible before natural, age-related losses begin; and they will minimize such losses in old age.

FIGURE E39-2 Osteoporosis (a) Cross section of (left) a normal bone compared with (right) a bone from a woman with osteoporosis. (b) The devastating effects of osteoporosis extend beyond a hunchbacked appearance; its victims are also at high risk for bone fractures.

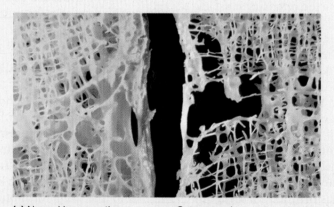

(a) Normal bone section   Osteoporosis   (b) Osteoporosis victim

CASE STUDY REVISITED HIDDEN HAZARDS OF SPACE TRAVEL

To prevent muscle loss during space flight, astronauts aboard the International Space Station exercise for about two hours daily, using a stationary bicycle and a treadmill. Bone loss, however, is more difficult to control. Researchers hypothesize that weightlessness (in unknown ways) favors the bone-dissolving activity of osteoclasts over the bone-strengthening actions of osteoblasts. Besides reducing bone

strength, there are other hidden hazards. As unused bone dissolves, it releases calcium into the bloodstream, which may eventually be deposited as kidney stones. Since weightlessness may remain a fact of life for space travelers, scientists are trying to determine whether bone production can be stimulated chemically, perhaps by boosting messenger proteins that favor osteoblast production and activity. Scientists have recently discovered and cloned the

gene for a bone-conserving protein called osteoprotegerin (OPG), which inhibits the activity of osteoclasts. Mice given an OPG injection and control mice were sent on a 12-day space shuttle mission. Compared to controls, the mice given OPG lost significantly less bone during their period of weightlessness, suggesting a potential aid for future astronauts.

To simulate prolonged weightlessness here on Earth, a team of international re-

Each type of vertebrate relies on a basic skeletal framework that first evolved in fishes but, over evolutionary time, has developed adaptations that support the animal's unique lifestyle. For a close-to-home example, simply find a dog and look carefully at its legs. Where is the dog's "ankle," "knee," "elbow," and "wrist"? Are all the bones of the human hand or foot contained within the dog's small paws? Compare the drawings in **FIGURE E39-3**—you might be surprised!

FIGURE E39-3 Dog and human limbs

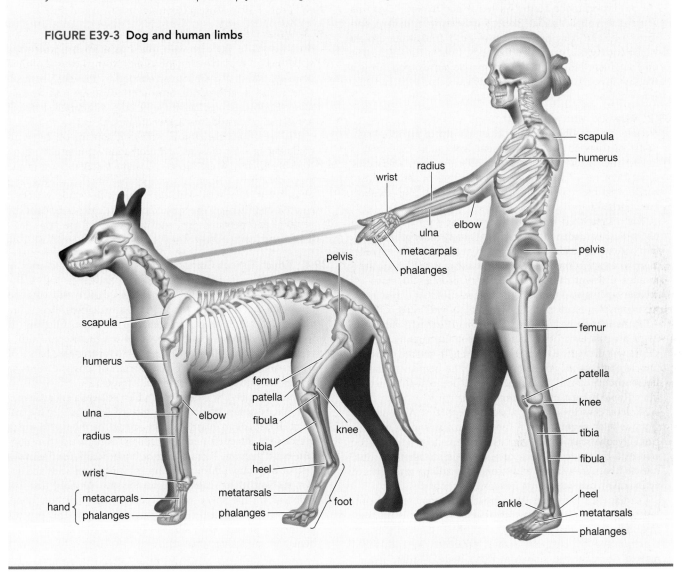

searchers convinced volunteers to spend three months confined to a bed with a slight head-down tilt (**FIG. 39-13**). One group was given an anti-osteoporosis drug; a second group was required to exercise in bed three times weekly; and the third group served as a control. The masses of data are still being analyzed, but preliminary findings show that even exercising three times weekly reduced muscle and bone loss, and the osteoporosis drug helped counteract loss of bone. As you might imagine, the study was tedious for the volunteers. Said one: "I really want to get out of this bed. I miss driving, I miss riding my bike. I miss taking a shower." Later, after slowly arising and beginning to wobble about on legs unaccustomed to loco-motion, he concluded: "It's an excellent experience. I'm glad I did it."

Consider This Some people believe that research into space travel is a waste of money and talent, since there are many problems and unmet needs on Earth. How might a NASA scientist justify research on bone loss in space to a skeptical audience? Would you volunteer for a bed-rest study like the one described here? Why or why not?

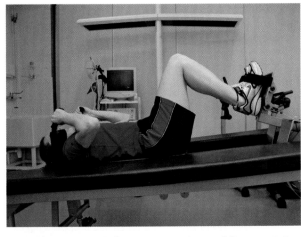

FIGURE 39-13 A volunteer exercises while remaining prone

CHAPTER REVIEW

SUMMARY OF KEY CONCEPTS

39.1 An Introduction to the Muscular and Skeletal Systems

Muscles can only contract. Skeletal muscles work in close harmony with the skeleton, which provides a framework against which contraction can move the body. Contraction of cardiac muscle pumps blood, and smooth muscle moves food through the digestive tract, influences blood vessel diameter, and contracts hollow organs such as the bladder and uterus.

39.2 How Do Muscles Work?

Skeletal muscle fibers (muscle cells) consist of myofibril subunits surrounded by sarcoplasmic reticulum. Sarcomeres linked end to end within each myofibril contain alternating thick filaments of myosin and thin filaments of actin and two accessory proteins (troponin and tropomyosin). Thin filaments are attached to proteins called Z lines.

When stimulated by a motor neuron at a synapse called a neuromuscular junction, each skeletal muscle fiber produces action potentials. Calcium is released from the sarcoplasmic reticulum when an action potential invades the muscle fiber via the T tubules. Calcium causes the accessory proteins to move off myosin binding sites on the actin molecules of thin filaments, allowing myosin heads of thick filaments to bind to them. Using ATP, the myosin heads bend, release, and reattach, sliding the filaments past each other and shortening the sarcomeres, thus shortening the muscle fiber. Both the strength and the degree of muscle contractions are determined by the number of muscle fibers stimulated and the frequency of action potentials in each fiber. Rapid firing causes maximum contraction.

Skeletal muscles rely on a steady supply of ATP derived from cellular respiration or, for bursts of strenuous activity, from glycolysis, which does not use oxygen and produces lactic acid as a by-product. In response to the stresses of exertion, muscle fibers grow larger and stronger by adding myofibrils of actin and myosin but do not increase in number.

Cardiac (heart) muscle also consists of sarcomeres that contain alternating thick and thin filaments. Its cells contract rhythmically and spontaneously, but these contractions are synchronized by electrical signals produced by specialized muscle fibers in the sinoatrial (SA) node. Cardiac muscle fibers are interconnected by intercalated disks that contain many gap junctions. These conduct electrical signals, producing coordinated contraction.

Smooth muscle lacks organized sarcomeres; but like cardiac muscle, its cells are electrically coupled by gap junctions. Smooth muscle surrounds hollow organs (uterus, digestive tract, bladder) and blood vessels, producing slow and sustained or rhythmic contractions, which are involuntary.

Web Tutorial 39.1 Muscle Structure

Web Tutorial 39.2 Muscle Contraction

39.3 What Does the Skeleton Do?

Three types of skeletons are found in animals. Hydrostatic skeletons (in cnidarians, mollusks, and worms) consist of fluid confined within a chamber surrounded by muscle. Exoskeletons, found in arthropods, are outer coverings with flexible joints. Endoskeletons, including the bony vertebrate skeleton, are found in echinoderms and chordates.

The vertebrate skeleton provides support for the body, attachment sites for muscles, and protection for internal organs. Red blood cells, white blood cells, and platelets form in the marrow of bones. Bone acts as a storage site for calcium and phosphorus. The axial skeleton includes the skull, vertebral column, and rib cage. The appendicular skeleton consists of the pectoral and pelvic girdles and the bones of the arms, legs, hands, and feet.

39.4 Which Tissues Comprise the Vertebrate Skeleton?

Cartilage is located at the ends of bones and forms pads in the knee joints and the intervertebral discs. It also supports the nose, ears, and respiratory passages. During embryological development, cartilage is the precursor of bone. Cartilage is formed by chondrocytes, which surround themselves with a matrix of fibrous collagen. Collagen also forms dense bands of tissue called ligaments that connect bones at movable joints, as well as tendons that connect muscles to bones.

Bone is formed by osteoblasts, which secrete a collagen matrix that becomes hardened by calcium phosphate. A typical bone consists of an outer shell of compact, hard bone, to which muscles are attached, and inner spongy bone, which may contain bone marrow. Remodeling of bone occurs continuously. Osteoclasts tunnel through the bone by means of acids and enzymes. Nourishing capillaries invade the tunnels, and osteoblasts fill the space with concentric layers of new bone, leaving a small central canal for each capillary. This process produces osteons. Osteoblasts become trapped within the bone and mature into osteocytes.

39.5 How Does the Body Move?

Skeletal muscles often form antagonistic pairs that move the skeleton. In the vertebrate skeleton, movement occurs around joints, where bones are joined by ligaments. Muscles attach to bones on either side of the joint by tendons. The contraction of one muscle bends the joint and stretches out its antagonistic muscle. At hinge joints, muscles are attached to the immovable bone at their origins and to the mobile bone at their insertions. Contraction of the flexor muscle bends the joint; contraction of its antagonistic extensor straightens it.

KEY TERMS

actin *page 800*
antagonistic muscles *page 805*
appendicular skeleton *page 806*
axial skeleton *page 806*

ball-and-socket joint *page 809*
bone *page 806*
cardiac muscle *page 799*
cartilage *page 806*
chondrocyte *page 806*

compact bone *page 806*
endoskeleton *page 805*
exoskeleton *page 804*
extensor *page 809*
flexor *page 809*
hinge joint *page 809*

hydrostatic skeleton *page 804*
insertion *page 809*
intervertebral disc *page 806*
joint *page 808*
ligament *page 806*

THINKING THROUGH THE CONCEPTS

1. Sketch a relaxed muscle fiber containing a myofibril, sarcomeres, and thick and thin filaments. How would a contracted muscle fiber look by comparison?

2. Describe the process of skeletal muscle contraction, beginning with an action potential in a motor neuron and ending with the relaxation of the muscle. Your answer should include the following words: *neuromuscular junction, T tubule, sarcoplasmic reticulum, calcium, thin filaments, binding sites, thick filaments, sarcomere, Z line,* and *active transport.*

3. Explain the following two statements: muscles can only actively contract; muscle fibers lengthen passively.

4. What are the three types of skeletons found in animals? For one of them, describe how the muscles are arranged around the skeleton and how contractions of the muscles result in movement of the skeleton.

5. Compare the structure and function of the following pairs: spongy and compact bone, smooth and skeletal muscle, and cartilage and bone.

6. Explain the functions of osteoblasts, osteoclasts, and osteocytes.

7. How is cartilage converted to bone during embryonic development? Where is cartilage located in the body, and what functions does it serve?

8. Describe a hinge joint and how it is moved by antagonistic muscles.

APPLYING THE CONCEPTS

1. Discuss some of the problems that would result if the human heart were made of skeletal muscle instead of cardiac muscle.

2. Myasthenia gravis is caused by the abnormal production of antibodies that bind to acetylcholine receptors on muscle cells and that eventually destroy the receptors. The disease causes muscles to become flaccid, weak, or paralyzed. Drugs, such as neostigmine, that inhibit the action of acetylcholinesterase (an enzyme that breaks down acetylcholine) are used to treat myasthenia gravis. How does neostigmine restore muscle activity?

3. Human muscle cells contain a mixture of slow-twitch and fast-twitch fibers. Slow-twitch muscle cells break down ATP slowly; they contain many mitochondria and large amounts of myoglobin, a dark pigment that acts as a reservoir for oxygen. All fast-twitch muscle cells break down ATP rapidly and possess smaller amounts of myoglobin. The relative numbers of these fibers in different muscles is under genetic control. Use this information to explain the location of dark (myoglobin-containing) and white meat in birds.

FOR MORE INFORMATION

Andersen, J. L., Schjerling, P., and Saltin, B. "Muscle, Genes, and Athletic Performance." *Scientific American*, September 2000. Muscle physiology is influenced by both genes and training, and it contributes to athletic ability.

Booth, F. W., and Neufer, P. D. "Exercise Controls Gene Expression." *American Scientist*, January–February 2005. As skeletal muscles are worked, they alter the way genes are expressed and produce healthy changes in the body.

Rosen, C. J. "Restoring Aging Bones." *Scientific American*, March 2003. New treatments for osteoporosis are helping aging women and men live more productive lives.

Ruff, C. B., "Gracilization of the Modern Human Skeleton." *American Scientist*, November–December, 2006. Over the last 2 million years, human skeletons have become less robust, as selection pressures on them have changed.

Sweeney, H. L. "Gene Doping." *Scientific American*, July 2004. Genetic engineering techniques may help people with muscle disorders, but will unscrupulous athletes use this technology to improve their performance?

Travis, J. "Boning Up." *Science News*, January 15, 2000. Researchers investigate a newly discovered gene whose protein shifts the balance between osteoblasts and osteoclasts.

Williams, C. "Don't Use It, Don't Lose It." *New Scientist*, September 2, 2006. New insights into why muscles atrophy with disuse may help in developing preventative treatments.

Animal Reproduction

The frozen zoo consists of tissue samples and sex cells, often of endangered species, preserved in liquid nitrogen. (**inset**) The last known po'ouli bird died in 2004, but its cells live on in the frozen zoo.

CASE STUDY THE FROZEN ZOO

IN JANUARY 2000, the last living bucardo, a mountain goat native to the Spanish Pyrenees, was killed by a falling tree. In 2004, the world's last known Hawaiian po'ouli bird died at the Maui Bird Conservation Center (inset). But there is a slender hope that future generations might yet see these extinct species. Thanks to San Diego's "frozen zoo," their cells, containing their genetic blueprints, live on, having been *cryopreserved* (kept alive in a deeply frozen state) before the species' untimely extinction. The San Diego Frozen Zoo houses collections of tissues, sperm, and eggs from over 5000 animals representing well over 300 species, all stored in canisters of liquid nitrogen at –320°F. Many of these species are endangered but, unlike the bucardo and po'ouli, most are not yet extinct. There are a few dozen frozen zoos on the planet, providing the raw material for a unique

brand of wildlife conservation using *assisted reproductive technology* (ART).

This approach to wildlife conservation involves techniques such as artificial insemination, in vitro fertilization, interspecies embryo transfer (the use of a surrogate mother from a different but related species), and even cloning. These techniques are surrounded by controversy and frustration, yet they have led to some inspiring success stories. Artificial insemination has become a cornerstone in efforts to save the black-footed ferret, giant panda, and cheetah. *In vitro fertilization* (IVF, in which sperm and egg are joined in a dish) is becoming well established as a conservation tool. A major advantage of IVF is that it allows sperm from an endangered species to be transported—between continents, if necessary—to fertilize an appropriate female. This eliminates the risk and trauma of transporting

the animals; it also circumvents the very real possibility that, once together, they will refuse to mate. Landmark successes of IVF include the first "test-tube" Siberian tiger in 1990, a gorilla in 1996, and the first panda in 2003. Ironically, the loss of species that makes ART so important is due to the enormous reproductive success of a single species—*Homo sapiens*. The burgeoning human population threatens wildlife throughout the world, as our energy use alters the global climate and we continue to usurp wildlife habitat for food, housing, and the extraction of natural resources.

How do animals, including people, reproduce? What options do we have for controlling reproduction? How does ART help infertile human couples have children? Are high-tech solutions for preserving endangered species justified?

40.1   HOW DO ANIMALS REPRODUCE?

Animals reproduce either sexually or asexually. **Sexual reproduction** requires the production of haploid gametes through meiosis. In a process called **fertilization**, two gametes—usually from separate parents—fuse and give rise to a diploid cell, which then divides mitotically to produce a diploid individual. Because the offspring receives genes from each of its parents, its genome is not identical to that of either parent. In contrast, **asexual reproduction** involves only a single animal that produces offspring through repeated mitoses of cells in some part of its body. The offspring are, therefore, genetically identical to the parent.

Because humans and most other animals reproduce sexually, we tend to regard sexual reproduction as the normal and best method. However, although sexual reproduction produces new combinations of genes, asexual reproduction is far more efficient because animals can do it alone. Not surprisingly, many animals reproduce asexually some of the time.

Asexual Reproduction Does Not Involve the Fusion of Sperm and Egg

Budding Produces a Miniature Version of the Adult

Many sponges and cnidarians, including hydra and some sea anemones, reproduce by **budding** (FIG. 40-1). *Buds*, which are miniature versions of the adult, grow directly from the parent's body and draw nourishment from it. When they are large enough, the buds break off and become independent individuals.

Fission Followed by Regeneration Can Produce a New Individual

Many animals are capable of **regeneration**, the ability to regrow lost body parts. For example, sea stars will regenerate a lost arm and lizards a tail that is lost to a predator. Regeneration is part of reproduction in species that reproduce by **fission**. Several annelid and flatworm species can reproduce

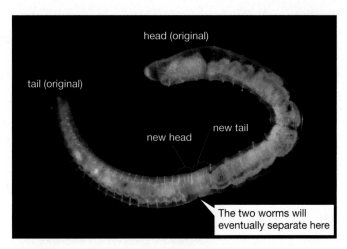

FIGURE 40-2 Fission followed by regeneration
This marine annelid (segmented) worm can reproduce by splitting its body and regenerating each half. QUESTION Which type of cell division gives rise to the cells of the offspring's body?

by dividing into two or more pieces, each of which regenerates an entire body (FIG. 40-2). A few brittle star species routinely reproduce in a similar fashion. Among cnidarians (anemones, sea jellies, and their relatives), some coral species and some sea jellies can split themselves in half lengthwise and regenerate two new individuals.

During Parthenogenesis, Eggs Develop Without Fertilization

The females of some animal species can reproduce by a process known as **parthenogenesis**, in which haploid egg cells divide mitotically and develop into adults without being fertilized. In some species, parthenogenetically produced offspring remain haploid. For example, male honeybees develop from unfertilized eggs and are haploid; their diploid sisters develop from fertilized eggs. Some fish, amphibians, and reptiles reproduce parthenogenetically, but restore the diploid number of chromosomes by duplicating all their chromosomes either before or after meiosis. All the resulting offspring are female.

Some species of fish, including relatives of the mollies and platys that are popular in tropical fish tanks, and some lizards, such as the whiptail of the southwestern United States and Mexico (FIG. 40-3), have populations consisting entirely of parthenogenetically reproducing females. Still other animals, such as some aphids, can reproduce either sexually or parthenogenetically, depending on environmental factors such as the season of the year and the availability of food (FIG. 40-4).

FIGURE 40-1 An anemone with many buds

FIGURE 40-3 The whiptail lizard

FIGURE 40-4 A female aphid gives birth
In spring and early summer, when food is abundant, some species of aphid females reproduce parthenogenetically; in fact, the females are born pregnant! In fall, they reproduce sexually. QUESTION Why do aphids switch to sexual reproduction in the fall?

FIGURE 40-5 Earthworms exchange sperm

Sexual Reproduction Requires the Union of Sperm and Egg

Given the obvious efficiency of asexual reproduction, no one is sure why sex evolved and became the dominant form of reproduction. Sex does have an important outcome: the genetic recombination that results from sexual reproduction creates novel genotypes—and therefore novel phenotypes—that are an important source of variation upon which natural selection may act.

In animals, sexual reproduction occurs when a haploid sperm fertilizes a haploid egg, generating a diploid offspring. In most animal species, an individual is either male or female. The sexes are distinguished by the type of gamete that each produces. Females produce **eggs**, which are large, nonmotile cells containing food reserves. Males produce small, motile **sperm**, with almost no cytoplasm or food reserves.

In some animals, such as earthworms and many snails, single individuals produce both sperm and eggs. Such individuals are commonly called **hermaphrodites** (after Hermaphroditos, a male Greek god whose body merged with that of a female water nymph, producing a half-male and half-female being). In most hermaphroditic species, reproduction involves a mutual exchange of sperm between individuals, as occurs between earthworms (**FIG. 40-5**). In some hermaphroditic species, however, individuals can fertilize their own eggs if a mate is unavailable. These animals, which include tapeworms and many pond snails, are relatively immobile and may find themselves isolated from other members of their species, making self-fertilization advantageous.

For species with two distinct sexes and hermaphrodites that cannot self-fertilize, successful reproduction requires that sperm and eggs from different animals be brought together for fertilization. The union of sperm and egg is accomplished in a variety of ways, depending on the animals' mobility and whether they breed in water or on land.

External Fertilization Occurs Outside the Parents' Bodies

In **external fertilization**, the union of the sperm and egg takes place outside the bodies of the parents. When animals breed in water, the parents release sperm and eggs into the water, through which the sperm swim to reach an egg. This process is called **spawning**. Because sperm and eggs are relatively short-lived, spawning animals must synchronize their reproductive behaviors, both *temporally* (male and female spawn at the same time) and *spatially* (male and female spawn in the same place). Synchronization may be achieved through chemical signals, courtship behaviors, environmental cues, or some combination of these factors.

Most spawning animals rely on environmental cues to some extent. Breeding usually occurs only during certain seasons of the year, and cues such as seasonal changes in day length typically stimulate the physiological changes that lead to readiness for breeding. More precise synchrony, however, is required to coordinate the actual release of sperm and egg. For example, many coral species synchronize spawning by the phase of the moon, simultaneously releasing blizzards of sperm and egg packets into the water (**FIG. 40-6**). Although many sperm and eggs make up each packet released by these hermaphroditic animals, they

FIGURE 40-6 Environmental cues may synchronize spawning
In the Great Barrier Reef of Australia, thousands of corals spawn simultaneously, creating a "blizzard" effect. Spawning in these corals is linked to the phase of the moon.

(a)

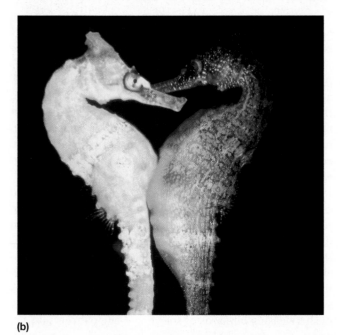

(b)

FIGURE 40-7 Courtship rituals synchronize release of sperm and eggs
(a) Courtship rituals among Siamese fighting fish (Betta splendens) ensure fertilization of the female's eggs, as male and female release sperm and eggs simultaneously. The male retrieves the eggs as they fall, spits them into his bubble nest floating above, and cares for the offspring during their first few weeks of life. (b) Spawning in the seahorse requires the male and female to orient their bodies so that the female can deposit her eggs in the male's pouch. QUESTION In addition to ensuring synchronized release of gametes, what other advantages do courtship rituals provide?

do not usually self-fertilize. The eggs are not ready for fertilization immediately, and the delay allows them to mix with sperm from other individuals of the same species.

Some animals communicate their sexual readiness to one another by sending visual, acoustic, or chemical signals. Chemical signals are especially common among immobile or sluggish invertebrates, such as mussels and sea stars. These animals release chemical signals called **pheromones**

into the water, where they are sensed by other members of the species. Usually, a female that is ready to spawn releases eggs and a sex pheromone into the water. Nearby males, detecting the mating pheromone, immediately release millions of sperm. The sperm themselves are lured by a chemical attractant released by the eggs in some, and likely most, animals. "Egg pheromones," which have been detected in animals as diverse as sea stars and humans, help ensure fertilization.

Synchronized timing alone does not guarantee efficient reproduction. Corals, sea stars, and mussels all waste enormous quantities of sperm and eggs because the gametes are released relatively far apart. In species of mobile animals, both temporal *and* spatial synchrony can be ensured by mating behaviors. Most fish, for example, have some form of courtship ritual in which the male and female come close together and release their gametes in the same place and at the same time. The courtship dances of Siamese fighting fish and the seahorse provide exquisite examples (**FIG. 40-7**). The female seahorse, laden with eggs, approaches the male and initiates an elaborate dance in which the partners approach, quiver, and nod heads before entwining their tails and lining up their bodies face to face. In an unusual reversal of sex roles, the female inserts a tube for depositing eggs into the pouch in the male's abdomen. As she injects her eggs into the pouch, the male releases a cloud of sperm from an opening just above; he then seals the fertilized eggs into his pouch. A few weeks later, the eggs have hatched and the male gives birth to perfect miniature seahorses. Frogs and toads assume a characteristic mating pose called *amplexus* (**FIG. 40-8**). In shallow water near the edges of ponds and lakes, the male mounts the female and prods the sides of her abdomen. This stimulates her to extrude her eggs, which he fertilizes by releasing sperm above them. The golden toads shown here in amplexus were once abundant in the cloud forests of Costa Rica, but they have not

FIGURE 40-8 Golden toads in amplexus
The smaller male clutches the female and stimulates her to release eggs. Golden toads are now believed to be extinct.

(a)

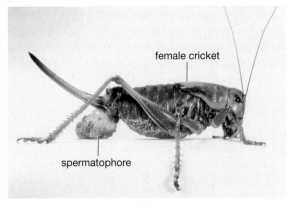

(b)

FIGURE 40-9 Internal fertilization is required for reproduction on land
(a) An endangered Sumatran rhino male mounts a female to mate. (b) A female mormon cricket takes a packet of sperm into her body.

been seen since 1989. Their disappearance, which scientists believe was caused by environmental changes related to land use and global warming, was so sudden that no one had thought to preserve this species' genetic heritage in a frozen zoo; it is lost forever.

Internal Fertilization Occurs Within the Female Body

During **internal fertilization**, sperm are placed within the female's body, where the egg is fertilized. Internal fertilization is an important adaptation to terrestrial life because sperm must remain bathed in fluid until they reach the eggs. Even in aquatic environments, internal fertilization increases the likelihood of success, because the sperm and eggs are confined together in a small space rather than relying on chance encounters in a large volume of water.

Internal fertilization usually occurs by **copulation**, in which the male deposits sperm directly into the female's reproductive tract (**FIG. 40-9a**). In a variation of internal fertilization, males of some species package their sperm in a container called a **spermatophore** (Greek for "sperm car-

FIGURE 40-10 Competition for females
Emerging from hibernation, masses of red-sided garter snake males compete to mate with females.

rier"). In many spermatophore-producing species, including some scorpions, grasshoppers, and salamanders, no copulation occurs. The male simply drops a spermatophore on the ground, and if a female finds it, she fertilizes herself by inserting it into her reproductive cavity, where the enclosed sperm are liberated (**FIG. 40-9b**).

Among animals that copulate to reproduce, males may compete for females. This has driven the evolution of a wide variety of sexually selected structures and reproductive behaviors. One spectacular example of competition for access to females occurs in the early spring of each year in the woods of western Canada. As the snows melt and the ground warms, male red-sided garter snakes emerge from the underground dens where they hibernate by the thousands. Later the females emerge, and a mating frenzy begins. In a sea of thousands of writhing snake bodies, each female attracts a crowd of dozens or even hundreds of males (**FIG. 40-10**). Only one will copulate successfully.

For fertilization to occur, a mature egg must be present. Many female snails and insects store sperm within their bodies for days or even months, ensuring a supply whenever eggs are ready. Among mammals, which do not store sperm, mating behavior must be synchronized. Often the female undergoes **ovulation** (release of the egg cell from the ovary) only during certain times of the year and signals her readiness to mate using both pheromones and behavior. In a few mammals, such as rabbits, copulation triggers ovulation. Zoo scientists attempting to breed a rare female Sumatran rhino (see Fig. 40-9a), which had not bred successfully in captivity for over a century, finally discovered that their ovulation is stimulated by courtship; two baby rhinos have now been born. Sumatran rhino sperm, stored at the frozen zoo in Cincinnati, Ohio, may one day help restore this critically endangered species. Only about 300 remain in the wild, partly because their horns are highly prized as a human aphrodisiac.

40.2 HOW DOES THE HUMAN REPRODUCTIVE SYSTEM WORK?

Humans and all other mammals have separate sexes, copulate, and fertilize their eggs internally. The **gonads** of mammals are paired organs that produce sex cells—sperm and eggs. Although most mammal species reproduce only during certain seasons of the year and consequently produce sperm and eggs only at that time, human reproduction is not restricted by season. Men produce sperm more or less continuously, and women *ovulate* (release a mature egg cell) about once a month.

The Ability to Reproduce Begins at Puberty

Sexual maturation occurs at **puberty**, a stage of development characterized by rapid growth and the appearance of secondary sexual characteristics in both sexes. Although puberty generally begins in the early teens, it may occasionally start as early as age 8 or as late as age 15. During puberty in both sexes, brain maturation causes the hypothalamus to increase **gonadotropin-releasing hormone (GnRH)**, which stimulates the anterior pituitary to produce

luteinizing hormone (LH) and follicle-stimulating hormone (FSH). These hormones stimulate the testes to produce more of the male sex hormone **testosterone** and the ovaries to produce more of the female sex hormone **estrogen**. In response to increases in testosterone, males develop secondary sexual characteristics: the **penis** (which deposits sperm in the female vagina) and *testes* enlarge; pubic, underarm, and facial hair appears; the larynx enlarges (causing a deepening voice); and muscular development increases. In response to increased estrogen (and other hormones that surge at puberty), females develop enlarged breasts and pubic and underarm hair, and begin menstruating. Changes also occur in the brain, which can make this an interesting stage for teenagers and their parents.

The Male Reproductive Tract Includes the Testes and Accessory Structures

The male reproductive system, summarized in Table 40-1 and shown in **FIGURE 40-11**, consists of the sperm-producing **testes** and accessory structures that secrete substances to activate and nourish the sperm, store it, and conduct it to the female reproductive tract.

Sperm Are Produced in the Testes

The testes, which produce both sperm and male sex hormones, are located in the **scrotum**, a pouch that hangs outside the main body cavity. This location keeps the testes about 4°C cooler than the core of the body, providing the

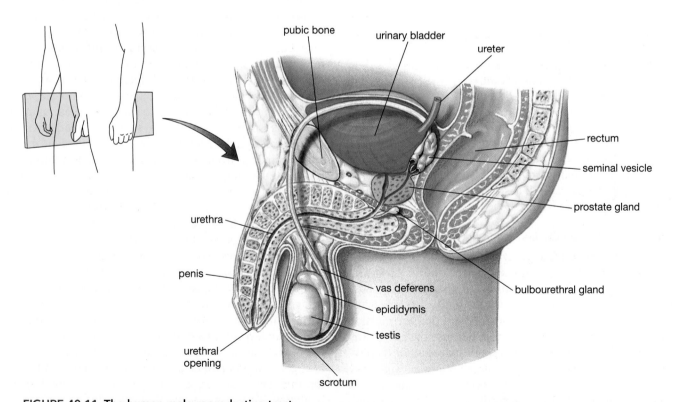

FIGURE 40-11 The human male reproductive tract
The male testes hang beneath the abdominal cavity in the scrotum. Sperm pass from the testis to the epididymis, through the vas deferens and urethra to the tip of the penis. Along the way, fluids are added from the seminal vesicles, the bulbourethral glands, and the prostate gland.

Table 40-1 The Human Male Reproductive Tract	
Structure	**Function**
Testes (male gonads)	Produce sperm and testosterone
Epididymis and vas deferens (ducts)	Store sperm; conduct sperm from testes to penis
Urethra (duct)	Conducts semen from vas deferens and urine from urinary bladder to the tip of the penis
Penis	Deposits sperm in female reproductive tract
Seminal vesicles (glands)	Secrete fluid into semen
Prostate gland	Secretes fluids into semen
Bulbourethral glands	Secrete fluid into semen

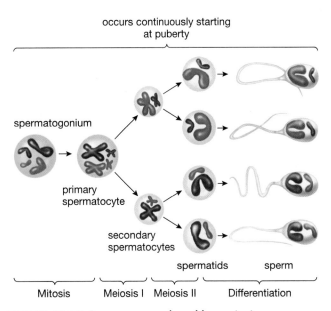

FIGURE 40-13 Sperm are produced by meiosis
Spermatogonia grow and differentiate to produce spermatocytes, which undergo meiosis and further differentiation to produce haploid sperm. Although 4 chromosomes are shown for simplicity, in humans, the diploid number is 46 and the haploid number is 23.

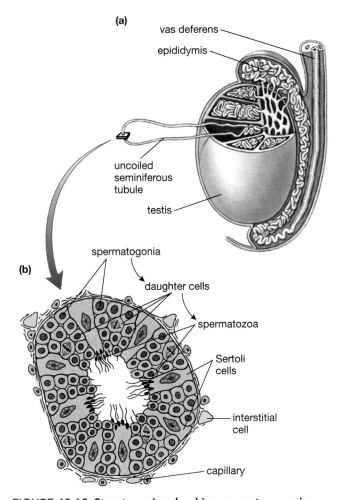

FIGURE 40-12 Structures involved in spermatogenesis
(a) A section of the testis, showing the seminiferous tubules, epididymis, and vas deferens. (b) Cross section of a seminiferous tubule. The walls of the seminiferous tubules are lined with nurturing Sertoli cells and spermatogonia undergoing meiosis. Mature sperm are freed into the central cavity. Testosterone is produced by interstitial cells.

optimal temperature for sperm development. Coiled, hollow **seminiferous tubules**, in which sperm are produced, nearly fill each testis (**FIG. 40-12a**). **Interstitial cells**, which synthesize the male hormone testosterone, are located in the spaces between the tubules (**FIG. 40-12b**).

Just inside the wall of each seminiferous tubule lie **spermatogonia** (singular, *spermatogonium*), the diploid cells from which the sperm eventually will arise, and the much larger *Sertoli cells* (Fig. 40-12b). Spermatogonia divide mitotically, replacing themselves so there is a continuing supply of spermatogonia, and forming cells that undergo **spermatogenesis** to produce haploid sperm (**FIG. 40-13**).

Spermatogenesis begins when spermatogonia grow and differentiate into **primary spermatocytes**, which are large diploid cells. The primary spermatocytes then undergo meiosis (described in Chapter 11). At the end of meiosis I, each primary spermatocyte gives rise to two haploid **secondary spermatocytes**. Each secondary spermatocyte divides again, during meiosis II, to produce two **spermatids**. Thus, each primary spermatocyte generates a total of four spermatids. Spermatids undergo radical rearrangements of their cellular components as they differentiate into sperm.

The spermatogonia, spermatocytes, and spermatids are enfolded in the **Sertoli cells**, which regulate the process of spermatogenesis and nourish the developing sperm. As spermatogenesis proceeds, the developing sperm migrate to the central cavity of the seminiferous tubule into which the mature sperm are released (see Fig. 40-12b).

A human sperm (**FIG. 40-14**) is unlike any other cell of the body. Most of the cytoplasm disappears, leaving a haploid nucleus that nearly fills the *head* of the sperm cell. Atop the nucleus lies a specialized lysosome called the **acrosome**. The acrosome contains enzymes that will dissolve protective layers around the egg and enable the sperm to enter and fertilize it. Behind the head is the *midpiece*, which is packed with mitochondria. These organelles provide the energy

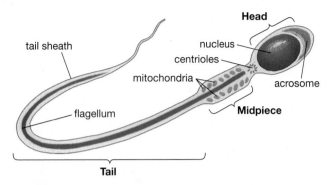

FIGURE 40-14 A human sperm cell
A mature sperm is a cell equipped with only the essentials: a haploid nucleus, the acrosome (containing enzymes that digest the barriers surrounding the egg), mitochondria for energy production, and a tail (a long flagellum) for locomotion.

needed to move the *tail*. Whiplike movements of the tail, which is actually a long flagellum, propel the sperm through the female reproductive tract.

In humans and other mammals, spermatogenesis does not begin until puberty, when GnRH from the hypothalamus stimulates the anterior pituitary to produce LH and FSH. LH stimulates the interstitial cells of the testes to produce testosterone (**FIG. 40-15**). Testosterone, in combination with FSH, stimulates the Sertoli cells to promote spermatogenesis. Like many physiological processes, sperm production is regulated by negative feedback. Testosterone, while stimulating spermatogenesis, also inhibits both GnRH release by the hypothalamus and LH and FSH release by the pituitary, limiting further testosterone production and sperm development. The Sertoli cells, when stimulated by FSH and testosterone, not only promote spermatogenesis, but secrete the hormone *inhibin* that also inhibits GnRH, LH, and FSH production (Fig. 40-15). This feedback process maintains sperm production at relatively constant levels throughout the male's reproductive life.

Accessory Structures Produce Semen and Conduct the Sperm Outside the Body

The seminiferous tubules merge to form the **epididymis**, a long, continuous, folded tube (see Fig. 40-12a). The epididymis leads into the **vas deferens**, a duct that carries sperm out of the scrotum. Most of the roughly hundred million sperm produced by a human male each day are stored in the vas deferens and epididymis. The vas deferens joins the **urethra**, which connects the bladder to the tip of the penis. This final common pathway is used, at different times, by both urine (during urination) and sperm (during ejaculation—a reflex caused by sexual stimulation that forces sperm out through the penis).

The fluid ejaculated from the penis, called **semen**, is only about 5% sperm; most of the semen fluid comes from three glands (see Fig. 40-11 and Table 40-1). Fluid pro-

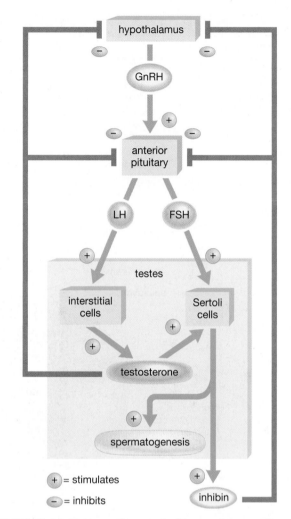

FIGURE 40-15 Hormonal control of spermatogenesis
GnRH from the hypothalamus stimulates the anterior pituitary to release LH and FSH. LH stimulates the interstitial cells to produce testosterone. Testosterone and FSH stimulate the Sertoli cells and the spermatogonia, causing spermatogenesis. Sertoli cells release inhibin, which, along with testosterone, inhibits further release of FSH and LH, forming a negative feedback loop that keeps the rate of spermatogenesis and the concentration of testosterone in the blood nearly constant. QUESTION   Why do injections of testosterone suppress sperm production?

duced by the **seminal vesicles** comprises about 60% of the semen. This fluid is rich in fructose that provides energy for the sperm; it also contains prostaglandins (see Chapter 37) that stimulate uterine contractions, which help transport the sperm up into the female reproductive tract. Its slightly alkaline pH protects the sperm from the vagina's acidic environment, which would inhibit sperm activity. The **prostate gland** produces a nutrient-rich secretion that comprises about 30% of the semen volume and includes enzymes that increase the fluidity of the semen after it is released into the vagina, allowing the sperm to swim more freely. **Bulbourethral glands** secrete mucus into the urethra that neutralizes remaining traces of acidic urine and helps lubricate the penis during intercourse.

Table 40-2 The Human Female Reproductive Tract

Structure	Function
Ovaries (female gonads)	Produce eggs, estrogen, and progesterone
Fimbria (opening of uterine tube)	Bear cilia that sweep egg into oviduct
Uterine tubes	Conduct egg to uterus; site of fertilization
Uterus	Muscular chamber where fetus develops
Cervix	Closes off lower end of uterus during pregnancy
Vagina	Receptacle for semen; birth canal

The Female Reproductive Tract Includes the Ovaries and Accessory Structures

The female reproductive tract is almost entirely contained within the abdominal cavity (Table 40-2 and **FIG. 40-16**). It consists of paired gonads—the **ovaries** (FIG. 40-17a)—and accessory structures that accept sperm, conduct the sperm to the egg, and nourish the developing **embryo**.

Eggs Are Produced in the Ovaries

Oogenesis, the formation of egg cells, begins during fetal development starting with the formation of precursor egg cells called **oogonia** (singular, *oogonium*). By the end of the third month of fetal development, all the oogonia have divided mitotically, producing **primary oocytes**. As fetal development continues, meiosis begins in all primary oocytes but is halted at prophase of meiosis I. By birth, a lifetime supply of primary oocytes is already in place. The ovaries start out with about 2 million primary oocytes. Many of these die; by puberty only about 400,000 remain. This is plenty, because only a few oocytes resume meiosis during each month of a woman's reproductive span, starting at puberty around age 13 and ending at *menopause* at about age 50.

Surrounding each oocyte is a layer of smaller cells that both nourish the developing oocyte and secrete female sex hormones. Together, the oocyte and these accessory cells make up a **follicle** (**FIG. 40-17b**). During the menstrual cycle, pituitary hormones stimulate the development of a dozen or more follicles, although usually only one follicle matures completely. At this time, the primary oocyte completes its first meiotic division (which was halted during development), dividing into a single **secondary oocyte** and a **polar body**, which is little more than a discarded set of

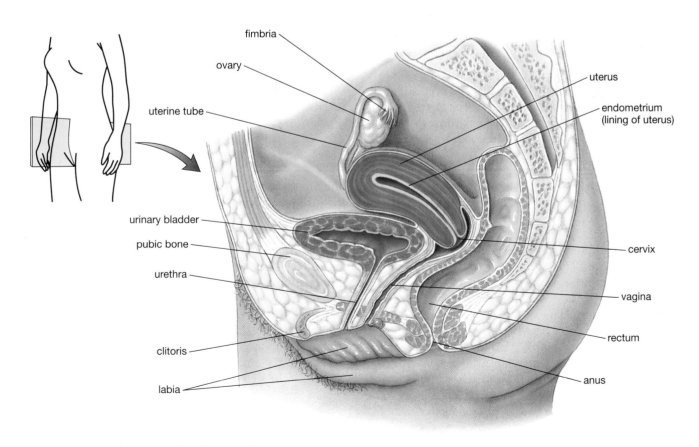

FIGURE 40-16 The human female reproductive tract
Eggs are produced in the ovaries and enter the uterine tube. Sperm and egg usually meet in the uterine tube, where fertilization and very early development occur. The early embryo embeds in the lining of the uterus, where development continues. The vagina receives sperm and serves as the birth canal.

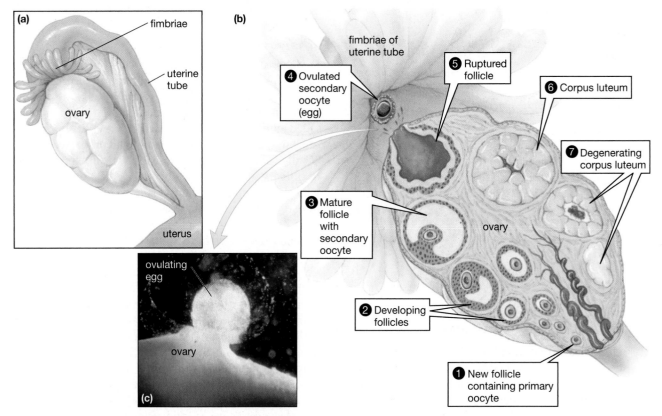

FIGURE 40-17 The structures involved in oogenesis
(a) External view of the ovary and uterine tube. **(b)** The development of follicles in an ovary, portrayed in a time sequence (clockwise from the lower right). ① A primary oocyte begins development within a follicle. ②, ③ The follicle grows, providing both hormones and nourishment for the enlarging oocyte. ④ At ovulation the egg, surrounded by follicle cells, bursts through the ovary wall. ⑤, ⑥, ⑦ The remaining follicle cells develop into the corpus luteum, which secretes hormones. If fertilization does not occur, the corpus luteum disintegrates after a few days. **(c)** An egg is released from a follicle within the ovary.

chromosomes (**FIG. 40-18**). Meanwhile, the accessory cells of the follicle multiply and secrete estrogen. As the follicle matures, it grows and eventually erupts through the surface of the ovary, releasing the secondary oocyte in a process called *ovulation* (**FIG. 40-17c**). The secondary oocyte then travels through the tube leading out of the ovary, called the **uterine tube** (sometimes called the *oviduct* or *Fallopian tube*). For convenience, we will refer to the ovulated secondary oocyte as the *egg*. If the egg is fertilized, this usually occurs in the uterine tube.

Some of the follicle cells accompany the egg; but most of them remain in the ovary, enlarge, and become glandular, forming the **corpus luteum** (see **FIG. 40-17b**). The corpus luteum secretes both estrogen and a second hormone, **progesterone**. If fertilization does not occur, the corpus luteum breaks down a few days later.

A human male is able to produce large numbers of sperm continuously. In contrast, a woman does not produce mature gametes (ovulate) unless her uterus is prepared to receive and nourish a fertilized egg. The **menstrual cycle** assures that ovulation is coordinated with preparation of the uterus. The menstrual cycle, regulated by interactions among hormones of the hypothalamus, anterior pituitary gland, and ovary, is described in detail in "A Closer Look: Hormonal Control of the Menstrual Cycle."

Accessory Structures Include the Uterine Tubes, Uterus, and Vagina

Each ovary nestles within the open end of its uterine tube (see **FIG. 40-17a**), which is fringed with ciliated "fingers" called *fimbriae* that nearly surround the ovary. The cilia create a current that sweeps the newly ovulated egg into the tube, where sperm may be swimming if intercourse has occurred recently. Fertilization usually happens within the tube. Beating cilia sweep the **zygote**, or fertilized egg, down the tube and into the **uterus** (sometimes called the *womb*). There it will develop for 9 months. The wall of the uterus has two layers that correspond to its dual functions of nourishing the developing embryo and bringing about childbirth. The inner lining, or **endometrium**, is richly supplied with blood vessels. This lining will form the mother's contribution to the **placenta**, the structure that transfers oxygen, carbon dioxide, nutrients, and wastes between mother and **fetus** (a term describing the later stages of mammalian development), as we will see in Chapter 41. The outer muscular wall of the uterus gradually expands as the developing child grows, then contracts strongly during delivery, expelling the infant out into the world.

Developing follicles secrete estrogen, which stimulates the uterine lining to grow an extensive network of blood vessels and nutrient-producing glands. After ovulation, estrogen and progesterone released by the corpus luteum stimulate

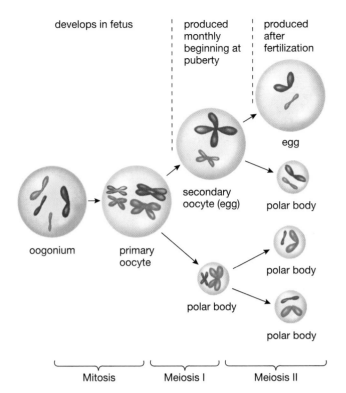

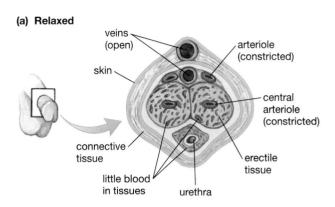

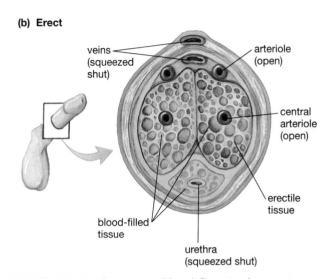

FIGURE 40-18 Egg cells are formed by meiosis
The oogonium undergoes mitosis and enlarges to form the primary oocytes. At meiosis I, most of the cytoplasm goes to the secondary oocyte, leaving a small polar body with chromosomes but little cytoplasm. At meiosis II, almost all the cytoplasm of the secondary oocyte goes to the egg, and a second small polar body discards the remaining "extra" chromosomes. The first polar body may also undergo the second meiotic division. In humans, meiosis II does not occur until a sperm penetrates the egg.

FIGURE 40-19 Changes in blood flow in the penis cause erection
(a) Smooth muscles encircling the arterioles leading into the penis are usually contracted, limiting blood flow. **(b)** During sexual excitement, these muscles relax, and blood flows into spaces within the penis. The swelling penis squeezes off the veins leaving the penis, increasing the blood pressure inside and causing the penis to become elongated and firm.

the endometrium to continue developing a thick bed for the embryo. Thus, if an egg is fertilized, it encounters a rich environment for growth. If the egg is not fertilized, however, the corpus luteum disintegrates, estrogen and progesterone levels fall, and the overgrown endometrium disintegrates as well. The uterus then contracts (sometimes causing menstrual cramps) and squeezes out the excess endometrial tissue. This causes a flow of tissue and blood called **menstruation** (from the Latin *mensis,* meaning "month").

The outer end of the uterus is nearly closed off by the **cervix**, a ring of connective tissue that encircles a tiny opening. The cervix holds the developing baby in the uterus and then relaxes at the onset of labor. This allows the central opening to expand, permitting birth of the child. Beyond the cervix is the **vagina**, which opens to the outside. The vagina maintains an acid pH to reduce infections, and it serves both as the receptacle for the penis during intercourse and as the birth canal (see Fig. 40-16).

Copulation Allows Internal Fertilization

As terrestrial mammals, humans use internal fertilization to deposit sperm into the moist environment of the female's reproductive tract. During intercourse, the penis is inserted into the vagina, where it releases sperm. The sperm swim from the vagina through the cervix and into the uterus, finally entering the uterine tubes. If the female has ovulated within the past day or so, the sperm will meet an egg in one of the uterine tubes. Only one sperm can succeed in fertilizing the egg and begin the development of a new human being.

During Copulation, Sperm Are Deposited in the Vagina

The male role in copulation begins with erection of the penis. Before erection, the penis is relaxed (flaccid), because the arterioles that supply it are constricted, allowing little blood flow (**FIG. 40-19a**). Under psychological and physical stimulation, the arterioles dilate and blood flows into spaces in the tissue within the penis (Viagra™ enhances dilation of these arterioles). As these *erectile tissues*

The menstrual cycle is controlled by hormones from the hypothalamus (GnRH), the anterior pituitary (FSH and LH), and the ovaries (estrogen and progesterone). It begins with the onset of menstruation, illustrated by the loss of the uterine lining as shown in the lowest panel of **FIGURE E40-1**. Hormonally, the menstrual cycle is initiated by the spontaneous release of gonadotropin-releasing hormone (GnRH) by cells in the hypothalamus (top panel). This release occurs continuously unless it is suppressed by other hormones, notably progesterone. The cycle starts on day 1 (which immediately follows day 28 of the cycle) and is stimulated by the increase in GnRH that begins around day 28. Follow the descriptions in the diagram by matching the numbers to Figure E40-1. (Numbers are duplicated in the figure when the description applies to multiple panels.)

① GnRH (top panel) stimulates the anterior pituitary (second panel) to release FSH (blue line) and LH (red line). These increases can first be observed around day 28. The endometrium of the uterus is shed during menstruation (lowest panel).

② FSH initiates the development of several follicles, which secrete estrogen, within the ovaries. Under the combined influences of FSH, LH, and estrogen, the follicles grow and the primary oocyte within each follicle begins developing. Usually, only one follicle completes development each month.

③ As the follicle grows, it secretes greater amounts of estrogen (purple line, fourth panel). This estrogen has three effects. First, it promotes the continued development of the follicle and its primary oocyte (third panel). Second, estrogen stimulates the growth of the endometrium of the uterus (lowest panel). Third, estrogen stimulates the hypothalamus to produce more GnRH (see top panel).

④ The GnRH stimulates a surge of LH (and a smaller increase in FSH) at about the 14th day of the cycle. The surge of LH has three important consequences. *First*, it triggers the resumption of meiosis I in the oocyte, producing the secondary oocyte and the first polar body.

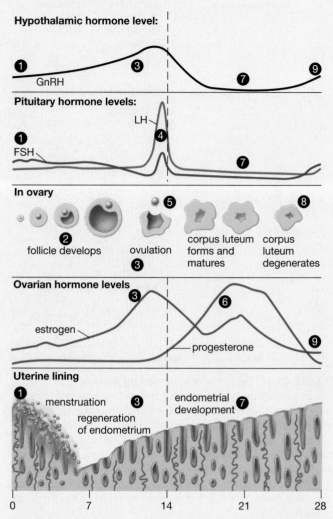

FIGURE E40-1 Hormonal control of the menstrual cycle The menstrual cycle is generated by interactions among the hormones of the hypothalamus, the anterior pituitary, and the ovaries. The circled numbers correspond to those in the text.

swell, they squeeze off the veins that drain the penis (**FIG. 40-19b**). Blood pressure increases, causing an erection. After the penis is inserted into the vagina, movements further stimulate touch receptors on the penis, triggering ejaculation. *Ejaculation* occurs when muscles encircling the epididymis, vas deferens, and urethra contract, forcing semen out through the penis and into the vagina. Although there is considerable normal variability, on average 3 or 4 milliliters of semen containing roughly 300 million sperm is ejaculated. *Male orgasm* causes ejaculation and feelings of intense pleasure and release of tension.

In the female, sexual arousal causes increased blood flow to the vagina, to the outer paired folds of tissue called the **labia** (singular, *labium*), and to the **clitoris**, a small structure just in front of the vagina (see Fig. 40-16). The clitoris (derived from the same embryological tissue as the tip of the penis) becomes engorged with blood. Stimulation by

the penis may result in *female orgasm*, a series of rhythmic contractions of the vagina and uterus accompanied by intensely pleasurable sensations. Female orgasm is not necessary for fertilization.

The intimate contact involved in copulation creates a situation in which disease-causing organisms can readily be transmitted, as described in "Health Watch: Sexually Transmitted Diseases."

During Fertilization, the Sperm and Egg Nuclei Unite

Sperm and eggs live for only a few days, so fertilization can succeed only if copulation occurs within a couple of days before or after ovulation. When leaving the ovary, the egg is surrounded by follicle cells. These cells, now called the **corona radiata**, and an inner jellylike layer, the **zona pellucida** ("clear area"), form a barrier between the sperm and the egg (**FIG. 40-20a**). Recent research supports the hypothesis that the

⑤ *Second*, the LH surge causes the final explosive growth of the follicle, culminating in ovulation, and *third*, it transforms the remnants of the follicle into the corpus luteum.

⑥ The corpus luteum secretes progesterone (green line) and estrogen (purple line).

⑦ Estrogen and progesterone together inhibit GnRH production and reduce FSH and LH, preventing the development of more follicles. Simultaneously, estrogen and progesterone stimulate the endometrium to develop a network of blood vessels and nutrient-producing glands. The endometrium eventually becomes about 4 millimeters thick.

⑧ If pregnancy does not occur, the corpus luteum starts to disintegrate about 12 days after ovulation. This disintegration is caused by the corpus luteum itself, which secretes progesterone that shuts down LH secretion. Because the corpus luteum can persist only while it is stimulated by LH (or by a similar hormone released by the developing embryo, as described later), it induces its own destruction, a form of negative feedback.

⑨ With the corpus luteum gone, estrogen and progesterone levels plummet. Deprived of stimulation by estrogen and progesterone, the endometrium of the uterus dies within few days; its blood and tissue form the menstrual flow that defines the first day of the new cycle. The reduced level of circulating progesterone no longer inhibits the hypothalamus, so the spontaneous release of GnRH resumes. GnRH stimulates the release of FSH and LH (cycling back to step ①), initiating the development of a new set of follicles and restarting the cycle.

During pregnancy, the embryo itself prevents these changes from occurring. Shortly after the ball of cells formed by the dividing fertilized egg embeds itself in the endometrium, it starts secreting an LH-like hormone called *chorionic gonadotropin (CG)*. This hormone travels in the bloodstream to the ovary, where it prevents the breakdown of the corpus luteum. The corpus luteum continues to secrete estrogen and progesterone for a few months, and the uterine lining continues to grow, nourishing the embryo. The embryo releases so much CG that the hormone is excreted in the mother's urine; most pregnancy tests use the presence of CG in urine to determine pregnancy.

Although negative feedback regulates the levels of most hormones, the hormones of the menstrual cycle are regulated by both positive and negative feedback. During the first half of the cycle, FSH and LH stimulate estrogen production by the follicles. High levels of estrogen then *stimulate* the mid-cycle surge of FSH and LH release (positive feedback). During the second half of the cycle, estrogen and progesterone together *inhibit* the release of FSH and LH (negative feedback). The early positive feedback causes hormone concentrations to reach high levels; later, negative feedback shuts the system down again unless pregnancy intervenes.

(a)

(b)

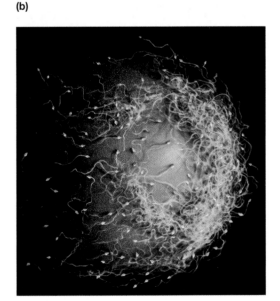

FIGURE 40-20 The secondary oocyte and fertilization
(a) A human secondary oocyte (egg) shortly after ovulation. Sperm must digest their way through the corona radiata and the zona pellucida to reach the oocyte. **(b)** Sperm surround the oocyte, attacking its defensive barriers. QUESTION Why is the oocyte so well protected by surrounding barriers?

HEALTH WATCH　　Sexually Transmitted Diseases

Sexually transmitted diseases (STDs) are transmitted primarily through sexual contact. Caused by viruses, bacteria, protists, or arthropods that infect the sexual organs and reproductive tract, STDs are a serious and growing health problem worldwide.

BACTERIAL INFECTIONS

Gonorrhea is a common STD often called "the clap." The bacteria can penetrate membranes lining the urethra, anus, cervix, uterus, uterine tubes, and throat. Males may experience painful urination and discharge of pus from the penis; female symptoms are often mild and include a vaginal discharge or painful urination. Although gonorrhea can be treated with antibiotics, because many infected individuals experience few or no symptoms, they go untreated and readily spread the disease. Gonorrhea can lead to infertility by blocking the uterine tubes with scar tissue. The bacterium attacks the eyes of newborns of infected mothers and was once a major cause of blindness. Today, most newborns are immediately given antibiotic eyedrops to prevent this.

Syphilis bacteria penetrate the mucous membranes of the genitals, lips, anus, or breasts. Because they don't survive prolonged exposure to air, they are spread almost entirely by intimate contact. Syphilis begins with a sore at the site of infection and can be cured with antibiotics. If untreated, syphilis bacteria spread through the body, damaging many organs including the skin, kidneys, heart, and brain, in some cases with fatal results. Syphilis can be transmitted to the fetus during pregnancy; and the skin, teeth, bones, liver, and central nervous system of the infant may be damaged.

Chlamydia causes inflammation of the urethra in males and of the urethra and cervix in females. In many cases, there are no obvious symptoms, so the infection goes untreated and spreads. The chlamydia bacterium can infect and block the uterine tubes, resulting in sterility. Chlamydial infection can cause eye inflammation in infants born to infected mothers and is a major cause of blindness in developing countries.

VIRAL INFECTIONS

Acquired immune deficiency syndrome, or **AIDS**, is caused by the human immunodeficiency virus (HIV) and was discussed in Chapter 36. It is spread primarily by sexual activity, contaminated blood and needles, and from mother to newborn. HIV attacks the immune system, leaving the victim vulnerable to a variety of infections. There is no cure, but drug combinations can prolong life considerably.

Genital herpes causes painful blisters on the genitals and surrounding skin and is transmitted primarily when blisters are present. Herpes remains in the body, emerging unpredictably, possibly in response to stress. Antiviral drugs can reduce the severity of outbreaks. A pregnant woman with an active case of genital herpes can transmit the virus to the developing fetus, in very rare cases causing mental or physical disability or stillbirth. Herpes can also be transmitted during childbirth.

Human papillomavirus (HPV) infects an estimated 50% of sexually active individuals at some time in their lives. Most show no symptoms and recover from the infection without knowing they had it. The virus may cause warts on the labia, vagina, cervix, or anus in females and on the penis, scrotum, groin, or thighs in males. The warts usually disappear, or they can be removed. HPV is of concern because it can cause cervical cancer, which kills nearly 4000 women each year in the U.S. In 2006, the U.S. FDA approved a vaccine against the forms of HPV that cause most cases of genital warts and cervical cancers. If administered to young women before they become sexually active, the vaccine could greatly reduce cervical cancer rates in the future.

PROTIST AND ARTHROPOD INFECTIONS

Trichomoniasis is caused by a protist that colonizes the mucous membranes lining the urinary tract and genitals of both males and females. Symptoms include a discharge caused by inflammation in response to the parasite. The protist is spread by intercourse but can also be acquired from contaminated clothing and toilet articles. Lengthy untreated infections can result in sterility.

Crab lice, also called *pubic lice*, are tiny arachnids (relatives of spiders) that live and lay their eggs in pubic hair. Their mouthparts are adapted for penetrating skin and sucking blood and body fluids, a process that causes severe itching. "Crabs" are not only irritating; they can also spread infectious diseases. They can be controlled through careful hygiene and chemical treatments.

---

human egg releases a chemical attractant that lures the sperm toward it.

In the uterine tube, hundreds of sperm reach the egg and encircle the corona radiata, each sperm releasing enzymes from its acrosome (**FIG. 40-20b**). These enzymes weaken both the corona radiata and the zona pellucida, allowing the sperm to wriggle through to the egg. If there aren't enough sperm, an insufficient amount of enzyme is released, and none of the sperm will reach the egg. This may be the reason that natural selection has led to the ejaculation of so many sperm. Perhaps 1 in 100,000 reach the uterine tube, and 1 in 20 of those encounters the egg, so

only a few hundred of the 300 million sperm that were ejaculated join in attacking the barriers around the egg.

When the first sperm finally contacts the egg's surface, the plasma membranes of egg and sperm fuse, and the sperm's head is drawn into the egg cytoplasm. As the sperm enters, it triggers two critical changes: first, vesicles near the surface of the egg release chemicals into the zona pellucida that reinforce it and prevent other sperm from entering. Second, the egg undergoes its second meiotic division, finally producing a haploid gamete. Fertilization occurs as the haploid nuclei of sperm and egg fuse, forming a diploid nucleus that contains all the genes of a new human being.

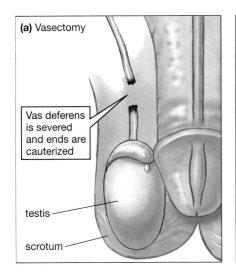

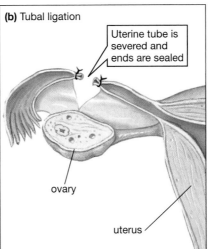

FIGURE 40-21 Sterilization
(a) Vasectomy involves removing a short segment of the vas deferens and cauterizing the cut ends, blocking the transmission of sperm. **(b)** Tubal ligation involves removing a short segment of the uterine tube and blocking off the cut ends, preventing sperm from reaching the oocyte and the oocyte from reaching the uterus.

Defects in the male or female reproductive system can prevent fertilization. For example, a blocked uterine tube can prevent sperm from reaching the egg. A man with a low sperm count (fewer than 20 million sperm per milliliter of semen) may be unable to impregnate a woman because too few sperm reach the egg. If the sperm are otherwise healthy, he can father children by *artificial insemination*, in which a large quantity of his sperm is injected directly into the vagina or uterus at the time of ovulation. Today, some couples seek high-technology help in the form of *in vitro* fertilization (see "Health Watch: High-Tech Reproduction").

40.3 HOW CAN PEOPLE LIMIT FERTILITY?

During most of human evolution, child mortality was high, and natural selection favored people who produced enough children to offset this high mortality rate. Today, although most people do not need to have many children to ensure that a few will reach adulthood, we still retain these reproductive drives. As a result, about 74 million new people are added to our overcrowded planet every year. Controlling birth rates has become an environmental necessity. On the individual level, birth control allows people to plan their families and provide the best opportunities for themselves and their children.

Historically, limiting fertility has not been easy. In the past, women in some cultures have tried such inventive, if bizarre, techniques as swallowing froth from the mouth of a camel or placing crocodile dung in the vagina. Since the 1970s, however, several effective techniques have been developed for **contraception**, the prevention of pregnancy. All forms of birth control have possible drawbacks, and hormonal forms have a range of potential side effects. The choice of a contraceptive should always be made in consultation with a health professional who can provide more complete information and advice.

Permanent Contraception Can Be Achieved Through Sterilization

In the long run, the most effortless method of contraception is **sterilization**, in which the pathways through which sperm or eggs must travel are interrupted (**FIG. 40-21**). In men, the vas deferens leading from each testis can be severed and the ends cauterized (heat-sealed) in an operation called a *vasectomy*. Sperm are still produced, but they cannot reach the penis during ejaculation. The surgery is performed under a local anesthetic, in some cases requires no stitches, and has no known effects on health or sexual performance. In a relatively new procedure, the vas deferens may be clamped shut using a small plastic clip.

The slightly more complex operation of *tubal ligation* renders a woman infertile by cutting and sealing off her uterine tubes. Ovulation still occurs, but sperm cannot travel to the egg, nor can the egg reach the uterus. An alternative consists of tiny springlike structures inserted into each uterine tube through the vagina and uterus. The procedure requires no incisions and only local anesthesia. The coil causes the uterine tube to form scar tissue that blocks passage of both sperm and eggs. Although sterilization is generally permanent, sometimes, in a delicate and expensive operation, a surgeon can reconnect the vas deferens or uterine tubes.

Temporary Contraception and Abortion Prevent or Terminate Pregnancy

Most temporary contraception techniques prevent ovulation or create a barrier between sperm and eggs. Table 40-3 summarizes these techniques.

As you now know, ovulation is triggered by a mid-cycle surge of LH. An obvious way to prevent ovulation is to suppress LH release by providing a continuous supply of estrogen and progesterone. This is the basis for birth control pills. Several other delivery systems for estrogen and progesterone (generally in synthetic form) are now available (see Table 40-3).

Table 40-3 Temporary Contraceptive Techniques

Method	Technique and Mechanism	Failure Rate[1]	STD Protection
Hormonal Methods: Prevent Ovulation			
Birth control pill	Pill containing either estrogen and synthetic progesterone (combination pill) or progesterone only (minipill). Taken daily.	0.1% to 3%	None
Contraceptive patch	Skin patch containing synthetic estrogen and progesterone. Replaced weekly.	< 1%[2]	None
Birth control shot	Injection of synthetic progesterone that blocks ovulation. Repeated at 3-month intervals.	0.3%	None
Vaginal ring	Flexible plastic ring impregnated with synthetic estrogen and progesterone. Inserted into vagina around the cervix, replaced every 4 weeks.	0.3% to 8%	None
Barrier Methods: Prevent Sperm and Egg from Meeting			
Abstinence	Deciding not to be sexually active.	0%	Excellent
Condom (male)	Thin latex sheath placed over penis just before intercourse. Prevents sperm from entering vagina. More effective with spermicide.	3% to 15%	Good
Condom (female)	Lubricated polyurethane pouch inserted into vagina; prevents sperm from entering cervix. More effective with spermicide.	5% to 21%	Probably good (little data available)
Sponge	Domed disposable sponge impregnated with spermicide. Inserted in vagina; works for 24 hours.	9% to 20% (failure rates double after giving birth)	Poor
Diaphragm/Cervical cap	Reusable, flexible, domed rubberlike barriers. Spermicide is placed within the dome, and device is fitted over the cervix before intercourse.	6% to 14%	Poor
Spermicide	Sperm-killing foam is placed in vagina prior to intercourse, forming a chemical barrier to sperm.	6% to 26%	Poor
Rhythm	Measuring body temperature and cervical mucus changes to estimate the time of ovulation and avoiding intercourse during the fertile period.	2% to 20% (rarely performed correctly)	None
Multiple Mechanisms of Action			
IUD (intrauterine device)[3]	Small plastic device treated with hormones or copper and inserted through the cervix into the uterus.	0.6% to 2%	None
"Morning after" pill (emergency contraception)[3]	Concentrated dose of the hormones in birth control pills, taken within 72 hours after intercourse.	25%	None

[1]Percentage of women becoming pregnant per year. The low and high numbers, respectively, indicate the differences between consistent, correct use and use in a more typical way that is not always consistent or correct.

[2]The patch is as effective as the pill, and more likely to be used properly; however, for women weighing more than 198 pounds, it becomes less effective.

[3]Although preventing fertilization seems to be the main mechanism, scientists cannot rule out that in some cases these last two methods may prevent implantation after fertilization.

Barrier methods are more effective when used with a *spermicide* (sperm-killing substance). The cervical cap and diaphragm block the opening of the cervix, preventing entry of sperm. Both male and female condoms, which also help protect against sexually transmitted diseases, prevent sperm from being deposited in the vagina. Less-reliable techniques include the use of spermicides alone and the *rhythm method* (abstinence from intercourse during ovulation). The rhythm method has a high failure rate because of inaccuracies in determining the menstrual cycle, which usually varies somewhat from month to month. Withdraw-al (removing the penis from the vagina before ejaculation) and douching (attempting to wash sperm out of the vagina before they have entered the uterus) are not reliable ways to avoid conception.

Contraception can also be achieved with an *intra-uterine device (IUD)*, a small T-shaped device inserted through the cervix and into the uterus by a physician. Research indicates that the main way in which IUDs work is by preventing fertilization. Copper or hormonal coatings on different IUDs and the reaction of the uterus to this foreign object create an environment that is hos-

HEALTH WATCH   High-Tech Reproduction

BIOETHICS

Although some assisted reproductive technologies may be used to save some animal species from extinction, most are used to help infertile couples have children. For women who do not ovulate regularly, fertility drugs (which cause release of extra FSH and LH) result in multiple ovulations. As a result of this procedure, the rate of multiple births, which pose far greater risks for the mother and babies, has dramatically increased (**FIG. E40-2**).

Through *intracytoplasmic sperm injection* (ICSI), even men whose sperm are unable to swim to and fertilize an egg may be able to father their own children. In ICSI, immature sperm cells are extracted from the testes and injected through a tiny needle directly into the egg's cytoplasm (**FIG. E40-3**).

Worldwide, well over 3 million people are now alive who were conceived in a shallow glass dish by *in vitro fertilization* (IVF; literally, "fertilization in glass"). A woman undergoing IVF is given drugs that stimulate multiple ovulations. Surgeons then insert a long needle into each ripe follicle and suck out the oocyte. Many oocytes are placed in a dish with freshly collected sperm, where they are incubated for one to seven days. A few healthy early embryos are then gently sucked into a tube and expelled into the uterus. Transplanting more than one embryo increases the success rate, but it also increases the probability of multiple births, which are much riskier than single births. For couples who carry serious genetic disorders, one cell of an IVF embryo may be removed and studied for defects in DNA (**FIG. E40-4**) before the embryo is implanted.

Using sperm-sorting technology, parents can now increase their odds of having a male or female child. Sperm carrying an X chromosome have 2.8% more DNA than sperm

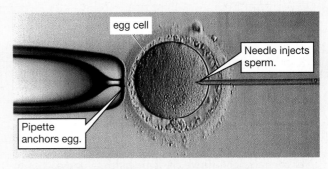

FIGURE E40-3 Injecting a sperm into an egg
An egg, held in place with a pipette, is injected with a single sperm cell that is placed directly into the egg's cytoplasm. Notice the corona radiata surrounding the egg. **QUESTION** Why is the sperm injected into the egg, instead of merely being placed in contact with it?

carrying a Y chromosome. This difference can be used as a basis for sorting a sperm sample and increasing the percentage of X or Y sperm, which are then placed directly into the woman's uterus. This can be important if the parents are carriers of sex-linked disorders, but some use sperm sorting in an attempt to balance their families. Enriching sperm samples in favor of X sperm has been most successful.

In the world of assisted reproduction, a widow could be impregnated by her dead husband's cryopreserved sperm. For example, one woman whose ovaries could not produce eggs gave birth to twins from donor eggs that had been cryopreserved for two years. A surrogate mother may bear a child for a woman who has had a hysterectomy or who simply does not want to go through pregnancy. The egg and sperm that produced the fetus within the surrogate mother could come from the couple who hired her; alternatively, they could come from unrelated individuals. Conceivably, a modern newborn could have up to five "parents"!

FIGURE E40-2 Septuplets

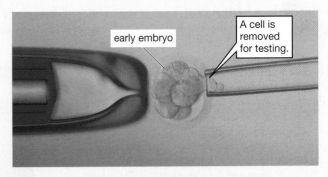

FIGURE E40-4 Removing a cell for genetic testing

tile to sperm and interferes with their progress into the uterine tube.

The "morning after" pill contains hormones similar to birth control pills but in larger dosages. This form of *emergency contraception* is most effective if taken within 72 hours after intercourse. These pills may work in several ways: by delaying or preventing ovulation, interfering with corpus luteum formation, interfering with the progress of sperm or egg through the uterine tube, or preventing implantation after fertilization.

SCIENTIFIC INQUIRY   Seeking a Male Contraceptive

The only temporary methods of birth control available to men are the condom and abstinence. One reason is that men produce about 100 million sperm daily, and even blocking 95% of these leaves enough to cause an unintended pregnancy. Still, research on male contraception has lagged behind that for women, partly because major drug companies believed that the market was too small to justify the enormous expense of researching and bringing these new drugs to market. But recent polls worldwide have shown that men are willing and even eager to shoulder far more responsibility for contraception. There are three different approaches to temporary male contraception under development.

Blocking the vas deferens: Tens of thousands of men in China already use silicone plugs that are injected into the vas deferens, blocking release of sperm. In India, a substance (RISUG) that partially blocks the vas deferens and damages the sperm that make it through is in advanced human clinical trials. Silicone plugs can be surgically removed, and RISUG can be dissolved by a solution injected into the vas deferens.

Hormonal methods: Administering testosterone prevents sperm formation by blocking both LH and FSH release through negative feedback (see Fig. 40-15). A promising male hormonal contraceptive (now in clinical trials) combines testosterone injections every 4–6 weeks with a synthetic progesterone implant that further suppresses FSH and LH.

Non-hormonal drugs: Animal testing is underway for a drug that blocks a protein in the epididymis that normally turns on the ability of sperm tails to swim. With their tails immobilized, sperm would be unable to swim to the egg. A second approach is a "vaccine" that causes the body to produce antibodies to a protein (*eppin*) that is crucial to producing functional sperm. In monkeys, this vaccine caused temporary sterility that could be maintained by periodic booster shots.

Although they won't be available in the U.S. for many years, these drugs and others under development promise to expand future contraceptive options for men.

Abortion Removes the Embryo from the Uterus

Abortion is not a form of contraception, because it terminates rather than prevents pregnancy. It generally involves dilating the cervix and removing the embryo by suction. Most abortions are performed during the first 3 months of pregnancy. Alternatively, abortion can be induced during the first 7 weeks of pregnancy by the drug RU-486 (mifepristone), which binds to progesterone receptors and blocks the actions of progesterone, which is essential to maintaining the uterine lining during pregnancy.

You may have noticed how birth control techniques are mainly designed for women. Why? Are male contraceptives under development? See "Scientific Inquiry: Seeking a Male Contraceptive."

CASE STUDY REVISITED   THE FROZEN ZOO

Although it is intended to save endangered species, assisted reproductive technology (ART) does not appeal to all conservationists. Some contend that the only appropriate way to preserve a species is to maintain enough natural habitat to support a breeding population large enough to sustain itself and maintain reasonable genetic diversity. Proponents of ART agree, but support high-tech efforts as a necessary adjunct to habitat preservation, particularly for critically endangered animals. Presiding over the frozen zoo in San Diego, geneticist Dr. Oliver Ryder explains: "[The frozen zoo] represents a genetic legacy— a DNA bank. In the future, scientists may have better tools, but they won't have access to more genes." Dr. Betsy Dresser, who heads the Audubon Center for Research of Endangered Species in New Orleans, describes ART as "a safety net." "If you freeze 200 or 300 embryos, that's enough to keep a population from going extinct." Dresser, who is working to develop interspecies embryo transfer techniques that will allow lions to serve as surrogate mothers for the endangered

FIGURE 40-22 Test-tube tiger

Siberian tiger (**FIG. 40-22**), adds: "I don't want to see tigers just in textbooks someday. Nor do I want people a hundred years from now to look back and say, 'God, they had this technology and they just let these animals go extinct.'" Advocates of ART look forward to a future when natural habitat is restored and protected so that populations of critically en-dangered species that have been bred in zoos (maintaining as much genetic diversity as possible) can be released to flourish and reproduce naturally in their native environments.

Consider This Frozen tissue from the world's last remaining bucardo and po'ouli provide the only hope that Earth may again harbor these unique species. But only cloning will produce a new bucardo or po'ouli. The animals produced by cloning will be genetically identical and may suffer from other problems that to date have plagued cloned animals (see Chapter 11). Should scientists put the money and effort into trying to resurrect these species? Defend your answer.

CHAPTER REVIEW

SUMMARY OF KEY CONCEPTS

40.1   How Do Animals Reproduce?

Animals reproduce either sexually or asexually. In sexual reproduction, haploid gametes, usually from two separate parents, unite and produce an offspring that is genetically different from either parent. Asexual reproduction, by budding, fission, or parthenogenesis, produces offspring that are genetically identical to the parent.

During sexual reproduction, the male gamete (a small, motile sperm) fertilizes a female gamete (a large, nonmotile egg). Some species are hermaphroditic, producing both sperm and eggs, but most have separate sexes. Fertilization can occur outside the bodies of the animals (external fertilization) or inside the body of the female (internal fertilization). External fertilization must occur in water so that the sperm can swim to meet the egg. Internal fertilization generally occurs by copulation, in which the male deposits sperm directly into the female's reproductive tract.

40.2   How Does the Human Reproductive System Work?

The human male reproductive tract consists of paired testes, which produce sperm and testosterone, and accessory structures that conduct the sperm to the female's reproductive tract and secrete fluids that activate swimming by the sperm and provide energy. In human males, spermatogenesis and testosterone production are stimulated by FSH and LH, secreted by the anterior pituitary. Spermatogenesis and testosterone production are nearly continuous, beginning at puberty and lasting until death.

The human female reproductive tract consists of paired ovaries, which produce eggs as well as the hormones estrogen and progesterone, and accessory structures that conduct sperm to the egg and receive and nourish the embryo during prenatal development. In human females, oogenesis, hormone production, and development of the lining of the uterus repeat in a monthly menstrual cycle. The cycle is controlled by hormones from the hypothalamus (GnRH), anterior pituitary (FSH and LH), and ovaries (estrogen and progesterone).

During copulation, the male ejaculates semen into the female's vagina. The sperm swim through the vagina and uterus into the uterine tube, where fertilization usually takes place. The unfertilized egg is surrounded by two barriers, the corona radiata and the zona pellucida. Enzymes released from the acrosomes in the heads of sperm digest these layers, permitting sperm to reach the egg. Only one sperm enters the egg and fertilizes it.

The ability to reproduce begins in puberty, when hypothalamic GnRH causes release of FSH and LH from the anterior pituitary. These, in turn, stimulate the sex glands to produce testosterone (male) and estrogen (female). These induce secondary sexual characteristics and the development of sperm and eggs.

Web Tutorial 40.1 The Male Reproductive System

Web Tutorial 40.2 The Female Reproductive System

40.3   How Can People Limit Fertility?

Contraception can be achieved by abstinence or by sterilization: severing the vas deferens in males (vasectomy) or the uterine tubes in females (tubal ligation). The uterine tubes may also be blocked by inserting a springlike device that causes scar tissue to form. Temporary contraception techniques include those that prevent ovulation by delivering estrogen and progesterone—for example, the pill, the contraceptive patch, the vaginal ring, and hormone injections. Barrier methods, which prevent sperm and egg from meeting, include the diaphragm, the cervical cap, the sponge, and the condom, accompanied by spermicide. Spermicide alone is less effective, while withdrawal and douching are unreliable. The rhythm method, which has a high failure rate, requires abstinence around the time of ovulation. Intrauterine devices primarily prevent sperm from reaching the egg. Emergency contraception (the "morning-after pill") has several possible mechanisms of action. Abortion causes expulsion of the developing embryo.

KEY TERMS

acquired immune deficiency syndrome (AIDS)   *page 828*
acrosome   *page 821*
asexual reproduction   *page 816*
budding   *page 816*

bulbourethral gland   *page 822*
cervix   *page 825*
chlamydia   *page 828*
clitoris   *page 826*
contraception   *page 829*

copulation   *page 819*
corona radiata   *page 826*
corpus luteum   *page 824*
crab lice   *page 828*
egg   *page 817*
embryo   *page 823*

endometrium   *page 824*
epididymis   *page 822*
estrogen   *page 820*
external fertilization   *page 817*
fertilization   *page 816*
fetus   *page 824*

THINKING THROUGH THE CONCEPTS

1. List the advantages and disadvantages of asexual reproduction, sexual reproduction, external fertilization, and internal fertilization, including an example of an animal that uses each type.

2. Compare the structures of the egg and sperm. What structural modifications do sperm have that facilitate movement, energy use, and gaining access to the egg?

3. What is the role of the corpus luteum in a menstrual cycle? In early pregnancy? What maintains its survival after ovulation?

4. Construct a chart of common sexually transmitted diseases. List the disease's name, cause (organism or virus), symptoms, and treatment.

5. List the structures, in order, through which a sperm passes, starting with the seminiferous tubules of the testis and ending in the uterine tube of the female.

6. Name the three accessory glands of the male reproductive tract. What are the functions of the secretions they produce?

7. Diagram the menstrual cycle, and describe the interactions among hormones secreted by the hypothalamus, pituitary gland, and ovaries that produce the cycle.

APPLYING THE CONCEPTS

1. Discuss the most effective or appropriate method of birth control for each of the following couples: Couple A, who have intercourse three times a week but never want to have children; Couple B, who have intercourse once a month and may want to have children someday; and Couple C, who have intercourse three times a week and want to have children someday.

2. Would a hypothetical contraceptive drug that blocked receptors for FSH and LH be useful in males? How would it work? What side effects might it have?

3. Think of all the ways a couple can obtain a child, including *in vitro* fertilization using the couple's eggs and sperm, *in vitro* fertilization using a donor's sperm or egg, and insemination of a surrogate mother with sperm from the couple's husband. Imagine some more. Do these options present ethical issues for you? What legal issues might possibly arise? What medical issues?

4. Fertility drugs have greatly increased the incidence of multiple births. When more than two embryos share the uterus, the incidence of premature birth and developmental problems increase substantially. The costs of caring for multiple premature infants are staggering. When fertility drugs produce multiple embryos, the physician can selectively eliminate some of these embryos early in development, so the remaining few have a better chance to develop fully and normally. Discuss the ethical implications of taking fertility drugs while taking these issues into consideration.

FOR MORE INFORMATION

Estabrook, B. "Staying Alive." *Wildlife Conservation*, June 2002. Assisted reproductive technology offers hope for saving endangered species.

Khamsi, R. "Sperm bounce Back After Male Contraception." *New Scientist*, April 28, 2006. Clinical trials show rapid recovery of sperm production after men stop taking hormonal contraceptives.

Kingsland, J. "Sperm Warfare." *New Scientist*, January 10, 2004. Male contraceptive research targets sperm.

Lanza, R. P., Dresser, B. L., and Damiani, P. "Cloning Noah's Ark." *Scientific American*, November 2000. For some endangered species, cloning may offer the best chance for survival.

Milius, S. "Battle of the Hermaphrodites." *Science News*, September 16, 2006. Two sexes in one body make for some interesting reproductive behaviors.

Ness, E. "How to Breed a 2,000-pound Rhino." *Discover*, November 2001. The endangered Sumatran rhino has given birth in captivity for the first time in over a century, and none too soon—only about 300 survive in the wild.

Ojcius, D. M., Darville, T., and Bavoil, P. M. "Can Chlamydia Be Stopped?" *Scientific American*, May 2005. Chlamydia is the leading cause of preventable blindness worldwide; new developments could help control it.

Riddle, J. M., and Estes, J. W. "Oral Contraceptives in Ancient and Medieval Times." *American Scientist*, May–June 1992. How did women control their fertility before modern medicine stepped in?

Whelan, J. "Reproduction Revolution: Sex for Fun, IVF for Children." *New Scientist*, October 20, 2006. This article explores high-tech reproductive options.

Wright, K. "Male Contraception." *Discover*, October 2002. The author explores challenges and progress in developing a male contraceptive.

41 Animal Development

John, who struggles with the effects of FAS as an adult, has assisted his adoptive mother Teresa Kellerman in educating people about the dangers of drinking during pregnancy. **(Inset)** Debbie now regrets drinking while she was pregnant with her daughter Sabrina, whose nervous system was damaged by alcohol.

(Photos courtesy of Tucson Citizen copyright 2001, and Teresa Kellerman).

CASE STUDY FACES OF FAS

"THE GUILT IS TREMENDOUS . . . I did it again and again. . . . I don't know how to tell them. It was something I could have prevented." Debbie, the young mother pictured here, has had seven children. Her daughter Cory has been diagnosed with **fetal alcohol syndrome (FAS)**, the most serious type of alcohol damage. At age three Cory was hyperactive and talked like a one-year-old. Doctors believe that Debbie's most recent child, Sabrina (inset), is almost certainly a victim as well. Her face bears the characteristic features of FAS. At seven months, she was weak, began having seizures, and could not eat solid food because she was unable to close her upper lip around the spoon. As this pregnant mother repeatedly got drunk, so did her developing

children. John (chapter opening photo) is a young adult with FAS; his mother drank while pregnant with him, and she was drunk when she delivered him.

The damage to children born to mothers who drink is irreparable. John was fortunate to have been adopted by a truly remarkable woman, Teresa Kellerman, who has become a leading parent advocate for alcohol-damaged children and their families. "Without intervention they end up homeless, jobless, addicted, arrested, pregnant or getting someone pregnant, living on the streets or dead, so many of them," says Kellerman, who founded the Fetal Alcohol Syndrome Community Resource Center in Tucson, Arizona. Sabrina's mother entered rehabilitation and has resolved to stay sober and

be a good parent to her many children. But even with the best parenting, neither John, Sabrina, nor the thousands of other children born each year with full-blown FAS are likely to ever live without supervision. Tens of thousands of others with milder alcohol damage may become functional but will never reach their full potential.

How does alcohol reach the developing child when a pregnant woman drinks? What features do doctors look for in diagnosing FAS? Is there any amount of alcohol that can be safely consumed during pregnancy? Is there a period of fetal development during which a pregnant woman can safely drink?

Developmental biologists continue to explore the amazing complexities of how a single cell—a zygote formed from the fusion of sperm and egg—transforms itself into a complex organism. Because the cells of the embryo proliferate by mitosis, each cell has an identical genome. What chemical commands transform genetically uniform cells into the different components of bones, blood, and the brain? As scientists learn more, optimism grows that we might harness the ability to direct cellular differentiation, eventually developing techniques to replace damaged cells in sick or disabled individuals. Here we explore the types and stages of animal development, a little of what is known about cell differentiation, and ways in which foreign substances can interfere with this delicate process.

41.1   HOW DO INDIRECT AND DIRECT DEVELOPMENT DIFFER?

When we think of development, images of a newborn infant may come to mind. Certainly, their proportions are different; but babies are, in all important ways, miniature versions of adult humans. People and other mammals—as well as birds and reptiles—are all born as "miniature adults," developing through a process called **direct development**. For the majority of animal species, however, *indirect development* is the norm.

During Indirect Development, Animals Undergo a Radical Change in Body Form

In **indirect development**, the juvenile animal differs significantly from the adult, undergoing radical changes during development, such as the transformation of a caterpillar into a butterfly. Indirect development occurs in most invertebrates, including insects and echinoderms, and amphibian vertebrates. Animals with indirect development typically produce huge numbers of eggs, and each egg has only a small amount of food reserve called **yolk**. The yolk nourishes the developing embryo during its transformation into a small, sexually immature form called a **larva** (**FIG. 41-1**). Because only a small amount of yolk is produced and the offspring usually fend for themselves after hatching, indirect development does not place great demands on the mother. This allows her to produce large numbers of offspring, most of which will not survive to adulthood. Use of this reproductive method is well illustrated by the spawning corals shown in Figure 40-6.

Some larval animals not only look very different from adult animals but also occupy entirely different habitats. In addition, most larvae feed on different organisms than they will as adults. This adaptation eliminates competition between adults and their offspring. For instance, the aquatic larva of the dragonfly feeds on aquatic organisms such as tadpoles; but the adult dragonfly, which is terrestrial, feeds on insects (**FIG. 41-1b**). Eventually, the larvae undergo a revolution in body form, or **metamorphosis**, and become sexually mature adults.

Although we tend to regard the adult form as the "real animal" and the larval stage as "preparatory," most of the life span of some animals, especially insects, is spent in the larval form. Some types of mayfly spend a year or more in an aquatic larval form, then metamorphose and emerge in huge swarms from freshwater streams, ponds,

FIGURE 41-1 Indirect development
(a) Many marine mollusks, such as this common whelk snail, undergo indirect development in which the nearly microscopic larva is very different from the adult in size, appearance, and lifestyle. **(b)** The larval dragonfly is aquatic and may feed on tadpoles (as shown here) and small fish, whereas the adult form is terrestrial and eats other insects. QUESTION List and explain some advantages and disadvantages of indirect development.

FIGURE 41-2 Direct development
The offspring of animals with direct development closely resemble their parents from the moment of birth. All are nurtured either by egg yolk or by nutrients from the mother's blood. **(a)** A male seahorse gives birth to young that have developed from yolk-rich eggs placed in his pouch by the female. **(b)** Lizards hatch from large, yolk-filled eggs. **(c)** Snails hatch from small, yolk-rich eggs. **(d)** Mammalian mothers nourish their developing young within their bodies. QUESTION List and explain some advantages and disadvantages of direct development.

and lakes. Adult mayflies live from a few hours to a few days. They do not feed; their sole occupation is to mate and lay eggs. Each female releases thousands of eggs over water, where they will eventually hatch into larvae, continuing the cycle.

Newborn Animals That Undergo Direct Development Resemble Miniature Adults

Other animals, including such diverse groups as land snails, reptiles, birds, and mammals, undergo direct development, in which the newborn animal is a miniature, but sexually immature, version of the adult (**FIG. 41-2**). As the young animal matures, it may grow much bigger, but it does not radically change its body form.

Juveniles of directly developing species are typically much larger than larvae, so they need much more nourishment before emerging into the world. Two strategies have evolved that meet the embryo's food requirement. Snails, reptiles, and birds produce eggs that contain relatively large amounts of yolk. Mammals, some snakes, and a few fish have relatively little yolk in their eggs; instead, developing embryos are nourished within the mother's body. Providing food for directly developing embryos places great demands on the mother. Many of these offspring, such as those of birds and mammals, require additional care and feeding after birth. This places additional demands on one, and often both, parents. In contrast to indirect development, relatively few offspring are produced, but more of them reach adulthood because the parents devote more resources to each individual.

Table 41-1 Vertebrate Embryonic Membranes

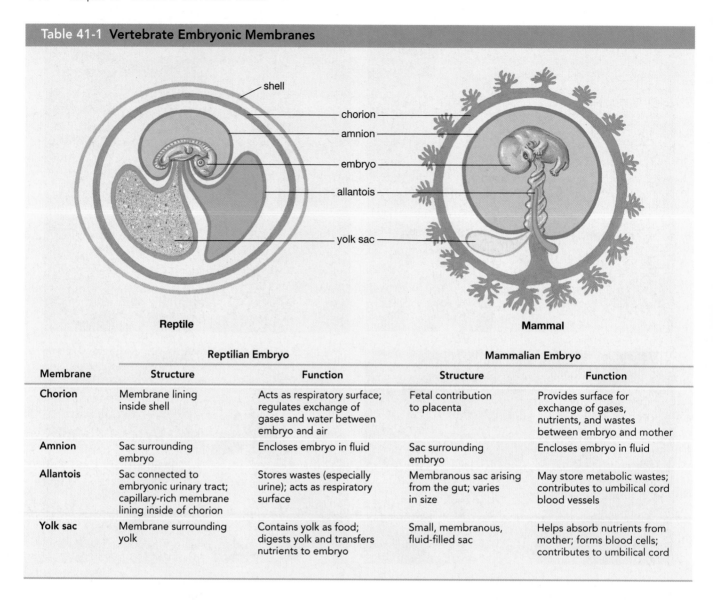

Reptile

Mammal

| Membrane | Reptilian Embryo | | Mammalian Embryo | |
	Structure	Function	Structure	Function
Chorion	Membrane lining inside shell	Acts as respiratory surface; regulates exchange of gases and water between embryo and air	Fetal contribution to placenta	Provides surface for exchange of gases, nutrients, and wastes between embryo and mother
Amnion	Sac surrounding embryo	Encloses embryo in fluid	Sac surrounding embryo	Encloses embryo in fluid
Allantois	Sac connected to embryonic urinary tract; capillary-rich membrane lining inside of chorion	Stores wastes (especially urine); acts as respiratory surface	Membranous sac arising from the gut; varies in size	May store metabolic wastes; contributes to umbilical cord blood vessels
Yolk sac	Membrane surrounding yolk	Contains yolk as food; digests yolk and transfers nutrients to embryo	Small, membranous, fluid-filled sac	Helps absorb nutrients from mother; forms blood cells; contributes to umbilical cord

Reptiles, Birds, and Mammals Produce Similar Extraembryonic Membranes

Amphibians were the first vertebrates to live on land, but their reproduction remains tied to water, where they deposit their eggs and their larval offspring grow and metamorphose into adults. Fully terrestrial vertebrate life was not possible until the evolution of the shelled **amniotic egg**. This innovation, which encases the embryo in a protected, liquid-filled space, arose first in reptiles and persists today in that group and its descendants: birds and mammals. It allows these groups to complete their development into the adult form in their own "private pond." The amniotic egg is characterized by four membranes, called **extraembryonic membranes**: the *chorion*, the *amnion*, the *allantois*, and the *yolk sac*. The **chorion** lines the shell and exchanges oxygen and carbon dioxide between the embryo and the egg's external environment. The **amnion** encloses the embryo in a watery environment; the **allantois** surrounds and isolates wastes; and (in nonmammalian vertebrates) the **yolk sac** contains the stored food, or "egg yolk." Although

eggs of most mammals contain almost no yolk, all four extraembryonic membranes persist, remnants of the reptilian genetic program for development. Table 41-1 compares the structures and functions of these extraembryonic membranes in reptiles and mammals.

41.2 HOW DOES ANIMAL DEVELOPMENT PROCEED?

The transformation from fertilized egg—a single cell—to a multicellular, differentiated embryo is a beautiful process. Actual development is smoothly continuous; the stages depicted are just convenient "snapshots." The initial stages of *cleavage*, *gastrulation*, *organogenesis*, and *growth* occur during embryonic life, in which nearly all the organs are formed. After birth, if the animal survives, it will grow further, achieve sexual maturity and reproduce, age, and finally die. Here, we examine the stages of embryonic development.

ectoderm   mesoderm   endoderm

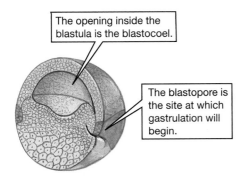

(a) The blastula just before gastrulation.
The three embryonic tissue types have not yet formed. Colors indicate the future fate of the cells after they begin differentiating in the gastrula.

The opening inside the blastula is the blastocoel.

The blastopore is the site at which gastrulation will begin.

(b) Cells migrate at the start of gastrulation.
Cells migrating in will form the endoderm and mesoderm layers of the gastrula; the cells remaining on the surface will form ectoderm.

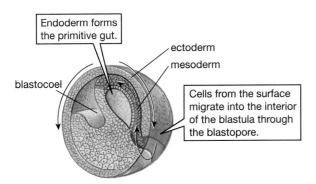

Endoderm forms the primitive gut.

ectoderm

mesoderm

blastocoel

Cells from the surface migrate into the interior of the blastula through the blastopore.

(c) Mesoderm differentiates.

primitive gut

remnant of blastocoel

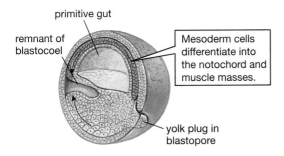

Mesoderm cells differentiate into the notochord and muscle masses.

yolk plug in blastopore

FIGURE 41-3 A frog blastula becomes a gastrula

Cleavage of the Zygote Begins Development

The formation of an embryo begins with **cleavage**, a series of mitotic divisions of the large fertilized egg cell or *zygote*. There is no growth between mitotic divisions, so as cleavage progresses, the available cytoplasm in the large zygote is split up into ever-smaller cells. Finally, a solid ball of small cells, the **morula**, is formed. As cleavage

continues, a cavity opens within the morula, and its cells become the outer covering of a hollow structure called the **blastula**. The space inside the blastula is called the *blastocoel* (**FIG. 41-3a**).

The details of cleavage differ by species. The pattern is largely determined by the amount of yolk present, because yolk hinders *cytokinesis* (cytoplasmic division). The almost yolkless eggs of sea urchins divide symmetrically, but eggs with extremely large yolks, such as a hen's egg, don't divide all the way through. Nevertheless, a hollow blastula is always produced; in reptiles and birds, it is flattened on top of the yolk.

Gastrulation Forms Three Tissue Layers

In the next step of development, an indentation called the **blastopore** forms on one side of the blastula. Blastula cells migrate inward through the blastopore, much as if you punched in an underinflated basketball (**FIG. 41-3b**). These cells form three embryonic tissue layers. The cell migration and differentiation that produces a three-layered embryo is called **gastrulation**, and the embryo that results is the **gastrula** (Table 41-2). Cells of the enlarging dimple, destined to become the digestive tract and associated organs, are now called **endoderm** (Greek for "inner skin"). The cells remaining on the outside, which will form the epidermis of the skin and the nervous system, are called **ectoderm** ("outer skin"). Meanwhile, some cells migrate between the endoderm and ectoderm, forming a third and final layer, the **mesoderm** ("middle skin"). Mesoderm gives rise to muscles, the skeleton (including the **notochord**, a supporting rod found at some stage in all chordates), and the circulatory system (**FIG. 41-3c**).

Adult Structures Develop During Organogenesis

Gradually, the ectoderm, mesoderm, and endoderm rearrange themselves into the organs characteristic of the animal's species in a process called **organogenesis** (see Table 41-2). In some cases, adult structures are, in effect, "sculpted" by the death of excess cells produced during embryonic development. Some cells are programmed to die at precise

Table 41-2 Derivation of Adult Tissues from Embryonic Cell Layers	
Embryonic Layer	**Adult Tissue**
Ectoderm	Epidermis of skin; hair; lining of mouth and nose; glands of skin; nervous system
Mesoderm	Dermis of skin; muscle, skeleton; circulatory system; gonads; kidneys; outer layers of digestive and respiratory tracts
Endoderm	Lining of digestive and respiratory tracts; liver; pancreas

FIGURE 41-4 A bullfrog loses its tail

times during development; this cell death is controlled by at least two mechanisms that function in different tissues. Some cells die during development unless they receive a "survival signal." Embryonic vertebrates, for example, have far more motor neurons for skeletal muscles in their spinal cords than do adults. These neurons survive only if they successfully form synapses on skeletal muscle cells and the extra ones die.

In other cases, embryonic structures form and then disappear because they receive a "death signal" at some stage of development. For example, all vertebrates pass through embryonic stages in which they have tails and webbed hands and feet. In humans, these stages can be seen clearly in the six-week-old human embryo (see Fig. 41-12). Two weeks later, the cells of the webbing are dying, to reveal separate fingers and toes; and the tail regresses as its cells die (see Fig. 41-13). In frogs, the tail is lost during metamorphosis from its tadpole larva. In this case, thyroid hormone, which triggers metamorphosis, also stimulates cells in the tail to produce enzymes that completely digest the tail (**FIG. 41-4**).

41.3   HOW IS DEVELOPMENT CONTROLLED?

Think for a moment about the biological miracle that transformed a single cell—a zygote—into the individual that is you. Biologists use dry terms for this incredible series of events. **Development** is the process by which an organism proceeds from fertilized egg through adulthood. **Differentiation** is the specialization of embryonic cells into different cell types, such as muscle cells, brain cells, and so on. How do cells differentiate from one another during development? We know that the zygote contains all the genes needed to direct the construction of the entire organism. Are any of these genes lost as cells differentiate?

Each Cell Contains the Entire Genetic Blueprint for the Organism

In the early 1950s, American embryologists Thomas King and Robert Briggs began pioneering experiments that were later continued by British embryologist John Gurdon. They transplanted the nucleus of a differentiated cell from the intestine of a tadpole into an unfertilized frog egg whose nucleus had been destroyed (**FIG. 41-5**). The intestinal nucleus directed the egg's development into a normal

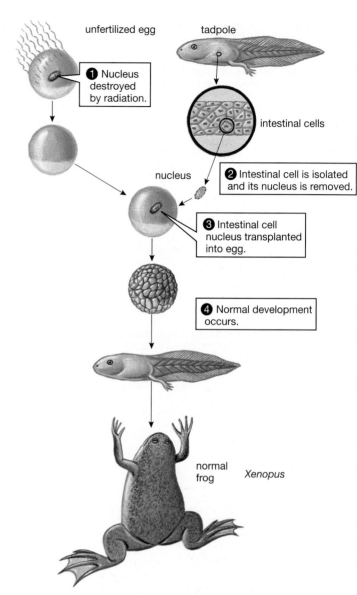

FIGURE 41-5 Cells retain all of their genes during differentiation
Researchers destroyed the nucleus of an unfertilized frog egg before transplanting the nucleus of a tadpole's intestinal cell into the egg. The resulting egg developed into a normal frog, demonstrating that intestinal cells retain all the genes necessary for the development of an entire organism. QUESTION In this experiment, would transplantation of a nucleus from any cell of an adult frog lead to normal development?

tadpole, a feat that would have been impossible if genes were lost during differentiation. The experiments supported the hypothesis that each differentiated cell in an animal contains its entire genetic blueprint. The knowledge that all cells retain the genes to produce an entire adult organism is used in stem cell technology, as described in "Scientific Inquiry: The Promise of Stem Cells." We now know that cells from different parts of an organism differ because of which genes they activate, transcribe into messenger RNA, and translate into proteins.

SCIENTIFIC INQUIRY   The Promise of Stem Cells

The ability of a single cell to give rise to the 200 or so different types of cells in an adult organism is one of the wonders of life. Every cell's nucleus contains the entire genetic blueprint for an organism, and whether a cell becomes muscle, bone, or brain is determined by complex factors in the cell's surroundings that determine which genes are active. These factors cause the cell to *differentiate*, that is, to assume a specialized form and function. A **stem cell** has not yet differentiated, continues to divide, and retains the potential to give rise to more than one cell type. People are excited by the medical implications of stem cell technology. Victims of heart attack, stroke, spinal cord injury, and degenerative diseases from arthritis to Parkinson's disease would benefit if their damaged tissues could be regenerated.

Embryonic stem cells (ESC) are derived from the inner cell mass of the blastocyst (see Fig. 41-10), a cluster of about 100 cells. In 1998, Dr. James Thompson and coworkers at the University of Wisconsin first isolated human ESCs, inducing them to grow in culture dishes and then to differentiate into a variety of human tissues, as illustrated in **FIGURE E41-1**. The appeal of ESCs is that they can produce any cell type in the body. However, because the blastocyst is an early human embryo, some U.S. legislators grapple with ethical concerns in funding ESC research.

Recent research has since shown that most adult tissues, including muscle, skin, liver, brain, heart, and blood, contain at least small numbers of stem cells, called **adult stem cells (ASC)**. In fact, stem cells from bone marrow, which produce both red and white blood cells, have been used for decades in transplants to treat diseases such as leukemia. Although scientists once thought that ASCs could differentiate into only a few cell types, researchers have been able to coax them into forming more different varieties than originally believed possible. The recent discovery that the placenta is rich in blood stem cells has generated excitement among researchers, who hope that the cells from this abundant source might be stimulated to form a variety of tissues in addition to blood.

Tissues derived from stem cells of one individual retain genetic markers that would cause the immune system of a different recipient to reject them without immune suppressant drugs. Anticipating new advances in stem cell therapy, some parents are having a sample of their newborn's placenta cryopreserved, so that stem cells with the child's exact genetic makeup will be available to repair damaged tissue later in the child's life. In the future, researchers hope to employ genetic engineering techniques to modify cell-surface proteins so that stem cells raised in culture from one individual could be used to treat others without rejection problems.

Therapeutic cloning, which would involve inserting a cell nucleus from an adult donor who needed tissue repair into an egg whose nucleus had been removed, would create embryonic stem cells that would not be rejected. Because it cannot be ruled out that this process, if sufficiently developed, could be used to produce a human clone of the donor, it remains controversial.

An ideal scenario would be to prod quiescent stem cells already residing in an injured tissue to reproduce. For example, heart muscle stem cells could be stimulated to replace muscle killed by a heart attack. Alternatively, doctors might harvest stem cells from an injured person, taking the cells from a tissue such as bone marrow where they are abundant, then treat the cells with specific differentiating factors and inject them into the damaged body part. Ideally, there they would reproduce and replace the lost tissue.

In addition to their tremendous potential for combating tissue damage and disease, stem cells might one day be cultured in large quantities and used to test new drugs. Medications that damage these embryonic cells would be likely to damage a developing embryo and should not be prescribed for pregnant women. Cultured stem cells could also be used to investigate the incredibly complex processes that control human development.

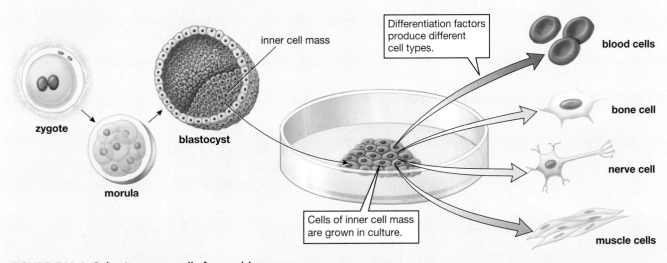

FIGURE E41-1 Culturing stem cells from a blastocyst

Gene Transcription Is Precisely Regulated During Development

How does a cell "decide" to become bone, muscle, or intestine? In any cell at any time, only a portion of the cell's genes are used, or transcribed. You may recall from Chapter 10 that *transcription* is the production of messenger RNA using a gene as a blueprint. The particular combination of genes that is transcribed in a cell largely determines the shape, structure, and biochemical activity of that cell. Differentiation during development is accomplished by a process called **induction**. Induction is the process by which specific cells are stimulated to follow a specific developmental pathway—for example, to become muscle or bone—under the influence of chemical messages produced by other cells. During induction, different sets of genes are selectively activated in different groups of cells, causing them to assume different forms and functions. In general, the molecules that control transcription are proteins (or proteins combined with substances such as steroid hormones) that bind to specific genes and either block or promote transcription.

In many invertebrates, various gene-regulating substances become concentrated in different places in the egg's cytoplasm as it develops. As the fertilized egg divides, each of its daughter cells receives different gene-regulating substances, driving them down different developmental paths. In contrast, cells produced by cleavage of vertebrate zygotes into the blastula stage can each give rise to a complete individual if they are separated, as occurs in the case of human identical twins.

The general developmental fate of most of an embryo's cells becomes sealed during gastrulation. In amphibian embryos, inducing cells form at the site of dimpling as the blastula is transformed into the gastrula. This area, called the *dorsal lip of the blastopore*, controls the developmental fate of the cells around it, as Hans Spemann and Hilde Mangold demonstrated using transplantation experiments in the 1920s (**FIG. 41-6**).

Cell Migration May Be Guided by Contact with Surface Proteins

Guided by chemical cues, cells migrate within the developing embryo (see Fig. 41-3). The process by which cells reach their appropriate positions—such as in the spinal cord or an arm muscle—is a topic of active research. Surface protein receptors associated with specific cell types may respond to specific chemical pathways produced by nearby cells. These chemical trails may lure cells with specific receptors to migrate along them. Although the exact mechanisms are not understood, the production of cell-type-specific proteins and of pathways along which cells bearing these specific proteins migrate depends on the transcription of specific genes as a result of induction.

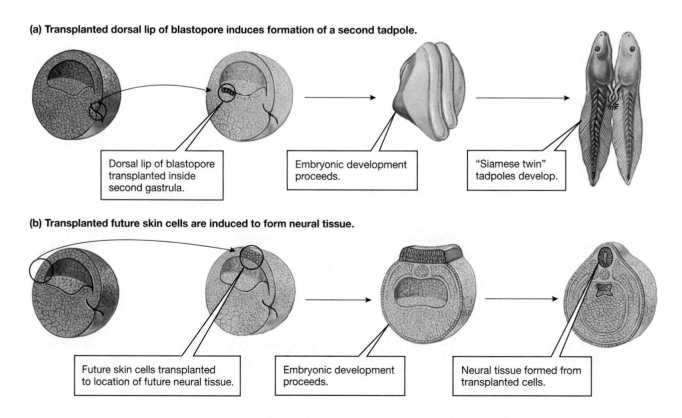

(a) Transplanted dorsal lip of blastopore induces formation of a second tadpole.

Dorsal lip of blastopore transplanted inside second gastrula.

Embryonic development proceeds.

"Siamese twin" tadpoles develop.

(b) Transplanted future skin cells are induced to form neural tissue.

Future skin cells transplanted to location of future neural tissue.

Embryonic development proceeds.

Neural tissue formed from transplanted cells.

FIGURE 41-6 Experiments demonstrating induction and its role in differentiation

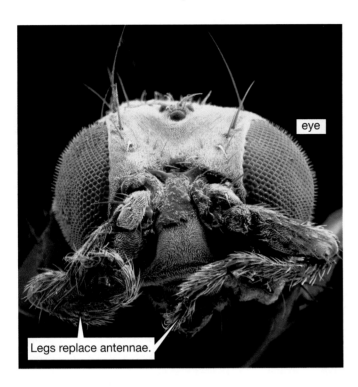

FIGURE 41-7 Homeobox gene segments regulate development
Due to a mutation in a homeobox gene, this fruit fly has perfectly formed legs where its antennae should be.

eye

Legs replace antennae.

Homeobox Gene Segments Are Important Developmental Regulators

How do different parts of an organism "know" which genes to express? The search for an answer to this seemingly simple question continues; but the *homeobox*, first discovered in fruit flies in the early 1980s, provides an important clue. **Homeoboxes** are short sequences of DNA found within larger genes. These DNA sequences code for sequences of amino acids within certain proteins. Hundreds of homeobox gene segments have been discovered, and many are involved in some way in early development. Slight differences among homeoboxes may endow them with different functions, such as directing the formation of different body parts. Scientists hypothesize that the sequence of amino acids coded by the homeobox segment of certain genes allows the proteins that these genes code for to bind to DNA. Research suggests that the DNA-binding proteins with homeobox segments are a special type of *transcription factor*, a chemical that binds to a gene and causes it to be "switched on" or transcribed. Transcription factors with homeobox segments are special because they act as master regulators, acting on all the genes needed to produce a specific body part, such as a leg. Further, their action is permanent, causing the affected genes to be continuously turned on in cells and the cell's offspring, committing them forever to be part of a leg, wing, or eye. If the homeobox segment is altered by a mutation, it can order a leg to appear where a fruit fly's antenna should be (**FIG. 41-7**).

Researchers have made the exciting discovery that very similar homeobox DNA sequences are present in animals including sponges, jellyfish, flatworms, insects, mice, and people. Related homeobox segments have also been found in fungi and plants. When sequences of genes are retained relatively unaltered over perhaps 500 million years of evolution, it suggests that they have an important role indeed, as you would predict for a master regulator gene that specifies the way an animal's body develops.

41.4 HOW DO HUMANS DEVELOP?

Human development is controlled by the same mechanisms that control the development of other animals. In fact, our development strongly reflects our evolutionary heritage. **FIGURE 41-8** summarizes human embryonic and fetal development.

During the First Two Months, Rapid Differentiation and Growth Occur

The egg emits chemicals that attract sperm, increasing its chances of being fertilized. A human egg is usually fertilized in the mother's uterine tube and undergoes a few cleavage divisions on its way to the uterus, a journey that takes about four days (**FIG. 41-9**). First a morula, a solid ball of cells, forms as the zygote begins to divide. By about six days after fertilization, the morula becomes a hollow ball of cells, the **blastocyst** (the mammalian version of a blastula; **FIG. 41-10**). Inside the blastocyst is a thickened region, the **inner cell mass** (Fig. 41-10; see also Fig. 41-9a). Within the uterus, the blastocyst burrows into the endometrium, a process called **implantation**. The outer cell layer will first become the chorion, which will form the embryonic contribution to the placenta; the inner cell mass develops into the embryo and the three other extraembryonic membranes.

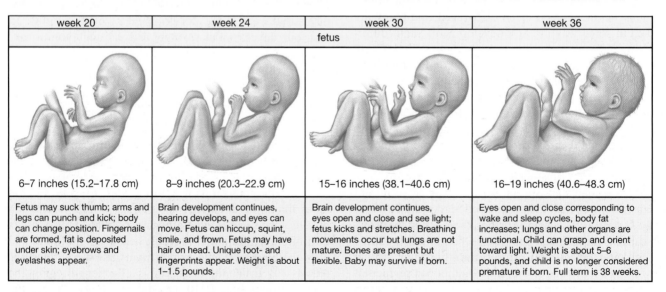

week 1	week 2	week 3	week 4	week 5	week 6
zygote to late blastula		embryo			
zygote, blastocyst, morula, late blastocyst		0.06 to 0.1 inches (1.5–2.5 mm)	0.12 to 0.20 inches (3–5 mm)	0.28 to 0.35 inches (7–9 mm)	0.32 to 0.43 inches (8–11 mm)
Cleavage of zygote forms the morula and then the blastula, which implants in the uterus.	Blastula burrows into endometrium; forms yolk sac, amnion, and embryonic disc.	Gastrulation occurs; notochord and beginning of neural tube form; heart beats.	Neural tube closes; arm buds, tail, and pharyngeal grooves form.	Eyes begin to form; leg buds form; brain enlarges.	External ears and webbed fingers form; pharyngeal grooves and tail are disappearing.

week 7	week 8	week 10	week 12	week 16
embryo		fetus		
0.67 to 0.79 inches (1.7–2.0 cm)	0.90 to 1.10 inches (2.3–2.8 cm)	1.25–1.75 inches (3.2–4.4 cm)	2–3 inches (5–7.6 cm)	4–5 inches (10.2–12.7 cm)
Webbed toes form; bones begin to stiffen; back straightens; eyelids begin forming.	All major organs and male genitals begin to form; arms can bend; fingers are distinct. Facial features and outer ears take shape.	After 8 weeks; the embryo is called a fetus. Red blood cells form; toes separate; eyelids have developed; major brain parts are present; hands can form fists.	Neck is well-defined; all organs are present; male and female genitals are present; arms and legs move; teeth begin to form; heartbeat can be detected electronically.	Sucking and swallowing movements occur; liver and pancreas begin functioning. Body has grown relative to the head; major organs continue developing. Mother may feel movement; weight is about 5 oz.

week 20	week 24	week 30	week 36
fetus			
6–7 inches (15.2–17.8 cm)	8–9 inches (20.3–22.9 cm)	15–16 inches (38.1–40.6 cm)	16–19 inches (40.6–48.3 cm)
Fetus may suck thumb; arms and legs can punch and kick; body can change position. Fingernails are formed, fat is deposited under skin; eyebrows and eyelashes appear.	Brain development continues, hearing develops, and eyes can move. Fetus can hiccup, squint, smile, and frown. Fetus may have hair on head. Unique foot- and fingerprints appear. Weight is about 1–1.5 pounds.	Brain development continues, eyes open and close and see light; fetus kicks and stretches. Breathing movements occur but lungs are not mature. Bones are present but flexible. Baby may survive if born.	Eyes open and close corresponding to wake and sleep cycles, body fat increases; lungs and other organs are functional. Child can grasp and orient toward light. Weight is about 5–6 pounds, and child is no longer considered premature if born. Full term is 38 weeks.

FIGURE 41-8 Human embryonic development

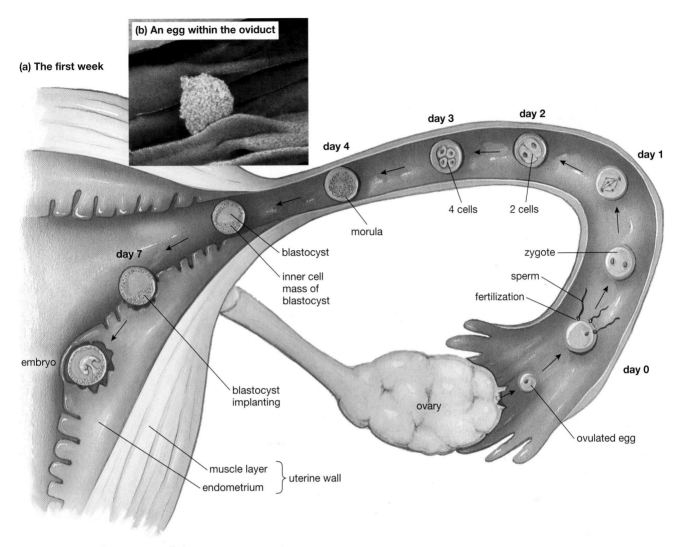

FIGURE 41-9 The journey of the egg
(a) The egg is fertilized in the uterine tube (oviduct) and, while dividing to form a morula, travels to the uterus; there, the hollow blastocyst implants in the endometrium, and development continues. **(b)** Photograph of the egg, surrounded by the follicle cells of the corona radiata, traveling within the oviduct toward the uterus.

After implantation, the inner cell mass grows and splits, forming two fluid-filled sacs that are separated by a double layer of cells called the **embryonic disc** (Fig. 41-10b). One sac, bounded by the amnion, forms the amniotic cavity. The amnion eventually encloses the embryo in the watery environment necessary for development. The yolk sac forms the second cavity, although in most mammals (those that form a placenta, called *placental mammals*), it contains no yolk. At this stage, the embryonic disc consists of an upper layer of future ectoderm cells and a lower layer of future endoderm cells. Gastrulation begins at the end of the second week after fertilization; many women first discover that they are pregnant at about this time when they fail to menstruate on schedule.

During the third week of development, the embryo, enclosed in its amniotic sac, curls to form the future head. The chorion, which will become the embryo's contribu-

tion to the placenta, extends tiny fingers called **chorionic villi** into the endometrium of the uterus. Embryonic blood vessels invade the chorionic villi, carrying blood pumped by the embryonic heart which begins to beat at about day 22. As the embryo grows during the fourth week (**FIG. 41-11**), the endoderm forms a tube—the embryonic gut—that will become the digestive tract. The umbilical cord forms from the yolk stalk and body stalk. The *yolk stalk* connects the yolk sac to the embryonic gut (an evolutionary "leftover" in mammals from the yolk's role of nourishing embryonic fish, birds, and reptiles). The embryo is connected to the chorion by the *body stalk* that includes the allantois and embryonic blood vessels. Within the embryo, ectoderm is forming structures that will become the brain and spinal cord. By the fourth week, the embryo bulges into the uterine cavity, completely surrounded by the amnion. It is linked to the placenta by the *umbilical*

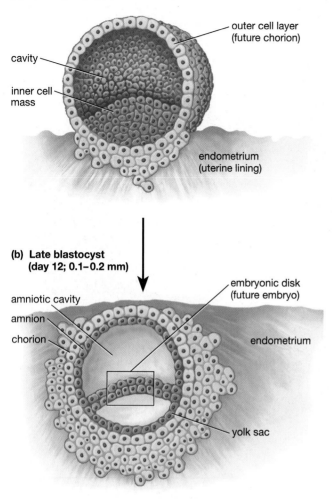

(a) Blastocyst (day 7; 0.1–0.2 mm)

outer cell layer
(future chorion)

cavity

inner cell
mass

endometrium
(uterine lining)

(b) Late blastocyst
(day 12; 0.1–0.2 mm)

embryonic disk
(future embryo)

amniotic cavity

amnion

chorion

endometrium

yolk sac

FIGURE 41-10 A blastocyst implants
(a) As it burrows into the uterine lining, the outer cell layer of
the blastocyst forms the chorion, the embryonic contribution
to the placenta. **(b)** The late blastocyst penetrates deeply into
the endometrium. The inner cell mass forms the amnion, yolk
sac, and the embryonic disc (future embryo).

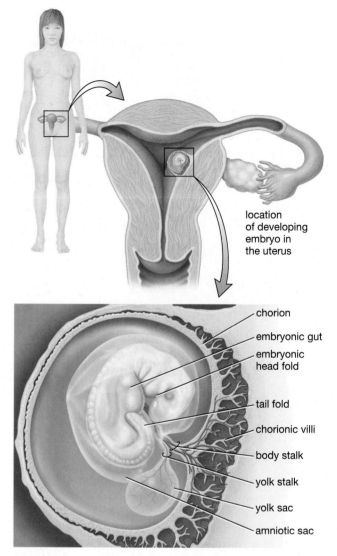

location
of developing
embryo in
the uterus

chorion

embryonic gut

embryonic
head fold

tail fold

chorionic villi

body stalk

yolk stalk

yolk sac

amniotic sac

FIGURE 41-11 A four-week human embryo
Endoderm forms the embryonic gut (future digestive tract),
which is connected to the yolk sac by the yolk stalk. The body
stalk carries embryonic blood into the chorionic villi.

stalk (the future umbilical cord; Fig. 41-11). A tail is clearly visible.

During the sixth week, the embryo clearly displays its
chordate ancestry (see Chapter 24 for more on chordates),
having developed a notochord, a prominent tail, and
pharyngeal grooves (indentations behind the head that are
homologous to the developing gills that fish and some amphibians retain as adults; **FIG. 41-12**). These structures will
disappear in the weeks that follow. Although it is only
about one inch in length, the embryo already has the rudimentary beginnings of eyes and a rapidly developing brain.

As the second month draws to an end, nearly all the
major organs have formed (**FIG. 41-13**). The gonads appear
and develop into testes or ovaries. Sex hormones—either
testosterone from the testes or estrogen from the ovaries—
are secreted. These hormones will affect the future development of embryonic organs—not only the reproductive

organs but also certain regions of the brain. After the second month of development, the embryo is called a **fetus**,
and it has taken on a generally human appearance.

The first two months of pregnancy are a time of extremely rapid differentiation and growth for the embryo,
and a time of considerable danger. Although vulnerable
throughout development, rapidly developing organs are
the most sensitive to toxic substances, such as drugs (including alcohol and nicotine) and certain medications
taken by the mother.

**The Placenta Secretes Hormones and Exchanges
Materials Between Mother and Embryo**

During the second week of pregnancy, the blastocyst burrows into the thickened lining of the uterus and obtains nutrients directly from the endometrium (see Fig. 41-10). The

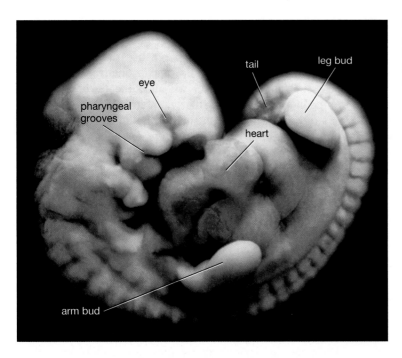

FIGURE 41-12 A six-week human embryo
As the sixth week begins, the head comprises about half of the human embryo. Arm and leg buds have formed, and a tail and pharyngeal grooves are clearly visible.

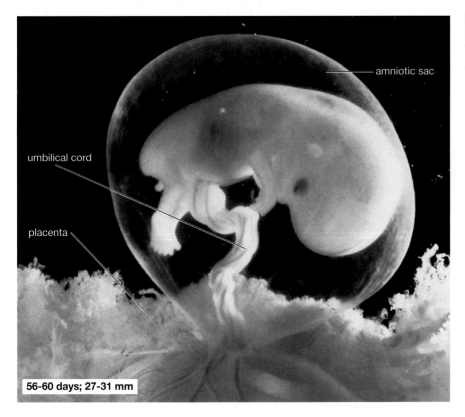

FIGURE 41-13 An eight-week human embryo
The embryo has acquired human features and is called a fetus after this time.

outer layer of the blastocyst forms the chorion, which begins to penetrate the endometrium with fingerlike chorionic villi. During the third week, the **placenta** begins to form from this complex interweaving of tissues from the embryo and the endometrium of the uterus. The placenta has two major functions: it secretes hormones and it allows the selective exchange of materials between the mother and the fetus.

As it develops during the first two months of pregnancy, the placenta begins secreting estrogen and progesterone. Estrogen stimulates the growth of the mother's uterus and mammary glands; progesterone also stimulates the mammary glands and inhibits premature contractions of the uterus. The placenta also regulates the exchange of materials between the blood of the mother and the blood of the fetus without allowing the two to mix. The chorionic villi contain a dense network of fetal capillaries and are bathed in pools of maternal blood (**FIG. 41-14**). Oxygen diffuses, and nutrients either diffuse or are actively transported from the maternal blood into the fetal capillaries, then carried into the fetus by the fetal umbilical vein. The umbilical arteries carry carbon dioxide and fetal wastes such as urea from the fetus to the mother. The carbon dioxide will be released from the mother's lungs, and the fetal urea will be excreted by her kidneys.

While allowing exchange by diffusion, the membranes of the capillaries and chorionic villi act as barriers to some large proteins and most cells. Despite this, some disease-causing organisms and many harmful chemicals, such as alcohol and nicotine, can penetrate the placental barrier, as described in "Health Watch: The Placenta Provides Only Partial Protection."

Growth and Development Continue During the Last Seven Months

The fetus continues to grow and develop for another seven months. Although the size of its body is "catching up" to the size of its head, the brain continues to develop rapidly and the head remains disproportionately large. Nearly every nerve cell ever formed during the human life span develops during embryonic life—one reason that the developing brain is such a sensitive target for alcohol and other drugs

ingested during pregnancy. As the brain and spinal cord grow, they begin to generate specific types of behavior. As early as the third month of pregnancy, the fetus begins to move and respond to stimuli. Some instinctive behaviors appear, such as sucking, which will allow the newborn to nurse immediately. Structures that the fetus will need when it emerges from the uterus, such as the lungs, stomach, intestine, and kidneys, gradually enlarge and become functional, although they will not be used until after birth. Nearly all fetuses 32 weeks or older can survive outside the womb with some medical assistance, and heroic measures can often save infants born as early as 26 weeks, but more mature fetuses have a much greater chance of healthy survival.

Development Culminates in Labor and Delivery

Usually, during the last months of pregnancy, the fetus becomes positioned head downward in the uterus, with the crown of the skull resting against (and being held up by) the cervix. The process of birth generally begins around the end of the ninth month (**FIG. 41-15**). Birth results from a complex interplay between uterine stretching caused by the growing fetus and fetal and maternal hormones that finally trigger **labor** (contractions of the uterus that result in *delivery*, the expulsion of the fetus from the uterus).

Unlike skeletal muscles, uterine muscles can contract spontaneously; and stretching enhances their tendency to contract. As the baby grows, it stretches the uterine muscles,

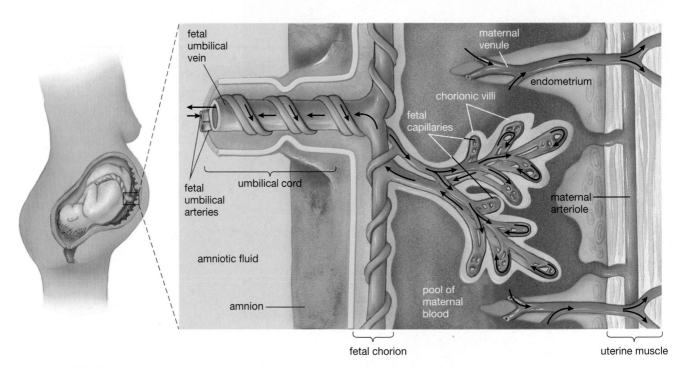

FIGURE 41-14 The placenta
The placenta is formed from the chorion of the embryo and the endometrium of the mother. Capillaries of the endometrium break down, releasing maternal blood that forms pools within the placenta. Chorionic villi containing embryonic capillaries extend into these pools. The placenta allows exchange of wastes and nutrients between the fetal capillaries and the maternal blood pools, while keeping the fetal and maternal blood supplies separate. The umbilical arteries carry deoxygenated blood from the fetus to the placenta, and the umbilical vein carries oxygenated blood back to the fetus. QUESTION A few types of mammals lack a placenta. What would you predict about the nature of development in nonplacental mammals?

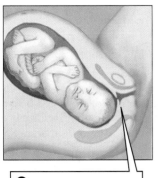

① The baby orients head downward, facing the mother's side. The cervix begins to thin and expand in diameter (dilate).

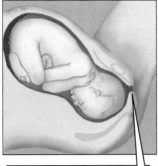

② The cervix dilates completely to 10 centimeters (almost 4 inches wide), and the baby's head enters the vagina, or birth canal. The baby rotates to face the mother's back.

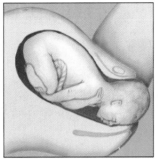

③ The baby's head emerges.

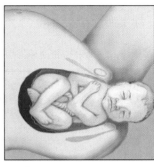

④ The baby rotates to the side once again as the shoulders emerge.

FIGURE 41-15 Delivery

which occasionally contract weeks before delivery. The exact stimuli that trigger human birth are unknown. The full-term fetus undoubtedly plays a major role by secreting substances that signal its readiness for birth. These trigger a a cascade of events that make the uterus even more likely to contract. When the combination of hormones and stretching activates the uterus beyond some critical point, strong contractions begin, signaling the onset of labor. As the contractions proceed, the baby's head pushes against the mother's cervix, making it expand in diameter (dilate). Stretch receptors in the walls of the cervix send signals to the hypothalamus, triggering oxytocin release. Under the dual stimulation of prostaglandin and oxytocin, the uterus contracts even more strongly. This positive feedback cycle is finally halted when the baby emerges from the vagina, or *birth canal.*

The infant's head is so large that it can barely fit through the mother's pelvis. The skull is compressed into a slightly conical shape as it passes through the vagina. We don't know if childbirth is painful for the infant, but it certainly is for the mother (see "Links to Life: Why Is Human Childbirth So Difficult?"). The infant is in for a rude awakening. The uterus was soft, fluid-cushioned, and warm. Suddenly, the baby must breathe on its own, regulate its own body temperature, and suckle to obtain food.

After a brief rest following childbirth, the uterus resumes its contractions and shrinks remarkably. During these contractions, the placenta is sheared from the uterine wall and expelled through the vagina as the *afterbirth.* The umbilical cord now releases prostaglandins that cause the muscles surrounding fetal blood vessels in the umbilical cord to contract and shut off blood flow. Although tying off the umbilical cord is standard practice, it is not usually necessary; if it were, other mammals would not survive birth.

Milk Secretion Is Stimulated by Hormones of Pregnancy

As the fetus grows, nourished by nutrients diffusing through the placenta, changes in the mother's breasts pre-pare her to continue nourishing her child after it is born. When pregnancy occurs, large quantities of estrogen and progesterone (acting together with several other hormones) stimulate **mammary glands**, milk-producing glands in the breasts, to grow, branch, and develop the capacity to secrete milk. The mammary glands are arranged in a circle around the nipple; each gland has a milk duct that leads to the *nipple*, which is a projection of epithelial tissue (**FIG. 41-16**). The actual secretion of milk, called **lactation**, is promoted by the pituitary hormone *prolactin.*

The level of prolactin rises steadily from about the fifth week of pregnancy and peaks when the baby is born,

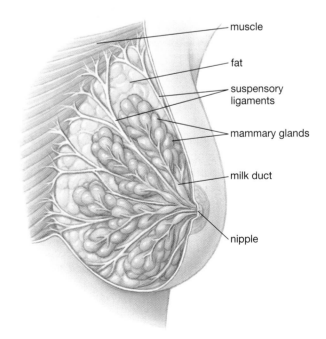

muscle

fat

suspensory ligaments

mammary glands

milk duct

nipple

FIGURE 41-16 **The structure of the mammary glands** During pregnancy, both fatty tissue and the milk-secreting glands and ducts increase in size.

HEALTH WATCH   The Placenta Provides Only Partial Protection

Through the first half of the twentieth century, physicians assumed that the placenta protected the developing fetus from most harmful substances in maternal blood. We now know that this is far from true. In fact, most drugs (both medicinal and "recreational") and even some disease-producing organisms readily penetrate the placental barrier and affect the fetus.

INFECTIONS CAN CROSS THE PLACENTA

Syphilis bacteria and the viruses that cause genital herpes and German measles occasionally cross the placenta and can damage the developing embryo. Infants may be born with HIV infections acquired from their mothers through the placenta.

DRUGS READILY CROSS THE PLACENTA

A tragic example of a drug that crosses the placenta is *thalidomide*, which was commonly prescribed in Europe in the late 1950s and early 1960s as a sedative and antinausea medication to combat morning sickness. Thalidomide's devastating effects on embryos were discovered only when many babies were born with missing or extremely abnormal limbs. In the late 1980s, the antiacne drug Accutane® was found to cause gross deformities in babies born to women using it. Researchers now know that Accutane contains a substance that acts in a manner similar to natural regulatory molecules that control body formation in the developing embryo. A more recent example is the drug valproic acid, an anticonvulsant commonly prescribed to control epilepsy. One study showed that children of women taking this drug during pregnancy were seven times as likely to have birth defects, and another showed that the average IQ scores of children exposed prenatally to valproic acid were significantly lower during childhood.

Although these are extreme examples, any drug, including aspirin, has the potential to harm the fetus. Any woman who thinks she may be pregnant should seek medical advice about any drugs that she takes.

THE EFFECTS OF SMOKING

Many women who smoke are so addicted that they don't stop smoking during pregnancy, thus exposing their developing children to nicotine, carbon monoxide, and a host of carcinogens. As a result, they have a higher incidence of miscarriages and tend to give birth to smaller infants, who have a higher risk of death shortly after birth. There is evidence that some children born to heavy smokers also suffer behavioral and intellectual impairment.

FETAL ALCOHOL SPECTRUM DISORDERS

The effects of alcohol on the developing fetus can be devastating. When a pregnant woman drinks, alcohol in the blood of her unborn child reaches a level as high as in her own blood and is not metabolized as rapidly. A woman who consumes one or two alcoholic drinks a day during the first three months of pregnancy significantly increases her chances of having a miscarriage. Despite warnings on every wine and beer bottle, over 12% of U.S. women drink during pregnancy, and each year they produce as many as 40,000 infants with *fetal alcohol spectrum disorders* (FASD; any of a range of disabilities caused by prenatal alcohol exposure). Women who drink even lightly during pregnancy on average have smaller babies, and a major study concluded that these children are more likely to be anxious, depressed, or aggressive. Many children born to mothers who drink heavily on a regular basis (4 to 5 drinks per day or more) or go on alcoholic binges during which they have 5 drinks (or more) at one time exhibit fetal alcohol syndrome. Such children are often of below-average intelligence and can be hyperactive and irritable, with poor impulse control. Certain facial abnormalities are used to diagnose FAS (**FIG. E41-2b**). In extreme cases, children afflicted with FAS may have small heads and small, improperly developed brains (**FIG. E41-2a**). FAS children may have inhibited growth, and a higher-than-normal incidence of defects of the heart and other organs. The damage is irreversible and can be fatal.

---

stimulating milk production. Milk is released when the infant's suckling stimulates nerve endings in the nipples, which signal the hypothalamus to cause the pituitary gland to release an extra surge of prolactin and oxytocin. Oxytocin causes muscles surrounding the mammary glands to contract, ejecting the milk into the ducts that lead to the nipples (see Fig. 41-16).

During the first few days after birth, the mammary glands secrete a thin, yellowish fluid called **colostrum**. Colostrum is high in protein and contains antibodies from the mother that help protect the infant against some diseases as its immune system is developing. Colostrum is gradually replaced by mature milk, which is higher in fat and milk sugar (lactose) and lower in protein.

Aging Is Inevitable

The aging process begins at the moment of birth. For thousands of years, people have attempted to delay aging and extend the human life span. The mythical "fountain of youth" has now been replaced by more modern elixirs. These pills, hormones, and diets, all claiming to delay aging, have only one thing in common: none of them have been demonstrated to work.

What is **aging** (**FIG. 41-17**)? Researchers have defined aging as a gradual accumulation of random damage to essential biological molecules—particularly DNA—that begins at an early age. Eventually, the body's ability to repair or compensate for the damage is exceeded, impairing function at all levels, from cells to tissues to organs. Aging is

FAS is the single most common known cause of mental retardation in the United States, with at least 4000 FAS infants born each year (about 1 in every 1000 births). Researchers have concluded that there is no safe level of alcohol consumption during any phase of pregnancy, and the U.S. Surgeon General advises pregnant women and those who are likely to become pregnant to completely avoid alcohol.

In summary, a pregnant woman should assume that any drugs she takes will reach her developing infant. Women who are likely to become pregnant need to know that crucial stages of development may occur before they even realize that they are pregnant, and so they should take the same precautions as they would if they were pregnant. The mother's choices during this critical nine-month period can strongly influence her child's future well-being.

(a)

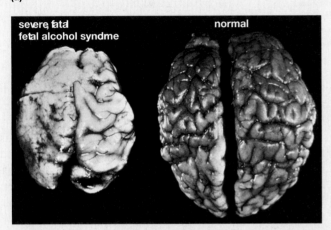

(b)

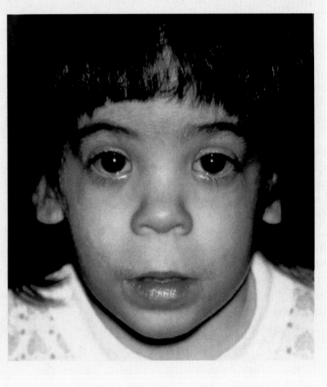

FIGURE E41-2 Alcohol impairs brain development
(a) The brain of a 6-week-old child with severe fetal alcohol syndrome (left) and the brain of a normal child of the same age (right) show the devastating effects of alcohol on the developing brain. (photo courtesy of Dr. Sterling Clarren, University of Washington, Seattle, WA). (b) Children with FAS usually have eyes that are narrow from side to side, a smooth philtrum (lacking a well-defined groove between the nose and upper lip), and a narrow upper lip (photo courtesy of Dr. Susan Astley, University of Washington, Seattle, WA).

FIGURE 41-17 Youth meets age

manifested in many ways: muscle and bone mass are lost, skin elasticity decreases, reaction time slows, and senses such as vision and hearing become less acute. A less-robust immune response renders the aging individual more vulnerable to disease. Eventually the individual can no longer fight off natural assaults, and death occurs.

Is aging, leading to death, a natural stage of development, programmed as a part of the life cycle? From an evolutionary perspective, aging is best considered to be a result of neglect, rather than genetic programming. Natural selection preserves only those mechanisms that keep an organism alive and healthy while it is producing and nurturing its young. Repair mechanisms that extend longevity beyond this period are not adaptive, and so have not been favored by natural selection. Further, deleterious mutations whose

LINKS TO LIFE   Why Is Human Childbirth So Difficult?

You've probably seen films of animals giving birth. Compared to the human ordeal, it seems painless, almost effortless. Why are we so different? Researchers Karen Rosenberg and Wenda Trevathan have found that assisted birth is universal throughout the enormous range of human cultures. They and others hypothesize that the need for help giving birth is a result of two human traits. First, we walk upright, and this *bipedalism* has altered the shape of the pelvis. Second, we have relatively enormous heads at birth, which barely fit through the pelvic opening and are usually deformed temporarily during birth. These human traits force the infant to emerge in an awkward position, somewhat backwards and sideways (see Fig. 41-15). The baby's large head makes childbirth difficult and painful, so the mother is somewhat incapacitated as the baby emerges. Monkeys, whose heads are relatively smaller, are born almost effortlessly with head and body facing forward, allowing the mother to reach down, help the infant emerge, and bring it up to her chest unassisted. Researchers postulate that as the brains and heads of our early ancestors increased dramatically in size, so did the difficulty and discomfort of human childbirth, necessitating the need for assistance.

effects manifest themselves only after an organism has finished reproducing are passed along to offspring, and so can readily spread and accumulate within a population.

One factor contributing to the cellular damage that accumulates with aging is the production of free radicals (see Chapter 2), which attack cellular components. Free radicals are generated as by-products of many crucial biochemical reactions, particularly those that harness energy. It is ironic that, while an organism would die almost instantly if its energy production were halted, the same metabolic reactions that produce usable energy may eventually—indirectly—lead to an organism's death. The ability to repair damage, particularly to DNA, relies on enzymes that are also encoded by DNA. These enzymes themselves become less functional as mutations accumulate in the genes that code for them, dooming the organism to ever-increasing numbers of genetic mistakes and metabolic malfunctions.

Researchers have bred genetically simple animals such as fruit flies and roundworms to have increased life spans, and they are studying the genetic differences between individuals with normal life spans and those that live longer. Whether genetic changes made as a result of this knowledge can make a difference in the human life span remains to be seen. It is likely that hundreds of genetically determined biochemical pathways all play interrelated roles in longevity. Would extending the human life span even be desirable, given the health problems that accompany old age and the resource constraints of an overpopulated planet? No one knows, but we can be certain it will not happen soon.

CASE STUDY REVISITED   FACES OF FAS

John Kellerman's birth mother arrived at a Denver hospital drunk. When her water broke (this refers to the release of amniotic fluid when the amnion breaks, usually during labor), the smell of alcohol suddenly filled the delivery room. Her newborn son was literally stewing in it. Although he has never had an alcoholic drink in his life, John was born drunk. His adoptive mother explains that without his medication, his behavior resembles that of a drunk person: silly, volatile, and lacking impulse control. The writings of young people with FAS poignantly convey the day-to-day difficulties that these innocent people face over a lifetime. In his poem "Help," John writes:

When my brain is not working right,
I need to let someone I trust
Help me
To stay safe and get calm again.
When my brain is not working right,
I feel like I am on a FASD Train
Going downhill,
And the engineer is asleep
And I can't wake him up,
And I can't put on the brakes.

CJ, an FAS victim from Canada, explains:

I am small, I have a different face . . .
    and Lots of Learning problems.
Moms do not do this on purpose
Do not be mean or mad or blame them
I am 17 years old and have had FAS
    all my life

I will have it forever
It will never go away
Don't be mean or mad or blame me.

Consider This There are about 20 states in the U.S. in which women using illegal drugs or abusing alcohol during pregnancy may be subject to legal action to protect fetal rights. How do you think that society should approach the dilemma of pregnant women who often unwittingly damage their unborn children? Is this child abuse? Based on what you now know about development, how would you deal with a friend who continued to smoke, drink, or take other drugs during pregnancy?

CHAPTER REVIEW

SUMMARY OF KEY CONCEPTS

41.1 How Do Indirect and Direct Development Differ?

Animals undergo either indirect or direct development. In indirect development, eggs (usually with relatively little yolk) hatch into larvae, which progress through a feeding stage and later undergo metamorphosis to become adults with notably different body forms. In direct development, the newborn animal is sexually immature but otherwise resembles a small adult. Animals with direct development tend to produce either large, yolk-filled eggs or nourish the developing embryo within the mother's body. In birds, reptiles, and mammals, extraembryonic membranes (the chorion, amnion, allantois, and yolk sac) encase the embryo in a fluid-filled space and regulate the exchange of nutrients and wastes between the embryo and its environment.

41.2 How Does Animal Development Proceed?

Animal development occurs in several stages. *Cleavage:* The fertilized egg undergoes cell divisions with little intervening growth, so the egg cytoplasm is partitioned into smaller cells. Cleavage divisions result in the formation of the morula, a solid ball of cells. A cavity then opens up within the morula, forming the blastula, a hollow ball of cells. *Gastrulation:* A dimple forms in the blastula, and cells migrate from the surface into the interior of the ball, eventually forming a three-layered gastrula. These three cell layers—ectoderm, mesoderm, and endoderm—give rise to all the adult tissues (see Table 41-2). *Organogenesis:* The cell layers of the gastrula form organs characteristic of the animal species. The juvenile animal increases in size and achieves sexual maturity. *Aging:* Cells begin to function less efficiently, as damage to DNA and other cellular components accumulates; the cell's self-repair abilities deteriorate and, eventually, the animal dies.

41.3 How Is Development Controlled?

All the cells of an animal body contain a full set of genetic information, yet cells are specialized for particular functions. During development, cells differentiate by stimulating and repressing the transcription of specific genes. Vertebrate cells retain the ability to form a complete individual if they are separated through the blastula stage, as occurs in the case of human identical twins.

During gastrulation, the developmental fate of most of the embryo's cells becomes sealed by a process called *induction* that is stimulated by chemical messages received from nearby cells. The differentiation of cells into specialized roles occurs as a result of differential gene expression caused by induction.

Which genes are expressed is determined by regulatory molecules, typically proteins (or proteins combined with substances such as steroid hormones) that bind to specific genes, either blocking or promoting transcription. Cells migrate within the developing embryo, by a process requiring chemical communication among cells. Surface proteins associated with specific cell types recognize chemical pathways laid out by other cells and cause cellular migration. Homeobox gene segments code for specific amino acid sequences that allow proteins called transcription factors to bind specific genes and cause them to be transcribed (used to produce messenger RNA). The special transcription factor proteins coded by homeobox-containing genes serve as "master regulators" of development, acting on all the genes needed to produce a specific body part, such as a leg.

41.4 How Do Humans Develop?

A fertilized egg (zygote) develops into a hollow blastocyst and implants in the endometrium. The outer wall will become the chorion and will form the embryonic contribution to the placenta; the inner cell mass develops into the embryo and the three other extraembryonic membranes. During gastrulation, cells migrate and differentiate into ectoderm, mesoderm, and endoderm. During the third week, the endoderm forms a tube that will become the digestive tract, the heart begins to beat, and the rudiments of a nervous system appear. By the end of the second month, the major organs have formed, and the embryo—now called a fetus—appears human. In the next seven months before birth, the fetus continues to grow; the lungs, stomach, intestine, kidneys, and nervous system enlarge, develop, and become more functional. Human embryonic development, the stages of which are summarized in Figure 41-8, follows the same principles as does development in other mammals.

During pregnancy, mammary glands in the mother's breasts grow under the influence of estrogen, progesterone, and other hormones. After about nine months, uterine contractions are triggered by a complex interplay of uterine stretch and prostaglandin and oxytocin release. As a result, the uterus expels the baby and then the placenta. After its birth, the infant begins suckling and activates the release of prolactin and oxytocin, which trigger milk secretion.

Aging is the gradual accumulation of cellular damage (particularly to genetic material) over time that leads to loss of functionality of an organism and, eventually, to death.

Web Tutorial 41.1 Overview of Animal Development

KEY TERMS

adult stem cell (ASC)
 page 843
aging *page 852*
allantois *page 840*
amnion *page 840*
amniotic egg *page 840*

blastocyst *page 845*
blastopore *page 841*
blastula *page 841*
chorion *page 840*
chorionic villus *page 847*
cleavage *page 841*

colostrum *page 852*
development *page 842*
differentiation *page 842*
direct development
 page 838
ectoderm *page 841*

embryonic disc *page 847*
embryonic stem cell (ESC)
 page 843
endoderm *page 841*
extraembryonic membrane
 page 840

THINKING THROUGH THE CONCEPTS

1. Distinguish between indirect and direct development, and give examples of each.

2. Describe the structure and function of the four extraembryonic membranes found in reptiles and birds. Are these four present in placental mammals? In what ways are their roles similar in reptiles and birds as compared to mammals? How do they differ?

3. What is yolk? How does it influence cleavage?

4. What is gastrulation? Describe gastrulation in frogs.

5. Name two structures derived from each of the three embryonic tissue layers—endoderm, ectoderm, and mesoderm.

6. How does cell death contribute to development?

7. Describe the process of induction, and give two examples.

8. Define differentiation. How do cells differentiate; that is, how is it that adult cells express some but not all genes of the fertilized egg?

9. In humans, where does fertilization occur, and what stages of development occur before the fertilized egg reaches the uterus?

10. Describe how the human blastocyst gives rise to the embryo and its extraembryonic membranes.

11. Explain how the structure of the placenta prevents mixing of fetal and maternal blood while allowing the exchange of substances between the mother and the fetus.

12. Is the placenta an effective barrier against substances that can harm the fetus? Describe two types of harmful agents that can cross the placenta, and discuss their effects on the fetus.

13. How do changes in the breast prepare a mother to nurse her newborn? How do hormones influence these changes and stimulate milk production?

14. Describe the events leading to the expulsion of the baby and the placenta from the uterus. Explain why this is an example of positive feedback.

APPLYING THE CONCEPTS

1. A researcher obtains two frog embryos at the gastrula stage of development. She carefully removes a cluster of cells from a location that she knows would normally become neural tube tissue and transplants it into the second gastrula in a location that would normally become skin. Does the second gastrula develop two neural tubes? Explain your answer.

2. Embryologists have used embryo fusion to produce *tetraparental* (four-parent) mice, and they have produced "geeps" from goat and sheep embryos. The resulting bodies are patchworks of cells from both animals. Why does fusion succeed with very early embryos (four-cell to eight-cell stages) and fail when much older embryos are used?

3. If the nucleus of an adult cell can be transplanted into an egg from which the nucleus has been removed to produce a clone of the adult, is it theoretically possible to produce human clones? Would such clones be exactly identical to the person who supplied the nucleus? Explain.

4. Based on your knowledge of cloning from this chapter and from Chapter 11, state as many arguments as possible for and against therapeutic cloning. Where do you stand?

FOR MORE INFORMATION

Cibelli, J. B., Lanza, R. P., and West, M. D. "The First Human Cloned Embryo." *Scientific American*, November 2001. Therapeutic cloning using the recipient's own cells can generate stem cells that will not undergo rejection by the immune system.

Hall, S. "The Good Egg." *Discover*, May, 2004. This article explains how the human zygote develops, with spectacular photographs.

Olshansky, S. J., Hayflick, L., and Carnes, B. A. "No Truth to the Fountain of Youth." *Scientific American,* June 2002. Anti-aging "remedies" abound—but none have been proven effective.

Park, A., and Cray, D. "The Politics of Stem Cells." *Time*, August 7, 2006. A beautifully illustrated layperson's introduction to stem cell research.

Rose, M. R. "Can Human Aging Be Postponed?" *Scientific American*, December 1999. Aging can theoretically be postponed, but this will not happen soon.

Rosenberg, K. W., and Trevathan, W. R. "The Evolution of Human Birth." *Scientific American*, November 2001. Explains why humans have such a difficult time delivering babies.

Plant Anatomy and Physiology

A prairie blanketed with Texas bluebonnets in the spring. Flowers such as these are actually elaborately modified leaves adapted to attract pollinators.

42

Plant Anatomy and Nutrient Transport

Each fall color has a purpose, but scientists are still investigating the function of red anthocyanin pigments, which are synthesized just before the leaf drops.

CASE STUDY WHY DO LEAVES TURN RED IN THE FALL?

EVERY AUTUMN, PEOPLE FLOCK to the deciduous forests of the northern United States—particularly New England—to enjoy the brilliant red, yellow, and orange colors of the leaves and the crisp, sunny weather that brings on this display. But these colors did not evolve to attract tourists! The yellow and orange pigments (carotenoids) are there all year, helping leaves trap sunlight for photosynthesis. As the leaves die and the predominant green chlorophyll pigment breaks down, these pigments are revealed. Of all the fall colors, the striking reds are the most mysterious. These are caused by anthocyanin pigments, which are synthesized just before the leaves drop in the fall and are not involved in photosynthesis.

Thanks to research that began over a century ago, scientists are finally obtaining hard evidence to explain why leaves synthesize a new red pigment just as they are dying. For the past decade, plant physiologists David Lee and Kevin Gould have sought answers to this question in Harvard Forest, a Massachusetts nature sanctuary with unrivaled fall colors. Their research follows up on the observations and hypotheses of German botanists of the late 1800s, who noticed that a combination of low temperatures and intense light seems to stimulate anthocyanin production. Perhaps, as these early investigators suggested, the red pigments help warm the leaf and protect it from the damaging effects of too much sun. But why protect a dying leaf?

It is easy to take plants for granted. We may admire a field of wildflowers or a giant sequoia, but we seldom stop to think about the intricate adaptations that allow even the most common plants to survive and prosper. Plants cannot move to seek food or water, to escape predators, to avoid winter, or to find a mate. Yet spruce perch atop the continuously frozen soil of the far north, mangroves stand immersed in Florida swamps, cacti thrive in the searing heat of the Mojave Desert, and bristlecone pines in the California mountains may live 4000 years. One of the best ways to appreciate plants is to consider how effectively they overcome the same environmental challenges encountered by all life-forms, including humans. To survive, organisms must:

- Obtain energy
- Obtain water and other nutrients
- Distribute water and nutrients throughout the body
- Exchange gases
- Support the body
- Grow and develop
- Reproduce

As we progress through this chapter introducing the plant body and how it works, think about how your own body performs the same basic functions. You may find yourself marveling at how natural selection, acting on different types of mutations, has produced these incredibly diverse forms of life.

42.1   HOW ARE PLANT BODIES ORGANIZED, AND HOW DO THEY GROW?

To understand how plants have evolved in ways that meet life's challenges, we must first explore the basic structure of the plant body. The enormous diversity of plant life was introduced in Chapter 21. In this chapter, we focus on the type of plant that dominates Earth's landscape: flowering plants or **angiosperms**. Angiosperm plant bodies consist of leaves, stems, and roots; and angiosperms produce flowers that generate seeds. **FIGURE 42-1** illustrates important angiosperm structures and functions.

There are two broad groups of angiosperms: *monocots* and *dicots*. Characteristic differences in each of these structures are shown in **FIGURE 42-2**. These two general groups are named after the *cotyledons* or *seed leaves* of the angiosperm embryo. **Monocots** have a single seed leaf ("mono" means "one," and "cot" is short for "cotyledon"). This group includes lilies, palms, orchids, and a large family collectively known as grasses. Although we think of grasses as growing in lawns or meadows, the grass family also includes wheat, rice, corn, barley, and oats. The seeds of these plants provide staple food for most of Earth's people, and the grasses also feed most domestic animals. **Dicots** have two seed leaves ("di" means "double"), and this group includes *deciduous* trees (those that drop their leaves in winter), bushes, and many wildflowers and garden plants.

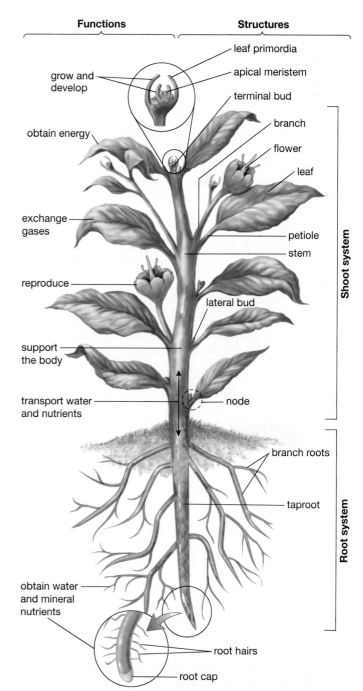

FIGURE 42-1 **Dicot flowering plant structures and functions**

Flowering Plants Consist of a Root System and a Shoot System

The bodies of angiosperms consist of two major regions, the *root system* and the *shoot system* (**FIG. 42-3**). The **root system** consists of all the roots of a plant. **Roots** are branched portions of the plant body, usually embedded in soil. Plant roots serve six major functions:

- They anchor the plant in the ground.
- They absorb water and minerals (plant nutrients) from the soil.

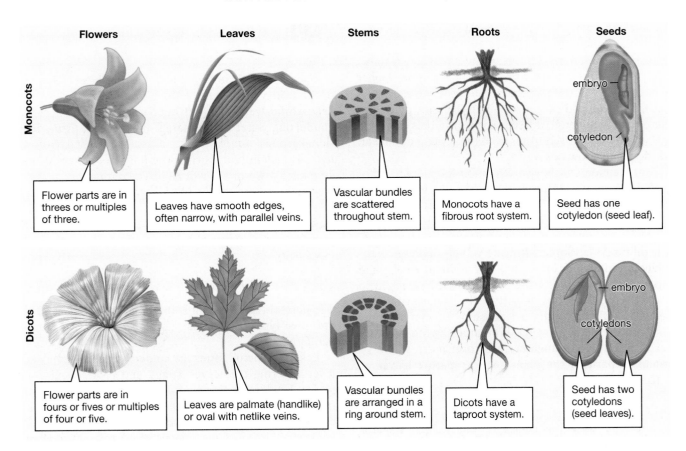

FIGURE 42-2 Monocots and dicots compared

- They store surplus food molecules manufactured during photosynthesis.
- They transport water, minerals, sugars, and hormones to and from the shoot.
- They produce some hormones.
- They interact with specific soil fungi and bacteria that provide nutrients to the plant.

The **shoot system** is usually located aboveground. In angiosperms, it consists of *stems, buds, leaves, flowers,* and *fruits.* Buds give rise to leaves and flowers. Leaves trap the energy of sunlight and capture it in high-energy molecules. Flowers are the plant's reproductive organs, producing male and female sex cells and helping them reach one another. Flowers later produce fruits that enclose and protect the seeds and often aid in seed dispersal. Stems, which are typically branched, provide support, and usually elevate the leaves, flowers, and fruit above the ground. This helps the leaves capture sunlight and the flowers attract animal pollinators or release their pollen to the winds. Elevating the fruit helps to disperse the seeds. Some parts of the shoot system are specialized to transport water, minerals, and food molecules; other parts produce hormones.

As Plants Grow, Meristem Cells Give Rise to Differentiated Cells

Animals and plants develop in dramatically different ways. One difference is the timing and distribution of growth. As

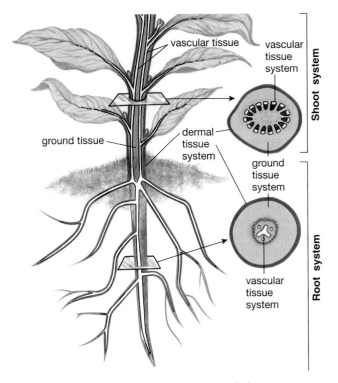

FIGURE 42-3 The structure of the root and shoot
Both the root and shoot of a flowering plant consist of three tissue systems: dermal, ground, and vascular tissue systems. EXERCISE On this drawing, circle all of the locations where primary growth occurs, and draw arrows pointing to some of the locations where secondary growth occurs.

animals grow, usually all parts of their bodies become larger until they reach their adult size. In contrast, flowering plants grow throughout their lives, never reaching a stable "adult" body form. Moreover, most plants grow longer only at the tips of their branches and roots, so the structures that developed earlier remain in the same place. For example, a swing tied to a tree branch, or initials carved in tree bark, do not move farther from the ground each year. Why do plants grow this way?

From the moment they sprout, plants are composed of two fundamentally different types of cells: *meristematic cells* and *differentiated cells*. **Meristematic cells** are capable of mitotic cell division. Like the stem cells of animals, they have not yet assumed an adult or "differentiated" form or function. Some offspring of the dividing meristem cells ("daughter cells") lose the ability to divide and become **differentiated cells**. These assume specialized structures and functions as part of the nongrowing portion of the plant. Continued divisions of meristematic cells keep the plant growing throughout its life, whereas their differentiated daughter cells form more stable or permanent parts of the plant, such as mature leaves or the trunks of trees.

Plants grow as a result of cell division and differentiation within *meristem tissue* which is composed of meristematic cells. **Apical meristem** ("tip meristems") tissue is located at the ends of roots and shoots (see Figs. 42-1 and 42-15). **Lateral meristem** ("side meristem" tissue; also called **cambium**), forms cylinders along the length of roots and stems (see Fig. 42-11).

Plants exhibit both *primary growth* and *secondary growth*. **Primary growth** occurs at the ends of roots and shoots as apical meristem cells divide and the resulting daughter cells differentiate. Primary growth is responsible for increases in length and for development of the specialized plant structures. The elongation of roots and shoots through primary growth allows them to enter new space from which to collect light, nutrients, and water. Primary growth also explains why your swing never gets any higher off the ground.

Secondary growth, which is responsible for increases in diameter, occurs by the division of lateral meristem cells and differentiation of their daughter cells. In most dicots as well as most *conifers* (cone-bearing evergreens; see Chapter 21) secondary growth causes the stems and roots to become thicker as they age. Although this chapter discusses secondary growth only in stems, keep in mind that secondary growth also occurs in roots.

Some angiosperms, described as *herbaceous plants*, exhibit only primary growth. As you might predict, herbaceous plants are soft-bodied with flexible stems, and they are usually *annuals* (living only one year). Herbaceous plants include lettuce, beans, lilies, and grasses. *Woody plants* exhibit both primary and secondary growth and are usually *perennials* (living for many years). Woody plants such as trees and bushes develop their hard, thickened stems and roots through secondary growth.

42.2 WHAT ARE THE TISSUES AND CELL TYPES OF PLANTS?

As meristem cells differentiate, they produce a wide variety of cell types. When one or more specialized types of cells work together to perform a specific function, such as conducting water and minerals, they form a *tissue*. Functional groups of more than one tissue are called *tissue systems*. The plant body is composed of three different tissue systems, shown in **FIGURE 42-3**. The **dermal tissue system** covers the outer surfaces of the plant body. The **ground tissue system** makes up most of the body of young plants. Its functions include photosynthesis, support, and storage. The **vascular tissue system** transports fluids throughout the plant body.

The Dermal Tissue System Covers the Plant Body

The dermal tissue system is the "skin" of the plant. There are two types of dermal tissue: *epidermal tissue* and *periderm*.

Epidermal tissue forms the **epidermis**, the outermost cell layer covering the leaves, stems, and roots of all young plants. Epidermal tissue also covers flowers, seeds, and fruit. In herbaceous plants, the epidermis forms the outer covering of the entire plant body throughout its life. The epidermal tissue of the aboveground parts of a plant is generally composed of tightly packed, thin-walled cells, covered with a waterproof, waxy **cuticle**, secreted by the epidermal cells. The cuticle reduces the evaporation of water from the plant and helps protect it from the invasion of disease microorganisms. In contrast, the epidermal cells of roots are not covered with cuticle, which would prevent them from absorbing water and minerals.

Some epidermal cells produce fine extensions called *hairs*. Many root epidermal cells bear **root hairs**, long projections of the individual cell that greatly enlarge the absorptive area of cell membrane. Epidermal hairs on the stems and leaves reduce evaporative water loss by reflecting sunlight and producing an unstirred layer of air near the plant's surface (**FIG. 42-4**). In contrast, some tropical plants use their hairy leaves to capture and hold water.

Periderm replaces epidermal tissue on the roots and stems of woody plants as they age. This dermal tissue is composed primarily of **cork cells**, which have thick, waterproof walls and are dead at maturity. Periderm also includes the **cork cambium** that gives rise to cork cells (see Fig. 42-12). Cork cells form the protective outer layer of the bark of trees and woody shrubs (see Fig. 42-13) as well as the tough covering of the older portions of their roots. Root segments that are covered with periderm anchor the plant but can no longer absorb water and minerals.

The Ground Tissue System Comprises Most of the Young Plant Body

The ground tissue system, which makes up the bulk of a young plant, consists of all nondermal and nonvascular (non "water and nutrient-conducting") tissues. There are

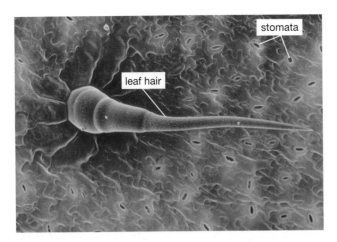

FIGURE 42-4 Dermal tissues cover plant surfaces
In shoot epidermis, such as in this zinnia leaf, the outer cell surfaces are covered with a transparent cuticle that reduces evaporation of water. The "leaf hair" also reduces evaporation by slowing the movement of air across the leaf surface. Stomata allow gases to move in and out of the leaf.

three types of ground tissue: *parenchyma, collenchyma,* and *sclerenchyma.*

Parenchyma tissue is the most abundant ground tissue. Its cells are thin walled and alive at maturity, and they typically carry out most of the plant's metabolic activities (**FIG. 42-5a**). Parenchyma cells have such diverse functions as photosynthesis, secretion of hormones, support, and food storage. Potatoes, seeds, fruits, and storage roots such as carrots are packed with parenchyma cells that store various types of sugars and starches.

Collenchyma tissue consists of elongated, polygonal (many-sided) cells with unevenly thickened cell walls (**FIG. 42-5b**). Collenchyma cells are alive at maturity but generally cannot divide. Although strong, the cell walls of collenchyma are still somewhat flexible. In herbaceous plants and in the leaf stalks and young growing stems of all plants, collenchyma tissue is an important source of support. The strings in celery stalks, for example, are mostly collenchyma cells in association with vascular tissue.

Sclerenchyma tissue is composed of cells with thick, hardened secondary cell walls (located between the outer primary wall and the plasma membrane) reinforced with

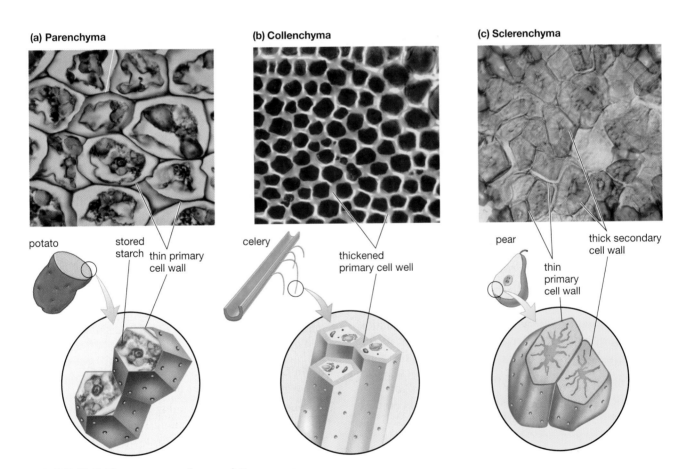

FIGURE 42-5 The structure of ground tissue
(a) Parenchyma cells are living and serve many functions. They have thin, flexible primary cell walls. These parenchyma cells are used for starch storage in a potato. **(b)** Collenchyma cells are living and have thickened, but somewhat flexible, primary walls. They help support the plant body (as seen in this celery stalk). **(c)** Sclerenchyma cells have thick, rigid secondary cell walls and die after they differentiate. Illustrated are "stone cells" that give pear fruit its slightly gritty texture.

(a)

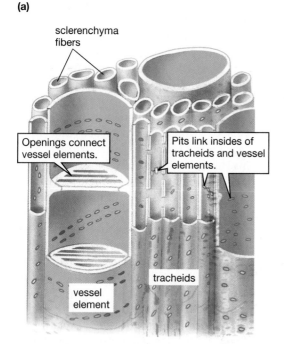

(b)

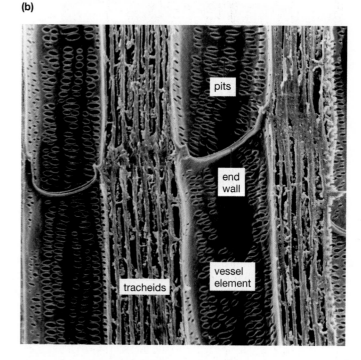

FIGURE 42-6 The structure of xylem
(a) Xylem includes sclerenchyma fibers for support and two types of conducting cells: tracheids and vessel elements. Tracheids are thin with ends and sides connected by pits. Pits in side walls interconnect tracheids and vessel elements. **(b)** A micrograph of xylem. Numerous pits are visible in the large vessel elements; the more abundant tracheids are much smaller tubes.

the stiffening substance *lignin* (**FIG. 42-5c**). Like collenchyma, sclerenchyma cells support and strengthen the plant body; however, unlike collenchyma, they die after they differentiate. Their thick, hardened cell walls then remain as a source of support. Sclerenchyma cells provide the fibers of hemp and jute, used for making rope. Other types of sclerenchyma cells form nut shells, the outer covering of peach pits, and the gritty texture of pears. Sclerenchyma cells also support the vascular tissues, described next.

The Vascular Tissue System Transports Water and Nutrients

The vascular tissue system of plants serves a function somewhat like that of the blood vessels in animals; it conducts water and dissolved substances throughout the body. As described later in this chapter, parts of the vascular system also contribute to the supportive "skeleton" of most plants. The vascular tissue system consists of two complex conducting tissues: *xylem* and *phloem*.

Xylem Conducts Water and Dissolved Minerals from the Roots to the Rest of the Plant

Xylem tissue in angiosperms includes supporting sclerenchyma fibers and two specialized cells types: *tracheids* and *vessel elements* (**FIG. 42-6**), which conduct water and dissolved mineral nutrients from roots up to the shoots. The final step in the formation of both tracheids and vessel

elements is cell death, leaving behind hollow tubes of nonliving cell wall.

Tracheids are thin, tubelike cells with thick cell walls, stacked atop one another (Fig. 42-6b). Their slanted, overlapping ends resemble the tips of hypodermic needles. The end walls contain **pits**, porous sections consisting of only a thin, water-permeable primary cell wall, which allows water and minerals to pass freely from one tracheid to the next, or from a tracheid to an adjacent vessel element.

Vessel elements, which are larger in diameter than tracheids, form relatively unobstructed pipelines called **vessels** from root to leaf (Fig. 42-6a). Vessel elements are stacked end to end, and their adjoining end walls may be connected by holes; or the walls may disintegrate, leaving an open tube.

Phloem Conducts Water, Sugars, Amino Acids, and Hormones Throughout the Plant Body

Phloem carries water containing dissolved substances synthesized by the plant, including sugars, amino acids, and hormones. In addition to sclerenchyma fibers, phloem contains *sieve-tube elements* and *companion cells*. **Sieve-tube elements** are the conductive cells of phloem (**FIG. 42-7**). As sieve-tube elements mature, most of their internal contents disintegrate, leaving behind only a thin layer of cytoplasm enclosed by the plasma membrane along the edge of the cell wall. At the ends of sieve-tube elements, membrane-lined pores connect adjacent cells, creating structures

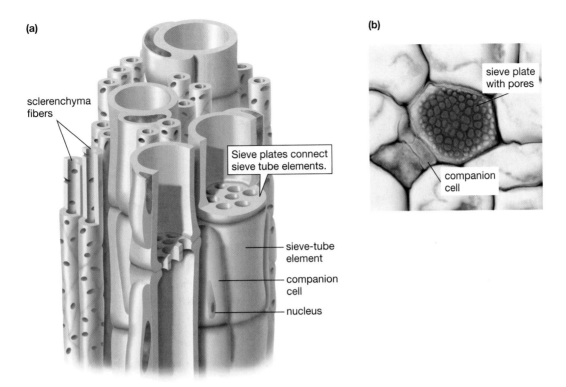

(a)

sclerenchyma fibers

Sieve plates connect sieve tube elements.

sieve-tube element

companion cell

nucleus

(b)

sieve plate with pores

companion cell

FIGURE 42-7 The structure of phloem
(a) Phloem includes sclerenchyma fibers for support, sieve-tube elements, and companion cells. Sieve-tube elements, stacked end to end, form the conducting system of phloem. Where they join, large membrane-lined pores allow fluid to pass freely between them. Each sieve-tube element has a companion cell that nourishes it and regulates its function. **(b)** A light micrograph of the end of a sieve tube element showing the sieve plate.

called *sieve plates*. Sieve plates allow fluid to move freely from one cell to the next. A continuous conducting system of *sieve tubes* is forged by a series of sieve-tube elements linked end to end.

Sieve-tube elements possess plasma membranes, a few mitochondria, and some endoplasmic reticulum, but they generally lack ribosomes, Golgi complexes, and nuclei. How, then, can sieve-tube elements remain alive? Each sieve-tube element is nourished by a smaller, adjacent **companion cell**. Companion cells are connected to sieve-tube elements by cytoplasm-filled channels called *plasmodesmata* (described in Chapter 5). Companion cells help maintain the integrity of the sieve-tube elements by donating high-energy compounds and directing the synthesis of important proteins.

42.3 WHAT ARE THE STRUCTURES AND FUNCTIONS OF LEAVES, ROOTS, AND STEMS?

Leaves Are Nature's Solar Cells

Leaves are the major photosynthetic structures of most plants; their rich green color arises from light-absorbing chlorophyll molecules. As chlorophyll breaks down in the fall, the yellows and oranges of additional light-trapping carotenoid pigments are revealed.

The shapes and structures of leaves have evolved in response to the environmental challenges that plants face in obtaining the essentials for photosynthesis: solar energy, carbon dioxide (CO_2), and water. Water is absorbed from the soil and transported to the leaf through the xylem, while CO_2 diffuses into the leaf from the air. Generally, then, a leaf will have a large light-gathering surface and openings to permit CO_2 entry. To prevent excessive evaporation, the leaf must be reasonably waterproof as well. The leaves of flowering plants display all of these features, as illustrated in **FIGURE 42-8**.

Leaves Have Two Major Parts: Blades and Petioles

A typical angiosperm leaf consists of a broad, flat portion, the **blade**, connected to the stem by a stalk called the **petiole** (see Fig. 42-1). The petiole positions the blade, usually orienting the leaf for maximum exposure to the sun. Inside the petiole and blade, vascular tissues provide a conducting system between the leaf and the rest of the plant body. Within the blade, these branch into **vascular bundles**, also called **veins**.

As shown in Figure 42-8, the structure of the leaf epidermis reflects the needs of the plant to conserve water while allowing CO_2 to enter. It consists of a layer of nonphotosynthetic, transparent epidermal cells that secrete a water-resistant waxy cuticle on their outer surfaces. The epidermis and its cuticle are pierced by adjustable pores,

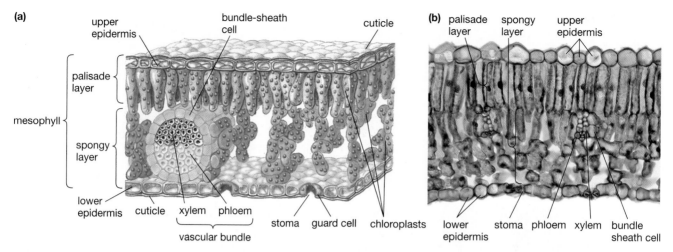

FIGURE 42-8 The structure of a typical dicot leaf
(a) The cells of the epidermis lack chloroplasts and are transparent, allowing sunlight to penetrate to the chloroplast-containing mesophyll cells beneath. The stomata that pierce the epidermis and the loose, open arrangement of the mesophyll cells ensure that CO_2 can diffuse into the leaf from the air and reach all the photosynthetic cells. **(b)** A light micrograph of a section from a lilac leaf.

the **stomata** (singular, stoma), which regulate the rate at which gases (CO_2, O_2, and water vapor) diffuse into or out of the leaf. Each stoma is surrounded by two sausage-shaped **guard cells**, which regulate the size of the opening into the interior of the leaf (see Fig. 42-8 and Fig. 42-22). Unlike the other epidermal cells, guard cells contain chloroplasts and can carry out photosynthesis. As we will see, photosynthesis in the guard cells contributes to their ability to adjust the size of the pore.

Beneath the epidermis lies the **mesophyll** ("middle of the leaf"), which consists of loosely packed parenchyma cells where most of the photosynthetic activity of the leaf occurs in chloroplasts. The spaces between these cells (see Fig. 42-8) allow CO_2 entering the stomata to diffuse easily among them, and oxygen (a photosynthetic by-product) to diffuse away. Many leaves possess two types of mesophyll cells—a layer of columnar *palisade cells* just beneath the upper epidermis and a layer of irregularly shaped *spongy cells* above the lower epidermis. Vascular bundles containing both xylem and phloem are embedded within the mesophyll, sending fine veins close to each photosynthetic cell. Thus, each mesophyll cell receives energy from sunlight transmitted through the clear epidermis, carbon dioxide from the air, and water from the xylem. The sugars it produces are carried to the rest of the plant by the phloem.

Stems Elevate and Support the Plant Body

Plant **stems** and their branches support and separate the leaves, lifting them into the sunlight and exposing them to the air. But in helping leaves obtain solar energy and CO_2 for photosynthesis, stems also separate leaves from the essential water and minerals that land plants obtain from soil. Further, stems separate the roots from the essential high-energy molecules that photosynthesizing leaves produce. So stems must not only conduct water and dissolved

minerals up to the leaves but also move high-energy molecules downward to sustain the roots.

The Stem Includes Four Types of Tissue

At the tip of the newly developing shoot sits a **terminal bud** consisting of apical meristem cells surrounded by developing leaves produced by the meristem (see Fig. 42-1). Some daughter cells of the terminal bud meristem differentiate into the specialized cell types of the stem. Most young stems are composed of four tissues: epidermis (dermal tissue), cortex (parenchyma), pith (parenchyma), and vascular tissues, described in the following sections. As Figure 42-2 illustrates, monocots and dicots differ somewhat in their arrangement of vascular tissues; here we discuss only dicot stems.

The Epidermis of the Stem Reduces Water Loss While Allowing Carbon Dioxide to Enter

The epidermis of the stem, like that of leaves, secretes a waxy cuticle that retards water loss. As in leaves, the stem epidermis is commonly perforated by stomata that regulate the entry of CO_2 and allow release of oxygen.

The Cortex and Pith Support the Stem, Store Food, and May Photosynthesize

The **cortex** is located between the epidermis and vascular tissues, while the **pith** lies central to the ring of vascular tissues, filling the inner portion of the young stem. In some young stems, it is difficult to tell where cortex ends and pith begins. Cortex and pith, which are parenchyma cells, perform three major functions: support, food storage, and (in some plants) photosynthesis.

- *Support*. In very young stems, water filling the central vacuoles of cortex and pith cells causes turgor pressure (see Chapter 5). Turgor pressure stiffens the cells, much as air inflates a tire. If you forget to water your houseplants, their drooping tips show the importance of turgor pressure in keeping young stems erect. Older stems also have collenchyma or sclerenchyma cells with thickened cell walls and do not rely on turgor pressure for support.

- *Storage.* Parenchyma cells in both cortex and pith convert sugar into starch and store the starch as a food reserve.
- *Photosynthesis.* In many stems, the outer layers of cortex cells contain chloroplasts and carry out photosynthesis. In plants such as cacti, in which the leaves are reduced to spines, the cortex of the stem is the only green photosynthetic part of the plant.

Vascular Tissues in the Stem Transport Water, Dissolved Nutrients, and Hormones

Vascular tissues, which transport water and its dissolved substances, interconnect all parts of the plant body. The vascular tissues produced by primary growth from the apical meristem are called primary xylem and primary phloem. In young dicot stems, the primary xylem and

phloem are separated by meristem tissue called vascular cambium, which is left behind by the apical meristem as the stem elongates. These three tissue layers extend along the length of the stem, initially forming rings of wedge-shaped bundles of xylem and phloem (**FIG. 42-9**; see Fig. 42-2). Secondary growth in dicot stems, discussed later, will result in concentric cylinders of xylem and phloem.

Stem Branches Arise from Meristem Cells of Lateral Buds

As the stem elongates, small clusters of apical meristem cells are "left behind" at discrete, characteristic locations, called **nodes**, on the stem (parts of the stem between nodes are **internodes**; see Fig. 42-9). The meristem cells at nodes give rise to **leaf primordia**, which will develop into the mature leaves unique to each species of plant. Just above the attachment point of these leaves, the meristem cells also produce **lateral buds**, which grow into branches.

When stimulated by the appropriate hormones (described in Chapter 44), the meristem cells of a lateral bud

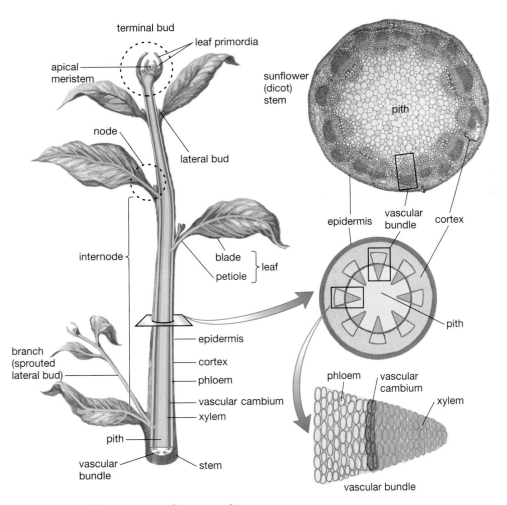

FIGURE 42-9 The structure of a young dicot stem
At the tip of the stem, the terminal bud includes the apical meristem and several leaf primordia produced by the meristem. Leaves and lateral buds are located at nodes, which are separated by internodes. In cross section, vascular tissue forms a ring of vascular bundles in dicots such as the sunflower shown in the photomicrograph. As the stem grows, leaf primordia develop into leaves, and internodes elongate. A lateral bud (meristem tissue) remains between each leaf and the stem, and may sprout into a branch.

FIGURE 42-10 How branches form

are activated and the bud sprouts, growing into a branch (**FIG. 42-10**). As the meristem cells divide, they release hormones. These hormones cause the parenchyma cells of the cortex between the bud and the vascular tissues of the stem to differentiate into xylem and phloem, which connects with the main vascular systems in the stem. As the branch grows, it duplicates the development of the stem. The apical meristem at its tip causes the branch to elongate and also produces leaf primordia and lateral buds.

Secondary Growth Produces Thicker, Stronger Stems

In conifers and perennial dicots, stems may survive up to hundreds of years, becoming thicker and stronger each year. This secondary growth in stems results from cell division in two lateral meristems: **vascular cambium** and *cork cambium* (**FIG. 42-11**).

Vascular Cambium Produces Secondary Xylem and Phloem

The vascular cambium is a cylinder of meristem cells located between the primary xylem and primary phloem. Daughter cells of the vascular cambium produced toward the inside of the stem differentiate into secondary xylem; those produced toward the outside of the stem differentiate into secondary phloem (see Fig. 42-11).

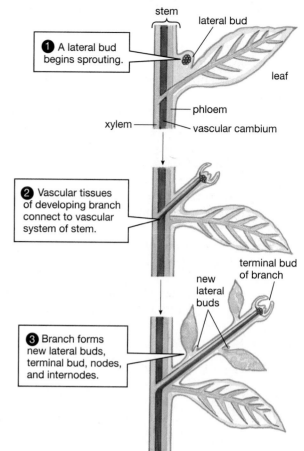

① A lateral bud begins sprouting.

stem
lateral bud
leaf
phloem
xylem
vascular cambium

② Vascular tissues of developing branch connect to vascular system of stem.

terminal bud of branch
new lateral buds

③ Branch forms new lateral buds, terminal bud, nodes, and internodes.

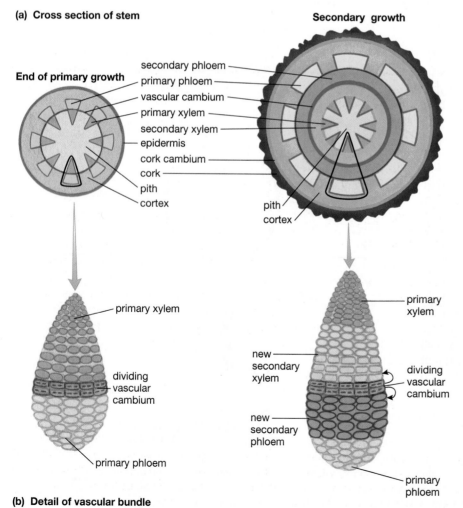

(a) Cross section of stem

Secondary growth

End of primary growth

secondary phloem
primary phloem
vascular cambium
primary xylem
secondary xylem
epidermis
cork cambium
cork
pith
cortex

pith
cortex

primary xylem

dividing vascular cambium

primary phloem

(b) Detail of vascular bundle

primary xylem

new secondary xylem

dividing vascular cambium

new secondary phloem

primary phloem

FIGURE 42-11 Secondary growth in a dicot stem
(a) Cross section of a dicot stem at the end of primary growth (left) and during early secondary growth (right). **(b)** A vascular bundle during secondary growth. Vascular cambium separates primary xylem and primary phloem. Division and differentiation of vascular cambium cells produce secondary xylem on the inside and secondary phloem on the outside. Cork cambium produces cork cells that cover the outside of the stem. QUESTION If a tree is "girdled" by removing a strip of bark completely around its trunk, it usually dies. Why?

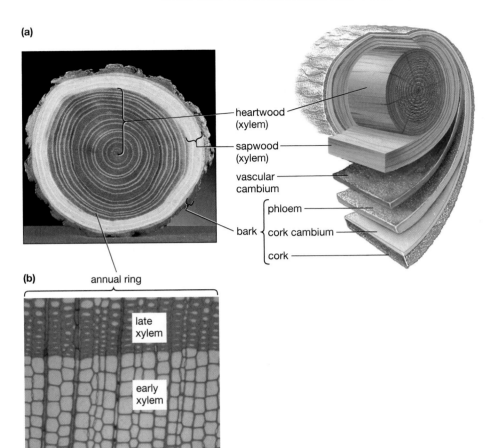

(a)

heartwood (xylem)

sapwood (xylem)

vascular cambium

phloem

bark {
cork cambium

cork

(b)

annual ring

late xylem

early xylem

FIGURE 42-12 How annual tree rings are formed

(a) Tree rings are clearly visible in this section of tree trunk. The ratio of cell wall to "hole" (the now-empty interior of the cell) determines the color of the wood: early wood, formed during the spring, with lots of hole, is pale; late wood, formed during the summer, with lots of wall, is dark. The water-transporting xylem of sapwood forms a lighter layer inside the bark. Xylem of the older heartwood, where the rings are most easily visible, no longer transports water and minerals. **(b)** As this micrograph shows, secondary xylem cells formed during the wet spring are large (early xylem), whereas secondary xylem cells formed during the hotter, drier summer are small (late xylem).

Secondary xylem, with its thicker cell walls, forms the wood in perennial woody plants, such as trees (the stems of trees are commonly called "trunks"). Young secondary xylem, called **sapwood**, transports water and minerals and is located just inside the vascular cambium. Older secondary xylem, the **heartwood**, fills the central portion of older stems or tree trunks (**FIG. 42-12**). Heartwood no longer carries water and solutes and provides only support and strength to the stem. Heartwood also serves as a collection site for metabolic wastes of the woody plant, such as gums, resins, and oils. These wastes increase the density of the heartwood, contribute to its darker color, and help the heartwood resist rotting.

Phloem cells are much weaker than xylem cells. As secondary phloem cells die over time, their sieve-tube elements and companion cells are crushed between the hard xylem on the inside of the trunk and the tough cork on the outside. Only a thin strip of recently formed phloem remains alive and functioning; secondary phloem contributes little to the increase in diameter of perennial plants.

In trees such as oaks and pines that are adapted to regions where there are pronounced seasons, cell division in the vascular cambium ceases during the cold of winter. In spring, the cambium cells divide, forming new xylem and phloem. The young cells grow by absorbing water and swelling while the newly formed cell walls are still soft. As the cells mature, their walls thicken and harden, preventing further growth. Because water is readily available in spring, young xylem cells swell considerably and are large at maturity. As summer progresses and water becomes scarcer, new xylem cells absorb less water and so are smaller when they mature. As a result, tree trunks in cross section show a pattern of alternating pale regions (large cells formed in spring) and dark regions (small cells formed in summer), as shown in Figure 42-12.

This pattern forms the familiar **annual rings** of growth. You can determine the approximate age of a tree that has been cut by counting the dark growth rings. Scientists can also use the width of rings to reconstruct past climate, because wet years produce more growth and wider rings. By studying the rings in ancient trees, including a 1000-year-old cypress, researchers have constructed an 800-year record of climate in Virginia, including a 7-year drought (from 1606 to 1612). They hypothesize that this drought was responsible for the mysterious disappearance of the Virginia colony of Jamestown, founded 1607.

Secondary Growth Causes the Epidermis to Be Replaced by Woody Cork

Recall that epidermal cells are mature, differentiated cells that can no longer divide. As new secondary xylem and phloem are added each year, enlarging the stem, the epidermis can't expand enough to keep up with the increasing circumference, and it splits off and dies. Apparently stimulated by hormones, some parenchyma cells in the cortex become rejuvenated and form a new lateral meristem

FIGURE 42-13 Cork forms the outer layer of bark
(a) An ancient sequoia in the Sierra Nevada of California. The cork of a sequoia eventually produces a thick, fire-resistant outer covering half a meter or more thick. This massive cork layer contributes to the sequoia's great longevity. Blackened areas on this cork are from past fires. **(b)** A layer of cork has been stripped from this cork oak. It will regenerate and can be harvested in about a decade.

layer, the cork cambium (see Fig. 42-11). These cells divide, forming daughter cells, called cork cells or simply cork, and develop tough, waterproof cell walls that protect the trunk both from drying out and from physical damage. Cork cells die as they mature and may form a protective layer up to a half meter thick in some large tree species, such as the sequoia (**FIG. 42-13a**). As the trunk expands from year to year, the outermost layers of cork split apart or peel off, accommodating the growth. Corks used to plug bottles are made from the outermost layer of cork from cork oaks, carefully peeled off by harvesters (**FIG. 42-13b**). The cork of the cork oak separates from the cork cambium, leaving this meristem layer behind, so the tree is not harmed. About 10 years later, harvesters will return and strip off the regenerated layer of cork.

The common term **bark** includes all the tissues outside the vascular cambium: phloem, cork cambium, and cork cells. Removing a strip of bark all the way around a tree, called *girdling*, kills the tree because it severs the phloem. Without phloem, sugars synthesized in the leaves cannot reach the roots. Deprived of energy, the roots can no longer take up minerals, and the tree dies.

Roots Anchor the Plant, Absorb Nutrients, and Store Food

Primary Growth Causes Roots to Elongate and to Branch

As a seed sprouts, the **primary root**—the first root to develop—grows down into the soil. Many dicots, such as carrots and dandelions, develop a taproot system. A **taproot system** consists of the primary root and many smaller roots that grow out from its sides (**FIG. 42-14a**). In contrast,

monocots such as grasses and palms produce a **fibrous root system** in which the primary root dies and is replaced by many new roots that emerge from the base of the stem (**FIG. 42-14b**). Root systems anchor the plant, absorb water and minerals from the soil, and store water and food molecules. Taproots are well-suited for storing large quantities of food (see "Evolutionary Connections: Interesting Adaptations of Roots, Stems, and Leaves").

(a)   (b)

FIGURE 42-14 Typical root systems in dicots and monocots
(a) Dicots typically have a taproot system, consisting of a long central root with many smaller, secondary roots branching from it. **(b)** Monocots usually have a fibrous root system, with many roots of equal size.

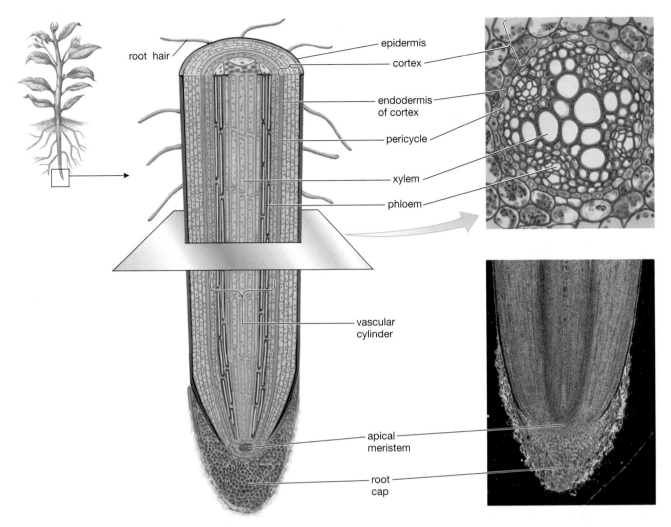

epidermis

cortex

root hair

endodermis
of cortex

pericycle

xylem

phloem

vascular
cylinder

apical
meristem

root
cap

FIGURE 42-15 Primary growth in roots
Primary growth in roots results from mitosis in the apical meristem near the tip. The root is composed of the
root cap, epidermis, vascular cylinder, and cortex.

In young roots of both taproot and fibrous root systems, divisions of the apical meristem give rise to four distinct regions (**FIG. 42-15**). At the very tip of the root, daughter cells produced on the lower portion of the apical meristem differentiate into the **root cap**. The root cap protects the apical meristem from being scraped off as the root pushes down between the rocky particles of the soil. Root-cap cells have thick cell walls and secrete a slimy lubricant that helps ease the root between soil particles. Nevertheless, root-cap cells wear away and must be continuously replaced by new cells from the meristem.

Daughter cells produced on the upper portion of the apical meristem differentiate into one of three parts: an outer epidermis, a layer of **cortex** beneath the epidermis, and a core called the *vascular cylinder* (see Fig. 42-15).

Under the influence of plant hormones, roots may branch. Root branches originate from the **pericycle** (the outermost layer of the vascular cylinder), a remnant of the apical meristem that retains the ability to divide. These pericycle cells divide and form the apical meristem of a **branch root** (**FIG. 42-16**). Branch root development is similar to

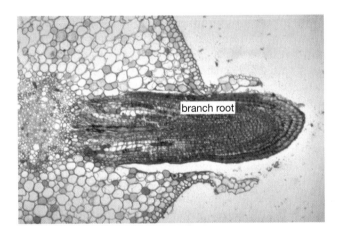

branch root

FIGURE 42-16 Branch roots
Branch roots emerge from the pericycle of a root. The center of this branch root is already differentiating into vascular tissue.

FIGURE 42-17 Root hairs
Root hairs, shown here in a sprouting radish, greatly increase a root's surface area, enhancing the absorption of water and minerals from the soil.

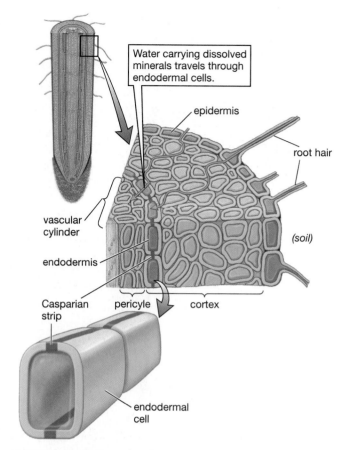

Water carrying dissolved minerals travels through endodermal cells.

epidermis

root hair

(soil)

vascular cylinder

endodermis

Casparian strip   pericycle   cortex

endodermal cell

FIGURE 42-18 The role of the Casparian strip
The Casparian strip is a band of waterproof material joining the walls of endodermal cells. Encircling each cell, the Casparian strip forces water to move by osmosis through the cell membranes instead of flowing through the porous cell walls. QUESTION What problem would arise in a root with no waterproof Casparian strip?

that of primary roots except that the branch must break out through the cortex and epidermis of the primary root. It does so both by crushing the cells that lie in its path and by secreting enzymes that digest them. The vascular tissues of the branch root connect with the vascular tissues of the primary root.

The Epidermis of the Root Is Permeable to Water and Actively Transports Mineral Nutrients

The root's outermost covering of cells is the epidermis, which is in contact with the soil and with the air and water that are trapped among the soil particles. Many epidermal cells grow root hairs into the surrounding soil (**FIG. 42-17**). By expanding the root's surface area, root hairs increase its ability to absorb water and minerals. Root hairs may add dozens of square meters of surface area to the roots of even small plants.

The root epidermal membrane actively transports minerals that the plant needs from the soil into the epidermal cells. The high concentration of dissolved minerals (solutes) then draws water into the cells by osmosis (as described in Chapter 5, osmosis is the movement of water through a membrane from a region of lower solute concentration to a region of higher solute concentration). The nonliving cell walls that surround plant cells are porous to water, so water also moves from soil into roots directly through the open spaces between the cellulose fibers that comprise the cell walls. This direct movement of water (including any dissolved substances it contains) is called **bulk flow**. In contrast to osmosis, bulk flow does not occur through a cell membrane.

Cortex Makes Up Much of the Interior of a Young Root

Cortex occupies most of the inside of a young root. Most of the cortex consists of a mass of large, loosely packed parenchyma cells just beneath the epidermis. Sugars produced in the shoot by photosynthesis are transported

down to the parenchyma cells of the cortex, where they are converted to energy-rich food compounds such as starch and stored. Roots of perennial plants store high-energy sugars and starches through the cold winter months, releasing them to power growth of the shoot system in the spring. Parenchyma cells are particularly abundant in roots specialized for carbohydrate storage, such as those of carrots and sweet potatoes.

The innermost layer of cortex consists of a ring of smaller, closely packed cells called the **endodermis** that encircles the vascular cylinder (see Fig. 42-15 and **FIG. 42-18**). The cell wall of each endodermal cell contains a band of fatty, waterproof material called the **Casparian strip**. The Casparian strip blocks water and dissolved minerals from traveling through the cell walls between endodermal cells, but it does not cover the cell surfaces facing the rest of the cortex or those facing the vascular cylinder. Water carrying dissolved substances can flow freely around both epidermal and cortex parenchyma cells by moving through the porous cell walls outside their plasma membranes. When water reaches the endodermis, however, the Casparian strip serves as a gatekeeper, forcing all the fluid that enters the vascular cylinder to move through the living membranes of the endodermal cells (see Fig. 42-18). Thus, the endodermal cells can regulate which substances enter the vascular cylinder, excluding some that might harm the plant. You'll learn another function of the Casparian strip in the next section.

The Root Vascular Cylinder Contains Conducting Tissues

The **vascular cylinder** contains the conducting tissues of xylem and phloem, which transport water and dissolved materials within the plant. The outermost layer of the vascular cylinder is the *pericycle*, located just inside the endoderm of the cortex. The pericycle cells receive water and minerals from the endodermal cells and actively transport the minerals into the interior of the vascular cylinder (water then follows by osmosis). Active transport of minerals out of the pericycle cells maintains the concentration gradient by lowering mineral concentrations in the pericycle cells. This keeps minerals moving by diffusion from the parenchyma of the cortex into the endoderm cells and then into the pericycle cells, with water always following by osmosis. The Casparian strip of the endodermal cells prevents the mineral-rich water from leaking back into the cortex through their porous cell walls.

Within the vascular cylinder, water carrying its high concentration of dissolved mineral nutrients flows through the cell walls of the nonliving xylem cells and up through the plant body (in a later section, you'll learn how water in plants seemingly defies gravity). Meanwhile, the phloem carries high-energy molecules derived from photosynthesis, such as sugars, down from the leaves. Some of this food is stored; the rest provides energy that allows the root cells to maintain themselves, grow, and actively transport minerals.

42.4 HOW DO PLANTS ACQUIRE MINERAL NUTRIENTS?

Roots Absorb Mineral Nutrients from Soil

Nutrients are chemicals found in the environment that organisms require for their growth and survival. About 20 mineral nutrients are essential or beneficial for plants. Plants require relatively large quantities of carbon (obtained from carbon dioxide), hydrogen (from water), oxygen (from air and water), phosphorus (from phosphate ions in soil), nitrogen (from nitrate and ammonium ions in soil), magnesium, calcium, and potassium (as ions in soil). Plants also require small quantities of nutrients such as iron, chlorine, copper, manganese, zinc, boron, and molybdenum. Carbon dioxide and oxygen usually enter plants by diffusion from the air into leaves, stems, and roots. Roots extract water and all other nutrients, collectively called **minerals**, from the soil.

Soil consists of bits of pulverized rock, air, water, and organic matter. Although the rock particles and the organic matter contain essential minerals, only those dissolved in the soil water are accessible to roots. The concentration of minerals in the soil water is very low, usually much lower than the concentration within plant cells and fluids. For example,

the concentration of potassium in root cells is at least 10 times greater than that in soil water, so diffusion cannot move potassium into the root. In general, roots use active transport to pump minerals into their epidermal cells against their concentration gradients. This requires ATP generated by root mitochondria, and this ATP production requires oxygen. To support plants, then, soil must have some oxygen-containing air spaces within it. Flooding (or overwatering) can kill plants by depriving their roots of oxygen. Mineral absorption by roots is explained in more detail in "A Closer Look: How Roots Absorb Water and Minerals."

Symbiotic Relationships Help Plants Acquire Nutrients

Many minerals are too scarce in soil water to support plant growth. One mineral—nitrogen—is almost always in short supply. Scientists hypothesize that red fall colors are actually synthesized to help the plant recover nitrogen-containing compounds in its leaves before they drop in the fall (see "Case Study Revisited: Why Do Leaves Turn Red in the Fall?"). Most plants have evolved beneficial relationships with either specialized fungi or bacteria that help the plants acquire scarce minerals such as nitrates and phosphates.

Fungal Mycorrhizae Help Plants Acquire Minerals

Most land plants form symbiotic relationships with fungi, producing root–fungus complexes called **mycorrhizae**, which help the plant extract and absorb minerals from soil. Microscopic fungal strands intertwine between the root cells and extend out into the soil (**FIG. 42-19**). The fungus renders nutrients such as phosphorus more accessible for uptake by the roots, perhaps by converting rock-bound minerals into simple soluble compounds that the root cell plasma membranes can transport. The fungus, in return, receives sugars, amino acids, and vitamins from the plant. In this way, both the fungus and the plant can grow in places where neither could survive alone, including deserts and high-altitude, nutrient-scarce, rocky soils.

Research has revealed that in some forests, mycorrhizae form an immense underground web that interlinks trees—even trees of different species. This web of fungi transfers carbon compounds produced by photosynthesis among the trees, causing those with access to abundant sunlight to subsidize their shaded neighbors. Thus, like an underground Robin Hood, the mycorrhizae transfer photosynthetic products from the rich to the poor. Researchers hypothesize that nutrient transfer among trees by mycorrhizae may be an important factor in the overall health of the forest.

Bacteria-Filled Nodules on the Roots of Legumes Help Those Plants Acquire Nitrogen

Since amino acids, nucleic acids, and chlorophyll all contain nitrogen, plants need large amounts of this element. Although nitrogen gas (N_2) makes up about 78% of the atmosphere, plants can take up nitrogen only through their roots, dissolved in water in the form of ammonium ions (NH_4^+) or nitrate ions (NO_3^-)

A CLOSER LOOK　How Roots Absorb Water and Minerals

Water enters the root freely through the porous, nonliving cell walls that surround each cell. This bulk flow continues until the water reaches the endodermis, where the Casparian strip forces water to move through the cell membrane by osmosis (black arrow in **FIG. E42-1**).

Most mineral absorption by roots occurs in a three-step process (see Fig. E42-1):

1. *Active transport into root hairs.* Root hairs projecting from the epidermal cells provide most of the surface area of the root and are in close contact with the soil water. The plasma membranes of the root hairs use the energy of ATP to transport minerals from the soil water, concentrating the minerals in the epidermal cytoplasm. Water then follows by osmosis.

2. *Diffusion through cytoplasm to pericycle cells.* The cytoplasm of adjacent living plant cells is interconnected by plasmodesmata. Minerals diffuse down their concentration gradients through these plasmodesmata, starting from the epidermal cells (where active transport has concentrated them) through the parenchyma cells of the cortex, then into the endodermal cells of the cortex, and finally into the pericycle cells that make up the outer layer of the vascular cylinder.

3. *Active transport into the extracellular space of the vascular cylinder.* Pericycle cells actively transport minerals out of their cytoplasm into the extracellular space within the vascular cylinder. Water follows by osmosis.

4. *Diffusion into the xylem.* Both the tracheids and vessel elements of mature xylem consist only of a "skeleton" of nonliving cell walls, which are full of holes (see Fig. 42-6). Water carrying dissolved minerals easily moves into the empty tubes of the xylem.

You can now appreciate one function of the waterproof Casparian strip that seals the spaces between the endodermal cells surrounding the vascular cylinder. If the water and dissolved minerals that enter the vascular cylinder could flow back through the cell walls outside the endodermal cells, minerals would leak back out of the vascular cylinder as fast as they were pumped in. The Casparian strip retains the mineral-rich water within the vascular cylinder, where it is pulled upward into the xylem.

FIGURE E42-1 Mineral and water uptake by roots
Dark blue arrow: Concentrated mineral solution follows a path through the interior of cells. Pericycle cells actively transport minerals into the interior of the vascular cylinder, maintaining a concentration gradient that keeps the minerals diffusing inward and water following by osmosis. **Pale blue arrow:** Water moves freely through cell walls until it reaches the Casparian strip and must travel by osmosis through the endodermal cell membranes to enter the vascular cylinder.

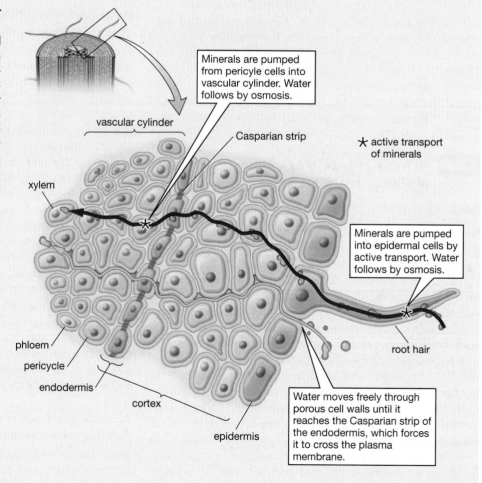

Minerals are pumped from pericyle cells into vascular cylinder. Water follows by osmosis.

vascular cylinder

Casparian strip

✳ active transport of minerals

xylem

Minerals are pumped into epidermal cells by active transport. Water follows by osmosis.

root hair

phloem

pericycle

endodermis

cortex

epidermis

Water moves freely through porous cell walls until it reaches the Casparian strip of the endodermis, which forces it to cross the plasma membrane.

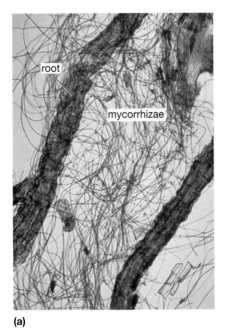

(a)

(b)

FIGURE 42-19 Mycorrhizae: a root–fungus symbiosis
(a) A tangled meshwork of fungal strands surrounds and penetrates into the root. (b) Seedlings growing under identical conditions with (right) and without (left) mycorrhizal fungi illustrate the importance of mycorrhizae in plant nutrition. QUESTION Based on what you've learned about root function, which part of the root system might you expect to be infected by mycorrhizal fungi?

Although N_2 diffuses from the atmosphere into the air spaces in the soil, it cannot be used by plants. Plants don't have the enzymes needed to carry out **nitrogen fixation**, the conversion of N_2 into NH_4^+ or NO_3. A variety of **nitrogen-fixing bacteria**, some of which live freely in the soil, do have these enzymes. However, nitrogen fixation is very costly (energetically speaking) using at least 12 ATPs per ammonium ion synthesized. Consequently, bacteria don't routinely manufacture a lot of extra NH_4^+ and liberate it into the soil.

Some plants, particularly **legumes** (such as peas, clover, alfalfa, and soybeans), enter into a mutually beneficial relationship with certain species of nitrogen-fixing bacteria. By secreting chemicals into the soil, legumes attract nitrogen-fixing bacteria to their roots. Once there, the bacteria enter the root hairs and make their way into cortex cells. Both the bacteria and their host cortex cells multiply, forming a swelling or **nodule** composed of bacteria-containing cortex cells (**FIG. 42-20**). In a cooperative relationship, the bacteria within the cortex cells use some of the plant's stored food molecules to power their metabolic processes, including nitrogen fixation. The bacteria obtain so much energy from the plant that they produce more NH_4^+ than they need. The surplus NH_4^+ diffuses into the cytoplasm of their host plant cells, providing the plant with a steady supply. Ammonium ions also diffuse into the surrounding soil, making it better able to support other types of plants. Farmers plant legumes not only for their commercial value but also to enrich the soil with NH_4^+ for future crops.

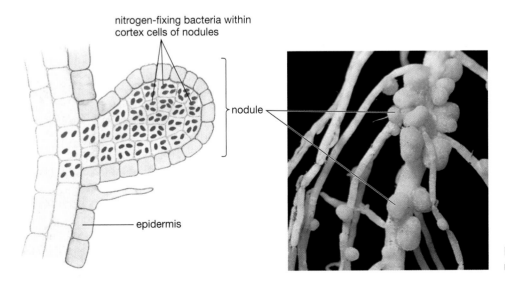

nitrogen-fixing bacteria within cortex cells of nodules

nodule

epidermis

FIGURE 42-20 Root nodules house nitrogen-fixing bacteria

FIGURE 42-21 The cohesion–tension theory of water flow from root to leaf in xylem
① As water molecules evaporate out of the leaves through transpiration, other water molecules replace them from the xylem of the leaf veins. ② As the top of the "water chain" is pulled up by evaporation, the rest of the chain, all the way down to the roots, comes along as well. ③ As the molecules of the water chain travel up the xylem in the roots, the decreased water pressure within the root xylem and the surrounding extracellular space causes water to enter from the soil, thus steadily replenishing the bottom of the chain.

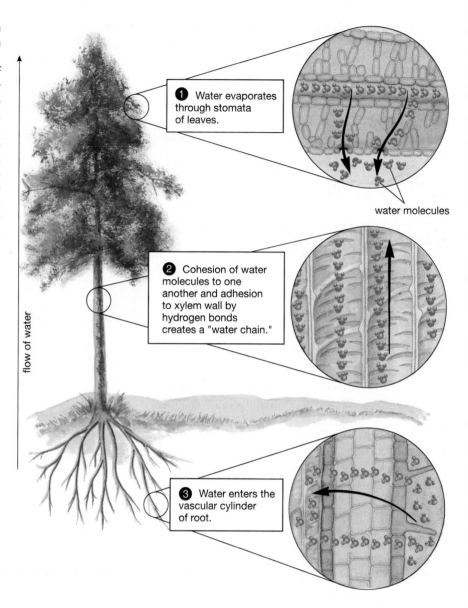

❶ Water evaporates through stomata of leaves.

water molecules

❷ Cohesion of water molecules to one another and adhesion to xylem wall by hydrogen bonds creates a "water chain."

flow of water

❸ Water enters the vascular cylinder of root.

42.5 HOW DO PLANTS MOVE WATER UPWARD FROM ROOTS TO LEAVES?

At least 90% of the water absorbed by the roots of plants evaporates through the stomata of leaves and, to a lesser extent, through the stomata of stems in a process called **transpiration**. As you will see, transpiration drives the movement of water upward through the plant body.

Water Movement in Xylem Is Explained by the Cohesion–Tension Theory

After entering the root xylem, water and minerals still must be moved to the uppermost reaches of the plant. In redwood trees, the distance may be more than 300 feet (about 100 meters). Bulk flow moves fluids up through the xylem from root to stem and leaf. Because minerals are dissolved in water, they are passively carried along as the water flows upward. But how do plants overcome the force of gravity and make water flow upward? The *cohesion–tension theory* provides an explanation.

According to the **cohesion–tension theory**, water moves upward through xylem primarily because it is actually pulled up by the leaves using the process of transpiration—evaporation of water from the leaves (**FIG. 42-21**). As its name suggests, this theory has two essential parts:

- *Cohesion.* Attraction among water molecules holds water together, forming a solid chain-like column within the xylem tubes.
- *Tension.* This "water chain" is pulled up the xylem; evaporation provides the necessary energy.

Let's briefly examine both.

Hydrogen Bonds Between Water Molecules Produce Cohesion

You may recall from Chapter 2 that water is a polar molecule; its ends carry small opposite charges. As a result, nearby water molecules attract one another, forming hydrogen bonds. Just as individually weak cotton threads together make the strong fabric of your jeans, the network of individually weak hydrogen bonds in water collectively produce a strong *cohesion*, or tendency to resist being separated. Experiments have demonstrated that the column of water within the xylem is at least as strong—and as unbreakable—as a steel wire of the same diameter. This is the "cohesion" part of the theory: hydrogen bonds among water molecules provide the cohesion that creates a chain of water extending the entire height of the plant within the xylem. Supplementing the cohesion between water molecules is adhesion between water molecules and the walls of xylem. Attraction of water molecules to the cell walls of the thin xylem tubes helps the water creep upward, just as water moves upward into a very narrow glass tube.

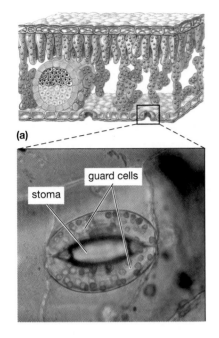

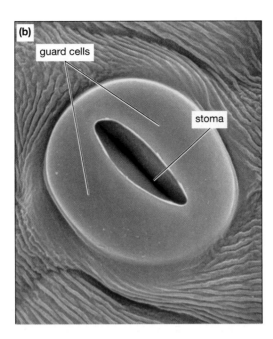

FIGURE 42-22 Stomata
Stomata seen through a **(a)** light microscope and **(b)** scanning electron microscope. In the light micrograph, note that the guard cells contain chloroplasts (the green ovals within the cells) but that the other epidermal cells do not. QUESTION When the stomata close, how is photosynthesis affected? How is the movement of water into the roots affected?

Transpiration Produces the Tension That Pulls Water Upward

Transpiration provides the force for water movement—the "tension" part of the theory. As a leaf transpires, the concentration of water in the mesophyll drops. This drop causes water to move by osmosis from the xylem into the dehydrating mesophyll cells. Water molecules leaving the xylem are linked to other water molecules in the same xylem tube by hydrogen bonds. Therefore, as water exits from the xylem and eventually evaporates, it pulls more water up the xylem. This process continues all the way to the roots, where water in the extracellular space around the xylem is pulled in through the holes in the walls of vessel elements and tracheids. The force generated by the evaporation of water from the leaves, transmitted down the xylem to the roots, is so strong that water can be absorbed even from quite dry soils. But can the cohesion–tension theory explain the movement of water from soil to the topmost leaves of giant redwoods? The answer is yes. Botanists have measured xylem water tensions strong enough to pull water up more than 600 feet (200 meters).

Scientists have calculated that a large maple tree may transpire about 250 gallons of water daily. How does a tree obtain the energy to lift about a ton of water about 45 feet above the ground every summer day? Think about transpiration a bit, and you will realize that pulling water up by transpiration only requires that the tree expose leaves with open stomata to sunlight. Solar energy provides the power directly by evaporating the water from the leaves. Also think about the impact of a forest of trees, each releasing hundreds of gallons of water into the air each day. As you might predict, transpiration can have a major impact on climate as described in "Earth Watch: Plants Help Regulate the Distribution of Water."

SUMMING UP

Water Transport in Xylem

Transpiration from leaves removes water from the top of a xylem tube. The transpired water is replaced by water from farther down the xylem tube, so water continues to move up the xylem tube by bulk flow. This upward flow removes water from the root xylem and the extracellular space surrounding it, promoting the movement of more water into the vascular cylinder of the root. The flow of water in xylem is unidirectional, from root system to shoot system, because only the shoot can transpire.

Adjustable Stomata Control the Rate of Transpiration

Although it provides the vital force that transports water and minerals up into the plant body, transpiration is also by far the largest source of water loss—a loss that can be fatal in hot, dry weather. Because most water lost through transpiration evaporates through the stomata of leaves and stems, you might think that a plant could prevent water loss simply by closing its stomata. Don't forget, however, that photosynthesis requires carbon dioxide from the air. Carbon dioxide diffuses in primarily through open stomata. Therefore, a plant must use its stomata to achieve a balance between acquiring carbon dioxide and losing water.

A stoma consists of a central opening surrounded by two sausage-shaped, photosynthetic guard cells that regulate the size of the opening (**FIG. 42-22**). With some exceptions, stomata open during the day, when sunlight allows photosynthesis, and close at night, conserving water. They will also close in the sunlight if the plant is losing too much water. Plants whose leaves are oriented horizontally generally have more stomata on the shaded, lower surface than on the sunny, upper surface in order to reduce evaporation.

The distribution of plants on Earth is limited by both environmental factors and plant adaptations. Probably the most important environmental factor influencing plant distribution is water: cacti inhabit deserts because they can withstand drought; orchids and mahogany trees need the frequent drenching rains of the rain forest. However, this relationship works both ways: plants, through transpiration, help regulate the amount and distribution of rainfall, soil water, and even river flow.

Consider the Amazon rain forest (**FIG. E42-2**). An acre of soil supports hundreds of towering trees, each bearing millions of leaves. The surface area of the leaves dwarfs the surface area of the soil, and up to 75% of all the water evaporating from the acre of forest is due to leaf transpiration. This transpiration raises the humidity of the air. About half of the water transpired from the leaves falls again as rain, so about one-third of the total rainfall is water recycled by transpiration. In a very real sense, the high humidity and frequent showers that the rain forest needs to survive are partly created by the trees themselves. Large-scale cutting

FIGURE E42-2 The Amazon rain forest
The rain-forest community helps produce and maintain its own environment.

of the Amazon rain forest continues. Trees are burned, releasing CO_2, and are replaced by crops to feed the growing population—and, recently, to grow soybeans for biodiesel fuels. Agricultural crops are not nearly as efficient at trapping and transpiring moisture as the trees they replace. Consequently, runoff increases, humidity falls, and rainfall declines. Decreased rainfall not only slows the regrowth of forest after the fields are abandoned but also harms adjacent forests, gradually changing their composition to non-rain-forest vegetation. Although portions of the rain forest can be protected from logging, the entire ecosystem is susceptible to the effects of climate change caused by the loss of transpiration in deforested areas.

The Monteverde cloud forest, which blankets the uppermost reaches of the Cordillera de Tilarán mountain range in Costa Rica, is entirely dependent on an almost constant shroud of fog. Transpiration from Costa Rica's lowland forests pumps moisture into the winds flowing up the mountain slopes; this moisture condenses and forms fog as the air cools. Scientists studying the Monteverde cloud forest have noticed an alarming trend: the clouds are lifting. Frog and toad populations have crashed in some areas, and birds not usually found in cloud forests are invading formerly clouded lowland areas. Why? A century of logging has eliminated over 80% of Costa Rica's lowland forests. Using satellite images, ecologist Robert Lawton and colleagues found that the deforested areas had relatively low cloud cover compared to nearby forests. Since the winds moving up the mountain slopes gain water from the lowlands, the researchers hypothesize that reduced transpiration has lowered the moisture content of the air, so it must rise farther before clouds form. Mist-free conditions are becoming more common, and scientists warn that if clouds disappear for days at a time, the fragile ecosystem could collapse.

These examples show that plants exert an enormous influence on the properties of the biosphere, such as humidity, rainfall, soil water, and river flow. These factors then feed back and influence all the life of the region. Human activities that alter plant cover can have far-reaching and unanticipated impacts on ecosystems.

Plants Regulate Their Stomata

How do plants control stomatal opening and closing? Stomata open when the guard cells take up water and elongate, bowing outward and increasing the space between them. Stomata close when guard cells lose water and shrink, reducing the space between them. The entry of water, in turn, is regulated by changes in the potassium content of the guard cells.

Several factors control the potassium concentration inside guard cells; the three most important are light, carbon dioxide, and water levels within the leaf. These help the plant achieve a balance between the need to photosynthesize and the need to conserve water.

- *Light.* When light strikes special pigments within the guard cells, it triggers a series of reactions that cause potassium to be actively transported into the guard cells. Water follows by osmosis, and the stomata swell and open. At night, the potassium pumping stops and potassium diffuses back out, causing the stomata to close and conserve water.

- *Carbon dioxide.* Low CO_2 concentrations (such as occur during daylight when photosynthesis exceeds cellular respiration) stimulate active transport of potassium into the guard cells. This transport causes stomata to open and allows CO_2 to diffuse into the leaf. At night, cellular respiration in the absence of photosynthesis raises CO_2 levels, halting the inward transport of potassium and allowing the guard cells to close.

- *Water.* If a leaf loses water faster than it can be replaced, it begins to wilt. Under these conditions, the leaf mesophyll cells release a hormone (abscisic acid, described in Chapter 44) that strongly inhibits active transport of potassium into the guard cells (even if light is adequate). As potassium diffuses out of the guard cells, water follows by osmosis, the guard cells shrink, and the stomata close. As you might guess, when your house or garden plants are wilted, they are unable to carry out normal levels of photosynthesis.

honeydew droplet

(a)

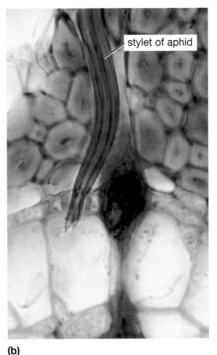

stylet of aphid

(b)

FIGURE 42-23 **Aphids feed on the sugary fluid in phloem sieve tubes**
(a) When an aphid pierces a sieve tube, pressure in the tube forces the fluid out of the phloem and into the aphid's digestive tract. The fluid is sometimes forced out its anus, as "honeydew." This fluid is collected by certain species of ants that, in turn, defend the aphids from predators. **(b)** The flexible stylet of an aphid passes through many layers of cells to penetrate a sieve-tube element.

The Pressure-Flow Theory Explains Sugar Movement in Phloem

The most widely accepted explanation for fluid transport in phloem of angiosperms is the **pressure-flow theory,** which states that differences in water pressure drive sap through phloem sieve tubes. These water pressure differences are actually created indirectly by the net production and use of sugar in different parts of the plant. Any portion of the plant that synthesizes more sugar than it uses is called a sugar **source**; a mature leaf is a good example. Conversely, any structure that uses up more sugar than it produces (this includes converting the sugar into starch for storage) is a sugar **sink**. Developing fruits are good examples of sugar sinks. Phloem sieve tubes carry sap away from sugar sources (which have excess sugar) and toward sugar sinks (where sugar is required).

The pressure-flow theory is illustrated in **FIGURE 42-24,** whose numbers correspond to those in the text. ① The sugar produced by a source cell (in the photosynthesizing

42.6 HOW DO PLANTS TRANSPORT SUGARS?

Sugars synthesized in the leaves must be moved to other parts of the plant, where they nourish non-photosynthetic structures such as roots or flowers, or are stored in cortex cells of roots and stems. Sugar transport is the function of phloem.

Botanists studying fluid in phloem have employed a most unlikely lab assistant: the aphid. *Aphids* are insects that feed on the sugary fluid called *phloem sap* contained in phloem sieve tubes. The aphid inserts its stylet, a pointed, hollow feeding tube, through the epidermis and cortex of a young stem into a sieve tube (**FIG. 42-23**). The aphid can then relax and let the fluid flow through the stylet into its digestive tract; the pressure driving the sap causes the aphid's body to expand like a balloon. By leaving the aphid's stylet in place while removing the rest of the aphid's body, botanists have been able to collect and then analyze the sap that flows out through the detached stylet. Phloem sap consists mostly of water containing about 10% to 25% dissolved sugar (mostly sucrose). The sap also distributes amino acids and plant hormones through the plant body. What drives the movement of this sugary solution?

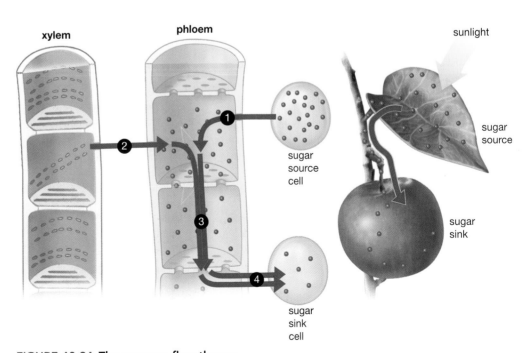

FIGURE 42-24 **The pressure-flow theory**
A photosynthesizing leaf is a sugar source, while a developing fruit is a sugar sink. Water pressure differences drive phloem sap containing sucrose from the leaf to the fruit. Numbers correspond to explanations within the text.

leaf) is actively transported into the phloem sieve tube. This raises the sugar (solute) concentration in the phloem sap in that portion of the sieve tube. ② Water (from xylem) follows the sugar into the sieve tube by osmosis. Because their rigid cell walls prevent sieve-tube cells from expanding, water entering the sieve tube increases the pressure of the phloem sap. ③ Water pressure drives the sugar-rich sap by bulk flow through the phloem sieve tubes into regions of lower pressure. ④ Cells of a sugar sink (the apple) actively transport sugar out of the phloem, and water follows by osmosis, creating a region of lower water pressure in this region of the sieve tube. The phloem sap thus moves from the source region where water pressure is high to the sink region where water pressure is lower (blue gradient in phloem), carrying the sugar with it.

Plant sugar sources and sinks can change with the seasons. For example, food storage structures, such as the taproots of carrots, are sugar sinks as they develop during the summer; but they are sugar sources in the spring, when they supply energy for a new plant to develop (if they haven't been harvested). The pressure-flow theory explains how phloem sap moves either up or down the plant, driven by pressure differences that are determined by the metabolic activities and demand for sugar of the various plant parts.

EVOLUTIONARY CONNECTIONS

Interesting Adaptations of Roots, Stems, and Leaves

Just as evolution has modified the basic shape of the vertebrate forelimb to suit the demands of running, swimming, and flying, so too have plant parts become modified in response to environmental demands. You may be surprised to learn that many familiar structures are derived from unlikely parts of a plant.

Specialized Roots Store Food; Others Photosynthesize

Roots have probably undergone fewer unusual modifications of their basic structure than have either stems or leaves. Some roots have extreme specializations for stor-

age, such as the beet, carrot, or radish (**FIG. 42-25a**). Some of the most unusual root adaptations occur in certain orchids that grow perched on trees. A few of these aerial orchids have green, photosynthetic roots (**FIG. 42-25b**).

Specialized Stems Produce New Plants, Store Water or Food, or Produce Thorns or Climbing Tendrils

Stems may perform functions very different from merely raising leaves up to the light. Strawberries, for example, grow horizontal *runners* that snake over the soil, sprouting new strawberry plants where nodes touch the soil (**FIG. 42-26a**). Meristem at the nodes touching the soil forms roots that allow the daughter plants to live independently.

The striking shape of the baobab tree (**FIG. 42-26b**) occurs because the tree uses its stem (trunk) as a massive water-storage organ, allowing it to thrive in dry climates. Cacti rely on their stems both to photosynthesize and to store water. The common white potato is actually an underground stem, specialized to store starch. Each "eye" is a lateral bud, ready to send up a branch when conditions become favorable, using the energy from its starch. If you store potatoes for too long, you may find them sprouting branches in your refrigerator. Irises have horizontal underground storage stems called rhizomes. They can be propagated by cutting up the rhizome; if it contains enough stored food, each piece with a node can generate a complete plant.

Many aboveground stems bear specialized branches, often in addition to "regular" branches. Some branches of grapes and Boston ivy are form grasping *tendrils*, which coil around trees, trellises, or buildings to give an otherwise prostrate plant better access to sunlight (**FIG. 42-26c**). One common branch adaptation is the *thorn*, such as that of the honey locust, which usually grows just above the attachment site of a leaf (**FIG. 42-26d**). Sharp thorns discourage animals from dining on the branches.

Specialized Leaves May Conserve and Store Water, Store Food, or Even Capture Insects

The most important environmental factors affecting the growth of leaves are temperature and availability of light and water. For example, plants growing on the floor of a tropical rain forest have plenty of water but very little

FIGURE 42-25 Root adaptations
(a) Dicot taproots modified for nutrient storage include (left to right) beets, carrots, and radishes. **(b)** This *Cattleya* orchid (a monocot) grows on a tree branch in the Amazon basin; its aerial roots dangle below the branch.

(a)

(b)

(c)

(d)

FIGURE 42-26 Stem adaptations
(a) The beach strawberry can reproduce using runners, which are horizontal stems. Where the node of a runner touches the soil, it sprouts roots and develops into a complete plant. (b) The enormously expanded water-storing trunk of the baobab tree allows it to thrive in a dry climate. (c) Tendrils are specialized stems that allow grape vines to cling to trees or trellises. (d) Honey locusts protect themselves with stems that form thorns.

light, owing to the deep shade cast by several layers of tree leaves above them. Consequently, their leaves tend to be extremely large—an adaptation demanded by the low light level and permitted by the abundant water.

At the other extreme, deserts receive bright sunlight virtually every day but have limited water and experience scorching temperatures. Some desert plants, called *succulents*, have thick leaves with large cells that store water from the infrequent rains against the inevitable long droughts (**FIG. 42-27a**). Such leaves are covered with a thick cuticle that reduces water evaporation. Cacti employ a very different strategy, reducing their leaves to thin spines that protect the plant from herbivores and provide almost no surface area for evaporation (**FIG. 42-27b**). Photosynthesis in cacti occurs in the cortex cells of their green, water-storing stems.

The common edible pea grasps fences or other plants with clinging tendrils. Unlike the modified branch tendrils

FIGURE 42-27 Leaf adaptations
(a) Succulent desert plants have fleshy leaves that store water from the infrequent rains. **(b)** Spines of desert cacti are nonphotosynthetic leaves whose surface area has been minimized, reducing evaporation and protecting the plant from grazing animals. **(c)** Daffodil bulbs consist of short central stems surrounded by thick water- and food-storing leaves. **(d)** A sundew grasps a lacewing fly with sticky, enzyme-laden hairs.

of grapes, pea tendrils are slender, supple leaflets. Some plants, such as onions, daffodils, and tulips, use thick, fleshy leaves as underground storage organs. A daffodil bulb consists of a short stem bearing thick, overlapping leaves that store nutrients during the winter (**FIG. 42-27c**). Finally, a few plants have turned the tables on the animals and have become predators. The protein-rich bodies of insects are excellent sources of nitrogen, if they can only be caught and digested. Carnivorous plants including Venus flytraps and sundews (**FIG. 42-27d**) have leaves that are modified into snares that can trap and digest unwary insects. These leaves have evolved in plants that colonize nitrogen-poor soils. Bogs are prime habitat for carnivorous plants because the acidic soil of bogs is hostile to nitrogen-fixing bacteria.

CASE STUDY REVISITED   WHY DO LEAVES TURN RED IN THE FALL?

At the onset of autumn, as temperatures cool and light levels remain high, the metabolic rate of the leaf slows, rendering it unable to use all the light it absorbs. The excess light energy can damage chloroplasts and leaf cells and further reduce photosynthesis. In the laboratory, Lee and Gould exposed both red and green dogwood leaves to intense light. They found that leaves containing more of the red anthocyanin pigment were much better protected from the effects of excess light energy than were those lacking it. Intense sunlight hitting leaves also causes the production of free radicals—highly reactive molecules that can damage cellular components. Scientists now have evidence that anthocyanins reduce the formation of free radicals by absorbing wavelengths of sun-

light energy that are not used in photosynthesis. What's more, these versatile red molecules act as antioxidants, reacting with any free radicals that are formed and rendering them harmless.

So why protect a dying leaf? Because both chlorophyll and carotenoids are rich in nitrogen, most of the plant's nitrogen is found in its leaves. To conserve this valuable nutrient, perennial plants salvage the nitrogen from dying leaves and pump it into woody tissues for storage over the winter. But this takes energy, which the plant derives from photosynthesis. Lee and Gould suggest that by protecting leaves during their final days, anthocyanin may allow the plant to continue photosynthesizing for as long as possible, thus acquiring the energy it needs to salvage nitrogen from chlorophyll and carotenoids for use in the coming spring.

Not all leaves turn red in the fall, and anthocyanins probably play various roles in the leaves of different plant species. As researchers probe deeper into the question of why leaves turn red, we can simply delight in the variety of autumn colors.

Consider This   Researcher William Hoch investigated the redness of leaves from 74 species of plants native to regions with cold winter climates (in the northern United States and Canada), where temperatures plunge in the fall, and milder climates (in coastal Europe). He found that the 41 species that produced the reddest leaves in autumn all came from the colder climates. What hypothesis does this support for the function of anthocyanins? Does it "prove" anything?

CHAPTER REVIEW

SUMMARY OF KEY CONCEPTS

42.1   How Are Plant Bodies Organized, and How Do They Grow?

The body of a land plant consists of root and shoot. Roots are usually underground, and their functions include anchoring the plant in the soil; absorbing water and minerals; storing surplus photosynthetic products; transporting water, minerals, photosynthetic products, and hormones; producing some hormones; and interacting with soil fungi and microorganisms that provide nutrients. Shoots are generally located aboveground and consist of stems, leaves, buds, and (in season) flowers and fruit. Shoot functions include photosynthesis, transport of materials, reproduction, and hormone synthesis.

Plant bodies are composed of two main classes of cells: meristematic cells and differentiated cells. Meristematic cells are undifferentiated and retain the capacity for mitotic cell division. Differentiated cells arise from divisions of meristem cells, become specialized for particular functions, and usually do not divide. Most meristem cells are located in apical meristems at the tips of roots and shoots and in lateral meristems in the shafts of roots and shoots. Primary growth (growth in length and differentiation of parts) results from the division and differentiation of cells from apical meristems; secondary growth (growth in diameter) results from the division and differentiation of cells from lateral meristems.

42.2   What Are the Tissues and Cell Types of Plants?

Plant bodies consist of three tissue systems: the dermal, ground, and vascular systems. The dermal tissue system forms the outer covering of the plant body. The dermal tissue system of leaves and of primary roots and stems is usually a single cell layer of epidermis. Dermal tissue after secondary growth is a multilayered covering of cork.

The ground tissue system consists of various cell types including parenchyma, collenchyma, and sclerenchyma. Most

are involved in photosynthesis, support, or storage. Ground tissue makes up most of a young plant during primary growth.

The vascular tissue system consists of xylem, which transports water and minerals from the roots to the shoots, and phloem, which transports water, sugars, amino acids, and hormones throughout the plant body.

42.3   What Are the Structures and Functions of Leaves, Roots, and Stems?

Leaves are the major photosynthetic organs of plants. The blade of a leaf consists of a water-resistant outer epidermis surrounding mesophyll cells, which have chloroplasts and carry out photosynthesis, and vascular bundles of xylem and phloem, which carry water, minerals, and photosynthetic products to and from the leaf. The epidermis is perforated by adjustable pores called stomata that regulate the exchange of gases and water.

Primary growth in dicot stems results in a structure consisting of an outer epidermis; supporting and photosynthetic cells of cortex beneath the epidermis; vascular tissues of xylem and phloem; and supporting and storage cells of pith at the center. Leaves and lateral buds are found at nodes along the surface of the stem. Under the proper hormonal conditions, lateral buds sprout into a branch. Secondary growth in stems results from cell divisions in the vascular cambium and cork cambium. Vascular cambium produces secondary xylem and secondary phloem, increasing the stem's diameter. Cork cambium produces waterproof cork cells that cover the outside of the stem.

Primary growth in roots results in a structure consisting of an outer epidermis, an inner vascular cylinder of xylem and phloem, and cortex between the two. The apical meristem near the tip of the root is protected by the root cap. Cells of the root epidermis absorb water and minerals from the soil. Root hairs are projections of epidermal cells that increase the surface area for absorption. Most cortex cells store surplus sugars (usually in the form of starch) produced by photosynthesis. The innermost layer of cortex cells is the endodermis, which controls the movement of water and minerals from the soil into the vascular cylinder.

Web Tutorial 42.1 Primary and Secondary Growth

Web Tutorial 42.2 Plant Transport Mechanisms

42.4 How Do Plants Acquire Mineral Nutrients?

Most minerals are taken up from the soil water by active transport into the root hairs. These minerals diffuse into the root through plasmodesmata to the pericycle, just inside the vascular cylinder. There they are actively transported into the extracellular space of the vascular cylinder. The minerals diffuse from the extracellular space into the tracheids and vessel elements of xylem.

Many plants have fungi called *mycorrhizae* associated with their roots that help absorb soil nutrients. Nitrogen can be absorbed only as ammonium or nitrate, both of which are scarce in most soils. Legumes have evolved a cooperative relationship with nitrogen-fixing bacteria that invade legume roots. The plant provides the bacteria with sugars, and the bacteria use some of the energy in those sugars to convert atmospheric nitrogen to ammonium, which the plant then absorbs.

42.5 How Do Plants Move Water Upward from Roots to Leaves?

The cohesion–tension theory explains xylem function: The cohesion of water molecules to one another by hydrogen bonds holds together the water within xylem tubes almost as if it were a solid chain. As water molecules evaporate from the leaves during transpiration, the hydrogen bonds pull other water molecules up the xylem to replace them. This movement is transmitted down the xylem to the root, where water loss from the vascular cylinder promotes water movement across the endodermis from the soil water by osmosis.

Water in the soil has a continuous, uninterrupted pathway through the porous cell walls of the outer layers of the root. The waterproof Casparian strip between endodermal cells forces water and dissolved minerals to move through selective cell membranes. Water moves by osmosis across the plasma membranes of the endodermal cells into the extracellular space of the vascular cylinder. The water pressure gradient caused by loss of water through transpiration is the primary force drawing water into the root.

42.6 How Do Plants Transport Sugars?

The pressure-flow theory explains sugar transport in phloem. Parts of the plant that synthesize sugar (for example, leaves) export sugar into the sieve tube. High sugar concentrations attract water to enter by osmosis, increasing the local hydrostatic pressure in the phloem. Other parts of the plant (for example, fruits) consume sugar, reducing hydrostatic pressure. Water and dissolved sugar move by bulk flow in the sieve tubes from areas of high to low pressure.

KEY TERMS

annual ring *page 869*
apical meristem *page 862*
bark *page 870*
blade *page 865*
branch root *page 871*
bulk flow *page 872*
cambium *page 862*
Casparian strip *page 872*
cohesion–tension theory
  page 876
collenchyma *page 863*
companion cell *page 865*
cork cambium *page 868*
cork cell *page 862*
cortex *page 871*
cuticle *page 862*
dermal tissue system
  page 862
dicot *page 860*
differentiated cell *page 862*
endodermis *page 872*

epidermal tissue *page 862*
epidermis *page 862*
fibrous root system *page 870*
ground tissue system
  page 862
guard cell *page 866*
heartwood *page 869*
internode *page 867*
lateral bud *page 867*
lateral meristem *page 862*
leaf *page 865*
leaf primordium *page 867*
legume *page 875*
meristematic cell *page 862*
mesophyll *page 866*
mineral *page 873*
monocot *page 860*
mycorrhizae *page 873*
nitrogen fixation *page 875*
nitrogen-fixing bacterium
  page 875

node *page 867*
nodule *page 875*
nutrient *page 873*
parenchyma *page 863*
pericycle *page 871*
periderm *page 862*
petiole *page 865*
phloem *page 864*
pit *page 864*
pith *page 866*
pressure-flow theory
  page 879
primary growth *page 862*
primary root *page 870*
root *page 860*
root cap *page 871*
root hair *page 862*
root system *page 860*
sapwood *page 869*
sclerenchyma *page 863*
secondary growth *page 862*

shoot system *page 861*
sieve-tube element
  page 864
sink *page 879*
source *page 879*
stem *page 866*
stoma *page 866*
taproot system *page 870*
terminal bud *page 866*
tracheid *page 864*
transpiration *page 876*
vascular bundle *page 865*
vascular cambium *page 868*
vascular cylinder *page 873*
vascular tissue system
  page 862
vein *page 865*
vessel *page 864*
vessel element *page 864*
xylem *page 864*

THINKING THROUGH THE CONCEPTS

1. Describe the locations and functions of the three tissue systems in land plants.

2. Distinguish between primary growth and secondary growth, and describe the cell types involved in each.

3. Distinguish between meristem cells and differentiated cells.

4. Diagram the internal structure of a dicot stem after primary growth, labeling and describing the function of epidermis, cortex, endodermis, pericycle, xylem, and phloem. What tissues are located in the vascular cylinder?

5. How do xylem and phloem differ?

6. What are the main functions of roots, stems, and leaves?

7. What types of cells form root hairs? What is the function of root hairs?

8. Diagram the internal structure of leaves. What structures regulate water loss and CO_2 absorption by a leaf?

9. Describe the daily cycle of the opening and closing of guard cells. How are various environmental conditions involved in this process?

10. A mutant form of aphid, the "klutzphid," inserts its stylet into the vessel elements of xylem. What materials are found in the fluids of xylem? Could an aphid live on xylem fluid? Would xylem fluid flow into the aphid? Explain your answer.

APPLYING THE CONCEPTS

1. An important goal of molecular botanists is to insert the genes for nitrogen fixation into crop plants such as corn or wheat (see Chapter 13). Why would the insertion of such genes be useful? What changes in farming practices would this technique allow?

2. Chapter 2 describes the unusual characteristics of water. Discuss several ways in which the evolution of vascular plants has been influenced by water's special characteristics.

3. A major environmental problem is desertification, in which overgrazing by cattle or other animals removes most of the vegetation in an area, and the region becomes drier and less able to support plants as a result. Explain this phenomenon based on your understanding of transpiration and how water moves through plants.

4. Desert grasses and wildflowers typically form fibrous root systems, while desert shrubs often form deep taproot systems. What advantages can you think of for each system? How does each type of root allow for survival in a desert environment?

5. Grasses (monocots) form their primary meristem near the ground surface rather than at the tips of branches the way dicots do. How does this feature allow you to grow a lawn and mow it every week in the summer? What would happen if you had a dicot lawn and tried to mow it?

6. Discuss the structures and adaptations that might occur in the leaves of plants living in (a) dry, sunny habitats; (b) wet, sunny habitats; (c) dry, shady habitats; and (d) wet, shady habitats. In which habitat would it be most difficult for a leaf to function well?

7. You and a friend carved your initials 5 feet above the ground on a tree on campus that was 40 feet tall. Now returning for your 25th reunion, you are ashamed of your actions; but you wonder if you will still find the damage on the tree, which is now 60 feet tall. How high above the ground should you look for your initials? Explain.

FOR MORE INFORMATION

Baskin, Y. "Forests in the Gas." *Discover*, October 1994. As carbon dioxide increases in the atmosphere, plant relationships will be altered.

Lee, D. W., and Gould, K. S. "Why Leaves Turn Red." *American Scientist*, November–December 2002. Researchers describe studies that led to a hypothesis that red pigments protect dying leaves so they can help the plant conserve valuable nutrients.

Milius, S. "Why Turn Red?" *Science News*, October 16, 2002. Why do leaves about to fall expend the energy to synthesize a new red pigment?

Perkins, S. "Lowland Tree Loss Threatens Cloud Forests." *Science News*, October 20, 2001. Costa Rica's Monteverde cloud forest relies on almost continuous cloud cover. This is now threatened by deforestation of lowland trees whose transpiration humidifies the air.

Zimmer, C. "The Web Below." *Discover*, November 1997. An underground web of mycorrhizae transfers nutrients between trees and helps maintain forest health.

CHAPTER 43

Plant Reproduction and Development

Amorphophallus titanium, also known as "corpse flower," has rarely been coaxed to bloom in the U.S. The central projection, called a *spadix*, is often over six feet tall and is filled with small male and female flowers.

CASE STUDY GORGEOUS? SURE! BUT HOT?

ROGER SEYMOUR of the University of Adelaide, Australia, recalls that he was first exposed to hot plants when a friend brought a philodendron blossom (actually a dense cluster of tiny flowers) to a party where everyone remarked at its mammal-like warmth—and its resemblance to a mammalian body part. In the 1970s, Seymour and coworkers reported that, like a mammal, the tree philodendron blossom maintains a relatively constant temperature of around 95°F (35°C) even in near-freezing air temperatures. In fact, their large blooms generate five times the heat of a comparably sized mammal. Recently, after inserting hair-thin temperature probes into flowers of the Asian sacred lotus (*Nelumbo nucifera*, **FIG. 43-1**), Seymour discovered that it also maintains a temperature of about 90°F, even when outside temperatures hover in the 40s. Warm flowers have been known for over 200 years, when French naturalist Jean-

Baptiste Lamarck first described self-heating blooms of the arum family. This group includes the aptly named dead-horse arum, whose warmth disperses its scent of decomposing flesh. The stench attracts blowflies, which eagerly dive inside the flower, mistaking it for the dead meat they consider a delicacy. The warm blossom of the eastern skunk cabbage (another member of the arum family) sometimes blooms within snowbanks, where its heat creates miniature ice caves (see Fig. 43-18). One of the world's most spectacular blooms, the "corpse flower" (also an arum), can reach a record 10 feet in height. It radiates heat and a stink that attracts carrion beetles in its native Sumatra. In the U.S. where only a dozen or so have ever bloomed, corpse flowers attract thousands of visitors (see chapter opener). One such event occurred at the University of Washington's botanical greenhouse, where the manager, Douglas

Ewing, grew the amazing plant from a seed. The stink, he enthused like a new father, "means it's doing what's natural, and I hope it continues to escalate and just drives us out of here." Why do some blooms mimic a warm, decaying corpse?

FIGURE 43-1 The Asian sacred lotus

43.1 WHAT ARE THE BASIC FEATURES OF PLANT LIFE CYCLES?

Plants Engage in Sex

Many plants can reproduce either sexually or asexually. During asexual reproduction, part of a single plant (such as a stem) produces a new plant using mitotic cell division. Asexually produced offspring, therefore, are genetically identical to the parent. In Chapter 42, you encountered several methods of asexual reproduction, including runners in strawberries, bulbs in daffodils, and rhizomes in irises. Asexual reproduction can be highly effective, allowing offspring to colonize an entire area where their original parent plant found optimal conditions.

However, if an offspring is genetically identical to its parent, then it is only as well adapted to the environment as its parent. What if the environment changes? Most sexually produced offspring combine genes from both parents, so they may be endowed with traits that differ from those of either parent. This new combination of traits may help the offspring cope with changing environments or invade slightly different habitats. As a result, most organisms, including plants, reproduce sexually at least some of the time.

You are probably somewhat familiar with sexual reproduction in animals—how do plants compare? During the animal life cycle, animals with diploid (2n) cells produce haploid (n) gametes (sperm or eggs) by the process of meiotic cell division (meiosis followed by cytoplasmic division). The gametes (a sperm and an egg) fuse to form a new diploid cell (the *zygote*) that develops into the adult organism through repeated mitotic cell divisions (mitosis followed by cytoplasmic division). The plant life cycle, however, is a bit

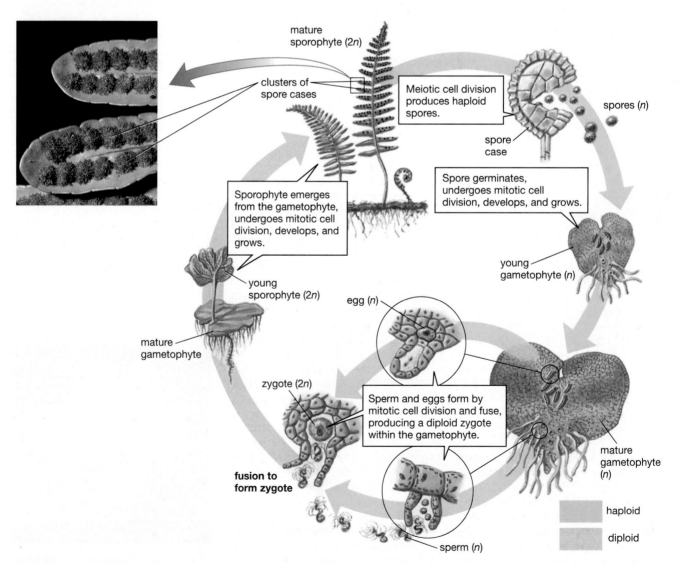

FIGURE 43-2 The life cycle of a fern, a nonflowering plant
Ferns illustrate the alternation-of-generations life cycle found in all plants, in which separate multicellular haploid and multicellular diploid organisms occur at different parts of the life cycle. The *n* refers to the haploid state, 2n to the diploid state. When you see ferns, look on the undersides of their leaves; occasionally you will find clusters of brown sporangia (inset).

more complex. Plants have two distinct, multicellular forms—one diploid and one haploid—that give rise to each other. For this reason, the plant life cycle exhibits **alternation of generations**: diploid plants (*sporophytes*) alternate with haploid plants (*gametophytes*).

Alternation of Generations Is Clearly Evident in Ferns and Mosses

Alternation of generations occurs in all plants. Evolutionarily ancient land plants, including mosses and ferns, have the most easily visible alternating generations, because the gametophyte is an independent plant. In ferns, it is smaller than the sporophyte, while in mosses the gametophyte is much larger (see Chapter 21, Fig. 21-4). As you learned in Chapter 21, these plants do not produce flowers. The gametophyte liberates motile sperm cells that reach an egg either by swimming through thin films of water that cover adjacent gametophytes or by being splashed by raindrops from one plant to the next. For this reason, ferns and mosses can reproduce only in moist habitats.

To illustrate alternation of generations, let's examine the life cycle of a fern, starting with the diploid adult plant (**FIG. 43-2**). This stage of the life cycle, the **sporophyte** ("spore plant" in Greek), bears reproductive cells that undergo meiotic cell division to produce haploid cells. These reproductive cells are **spores**, rather than gametes. Unlike gametes, spores do not fuse together to form a diploid cell. Instead, the spore is carried by wind or water to the soil. There, the spore **germinates** (begins to grow and develop), undergoing repeated mitotic cell divisions to form a multicellular, haploid organism. This organism produces gametes and so is called the **gametophyte** ("gamete plant" in Greek). Because its cells are already haploid, the gametophyte can produce sperm and eggs without further meiosis. A single gametophyte usually produces both sperm and eggs, but typically at different times, thereby preventing self-fertilization. Sperm and egg from different gametophytes fuse to form a fertilized egg, or **zygote**, that develops into a new diploid sporophyte plant.

43.2 HOW IS REPRODUCTION IN SEED PLANTS ADAPTED TO DRIER ENVIRONMENTS?

Many terrestrial habitats are relatively dry, and sperm do not have much opportunity to swim to eggs. Yet the sperm, the egg, the zygote that forms when they merge, and the embryo that develops from the zygote must all be kept moist to survive. Both flowering and nonflowering seed plants have been successful in colonizing dry land. During the course of evolution, their male and female gametophytes have become microscopic in size. A male gametophyte surrounded by a protective coat is called a **pollen grain**. The pollen grain encloses sperm cells in a watertight packet that can be easily transported to another plant. The egg-producing female gametophyte remains moist and protected within the flower, and the pollen grain ensures

FIGURE 43-3 Conifers are wind pollinated
Even slight breezes blow thick clouds of pollen from ripe male cones. Look for these "soft cones" in clusters near the ends of branches of pine, spruce, and fir trees, often in late spring. The cones disintegrate after releasing their pollen. The larger, woody cones are female and produce seeds at the base of each scale. QUESTION Compared to flowering plants pollinated by animals, what advantages do wind-pollinated plants have? What disadvantages?

that the sperm are delivered directly to the egg, as described later. The fertilized egg becomes enclosed in a drought-resistant seed. The **seed**, which consists of an embryonic plant and a food reserve encased within a protective outer coating, may lie dormant (in a resting state) for months or years, waiting for conditions favorable for germination and growth.

The earliest seed plants were the gymnosperms, represented today mainly by conifers, a group that includes pines, firs, and spruces. As described in Chapter 21, conifers do not produce flowers; instead, they bear male and female gametophytes on separate reproductive structures called *cones*. During early spring, the small male cones release millions of pollen grains that are carried far and wide by the wind (**FIG. 43-3**). So many grains are floating around that some, by chance, enter the pollen chambers located on the scales of the female cones, where they are captured by sticky coatings of sugars and resins. The pollen grains then send out pollen tubes that tunnel to the female gametophytes at the base of each of the *scales* (woody plates that make up the cone). Sperm travel through the pollen tubes and fertilize eggs within each female gametophyte, forming a diploid zygote that begins the next generation. In Chapter 21, the life cycle of a conifer (a pine) is illustrated in Figure 21-9.

43.3 WHAT IS THE FUNCTION AND STRUCTURE OF THE FLOWER?

Most Flowers Lure Animals That Pollinate Them

Clearly, wind pollination is successful; pines and other conifers dominate northern forests. Some flowering plants also use the wind to carry their pollen. These include many

HEALTH WATCH   Are You Allergic to Pollen?

Wind pollination can succeed only if plants release huge quantities of pollen into the air. Unfortunately for allergy sufferers, it is easy to inhale these microscopic male gametophytes. Proteins in pollen coats activate the immune systems of sensitive individuals, causing itchy, watery eyes, runny noses, burning throats, coughing, and sneezing. If you are among these unfortunate people, your own immune system is creating all these symptoms in an attempt to rid you of the harmless pollen, which it mistakes for dangerous disease organisms. People who suffer from "hay fever" are usually sensitive only to specific types of pollen. In temperate climates, springtime sufferers may be allergic to tree pollen, while summertime allergies may be caused by grasses. In the U.S., however, the worst cause of hay fever is not hay, but ragweed, which pollinates during late summer and fall (**FIG. E43-1**). Ragweed flowers, like those of most wind-pollinated plants, are inconspicuous because they are not adapted to attract animal pollinators. A single plant can release a million pollen grains daily; collectively, ragweed is estimated to release 100 millions tons of pollen in the U.S. each year. Ragweed pollen has been collected 400 miles out to sea and 2 miles up in the atmosphere. The small size and copious quantities of ragweed pollen grain and the type of proteins in their coats make them a particular menace for allergy suffers.

Plants pollinated by bees and other animals are rarely a cause of allergies, because their pollen is sticky and produced in small amounts. Goldenrod, which blooms during the ragweed season and is conspicuously yellow, has often been blamed for allergies that are actually caused by ragweed. In fact, its yellow blooms attract bee and butterfly pollinators, and most people can enjoy it in perfect comfort (**FIG. E43-2**).

FIGURE E43-1 Inconspicuous ragweed flowers and their pollen

FIGURE E43-2 Goldenrod

trees—oak, maple, birch, poplar, cottonwood, and aspen. Wind-pollinated flowers are also produced by grasses, corn, and ragweed, the bane of allergy sufferers (see "Health Watch: Are You Allergic to Pollen?").

Wind pollination is an inefficient process, because most pollen grains do not reach their target. In a world of stationary plants and mobile animals, an early gymnosperm that enticed an animal to carry its pollen from male to female cone would greatly enhance its reproductive rate and hence its evolutionary success. As it happens, gymnosperms and insects established just such a relationship about 150 million years ago.

Insects, especially beetles, are among the most abundant animals on Earth. Today, they exploit nearly every possible food resource on land, including the reproductive parts of gymnosperms. About 150 million years ago, some beetles fed on both the protein-rich pollen of male cones and the sugar-rich secretions of female cones. Beetles can make quite a mess when they feed, and pollen feeders often wind up with pollen dusted all over their bodies. If the same beetle visited one plant and ate pollen, then wandered over to another plant of the same species to dine on the sugary secretions of a female cone, some of the loose pollen would quite likely rub off on the female cone. This set the stage for the evolution of flowering plants.

To pollinate efficiently, the same insect must visit several plants of the same species, picking up and depositing pollen along the way. For plants, two key adaptations were necessary to promote this. First, enough pollen or nectar (a sugary fluid) had to be produced within the reproductive structures so that insects would regularly visit them to feed. Second, the location and richness of these storehouses of pollen and nectar had to be advertised to the insects, both to show them where to go and to entice them to specialize on that particular plant species. Any mutation that contributed to these adaptations would enhance the reproductive success of the plant that carried the mutation and would be favored by natural selection. By about 130 million years ago, flowers had evolved with exactly these adaptations. The advantages of flowers are so great that flowering plants, or *angiosperms*, dominate today's *temperate* (middle latitude) and *tropical* (low latitude, near the equator) regions. Flowers are pollinated by animals that include bees, moths, butterflies, hummingbirds, a few

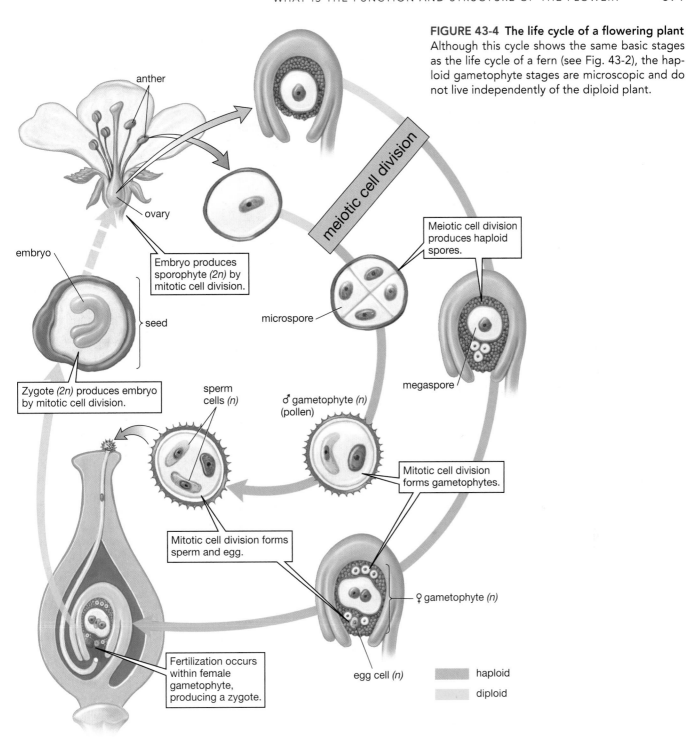

FIGURE 43-4 The life cycle of a flowering plant Although this cycle shows the same basic stages as the life cycle of a fern (see Fig. 43-2), the haploid gametophyte stages are microscopic and do not live independently of the diploid plant.

types of mammals, and beetles (as described in the Case Study). Although flowers did not evolve to attract people, for reasons that remain mysterious we often respond with pleasure to their scents, shapes, and colors.

Flowers Are the Reproductive Structures of Angiosperms

Flowers are the reproductive structures of flowering plants, produced by the sporophyte generation. Within flowers, two types of spores are formed by meiotic cell division (**FIG. 43-4**). These haploid spores develop into microscopic

gametophytes that never assume an independent existence. The larger spore type, the *megaspore* ("mega" is from the Greek meaning "large"), undergoes a few mitotic divisions and develops into the female gametophyte, a small cluster of cells permanently retained within the flower. The other type of spore, the *microspore* ("micro" is from the Greek meaning "small"), develops into the male gametophyte that contains two sperm. The male gametophyte, surrounded by a protective coat, becomes the pollen grain, which drifts on the wind or is carried by an animal from one flower to another. On the recipient flower, the pollen grain produces a tube that tunnels through the

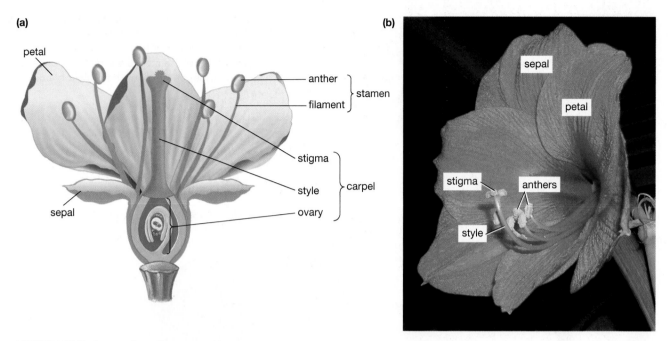

FIGURE 43-5 A complete flower
(a) A complete flower has four parts: sepals, petals, stamens (the male reproductive structures), and at least one carpel (the female reproductive structure). This drawing shows a complete dicot flower. **(b)** The amaryllis is a complete monocot flower, with 3 sepals (virtually identical to the petals), 3 petals, 6 stamens, and 3 carpels (fused into a single structure). The anthers are well below the stigma, making self-pollination unlikely.

flower's tissues to the female gametophyte within. Sperm travel down this tube and enter the female gametophyte, where fertilization occurs. In the sections that follow, we explore the intimate details of sexual reproduction in flowering plants.

Complete Flowers Have Four Major Parts

Evolution commonly produces new structures by modifying old ones; flower parts are actually highly modified leaves, shaped by mutation and natural selection into a form that enhances pollination. A **complete flower**, such as that of a petunia, rose, or lily, consists of a central axis supporting four successive sets of modified leaves (**FIG. 43-5**). These modified leaves form the *sepals, petals, stamens,* and *carpels*. The **sepals** are located at the base of the flower. In dicots, sepals are typically green and leaflike (Fig. 43-5a); in monocots, most sepals resemble the petals (Fig. 43-5b). In either case, sepals surround and protect the flower bud as the remaining three structures develop. Just above the sepals are the **petals**, which are often brightly colored and fragrant, advertising the location of the flower.

The male reproductive structures, the **stamens**, are attached just above the petals. Each stamen typically consists of a slender **filament** that supports an **anther**, the structure that produces pollen. The female reproductive structure, the **carpel**, frequently occupies a central position in the flower. A generalized carpel is somewhat vase-shaped, with a sticky **stigma** for catching pollen mounted atop an elongated **style**. The style connects the stigma with the bulbous **ovary** (Fig. 43-5a). Inside the ovary are one or more **ovules**, in which the female gametophytes develop. When

mature, each ovule will become a seed, and the ovary will develop into a protective, adhesive, and/or edible enclosure, the **fruit**.

Incomplete flowers lack one or more of the four floral parts. For example, grass flowers (see Fig. 43-9) lack both petals and sepals. Other incomplete flowers lack either the male stamens or the female carpels. In such cases, the flowers are described as *imperfect*, as well as incomplete. This is not a value judgment; plant species with imperfect flowers are highly successful. They produce separate male and female flowers, sometimes on the same plant, as in the case of garden squash such as zucchini (**FIG. 43-6**) or the corpse flower and philodendron. Both of these contain a central projection, called a *spadix*, that bears many tiny male and female flowers. Other plants with imperfect flowers generate male and female flowers on separate male and female plants. An example is the American holly, whose familiar red berries are produced only by female plants.

Pollen Contains the Male Gametophyte

Pollen develops within the anther of the sporophyte plant. Each anther consists of four chambers called pollen sacs (**FIG. 43-7**). Within each pollen sac, hundreds to thousands of diploid **microspore mother cells** develop. Each microspore mother cell undergoes meiotic cell division (described in Chapter 11) to produce four haploid **microspores**. Each microspore then undergoes one mitotic cell division to produce a haploid male gametophyte. In many species, the immature male gametophyte consists of only two cells: a large **tube cell** and a smaller **generative cell** that resides within the cytoplasm of the tube cell (Fig. 43-7 photo inset).

FIGURE 43-6 Male and female flowers
Plants of the squash family, such as these zucchinis, bear separate female (left) and male (right) flowers. Each plant initially produces only male flowers, ensuring some cross-pollination between plants that flower at slightly different times. Note the small zucchini (actually a fruit) forming at the base of the female flower. Zucchini fruits are produced only by female flowers. QUESTION In species with separate male and female flowers on the same plant, why would natural selection favor individuals whose male and female flowers bloom at different times?

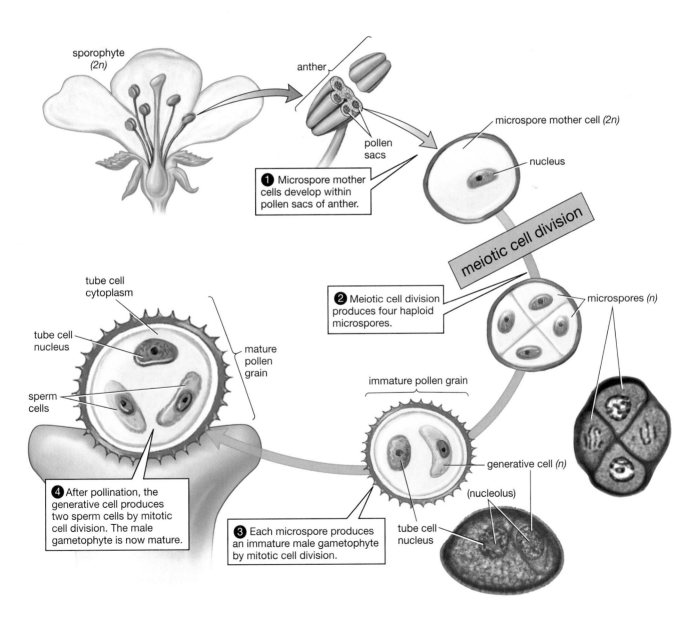

FIGURE 43-7 Male gametophyte (pollen) development

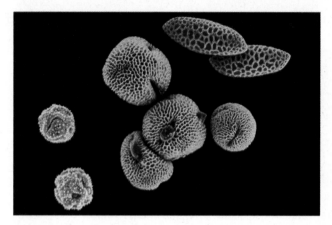

FIGURE 43-8 Pollen grains
The tough outer coats of many pollen grains are elaborately sculptured in species-specific shapes and patterns. The pollen grains in this false-color SEM photo are from a geranium (orange), a tiger lily (fuschia), and a dandelion (yellow).

As the gametophyte matures, the generative cell undergoes mitotic cell division and produces two haploid sperm cells (Fig. 43-7). A tough surface coat develops around it, often with an elaborate pattern of pits and protrusions characteristic of the plant species (**FIG. 43-8**). This coat protects the

FIGURE 43-9 Wind-pollinated flowers
The flowers of grasses and many deciduous trees are wind pollinated, with anthers (yellow structures hanging beneath the flowers) exposed to the wind. Petals are reduced or absent.

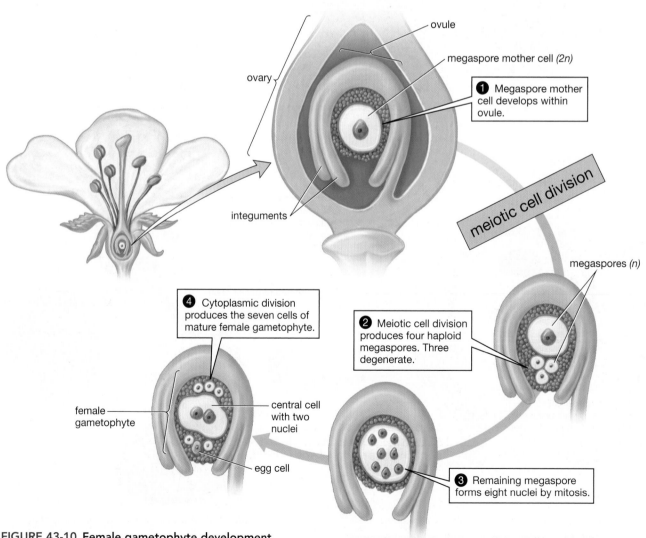

ovule

megaspore mother cell *(2n)*

ovary

❶ Megaspore mother cell develops within ovule.

integuments

meiotic cell division

megaspores *(n)*

❷ Meiotic cell division produces four haploid megaspores. Three degenerate.

❹ Cytoplasmic division produces the seven cells of mature female gametophyte.

female gametophyte

central cell with two nuclei

egg cell

❸ Remaining megaspore forms eight nuclei by mitosis.

FIGURE 43-10 Female gametophyte development

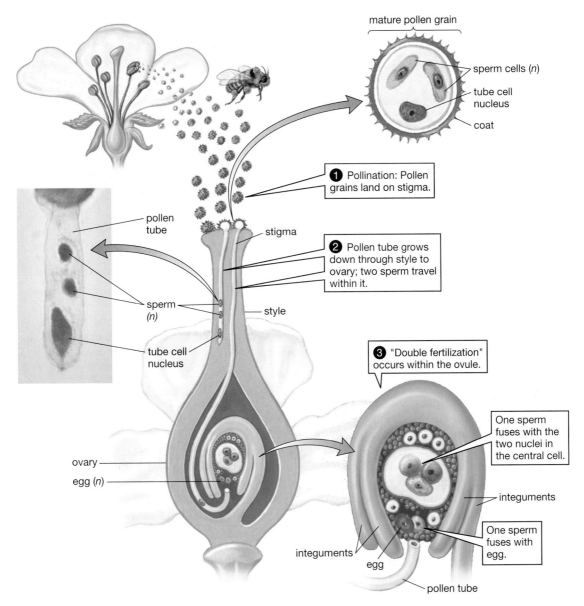

mature pollen grain

sperm cells (n)

tube cell nucleus

coat

1 Pollination: Pollen grains land on stigma.

pollen tube

stigma

2 Pollen tube grows down through style to ovary; two sperm travel within it.

sperm (n)

style

tube cell nucleus

3 "Double fertilization" occurs within the ovule.

One sperm fuses with the two nuclei in the central cell.

ovary

egg (n)

integuments

One sperm fuses with egg.

integuments

egg

pollen tube

FIGURE 43-11 Pollination and fertilization of a flower

cells during their journey to the sometimes-distant female carpel. The male gametophyte and its protective coat together form the pollen grain.

When the pollen has matured, the pollen sacs of the anther split open. In wind-pollinated flowers, such as those of grasses (**FIG. 43-9**) and oaks, the pollen grains spill out and are widely distributed by wind currents; a few of those grains reach and pollinate other flowers of the same species. In animal-pollinated flowers, the pollen adheres weakly to the anther case until the pollinator comes along and brushes or picks it off.

The Female Gametophyte Forms Within the Ovule of the Ovary

Within an ovary, masses of cells differentiate into ovules. Each young ovule consists of protective outer layers of cells called **integuments**, which surround a single, diploid **megaspore mother cell** (**FIG. 43-10**). This large cell produces the female gametophyte. First, it undergoes one meiotic cell division that produces four large haploid **megaspores**. Three

of these degenerate, and one survives. The nucleus of this megaspore divides by mitosis three times, producing eight haploid nuclei. Plasma membranes then form, dividing the cytoplasm into seven (not eight) cells. There are three small cells at each end, each with one nucleus, and one large central cell with two nuclei. This seven-celled organism is the haploid female gametophyte. The **egg cell** is one of the three at the lower end, located near an opening in the integuments of the ovule (Fig. 43-10).

Pollination of the Flower Leads to Fertilization

Pollination is necessary for *fertilization*, but these are two distinct events. **Pollination** occurs when a pollen grain lands on the stigma of a flower of the same plant species, beginning a remarkable series of events (**FIG. 43-11**). The pollen grain absorbs water from the stigma. The generative cell undergoes mitotic cell division to form two sperm cells. Meanwhile, the tube cell elongates, burrowing into the style and producing a tube that will conduct sperm down the style and into an ovule within the ovary.

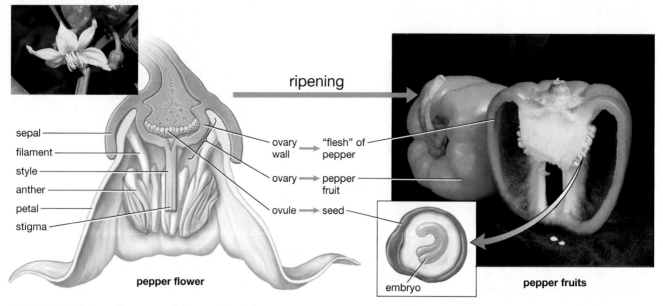

FIGURE 43-12 Development of fruit and seeds in a pepper
Fruits and seeds develop from flower parts. The ovary wall ripens into the fruit flesh. Each pepper ovary houses many ovules, which develop into seeds. The zygote within each seed develops into the embryo.

If all goes well, the pollen tube reaches the opening in the integument of an ovule and breaks into the female gametophyte. The tube's tip ruptures, releasing the two sperm. One sperm merges with the egg cell, a process called **fertilization**. Fertilization produces the diploid zygote that will develop into the embryo and eventually into a new sporophyte. The second sperm enters the large central cell, where its nucleus fuses with the two nuclei already present, forming a *triploid nucleus* (containing three sets of chromosomes). Through repeated mitotic divisions, this cell will develop into the triploid (*3n*) **endosperm**, a food-storage tissue within the seed. **Double fertilization** describes the fusion of the egg with one sperm and the fusion of the two central nuclei with the second sperm, a process unique to flowering plants. The other five cells of the female gametophyte degenerate soon after fertilization.

43.4 HOW DO FRUITS AND SEEDS DEVELOP?

Drawing on the resources of the parent plant, the female gametophyte and the surrounding integuments of the ovule develop into a seed. The seed is surrounded by the ovary, which will form a fruit (**FIG. 43-12**). Having already served their functions of attracting pollinators and producing pollen, petals and stamens shrivel and fall away as the fruit enlarges.

The Fruit Develops from the Ovary

When you eat a fruit, you are consuming the plant's ripened ovary (sometimes accompanied by other flower parts). The foods we commonly call fruits (apples, berries, peaches, oranges, bananas) are usually sweet and often juicy, but many that we call "vegetables," including avocados, zucchini, tomatoes, and peppers (see Fig. 43-12), are fruits as well.

Fruits may also be hard, winged, or sharp and spiked. The various shapes, colors, and textures of fruits all serve the same function; they help disperse seeds away from the parent plant, in many cases taking advantage of the mobility of animals (see "Earth Watch: On Dodos, Bats, and Disrupted Ecosystems"). For example, the burrs that attach to your socks during a walk in a fall meadow are likely to be specialized fruits, hitchhiking on you to disperse their seeds.

The Seed Develops from the Ovule

Inside the ovule, two distinct developmental processes occur to produce the seed (**FIG. 43-13**). First, the triploid central cell divides rapidly; the resulting daughter cells absorb nutrients from the parent plant, forming a food-filled endosperm. Second, the zygote develops into the embryo (Fig. 43-13a, b) while the other five cells of the female gametophyte degenerate. Both dicot and monocot embryos consist of an embryonic root and embryonic shoot (Fig. 43-13c). The shoot portion includes one or two **cotyledons**, or seed leaves, which absorb food molecules from the endosperm and transfer them to other parts of the embryo. When we eat peas, beans, corn, rice, or wheat, we are benefiting from the food that these plants stored in their seeds for their own embryos. Meanwhile, the outer coverings or integuments of the ovule thicken, harden, and become the **seed coat** that surrounds and protects the seed.

In monocots ("one cotyledon"), the cotyledon typically absorbs some endosperm during development, but most of it remains in the mature seed to be used directly by the germinating seedling, as illustrated by the corn kernel in **FIGURE 43-14a**. Rice, barley, and wheat are also monocots. Flour made from these grains is ground-up endosperm, and we sometimes consume the wheat embryo separately as "wheat germ."

In many dicots ("two cotyledons"), the cotyledons absorb most of the endosperm during seed development, so

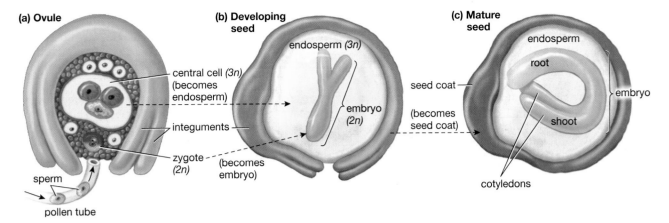

FIGURE 43-13 Seed development
(a) Seed formation begins after one sperm fuses with the egg, forming a diploid zygote, and the second sperm fuses with the central cell. **(b)** The endosperm develops from the triploid central cell, which undergoes many mitotic cell divisions as it absorbs nutrients from the parent plant. Then the embryo develops, absorbing nutrients from the endosperm. **(c)** The two cotyledons of dicots (such as this pepper) absorb endosperm as the seed develops.

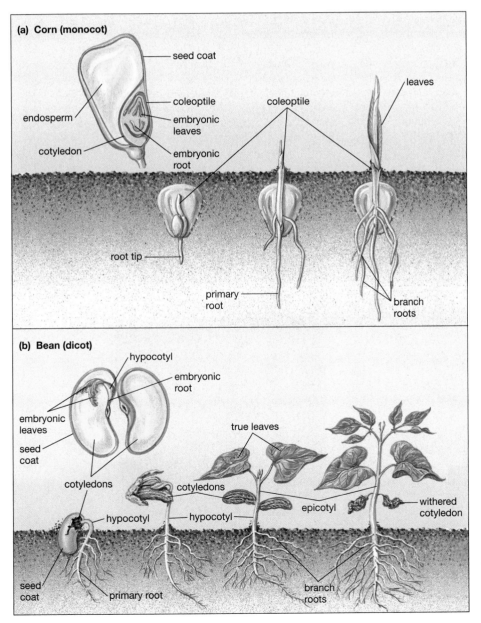

FIGURE 43-14 Seed germination
First the root grows rapidly, absorbing water and nutrients. **(a)** In monocots such as corn, the shoot tip is protected within a tough coleoptile. **(b)** In dicots such as the bean, the hypocotyl (shown) or the epicotyl bends, forming a hook that emerges from the soil first, protecting the shoot tip.

897

EARTH WATCH   On Dodos, Bats, and Disrupted Ecosystems

Flowering plants dominate terrestrial ecosystems largely because of mutually beneficial relationships with animals that pollinate their flowers and disperse their seeds. Within complex ecosystems, these relationships sustain both plant and animal populations and ultimately the ecosystem itself. As native animals are destroyed, plants that coevolved with them are also threatened, and natural ecosystems that evolved over millennia may collapse.

The tambalacoque tree, like most of the native plants on the island of Mauritius in the Indian Ocean, is threatened. Tambalacoque trees produce large, edible fruit similar to peaches, with a pulpy outside surrounding a stone-hard pit. Today, the remaining trees produce healthy fruits that fall to the ground and rapidly rot in the tropical climate. Before humans arrived, the island was home to the dodo (**FIG. E43-3a**). Early sailors found the large, slow dodos easy prey, and by 1681 they had hunted the dodo to extinction. Other native animals, including giant tortoises, large-billed parrots, and the giant skink (a large reptile), were also driven to extinction as humans destroyed natural habitats by clearing the land for farming and introducing monkeys, pigs, deer, and various plants to Mauritius. All of these factors threatened the tambalacoque forests. Scientists believe that some of the now-extinct animals ate the tambalacoque fruit before it had a chance to rot, thoroughly cleaning the seeds and thus protecting them from attack by fungi. They also dispersed the seeds, ensuring that some reached favorable habitats.

In another example, on the island of Madagascar off the coast of Africa, researchers have identified more than 20 tree species that depend primarily on lemurs (tree-dwelling primates) for seed dispersal. But a burgeoning human population is destroying lemur habitats, and the animals are rapidly disappearing. Where lemurs are disappearing, so are these trees.

In many tropical forests, fruit-eating bats are the most important agents of seed dispersal (**FIG. E43-3b**). These bats may fly more than 20 miles each night, consuming up to twice their weight in fruit and defecating the seeds in flight. Biologist Donald Thomas discovered that after passing through a bat's digestive tract, nearly all the seeds germinated; in contrast, seeds planted directly from fruit had only a 10% germination rate. Today, in the tropical forests of southern Mexico, fruit-eating and dispersing animals such as monkeys, deer, and tapir have been overhunted, and fruit-eating bats are threatened by habitat destruction as land is cleared for agriculture. Tropical fruits are rotting on the forest floor or sending up doomed sprouts under the shade of their parents; dispersal has been dramatically reduced. As Alejandro Estrada of the University of Mexico put it, "The continued existence of tropical forests whose primates and . . . birds and bats have been shot is just as precarious as if their trees had been chain-sawed and bulldozed." The web of interdependent life-forms linked by interactions forged over millennia of coevolution is fragile and easily disrupted. Only by understanding and preserving the complex and crucial interactions among plants and animals can we hope to conserve diverse, functioning ecosystems.

(a)

(b)

FIGURE E43-3 Animal seed dispersers are crucial to some ecosystems
(a) The dodo and other now-extinct animals probably helped disperse and promote germination of the tambalacoque trees of Mauritius. The trees are now endangered and rarely germinate in the wild. **(b)** A bat eats a ripe fig in Kenya. Without bats and other seed-dispersing animals, some types of tropical forest communities might not survive.

the mature seed is virtually filled with embryo, as illustrated by the bean in **FIGURE 43-14b**. If you strip the thin seed coat from a bean or a pea, you will find that the inside splits easily into two halves; each is a cotyledon.

43.5 HOW DO SEEDS GERMINATE AND GROW?

Seeds need warmth and moisture to germinate. But even under ideal conditions, many newly matured seeds do not germinate immediately. Instead, they enter a period of **dormancy** in which their metabolic activity is low and they are able to resist adverse environmental conditions.

Seed Dormancy Helps Ensure Germination at an Appropriate Time

Seed dormancy solves two problems. First, it prevents seeds from germinating within a moist fruit, where the emerging plant might be consumed by a fruit-eating animal or attacked by mold as the fruit decomposed. If they survived, multiple seedlings germinating within a fruit would grow in a dense cluster, competing for nutrients and light. Second, environmental conditions that are suitable for seedling growth (such as warmth and moisture) may not coincide with seed maturation. For example, seeds in *temperate climates* (in middle latitudes where there are four distinct seasons), which mature in the late summer, face the harsh winter to come. These avoid freezing as tender young sprouts by spending the winter as dormant seeds. In warm, moist, tropical regions, where environmental conditions are suitable for germination throughout the year, seed dormancy is much less common.

Besides warmth and water, many plant species have additional requirements for seed germination. These are finely tuned to the plant's native environment and the mechanisms it uses for dispersal. The three most common requirements to break seed dormancy are drying, exposure to cold, and disruption of the seed coat.

- *Drying.* This prevents seeds from germinating while they are still within a moist fruit. Seeds that require drying often are dispersed by fruit-eating animals, which cannot digest the seeds. The seeds are excreted and exposed to air, where they dry out. Later, when temperature and moisture levels are favorable, they germinate.

- *Cold.* Seeds of many temperate and arctic plants will not germinate unless they are exposed to prolonged subfreezing temperatures, followed by sufficient warmth and moisture. This ensures that seeds released in mild autumn weather do not immediately germinate, only to succumb to winter cold. Requiring a substantial cold spell keeps them from sprouting until the following spring.

- *Seed coat disruption.* The seed coat itself may need to be weathered or partially digested before germination can

occur, and some coats contain chemicals that inhibit germination (as described in Chapter 29). In deserts, for example, years may go by without enough water for a plant to complete its life cycle. The seed coats of many desert plants have water-soluble chemicals that inhibit germination, and only a hard rainfall can wash away enough of the inhibitors to allow sprouting. A seed of the Asian sacred lotus (see Fig. 43-1), found buried in a dry lakebed in China, sprouted four days after its seed coat was filed away. Radiocarbon dating revealed it to be nearly 1300 years old.

Upon Germination, the Root Emerges First, Followed by the Shoot

During germination, the embryo absorbs water, which makes it swell and burst its seed coat. The root usually emerges first and grows rapidly, absorbing water and minerals from the soil. As it tunnels downward, the apical meristem at the root tip is protected by a root cap (see Fig. 42-15 in Chapter 42). Much of the water is transported to the shoot, and its cells elongate and push upward through the soil toward the light.

The embryonic shoot usually consists of two regions. Below the cotyledons, but above the root, is the **hypocotyl** ("hypo" is from the Greek, meaning "beneath" or "lower"); above the cotyledons, the shoot is called the **epicotyl** ("epi" means "above"). At the tip of the epicotyl lies the apical meristem of the shoot; its daughter cells will later differentiate into the specialized cell types of stem, leaves, and flowers. In some embryos, one or two developing leaves may be beginning to form.

As it forces its way up through the soil, the germinating shoot must protect its delicate apical meristem and tender leaflets from abrasion. In monocots, the **coleoptile**, a tough sheath, encloses the shoot tip like a glove around a finger (see Fig. 43-14a), pushing aside soil particles as the tip grows. Once out in the air, the coleoptile degenerates, allowing the tender shoot to emerge. In dicots, which lack coleoptiles, the shoot forms a hook in the hypocotyl or epicotyl (see Fig. 43-14b). The bend of the hook, encased in thick-walled epidermal cells, forces its way up through the soil, clearing the path for the downward-pointing apical meristem and its delicate new leaves. The shoot immediately begins to straighten after it emerges, orienting its leaves toward the sunlight.

Cotyledons Nourish the Sprouting Seed

Food stored in the seed provides the energy for sprouting. Recall that the two cotyledons of many dicots absorb the endosperm during seed developments, so they are swollen with stored food. In dicots with hypocotyl hooks, such as beans and squash, the elongating shoot carries the cotyledons out of the soil into the air. These aboveground cotyledons typically become green and photosynthetic and transfer both previously stored food and newly synthesized

human vision

"bee vision"

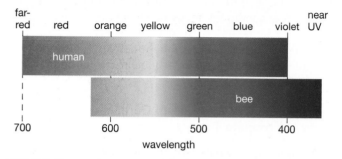

far-red | red | orange | yellow | green | blue | violet | near UV

human

bee

700 600 500 400

wavelength

FIGURE 43-15 Cotyledons nourish the developing plant
In dicots such as beans and the squash shown here, the seed leaves or cotyledons expand and photosynthesize. The first "true leaf" comes out a bit later. Eventually, the cotyledons shrivel up and die.

FIGURE 43-16 Ultraviolet patterns guide bees to nectar
(Bottom) The spectra of color vision for humans and bees overlap considerably but are shifted on the edges. Humans are sensitive to red, which bees do not perceive; bees can see near-UV light, which is invisible to the human eye. (Top) Many flowers photographed under (left) ordinary daylight and (right) under UV light show striking differences in color patterns. Bees can see the UV patterns that presumably direct them to the nectar- and pollen-containing centers of the flowers.

sugars to the shoot (**FIG. 43-15**). In dicots with epicotyl hooks, the cotyledons remain below the ground, shriveling up as the embryo absorbs their stored food. Many monocots (see the corn in Fig. 43-14a) retain most of their food reserve in the endosperm until germination, when it is digested and absorbed by the cotyledon as the embryo grows. The cotyledon stays belowground in the remnants of the seed.

Once out in the air, the shoot rapidly spreads its leaves to the sun. Simultaneously, the root system delves into the soil. The apical meristem cells of shoot and root divide, giving rise to the mature structures discussed in Chapter 42. Eventually this plant will mature, flower, and set seed, renewing the cycle of life. The regulation of this cycle, including why shoots grow upward while roots grow downward and how plants produce flowers at the proper time of year, is described in Chapter 44.

43.6 WHAT ARE SOME ADAPTATIONS FOR POLLINATION AND SEED DISPERSAL?

Coevolution Matches Plants and Pollinators

In many instances, plants and their pollinators have coevolved; that is, each has acted as an agent of natural selection on the other. Animal-pollinated flowers must attract useful pollinators and frustrate undesirable visitors that might eat nectar or pollen without fertilizing the flower. The animals must be able to locate and identify nutritious flowers and extract the nectar or pollen. Animal-pollinated flowers can be loosely grouped into three categories, depending on the benefits (real or deceptive) that they offer to potential pollinators: food, sex, or a nursery. As you will see, some plants have evolved in ways that trick insects into pollinating them and can thus avoid expending energy to provide the insects with food.

(a)

(b)

stamens

FIGURE 43-17 "Pollinating" a pollinator
(a) In Scotch broom flowers, the bee finds nectar near the junction of top and bottom petals. (b) The bee's weight deflects the bottom petals downward, causing pollen-laden stamens to pop up and cover the bee's hairy back with pollen. The bee will carry the pollen to other Scotch broom flowers, leaving some on ready stigmas.

FIGURE 43-18 A skunk cabbage
Scavenging insects are attracted by foul odors emitted by flowers such as the skunk cabbage, whose warmth may melt surrounding snow.

Some Flowers Provide Food for Pollinators

Many flowers provide food for foraging animals such as beetles, bees, moths and butterflies, or hummingbirds. In return, the animals unwittingly distribute pollen from flower to flower. We can thank the bees for most of the sweet-smelling flowers because sweet "flowery" odors attract these pollinators (flowers produce scents in petals or other flower parts, depending on the species). Bees have good color vision, but they do not see exactly the same range of colors that humans do (**FIG. 43-16**). To attract a bee from afar, bee-pollinated flowers must look brightly colored to a bee. Typically, these flowers are white, blue, yellow, or orange; many have other markings that reflect UV light, including central spots or lines pointing toward the center.

Bee-pollinated flowers have structural adaptations that help ensure pollen transfer. In the Scotch broom flower, for example, nectar forms in a crevice between enclosing petals. In newly opened flowers, pollen-laden stamens lurk within the crevice. When a bee visits a young flower, the stamens emerge, brushing pollen onto its back as the bee's weight deflects the petals downward (**FIG. 43-17**). In older flowers the sticky stigma of the carpel protrudes from the crevice, so when a pollen-coated bee delves in for nectar, she leaves pollen behind on the stigma.

Many flowers adapted for moth and butterfly pollinators have nectar-containing tubes that accommodate the long tongues of these insects. Flowers pollinated by night-flying moths open only in the evening; most are white and some give off strong, musky odors that attract moths through the darkness. Beetles and flies often prefer to feed on animal material, and beetle-pollinated flowers often smell like dung or rotting flesh, which attracts these scavenging insects. Some of these, including the corpse flower (see the chapter opening photo) and the skunk cabbage (**FIG. 43-18**) also heat up. Their warmth attracts pollinators and helps broadcast their foul scents (see the Case Study Revisited). These particular flowers deceive their pollinators by smelling like nutrient-rich rotting meat but offering no food at all.

Hummingbirds are one of the few important vertebrate pollinators (**FIG. 43-19a**), although several other mammals also pollinate flowers (**FIG. 43-19b**). Since hummingbirds have a poor sense of smell, hummingbird-pollinated flowers

FIGURE 43-19 Some vertebrate pollinators
(a) A hummingbird feeds at a hibiscus flower. Notice how the anthers are positioned to deposit pollen on its head. (b) As the honey possum of Australia stuffs its face into this flower, pollen adheres to its muzzle and whiskers. A visit to another flower will transfer the pollen. QUESTION Why have many plants that are pollinated by hummingbirds evolved flowers shaped like long, narrow tubes?

FIGURE 43-20 Sexual deception promotes pollination
This male wasp is trying to copulate with an orchid flower. The result is successful reproduction—by the orchid, not the wasp!

Sexy Deceptions Attract Pollinators

To pollinate their flowers, a few plants, most notably the orchids, take advantage of the mating drive and stereotyped behaviors of male wasps. Some orchid flowers mimic female wasps or bees both in scent (the orchids release a sexual attractant similar to that produced by the female insect) and shape (**FIG. 43-20**). The males land atop these "fake females" and attempt to copulate, but get only a packet of pollen for their efforts. As they repeat their attempts on other orchids of the same species, the pollen packet is transferred.

Some Plants Provide Nurseries for Pollinators

Perhaps the most elaborate relationships between plants and pollinators occur in a few cases in which insects fertilize a flower and then lay their eggs in the flower's ovary. This arrangement occurs between milkweeds and milkweed bugs, figs and certain wasps, and yuccas and yucca moths (**FIG. 43-21**). The yucca moth's remarkable behavior results in the pollination of yuccas and a well-stocked pantry for its own offspring. A female moth visits a yucca flower, collects pollen, and rolls it into a compact ball. She carries the pollen ball to another yucca flower, drills a hole in the ovary wall, and lays her eggs inside the ovary. Then she smears the pollen ball all over the stigma of the flower, performing this genetically programmed behavior flawlessly. By pollinating the yucca, the moth ensures that the plant will provide a supply of developing seeds for its caterpillar offspring. Because the caterpillars eat only a

seldom synthesize fragrant chemicals. However, they often produce more nectar than other flowers, because hummingbirds need more energy than insects do and will favor flowers that provide it. Hummingbird-pollinated flowers may have a deep, tubular shape that accommodates the birds' long bills and tongues. In addition, they are often red, which is attractive to hummingbirds but is not seen well by bees (see Fig. 43-16).

stamen

carpel

FIGURE 43-21 A mutually dependent relationship
(a) Yuccas bloom on the dry plains of eastern Colorado in early summer. (b) A yucca moth places pollen on the stigma of a yucca flower.

(a)

(b)

(a)   (b)

FIGURE 43-22 **Wind-dispersed fruits**
(a) Dandelions have filamentous tufts that catch the breezes.
(b) Maple fruits resemble miniature glider–helicopters, whirling away from the tree as they fall. EXERCISE To see how the wings aid in seed dispersal, take two maple fruits and pluck the wings off one. Hold both fruits over your head and drop them. Compare where they land.

FIGURE 43-23 **Water-dispersed fruit**
This coconut may have been washed ashore after a long journey at sea. Coconut "meat" and coconut "milk" are two different types of endosperm. The large size and massive food reserves of coconuts are probably adaptations for successful germination and growth on barren, sandy beaches.

fraction of the seeds, the yucca also reproduces successfully. The mutual adaptation of yucca and yucca moth is so complete that neither can reproduce without the other.

Fruits Help Disperse Seeds

A plant benefits if its seeds are dispersed far enough away so that its offspring don't compete with it for light and nutrients. Plant species will also be more successful and widespread if they can disperse their seeds both locally and to distant habitats. In flowering plants, fruits use a fascinating variety of mechanisms to disperse seeds.

Explosive Fruits Allow Shotgun Dispersal

A few plants develop explosive fruits that eject their seeds meters away from the parent plant. Mistletoes, common parasites of trees, produce fruits that shoot out sticky seeds. If one seed strikes a nearby tree, it sticks to the bark and germinates, sending rootlike fibers into the vascular tissues of its host, from which it draws its nourishment. Because the proper germination site for a mistletoe seed is not the ground but a tree limb, shooting the seeds, not dropping them, is clearly useful.

Lightweight Fruits Allow Wind Dispersal

Dandelions and maples (**FIG. 43-22**) produce lightweight fruits with large wind-catching surfaces. Each hairy tuft on a dandelion puff is a separate fruit bearing a single small seed that it can carry for miles if the winds cooperate. The single wing of the maple fruit, in contrast, causes its large seed to spiral as it falls, usually taking it only a few meters from its parent tree.

Floating Fruits Allow Water Dispersal

Many fruits can float on water for a time and may be dispersed by streams or rivers. The coconut fruit, however, is a champion floater. Round, buoyant, and waterproof, the coconut drops from its parent palm, often near a sandy shore. It may germinate in place, or it may be washed out to sea and float for weeks or months until it washes ashore on some distant isle (**FIG. 43-23**). There it may germinate, possibly establishing a new coconut colony where none previously existed.

Clingy or Tasty Fruits Allow Animal Dispersal

Clingy fruits, such as burdocks, burr clover, foxtails, and sticktights, grasp animal fur (or human clothing) with prongs, hooks, spines, or adhesive hairs (**FIG. 43-24**). Their

FIGURE 43-24 **The cocklebur fruit uses hooked spines to hitch a ride on furry animals**

parent plant holds its ripe fruit very loosely, so that even slight contact with fur pulls the fruit off the plant and leaves it stuck to the animal. Some of these fruits later fall off as the animal rolls on the ground, brushes against objects, or grooms and sheds its fur.

Unlike clingy hitchhiker fruits, edible fruits benefit both the plant and the animal disperser. The plant stores sugars, starches, and appealing flavors in a fleshy fruit that surrounds the seeds, enticing hungry animals (**FIG. 43-25**). Some such fruits, including peaches, plums, and avocados, contain large, hard seeds that animals usually do not eat. Other fruits, such as blackberries, raspberries, strawberries, tomatoes, and peppers, have small seeds that animals swallow. These seeds are eventually excreted unharmed. Some have seed coats that must be scraped or digested by passing through an animal's digestive tract before they will germinate (see "Earth Watch: On Dodos, Bats, and Disrupted Ecosystems"). An ecology graduate student studying the very spicy seeds of chili peppers found that their burning flavor discourages local mammals from eating the fruit but does not deter birds, who are insensitive to it. He further discovered that the digestive tracts of mammals destroy chili seeds, but seeds that have passed through a bird's digestive tract germinate at three times the rate of those that just fall to the ground. Besides being transported away from its parent plant, a seed that is swallowed and excreted benefits in another way: it ends up with its own supply of fertilizer!

FIGURE 43-25 The colors of ripe fruits attract animals A bright red raspberry fruit has attracted a resplendent quetzal in Costa Rica. Only ripe fruits containing mature seeds are sweet and brightly colored, appealing to animals that feed on them and disperse their seeds.

CASE STUDY REVISITED GORGEOUS? SURE! BUT HOT?

You have learned by now that most heat-producing flowers or flower clusters are pollinated by beetles, flies, and sometimes bees, and their scent mimics the pollinator's preferred food. The corpse flower, dead-horse arum, and skunk cabbage all attract beetles or flies that feed on decaying meat or animal wastes. In contrast, philodendrons have faint, pleasant aromas, and attract beetles that feed on their pollen and other flower parts. The sacred lotus, pollinated by bees and beetles smells sweet and fruity. But why heat up?

Heat causes odorous chemicals to evaporate, spreading the scent into the surrounding air and attracting pollinators, sometimes from considerable distances. Some blossoms, such as those of the dead-horse arum, deceive their pollinators. Lured by the odor of warm, rotting flesh, flies are enticed into a chamber surrounding the central spadix where a hairy barricade traps them overnight (**FIG. 43-26**). As they struggle to escape, the flies shower the female flowers with pollen picked up from an earlier close encounter

with an arum. By the next day, the hairs have wilted and the male flowers have matured, dusting pollen onto the flies as they escape. Being slow learners, the flies are deceived by, and pollinate, many more dead-horse arums.

Roger Seymour hypothesizes that another important function of some warm flowers is to reward their cold-blooded pollinators, which must achieve a body temperature of at least 85°F (29.4°C) before they can fly. The pleasant aroma of

FIGURE 43-26 Flies on a dead-horse arum

philodendron flowers attracts beetles that feed on their pollen and other flower parts. In addition to providing real food for the beetles, these benevolent arum flowers also provide a brothel. Beetles crawl down around the spadix during the day, where they settle into an evening of feasting and mating within the blossom's warm confines (**FIG. 43-27**). Measurements of beetle energy consumption reveal that they use 50% less energy by hanging out in the philodendron "nightclub" than had they remained active outside during the cool night.

Heat-producing flowers are rare, and many are members of evolutionarily ancient groups. Some botanists hypothesize that warmth was an early innovation to attract beetle pollinators. Today most flowers supply their pollinators with a sip of nectar that provides a little energy and then send them on their way with a quick dusting of pollen—a much faster and less energy-consuming process than warming the flower. Perhaps over evolutionary time, as Seymour puts it, "nightclubs were replaced by fast food."

Consider This Think about the pollination strategies used by the dead-horse arum and the philodendron, mentioned in this essay, and the "fast food" flowers emphasized in the chapter. In air of the same temperature (say 70°F), which flower is likely to use the most energy? The least energy? Suggest evolutionary reasons for the fast-food flowers that predominate today.

FIGURE 43-27 A beetle brothel
This philodendron flower (from French Guiana) produces scent on the first day of blooming (left). Then it heats up, attracting beetles that congregate, feed, mate, and conserve energy on the warm blossom (right).

CHAPTER REVIEW

SUMMARY OF KEY CONCEPTS

43.1 What Are the Basic Features of Plant Life Cycles?

The sexual life cycle of plants—alternation of generations—includes both a multicellular diploid form (the sporophyte generation) and a multicellular haploid form (the gametophyte generation).

43.2 How Is Reproduction in Seed Plants Adapted to Drier Environments?

In seed plants, the gametophyte stage is microscopic and does not live independently. The male gametophyte encased in a coat is the pollen grain, carried from plant to plant by wind or animals. The female gametophyte is retained within the body of the sporophyte plant. These adaptations allow seed plants to reproduce in relatively dry environments.

43.3 What Is the Function and Structure of the Flower?

Flowering plants evolved from gymnosperms. In gymnosperms, wind blows pollen from male cones to female cones, but wind pollination is inefficient. In many habitats, flowering plants enjoy a selective advantage over gymnosperms because many types of flowers attract insects that carry pollen from plant to plant.

Complete flowers consist of four parts: sepals, petals, stamens (male reproductive structures), and carpels (female reproductive structures). The sepals form the outer covering of the flower bud. Most petals (and in some cases, the sepals) are brightly colored and attract pollinators to the flower. The stamen consists of a filament that bears an anther, in which pollen develops. The carpel consists of the ovary, in which one or more female gametophytes develop, and a style that bears a sticky stigma to which pollen adheres during pollination.

Pollen, which consists of the male gametophyte encased in a tough coat, develops in the anthers. The diploid microspore mother cell undergoes meiotic cell division to produce four haploid microspores. Each undergoes mitotic cell division to form the haploid male gametophyte. The immature pollen grain consists of a tube cell and a generative cell that will later divide to produce two sperm cells.

The female gametophyte develops within the ovules of the ovary. A diploid megaspore mother cell undergoes meiotic cell division to form four haploid megaspores. Three degenerate; the fourth undergoes mitotic divisions to produce the eight nuclei of the female gametophyte. One of these becomes the egg; another becomes a large central cell with two nuclei, and the rest degenerate.

Pollination is the transfer of pollen from anther to stigma. When a pollen grain lands on a stigma, its tube cell grows through the style to the female gametophyte. The generative cell divides to form two sperm cells that travel down the style within the tube cell, eventually entering the female gametophyte. One sperm fuses with the egg to form a diploid zygote, which will give rise to the embryo. The other sperm fuses with the two nuclei of the central cell, producing a triploid cell that will give rise to the endosperm, a food-storage tissue within the seed.

Web Tutorial 43.1 Reproduction in Flowering Plants

43.4 How Do Fruits and Seeds Develop?

Fruits may be juicy and edible, enclosing seeds that are adapted for passage through an animal digestive tract. Alternatively, fruits may have hooks that attach to animal fur, or wings that promote wind dispersal. The function of the fruit is to disperse the seed. The embryo consists of an embryonic root and embryonic shoot, including the cotyledon (one in monocots, two in dicots). Cotyledons absorb food from the endosperm and transfer it to the growing embryo. The seed is enclosed within a fruit, which develops from the ovary wall.

43.5 How Do Seeds Germinate and Grow?

Seed germination requires warmth and moisture. Energy for germination comes from food stored in the endosperm, which is transferred to the embryo by the cotyledons. Seeds may remain dormant for some time after fruit ripening, particularly in temperate climates. To germinate, some seeds also require drying, exposure to cold, or disruption of the seed coat. The root emerges first from the germinating seed, absorbing water and nutrients for the shoot. Monocot shoots are often protected with a coleoptile, while many dicots use epicotyl or hypocotyl hooks to break through the soil. Cotyledons supply the young shoot with food energy for growth.

43.6 What Are Some Adaptations for Pollination and Seed Dispersal?

Plants and their animal pollinators and seed dispersers have co-evolved, acting as agents of natural selection on one another over evolutionary time. Flowers attract animals with scent, food such as nectar, and appropriate colors and shapes that correspond to the animal's body and senses. Some flowers deceive pollinators, attracting insects with food scents or the shape of a mate. Some plants and their pollinators, such as the yucca plant and yucca moth, are completely dependent on one another. Fruits disperse seeds in many ways. Some are adapted to be transported by wind or water. Some cling to animal fur; others entice animals to eat the fruit without harming the seeds.

KEY TERMS

alternation of generations *page 889*
anther *page 892*
carpel *page 892*
coleoptile *page 899*
complete flower *page 892*
cotyledon *page 896*
dormancy *page 899*
double fertilization *page 896*
egg cell *page 895*
endosperm *page 896*

epicotyl *page 899*
fertilization *page 895*
filament *page 892*
flower *page 891*
fruit *page 892*
gametophyte *page 889*
generative cell *page 892*
germinate *page 889*
hypocotyl *page 899*
incomplete flower *page 892*
integument *page 895*

megaspore *page 895*
megaspore mother cell *page 895*
microspore *page 892*
microspore mother cell *page 892*
ovary *page 892*
ovule *page 892*
petal *page 892*
pollen grain *page 889*
pollination *page 895*

seed *page 889*
seed coat *page 896*
sepal *page 892*
spore *page 889*
sporophyte *page 889*
stamen *page 892*
stigma *page 892*
style *page 892*
tube cell *page 892*
zygote *page 889*

THINKING THROUGH THE CONCEPTS

1. Diagram the plant life cycle, comparing ferns with flowering plants. Which stages are haploid, and which are diploid? At which stage are gametes formed?

2. What are the advantages of the reduced gametophyte stages in seed plants, compared with the more substantial gametophytes of ferns?

3. Diagram a complete flower. Where are the male and female gametophytes formed?

4. How does an egg develop within the female gametophyte? How does this structure allow double fertilization to occur?

5. What is a pollen grain, and how is it formed?

6. What are the parts of a seed, and how does each part contribute to the development of a seedling?

7. Describe the characteristics you would expect to find in flowers that are pollinated by the wind, beetles, bees, and hummingbirds, respectively. In each case, explain why the traits have developed.

8. What is the endosperm? From which cell of the female gametophyte is it derived? Is endosperm usually more abundant in the mature seed of a dicot or of a monocot?

9. Describe three mechanisms whereby seed dormancy is broken in different types of seeds. How are these mechanisms related to the typical environment of the plant?

10. How do monocot and dicot seedlings protect the delicate shoot tip during seed germination?

11. Describe three types of fruits and the mechanisms whereby these fruit structures help disperse their seeds.

APPLYING THE CONCEPTS

1. A friend gives you some seeds to grow in your yard. When you plant some, nothing happens. What might you try to get the seed to germinate?

2. Charles Darwin once described a flower that produced nectar at the bottom of a tube 43 centimeters (10.5 inches) deep. He predicted that there must be a moth or other animal with a 43-centimeter-long "tongue" to match; he was right. Such specialization almost certainly means that this particular flower could be pollinated only by that specific moth. What are the advantages and disadvantages of such specialization?

3. Many plants that we call weeds were brought from another continent either accidentally or purposefully. In their new environment, they have few competitors or animal predators, so they tend to grow in such large numbers that they displace native plants. Think of several ways in which humans become involved in plant dispersal. To what degree do you think humans have changed the distributions of plants? In what ways is this change helpful to humans? In what ways is it a disadvantage?

4. In the Tropics, there are a number of plant–animal evolutionary relationships in which each depends on the other for survival. Given the rapid rate of destruction of tropical ecosystems, how does this type of relationship leave both organisms particularly vulnerable to extinction?

FOR MORE INFORMATION

Brown, Kathryn. "Patience Yields Secrets of Seed Longevity." *Science*, March 9, 2001. William Beal's plant germination study continues over 120 years later.

Eiseley, L. "How Flowers Changed the World." *National Wildlife*, April–May 1996. The late philosopher/naturalist Loren Eiseley eloquently explains how the evolution of flowers has changed the history of life on Earth. Beautifully written and illustrated.

Milius, S. "The Science of Big, Weird Flowers." *Science News*, September 11, 1999. Many giant flowers deceive flies and beetles into pollinating them by smelling like carrion.

Milius, S. "Warm-Blooded Plants." *Science News*, December 13, 2003. Some plants heat up, attracting and sometimes benefiting pollinators.

Milius, S. "Moss Express: Insects and Mites Tote Mosses' Sperm." *Science News*, September 2, 2006. New research suggests that some mosses may not rely entirely on water to transfer sperm but may rely on tiny arthropods as well.

Mlot, C. "Where There's Smoke, There's Germination." *Science News*, May 31, 1997. Researchers have discovered that nitrogen dioxide, such as that released in smoke, is a potent stimulus for seed germination in some species.

Moore, P. D. "The Buzz About Pollination." *Nature*, November 7, 1996. The buzzing of bees may actually shake loose the pollen of certain specially adapted flowers, in a kind of "sonic pollination."

Pichersky, E. "Plant Scents." *American Scientist*, November–December, 2004. Floral scents may not only attract pollinators but also deter attack from predators and disease organisms.

Seymour, R. S., and Schultz-Motel, P. "Thermoregulating Lotus Flowers." *Nature*, September 26, 1996. The lotus flower generates a significant amount of heat and effectively regulates its own temperature. The heat may serve as an attractant to pollinators.

44 Plant Responses to the Environment

A fly enters a Venus flytrap. (Inset) The flytrap plant in its natural environment.

CASE STUDY PREDACIOUS PLANTS

IN A MARSH, PLANTS ARE "HUNGRY"—not for sunlight, but for nitrogen. Marshes and bogs tend to be acidic, and the acid discourages growth of nitrogen-fixing bacteria. Although it is scarce in this watery environment, nitrogen is abundant in the proteins of animal bodies, and some bog-dwelling plants have evolved carnivorous lifestyles that satisfy their nitrogen needs.

In a bog somewhere in South Carolina, an unsuspecting fly lands on a seemingly harmless plant (see the chapter opener inset), attracted by nectar lining the edges of the plant's clamshell-like leaves. Suddenly the leaves snap together, pressing the insect's body against digestive glands that line their inner surfaces. Their spiked edges mesh, trapping the hapless insect. During the next four or five days, enzymes will digest the fly, and the leaf will absorb the nitrogen-containing molecules before the trap opens to attract its next victim.

Nearby, a lacewing insect lands on a cluster of glistening, sweet droplets—only to find itself struggling helplessly in the sticky mass (FIG. 44-1). To make matters worse, red tentacles bearing more balls of sweet glue bend toward it, miring it hopelessly. Enzymes in the secretion will digest the insect's body and the sundew will "feast," absorbing the nitrogen-rich nutrients.

Beneath the surface of the marsh, still another drama unfolds. A bladderwort dangles hundreds of pear-shaped chambers in the water, which—depending on the species—range in size from pinheads to peas. Each of these "bladders" is sealed by a watertight trapdoor whose lower edge is fringed with bristles. A tiny crustacean (related to shrimp, but barely visible to us) swims by, brushing the hairs. Within one-sixtieth of a second, the bladder traps the animal inside, where enzymes gradually kill and digest it. But how do these predacious plants move quickly enough to trap flying or swimming animals?

FIGURE 44-1 A sundew traps a lacewing

44.1 WHAT ARE PLANT HORMONES, AND HOW DO THEY ACT?

Plant cells, like those of animals, are miniature factories filled with diverse chemicals that allow plants to respond appropriately to their environment. Some convey messages within the plant, and others even communicate among individual plants. Animal physiologists have long recognized that chemicals called **hormones** are produced by cells in one location and transported to other parts of the body, where they exert specific effects. Comparably, plant-regulating chemicals are called **plant hormones**. So far, plant physiologists have identified five major classes of plant hormones: *auxins, gibberellins, cytokinins, ethylene,* and *abscisic acid* (Table 44-1). Several more hormones have recently been identified; one resembles aspirin, and another is a gas that is also used as a messenger molecule in animals. You'll learn more about these later in this chapter.

Each hormone can elicit a variety of responses from plant cells, depending on factors such as the type of cell it reaches, the developmental stage of the plant, the concentration of the hormone, and the presence of other hormones. Further, some plant hormones play different roles in different plant species. So a plant physiologist might study the role of a hormone in a plant species that grows easily and rapidly in the lab, only to find that it doesn't work the same way in other plants. This should not be too surprising, since animal hormones, which have been more intensely studied, also have differing effects among species. The hormone thyroxin, for example, helps a salmon make the transition from its freshwater hatching stream to the ocean, a frog to metamorphose from tadpole to adult, and a snake to shed its skin.

FIGURE 44-2 Tendrils show thigmotropism
In response to contact, auxin stimulates unequal elongation of cells in these cucumber tendrils, causing them to bend around nearby objects such as garden stakes, fences, or this grass stem.

In this chapter, we focus on the five major categories of plant hormone, whose effects are summarized here.

- **Auxins** are a group of chemically related hormones. They promote the elongation of cells in coleoptiles and other parts of the shoot; high concentrations cause cells to elongate (see "Scientific Inquiry: How Were Plant Hormones Discovered?"). In roots, which differ from stems in their response to auxin, low concentrations stimulate elongation, while slightly higher concentrations inhibit elongation. Both light and gravity affect the distribution of auxin in roots and shoots, so auxin plays a major role in both **phototropism** (growth toward light; "tropism" refers to an involuntary orienting response to a stimulus) and **gravitropism** (directional growth with respect to gravity). **Thigmotropism** (orientation in response to touch), in which some plants wrap tendrils (modified leaves or stems) around supporting structures, is stimulated by auxin (**FIG. 44-2**). Auxin also influences the differentiation of conducting tissues (xylem and phloem) and the development of fruits. Auxin may prevent sprouting of lateral buds on a stem. It stimulates root branching and can be used to cause stems of plants to grow roots. Synthetic auxin (2,4-D) is widely used to kill dicot plants by disrupting the normal balance among auxin and other plant hormones.

- **Gibberellins** are a group of chemically similar molecules that, like auxin, promote the elongation of cells in stems. In some plants, gibberellins stimulate flowering, fruit development, seed germination, and bud sprouting.

Table 44-1 Hormone Actions in Plants

Hormone	Functions
Abscisic acid	Closing of stomata; inhibited plant growth; seed dormancy; bud dormancy
Auxins	Cell elongation in coleoptiles and shoots; phototropism; gravitropism in shoots and roots; root growth and branching; apical dominance; development of vascular tissue; fruit development; retarding senescence in leaves and fruit; ethylene production in fruit
Cytokinins	Promotes lateral bud sprouting; delays senescence of leaves; promotes cell division; stimulates development of fruit, endosperm, and embryo
Ethylene	Ripening of fruit; abscission of fruits, flowers, and leaves; inhibition of stem elongation; formation of hook in dicot seedlings
Gibberellins	Germination of seeds and sprouting of buds; elongation of stems; stimulation of flowering; development of fruit

- **Cytokinins** promote cell division in many plant tissues; consequently, they stimulate the sprouting of buds and the development of the fruit, seed endosperm, and embryo. Cytokinins also stimulate plant metabolism, delaying the aging of plant parts, especially leaves.
- **Ethylene** is an unusual plant hormone because it is a gas at typical environmental temperatures. Ethylene is best known, and most commercially valuable, for its ability to cause fruit to ripen. It also stimulates the formation of weak-celled *abscission layers*, which allow leaves, flowers, and fruit to drop off at the appropriate times of year.
- **Abscisic acid** helps plants withstand unfavorable environmental conditions. As you learned in Chapter 42, abscisic acid causes stomata to close when water is scarce. It decreases plant growth in response to stressful conditions. It helps maintain dormancy in buds and seeds at times when germination would lead to death. Abscisic acid was named based on the mistaken hypothesis that it caused leaf abscission.

In the rest of this chapter we illustrate how hormones regulate growth and development during plant life cycles.

44.2 HOW DO HORMONES REGULATE THE PLANT LIFE CYCLE?

The life cycle of a plant results from a complex interplay between its genetic information and its environment. Many hormones act by influencing the activity of genes, which in turn regulate growth and development as well as nearly all responses to environmental factors. At each stage in their life cycles, often in response to environmental stimuli, plants produce distinctive sets of hormones. These hormones interact with one another and with genes to control the growth, maturation, reproduction, and senescence that comprise the stages of the plant life cycle.

The Plant Life Cycle Begins with a Seed

Abscisic Acid Maintains Seed Dormancy

As we pointed out in Chapter 42, a warm autumn day provides ideal germination conditions for a seed maturing within a juicy fruit, yet the seed remains dormant until the following spring. In many seeds, abscisic acid enforces dormancy. Abscisic acid slows down the metabolism of the embryo within the seed, preventing its growth. It is common for the seeds of temperate and high-latitude plants to require a period of cold, such as a full winter season, to break dormancy and allow the seed to germinate. In these seeds, chilling seems to cause a gradual reduction of abscisic acid levels. In contrast, the seed coats of some desert plants contain high concentrations of abscisic acid that must be washed away to allow germination. This requires a hard rain, and helps assure that germination will occur only when enough moisture is available for growth (**FIG. 44-3**).

Gibberellin Stimulates Germination

Germination is stimulated by other hormones, especially gibberellin. In germinating seeds, gibberellin initiates the synthesis of enzymes that digest the food reserves of the endosperm and cotyledons, making sugars, lipids, and amino acids available to the growing embryo.

For Some Plants, Where There's Smoke, There's Germination

Even a fire-blackened landscape will look quite green in a few years. Some plant species thrive in ecosystems maintained by frequent fires. Some produce seeds coated with resins that require the high temperature caused by fire to open. Recently, plant physiologists discovered that in some species, chemical signals from smoke can stimulate germination. Studying whispering bells, a plant found in chaparral ecosystems that are maintained by frequent fires, the researchers found that 30 seconds of exposure to smoke triggered seed germination. Although different smoke chemicals stimulate germination in different species, for whispering bells, nitrogen dioxide seems to be the "active ingredient." Does nitrogen dioxide promote gibberellin synthesis in these seeds? Future research will tell.

FIGURE 44-3 A desert in bloom
The yellow flowers are desert marigolds, which are stimulated by rain to germinate in large numbers. Abscisic acid helps maintain dormancy in many desert plants such as these, ensuring that they germinate when water is available.

SCIENTIFIC INQUIRY How Were Plant Hormones Discovered?

Anyone who keeps houseplants on a windowsill knows that the plants bend toward the window in response to the sunlight streaming in. More than a hundred years ago, Charles Darwin and his son Francis studied this phenomenon of growth toward the light, or phototropism.

FIRST, THE DARWINS DETERMINED THE DIRECTION OF INFORMATION TRANSFER

The Darwins illuminated grass coleoptiles (protective sheaths surrounding monocot seedlings) from various angles. They noted that a region of the coleoptile a few millimeters below the tip bent toward the light, causing the tip to point toward the light source. When they covered the tip of the coleoptile with an opaque cap, the coleoptile didn't bend.

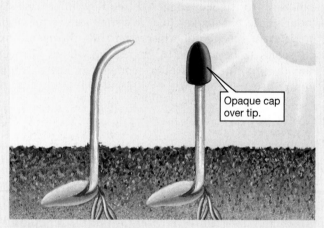

FIGURE E44-1 Tip doesn't bend in dark

A clear cap allowed the stem to bend, and it still bent if the bending region below the tip was covered with an opaque sleeve.

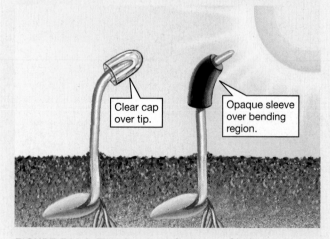

FIGURE E44-2 Tip perceives direction of light

The Darwins concluded that the tip of the coleoptile perceives the direction of light and that bending occurs farther down the coleoptile. This suggests that the tip transmits information about light direction down to the bending region. How does the coleoptile bend? Although the Darwins didn't know this, the coleoptile bends due to unequal elongation of cells on opposite sides of the shaft. The cells on the darker side elongate faster than those facing the light, bending the shaft toward the light. This indicates that information transmitted from the tip to the bending region causes unequal cell elongation.

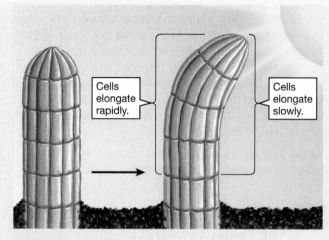

FIGURE E44-3 Cells elongate on shaded side

LATER, PETER BOYSEN-JENSEN DEMONSTRATED THAT THE INFORMATION IS CHEMICAL IN NATURE

About 30 years later, in 1913, Peter Boysen-Jensen cut the tips off coleoptiles and found that the remaining stump neither elongated nor bent toward the light. If he replaced the tip and placed the patched-together coleoptile in the dark, it elongated straight up. In the light, it showed normal phototropism. When he inserted a thin layer of porous gelatin that prevented direct contact but permitted diffusion of substances between the severed tip and the stump, he still observed elongation and bending. In contrast, an impenetrable barrier eliminated these responses.

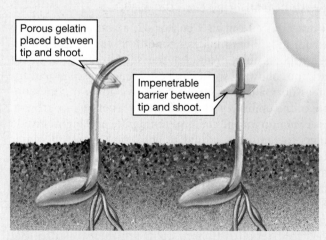

FIGURE E44-4 A diffusing substance promotes bending

Boysen-Jensen concluded that a chemical is produced in the tip and moves down the shaft, causing cell elongation. In the dark, the chemical that causes the cells to elongate diffuses straight down from the tip and causes the coleoptile to elongate straight up. Presumably, light causes the chemical to become more concentrated on the "shady" far side of the shaft, so cells on the shady side elongate faster than do cells on the "sunny" near side, causing the shaft to bend toward the light.

FINALLY, THE CHEMICAL AUXIN WAS IDENTIFIED

The next step was to isolate and identify the chemical. In the 1920s, Frits Went devised a way to collect the elongation-promoting chemical. He cut off the tips of oat coleoptiles and placed them on a block of agar (a porous, gelatinous material) for a few hours. Went hypothesized that the chemical would migrate out of the coleoptiles into the agar.

He then cut up the agar, now presumably loaded with the chemical, and placed small pieces on the tops of coleoptile stumps growing in darkness. When he put a piece of agar squarely atop a stump, the stump elongated straight up. All the stump cells received equal amounts of the chemical and elongated at the same rate. If he placed a piece on one side of a cut stump, the stump would invariably bend away from the side with the agar.

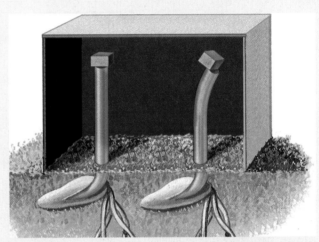

FIGURE E44-6 Diffusing substance causes bending in dark

It was apparent that cells on the side under the agar received more of the chemical and were more stimulated to elongate. Went called the chemical *auxin*, from a Greek word meaning "to increase." Kenneth Thimann later purified auxin and determined its molecular structure.

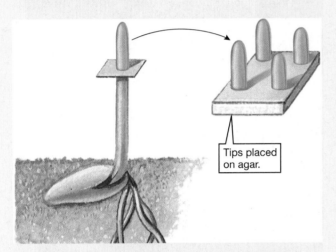

Tips placed on agar.

FIGURE E44-5 Diffusing substance is collected in agar

Auxin Controls the Orientation of the Sprouting Seedling

When the embryo breaks out of the seed coat, it immediately faces a crucial problem: Which way is up? Whether the seed was buried by a squirrel or fell randomly to the ground, it faces a high probability of being oriented upside down. Auxin controls the responses of both roots and shoots to light and gravity. As you may recall, embryonic roots emerge first, followed by the shoot.

Auxin Stimulates Shoot Elongation Away from Gravity and Toward Light

Auxin is synthesized in the embryonic shoot tip. As the tip emerges from a buried seed, the auxin moves downward and stimulates the cells of the stem to elongate. If the stem is horizontal, cells within the stem detect the direction of gravity and cause auxin to accumulate on the lower side

(FIG. 44-4a). Therefore, the lower cells elongate rapidly and cause the stem to bend upward. When the shoot tip becomes vertical, auxin becomes evenly distributed. The stem then grows straight up, emerging from the soil into the light. The same effect can be observed in older plants as well (FIG. 44-4c).

In addition to gravitropism, auxin also mediates phototropism. Ordinarily, the distribution of auxin caused by light is the same as the distribution caused by gravity because the direction of brightest light (the sun) is usually roughly opposite that of gravity. If a young shoot buried horizontally underground is close enough to the surface that some light penetrates down to it, both light and gravity cause auxin to be transported to the lower side of the shoot and promote upward bending. Thus, gravitropism and phototropism usually work together to bring the shoot upward into the light. Shoots placed on a windowsill clearly demonstrate phototropism (FIG. 44-4d).

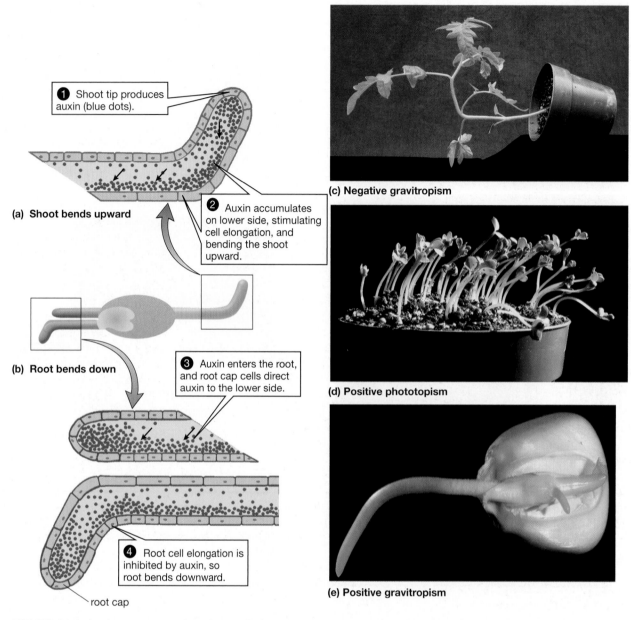

(a) Shoot bends upward

❶ Shoot tip produces auxin (blue dots).

❷ Auxin accumulates on lower side, stimulating cell elongation, and bending the shoot upward.

(b) Root bends down

❸ Auxin enters the root, and root cap cells direct auxin to the lower side.

❹ Root cell elongation is inhibited by auxin, so root bends downward.

root cap

(c) Negative gravitropism

(d) Positive phototopism

(e) Positive gravitropism

FIGURE 44-4 Auxin causes gravitropism and phototropism
In **(a)** and **(b)**, auxin on the lower side of the shoot and root is distributed similarly, but with opposite effects. **(c)** This tomato plant grows away from gravity (negative gravitropism) after resting in the dark on its side for less than a day. **(d)** These radish seedlings bend toward light (positive phototropism). **(e)** The developing root of this germinating corn seed grows downward (positive gravitropism) after the seed is placed on its side.

Auxin Helps Control the Direction of Root Growth

Auxin is transported down from the shoot into the root cap. From there it is distributed in response to gravity and helps control the direction of root growth. If the root is not vertical, the root cap senses the direction of gravity and causes the auxin to accumulate on the lower side of the root. In roots, moderate concentrations of auxin inhibit cell elongation. Therefore, cell elongation along the lower side of the root, where auxin accumulates, is inhibited, whereas cell elongation along the upper side of the root remains unaffected. As a result, the root bends downward (**FIG. 44-4b** and **44-4e**). When the root tip points directly downward, the auxin is distributed equally on all sides, and the root continues to grow straight downward. Because auxin slows but does not eliminate root cell elongation, the root will continue to grow downward.

How Do Plants Sense Gravity?

The simple answer to this question is that no one is certain. Many researchers hypothesize that starch-filled plastids in some stem and root-cap cells inform the plant which way is up and down. By staining these plastids and observing them under the microscope, plant physiologists have discovered that they rest on the lower part of the cell, rapidly settling to the "new" downward side of the cell when the plant is laid on its side (**FIG. 44-5**).

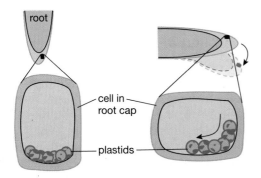

FIGURE 44-5 Plastids may be gravity detectors

How might falling plastids initiate the response to gravity? One hypothesis is that the plastids are enmeshed in fibers of the cytoskeleton (see Chapter 4), and these fibers connect the plastids to ion channels. As the plastids are pulled downward by gravity, they pull the channels open, allowing a flow of ions (probably calcium ions) into the cell. In this way, a mechanical stimulus (plasmid movement) could be converted to a chemical stimulus (increased calcium). This chemical change might then start a series of reactions that cause auxin to accumulate on the downward side of the shoot or root. Plant physiologists are still testing these hypotheses.

The Genetically Determined Shape of the Mature Plant Is the Result of Interactions Among Hormones

As a plant grows, both its root and shoot develop branching patterns that are mostly determined by their genes. For example, the stems of some plants such as sunflowers hardly branch at all. Others, such as oaks and cottonwoods, branch profusely with no clear pattern, and still others branch in very regular patterns, producing the conical shapes of firs and spruces.

The amount of growth in shoot and root systems must also be kept in balance. The shoot must be large enough to supply the roots with sugars, and the roots must be large enough to provide the shoot with water and minerals. Interactions between auxin and cytokinin regulate root and stem branching, thereby regulating the relative sizes of root and shoot systems.

The Shoot Tip Produces Auxin That Inhibits Stem Branching

Gardeners know that pinching back the tip of a growing plant causes the plant to become bushier. This happens because the growing tip suppresses the sprouting of lateral buds to form branches, a phenomenon known as **apical dominance**; the shoot tip is the "apex" of the plant, and it "dominates" the lower lateral buds (**FIG. 44-6**). The control mechanism for the sprouting of lateral buds remains a subject of research, but it appears that proper levels and ratios of auxin and cytokinin must be present. Auxin is produced by the shoot tip (where it is found in highest concentration) and transported down the stem, gradually decreasing in concentration. Cytokinin is produced by the root tips (where its concentration is highest) and is transported up through the roots and into the stem. Therefore, the relative concentrations of these two hormones will vary along the length of the stem and roots. Buds at different positions will experience differing hormonal influences.

Auxin by itself appears to inhibit the sprouting of lateral buds, whereas auxin and cytokinin together stimulate bud sprouting. The lateral buds closest to the shoot tip receive enough auxin to inhibit their growth, but they receive very little cytokinin because they are so far from the roots. Therefore, they remain dormant. Lower buds receive less auxin, but more cytokinin. They are stimulated to

FIGURE 44-6 Apical dominance
The plant on the right is growing naturally, while the plant on the left has had its growing tip cut off, reducing auxin levels in the stem and allowing lateral buds to sprout just below the cut stem.

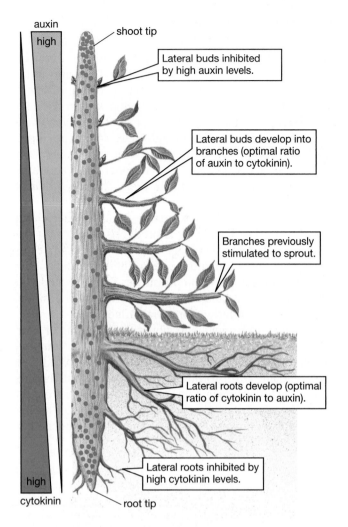

FIGURE 44-7 The role of auxin and cytokinin in lateral bud sprouting
The relative amounts of auxin (blue dots) and cytokinin (red dots) control the sprouting of lateral buds and branching and development of lateral roots. Auxin is produced by shoot tips and moves downward; cytokinin is produced by root tips and moves upward. QUESTION If you removed a plant's shoot tip and applied auxin to the cut surface, how would you expect the lateral buds to respond?

sprout by optimal concentrations of both hormones (**FIG. 44-7**). In many types of plants, this gradient in auxin and cytokinin ratios produces an orderly progression of bud sprouting from the bottom to the top of the shoot. The exact ratio of cytokinin to auxin that promotes sprouting varies among species.

Auxin Stimulates and Cytokinin Inhibits Root Branching

Even in extremely low concentrations, auxin, produced in the shoot tip, stimulates root branching. As described in Chapter 42, branch roots arise from the pericycle layer of the vascular cylinder. Auxin, transported down from the stem, stimulates pericycle cells to divide and form a branch root. Auxin also stimulates the growth of new roots from stems. Commercial auxin powder allows people to produce

a new plant by dipping the cut end of a stem into the auxin and placing it in soil or water, where it will develop roots. In contrast, cytokinin produced in the root tip inhibits roots from branching. Lower roots receive more cytokinin and less auxin from the shoot tip, while roots closer to the soil surface receive more auxin and less cytokinin (see Fig. 44-7). Because roots and the shoots supply complementary nutrients (sugar from the photosynthetic shoots, and water and minerals from the roots), it is important that their development remain balanced. Through the competing and differing actions of auxin and cytokinin in shoots and roots, the two systems may regulate each other's growth.

Day Length Controls Flowering

The timing of flowering and seed production is finely tuned to the physiology of the plant and the rigors of its environment. In temperate climates, plants must flower early enough so that their seeds can mature before the deadly frosts of autumn. Depending on how quickly the seed and fruit develop, flowering may occur in spring, as it does in oaks; in summer, as in lettuce; or even in autumn, as in asters.

What environmental cues do plants use to determine the seasons? Most cues, such as temperature or water availability, are quite variable: October can be warm, a late snow could fall in May, or the summer might be unusually cool and wet. The only reliable cue is day length: longer days always mean that spring and summer are coming; shorter days foretell the onset of autumn and winter.

With respect to flowering, botanists classify plants as *day-neutral*, *long-day*, or *short-day*. A **day-neutral plant** flowers as soon as it has sufficiently grown and developed, regardless of the length of the day. Day-neutral plants include tomatoes, corn, snapdragons, and roses. Although the naming is traditional, long-day and short-day plants are better described as *short-night* and *long-night* plants because their flowering actually depends on the duration of continuous darkness rather than on day length (**FIG. 44-8**). **Short-night plants** (which include lettuce, spinach, iris, clover, and petunias) flowers when the length of darkness is shorter than a species-specific critical period. **Long-night plants** (including asters, potatoes, soybeans, goldenrod, and cockleburs) flower when the length of uninterrupted darkness is longer than a species-specific critical period. Thus, spinach is classified as a short-night plant because it flowers only if the night is shorter than 11 hours (its critical period), and the cocklebur is a long-night plant because it flowers only if uninterrupted darkness lasts more than 8.5 hours. Both of these plants will flower with 10-hour nights.

Plant scientists can induce flowering in the cocklebur by exposing a single leaf to long nights (longer than its 8.5-hour critical period) in a special chamber, while the rest of the plant continues to experience short nights. Clearly, a signal that induces flowering must travel from the leaf to the flower bud. Plant physiologists have been attempting for decades to isolate this elusive signaling molecule, often

Darkness longer than critical length	Darkness shorter than critical length

FIGURE 44-8 The effects of night length on flowering

day-neutral   rose

long-night   aster

short-night   petunia

Phytochrome changes from one form to the other when the pigment absorbs light of the appropriate color; upon absorbing red light, P_r is converted into active P_{fr}; upon absorbing far-red light, P_{fr} is transformed back into inactive P_r. In the dark, P_{fr} spontaneously reverts to inactive P_r. Sunlight consists of all wavelengths of visible light, including both red and far-red. Therefore, during the day, a sunlit leaf will contain both forms of phytochrome. Because a reasonable amount of P_{fr} will be present in sunlight, this active form of phytochrome will control the plant's responses.

Plants seem to use the phytochrome system in combination with their internal biological clocks to detect the duration of continuous darkness. Cockleburs, for example, flower under a schedule of 16 hours of darkness and 8 hours of light. However, interrupting the middle of the dark period with just a minute or two of light prevents flowering. Thus, their flowering is controlled by the length of continuous darkness. It is evident that even brief exposure to sunlight or white light will reset their biological clocks. The color of the light used for the night exposure is also important. A nighttime flash of pure red light inhibits flowering, presumably by converting P_r into

called *florigen* (literally, "flower maker"). Some researchers believe that they are close to demonstrating a flower-stimulating substance for a specific type of plant, using genetic manipulation. It is likely, however, that interactions among multiple, as yet unidentified plant hormones stimulate or inhibit flowering, and that these chemicals may differ among plant species. Researchers have had more success in determining how plants measure the length of uninterrupted darkness, which is a crucial stimulus for producing whatever substances control flowering, as described next.

Pigments Called Phytochromes Detect Light and Reset the Biological Clock

To measure continuous darkness, a plant needs two things: some sort of metabolic clock to measure time (the duration of darkness) and a light-detecting system to set the clock. Virtually all organisms have an internal **biological clock** that measures time even without environmental cues. In most organisms, including plants, the biological clock is poorly understood. But we know that environmental cues, particularly light, can reset the clock. How do plants detect light?

The light-detecting system of plants is a pigment in leaves called **phytochrome** (literally, "plant color"). Phytochrome occurs in two forms that can be interconverted. One form, called P_r, is inactive and strongly absorbs red light (r). The other form, called P_{fr}, is the active form, either stimulating or inhibiting a response to light in the plant. This active form absorbs far-red light (fr; almost infrared) and so is called P_{fr} (FIG. 44-9).

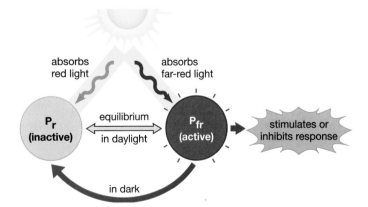

FIGURE 44-9 The light-sensitive phytochrome pigment
Phytochrome exists in inactive (P_r) and active (P_{fr}) forms. In daylight, the two forms of the phytochrome reach an equilibrium, and the active P_{fr} form determines the response. In darkness, P_{fr} spontaneously converts to inactive P_r. Upon absorbing red light, P_r is converted into active P_{fr}; upon absorbing far-red light, P_{fr} is transformed back into inactive P_r.

the active P_{fr} that inhibits flowering. In contrast, a far-red flash that produces inactive P_r has no effect on flowering, as if no light were detected. This observation implicates phytochrome in the control of flowering. However, scientists are still researching how the response of phytochrome to light determines whether a plant will flower.

Phytochrome Influences Other Responses of Plants to Their Environment

Phytochrome is involved in other plant responses to light. One action of phytochrome is to regulate seedling elongation after germination. Scientists have discovered that P_{fr} inhibits the elongation of seedlings. Because P_{fr} reverts to P_r in the dark, seedlings germinating in the darkness of the soil contain no inhibiting P_{fr} and consequently elongate very rapidly. This brings them out of the soil and into the sunlight (**FIG. 44-10**).

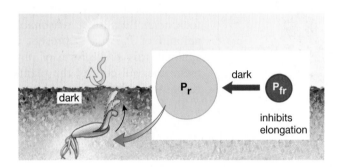

FIGURE 44-10 Decreased P_{fr} allows elongation

Once out in the sunlight, the phytochrome is exposed to all wavelengths of light, including red light, converting some P_r into P_{fr}. The P_{fr} inhibits elongation and prevents the seedlings from growing too rapidly and becoming weak and spindly (**FIG. 44-11**).

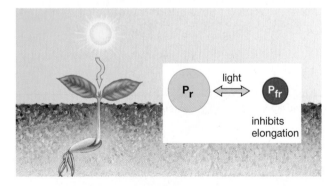

FIGURE 44-11 Increased P_{fr} inhibits elongation

Seedlings growing beneath other plants will be exposed largely to far-red light, because the green chlorophyll in the leaves above them absorbs most of the red light but transmits far-red. Far-red light converts the inhibitory P_{fr} into inactive P_r, so shaded seedlings also grow rapidly, which may bring them out of the shade.

Other plant responses that are stimulated by P_{fr} include leaf growth, chlorophyll synthesis, and the straightening of

the epicotyl or hypocotyl hook of dicot seedlings (see Chapter 43). As in the case of stem elongation, these responses are adaptations related to burial in the soil or shading by the leaves of other plants. For example, a newly germinating shoot needs to retain its protective bend while still in the soil (that is, in the dark) and straighten out only in the open air, where sunlight converts P_r to P_{fr}.

Hormones Coordinate the Development of Seeds and Fruit

After fertilization, the developing seeds release auxin or gibberellin (or both) into the surrounding ovary tissues. Cytokinins, under the influence of auxin, promote cell division within the ovary and seeds. Cells of the ovary multiply and grow larger, storing starches and other food materials, and produce a mature fruit. In this way, the plant coordinates the development of seeds and fruit. The most important commercial use of gibberellin is in grape production. In the U.S., gibberellin is applied heavily to green grapes, which grow larger and form looser clusters as a result (**FIG. 44-12a**).

FIGURE 44-12 Commercial uses for plant hormones
(a) The grapes on the right have been sprayed with gibberellin, producing larger, looser clusters. **(b)** Bananas that ripen naturally (left) remain green longer than those exposed to ethylene (right).

FIGURE 44-13 Ripe fruit becomes attractive to animal seed dispersers
The prickly pear cactus fruit is green, hard, and bitter before it ripens, which discourages animals from eating it. After the seeds mature, the fruit becomes soft, red, and tasty, attracting animals such as this desert tortoise. The mature seeds are not harmed by the animal's digestive tract and are dispersed in the animal's feces. QUESTION Agricultural engineers have developed genetically modified tomato plants in which ethylene production is blocked. Why might such a plant be valuable to tomato growers?

Seeds and fruits acquire nutrients for growth and development from their parent plant. In nature, not surprisingly, seed maturation and fruit ripening are closely coordinated. Most unripe fruits are inconspicuously colored (usually green, like the rest of the plant), hard, bitter, and in some cases even poisonous. As a result, animals seldom eat unripe fruit.

As the seeds mature, the fruit ripens; it becomes softer as enzymes weaken its cell walls, sweeter as starches are converted to sugar, and more brightly colored as green chlorophyll is broken down and more yellow, orange, and red pigments are synthesized. Bright colors make fruits attractive to animals (**FIG. 44-13**). Look around the produce section of your supermarket at all the brightly colored fruits—most of these colors are adaptations that attract animal seed dispersers.

In fruits such as bananas, apples, pears, tomatoes, and avocados, the changes that accompany ripening are stimulated by ethylene. Ethylene is synthesized by fruit cells in response to a surge of auxin that is released by the seeds—an important mechanism by which seed and fruit development are coordinated. Because ethylene is a gas, many ripe fruits continually leak ethylene into the air. In nature, this probably doesn't make much difference. But if fruit is stored in a closed container, ethylene released from one fruit often hastens the ripening of the rest. Discovering the role of ethylene in ripening has revolutionized modern fruit and vegetable marketing. Bananas grown in Central America can be picked green and tough and shipped to North American markets. By exposing them to ethylene at their destination,

grocers can market perfectly ripe bananas (see **FIG. 44-12b**). Not all fruits ripen properly when separated from the plant. Strawberries, for example, do not ripen in response to ethylene and must be allowed to vine ripen; shipping these ripe, soft fruits to markets without damage is a challenge. Although green tomatoes do ripen when gassed with ethylene, they never taste the same as those that ripen on the vine.

Senescence and Dormancy Prepare the Plant for Winter

In autumn, fruit ripens and drops to the ground, making it more available to animals that will eat it and disperse its seeds. Perennial broadleaf plants also shed their leaves in autumn because leaves would be a liability in winter, unable to photosynthesize but still allowing water to evaporate. Leaves, fruits, and flowers undergo a rapid aging called **senescence**.

Senescence is a complex process controlled by several different hormones. In most plants, healthy leaves and developing seeds produce auxin, which in turn helps maintain the health of the leaf or fruit. Simultaneously, the roots synthesize cytokinin, which is transported up the stem and out to the branches. Cytokinin also prevents senescence. (Cut flowers purchased from a florist have often been sprayed with cytokinin to keep them fresher longer.) But as winter approaches, cytokinin production in roots slows, and fruits and leaves produce less auxin. Meanwhile, ethylene is released by both aging leaves and ripening fruit. Ethylene stimulates leaf senescence, during which proteins, starches, and chlorophyll are broken down to simple molecules. These are transported to the roots and other permanent tissues of the plant for winter storage. Senescence ends with the formation of the **abscission layer** at the base of the petiole connecting the leaf to the stem (**FIG. 44-14**). Ethylene stimulates this

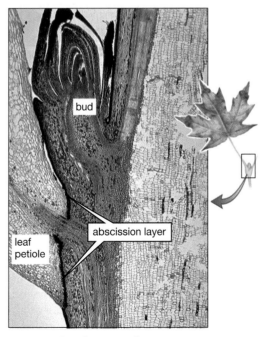

FIGURE 44-14 The abscission layer
This cross section shows the abscission layer forming at the base of a maple leaf. A new leaf bud is visible above the dying leaf.

layer of thin-walled cells to produce an enzyme that digests their cell walls. When the petiole attachment site weakens sufficiently, the leaf or fruit falls. Ethylene's role in leaf dropping led to its discovery. In the 1800s, gas lamps were installed in German cities, and residents soon observed that plants growing near leaky gas mains supplying the lamps grew abnormally and lost their leaves prematurely. In the early 1900s, a Russian plant physiologist tested all the components of the "illuminating gas" on plants and discovered that ethylene was responsible for these effects.

Meanwhile, other changes also occur that prepare the plant for winter. New buds, rather than developing into leaves and branches as they would have during spring and summer, now become tightly wrapped up and dormant, waiting out the winter. Dormancy in buds, as in seeds, is enforced by abscisic acid. Metabolism slows to a crawl, and the plant enters its long winter "sleep," awaiting signals of warmth and longer spring days before "awakening" once again.

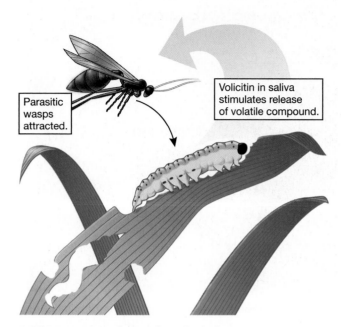

FIGURE 44-15 A chemical cry for help

44.3 CAN PLANTS COMMUNICATE AND MOVE RAPIDLY?

Plants May Summon "Bodyguards" When Attacked

The hundred-million-year war between plants and their parasites and predators has led to the evolution of sophisticated plant defenses that surprise people who are accustomed to thinking of plants as passive, helpless organisms. Researchers studying how plants respond to attack by predators or disease-causing viruses have recently discovered that plants under attack help protect themselves—and sometimes neighboring plants as well—by releasing volatile chemicals into the air around them in what could be called a chemical "cry for help." Working with maize (a relative of corn), researchers discovered that in response to attack by hungry caterpillars, maize plants release a mixture of volatile chemicals. A wasp has evolved an attraction to these volatile chemicals, probably because they signal the presence of the caterpillars, which provide food for the wasp's offspring. The wasp lays its eggs in a caterpillar's body, where the larvae hatch and consume their host from the inside out. Scientists discovered that merely tearing the leaves of the maize plant will not elicit the chemical alarm signal; the attack on the plant must come from an actual caterpillar. A chemical called *volicitin* in caterpillar saliva causes maize to release the volatile chemicals (FIG. 44-15). Similarly, when spider mites attack lima bean plants, the beans release a chemical that attracts another type of mite that preys on the spider mite. Wild tobacco plants attacked by hornworms (hawk moth larvae) actually time their chemical defenses, releasing wasp-attracting chemicals during the day but releasing a blend of chemicals at night that seems to deter the night-flying hawk moths from laying eggs on the plants. As you contemplate these apparently ingenious strategies, keep in mind that they evolved gradually because natural selection favored them.

Plants May Warn Their Neighbors and Offspring of Attacks

If attacked plants can summon help, might their neighbors sense the message and prepare themselves for attack as well? Evidence is accumulating that many types of plants (including barley and willow, alder, and birch trees) warn members of their own species of attack; the warning boosts defenses in healthy individuals, making them more able to defend themselves. This communication may also cross species lines. Wounded sagebrush release large amounts of volatile chemicals; researchers discovered that wild tobacco plants planted downwind from wounded sagebrush are less prone to insect damage than are those growing downwind of healthy sagebrush, supporting the hypothesis that airborne chemical signals from injured sagebrush activate defense mechanisms in the tobacco plants. Likewise, injured lima bean plants seem to trigger defenses in nearby cucumber plants.

Researchers studying radish plants attacked by butterfly larvae found that they produce more of a bitter-tasting chemical and more defensive hairs on their leaves. When these better-defended plants reproduced, their seedlings were less attractive to predators than were seedlings from uninjured parent plants. This supports the hypothesis that plants not only defend themselves but also pass on a chemical signal within their seeds that triggers development of more defenses in their offspring.

Many plants produce *salicylic acid*, the compound from which aspirin is derived. Native Americans used tea made from willow bark (which contains large amounts of this substance) as a painkiller. Ilya Raskin and colleagues at Rutgers University found that tobacco plants infected with a plant virus produce large quantities of salicylic acid. The salicylic acid, in turn, activates an immune response in the plants, helping them fight off the viral attack. The plant

FIGURE 44-16 A rapid response to touch
(a) A leaf of the sensitive plant (*Mimosa*) consists of an array of leaflets emerging from a central stalk, which is attached to the main stem by a short petiole. (b) Touching causes the leaflets to close together and the petiole to droop.

also converts some of the salicylic acid to *methyl salicylate* (used to flavor "wintergreen" candy). This highly volatile compound diffuses into the air from virus-infected plant tissues and is absorbed by nearby plants. The plant's healthy neighbors reconvert the methyl salicylate to salicylic acid, enhancing their immune defenses and making them better able to resist the viral infection.

As Ilya Raskin explains, "Plants can't run away and they can't make…noises. But they are wonderful chemists." Hopefully, human chemists, learning from the plants, will design chemicals to enable farmers of the future to protect their crops from predators and diseases using natural signaling substances instead of toxic pesticides.

Some Plants Move Rapidly

All plants are alive, but (as described in the Case Study) some are livelier than others. A few plants exhibit rapid movements coordinated by electrical signals that resemble the nerve impulses of animals. These include the sundew (see Fig. 44-1), the sensitive plant, and the Venus flytrap (described in "Case Study Revisited: Predacious Plants?"; see the chapter opener photo).

Sundew leaves are covered with long, glandular hairs whose tips secrete globs of a complex brew containing sweet nectar, a sticky substance, and protein-digesting enzymes. An insect attracted by the nectar becomes entrapped in the gluey globs and struggles to escape. Vibrations from the struggle open channels in the hairs, allowing a flow of ions that produces an electric current. This electrical signal, by an unknown mechanism, causes the hairs to bend toward the source of vibrations. The insect is then further entangled, digested, and absorbed by these amazing appendages. If you touch a sensitive plant (*Mimosa*), its leaflets immediately fold together, and its petioles droop (FIG. 44-16). This rapid and dramatic movement probably discourages leaf eaters. In the sensitive plant, electrical signals travel from the touched leaf to the petiole that attaches the leaf to the stem. As the signal travels, it causes specialized "motor cells" to increase their permeability to certain ions, including potassium. The motor cells are located at the base of each leaflet and in the petiole where it joins the stem. As ions flow out of the motor cells, water follows by osmosis. As the cells shrink from water loss, both the leaflets and the petioles rapidly droop.

The world's speediest plant is the aquatic bladderwort, a predator that traps fast-swimming aquatic creatures that provide it with nitrogen. The bladderwort's trap is an evolutionary marvel that doesn't rely on electrical signals to capture prey. The door to its hollow bladder is hinged at the top, opens inward, and is sealed shut by sticky secretions. Any water that leaks in is extruded by glandular cells lining the bladder. This produces a slightly lower water pressure inside that pulls the walls inward, giving them a concave shape. Stiff bristles guard the sealed trapdoor. If a small prey animal bumps the bristles, they act as levers, pushing the flexible door inward and breaking the seal (FIG. 44-17). The lower pressure inside instantly sucks in water, carrying the prey to its death and eventual digestion.

FIGURE 44-17 A bladderwort snares its prey
The pale bladders on this bladderwort (genus *Utricularia*) are clearly visible. One bladder is enlarged in the inset, showing three small crustaceans trapped inside.

CASE STUDY REVISITED PREDACIOUS PLANTS

The Venus flytrap (see the chapter opener photo) also uses electrical impulses to trap prey. Each of the fringed trapping leaves bears three sensory "hairs" on its inside surface. Nectar attracts foraging insects that bump into the hairs, stimulating a flow of ions and an electrical signal similar to that used by the sundew, the sensitive plant, and animal nerve cells. This signal sets off a rapid chain of events that causes the trap to close in about half a second. The question is, how?

This plant engineering problem recently caught the attention of Lakshminarayanan Mahadevan, a Harvard researcher in applied mathematics and mechanics whose studies have led to new hypotheses about how flytrap leaves snap shut. His research team painted dots of fluorescent paint on many locations on the flytrap leaves and then tracked the spots with a high-speed video camera as the leaves closed. Using these data, the researchers designed a computer simulation of the process and concluded that a pair of flytrap leaves somewhat resemble a tennis ball that has been split nearly in half, with each half turned inside out. Slight pressure can cause the tennis ball halves to suddenly resume their concave shapes, snapping back together. Open flytrap leaves, the researchers hypothesize, are under a similar type of tension, probably due to compression of cells in the central mesophyll layer. If an insect bumps the hairs, it causes these cells to rapidly absorb water and swell, effectively "popping" each leaf from a slightly convex to a slightly concave shape that closes them around the insect (**FIG. 44-18**).

This suggests that the open leaves store potential energy that is released as the leaves snap shut. You might predict, then, that opening the leaves would be a slow, energy-intensive process—and you would be right. Reopening the trap takes several hours and consumes large quantities of ATP. So it is very important that something nutritious actually be in the leaf before it closes. Amazingly, the plant has a built-in "fail-safe" mechanism that usually prevents it from closing on an inanimate object. To set off the trap, one hair must be touched twice in rapid succession or two hairs must be touched almost simultaneously.

But the flytrap still holds mysteries. How do the sensory hairs "know" which ones have been touched and how quickly? How do they then transform the touch stimulus into an electrical signal? How does the electrical signal cause cells to absorb water? As so often happens in biology, the answer to one question immediately suggests several new ones; science is a never-ending quest for deeper understanding.

Consider This Many wetlands in the United States are threatened by runoff water from nearby farms, which may be heavily fertilized or rich in animal wastes. Carnivorous plants thrive in nitrogen-poor bogs partly because other species, which can't trap nitrogen-rich food, cannot compete with them. Explain why runoff from farms poses a threat to carnivorous plants in nearby wetlands.

FIGURE 44-18 Success!

CHAPTER REVIEW

SUMMARY OF KEY CONCEPTS

44.1 What Are Plant Hormones, and How Do They Act?
Plant hormones are chemicals produced by cells in one part of a plant body and transported to other parts of the plant, where they exert specific effects. The five major classes of plant hormones are auxins, gibberellins, cytokinins, ethylene, and abscisic acid. The functions of these hormones are summarized in Table 44-1.

44.2 How Do Hormones Regulate the Plant Life Cycle?
Dormancy in seeds is enforced by abscisic acid. Falling levels of abscisic acid and rising levels of gibberellin trigger germination. As the seedling grows, it shows differential growth in response to the direction of light (phototropism) and gravity (gravitropism). Auxin mediates phototropism and gravitropism in shoots and gravitropism in roots. Similar concentrations of auxin stimulate cell elongation in shoots while inhibiting it in roots. Plants may detect gravity by means of starch-containing plastids. Auxin causes some plants to wrap around nearby objects (thigmotropism).

Branching in stems and roots results from the interplay of auxin (produced in shoot tips and transported downward) and cytokinin (synthesized in roots and transported up the shoot). An optimum concentration of both auxin and cytokinin stimulates the growth of lateral buds and branching of roots.

The timing of flowering is often controlled by duration of darkness. Flowering is probably both stimulated and inhibited by a variety of undescribed hormones called *florigens*. Plants can detect light and darkness by changes in phytochrome, a leaf pigment. Phytochrome influences plant responses to light including flowering, straightening the epicotyl or hypocotyl hook, seedling elongation, leaf growth, and chlorophyll synthesis.

Developing seeds produce auxin and/or gibberellin, which diffuse into the surrounding ovary tissues and cause the production of a fruit. A surge of auxin as the seed matures stimulates cells of some fruits to release another hormone, ethylene, causing ripening. Ripening includes the conversion of starches to sugars, softening of the fruit, development of bright colors, and, commonly, the formation of an abscission layer at the base of the petiole.

Several changes prepare perennial plants of temperate zones for winter. Leaves and fruits undergo a rapid aging process called senescence, including the formation of an abscission layer. Senescence occurs as a result of a fall in levels of auxin and cytokinin and, perhaps, a rise in ethylene concentrations. Other parts of the plant, including buds, become dormant. Dormancy in buds is enforced by high concentrations of abscisic acid.

Web Tutorial 44.1 Plant Hormones

Web Tutorial 44.2 The Light-Sensitive Pigment Phytochrome

Web Tutorial 44.3 Plant Responses to Stimuli

44.3 Can Plants Communicate and Move Rapidly?

Some plants under attack by insects release volatile chemicals into the air around them that attract insect predators. Volatile compounds released by injured or infected plants may stimulate neighboring plants to produce substances that help protect them from predation or infection.

Some plants can move rapidly. In the Venus flytrap and sensitive plant, touch sensors in leaves generate electrical signals that lead to a flow of ions. This ion flow causes specialized cells to quickly absorb or lose water. Changes in the size of these cells cause leaves to move or the petiole to droop. The aquatic bladderwort generates lower pressure within its bladder and uses it to suck in prey.

KEY TERMS

abscisic acid *page 911*
abscission layer *page 919*
apical dominance *page 915*
auxin *page 910*
biological clock *page 917*

cytokinin *page 911*
day-neutral plant *page 916*
ethylene *page 911*
gibberellin *page 910*
gravitropism *page 910*

hormone *page 910*
long-night plant *page 917*
phototropism *page 910*
phytochrome *page 917*
plant hormone *page 910*

senescence *page 918*
short-night plant *page 917*
thigmotropism *page 910*

THINKING THROUGH THE CONCEPTS

1. What did the Darwins, Boysen-Jensen, and Went each contribute to our understanding of phototropism? Do their experiments truly prove that auxin is the hormone that controls phototropism? What other experiments would you like to see?

2. Which hormone maintains apical dominance? Which hormone maintains seed dormancy?

3. How can auxin cause shoots to grow up and roots to grow down?

4. What is apical dominance? How do auxin and cytokinin interact in determining the growth of lateral buds?

5. What is a biological clock?

6. What is the phytochrome system? How do the two forms of phytochrome interact to help control the plant life cycle?

7. Describe the role of phytochrome in shoot elongation of a seed germinating underground. What is the adaptive significance of this response?

8. Which hormones cause fruit development? Which hormone causes fruit ripening?

9. Which hormone is involved in maintaining bud dormancy? In leaf and fruit drop?

10. What is one major commercial use of gibberellin? Of ethylene?

11. Describe one example of a chemical defense mechanism in plants.

12. Describe how a sensitive plant closes its leaves. Why might this behavior have evolved?

13. What is the advantage of predatory behavior in plants? What habitats favor this behavior, and why?

APPLYING THE CONCEPTS

1. Suppose you got a job in a greenhouse in which the owner was trying to start the flowering of chrysanthemums (a long-night plant) for Mother's Day. You accidentally turned on the light in the middle of the night. Would you be likely to lose your job? Why or why not? What would happen if you turned on the lights in the day?

2. A student reporting on a project said that one of her seeds did not grow properly because it was planted upside down so that it got confused and tried to grow down. Do you think the teacher accepted this explanation? Why or why not?

3. Bean sprouts such as those you might eat in a salad need to be grown in the dark to form their long stems. If they are grown in the light, they will be short and green. Why do seedlings grow long and spindly in the dark? What advantages would growing this way in darkness have in nature?

4. Suppose that on July 4, you discover that both a long-night plant and a short-night plant have bloomed in your garden. Discuss how it is possible for both plants to bloom.

5. Suppose you work in a lab with a well-equipped greenhouse, healthy tomato plants, tomato hornworms by the dozen, and a large supply of the parasitic wasps that attack tomato hornworms. Design a controlled study that will support or refute the hypothesis that tomato plants, like maize, can summon the wasps when attacked by a hornworm. Be sure to control for other types of attack.

FOR MORE INFORMATION

Farmer, E. E. "New Fatty Acid-Based Signals: A Lesson from the Plant World." *Science*, May 9, 1997. The author describes research leading to the discovery of volicitin, which attracts parasitic wasps to plants under attack by caterpillars.

Hansen, E. "Where Rocks Sing, Ants Swim, and Plants Eat Animals." *Discover*, October 2001. Researchers explore carnivorous plants in the wilds of Borneo.

Mlot, C. "Where There's Smoke, There's Germination." *Science News*, May 31, 1997. Researchers have discovered that nitrogen dioxide produced by fires can induce germination in plants that live in ecosystems where fires are common.

Moffatt, A. S. "How Plants Cope with Stress." *Science*, November 1, 1994. The hormone "systemin," similar to animal hormones, enables plants to respond to stress.

Russell, S. A. "Talking Plants." *Discover*, April 2002. This article is a clear and engaging summary of research documenting chemical communication among plants.

Saunders, F. "Keep the Aspirin Flying." *Discover*, January 1998. The author describes how plants use methyl salicylate to help nearby plants resist infection.

Selim, J. "Snap, Crackle, and Pop!" *Discover*, May, 2005. Computers provide new insights into how Venus flytrap leaves snap shut.

APPENDIX I
Metric System Conversions

To Convert Metric Units:	Multiply by:	To Get English Equivalent:
Length		
Centimeters (cm)	0.3937	Inches (in.)
Meters (m)	3.2808	Feet (ft)
Meters (m)	1.0936	Yards (yd)
Kilometers (km)	0.6214	Miles (mi)
Area		
Square centimeters (cm^2)	0.155	Square inches (in.^{2})
Square meters (m^2)	10.7639	Square feet (ft^2)
Square meters (m^2)	1.1960	Square yards (yd^2)
Square kilometers (km^2)	0.3831	Square miles (mi^2)
Hectare (ha) (10,000 m^2)	2.4710	Acres (a)
Volume		
Cubic centimeters (cm^3)	0.06	Cubic inches (in.^{3})
Cubic meters (m^3)	35.30	Cubic feet (ft^3)
Cubic meters (m^3)	1.3079	Cubic yards (yd^3)
Cubic kilometers (km^3)	0.24	Cubic miles (mi^3)
Liters (L)	1.0567	Quarts (qt), U.S.
Liters (L)	0.26	Gallons (gal), U.S.
Mass		
Grams (g)	0.03527	Ounces (oz)
Kilograms (kg)	2.2046	Pounds (lb)
Metric ton (tonne) (t)	1.10	Ton (tn), U.S.
Speed		
Meters/second (mps)	2.24	Miles/hour (mph)
Kilometers/hour (kmph)	0.62	Miles/hour (mph)

To Convert English Units:	Multiply by:	To Get Metric Equivalent:
Length		
Inches (in.)	2.54	Centimeters (cm)
Feet (ft)	0.3048	Meters (m)
Yards (yd)	0.9144	Meters (m)
Miles (mi)	1.6094	Kilometers (km)
Area		
Square inches (in.^{2})	6.45	Square centimeters (cm^2)
Square feet (ft^2)	0.0929	Square meters (m^2)
Square yards (yd^2)	0.8361	Square meters (m^2)
Square miles (mi^2)	2.5900	Square kilometers (km^2)
Acres (a)	0.4047	Hectare (ha) (10,000 m^2)
Volume		
Cubic inches (in.^{3})	16.39	Cubic centimeters (cm^3)
Cubic feet (ft^3)	0.028	Cubic meters (m^3)
Cubic yards (yd^3)	0.765	Cubic meters (m^3)
Cubic miles (mi^3)	4.17	Cubic kilometers (km^3)
Quarts (qt), U.S.	0.9463	Liters (L)
Gallons (gal), U.S.	3.8	Liters (L)
Mass		
Ounces (oz)	28.3495	Grams (g)
Pounds (lb)	0.4536	Kilograms (kg)
Ton (tn), U.S.	0.91	Metric ton (tonne) (t)
Speed		
Miles/hour (mph)	0.448	Meters/second (mps)
Miles/hour (mph)	1.6094	Kilometers/hour (kmph)

Metric Prefixes

Prefix			Meaning
giga-	G	$10^9 =$	1,000,000,000
mega-	M	$10^6 =$	1,000,000
kilo-	k	$10^3 =$	1000
hecto-	h	$10^2 =$	100
deka-	da	$10^1 =$	10
		$10^0 =$	1
deci-	d	$10^{-1} =$	0.1
centi-	c	$10^{-2} =$	0.01
milli-	m	$10^{-3} =$	0.001
micro-	μ	$10^{-6} =$	0.000001

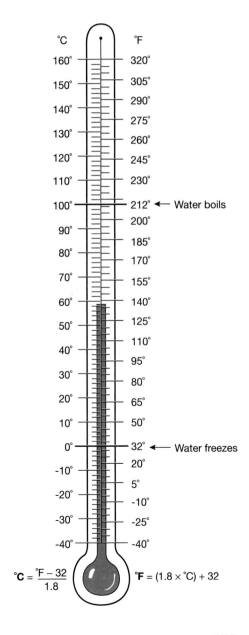

°C   °F

160° — 320°
150° — 305°
        290°
140° — 275°
130° — 260°
120° — 245°
110° — 230°
100° — 212° ← Water boils
        200°
90° — 185°
80° — 170°
70° — 155°
60° — 140°
50° — 125°
        110°
40° — 95°
30° — 80°
20° — 65°
10° — 50°
0° — 32° ← Water freezes
        20°
-10° — 5°
-20° — -10°
-30° — -25°
-40° — -40°

$$°C = \frac{°F - 32}{1.8}$$

$$°F = (1.8 \times °C) + 32$$

Classification of Major Groups of Organisms*

Domain	Kingdom	Phylum	Common Name
Bacteria (prokaryotic, peptidoglycan in cell wall)			**bacteria**
Archaea (prokaryotic, no peptidoglycan in cell wall)			**archaeans**
Eukarya (eukaryotic)		Rhodophyta	red algae
		Chlorophyta	green algae
		Euglenida	euglenids
		Foraminifera	forams
	Excavata		**excavates**
		Parabasalia	parabasalids
		Diplomonadida	diplomonads
	Amoebozoa		**amoebozoids**
		Gymnamoebae	lobose amoebae
		Acrasiomycota	cellular slime molds
	Alveolata		**alveolates**
		Apicomplexa	sporozoans
		Pyrrophyta	dinoflagellates
		Ciliophora	ciliates
	Stramenopila		**stramenopiles**
		Oomycota	egg fungi
		Phaeophyta	brown algae
		Bacillariophyta	diatoms
	Plantae (multicellular, photosynthetic)		**plants**
		Bryophyta	mosses
		Pteridophyta	ferns
		Coniferophyta	evergreens
		Anthophyta	flowering plants
	Fungi (multicellular, heterotrophic, absorb nutrients)		**fungi**
		Chytridiomycota	chytrids
		Zygomycota	zygote fungi
		Ascomycota	sac fungi
		Basidiomycota	club fungi
	Animalia (multicellular, heterotrophic, ingest nutrients)		**animals**
		Porifera	sponges
		Cnidaria	hydras, sea anemones, jellyfish, corals
		Ctenophora	comb jellies
		Platyhelminthes	flatworms
		Nematoda	roundworms
		Annelida	segmented worms
		Oligochaeta	earthworms
		Polychaeta	tube worms
		Hirudinea	leeches
		Arthropoda	arthropods ("jointed legs")
		Insecta	insects
		Arachnida	spiders, ticks
		Myriapoda	millipedes and centipedes
		Crustacea	crabs, lobsters
		Mollusca	mollusks ("soft-bodied")
		Gastropoda	snails
		Pelecypoda	mussels, clams
		Cephalopoda	squid, octopuses
		Echinodermata	sea stars, sea urchins, sea cucumbers
		Chordata	chordates
		Urochordata	tunicates
		Cephalochordata	lancelets
		Myxini	hagfishes
		Vertebrata	vertebrates
		Pertromyzontiformes	lampreys
		Chondrichthyes	sharks, rays
		Actinopterygii	ray-finned fishes
		Actinistia	coelacanths
		Dipnoi	lungfishes
		Amphibia	frogs, salamanders
		Reptilia	turtles, snakes, lizards, crocodilians, birds
		Mammalia	mammals

*This table lists only those taxonomic categories described in the textbook.

Biological Vocabulary: Common Roots, Prefixes, and Suffixes

Biology has an extensive vocabulary often based on Greek or Latin rather than English words. Rather than memorizing every word as if it were part of a new, foreign language, you can figure out the meaning of many new terms if you learn a much smaller number of word roots, prefixes, and suffixes. We have provided common meanings in biology rather than literal translations from Greek or Latin. For each item in the list, the following information is given: meaning; part of word (prefix, suffix, or root); example from biology.

a–, an–, e–: without, lack of (prefix); *abiotic*, without life

acro–: top, highest (prefix); *acrosome*, vesicle of enzymes at the tip of a sperm

ad–: to (prefix); *adhesion*, property of sticking to something else

allo–: other (prefix); *allopatric* (literally, "different fatherland), restricted to different regions

amphi–: both, double, two (prefix); *amphibian*, a class of vertebrates that usually has two life stages (aquatic and terrestrial; e.g., a tadpole and an adult frog)

andro: man, male (root); *androgen*, a male hormone such as testosterone

antero–: front (prefix or root); *anterior*, toward the front of

anti–: against (prefix); *antibiotic* (literally "against life"), a substance that kills bacteria

apic–: top, highest (prefix); *apical meristem*, the cluster of dividing cells at the tip of a plant shoot or root

arthr–: joint (prefix); *arthropod*, animals such as spiders, crabs, and insects, with exoskeletons that include jointed legs

–ase: enzyme (suffix); *protease*, an enzyme that digests protein

auto–: self (prefix); *autotrophic*, self–feeder (e.g., photosynthetic)

bi–: two (prefix); *bipedal*, having two legs

bio: life (root); *biology*, the study of life

blast: bud, precursor (root); *blastula*, embryonic stage of development, a hollow ball of cells

bronch–: windpipe (root); *bronchus*, a branch of the trachea (windpipe) leading to a lung

carcin, –o: cancer (root); *carcinogenesis*, the process of producing a cancer

cardi–, –a–, –o–: heart (root); *cardiac*, referring to the heart

carn–, –i–, –o–: flesh (prefix or root); *carnivore*, an animal that eats other animals

centi–: one hundredth (prefix); *centimeter*, a unit of length, 1 one-hundredth of a meter

cephal–, –i–, –o–: head (prefix or root); *cephalization*, the tendency for the nervous system to be located principally in the head

chloro–: green (prefix or root); *chlorophyll*, the green, light-absorbing pigment in plants

chondr–: cartilage (prefix); *Chondrichthyes*, class of vertebrates including sharks and rays, with a skeleton made of cartilage

chrom–: color (prefix or root); *chromosome*, a threadlike strand of DNA and protein in the nucleus of a cell (*Chromosome* literally means "colored body," because chromosomes absorb some of the colored dyes commonly used in microscopy.)

–cide: killer (suffix); *pesticide*, a chemical that kills "pests" (usually insects)

–clast: break down, broken (root or suffix); *osteoclast*, a cell that breaks down bone

co–: with or together with (prefix); *cohesion*, property of sticking together

coel–: hollow (prefix or root); *coelom*, the body cavity that separates the internal organs from the body wall

contra–: against (prefix); *contraception*, acting to prevent conception (pregnancy)

cortex: bark, outer layer (root); *cortex*, outer layer of kidney

crani–: skull (prefix or root); *cranium*, the skull

cuti: skin (root); *cuticle*, the outermost covering of a leaf

–cyte, cyto–: cell (root or prefix); *cytokinin*, a plant hormone that promotes cell division

de–: from, out of, remove (prefix); *decomposer*, an organism that breaks down organic matter

dendr–: treelike, branching (root); *dendrite*, highly branched input structures of nerve cells

derm: skin, layer (root); *ectoderm*, the outer embryonic germ layer of cells

deutero–: second (prefix); *deuterostome* (literally, "second opening"), an animal in which the coelom is derived from the gut

di–: two (prefix); *dicot*, an angiosperm with two cotyledons in the seed

diplo–: both, double, two (prefix or root); *diploid*, having paired homologous chromosomes

dys–: difficult, painful (prefix); *dysfunction*, an inability to function properly

ecto–: outside (prefix); *ectoderm*, the outermost tissue layer of animal embryos

–elle: little, small (suffix); *organelle* (literally, "little organ"), a subcellular structure that performs a specific function

endo– (or ento–): inside, inner (prefix); *endocrine*, pertaining to a gland that secretes hormones inside the body

epi–: outside, outer (prefix); *epidermis*, outermost layer of skin

equi–: equal (prefix); *equidistant*, the same distance

erythro–: red (prefix); *erythrocyte*, red blood cell

eu–: true, good (prefix); *eukaryotic*, pertaining to a cell with a true nucleus

ex– (or exo–): out of (prefix); *exocrine*, pertaining to a gland that secretes a substance (e.g., sweat) outside of the body

extra–: outside of (prefix); *extracellular*, outside of a cell

–fer: to bear, to carry (suffix); *conifer*, a tree that bears cones

gastr–: stomach (prefix or root); *gastric*, pertaining to the stomach

gen: to produce (prefix, suffix, or root); *antigen*, a substance that causes the body to produce antibodies

gyn, –o: female (prefix or root); *gynecology*, the study of the female reproductive tract

haplo–: single (prefix); *haploid*, having a single copy of each type of chromosome

hem– (or hemato–): blood (prefix or root); *hemoglobin*, the molecule in red blood cells that carries oxygen

hemi–: half (prefix); *hemisphere*, one of the halves of the cerebrum

hetero–: other (prefix); *heterotrophic*, an organism that feeds on other organisms

hom–, homo–, homeo–: same (prefix); *homeostasis*, to maintain constant internal conditions in the face of changing external conditions

hydro–: water (usually prefix); *hydrophilic*, being attracted to water

hyper–: above, greater than (prefix); *hyperosmotic*, having a greater osmotic strength (usually higher solute concentration)

hypo–: below, less than (prefix); *hypodermic*, below the skin

inter–: between (prefix); *interneuron*, a neuron that receives input from one (or more) neuron and sends output to another neuron (or many neurons)

intra–: within (prefix); *intracellular*, pertaining to an event or substance that occurs within a cell

iso–: equal (prefix); *isotonic*, pertaining to a solution that has the same osmotic strength as another solution

–itis: inflammation (suffix); *hepatitis*, an inflammation (or infection) of the liver

leuc–, leuco– leuk–, leuko–: white (prefix); *leukocyte*, a white blood cell

lip–: fat (prefix or root); *lipid*, the chemical category to which fats, oils, and steroids belong

–logy: study of (suffix); *biology*, the study of life

–lysis: loosening, split apart (root or suffix); *lysis*, to break open a cell

macro–: large (prefix); *macrophage*, a large white blood cell that destroys invading foreign cells

medulla: marrow, middle substance (root); *medulla*, inner layer of kidney

mere: segment, body section (suffix); *sarcomere*, the functional unit of a vertebrate skeletal muscle cell

meso–: middle (prefix); *mesophyll*, middle layers of cells in a leaf

meta–: change, after (prefix); *metamorphosis*, to change body form (e.g., developing from a larva to an adult)

micro–: small (prefix); *microscope*, a device that allows one to see small objects

milli–: one thousandth (prefix); *millimeter*, a unit of measurement of length, 1 one-thousandth of a meter

mito–: thread (prefix); *mitosis*, cell division (in which chromosomes appear as threadlike bodies)

mono–: single (prefix); *monocot*, a type of angiosperm with one cotyledon in the seed

morph–: shape, form (prefix or root); *polymorphic*, having multiple forms

multi–: many (prefix); *multicellular*, pertaining to a body composed of more than one cell

myo–: muscle (prefix); *myofibril*, protein strands in muscle cells

neo–: new (prefix); *neonatal*, relating to or affecting a newborn child

neph–: kidney (prefix or root); *nephron*, functional unit of mammalian kidney

neur–, neuro–: nerve (prefix or root); *neuron*, a nerve cell

oligo–: few (prefix); *oligomer*, a molecule made up of a few subunits (see also *poly*)

omni–: all (prefix); *omnivore*, an animal that eats both plants and animals

oo–, ov–, ovo–: egg (prefix); *oocyte*, one of the stages of egg development

opsi–: sight (prefix or root); *opsin*, protein part of light-absorbing pigment in eye

opso–: tasty food (prefix or root); *opsonization*, process whereby antibodies and/or complement render bacteria easier for white blood cells to engulf

–osis: a condition, disease (suffix); *atherosclerosis*, a disease in which the artery walls become thickened and hardened

oss–, osteo–: bone (prefix or root); *osteoporosis*; a disease in which the bones become spongy and weak

pater, patr–: father (usually root); *paternal*, from or relating to a father

path, –i, –o–: disease (prefix or root); *pathology*, the study of disease and diseased tissue

–pathy: disease (suffix); *neuropathy*, a disease of the nervous system

peri–: around (prefix); *pericycle*, the outermost layer of cells in the vascular cylinder of a plant root

phago–: eat (prefix or root); *phagocyte*, a cell that eats other cells (e.g., some types of white blood cell)

–phil, philo–: to love (prefix or suffix); *hydrophilic* (literally, "water-loving"), pertaining to a watersoluble molecule

–phob, phobo–: to fear (prefix or suffix); *hydrophobic* (literally, "water-fearing"), pertaining to a water-insoluble molecule

phyll: leaf (root or suffix); *chlorophyll*, the green, light-absorbing pigment in a leaf

phyte: plant (root or suffix); *gametophyte* (literally, "gamete plant"), the gamete-producing stage of a plant life cycle

–plasm: formed substance (root or suffix); *cytoplasm*, the material inside a cell

ploid: chromosomes (root); *diploid*, having paired chromosomes

pneumo–: lung (root); *pneumonia*, a disease of the lungs

–pod: foot (root or suffix); *gastropod* (literally, "stomach-foot"), a class of molluscs, principally snails, that crawl on their ventral surfaces

poly–: many (prefix); *polysaccharide*, a carbohydrate polymer composed of many sugar subunits

post–, postero–: behind (prefix); *posterior*, pertaining to the hind part

pre–, pro–: before, in front of (prefix); *premating isolating mechanism*, a mechanism that prevents gene flow between species, acting to prevent mating (e.g., having different courtship rituals or different mating seasons)

prim–: first (prefix); *primary cell wall*, the first cell wall laid down between plant cells during cell division

pro–: before (prefix); *prokaryotic*, pertaining to a cell without (that evolved before the evolution of) a nucleus

proto–: first (prefix); *protocell*, a hypothetical evolutionary ancestor to the first cell

pseudo–: false (prefix); *pseudopod* (literally, "false foot"), the extension of the plasma membrane by which some cells, such as *Amoeba*, move and capture prey

quad–, quat–: four (prefix); *quaternary structure*, the "fourth level" of protein structure, in which multiple peptide chains form a complex three-dimensional structure

ren: kidney (root); *adrenal*, gland attached to the mammalian kidney

retro–: backward (prefix); *retrovirus*, a virus that uses RNA as its genetic material; this RNA must be copied "backward" to DNA during infection of a cell by the virus

sarco–: muscle (prefix); *sarcoplasmic reticulum*, a calcium-storing, modified endoplasmic reticulum found in muscle cells

scler–: hard, tough (prefix); *sclerenchyma*, a type of plant cell with a very thick, hard cell wall

semi–: one-half (prefix); *semiconservative replication*, the mechanism of DNA replication, in which one strand of the original DNA double helix becomes incorporated into the new DNA double helix

–some, soma–, somato–: body (prefix or suffix); *somatic nervous system*, part of the peripheral nervous system that controls the skeletal muscles that move the body

sperm, sperma–, spermato–: seed (usually root); *gymnosperm*, a type of plant producing a seed not enclosed within a fruit

stasis, stat–: stationary, standing still (suffix or prefix); *homeostasis*, the physiological process of maintaining constant internal conditions despite a changing external environment

stoma, –to: mouth, opening (prefix or root); *stoma*, the adjustable pore in the surface of a leaf that allows carbon dioxide to enter the leaf

sub–: under, below (prefix); *subcutaneous*, beneath the skin

sym–: same (prefix); *sympatric* (literally "same father"), found in the same region

testis: witness (root); *testis*, male reproductive organ (derived from the custom in ancient Rome that only males had standing in the eyes of the law; *testimony* has the same derivation)

therm–: heat (prefix or root); *thermoregulation*, the process of regulating body temperature

trans–: across (prefix); *transgenic*, having genes from another organism (usually another species); the genes have been moved "across" species

tri–: three (prefix); *triploid*, having three copies of each homologous chromosome

troph: food, nourishment (root); *autotrophic*, self-feeder (e.g., photosynthetic)

–tropic: change, turn (suffix); *phototropism*, the process by which plants orient toward the light

ultra–: beyond (prefix); *ultraviolet*, light of wavelengths beyond the violet

uni–: one (prefix); *unicellular*, referring to an organism composed of a single cell

vita: life (root); *vitamin*, a molecule required in the diet to sustain life

–vor: eat (usually root); *herbivore*, an animal that eats plants

zoo–, zoa–: animal (usually root); *zoology*, the study of animals

Glossary

abdomen: the body segment at the posterior end of an animal with segmentation; contains most of the digestive structures.

abiotic (ā-bī-ah′-tik): nonliving; the abiotic portion of an ecosystem includes soil, rock, water, and the atmosphere.

abortion: the procedure for terminating pregnancy; the cervix is dilated, and the embryo and placenta are removed.

abscisic acid (ab-sis′-ik): a plant hormone that generally inhibits the action of other hormones, enforcing dormancy in seeds and buds and causing the closing of stomata.

abscission layer: a layer of thin-walled cells, located at the base of the petiole of a leaf, that produces an enzyme that digests the cell wall holding leaf to stem, allowing the leaf to fall off.

absorption: the process by which nutrients are taken into cells.

accessory pigments: colored molecules other than chlorophyll that absorb light energy and pass it to chlorophyll.

acellular slime mold: a type of funguslike protist that forms a multinucleate structure that crawls in amoeboid fashion and ingests decaying organic matter; also called *plasmodial slime mold*.

acetylcholine (ah-sēt′-il-kō′-lēn): a neurotransmitter in the brain and in synapses of motor neurons that innervate skeletal muscles.

acid: a substance that releases hydrogen ions (H^+) into solution; a solution with a pH of less than 7.

acid deposition: the deposition of nitric or sulfuric acid, either in rain (acid rain) or in the form of dry particles, as a result of the production of nitrogen oxides or sulfur dioxide through burning, primarily of fossil fuels.

acidic: with an H^+ concentration exceeding that of OH^-; releasing H^+.

acquired immune deficiency syndrome (AIDS): an infectious disease caused by the human immunodeficiency virus (HIV); attacks and destroys T cells, thus weakening the immune system.

acrosome (ak′-rō-sōm): a vesicle, located at the tip of an animal sperm, that contains enzymes needed to dissolve protective layers around the egg.

actin (ak′-tin): a major muscle protein whose interactions with myosin produce contraction; found in the thin filaments of the muscle fiber; see also *myosin*.

action potential: a rapid change from a negative to a positive electrical potential in a nerve cell. This signal travels along an axon without a change in amplitude.

activation energy: in a chemical reaction, the energy needed to force the electron shells of reactants together, prior to the formation of products.

active site: the region of an enzyme molecule that binds substrates and performs the catalytic function of the enzyme.

active transport: the movement of materials across a membrane through the use of cellular energy, normally against a concentration gradient.

adaptation: a trait that increases the ability of an individual to survive and reproduce compared to individuals without the trait.

adaptive radiation: the rise of many new species in a relatively short time as a result of a single species that invades different habitats and evolves under different environmental pressures in those habitats.

adenine: a nitrogenous base found in both DNA and RNA; abbreviated as *A*.

adenosine diphosphate (a-den′-ō-sēn dī-fos′-fāt; ADP): a molecule composed of the sugar ribose, the base adenine, and two phosphate groups; a component of ATP.

adenosine triphosphate (a-den′-ō-sēn trī-fos′-fāt; ATP): a molecule composed of the sugar ribose, the base adenine, and three phosphate groups; the major energy carrier in cells. The last two phosphate groups are attached by "high-energy" bonds.

adipose tissue (a′-di-pōs): tissue composed of fat cells.

adrenal cortex: the outer part of the adrenal gland, which secretes steroid hormones that regulate metabolism and salt balance.

adrenal gland: a mammalian endocrine gland, adjacent to the kidney; secretes hormones that function in water regulation and in the stress response.

adrenal medulla: the inner part of the adrenal gland, which secretes epinephrine (adrenaline) and norepinephrine (noradrenaline) in the stress response.

adrenocorticotropic hormone (a-drēn-ō-kor-tik-ō-trō′-pik; ACTH): a hormone, secreted by the anterior pituitary, that stimulates the release of hormones by the adrenal cortex, especially in response to stress.

aerobic: using oxygen.

age structure: the distribution of males and females in a population according to age groups.

agglutination (a-gloo-tin-ā′-shun): the clumping of foreign substances or microbes, caused by binding with antibodies.

aggression: antagonistic behavior, normally among members of the same species, often resulting from competition for resources.

aggressive mimicry (mim′ik-rē): the evolution of a predatory organism to resemble a harmless animal or part of the environment, thus gaining access to prey.

aging: a gradual accumulation of random damage to essential biological molecules, particularly DNA, that begins at an early age. Eventually, the body's ability to repair the damage is exceeded, impairing function at all levels, from cells to tissues to organs.

aldosterone: a hormone, secreted by the adrenal cortex, that helps regulate ion concentration in the blood by stimulating the reabsorption of sodium by the kidneys and sweat glands.

alga (al′-ga; pl., algae, al′-jē): any photosynthetic member of the eukaryotic kingdom Protista.

allantois (al-an-tō′-is): one of the embryonic membranes of reptiles, birds, and mammals; in reptiles and birds, serves as a waste-storage organ; in mammals, forms most of the umbilical cord.

allele (al-ēl′): one of several alternative forms of a particular gene.

allele frequency: for any given gene, the relative proportion of each allele of that gene in a population.

allergy: an inflammatory response produced by the body in response to invasion by foreign materials, such as pollen, that are themselves harmless.

allopatric speciation (al-ō-pat′-rik): speciation that occurs when two populations are separated by a physical barrier that prevents gene flow between them (geographical isolation).

allosteric regulation: the process by which enzyme action is enhanced or inhibited by small organic molecules that act as regulators by binding to the enzyme and altering its active site.

alternation of generations: a life cycle, typical of plants, in which a diploid sporophyte (spore-producing) generation alternates with a haploid gametophyte (gamete-producing) generation.

altruism: a type of behavior that may decrease the reproductive success of the individual performing it but benefits that of other individuals.

alveolate (al-vē′-ō-lāt): a member of the Alveolata, a large assemblage of protists that is assigned kingdom status by many systematists. The alveolates, which are characterized by a system of sacs beneath the cell membrane, include ciliates, foraminiferans, dinoflagellates, and apicomplexans.

alveolus (al-vē′-ō-lus; pl., alveoli): a tiny air sac within the lungs, surrounded by capillaries, where gas exchange with the blood occurs.

amino acid: the individual subunit of which proteins are made, composed of a central carbon atom bonded to an amino group ($—NH_2$), a carboxyl group ($—COOH$), a hydrogen atom, and a variable group of atoms denoted by the letter *R*.

amino acid derived hormone: a class of hormone that is synthesized by the body from single amino acids. Examples include epinephrine and thyroxine.

ammonia: NH_3; a highly toxic nitrogen-containing waste product of amino acid breakdown. In the mammalian liver, it is converted to urea.

amniocentesis (am-nē-ō-sen-tē′-sis): a procedure for sampling the amniotic fluid surrounding a fetus: A sterile needle is inserted through the abdominal wall, uterus, and amniotic sac of a pregnant woman; 10 to 20 milliliters of amniotic fluid are withdrawn. Various tests may be performed on the fluid and the fetal cells suspended in it to provide information on the developmental and genetic state of the fetus.

amnion (am′-nē-on): one of the embryonic membranes of reptiles, birds, and mammals; encloses a fluid-filled cavity that envelops the embryo.

amniotic egg (am-nē-ōt′-ik): the egg of reptiles and birds; contains an amnion that encloses the embryo in a watery environment, allowing the egg to be laid on dry land.

amoeba: a type of animal-like protist that uses a characteristic streaming mode of locomotion by extending a cellular projection called a *pseudopod*.

amoeboid cell: a protist or animal cell that moves by extending a cellular projection called a pseudopod.

amphibian: a member of the chordate class Amphibia, which includes the frogs, toads, and salamanders, as well as the limbless caecelians.

amplexus (am-plek´-sus): in amphibians, a form of external fertilization in which the male holds the female during spawning and releases his sperm directly onto her eggs.

ampulla: a muscular bulb that is part of the water-vascular system of echinoderms; controls the movement of tube feet, which are used for locomotion.

amygdala (am-ig´-da-la): part of the forebrain of vertebrates that is involved in the production of appropriate behavioral responses to environmental stimuli.

amylase (am´-i-lās): an enzyme, found in saliva and pancreatic secretions, that catalyzes the breakdown of starch.

anaerobe: an organism whose respiration does not require oxygen.

anaerobic: not using oxygen.

analogous structures: structures that have similar functions and superficially similar appearance but very different anatomies, such as the wings of insects and birds. The similarities are due to similar environmental pressures rather than to common ancestry.

anaphase (an´-a-fāz): in mitosis, the stage in which the sister chromatids of each chromosome separate from one another and are moved to opposite poles of the cell; in meiosis I, the stage in which homologous chromosomes, consisting of two sister chromatids, are separated; in meiosis II, the stage in which the sister chromatids of each chromosome separate from one another and are moved to opposite poles of the cell.

androgen: a male sex hormone.

androgen insensitivity: a rare condition in which an individual with XY chromosomes is female in appearance because the body's cells don't respond to the male hormones that are present.

angina (an-jī´-nuh): chest pain associated with reduced blood flow to the heart muscle, caused by the obstruction of coronary arteries.

angiosperm (an´-jē-ō-sperm): a flowering vascular plant.

angiotensin (an-jē-ō-ten´-sun): a hormone that functions in water regulation in mammals by stimulating physiological changes that increase blood volume and blood pressure.

annual ring: a pattern of alternating light (early) and dark (late) xylem of woody stems and roots, formed as a result of the unequal availability of water in different seasons of the year, normally spring and summer.

antagonistic muscles: a pair of muscles, one of which contracts and in so doing extends the other; an arrangement that makes possible movement of the skeleton at joints.

anterior: the front, forward, or head end of an animal.

anterior pituitary: a lobe of the pituitary gland that produces prolactin and growth hormone as well as hormones that regulate hormone production in other glands.

anther (an´-ther): the uppermost part of the stamen, in which pollen develops.

antheridium (an-ther-id´-ē-um): a structure in which male sex cells are produced, found in the bryophytes and certain seedless vascular plants.

antibiotic resistance: the ability of a mutated pathogen to resist the effects of an antibiotic that normally kills it.

antibody: a protein, produced by cells of the immune system, that combines with a specific antigen and normally facilitates the destruction of the antigen.

anticodon: a sequence of three bases in transfer RNA that is complementary to the three bases of a codon of messenger RNA.

antidiuretic hormone (an-tē-di-ūr-et´-ik; ADH): a hormone produced by the hypothalamus and released into the bloodstream by the posterior pituitary when blood volume is low; increases the permeability of the distal tubule and the collecting duct to water, allowing more water to be reabsorbed into the bloodstream.

antigen: a complex molecule, normally a protein or polysaccharide, that stimulates the production of a specific antibody.

antioxidant: any molecule that reacts with free radicals, neutralizing their ability to damage biological molecules. Vitamins C and E are examples of dietary antioxidants.

aphotic zone: the region of the ocean below 200 m, where sunlight does not penetrate.

apical dominance: the phenomenon whereby a growing shoot tip inhibits the sprouting of lateral buds.

apical meristem (āp´-i-kul mer´-i-stem): the cluster of meristematic cells at the tip of a shoot or root (or one of their branches).

apicomplexan (ā-pē-kom-pleks´-an): a member of the protist phylum Apicomplexa, which includes mostly parasitic, single-celled eukaryotes such as *Plasmodium*, which causes malaria in humans. The apicomplexans are part of a larger group known as the alveolates.

appendicular skeleton (ap-pen-dik´-ū-lur): the portion of the skeleton consisting of the bones of the extremities and their attachments to the axial skeleton; the pectoral and pelvic girdles, the arms, legs, hands, and feet.

aqueous humor (ā´-kwē-us): the clear, watery fluid between the cornea and lens of the eye.

Archaea: one of life's three domains; consists of prokaryotes that are only distantly related to members of the domain Bacteria.

archegonium (ar-ke-gō´-nē-um): a structure in which female sex cells are produced; found in the bryophytes and certain seedless vascular plants.

arteriole (ar-tēr´-ē-ōl): a small artery that empties into capillaries. Constriction of the arteriole regulates blood flow to various parts of the body.

artery (ar´-tuh-rē): a vessel with muscular, elastic walls that conducts blood away from the heart.

arthropod: a member of the animal phylum Arthropoda, which includes the insects, spiders, ticks, mites, scorpions, crustaceans, millipedes, and centipedes.

artificial selection: a selective breeding procedure in which only those individuals with particular traits are chosen as breeders; used mainly to enhance desirable traits in domestic plants and animals; may also be used in evolutionary biology experiments.

ascus (as´-kus): a saclike case in which sexual spores are formed by members of the fungal division Ascomycota.

asexual reproduction: reproduction that does not involve the fusion of haploid sex cells. The parent body may divide and new parts regenerate, or a new, smaller individual may form as an attachment to the parent, to drop off when complete.

association neuron: in a neural network, a nerve cell that is postsynaptic to a sensory neuron and presynaptic to a motor neuron. In actual circuits, there may be many association neurons between individual sensory and motor neurons.

atherosclerosis (ath´-er-ō-skler-ō´-sis): a disease characterized by the obstruction of arteries by cholesterol deposits and thickening of the arterial walls.

atom: the smallest particle of an element that retains the properties of the element.

atomic nucleus: the central part of an atom that contains protons and neutrons.

atomic number: the number of protons in the nuclei of all atoms of a particular element.

atrial natriuretic peptide (ā´-trē-ul nā-trē-ū-ret´-ik; ANP): a hormone, secreted by cells in the mammalian heart, that reduces blood volume by inhibiting the release of ADH and aldosterone.

atrioventricular (AV) node (ā´-trē-ō-ven-trik´-ū-lar nōd): a specialized mass of muscle at the base of the right atrium through which the electrical activity initiated in the sinoatrial node is transmitted to the ventricles.

atrioventricular valve: a heart valve that separates each atrium from each ventricle, preventing the backflow of blood into the atria during ventricular contraction.

atrium (ā´-trē-um): a chamber of the heart that receives venous blood and passes it to a ventricle.

auditory canal (aw´-di-tor-ē): a canal within the outer ear that conducts sound from the external ear to the tympanic membrane.

auditory nerve: the nerve leading from the mammalian cochlea to the brain, carrying information about sound.

auditory tube: a tube connecting the middle ear with the pharynx that allows pressure to equilibrate between the middle ear and the outside air (also called the Eustachian tube).

autoimmune disease: a disorder in which the immune system produces antibodies against the body's own cells.

autonomic nervous system: the part of the peripheral nervous system of vertebrates that synapses on glands, internal organs, and smooth muscle and produces largely involuntary responses.

autosome (aw´-tō-sōm): a chromosome that occurs in homologous pairs in both males and females and that does not bear the genes determining sex.

autotroph (aw´-tō-trof): "self-feeder"; normally, a photosynthetic organism; a producer.

auxin (awk´-sin): a plant hormone that influences many plant functions, including phototropism, apical dominance, and root branching; generally stimulates cell elongation and, in some cases, cell division and differentiation.

axial skeleton: the skeleton forming the body axis, including the skull, vertebral column, and rib cage.

axon: a long extension of a nerve cell, extending from the cell body to synaptic endings on other nerve cells or on muscles.

bacillus (**buh-sil´-us;** pl., **bacilli**): a rod-shaped bacterium.

Bacteria: one of life's three domains; consists of prokaryotes that are only distantly related to members of the domain Archaea.

bacterial conjugation: the exchange of genetic material between two bacteria.

bacteriophage (**bak-tir´-ē-ō-faj**): a virus specialized to attack bacteria.

bacterium (**bak-tir´-ē-um;** pl., **bacteria**): an organism consisting of a single prokaryotic cell surrounded by a complex polysaccharide coat.

balanced polymorphism: the prolonged maintenance of two or more alleles in a population, normally because each allele is favored by a separate environmental pressure.

ball-and-socket joint: a joint in which the rounded end of one bone fits into a hollow depression in another, as in the hip; allows movement in several directions.

bark: the outer layer of a woody stem, consisting of phloem, cork cambium, and cork cells.

Barr body: an inactivated X chromosome in cells of female mammals, which have two X chromosomes; normally appears as a dark spot in the nucleus.

basal body: a structure resembling a centriole that produces a cilium or flagellum and anchors this structure within the plasma membrane.

base: (1) a substance capable of combining with and neutralizing H$^+$ ions in a solution; a solution with a pH of more than 7; (2) in molecular genetics, one of the nitrogen-containing, single- or double-ringed structures that distinguish one nucleotide from another. In DNA, the bases are adenine, guanine, cytosine, and thymine.

basic: with an H$^+$ concentration less than that of OH$^-$; combining with H$^+$.

basidiospore (**ba-sid´-ē-ō-spor**): a sexual spore formed by members of the fungal division Basidiomycota.

basidium (**bas-id´-ē-um**): a diploid cell, typically club-shaped, formed by members of the fungal division Basidiomycota; produces basidiospores by meiosis.

basilar membrane (**bas´-eh-lar**): a membrane in the cochlea that bears hair cells that respond to the vibrations produced by sound.

basophil (**bas´-ō-fil**): a type of white blood cell that releases both substances that inhibit blood clotting and chemicals that participate in allergic reactions and in responses to tissue damage and microbial invasion.

B cell: a type of lymphocyte that participates in humoral immunity; gives rise to plasma cells, which secrete antibodies into the circulatory system, and to memory cells.

behavior: any observable activity of a living animal.

behavioral isolation: the lack of mating between species of animals that differ substantially in courtship and mating rituals.

bilateral symmetry: a body plan in which only a single plane through the central axis will divide the body into mirror-image halves.

bile (**bīl**): a liquid secretion, produced by the liver, that is stored in the gallbladder and released into the small intestine during di-

gestion; a complex mixture of bile salts, water, other salts, and cholesterol.

bile salt: a substance that is synthesized in the liver from cholesterol and amino acids and that assists in the breakdown of lipids by dispersing them into small particles on which enzymes can act.

binary fission: the process by which a single bacterium divides in half, producing two identical offspring.

binocular vision: the ability to see objects simultaneously through both eyes, providing greater depth perception and more-accurate judgment of the size and distance of an object from the eyes.

biocapacity: an estimate of the sustainable resources and waste-absorbing capacity actually available on Earth. While related to the concept of carrying capacity explained in Chapter 26, both the footprint and biocapacity calculations are subject to change as new technologies change the way people use resources.

biodegradable: able to be broken down into harmless substances by decomposers.

biodiversity: the total number of species within an ecosystem and the resulting complexity of interactions among them.

biogeochemical cycle: also called a *nutrient cycle*, the process by which a specific nutrient in an ecosystem is transferred between living organisms and the nutrient's reservoir in the nonliving environment.

biological clock: a metabolic timekeeping mechanism found in most organisms, whereby the organism measures the approximate length of a day (24 hours) even without external environmental cues such as light and darkness.

biological magnification: the increasing accumulation of a toxic substance in progressively higher trophic levels.

biomass: the dry weight of organic material in an ecosystem.

biome (**bī´-ōm**): a terrestrial ecosystem that occupies an extensive geographical area and is characterized by a specific type of plant community; for example, deserts.

biosphere (**bī´-ō-sfēr**): that part of Earth inhabited by living organisms; includes both living and nonliving components.

biosphere reserves: Designated by the UN, these are regions intended to maintain biodiversity and evaluate techniques for sustainable human development while maintaining local cultural values.

biotechnology: any industrial or commercial use or alteration of organisms, cells, or biological molecules to achieve specific practical goals.

biotic (**bī-ah´-tik**): living.

biotic potential: the maximum rate at which a population could increase, assuming ideal conditions that allow a maximum birth rate and minimum death rate.

birth control pill: a temporary contraceptive method that prevents ovulation by providing a continuing supply of estrogen and progesterone, which in turn suppresses LH release; must be taken daily, normally for 21 days of each menstrual cycle.

birth rate: the number of births per individual in a specified unit of time, such as a year.

bladder: a hollow muscular storage organ for storing urine.

blade: the flat part of a leaf.

blastocyst (**blas´-tō-sist**): an early stage of mammalian embryonic development, consisting of a hollow ball of cells, enclosing a mass of cells attached to its inner surface, which becomes the embryo.

blastopore: the site at which a blastula indents to form a gastrula.

blastula (**blas´-tū-luh**): in animals, the embryonic stage attained at the end of cleavage, in which the embryo normally consists of a hollow ball with a wall one or several cell layers thick.

blind spot: the area of the retina at which the axons of the ganglion cells merge to form the optic nerve.

blood: a fluid consisting of plasma in which blood cells are suspended; carried within the circulatory system.

blood–brain barrier: relatively impermeable capillaries of the brain that protect the cells of the brain from potentially damaging chemicals that reach the bloodstream.

blood clotting: a complex process by which platelets, the protein fibrin, and red blood cells block an irregular surface in or on the body, such as a damaged blood vessel, sealing the wound.

blood vessel: a channel that conducts blood throughout the body.

body mass index (BMI): a number derived from an individual's weight and height used to estimate body fat. The formula is: weight (in kg)/height^2 (in meters^2).

bone: a hard, mineralized connective tissue that is a major component of the vertebrate endoskeleton; provides support and sites for muscle attachment.

book lung: a structure composed of thin layers of tissue, resembling pages in a book, that are enclosed in a chamber and used as a respiratory organ by certain types of arachnids.

boom-and-bust cycle: a population cycle characterized by rapid exponential growth followed by a sudden massive die-off, seen in seasonal species and some populations of small rodents, such as lemmings.

Bowman's capsule: the cup-shaped portion of the nephron in which blood filtrate is collected from the glomerulus.

bradykinin (**brā´-dē-ki´-nin**): a chemical, formed during tissue damage, that binds to receptor molecules on pain nerve endings, giving rise to the sensation of pain.

brain: the part of the central nervous system of vertebrates that is enclosed within the skull.

branch root: a root that arises as a branch of a preexisting root, through divisions of pericycle cells and subsequent differentiation of the daughter cells.

bronchiole (**bron´-kē-ōl**): a narrow tube, formed by repeated branching of the bronchi, that conducts air into the alveoli.

bronchus (**bron´-kus**): a tube that conducts air from the trachea to each lung.

bryophyte (**brī´-ō-fīt**): a simple nonvascular plant of the division Bryophyta, including mosses and liverworts.

bud: in animals, a small copy of an adult that develops on the body of the parent and eventually breaks off and becomes independent; in plants, an embryonic shoot, normally very short and consisting of an apical meristem with several leaf primordia.

budding: asexual reproduction by the growth of a miniature copy, or bud, of the adult animal on the body of the parent.

The bud breaks off to begin independent existence.

buffer: a compound that minimizes changes in pH by reversibly taking up or releasing H^+ ions.

bulbourethral gland (bul-bō-ū-rē´-thrul): in male mammals, a gland that secretes a basic, mucus-containing fluid that forms part of the semen.

bulk flow: the movement of many molecules of a gas or fluid in unison from an area of higher pressure to an area of lower pressure.

bundle-sheath cell: one of a group of cells that surround the veins of plants; in C_4 (but not in C_3) plants, bundle-sheath cells contain chloroplasts.

C_3 cycle: the cyclic series of reactions whereby carbon dioxide is fixed into carbohydrates during the light-independent reactions of photosynthesis; also called *Calvin-Benson cycle*.

C_4 pathway: the series of reactions in certain plants that fixes carbon dioxide into oxaloacetic acid, which is later broken down for use in the C_3 cycle of photosynthesis.

calcitonin (kal-si-tōn´-in): a hormone, secreted by the thyroid gland, that inhibits the release of calcium from bone.

calorie (kal´-ō-rē): the amount of energy required to raise the temperature of 1 gram of water by 1 degree Celsius.

Calorie: a unit of energy, in which the energy content of foods is measured; the amount of energy required to raise the temperature of 1 liter of water 1 degree Celsius; also called a *kilocalorie*, equal to 1000 calories.

Calvin-Benson cycle: see C_3 *cycle*.

cambium (kam´-bē-um; pl., **cambia**): a lateral meristem, parallel to the long axis of roots and stems, that causes secondary growth of woody plant stems and roots. See *cork cambium; vascular cambium*.

camouflage (cam´-a-flaj): coloration and/or shape that renders an organism inconspicuous in its environment.

cancer: a disease in which some of the body's cells escape from normal regulatory processes and divide without control.

capillary: the smallest type of blood vessel, connecting arterioles with venules. Capillary walls, through which the exchange of nutrients and wastes occurs, are only one cell thick.

capsule: a polysaccharide or protein coating that some disease-causing bacteria secrete outside their cell wall.

carbohydrate: a compound composed of carbon, hydrogen, and oxygen, with the approximate chemical formula $(CH_2O)_n$; includes sugars and starches.

carbon fixation: the initial steps in the C_3 cycle, in which carbon dioxide reacts with ribulose bisphosphate to form a stable organic molecule.

cardiac cycle (kar´-dē-ak): the alternation of contraction and relaxation of the heart chambers.

cardiac muscle (kar´-dē-ak): the specialized muscle of the heart, able to initiate its own contraction, independent of the nervous system.

carnivore (kar´-neh-vor): literally, "meat eater"; a predatory organism that feeds on herbivores or on other carnivores; a secondary (or higher) consumer.

carotenoid (ka-rot´-en-oid): a red, orange, or yellow pigment, found in chloroplasts, that serves as an accessory light-gathering molecule in thylakoid photosystems.

carpel (kar´pel): the female reproductive structure of a flower, composed of stigma, style, and ovary.

carrier: an individual who is heterozygous for a recessive condition; displays the dominant phenotype but can pass on the recessive allele to offspring.

carrier protein: a membrane protein that facilitates the diffusion of specific substances across the membrane. The molecule to be transported binds to the outer surface of the carrier protein; the protein then changes shape, allowing the molecule to move across the membrane through the protein.

carrying capacity: the maximum population size that an ecosystem can support indefinitely; determined primarily by the availability of space, nutrients, water, and light.

cartilage (kar´-teh-lij): a form of connective tissue that forms portions of the skeleton; consists of chondrocytes and their extracellular secretion of collagen; resembles flexible bone.

Casparian strip (kas-par´-ē-un): a waxy, waterproof band, located in the cell walls between endodermal cells in a root, that prevents the movement of water and minerals into and out of the vascular cylinder through the extracellular space.

catalyst (kat´-uh-list): a substance that speeds up a chemical reaction without itself being permanently changed in the process; lowers the activation energy of a reaction.

catastrophism: the hypothesis that Earth has experienced a series of geological catastrophes, probably imposed by a supernatural being, that accounts for the multitude of species, both extinct and modern, and preserves creationism.

cell: the smallest unit of life, consisting, at a minimum, of an outer membrane that encloses a watery medium containing organic molecules, including genetic material composed of DNA.

cell body: the part of a nerve cell in which most of the common cellular organelles are located; typically a site of integration of inputs to the nerve cell.

cell cycle: the sequence of events in the life of a cell, from one division to the next.

cell division: splitting of one cell into two; the process of cellular reproduction.

cell-mediated immunity: an immune response in which foreign cells or substances are destroyed by contact with T cells.

cell plate: in plant cell division, a series of vesicles that fuse to form the new plasma membranes and cell wall separating the daughter cells.

cellular respiration: the oxygen-requiring reactions, occurring in mitochondria, that break down the end products of glycolysis into carbon dioxide and water while capturing large amounts of energy as ATP.

cellular slime mold: a funguslike protist consisting of individual amoeboid cells that can aggregate to form a sluglike mass, which in turn forms a fruiting body.

cellulase: an enzyme that catalyzes the breakdown of the carbohydrate cellulose into its component glucose molecules; almost entirely restricted to microorganisms.

cellulose: an insoluble carbohydrate composed of glucose subunits; forms the cell wall of plants.

cell wall: a layer of material, normally made up of cellulose or cellulose-like materials, that is outside the plasma membrane of plants, fungi, bacteria, and some protists.

central nervous system: in vertebrates, the brain and spinal cord.

central vacuole: a large, fluid-filled vacuole occupying most of the volume of many plant cells; performs several functions, including maintaining turgor pressure.

centriole (sen´-trē-ol): in animal cells, a short, barrel-shaped ring consisting of nine microtubule triplets; a microtubule-containing structure at the base of each cilium and flagellum; gives rise to the microtubules of cilia and flagella and is involved in spindle formation during cell division.

centromere (sen´-trō-mer): the region of a replicated chromosome at which the sister chromatids are held together until they separate during cell division.

cephalization (sef-ul-i-zā´-shun): the tendency of sensory organs and nervous tissue to become concentrated in the head region over evolutionary time.

cerebellum (ser-uh-bel´-um): the part of the hindbrain of vertebrates that is concerned with coordinating movements of the body.

cerebral cortex (ser-ē´-brul kor´-tex): a thin layer of neurons on the surface of the vertebrate cerebrum, in which most neural processing and coordination of activity occurs.

cerebral hemisphere: one of two nearly symmetrical halves of the cerebrum, connected by a broad band of axons, the corpus callosum.

cerebrospinal fluid: a clear fluid, produced within the ventricles of the brain, that fills the ventricles and cushions the brain and spinal cord.

cerebrum (ser-ē´-brum): the part of the forebrain of vertebrates that is concerned with sensory processing, the direction of motor output, and the coordination of most bodily activities; consists of two nearly symmetrical halves (the hemispheres) connected by a broad band of axons, the corpus callosum.

cervical cap: a birth control device consisting of a rubber cap that fits over the cervix, preventing sperm from entering the uterus.

cervix (ser´-viks): a ring of connective tissue at the outer end of the uterus, leading into the vagina.

channel protein: a membrane protein that forms a channel or pore completely through the membrane and that is usually permeable to one or to a few water-soluble molecules, especially ions.

chaparral: a biome located in coastal regions, with very low annual rainfall, characterized by shrubs and small trees.

chemical bond: the force of attraction between neighboring atoms that holds them together in a molecule.

chemical equilibrium: the condition in which the "forward" reaction of reactants to products proceeds at the same rate as the "backward" reaction from products to reactants, so that no net change in chemical composition occurs.

chemical reaction: the process that forms and breaks chemical bonds that hold atoms together.

chemiosmosis (ke-me-oz-mo'-sis): a process of ATP generation in chloroplasts and mitochondria. The movement of electrons down an electron transport system is used to pump hydrogen ions across a membrane, thereby building up a concentration gradient of hydrogen ions across the membrane; the hydrogen ions diffuse back across the membrane through the pores of ATP-synthesizing enzymes; the energy of their movement down their concentration gradient drives ATP synthesis.

chemoreceptor: a sensory receptor that responds to chemicals from the environment; used in the chemical senses of taste and smell.

chemosynthetic (kem'-o-sin-the-tik): capable of oxidizing inorganic molecules to obtain energy.

chemotactic (kem-o-tak'-tik): moving toward chemicals given off by food or away from toxic chemicals.

chiasma (ki-as'-muh; pl., **chiasmata**): a point at which a chromatid of one chromosome crosses with a chromatid of the homologous chromosome during prophase I of meiosis; the site of exchange of chromosomal material between chromosomes.

chitin (ki'-tin): a compound found in the cell walls of fungi and the exoskeletons of insects and some other arthropods; composed of chains of nitrogen-containing, modified glucose molecules.

chlamydia (kla-mid'-e-uh): a sexually transmitted disease, caused by a bacterium, that causes inflammation of the urethra in males and of the urethra and cervix in females.

chlorophyll (klor'-o-fil): a pigment found in chloroplasts that captures light energy during photosynthesis; absorbs violet, blue, and red light but reflects green light.

chloroplast (klor'-o-plast): the organelle in plants and plantlike protists that is the site of photosynthesis; surrounded by a double membrane and containing an extensive internal membrane system that bears chlorophyll.

cholecystokinin (ko'-le-sis-to-ki'-nin): a digestive hormone, produced by the small intestine, that stimulates the release of pancreatic enzymes.

chondrocyte (kon'-dro-sit): a living cell of cartilage. With their extracellular secretions of collagen, chondrocytes form cartilage.

chorion (kor'-e-on): the outermost embryonic membrane in reptiles, birds, and mammals; in birds and reptiles, functions mostly in gas exchange; in mammals, forms most of the embryonic part of the placenta.

chorionic gonadotropin (CG): a hormone, secreted by the chorion (one of the fetal membranes), that maintains the integrity of the corpus luteum during early pregnancy.

chorionic villus (kor-e-on-ik; pl., **chorionic villi**): in mammalian embryos, a fingerlike projection of the chorion that penetrates the uterine lining and forms the embryonic portion of the placenta.

chorionic villus sampling (CVS): a procedure for sampling cells from the chorionic villi produced by a fetus: A tube is inserted into the uterus of a pregnant woman, and a small sample of villi are suctioned off for genetic and biochemical analyses.

choroid (kor'-oid): a darkly pigmented layer of tissue, behind the retina, that contains blood vessels and pigment that absorbs stray light.

chromatid (kro'-ma-tid): one of the two identical strands of DNA and protein that forms a replicated chromosome. The two sister chromatids are joined at the centromere.

chromatin (kro'-ma-tin): the complex of DNA and proteins that makes up eukaryotic chromosomes.

chromist: a member of the Chromista, a large assemblage of protists that is assigned kingdom status by many systematists. The chromists include the diatoms, brown algae, and water molds.

chromosome (kro'-mo-som): a single DNA double helix together with proteins that help to organize the DNA.

chronic bronchitis: a persistent lung infection characterized by coughing, swelling of the lining of the respiratory tract, an increase in mucus production, and a decrease in the number and activity of cilia.

chyme (kim): an acidic, souplike mixture of partially digested food, water, and digestive secretions that is released from the stomach into the small intestine.

ciliate (sil'-e-et): a protozoan characterized by cilia and by a complex unicellular structure, including harpoonlike organelles called trichocysts. Members of the genus *Paramecium* are well-known ciliates.

cilium (sil'-e-um; pl., **cilia**): a short, hairlike projection from the surface of certain eukaryotic cells that contains microtubules in a 9 + 2 arrangement. The movement of cilia may propel cells through a fluid medium or move fluids over a stationary surface layer of cells.

circadian rhythm (sir-ka'-de-un): an event that recurs with a period of about 24 hours, even in the absence of environmental cues.

citric acid cycle: see *Krebs cycle*.

class: the taxonomic category composed of related orders. Closely related classes form a division or phylum.

cleavage: the early cell divisions of embryos, in which little or no growth occurs between divisions; reduces the cell size and distributes gene-regulating substances to the newly formed cell.

climate: patterns of weather that prevail from year to year and even from century to century in a given region.

climax community: a diverse and relatively stable community that forms the endpoint of succession.

clitoris: an external structure of the female reproductive system; composed of erectile tissue; a sensitive point of stimulation during sexual response.

clonal selection: the mechanism by which the immune response gains specificity; an invading antigen elicits a response from only a few lymphocytes, which proliferate to form a clone of cells that attack only the specific antigen that stimulated their production.

clone: offspring that are produced by mitosis and are therefore genetically identical to each other.

cloning: the process of producing many identical copies of a gene; also the production of many genetically identical copies of an organism.

closed circulatory system: the type of circulatory system, found in certain worms and vertebrates, in which the blood is always confined within the heart and vessels.

club fungus: a fungus of the division Basidiomycota, whose members (which include mushrooms, puffballs, and shelf fungi) reproduce by means of basidiospores.

clumped distribution: the distribution characteristic of populations in which individuals are clustered into groups; may be social or based on the need for a localized resource.

cnidocyte (nid'-o-sit): in members of the phylum Cnidaria, a specialized cell that houses a stinging apparatus.

cochlea (kahk'-le-uh): a coiled, bony, fluid-filled tube found in the mammalian inner ear; contains receptors (hair cells) that respond to the vibration of sound.

codominance: the relation between two alleles of a gene, such that both alleles are phenotypically expressed in heterozygous individuals.

codon: a sequence of three bases of messenger RNA that specifies a particular amino acid to be incorporated into a protein; certain codons also signal the beginning or end of protein synthesis.

coelom (se'-lom): a space or cavity that separates the body wall from the inner organs.

coenzyme: an organic molecule that is bound to certain enzymes and is required for the enzymes' proper functioning; typically, a nucleotide bound to a water-soluble vitamin.

coevolution: the evolution of adaptations in two species due to their extensive interactions with one another, such that each species acts as a major force of natural selection on the other.

cohesion: the tendency of the molecules of a substance to stick together.

cohesion–tension theory: a model for the transport of water in xylem, by which water is pulled up the xylem tubes, powered by the force of evaporation of water from the leaves (producing tension) and held together by hydrogen bonds between nearby water molecules (cohesion).

coleoptile (ko-le-op'-til): a protective sheath surrounding the shoot in monocot seeds, allowing the shoot to push aside soil particles as it grows.

collagen (kol'-uh-jen): a fibrous protein in connective tissue such as bone and cartilage.

collar cell: a specialized cell lining the inside channels of sponges. Flagella extend from a sievelike collar, creating a water current that draws microscopic organisms through the collar and into the body, where they become trapped.

collecting duct: a conducting tube, within the kidney, that collects urine from many nephrons and conducts it through the renal medulla into the renal pelvis. Urine may become concentrated in the collecting ducts if ADH is present.

collenchyma (kol-en'-ki-muh): an elongated, polygonal plant cell type with irregularly thickened primary cell walls that is alive at maturity and that supports the plant body.

colon: the longest part of the large intestine, exclusive of the rectum.

colostrum (ko-los'-trum): a yellowish fluid, high in protein and containing antibodies, that is produced by the mammary glands before milk secretion begins.

commensalism (kum-en'-sal-iz-um): a symbiotic relationship in which one species benefits while another species is neither harmed nor benefited.

communication: the act of producing a signal that causes another animal, normally of the same species, to change its behavior in a way that is beneficial to one or both participants.

community: all the interacting populations within an ecosystem.

compact bone: the hard and strong outer bone; composed of osteons.

companion cell: a cell adjacent to a sieve-tube element in phloem, involved in the control and nutrition of the sieve-tube element.

competition: interaction among individuals who attempt to utilize a resource (for example, food or space) that is limited relative to the demand for it.

competitive exclusion principle: the concept that no two species can simultaneously and continuously occupy the same ecological niche.

competitive inhibition: the process by which two or more molecules that are somewhat similar in structure compete for the active site of an enzyme.

complement: a group of blood-borne proteins that participate in the destruction of foreign cells to which antibodies have bound.

complementary base pair: in nucleic acids, bases that pair by hydrogen bonding. In DNA, adenine is complementary to thymine and guanine is complementary to cytosine; in RNA, adenine is complementary to uracil, and guanine to cytosine.

complement reaction: an interaction among foreign cells, antibodies, and complement proteins that results in the destruction of the foreign cells.

complement system: a series of reactions in which complement proteins bind to antibody stems, attracting to the site phagocytic white blood cells that destroy the invading cell that triggers the reactions.

complete flower: a flower that has all four floral parts (sepals, petals, stamens, and carpels).

compound: a substance whose molecules are formed by different types of atoms; can be broken into its constituent elements by chemical means.

compound eye: a type of eye, found in arthropods, that is composed of numerous independent subunits called *ommatidia.* Each ommatidium apparently contributes a piece of a mosaiclike image perceived by the animal.

concentration: the number of particles of a dissolved substance in a given unit of volume.

concentration gradient: the difference in concentration of a substance between two parts of a fluid or across a barrier such as a membrane.

conclusion: in the scientific method, the decision about the validity of a hypothesis on the basis of experimental evidence.

condensation: compaction of eukaryotic chromosomes into discrete units in preparation for mitosis or meiosis.

condom: a contraceptive sheath worn over the penis during intercourse to prevent sperm from being deposited in the vagina.

conducting portion: the portion of the respiratory system in lung-breathing vertebrates that carries air to the alveoli.

cone: a cone-shaped photoreceptor cell in the vertebrate retina; not as sensitive to light as are the rods. The three types of cones are most sensitive to different colors of light and provide color vision; see also *rod.*

conifer (**kon′-eh-fer**): a member of a class of tracheophytes (Coniferophyta) that reproduces by means of seeds formed inside cones and that retains its leaves throughout the year.

conjugation: in prokaryotes, the transfer of DNA from one cell to another via a temporary connection; in single-celled eukaryotes, the mutual exchange of genetic material between two temporarily joined cells.

connective tissue: a tissue type consisting of diverse tissues, including bone, fat, and blood, that generally contain large amounts of extracellular material.

conservation biology: the application of knowledge from ecology and other areas of biology to conserve biodiversity.

constant region: the part of an antibody molecule that is similar in all antibodies of a given class.

consumer: an organism that eats other organisms; a heterotroph.

contest competition: a mechanism for resolving intraspecific competition by using social or chemical interactions.

contraception: the prevention of pregnancy.

contractile vacuole: a fluid-filled vacuole in certain protists that takes up water from the cytoplasm, contracts, and expels the water outside the cell through a pore in the plasma membrane.

control: that portion of an experiment in which all possible variables are held constant; in contrast to the "experimental" portion, in which a particular variable is altered.

convergence: a condition in which a large number of nerve cells provide input to a smaller number of cells.

convergent evolution: the independent evolution of similar structures among unrelated organisms as a result of similar environmental pressures; see *analogous structures.*

convolution: a folding of the cerebral cortex of the vertebrate brain.

copulation: reproductive behavior in which the penis of the male is inserted into the body of the female, where it releases sperm.

coral reef: a biome created by animals (reef-building corals) and plants in warm tropical waters.

core reserves: natural areas protected from most human uses that encompass enough space to preserve ecosystems with all their biodiversity.

cork cambium: a lateral meristem in woody roots and stems that gives rise to cork cells.

cork cell: a protective cell of the bark of woody stems and roots; at maturity, cork cells are dead, with thick, waterproofed cell walls.

cornea (**kor′-nē-uh**): the clear outer covering of the eye, in front of the pupil and iris.

corona radiata (**kuh-rō′-nuh rā-dē-a′-tuh**): the layer of cells surrounding an egg after ovulation.

corpus callosum (**kor′pus kal-ō′-sum**): the band of axons that connect the two cerebral hemispheres of vertebrates.

corpus luteum (**kor′-pus loo′-tē-um**): in the mammalian ovary, a structure that is derived from the follicle after ovulation and that secretes the hormones estrogen and progesterone.

cortex: the part of a primary root or stem located between the epidermis and the vascular cylinder.

cortisol (**kor′-ti-sol**): a steroid hormone released into the bloodstream by the adrenal cortex in response to stress. Cortisol helps the body cope with short-term stressors by raising blood glucose levels, and also inhibits the immune response.

cotyledon (**kot-ul-ē′don**): a leaflike structure within a seed that absorbs food molecules from the endosperm and transfers them to the growing embryo; also called *seed leaf.*

coupled reaction: a pair of reactions, one exergonic and one endergonic, that are linked together such that the energy produced by the exergonic reaction provides the energy needed to drive the endergonic reaction.

covalent bond (**kō-vā′-lent**): a chemical bond between atoms in which electrons are shared.

crab lice: an arthropod parasite that can infest humans; can be transmitted by sexual contact.

creationism: the hypothesis that all species on Earth were created in essentially their present form by a supernatural being and that significant modification of those species—specifically, their transformation into new species—cannot occur by natural processes.

crista (**kris′-tuh;** pl., **cristae**): a fold in the inner membrane of a mitochondrion.

critically endangered species: a species that faces an extreme risk of extinction in the wild in the immediate future.

crop: an organ, found in both earthworms and birds, in which ingested food is temporarily stored before being passed to the gizzard, where it is pulverized.

cross-bridge: in muscles, an extension of myosin that binds to and pulls on actin to produce muscle contraction.

cross-fertilization: the union of sperm and egg from two individuals of the same species.

crossing over: the exchange of corresponding segments of the chromatids of two homologous chromosomes during meiosis.

cultural evolution: changes in the behavior of a population of animals, especially humans, by learning behaviors acquired by members of previous generations.

cuticle (**kū′-ti-kul**): a waxy or fatty coating on the exposed surfaces of epidermal cells of many land plants, which aids in the retention of water.

cyanobacterium: a photosynthetic prokaryotic cell that utilizes chlorophyll and releases oxygen as a photosynthetic by-product; sometimes called *blue-green algae.*

cyclic AMP: a cyclic nucleotide, formed within many target cells as a result of the reception of amino acid derivatives or peptide hormones, that causes metabolic changes in the cell; often called a *second messenger.*

cyclic nucleotide (**sik′-lik noo′-klē-ō-tid**): a nucleotide in which the phosphate group is bonded to the sugar at two points, forming a ring; serves as an intracellular messenger.

cyst (sist): an encapsulated resting stage in the life cycle of certain invertebrates, such as parasitic flatworms and roundworms.

cystic fibrosis: an inherited disorder characterized by the buildup of salt in the lungs and the production of thick, sticky mucus that clogs the airways, restricts air exchange, and promotes infection.

cytokine (si′-to̅-kin): any of several chemical messenger molecules released by cells that facilitate communication with other cells and transfer signals within and between the various systems of the body. Cytokines are important in cellular differentiation and the immune system.

cytokinesis (si-to̅-ki-ne̅′-sis): the division of the cytoplasm and organelles into two daughter cells during cell division; normally occurs during telophase of mitosis.

cytokinin (si-to̅-ki′-nin): a plant hormone that promotes cell division, fruit growth, and the sprouting of lateral buds and prevents the aging of plant parts, especially leaves.

cytoplasm (si′-to̅-plaz-um): the material contained within the plasma membrane of a cell, exclusive of the nucleus.

cytosine: a nitrogenous base found in both DNA and RNA; abbreviated as C.

cytoskeleton: a network of protein fibers in the cytoplasm that gives shape to a cell, holds and moves organelles, and is typically involved in cell movement.

cytotoxic T cell: a type of T cell that, upon contacting foreign cells, directly destroys them.

day-neutral plant: a plant in which flowering occurs as soon as the plant has grown and developed, regardless of day length.

death rate: the number of deaths per individual in a specified unit of time, such as a year.

decomposer: an organism, normally a fungus or bacterium, that digests organic material by secreting digestive enzymes into the environment, in the process liberating nutrients into the environment.

deductive reasoning: the process of generating hypotheses about how a specific experiment or observation will turn out.

deforestation: the excessive cutting of forests, primarily rain forests in the Tropics, to clear space for agriculture.

dehydration synthesis: a chemical reaction in which two molecules are joined by a covalent bond with the simultaneous removal of a hydrogen from one molecule and a hydroxyl group from the other, forming water; the reverse of hydrolysis.

deletion mutation: a mutation in which one or more pairs of nucleotides are removed from a gene.

demographic transition: a change in population dynamic in which a stable population experiences rapid growth and then returns to a stable (although much larger) size.

demography: the study of the changing human population. Using complex life tables, demographers measure and compare many aspects of human populations in different countries and world regions.

denature: to disrupt the secondary and/or tertiary structure of a protein while leaving its amino acid sequence intact. Denatured proteins can no longer perform their biological functions.

dendrite (den′-drit): a branched tendril that extends outward from the cell body of a neuron; specialized to respond to signals from the external environment or from other neurons.

denitrifying bacterium (de̅-ni′-treh-fi̅-ing): a bacterium that breaks down nitrates, releasing nitrogen gas to the atmosphere.

density-dependent: referring to any factor, such as predation, that limits population size more effectively as the population density increases.

density-independent: referring to any factor that limits a population's size and growth regardless of its density.

deoxyribonucleic acid (de̅-ox-e̅-ri-bo̅-noo-kla̅′-ik; DNA): a molecule composed of deoxyribose nucleotides; contains the genetic information of all living cells.

dermal tissue system: a plant tissue system that makes up the outer covering of the plant body.

dermis (dur′-mis): the layer of skin beneath the epidermis; composed of connective tissue and containing blood vessels, muscles, nerve endings, and glands.

desert: a biome in which less than 25 to 50 centimeters (10 to 20 inches) of rain falls each year.

desertification: the spread of deserts by human activities.

desmosome (dez′-mo̅-so̅m): a strong cell-to-cell junction that attaches adjacent cells to one another.

detritus feeder (de-tri̅′-tus): one of a diverse group of organisms, ranging from worms to vultures, that live off the wastes and dead remains of other organisms.

deuterostome (doo′-ter-o̅-sto̅m): an animal with a mode of embryonic development in which the coelom is derived from outpocketings of the gut; characteristic of echinoderms and chordates.

development: the process by which an organism proceeds from fertilized egg through adulthood to eventual death.

diabetes mellitus (di-uh-be̅′-te̅s mel-i̅′-tus): a disease characterized by defects in the production, release, or reception of insulin; characterized by high blood glucose levels that fluctuate with sugar intake.

dialysis (di̅-al′-i-sis): the passive diffusion of substances across an artificial semipermeable membrane.

diaphragm (di̅′-uh-fram): in the respiratory system, a dome-shaped muscle forming the floor of the chest cavity that, when it contracts, pulls itself downward, enlarging the chest cavity and causing air to be drawn into the lungs; in a reproductive sense, a contraceptive rubber cap that fits snugly over the cervix, preventing the sperm from entering the uterus and thereby preventing pregnancy.

diatom (di̅′-uh-tom): a protist that includes photosynthetic forms with two-part glassy outer coverings; important photosynthetic organisms in fresh water and salt water.

dicot (di̅′-kaht): short for dicotyledon; a type of flowering plant characterized by embryos with two cotyledons, or seed leaves, modified for food storage.

differentially permeable: referring to a membrane through which some substances can pass more easily than can other substances.

differential reproduction: differences in reproductive output among individuals of a population, normally as a result of genetic differences.

differentiated cell: a mature cell specialized for a specific function; in plants, differentiated cells normally do not divide.

differentiation: the process whereby relatively unspecialized cells, especially of embryos, become specialized into particular tissue types.

diffusion: the net movement of particles from a region of high concentration of that particle to a region of low concentration, driven by the concentration gradient; may occur entirely within a fluid or across a barrier such as a membrane.

digestion: the process by which food is physically and chemically broken down into molecules that can be absorbed by cells.

digestive system: a group of organs responsible for ingesting and then digesting food substances into simple molecules that can be absorbed and then expelling undigested wastes from the body.

dinoflagellate (di-no̅-fla′-jel-et): a protist that includes photosynthetic forms in which two flagella project through armorlike plates; abundant in oceans; can reproduce rapidly, causing "red tides."

dioecious (di̅-e̅′-shus): pertaining to organisms in which male and female gametes are produced by separate individuals rather than in the same individual.

diploid (dip′-loid): referring to a cell with pairs of homologous chromosomes.

direct development: a developmental pathway in which the offspring is born as a miniature version of the adult and does not radically change in body form as it grows and matures.

directional selection: a type of natural selection in which one extreme phenotype is favored over all others.

disaccharide (di̅-sak′-uh-ri̅d): a carbohydrate formed by the covalent bonding of two monosaccharides.

disruptive selection: a type of natural selection in which both extreme phenotypes are favored over the average phenotype.

distal tubule: in the nephrons of the mammalian kidney, the last segment of the renal tubule through which the filtrate passes just before it empties into the collecting duct; a site of selective secretion and reabsorption as water and ions pass between the blood and the filtrate across the tubule membrane.

disturbance: any event that disrupts the ecosystem by altering its community, its abiotic structure, or both; disturbance precedes succession.

disulfide bridge: the covalent bond formed between the sulfur atoms of two cysteines in a protein; typically causes the protein to fold by bringing otherwise distant parts of the protein close together.

divergence: a condition in which a small number of nerve cells provide input to a larger number of cells.

divergent evolution: evolutionary change in which the differences between two lineages become more pronounced with the passage of time.

division: the taxonomic category contained within a kingdom and consisting of related classes of plants, fungi, bacteria, or plantlike protists.

DNA–DNA hybridization: a technique by which DNA from two species is separated into single strands and then allowed to reform; hybrid double-stranded DNA from the two species can occur where the sequence of nucleotides is complementary. The greater the degree of hybridization, the closer the evolutionary relatedness of the two species.

DNA profiling: the pattern of short tandem repeats of specific DNA segments; using 13 short tandem repeats, the DNA profile of one person is different from that of any other person on Earth.

DNA helicase: an enzyme that helps unwind the DNA double helix during DNA replication.

DNA library: a readily accessible, easily duplicable complete set of all the DNA of a particular organism, normally cloned into bacterial plasmids.

DNA ligase: an enzyme that joins the sugars and phosphates in a DNA strand to create a continuous sugar-phosphate backbone.

DNA polymerase: an enzyme that bonds DNA nucleotides together into a continuous strand, using a preexisting DNA strand as a template.

DNA probe: a sequence of nucleotides that is complementary to the nucleotide sequence in a gene under study; used to locate a given gene during gel electrophoresis or other methods of DNA analysis.

DNA replication: the copying of the double-stranded DNA molecule, producing two identical DNA double helices.

DNA sequencing: the process of determining the order of nucleotides in a DNA molecule.

domain: the broadest category for classifying organisms; organisms are classified into three domains: Bacteria, Archaea, and Eukarya.

dominance hierarchy: a social arrangement in which a group of animals, usually through aggressive interactions, establishes a rank for some or all of the group members that determines access to resources.

dominant: an allele that can determine the phenotype of heterozygotes completely, such that they are indistinguishable from individuals homozygous for the allele; in the heterozygotes, the expression of the other (recessive) allele is completely masked.

dopamine (dōp′-uh-mēn): a transmitter in the brain whose actions are largely inhibitory. The loss of dopamine-containing neurons causes Parkinson's disease.

dormancy: a state in which an organism does not grow or develop; usually marked by lowered metabolic activity and resistance to adverse environmental conditions.

dorsal (dor′-sul): the top, back, or uppermost surface of an animal oriented with its head forward.

dorsal root ganglion: a ganglion, located on the dorsal (sensory) branch of each spinal nerve, that contains the cell bodies of sensory neurons.

double covalent bond: a covalent bond in which two atoms share two pairs of electrons.

double fertilization: in flowering plants, the fusion of two sperm nuclei with the nuclei of two cells of the female gametophyte. One sperm nucleus fuses with the egg to form the zygote; the second sperm nucleus fuses with the two haploid nuclei of the primary endosperm cell, forming a triploid endosperm cell.

double helix (hē′-liks): the shape of the two-stranded DNA molecule; like a ladder twisted lengthwise into a corkscrew shape.

doubling time: the time it would take a population to double in size at its current growth rate.

douching: washing the vagina; after intercourse, an attempt to wash sperm out of the vagina before they enter the uterus; an ineffective contraceptive method.

Down syndrome: a genetic disorder caused by the presence of three copies of chromosome 21; common characteristics include mental retardation, distinctively shaped eyelids, a small mouth with protruding tongue, heart defects, and low resistance to infectious diseases; also called *trisomy 21*.

duct: a tube or opening through which exocrine secretions are released.

duplicated chromosome: a eukaryotic chromosome following DNA replication; consists of two sister chromatids joined at the centromeres.

ecdysone: a steroid hormone that triggers molting in insects and other arthropods.

echolocation: the use of ultrasonic sounds, which bounce back from nearby objects, to produce an auditory "image" of nearby surroundings; used by bats and porpoises.

ecological footprint: an estimate of the area of Earth's surface required to produce the resources we use and to absorb the wastes we generate, expressed in acres of average productivity.

ecological isolation: the lack of mating between organisms belonging to different populations that occupy distinct habitats within the same general area.

ecological niche (nitch): the role of a particular species within an ecosystem, including all aspects of its interaction with the living and nonliving environments.

ecology (ē-kol′-uh-jē:) the study of the interrelationships of organisms with each other and with their nonliving environment.

ecosystem (ē′-kō-sis-tem): all the organisms and their nonliving environment within a defined area.

ecosystem services: the processes through which natural ecosystems and their living communities sustain and fulfill human life. Ecosystem services include purifying air and water, replenishing oxygen, pollinating plants, flood control, wildlife habitat, and many more.

ectoderm (ek′-tō-derm): the outermost embryonic tissue layer, which gives rise to structures such as hair, the epidermis of the skin, and the nervous system.

ectotherm: an animal that obtains most of its body warmth from its environment. Body temperatures of ectotherms vary with the temperature of their surroundings.

effector (ē-fek′-tor): a part of the body (normally a muscle or gland) that carries out responses as directed by the nervous system.

egg: the haploid female gamete, normally large and nonmotile, containing food reserves for the developing embryo.

electrocardiogram (ECG): the read-out of an instrument that records the electrical activity generated by cardiac muscle action potentials. These electrical events are measured by electrodes placed at specific sites on the surface of the body.

electrolocation: the production of high-frequency electrical signals from an electric organ in front of the tail of weak electrical fish; used to detect and locate nearby objects.

electron: a subatomic particle, found in an electron shell outside the nucleus of an atom, that bears a unit of negative charge and very little mass.

electron carrier: a molecule that can reversibly gain or lose electrons. Electron carriers generally accept high-energy electrons produced during an exergonic reaction and donate the electrons to acceptor molecules that use the energy to drive endergonic reactions.

electron shell: a region within which electrons orbit that corresponds to a fixed energy level at a given distance from the atomic nucleus of an atom.

electron transport system: a series of electron carrier molecules, found in the thylakoid membranes of chloroplasts and the inner membrane of mitochondria, that extract energy from electrons and generate ATP or other energetic molecules.

element: a substance that cannot be broken down, or converted, to a simpler substance by ordinary chemical means.

El Niño (el nēn′-yō): literally "boy child"; a reduction in intensity of Northeast Tradewinds that causes widespread disruption of weather patterns.

embryo: in animals, the stages of development that begin with the fertilization of the egg cell and end with hatching or birth; in mammals in particular, the early stages in which the developing animal does not yet resemble the adult of the species.

embryonic disc: in human embryonic development, the flat, two-layered group of cells that separates the amniotic cavity from the yolk sac.

embryonic stem cell: a cell derived from an early embryo that is capable of differentiating into any of the adult cell types.

embryo sac: the haploid female gametophyte of flowering plants.

emergent property: an intangible attribute that arises as the result of complex ordered interactions among individual parts.

emigration (em-uh-grā′-shun): migration of individuals out of an area.

emphysema (em-fuh-sē′-muh): a condition in which the alveoli of the lungs become brittle and rupture, causing decreased area for gas exchange.

endangered species: a species that faces a very high risk of extinction in the wild in the near future.

endergonic (en-der-gon′-ik): pertaining to a chemical reaction that requires an input of energy to proceed; an "uphill" reaction.

endocrine disruptors: environmental pollutants that interfere with endocrine function, often by disrupting the action of sex hormones.

endocrine gland: a ductless, hormone-producing gland consisting of cells that release their secretions into the extracellular fluid from which the secretions diffuse into nearby capillaries.

endocrine hormones: chemical messages produced by specialized cells and released into the circulatory system. They cause a prolonged or temporary change in target cells bearing specific receptors for these hormones.

endocrine system: an animal's organ system for cell-to-cell communication, composed of hormones and the cells that secrete them and receive them.

endocytosis (en-dō-sī-tō'-sis): the process in which the plasma membrane engulfs extracellular material, forming membrane-bound sacs that enter the cytoplasm and thereby move material into the cell.

endoderm (en'-dō-derm): the innermost embryonic tissue layer, which gives rise to structures such as the lining of the digestive and respiratory tracts.

endodermis (en-dō-der'-mis): the innermost layer of small, close-fitting cells of the cortex of a root that form a ring around the vascular cylinder.

endogenous pyrogen: a chemical, produced by the body, that stimulates the production of a fever.

endometrium (en-dō-mē'-trē-um): the nutritive inner lining of the uterus.

endoplasmic reticulum (ER) (en-dō-plaz'-mik re-tik'-ū-lum): a system of membranous tubes and channels in eukaryotic cells; the site of most protein and lipid syntheses.

endorphin (en-dor'-fin): one of a group of peptide neuromodulators in the vertebrate brain that, by reducing the sensation of pain, mimics some of the actions of opiates.

endoskeleton (en'-dō-skel'-uh-tun): a rigid internal skeleton with flexible joints to allow for movement.

endosperm: a triploid food storage tissue in the seeds of flowering plants that nourishes the developing plant embryo.

endospore: a protective resting structure of some rod-shaped bacteria that withstands unfavorable environmental conditions.

endosymbiont hypothesis: the hypothesis that certain organelles, especially chloroplasts and mitochondria, arose as mutually beneficial associations between the ancestors of eukaryotic cells and captured bacteria that lived within the cytoplasm of the pre-eukaryotic cell.

endotherm: an animal that obtains most of its body heat from metabolic activates. Endotherm body temperature can remain relatively constant within a range of environmental temperatures.

energy: the capacity to do work.

energy-carrier molecule: a molecule that stores energy in "high-energy" chemical bonds and releases the energy to drive coupled endothermic reactions. In cells, ATP is the most common energy-carrier molecule.

energy level: the specific amount of energy characteristic of a given electron shell in an atom.

energy pyramid: a graphical representation of the energy contained in succeeding trophic levels, with maximum energy at the base (primary producers) and steadily diminishing amounts at higher levels.

entropy (en'-trō-pē): a measure of the amount of randomness and disorder in a system.

environmental estrogens: chemicals in the environment that mimic some of the effects of estrogen in animals.

environmental resistance: any factor that tends to counteract biotic potential, limiting population size.

enzyme (en'zīm): a protein catalyst that speeds up the rate of specific biological reactions.

eosinophil (ē-ō-sin'-ō-fil): a type of white blood cell that converges on parasitic invaders and releases substances to kill them.

epicotyl (ep'-ē-kot-ul): the part of the embryonic shoot located above the cotyledons but below the tip of the shoot.

epidermal tissue: dermal tissue in plants that forms the epidermis, the outermost cell layer that covers young plants.

epidermis (ep-uh-der'-mis): in animals, specialized epithelial tissue that forms the outer layer of skin; in plants, the outermost layer of cells of a leaf, young root, or young stem.

epididymis (e-pi-di'-di-mus): a series of tubes that connect with and receive sperm from the seminiferous tubules of the testis.

epiglottis (ep-eh-glah'-tis): a flap of cartilage in the lower pharynx that covers the opening to the larynx during swallowing; directs food down the esophagus.

epinephrine (ep-i-nef'-rin): a hormone, secreted by the adrenal medulla, that is released in response to stress and that stimulates a variety of responses, including the release of glucose from liver and an increase in heart rate.

epithelial cell (eh-puh-thē'-lē-ul): the cell type that forms epithelial tissue.

epithelial tissue (eh-puh-thē'-lē-ul): a tissue type that forms membranes that cover the body surface and line body cavities, and that also gives rise to glands.

equilibrium population: a population in which allele frequencies and the distribution of genotypes do not change from generation to generation.

erythroblastosis fetalis (eh-rith'-rō-blas-tō'-sis fē-tal'-is): a condition in which the red blood cells of a newborn Rh-positive baby are attacked by antibodies produced by its Rh-negative mother, causing jaundice and anemia. Retardation and death are possible consequences if treatment is inadequate.

erythrocyte (eh-rith'-rō-sīt): a red blood cell, active in oxygen transport, that contains the red pigment hemoglobin.

erythropoietin (eh-rith'-rō-pō-ē'-tin): a hormone produced by the kidneys in response to oxygen deficiency that stimulates the production of red blood cells by the bone marrow.

esophagus (eh-sof'-eh-gus): a muscular passageway that conducts food from the pharynx to the stomach in humans and other mammals.

essential amino acid: an amino acid that is a required nutrient; the body is unable to manufacture essential amino acids, so they must be supplied in the diet.

essential fatty acid: a fatty acid that is a required nutrient; the body is unable to manufacture essential fatty acids, so they must be supplied in the diet.

estrogen: in vertebrates, a female sex hormone, produced by follicle cells of the ovary, that stimulates follicle development, oogenesis, the development of secondary sex characteristics, and growth of the uterine lining.

estuary: a wetland formed where a river meets the ocean; the salinity there is quite variable but lower than in sea water and higher than in fresh water.

ethology (ē-thol'-ō-jē): the study of animal behavior in natural or near-natural conditions.

ethylene: a plant hormone that promotes the ripening of fruits and the dropping of leaves and fruit.

euglenoid (ū'-gle-noid): a protist characterized by one or more whiplike flagella that are used for locomotion and by a photoreceptor that detects light. Euglenoids are photosynthetic, but if deprived of chlorophyll, some are capable of heterotrophic nutrition.

Eukarya (ū-kar'-ē-a): one of life's three domains; consists of all eukaryotes (plants, animals, fungi, and protists).

eukaryote (ū-kar'-ē-ōt): an organism whose cells are eukaryotic; plants, animals, fungi, and protists are eukaryotes.

eukaryotic (ū-kar-ē-ot'-ik): referring to cells of organisms of the domain Eukarya (plants, animals, fungi, and protists). Eukaryotic cells have genetic material enclosed within a membrane-bound nucleus and contain other membrane-bound organelles.

Eustachian tube (ū-stā'-shin): a tube connecting the middle ear with the pharynx; allows pressure between the middle ear and the atmosphere to equilibrate.

eutrophic lake: a lake that receives sufficiently large inputs of sediments, organic material, and inorganic nutrients from its surroundings to support dense communities; murky with poor light penetration.

evergreen: a plant that retains green leaves throughout the year.

evolution: the descent of modern organisms with modification from preexisting lifeforms; strictly speaking, any change in the proportions of different genotypes in a population from one generation to the next.

excretion: the elimination of waste substances from the body; can occur from the digestive system, skin glands, urinary system, or lungs.

excretory pore: an opening in the body wall of certain invertebrates, such as the earthworm, through which urine is excreted.

exergonic (ex-er-gon'-ik): pertaining to a chemical reaction that liberates energy (either as heat or in the form of increased entropy); a "downhill" reaction.

exhalation: the act of releasing air from the lungs, which results from a relaxation of the respiratory muscles.

exocrine gland: a gland that releases its secretions into ducts that lead to the outside of the body or into the digestive tract.

exocytosis (ex-ō-sī-tō'-sis): the process in which intracellular material is enclosed within a membrane-bound sac that moves to the plasma membrane and fuses with it, releasing the material outside the cell.

exon: a segment of DNA in a eukaryotic gene that codes for amino acids in a protein (see also *intron*).

exoskeleton (ex'-ō-skel'-uh-tun): a rigid external skeleton that supports the body, protects the internal organs, and has flexible joints that allow for movement.

exotic/exotic species: a foreign species introduced into an ecosystem where it did not evolve; such species may flourish and outcompete native species.

experiment: in the scientific method, the testing of a hypothesis by carefully controlled observations, leading to a conclusion.

exponential growth: a continuously accelerating increase in population size.

extensor: a muscle that straightens a joint.

external ear: in mammals, the parts of the ear outside the tympanic membrane; consists of the pinna and auditory canal.

external fertilization: the union of sperm and egg outside the body of either parent.

extinction: the death of all members of a species.

extracellular digestion: the physical and chemical breakdown of food that occurs outside a cell, normally in a digestive cavity.

extraembryonic membrane: in the embryonic development of reptiles, birds, and mammals, either the chorion, amnion, allantois, or yolk sac; functions in gas exchange, provision of the watery environment needed for development, waste storage, and storage of the yolk, respectively.

eyespot: a simple, lensless eye found in various invertebrates, including flatworms and jellyfish. Eyespots can distinguish light from dark and sometimes the direction of light, but they cannot form an image.

facilitated diffusion: the diffusion of molecules across a membrane, assisted by protein pores or carriers embedded in the membrane.

fairy ring: a circular pattern of mushrooms formed when reproductive structures erupt from the underground hyphae of a club fungus that has been growing outward in all directions from its original location.

family: the taxonomic category contained within an order and consisting of related genera.

farsighted: the inability to focus on nearby objects, caused by the eyeball being slightly too short or the cornea too flat.

fat (molecular): a lipid composed of three saturated fatty acids covalently bonded to glycerol; solid at room temperature.

fat (tissue): adipose tissue; connective tissue that stores fat; composed of cells packed with triglycerides.

fatty acid: an organic molecule composed of a long chain of carbon atoms, with a carboxylic acid (COOH) group at one end; may be saturated (all single bonds between the carbon atoms) or unsaturated (one or more double bonds between the carbon atoms).

feces: semisolid waste material that remains in the intestine after absorption is complete and is voided through the anus. Feces consist of indigestible wastes and bacteria.

feedback inhibition: in enzyme-mediated chemical reactions, the condition in which the product of a reaction inhibits one or more of the enzymes involved in synthesizing the product.

fermentation: anaerobic reactions that convert the pyruvic acid produced by glycolysis into lactic acid or alcohol and CO_2.

fertilization: the fusion of male and female haploid gametes, forming a zygote.

fetal alcohol syndrome (FAS): a cluster of symptoms, including retardation and physical abnormalities, that occur in infants born to mothers who consumed large amounts of alcoholic beverages during pregnancy.

fetus: the later stages of mammalian embryonic development (after the second month for humans), when the developing animal has come to resemble the adult of the species.

fever: an elevation in body temperature caused by chemicals (pyrogens) that are released by white blood cells in response to infection.

fibrillation: rapid, uncoordinated, and ineffective contractions of heart muscle cells.

fibrin (fī'-brin): a clotting protein formed in the blood in response to a wound; binds with other fibrin molecules and provides a matrix around which a blood clot forms.

fibrinogen (fī-brin'-ō-jen): the inactive form of the clotting protein fibrin. Fibrinogen is converted into fibrin by the enzyme thrombin, which is produced in response to injury.

fibrous root system: a root system, commonly found in monocots, characterized by many roots of approximately the same size arising from the base of the stem.

filament: in flowers, the stalk of a stamen, which bears an anther at its tip.

filtrate: the fluid produced by filtration; in the kidneys, the fluid produced by the filtration of blood through the glomerular capillaries.

filtration: within Bowman's capsule in each nephron of a kidney, the process by which blood is pumped under pressure through permeable capillaries of the glomerulus, forcing out water, dissolved wastes, and nutrients.

fimbria (fim'-brē-uh; pl., **fimbriae**): in female mammals, the ciliated, fingerlike projections of the oviduct that sweep the ovulated egg from the ovary into the oviduct.

first law of thermodynamics: the principle of physics that states that within any isolated system, energy can be neither created nor destroyed but can be converted from one form to another.

fission: asexual reproduction by dividing the body into two smaller, complete organisms.

fitness: the reproductive success of an organism, usually expressed in relation to the average reproductive success of all individuals in the same population.

flagellum (fla-jel'-um; pl., **flagella**): a long, hairlike extension of the plasma membrane; in eukaryotic cells, it contains microtubules arranged in a 9 + 2 pattern. The movement of flagella propel some cells through fluids.

flame cell: in flatworms, a specialized cell, containing beating cilia, that conducts water and wastes through the branching tubes that serve as an excretory system.

flexor: a muscle that flexes (decreases the angle of) a joint.

florigen: one of a group of plant hormones that can both trigger and inhibit flowering; daylength is a stimulus.

flower: the reproductive structure of an angiosperm plant.

fluid: a liquid or gas.

fluid mosaic model: a model of membrane structure; according to this model, membranes are composed of a double layer of phospholipids in which various proteins are embedded. The phospholipid bilayer is a somewhat fluid matrix that allows the movement of proteins within it.

follicle: in the ovary of female mammals, the oocyte and its surrounding accessory cells.

follicle-stimulating hormone (FSH): a hormone, produced by the anterior pituitary, that stimulates spermatogenesis in males and the development of the follicle in females.

food chain: a linear feeding relationship in a community, using a single representative from each of the trophic levels.

food vacuole: a membranous sac, within a single cell, in which food is enclosed. Digestive enzymes are released into the vacuole, where intracellular digestion occurs.

food web: a representation of the complex feeding relationships (in terms of interacting food chains) within a community, including many organisms at various trophic levels, with many of the consumers occupying more than one level simultaneously.

foraminiferan (for-am-i-nif'-er-un): an aquatic (largely marine) protist characterized by a typically elaborate calcium carbonate shell.

forebrain: during development, the anterior portion of the brain. In mammals, the forebrain differentiates into the thalamus, the limbic system, and the cerebrum. In humans, the cerebrum contains about half of all the neurons in the brain.

fossil: the remains of a dead organism, normally preserved in rock; may be petrified bones or wood; shells; impressions of body forms, such as feathers, skin, or leaves; or markings made by organisms, such as footprints.

fossil fuel: a fuel such as coal, oil, and natural gas, derived from the remains of ancient organisms.

founder effect: a type of genetic drift in which an isolated population founded by a small number of individuals may develop allele frequencies that are very different from those of the parent population as a result of chance inclusion of disproportionate numbers of certain alleles in the founders.

fovea (fō'-vē-uh): in the vertebrate retina, the central region on which images are focused; contains closely packed cones.

free-living: not parasitic.

free nerve ending: on some receptor neurons, a finely branched ending that responds to touch and pressure, to heat and cold, or to pain; produces the sensations of itching and tickling.

free nucleotides: nucleotides that have not been joined together to form a DNA or RNA strand.

free radical: a molecule with an unpaired electron, which makes it highly unstable and reactive with nearby molecules. By stealing an electron from the molecule it attacks, it creates a new free radical and begins a chain reaction that can lead to the destruction of biological molecules crucial to life.

fruit: in flowering plants, the ripened ovary (plus, in some cases, other parts of the flower), which contains the seeds.

fruiting body: a spore-forming reproductive structure of certain protists, bacteria, and fungi.

functional group: one of several groups of atoms commonly found in an organic molecule, including hydrogen, hydroxyl, amino, carboxyl, and phosphate groups, that determine the characteristics and chemical reactivity of the molecule.

gallbladder: a small sac, next to the liver, in which the bile secreted by the liver is stored and concentrated. Bile is released from the gallbladder to the small intestine through the bile duct.

gamete (gam'-ēt): a haploid sex cell formed in sexually reproducing organisms.

gametic incompatibility: the inability of sperm from one species to fertilize eggs of another species.

gametophyte (ga-mēt′-ō-fīt): the multicellular haploid stage in the life cycle of plants.

ganglion (gang′-lē-un): a cluster of neurons.

ganglion cell: a type of cell, of which the innermost layer of the vertebrate retina is composed, whose axons form the optic nerve.

gap junction: a type of cell-to-cell junction in animals in which channels connect the cytoplasm of adjacent cells.

gas-exchange portion: the portion of the respiratory system in lung-breathing vertebrates where gas is exchanged in the alveoli of the lungs.

gastric inhibitory peptide: a hormone, produced by the small intestine, that inhibits the activity of the stomach.

gastrin: a hormone, produced by the stomach, that stimulates acid secretion in response to the presence of food.

gastrovascular cavity: a saclike chamber with digestive functions, found in simple invertebrates; a single opening serves as both mouth and anus, and the chamber provides direct access of nutrients to the cells.

gastrula (gas′-troo-luh): in animal development, a three-layered embryo with ectoderm, mesoderm, and endoderm cell layers. The endoderm layer normally encloses the primitive gut.

gastrulation (gas-troo-la′-shun): the process whereby a blastula develops into a gastrula, including the formation of endoderm, ectoderm, and mesoderm.

gel electrophoresis: a technique in which molecules (such as DNA fragments) are placed on restricted tracks in a thin sheet of gelatinous material and exposed to an electric field; the molecules then migrate at a rate determined by certain characteristics, such as size.

gene: the unit of heredity; a segment of DNA located at a particular place on a chromosome that encodes the information for the amino acid sequence of a protein, and hence particular traits.

gene flow: the movement of alleles from one population to another owing to the migration of individual organisms.

gene pool: the total of all alleles of all genes in a population; for a single gene, the total of all the alleles of that gene that occur in a population.

generative cell: in flowering plants, one of the haploid cells of a pollen grain; undergoes mitosis to form two sperm cells.

genetically modified organism (GMO): an organism that has been produced through the techniques of genetic engineering.

genetic code: the collection of codons of mRNA, each of which directs the incorporation of a particular amino acid into a protein during protein synthesis.

genetic drift: a change in the allele frequencies of a small population purely by chance.

genetic engineering: the modification of genetic material to achieve specific goals.

genetic equilibrium: a state in which the allele frequencies and the distribution of genotypes of a population do not change from generation to generation.

genetic recombination: the generation of new combinations of alleles on homologous chromosomes due to the exchange of DNA during crossing over.

genital herpes: a sexually transmitted disease, caused by a virus, that can cause painful blisters on the genitals and surrounding skin.

genital warts: a sexually transmitted disease, caused by a virus, that forms growths or bumps on the external genitalia, in or around the vagina or anus, or on the cervix in females or penis, scrotum, groin, or thigh in males.

genome (jē′-nōm): the entire set of genes carried by a member of any given species.

genotype (jēn′-ō-tip): the genetic composition of an organism; the actual alleles of each gene carried by the organism.

genus (jē-nus): the taxonomic category contained within a family and consisting of very closely related species.

geographical isolation: the separation of two populations by a physical barrier.

germ layer: a tissue layer formed during early embryonic development.

germination: the growth and development of a seed, spore, or pollen grain.

gherelin: a hormone produced by the stomach and upper small intestine when food is absent, stimulating hunger.

gibberellin (jib-er-el′-in): a plant hormone that stimulates seed germination, fruit development, and cell division and elongation.

gill: in aquatic animals, a branched tissue richly supplied with capillaries around which water is circulated for gas exchange.

gizzard: a muscular organ, found in earthworms and birds, in which food is mechanically broken down prior to chemical digestion.

gland: a cluster of cells that are specialized to secrete substances such as sweat or hormones.

glial cell: a cell of the nervous system that provides support and insulation for neurons.

global warming: a gradual rise in global atmospheric temperature as a result of an amplification of the natural greenhouse effect due to human activities.

glomerulus (glō-mer′-ū-lus): a dense network of thin-walled capillaries, located within the Bowman's capsule of each nephron of the kidney, where blood pressure forces water and dissolved nutrients through capillary walls for filtration by the nephron.

glucagon (gloo′-ka-gon): a hormone, secreted by the pancreas, that increases blood sugar by stimulating the breakdown of glycogen (to glucose) in the liver.

glucocorticoid (gloo-kō-kor′-tik-oid): a class of hormones, released by the adrenal cortex in response to the presence of ACTH, that make additional energy available to the body by stimulating the synthesis of glucose.

glucose: the most common monosaccharide, with the molecular formula $C_6H_{12}O_6$; most polysaccharides, including cellulose, starch, and glycogen, are made of glucose subunits covalently bonded together.

glycerol (glis′-er-ol): a three-carbon alcohol to which fatty acids are covalently bonded to make fats and oils.

glycogen (gli′-kō-jen): a long, branched polymer of glucose that is stored by animals in the muscles and liver and metabolized as a source of energy.

glycolysis (gli-kol′-i-sis): reactions, carried out in the cytoplasm, that break down glucose into two molecules of pyruvic acid, producing two ATP molecules; does not require oxygen but can proceed when oxygen is present.

glycoprotein: a protein to which a carbohydrate is attached.

goiter: a swelling of the neck caused by iodine deficiency, which affects the functioning of the thyroid gland and its hormones.

Golgi complex (gōl′-jē): a stack of membranous sacs, found in most eukaryotic cells, that is the site of processing and separation of membrane components and secretory materials.

gonad: an organ where reproductive cells are formed; in males, the testes, and in females, the ovaries.

gonadotropin-releasing hormone (GnRH): a hormone produced by the neurosecretory cells of the hypothalamus, which stimulates cells in the anterior pituitary to release FSH and LH. GnRH is involved in the menstrual cycle and in spermatogenesis.

gonorrhea (gon-uh-rē′-uh): a sexually transmitted bacterial infection of the reproductive organs; if untreated, can result in sterility.

gradient: a difference in concentration, pressure, or electrical charge between two regions.

Gram stain: a stain that is selectively taken up by the cell walls of certain types of bacteria (gram-positive bacteria) and rejected by the cell walls of others (gram-negative bacteria); used to distinguish bacteria on the basis of their cell wall construction.

granum (gra′-num; pl., grana): a stack of thylakoids in chloroplasts.

grassland: a biome, located in the centers of continents, that supports grasses; also called *prairie*.

gravitropism: growth with respect to the direction of gravity.

gray crescent: in frog embryonic development, an area of intermediate pigmentation in the fertilized egg; contains gene-regulating substances required for the normal development of the tadpole.

gray matter: the outer portion of the brain and inner region of the spinal cord; composed largely of neuron cell bodies, which give this area a gray color.

greenhouse effect: the process in which certain gases such as carbon dioxide and methane trap sunlight energy in a planet's atmosphere as heat; the glass in a greenhouse does the same. The result, global warming, is being enhanced by the production of these gases by humans.

greenhouse gas: a gas, such as carbon dioxide or methane, that traps sunlight energy in a planet's atmosphere as heat; a gas that participates in the greenhouse effect.

ground tissue system: a plant tissue system consisting of parenchyma, collenchyma, and sclerenchyma cells that makes up the bulk of a leaf or young stem, excluding vascular or dermal tissues. Most ground tissue cells function in photosynthesis, support, or carbohydrate storage.

growth hormone: a hormone, released by the anterior pituitary, that stimulates growth, especially of the skeleton.

growth rate: a measure of the change in population size per individual per unit of time.

guanine: a nitrogenous base found in both DNA and RNA; abbreviated as *G*.

guard cell: one of a pair of specialized epidermal cells surrounding the central opening of a stoma of a leaf, which regulates the size of the opening.

gymnosperm (jim′-nō-sperm): a nonflowering seed plant, such as a conifer, cycad, or gingko.

gyre (jīr): a roughly circular pattern of ocean currents, formed because continents interrupt the currents' flow; rotates clockwise in the Northern Hemisphere and counterclockwise in the Southern Hemisphere.

habitat fragmentation: the process by which human development and activities produce patches of wildlife habitat that may not be large enough to sustain viable populations.

habituation (heh-bich-oo-ā′-shun): simple learning characterized by a decline in response to a harmless, repeated stimulus.

hair cell: a type of mechanoreceptor cell found in the inner ear that produces an electrical signal when stiff "hairlike" cilia projecting from the surface of the cell are bent. Hair cells in the cochlea respond to sound vibrations; those in the vestibular system respond to motion and gravity.

hair follicle: a gland in the dermis of mammalian skin, formed from epithelial tissue, that produces a hair.

halophile (hā′-lō-fīl): literally, "salt-loving"; an organism that thrives in salty conditions.

haploid (hap′-loid): referring to a cell that has only one member of each pair of homologous chromosomes.

Hardy-Weinberg principle: a mathematical model proposing that, under certain conditions, the allele frequencies and genotype frequencies in a sexually reproducing population will remain constant over generations.

Haversian system (ha-ver′-sē-un): see *osteon*.

head: the anteriormost segment of an animal with segmentation.

heart: a muscular organ responsible for pumping blood within the circulatory system throughout the body.

heart attack: a severe reduction or blockage of blood flow through a coronary artery, depriving some of the heart muscle of its blood supply.

heartwood: older xylem that contributes to the strength of a tree trunk.

heat of fusion: the energy that must be removed from a compound to transform it from a liquid into a solid at its freezing temperature.

heat of vaporization: the energy that must be supplied to a compound to transform it from a liquid into a gas at its boiling temperature.

heliozoan (hē-lē-ē-ō-zō′-un): an aquatic (largely freshwater) animal-like protist; some have elaborate silica-based shells.

helix (hē′-liks): a coiled, springlike secondary structure of a protein.

helper T cell: a type of T cell that helps other immune cells recognize and act against antigens.

hemocoel (hē′-mō-sēl): a blood cavity within the bodies of certain invertebrates in which blood bathes tissues directly; part of an open circulatory system.

hemodialysis (hē-mō-di-al′-luh-sis): a procedure that simulates kidney function in individuals with damaged or ineffective kidneys; blood is diverted from the body, artificially filtered, and returned to the body.

hemoglobin (hē′-mō-glō-bin): the iron-containing protein that gives red blood cells their color; binds to oxygen in the lungs and releases it to the tissues.

hemophilia: a recessive, sex-linked disease in which the blood fails to clot normally.

herbivore (erb′-i-vor): literally, "plant-eater"; an organism that feeds directly and exclusively on producers; a primary consumer.

hermaphrodite (her-maf′-ruh-dīt′): an organism that possesses both male and female sexual organs.

hermaphroditic (her-maf′-ruh-dīt′-ik): possessing both male and female sexual organs. Some hermaphroditic animals can fertilize themselves; others must exchange sex cells with a mate.

heterotroph (het′-er-ō-trōf′): literally, "other-feeder"; an organism that eats other organisms; a consumer.

heterozygous (het-er-ō-zī′-gus): carrying two different alleles of a given gene; also called *hybrid*.

hindbrain: the posterior portion of the brain, containing the medulla, pons, and cerebellum.

hinge joint: a joint at which one bone is moved by muscle and the other bone remains fixed, such as in the knee, elbow, or fingers; allows movement in only two dimensions.

hippocampus (hip-ō-kam′-pus): the part of the forebrain of vertebrates that is important in emotion and especially learning.

histamine: a substance released by certain cells in response to tissue damage and invasion of the body by foreign substances; promotes the dilation of arterioles and the leakiness of capillaries and triggers some of the events of the inflammatory response.

homeobox (hō′-mē-ō-boks): a sequence of DNA coding for special, 60-amino-acid proteins, which activate or inactivate genes that control development; these sequences specify embryonic cell differentiation.

homeostasis (hōm-ē-ō-stā′sis): the maintenance of a relatively constant environment required for the optimal functioning of cells, maintained by the coordinated activity of numerous regulatory mechanisms, including the respiratory, endocrine, circulatory, and excretory systems.

hominid: a human or a prehistoric relative of humans, beginning with the Australopithecines, whose fossils date back at least 4.4 million years.

homologous structures: structures that may differ in function but that have similar anatomy, presumably because the organisms that possess them have descended from common ancestors.

homologue (hō-′mō-log): a chromosome that is similar in appearance and genetic information to another chromosome with which it pairs during meiosis; also called *homologous chromosome*.

homozygous (hō-mō-zī′-gus): carrying two copies of the same allele of a given gene; also called *true-breeding*.

hormone: a chemical that is synthesized by one group of cells, secreted, and then carried in the bloodstream to other cells, whose activity is influenced by reception of the hormone.

host: the prey organism on or in which a parasite lives; is harmed by the relationship.

human immunodeficiency virus (HIV): a pathogenic virus that causes acquired immune deficiency syndrome (AIDS) by attacking and destroying the immune system's helper T cells.

humoral immunity: an immune response in which foreign substances are inactivated or destroyed by antibodies that circulate in the blood.

Huntington disease: an incurable genetic disorder, caused by a dominant allele, that produces progressive brain deterioration, resulting in the loss of motor coordination, flailing movements, personality disturbances, and eventual death.

hybrid: an organism that is the offspring of parents differing in at least one genetically determined characteristic; also used to refer to the offspring of parents of different species.

hybrid infertility: reduced fertility (typically, complete sterility) in the hybrid offspring of two species.

hybrid inviability: the failure of a hybrid offspring of two species to survive to maturity.

hybridoma: a cell produced by fusing an antibody-producing cell with a myeloma cell; used to produce monoclonal antibodies.

hydrogen bond: the weak attraction between a hydrogen atom that bears a partial positive charge (due to polar covalent bonding with another atom) and another atom, normally oxygen or nitrogen, that bears a partial negative charge; hydrogen bonds may form between atoms of a single molecule or different molecules.

hydrologic cycle: the water cycle, driven by solar energy; a nutrient cycle in which the main reservoir of water is the ocean and most of the water remains in the form of water throughout the cycle (rather than being used in the synthesis of new molecules).

hydrolysis (hī-drol′-i-sis): the chemical reaction that breaks a covalent bond by means of the addition of hydrogen to the atom on one side of the original bond and a hydroxyl group to the atom on the other side; the reverse of dehydration synthesis.

hydrophilic (hī-drō-fil′-ik): pertaining to a substance that issolves readily in water or to parts of a large molecule that form hydrogen bonds with water.

hydrophobic (hī-drō-fō′-bik): pertaining to a substance that does not dissolve in water.

hydrophobic interaction: the tendency for hydrophobic molecules to cluster together when immersed in water.

hydrostatic skeleton (hī-drō-stat′-ik): a body type that uses fluid contained in body compartments to provide support and mass against which muscles can contract.

hydrothermal vent community: a community of unusual organisms, living in the deep ocean near hydrothermal vents, that depends on the chemosynthetic activities of sulfur bacteria.

hypertension: arterial blood pressure that is chronically elevated above the normal level.

hypertonic (hī-per-ton′-ik): referring to a solution that has a higher concentration of dissolved particles (and therefore a lower concentration of free water) than has the cytoplasm of a cell.

hypha (hī'-fuh; pl., **hyphae**): a threadlike structure that consists of elongated cells, typically with many haploid nuclei; many hyphae make up the fungal body.

hypocotyl (hī'-pō-kot-ul): the part of the embryonic shoot located below the cotyledons but above the root.

hypothalamus (hī-pō-thal'-a-mus): a region of the brain that controls the secretory activity of the pituitary gland; synthesizes, stores, and releases certain peptide hormones; directs autonomic nervous system responses.

hypothesis (hī-poth'-eh-sis): in the scientific method, a supposition based on previous observations that is offered as an explanation for the observed phenomenon and is used as the basis for further observations, or experiments.

hypotonic (hī-pō-ton'-ik): referring to a solution that has a lower concentration of dissolved particles (and therefore a higher concentration of free water) than has the cytoplasm of a cell.

immigration (im-uh-grā'-shun): migration of individuals into an area.

immune response: a specific response by the immune system to the invasion of the body by a particular foreign substance or microorganism, characterized by the recognition of the foreign substance by immune cells and its subsequent destruction by antibodies or by cellular attack.

immune system: cells such as macrophages, B cells, and T cells and molecules such as antibodies that work together to combat microbial invasion of the body.

imperfect fungus: a fungus of the division Deuteromycota; no species in this division has been observed to form sexual reproductive structures.

implantation: the process whereby the early embryo embeds itself within the lining of the uterus.

imprinting: the process by which an animal forms an association with another animal or object in the environment during a sensitive period of development.

inclusive fitness: the reproductive success of all organisms that bear a given allele, normally expressed in relation to the average reproductive success of all individuals in the same population; compare with *fitness*.

incomplete dominance: a pattern of inheritance in which the heterozygous phenotype is intermediate between the two homozygous phenotypes.

incomplete flower: a flower that is missing one of the four floral parts (sepals, petals, stamens, or carpels).

independent assortment: see *law of independent assortment*.

indirect development: a developmental pathway in which an offspring goes through radical changes in body form as it matures.

induction: the process by which a group of cells causes other cells to differentiate into a specific tissue type.

inductive reasoning: the process of creating a generalization based on many specific observations that support the generalization, coupled with an absence of observations that contradict it.

inflammatory response: a nonspecific, local response to injury to the body, characterized by the phagocytosis of foreign substances and tissue debris by white blood cells and by the walling off of the injury site by the clotting of fluids that escape from nearby blood vessels.

inhalation: the act of drawing air into the lungs by enlarging the chest cavity.

inheritance: the genetic transmission of characteristics from parent to offspring.

inheritance of acquired characteristics: the hypothesis that organisms' bodies change during their lifetimes by use and disuse and that these changes are inherited by their offspring.

inhibiting hormone: a hormone, secreted by the neurosecretory cells of the hypothalamus, that inhibits the release of specific hormones from the anterior pituitary.

innate (in-āt'): inborn; instinctive; determined by the genetic makeup of the individual.

inner cell mass: in human embryonic development, the cluster of cells, on one side of the blastocyst, that will develop into the embryo.

inner ear: the innermost part of the mammalian ear; composed of the bony, fluid-filled tubes of the cochlea and the vestibular apparatus.

inorganic: describing any molecule that does not contain both carbon and hydrogen.

insertion: the site of attachment of a muscle to the relatively movable bone on one side of a joint.

insertion mutation: a mutation in which one or more pairs of nucleotides are inserted into a gene.

insight learning: a complex form of learning that requires the manipulation of mental concepts to arrive at adaptive behavior.

instinctive: innate; inborn; determined by the genetic makeup of the individual.

insulin: a hormone, secreted by the pancreas, that lowers blood sugar by stimulating many cells to take up glucose and by stimulating the liver to convert glucose to glycogen.

integration: in nerve cells, the process of adding up electrical signals from sensory inputs or other nerve cells to determine the appropriate outputs.

integument (in-teg'-ū-ment): in plants, the outer layers of cells of the ovule that surround the embryo sac; develops into the seed coat.

intensity: the strength of stimulation or response.

interferon: a protein released by certain virus-infected cells that increases the resistance of other, uninfected, cells to viral attack.

intermediate filament: part of the cytoskeleton of eukaryotic cells that probably functions mainly for support and is composed of several types of proteins.

intermembrane compartment: the fluid-filled space between the inner and outer membranes of a mitochondrion.

internal fertilization: the union of sperm and egg inside the body of the female.

internode: the part of a stem between two nodes.

interphase: the stage of the cell cycle between cell divisions; the stage in which chromosomes are replicated and other cell functions occur, such as growth, movement, and acquisition of nutrients.

interspecific competition: competition among individuals of different species.

interstitial cell (in-ter-sti'-shul): in the vertebrate testis, a testosterone-producing cell located between the seminiferous tubules.

interstitial fluid (in-ter-sti'-shul): fluid, similar in composition to plasma (except lacking large proteins), that leaks from capillaries and acts as a medium of exchange between the body cells and the capillaries.

intertidal zone: an area of the ocean shore that is alternately covered and exposed by the tides.

intervertebral disc (in-ter-ver-tē'-brul): a pad of cartilage between two vertebrae that acts as a shock absorber.

intracellular digestion: the chemical breakdown of food within single cells.

intraspecific competition: competition among individuals of the same species.

intrauterine device (IUD): a small copper or plastic loop, squiggle, or shield that is inserted in the uterus; a contraceptive method that works by irritating the uterine lining so that it cannot receive the embryo.

intron: a segment of DNA in a eukaryotic gene that does not code for amino acids in a protein.

invasive species: organisms with a high biotic potential that are introduced (deliberately or accidentally) into ecosystems where they did not evolve and where they encounter little environmental resistance and tend to displace native species.

invertebrate (in-vert'-uh-bret): an animal that never possesses a vertebral column.

ion (ī'-on): a charged atom or molecule; an atom or molecule that has either an excess of electrons (and hence is negatively charged) or has lost electrons (and is positively charged).

ionic bond: a chemical bond formed by the electrical attraction between positively and negatively charged ions.

iris: the pigmented muscular tissue of the vertebrate eye that surrounds and controls the size of the pupil, through which light enters.

islet cell: a cell in the endocrine portion of the pancreas that produces either insulin or glucagon.

isolating mechanism: a morphological, physiological, behavioral, or ecological difference that prevents members of two species from interbreeding.

isotonic (ī-sō-ton'-ik): referring to a solution that has the same concentration of dissolved particles (and therefore the same concentration of free water) as has the cytoplasm of a cell.

isotope: one of several forms of a single element, the nuclei of which contain the same number of protons but different numbers of neutrons.

J-curve: the J-shaped growth curve of an exponentially growing population in which increasing numbers of individuals join the population during each succeeding time period.

joint: a flexible region between two rigid units of an exoskeleton or endoskeleton, allowing for movement between the units.

karyotype: a preparation showing the number, sizes, and shapes of all chromosomes within a cell and, therefore, within the individual or species from which the cell was obtained.

keratin (ker'-uh-tin): a fibrous protein in hair, nails, and the epidermis of skin.

keystone species: a species whose influence on community structure is greater than its abundance would suggest.

kidney: one of a pair of organs of the excretory system that is located on either side of the spinal column and filters blood, removing wastes and regulating the composition and water content of the blood.

kinetic energy: the energy of movement; includes light, heat, mechanical movement, and electricity.

kinetochore (ki-net′-ō-kor): a protein structure that forms at the centromere regions of chromosomes; attaches the chromosomes to the spindle.

kingdom: the second broadest taxonomic category, contained within a domain and consisting of related phyla or divisions.

kin selection: a type of natural selection that favors a certain allele because it increases the survival or reproductive success of relatives that bear the same allele.

Klinefelter syndrome: a set of characteristics typically found in individuals who have two X chromosomes and one Y chromosome; these individuals are phenotypically males but are sterile and have several femalelike traits, including broad hips and partial breast development.

Krebs cycle: a cyclic series of reactions, occurring in the matrix of mitochondria, in which the acetyl groups from the pyruvic acids produced by glycolysis are broken down to CO_2, accompanied by the formation of ATP and electron carriers; also called *citric acid cycle*.

kuru: a degenerative brain disease, first discovered in the cannibalistic Fore tribe of New Guinea, that is caused by a prion.

labium (pl., labia): one of a pair of folds of skin of the external structures of the mammalian female reproductive system.

labor: a series of contractions of the uterus that result in birth.

lactation: the secretion of milk from the mammary glands.

lacteal (lak-tēl′): a single lymph capillary that penetrates each villus of the small intestine.

lactose (lak′-tōs): a disaccharide composed of glucose and galactose; found in mammalian milk.

lactose intolerance: inadequate ability to break down milk sugar caused by low secretion of lactase. Symptoms include bloating, cramps, and diarrhea after consuming many milk products.

La Niña (la nēn′-ya): literally "girl child"; a reversal of the El Niño weather pattern.

large intestine: the final section of the digestive tract; consists of the colon and the rectum.

larva (lar′-vuh): an immature form of an organism with indirect development prior to metamorphosis into its adult form; includes the caterpillars of moths and butterflies and the maggots of flies.

larynx (lar′-inks): that portion of the air passage between the pharynx and the trachea; contains the vocal cords.

lateral bud: a cluster of meristematic cells at the node of a stem; under appropriate conditions, it grows into a branch.

lateral meristem: a meristematic tissue that forms cylinders parallel to the long axis of roots and stems; normally located between the primary xylem and primary phloem (vascular cambium) and just outside the phloem (cork cambium); also called *cambium*.

law of independent assortment: the independent inheritance of two or more distinct traits; states that the alleles for one trait may be distributed to the gametes independently of the alleles for other traits.

law of segregation: Gregor Mendel's conclusion that each gamete receives only one of each parent's pair of genes for each trait.

laws of thermodynamics: the physical laws that define the basic properties and behavior of energy.

leaf: an outgrowth of a stem, normally flattened and photosynthetic.

leaf primordium (pri-mor′-dē-um; pl., primor-dia): a cluster of meristem cells, located at the node of a stem, that develops into a leaf.

learning: an adaptive change in behavior as a result of experience.

legume (leg′-ūm): a member of a family of plants characterized by root swellings in which nitrogen-fixing bacteria are housed; includes soybeans, lupines, alfalfa, and clover.

lens: a clear object that bends light rays; in eyes, a flexible or movable structure used to focus light on a layer of photoreceptor cells.

leptin: a peptide hormone. One of the functions of leptin, which is released by fat cells, is to help the body monitor its fat stores and regulate weight.

leukocyte (loo′-kō-sit): any of the white blood cells circulating in the blood.

lichen (li′-ken): a symbiotic association between an alga or cyanobacterium and a fungus, resulting in a composite organism.

life cycle: the events in the life of an organism from one generation to the next.

life table: a data table that groups organisms born at the same time and tracks them throughout their life span, recording how many continue to survive in each succeeding year (or other unit of time). Various parameters such as sex may be used in the groupings. Human life tables may include many other parameters (such as socioeconomic status) used by demographers.

ligament: a tough connective tissue band connecting two bones.

light-dependent reactions: the first stage of photosynthesis, in which the energy of light is captured as ATP and NADPH; occurs in thylakoids of chloroplasts.

light-harvesting complex: in photosystems, the assembly of pigment molecules (chlorophyll and accessory pigments) that absorb light energy and transfer that energy to electrons.

light-independent reactions: the second stage of photosynthesis, in which the energy obtained by the light-dependent reactions is used to fix carbon dioxide into carbohydrates; occurs in the stroma of chloroplasts.

lignin: a hard material that is embedded in the cell walls of vascular plants and provides support in terrestrial species; an early and important adaptation to terrestrial life.

limbic system: a diverse group of brain structures, mostly in the lower forebrain, that includes the thalamus, hypothalamus, amygdala, hippocampus, and parts of the cerebrum and is involved in basic emotions, drives, behaviors, and learning.

limnetic zone: a lake zone in which enough light penetrates to support photosynthesis.

linkage: the inheritance of certain genes as a group because they are parts of the same chromosome. Linked genes do not show independent assortment.

lipase (li′-pās): an enzyme that catalyzes the breakdown of lipids such as fats.

lipid (li′-pid): one of a number of organic molecules containing large nonpolar regions composed solely of carbon and hydrogen, which make lipids hydrophobic and insoluble in water; includes oils, fats, waxes, phospholipids, and steroids.

littoral zone: a lake zone, near the shore, in which water is shallow and plants find abundant light, anchorage, and adequate nutrients.

liver: an organ with varied functions, including bile production, glycogen storage, and the detoxification of poisons.

lobefin: a member of the fish order Sarcopterygii, which includes coelacanths and lungfishes. Ancestors of today's lobefins gave rise to the first amphibians, and thus ultimately to all tetrapod vertebrates.

local hormones: a general term for messenger molecules produced by most cells and released into the cells' immediate vicinity. Local hormones, which include prostaglandins and cytokines, influence nearby cells bearing appropriate receptors.

locus: the physical location of a gene on a chromosome.

long-day plant: a plant that will flower only if the length of daylight is greater than some species-specific duration.

long-night plant: a plant that will flower only if the duration of uninterrupted darkness is longer than some species-specific duration (sometimes called a *short-day plant*).

long-term memory: the second phase of learning; a more-or-less permanent memory formed by a structural change in the brain, brought on by repetition.

loop of Henle (hen′-lē): a specialized portion of the tubule of the nephron in birds and mammals that creates an osmotic concentration gradient in the fluid immediately surrounding it. This gradient in turn makes possible the production of urine more osmotically concentrated than blood plasma.

lung: a paired respiratory organ in which gas exchange occurs, consisting of inflatable chambers within the chest cavity.

luteinizing hormone (LH): a hormone, produced by the anterior pituitary, that stimulates testosterone production in males and the development of the follicle, ovulation, and the production of the corpus luteum in females.

lymph (limf): a pale fluid, within the lymphatic system, that is composed primarily of interstitial fluid and lymphocytes.

lymphatic system: a system consisting of lymph vessels, lymph capillaries, lymph nodes, and the thymus and spleen; helps protect the body against infection, absorbs fats, and returns excess fluid and small proteins to the blood circulatory system.

lymph node: a small structure that filters lymph; contains lymphocytes and macrophages, which inactivate foreign particles such as bacteria.

lymphocyte (lim′-fō-sit): a type of white blood cell important in the immune response.

lysosome (li′-sō-sōm): a membrane-bound organelle containing intracellular digestive enzymes.

macronutrient: a nutrient needed in relatively large quantities (often defined as making up more than 0.1% of an organism's body).

macrophage (mak′-rō-fāj): a type of white blood cell that engulfs microbes and destroys them by phagocytosis; also presents microbial antigens to T cells, helping stimulate the immune response.

magnetotactic: able to detect and respond to Earth's magnetic field.

major histocompatibility complex (MHC): proteins, normally located on the surfaces of body cells, that identify the cell as "self"; also important in stimulating and regulating the immune response.

maltose (mal′-tōs): a disaccharide composed of two glucose molecules.

mammal: a member of the chordate class Mammalia, which includes vertebrates with hair and mammary glands.

mammary gland (mam′-uh-rē): a milk-producing gland used by female mammals to nourish their young.

mantle (man′-tul): an extension of the body wall in certain invertebrates, such as mollusks; may secrete a shell, protect the gills, and, as in cephalopods, aid in locomotion.

marsupial (mar-soo′-pē-ul): a mammal whose young are born at an extremely immature stage and undergo further development in a pouch while they remain attached to a mammary gland; includes kangaroos, opossums, and koalas.

mass extinction: a relatively sudden loss of many forms of life as a result of an environmental change. The fossil record reveals five mass extinctions over geologic time.

mast cell: a cell of the immune system that synthesizes histamine and other molecules used in the body's response to trauma and that are a factor in allergic reactions.

matrix: the fluid contained within the inner membrane of a mitochondrion.

mechanical incompatibility: the inability of male and female organisms to exchange gametes, normally because their reproductive structures are incompatible.

mechanoreceptor: a receptor that responds to mechanical deformation, such as that caused by pressure, touch, or vibration.

medulla (med-ū′-luh): the part of the hindbrain of vertebrates that controls automatic activities such as breathing, swallowing, heart rate, and blood pressure.

medusa (meh-doo′-suh): a bell-shaped, typically free-swimming stage in the life cycle of many cnidarians; includes jellyfish.

megakaryocyte (meg-a-kar′-ē-ō-sit): a large cell type that remains in the bone marrow, pinching off pieces of itself that then enter the circulation as platelets.

megaspore: a haploid cell formed by meiosis from a diploid megaspore mother cell; through mitosis and differentiation, develops into the female gametophyte.

megaspore mother cell: a diploid cell, within the ovule of a flowering plant, that undergoes meiosis to produce four haploid megaspores.

meiosis (mī-ō′-sis): in eukaryotic organisms, a type of nuclear division in which a diploid nucleus divides twice to form four haploid nuclei.

meiotic cell division: meiosis followed by cytokinesis.

melanocyte-stimulating hormone (me-lan′-ō-sit): a hormone, released by the anterior pituitary, that regulates the activity of skin pigments in some vertebrates.

melatonin (mel-uh-tōn′-in): a hormone, secreted by the pineal gland, that is involved in the regulation of circadian cycles.

membrane: in multicellular organisms, a continuous sheet of epithelial cells that covers the body and lines body cavities; in a cell, a thin sheet of lipids and proteins that surrounds the cell or its organelles, separating them from their surroundings.

memory B cell: a type of white blood cell that is produced as a result of the binding of an antibody on a B cell to an antigen on an invading microorganism. Memory B cells persist in the bloodstream and provide future immunity to invaders bearing that antigen.

memory T cell: a type of white blood cell that is produced as a result of the binding of a receptor on a T cell to an antigen on an invading microorganism. Memory T cells persist in the bloodstream and provide future immunity to invaders bearing that antigen.

meninges (men-in′-jēz): three layers of connective tissue that surround the brain and spinal cord.

menstrual cycle: in human females, a complex 28-day cycle during which hormonal interactions among the hypothalamus, pituitary gland, and ovary coordinate ovulation and the preparation of the uterus to receive and nourish the fertilized egg. If pregnancy does not occur, the uterine lining is shed during menstruation.

menstruation: in human females, the monthly discharge of uterine tissue and blood from the uterus.

meristem cell (mer′-i-stem): an undifferentiated cell that remains capable of cell division throughout the life of a plant.

mesoderm (mēz′-ō-derm): the middle embryonic tissue layer, lying between the endoderm and ectoderm, and normally the last to develop; gives rise to structures such as muscle and skeleton.

mesoglea (mez-ō-glē′-uh): a middle, jelly-like layer within the body wall of cnidarians.

mesophyll (mez′-ō-fil): loosely packed parenchyma cells under the epidermis of a leaf.

messenger RNA (mRNA): a strand of RNA, complementary to the DNA of a gene, that conveys the genetic information in DNA to the ribosomes to be used during protein synthesis; sequences of three bases (codons) in mRNA specify particular amino acids to be incorporated into a protein.

metabolic pathway: a sequence of chemical reactions within a cell, in which the products of one reaction are the reactants for the next reaction.

metabolism: the sum of all chemical reactions that occur within a single cell or within all the cells of a multicellular organism.

metamorphosis (met-a-mor′-fō-sis): in animals with indirect development, a radical change in body form from larva to sexually mature adult, as seen in amphibians (tadpole to frog) and insects (caterpillar to butterfly).

metaphase (met′-a-fāz): the stage of mitosis in which the chromosomes, attached to spindle fibers at kinetochores, are lined up along the equator of the cell.

methanogen (me-than′-ō-jen): a type of anaerobic archaean capable of converting carbon dioxide to methane.

microbe: a microorganism.

microevolution: change over successive generations in the composition of a population's gene pool.

microfilament: part of the cytoskeleton of eukaryotic cells that is composed of the proteins actin and (in some cases) myosin; functions in the movement of cell organelles and in locomotion by extension of the plasma membrane.

micronutrient: a nutrient needed only in small quantities (often defined as making up less than 0.01% of an organism's body).

microsphere: a small, hollow sphere formed from proteins or proteins complexed with other compounds.

microspore: a haploid cell formed by meiosis from a microspore mother cell; through mitosis and differentiation, develops into the male gametophyte.

microspore mother cell: a diploid cell contained within an anther of a flowering plant, which undergoes meiosis to produce four haploid microspores.

microtubule: a hollow, cylindrical strand, found in eukaryotic cells, that is composed of the protein tubulin; part of the cytoskeleton used in the movement of organelles, cell growth, and the construction of cilia and flagella.

microvillus (mī-krō-vi′-lus; pl., microvilli): a microscopic projection of the plasma membrane of each villus; increases the surface area of the villus.

midbrain: during development, the central portion of the brain; contains an important relay center, the reticular formation.

middle ear: the part of the mammalian ear composed of the tympanic membrane, the Eustachian tube, and three bones (hammer, anvil, and stirrup) that transmit vibrations from the auditory canal to the oval window.

middle lamella: a thin layer of sticky polysaccharides, such as pectin, and other carbohydrates that separates and holds together the primary cell walls of adjacent plant cells.

mimicry (mim′-ik-rē): the situation in which a species has evolved to resemble something else—typically another type of organism.

mineral: an inorganic substance, especially one in rocks or soil.

minimum viable population (MVP): the smallest isolated population that can persist indefinitely and survive natural events such as fires and floods.

mitochondrion (mī-tō-kon′-drē-un): an organelle, bounded by two membranes, that is the site of the reactions of aerobic metabolism.

mitosis (mī-tō′-sis): a type of nuclear division, used by eukaryotic cells, in which one copy of each chromosome (already duplicated during interphase before mitosis) moves into each of two daughter nuclei; the daughter nuclei are therefore genetically identical to each other.

mitotic cell division: mitosis followed by cytokinesis.

molecule (mol′-e-kūl): a particle composed of one or more atoms held together by chemical bonds; the smallest particle of a compound that displays all the properties of that compound.

molt: to shed an external body covering, such as an exoskeleton, skin, feathers, or fur.

monoclonal antibody: an antibody produced in the lab by the cloning of hybridoma cells; each clone of cells produces a single antibody.

monocot: short for monocotyledon; a type of flowering plant characterized by embryos with one seed leaf, or cotyledon.

monoecious (**mon-ē´-shus**): pertaining to organisms in which male and female gametes are produced in the same individual.

monomer (**mo´-nō-mer**): a small organic molecule, several of which may be bonded together to form a chain called a *polymer.*

monophyletic: referring to a group of species that contains all the known descendants of an ancestral species.

monosaccharide (**mo-nō-sak´-uh-rīd**): the basic molecular unit of all carbohydrates, normally composed of a chain of carbon atoms bonded to hydrogen and hydroxyl groups.

monotreme: a mammal that lays eggs; for example, the platypus.

morula (**mor´-ū-luh**): in animals, an embryonic stage during cleavage, when the embryo consists of a solid ball of cells.

motor neuron: a neuron that receives instructions from sensory neurons or interneurons and activates effector organs, such as muscles or glands.

motor unit: a single motor neuron and all the muscle fibers on which it forms synapses.

mouth: the opening of a tubular digestive system into which food is first introduced.

mucous membrane: the lining of the inside of the respiratory and digestive tracts.

multicellular: many-celled; most members of the kingdoms Fungi, Plantae, and Animalia are multicellular, with intimate cooperation among cells.

multiple alleles: many alleles of a single gene, perhaps dozens or hundreds, as a result of mutations.

muscle fiber: an individual muscle cell.

mutation: a change in the base sequence of DNA in a gene; normally refers to a genetic change significant enough to alter the appearance or function of the organism.

mutualism (**mū´-choo-ul-iz-um**): a symbiotic relationship in which both participating species benefit.

mycelium (**mi-sēl´-ē-um**): the body of a fungus, consisting of a mass of hyphae.

mycorrhiza (**mi-kō-rī´zuh;** pl., **mycorrhizae**): a symbiotic relationship between a fungus and the roots of a land plant that facilitates mineral extraction and absorption.

myelin (**mī´-uh-lin**): a wrapping of insulating membranes of specialized nonneural cells around the axon of a vertebrate nerve cell; increases the speed of conduction of action potentials.

myofibril (**mi-ō-fī´-bril**): a cylindrical subunit of a muscle cell, consisting of a series of sarcomeres; surrounded by sarcoplasmic reticulum.

myometrium (**mi-ō-mē´-trē-um**): the muscular outer layer of the uterus.

myosin (**mi´-ō-sin**): one of the major proteins of muscle, the interaction of which with the protein actin produces muscle contraction; found in the thick filaments of the muscle fiber; see also *actin.*

natural causality: the scientific principle that natural events occur as a result of preceding natural causes.

natural killer cell: a type of white blood cell that destroys some virus-infected cells and cancerous cells on contact; part of the immune system's nonspecific internal defense against disease.

natural selection: the unequal survival and reproduction of organisms due to environmental forces, resulting in the preservation of favorable adaptations. Usually, natural selection refers specifically to differential survival and reproduction on the basis of genetic differences among individuals.

near-shore zone: the region of coastal water that is relatively shallow but constantly submerged; includes bays and coastal wetlands and can support large plants or seaweeds.

nearsighted: the inability to focus on distant objects caused by an eyeball that is slightly too long or a cornea that is too curved.

negative feedback: a situation in which a change initiates a series of events that tend to counteract the change and restore the original state. Negative feedback in physiological systems maintains homeostasis.

nephridiopore: the opening of the simple kidney (nephridium) of earthworms to the outside.

nephridium (**nef-rid´-ē-um**): an excretory organ found in earthworms, mollusks, and certain other invertebrates; somewhat resembles a single vertebrate nephron.

nephron (**nef´-ron**): the functional unit of the kidney; where blood is filtered and urine formed.

nephrostome (**nef´-rō-stōm**): the funnel-shaped opening of the nephridium of some invertebrates such as earthworms; coelomic fluid is drawn into the nephrostome for filtration.

nerve: a bundle of axons of nerve cells, bound together in a sheath.

nerve cord: a paired neural structure in most animals that conducts nervous signals to and from the ganglia; in chordates, a nervous structure lying along the dorsal side of the body; also called spinal cord.

nerve net: a simple form of nervous system, consisting of a network of neurons that extend throughout the tissues of an organism such as a cnidarian.

nerve tissue: the tissue that make up the brain, spinal cord, and nerves; consists of neurons and glial cells.

net primary productivity: the energy stored in the autotrophs of an ecosystem over a given time period.

neural tube: a structure, derived from ectoderm during early embryonic development, that later becomes the brain and spinal cord.

neuromuscular junction: the synapse formed between a motor neuron and a muscle fiber.

neuron (**noor´-on**): a single nerve cell.

neuropeptide: a small protein molecule with neurotransmitter-like actions.

neurosecretory cell: a specialized nerve cell that synthesizes and releases hormones.

neurotransmitter: a chemical that is released by a nerve cell close to a second nerve cell, a muscle, or a gland cell and that influences the activity of the second cell.

neutral mutation: a mutation that has little or no effect on the function of the encoded protein.

neutralization: the process of covering up or inactivating a toxic substance with antibody.

neutron: a subatomic particle that is found in the nuclei of atoms, bears no charge, and has a mass approximately equal to that of a proton.

neutrophil (**nū´-trō-fil**): a type of white blood cell that engulfs invading microbes and contributes to the nonspecific defenses of the body against disease.

nitrogen fixation: the process that combines atmospheric nitrogen with hydrogen to form ammonium (NH_4^+).

nitrogen-fixing bacterium: a bacterium that possess the ability to remove nitrogen (N_2) from the atmosphere and combine it with hydrogen to produce ammonium (NH_4^+).

node: in plants, a region of a stem at which leaves and lateral buds are located; in vertebrates, an interruption of the myelin on a myelinated axon, exposing naked membrane at which action potentials are generated.

nodule: a swelling on the root of a legume or other plant that consists of cortex cells inhabited by nitrogen-fixing bacteria.

nondisjunction: an error in meiosis in which chromosomes fail to segregate properly into the daughter cells.

nonpolar covalent bond: a covalent bond with equal sharing of electrons.

norepinephrine (**nor-ep-i-nef-rin´**): a neurotransmitter, released by neurons of the parasympathetic nervous system, that prepares the body to respond to stressful situations; also called *noradrenaline.*

northern coniferous forest: a biome with long, cold winters and only a few months of warm weather; populated almost entirely by evergreen coniferous trees; also called taiga.

no-till: a method of growing crops that leaves the remains of harvested crops in place to form mulch for the next year's crops.

notochord (**nōt´-ō-kord**): a stiff but somewhat flexible, supportive rod found in all members of the phylum Chordata at some stage of development.

nuclear envelope: the double-membrane system surrounding the nucleus of eukaryotic cells; the outer membrane is typically continuous with the endoplasmic reticulum.

nucleic acid (**noo-klā´-ik**): an organic molecule composed of nucleotide subunits; the two common types of nucleic acids are ribonucleic acid (RNA) and deoxyribonucleic acid (DNA).

nucleoid (**noo-klē-oid**): the location of the genetic material in prokaryotic cells; not membrane-enclosed.

nucleolus (**noo-klē´-ō-lus**): the region of the eukaryotic nucleus that is engaged in ribosome synthesis; consists of the genes encoding ribosomal RNA, newly synthesized ribosomal RNA, and ribosomal proteins.

nucleotide: a subunit of which nucleic acids are composed; a phosphate group bonded to a sugar (deoxyribose in DNA), which is in turn bonded to a nitrogen-containing base (adenine, guanine, cytosine, or thymine in DNA). Nucleotides are linked together, forming a strand of nucleic acid, as follows: Nucleotides are linked together, forming a strand of nucleic acid, by bonds between the phosphate of one nucleotide and the sugar of the next nucleotide.

nucleotide substitution: a mutation that replaces one nucleotide in a DNA molecule with another; for example, a change from an adenine to a guanine.

nucleus (atomic): the central region of an atom, consisting of protons and neutrons.

nucleus (cellular): the membrane-bound organelle of eukaryotic cells that contains the cell's genetic material.

nutrient: a substance acquired from the environment and needed for the survival, growth, and development of an organism.

nutrient cycle: a description of the pathways of a specific nutrient (such as carbon, nitrogen, phosphorus, or water) through the living and nonliving portions of an ecosystem. Also called biogeochemical cycle.

nutrition: the process of acquiring nutrients from the environment and, if necessary, processing them into a form that can be used by the body.

observation: in the scientific method, the noting of a specific phenomenon, leading to the formulation of a hypothesis.

oil: a lipid composed of three fatty acids, some of which are unsaturated, covalently bonded to a molecule of glycerol; liquid at room temperature.

olfaction (ōl-fak′-shun): a chemical sense, the sense of smell; in terrestrial vertebrates, the result of the detection of airborne molecules.

oligotrophic lake: a lake that is very low in nutrients and hence clear with extensive light penetration.

ommatidium (ōm-ma-tid′-ē-um): an individual light-sensitive subunit of a compound eye; consists of a lens and several receptor cells.

omnivore: an organism that consumes both plants and other animals.

one gene, one protein rule: the premise that each gene encodes the information for the synthesis of a single protein.

oogenesis: the process by which egg cells are formed.

oogonium (ō-ō-gō′-nē-um; pl., **oogonia**): in female animals, a diploid cell that gives rise to a primary oocyte.

open circulatory system: a type of circulatory system found in some invertebrates, such as arthropods and mollusks, that includes an open space (the hemocoel) in which blood directly bathes body tissues.

operant conditioning: a laboratory training procedure in which an animal learns to perform a response (such as pressing a lever) through reward or punishment.

operculum: an external flap, supported by bone, that covers and protects the gills of most fish.

opioid (ōp′-ē-oid): one of a group of peptide neuromodulators in the vertebrate brain that mimic some of the actions of opiates (such as opium) and also seem to influence many other processes, including emotion and appetite.

optic nerve: the nerve leading from the eye to the brain, carrying visual information.

order: the taxonomic category contained within a class and consisting of related families.

organ: a structure (such as the liver, kidney, or skin) composed of two or more distinct tissue types that function together.

organelle (or-guh-nel′): a structure, found in the cytoplasm of eukaryotic cells, that performs a specific function; sometimes refers specifically to membrane-bound structures, such as the nucleus or endoplasmic reticulum.

organic/organic molecule: describing a molecule that contains both carbon and hydrogen.

organism (or′-guh-niz-um): an individual living thing.

organogenesis (or-gan-ō-jen′-uh-sis): the process by which the layers of the gastrula (endoderm, ectoderm, mesoderm) rearrange into organs.

organ system: two or more organs that work together to perform a specific function; for example, the digestive system.

origin: the site of attachment of a muscle to the relatively stationary bone on one side of a joint.

osmoregulation: homeostatic maintenance of the water and salt content of the body within a limited range.

osmosis (oz-mō′-sis): the diffusion of water across a differentially permeable membrane, normally down a concentration gradient of free water molecules. Water moves into the solution that has a lower concentration of free water from a solution with the higher concentration of free water.

osmotic pressure: the pressure required to counterbalance the tendency of water to move from a solution with a higher concentration of free water molecules into a solution with a lower concentration of free water molecules.

osteoblast (os′-tē-ō-blast): a cell type that produces bone.

osteoclast (os′-tē-ō-klast): a cell type that dissolves bone.

osteocyte (os′-tē-ō-sīt): a mature bone cell.

osteon: a unit of hard bone consisting of concentric layers of bone matrix, with embedded osteocytes, surrounding a small central canal that contains a capillary.

osteoporosis (os′-tē-ō-por-ō′-sis): a condition in which bones become porous, weak, and easily fractured; most common in elderly women.

outer ear: the outermost part of the mammalian ear, including the external ear and auditory canal leading to the tympanic membrane.

oval window: the membrane-covered entrance to the inner ear.

ovary: in animals, the gonad of females; in flowering plants, a structure at the base of the carpel that contains one or more ovules and develops into the fruit.

overexploitation: hunting or harvesting natural populations at a rate that exceeds their ability to replenish their numbers.

oviduct: in mammals, the tube leading from the ovary to the uterus.

ovulation: the release of a secondary oocyte, ready to be fertilized, from the ovary.

ovule: a structure within the ovary of a flower, inside which the female gametophyte develops; after fertilization, develops into the seed.

oxytocin (oks-ē-tō′-sin): a hormone, released by the posterior pituitary, that stimulates the contraction of uterine and mammary gland muscles.

ozone layer: the ozone-enriched layer of the upper atmosphere that filters out some of the sun's ultraviolet radiation.

pacemaker: a cluster of specialized muscle cells in the upper right atrium of the heart that produce spontaneous electrical signals at a regular rate; the sinoatrial node.

pain receptor: a receptor cell that responds to certain chemicals that are produced as a result of tissue damage, such as potassium ions or bradykinin, and is responsible for the sensation of pain.

palisade cell: a columnar mesophyll cell, containing chloroplasts, just beneath the upper epidermis of a leaf.

pancreas (pan′-krē-us): a combined exocrine and endocrine gland located in the abdominal cavity next to the stomach. The endocrine portion secretes the hormones insulin and glucagon, which regulate glucose concentrations in the blood. The exocrine portion secretes enzymes for fat, carbohydrate, and protein digestion into the small intestine and neutralizes the acidic chyme.

pancreatic juice: a mixture of water, sodium bicarbonate, and enzymes released by the pancreas into the small intestine.

parasite (par′-uh-sit): an organism that lives in or on, and feeds on, a larger organism called a *host*, weakening it.

parasitism: a symbiotic relationship in which one organism (commonly smaller and more numerous than its host) benefits by feeding on the other, which is normally harmed but not immediately killed.

parasympathetic division: the division of the autonomic nervous system that produces largely involuntary responses related to the maintenance of normal body functions, such as digestion.

parathormone: a hormone, secreted by the parathyroid gland, that stimulates the release of calcium from bones.

parathyroid gland: one of four small endocrine glands, embedded in the surface of the thyroid gland, that produces parathyroid hormone, which (with calcitonin from the thyroid gland) regulates calcium ion concentration in the blood.

parathyroid hormone: a hormone released by the parathyroid gland that works in conjunction with calcitonin to regulate calcium ion concentration in the blood.

parenchyma (par-en′-ki-muh): a plant cell type that is alive at maturity, normally with thin primary cell walls, that carries out most of the metabolism of a plant. Most dividing meristem cells in a plant are parenchyma.

parthenogenesis (par-the-nō-jen′uh-sis): a specialization of sexual reproduction, in which a haploid egg undergoes development without fertilization.

passive transport: the movement of materials across a membrane down a gradient of concentration, pressure, or electrical charge without using cellular energy.

pathogen: an organism (or a toxin) capable of producing disease.

pathogenic (path′-ō-jen-ik): capable of producing disease; refers to an organism with such a capability (a pathogen).

pedigree: a diagram showing genetic relationships among a set of individuals, normally with respect to a specific genetic trait.

pelagic (puh-la′-jik): free-swimming or floating.

penis: an external structure of the male reproductive and urinary systems; serves to deposit sperm into the female reproductive system and delivers urine to the exterior.

peptide (pep′-tid): a chain composed of two or more amino acids linked together by peptide bonds.

peptide bond: the covalent bond between the amino group's nitrogen of one amino acid and the carboxyl group's carbon of a second amino acid, joining the two amino acids together in a peptide or protein.

peptide hormone: a hormone consisting of a chain of amino acids; includes small proteins that function as hormones.

peptidoglycan (pep-tid-ō-glī′-kan): a component of prokaryotic cell walls that consists of chains of sugars cross-linked by short chains of amino acids called peptides.

pericycle (per′-i-si-kul): the outermost layer of cells of the vascular cylinder of a root.

periderm: the outer cell layers of roots and stems that have undergone secondary growth, consisting primarily of cork cambium and cork cells.

peripheral nerve: a nerve that links the brain and spinal cord to the rest of the body.

peripheral nervous system: in vertebrates, the part of the nervous system that connects the central nervous system to the rest of the body.

peristalsis: rhythmic coordinated contractions of the smooth muscles of the digestive tract that move substances through the digestive tract.

permafrost: a permanently frozen layer of soil in the arctic tundra that cannot support the growth of trees.

petal: part of a flower, typically brightly colored and fragrant, that attracts potential animal pollinators.

petiole (pet′-ē-ōl): the stalk that connects the blade of a leaf to the stem.

phagocytic cell (fa-gō-sit′-ik): a type of immune system cell that destroys invading microbes by using phagocytosis to engulf and digest the microbes.

phagocytosis (fa-gō-si-tō′-sis): a type of endocytosis in which extensions of a plasma membrane engulf extracellular particles and transport them into the interior of the cell.

pharyngeal gill slit (far-in′-jē-ul): an opening, located just posterior to the mouth, that connects the digestive tube to the outside environment; present (as some stage of life) in all chordates.

pharynx (far′-inks): in vertebrates, a chamber that is located at the back of the mouth and is shared by the digestive and respiratory systems; in some invertebrates, the portion of the digestive tube just posterior to the mouth.

phenotype (fēn′-ō-tīp): the physical characteristics of an organism; can be defined as outward appearance (such as flower color), as behavior, or in molecular terms (such as glycoproteins on red blood cells).

pheromone (fer′-uh-mōn): a chemical produced by an organism that alters the behavior or physiological state of another member of the same species.

phloem (flō′-um): a conducting tissue of vascular plants that transports a concentrated sugar solution up and down the plant.

phospholipid (fos-fō-li′-pid): a lipid consisting of glycerol bonded to two fatty acids and one phosphate group, which bears another group of atoms, typically charged and containing nitrogen. A double layer of phospholipids is a component of all cellular membranes.

phospholipid bilayer: a double layer of phospholipids that forms the basis of all cellular membranes. The phospholipid heads, which are hydrophilic, face the water of extracellular fluid or the cytoplasm; the tails, which are hydrophobic, are buried in the middle of the bilayer.

photic zone: the region of the ocean where light is strong enough to support photosynthesis.

photon (fō′-ton): the smallest unit of light energy.

photopigment (fō′-tō-pig-ment): a chemical substance in photoreceptor cells that, when struck by light, changes in molecular conformation.

photoreceptor: a receptor cell that responds to light; in vertebrates, rods and cones.

photorespiration: a series of reactions in plants in which O_2 replaces CO_2 during the C_3 cycle, preventing carbon fixation; this wasteful process dominates when C_3 plants are forced to close their stomata to prevent water loss.

photosynthesis: the complete series of chemical reactions in which the energy of light is used to synthesize high-energy organic molecules, normally carbohydrates, from low-energy inorganic molecules, normally carbon dioxide and water.

photosystem: in thylakoid membranes, a light-harvesting complex and its associated electron transport system.

phototactic: capable of detecting and responding to light.

phototropism: growth with respect to the direction of light.

pH scale: a scale, with values from 0 to 14, used for measuring the relative acidity of a solution; at pH 7 a solution is neutral, pH 0 to 7 is acidic, and pH 7 to 14 is basic; each unit on the scale represents a tenfold change in H^+ concentration.

phycocyanin (fī-kō-si′-uh-nin): a blue or purple pigment that is located in the membranes of chloroplasts and is used as an accessory light-gathering molecule in thylakoid photosystems.

phylogeny (fī-lah′-jen-ē): the evolutionary history of a group of species.

phylum (fī-lum): the taxonomic category of animals and animal-like protists that is contained within a kingdom and consists of related classes.

phytochrome (fī′-tō-krōm): a light-sensitive plant pigment that mediates many plant responses to light, including flowering, stem elongation, and seed germination.

phytoplankton (fī′-tō-plank-ten): photosynthetic protists that are abundant in marine and freshwater environments.

pilus (pil′-us; pl., pili): a hairlike projection that is made of protein, located on the surface of certain bacteria, and is typically used to attach a bacterium to another cell.

pineal gland (pi-nē′-al): a small gland within the brain that secretes melatonin; controls the seasonal reproductive cycles of some mammals.

pinocytosis (pi-nō-si-tō′-sis): the nonselective movement of extracellular fluid, enclosed within a vesicle formed from the plasma membrane, into a cell.

pioneer: an organism that is among the first to colonize an unoccupied habitat in the first stages of succession.

pit: an area in the cell walls between two plant cells in which secondary walls did not form, such that the two cells are separated only by a relatively thin and porous primary cell wall.

pith: cells forming the center of a root or stem.

pituitary gland: an endocrine gland, located at the base of the brain, that produces several hormones, many of which influence the activity of other glands.

placenta (pluh-sen′-tuh): in mammals, a structure formed by a complex interweaving of the uterine lining and the embryonic membranes, especially the chorion; functions in gas, nutrient, and waste exchange between embryonic and maternal circulatory systems and secretes hormones.

placental (pluh-sen′-tul): referring to a mammal possessing a placenta (that is, species that are not marsupials or monotremes).

plankton: microscopic organisms that live in marine or freshwater environments; includes phytoplankton and zooplankton.

plant hormone: the plant-regulating chemicals auxin, gibberellins, cytokinins, ethylene, and abscisic acid; somewhat resemble animal hormones in that they are chemicals produced by cells in one location that influence the growth or metabolic activity of other cells, typically some distance away in the plant body.

plaque (plak): a deposit of cholesterol and other fatty substances within the wall of an artery.

plasma: the fluid, noncellular portion of the blood.

plasma cell: an antibody-secreting descendant of a B cell.

plasma membrane: the outer membrane of a cell, composed of a bilayer of phospholipids in which proteins are embedded.

plasmid (plaz′-mid): a small, circular piece of DNA located in the cytoplasm of many bacteria; normally does not carry genes required for the normal functioning of the bacterium but may carry genes that assist bacterial survival in certain environments, such as a gene for antibiotic resistance.

plasmodesma (plaz-mō-dez′-muh; pl., plasmo-desmata): a cell-to-cell junction in plants that connects the cytoplasm of adjacent cells.

plasmodial slime mold: see *acellular slime mold.*

plasmodium (plaz-mō′-dē-um): a sluglike mass of cytoplasm containing thousands of nuclei that are not confined within individual cells.

plastid (plas′-tid): in plant cells, an organelle bounded by two membranes that may be involved in photosynthesis (chloroplasts), pigment storage, or food storage.

platelet (plāt′-let): a cell fragment that is formed from megakaryocytes in bone marrow and lacks a nucleus; circulates in the blood and plays a role in blood clotting.

plate tectonics: the theory that Earth's crust is divided into irregular plates that are converging, diverging, or slipping by one another; these motions cause continental drift, the movement of continents over Earth's surface.

pleated sheet: a form of secondary structure exhibited by certain proteins, such as silk, in which many protein chains lie side-by-side, with hydrogen bonds holding adjacent chains together.

pleiotropy (ple'-ō-trō-pē): a situation in which a single gene influences more than one phenotypic characteristic.

pleural membrane: a membrane that lines the chest cavity and surrounds the lungs.

point mutation: a mutation in which a single base pair in DNA has been changed.

polar body: in oogenesis, a small cell, containing a nucleus but virtually no cytoplasm, produced by the first meiotic division of the primary oocyte.

polar covalent bond: a covalent bond with unequal sharing of electrons, such that one atom is relatively negative and the other is relatively positive.

polar nucleus: in flowering plants, one of two nuclei in the primary endosperm cell of the female gametophyte; formed by the mitotic division of a megaspore.

pollen/pollen grain: the male gametophyte of a seed plant.

pollination: in flowering plants, when pollen grains land on the stigma of a flower of the same species; in conifers, when pollen grains land within the pollen chamber of a female cone of the same species.

polygenic inheritance: a pattern of inheritance in which the interactions of two or more functionally similar genes determine phenotype.

polymer (pah'-li-mer): a molecule composed of three or more (perhaps thousands) smaller subunits called *monomers*, which may be identical (for example, the glucose monomers of starch) or different (for example, the amino acids of a protein).

polymerase chain reaction (PCR): a method of producing virtually unlimited numbers of copies of a specific piece of DNA, starting with as little as one copy of the desired DNA.

polyp (pah'-lip): the sedentary, vase-shaped stage in the life cycle of many cnidarians; includes hydra and sea anemones.

polypeptide: a short polymer of amino acids; often used as a synonym for protein.

polyploidy (pahl'-ē-ploid-ē): having more than two homologous chromosomes of each type.

polysaccharide (pahl-ē-sak'-uh-rid): a large carbohydrate molecule composed of branched or unbranched chains of repeating monosaccharide subunits, normally glucose or modified glucose molecules; includes starches, cellulose, and glycogen.

pons: a portion of the hindbrain, just above the medulla, that contains neurons that influence sleep and the rate and pattern of breathing.

population: all the members of a particular species within an ecosystem, found in the same time and place and actually or potentially interbreeding.

population bottleneck: a form of genetic drift in which a population becomes extremely small; may lead to differences in allele frequencies as compared with other populations of the species and to a loss in genetic variability.

population cycle: regularly recurring, cyclic changes in population size.

population genetics: the study of the frequency, distribution, and inheritance of alleles in a population.

positive feedback: a situation in which a change initiates events that tend to amplify the original change.

post-anal tail: a tail that extends beyond the anus; exhibited by all chordates at some stage of development.

posterior: the tail, hindmost, or rear end of an animal.

posterior pituitary: a lobe of the pituitary gland that is an outgrowth of the hypothalamus and that releases antidiuretic hormone and oxytocin.

postmating isolating mechanism: any structure, physiological function, or developmental abnormality that prevents organisms of two different populations, once mating has occurred, from producing vigorous, fertile offspring.

postsynaptic neuron: at a synapse, the nerve cell that changes its electrical potential in response to a chemical (the neurotransmitter) released by another (presynaptic) cell.

postsynaptic potential (PSP): an electrical signal produced in a postsynaptic cell by transmission across the synapse; it may be excitatory (EPSP), making the cell more likely to produce an action potential, or inhibitory (IPSP), tending to inhibit an action potential.

potential energy: "stored" energy, normally chemical energy or energy of position within a gravitational field.

prairie: a biome, located in the centers of continents, that supports grasses; also called *grassland*.

preadaptation: a feature evolved under one set of environmental conditions that, purely by chance, helps an organism adapt to new environmental conditions.

prebiotic evolution: evolution before life existed; especially, the abiotic synthesis of organic molecules.

precapillary sphincter (sfink'-ter): a ring of smooth muscle between an arteriole and a capillary that regulates the flow of blood into the capillary bed.

predation (pre-dā'-shun): the act of killing and eating another living organism.

predator: an organism that kills and eats other organisms.

premating isolating mechanism: any structure, physiological function, or behavior that prevents organisms of two different populations from exchanging gametes.

pressure-flow theory: a model for the transport of sugars in phloem, by which the movement of sugars into a phloem sieve tube causes water to enter the tube by osmosis, while the movement of sugars out of another part of the same sieve tube causes water to leave by osmosis; the resulting pressure gradient causes the bulk movement of water and dissolved sugars from the end of the tube into which sugar is transported toward the end of the tube from which sugar is removed.

presynaptic neuron: a nerve cell that releases a chemical (the neurotransmitter) at a synapse, causing changes in the electrical activity of another (postsynaptic) cell.

prey: organisms that are killed and eaten by another organism.

primary cell wall: cellulose and other carbohydrates secreted by a young plant cell between the middle lamella and the plasma membrane.

primary consumer: an organism that feeds on producers; an herbivore.

primary endosperm cell: the central cell of the female gametophyte of a flowering plant, containing the polar nuclei (normally two); after fertilization, undergoes repeated mitotic divisions to produce the endosperm of the seed.

primary growth: growth in length and development of the initial structures of plant roots and shoots, due to the cell division of apical meristems and differentiation of the daughter cells.

primary oocyte (ō'-ō-sit): a diploid cell, derived from the oogonium by growth and differentiation, that undergoes meiosis, producing the egg.

primary phloem: phloem in young stems produced from an apical meristem.

primary root: the first root that develops from a seed.

primary spermatocyte (sper-ma'-tō-sit): a diploid cell, derived from the spermatogonium by growth and differentiation, that undergoes meiosis, producing four sperm.

primary structure: the amino acid sequence of a protein.

primary succession: succession that occurs in an environment, such as bare rock, in which no trace of a previous community was present.

primary xylem: xylem in young stems produced from an apical meristem.

primate: a mammal characterized by the presence of an opposable thumb, forward-facing eyes, and a well-developed cerebral cortex; includes lemurs, monkeys, apes, and humans.

primitive streak: in reptiles, birds, and mammals, the region of the ectoderm of the two-layered embryonic disc through which cells migrate, forming mesoderm.

prion (prē'-on): a protein that, in mutated form, acts as an infectious agent that causes certain neurodegenerative diseases, including kuru and scrapie.

producer: a photosynthetic organism; an autotroph.

product: an atom or molecule that is formed from reactants in a chemical reaction.

profundal zone: a lake zone in which light is insufficient to support photosynthesis.

progesterone (prō-ge'-ster-ōn): a hormone, produced by the corpus luteum, that promotes the development of the uterine lining in females.

prokaryote (prō-kar'-ē-ōt): an organism whose cells are prokaryotic; bacteria and archaea are prokaryotes.

prokaryotic (prō-kar-ē-ot'-ik): referring to cells of the domains Bacteria or Archaea. Prokaryotic cells have genetic material that is not enclosed in a membrane-bound nucleus; they also lack other membrane-bound organelles.

prolactin: a hormone, released by the anterior pituitary, that stimulates milk production in human females.

promoter: a specific sequence of DNA to which RNA polymerase binds, initiating gene transcription.

prophase (prō'-fāz): the first stage of mitosis, in which the chromosomes first become visible in the light microscope as thickened, condensed threads and the spindle begins to form; as the spindle is completed, the nuclear envelope breaks apart, and the spindle fibers invade the nuclear region and attach to the kinetochores of the chromosomes. Also, the first stage of meiosis: In meiosis I, the homologous chromosomes pair up and exchange parts at chiasmata; in meiosis II, the spindle re-forms and chromosomes attach to the microtubules.

prostaglandin (pro-stuh-glan´-din): a family of modified fatty acid hormones manufactured by many cells of the body.

prostate gland (pros´-tāt): a gland that produces part of the fluid component of semen; prostatic fluid is basic and contains a chemical that activates sperm movement.

protease (prō´-tē-ās): an enzyme that digests proteins.

protein: polymer of amino acids joined by peptide bonds.

protist: a eukaryotic organism that is not a plant, animal, or fungus. The term encompasses a diverse array of organisms and does not represent a monophyletic group.

protocell: the hypothetical evolutionary precursor of living cells, consisting of a mixture of organic molecules within a membrane.

proton: a subatomic particle that is found in the nuclei of atoms, bears a unit of positive charge, and has a relatively large mass, roughly equal to the mass of the neutron.

protonephridium (prō-tō-nef-rid´-ē-um; pl., **protonephridia**): an excretory system consisting of tubules that have external opening but lack internal openings; for example, the flame-cell system of flatworms.

protostome (prō´-tō-stōm): an animal with a mode of embryonic development in which the coelom is derived from splits in the mesoderm; characteristic of arthropods, annelids, and mollusks.

protozoan (prō-tuh-zō´-an; pl., **protozoa**): a nonphotosynthetic or animal-like protist.

proximal tubule: in nephrons of the mammalian kidney, the portion of the renal tubule just after the Bowman's capsule; receives filtrate from the capsule and is the site where selective secretion and reabsorption between the filtrate and the blood begins.

pseudocoelom (soo´-dō-sēl´-ōm): "false coelom"; a body cavity that has a different embryological origin than a coelom but serves a similar function; found in roundworms.

pseudoplasmodium (soo´-dō-plaz-mō´-dē-um): an aggregation of individual amoeboid cells that form a sluglike mass.

pseudopod (sood´-ō-pod): an extension of the plasma membrane by which certain cells, such as amoebae, locomote and engulf prey.

puberty: a stage of development (in humans, usually beginning in the early teenage years) characterized by rapid growth and the appearance of secondary sexual characteristics in response to increased secretion of testosterone in males and estrogen in females.

Punnett square method: an intuitive way to predict the genotypes and phenotypes of offspring in specific crosses.

pupa: a developmental stage in some insect species in which the organism stops moving and feeding and may be encased in a cocoon; occurs between the larval and the adult phases.

pupil: the adjustable opening in the center of the iris, through which light enters the eye.

Purkinje fibers: specialized cardiac muscle cells that rapidly conduct electrical signals from the AV bundle branches up into both ventricles, causing them to contract simultaneously.

pyloric sphincter (pī-lor´-ik sfink´-ter): a circular muscle, located at the base of the stomach, that regulates the passage of chyme into the small intestine.

pyruvate: a three-carbon molecule that is formed by glycolysis and then used in fermentation or cellular respiration.

quaternary structure (kwat´-er-nuh-rē): the complex three-dimensional structure of a protein composed of more than one peptide chain.

queen substance: a chemical, produced by a queen bee, that can act as both a primer and a pheromone.

radial symmetry: a body plan in which any plane along a central axis will divide the body into approximately mirror-image halves. Cnidarians and many adult echinoderms have radial symmetry.

radioactive: pertaining to an atom with an unstable nucleus that spontaneously disintegrates, with the emission of radiation.

radiolarian (rā-dē-ō-lar´-ē-un): an aquatic protist (largely marine) characterized by typically elaborate silica shells.

radula (ra´-dū-luh): a ribbon of tissue in the mouth of gastropod mollusks; bears numerous teeth on its outer surface and is used to scrape and drag food into the mouth.

rain shadow: a local dry area created by the modification of rainfall patterns by a mountain range.

random distribution: distribution characteristic of populations in which the probability of finding an individual is equal in all parts of an area.

reactant: an atom or molecule that is used up in a chemical reaction to form a product.

reaction center: in the light-harvesting complex of a photosystem, the chlorophyll molecule to which light energy is transferred by the antenna molecules (light-absorbing pigments); the captured energy ejects an electron from the reaction center chlorophyll, and the electron is transferred to the electron transport system.

receptor: a cell that responds to an environmental stimulus (chemicals, sound, light, pH, and so on) by changing its electrical potential; also, a protein molecule in a plasma membrane that binds to another molecule (hormone, neurotransmitter), triggering metabolic or electrical changes in a cell.

receptor-mediated endocytosis: the selective uptake of molecules from the extracellular fluid by binding to a receptor located at a coated pit on the plasma membrane and pinching off the coated pit into a vesicle that moves into the cytoplasm.

receptor potential: an electrical potential change in a receptor cell, produced in response to the reception of an environmental stimulus (chemicals, sound, light, heat, and so on). The size of the receptor potential is proportional to the intensity of the stimulus.

receptor protein: a protein, located on a membrane (or in the cytoplasm), that recognizes and binds to specific molecules. Binding by receptor proteins typically triggers a response by a cell, such as endocytosis, increased metabolic rate, or cell division.

recessive: an allele that is expressed only in homozygotes and is completely masked in heterozygotes.

recognition protein: a protein or glycoprotein protruding from the outside surface of a plasma membrane that identifies a cell as belonging to a particular species, to a specific individual of that species, and in many cases to one specific organ within the individual.

recombinant DNA: DNA that has been altered by the recombination of genes from a different organism, typically from a different species.

recombination: the formation of new combinations of the different alleles of each gene on a chromosome; the result of crossing over.

rectum: the terminal portion of the vertebrate digestive tube, where feces are stored until they can be eliminated.

reflex: a simple, stereotyped movement of part of the body that occurs automatically in response to a stimulus.

regeneration: the regrowth of a body part after loss or damage; also, asexual reproduction by means of the regrowth of an entire body from a fragment.

releasing hormone: a hormone, secreted by the hypothalamus, that causes the release of specific hormones by the anterior pituitary.

renal artery: the artery carrying blood to each kidney.

renal cortex: the outer layer of the kidney; where nephrons are located.

renal medulla: the layer of the kidney just inside the renal cortex; where loops of Henle produce a highly concentrated interstitial fluid, important in the production of concentrated urine.

renal pelvis: the inner chamber of the kidney; where urine from the collecting ducts accumulates before it enters the ureters.

renal vein: the vein carrying blood away from each kidney.

renin: an enzyme that is released (in mammals) when blood pressure and/or sodium concentration in the blood drops below a set point; initiates a cascade of events that restores blood pressure and sodium concentration.

replacement-level fertility (RLF): the average birthrate at which a reproducing population exactly replaces itself during its lifetime.

replication bubble: in DNA replication, the unwound portion of the two parental DNA strands that have been separated by DNA helicase.

reproductive isolation: the failure of organisms of one population to breed successfully with members of another; may be due to premating or postmating isolating mechanisms.

reptile: a member of the chordate group that includes the snakes, lizards, turtles, alligators, and crocodiles; not a monophyletic group.

reservoir: the major source and storage site of a nutrient in an ecosystem, normally in the abiotic portion.

resource partitioning: the coexistence of two species with similar requirements, each occupying a smaller niche than either would if it were by itself; a means of minimizing their competitive interactions.

respiratory center: a cluster of neurons, located in the medulla of the brain, that sends rhythmic bursts of nerve impulses to the respiratory muscles, resulting in breathing.

resting potential: a negative electrical potential in unstimulated nerve cells.

restriction enzyme: an enzyme, normally isolated from bacteria, that cuts double-stranded DNA at a specific nucleotide sequence; the nucleotide sequence that is cut differs for different restriction enzymes.

restriction fragment: a piece of DNA that has been isolated by cleaving a larger piece of DNA with restriction enzymes.

restriction fragment length polymorphism (RFLP): a difference in the length of restriction fragments, produced by cutting samples of DNA from different individuals of the same species with the same set of restriction enzymes; the result of differences in nucleotide sequences among individuals of the same species.

reticular formation (reh-tik′-ū-lar): a diffuse network of neurons extending from the hindbrain, through the midbrain, and into the lower reaches of the forebrain; involved in filtering sensory input and regulating what information is relayed to conscious brain centers for further attention.

retina (ret′-in-uh): a multilayered sheet of nerve tissue at the rear of camera-type eyes, composed of photoreceptor cells plus associated nerve cells that refine the photoreceptor information and transmit it to the optic nerve.

retrovirus: a virus that uses RNA as its genetic material. When it invades a eukaryotic cell, a retrovirus "reverse transcribes" its RNA into DNA, which then directs the synthesis of more viruses, using the transcription and translation machinery of the cell.

reverse transcriptase: an enzyme found in retroviruses that catalyzes the synthesis of DNA from an RNA template.

Rh factor: a protein on the red blood cells of some people (Rh-positive) but not others (Rh-negative); the exposure of Rh-negative individuals to Rh-positive blood triggers the production of antibodies to Rh-positive blood cells.

rhizoid (rī′-zoid): a rootlike structure found in bryophytes that anchors the plant and absorbs water and nutrients from the soil.

rhizome (rī′-zōm): an undergound stem, usually horizontal, that stores food.

rhythm method: a contraceptive method involving abstinence from intercourse during ovulation.

ribonucleic acid (rī-bō-noo-klā′-ik; RNA): a molecule composed of ribose nucleotides, each of which consists of a phosphate group, the sugar ribose, and one of the bases adenine, cytosine, guanine, or uracil; involved in converting the information in DNA into protein; also the genetic material of some viruses.

ribosomal RNA (rRNA): a type of RNA that combines with proteins to form ribosomes.

ribosome: an organelle consisting of two subunits, each composed of ribosomal RNA and protein; the site of protein synthesis, during which the sequence of bases of messenger RNA is translated into the sequence of amino acids in a protein.

ribozyme: an RNA molecule that can catalyze certain chemical reactions, especially those involved in the synthesis and processing of RNA itself.

RNA polymerase: in RNA synthesis, an enzyme that catalyzes the bonding of free RNA nucleotides into a continuous strand, using RNA nucleotides that are complementary to those of a strand of DNA.

rod: a rod-shaped photoreceptor cell in the vertebrate retina, sensitive to dim light

but not involved in color vision; see also *cone.*

root: the part of the plant body, normally underground, that provides anchorage, absorbs water and dissolved nutrients and transports them to the stem, produces some hormones, and in some plants serves as a storage site for carbohydrates.

root cap: a cluster of cells at the tip of a growing root, derived from the apical meristem; protects the growing tip from damage as it burrows through the soil.

root hair: a fine projection from an epidermal cell of a young root that increases the absorptive surface area of the root.

root system: all of the roots of a plant.

rough endoplasmic reticulum: endoplasmic reticulum lined on the outside with ribosomes.

runner: a horizontally growing stem that may develop new plants at nodes that touch the soil.

sac fungus: a fungus of the division Ascomycota, whose members form spores in a saclike case called an *ascus.*

sapwood: young xylem that transports water and minerals in a tree trunk.

saprobe (sap′-rōb): an organism that derives its nutrients from the bodies of dead organisms.

sarcodine (sar-kō′-dīn): a nonphotosynthetic protist (protozoan) characterized by the ability to form pseudopodia; some sarcodines, such as amoebae, are naked, whereas others have elaborate shells.

sarcomere (sark′-ō-mer): the unit of contraction of a muscle fiber; a subunit of the myofibril, consisting of actin and myosin filaments and bounded by Z lines.

sarcoplasmic reticulum (sark′-ō-plas′-mik re-tik′-ū-lum): the specialized endoplasmic reticulum in muscle cells; forms interconnected hollow tubes. The sarcoplasmic reticulum stores calcium ions and releases them into the interior of the muscle cell, initiating contraction.

saturated: referring to a fatty acid with as many hydrogen atoms as possible bonded to the carbon backbone; a fatty acid with no double bonds in its carbon backbone.

savanna: a biome that is dominated by grasses and supports scattered trees and thorny scrub forests; typically has a rainy season in which all the year's precipitation falls.

scientific method: a rigorous procedure for making observations of specific phenomena and searching for the order underlying those phenomena.

scientific name: the name of an organism formed from the two smallest major taxonomic categories—the genus and the species.

scientific theory: a general explanation of natural phenomena developed through extensive and reproducible observations; more general and reliable than a hypothesis.

sclera: a tough, white connective tissue layer that covers the outside of the eyeball and forms the white of the eye.

sclerenchyma (skler-en′-ki-muh): a plant cell type with thick, hardened secondary cell walls that normally dies as the last stage of differentiation and both supports and protects the plant body.

scramble competition: a free-for-all scramble for limited resources among individuals of the same species.

scrotum (skrō′-tum): the pouch of skin containing the testes of male mammals.

S-curve: the S-shaped growth curve that describes a population of long-lived organisms introduced into a new area; consists of an initial period of exponential growth, followed by decreasing growth rate, and, finally, relative stability around a growth rate of zero.

sebaceous gland (se-bā′-shus): a gland in the dermis of skin, formed from epithelial tissue, that produces the oily substance sebum, which lubricates the epidermis.

secondary cell wall: a thick layer of cellulose and other polysaccharides secreted by certain plant cells between the primary cell wall and the plasma membrane.

secondary consumer: an organism that feeds on primary consumers; a carnivore.

secondary growth: growth in the diameter of a stem or root due to cell division in lateral meristems and differentiation of their daughter cells.

secondary oocyte (ō′-ō-sīt): a large haploid cell derived from the first meiotic division of the diploid primary oocyte.

secondary phloem: phloem produced from the cells that arise toward the outside of the vascular cambium.

secondary spermatocyte (sper-ma′-tō-sīt): a large haploid cell derived by meiosis I from the diploid primary spermatocyte.

secondary structure: a repeated, regular structure assumed by protein chains held together by hydrogen bonds; for example, a helix.

secondary succession: succession that occurs after an existing community is disturbed—for example, after a forest fire; much more rapid than primary succession.

secondary xylem: xylem produced from cells that arise at the inside of the vascular cambium.

second law of thermodynamics: the principle of physics that states that any change in an isolated system causes the quantity of concentrated, useful energy to decrease and the amount of randomness and disorder (entropy) to increase.

second messenger: an intracellular chemical, such as cyclic AMP, that is synthesized or released within a cell in response to the binding of a hormone or neurotransmitter (the first messenger) to receptors on the cell surface; brings about specific changes in the metabolism of the cell.

secretin: a hormone, produced by the small intestine, that stimulates the production and release of digestive secretions by the pancreas and liver.

seed: the reproductive structure of a seed plant; protected by a seed coat; contains an embryonic plant and a supply of food for it.

seed coat: the thin, tough, and waterproof outermost covering of a seed, formed from the integuments of the ovule.

segmentation (seg-men-tā′-shun): an animal body plan in which the body is divided into repeated, typically similar units.

segmentation movement: a contraction of the small intestine that results in the mixing of partially digested food and digestive enzymes. Segmentation movements also bring nutrients into contact with the absorptive intestinal wall.

segregation: see *law of segregation.*

selectively permeable: the quality of a membrane that allows certain molecules or ions to move through it more readily than others.

self-fertilization: the union of sperm and egg from the same individual.

selfish gene: the concept that genes, rather than organisms, are the unit of natural selection.

semen: the sperm-containing fluid produced by the male reproductive tract.

semiconservative replication: the process of replication of the DNA double helix; the two DNA strands separate, and each is used as a template for the synthesis of a complementary DNA strand. Consequently, each daughter double helix consists of one parental strand and one new strand.

semilunar valve: a valve located between the right ventricle of the heart and the pulmonary artery or between the left ventricle and the aorta; prevents the backflow of blood into the ventricles when they relax.

seminal vesicle: in male mammals, a gland that produces a basic, fructose-containing fluid that forms part of the semen.

seminiferous tubule (sem-i-ni′-fer-us)**:** in the vertebrate testis, a series of tubes in which sperm are produced.

senescence: in plants, a specific aging process, typically including deterioration and the dropping of leaves and flowers.

sensitive period: the particular stage in an animal's life during which it imprints.

sensory neuron: a nerve cell that responds to a stimulus from the internal or external environment.

sensory receptor: a cell (typically, a neuron) specialized to respond to particular internal or external environmental stimuli by producing an electrical potential.

sepal (se̅′-pul)**:** the set of modified leaves that surround and protect a flower bud; in dicots, usually opening into green, leaflike structures when the flower blooms.

septum (pl., septa): a partition that separates the fungal hypha into individual cells; pores in septa allow the transfer of materials between cells.

serotonin (ser-uh-to̅′-nin)**:** in the central nervous system, a neurotransmitter that is involved in mood, sleep, and the inhibition of pain.

Sertoli cell: in the seminiferous tubule, a large cell that regulates spermatogenesis and nourishes the developing sperm.

sessile (ses′-ul)**:** not free to move about, usually permanently attached to a surface.

severe combined immune deficiency (SCID): a disorder in which no immune cells, or very few, are formed; the immune system is incapable of responding properly to invading disease organisms, and the individual is very vulnerable to common infections.

sex chromosomes: the pair of chromosomes that usually determines the sex of an organism; for example, the X and Y chromosomes in mammals.

sex-linked: referring to a pattern of inheritance characteristic of genes located on one type of sex chromosome (for example, X) and not found on the other type (for example, Y); also called X-linked. In sex-linked inheritance, traits are controlled by genes carried on the X chromosome; females show the dominant trait unless they are homozygous recessive, whereas males express whichever allele is on their single X chromosome.

sexually transmitted disease (STD): a disease that is passed from person to person by sexual contact.

sexual recombination: during sexual reproduction, the formation of new combinations of alleles in offspring as a result of the inheritance of one homologous chromosome from each of two genetically distinct parents.

sexual reproduction: a form of reproduction in which genetic material from two parent organisms is combined in the offspring; normally, two haploid gametes fuse to form a diploid zygote.

sexual selection: a type of natural selection in which the choice of mates by one sex is the selective agent.

shoot system: all the parts of a vascular plant exclusive of the root; normally aboveground, consisting of stem, leaves, buds, and (in season) flowers and fruits; functions include photosynthesis, transport of materials, reproduction, and hormone synthesis.

short-day plant: a plant that will flower only if the length of daylight is shorter than some species-specific duration.

short-night plant: a plant that will flower only if the duration of darkness is shorter than some species-specific duration (sometimes called a *long-day plant*).

sickle-cell anemia: a recessive disease caused by a single amino acid substitution in the hemoglobin molecule. Sickle-cell hemoglobin molecules tend to cluster together, distorting the shape of red blood cell shape and causing them to break and clog capillaries.

sieve plate: in plants, a structure between two adjacent sieve-tube elements in phloem, where holes formed in the primary cell walls interconnect the cytoplasm of the elements; in echinoderms, the opening through which water enters the water-vascular system.

sieve tube: in phloem, a single strand of sieve-tube elements that transports sugar solutions.

sieve-tube element: one of the cells of a sieve tube, which form the phloem.

simple diffusion: the diffusion of water, dissolved gases, or lipid-soluble molecules through the phospholipid bilayer of a cellular membrane.

single covalent bond: a covalent bond in which two atoms share one pair of electrons.

sink: in plants, any structure that uses up sugars or converts sugars to starch and toward which phloem fluids will flow.

sinoatrial (SA) node (si′-no̅-a̅t′-re̅-ul)**:** a small mass of specialized muscle in the wall of the right atrium; generates electrical ignals rhythmically and spontaneously and serves as the heart's pacemaker.

skeletal muscle: the type of muscle that is attached to and moves the skeleton and is under the direct, normally voluntary, control of the nervous system; also called *striated muscle*.

skeleton: a supporting structure for the body, on which muscles act to change the body configuration; may be external or internal.

skin: the tissue that makes up the outer surface of an animal body.

slime layer: a sticky polysaccharide or protein coating that some disease-causing bacteria secrete outside their cell wall; helps the cells aggregate and stick to smooth surfaces.

small intestine: the portion of the digestive tract, located between the stomach and large intestine, in which most digestion and absorption of nutrients occur.

smell: the olfactory sense that allows animals to respond to odorous, airborne chemicals in their external environment.

smooth endoplasmic reticulum: endoplasmic reticulum without ribosomes.

smooth muscle: the type of muscle that surrounds hollow organs, such as the digestive tract, bladder, and blood vessels; normally not under voluntary control.

sodium–potassium pump: a set of active-transport molecules that use the energy of ATP to pump sodium ions out of the cell and potassium ions in, maintaining the concentration gradients of these ions across the membrane.

solvent: a liquid capable of dissolving (uniformly dispersing) other substances in itself.

somatic nervous system: that portion of the peripheral nervous system that controls voluntary movement by activating skeletal muscles.

source: in plants, any structure that actively synthesizes sugar and away from which phloem fluid will be transported.

spawning: a method of external fertilization in which male and female parents shed gametes into the water, and sperm must swim through the water to reach the eggs.

speciation: the process of species formation, in which a single species splits into two or more species.

species (spe̅′-se̅s)**:** the basic unit of taxonomic classification, consisting of a population or series of populations of closely related and similar organisms. In sexually reproducing organisms, a species can be defined as a population or series of populations of organisms that interbreed freely with one another under natural conditions but that do not interbreed with members of other species.

specific heat: the amount of energy required to raise the temperature of 1 gram of a substance by 1 °C.

sperm: the haploid male gamete, normally small, motile, and containing little cytoplasm.

spermatid: a haploid cell derived from the secondary spermatocyte by meiosis II; differentiates into the mature sperm.

spermatogenesis: the process by which sperm cells form.

spermatogonium (pl., spermatogonia): a diploid cell, lining the walls of the seminiferous tubules, that gives rise to a primary spermatocyte.

spermatophore: in a variation on internal fertilization in some animals, the males package their sperm in a container that can be inserted into the female reproductive tract.

spermicide: a sperm-killing chemical; used for contraceptive purposes.

spicule (spik′-ul)**:** a subunit of the endoskeleton of sponges that is made of protein, silica, or calcium carbonate.

spinal cord: the part of the central nervous system of vertebrates that extends from the base of the brain to the hips and is protected by the bones of the vertebral column; contains the cell bodies of motor neurons that form synapses with skeletal muscles, the circuitry for some simple reflex behaviors, and axons that communicate with the brain.

spindle microtubules: microtubules organized in a spindle shape that separate chromosomes during mitosis or meiosis.

spiracle (spi'-ruh-kul): an opening in the abdominal segment of insects through which air enters the tracheae.

spirillum (spi'-ril-um; pl., spirilla): a spiral-shaped bacterium.

spleen: an organ of the lymphatic system in which lymphocytes are produced and blood is filtered past lymphocytes and macrophages, which remove foreign particles and aged red blood cells.

spongy bone: porous, lightweight bone tissue in the interior of bones; the location of bone marrow.

spongy cell: an irregularly shaped mesophyll cell, containing chloroplasts, located just above the lower epidermis of a leaf.

spontaneous generation: the proposal that living organisms can arise from nonliving matter.

sporangium (spor-an'-jē-um; pl., sporangia): a structure in which spores are produced.

spore: in plants and fungi, a haploid cell capable of developing into an adult without fusing with another cell (without fertilization). In bacteria and some other organisms, a stage in the life cycle that is resistant to extreme environmental conditions.

sporophyte (spor'-ō-fīt): the diploid form of a plant that produces haploid, asexual spores through meiosis.

sporozoan (spor-ō-zō'-un): a parasitic protist with a complex life cycle, typically involving more than one host; named for their ability to form infectious spores. A well-known sporozoan (genus *Plasmodium*) causes malaria.

stabilizing selection: a type of natural selection in which those organisms that display extreme phenotypes are selected against.

stamen (stā'-men): the male reproductive structure of a flower, consisting of a filament and an anther, in which pollen grains develop.

starch: a polysaccharide that is composed of branched or unbranched chains or glucose molecules; used by plants as a carbohydrate-storage molecule.

start codon: the first AUG codon in a messenger RNA molecule.

startle coloration: a form of mimicry in which a color pattern (in many cases resembling large eyes) can be displayed suddenly by a prey organism when approached by a predator.

stem: the portion of the plant body, normally located above ground, that bears leaves and reproductive structures such as flowers and fruit.

stem cell: an undifferentiated cell that is capable of dividing and giving rise to one or more distinct types of differentiated cell(s).

sterilization: a generally permanent method of contraception in which the pathways through which the sperm (vas deferens) or egg (oviducts) must travel are interrupted.

steroid: a lipid consisting of four fused carbon rings, with various functional groups attached.

steroid hormone: a class of hormone whose chemical structure (four fused carbon rings with various functional groups) re-

sembles cholesterol; steroids, which are lipids, are secreted by the ovaries and placenta, the testes, and the adrenal cortex.

stigma (stig'-muh): the pollen-capturing tip of a carpel.

stoma (stō'-muh; pl., stomata): an adjustable opening in the epidermis of a leaf, surrounded by a pair of guard cells, that regulates the diffusion of carbon dioxide and water into and out of the leaf.

stomach: the muscular sac between the esophagus and small intestine where food is stored and mechanically broken down and in which protein digestion begins.

stop codon: a codon in messenger RNA that stops protein synthesis and causes the completed protein chain to be released from the ribosome.

strand: a single polymer of nucleotides; DNA is composed of two strands.

striated muscle: see *skeletal muscle*.

stroke: an interruption of blood flow to part of the brain caused by the rupture of an artery or the blocking of an artery by a blood clot. Loss of blood supply leads to rapid death of the area of the brain affected.

stroma (strō'-muh): the semi-fluid material inside chloroplasts in which the grana are embedded.

style: a stalk connecting the stigma of a carpel with the ovary at its base.

subatomic particle: the particles of which atoms are made: electrons, protons, and neutrons.

subclimax: a community in which succession is stopped before the climax community is reached and is maintained by regular disturbances—for example, tallgrass prairie maintained by periodic fires.

substrate: the atoms or molecules that are the reactants for an enzyme-catalyzed chemical reaction.

subunit: a small organic molecule, several of which may be bonded together to form a larger molecule. See also *monomer*.

succession (suk-seh'-shun): a structural change in a community and its nonliving environment over time. During succession, species replace one another in a somewhat predictable manner until a stable, self-sustaining climax community is reached.

sucrose: a disaccharide composed of glucose and fructose.

sugar: a simple carbohydrate molecule, either a monosaccharide or a disaccharide.

sugar-phosphate backbone: a major feature of DNA structure, formed by attaching the sugar of one nucleotide to the phosphate from the adjacent nucleotide in a DNA strand.

surface tension: the property of a liquid to resist penetration by objects at its interface with the air, due to cohesion between molecules of the liquid.

survivorship curve: a curve resulting when the number of individuals of each age in a population is graphed against their age, usually expressed as a percentage of their maximum life span.

sustainable development: human activities that meet present needs for a reasonable quality of life without exceeding nature's limits and without compromising the ability of future generations to meet their needs.

symbiosis (sim'-bi-ō'sis): a close interaction between organisms of different species

over an extended period. Either or both species may benefit from the association, or (in the case of parasitism) one of the participants is harmed. Symbiosis includes parasitism, mutualism, and commensalism.

symbiotic: referring to an ecological relationship based on symbiosis.

sympathetic division: the division of the autonomic nervous system that produces largely involuntary responses that prepare the body for stressful or highly energetic situations.

sympatric speciation (sim-pat'-rik): speciation that occurs in populations that are not physically divided; normally due to ecological isolation or chromosomal aberrations (such as polyploidy).

synapse (sin'-aps): the site of communication between nerve cells. At a synapse, one cell (presynaptic) normally releases a chemical (the neurotransmitter) that changes the electrical potential of the second (postsynaptic) cell.

synaptic terminal: a swelling at the branched ending of an axon; where the axon forms a synapse.

syphilis (si'-ful-is): a sexually transmitted bacterial infection of the reproductive organs; if untreated, can damage the nervous and circulatory systems.

systematics: the branch of biology concerned with reconstructing phylogenies and with naming and classifying species.

taiga (ti'-guh): a biome with long, cold winters and only a few months of warm weather; dominated by evergreen coniferous trees; also called *northern coniferous forest*.

taproot system: a root system, commonly found in dicots, that consists of a long, thick main root and many smaller lateral roots, all of which grow from the primary root.

target cell: a cell on which a particular hormone exerts its effect.

taste: a chemical sense for substances dissolved in water or saliva; in mammals, perceptions of sweet, sour, bitter, salty, and umami produced by the stimulation of receptors on the tongue.

taste bud: a cluster of taste receptor cells and supporting cells that is located in a small pit beneath the surface of the tongue and that communicates with the mouth through a small pore. The human tongue has about 10,000 taste buds.

taxis (taks'-is; pl., taxes): an innate behavior that is a directed movement of an organism toward or away from a stimulus such as heat, light, or gravity.

taxonomy (tax-on'-uh-mē): the science by which organisms are classified into hierarchically arranged categories that reflect their evolutionary relationships.

Tay-Sachs disease: a recessive disease caused by a deficiency in enzymes that regulate lipid breakdown in the brain.

T cell: a type of lymphocyte that recognizes and destroys specific foreign cells or substances or that regulates other cells of the immune system.

T-cell receptor: a protein receptor, located on the surface of a T cell, that binds a specific antigen and triggers the immune response of the T cell.

tectorial membrane (tek-tor'-ē-ul): one of the membranes of the cochlea in which the hairs of the hair cells are embedded. In sound reception, movement of the basilar membrane relative to the tectorial membrane bends the cilia.

telomere (tē′-le-mēr): the nucleotides at the end of a chromosome that protect the chromosome from damage during condensation, and prevent the end of one chromosome from attaching to the end of another chromosome.

telophase (tēl′-ō-fāz): in mitosis and both divisions of meiosis, the final stage, in which the spindle fibers usually disappear, nuclear envelopes re-form, and cytokinesis generally occurs. In mitosis and meiosis II, the chromosomes also relax from their condensed form.

temperate deciduous forest: a biome in which winters are cold and summer rainfall is sufficient to allow enough moisture for trees to grow and shade out grasses.

temperate rain forest: a biome in which there is no shortage of liquid water year-round and that is dominated by conifers.

template strand: the strand of the DNA double helix from which RNA is transcribed.

temporal isolation: the inability of organisms to mate if they have significantly different breeding seasons.

tendon: a tough connective tissue band connecting a muscle to a bone.

tendril: a slender outgrowth of a stem that coils about external objects and supports the stem; normally a modified leaf or branch.

tentacle (ten′-te-kul): an elongate, extensible projection of the body of cnidarians and cephalopod mollusks that may be used for grasping, stinging, and immobilizing prey, and for locomotion.

terminal bud: meristem tissue and surrounding leaf primordia that are located at the tip of the plant shoot.

territoriality: the defense of an area in which important resources are located.

tertiary consumer (ter′-shē-er-ē): a carnivore that feeds on other carnivores (secondary consumers).

tertiary structure (ter′-shē-er-ē): the complex three-dimensional structure of a single peptide chain; held in place by disulfide bonds between cysteines.

test cross: a breeding experiment in which an individual showing the dominant phenotype is mated with an individual that is homozygous recessive for the same gene. The ratio of offspring with dominant versus recessive phenotypes can be used to determine the genotype of the phenotypically dominant individual.

testis (pl., testes): the gonad of male mammals.

testosterone: in vertebrates, a hormone produced by the interstitial cells of the testis; stimulates spermatogenesis and the development of male secondary sex characteristics.

thalamus: the part of the forebrain that relays sensory information to many parts of the brain.

theory: in science, a general explanation of natural phenomena developed through extensive and reliable observations; more general and reliable than a hypothesis.

thermoacidophile (ther-mō-a-sid′-eh-fil): an archaean that thrives in hot, acidic environments.

thermoreceptor: a sensory receptor that responds to changes in temperature.

thick filament: in the sarcomere, a bundle of myosin that interacts with thin filaments, producing muscle contraction.

thin filament: in the sarcomere, a protein strand that interacts with thick filaments, producing muscle contraction; composed primarily of actin, with accessory proteins.

thorax: the segment between the head and abdomen in animals with segmentation; the segment to which structures used in locomotion are attached.

thorn: a hard, pointed outgrowth of a stem; normally a modified branch.

threatened species: all species classified as critically endangered, endangered, or vulnerable.

threshold: the electrical potential (less negative than the resting potential) at which an action potential is triggered.

thrombin: an enzyme produced in the blood as a result of injury to a blood vessel; catalyzes the production of fibrin, a protein that assists in blood clot formation.

thylakoid (thī′-luh-koid): a disk-shaped, membranous sac found in chloroplasts, the membranes of which contain the photosystems and ATP-synthesizing enzymes used in the light-dependent reactions of photosynthesis.

thymine: a nitrogenous base found only in DNA; abbreviated as *T*.

thymosin: a hormone, secreted by the thymus, that stimulates the maturation of cells of the immune system.

thymus (thī′-mus): an organ of the lymphatic system that is located in the upper chest in front of the heart and that secretes thymosin, which stimulates lymphocyte maturation.

thyroid gland: an endocrine gland, located in front of the larynx in the neck, that secretes the hormones thyroxine (affecting metabolic rate) and calcitonin (regulating calcium ion concentration in the blood).

thyroid-stimulating hormone (TSH): a hormone, released by the anterior pituitary, that stimulates the thyroid gland to release hormones.

thyroxine (thī-rox′-in): a hormone, secreted by the thyroid gland, that stimulates and regulates metabolism.

tight junction: a type of cell-to-cell junction in animals that prevents the movement of materials through the spaces between cells.

tissue: a group of (normally similar) cells that together carry out a specific function; for example, muscle; may include extracellular material produced by its cells.

tonsil: a patch of lymphatic tissue consisting of connective tissue that contains many lymphocytes; located in the pharynx.

trachea (trā′-kē-uh): in birds and mammals, a flexible tube, supported by rings of cartilage, that conducts air between the larynx and the bronchi; in insects, an elaborately branching tube that carries air from openings called *spiracles* near each body cell.

tracheid (trā′-kē-id): an elongated xylem cell with tapering ends that contains pits in the cell wall; forms tubes that transport water.

tracheophyte (trā′-kē-ō-fīt): a plant that has conducting vessels; a vascular plant.

transcription: the synthesis of an RNA molecule from a DNA template.

transducer: a device that converts signals from one form to another. Sensory receptors are transducers that convert environmental stimuli, such as heat, light, or vibration, into electrical signals (such as action potentials) recognized by the nervous system.

transfer RNA (tRNA): a type of RNA that binds to a specific amino acid, carries it to a ribosome, and positions it for incorporation into the growing protein chain during protein synthesis. A set of three bases in tRNA (the anticodon) is complementary to the set of three bases in mRNA (the codon) that codes for that specific amino acid in the genetic code.

transformation: a method of acquiring new genes, whereby DNA from one bacterium (normally released after the death of the bacterium) becomes incorporated into the DNA of another, living, bacterium.

transgenic: referring to an animal or a plant that expresses DNA derived from another species.

translation: the process whereby the sequence of bases of messenger RNA is converted into the sequence of amino acids of a protein.

transpiration (trans′-per-ā-shun): the evaporation of water through the stomata of a leaf.

transport protein: a protein that regulates the movement of water-soluble molecules through the plasma membrane.

trial-and-error learning: a process by which adaptive responses are learned through rewards or punishments provided by the environment.

trichomoniasis (trik-ō-mō-nī′-uh-sis): a sexually transmitted disease, caused by the protist *Trichomonas*, that causes inflammation of the mucous membranes that line the urinary tract and genitals.

tricuspid valve: the valve between the right ventricle and the right atrium of the heart.

triglyceride (trī-glis′-er-id): a lipid composed of three fatty-acid molecules bonded to a single glycerol molecule.

triple covalent bond: a covalent bond that occurs when two atoms share three pairs of electrons.

trisomy 21: see *Down syndrome*.

trisomy X: a condition of females who have three X chromosomes instead of the normal two; most such women are phenotypically normal and are fertile.

trophic level: literally, "feeding level"; the categories of organisms in a community, and the position of an organism in a food chain, defined by the organism's source of energy; includes producers, primary consumers, secondary consumers, and so on.

tropical deciduous forest: a biome with pronounced wet and dry seasons and plants that shed their leaves during the dry season to minimize water loss.

tropical rain forest: a biome with evenly warm, evenly moist conditions; dominated by broadleaf evergreen trees; the most diverse biome.

true-breeding: pertaining to an individual all of whose offspring produced through self-fertilization are identical to the parental type. True-breeding individuals are homozygous for a given trait.

T tubule: a deep infolding of the muscle plasma membrane; conducts the action potential inside a cell.

tubal ligation: a surgical procedure in which a woman's oviducts are cut and tied so that the egg cannot reach the uterus, making her infertile.

tube cell: the outermost cell of a pollen grain; digests a tube through the tissues of the carpel, ultimately penetrating into the female gametophyte.

tube foot: a cylindrical extension of the water-vascular system of echinoderms; used for locomotion, grasping food, and respiration.

tubular reabsorption: the process by which cells of the tubule of the nephron remove water and nutrients from the filtrate within the tubule and return those substances to the blood.

tubular secretion: the process by which cells of the tubule of the nephron remove additional wastes from the blood, actively secreting those wastes into the tubule.

tubule (toob′-ūl): the tubular portion of the nephron; includes a proximal portion, the loop of Henle, and a distal portion. Urine is formed from the blood filtrate as it passes through the tubule.

tumor: a mass that forms in otherwise normal tissue; caused by the uncontrolled growth of cells.

tundra: a biome with severe weather conditions (extreme cold and wind and little rainfall) that cannot support trees.

turgor pressure: pressure developed within a cell (especially the central vacuole of plant cells) as a result of osmotic water entry.

Turner syndrome: a set of characteristics typical of a woman with only one X chromosome: sterile, with a tendency to be very short and to lack normal female secondary sexual characteristics.

tympanic membrane (tim-pan′-ik): the eardrum; a membrane, stretched across the opening of the middle ear, that transmits vibrations to the bones of the middle ear.

unicellular: single-celled; most members of the domains Bacteria and Archaea and the kingdom Protista are unicellular.

uniform distribution: the distribution characteristic of a population with a relatively regular spacing of individuals, commonly as a result of territorial behavior.

uniformitarianism: the hypothesis that Earth developed gradually through natural processes, similar to those at work today, that occur over long periods of time.

unsaturated: referring to a fatty acid with fewer than the maximum number of hydrogen atoms bonded to its carbon backbone; a fatty acid with one or more double bonds in its carbon backbone.

upwelling: an upward flow that brings cold, nutrient-laden water from the ocean depths to the surface; occurs along western coastlines.

uracil: a nitrogenous base found in RNA; abbreviated as U.

urea (ū-rē′-uh): a water-soluble, nitrogen-containing waste product of amino acid breakdown; one of the principal components of mammalian urine.

ureter (ū′-re-ter): a tube that conducts urine from each kidney to the bladder.

urethra (ū-rē′-thruh): the tube leading from the urinary bladder to the outside of the body; in males, the urethra also receives sperm from the vas deferens and conducts both sperm and urine (at different times) to the tip of the penis.

uric acid (ūr′-ik): a nitrogen-containing waste product of amino acid breakdown; a relatively insoluble white crystal excreted by birds, reptiles, and insects.

urinary system: the organ system that produces, stores, and eliminates urine, which contains cellular wastes, excess water and nutrients, and toxic or foreign substances. The urinary system is critical for maintaining homeostatic conditions within the bloodstream. It includes the kidneys, ureters, bladder, and urethra.

urine: the fluid produced and excreted by the urinary system of vertebrates; contains water and dissolved wastes, such as urea.

uterine tube: also called the oviduct, the tube leading from the ovary to the uterus, into which the secondary oocyte (egg cell) is released.

uterus: in female mammals, the part of the reproductive tract that houses the embryo during pregnancy.

vaccination: an injection into the body that contains antigens characteristic of a particular disease organism and that stimulates an immune response.

vacuole (vak′-ū-ōl): a vesicle that is typically large and consists of a single membrane enclosing a fluid-filled space.

vagina: the passageway leading from the outside of a female mammal's body to the cervix of the uterus.

variable: a factor in a scientific experiment that is deliberately manipulated in order to test a hypothesis.

variable region: the part of an antibody molecule that differs among antibodies; the ends of the variable regions of the light and heavy chains form the specific binding site for antigens.

vascular (vas′-kū-lar): describing tissues that contain vessels for transporting liquids.

vascular bundle a strand of xylem and phloem in leaves and stems; in leaves, commonly called a *vein*.

vascular cambium: a lateral meristem that is located between the xylem and phloem of a woody root or stem and that gives rise to secondary xylem and phloem.

vascular cylinder: the centrally located conducting tissue of a young root, consisting of primary xylem and phloem.

vascular tissue system: a plant tissue system consisting of xylem (which transports water and minerals from root to shoot) and phloem (which transports water and sugars throughout the plant).

vas deferens (vaz de′-fer-enz): the tube connecting the epididymis of the testis with the urethra.

vasectomy: a surgical procedure in which a man's vas deferens are cut, preventing sperm from reaching the penis during ejaculation, thereby making him infertile.

vector: a carrier that introduces foreign genes into cells.

vein: in vertebrates, a large-diameter, thin-walled vessel that carries blood from venules back to the heart; in vascular plants, a vascular bundle, or a strand of xylem and phloem in leaves.

ventral (ven′-trul): the lower side or underside of an animal whose head is oriented forward.

ventricle (ven′-tre-kul): the lower muscular chamber on each side of the heart, which pumps blood out through the arter-

ies. The right ventricle sends blood to the lungs; the left ventricle pumps blood to the rest of the body.

venule (ven′-ūl): a narrow vessel with thin walls that carries blood from capillaries to veins.

vertebral column (ver-tē′-brul): a column of serially arranged skeletal units (the vertebrae) that enclose the nerve cord in vertebrates; the backbone.

vertebrate: an animal that possesses a vertebral column.

vesicle (ves′-i-kul): a small, membrane-bound sac within the cytoplasm.

vessel: a tube of xylem composed of vertically stacked vessel elements with heavily perforated or missing end walls, leaving a continuous, uninterrupted hollow cylinder.

vessel element: one of the cells of a xylem vessel; elongated, dead at maturity, with thick, lignified lateral cell walls for support but with end walls that are either heavily perforated or missing.

vestigial structure (ves-tij′-ē-ul): a structure that serves no apparent purpose but is homologous to functional structures in related organisms and provides evidence of evolution.

villus (vi′-lus; pl., **villi**): a fingerlike projection of the wall of the small intestine that increases the absorptive surface area.

viroid (vi′-roid): a particle of RNA that is capable of infecting a cell and of directing the production of more viroids; responsible for certain plant diseases.

virus (vi′-rus): a noncellular parasitic particle that consists of a protein coat surrounding genetic material; multiplies only within a cell of a living organism (the host).

vitamin: one of a group of diverse chemicals that must be present in trace amounts in the diet to maintain health; used by the body in conjunction with enzymes in a variety of metabolic reactions.

vitreous humor (vit′-rē-us): a clear, jelly-like substance that fills the large chamber of the eye between the lens and the retina.

vocal cord: one of a pair of bands of elastic tissue that extend across the opening of the larynx and produce sound when air is forced between them. Muscles alter the tension on the vocal cords and control the size and shape of the opening, which in turn determines whether sound is produced and what its pitch will be.

vulnerable species: a species that faces a high risk of extinction in the medium-term future.

waggle dance: a symbolic form of communication used by honeybee foragers to communicate the location of a food source to their hivemates.

warning coloration: bright coloration that warns predators that the potential prey is distasteful or even poisonous.

water mold: a funguslike protist that includes some pathogens, such as the downy mildew, which attacks grapes.

water-vascular system: a system in echinoderms that consists of a series of canals through which seawater is conducted and is used to inflate tube feet for locomotion, grasping food, and respiration.

wax: a lipid composed of fatty acids covalently bonded to long-chain alcohols.

weather: short-term fluctuations in temperature, humidity, cloud cover, wind, and precipitation in a region over periods of hours to days.

Werner syndrome: a rare condition in which a defective gene causes premature aging; caused by a mutation in the gene that codes for DNA replication/repair enzymes.

white matter: the portion of the brain and spinal cord that consists largely of myelin-covered axons and that give these areas a white appearance.

wildlife corridors: strips of protected land linking larger areas. They allow animals to move freely and safely between habitats that would otherwise be isolated by human activities.

withdrawal: the removal of the penis from the vagina just before ejaculation in an attempt to avoid pregnancy; an ineffective contraceptive method.

working memory: the first phase of learning; short-term memory that is electrical or biochemical in nature.

xylem ($\bar{\text{zi}}$-**lum**): a conducting tissue of vascular plants that transports water and minerals from root to shoot.

yolk: protein-rich or lipid-rich substances contained in eggs that provide food for the developing embryo.

yolk sac: one of the embryonic membranes of reptilian, bird, and mammalian embryos; in birds and reptiles, a membrane surrounding the yolk in the egg; in mammals, forms part of the umbilical cord and the digestive tract but contains no yolk.

Z line: a fibrous protein structure to which the thin filaments of skeletal muscle are attached; forms the boundary of a sarcomere.

zona pellucida (**pel-oo′-si-duh**): a clear, noncellular layer between the corona radiata and the egg.

zooflagellate ($\bar{\text{zo}}$-$\bar{\text{o}}$-**fla′-jel-et**): a nonphotosynthetic protist that moves by using flagella.

zooplankton: nonphotosynthetic protists that are abundant in marine and freshwater environments.

zoospore ($\bar{\text{zo}}$′-$\bar{\text{o}}$-**spor**): a nonsexual reproductive cell that swims by using flagella; formed by members of the protistan division Oomycota.

zygospore ($\bar{\text{zi}}$′-**go**-**spor**): a fungal spore, produced by the division Zygomycota, that is surrounded by a thick, resistant wall and forms from a diploid zygote.

zygote ($\bar{\text{zi}}$′-**got**): in sexual reproduction, a diploid cell (the fertilized egg) formed by the fusion of two haploid gametes.

zygote fungus: a fungus of the division Zygomycota, which includes the species that cause fruit rot and bread mold.

Answers to Figure Caption Questions

CHAPTER 1

Figure 1-1
Some examples: Answerable at the cell level but not at the tissue level: How are signals transmitted in a neuron? How do white blood cells move to the site of wounds? How do chromosomes move during cell division? How do bacteria stick to surfaces? Answerable at the tissue level but not at the cell level: Which part of the brain controls speech? How does the kidney help maintain the body's water balance? What are the functions of skin? How does water get from a plant's roots to its leaves?

Figure 1-5
The antibacterial chemicals produced by fungi probably evolved because they improved the fungi's ability to compete with bacteria for access to resources such as food and space (by excluding bacteria from areas where the fungi are present).

Figure 1-9
Sweating also reduces the body's water content and its content of dissolved salts and other ions. Cooling the body to restore temperature homeostasis can thus have side effects that disrupt optimal water levels and dissolved salt concentrations. These disruptions in turn stimulate mechanisms to restore homeostasis in those features.

Figure E1-1
Redi's experiment demonstrated that the maggots were caused by something that was excluded by a gauze cover, but the possibility remained that some agent other than flies produced the maggots. An effective follow-up experiment might involve a series of closed, meat-containing systems that were identical in all respects other than the addition of a single possible causal element. Perhaps one container would have flies added, one roaches, one dust or soot, and so on. And, of course, a control would have nothing added.

CHAPTER 2

Figure 2-2
Atoms with non-full outer shells are reactive, with a strong tendency to form bonds with other atoms, and thus suitable for roles in the myriad chemical reactions of metabolism and for forming the complex molecules from which living things are constructed. Life's most prevalent molecules are notable for their tendency to participate in covalent bonds.

Figure 2-6
Oxygen's nucleus has eight protons, while hydrogen's has only one proton. So the positive charge of the oxygen nucleus is far stronger than that of the hydrogen nucleus.

Figure 2-9
Free radicals have atoms (often oxygens) with one or more unpaired electrons in their outer shells, making them very unsta-ble and causing them to capture electrons from nearby molecules to complete their outer shells. This can lead to changes in biological molecules, including DNA, which are critical to proper cellular functioning.

CHAPTER 3

Figure 3-9
The main drawback of widespread use of plastics is their resistance to natural degradation and consequent problems with long-term persistence of plastic waste in the environment. Starch is easily digestible by decomposer microbes, so starch-based plastics could potentially be far more biodegradable than plastics based on microbe-resistant molecules such as cellulose. The main challenge of designing starch-based plastics is making them sufficiently durable and strong.

Figure 3-16
As lipids, steroids are soluble in the lipid-based cell membrane and can cross it (and the nuclear membrane) to act inside the cell. Other types of hormones (mostly peptides) are not lipid soluble and cannot easily cross the cell membrane.

Figure 3-21
Heat energy can break chemical bonds, and the hydrogen bonds that account for secondary (and higher level) protein structure are especially susceptible to heat. Because a protein's functional ability usually depends on its shape, breaking shape-controlling hydrogen bonds disrupts function.

CHAPTER 4

Figure 4-4
Of the four structures listed, only the ribosome is found in all of the main branches of life (i.e., found in bacteria, archaea, and all eukaryotes). Thus, ribosomes must have been present in the common ancestor of all living cells, and nuclei, mitochondria, and chloroplasts must have arisen later.

Figure 4-10
Key processes such as DNA replication and transcription require that enzyme molecules have access to the DNA strand. Condensation of the genetic material restricts that access because it leaves little space around individual strands.

CHAPTER 5

Figure 5-7
In simple diffusion (left side of figure to right), the initial diffusion rate increases as the initial concentration gradient increases. In facilitated diffusion (right), the initial diffusion rate also increases with the initial concentration gradient but eventually reaches an upper limit when the carrier proteins are saturated.

Figure 5-10
Freshwater fishes must (and do) have physiological mechanisms that constantly export water to the environment to compensate for the water that flows into their bodies by osmosis.

Figure 5-11
The rigid cell wall of plant cells counteracts the pressure exerted by water entering by osmosis. (Water will continue entering the cell only until the osmotic pressure is balanced by the mechanical pressure of the elastic cell wall.) Animal cells lack a cell wall, and when placed in a highly hypotonic solution will absorb water by osmosis until the cell membrane bursts.

Figure 5-16
Exocytosis uses cellular energy, while diffusion is passive. In addition, while materials move through a membrane during diffusion out of a cell, in exocytosis, materials are expelled without passing directly through the plasma membrane. Exocytosis allows materials that are too large to pass through membranes to be eliminated from the cell.

CHAPTER 6

Figure 6-6
Other possibilities include mechanical energy (e.g., shaking), electricity, and radiation.

Figure 6-8
The conversion breaks a "high-energy" phosphate bond, and the energy stored in that bond can be transferred to a molecule involved in a reaction.

Figure 6-14
No. A catalyst lowers a reaction's activation energy, but does not eliminate it. The reaction energy must still be overcome in order for the reaction to proceed.

Figure 6-15
The best bet would be to increase the concentration of enzyme, because reaction rate is most often limited by the number of available enzyme molecules. Other things that could be helpful include increasing the reaction temperature (but not so much as to denature the enzyme) and adjusting the pH to the level at which the enzyme's activity is highest (though this last modification would require specific knowledge about the enzyme).

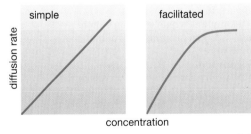

CHAPTER 7

Figure 7-7
Almost all of the ATP and NADPH produced in the chloroplast is used for the production of sugar in the Calvin-Benson cycle. Mitochondria are needed to extract the energy stored in the sugar molecules.

Figure 7-8
The H ions cross the membrane through the H^+ channel coupled to an ATP-synthesizing enzyme.

Figure 7-12
The C_4 pathway is less efficient than the C_3 pathway; C_4 uses one extra ATP per CO_2 molecule (for regenerating PEP). Thus, when CO_2 is abundant and photorespiration is not a problem, C_3 plants produce sugar at lower energy cost, and they outcompete C_4 plants.

CHAPTER 8

Figure 8-3
In oxygen-rich environments, both types of bacteria can survive, but aerobic bacteria prevail because their cellular respiration is far more efficient (produces more ATP per glucose molecule) than is glycolysis. In oxygen-poor environments, however, aerobic bacteria are limited by the oxygen shortage, and anaerobes would prevail despite their inefficiency.

Figure 8-8
In the absence of oxygen, ATP production halts. Oxygen is the final acceptor in the electron transport chain, and if it is not present, electrons cannot proceed along the chain (they "stack up" in the chain) and production of ATP by chemiosmosis comes to a halt.

CHAPTER 9

Figure 9-5
It takes more energy to break apart a C—G base pair, because these are held together by three hydrogen bonds, compared with the two hydrogen bonds that bind A to T.

Figure 9-7
DNA polymerase always moves in the 3′ to 5′ direction on a parental strand. Because the two strands of a DNA double helix are oriented in opposite directions, the 5′ direction on one strand leads toward the replication fork and the 5′ direction on the other strand leads away from the fork. Therefore, DNA polymerase must move in opposite directions on the two strands.

CHAPTER 10

Figure 10-1
This mutant would grow in simple medium if either ornithine, citrulline, or arginine were added.

Figure 10-4
RNA polymerase always travels in the 3′ to 5′ direction. Because the two DNA strands run in opposite directions, if the other DNA strand were the template strand, then RNA polymerase must travel in the opposite direction (that is, right to left in this illustration).

Figure 10-8
Grouped in codons, the original mRNA sequence is CGA AUC UAG UAA. Changing all G to U would produce the sequence CUA AUC UAU UAA. The two changes are in the first codon (CGA to CUA) and the third codon (UAG to UAU). Refer to the genetic code illustrated in Table 10-3. First, CGA encodes arginine, while CUA encodes leucine, so the first G → U change would substitute leucine for arginine in the protein. Second, UAG is a stop codon, but UAU encodes tyrosine. Therefore, the second G → U change would add tyrosine to the protein instead of stopping translation. The final codon in the illustration, UAA, is a stop codon, so the new protein would end with tyrosine.

CHAPTER 11

Figure 11-10
If the sister chromatids of one replicated chromosome failed to separate, then one daughter cell would not receive any copy of that chromosome, while the other daughter cell would receive both copies.

Figure 11-21
If one pair of homologues failed to separate at anaphase I, one of the resulting daughter cells (and the gametes produced from it) would have both homologues and the other daughter cell (and the gametes produced from it) would not have any copies of that homologue.

CHAPTER 12

Figure 12-14
Use Punnett squares to determine if the genotype of a plant bearing smooth yellow seeds can be revealed by a test cross with a plant bearing wrinkled green seeds. A plant with wrinkled green seeds has the genotype *ssyy*. A plant with smooth yellow seeds could be *SSYY, SsYY, SSYy,* or *SsYy.* Set up four Punnett squares to see if the smooth yellow plant's genotype can be revealed by a test cross.

CHAPTER 13

Figure 13-3
Primers direct DNA polymerase to begin synthesizing new DNA at a specific location on the double helix. Therefore, they are essential to producing multiple copies of specific segments of DNA (genes), rather than copying huge sections of unwanted DNA.

Figure 13-7
As they do for other genes, each person normally has two copies of each STR gene, one on each of a pair of homologous chromosomes. A person may be homozygous (two copies of the same allele) or heterozygous (one copy of each of two alleles) for each STR. The bands on the gel represent individual alleles of an STR gene. Therefore, a single person can have one band (if homozygous) or two bands (if heterozygous). If a person is homozygous for an STR allele, then he has two copies of the same allele. The DNA from both (identical) alleles will run in the same place on the gel, and therefore that (single) band will have twice as much DNA as each of the two bands of DNA from a heterozygote. The more DNA, the brighter the band.

Figure 13-8
Fields are often plowed or harrowed (the soil cut through by rotating blades) to uproot and kill weeds. The disturbed topsoil is susceptible to being blown away by the wind or washed away by hard rains. In principle, herbicide-resistant crops could be planted directly into fields without plowing. Weeds could be controlled, not by plowing or harrowing, but by herbicides that would kill weeds but not harm the crops. If the soil remains undisturbed, with a more or less continuous cover of plants (weeds at first, and then crops), it will erode less readily.

CHAPTER 14

Figure 14-6
No. Evolution can include changes in traits that are not revealed in external morphology, such as physiological systems and metabolic pathways. More generally, evolution in the sense of changes in a species' gene pool is inevitable in all lineages; genetic evolution is not necessarily reflected in morphological change.

Figure 14-8
Possibilities include coccyx (tailbone)—homologous with tail bones of a cat (or any tetrapod); goose bumps—homologous with erectile hairs of chimp (or any mammal; used for aggressive displays and insulation); appendix—homologous with cecum of a rabbit (and other herbivorous mammals; extension of large intestine used for storage); wisdom teeth—homologous with grinding rear molars of leaf-eating monkey (or other mammals); ear-wiggling muscles (the muscles around the ear that some people can use to make their ears move)—homologous with muscles in a dog (and other mammals) that enable the external ear to orient toward sound.

Figure 14-9
Analagous. Both peacock tail feathers and dog tails are used for communication, so they have a common function, but very different structures.

CHAPTER 15

Figure 15-3
If the antibiotic consistently induced mutations for antibiotic resistance, you would predict the same pattern of colonies on the streptomycin-containing plates as on the original plate.

Figure 15-5
For a locus with two alleles, one dominant and one recessive, there are two possible phenotypes. A mating between a heterozygote and a homozygote-recessive

yields offspring with a 50:50 ratio of the two phenotypes.

	B	**b**
b	Bb (black)	bb (brown)
B	Bb (black)	bb (brown)

Figure 15-6
Allele A should behave roughly as it does in the size 4 population, its frequency drifting to fixation or loss in almost all cases. But, because the population is a bit larger, the allele should, on average, take a longer time (greater number of generations) to reach fixation or loss. The longer period of drift should also allow for more reversals of direction (e.g., frequency drifting down, then up, then down again, etc.) than in the size 4 population.

Figure 15-7
Mutations inevitably and continually add variability to a population and, after the population becomes larger, the counteracting, diversity-reducing effects of drift decrease. The net result is an increase in genetic diversity.

Figure 15-11
Greater for males. A female's reproductive success is limited by her maximum litter size, but a male's potential reproductive success is limited only by the number of available females. When, as in bighorn sheep, males battle for access to females, the most successful males can impregnate many females, while unsuccessful males may not fertilize any females at all. Thus, the difference between the most and least successful male can be very large. In contrast, even the most successful female can have only one litter of offspring per breeding season, which is not that many more offspring than for a female who fails to reproduce.

Figure 15-13
There is always a limit to directional selection. As a trait becomes more extreme, eventually the cost of increasing it further outweigh the benefits (for example, the cost of obtaining extra food may outweigh the benefit of larger size).

CHAPTER 16

Figure 16-9
Possibilities include continental drift, climate changes (especially glacial advances) that cause habitat fragmentation, formation of islands by volcanic activity or rising sea level, movements of organisms to existing islands (including "islands" of isolated habitats such as lakes, mountaintops, deep-ocean vents), formation of barriers to movement (e.g., new mountain ranges, deserts, rivers). These processes are indeed sufficiently common and widespread to account for a multitude of speciation events over the history of life.

Figure 16-10
The key question is whether the two populations (apple and hawthorn) interbreed. Tests might involve careful observation of flies under natural conditions, lab experiments in which captive flies of the two types are provided with opportunities to interbreed, or genetic comparisons to determine the degree of gene flow between the two types of flies.

Figure 16-12
Presumably sympatric. The species in Lake Malawi are found only there and are all more closely related to one another than to any species from outside the lake. This pattern suggests that the species all arose from a single common ancestor that was present in the lake, and that all the speciation events that led to the current array of species took place in a single geographic location, Lake Malawi.

Figure 16-14
Natural selection cannot look forward and ensure that the only traits that evolve are those that ensure survival of the species as a whole. Instead, natural selection ensures only the preservation of traits that help individuals survive and reproduce more successfully than individuals lacking the trait. So if, in a particular species, highly specialized individuals survive and reproduce better than less-specialized individuals, the specialized phenotype will come to predominate, even if it ultimately puts the species at greater risk of extinction.

CHAPTER 17

Figure 17-2
The presence of oxygen would prevent the accumulation of organic compounds by quickly oxidizing them or their precursors. All of the successful abiotic synthesis experiments used oxygen-free "atmospheres."

Figure 17-4
The bacterial sequence would be most similar to that of the plant mitochondrion, because (as the descendant of the immediate ancestor of the mitochondrion) the bacterium shares with the mitochondrion a more recent common ancestor than with the chloroplast or the nucleus.

Figure 17-7
Most likely because of competition with seed plants, which had not yet arisen during the period when ferns and club mosses reached large sizes. After seed plants arose, competition from them eventually eliminated other types of plants from many ecological niches, presumably including those niches that favored evolution of large size.

Figure 17-8
No. The mudskipper merely demonstrates the plausibility of a hypothetical intermediate step in the proposed scenario for the origin of land-dwelling tetrapods. But the existence of a modern example similar in form to the hypothetical intermediate form does not provide information about the actual identity of that intermediate form.

Figure 17-10
On average, the rate at which new species arise has been larger than the rate at which species have gone extinct.

Figure 17-18
The African replacement hypothesis. These fossils are the oldest modern humans found so far, and their presence in Africa suggests that modern humans were present in Africa before they were present anywhere else, which, if true, would mean that they originated in Africa.

Figure E17-1
356.5 million years old. (3:1 ratio means that $3/4$ of the original uranium-235 is left, so $1/2$ of its half-life has passed.)

CHAPTER 18

Figure E18-3
A monophyletic group includes all of the descendants of a common ancestor. Fungi and animals are monophyletic; protists, great apes, seedless plants, and prokaryotes are not. (Descendants of the most recent common ancestor of protists include all the other eukaryotes; descendants of the most recent common ancestor of great apes include humans; descendants of the most recent common ancestor of prokaryotes include eukaryotes; and the descendants of the most recent common ancestor of seedless plants include seed plants.)

CHAPTER 19

Figure 19-4
Protective structures like endospores are most likely to evolve in environments in which protection is especially advantageous. Compared to other environments inhabited by bacteria, soils are especially vulnerable to drying out, which can be fatal to unprotected bacteria. Bacteria that could resist long dry periods would gain an evolutionary advantage.

Figure 19-5
Enzymes from bacteria that live in hot environments are active at high temperatures (temperatures that usually denature enzymes in organisms from more temperate environments). This ability to function at high temperatures makes the enzymes useful in test-tube reactions (such as PCR) that are run at high temperatures.

Figure 19-7
Binary fission eliminates the need to find a mate, and is useful under relatively constant environments because a well-adapted individual passes on all its genes and all its traits to its offspring.

Figure 19-9
The concentration of nitrogen gas would increase, because the major process for removing atmospheric nitrogen would end, while the processes that add nitrogen gas to the atmosphere would continue.

Figure 19-12
Viruses lack ribosomes and the rest of the "machinery" required to manufacture proteins.

Figure 19-13
Viruses replicate by integrating their genetic material into the host cell's genome. Thus, if biotechnologists can insert foreign

genetic material into a virus, the virus will naturally tend to transfer the foreign genes to cells they infect.

CHAPTER 20
Figure 20-2
Sex is the process that combines the genetic material from two individuals. In plants and animals, this occurs only during reproduction. But in many protists (and prokaryotes), gene exchange occurs independently of reproduction, which is often asexual.
Figure 20-7
Convergent evolution. Water molds and fungi live in similar environments and acquire nutrients in similar ways. These ecological similarities have fostered the evolution of superficially similar structures, even though the two taxa are only distantly related.

CHAPTER 21
Figure 21-3
Bryophytes lack lignin (which provides stiffness and support) and conducting vessels (which transport materials to distant parts of the body). Vessels and stiff stems seem to be required to achieve more than minimal height.
Figure 21-5
All the pictured structures are sporophytes. In ferns, horsetails, and club mosses, the gametophyte is small and inconspicuous.
Figure 21-7
The most common adaptations are hard, protective shells and incorporation of toxic and/or distasteful chemicals.
Figure 21-10
Pollination

Type	Advantages	Disadvantages
Wind	Not dependent on presence of animals; no investment in nectar or showy flowers; pollen can disperse over large distances	Larger investment in pollen because most fails to reach an egg; higher chance of failure to fertilize any egg
Animal	Each pollen grain has much greater chance of reaching suitable egg	Depends on presence of animals; must invest in nectar and showy flowers

Both types of pollination persist in angiosperms because the cost-benefit balance, and therefore the most adaptive pollination system, differs depending on the ecological circumstances of a species.

CHAPTER 22
Figure 22-1
Its filamentous shape helps the fungal body to penetrate and extend into its food sources; its shape also maximizes the ratio of surface area to interior volume (which

maximizes the area available for absorbing nutrients). The extreme thinness of the filaments ensures that no cell is very far from the surface at which nutrients are absorbed.
Figure 22-8
No. The common source of the two hyphae means that they will both inherit the same mating type. Hyphae must be of different mating types in order to reproduce sexually.
Figure 22-9
Haploid (although there might be some diploid cells in the basidia concealed in the gills of these haploid mushrooms).
Figure 22-17
In nature, bacteria compete with fungi for access to food and living space. The antibiotic chemicals produced by fungi serve as a defense against competition from bacteria.

CHAPTER 23
Figure 23-4
Sponges are "primitive" only in the sense that their lineage arose early in the evolutionary history of animals and their body plan is comparatively simple. Nonetheless, the sponge is well adapted to its marine habitats and highly successful.
Figure 23-6
(a) polyp, (b) medusa, (c) polyp, (d) medusa
Figure 23-10
Parasitic tapeworms have no gut and absorb nutrients across their body surfaces. Their ribbonlike shape maximizes surface area for absorption and allows the worm body to extend through the greatest possible area of the host's body (to be in contact with as many nutrients as possible).
Figure 23-11
Two openings allow one-way travel of food through the gut. One-way movement allows more efficient digestion than two-way movement; digestive waste from which all nutrients have been extracted can be excreted quickly without the need for reverse travel back along the gut, and food can be processed more quickly.
Figure 23-12
Water travels easily through the moist epidermis of a leech. When a high concentration of salt is dissolved in the moisture on the outside of a leech's body, water moves rapidly out of the leech's body by osmosis, dehydrating and ultimately killing the animal.

CHAPTER 24
Figure 24-6
A freshwater fish's body is immersed in a hypotonic solution, so water tends to continuously enter the body by osmosis. The physiological challenge is to get rid of all this excess water. For a saltwater fish, the challenge is reversed. The surrounding solution is hypertonic, so water tends to leave the body. The physiological challenge is to retain sufficient water.

Figure 24-8
One advantage is that adults and juveniles occupy different habitats and therefore do not compete with one another for resources (the niche occupied by an individual over its lifetime is broadened).
Figure 24-11
Flight is a very expensive trait (consumes a lot of energy, requires many special structures). In circumstances in which the benefits of flight are low, such as in habitats without predators or in species whose size is very large, natural selection may favor individuals that forgo an investment in flight, and flightlessness can arise.

CHAPTER 25
Figure 25-1
One possibility is that the variation necessary for selection has never arisen. (If no members of the species by chance gain the ability to discriminate between their own chicks and cuckoo chicks, then selection has no opportunity to favor the novel behavior.) Another possibility is that the cost of the behavior is relatively low. (If parasitism by cuckoos is rare, a parent that feeds any begging chick in its nest is, on average, much more likely to benefit than suffer.)
Figure 25-7
Because the crossbreeding experiment shows that differences in orientation direction between the two populations stem from genetic differences, birds from the western population should orient in a southwesterly direction, regardless of the environment in which they are raised.
Figure 25-21
The male bowerbird's genetic fitness is communicated through his bower-building ability. The female instinctively recognizes this and seeks out the best bower-builder to sire her offspring. These offspring will inherit at least some of his desirable genes, which will enhance the survival and reproduction of the offspring.
Figure 25-24
Canids forage mainly by smell; apes are mostly visual foragers. Modes of sexual signaling are affected not only by the nature of the information to be encoded, but also by sensory biases and sensitivities of the species involved. Communication systems may evolve to take advantage of traits that originally evolved for other functions.

CHAPTER 26
Figure 26-2
The death rate would need to equal the number of new bacteria produced; that is, a number of bacteria equal to the pre-doubling population size would need to die with each doubling.
Figure 26-4
Many variables interact in complex ways to produce real population cycles. Weather, for example, affects the lemmings' food supply and thus their ability to survive and reproduce. Predation of

lemmings is influenced by both the number of predators and the availability of other prey, which in turn is influenced by multiple environmental variables.

Figure 26-12
Emigration relieves population pressure in an overpopulated area, spreading the migrating animals into new habitats that may have more resources. Human emigration within and between countries is often driven by the desire or need for more resources, although social factors—such as wars and religious or racial persecution—also fuel human emigration. (This is a subjective question that can lead to discussion of the extent to which overpopulation drives human emigration.)

Figure 26-15
Many scenarios are possible based on different assumptions about increases in technology, changes in birth and death rates, and the resiliency of the ecosystems that sustain us. (This is a subjective question.)

Figure 26-19
In positive feedback, a change creates a situation that amplifies the change. When fertility exceeds RLF, there are more children than parents. As the additional children mature and become parents themselves, this more numerous generation produces still more additional children, and so on.

Figure 26-20
High birth rates in developing countries are sustained by cultural expectations, lack of health education, and lack of access to contraception. Lower birth rates in developed countries are encouraged by easy access to contraception, the relatively high cost of raising children, and more varied career opportunities for women. (Students should be able to expand on, or add to, these factors.)

Figure 26-21
U.S. population growth is in the rapidly rising "exponential" phase of the S-curve. Stabilization will require some combination of reduction in immigration rates and birth rates. An increase in death rates is less likely, but cannot be ruled out entirely in any future scenario. (Students should be encouraged to speculate about the timing and the reason for various time frames.)

CHAPTER 27

Figure 27-1
Invasive species, because they did not evolve in the habitat to which they were introduced, may occupy a niche that is nearly identical to a native species (for example, zebra mussels compete with other freshwater mussel and clam species). A successful invasive species may have adaptations (for example, a higher growth rate or reproductive rate) that allow it to outcompete the native species. Further, all native species are likely to have local predators, while the introduced (invasive) species may not. The absence of predators would also help

the invasive species outcompete a native species if they occupy very similar niches.

Figure 27-3
This is an open-ended question; a few examples follow: Keen eyesight of hawk and camouflage color of mouse. Forward-pointing eyes of predators allowing binocular vision and good localization of prey, and side-situated eyes of many prey that afford nearly 360-degree vision, thus allowing them to spot predators from almost any angle. Grasses have evolved tough silicon embedded in their leaves, and grazing herbivores have teeth that grow continuously throughout life so that they are not completely worn down by the abrasive grasses.

Figure 27-5
Although scientists have not observed these examples of evolution as they occurred, a likely scenario is as follows: Chance mutation caused some animals to more closely resemble their surroundings than did other members of their species. These individuals were less likely to be seen and eaten by predators. As a result, more survived and reproduced, passing their genetically determined appearance on to their offspring. Over large spans of time, more chance mutations that further enhanced the resemblance to the environment further reduced predation, selecting for individuals bearing these genetic traits.

Figure 27-11
While tasty prey has often evolved to blend in with its surroundings, poisonous prey species (such as the monarch caterpillar that stores milkweed toxin in its body) frequently advertise their presence with bright "warning colors." These bright colors make it easy for predators to learn to actively avoid animals bearing them.

Figure 27-13
People whose activities influence communities can make a special effort to preserve or restore those species on which the structure of the community depends.

Figure 27-14
Forests in which fires are suppressed grow extremely dense stands of trees, which tend to be less healthy because they are competing with one another for light, nutrients, and water. A fire under these conditions will burn hotter and more extensively, leaving fewer living plants and seeds to recolonize the area, slowing succession. Smaller, more frequent fires open spaces within a forest and allow patches of localized succession to occur, increasing the habitat for a variety of animals and increasing the species diversity of plants.

CHAPTER 28

Figure 28-3
On land, high productivity is supported by optimal temperatures for plant growth, a long growing season, and plenty of moisture, such as found in rain forests. Lack of water limits desert productivity. In aquatic

ecosystems, high productivity is supported by an abundance of nutrients and adequate light, such as found in estuaries. Lack of nutrients limits open ocean productivity.

Figure 28-6
Whenever energy is utilized, the second law of thermodynamics applies: When energy is converted from one form to another, the amount of useful energy decreases. Much of this energy is lost as heat. Because relatively little energy is transferred from one trophic level to the next, in order to maintain the tremendously organized state of life, animals must consume a large number of calories from the trophic level below them. The availability of energy is highest in photosynthetic organisms, which trap it directly from sunlight. Energy availability decreases with each successive energy transfer to higher trophic levels.

Figure 28-9
Humanity's need to grow crops to feed our growing population has led to the trapping of nitrogen using industrial processes; it is then used as fertilizer. Additionally, large-scale livestock feedlots generate enormous amounts of nitrogenous waste. Nitrogen oxides are also generated when fossil fuels are burned in power plants, vehicles, and factories, and when forests are burned. Consequences include the overfertilization of lakes, rivers, and portions of the ocean that receive runoff from land. Another important consequence is acid deposition, in which nitrogen oxides formed by combustion produce nitric acid in the atmosphere; this acid is then deposited on land.

CHAPTER 29

Figure 29-1
The equator-to-pole temperature gradient would still remain, but there would be no change in day length or seasons throughout the year.

Figure 29-6
Both live in arid environments, so their evolution has selected for fleshy, water-storing bodies. In each case, leaf area has been reduced, limiting evaporation. Both of these plants would be attractive to desert animals because they store water; as a result of this predation pressure, both carry defensive spines.

Figure 29-9
Nutrients are indeed abundant in tropical rain forests, but they are not stored in the soil. The optimal temperature and moisture of tropical climates allow plants to make such efficient use of nutrients that nearly all nutrients are stored in plant bodies, and to a lesser extent, in the bodies of the animals they support. These growing conditions support such a vast array of plants that these, in turn, provide a wealth of habitats and food sources for diverse animals.

Figure 29-14
Seeds adapted to germinate only when water is adequate often have chemicals

that inhibit germination in their seed coats. These chemicals are water soluble, washing away if enough rain falls to allow the plant a good chance of completing its life cycle.

Figure 29-17
Tallgrass prairie is vulnerable to two major encroachments. First, if humans suppress natural wildfires, the adequate rainfall amounts allow forests to take over. Second, these biomes provide the world's most fertile soils and have excellent growing conditions for farming, which has now displaced native vegetation.

Figure 29-29
Nearshore ecosystems have an abundance of the two limiting factors for life in water: nutrients and light to support photosynthetic organisms. Both upwelling from ocean depths and runoff from the land can provide nutrients, depending on the location of the ecosystem. The shallow water in these areas allows adequate light to penetrate to support rooted plants and/or anchored algae, which in turn provides food and shelter for a wealth of marine life.

Figure 29-30
Bleaching refers to the loss of symbiotic algae that normally inhabit the corals' tissues, providing them with energy captured during photosynthesis and calcium carbonate used in coral skeletons. Loss of these resources can eventually kill the coral. Bleaching is a common response to water that is excessively warm, and thus global warming may contribute to the demise of coral reefs.

CHAPTER 31

Figure 31-2
If the mammal's heat-sensing nerve endings were rendered nonfunctional, the nervous system would not send a signal to the hypothalamus when the body reached the set-point temperature. Consequently, the hypothalamus would send continuous "turn on" signals to the body's heat-generating and heat-retention mechanisms, which would continue to increase body temperature. This could lead to death by impairing the function of enzymes and disrupting biochemical pathways crucial to life.

Figure 31-7
In smaller bodies, the ratio of surface area to volume (and therefore to body mass) is greater than in larger bodies, so rate of heat loss (per unit of body mass) is greater. Therefore, younger, smaller individuals require additional insulation to maintain body temperature. (The relationship between volume and surface area is covered in Chapter 5.)

Figure 31-8
Blood is considered a specialized form of connective tissue because it is composed largely of a matrix of extracellular fluid (plasma) in which individual cells are suspended.

Figure 31-11
Skin is a functional unit composed of several tissue types (connective, epithelial, muscle, nerve), while blood is a more homogeneous functional unit composed of fairly similar cells and their surrounding fluid matrix, the plasma. One cannot identify two or more tissue types in blood, as is required for a structure to be an organ.

CHAPTER 32

Figure 32-4
Due to exercise-induced increase in muscle size, the hearts of well-conditioned athletes are larger than those of sedentary people. The heart's growth makes the volume of the ventricles larger, so a greater volume of blood is pumped with each heartbeat. Since the body's resting demand for oxygen remains unchanged, the demand can be met with fewer heartbeats per unit time.

Figure 32-6
Graph for Figure 32-6 exercise is shown here:

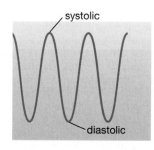

Figure 32-8
Iron is a key component of hemoglobin, which is necessary for building red blood cells capable of transporting oxygen. Consumption of iron in the diet is necessary to replace iron that is excreted with body waste because red blood cells are continually broken down after they die, and the iron recycling in the body is not 100% efficient.

Figure 32-10
The hormone erythropoietin stimulates production of additional red blood cells. These extra cells increase the blood's capacity to carry oxygen to muscles, thereby increasing the amount of time that the muscles can work without becoming fatigued (i.e., the time until a muscle's oxygen supply is depleted).

Figure 32-15
The principles of diffusion (see Chapter 5) tell us that molecules move down gradients from high to low concentration. Because the tissues through which capillaries pass generally consume oxygen, the concentration of oxygen is higher inside the capillaries than out, and oxygen diffuses along its concentration gradient from the inside of the capillary to the outside. Conversely, tissues generally produce carbon dioxide, so the concentration of carbon dioxide is higher outside the capillaries

than in, and carbon dioxide diffuses along its concentration gradient from the outside of the capillary to the inside.

CHAPTER 33

Figure 33-4
The open circulatory systems of insects are not well suited for transporting gases long distances within the body (compared to closed circulatory systems). Instead, respiratory organs (and their opening to the environment) must be situated close to all of the different organs and tissues of the insect body.

Figure 33-5
Because juveniles of many amphibian species breathe through gills, and because both juveniles and adults often depend on respiration through the skin, amphibians live in moist habitats, such as marshes, swamps, ponds, and rain forests.

Figure 33-11
The woman's high-altitude inhalation contains a smaller volume of air than an inhalation at sea level does. When the diaphragm contracts to expand the chest cavity, the volume of air that enters the lungs depends on the air pressure outside the body. At high elevation, atmospheric pressure is lower and less air is pushed into the lungs. In general, people traveling from low to high elevation compensate by taking deeper breaths (enlarging the chest cavity more).

CHAPTER 34

Figure 34-6
Because the microorganism populations live in the first chamber of the digestive system, many of them are passed to the other chambers, where they are digested and become an additional source of protein, carbohydrate, and other nutrients.

Figure 34-8
Lack of teeth is an adaptation for flight. The demands of flight have favored the evolution of adaptations that reduce body weight, including the loss of teeth (and the heavy jaws and jaw muscles associated with teeth).

Figure 34-14
A likely outcome is that the requisite absorptive surface area would have been achieved by some other sort of adaptation, perhaps a much longer intestine and a correspondingly large abdominal cavity to hold it. One could speculate about other features that might have evolved in concert with a superlong intestine (e.g., skeletal and muscular features to support it).

CHAPTER 35

Figure 35-2
Excretion of nitrogenous wastes as ammonia is restricted to animals that live in water. Despite the toxicity of ammonia, animals that live surrounded by water can excrete it more or less continuously and thus escape its toxic effects (and also

avoid expending energy to convert ammonia to urea or uric acid).

Figure 35-8
Consumption of alcohol tends to cause dehydration. Drinking large amounts of liquids (including alcoholic ones) decreases the osmolarity of blood below its set point, halting ADH release and stimulating urine flow and water loss. But the effect of the resulting negative feedback (i.e., increased blood osmolarity) is dampened, because ADH release is disrupted by the effects of alcohol. Thus, water loss continues even after blood osmolarity rises above its set point.

CHAPTER 36
Figure 36-5
Certain white blood cells (e.g., monocytes) are capable of amoeba-like locomotion in which the cell changes shape to extend protuberances that help it move. These amoeboid properties allow white blood cells to ooze through tiny openings in the capillary wall. Red blood cells are not amoeboid; they retain a single shape and thus cannot move through the tiny spaces in the capillary wall.

Figure 36-6
The combination of constant and variable components allows an antibody to accomplish not only functions that are unique to a particular antibody molecule (i.e., recognize and bind to a specific antigen) but also functions that are common to all antibodies in a given class (i.e., carry out the appropriate response to neutralize the invader). Variable regions provide the unique properties; constant regions provide the universal properties.

Figure 36-10
Cytotoxic T cells are activated by antigens on the target cell's surface, so the membranes of cancer cells might be expected to contain distinctive, abnormal molecules. This is in fact the case; cancer cells present tumor-specific antigens.

CHAPTER 37
Figure 37-6
The involvement of nerve cells in feedback loops allows sensory nerve signals to function in the loops. Thus, events such as temperature changes, glucose level changes, and pressure changes can turn hormone release on or off; information about such events cannot be readily communicated by chemical messages.

Figure 37-4
Dwarfism can be treated by administering growth hormone to affected individuals; it is much more difficult to halt the production of excess hormone (especially of a pituitary hormone such as growth hormone). Disruption of the pituitary is likely to disrupt the release of many other hormones as well. (Many cases of gigantism are caused by pituitary tumors, which can be treated by surgery.)

Figure 37-10
A high level of TSH indicates that the thyroid gland is producing an insufficient amount of thyroxine (hypothyroidism). With too little thyroxine circulating in the blood, the anterior pituitary does not receive a "turn off" signal and churns out a continuous stream of TSH.

Figure 37-12
Blood glucose levels would be low, and they would plummet rapidly even after being replenished by eating. After a meal, the person's elevated glucose would stimulate insulin production, which would reduce glucose levels, thus stimulating glucagon production. But body cells would not receive the glucagon messages that stimulate them to release glucose. Persistent low levels of glucose would stimulate production of additional glucagon, so levels of circulating glucagon would be high but have no effect.

Figure 37-15
They would likely remain in tadpole form and fail to metamorphose into frogs.

CHAPTER 38
Figure 38-4
Based on the evidence, the toxin must act on the postsynaptic part of the synapse, by blocking neurotransmitter receptors. If the toxin acted by blocking release of neurotransmitters from the presynaptic neuron, then adding neurotransmitter to the synapse would have generated a postsynaptic action potential. If the toxin acted by preventing sodium channels from opening, then it would not have been possible to stimulate an action potential in the presynaptic neuron.

Figure 38-5
Sensitive areas have a higher density of touch-sensitive sensory neurons. This would allow two types of increased sensitivity: (1) The higher density of receptors would mean that touching any particular place on the sensitive skin area would be more likely to stimulate at least one touch receptor, allowing the brain to determine the location of the stimulus more precisely. (2) A higher density of receptors would also mean that a touch of a particular force would stimulate a larger number of neurons, which the central nervous system would likely perceive as a stronger stimulus.

Figure 38-10
A severed or damaged spinal cord prevents the sensation of pain from being relayed to the brain but does not disrupt the reflex circuit, which requires only transmission within a small portion of the spinal cord.

Figure 38-22
For nearsightedness, the cornea is flattened, which diverges incoming light rays before they reach the lens. For farsightedness, the edges of the cornea are reduced, which leaves the cornea with a more rounded shape that converges incoming light rays.

Figure 38-25
Non-predators with binocular vision tend to be species whose lifestyles make depth perception especially important. Examples include monkeys that must make accurate leaps from tree to tree and bats that must fly quickly and accurately to flowers or fruit. In such species, the advantages of binocular vision presumably outweigh the increased predation risk from the resulting narrower field of view.

Figure 38-29
There are several reasons that porpoises do not have large external ears: (1) Although large external ears would probably enhance porpoises' ability to detect the existence and direction of a returning echolocation call, the hydrodynamic drag exerted by external ears would be costly for an animal that must swim swiftly in the sea, and would be selected against. (2) Sound travels much better in water than in air, so the returning echoes of porpoise echolocation calls would not be as faint as the returning echoes of bat echolocation calls. (3) Porpoises hunt much larger prey than bats do, which would also contribute to the returning echoes being louder. (4) The head of a porpoise is much larger than the head of a bat, so the separation between its ears, even without external ears, is still relatively large, allowing effective location of sounds.

CHAPTER 39
Figure 39-4
Neither the thick filaments nor the thin filaments shorten. Both remain the same length throughout contraction, but the sarcomere as a whole shortens as the thin filaments slide toward the center of the sarcomere.

Figure 39-8
Thick armor offers good protection against predators but is very heavy. Water helps support the exoskeleton and reduces the energy required to move it around. On land, the energy cost of armor is prohibitively high, especially for flying animals (such as most insects).

CHAPTER 40
Figure 40-2
Mitosis. (In animals, meiosis occurs only as part of sexual reproduction.)

Figure 40-4
The best hypothesis is that having genetically diverse offspring provides an advantage primarily during the unpredictable environmental stresses that arise during the overwintering period. During the summer, it's safe to pursue a reproductive strategy that emphasizes sheer numbers of offspring, but a female that spends her autumn producing genetically diverse offspring is more likely to have some that survive the winter.

Figure 40-7

Courtship rituals help ensure that animals mate with other individuals of their own species. Courtship rituals are reproductive isolating mechanisms that help prevent potentially disadvantageous hybrid mating. The courtship activities in some species (such as the Sumatran rhino) stimulates ovulation.

Figure 40-15

Circulating testosterone suppresses the release of FSH and LH from the anterior pituitary, thus suppressing sperm production. (Testosterone-based male contraceptives are under development.)

Figure 40-20

One reason is that these barriers can be penetrated only by a large number of healthy sperm. Males producing large numbers of healthy sperm have the highest probability of producing healthy offspring.

Figure E40-2

By injecting the sperm, the complex chemical interaction between the sperm and egg that accompanies normal fertilization is avoided. It is difficult to predict the outcome of the interaction between a sperm and the egg's protective layers, or to know if a given sperm cell will be able to successfully penetrate an egg. So injecting sperm improves the likelihood of successful fertilization.

CHAPTER 41

Figure 41-1

Indirect development reduces competition between the adult and its offspring. Indirectly reproducing animals usually produce large numbers of eggs. Because only a small amount of yolk is produced and the offspring usually fend for themselves after hatching, indirect development does not place great demands on the mother. This allows her to produce large numbers of offspring. Although most will not survive to adulthood, they may disperse widely and enough are likely to survive to maintain the population.

Figure 41-2

Animals using direct development generally produce relatively small numbers of eggs or offspring, which are nourished inside the mother's body or develop within eggs with large amounts of yolk. Parents may care for the young, who have a much greater chance of survival than do the larvae of indirectly reproducing animals.

Figure 41-5

No, a haploid cell (gametes and their precursors) would not work, because haploid vertebrate cells contain only half the normal number of chromosomes.

Figure 41-14

Because there is no placenta for exchange of nutrients, wastes, and gases, the embryo is retained in the uterus for a much shorter time. Offspring are born in a much less developed state (marsupials) or even leave the mother as eggs (monotremes). In marsupials, the embryo is enclosed within an amnion that segregates it from the mother's blood (and thus from attack by the mother's immune system); but nutrients cannot pass across the amnion, and the embryo must be ejected from the uterus as soon as its yolk is exhausted. Development continues outside the uterus (typically inside an external pouch).

CHAPTER 42

Figure 42-3

Primary growth occurs at the tip of the primary root, at the tips of all lateral roots, at the terminal bud, at all lateral buds, and at the tips of lateral branches. Secondary growth will occur all along the margins of the primary and lateral roots and shoots.

Figure 42-11

As shown in the figure, the phloem layer lies close to the outside of the stem and is part of the bark. Removing a strip of bark entirely around the main stem (trunk) creates an unbridgeable break in the phloem vessels connecting the roots to the rest of the plant. Because phloem carries nutrients from their point of acquisition to their point of use, the plant cannot survive without a phloem connection between (for example) its sugar sources (e.g., leaves) and its sugar sinks (e.g., growing root tips).

Figure 42-18

One major problem would be that dissolved nutrients could leak out of the vascular cylinder through spaces between the cells of the endodermis. If that happened, the nutrients could diffuse down their concentration gradient into the surrounding soil and be lost to the plant. The Casparian strip ensures that the movement of water and ions in and out of the vascular cylinder is regulated by the cell membranes of the endodermal cells.

Figure 42-19

Only the growing tips of roots are infected because this is where most absorption of water and nutrients takes place.

Figure 42-22

Photosynthesis in the leaf slows and eventually stops because its rate is limited by the availability of CO_2 and closed stomata prevent air from reaching the sites of photosynthesis. When stomata are closed, transpiration slows or stops, transpirational pull of water up the xylem column slows or stops, and bulk flow of water into the roots slows or stops. But some water still enters the roots by osmosis (i.e., ions are still pumped into root cells, and water still follows by osmosis).

CHAPTER 43

Figure 43-3

Advantages: Wind-pollinated individuals can potentially fertilize other plants over a much larger geographic area. Fertilization is not dependent on success of pollinators (i.e., not dependent on animals whose populations may decline or disappear). There is no need to invest energy in structures and/or chemicals that attract pollinators. *Disadvantages*: The plant must produce enormous quantities of pollen because so much is wasted due to its random dispersal and because most pollen does not reach the female gametophyte.

Figure 43-6

Separate bloom times reduce the chances of self-fertilization, which would result in inbreeding. Plants that do not inbreed in many cases have more successful offspring. Inbred individuals are more likely to have homozygous recessive alleles that are deleterious.

Figure 43-19

When nectar is at the base of a long, tubular flower, it is accessible to hummingbirds (which have long bills and tongues) but not to insects and other small animals that might like to eat it. A flower whose nectar is inaccessible to insects is more likely to have nectar available for a visiting hummingbird; it is thus more likely that hummingbirds will learn to repeatedly visit this type of flower, ensuring that the flowers will be pollinated and reproduce.

CHAPTER 44

Figure 44-7

The auxin would replace the tip in creating apical dominance, and lateral branches would be inhibited from sprouting.

Figure 44-13

Ripe tomatoes are often damaged in shipping. Blocking ethylene production would prevent ripening. Green tomatoes could be ripened after they reach the market by exposing them to ethylene.

Photo Credits

Animals Animals/Earth Scenes; 16-12UR: Jack Jeffrey/Photo Resource Hawaii/Alamy Images; 16-12WL: A.C. Medeiros/Dr. Gerald D. Carr, PhD; 16-12WR: Dr. Gerald D. Carr, PhD; 16-13: Tom McHugh/Steinhart Aquarium/Photo Researchers, Inc.; 16-14: Allen Blake Sheldon/Allen Blake Sheldon Nature Photography; E16-1: Martin Harvey/Peter Arnold, Inc.; E16-3: Mileniusz Spanowicz

Chapter 17 opener: National Geographic Image Collection; 17-3: Hybrid painting, 204, for Scientific American. ©206 by Don Dixon/cosmographica.com; 17-5: Michael Abbey/Visuals Unlimited; 17-6a: Chase Studio; 17-6b: James L. Amos/Photo Researchers, Inc.; 17-6c: Chris Howes/Wild Places Photography/Alamy Images; 17-6d: Douglas Faulkner/Photo Researchers, Inc.; 17-7: Illustration by Ludek Pesek/Science Photo Library/Photo Researchers, Inc.; 17-8: Terry Whittaker/Photo Researchers, Inc.; 17-9: Illustration by Chris Butler/Science Photo Library/Photo Researchers, Inc.; 17-12a: Tom McHugh/Chicago Zoological Park/Photo Researchers, Inc.; 17-12b: Frans Lanting/Minden Pictures; 17-12c: Nancy Adams/Tom Stack & Associates, Inc.; 17-13: ©Michel Brunet/M.P.F.T.; 17-16: David Frayer, Dept. of Anthropology, University of Kansas; 17-17: Jerome Chatin/Gamma Press USA, Inc.

Chapter 18 opener: Tom Brakefield/DRK Photo; 18-1a: Wayne Lankinen/Bruce Coleman Inc.; 18-1b: M.C. Chamberlain/DRK Photo; 18-1c: Maslowski/Photo Researchers, Inc.; 18-2a: C. Steven Murphree/Biological Photo Service; 18-2b: Dr. Greg Rouse, Department of Invertebrate Zoology, National Museum of Natural History, Smithsonian Institution; 18-2c: Dr. Jeremy Burgess/Science Photo Library/Photo Researchers, Inc.; 18-4a: Hans Gelderblom/Getty Images Inc. - Stone Allstock; 18-4b: Reprinted by permission of Springer-Verlag from W.J. Jones, J.A. Leigh, F. Mayer, C.R. Woese, and R.S. Wolfe, Methanococcus jannaschii sp. nov., an extremely thermophilic methanogen from a submarine hydrothermal vent. Archives of Microbiology 136:254-261 (1983). © 1983 by Springer-Verlag GmbH & Co KG. Image courtesy of W. Jack Jones.; 18-7: Zig Koch/Kino Fotoarqiovo

Chapter 19 opener: AP Wide World Photos; 19-1a: David M. Phillips/Visuals Unlimited; 19-1b: SPL/Photo Researchers, Inc.; 19-1c: © Scott Camazine/Photo Researchers, Inc.; 19-3: © Eye of Science/Photo Researchers, Inc.; 19-4: A.B. Dowsett/Science Photo Library/Photo Researchers, Inc.; 19-5: Alan L. Detrick/Photo Researchers, Inc.; 19-6: Biophoto Associates/Photo Researchers, Inc.; 19-7: CNRI/Science Photo Library/Photo Researchers, Inc.; 19-8: Dennis Kunkel/Phototake NYC; 19-9a,b: C.P. Vance/Visuals Unlimited; 19-11UL: MANFRED KAGE/Peter Arnold, Inc.; 19-11UR: Dept. of Microbiology, Biozentrum/Photo Researchers, Inc.; 19-11WL,R: Dr. Linda Stannard, UCT/Photo Researchers, Inc.; 19-13: Oliver Meckes/Ottawa/Photo Researchers, Inc.; 19-14: © EM Unit, VLA/Photo Researchers, Inc.; E19-1: Paul Poplis/FoodPix/Jupiter Images - FoodPix - Creatas - Brand X - Banana Stock - PictureQuest

Chapter 20 opener: Olivier Digoit/Alamy Images; 20-1: M.I. Walker/Photo Researchers, Inc.; 20-2a: Carolina Biological Supply Company/Phototake NYC; 20-2b: Eric Grave/Science Source/Photo Researchers, Inc.; 20-3: P.M. Motta and F.M. Magliocca/Science Photo Library/Photo Researchers, Inc.; 20-4: David M. Phillips/The Population Control/Photo Researchers, Inc.; 20-6: Oliver Meckes/Photo Researchers, Inc.; 20-7: William Merrill, Penn State University; 20-8: Manfred Kage/Peter Arnold, Inc.; 20-9a: D.P. Wilson/Eric and David Hosking/Photo Researchers, Inc.; 20-9b: Minden Pictures; 20-10: David M. Phillips/Visuals Unlimited; 20-11: Peter J.S. Franks, Scripps Institution of Oceanography; 20-14: Oliver Meckes & Nicole Ottawa/Eye of Science/Photo Researchers, Inc.; 20-15a: Ed Degginger/Color-Pic, Inc.; 20-15b: Manfred Kage/Peter Arnold, Inc.; 20-16: Dennis

Kunkel/Phototake NYC; 20-17a: P.W. Grace/Science Source/Photo Researchers, Inc.; 20-17b: Cabisco/Visuals Unlimited; 20-18W: Cabisco/Visuals Unlimited; 20-19: Mark Conlin Photography; 20-20a: © Ray Simons/Photo Researchers, Inc.; 20-20b: SeaPics.com

Chapter 21 opener: Frans Lanting/Minden Pictures; 21-3a: Lee W. Wilcox; 21-3b: John Gerlach/Tom Stack & Associates, Inc.; 21-3c: John Shaw/Tom Stack & Associates, Inc.; 21-3d: Glen Allison/Photographer's Choice/Getty Images, Inc.; 21-5a: Dwight R. Kuhn Photography; 21-5b: Milton Rand/Tom Stack & Associates, Inc.; 21-5c: Larry Ulrich/DRK Photo; 21-5d: Konrad Wothe/Minden Pictures; 21-6UL: Milton Rand/Tom Stack & Associates, Inc.; 21-7C: Andy Roberts/Getty Images Inc. - Stone Allstock; 21-7d: John Kaprielian/Photo Researchers, Inc.; 21-8a: Maurice Nimmo/SPL/A-Z Botanical Collection, Ltd./Photo Researchers, Inc.; 21-8b: Teresa and Gerald Audesirk; 21-8c: © Peter Johnson / CORBIS All Rights Reserved; 21-8d: Image100/Alamy Images; 21-9a: Dr. William M. Harlow/Photo Researchers, Inc.; 21-9b: Gilbert S. Grant/Photo Researchers, Inc.; 21-10a: Dwight R. Kuhn Photography; 21-10b: David Dare Parker/OnAsia; 21-10bi: Matt Jones/Auscape International Proprietary Ltd.; 21-10c: Dwight R. Kuhn Photography; 21-10d: Teresa and Gerald Audesirk; 21-10e: Larry West/Photo Researchers, Inc.

Chapter 22 opener: Michael W. Beug; 22-1a: Robert & Linda Mitchell Photography; 22-1b: Elmer Koneman/Visuals Unlimited; 22-2: Jeff Lepore/Photo Researchers, Inc.; 22-4: Thomas J. Volk, TomVolkFungi.net; 22-5a: Breck P. Kent; 22-5b: Carolina Biological Supply Company/Phototake NYC; 22-6ins: Andrew Syred/Photo Researchers, Inc.; 22-7a: W.K. Fletcher/Photo Researchers, Inc.; 22-7b: David Dvorak, Jr.; 22-8ins: S Lowry/Univ.Ulster/Getty Images Inc. - Stone Allstock; 22-9a: Scott Camazine/Photo Researchers, Inc.; 22-9b: David M. Dennis/Tom Stack & Associates, Inc.; 22-9c: Hans Reinhard/Bruce Coleman Inc.; 22-10: Darrell Hensley, Ph.D., University of Tennessee, Entomology & Plant Pathology; 22-12a: Jeff Foott Productions; 22-12b: Robert & Linda Mitchell Photography; 22-13: Stanley L. Flegler/Visuals Unlimited; 22-14: David M. Dennis/Tom Stack & Associates, Inc.; 22-15a: Michael Fogden/DRK Photo; 22-15b: Hugh Sturrock/University of Edinburgh; 22-16: David M. Phillips/Visuals Unlimited; 22-17: Teresa and Gerald Audesirk; 22-18: M. Viard/Jacana/Photo Researchers, Inc.; 22-19: G.L. Barron/Biological Photo Service; 22-20: Cabisco/Visuals Unlimited; E22-1: Matt Meadows/Peter Arnold, Inc.

Chapter 23 opener: Eloy Alonso Gonzalez/Reuters Limited; 23-4a: Larry Lipsky/DRK Photo; 23-4b: Brian Parker/Tom Stack & Associates, Inc.; 23-4c: © Charles Seaborn/Odyssey/Chicago; 23-6a: Gregory Ochocki/Photo Researchers, Inc.; 23-6b: Mark Webster/Photolibrary.Com; 23-6c: Teresa and Gerald Audesirk; 23-6d: David B. Fleetham/SeaPics.com; 23-9a: Dr. Richard Kessel & Dr. Gene Shih/Visuals Unlimited/Getty Images, Inc.; 23-9b: M.I. (Spike) Walker/Alamy Images; 23-9c: Dr. Wolfgang Seifarth; 23-10UR1: Martin Rotker/Phototake NYC; 23-10UR2: Stanley Flegler/Visuals Unlimited; 23-12a: Kjell B. Sandved/Butterfly Alphabet, Inc.; 23-12b: ©204Peter Batson/Image Quest Marine; 23-12c: J.H. Robinson/Photo Researchers, Inc.; 23-14a: Ray Coleman/Photo Researchers, Inc.; 23-14b: Alex Kerstitch/Estate of Alex Kerstitch; 23-15a: Fred Bavendam/Peter Arnold, Inc.; 23-15b: Ed Reschke/Peter Arnold, Inc.; 23-16a: Fred Bavendam/Peter Arnold, Inc.; 23-16b: Kjell B. Sandved/Photo Researchers, Inc.; 23-16c: Alex Kerstitch/Estate of Alex Kerstitch; 23-17: © Reg Morrison/Auscape/Minden Pictures; 23-18: Dwight R. Kuhn Photography; 23-20: © David Scharf/Peter Arnold, Inc.; 23-21a: Carolina Biological Supply Company/Phototake NYC; 23-21b: Peter J. Bryant/Biological Photo Service; 23-21c: Stephen Dalton/Photo Researchers, Inc.; 23-21d: Werner H. Muller/Peter Arnold, Inc.; 23-21e: Stanley Breeden/DRK Photo;

23-22a: Dwight Kuhn/Dwight R. Kuhn Photography; 23-22b: Tim Flach/Getty Images Inc. - Stone Allstock; 23-22c: Teresa and Gerald Audesirk; 23-23a: Marty Cordano/DRK Photo; 23-23b: ©Tom Brakefield/CORBIS; 23-24a: Tom Branch/Photo Researchers, Inc.; 23-24b: Peter J. Bryant/Biological Photo Service; 23-24c: Carolina Biological Supply Company/Phototake NYC; 23-24d: Alex Kerstitch/Estate of Alex Kerstitch; 23-25: Tom E. Adams/Peter Arnold, Inc.; 23-26a: Carolina Biological Supply Company/Phototake NYC; 23-26b: Reproduced by permission from Howard Shiang, D.V.M., Journal of the American Veterinary Medical Association 163:981, Oct. 1973.; 23-27a: Teresa and Gerald Audesirk; 23-27b: Jeff Foott Productions; 23-27c: Chris Newbert/Bruce Coleman Inc.; 23-28b: Michael Male/Photo Researchers, Inc.; 23-29: Dr. Tsunemi Kuboders/National Science Museum/AP Wide World Photos; E23-1: Volker Steger/Photo Researchers, Inc.

Chapter 24 opener: Getty Images, Inc.; 24-2: John Giannicchi/Science Source/Photo Researchers, Inc.; 24-3ins: Runk/Shoenberger/Grant Heilman Photography, Inc.; 24-3R: Tom McHugh/Photo Researchers, Inc.; 24-4a: Tom McHugh/Steinhart Aquarium/Photo Researchers, Inc.; 24-4b: Tom Stack & Associates, Inc.; 24-4bI: Breck P. Kent; 24-5a: Jeffrey L. Rotman Photography; 24-5b: David Hall/Photo Researchers, Inc.; 24-6a: Peter David/Getty Images Inc. - Hulton Archive Photos; 24-6b: Mike Neumann/Photo Researchers, Inc.; 24-6c: Stephen Frink/Getty Images Inc. - Stone Allstock; 24-7a: Tom McHugh/Photo Researchers, Inc.; 24-7b: Alan Root/Survival Anglia/OSF/Photolibrary.Com; 24-8a: Breck P. Kent/Animals Animals/Earth Scenes; 24-8b: Joe McDonald/Tom Stack & Associates, Inc.; 24-8c: Cosmos Blank/National Audubon Society/Photo Researchers, Inc.; 24-9a: David G. Barker/Tom Stack & Associates, Inc.; 24-9b: Roger K. Burnard/Biological Photo Service; 24-9c: Frans Lanting/Minden Pictures; 24-10: Carolina Biological Supply Company/Phototake NYC; 24-11a: Walter E. Harvey/Photo Researchers, Inc.; 24-11b: Carolina Biological Supply Company/Phototake NYC; 24-11c: Ray Ellis/Photo Researchers, Inc.; 24-12: Tom McHugh/Photo Researchers, Inc.; 24-13a: Dave Watts/Nature Picture Library; 24-13b: Craig Ingram/Alamy Images; 24-14a: Mark Newman/Index Stock Imagery, Inc.; 24-14ai: D. Parer and E. Parer-Cook/Auscape International Proprietary Ltd.; 24-14b: Dave Watts/Alamy Images; 24-14c: Klein/Peter Arnold, Inc.; 24-15a: Flip Nicklin/Minden Pictures; 24-15b: Jonathan Watts/Science Photo Library/Photo Researchers, Inc.; 24-15c: C. and M. Denis-Huot/Peter Arnold, Inc.; 24-15d: S.R. Maglione/Photo Researchers, Inc.; E24-1: Stanley Breeden/DRK Photo; E24-2: © Richard T. Nowitz / CORBIS All Rights Reserved

Chapter 25 opener: Jennifer Graylock/AP Wide World Photos; Opener inset: Joe McDonald/DRK Photo; 25-1a,b: Eric and David Hosking/Frank Lane Picture Agency Limited; 25-3a-e: Boltin Picture Library; 25-4: Frans Lanting/Minden Pictures; 25-6: Thomas McAvoy/Getty Images/Time Life Pictures; 25-8: © Renee Lynn / CORBIS All Rights Reserved; 25-9: Ken Cole/Animals Animals/Earth Scenes; 25-10: Richard K. LaVal/Animals Animals/Earth Scenes; 25-11: Nuridsany et Perennou/Photo Researchers, Inc.; 25-12: Robert & Linda Mitchell/Robert & Linda Mitchell Photography; 25-13a: Stephen J. Krasemann/Photo Researchers, Inc.; 25-13b: Hans Pfletschinger/Peter Arnold, Inc.; 25-14a: M.P. Kahl/DRK Photo; 25-14b: Marc Chamberlain/Marc C. Chamberlain; 25-15: Ray Dove/Visuals Unlimited; 25-16: William Ervin/Natural Imagery; 25-17: Michael K. Nichols/National Geographic Image Collection; 25-18: John D. Cunningham/Visuals Unlimited; 25-20: Joe McDonald/Visuals Unlimited; 25-21a: © Konrad Wothe/Minden Pictures; 25-21b: Frans Lanting/Minden Pictures; 25-22: Anne et Jacques Six; 25-24a: Dwight R. Kuhn Photography; 25-24b: Teresa and Gerald Audesirk; 25-25: Fred Bruemmer/Peter Arnold, Inc.; 25-27: © Raymond Mendez/Animals Animals/Earth Scenes; 25-28: Photo Lennart Nilsson/Albert Bonniers Forlag; 25-29: William P. Fifer, New York State Psychiatric

Index

126 INDEX